Collected Works of J. D. Eshelby

The Mechanics of Defects and Inhomogeneities

Edited by

XANTHIPPI MARKENSCOFF
University of California, San Diego, CA, U.S.A.

and

ANURAG GUPTA
University of California, Berkeley, CA, U.S.A.

Springer

A C.I.P. Catalogue record for this book is available from the Library of Congress.

ISBN-10 1-4020-4416-X (HB)
ISBN-13 978-1-4020-4416-8 (HB)

Published by Springer,
P.O. Box 17, 3300 AA Dordrecht, The Netherlands.

www.springer.com

Every effort has been made to contact the copyright holders of the articles which have been reproduced from other sources. Anyone who has not been properly credited is requested to contact the publishers, so that due acknowledgements may be made in subsequent editions.

Printed on acid-free paper

Printed in the Netherlands.

Table of Contents

Biographical Data

1916 December 21: John Douglas Eshelby born in Puddington, Cheshire, United Kingdom.

1937 First Class Honours in Physics, University of Bristol, UK.

1950 Ph.D. in Physics, University of Bristol, with a thesis on "Stationary and Moving Dislocations".

1953 Research Associate in the Physics Department at the University of Illinois at Urbana, USA.

1954 Lecturer in the Department of Physical Metallurgy at the University of Birmingham, UK.

1963 Visiting Professor at the Technische Hochschule and the Max Planck Institute in Stuttgart, Germany.

1964 Senior Visiting Fellow in the Metal Physics group at the Cavendish Laboratory, Cambridge University, and thereafter Fellow and Lecturer at Churchill College, Cambridge, UK.

1966 Reader in the Department of the Theory of Materials at the University of Sheffield.

1971 Elected to a Personal Chair in the Department of the Theory of Materials, University of Sheffield.

1974 Elected Fellow of the Royal Society for being "distinguished for his theoretical studies of the micromechanics of crystalline imperfections and material inhomogeneities".

1977 Awarded the Timoshenko Medal by the American Society of Mechanical Engineers.

1981 December 10: Dies in Sheffield, UK.

For a detailed biography the reader is referred to: "John Douglas Eshelby. 21 December 1916–10 December 1981", by B. A. Bilby, *Biographical Memoirs of Fellows of the Royal Society*, Vol. 36, 1990, pp. 126–150.

Preface

My task here is to guide the reader as to the contents of this volume, *Eshelby's Collected Works* rather than to try to evaluate the significance of Eshelby's work, which I could not properly do anyway (this being left to history).

Because people have always been interested in the lives and personalities of those whose originality and creativity were outstanding, we have included in this volume Eshelby's handwritten notes and the poem *Albert and the Lion*, and asked David Barnett, Bruce Bilby, Alfred Seeger and John Willis to write a foreword to this volume. They all wrote their reminiscences of Eshelby as a scientist and as a person, and, as I am reading them, Eshelby's spirit is coming alive.

I was very fortunate that I had the chance to meet Eshelby during his visit to Brown University in the fall of 1977. I was too young at the time to fully appreciate his remarkable lecture on the energy-momentum tensor, the handwritten distributed notes of which I enclosed here. I had just started (what developed into a life-long endeavor toward the effective mass of a moving dislocation) to look on dislocations moving from rest nonuniformly, and I consulted with him as to the approach to follow. He was generous with his time and very kind to me. He sketched his thoughts with pencil on several pages, asserting each piece of the superposition firmly on a figure (the scanned pages are included below). In the years to follow, I spent long hours reading and re-reading his papers, mainly those on the forces on elastic singularities, inclusions, and moving dislocations, deriving immense pleasure. Intrigued by his ellipsoidal inclusion problem, I was able to prove that the ellipsoid perturbs into another ellipsoid if it has to maintain the (constant stress) Eshelby property, and that the shape with the Eshelby property has to be a nine-dimensional manifold.

Eshelby has brought his own thinking to the mechanics of defects and inhomogeneities. Understanding Eshelby's papers was for me like learning a new language. The purpose of this book is to bring his work together, because it has a unique cohesiveness of thought and approach with application to a wide range of phenomena. His "force on an elastic singularity" concept has led to a whole new configurational mechanics of continua concerned with driving forces on moving defects, boundaries, phase transformations, etc. (e.g., conference proceedings on *Mechanics of Material Forces*, Springer, 2005), and his ellipsoidal inclusion solution had immense impact on the modeling of the mechanical properties of metals and composite materials (e.g., T. Mura, *Micromechanics of Defects in Solids*, 1987; S. Nemat-Nasser and M. Hori, *Micromechanics: Overall Properties of Heterogeneous Materials*, 1999). As written in his London *Times* obituary on January 2, 1982, "he had no time for useless erudition". His papers go to the heart of the physics of the problem with a depth of insight that is of immense beauty.

Eshelby has not written a book, but several review articles are like monographs and take the place of the book(s) he has not written. These review articles are references

#17: The continuum theory of lattice defects (1956);
#26: Elastic inclusions and inhomogeneities (1961);
#36: Stress analysis: theory of elasticity (1968);
#37: Stress analysis: fracture mechanics (1968);

#39: Dislocations and the theory of fracture (1968), with B.A. Bilby;
#43: Energy relations and the energy-momentum tensor in continuum mechanics (1970);
#47: The calculation of energy release rates (1975);
#48: Point defects (1975);
#53: Boundary problems, in *Dislocations in Solids* (1979);
#56: Aspects of the theory of dislocations (1982).

Eshelby's scientific contributions can be grouped into four main areas of solid mechanics and materials:

- Forces on elastic singularities and the energy momentum tensor,
- Fracture mechanics,
- Inclusions and inhomogeneities,
- Dislocations and point defects.

A brief description of each of these contributions follows.

Forces on elastic singularities and energy momentum tensor (references: #6, 17, 24, 43, 47, 50, 55)

In these papers is presented the energy momentum tensor and the conserved integrals that lead to "a force on an elastic singularity". Eshelby created an ingenious thought experiment that constructs the difference in the energy (of the whole system), when the defect is displaced infinitesimally, as the difference in the work of the tractions on a surface surrounding the defect that is cut, deformed in the replica, and re-inserted. This difference in the work of the tractions on the surface of the cut-reweld experiment equals the change of energy of the whole system when the defect moves infinitesimally, and is expressed as a surface integral of the energy momentum tensor. If it is zero, there is no elastic singularity, if it is not, then it can be interpreted as a configurational force acting on the defect. The power of this is that any singularities in the neighborhood of the defect are circumvented, and that the surface integral is independent of the surface chosen. The expression for the force on a defect given in #24 in 1959 is in the form of the J integral independently discovered by Rice in 1968 for notches and cracks, which had sweeping applications to fracture mechanics.

Fracture mechanics (references #37–45)

In fracture mechanics Eshelby considered the crack both as a singularity in itself, and as a distribution of dislocations, applying to it the energetics that he developed for energy-release rates. In the review article with Bilby we have an excellent exposition of fracture mechanics in general, including the Barenblatt crack, and the Peach–Koehler force on a dislocation. In dynamic fracture mechanics, Eshelby (with Atkinson) derived the energy flux into a moving crack tip (#38) in order to establish an energy balance equation from which the motion of the crack can be determined (#42, 43). He did this for dislocations too, as early as 1953 (#10, 43) which is a more difficult problem, since the dislocation has an effective mass while the crack does not, stating that "the dislocation is haunted by its past". In #40, he derived the elastic fields for a mode III propagating crack, a detailed account of which is given in Willis' foreword.

Inclusions and inhomogeneities (references #18, 25, 26, 31, 32, 49, 51)

In references #18 and #25 Eshelby attacked the problem of determining the elastic fields in a material containing a region (called inclusion) of "transformation strain", or "stress-free strain", due to thermal stresses or phase transformation, which would produce no stress if the deformation were unconstrained, but it does when constrained inside a matrix. Eshelby's thought follows similar steps of imaginary cutting, straining and rewelding operations, essentially bringing the inclusion to an undeformed shape by the application of a layer of surface body forces, which are cancelled by adding the opposite when the inclusion is re-fitted in the matrix. Based on the properties of potential functions, Eshelby showed that the strain is constant inside an ellipsoidal inclusion with a uniform eigenstrain, the two of them being related by the Eshelby tensor that carries the geometric information. The importance of the constant strain inside an ellipsoidal inclusion is that it allows finding the eigenstrains in an "equivalent inclusion" problem, which is equivalent to the one of a far field being perturbed by an ellipsoidal region of different elastic constants, an "inhomogeneity". Knowledge of the outside stress field (#25) allows accounting for the interaction of inclusions (#32), where the concept of the force on an elastic singularity, and the corresponding energetics, is applied to the defect being the inclusion/inhomogeneity. Important applications of the ellipsoidal inclusion solution are the limiting cases of penny-shaped cracks and dislocation loops, for which the eigenstrain is a delta function normal to the slip-plane. Publication #18 in 1957 is the most cited reference in solid mechanics in the last fifty years and has been called by many the elasticity solution of the century. Indeed, the constant strain/stress property, coupled with the generality of the ellipsoidal geometry to approximate a variety of shapes, had far reaching applications to the modeling of the effective elastic properties of composites containing fibers, reinforced particles, voids, phase transformations, etc.

Dislocations and point defects (references #1–3, 5, 7–16, 19–22, 27–31, 33–35, 39, 46, 48, 52, 53, 56)

In the area of dislocations and lattice defects Eshelby contributed also significantly. He provided some fundamental elasticity solutions for single dislocations in isotropic and anisotropic solids (#3, 8, 33, 56), and in bounded media including whiskers (#7, 9, 11, 19, 21, 53), for single moving dislocations (#2, 10, 16, 28, 46, 56), for the interaction of dislocations with external stress fields (#17, 56) and other physical phenomena (such as heat, elastic waves, ionic charges, etc., #15, 20, 30, 34), as well as fundamental solutions for equilibrium positions of arrays of dislocations and the forces in dislocation pile-ups. The review article #17 (1956) gives an exposition of single dislocation theory and the Kröner theory of continuously distributed dislocations, presenting the energetics of dislocation interactions with external fields and boundaries in the more general context of interaction energies between stress systems. Eshelby revisited dislocation theory (#56, 1982), giving to it his own characteristic perspective of geometry and energetics and extended the Peach–Koehler force to account for the self-force that a dislocation loop exerts on itself. For terminating dislocations (#35), borrowing from electromagnetism again, he modeled them as Kröner incompatibilities. For moving dislocations, Eshelby formulated the equation of motion of a dislocation under an external stress field (#10), concluding that because of the logarithmic singularity induced by the acceleration, a cut-off radius implies the need for atomistics (#43). In the area of lattice point defects Eshelby contributed the fundamental solution of the change of volume of the body due to interstitial atoms (or a uniform distribution of point defects) accounting for surface image effects, and also providing a study of their interactions (#12, 13, 14, 17, 48, 52).

In the fall of 2003, while I was visiting the University of California, Berkeley as a Miller Professor, for which I am very grateful (this volume being one of the reasons), I met and collaborated with Mr. Anurag Gupta, a young PhD student in Solid Mechanics. We worked on higher order conservation laws based on the Beltrami–Michell equations, and Anurag became truly enamored with Eshelbian mechanics and configurational forces. He proposed that we edit together *Eshelby's Collected Works*, and, when initially we were not sure we could find a publisher, he suggested that we just publish them on the web anyway.

His enthusiasm has been unbounded, and I am very happy to see the young generation thrilled about solid mechanics, a sign that it will endure in time because its appeal, beauty and power are eternal.

Xanthippi Markenscoff
La Jolla, California
March 12, 2006

On x-axis

$t<0$ $w=0$

$t>0$ $\Delta u =$

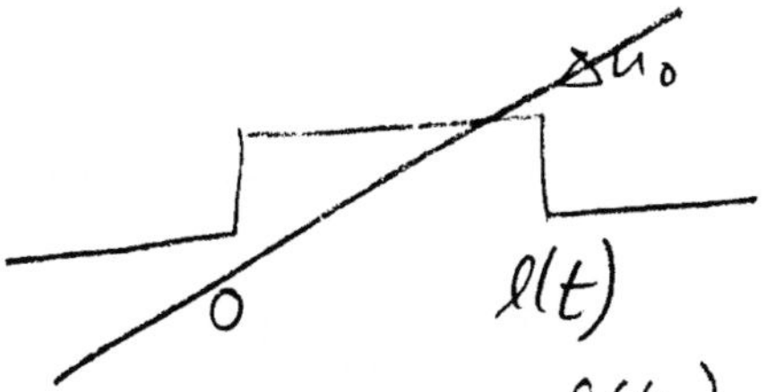

$\Delta u_1 =$

$\Delta u =$

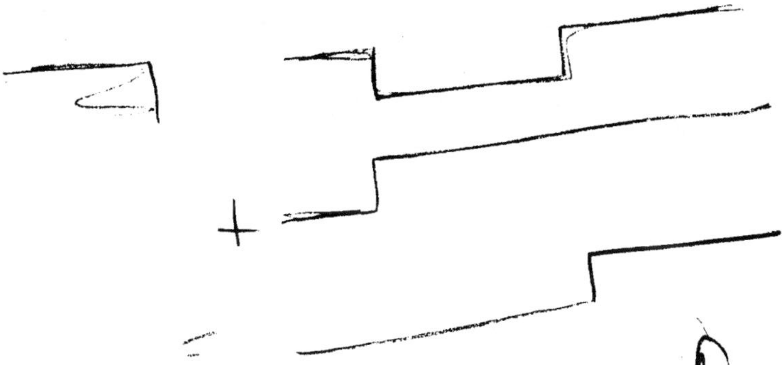

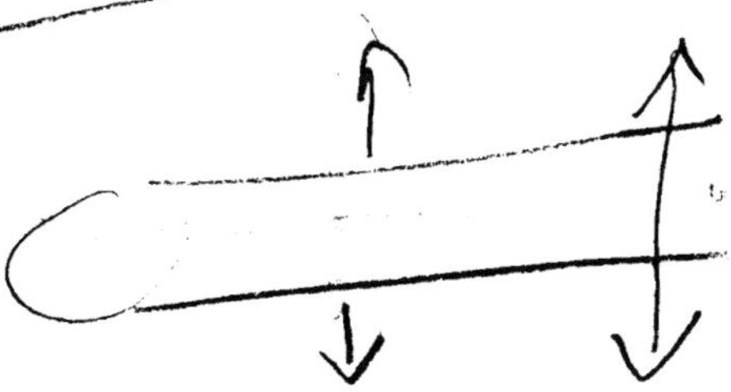

$x \gg ct$

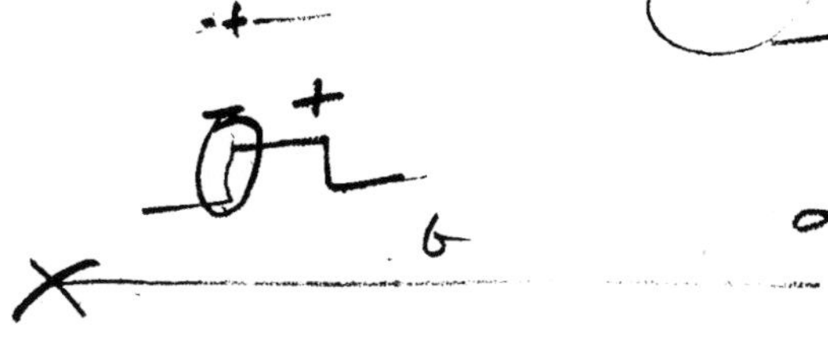

∞ ~~$u_y = f($~~

$u_y = H(ct-y)$

$f(y,t)$

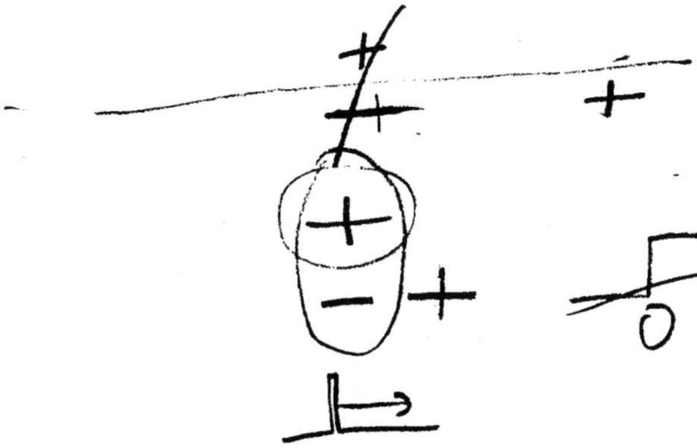

$\Delta u = H(-x)\,H(x-l(t))$

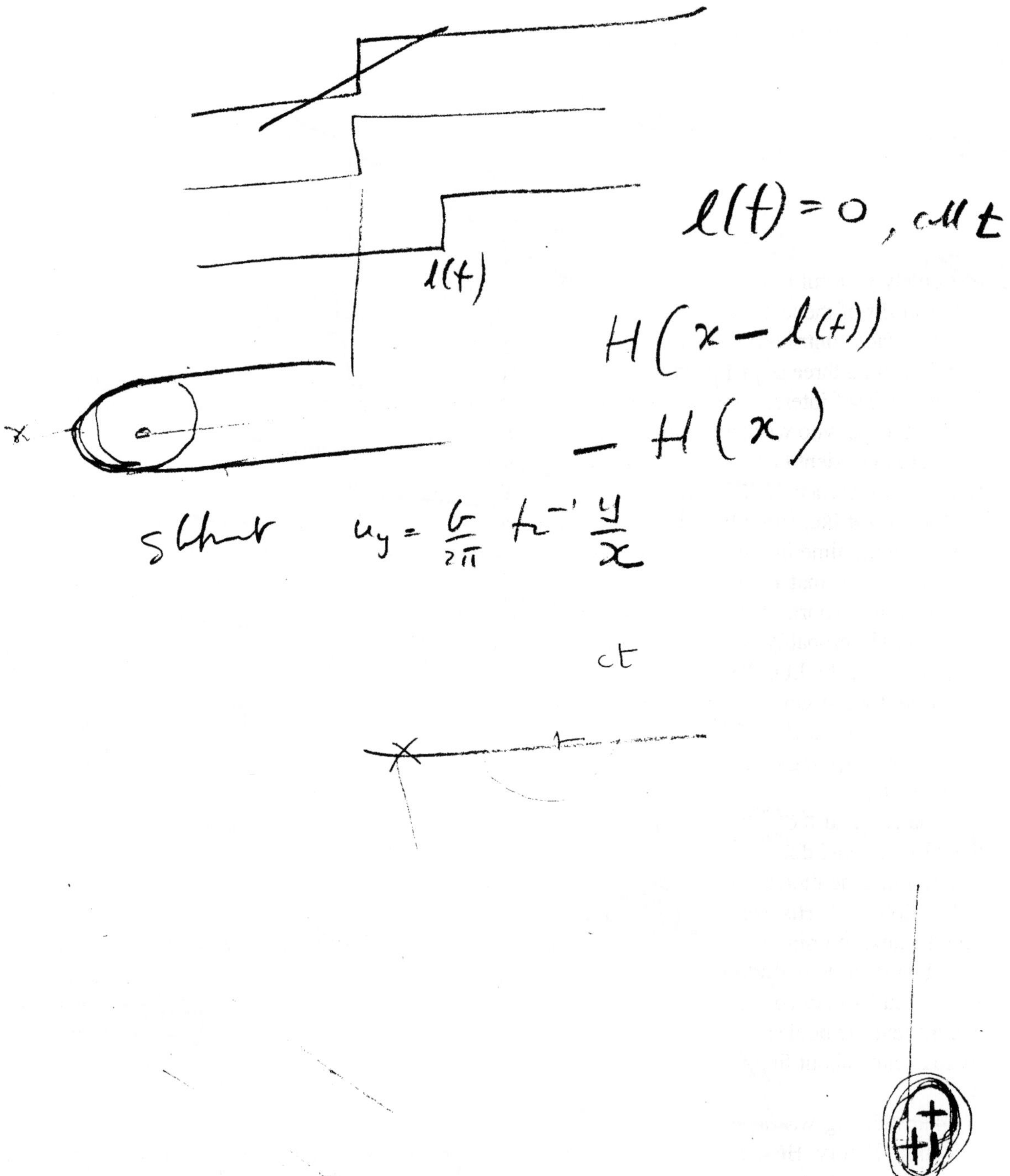
l(t)
l(t) = 0, all t
H(x − l(t))
− H(x)
ct

Foreword by D. M. Barnett

I am extremely grateful to Professor Markenscoff and Mr. Gupta for collaborating to produce this collection of the works of the late John Douglas (Jock) Eshelby, and I am very flattered to have been asked to be one of the preface contributors for the collection. I cannot pretend to have known Professor Eshelby well, but we did meet on three occasions and corresponded on one. Given space limitations, I would simply like to describe my brief interactions with him and try to communicate the tremendous influence he had upon me and upon those who work at the materials science/solid mechanics interface.

As a doctoral student I had traveled from Stanford to Ottawa, Canada in the summer of 1966 to attend an international conference on the deformation of crystalline solids. All major US airlines except American Airlines were on strike, and it had taken me almost two days to reach Ottawa via LA, Dallas, and Chicago including sleeping time in the shower rooms at Love Field and O'Hare airport (which one could rent for 25 cents an hour at that time). I was quite bleary-eyed upon arrival at the conference registration and sat down next to a short, reddish-haired gentleman who glanced at my name tag and realized he should affix his own. He probably sensed my mouth preparing to drop when he said, "I hope you aren't going to say, 'You're not *the* J.D. Eshelby!'." I allowed as to how I was, but that, having been properly warned, I wouldn't, and we struck up a nice conversation. I still remember the most beautiful talk he gave about work done by Theo Laub and himself on terminating dislocations. Years later, after reading Hardy's "A Mathematician's Apology", I recalled this first encounter and realized what Hardy meant by "people in the Bradman class".

A year later I had the pleasure of meeting Eshelby again at a 1967 meeting on generalized continua organized by my post-doctoral host, the late Professor Ekkehart Kroner, partly in Freudenstadt and partly in Stuttgart. I had mentioned to Eshelby my fondness for an old cartoon about Albert and the Lion narrated by Stanley Holloway. His eyes lit up as he recounted how his father used to recite that particular piece to him, and the next morning he brought me a written version of the poem, as he recalled it. I still have it on my office board, an Eshelby original written on stationery from the Hotel and Kurhaus Max Laufer zum Rappen in Freudenstadt (presented here at the end of this forward). Later, in Stuttgart, I accompanied him to one of his favorite bookstores, where he bought a copy of Lie Group Theory for Pedestrians (in Russian). He was so excited about finding the book that I had to remind him to pay for it before accidentally leaving with it.

Our final meeting was at the 1969 National Bureau of Standards meeting on Fundamental Aspects of Dislocation Theory. He did not present a paper there, but he was involved in a delightful interchange about the reality (or lack thereof) of a so-called Rosenstock–Newell solid. (The interchange appears in the conference proceedings and is certainly worth perusing.) I should add that Eshelby had been invited to Stanford by George Herrmann, my chairman in applied mechanics, to visit us and give us some lectures in 1982. Unfortunately, that visit was never to take place, as Jock died weeks before its scheduled occurrence.

It is virtually impossible to describe the legacy of Jock Eshelby, but I can try to

relate what I believe to be his influence on me. The first archival paper I really delved into was by Eshelby, Read, and Shockley on straight dislocations in anisotropic linear elastic media; I still have my considerable notes on that paper in my office. I think, however, it was his work on energies of and forces on elastic singularities which stuck with me the most. Firstly, having been trained as a mechanical engineer, our courses on elasticity always emphasized generating pointwise solutions for elastic fields. Eshelby's work showed me solutions in this depth were often not necessary if, as was usually the case in physics and in materials science, energies and their variations when defects underwent virtual displacements were really the objects of interest. I always marveled at his ability to extract the maximum amount of useful information with the minimum expenditure of effort. More importantly, his work showed me, and virtually everyone else, that we must never forget to include the loss of potential energy of the loading mechanism in reckoning the total potential energy of a deformed solid. An understanding of his treatment of elastic interaction energies is essential for teaching young students how one computes these quantities correctly (and with the correct sign).

Of the research I have been involved with, I must admit to a certain degree of pleasure with a very short note I drafted with Steve Gavazza on the "image" part of the Peach–Koehler force on a dislocation loop element in a linear elastic solid with a free surface. If one is not careful, one obtains one-half the required result (the result is well-known, but the derivation is non-trivial); we published the work, in part to show that we had understood what Jock had taught us. Eshelby referred to our manuscript in his paper on Boundary Value Problems, and I think he liked our efforts. Steve and I took much satisfaction with that.

Unlike many of my colleagues in solid mechanics, I found his "cutting and welding" arguments far preferable to the more formal approaches common to the solution of boundary value problems. His thinking struck me as being more akin to that of a physicist than of a solid mechanician, and I guess I simply preferred his approach. One of my biggest fears is watching modern materials science education moving away from familiarizing students with analysis of the type pioneered by Jock. I have no doubts that defect research will persist, but I think it will occur within the solid mechanics community, and perhaps that is good news; the possible bad news is that Eshelby's unique brand of reasoning may disappear. Perhaps this volume of his collected works will play some role in reviving "Eshelby-think". If so, the volume will have served its purpose.

HOTEL UND KURHAUS MAX LAUFER ZUM RAPPEN
SEIT ÜBER 125 JAHREN IM FAMILIENBESITZ · DIREKTION: HANS KARR

7290 FREUDENSTADT,

<u>Albert and the Lion.</u>

There's a famous seaside place
called Blackpool,
Which is famous for fresh air and fun,
And Mr & Mrs Ramsbottom were
there with young Albert, their son.
A fine little lad was young Albert, all
dressed up as a swell, with a stick with
an 'orses 'ead 'andle, the finest
that Woolworths could sell.
They didn't think much to the ocean, the
waves they was fiddlin' and small,
there were no wrecks, nobody drownded, 'fact
nothing to laff at at all.
So seeking for further amusement they paid
and went in to the Zoo, where there's lions and
tigers and camels, and old ale and sandwiches too.
There were one gurt big lion called Albert,
'is face were all covered with scars,
and he lay in a somnolent

Telefon (07441) 35 03 / 04 · Drahtanschrift: Rappenhotel · Postscheckkonto: Stuttgart 9680 · Kreissparkasse Freudenstadt 495 · Volksbank Freudenstadt 851

2

HOTEL UND KURHAUS MAX LAUFER ZUM RAPPEN

SEIT ÜBER 125 JAHREN IM FAMILIENBESITZ · DIREKTION: HANS KARR

7290 FREUDENSTADT,

posture with 'is face 'gainst t' bars.

Now Albert had heard about lions,
how they was ferocious and wild,
and to see Wallace sleeping so
peaceful, well, it didn't seem
right to the child.

So straightway the brave little
fellow, not showing a morsel of fear,
took 'is stick with the 'orses 'ead 'andle,
and shoved it in Wallace's ear.

You could see the lion didn't
like it, for giving a kind of a roll,
'e pulled Albert inside the cage with
'im, and swallowed the little lad whole.

Now Pa, who had seen the occurrence,
and didn't know what to do next,

3

HOTEL UND KURHAUS MAX LAUFER ZUM RAPPEN

SEIT ÜBER 125 JAHREN IM FAMILIENBESITZ · DIREKTION: HANS KARR

7290 FREUDENSTADT,

Said "Mother, yon lion's 'et Albert",
and Mother said "Ee, I am vexed".
The Keeper was quite nice about
it, he said "What a nasty mishap.
Are you sure its your ~~[illegible]~~ boy 'e's eaten",
and Pa said "Am I sure, 'ere's is cap".

4

HOTEL UND KURHAUS MAX LAUFER ZUM RAPPEN
SEIT ÜBER 125 JAHREN IM FAMILIENBESITZ · DIREKTION: HANS KARR

7290 FREUDENSTADT,

~~Said "Mother, yon lion's eat Albert",~~
~~and Mother said "~~Well~~, I am vexed",~~
~~are you sure it's our Albert he's eaten",~~
~~and Pa said "~~Do~~ Am I sure, there's~~
~~'is cap.~~

The Manager had to be sent for,
he came and he said "What's to do,"
Pa said "Yon lion's eat Albert, and
'im in 'is Sunday best, too".

The Manager wanted no nonsense,
he took out his purse right away,
Said "How much to settle the matter?"

Pa said "How much do you
usually pay".

But Ma had turned a bit
~~awkward~~ awkward when she thought
where her Albert had gone. She

HOTEL UND KURHAUS MAX LAUFER ZUM RAPPEN

SEIT ÜBER 125 JAHREN IM FAMILIENBESITZ · DIREKTION: HANS KARR

7290 FREUDENSTADT,

said "Someone's got to be summonsed",
so that was decided upon.

So off they went to the police court,
in front of the Magistrate chap,
and explained what had happened
to Albert, and proved it by showing
his cap.

The Magistrate gave his opinion,
that no-one was really to blame,
and he hoped the Ramsbottoms
would have further sons to their name.

At this Mother got proper blazin';
~~[illegible]~~ "thank you kindly" said she
"what, spend all our lives raisin'
children to feed ruddy lions, —
not me!"

Foreword by B. Bilby

The writer was pleased to learn of this project and to be invited with others to write a brief foreword to it. He has already written twice at some length about Jock and his work (Bilby 1985, 1990) and will, as requested, talk here mainly about the man himself. Jock's work, which speaks for itself, was a great part of his life and is his true monument. It will certainly be very useful to have his papers collected together in a handy form. These papers will remain essential reading for any serious researcher entering the field for some time to come. For they are of a very special kind. He took great care over their writing and his hope of each was that it might be "a little gem". Indeed, as the writer has said, many are treasure houses, to be studied in depth: full of asides and conjectures, with hints for future research. A characteristic of his work was his frequent use of concepts in one field to guide him to the solution of problems in another, so that his papers abound in interesting analogies. Jock regarded himself as "a humble supplier of tools for the trade" and would often, in a few sentences, outline extensions or consequences of the ideas he had formulated, leaving their detailed development and application to other craftsmen. In the years since his death there has been a steady flow of references to his work and of elaborations to it. There is little doubt that this process will continue.

The writer's first encounter with Jock was at Bristol in 1947 or 1948: he sat at a table in an airy room, studying a paper in a thick volume of the Philosophical Transactions propped efficiently before him on a large book-rest. Thereafter we met at conferences and exchanged papers until 1966, when he joined the Department of the Theory of Materials at the University of Sheffield. In those days there was more time in Universities and Jock's room soon became a centre for discussion before the blackboard at coffee and tea time, when much valuable work was done. Here he would often enliven the argument with his dry wit. Typical of his humour were comments such as "When I was young I used to think I was indecisive, but now I'm not so sure" and "I would go to any lengths to avoid a bit of trouble". Uninteresting matters might raise him "to a high pitch of indifference". Of some obscure result hindering the main thrust of debate he would say, "It's obvious! I forget exactly why".

Discussing a problem, he often urged "Look at it my way – the right way". He believed, like Willard Gibbs, that the objective of theoretical physics was to make things look as simple as possible. Indeed he was very aware of the importance of consulting the right books and learning from the right teachers, perhaps because of the way he himself had been obliged to learn. For when he was about 13 he had rheumatic fever and never went to school again. His education was undertaken privately by tutors – the village parson and schoolmaster – and so he had, in many ways, to "find things out for himself". He displayed an exceptional ability to do so. It is interesting to speculate that the fact that he had largely to educate himself until he went to study Physics at the University of Bristol, may well have contributed to his becoming such an original and creative thinker. Another characteristic, shown when grappling with a thorny problem before the blackboard, was his irritation with any interruption foreign to the discourse. "Don't stop the flow", he would cry, well knowing that the atmosphere and conditions enabling one to grasp some difficult point would not easily recur.

He chose with care the problems on which he worked, selecting those which were topical and difficult, but just sufficiently tractable for solution at the time. He liked to think deeply about the implications of his calculations and his general advice about any problem was not to spoil it by undue haste: "Don't rush it!" he would say. He was very conscious of the power of analytical methods in revealing relationships and it pleased him to puzzle out solutions; with tongue in cheek he would say that a numerical attack was "not quite fair". One can only speculate about what he might have done with the computing power now available. Anticipating with some disquiet the consequences of the chip, he "wondered what all the worthy clerks would do for a living" and jested that he "was rather relieved that the only paper he had published with an inventor of the transistor was not about semiconductors". But he was eager to use the early electronic calculators – and, characteristically, discovered an error in the arctan function of one of them.

Jock was a great collector of secondhand books and liked to snap up a bargain – "I got it in a sale" he would say. He had a comprehensive library of scientific and technical works in his field (where he generally had 'the' book on a subject – perhaps because of the nature of his own early learning). An amusing story concerns his own copy of Mott and Gurney, which he lent to some student and then, as often happens, could not recall to whom he had lent it. So in his prowling round the bookshops he kept an eye open for another copy. Sure enough, one day when he was in Poole's shop in Charing Cross Road he spotted a copy on the top shelf. Retrieving it with some difficulty, he was somewhat put out to find his own signature on the fly-leaf! Jock's academic interests ranged over a wide field and he amazed his arts colleagues with his general knowledge and his erudition in their own subjects. He was very interested in philology and had a collection of books on this topic, as well as classic texts, readers, dictionaries and grammars, covering many languages. Another interest was archaeology and the Ancient World. He once said that he might well have been an assyriologist, if he had not found his way into physics.

Eshelby was a retiring man, with a shyness sometimes concealed by a gruff exterior. But he was always willing to help the genuine enquirer. A great admirer of Lord Rayleigh, he was something of a gentleman scientist himself. A bachelor and a scholar, his needs were simple and he was not much concerned with personal advancement.

Although he was steeped in the classical physics of the nineteenth century and most of his papers are mainly concerned with classical concepts, the paper on the interaction of kinks and elastic waves contains a masterly example of the application of quantum theory ("All physics is there", a distinguished colleague once said of this paper). Jock will be remembered as a master of the theory of dislocations and point defects; and for his studies on the energy-momentum tensor of elastic and other continua, on the theory of inclusions and inhomogeneities and on the equations of motion of elastic cracks. He liked to think of his most important works as "amusing applications of the theorem of Gauss" and once introduced himself as "a theoretical physicist – one who does no useful work". Despite these sayings he was very critical of useless erudition; "This result has been obtained before but is derived more elaborately here": and also of 'bandwagons'; "Whenever I see a bandwagon my first instinct is to let its tyres down". His first demand of a theory was "Make it do something!" He has always written with an eye on practical ends, providing a sure way for scientific advance by creating incisive papers on the topics that interested him. This is useful work indeed, for these papers are a scholarly foundation on which others can build, and have already led to wide application of his results.

It may well be that this volume will encourage more craftsmen to further hone and use the tools he has provided; and even, perhaps, will act as a spur to other toolmakers.

Bruce Bilby
Sheffield, 27 March 2006

Bilby, B.A., 1985, Introductory Lecture, in *Fundamentals of Deformation and Fracture*, International Union of Theoretical & Applied Mechanics Eshelby Memorial Symposium, Sheffield, 2–5 April 1984, B.A. Bilby, K.J. Miller & J.R. Willis (Eds), Cambridge University Press, pp. 11–32.

Bilby, B.A., 1990, *John Douglas Eshelby. Biographical Memoirs of Fellows of the Royal Society*, Vol. 36, 127–150.

Foreword by A. Seeger

John Douglas Eshelby and the Elasticity Theory of Crystal Defects

In the nineteenth century, from about 1820 onwards, the theory of elastic solids was the *pièce de résistance* of theoretical physics, as evidenced by the work of Louis Marie Henri Navier, Siméon-Denis Poisson, Augustin-Louis Cauchy, Rudolf Emanuel Clausius, Gabriel Lamé, Adhémar Barre de Saint Venant, George Gabriel Stokes, George Green, Gustav Robert Kirchhoff, William Thomson (Lord Kelvin), Valentin-Joseph Boussinesq, and many others. The early work, up to about 1830, was based on the assumption that solids consisted of a regular arrangement of molecules (called "molecular hypothesis"). When the difficulties of making the transition to isotropy were realized, the emphasis shifted to the continuum description. In the second half of the century the continuum mechanics of *crystals* was particularly advanced by Franz Neumann and Woldemar Voigt.

In the early twentieth century, the mainstream of theoretical physics turned to quantum and relative theory and to their application to molecules, atoms, and nuclei. Elasticity became part of mechanics or applied mathematics, which was now considered as a neighbouring field rather than a central part of physics. Only very gradually was it realized that many mechanical properties of solids, such as the wide range of the phenomena of plastic deformation, of internal friction, and of their recovery during heat treatment, could not be understood without considering the atomic structure of crystalline materials, which had become accessible in 1912 through the X-ray work of Max von Laue and his associates. Simple atomistic models for the non-elastic behaviour of crystalline materials were subsequently proposed, in an ad-hoc manner, by Ludwig Prandtl, Ulrich Dehlinger, Jakov Il'ič Frenkel', and others. The key idea was that real materials contained localized perturbations of the regular crystal structure and that these "lattice defects" were able to move through the crystals under the influence of thermal agitation and external forces.

The link between the atomistic models and the classical work on continuum mechanics of elastic bodies was established by G.I. Taylor (1934) and J.M. Burgers (1939). The underlying concept may be described as follows. The perturbations held responsible for the non-elastic behaviour of crystalline materials may be two-dimensional (e.g. grain boundaries separating adjacent crystallites), one-dimensional (e.g. dislocations), or zero-dimensional (so-called atomic defects such as atoms inserted into interstices of the crystal structure [interstitials] or vacant lattice sites [vacancies]). They are embedded in a matrix in which the ideal crystal structure is preserved apart from small strains created by the defects. Therefore, in the bulk of the material the linearized theory of elasticity remains applicable. The attractive consequence of this concept is that quantities pertaining to the ideal crystal, such as elastic stiffnesses and compliances, can be brought into the quantitative descriptions of the defects. However, it is obvious that the continuum mechanics approach is bound to fail at and in the immediate neighbourhood of the defects. A characteristic of theoretical physics is that when a simplified theory fails (here the theory of elastic continua) or an adequate general theory is not known (a striking example is the "big bang" in cosmology), the untreatable or unknown is replaced by a singularity in the simpler theory. In the present case this raised the question of how intuitive descriptions of lattice defects may be quantitatively related to properties of the singularities appearing in the continuum description.

In 1946, John Eshelby entered the field through his Ph.D. work at Bristol University after military service in World War II. He soon became one of the leading figures, if not *the* leading one. He distinguished himself by searching for general answers without losing the contact with the practical problems to be solved and by very lucid expositions of his results. Many of the new concepts in continuum mechanics that emerged in the second half of the twentieth century were due to John Eshelby or at least influenced by him. Quite a few of them have proved to be of great importance to material science and should be taught in university courses. Examples are what Eshelby called the "Maxwell tensor of elasticity" and its relationship to the forces acting on the elastic singularities, the relationship between the lattice parameters and the macroscopic dimensions of distorted crystals, the *Eshelby factor* connecting A.E.H. Love's "centres of compression or dilatation" to the change of the volume of a sample of finite size into which they have been introduced, and – perhaps most important of them all – Eshelby's by now classical results on ellipsoidal inclusions whose elastic constants differ from those of the matrix. These results form the theoretical basis of much of the modern work on laminated structures and precipitation-hardened materials.

The scientific community is still waiting for the book in which the developments sketched above are collected and made accessible to a wider readership (including students of materials science), particularly now that eigenstress and eigenstrain problems are of growing importance in the miniturization of electronic devices. With his sense for the essential, John Eshelby would have been the right person to write such a book. During our contacts, about which I will more presently, I urged him repeatedly to do so, perhaps in the style of the fourth edition (Oxford 1927) of A.E.H. Love's *A Treatise on the Mathematical Theory of Elasticity*, which we both valued very highly. Jock's untimely death put an end to my hopes that in this way his unparalleled insight into the subtleties of the application of continuum mechanics to solid state physics might be passed on for use by future generations. Fortunately, on several topics he published excellent reviews. I like to consider the present collection, which contains these reviews together with Eshelby's original papers, as a substitute for a book that was never written.

When John Eshelby's first papers on dislocation theory appeared in 1949, I had been working in this field for about a year. Anxious to know where I stood with my own thinking, I studied the Eshelby papers carefully. Shortly before I took up a British Council scholarship at the H.H. Wills Laboratory of Bristol University on 1 October 1951, I was most impressed by the elegant mathematics of "The equilibrium of linear arrays of dislocations", by J.D. Eshelby, F.C. Frank and F.R.N. Nabarro (*Philos. Mag.* **42**, 1951, 351–364), now a classic. Naturally, I was looking forward to meeting the authors. By the time I arrived in Bristol, Frank Nabarro had already left for Birmingham University, but John Eshelby and Charles Frank were there.

In Bristol, the theoretical Ph.D. students and post-docs under the general guidance of Nevill Mott were housed on the first floor of the Royal Fort House, a separate building next to Stuart House, where Professor Mott and his family lived. A desk was assigned to me next to the office in which John Eshelby and Jacques Friedel worked. Jock (as Eshelby was called by this friends) and I soon became close friends. Not only were we interested in similar scientific questions (though usually using different approaches to the same problem, as for instance when treating dislocations in elastically anisotropic media or in the kink mobility problem) but we also had several hobbies in common, such as modern languages and history. We also discovered a common feature in our background. Due to circumstances (ill health in Jock's case, war and post-war conditions in mine), we had had to rely on our curiosity for filling the gaps in our formal education by self-teaching.

The post-docs, Ph.D. students, and visitors of the theoretical group in Bristol came from all over the world. This created an immensely stimulating intellectual atmosphere. For example, the initial stimulus for Eshelby's masterful solution of the inclusion problem was brought by Jacques Friedel from the lecture course on mechanics at the École Polytechnique at Paris (one of the centres of higher education in which the classical theory of elasticity was still held in high esteem). If my memory serves me right, Jock was the only member of the group who had graduated from Bristol University. (On one occasion Nevill Mott confided to me that in his opinion Bristol was not the right place to breed good theoretical physicists, Jock being the exception that proves the rule.) Moreover, Jock had grown up in the area (in Northern Somerset). He introduced me not only to English customs and to the niceties of the English language but also to the country around Bristol, including Somerset cider, to which I might have become addicted had I stayed in Bristol much longer.

Jock was my senior by almost eleven years. Having grown up in intellectual isolation during the war and having lost my father early, I considered Jock a fatherly friend in some respects. His advice was always unselfish and generously given. When I presented him with a result of which I was proud because of the elegant mathematics I had made use of, he would bring me down to earth by dryly asking "and what does it do for you?" It came as a shock to me when sometime in 1952 Jock told me that at the invitation of Frederick Seitz he was going to depart for a temporary position at the University of Illinios. In addition to personal reasons, this contributed to my declining Nevill Mott's offer to prolong my stay in Bristol for another year.

For several years, our contacts remained confined to meeting at conferences and exchanging letters and papers. I was therefore very glad when, after having been appointed to a chair at Stuttgart University, in 1963 I could arrange a visiting professorship for Jock. In addition to letting us profit from his deep insight into the physics of elastic singularities, during his stay in Stuttgart Jock did a lot to improve the English of our research students. Having somehow managed to carry over his dry British humour to his German, he soon became the central figure of our five o'clock tea. Never before or afterwards was the attendance as good as during Jock's time. More than four decades later, some of his sayings – often reflecting deep wisdom – are still remembered. It is a pity that a collection of Eshelby jokes that might have been appended to the present work does not appear to exist.

Stuttgart, 4 January 2006

Foreword by J. R. Willis

I first encountered the work of J.D. Eshelby (henceforth, Jock) when I was a graduate student, in 1962. My advisor, Maurice Jaswon, had solved the problem of a cylindrical inclusion of elliptic cross-section in an isotropic medium, and he proposed that I should generalise this to anisotropy. That introduced me to Jock's wonderful ellipsoidal inclusion paper of 1957. I recall feeling some concern, that I was working on something that was already five years old – I had, of course, no appreciation at the time that the paper was a classic, destined to be influential for the next fifty years (so far)! I also had to work hard, for a very long time, before I really absorbed Jock's beautiful "cut and weld" arguments. The effort was well worthwhile, of course, because the insight gained thereby has supported perhaps half of my subsequent career. I solved, in fact, in 1964, the ellipsoidal inclusion problem for an anisotropic medium, and was stupid enough not to publish it (Rodney Hill referred to it in a footnote in one of his papers, in 1965. The solution has been found – and published – by several others and appears, for example, in Toshio Mura's book on Micromechanics). The property that I exploited was the homogeneity of the Green's function. Eshelby himself had recognised that this property would ensure that the strain inside the inclusion would be uniform but he did not complete the development. I did nevertheless build on the insight so gained, in solving quite a variety of problems for an anisotropic medium, so again I have Jock Eshelby to thank. Naturally, I studied also several others of Eshelby's classic papers, including in particular the paper of 1951 on the force on an elastic singularity, upon which, it seems to me, at least a whole generation of research in metal physics, relating to point defects, has been based.

I was privileged to know Jock Eshelby, from 1965 onwards, when I took a post in Cambridge. I experienced at first hand his modesty, his dry humour and his kindness, and like so many others developed great affection for him, to combine with respect for his scientific accomplishments. In fact, he went through what I believe was a somewhat lean patch around that time: he devoted a lot of effort to a search for explicit solutions for special anisotropic media. Of course he was fully aware of the theorem of Galois but still he hoped to find interesting special cases, and he was at best partially successful. He "recovered", as it were, when he took up the problem of an accelerating crack under "Mode III" loading, exploiting remarkable insight, as well as background knowledge based on earlier work on moving dislocations as well as moving electrons. He found the general solution that he sought, but the problem had in fact been solved previously by B.V. Kostrov. It seemed to me to be a rare lapse for Jock, that he believed for a while that his solution, which assumed a stationary crack under pre-existing load then started to extend, was essentially different from Kostrov's, in which the crack started to move the instant that the load was applied. One solution can be obtained from the other by superposition – a procedure that Jock applied with tremendous effect in much of his other work. I do believe, however, that Jock's solution was crucial to the next development. It was clear, from his exposition, that the solution had a simple property: if the loading on the crack was time-independent, then the field radiated from the moving crack tip was also time-independent, behind the wave front. It was Ben Freund who had the insight (or perhaps the courage) to investigate whether any similar property would be true, for a crack under "Mode I" loading. Ben solved the problem of a crack which was stationary up to a given time, and then moved with constant speed: it turned out that the traction on the line ahead of the crack was indeed time-independent. Hence, it was possible then to change the speed, to a different constant value, and so, by a limiting procedure, to solve the problem of a Mode I crack in arbitrary motion, under time- independent loading. Ben studied also motion under other simple patterns of loading; Kostrov then succeeded in finding the general solution. Thus, Jock had a major influence in the development of the theory of the dynamics of cracks.

I have so far not alluded to one other of Jock's great contributions: his introduction of the elastic energy-momentum tensor and its associated path-independent integral. In its manifestation as the J-integral of J.R. Rice, it is one of the most important entities in fracture mechanics, linear and nonlinear. It is a significant tool in the analysis of phase transformations and related problems. It has a resonance with mathematical concepts related to invariance under groups of transformations (Noether's theorem) and it has spawned a whole new description of mechanics, formulated in "material space". These developments extend far beyond my expertise but surely comment will be made elsewhere.

Considering recent developments, it is clear as well as remarkable that Jock's scientific legacy is even greater than was apparent 25 years ago, when we lost a mentor, an inspiration and a friend.

Full List of J. D. Eshelby's Publications and Acknowledgements

The numbering of the articles below is the same as in the Table of Contents.

#1 Dislocations as a cause of mechanical damping in metals (1949). *Proc. Roy. Soc. London A* **197**, 396–416.
Reprinted with kind permission from the Royal Society, London.

#2 Uniformly moving dislocations (1949). *Proc. Phys. Soc. London A* **62**, 307–314.
Reprinted with kind permission from the Institute of Physics Publishing (http://www.iop.org/journals/).

#3 Edge dislocations in anisotropic materials (1949). *Phil. Magazine* **40**, 903–912.
Reprinted with kind permission from Taylor & Francis Ltd. (http://www.tandf.co.uk/journals/).

#4 The fundamental physics of heat conduction (1951). In *Proceedings of a General Discussion on Heat Transfer*, pp. 267–270, Institution of Mechanical Engineers & ASME.
Reprinted with kind permission from the Council of the Institution of Mechanical Engineers.

#5 The equilibrium of linear arrays of dislocations (1951), with F. C. Frank and F. R. N. Nabarro. *Phil. Magazine* **42**, 351–364.
Reprinted with kind permission from Taylor & Francis Ltd. (http://www.tandf.co.uk/journals/).

#6 The force on an elastic singularity (1951). *Phil. Trans. Roy. Soc. London A* **244**, 87–111.
Reprinted with kind permission from the Royal Society, London.

#7 Dislocations in thin plates (1951), with A. N. Stroh. *Phil. Magazine* **42**, 1401–1405.
Reprinted with kind permission from Taylor & Francis Ltd. (http://www.tandf.co.uk/journals/).

#8 Anisotropic elasticity with applications to dislocation theory (1953), with W. T. Read and W. Shockley. *Acta metall.* **1**, 251–259.
Reprinted from Acta Metallurgica, Copyright 1953, with kind permission from Elsevier.

#9 Screw dislocations in thin rods (1953). *J. Appl. Phys.* **24**, 176–179. Reprinted from Journal of Applied Physics, Copyright 1953, with kind permission from American Institute of Physics.

#10 The equation of motion of a dislocation (1953). *Phys. Rev.* **90**, 248–255.
Reprinted from Physical review, Copyright 1953, with kind permission from the American Physical Society.

#11 A tentative theory of metallic whisker growth (1953). *Phys. Rev.* **91**, 755–756.
Reprinted from Physical review, Copyright 1953, with kind permission from the American Physical Society.

#12 Geometrical and apparent x-ray expansions of a crystal containing lattice defects (1953). *J. Appl. Phys.* **24**, 1249.
Reprinted from Journal of Applied Physics, Copyright 1953, with kind permission from the American Institute of Physics.

#13 Distortion of a crystal by point imperfections (1954). *J. Appl. Phys.* **25**, 255–261. Reprinted from Journal of Applied Physics, Copyright 1954, with kind permission from the American Institute of Physics.

#14 The elastic interaction of point defects (1955). *Acta metall.* **3**, 487–490.
Reprinted from Acta Metallurgica, Copyright 1955, with kind permission from Elsevier.

#15 Note on the heating effect of moving dislocations (1956), with P. L. Pratt. *Acta metall.* **4**, 560–562.
Reprinted from Acta Metallurgica, Copyright 1956, with kind permission from Elsevier.

#16 Supersonic dislocations and dislocations in dispersive media (1956). *Proc. Phys. Soc. London B* **69**, 1013–1019.
Reprinted with kind permission from the Institute of Physics Publishing (http://www.iop.org/Journals/).

#17 The continuum theory of lattice defects (1956). *Solid State Physics* **3**, 79–144. Academic Press.
Reprinted from Solid State Physics, Copyright 1956, with kind permission from Elsevier.

#18 The determination of the elastic field of inclusion and related problems (1957). *Proc. Roy. Soc. London A* **241**, 376–396.
Reprinted with kind permission from the Royal Society, London.

#19 A note on the Gomer effect. Discussion to "Some observations on field emission from mercury whiskers" by R. Gomer. (1958). In *Growth and Perfection of Crystals*, eds. R. H. Doremus, B. W. Roberts, and D. Turnbull, pp. 130–132, Wiley, New York.

#20 Charged dislocations and the strength of ionic crystals (1958), with C. W. A. Newey, P. L. Pratt, and A. B. Lidiard. *Phil. Magazine* **3**, 75–89.
Reprinted with kind permission from Taylor & Francis Ltd. (http://www.tandf.co.uk/journals/).

#21 The twist in a crystal whisker containing a dislocation (1958). *Phil. Magazine* **3**, 440–447.
Reprinted with kind permission from Taylor & Francis Ltd. (http://www.tandf.co.uk/journals/).

#22 The elastic model of lattice defects (1958). *Ann. der Phys.* **1**, 116–121.
Reprinted with kind permission from Wiley-VCH.

#23 Stress-induced ordering and strain ageing in low carbon steels (1959), with D. V. Wilson and B. Russell. *Acta metall.* **7**, 628–631.
Reprinted from Acta Metallurgica, Copyright 1959, with kind permission from Elsevier.

#24 Scope and limitations of the continuum approach (1959). In *Internal Stresses and Fatigue in Metals*, eds. G. N. Rassweiler and W. I. Grube, pp. 41–58, Elsevier, Amsterdam.
Reprinted with kind permission from the University of Sheffield.

#25 The elastic field outside an ellipsoidal inclusion (1959). *Proc. Roy. Soc. London A* **252**, 561–569.
Reprinted with kind permission from the Royal Society, London.

#26 Elastic inclusions and inhomogeneities (1961). *Prog. Solid Mech.* **2**, 89–140.
Reprinted with kind permission from the University of Sheffield.

#27 Dislocations in visco-elastic materials (1961). *Phil. Magazine* **6**, 953–963.
Reprinted with kind permission from Taylor & Francis Ltd.
(http://www.tandf.co.uk/journals/).

#28 The interaction of kinks and elastic waves (1962). *Proc. Roy. Soc. London A* **266**, 222–246.
Reprinted with kind permission from the Royal Society, London.

#29 The energy and line tension of a dislocation in a hexagonal crystal (1962), with Y. T. Chou. *J. Mech. Phys. Solids* **10**, 27–34.
Reprinted from Journal of Mechanics and Physics of Solids, Copyright 1962, with kind permission from Elsevier.

#30 The distortion and electrification of plates and rods by dislocations (1962). *Phys. Stat. Sol.* **2**, 1021–1028.
Reprinted with kind permission from Wiley-VCH.

#31 The distribution of dislocations in an elliptical glide zone (1963). *Phys. Stat. Sol.* **3**, 2057–2060.
Reprinted with kind permission from Wiley-VCH.

#32 On the elastic interaction between inclusions (1966). *Acta metall.* **14**, 1306–1309. (Appendix to a paper "On the modulated structure of aged Ni-Al alloys" by A. J. Ardell and R. B. Nicholson.)
Reprinted from Acta Metallurgica, Copyright 1966, with kind permission from Elsevier.

#33 A simple derivation of the field of an edge dislocation (1966). *Brit. J. Appl. Phys.* **17**, 1131–1135.
Reprinted with kind permission from the Institute of Physics Publishing
(http://www.iop.org/Journals/).

#34 The velocity of a wave along a dislocation (1966), with T. Laub. *Phil. Magazine* **14**, 1285–1293.
Reprinted with kind permission from Taylor & Francis Ltd.
(http://www.tandf.co.uk/journals/).

#35 The interpretation of terminating dislocations (1967), with T. Laub. *Canada J. Phys.* **45**, 887–892.
Reprinted with kind permission from the NRC Research Press.

#36 Stress analysis: theory of elasticity (1968). In *Fracture Toughness*, ISI Publication 121, Chapter 2, pp. 13–29, The Iron and Steel Institute, London.
Reprinted with kind permission from Maney Publishing.

#37 Stress analysis: fracture mechanics (1968). In *Fracture Toughness*, ISI Publication 121, Chapter 3, pp. 30–48, The Iron and Steel Institute, London.
Reprinted with kind permission from Maney Publishing.

#38 The flow of energy into the tip of a moving crack (1968), with C. Atkinson. *Int. J. Fracture Mech.* **4**, 3–8.
Reprinted with kind permission from Springer.

#39 Dislocations and the theory of fracture (1968), with B. A. Bilby. In *Fracture, An Advanced Treatise*, Vol. 1, ed. H. Liebowitz, pp. 99–182, Academic Press, New York.
Reprinted from Fracture, An Advanced Treatise, Copyright 1968, with kind permission from Elsevier.

#40 The elastic field of a crack extending non-uniformly under general anti-plane loading (1969). *J. Mech. Phys. Solids* **17**, 177–199.
Reprinted from Journal of Mechanics and Physics of Solids, Copyright 1969, with kind permission from Elsevier.

#41 Axisymmetric stress field around spheroidal inclusions and cavities in a transversely isotropic material (1969). *J. Appl. Mech.* **36**, 652.
Reprinted with kind permission from ASME.

#42 The starting of a crack (1969). In *Physics of Strength and Plasticity – The Orowan 65th Anniversary Volume*, ed. A. S. Argon, pp. 263–275, MIT Press, Massachusetts.
Reprinted with kind permission from MIT Press.

#43 Energy relations and the energy-momentum tensor in continuum mechanics (1970). In *Inelastic Behaviour of Solids*, eds. M. F. Kanninen, W. F. Adler, A. R. Rosenfield, and R. I. Jaffee, pp. 77–115, McGraw-Hill, New York.
Reprinted with kind permission from the McGraw-Hill Companies.

#44 The fracture mechanics of flint-knapping and allied processes (1971), with J. G. Fonseca and C. Atkinson. *Int. J. Fracture Mech.* **7**, 421–433.
Reprinted with kind permission from Springer.

#45 Fracture Mechanics (1971). *Science Progress* **59**, 161–179.
Reprinted with kind permission from Blackwell Publishing.

#46 Dislocation theory for geophysical applications (1973). *Phil. Trans. Roy. Soc. London A* **274**, 331–338.
Reprinted with kind permission from the Royal Society, London.

#47 The calculation of energy release rates (1975). In *Prospects of Fracture Mechanics*, eds. G. C. Sih, H. C. van Elst, and D. Broek, pp. 69–84, Noordhoff, Leyden.
Reprinted with kind permission from Springer.

#48 Point defects (1975). In *The Physics of Metals – Sir Nevill Mott 60th Anniversary Volume*, ed. P. B. Hirsch, pp. 1–42, Cambridge University Press, Cambridge.
Reprinted with kind permission from the Cambridge University Press.

#49 The change of shape of a viscous ellipsoidal region embedded in a slowly deforming matrix having a different viscosity (1975), with B. A. Bilby and A. K. Kundu. *Tectonophysics* **28**, 265–274.
Reprinted from Tectonophysics, Copyright 1975, with kind permission from Elsevier.

#50 The elastic energy-momentum tensor (1975). *J. Elasticity* **5**, 321–335.
Reprinted with kind permission from Springer.

#51 The change of shape of a viscous ellipsoidal region embedded in a slowly deforming matrix having a different viscosity – Some comments on a discussion by N. C. Gay (1976), with B. A. Bilby, M. L. Kolbuszewski, and A. K. Kundu. *Tectonophysics* **35**, 408–409.
Reprinted from Tectonophysics, Copyright 1976, with kind permission from Elsevier.

#52 Interaction and diffusion of point defects (1977). In *Point Defect Behaviour and Diffusional Processes*, eds. R. E. Smallman and E. Harris, pp. 3–10, The Metals Society, London.
Reprinted with kind permission from Maney Publishing.

#53 Boundary problems (1979). In *Dislocations in Solids*, Vol. 1, ed. F. R. N. Nabarro, pp. 167–221, North-Holland, Amsterdam.
Reprinted with kind permission from the University of Sheffield.

#54 The force on a disclination in a liquid crystal (1980). *Phil. Magazine A* **42**, 359–367.
Reprinted with kind permission from Taylor & Francis Ltd.
(http://www.tandf.co.uk/journals/).

#55 The energy-momentum tensor of complex continua (1980). In *Continuum Models of Discrete Systems*, eds. E. Kröner and K.-H. Anthony, pp. 651–665, University of Waterloo Press, Waterloo.
Reprinted with kind permission from the University of Waterloo Press.

#56 Aspects of the theory of dislocations (1982). In *Mechanics of Solids – The Rodney Hill 60th Anniversary Volume*, eds. H. G. Hopkins and M. J. Sewell, pp. 185–225, Pergamon Press, Oxford.
Reprinted with kind permission from the University of Sheffield.

#57 The stresses on and in a thin inextensible fibre in a stretched elastic medium (1982). *Engineering Fracture Mechanics* **16**, 453–455.
Reprinted from Engineering Fracture Mechanics, Copyright 1982, with kind permission from Elsevier.

The photograph of J. D. Eshelby has been reprinted with kind permission from the University of Sheffield. We would like to acknowledge Mr. Jacky Hodgson, Head of special collections at the main library of

University of Sheffield for providing us with the photograph. We are very thankful to Nathalie Jacobs, Publishing Editor of Mechanical Engineering at Springer for the enthusiastic support with which she embraced every aspect of the project, and Jolanda Karada for the marvelous reproduction of articles from poor quality photocopies.

Collected Works of J. D. Eshelby

Dislocations as a cause of mechanical damping in metals

By J. D. Eshelby, *H. H. Wills Physics Laboratory, University of Bristol*

(*Communicated by N. F. Mott, F.R.S.—Received* 1 *December* 1948)

Zener has shown how thermoelastic effects give rise to damping of the mechanical vibrations of a solid. For example, in a vibrating reed opposite sides are alternately compressed and extended. This gives rise to an alternating temperature-difference across the width of the reed, and the resulting flow of heat leads to dissipation of mechanical energy.

In a vibrating single crystal of a metal an additional energy loss is observed which is usually attributed to the motion of dislocations. In the present paper the following mechanism is proposed. Dislocations are trapped in 'potential troughs' at the minima of the internal stress in the crystal. When the crystal vibrates the dislocations oscillate in their potential troughs and the moving stress-system associated with them produces a fluctuating temperature distribution in the material; this leads to damping as in Zener's case. The rate of loss of energy produced by a dislocation oscillating with given amplitude is calculated and the effect of a collection of them is discussed. An actual estimate of the damping in a vibrating crystal requires (i) a knowledge of the relation between the amplitude of oscillation of a dislocation and the vibrational stress causing it to move, and (ii) a knowledge of the density of dislocations in the material. A tentative discussion of (i) is given. The quantity (ii) is unknown; however, it is shown that the damping depends only on the ratio of the number of dislocations per unit area to the number of potential troughs per unit area. If this ratio is calculated from the theoretical result and the observed damping in copper single crystals, it is found to be of the order of unity. The present theory predicts that the damping should increase with frequency. This is in disagreement with the limited experimental data available.

Two subsidiary effects are also investigated, the thermoelastic damping arising from the interaction between the vibrational stresses and the stresses surrounding *stationary* dislocations, and the damping due to the emission of elastic waves from an oscillating dislocation. Both these effects are shown to be small compared with the thermoelastic damping caused by moving dislocations.

1. Introduction

A number of workers have measured the damping of mechanical vibrations in metal single crystals for low strain-amplitudes (about 10^{-7} to 10^{-5}) at various frequencies between 10 and 100 kc./sec. The damping is usually defined as the quantity Δ representing the energy dissipated per half-cycle divided by the total strain energy of the material. When the damping arising from large-scale thermal currents (Zener 1940) has been eliminated or allowed for, there still remains a loss which is generally supposed to be due to the motion of dislocations. For low amplitudes of vibration this loss is nearly independent of amplitude ('region of level decrement', in Read & Tyndall 1946), but increases at greater amplitudes. Only the low-amplitude damping is considered in this paper.

Attempts to explain this damping in terms of the inelastic strain (creep) observed in static tests (Seitz & Read 1941; Read & Tyndall 1946) have not been very successful; the observed values of Δ correspond to a much greater creep rate than is actually observed. The main object of this paper is to consider the contribution to the damping by 'elastically bound' dislocations, i.e. dislocations normally resting at the bottom of troughs in the potential distribution representing the internal stresses in the crystal lattice, and displaced from their equilibrium positions by an applied shear stress. Static stresses insufficient to lift a dislocation over the barrier separating two potential troughs, even with the aid of thermal fluctuations, will produce no permanent inelastic deformation. Alternating stresses of the same magnitude, arising from the vibration of the specimen in a damping test, will make the dislocation oscillate in its potential trough. This motion gives rise to energy loss by the mechanism discussed below, and thus a source of damping is provided which is not associated with large creep rates in static tests.

Several mechanisms can be imagined whereby oscillating dislocations could contribute to the dissipation of energy. The main one considered in this paper is a development of Zener's (1940) theory of thermoelastic damping. The movement of a dislocation alters the stresses at any point. This alteration of stress is associated with a change of temperature. The latter varies from point to point, so that temperature gradients exist in the material. The resulting flow of heat leads to dissipation of mechanical energy.

As a preliminary to the investigation of the mechanism of the last paragraph, the damping due to the interaction between the vibrational stress and the stresses surrounding *stationary* dislocations is investigated. Finally, the damping of a moving dislocation due to emission of elastic waves (analogous to radiation damping) is briefly considered.

The whole treatment is two-dimensional, i.e. the dislocations are all considered to be parallel and of infinite length.

2. Thermoelastic effects

Consider a material with zero thermal conductivity. It is shown in the Appendix (equations (a 1) and (a 3)) that the relation between the rate of change of temperature T at any point and the stresses there is

$$\frac{dT}{dt} = \frac{T}{c_p}\left\{\frac{1}{4}\left(\frac{\partial G^{-1}}{\partial T}\right)_p \frac{d(\sigma_0^2)}{dt} + \frac{9}{2}\left(\frac{\partial}{\partial T}\left(\frac{\nu}{E}\right)\right)_p \frac{d(p^2)}{dt} + \alpha\frac{dp}{dt}\right\}, \tag{1}$$

where p is the hydrostatic pressure (one-third of the sum of the diagonal terms of the stress tensor), σ_0^2 is the sum of the squares of the nine stress components, G, E and ν are the shear modulus, Young's modulus and Poisson's ratio, α is the volume expansion coefficient and T the absolute temperature. When the thermal conductivity is not zero the temperature can be found as a function of position and time by solving the heat-conduction equation with a distribution of heat sources of strength $c_p dT/dt$ per unit volume with dT/dt given by (1). If the resulting temperature varies from point to point heat currents will flow and the mechanical energy of the system will be dissipated.

If the material is initially unstressed a homogeneous stress varying harmonically with time will produce no loss, since there will be no temperature gradient. If the material contains an initial non-homogeneous stress there will be a non-homogeneous temperature distribution and hence energy loss, since p^2 and σ_0^2 will contain products of the initial and applied stress components.

This is the basis of the first damping mechanism considered below. The initial inhomogeneous stress is taken to be that of stationary dislocations, and the homogeneous stress is the stress due to the vibration of the specimen. The vibrational

26-2

stress is not strictly homogeneous, but will be approximately so over the region in which its interaction with a dislocation is important.

The above mechanism postulates that the dislocations remain stationary in the material. If they move under the influence of the vibrational stresses the stress at any point will vary and give rise to additional temperature changes and consequently damping. This is the second mechanism considered.

3. Damping by stationary dislocations

Swift & Richardson (1947) have already suggested that the amplitude-independent damping which they observed for zinc crystals might arise from the interaction between the applied stress and the stresses surrounding the dislocations.

It is shown in the Appendix that a single dislocation in an infinite medium will, under the influence of an applied shear $\sigma_0 \sin \omega t$, produce an energy loss

$$w \sim \frac{a^2}{c_p T}\left(\frac{d \log G}{d \log T}\right)^2 \sigma_0^2 \tag{2}$$

per cycle per unit length. This holds apart from a numerical factor both for edge and screw dislocations.

Damping processes are usually discussed in terms of a set of relaxation times each associated with a certain oscillator strength. From this point of view the fact that w is independent of frequency has the following interpretation: the dislocation has a continuous relaxation spectrum, and the oscillator strength is the same for all relaxation times. Actually the spectrum will be cut off for relaxation times less l^2/κ, where l is the distance below which the elastic solution for the stresses surrounding the dislocation breaks down and κ is the thermal diffusivity of the material. Since $l \sim 10^{-7}$ cm. and $\kappa \sim 1$ cm.^{2}/sec. this could produce no observable effect at kilocycle frequencies.

The elastic energy per unit volume due to the vibrations is of the order of σ_0^2/G. If there are n dislocations threading unit area, and it is assumed that the contribution of each dislocation to the total loss is given by (2), then

$$\Delta \sim \frac{a^2 G}{c_p T}\left(\frac{d \log G}{d \log T}\right)^2 n. \tag{3}$$

With the following values for copper,

$$a = 3 \times 10^{-8}\,\text{cm.}, \qquad G = 4{\cdot}5 \times 10^{11}\,\text{dynes/cm.}^2,$$

$$c_p = 3{\cdot}4 \times 10^7\,\text{ergs/cm.}^3\,{}^\circ\text{C}, \qquad T = 300^\circ\,\text{K},$$

$$(d \log G/d \log T)^2 = 0{\cdot}02,$$

Δ is found to be of the order of $10^{-16}n$. The observed value of Δ, about 10^{-5}, would require about 10^{11} dislocations per cm.^{2}. Similar results hold for other materials.

Provided the mean distance between dislocations is appreciably greater than $(\kappa/\omega)^{\frac{1}{2}}$, the assumption that the total loss is the sum of the losses which each dislocation would produce in the absence of the others is justified. When they are closer together than this the calculation is rendered difficult for reasons discussed in the

Appendix. The correct value is not, however, likely to exceed (3). Since a density of 10^{11} dislocations per cm.^{2} seems excessive for an undistorted single crystal this mechanism will not be discussed further.

4. Damping by oscillating dislocation

An expression is derived in the Appendix for the energy loss per cycle due to an oscillating dislocation. For a screw dislocation the result is

$$w = \frac{1}{2^7\pi^2}\frac{d^2\omega G a^2}{\kappa}\frac{a^2}{l^2}\frac{G}{c_p T}\left(\frac{d\log G}{d\log T}\right)^2 \text{ ergs per unit length.} \tag{3a}$$

The corresponding expression for the edge dislocation is rather complicated; it may be found from equation (a 24) in the Appendix. If the round value $\frac{1}{3}$ is taken for Poisson's ratio and it is assumed that

$$\frac{1}{G}\frac{dG}{dT} \doteqdot \frac{1}{E}\frac{dE}{dT} \gg \frac{1}{\nu}\frac{d\nu}{dT},$$

which is approximately correct for metals, the expression for the edge dislocation becomes, in ergs per unit length,

$$w \doteqdot \frac{1}{10}\frac{d^2\omega G a^2}{\kappa}\left\{\frac{1}{10}\frac{a^2}{l^2}\frac{G}{c_p T}\left(\frac{d\log G}{d\log T}\right)^2 + \frac{c_p - c_v}{c_p}\log\frac{\kappa}{\omega l^2}\right\}. \tag{4}$$

The symbols have the following meaning:

$\omega/2\pi$ frequency of oscillation.
d amplitude of oscillation.
a interatomic spacing.
G, E, ν shear modulus, Young's modulus and Poisson's ratio.
c_p, c_v specified heats per unit volume at constant pressure and volume respectively.
κ thermal diffusivity (thermal conductivity divided by density and specific heat).
l 'cut-off' length.

The length l is a parameter introduced to allow for the fact that the expression derived from elasticity theory for the stresses surrounding a dislocation ceases to hold in its immediate neighbourhood. It represents approximately the distance from the dislocation below which the elastic solution ceases to hold good.

The main assumptions made in deriving these results are the following:

(i) The dislocation is of infinite length and oscillates with simple harmonic motion along its slip-plane. The medium in which it exists is of infinite extent, and there are no other dislocations present.

(ii) It carries with it the stress-system which surrounds it when stationary. This will be a good approximation if its maximum velocity is much less than that of sound in the material. That the velocity reached in damping experiments must be much less than that of sound can be seen as follows. The specimen used is about one-quarter wave-length long. A dislocation moving with the velocity of sound would therefore more than cover the whole length of the specimen in a half-cycle. Since the movement is presumably a very small fraction of the length of the specimen the velocity must be much smaller than that of sound.

(iii) The cut-off length l and amplitude of oscillation d are both much smaller than the length $(\kappa/\omega)^{\frac{1}{2}}$, about 10^{-3} cm., for metals in the range 10 to 100 kc./sec.

Approximate values of the constants occurring in (4) are, for copper at 300° K,

$$\frac{G}{c_p T} = 42, \quad \left(\frac{d \log G}{d \log T}\right)^2 = 0{\cdot}020, \quad \frac{c_p - c_v}{c_p} = 0{\cdot}027, \quad \kappa = 1{\cdot}2\,\text{cm.}^2\text{sec.}^{-1}.$$

With these values the { } of (4) becomes approximately

$$2{\cdot}7 \times 10^{-2}\left(3\frac{a^2}{l^2} + \log\frac{1{\cdot}2}{\omega l^2}\right).$$

The exact choice of l in the second term is not critical. For $l \sim 10^{-7}$ cm. and a frequency in the neighbourhood of 30 to 50 kc./sec. the logarithm will be of the order of 20. The elastic solution for the stresses surrounding the dislocation is generally supposed to hold to within a few atomic distances of its centre. 10^{-1} would be a good estimate for a^2/l^2; unity is certainly an over-estimate. It is probably justifiable to neglect the first term in comparison with the second, and this will be done in what follows. In the expression for the screw dislocation the logarithmic term is missing. It arises from the hydrostatic-pressure component of the stresses which is zero for a screw dislocation. The contribution from the screw dislocations will also be neglected.

According to the discussion in the Appendix the loss due to a number of dislocations situated at random in the material is approximately the sum of the losses they would produce separately. If the vibration shear stress in the material is σ_0 the elastic energy of the material per unit volume is of the order of $\frac{1}{2}\sigma_0^2/G$. If there are n dislocations threading unit area each giving the contribution (4) the damping will be

$$\Delta = \frac{1}{10\kappa}\frac{c_p - c_v}{c_p} G^2 a^2 \omega \log\left(\frac{\kappa}{\omega l^2}\right)\left(\frac{d}{\sigma_0}\right)^2 n. \tag{5}$$

If the amplitudes of the various dislocations differ, d^2 will represent an average value.

To find Δ explicitly it is necessary to know the value of d/σ_0. There is another phenomenon which should depend on this ratio, the reduction of shear modulus produced by the dislocations. A dislocation in moving a distance d across a unit cube produces a shear ad. If there are n dislocations threading unit area and they each move a distance d under the influence of a shear, the elastic strain σ_0/G will be accompanied by an additional strain nad, so that the apparent shear modulus G' will be given by

$$\frac{\sigma_0}{G'} = \frac{\sigma_0}{G} + nad.$$

The change in the inverse of the shear modulus on introducing the dislocations into a material previously containing none will be

$$\frac{\delta G^{-1}}{G^{-1}} = naG\frac{d}{\sigma_0}. \tag{5a}$$

Combined with (5) this gives

$$n = \frac{1}{10\Delta\kappa}\frac{c_p - c_v}{c_p}\omega\left(\frac{\delta G^{-1}}{G^{-1}}\right)^2 \log\frac{\kappa}{\omega l^2}. \tag{6}$$

This relation should be independent of any variation of d/σ_0 with frequency, provided G and Δ are measured at the same frequency. Since d^2 and d in (5) and (5*a*) represent averages, (6) should strictly be multiplied by $\overline{d^2}/\bar{d}^2$. There seem to be no measurements on single crystals which would serve to check these results.

Lawson (1941) investigated the change of damping and Young's modulus when compressive stresses of the order of 1 kg./mm.² were applied to polycrystalline specimens of very pure copper. The crystal grains were several millimetres across, and Lawson considers that the effect of intercrystallite thermal currents can be neglected. The following data refer to the only case in which he gives numerical values both for damping and Young's modulus:

Δ before compression 2×10^{-3},
Δ after compression 4×10^{-3},
change of Young's modulus 6 to 7 %,
frequency *c.* 50 kc./sec.

It will be assumed that the proportional charges in G and Young's modulus are the same. The δG^{-1} in (6) refers to the total change when dislocations are introduced into a material originally free from them. If the comparison is to be made for the higher value of Δ, it is easy to see that the observed change of inverse modulus must be multiplied by 2. Inserted in (6) these figures give

$$n \doteqdot 5 \times 10^4 \text{ dislocations/cm.}^2.$$

Though there is no direct evidence for the number of dislocations in single crystals, this value is rather low compared with the value of about 10^6 generally assumed. Lawson's value of Δ is large compared to that found for copper single crystals (Reid 1941), about 10^{-4} to 10^{-5}. If some effect not present in single crystals is operative in his case the value 10^4 to 10^5 will be an underestimate.

According to a model which has been used successfully in discussing transient creep (Mott & Nabarro 1948) the dislocations are restrained from moving freely by stresses in the lattice. The variation of the stress with position is represented roughly by the expression

$$\sigma = \sigma_i \sin\frac{2\pi x}{\Lambda},$$

where σ_i is the maximum internal strain and Λ is the distance between successive maxima. The dislocations occupy the points where $\sigma = 0$. When an external shear stress σ_0 is applied these points move a distance $d = \Lambda\sigma_0/2\pi\sigma_i$. If the dislocation is situated in a unit cube it produces a shear ad in moving the distance d. The effective force on the dislocation is given by $Fd = \sigma_0 ad$ or $F = \sigma_0 a$, and the dislocation can be regarded as restrained by an elastic restoring force $2\pi\sigma_i a/\Lambda$ per unit displacement per unit length, or as occupying the bottom of a trough in a potential field

$$V = (\sigma_i a\Lambda/2\pi)\cos 2\pi x/\Lambda. \tag{7}$$

For slow movement of the dislocation d/σ_0 should thus have the value $\Lambda/2\pi\sigma_i$. However, the inertia of the material set in motion by the movement of the dislocation cannot in general be neglected. A dislocation can be shown to have an equivalent mass of the order of ρa^2 per unit length, where ρ is the density of the material. Consider the normal equation of forced harmonic motion

$$m\ddot{x}+r\dot{x}+kx = F\sin\omega t. \tag{8}$$

The energy dissipated per cycle is $2\pi r d^2\omega$, where d is the amplitude of the oscillation. This is just of the form (4) apart from the logarithmic term which varies only slowly with ω. It is reasonable to suppose that the motion of the dislocation is governed by (8) with

$$m = \rho a^2, \quad r = \beta G a^2/2\pi, \quad k = 2\pi\sigma_i a/\Lambda,$$

where

$$\beta = \frac{1}{10\kappa}\frac{c_p-c_v}{c_p}\log\frac{\kappa}{\omega l^2}.$$

The natural undamped frequency of the motion is given by

$$\omega_0 = (k/m)^{\frac{1}{2}} = (2\pi\sigma_i/\rho\Lambda a)^{\frac{1}{2}}.$$

At very low frequencies $(d/\sigma_0)^2 = (\Lambda/2\pi\sigma_i)^2$. At other frequencies this value must be multiplied by the factor

$$\{(1-\omega^2/\omega_0^2)^2+r^2\omega^2/k^2\}^{-1}. \tag{9}$$

In terms of the quantity

$$z = G\Lambda/2\pi\sigma_i,$$

the two parameters in (9) have the values

$$\omega_0 = (G/\rho a)^{\frac{1}{2}} z^{-\frac{1}{2}} \sim 10^{10} z^{-\frac{1}{2}}, \quad r/k = (\beta a/2\pi)\, z \sim 10^{-9} z. \tag{10}$$

According to the model being used σ_i represents the critical shear stress of the crystal. G/σ_i usually lies between 10^2 and 10^5 for single crystals. Presumably Λ lies between 10^{-2} and 10^{-6} cm. Then z will lie between 10^3 and 10^{-4}. Thus for frequencies in the kilocycle range ω/ω_0 and $r\omega/k$ will always be small compared with unity and the low-frequency value $d/\sigma_0 = \Lambda/2\pi\sigma_i$ may be used. Equation (5) therefore becomes

$$\Delta = \frac{\omega a^2}{10\kappa}\frac{c_p-c_v}{c_p}\log\frac{\kappa}{\omega l^2}\left(\frac{G\Lambda}{2\pi\sigma_i}\right)^2 n. \tag{11}$$

Of the quantities in (11) σ_i can be found from the yield strength and the exact value of l is unimportant. The remainder are known constants except for Λ and n which are quite unknown. The combination $n' = \Lambda^2 n$, however, represents the proportion of potential troughs occupied by dislocations. The value of n' calculated from

$$n' = \Delta\left[\frac{\omega a^2}{10\kappa}\frac{c_p-c_v}{c_p}\log\frac{\kappa}{\omega l^2}\left(\frac{G}{2\pi\sigma_i}\right)^2\right]^{-1} \tag{12}$$

should give some idea of the reasonableness of the assumptions on which (11) is based.

For a single crystal of copper at 33·5 kc./sec. Read (1941) found $\Delta \sim 10^{-4}$ for the freshly mounted specimen, falling to $\Delta \sim 10^{-5}$ on annealing. Seitz (1943) gives the yield stress 0·1 kg./mm.^{2} for a copper crystal of comparable purity. Using these figures, together with the values already quoted, for the remaining constants gives

$$n' = 2 \times 10^3 \Delta,$$

or $\frac{1}{5}$ for the unannealed and $\frac{1}{50}$ for the annealed specimen. A result of the order of one dislocation in each potential trough in the unannealed material seems reasonable. A value greatly in excess of unity is unlikely, since the dislocations would then be substantially controlled by each other's stresses so that effectively $\sigma_i \sim Ga/\Lambda$ and $n \sim \Lambda^{-2}$; this is easily seen to lead to $\Delta \propto n^{-1}$, or a decrease of damping with number of dislocations. Read found, on the contrary, that Δ increased on subjecting his specimens to static stresses and which presumably increased the number of dislocations.

A large number of measurements have been made on zinc (Read 1941; Read & Tyndall 1946; Swift & Richardson 1947). This material shows strong 'self-annealing' even at room temperature; for example, a specimen had a Δ of 10^{-3} on first mounting in the apparatus, 7×10^{-5} after 30 hr. and 6×10^{-6} after 7 days. Because of its great elastic and thermal anisotropy and the fact that the stresses surrounding a dislocation in a hexagonal crystal have not been calculated, it would be rash to assume that the present theory can be applied. However, it may be of interest to compare the values of n' obtained from (12) using bulk values of the constants involved. Read & Tyndall found a departure from Hooke's law at a shear strain of about 8×10^{-6}; Swift & Richardson adopt the same value in discussing their results. If this strain is identified with σ_i/G, (12) gives $n' = 45\Delta$. The observed Δ varies between about 3×10^{-3} and 3×10^{-6}, giving an n' between 10^{-1} and 10^{-4}.

According to (11) Δ is approximately proportional to frequency. No account has been taken of this when comparing theory and experiment in the last few paragraphs, since most measurements have in fact been made at isolated frequencies.

Read (1941) has published curves for a zinc crystal at 39 and 78 kc./sec., showing that Δ is inversely proportional to the frequency. In view of the great differences between specimen and specimen, both in the value of Δ and its variation with amplitude, this relation cannot be assumed to hold universally. On the other hand, according to Seitz (1943) Δ is independent of frequency for copper; there seems to be no reference to this in the literature. Clearly more experimental results are needed, but it may be of interest to consider what modifications (apart from dropping the limitation to a two-dimensional problem) would be needed to bring the present theory into line with these results if confirmed.

The frequency dependence in (11) arises essentially from two facts. First, the thermoelastic effects are equivalent to a resistive force on the dislocation approximately proportional to its velocity; this is confirmed by calculations for a dislocation moving with uniform velocity. Secondly, the natural frequency of oscillation of a dislocation in its potential trough is much higher than the applied frequency.

If ω_0 could be taken to be of the order of the frequency used in damping measurements Δ would no longer be proportional to frequency, being given by the product of (11) and the expression (9). Some rather indirect evidence for such a low value of ω_0 is provided by a comparison of observed creep rates with theory. The unit process in slip is considered to be the movement of a dislocation from trough to trough of the potential (7) under the influence of thermal fluctuations and an applied stress. The variation of creep rate with temperature might therefore be expected to be given by a rate-process formula of the type $f\exp(-U/kT)$. If experimental data are fitted to such an expression (Kauzmann 1941) the frequency factor f is of an altogether lower order of magnitude than the ω_0 estimated from (10).

5. Radiation damping

An isolated screw dislocation oscillating in an infinite medium dissipates energy at a rate

$$w = \pi\rho a^2 d^2 \omega^2/16$$

per cycle per unit length as a result of the radiation of elastic waves (Appendix, § A 6). ρ is the density of the medium, and the other symbols are as in § 4. Compared with the thermoelastic dissipation (3 a) this is extremely small at kilocycle frequencies. The radiation damping for an edge dislocation will be of the same order and hence will be small in comparison with the thermoelastic damping (4). In actual damping experiments the wave-length of the emitted elastic waves is greater than the dimensions of the specimen, and the calculation for the infinite specimen does not apply. Reflexion at the surface of the specimen would not produce energy loss but a reactive effect on the motion of the dislocation. Scattering by lattice imperfections (e.g. the centres of other dislocations where the stress-strain relation is non-linear) might ultimately lead to dissipation.

6. Conclusion

The thermoelastic effect arising from stationary dislocations is inadequate to explain the damping observed in single crystals.

The corresponding effect for moving dislocations seems capable of giving an effect of the right order of magnitude, but the frequency dependence needs more detailed investigation both theoretically and experimentally.

The experimental methods hitherto used do not lend themselves to measurements of damping and elastic constants over a range of frequencies on the same specimen without removal from the apparatus. A technique which could give such results might throw valuable light on the dynamics of moving dislocations.

Appendix

A 1. *Method of obtaining energy loss*

The first step is the calculation of the temperature changes associated with the varying elastic conditions in the medium. The state of the medium is determined by the temperature T and the stress tensor p_{ij}. For the entropy change of unit volume we have

$$dS = \left(\frac{\partial S}{\partial T}\right)_p dT + \left(\frac{\partial S}{\partial p_{ij}}\right)_T dp_{ij}.$$

The convention that a repeated index is summed for the values 1, 2, 3 will be used. The second term on the right therefore represents nine terms and in the calculations below

$$(p_{ij})^2 = p_{ij}p_{ij} = p_{11}^2 + \ldots + 2p_{12}^2 + \ldots$$

and

$$p_{mm} = p_{11} + p_{22} + p_{33}.$$

The coefficient of dT is by definition c_p/T. The coefficient of dp_{ij} can be transformed by considering the Gibbs function, the change of which is given by

$$d\phi = -e_{ij}dp_{ij} - S\,dT.$$

Here e_{ij} is the strain tensor defined in terms of the displacement vector u_i by the relation

$$e_{ij} = \frac{1}{2}\left(\frac{\partial u_i}{\partial x_j} + \frac{\partial u_j}{\partial x_i}\right).$$

Since $d\phi$ is a perfect differential

$$\left(\frac{\partial e_{ij}}{\partial T}\right)_p = \left(\frac{\partial S}{\partial p_{ij}}\right)_T.$$

e_{ij} and p_{ij} are related by

$$e_{ij} = \epsilon p_{ij} + \zeta p_{mm}\delta_{ij} + \tfrac{1}{3}\alpha(T - T_0)\,\delta_{ij},$$

where

$$\epsilon = \tfrac{1}{2}G, \quad \zeta = \nu/E. \tag{a 1}$$

G and E are the shear modulus and Young's modulus and ν is Poisson's ratio. The final term represents the thermal expansion; α is the coefficient of volume expansion and T_0 some arbitrary temperature at which the material is considered to be unstrained in the absence of stress. Hence

$$dS = \frac{c_p}{T}dT + \{\epsilon' p_{ij} + \zeta' p_{mm}\delta_{ij} + \tfrac{1}{3}\alpha\delta_{ij}\}\,dp_{ij}, \tag{a 2}$$

where

$$\epsilon' = \frac{1}{2}\frac{dG^{-1}}{dT}, \quad \zeta' = \frac{d}{dt}\left(\frac{\nu}{E}\right).$$

If the second term in (a 2) is integrated from an initial state p_{ij}^0 to a final state p_{ij} and multiplied by T/c_p, we obtain a quantity

$$v_0 = \frac{T}{c_p}\{\tfrac{1}{2}\epsilon'(p_{ij}^2 - p_{ij}^{02}) + \tfrac{1}{2}\zeta'[(p_{mm})^2 - (p_{mm}^0)^2] + \tfrac{1}{3}\alpha(p_{mm} - p_{mm}^0)\}, \tag{a 3}$$

and the entropy equation can be written in the form

$$T\,dS = c_p(dv + dv_0),$$

where dv has been written for dT. v will be used to denote the change of temperature from the mean.

If the thermal conductivity is k then $k\nabla^2 v$ is the rate of flow of heat into unit volume. If we identify this with $T\,dS$ the last equation can be written as

$$\kappa\nabla^2 v - \dot{v} + \dot{v}_0 = 0, \tag{a 4}$$

where $\kappa = k/c_p$ is the thermal diffusivity. Clearly v_0 represents the value of v when the conductivity is zero. When v is known the rate of dissipation of energy can be found from the increase of entropy. If T is the mean absolute temperature the rate of increase of entropy per unit volume is

$$\frac{k^2\nabla^2 v}{T+v} = \frac{k\nabla^2 v}{T} - \frac{1}{T^2} kv\nabla^2 v,$$

if higher terms are neglected. The total change of entropy is found by integration over the whole volume. Application of Green's theorem converts the first term into an integral over the boundary of the solid. If grad $v = 0$ over the boundary this term gives no contribution to the energy loss. We shall assume that this is the case. The rate of dissipation of energy is thus

$$-\frac{k}{T}\int_{\mathbf{r}} v\nabla^2 v\, d\mathbf{r} = \frac{c_p}{T}\int_{\mathbf{r}} v(\dot{v}_0 - \dot{v})\, d\mathbf{r}$$

by (a 4). For a cyclic process with period $2\pi/\omega$ the energy loss per cycle is

$$w = \frac{c_p}{T}\int_0^{2\pi/\omega}\int_{\mathbf{r}} v(\dot{v}_0 - \dot{v})\, d\mathbf{r}\, dt = \frac{c_p}{T}\int_0^{2\pi/\omega}\int_{\mathbf{r}} v\dot{v}_0\, d\mathbf{r}\, dt. \tag{a 5}$$

A more rigorous discussion is given by Päsler (1944).

For a two-dimensional problem in which the medium is of infinite extent w is most easily calculated by making a two-dimensional Fourier analysis of v and v_0. w can then be found without calculating v explicitly. For any function of position $f(\mathbf{r})$ we define the Fourier transform $\bar{f}(\mathbf{k})$ by

$$\bar{f}(\mathbf{k}) = \frac{1}{2\pi}\int_{\mathbf{r}} f(\mathbf{r})\, e^{i\mathbf{k}.\mathbf{r}}\, d\mathbf{r}, \tag{a 6}$$

the integral extending over all space. If f depends on the time t as a parameter so will $\bar{f}$. We need two results from the theory of Fourier transforms (cf. Titchmarsh 1937). The first is the Fourier integral theorem which states that

$$f(\mathbf{r}) = \frac{1}{2\pi}\int_{\mathbf{k}} \bar{f}(\mathbf{k})\, e^{-i\mathbf{k}.\mathbf{r}}\, d\mathbf{k} = \frac{1}{(2\pi)^2}\int_{\mathbf{k}} d\mathbf{k}\int_{\mathbf{r}} f(\mathbf{r}')\, e^{i\mathbf{k}.(\mathbf{r}'-\mathbf{r})}\, d\mathbf{r}', \tag{a 7}$$

and the second is Parseval's theorem

$$\int_{\mathbf{k}} f_1(\mathbf{r}) f_2^*(\mathbf{r})\, d\mathbf{r} = \int_{\mathbf{k}} \bar{f}_1(\mathbf{k})\, \bar{f}_2^*(\mathbf{k})\, d\mathbf{k}, \tag{a 8}$$

where * denotes the complex conjugate.

Since $\nabla^2 e^{i\mathbf{k}.\mathbf{r}} = -k^2 e^{i\mathbf{k}.\mathbf{r}}$, (a 4) in conjunction with (a 7) gives

$$\kappa k^2 \bar{v} + \dot{\bar{v}} = \dot{\bar{v}}_0. \tag{a 9}$$

As v and v_0 are real (a 8) applied to (a 5) gives

$$w = \frac{c_p}{T}\int_0^{2\pi/\omega}\int_{\mathbf{k}} \bar{v}\dot{\bar{v}}_0^*\, d\mathbf{k} = \frac{c_p\kappa}{T}\int_0^{2\pi/\omega}\int_{\mathbf{k}} |\,\bar{v}\,|^2 k^2\, d\mathbf{k}; \tag{a 10}$$

the second equation follows from (a 9).

The following result will also be needed. If the function $f(\mathbf{r})$ is shifted as a whole by a vector displacement $\mathbf{d}$, becoming $f(\mathbf{r}-\mathbf{d})$, then the new transform is given by

$$\begin{aligned}\bar{f}'(\mathbf{k}) &= \frac{1}{2\pi}\int_{\mathbf{r}} f(\mathbf{r}-\mathbf{d})\,\mathrm{e}^{i\mathbf{k}.\mathbf{r}}\,d\mathbf{r} \\ &= \frac{1}{2\pi}\int_{\mathbf{r}'} f(\mathbf{r}')\,\mathrm{e}^{i\mathbf{k}.(\mathbf{r}'+\mathbf{d})}\,d\mathbf{r}' \\ &= \mathrm{e}^{i\mathbf{k}.\mathbf{d}}\bar{f}(\mathbf{k}).\end{aligned} \qquad \text{(a 11)}$$

A 2. *Stresses around edge and screw dislocations*

For convenience the stresses round a dislocation are given below, together with the combinations of them which will be required later.

(i) *Edge dislocation.* The components in the xy plane have been given by Koehler (1941). The remaining components have been calculated on the assumption of plane strain:

$$\left.\begin{aligned} p_{11} &= -D(\sin 3\theta + 3\sin\theta)/2r, & p_{33} &= -D(4\nu\sin\theta)/2r, \\ p_{22} &= D(\sin 3\theta - \sin\theta)/2r, & p_{13} &= p_{23} = 0, \\ p_{12} &= D(\cos 3\theta + \cos\theta)/2r, & p_{mm} &= -D\{4(1+\nu)\sin\theta\}/2r, \\ & p_{ij}^2 = 2D^2\{(1+\nu^2) - \nu^2\cos 2\theta\}/r^2, \end{aligned}\right\} \qquad \text{(a 12)}$$

where $D = Ga/2\pi(1-\nu)$ and a is the interatomic spacing.

(ii) *Screw dislocation.* The displacement has the simple form (Burgers 1939)

$$u_3 = \frac{a}{2\pi}\arctan\frac{y}{x} = \frac{a}{2\pi}\theta, \quad u_1 = 0, \quad u_2 = 0. \qquad \text{(a 13)}$$

The stresses can easily be derived; all vanish but two:

$$\left.\begin{aligned} p_{23} &= \frac{aG}{2\pi}\frac{\cos\theta}{r}, & p_{13} &= \frac{aG}{2\pi}\frac{\sin\theta}{r}, \\ p_{mm} &= 0, & p_{ij}^2 &= \frac{a^2G^2}{2\pi^2r^2}. \end{aligned}\right\} \qquad \text{(a 14)}$$

A 3. *Energy loss due to stationary dislocations*

If a material has a large initial strain p_{ij}^0 and a small additional strain p'_{ij} is applied' (a 3) gives

$$v_0 = \frac{T}{c_p}\{\tfrac{1}{2}\epsilon' p_{ij}^0 p'_{ij} + \tfrac{1}{2}\zeta'(p_{mm}^0)(p'_{mm}) + \tfrac{1}{3}\alpha(p'_{mm})\}.$$

In our application p_{ij}^0 represents the stresses surrounding a dislocation and p'_{ij} is the stress arising from the vibration of the material. As the simplest case consider a screw dislocation subject to the alternating shear $p'_{23} = p'_{32} = S\cos\omega t$. Then by (a 14)

$$v_0 = A\frac{\cos\theta}{r}\cos\omega t, \quad A = \frac{\epsilon' aGST}{2\pi c_p}.$$

From (a 6) $$\bar{v}_0 = \frac{A}{2\pi}\cos\omega t\int_0^\infty dr\int_0^{2\pi}\mathrm{e}^{i\mathbf{kr}}\cos\theta\,d\theta.$$

If we introduce polar co-ordinates (k, θ') in **k**-space and $\phi = \theta - \theta'$,

$$\bar{v}_0 = \frac{A}{2\pi}\cos\omega t \int_0^\infty dr \int_0^{2\pi} e^{ikr\cos\phi}\cos(\theta'+\phi)\,d\phi$$

$$= A\cos\omega t\cos\theta' \int_0^\infty J_1(kr)\,dr = A\frac{\cos\theta'}{k}\cos\omega t.$$

The steady-state solution of (a 9) is

$$\bar{v} = \frac{A\omega\cos\theta'}{k}\frac{\cos(\omega t+\delta)}{(\kappa^2k^4+\omega^2)^{\frac{1}{2}}}, \quad \tan\delta = \frac{\omega}{\kappa k^2}.$$

Hence from (a 10)

$$w = \frac{A^2\omega^2c_p\kappa}{T}\int_0^{2\pi/\omega}\cos^2\omega t\,dt\int_0^{2\pi}\cos^2\theta'\,d\theta'\int_0^\infty\frac{k\,dk}{\kappa^2k^4+\omega^2} = \frac{\pi^3A^2c_p}{4T}. \qquad \text{(a 15)}$$

It is easy to take account of the fact that v_0 does not in fact increase as $1/r$ at small distances from the origin. For example, if $1/r$ is replaced by $(1-e^{-r/l})/r$, the expression for $\bar{v}_0$ becomes multiplied by $(1+k^2l^2)^{-\frac{1}{2}}$. w is then equal to (a 15) multiplied by the factor

$$(1+4x^2\pi^{-1}\log x)(1+x^4)^{-1}, \quad \text{where} \quad x = (l^2\omega/\kappa)^{\frac{1}{2}}.$$

For frequencies appreciably less than κ/l^2 the expression for w is practically unaltered. Since for a metal $\kappa \sim 1$ and l (the distance from the centre of the dislocation within which the elastic solution breaks down) is certainly less than 10^{-6} cm., no error will be made by taking $l = 0$ for kilocycle frequencies. Essentially the same result follows if $1/r$ is replaced by any function which remains finite at the origin and approaches $1/r$ for $r > l$.

In full (a 15) reads

$$w = \frac{\pi}{64}\left(\frac{d\log G}{d\log T}\right)^2\frac{a^2S^2}{Tc_p}. \qquad \text{(a 16)}$$

Other cases in which an edge or screw dislocation is subjected to various kinds of shear can be treated similarly. For example, if an edge dislocation is subjected to a shear of magnitude S and any direction in the xy plane so that

$$p'_{12} = S\cos\omega t\sin 2\phi, \quad p'_{11} = -p'_{22} = S\cos\omega t\cos 2\phi,$$

where ϕ is an arbitrary angle, then the loss per cycle is obtained from (a 16) by replacing $\pi/64$ by $\pi/32(1-\sigma)^2$. The results are of the same order of magnitude in all cases. If the diagonal terms p_{mm} are involved the factor $d\log(E/\sigma)/d\log T$ will occur; it is of the same order of magnitude as $d\log G/d\log T$. The actual value of v can be found with the aid of (a 7). The result is

$$v = \frac{A\cos\theta}{r}\{(1+x\,\mathrm{ker}'\,x)\cos\omega t - x\,\mathrm{kei}'\,x\sin\omega t\}, \quad x^2 = \frac{r^2\omega}{\kappa}.$$

From the properties of the ker and kei functions it is easy to show that as $\kappa \to 0$, $v \to v_0$ and as $\kappa \to \infty$, $v \to 0$ as might be expected. As $\mathrm{ker}\,x$ and $\mathrm{kei}\,x$ fall off exponentially for $x > 1$, it follows that $(v_0 - v)$ is only appreciable within a cylinder of radius $\sim (\kappa/\omega)^{\frac{1}{2}}$ about the axis of the dislocation.

A 4. *Energy loss due to a moving dislocation*

Consider a dislocation moving along the X-axis as slip plane and carrying with it the stress-system it has when stationary. Let the position of its centre be $x = d\sin\omega t$. Then by (a 11)

$$\bar{v}_0(t) = \bar{v}_0(0)\exp\{ikd\sin\omega t\cos\theta'\}. \tag{a 17}$$

For a periodic (but not necessarily sinusoidal) $\bar{v}_0$ the general periodic solution of (a 9) is

$$\bar{v}(t) = \int_{-\infty}^{t} \dot{\bar{v}}_0(\tau)\exp\{-\kappa k^2(t-\tau)\}\,d\tau.$$

From (a 17)

$$\dot{\bar{v}}_0(\tau) = ikd\omega\cos\omega\tau\cos\theta'\bar{v}_0(0)\exp\{ikd\sin\omega\tau\cos\theta'\},$$

and hence

$$\bar{v}\dot{\bar{v}}_0^* = d^2\omega^2k^2\cos\omega t\cos^2\theta'\,|\,\bar{v}_0(0)\,|^2\int_{-\infty}^{t}\cos\omega\tau\exp\{-\kappa k^2(t-\tau)\}(\cos kq + i\sin kq)\,d\tau, \tag{a 18}$$

where

$$q = d(\sin\omega\tau - \sin\omega t)\cos\theta', \tag{a 19}$$

from which the energy loss per cycle can be found by (a 10).

For an edge dislocation (a 12) gives, neglecting the effect of any applied forces,

$$v_0(0) = Ar^{-2} + Br^{-2}\cos 2\theta + Cr^{-1}\sin\theta, \tag{a 20}$$

where

$$A = D^2\{(1+\sigma^2)\,\epsilon' + (1+\sigma)^2\zeta'\}\,T/c_p,$$
$$B = -D^2\{\epsilon'\sigma^2 + \zeta'(1+\sigma)^2\}\,T/c_p,$$
$$C = \tfrac{2}{3}\alpha D(1+\sigma)\,T/c_p.$$

As the simplest way of taking into account the fact that the stresses are finite at the origin we replace r by $\rho = (r^2+l^2)^{\frac{1}{2}}$. Then (a 6) gives

$$\begin{aligned}\bar{v}_0(0) &= \frac{1}{2\pi}\int_0^{\infty} r\,dr\int_0^{2\pi}\exp\{ikr\cos\phi\}\,d\phi\,\{A\rho^{-2} + B\rho^{-2}\cos 2(\theta'+\phi) + C\rho^{-1}\sin(\theta'+\phi)\}\\ &= \int_0^{\infty} r\,dr\{A\rho^{-2}J_0(kr) - B\rho^{-2}J_2(kr)\cos 2\theta' - iC\rho^{-1}J_1(kr)\sin\theta'\}\\ &= AH_1 - BH_2\cos 2\theta' - iCH_3\sin\theta',\end{aligned}$$

where

$$H_1 = \int_0^{\infty}\frac{J_0(kr)}{r^2+l^2}\,r\,dr = K_0(kl)$$

by W. B. F., p. 425, eqn. (5),†

$$H_2 = \int_0^{\infty}\frac{J_2(kr)}{r^2+l^2}\,r\,dr = 2(kl)^{-2} - K_2(kl),$$

by a modification of the method of W. B. F., § 13·51, taking into account the singularity at the origin, and

$$H_3 = \int_0^{\infty}\frac{J_1(kr)}{(r^2+l^2)^{\frac{1}{2}}}\,r\,dr = \tfrac{1}{2}l\{I_0K_1 - I_1K_0\},$$

† Here and elsewhere W. B. F. refers to the book by Watson (1922).

where the argument of the I's and K's is $\frac{1}{2}kl$. This last result is most easily found from W. B. F., p. 435, eqn. (3), with $\nu = 0$ on differentiating with respect to a.

The integrations required by (a 10) will now be carried out on the assumption that

$$d^2\omega/\kappa \ll 1 \tag{a 21}$$

and

$$l^2\omega/\kappa \ll 1, \tag{a 22}$$

but with no assumption about the relative size of d and l. We have

$$|\bar{v}_0(0)|^2 = A^2H_1^2 - 2ABH_1H_2\cos 2\theta' + B^2H_2^2\cos^2 2\theta' + C^2H_3^2\sin^2\theta'.$$

In (a 18) this factor and $\cos^2\theta'$ are even functions of θ', whilst the term $i\sin kq$ is odd, so that this term will not contribute to the θ'-integration and can be omitted. Consider the ratio of the coefficients of k in the factors $\cos kq$ and $\exp\{-\kappa k^2(t-\tau)\}$. It is

$$x = d(\sin\omega\tau - \sin\omega t)\cos\theta'/\kappa^{\frac{1}{2}}(t-\tau)^{\frac{1}{2}}.$$

The values of τ and t which occur satisfy

$$-\infty < \tau \leqslant t \leqslant 2\pi/\omega.$$

For the critical case $\tau = t$, $x = 0$. In other cases x is of the order of $d(\omega/\kappa)^{\frac{1}{2}}$. If (a 21) holds the argument of the cosine is always much smaller than that of the exponential. When $\bar{v}\bar{v}_0^*$ is integrated with respect to k the contributions from k-values for which $\kappa k^2(t-\tau) \gg 1$ will be small. Hence in the k-integration $\cos kq$ can be replaced by unity. Thus

$$\begin{aligned} w &= \frac{c_p d^2\omega^2}{T}\int_0^{2\pi/\omega}\exp\{-\kappa k^2 t\}\cos\omega t\,dt\int_0^{2\pi}\cos^2\theta'\,d\theta'\int_0^\infty|\bar{v}_0(0)|^2k^3\,dk \\ &\qquad\qquad \times\int_{-\infty}^{t}\exp\{\kappa k^2\tau\}\cos\omega\tau\,d\tau \\ &= \frac{\pi c_p d^2\omega\kappa}{T}\int_0^{2\pi}\cos^2\theta'\,d\theta'\int_0^\infty|\bar{v}_0(0)|^2\frac{k^5\,dk}{\kappa^2k^4+\omega^2} \\ &= \frac{\pi^2 c_p d^2\omega\kappa}{T}\int_0^\infty\{A^2H_1^2 - ABH_1H_2 + \tfrac{1}{2}B^2H_2^2 + \tfrac{1}{4}C^2H_3^2\}\frac{k^5\,dk}{\kappa^2k^4+\omega^2} \\ &= \frac{\pi^2 c_p d^2\omega}{Tl^2\kappa}\{A^2L_1 - ABL_2 + \tfrac{1}{2}B^2L_3 + \tfrac{1}{4}C^2L_4\}, \end{aligned} \tag{a 23}$$

say. If $X = (x^4+\xi^4)^{-1}$ with $\xi = (l^2\omega/\kappa)^{\frac{1}{2}}$ then

$$L_1 = \int_0^\infty xK_0^2(x)\,x^4X\,dx = \int_0^\infty xK_0^2(x)\,dx - \xi^4\int_0^\infty xK_0^2(x)\,X\,dx.$$

The first integral has the value $\frac{1}{2}$. The second must be less than

$$\xi^4\int_0^\infty X\,dx = \pi\xi/\surd 2,$$

since $0 \leqslant xK_0(x) < 1$ for $0 \leqslant x$. Thus

$$L_1 = \tfrac{1}{2} + O(\xi).$$

Similarly
$$L_2 = \int_0^\infty xK_0(x)\{2x^{-2} - K_2(x)\}\,dx + O(\xi) = \frac{\pi^2}{12} - \tfrac{1}{2} + O(\xi)$$

and $$L_3 = \int_0^\infty x\{K_2(x) - 2x^{-2}\}^2\,dx + O(\xi) = \tfrac{1}{2} + O(\xi).$$

The relation $$I_0(x)\,K_1(x) + I_1(x)\,K_0(x) = x^{-1}$$

gives $$L_4/l^2 = \int_0^\infty x\{x^{-2} - 4I_0K_0I_1K_1\}\,x^4Y\,dx,$$

where $Y = (x^4 + \xi^4/16)^{-1}$. This can be split up to give

$$L_4/l^2 = \int_0^1 x^3Y\,dx - 4\int_0^1 xI_0K_0I_1K_1Y\,dx + \int_1^\infty x\{x^{-2} - 4I_0K_0I_1K_1\}\,dx$$
$$+\tfrac{1}{4}\xi^4\int_0^1 xI_0K_0I_1K_1Y\,dx - \tfrac{1}{4}\xi^4\int_1^\infty x\{x^{-2} - 4I_0K_0I_1K_1\}\,Y\,dx.$$

The first term is a standard form, the second and third can be evaluated numerically and the last two can be shown to be of the order of ξ. In this way it is found that

$$L_4/l^2 = -\log\xi + 0{\cdot}109 + O(\xi).$$

Hence if (a 22) holds

$$w = \frac{\pi^2c_p d^2\omega}{2\kappa T}\left\{l^{-2}(A^2 - 0{\cdot}645AB + \tfrac{1}{2}B^2) + \tfrac{1}{4}C^2\left(\log\frac{\kappa}{\omega l^2} + 0{\cdot}218\right)\right\}. \qquad \text{(a 24)}$$

For a screw dislocation (a 14) gives

$$v_0(0) = Ar^{-1}, \quad \text{where now} \quad A = T\epsilon' a^2G^2/4\pi^2c_p.$$

The loss per cycle can be found from (a 24) by giving A this value and setting $B = C = 0$.

Other choices of the 'cut-off' function ρ lead to similar results. For the term in C^2 if

$$\rho^{-1} = r^{-1}\{1 - \exp(-r/l)\} \quad \text{or} \quad \rho^{-1} = \begin{cases} r^{-1} & (r \geqslant l), \\ 0 & (r < l), \end{cases}$$

it can be shown that the quality $0{\cdot}218$ in (a 24) is replaced by zero or $0{\cdot}232$ respectively. If $\log(\kappa/\omega l^2)$ is of the order of 10 the loss arising from the C^2 term is very insensitive to the choice of the form of ρ or the value of l. The terms involving A and B are naturally more sensitive, since l occurs in the form l^{-2}. It is also difficult to find analytical expressions for ρ which are easy to deal with. A rather artificial choice is

$$\rho^{-2} = r^{-2}\{1 - J_0(r/l)\},$$

which is finite at the origin and approaches the value r^{-2} when r is a few multiples of l. Using W.B.F., p. 406, eqn. (11), it is easy to show that in the A^2 term the effect of this choise of ρ is to replace the l of (a 24) by $l\sqrt{2}$. The terms in A and B will be found to be smaller than the C^2 term, so that more detailed calculations are unnecessary.

A 5. *The effect of a large number of dislocations*

If an infinite block of material contains a large number of dislocations, the temperature variation is given by the sum of the v's of the individual dislocations. Cross-terms between these will occur when w is found from (a 5) or (a 10).

The method used so far can be summed up as follows. Instead of dealing with the temperature variation v directly we construct the transformed function $\bar{v}$ in reciprocal $\mathbf{k}$-space. From this w is derived by integration according to (a 10). This is analogous to the method used in X-ray analysis, where the transform of the electron density of the scattering material is plotted in $\mathbf{k}$-space and the distribution of scattered radiation is determined by calculation in $\mathbf{k}$-space. The problem of a more or less random distribution of dislocations is thus analogous to the problem of X-ray scattering from an amorphous solid. The calculation below of the damping produced by a large number of dislocations is based on the averaging process used in the X-ray case (cf. Debye 1927).

In the case of stationary dislocations producing damping by interaction with an applied stress there is a fundamental difficulty. If a dislocation is reversed in sign so is its temperature variation. A macroscopically undistorted material will contain about equal numbers of dislocations of either sign. This makes any calculation of the total damping difficult. Since the effect of stationary dislocations seems to be unimportant this case will not be pursued. This much can be said, however. It has been seen that the expression $(v-v_0)$ extends only to a distance of the order of $(\kappa/\omega)^{\frac{1}{2}}$ about each dislocation; provided, therefore, the mean distance between dislocations is much greater than this distance the loss will be effectively the sum of the losses produced by each dislocation in the absence of the others.

We next consider a set of edge dislocations oscillating under the influence of an applied alternating shear. For simplicity we assume that they have a choice of two slip-planes at right angles and consider only the contribution to v from the hydrostatic pressure term, so that from (a 20)

$$v_0(0) = \frac{c}{r}\sin\theta \propto \frac{a}{r}\sin\theta, \tag{a 25}$$

where a is the 'strength' of the dislocation, i.e. the interatomic spacing provided with a $\pm$ sign according to the sign of the dislocation. If the dislocation is oscillating along its slip-plane (the x-axis) with amplitude d, then at points where $r \gg d$

$$v_0(t) \propto \frac{a}{r}\sin\theta + d\sin\omega t\frac{\partial}{\partial x}\frac{a}{r}\sin\theta = \frac{a}{r}\sin\theta - \frac{ad}{r^2}\sin 2\theta\,\sin\omega t. \tag{a 26}$$

The constant first term does not contribute to the damping. When the sign of a is changed the direction of motion under a given applied shear is reversed; in other words, the sign of d is also changed, so that the second term in (a 26) is unaltered. This is illustrated in figure 1 (a) and (b), where the T's represent the slip-plane and extra half-plane of atoms of an edge dislocation. The arrow denotes the direction of motion under the shear σ. The rule is that the dislocation moves in the direction of the σ-arrow to which the stem of its T points. The $\pm$ signs give the sign of the second term in (a 26) in the four quadrants. If the slip-plane of the dislocation is turned through a right angle, (c) or (d), the second term of (a 26) is unaltered except possibly in sign. The stresses σ are accompanied by the stresses σ' on account of the symmetry of the stress-tensor. The diagrams (c) and (d) can now be filled in as were (a) and (b). It is clear that at large distances from their centres the four dislocations give indistinguishable temperature variations. In this sense they are all of the same sign. Of course at distances of the order of d from their centres the temperature distribution about (*a*), (*b*), (*c*) and (*d*) are different. If the damping due to a pair of dislocations in a given relative position is calculated in detail the result is found to be independent of whether the pair is of type (*aa*), (*ab*), (*ac*), ..., provided the distance between them is large compared with d.

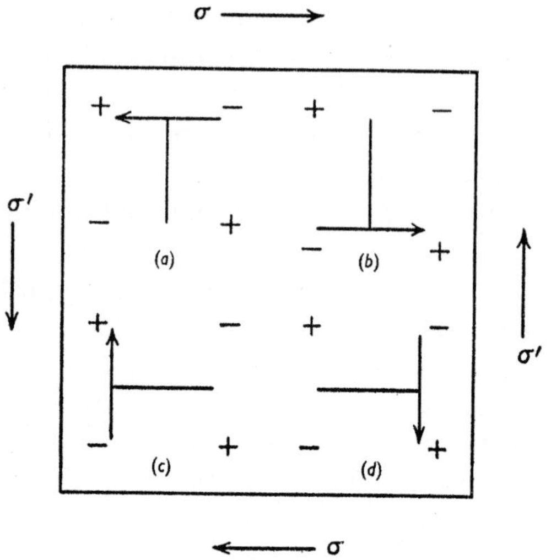

FIGURE 1.

We shall therefore assume that a good approximation to the actual case of a large number of edge dislocations will be obtained by attributing to them the $v_0(0)$ of (a 25). In addition to taking them all of the same sign we also assume that d is the same for all of them.

For a number N of identical dislocations in an infinite medium, situated at the points $\mathbf{r}_n$, we have by (a 11)

$$\overline{V}_0 = \overline{v}_0 \sum_0^N e^{i\mathbf{k}.\mathbf{r}_n}, \tag{a 27}$$

where $\overline{V}_0$ is the transform for the set of dislocations and $\overline{v}_0$ refers to a single one at the origin. Then

$$\begin{aligned} \overline{V}\dot{\overline{V}}_0^* &= \overline{v}\dot{\overline{v}}_0^* \sum_{m,n} e^{i\mathbf{k}.(\mathbf{r}_m-\mathbf{r}_n)} \\ &= \overline{v}\dot{\overline{v}}_0^* \{N + 2\sum_{m>n} \cos \mathbf{k}.(\mathbf{r}_m - \mathbf{r}_n)\}, \end{aligned} \tag{a 28}$$

where there are $\frac{1}{2}N(N-1)$ terms in the summation. Suppose that the N dislocations are scattered over a circle of radius r_1. We now assume that the summation in (a 27) may be replaced by the integral

$$\tfrac{1}{2}N(N-1)\int_{\mathbf{r}} \frac{d\mathbf{r}}{S}\int_{\mathbf{r}'} \frac{d\mathbf{r}'}{S} \cos \mathbf{k}.(\mathbf{r}-\mathbf{r}')\, P(\mathbf{r}-\mathbf{r}'), \tag{a 29}$$

where $S = \pi r_1^2$ is the area of the circle and $P(\mathbf{r}-\mathbf{r}')$ is a density function expressing the relative probability of finding a dislocation at a vector distance $\mathbf{r}-\mathbf{r}'$ from any

27-2

particular dislocation. For a completely random distribution $P = 1$ everywhere inside r_1. If the dislocations tend to avoid one another and form an irregular lattice, P will be less than unity for small $|\mathbf{r}-\mathbf{r}'|$ and will approach unity for large $|\mathbf{r}-\mathbf{r}'|$. As a rough approximation to a more complicated distribution function we take

$$P = 1, \quad |\mathbf{r}-\mathbf{r}'| \geqslant r_0,$$
$$P = 0, \quad |\mathbf{r}-\mathbf{r}'| < r_0,$$

i.e. each dislocation is surrounded by a circle of radius r_0 inside which another cannot lie. It is easiest to carry out the integration first assuming that $P = 1$ everywhere and then subtract a correction. The integral in (a 29) then becomes

$$\int_0^{2\pi}\int_0^{r_1}\int_0^{2\pi}\int_0^{r_1} \cos \mathbf{k}.(\mathbf{r}-\mathbf{r}')\, r\,dr\,d\theta\, r'\,dr'\,d\theta', \tag{a 30}$$

less the correction $$\int_0^{2\pi}\int_0^{r_1}\int_0^{2\pi}\int_0^{r_0} \cos \mathbf{k}.(\mathbf{r}-\mathbf{r}')\, r\,dr\,d\theta\, r'\,dr'\,d\theta', \tag{a 31}$$

with the understanding that in the integration from 0 to r_0 only those parts of the circle of radius r_0 about $\mathbf{r}$ which lie within r_1 are to contribute. If

$$I(x) = \int_0^{2\pi}\int_0^{x} \cos(kr\cos\theta)\, r\,dr\,d\theta = 2\pi x^2 \frac{J_1(kx)}{kx},$$

the expression (a 30) is $I^2(r_1)$. In (a 31) the integration over the dashed co-ordinates is equivalent to multiplication by πr_1^2 with a small error of order $r_0 r_1/r_1^2$ due to the intersection of some of the circles of radius r_0 with the circumference of r_1. Integration over the undashed co-ordinates yields $I(r_0)$. Hence if we replace $N(N-1)$ by N^2 (a 28) becomes

$$\overline{V\dot{V}_0^*} = \overline{v\dot{v}_0^*}\, N\left\{1 + 4N\frac{J_1^2(kr_1)}{(kr_1)^2} - 2N\frac{r_0^2}{r_1^2}\frac{J_1(kr_0)}{kr_0}\right\}.$$

The energy loss will be given by (a 23) with the factor $N\{\,\}$ inserted in the integrand. The second term in $\{\,\}$ represents, for large r_1, a peak at the origin of height N and width of the order of $1/r_1$. It can easily be shown to give a negligible contribution to w. Since $|J_1(x)/x| \leqslant \frac{1}{2}$ the w of (a 23) will be altered by a factor lying between the limits $N(1 \pm Nr_0^2/r_1^2)$. $\pi r_1^2/N$ is the mean area occupied per dislocation and πr_0^2 is the forbidden area surrounding each dislocation; Nr_0^2/r_1^2 will therefore be a small fraction of unity for small departures from complete randomness, and the loss per cycle will be approximately N times that of a single dislocation. In the extreme case where the circles r_0 round each dislocation form a two-dimensional close-packed array $Nr_0^2/r_1^2 = \frac{1}{2}\pi\sqrt{3} \doteqdot 0{\cdot}9$, and a detailed calculation is necessary. If $N\{\,\}$ is inserted in the integrand of (a 23) only the first term of the expansion of H_3 need be retained if $r_0 \gg l$. Using the result

$$\int_0^\infty \frac{J_1(x)}{x^4+a^4} x^2\,dx = (\text{kei}'\,a)/a,$$

we easily find $$w = N\frac{\pi^2 c_p d'\omega C^2}{2T\kappa}\left\{\frac{1}{4}\log\frac{\kappa}{\omega l^2} - N\frac{r_0^2}{r_1^2}\frac{\text{kei}'\,[r_0(\omega/\kappa)^{\frac{1}{2}}]}{r_0(\omega/\kappa)^{\frac{1}{2}}}\right\}.$$

For $x>1$ kei$'x$ falls off exponentially with x, whilst for small x, $x^{-1}\text{kei}'x \sim -\frac{1}{2}\log x$. Thus when $r_0 > (\omega/\kappa)^{\frac{1}{2}}$, w is effectively N times that of a single dislocation, and when $r_0 < (\omega/\kappa)^{\frac{1}{2}}$ the same result holds with $\log(\kappa/\omega l^2)$ replaced by $\log(r_0^2/l^2)$. For kilocycle frequencies and $l \sim 10^{-7}$ cm., $\log(\kappa/\omega l^2) \sim 20$, and for any reasonable value of r_0, $\log(r_0^2/l^2)$ will be of the same order.

The above calculations refer to a finite collection of dislocations in an infinite medium. The physical case is that of a finite body filled with dislocations. Imagine the finite body inserted into a suitable cavity in an infinite medium. The oscillating heat-sources representing the dislocations near the surface of the original body will only affect the new material to a depth of the order of $(\kappa/\omega)^{\frac{1}{2}}$ (cf. Carslaw & Jaeger 1947). Since the process of energy-loss is essentially a volume effect the damping will only be altered by a fraction of the order of $L^2(\kappa/\omega)^{\frac{1}{2}}/L^3$, where L is some linear dimension of the body. Provided $(\kappa/\omega)^{\frac{1}{2}} \ll L$ the results for the infinite case may be applied to the finite body.

A 6. *Radiation damping of dislocations*

For simplicity consider a screw dislocation. The stress around it is everywhere a pure shear. If the centre of the dislocation is at $(\xi, 0)$, then by (a 13) the displacement is

$$u_3 = \frac{a}{2\pi}\text{arc tan}\frac{y}{x-\xi} \doteqdot \frac{a}{2\pi}\theta + \frac{a}{2\pi}\frac{\sin\theta}{r}\xi \tag{a 32}$$

for $r \gg \xi$. If we put $\xi = d\cos\omega t$ in (a 32), u_3 will not be a solution of the elastic wave-equation. It will be approximately correct for r much less than the wave-length of sound of frequency $\omega/2\pi$ in the material. The problem is to find a solution of the wave-equation representing outgoing waves for large r and approximating to (a 32) with $\xi = d\cos\omega t$ for small r. Since the term $a\theta/2\pi$ already satisfies the wave-equation the problem reduces to solving

$$c^2\nabla^2 u - \ddot{u} = 0, \quad c = (G/\rho)^{\frac{1}{2}},$$

with

$$u = \frac{ad}{2\pi}\frac{\sin\theta}{r}\mathrm{e}^{-i\omega t} \quad (d \ll r \ll 1/k)$$

and

$$u = F(r,\theta)\,\mathrm{e}^{i(kr-\omega t)} \quad (r \gg 1/k),$$

where c is the velocity of shear waves and $k = \omega/c$. The solution must be of the form

$$u = A\,\mathrm{e}^{-i\omega t}\sin\theta\, Z_1(kr),$$

where Z_1 is a Bessel function of order unity. The Bessel function which behaves like e^{ix} for large x is $H_1^{(1)}(x)$, which has for large and small x the forms $2i/\pi x$ and $(2/\pi x)^{\frac{1}{2}}\mathrm{e}^{ix}$ respectively. This easily leads to the following form for u at large r, expressed now in real form:

$$u = \frac{akd}{4}\left(\frac{2}{\pi kr}\right)^{\frac{1}{2}}\cos[\phi(r) - \omega t]\sin\theta.$$

The rate of flux of energy in the direction of propagation (radially outwards) is $\frac{1}{2}\rho u^2 c$ per unit area (Rayleigh 1894). If this is integrated over a cylinder of radius r and over a period of the oscillation it is easily shown that

$$w = \frac{\pi}{16}\rho a^2 d^2 \omega^2 \tag{a 33}$$

is the energy radiated away per cycle per unit length of dislocation.

References

Burgers, J. M. 1939 *Proc. K. Akad. Wet. Amst.* **42**, 293.
Carslaw, H. S. & Jaeger, J. C. 1947 *Conduction of heat in solids.* Oxford: Clarendon Press.
Debye, P. 1927 *Phys. Z.* **28**, 135.
Kauzmann, W. 1941 *Trans. Amer. Inst. Min. (Metall.) Engrs*, **143**, 57.
Koehler, J. S. 1941 *Phys. Rev.* **60**, 397.
Lawson, A. W. 1941 *Phys. Rev.* **60**, 330.
Mott, N. F. & Nabarro, F. R. N. 1948 *Phys. Soc. Report of Conference on the Strength of Solids*, p. 1.
Päsler, M. 1944 *Z. Phys.* **122**, 357.
Rayleigh, Lord 1894 *Theory of sound.* London: Macmillan.
Read, T. A. 1941 *Trans. Amer. Inst. Min. (Metall.) Engrs*, **143**, 30.
Read, T. A. & Tyndall, P. T. 1946 *J. Appl. Phys.* **17** (9), 720.
Seitz, F. 1943 *The physics of metals.* New York: McGraw-Hill.
Seitz, F. & Read, T. A. 1941 *J. Appl. Phys.* **12** (6), 475.
Swift, L. M. & Richardson, J. E. 1947 *J. Appl. Phys.* **18** (4), 417.
Titchmarsh, E. C. 1937 *Fourier integrals.* Oxford: Clarendon Press.
Watson, G. N. 1922 *A treatise on Bessel functions.* Cambridge University Press.
Zener, C. 1940 *Proc. Phys. Soc.* **52**, 152.

Uniformly Moving Dislocations

By J. D. ESHELBY
H. H. Wills Physical Laboratory, University of Bristol

Communicated by N. F. Mott; MS. received 2nd December 1948

ABSTRACT. An expression is derived for the displacements in an isotropic elastic medium which contains an edge dislocation moving with uniform velocity c. When $c=0$ the solution reduces to that given by Burgers for a stationary edge dislocation. The energy density in the medium becomes infinite as c approaches c_2, the velocity of shear waves in the medium; this velocity therefore sets a limit beyond which the dislocation cannot be accelerated by applied stresses. The atomic structure of the medium is next partly taken into account, following the method already used by Peierls and Nabarro for the stationary dislocation. The solution found in this way differs from the one in which the atomic structure is neglected only within a region of width ζ which extends not more than a few atomic distances from the centre. ζ varies with c and vanishes when $c=c_r$, the velocity of Rayleigh waves. It becomes negative when $c_r< c< c_2$. Thus c_r rather than c_2 appears to be the limiting velocity when the atomic nature of the medium is taken into account. Since $c_r \simeq 0{\cdot}9c_2$ the difference is not of much importance.

The same method applied to a screw dislocation gives, in the purely elastic case, the expression already derived by Frank. The corresponding Peierls–Nabarro calculation shows that the width ζ is proportional to $(1-c^2/c_2^2)^{\frac{1}{2}}$. This "relativistic" behaviour is analogous to Frenkel and Kontorowa's results for their one-dimensional dislocation model.

§ 1. INTRODUCTION

FRANK (1949) has discussed the motion of screw dislocations: he shows that the displacement system suffers a contraction in the direction of motion as $(1-c^2/c_2^2)^{\frac{1}{2}}$ where c is the velocity of motion and c_2 is the velocity of shear waves in the medium. He has also discussed the case of edge dislocations approximately. In the present paper an exact solution within the limits of isotropic elasticity theory is given for the motion of an edge dislocation. The material is supposed to be cut at the slip plane, and the displacements due to uniformly moving distributions of force on the cut surfaces are derived. These travelling displacements may be so chosen that the two surfaces can be stuck together again leaving a moving dislocation in the material. This method lends itself (§ 3) to an extension of the calculation of Peierls (1940) and Nabarro (1947) in which the shape of the dislocation near its centre is determined by considering the balance between the elastic forces and the interatomic forces acting across the slip plane. In § 4 the same method is applied to the screw dislocation. Frank's results are reproduced and in addition the Peierls–Nabarro calculation for this case shows that the width of the screw dislocation behaves relativistically just as it does in the one-dimensional model of Frenkel and Kontorowa.

§ 2. MOVING EDGE DISLOCATION

Imagine the infinite unstressed elastic body of Figure 1 to be cut in two along the plane $y=0$ and the two halves slightly separated. Let tangential forces $p(x)$ per unit area be applied to the surface AB, independent of z and everywhere parallel to the x-axis, and producing horizontal and vertical displacements u, v

at the surface. Now let exactly equal opposite forces $-p(x)$ be applied at corresponding points on CD producing displacements $\bar{u}=-u$, $\bar{v}=v$. The two halves can be welded together and the forces p removed, leaving the body in a "self-stressed" state. A knowledge of u along the x-axis is sufficient to determine u and v throughout the solid.

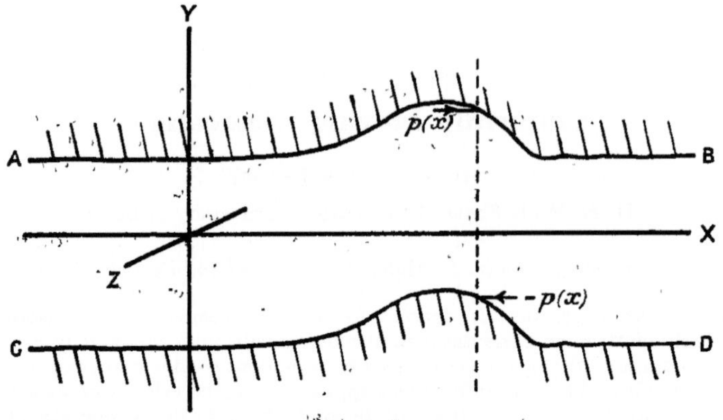

Figure 1.

The particular choice

$$u=-\bar{u}=0,\quad x>0;\qquad u=-\bar{u}=\lambda_0/2,\quad x<0 \qquad \ldots\ldots(1)$$

for the surface displacements gives the edge-type dislocation of Burgers (1939) for which

$$\left.\begin{aligned} u&=\frac{\lambda_0}{2\pi}\tan^{-1}\frac{y}{x}+\frac{\lambda_0}{2\pi}\frac{\lambda+\mu}{\lambda+2\mu}\frac{xy}{x^2+y^2},\\ v&=-\frac{\lambda_0}{2\pi}\frac{\mu}{\lambda+2\mu}\ln(x^2+y^2)^{\frac{1}{2}}+\frac{\lambda_0}{2\pi}\frac{\lambda+\mu}{\lambda+2\mu}\frac{y^2}{x^2+y^2},\end{aligned}\right\} \qquad \ldots\ldots(2)$$

where λ and μ are the Lamé constants and λ_0 is the "strength" of the dislocation. u increases by λ_0 on describing a circuit around the origin.

Suppose now that the distributions of force p and $-p$ are moving uniformly along the x-axis with velocity c, so that they depend on x and t only through $x'=x-ct$. The corresponding displacements will be of the form $u(x',y)$, $v(x',y)$; their connection with each other and with $p(x')$ can be found by an extension of the analysis used in discussing Rayleigh surface waves (Love 1944).

Consider the displacements

$$u_1=Ae^{-\gamma sy}\sin sx',\qquad v_1=A\gamma e^{-\gamma sy}\cos sx',$$
$$u_2=B\beta e^{-\beta sy}\sin sx',\qquad v_2=Be^{-\beta sy}\cos sx',$$

where

$$\beta=(1-c^2/c_2^2)^{\frac{1}{2}},\qquad \gamma=(1-c^2/c_1^2)^{\frac{1}{2}},\qquad x'=x-ct,$$

and $c_1=\{(\lambda+2\mu)/\rho\}^{\frac{1}{2}}$, $c_2=(\mu/\rho)^{\frac{1}{2}}$ are the velocities of longitudinal and shear waves in the medium, whose density is ρ. u_1 and v_1 correspond to an irrotational disturbance and satisfy

$$(c_1^2\nabla^2-\partial^2/\partial t^2)(u_1,\,v_1)=0,\qquad \partial u_1/\partial y-\partial v_1/\partial x=0,\qquad \ldots\ldots(3)$$

while u_2 and v_2 correspond to a distortional disturbance and satisfy

$$(c_2^2\nabla^2 - \partial^2/\partial t^2)(u_2, v_2) = 0, \qquad \partial u_2/\partial x + \partial v_2/\partial y = 0,$$

and consequently the combination $u = u_1 + u_2$, $v = v_1 + v_2$ satisfies the equations governing the motion of an elastic solid, namely

$$(\lambda + \mu)\frac{\partial}{\partial x}\left(\frac{\partial u}{\partial x} + \frac{\partial v}{\partial y}\right) + \mu\nabla^2 u - \rho\frac{\partial^2 u}{\partial t^2} = 0$$

and a similar equation for v. The condition that the normal stress

$$p_{yy} = \lambda\partial u/\partial x + (\lambda + 2\mu)\partial v/\partial y$$

should vanish at the surface $y = 0$ is easily found to be $B/A = -\alpha^2/\beta$, where $\alpha = (1 - c^2/2c_2^2)^{\frac{1}{2}}$. Corresponding u and v are then

$$u = (2c_2^2/c^2)(e^{-\gamma sy} - \alpha^2 e^{-\beta sy})\sin sx', \qquad \ldots\ldots(4)$$

$$v = (2c_2^2/c^2)(\gamma e^{-\gamma sy} - \alpha^2 e^{-\beta sy}/\beta)\cos sx', \qquad \ldots\ldots(5)$$

the amplitude of u being chosen to be unity for $y = 0$. The shear stress at the surface must be equal to the applied force $p(x')$; it is easily found to be

$$p_{xy}(x', 0) = p(x') = D(c)s\sin sx', \qquad \ldots\ldots(6)$$

where $D(c) = -2\mu(2c_2^2/c^2)(\gamma - \alpha^4/\beta)$. Solving $D(c) = 0$ gives the velocity of Rayleigh waves.

If a more general travelling surface wave is built up in the form

$$u(x', 0) = \int_0^\infty f(s)\sin sx'\,ds,$$

then

$$u(x', y) = (2c_2^2/c^2)\int_0^\infty (e^{-\gamma sy} - \alpha^2 e^{-\beta sy})f(s)\sin sx'\,ds, \qquad \ldots\ldots(7)$$

$$v(x', y) = (2c_2^2/c^2)\int_0^\infty (\gamma e^{-\gamma sy} - \alpha^2 e^{-\beta sy}/\beta)f(s)\cos sx'\,ds, \qquad \ldots\ldots(8)$$

$$p(x')/D(c) = \int_0^\infty f(s)s\sin sx'\,ds. \qquad \ldots\ldots(9)$$

If u, v are split up into u_1, v_1 and u_2, v_2, the suffix 1 referring to the parts containing γ and 2 to those containing β, then γu_1, v_1 are Hilbert transforms (Titchmarsh 1937) and so are u_2, βv_2. A more useful relation between them is the following. Equations (3), (4) and (5) give

$$\left(\frac{\partial^2}{\partial x'^2} + \frac{\partial^2}{\partial(\gamma y)^2}\right)(\gamma u_1, v_1) = 0, \qquad -\frac{\partial(\gamma u_1)}{\partial x'} = \frac{\partial v_1}{\partial(\gamma y)}, \qquad \frac{\partial(\gamma u_1)}{\partial(\gamma y)} = \frac{\partial v_1}{\partial x'},$$

and this will continue to hold for the more general u_1, v_1. By the usual theory of conjugate functions it follows that if

$$\gamma u_1 = \operatorname{Im} P(x + i\gamma y) \quad \text{then} \quad v_1 = \operatorname{Re} P(x + i\gamma y), \qquad \ldots\ldots(10)$$

and similarly if

$$u_2 = \operatorname{Im} Q(x + i\beta y) \quad \text{then} \quad \beta v_2 = \operatorname{Re} Q(x + i\beta y), \qquad \ldots\ldots(11)$$

where P and Q are analytic functions. (The relation between Fourier transforms, Hilbert transforms and the complex variable are fully discussed by Titchmarsh).

Suppose that at the surfaces AB and CD u and $\bar{u}$ are given by (1) with x replaced by x'. This may be written

$$u = -\bar{u} = \frac{\lambda_0}{4} - \frac{\lambda_0}{4}\frac{2}{\pi}\int_0^\infty \frac{\sin sx'}{s}\,ds.$$

The first term clearly represents a constant displacement for all x', y. Thus by (7) for $y>0$

$$\begin{aligned} u(x',y) &= \frac{\lambda_0}{4} - \frac{\lambda_0}{2\pi}\left(\frac{2c_2^2}{c^2}\right)\int_0^\infty (e^{-\gamma sy} - \alpha^2 e^{-\beta sy})\frac{\sin sx'}{s}\,ds \\ &= \frac{\lambda_0}{2\pi}\frac{2c_2^2}{c^2}\left[\tan^{-1}\frac{\gamma y}{x'} - \alpha^2\tan^{-1}\frac{\beta y}{x'}\right]. \end{aligned} \qquad \ldots\ldots(12)$$

Since

$$\ln(x+iy) = \ln(x^2+y^2)^{\frac{1}{2}} + i\tan^{-1}(y/x),$$

it follows from (10) and (11) that

$$v(x',y) = \frac{\lambda_0}{2\pi}\frac{2c_2^2}{c^2}\left[\gamma\ln(x'^2+\gamma^2y^2)^{\frac{1}{2}} - \frac{\alpha^2}{\beta}\ln(x'^2+\beta^2y^2)^{\frac{1}{2}}\right]. \qquad \ldots\ldots(13)$$

There is a corresponding expression for (u, v) when $y<0$ derived from $\bar{u}$. Except at the singularity at $x=0$ there is no relative motion between the planes AB and CD and they can be imagined to be welded together. It is easy to see that (12) and (13) give the displacements for $y<0$ as well as for $y>0$.

On passing to the limit $c=0$, (12) and (13) become (2). Independently of the method of derivation the displacements (12) and (13) have the following properties: (i) they satisfy the wave equation; (ii) u increases by λ_0 on describing any circuit round the point $x'=x-ct=0$; (iii) they reduce to the expressions for a stationary edge dislocation when $c=0$.

The irrotational and distortional terms behave relativistically with respective limiting velocities c_1 and c_2, but the proportion between them also alters with velocity. On account of the factor $1/\beta$ in the second term of v the energy becomes infinite as c approaches c_2 even if the singularities at $x'=0$ and infinity are excluded in the usual way. If solutions exist for $c>c_2$ it would be difficult to connect them continuously with the solution for $c<c_2$ and so identify them as moving edge dislocations; c_2 is thus the limiting velocity.

If (1) is replaced by

$$u = -\bar{u} = \frac{\lambda_0}{4}\left(1 + \frac{2}{\pi}\tan^{-1}\frac{x'}{\zeta}\right), \qquad \ldots\ldots(14)$$

to which it reduces when $\zeta\to 0$, a corresponding calculation shows that u and v in the body of the material are given by (12) and (13) on replacing y by $y\pm\zeta$, the sign of ζ to be chosen to be the same as that of y. Formally this "spread" dislocation is equivalent to a distribution of elementary dislocation of the type (12), (13), the total strength between x' and $x'+dx'$ being

$$d\lambda_1 = \frac{\lambda_0}{\pi}\frac{\zeta}{x'^2+\zeta^2}\,dx'. \qquad \ldots\ldots(15)$$

The change of u round a circuit depends on the size and position ot the circuit but approaches λ_0 for a large circuit surrounding $x'=0$.

At first sight it is surprising that the method used should lead to a moving form of the Burgers expression (2) rather than to a moving form of the original Taylor expression

$$u=\frac{\lambda_0}{2\pi}\tan^{-1}\frac{y}{x}, \qquad v=\frac{\lambda_0}{2\pi}\ln(x^2+y^2)^{\frac{1}{2}}, \qquad \ldots\ldots(16)$$

or something similar. This behaviour can be traced to the condition $B/A=-\alpha^2/\beta$ imposed to make p_{yy} vanish over the plane $y=0$. The actual condition used was that the Fourier transform of $p_{yy}(x',0)$ should vanish for all x'. Strictly this implies (Doetsch 1937) that $p_{yy}(x',0)$ vanishes "almost everywhere" and that any singularities are of such a feeble kind that $\int_0^X p_{yy}(x',0)\,dx'=0$ for any X. A singularity of the type of a Dirac δ-function is thus excluded. The last equation implies that $\int_{-\infty}^{\infty} p_{yy}(x',0)\,dx=0$ which in turn means that

$$\int_{-\infty}^{\infty} p_{yy}(x',y)\,dx=0, \qquad \ldots\ldots(17)$$

as can be seen by considering the equilibrium of a slab bounded by the planes $y=0$ and $y=$const., or analytically by the method of Love (1944, §117). As is well-known (16) does not satisfy (17), which accounts for its non-appearance in the previous analysis.

The whole argument could in fact be repeated without requiring that $p_{yy}(x',0)=0$ everywhere, and the two halves of the material could be welded together as before and would be in equilibrium provided suitable external forces could be applied to the junction plane. Among other solutions (12), or the corresponding form with y replaced by $y\pm\zeta$, would reappear but with the coefficients of the inverse tangents arbitrary. The first or second terms of (12) and (13) taken separately do in fact satisfy the wave equation and give rise to the two following types of dislocations which could exist with suitable forces applied to the plane $y=0$:

$$u=\frac{\lambda_0}{2\pi}\tan^{-1}\frac{\gamma(y\pm\zeta)}{x'}, \qquad v=\frac{\lambda_0}{2\pi}\ln\{x'^2+\gamma^2(y\pm\zeta)^2\}^{\frac{1}{2}} \qquad \ldots\ldots(18)$$

and

$$u=\frac{\lambda_0}{2\pi}\tan^{-1}\frac{\beta(y\pm\zeta)}{x'}, \qquad v=\frac{\lambda_0}{2\pi}\frac{1}{\beta}\ln\{x'^2+\beta^2(y\pm\zeta)^2\}^{\frac{1}{2}}. \qquad \ldots\ldots(19)$$

Their limiting velocities are c_1 and c_2 respectively. For (18)

$$p_{yy}=\pm\frac{\lambda_0}{\pi}\mu\gamma\alpha^2\frac{\zeta}{x'^2+\zeta^2}$$

on the surfaces AB and BC (Figure 1) respectively. Thus to support it the dislocation requires a total external force

$$F_1=2\int_{-\infty}^{\infty} p_{yy}\,dx=2\lambda_0\mu\alpha^2$$

parallel to the y-axis distributed over a distance of the order ζ along the x-axis. When $\zeta=0$ the force is concentrated at the origin. For a velocity $c=\sqrt{2}c_2$, F_1 vanishes and the dislocation could exist without external forces provided $\sqrt{2}c_2$ is less than the limiting velocity c_1: the condition for this is $\lambda>0$, which is satisfied.

in normal physical cases. This agrees with (12) and (13) on putting $c=\sqrt{2}c_2$. This isolated solution cannot be reached from the solutions for $c<c_2$. On the other hand (19) requires a total force $F_2=2\lambda_0\mu\beta$ which does not vanish below the limiting velocity. At any velocity less than c_2 a linear combination of (18) and (19) can exist provided the proportion between them is such that the external forces required to support each separately are equal and opposite. This is equivalent to the condition $B/A=-\alpha^2/\beta$ and leads back to (12) and (13).

When $c=0$ and $\zeta=0$ both (18) and (19) reduce to (16). The Taylor expression is thus both irrotational and dilatationless and requires for its maintenance a distribution of force along the z-axis of linear density $2\lambda_0\mu$ directed parallel to the y-axis. If it is to be in equilibrium in the slab defined by the planes $y=\pm Y$ there must be a normal pressure of total amount $\lambda_0\mu$ suitably distributed over the surface Y and a corresponding normal tension of total amount $\lambda_0\mu$ distributed over the surface $-Y$.

§ 3. THE WIDTH OF A MOVING EDGE DISLOCATION

By considering the surfaces AB and CD of Figure 1 not as welded together but as kept in place by interatomic forces across the slip-plane, Peierls (1940) and Nabarro (1947) were led to an expression of the form (14) (with x replacing x') for a stationary edge dislocation with a width $\zeta=\lambda_0/2(1-\sigma)$ depending on σ, Poisson's ratio. Briefly their argument was the following. Starting from the relation

$$u=\sin sx, \qquad p_{xy}=-\{\mu/(1-\sigma)\}s\sin sx, \qquad \ldots\ldots(20)$$

connecting a sinusoidal displacement of the surface AB with the tangential force necessary to produce it, and from an assumed law

$$p_{xy}=-(\mu/2\pi)\sin 2\pi(u-\bar{u})/\lambda_0, \qquad \ldots\ldots(21)$$

giving the atomic forces arising from the relative horizontal displacement of AB and CD at any point, they reach an integral equation for u which has (18) as a solution. The atomic nature of the medium only enters the problem through the periodicity of (21) and so the origin of x in (14) is arbitrary.

In so far as (21) is a good approximation in the stationary case it will also be nearly correct if du/dt is small compared with the natural velocity of vibration of the atoms, i.e. if the velocity of the dislocation is small compared with c_2. Even if c is of this order du/dt will only approach c near the centre of the dislocation where (21) is least accurate even in the static case. A repetition of the Peierls–Nabarro calculation for the moving dislocation may therefore have some significance. According to (4) and (6)

$$u=\sin sx', \qquad p_{xy}=D(c)s\sin sx' \qquad \ldots\ldots(22)$$

replaces (20), to which it reduces when $c=0$. This leads to (14) (with x replaced by x') where now

$$\zeta=\delta\lambda_0/2(1-\sigma), \qquad \delta=D(c)/D(0).$$

In Figure 2, δ is shown for $\sigma=\frac{1}{3}$, compared with the simple relativistic contraction factor β occurring in the case of the screw dislocation (§ 4). δ vanishes when $c=c_r$, the Rayleigh wave velocity.

According to the present model the isolated solution of § 2 with $c=\sqrt{2}c_2$ would have a complex ζ unless $\lambda=0$. In that case $\zeta=0$ and $\gamma=\alpha=0$ so that it would exist in the completely contracted form $v=0$, $u=0$, $x'>0$; $u=\lambda_0/2$, $x'<0$, $y>0$; $u=-\lambda_0/2$, $x'<0$, $y<0$.

It is hardly worth while discussing what happens when $c_r<c<c_2$ since the approximations made will be very poor in that region. The statement at the end of §2 should be qualified by saying that the limiting velocity is of the order of $0{\cdot}9c_2$.*

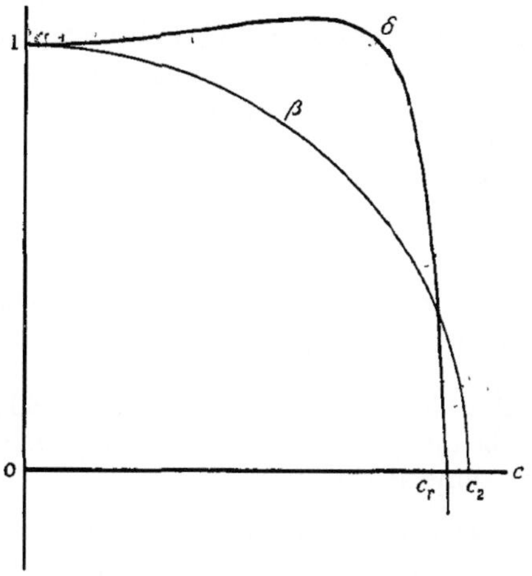

Figure 2. β and δ for $\sigma=\frac{1}{3}$ as functions of c.

Apart from any considerations of interatomic forces the dislocation of this section when ζ is any function of c which vanishes for $c=0$ could be considered as the moving form of (1). However, if it is required that the change of u shall be the same for any circuit surrounding $x'=0$ the form (12), (13) is unique.

§ 4. MOVING SCREW DISLOCATION

The screw dislocation already discussed by Frank may be treated in the same way, or more briefly by the following equivalent method.

If $\phi(x, y)$ satisfies $(\partial^2\phi/\partial x^2)+(\partial^2\phi/\partial y^2)=0$ then (Jeffreys and Jeffreys 1946, Titchmarsh 1937)

$$\phi(x, y)=\frac{1}{\pi}\int_{-\infty}^{\infty} d\xi \int_0^{\infty} \phi(\xi, 0)e^{-sy}\cos s(x-\xi)\,ds.$$

A purely distortional elastic disturbance in which the displacement w is everywhere in the z-direction must satisfy

$$\frac{\partial^2 w}{\partial x^2}+\frac{\partial^2 w}{\partial y^2}-\frac{1}{c_2^2}\frac{\partial^2 w}{\partial t^2}=0.$$

If it is of the form $w(x', y)$ this becomes

$$\frac{\partial^2 w}{\partial x'^2}+\frac{\partial^2 w}{\partial(\beta y)^2}=0,$$

* *Note added in proof.* The limiting velocity is exactly c_r only because we have confined the departure from linear elasticity to a plane.

so that if
$$w(x',0)=\int_0^\infty f(s)\sin sx'\,ds, \qquad ......(23)$$

then
$$w(x',y)=\int_0^\infty f(s)e^{-\beta sy}\sin sx'\,ds$$

and
$$p_{yz}(x',0)=\mu\frac{\partial w}{\partial y}=-\mu\beta\int_0^\infty f(s)s\sin sx'\,ds. \qquad ......(24)$$

If $w(x',0)$ is given by the right hand side of (1) (with x replaced by x') then

$$w(x',y)=\frac{\lambda_0}{2\pi}\tan^{-1}\frac{\beta y}{x'} \qquad ......(25)$$

in agreement with Frank.

If the law

$$p_{yz}=-(\mu/2\pi)\sin 2\pi(w-\bar{w})\lambda_0 \qquad ......(26)$$

is assumed for the interatomic forces across the slip-plane a repetition of the Peierls–Nabarro calculation using (23), (24) and (26) in place of (22) and (21) gives $w=-\bar{w}=(\lambda_0/2\pi)\tan^{-1}(\beta\zeta/x')$ where now $\zeta=\lambda_0/2$ is the width of the stationary dislocation which is thus less than the width of the edge dislocation by the factor $(1-\sigma)\simeq 2/3$. The displacement is obtained by replacing y by $y\pm\zeta$ in (25), the sign of ζ being the same as that of y, i.e.

$$w=(\lambda_0/2\pi)\tan^{-1}\beta y(1+\zeta/|y|)/x'. \qquad ......(27)$$

It corresponds to a distribution of dislocations of the type (25) along the x-axis according to the law (19). (23) can be obtained from (25) by cutting out the material between the planes $y=\pm\zeta$ and sticking together the remainder.

According to (27) the width of a screw dislocation suffers a simple relativistic contraction (with limiting velocity c_2) in complete analogy with the one-dimensional model of Frenkel and Kontorowa (1938).

The calculation leading to (27) is not very significant for a cubic material since there is not a unique slip-plane as there is for the edge dislocation. However (27) may be considered as a crude approximation to the case of a screw dislocation in a hexagonal material. Here there is a unique slip-plane perpendicular to the hexagonal axis. The problem should, of course, be solved using the equations of non-isotropic elastic theory.

ACKNOWLEDGMENTS

The author wishes to thank Dr. F. C. Frank and Mr. F. R. N. Nabarro for helpful discussion and suggestions.

REFERENCES

Burgers, J. M., 1939, *Proc. Kon. Ned. Akad. v. Wet.*, **42**, 293.
Doetsch, G., 1937, *Laplace-Transformation* (Berlin: Springer), p. 102.
Frank, F. C., 1949, *Proc. Phys. Soc.* A, **62**, 131.
Frenkel, J., and Kontorowa, T., 1938, *Phys. Z. Sowjet*, **13**, 1.
Jeffreys, H. and B. S., 1946, *Methods of Mathematical Physics* (Cambridge: University Press), p. 426.
Love, A. E. H., 1944, *Elasticity* (Cambridge: University Press), p. 309.
Nabarro, F. R. N., 1947, *Proc. Phys. Soc.*, **59**, 256.
Peierls, R., 1940, *Proc. Phys. Soc.*, **52**, 34.
Titchmarsh, B. C., 1937, *Fourier Integrals* (Oxford: University Press), Chap. V.

LXXXII. *Edge Dislocations in Anisotropic Materials.*

J. D. Eshelby,
H. H. Wills Physical Laboratory, University of Bristol*.

[Received May 18, 1949.]

Abstract.

Burgers has discussed dislocations in materials with cubic symmetry for the general case in which the dislocation axis is an arbitrary curve. The results of the present paper are limited to dislocations of edge type with an infinite straight line as axis, but apply to a material with the symmetry of any of the crystal classes. The axis of the dislocation may be arbitrarily inclined to the symmetry axes of the material. Expressions are given for the elastic displacements and the energy of the dislocation. Nabarro's calculation of the width of a dislocation is extended to the anisotropic case.

§1. Introduction.

Burgers (1939) has discussed elastic dislocations in both isotropic materials and crystals with cubic symmetry for the general case in which the axis of a dislocation is allowed to be an arbitrary curve. It would be difficult to extend his analysis to cover crystals of lower symmetry. However, for a dislocation of edge type whose axis is a straight line of infinite extent the problem is one of plane strain. Green (1945) has already solved the problem of generalized plane stress in an anisotropic plate. In the present paper Green's results are adapted to the case of plane strain (§2) and used to find the elastic displacements surrounding the dislocation (§3). In §4 the elastic energy of the dislocation is calculated and in §5 Nabarro's calculation of the width of a dislocation is repeated for the anisotropic case. §6 contains some illustrative numerical results.

§2. The Solution of Problems of Plane Strain in Anisotropic Materials.

A problem in generalized plane stress in an isotropic plate can be solved by regarding it as a problem in plane strain and in the results replacing the stresses by their mean values across the plate and also replacing λ by $\lambda'=2\lambda\mu/(\lambda+2\mu)$, where λ and μ are the Lamé constants (Love 1944). Conversely a problem in plane strain could be replaced by one in plane stress. This latter process can be extended to the anisotropic case.

* Communicated by Prof. N. F. Mott, F.R.S.

For plane strain in which the z-displacement w vanishes and the x- and y-displacements u, v are independent of z we have, with the usual notation :

$$
\begin{aligned}
e_{zz}&=s_{31}\widehat{xx}+s_{32}\widehat{yy}+s_{33}\widehat{zz}+s_{34}\widehat{yz}+s_{35}\widehat{xz}+s_{36}\widehat{xy}=0,\\
e_{yz}&=s_{41}'\widehat{xx}+s_{42}\widehat{yy}+s_{43}\widehat{zz}+s_{44}\widehat{yz}+s_{45}\widehat{xz}+s_{46}\widehat{xy}=0,\\
e_{xz}&=s_{51}\widehat{xx}+s_{52}\widehat{yy}+s_{53}\widehat{zz}+s_{54}\widehat{yz}+s_{55}\widehat{xz}+s_{56}\widehat{xy}=0.
\end{aligned}
$$

These equations can be used to eliminate $\widehat{zz}$, $\widehat{yz}$, $\widehat{xz}$ from the expressions for e_{xx}, e_{yy}, e_{xy}, which become

$$
\begin{aligned}
e_{xx}&=s_{11}'\widehat{xx}+s_{12}'\widehat{yy}+s_{16}'\widehat{xy},\\
e_{yy}&=s_{21}'\widehat{xx}+s_{22}'\widehat{yy}+s_{26}'\widehat{xy},\\
e_{xy}&=s_{61}'\widehat{xx}+s_{62}'\widehat{yy}+s_{66}'\widehat{xy},
\end{aligned}
$$

where

$$
s_{ij}'=\begin{vmatrix} s_{ij} & s_{i3} & s_{i4} & s_{i5}\\ s_{3j} & s_{33} & s_{34} & s_{35}\\ s_{4j} & s_{43} & s_{44} & s_{45}\\ s_{5j} & s_{53} & s_{54} & s_{55}\end{vmatrix} \div \begin{vmatrix} s_{33} & s_{34} & s_{35}\\ s_{43} & s_{44} & s_{45}\\ s_{53} & s_{54} & s_{55}\end{vmatrix},\ i,j=1,2,6.
$$

Clearly $s_{ij}'=s_{ji}'$.

The following form is convenient for calculation (Schwein's expansion, *cf.* Aitken (1946)) :

$$
s_{ij}'=s_{ij}-s_{3j}s_{i3}/s_{33}-\begin{vmatrix} s_{3j} & s_{33}\\ s_{4j} & s_{43}\end{vmatrix}\cdot\begin{vmatrix} s_{i3} & s_{i4}\\ s_{33} & s_{34}\end{vmatrix}\Big/ s_{33}\begin{vmatrix} s_{33} & s_{34}\\ s_{43} & s_{44}\end{vmatrix}
$$
$$
-\begin{vmatrix} s_{3j} & s_{33} & s_{34}\\ s_{4j} & s_{43} & s_{44}\\ s_{5j} & s_{53} & s_{54}\end{vmatrix}\cdot\begin{vmatrix} s_{i3} & s_{i4} & s_{i5}\\ s_{33} & s_{34} & s_{35}\\ s_{43} & s_{44} & s_{45}\end{vmatrix}\Big/\begin{vmatrix} s_{33} & s_{34}\\ s_{43} & s_{44}\end{vmatrix}\cdot\begin{vmatrix} s_{33} & s_{34} & s_{35}\\ s_{43} & s_{44} & s_{45}\\ s_{53} & s_{54} & s_{55}\end{vmatrix} \quad . \ . \ . \ (1)
$$

Comparison with the results given by Green (1945) for the corresponding problem in generalized plane stress shows that it is only necessary to replace his s_{ij} by s_{ij}' to obtain solutions of a problem in plane strain. With this change the relevant part of his analysis is given below.

Introduce the two complex variables

$$
z_1=x+i\lambda_1 y,\quad z_2=x+i\lambda_2 y
$$

where λ_1, λ_2 are complex constants given implicitly in terms of the elastic constants by the following relation with $n=1$ or 2 :

$$
\lambda_n=\frac{1-\gamma_n-i\delta_n}{1+\gamma_n+i\delta_n},\quad \gamma_n=\frac{\alpha_n-1}{\alpha_n+1+2(\alpha_n-\frac{1}{4}k_n^2)^{\frac{1}{2}}},\quad \delta_n=\frac{-k_n}{\alpha_n+1+2(\alpha_n-\frac{1}{4}k_n^2)^{\frac{1}{2}}} \quad . \ . \ . \ (2)
$$

where

$$\alpha_1+\alpha_2+k_1k_2=(2s'_{12}+s'_{66})/s'_{22}, \quad \alpha_1\alpha_2=s'_{11}/s'_{22}$$

$$k_1+k_2=-2s'_{26}/s'_{22}, \quad k_1\alpha_2+k_2\alpha_1=-2s'_{16}/s'_{22}.$$

$\alpha_n-\frac{1}{4}k_n^2$ and α_n are real and positive and k_n, γ_n, δ_n are real.

For convenience we replace Green's f, g by f_1, f_2. In the expressions below, Σ represents summation over the values $n=1$, 2 and c.c. the complex conjugate of the quantity preceding it. To the stress function

$$\chi=\Sigma f_n(z_n)+\text{c.c.}$$

there correspond the displacements

$$u=-\Sigma C_n f'_n(z_n)+\text{c.c.}, \quad v=i\Sigma D_n f'_n(z_n)+\text{c.c.}$$

where

$$C_n=s'_{11}\lambda_n^2-s'_{12}+i\lambda_n s'_{16},$$

$$D_n=\lambda_n^{-1}(s'_{12}\lambda_n^2-s'_{22}+i\lambda_n s'_{26}),$$

and the stresses

$$\widehat{xx}=-\Sigma\lambda_n^2 f''_n(z_n)+\text{c.c.}$$

$$\widehat{yy}=\Sigma f''_n(z_n)+\text{c.c.}$$

$$\widehat{xy}=-i\Sigma f''_n(z_n)+\text{c.c.}$$

The couple on a circuit in the material is the change of

$$M=\Sigma\{z_n f'_n(z_n)-f_n(z_n)\}+\text{c.c.}$$

on going round the circuit. Similarly the x- and y-components of the total force acting on the circuit are given by the changes of

$$X=-i\Sigma\lambda_n f'_n(z_n)+\text{c.c.}$$

$$Y=\Sigma f'_n(z_n)+\text{c.c.}$$

on going round it.

§3. Edge Dislocations in Anistropic Bodies.

To represent an edge dislocation we need, following Burgers (1939), a solution for which (u, v) increases by a constant vector $\mathbf{b}$ on describing a circuit about the origin. The total force and couple on any circuit must vanish. Let $f_n(z_n)=\frac{1}{2}A_n \log z_n$, where A_n is a complex constant.

If the suffixes r, i denote the real and imaginary parts of a quantity then

$$\left.\begin{aligned} u&=-\Sigma\{(C_{nr}A_{nr}-C_{ni}A_{ni})\log\sqrt{(x_n^2+y_n^2)}-(C_{ni}A_{nr}+C_{nr}A_{ni})\tan^{-1}(y_n/x_n)\},\\ v&=-\Sigma\{(D_{ni}A_{nr}+D_{nr}A_{ni})\log\sqrt{(x_n^2+y_n^2)}+(D_{nr}A_{nr}-D_{ni}A_{ni})\tan^{-1}(y_n/x_n)\},\end{aligned}\right\} \quad (3)$$

where

$$x_n=x-\lambda_{ni}y, \qquad y_n=\lambda_{nr}y.$$

The couple given by M vanishes for any choice of A_1, A_2. The condition that the force given by Y should vanish is $A_{2i}=-A_{1i}$, and the corresponding condition for X is

$$\lambda_{1r}A_{1r}+\lambda_{2r}A_{2r}-(\lambda_{1i}-\lambda_{2i})A_{1i}=0. \quad \quad (4)$$

The change in (u, v) round a circuit about the origin arises from the change of 2π in the inverse tangents in (3). If the Burgers vector is required to be (b_x, b_y) we must have

$$C_{1i}A_{1r}+C_{2i}A_{2r}+(C_{1r}-C_{2r})A_{1i}=b_x/2\pi, \quad \quad (5)$$

$$D_{1r}A_{1r}+D_{2r}A_{2r}-(D_{1i}-D_{2i})A_{1i}=b_y/2\pi. \quad \quad (6)$$

Equations (4), (5) and (6) fix the values of the constants A_{1r}, A_{1i}, A_{2i}. In what follows we shall put $b_x=b$, $b_y=0$, so that the Burgers vector lies along the x-axis : the slip-plane is then the plane $y=0$. The expressions for (u, v) become fairly simple if $s'_{16}=s'_{26}=0$. This covers a number of cases in which the slip-plane and Burgers vector are simply related to the symmetry axes of the material. Several cases have to be distinguished. If

$$s'_{16}=s'_{26}=0 \text{ and } (2s'_{12}+s'_{66})^2>4s'_{11}s'_{22} \quad \quad (7)$$

λ_1 and λ_2 are real and

$$\left.\begin{aligned} u&=\frac{b}{2\pi}\frac{s'_{11}\lambda_1^2-s'_{12}}{s'_{11}(\lambda_1^2-\lambda_2^2)}\tan^{-1}\frac{\lambda_1 y}{x}+\frac{b}{2\pi}\frac{s'_{11}\lambda_2^2-s'_{12}}{s'_{11}(\lambda_2^2-\lambda_1^2)}\tan^{-1}\frac{\lambda_2 y}{x},\\ v&=-\frac{b}{2\pi}\frac{s'_{12}\lambda_1-s'_{22}\lambda_1^{-1}}{s'_{11}(\lambda_1^2-\lambda_2^2)}\log\sqrt{(x^2+y^2)}-\frac{b}{2\pi}\frac{s'_{22}\lambda_2-s'_{22}\lambda_1^{-1}}{s'_{11}(\lambda_2^2-\lambda_1^2)}\log\sqrt{(x^2+y^2)},\end{aligned}\right\} \quad (8)$$

where

$$\lambda_1=\lambda_{1r}=\alpha_1^{-\frac{1}{2}}, \quad \lambda_2=\lambda_{2r}=\alpha_2^{-\frac{1}{2}}$$

and α_1, α_2 are the roots of

$$s'_{22}\alpha^2-(2s'_{12}+s'_{66})\alpha+s'_{11}=0.$$

On the other hand, if

$$s'_{16}=s'_{26}=0 \quad \text{and} \quad (2s'_{12}+s'_{66})^2<4s'_{11}s'_{22}, \quad \quad (9)$$

λ_1 and λ_2 are complex conjugates and

$$\left.\begin{aligned} u&=\frac{b}{4\pi}\left\{\tan^{-1}\frac{\lambda_{1r}y}{x-\lambda_{1i}y}+\tan^{-1}\frac{\lambda_{1r}y}{x+\lambda_{1i}y}+\frac{s'_{11}(\lambda_{1r}^2-\lambda_{1i}^2)-s'_{12}}{2s'_{11}\lambda_{1r}\lambda_{1i}}\right.\\ &\qquad\left.\times[-\log\sqrt{((x-\lambda_{1i}y)^2+\lambda_{1r}^2y^2)}+\log\sqrt{((x+\lambda_{1i}y)^2+\lambda_{1r}^2y^2)}]\right\}\\ v&=-\frac{b}{4\pi}\left\{\left[\frac{s'_{12}}{2s'_{11}\lambda_{1r}}+\frac{s'_{22}}{2s'_{11}\lambda_{1r}(\lambda_{1r}^2+\lambda_{1i}^2)}\right]\left[\log\sqrt{((x-\lambda_{1i}y)^2+\lambda_{1r}^2y^2)}\right.\right.\\ &\qquad\left.+\log\sqrt{((x+\lambda_{1i}y)^2+\lambda_{1r}^2y^2)}\right]+\left[\frac{s'_{12}}{2s'_{11}\lambda_{1i}}-\frac{s'_{22}}{2s'_{11}\lambda_{1i}(\lambda_{1r}^2+\lambda_{1i}^2)}\right]\\ &\qquad\left.\times\left[\tan^{-1}\frac{\lambda_{1r}y}{x-\lambda_{1i}y}-\tan^{-1}\frac{\lambda_{1r}y}{x+\lambda_{1i}y}\right]\right\}\end{aligned}\right\}, \quad (10)$$

where $\lambda_1=\lambda_{1r}+i\lambda_{1i}$ can be found from (2) with

$$\alpha_1=(s'_{11}/s'_{22})^{\frac{1}{2}}, \quad k_1^2=2(s'_{11}/s'_{22})^{\frac{1}{2}}-(2s'_{12}+s'_{66})/s'_{22}.$$

When
$$s'_{16}=s'_{26}=0 \quad \text{and} \quad (2s'_{12}+s'_{66})^2=4s'_{11}s'_{22} \quad . \quad . \quad . \quad . \quad (11)$$
$\lambda_{1r}=\lambda_{2r}=1$ and $\lambda_{1i}=\lambda_{2i}=0$. The displacements can be obtained by passing to the limit in (8) or (10). The result is

$$\left.\begin{aligned} u&=\frac{b}{2\pi}\tan^{-1}\frac{y}{x}+\frac{b}{2\pi}\frac{s'_{11}-s'_{12}}{2s'_{11}}\frac{xy}{x^2+y^2}, \\ v&=-\frac{b}{2\pi}\frac{s'_{12}+s'_{22}}{2s'_{11}}\log\sqrt{(x^2+y^2)}-\frac{b}{2\pi}\frac{s'_{12}-s'_{22}}{2s'_{11}}\frac{y^2}{x^2+y^2}. \end{aligned}\right\} \quad . \quad (12)$$

For an isotropic substance

$$s'_{11}/s'_{22}=1, \quad s'_{12}/s'_{11}=-\lambda/(\lambda+2\mu)$$

and (12) agrees with the expression given by Burgers (1939).

In the general case when $s'_{16}\neq 0$, $s'_{26}\neq 0$ the expressions obtained by solving (4), (5) and (6) for A_{1r}, A_{2r}, A_{1i} and substituting them in (3) are rather clumsy. In the application in § 4, 5 only the values of (u, v) and $\widehat{xy}$ at the slip-plane $y=0$ will be required. There is a discontinuity in u on crossing the x-axis. If

$$u_1(x)=\lim_{\epsilon\to 0} u(x, \epsilon), \quad u_2(x)=\lim_{\epsilon\to 0} u(x, -\epsilon)$$

and similarly for v_1, v_2, then

$$\left.\begin{aligned} &u_1=u_2=P\log x, \quad x>0, \\ &u_1=P\log x+\tfrac{1}{2}b, \quad u_2=P\log x-\tfrac{1}{2}b, \quad x<0, \\ &v_1=v_2=Q\log x, \quad x\gtrless 0, \end{aligned}\right\} \quad . \quad . \quad (13)$$

where P, Q are functions of the s'_{ij} whose values will not be needed. If ϕ represents the discontinuity in u on crossing the x-axis then

$$\left.\begin{aligned} \phi=u_1-u_2&=0, \quad x>0, \\ &=b, \quad x<0. \end{aligned}\right\} \quad . \quad . \quad . \quad . \quad . \quad . \quad . \quad . \quad (14)$$

The shear stress at the slip-plane is

$$\widehat{xy}_0=\frac{b}{2\pi}\frac{K}{x} \quad . \quad . \quad . \quad . \quad . \quad . \quad . \quad . \quad (15)$$

where

$$K=\begin{vmatrix} \lambda_{1r} & \lambda_{2r} & -(\lambda_{1i}-\lambda_{2i}) \\ \lambda_{1i} & \lambda_{2i} & (\lambda_{1r}-\lambda_{2r}) \\ D_{1r} & D_{2r} & (D_{1i}-D_{2i}) \end{vmatrix} \div \begin{vmatrix} \lambda_{1r} & \lambda_{2r} & -(\lambda_{1i}-\lambda_{2i}) \\ C_{1i} & C_{2i} & (C_{1r}-C_{2r}) \\ D_{1r} & D_{2r} & (D_{1i}-D_{2i}) \end{vmatrix}.$$

When conditions (6) or (8) hold

$$K=[s'_{11}\{(2s'_{12}+s'_{66})+2(s'_{11}s'_{22})^{\frac{1}{2}}\}]^{-\frac{1}{2}} \quad . \quad . \quad . \quad . \quad . \quad (16)$$

This reduces to $K=1/(2s'_{11})$ for the condition (10) and to $K=\mu/(1-\sigma)$ for isotropy, where σ is Poisson's ratio.

§4. Elastic Energy of the Dislocation.

If the elastic energy density of a dislocation is integrated throughout the material the result diverges at the origin and infinity. The integration will therefore be confined to the space between a small cylinder of radius r_0 and a large cylinder of radius r_1, both concentric with the z-axis. We can avoid having to know the stresses at all points of the medium by an artifice. Imagine an unstrained block of material to be cut along the half-plane $y=0$, $x \geqslant 0$. A dislocation may be formed by applying suitable forces to the cut surfaces and then welding them together again. The work done in this process is

$$\mathrm{W}=\tfrac{1}{2}\int_{r_0}^{r_1} \{\widehat{xy}_0(u_1-u_2)+\widehat{yy}_0(v_1-v_2)\}\, dx = \frac{b^2\mathrm{K}}{4\pi}\log\frac{r_1}{r_0} \quad . \quad (16)$$

per unit length of dislocation in the z-direction, by (14) and (15). This must be the energy of the dislocation per unit length. For isotropy it reduces to the value given by Koehler (1941, equation (15)).

§5. The Width of the Dislocation.

Near the origin the strains corresponding to the displacements (3) become large and Hooke's law cannot be expected to hold good. Nabarro (1940) has discussed the form of a dislocation near its centre assuming that isotropic elastic theory is valid except in the neighbourhood of the slip plane. The results of the present paper may be used to find the influence of anisotropy on his results.

Suppose that an infinite block of anisotropic material is cut in two along the plane $y=0$. Let arbitrary forces be applied to the surface of the upper half-block, uniformly distributed in the z-direction, so that the problem is a two-dimensional one. If exactly opposite forces are applied to corresponding points of the surface of the lower half-block the blocks can be welded together to give a dislocation of a more general kind than that already discussed : the discontinuity $\phi=u_1-u_2$ varies from point to point along the x-axis.

Imagine a number of dislocations of the type (13) distributed along the x-axis. As we move along the axis in the negative direction ϕ changes by b each time a dislocation is passed. An arbitrary variation of ϕ with x can thus be considered as due to a continuous distribution of elementary dislocations along the x-axis. If the sum of the Burgers vectors of the elementary dislocations lying between x' and $x'+dx'$ is $b'\,dx'$ their contribution to the shear stress at $(x, 0)$ is, by (15)

$$\frac{b'\mathrm{K}dx'}{2\pi(x-x')},$$

whilst

$$\frac{d\phi}{dx'}=-\frac{db'}{dx'}:$$

By integration we find

$$\widehat{xy}_0=\frac{\mathrm{K}}{2\pi}\int_{-\infty}^{\infty}\frac{1}{x'-x}\frac{d\phi}{dx'}\,dx' \quad . \quad . \quad . \quad . \quad . \quad . \quad (17)$$

as the relation between the shear stress at any point and the relative displacements at all points of the slip-plane. Equation (17) corresponds to Nabarro's equation (10).

We now have to find a relation between ϕ and $\widehat{xy}_0$ at any point expressing the departure from Hooke's law. Suppose that the material is given a shear in which the atomic planes parallel to the slip-plane slide over each other without change of the spacing a between them. The only non-vanishing strain component is $e_{xy}=\partial u/\partial y$, so that if Hooke's law holds, $\widehat{xy}=s_{66}^{-1}\partial u/\partial y$. At the slip-plane this may be written

$$\widehat{xy}=\phi/s_{66}a. \quad . \quad . \quad . \quad . \quad . \quad . \quad . \quad . \quad . \quad (18)$$

For large ϕ/a this will cease to be true. Clearly $\widehat{xy}$ must be periodic with period equal to the Burgers vector b which represents slip by one lattice spacing. If the relation between $\widehat{xy}$ and ϕ is assumed to be sinusoidal and is required to reduce to (18) for small ϕ it must be

$$\widehat{xy}=\frac{1}{2\pi s_{66}}\frac{b}{a}\sin\frac{2\pi\phi}{b}. \quad . \quad . \quad . \quad . \quad . \quad . \quad . \quad (19)$$

This corresponds to Nabarro's equation (1). An integral equation for ϕ can be found by equating the $\widehat{xy}_0$ of (17) to minus the $\widehat{xy}$ of (19). (The normal to the surface of the upper half-block points in the negative direction.) The equation corresponding to Nabarro's (11) is thus

$$\int_{-\infty}^{\infty}\frac{1}{x-x'}\frac{d\phi}{dx'}dx'=\frac{b}{\mathrm{K}as_{66}}\sin\frac{2\pi\phi}{b}.$$

By comparison with Nabarro's equation (12),

$$\frac{\phi}{b}=-\frac{1}{\pi}\tan^{-1}\frac{x}{\zeta} \quad \text{with} \quad \zeta=\frac{1}{2}\mathrm{K}s_{66}a. \quad . \quad . \quad . \quad . \quad (20)$$

ζ is a measure of the distance along the x-axis within which the departure from Hooke's law is important. It gives the " width " of the dislocation.

If $|x|\gg\zeta$, ϕ is $-\frac{1}{2}b$ when $x>0$ and $\frac{1}{2}b$ when $x<0$: for large x it thus differs from the ϕ of (14) by an inessential constant relative displacement of the upper and lower half blocks. The distribution of elementary dislocations which corresponds to (20) is

$$\frac{db'}{dx}=\frac{b}{\pi}\frac{\zeta}{x^2+\zeta^2},$$

Hence from (13) we have

$$u_1, u_2=\pm\frac{1}{2}\phi+\frac{\mathrm{P}b\zeta}{\pi}\int_{-\infty}^{\infty}\frac{\log|x-x'|}{x'^2+\zeta^2}dx',$$

$$=\mp\frac{b}{2\pi}\tan^{-1}\frac{x}{\zeta}+\mathrm{P}b\log\sqrt{(x^2+\zeta^2)}.$$

Similarly v_1 and v_2 are given by the second term of this expression with P replaced by Q.

The other results in Nabarro's paper which depend only on the displacement and shear stress at the slip-plane may similarly be found by comparison with his formulæ. For example, the ratio of the stress required to move a dislocation to the stress necessary to make the atomic planes move rigidly over one another is

$$4\pi(\zeta/b)\exp(-2\pi\zeta/b). \qquad (21)$$

§ 6. Discussion.

According to (16) the energy of a dislocation is proportional to the square of its Burgers vector and to the constant K. The logarithmic term will be neglected. If the body is treated as isotropic, K has the value $\mu/(1-\sigma)$ for all orientations and the energy depends only on the

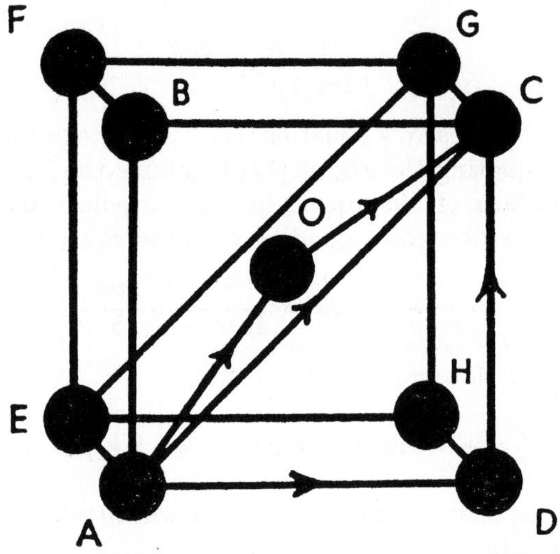

Burgers vector. When anisotropy is taken into account K varies with the direction of the dislocation, since the s_{ij} change with the orientation of the coordinate axes.

The magnitude and direction of the Burgers vector and the direction of the slip-plane have to be introduced by considering the lattice structure of the crystal. The energy for various choices of Burgers vector and slip-plane can then be compared by using elastic theory. The quantity $\zeta/b=\frac{1}{2}Ks_{66}a/b$, which determines the case of slipping, can also be calculated: K and s_{66} vary with orientation and a/b has values determined by the lattice geometry which may vary for different choices of Burgers vector and slip-plane.

As an example consider a crystal with body-centred cubic structure. The three shortest lattice displacements are AO, AD, AC. They are possible Burgers vectors. Three possible combinations of Burgers vectors and slip-plane are: (i) AD, ADHE; (ii) AC, ACGE; (iii) AO, ACGE.

It is convenient to write s_{ij}^0 for the values of the elastic constants when the x-, y- and z-axis are parallel to the fourfold axes of the material, and use s_{ij} to refer to an arbitrary coordinate system. The s_{ij}^0 are the elastic constants as usually tabulated.

For case (i)

$$K^{-2}=2[(s_{11}^0)^2-(s_{12}^0)^2]\{1+(s_{12}^0/s_{11}^0)-2(s_{12}^0/s_{11}^0)^2+\tfrac{1}{2}(s_{44}^0/s_{11}^0)\} \quad . \quad (22)$$

$$s_{66}=s_{44}^0, \quad a/b=\tfrac{1}{2}AB/AD=\tfrac{1}{2}.$$

For case (ii) the s_{ij} have to be obtained from the s_{ij}^0 by the transformation corresponding to a rotation of 45° about AE. It will be found that K is still given by (22) whilst

$$s_{66}=2(s_{11}^0-s_{12}^0), \quad a/b=\tfrac{1}{2}BD/AC=\tfrac{1}{2}.$$

The non-vanishing s_{ij} for case (iii) are given in terms of the s_{ij}^0 by the relations

$$s_{11}=s_{11}^0-\tfrac{2}{3}S, \quad s_{22}=s_{33}=s_{11}^0-\tfrac{1}{2}S, \quad s_{44}=s_{44}^0+\tfrac{2}{3}S, \quad s_{55}=s_{66}=s_{44}^0+\tfrac{4}{3}S,$$

$$s_{12}=s_{13}=s_{12}^0+\tfrac{1}{3}S, \quad s_{23}=s_{12}^0+\tfrac{1}{6}S, \quad s_{25}=-s_{35}=\tfrac{1}{2}s_{46}=-\tfrac{1}{3}\sqrt{2}S,$$

where $S=s_{11}^0-s_{12}^0-\tfrac{1}{2}s_{44}^0$.

K can be found from (1) and (16). The value of a/b is $\tfrac{1}{2}BD/AD=\sqrt{(\tfrac{2}{3})}$.

Numerical values for α-iron are given in Table I., using the elastic constants given by Schmid and Boas (1935). The isotropic values of K and s_{66} are calculated from the elastic constants found by Schmid and Boas by averaging the anisotropic constants : this gives a fairer comparison than would the observed bulk values.

TABLE I.

Energy per unit length of a dislocation in α-iron.

	Slip-plane	Slip direction	$K\times10^{-11}$ dyne/cm.^{2}	$s_{66}\times10^{13}$ cm.^{2}/dyne	Energy $\times10^5$ erg/cm.	a/b	ζ/b
(i)	(100)	[010]	10·5	8·6	6·82	$\tfrac{1}{2}$	0·23
(ii)	(110)	[010]	10·5	20·8	13·65	$\tfrac{1}{2}$	0·55
(iii)	(110)	[111]	12·2	16·7	5·95	$\sqrt{(\tfrac{2}{3})}$	0·83
(iv)	(Isotropy)		11·4	13·1	—	—	—

Frank (1949) has suggested that a dislocation may "dissociate" into two or more dislocations whose Burgers vectors added vectorially are equivalent to that of the original dislocation ; conversely, several dislocations might "associate" to form a single one. If we use D(**b**) to denote the dislocation with Burgers vector **b** we may write the symbolical reaction

$$D(\mathbf{b}_1)+D(\mathbf{b}_2)\rightleftarrows D(\mathbf{b}_3)$$

where the **b**s satisfy

$$\mathbf{b}_1+\mathbf{b}_2=\mathbf{b}_3$$

and the axes of the dislocations are perpendicular to the plane in which $\mathbf{b}_1$, $\mathbf{b}_2$, $\mathbf{b}_3$ lie. According to (16) the energy of dissociation is

$$E=(K_1b_1^2+K_2b_2^2-K_3b_3^2)/4\pi$$

per unit length. The three Ks have to be calculated for the orientation of the corresponding **b**, as explained above. The reaction will go in the direction of dissociation or association according as E is positive or negative. If the medium is isotropic the condition becomes

$$b_1^2+b_2^2 \gtrless b_3^2,$$

i. e. there will be dissociation or association according as the angle between $\mathbf{b}_1$ and $\mathbf{b}_2$ is acute or obtuse.

For a body-centred material the reaction

$$D(AD)+D(DC) \rightleftarrows D(AC)$$

has E=0 in the case of isotropy, since ADC is a right angle. The same is true when anisotropy is taken into account, since according to the previous discussion all three Ks are given by (22). The reaction

$$D(AO)+D(OC) \rightleftarrows D(AC)$$

has a negative E in the isotropic case, since AOC is obtuse. Whether E is positive or negative in the anisotropic case will depend on the elastic constants of the particular material. For example, in the case of α-iron taking $K=11{\cdot}4\times10^{11}$ (from the entry (iv) of Table I.) and AD=2·86 A gives E (isotropic)$=-3{\cdot}7\times10^5$ erg/cm., whilst entries (ii) and (iii) give directly E (anisotropic)$=(2\times5{\cdot}95-13{\cdot}65)\times10^{-5}=-1{\cdot}75\times10^{-5}$ erg/cm. In this case taking into account anisotropy reduces the magnitude of E, but is insufficient to reverse its sign.

The last column of Table I. implies, in conjunction with equation (21), that of the three cases considered dislocations of type (iii) (with Burgers vectors along cube diagonals) should move much more readily under the influence of an applied shear stress than types (i) and (ii).

Similar calculations could be made for other materials and other crystal structures. The necessary calculation of the s_{ij} from the s_{ij}^0 and the s'_{ij} from the s_{ij} are, however, rather tedious.

References.

Aitken, A. C., 1946, *Determinants and Matrices* (Edinburgh : Oliver and Boyd), p. 109.

Burgers, J. M., 1939, *Proc. Kon. Ned. Akad. v. Wet.*, **42**, 293, 378.

Frank, F. C., 1949, Private communication.

Green, A. E., 1945, *Proc. Camb. Phil. Soc.*, **41**, 224.

Koehler, J. S., 1941, *Phys. Rev.*, **60**, 397.

Love, A. E. H., 1944, *Elasticity* (Cambridge : University Press).

Nabarro, F. R. N., 1940, *Proc. Phys. Soc.*, **52**, 34.

Schmid, E., and Boas, W., 1935, *Kristallplastizität* (Berlin : Springer), pp. 200, 326.

The Fundamental Physics of Heat Conduction

By J. D. Eshelby, Ph.D.

INTRODUCTION

This paper sketches the picture which theoretical physics gives of the mechanisms of heat conduction in metals and insulators for the temperature range used in normal engineering practice.

The chief points to be explained are :—

(1) The marked difference between metals and non-metals.

(2) The temperature dependence of thermal conductivity. (Roughly speaking, the conductivity is independent of temperature for metals, and is inversely proportional to the absolute temperature for non-metallic crystals.)

(3) The close connexion between thermal and electrical conductivity in metals (Wiedemann–Franz law).

(4) The effect on thermal conductivity of impurities, cold work, and alloying.

Thermal conductivity is a non-equilibrium property and requires for its evaluation a knowledge of the "time of relaxation", that is, the time required for a disturbed system to reach statistical equilibrium. This can be replaced by the more vivid concept of the mean free path of the carriers which are responsible for the conduction. The calculation of the mean free path is difficult, and it is impossible in a short paper to develop the theory to the point where numerical results can be obtained.

The MS. of this paper was received at the Institution of Mechanical Engineers on 28th February 1951.

THE BASIC EQUATION

Physical insight into the process of conduction first became possible for the simplest state of matter—a gas—when the Kinetic Theory was developed by Maxwell and Boltzmann. It is convenient to recall the derivation of the expression for the thermal conductivity of a gas, since the same formula plays an important part in the theory of solids.

It is assumed that the gas consists of a large number of identical molecules moving randomly with mean velocity V and mean free path (mean distance between collisions) l, and that at each collision a molecule takes up the thermal energy appropriate to the temperature at the point of collision, and retains it until its next collision. To take into account the random directions of motion of the molecules, use is made of the well-known artifice of assuming that the molecules may be divided into six equal interpenetrating streams, three moving parallel to the directions

OX, OY, OZ of a rectangular co-ordinate system and three in the reverse directions.

If it is supposed that there is a temperature-gradient dT/dx along the x-axis, and a unit area at the origin perpendicular to the x-axis where the temperature, say, T_0, is considered, the stream of molecules moving through it from right to left have on the average come from a point distant l from the origin, and the temperature there is $T_0+l\,.\,dT/dx$. The heat energy which they transport through the area per unit time is thus $\frac{1}{6}sV(T_0+l\,.\,dT/dx)$. The stream moving in the opposite direction similarly transports an energy $\frac{1}{6}sV(T_0-l\,.\,dT/dx)$ through the surface in unit time. The total rate of flux of energy is thus $\frac{1}{3}lVs\,.\,dT/dx$. By definition this is equal to kdT/dx. Thus

$$k = \tfrac{1}{3}lVs \qquad (4)$$

THE EARLY ELECTRON THEORY

At the turn of the century the development of the electron theory made possible a theoretical approach to the problem of metallic conduction. The theory regarded a metal as a lattice of positive ions immersed in a "gas" of electrons which moved freely between the ions with thermal velocities. Equation (4) can be applied to this model. According to the ideas of the time, the specific heat of the electron gas should be $\frac{3}{2}\sigma N$ erg/°C. cm.^{3}, where σ is Boltzmann's constant, N is the number of electrons per unit volume, and V can be found from the relation $\frac{1}{2}mV^2 = \frac{3}{2}\sigma T$, where m is the mass of an electron. The theory could give no account of l. The natural suggestion that it was of the order of the inter-ionic distance and independent of T gave incorrect results.

There is, however, another quantity which also depends on l, namely, the electrical conductivity, K. There should therefore be a relation between k and K which does not involve l. An electric field **E** acting on an electron of charge e gives it an acceleration $e\mathbf{E}/m$. By the time it has covered a distance l it has gained a velocity $e\mathbf{E}l/mV$ in addition to its thermal velocity V. If it is assumed that the electron loses this additional velocity at each collision the mean additional velocity (drift velocity) is $\frac{1}{2}e\mathbf{E}l/mV$. The electric current per unit area is the product of the drift velocity, the electronic charge and the number of electrons per unit volume, and according to Ohm's law is equal to $K\mathbf{E}$. The electrical conductivity is therefore given by

$$K = \tfrac{1}{2}e^2lN/mV \qquad (5)$$

Elimination of l from equations (4) and (5) gives

$$\frac{k}{K} = 2\frac{\sigma}{e^2}(\tfrac{1}{2}mV^2) = 3\frac{\sigma^2 T}{e^2}$$

or

$$k/KT = 2{\cdot}2\times10^{-8} \text{ watt ohm/°C.}^2$$

According to this, the ratio of the thermal to the electrical conductivity should be the same for all metals at the same temperature (Wiedemann–Franz law), and should be proportional to the absolute temperature (Lorenz law). Experimentally these laws are found to be obeyed quite closely by normal metals above a temperature of, say, −150° C. (−238° F.). This relation, which survives almost unchanged in more recent theories, is important because it relates the thermal conductivity to the electrical conductivity, about which there is much more experimental and theoretical information.

Despite its success with the Wiedemann–Franz law the early theory had several defects. The chief of these was that the electrons should contribute to the specific heat half as much as did the ions. In fact, at normal temperatures the contribution of the ions alone accounts for the observed values.

THE MODERN THEORY

According to the wave-mechanical interpretation of quantum theory the behaviour of a particle is described by a wave-function ψ. For a swarm of particles ψ^2 is a measure of the number of particles per unit volume. For one particle ψ^2dv is the probability of finding the particle in a particular volume-element dv. The ψ-function satisfies a wave-equation similar to that for light in a medium of varying refractive index. The place of the varying refractive index is taken by the potential of the field of force in which the electron finds itself.

When the potential is that of an atomic nucleus the wave-equation is found to have solutions for certain values only of the energy of the electron. The atom thus has a discrete set of energy-levels. It can accommodate a number of electrons in these levels subject to the important limitation that not more than two electrons may occupy the same level (Pauli principle). By inserting electrons successively in the energy levels, beginning with the lowest, and observing Pauli's rule, the periodic table can be built up just as Bohr built it before the advent of the newer quantum theory.

A simple mechanical model will be used to illustrate the effect of bringing a group of identical atoms together to form a crystal. Three identical and independent simple harmonic oscillators, are considered each consisting of a weight attached to a spring. Each is capable of vibrating with the same frequency, and if one is set in motion the other two will not be affected. Now if the oscillators are coupled by additional springs (Fig. 9), the

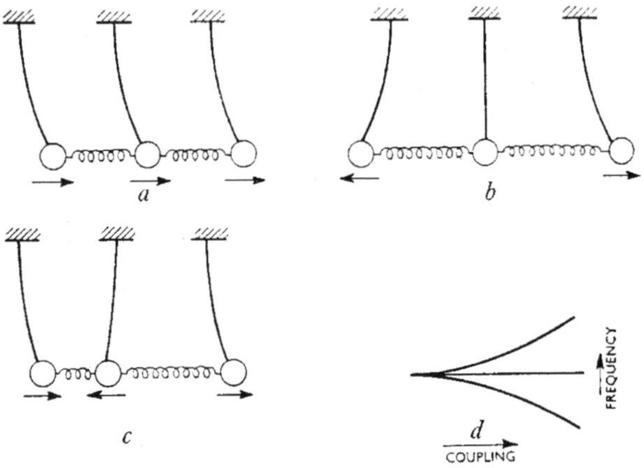

Fig. 9. Oscillations of a Coupled System

a Equal amplitude and phase.
b Second normal mode.
c Less obvious normal mode.
d Relation between frequencies and strength of coupling.

oscillations of the system will be more complicated. Certain simple types of motion can, however, be considered in which the deflexion of each of the three masses is given by one and the same simple harmonic function of time. The simplest of these "normal modes" is that in which the three particles vibrate with equal amplitude and phase (Fig. 9*a*). The coupling springs have no effect and the frequency is the same as that of one of the oscillators before being coupled. A second normal mode is shown in Fig. 9*b*. The outer masses vibrate 180 deg. out of phase and the central one is stationary. Clearly, the outer masses are acted upon by equal but opposite restoring forces somewhat greater than those they experienced when uncoupled, and the central mass is acted on by zero net force. The frequency is higher than that of the uncoupled oscillators.

A rather less obvious normal mode is shown in Fig. 9*c*. The outer masses vibrate in phase with each other and the central mass is 180 deg. out of phase with respect to them. It is clear that the restoring force on each mass is less than when they are uncoupled, so that the frequency is smaller than in the uncoupled state. It can be shown that there are no other normal modes. Thus the three equal vibrational frequencies of the separate oscillators are transformed into three unequal vibrational frequencies of the coupled system: the separation between the highest and lowest frequencies is proportional to the strength of the coupling (Fig. 9*d*). Any state of motion of the system can be regarded as a superposition of normal modes of vibration. Thus, if at a certain instant the first mass is vibrating and the other two are stationary, then this must be because the appropriate combination of normal modes happens temporarily to have zero resultant for the second and third masses. Because the three normal frequencies are different this state of affairs cannot be permanent: the normal modes will get out of step and all three particles will in general be in motion, though the system may be instantaneously in a state in which, for example, the first and second masses are stationary and the third is in motion.

The point to be emphasized is that it is impossible to confine motion entirely to one of the three coupled oscillators. A similar argument can be applied to the coupling of N identical oscillators. These will then be N normal modes spread over a range of values proportional to the degree of coupling.

The behaviour of N atoms is similar. (N is of the order of 10^{23} for a cubic centimetre of material.) The frequency of an uncoupled oscillator is replaced by the frequency ν of the ψ-wave, which according to Bohr's condition is related to the energy $\mathbf{E}$ of the corresponding state by the equation $\mathbf{E} = \mathbf{h}\nu$, where $\mathbf{h}$ is Planck's constant. As the atoms are brought together the N identical frequencies of one particular state of the N atoms is split into N different frequencies. That is, each energy-level splits into N energy-levels covering a range proportional to the coupling between the atoms, measured by the degree of overlap of their electrostatic potentials. The same splitting takes place for each of the energy levels of the original atoms, but it is most marked for the higher energy-levels (Fig. 10*a*). According

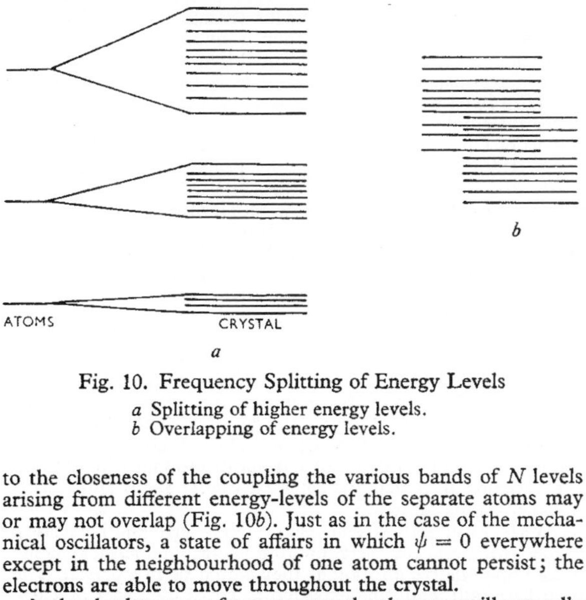

Fig. 10. Frequency Splitting of Energy Levels
a Splitting of higher energy levels.
b Overlapping of energy levels.

to the closeness of the coupling the various bands of N levels arising from different energy-levels of the separate atoms may or may not overlap (Fig. 10*b*). Just as in the case of the mechanical oscillators, a state of affairs in which $\psi = 0$ everywhere except in the neighbourhood of one atom cannot persist; the electrons are able to move throughout the crystal.

At the absolute zero of temperature the electrons will naturally occupy the lowest levels. First, the case is considered where there are enough electrons to fill the lower bands and the uppermost band is left half-filled. (The upper band may be simple or formed from the overlapping of two bands.) The "distribution function" $f(\mathbf{E})$, showing the probability of a state of energy $\mathbf{E}$ being occupied, will have the form of the stepped curve (1) in Fig. 11*a*, corresponding to $f = 1$ for $\mathbf{E} < \mathbf{E}_0$, and $f = 0$ for $\mathbf{E} > \mathbf{E}_0$. The calculation of f for a temperature T other than absolute zero follows the classical method of Maxwell and Boltzmann, subject to the proviso that not more than two electrons may occupy a given energy-state. The result for normal temperatures is shown by curve 2. A few states above $\mathbf{E}_0$ are populated with electrons, and an equal number of previously occupied states just below $\mathbf{E}_0$ are unoccupied. The distribution function differs appreciably from the curve for zero temperature only in a range where σT is of the order of $\mathbf{E}_0$.

It is supposed next that for $T = 0$ there are enough electrons just to fill a number of bands completely (Fig. 11*b*). When the temperature is raised, a few electrons will be found in the lower levels of the next higher band. It can be shown that at $\mathbf{E}_1$ (Fig. 11*b*) the distribution function is approximately $f = \exp.\{-\frac{1}{2}(\mathbf{E}_1 - \mathbf{E}_0)/\sigma T\}$. Hence if the gap $\mathbf{E}_1 - \mathbf{E}_0$ is large in comparison with σT the number of occupied levels at $\mathbf{E}_1$ and the corresponding depopulation of levels near $\mathbf{E}_0$ will be negligible.

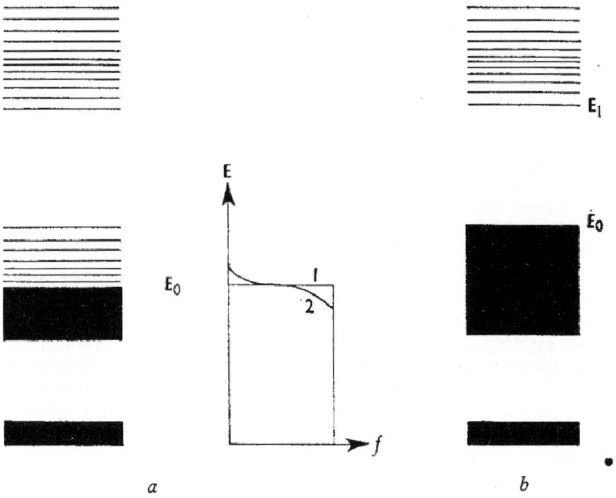

Fig. 11. Distribution Function

METALS AND NON-METALS

The foregoing results can now be applied to the process of heat conduction. If a specimen is hotter at the left side than at the right, an electron which wanders from left to right will have greater energy than its surroundings. By some collision process (which has still to be determined) it loses this excess and drops to a state of lower energy. Eventually it may wander to the left again where it will be raised to a higher energy-level by collision, and is then ready to repeat the cycle, provided that there are not more than two electrons occupying the same energy-level. Unless it can find a suitable level occupied by one or no electrons into which it can jump, an electron cannot leave its original level. Consequently, in the case shown in Fig. 11*a* only the electrons within an energy-range of the order of σT centred on $\mathbf{E}_0$ can suffer collisions and contribute to the thermal conductivity. In the case shown in Fig. 11*b* there are virtually no vacant levels into which an electron could jump and, consequently, the electronic contribution to the thermal conductivity is negligible. In this way the modern theory of the solid state accounts for the marked difference between metallic and non-metallic conduction: metallic conductors contain an incompletely filled band, whilst non-metals contain only completely filled bands.

Detailed calculations are necessary to decide whether any particular substance has completely or partially filled bands. A few general deductions can, however, be made. Atoms with a single valency electron (and hence an odd number of electrons) are likely to have a half-filled band, since each band can accommodate two electrons per atom (i.e. $2N$ per band). The good conductivity of the alkali metals and noble metals is to be accounted for in this way. By the same argument, divalent elements should be non-metallic unless the highest band is really a pair of overlapping bands which can contain $4N$ electrons in all. Calculation verifies that this is so for the divalent metals.

Metallic Conduction. The derivation of the absolute magnitudes of the quantities V, s, and l, which according to equation (4) determine k, would be a formidable task, so that the discussion is limited to their temperature-dependence. It has been shown that only electrons with energy in the neighbourhood of $\mathbf{E}_0$ take part in conduction. The appropriate value of V is that derived from the relation $\frac{1}{2}mV^2 = \mathbf{E}_0$ which is independent of T. Again, only electrons in the range where σT is of the order of $\mathbf{E}_0$ are free to receive additional thermal energy when the temperature is slightly raised. Hence, the electronic specific heat s is proportional to T. (This overcomes one of the chief difficulties of the older theories.)

If the values of V and s are substituted in equation (4) and the result is compared with observed values of k for metals, l is found to be of the order of a hundred inter-atomic spacings at room temperature. This is a surprising result. Since the size of the positive ions is of the order of the spacing between them, an electron would not be expected to be able to travel more than one inter-atomic distance before colliding with an ion. This paradox can be resolved with the aid of wave-mechanics. Light-waves can traverse the regular lattice of a perfect transparent crystal without scattering, and only deviations from regularity cause scattering. Similarly the ψ-waves of an electron are not scattered by a perfectly regular ionic lattice. In particle language this means that an electron can move through a regular lattice without colliding with it, and consequently l would be infinite.

Because of thermal agitation the ionic lattice is not perfect, but each lattice point has a mean displacement from its normal position proportional to the square root of the absolute temperature. Wave-mechanical calculation shows that the scattering of a ψ-wave is proportional to this mean displacement. The probability of an electron being scattered is proportional to the square of the scattered part of the ψ-wave and is therefore proportional to T. In particle language the probability of an electron undergoing a collision is proportional to T and, consequently, its mean free path is proportional to $1/T$. This analysis gives the results

$$V = \text{constant}, \quad s \propto T, \quad l \propto 1/T$$

Hence, from equation (4) k is independent of T, which is in fair agreement with observation for metals.

Impurities in a metal produce disturbances in the lattice analogous to those produced by thermal motion, except that they are independent of temperature. If thermal fluctuations were absent the mean free path would be proportional to the concentration of impurities, and independent of temperature, and the resulting conductivity would be proportional to the temperature. If l_t and l_i are mean free paths for thermal and impurity scattering, the net mean free path will clearly be less than either l_t or l_i. In fact, the inverses of the free paths may be added: $1/l = 1/l_t + 1/l_i$. Therefore

$$\frac{1}{k} = \frac{1}{k_t} + \frac{1}{k_i} = \frac{1}{k_t} + \frac{AC}{T} \quad . \quad . \quad . \quad . \quad . \quad (6)$$

where A is a constant and C is the concentration of impurities. For a metallic impurity in a metallic matrix A has a value of the order of unity, when k is measured in watts/cm./sec./deg., and C is in atomic per cent and is small. For example, A is about 1·3 for silver in magnesium and about 0·13 for silver in gold. It is clear that A is determined by the degree of "misfit" of the impurity. When C approaches 100 per cent, the roles of impurity and matrix are interchanged and $1/k_i$ is proportional to $(100-C)$. Consequently, if the thermal conductivity of an alloy is plotted as a function of the concentration of one component the curve has a minimum somewhere between 0 and 100 per cent.

Measurements of thermal conductivity are not very suitable as a means of detecting impurities, since no distinction can be made between a large amount of impurity with small scattering power and a small amount with large scattering power; moreover, the same information can be more easily obtained from the electrical conductivity.

Plastic deformation of a metal produces disturbance of the lattice which may be regarded as "frozen heat motion". The thermal conductivity is again given by equation (6) where AC is now a measure of the disturbance of the lattice.

Non-metallic Conduction. In non-metallic solids the electronic conductivity is negligible. The thermal vibrations of a crystal lattice can be regarded as a superposition of elastic (sound) waves. If a crystal has two faces at different temperatures, heat energy will be carried from one to the other by "sound radiation" in much the same way as radiant energy is transferred across a vacuum by electromagnetic waves. Since the fundamental carriers of heat energy are sound waves, V in equation (4) is to be identified with the velocity of sound in the material. According to Debye's theory, the specific heat of a body is simply the rate of change with temperature of the energy of these thermal sound waves. Therefore, s in equation (4) is to be equated to the ordinary specific heat of the body. At normal temperatures V and s are approximately constant.

A value for the mean free path l has yet to be found. The heat transfer may be imagined as taking place by the movement of discrete packets of sound energy and, if these packets suffered no attenuation, the thermal conductivity would be infinite. If, on the other hand, they undergo an attenuation such that in traversing a distance x their intensity is reduced by a factor exp. $(-x/l)$, then l will be the effective mean free path.

For the metallic conduction the sound waves were regarded as the scatterers of the electronic ψ-waves, but for non-metallic conduction nothing seems to be left to account for the scattering of the sound waves themselves. The way out of the difficulty was found by Debye—the sound waves scatter each other.

In a lattice in which the force on an atom is proportional to its displacement from the equilibrium position, any number of waves can be superimposed without interacting with one another. In an actual crystal, however, the force is not strictly proportional to displacement. This is shown, for example, by the fact that the contraction of a body under a hydrostatic pressure is less than its expansion under an equal hydrostatic tension. The linear force-displacement relation $F = Bx$ is replaced by $F = (B+Cx)x$, where C is a measure of the "anharmonicity". Hence the amplitude of mutual scattering of the elastic waves is proportional to the root-mean-square displacement of the atoms. The scattering intensity is therefore proportional to the mean square displacement which has already been shown to be proportional to T. Hence the mean free path is inversely proportional to T, just as in the scattering of electrons by the lattice. Combined with the constancy of s and V, this leads to a thermal conductivity inversely proportional to T. Experiment verifies this for a number of non-metallic crystalline substances.

In a similar manner to the ψ-waves, the sound waves will be scattered by imperfections of the crystal. The mean free path for this process will be independent of temperature. The conductivity will be given by taking, for the mean free path, the reciprocal of the sum of the reciprocals of the mean free paths for scattering of wave by wave and by impurities. This can be applied, for example, to the variation of conductivity with composition for solid solutions of one ionic substance in another. The variation is similar to that for metallic alloys and can be explained in the same way.

Glasses form an extreme case. They have no long-range crystal structure, but neighbouring atoms are approximately in the same relative position as they would be in a crystal. The mean free path would be expected to be of the order of a few atomic distances and, consequently, the conductivity to be nearly independent of temperature, as indeed it is. For fused quartz the value of l at room temperature, calculated from the observed conductivity, is 7 Ångström units. This is just the size of the silicon oxide tetrahedra from which the glass is built up.

Contemporary theoretical work on thermal conductivity is chiefly concerned with effects occurring at extremely low temperatures (liquid hydrogen and helium). Though not of immediate practical value to the heat engineer this work is helping to add to basic knowledge of the solid state.

XLI. *The Equilibrium of Linear Arrays of Dislocations.*

By J. D. Eshelby, F. C. Frank,
H. H. Wills Physical Laboratory, University of Bristol,
and F. R. N. Nabarro,
Department of Metallurgy, Uuiversity of Birmingham *.

[Received February 1, 1951.]

Abstract.

A method is given for finding the equilibrium positions of a set of like dislocations in a common slip-plane under the influence of a given applied stress. Their positions are given by the roots of a certain set of orthogonal polynomials. The case of a set of free dislocations piled up against a fixed dislocation by a constant applied stress is discussed in detail and the resulting stress-distribution is compared with that produced by a crack with freely slipping surfaces.

§ 1. Introduction.

The following problem (Frank 1951, Kuhlmann 1951, Nabarro 1951) arises in interpreting the plastic behaviour of solids in terms of dislocations. A set of identical dislocations lie in the same slip-plane. What positions will they take up under the combined action of their mutual repulsions and the force exerted on them by a given applied shear stress, in general a function of position along the plane?

Since the dislocations (assumed to be of infinite length and parallel to one another) repel each other inversely as the distance between them the problem is unchanged if they are replaced by a set of line-charges and the applied stress is replaced by an electric field. This electrostatic problem was used by Stieltjes (1885) (*cf.* also Szegö 1939) to illustrate the properties of the zeros of orthogonal polynomials. In the present paper we show how the properties of the classical orthogonal polynomials may be used to discuss the dislocation problem.

Stieltjes solved the problem by minimizing the potential energy of the charges. We start from the idea of the force acting on a dislocation. The two methods are equivalent.

§ 2.

Suppose that there is an infinite straight dislocation parallel to the z-axis, having the plane $y=0$ as slip plane, and passing through the point $x=x_i$. Consider the shear stresses p_{xy}, p_{yz} in the slip-plane. If the

* Communicated by the Authors.

dislocation is of pure screw type $p_{xy}=0$, $p_{yz}\neq 0$. If it is of pure edge type $p_{xy}\neq 0$, $p_{yz}=0$. In either case the non-vanishing stress component which it produces at the point x in the slip-plane is

$$p=\frac{A}{x-x_i}. \qquad (1)$$

For a screw dislocation $A=A_s=\mu b/2\pi$ and for an edge dislocation

$$A=A_e=\mu b/2\pi(1-\sigma)$$

(Burgers 1939). Here μ is the shear modulus, σ is Poisson's ratio and b is the magnitude of the Burgers vector **b** which gives the change of the displacement vector on passing once round the dislocation line. For an edge dislocation in an anisotropic material $A=Kb/2\pi$ where K is a certain function of the elastic constants (Eshelby 1949).

A screw dislocation is acted on by a force bp_{yz} per unit length, where p_{yz} is the total stress acting at its centre, excluding that produced by itself. Similarly an edge dislocation is acted on by a force bp_{xy}. If there are n dislocations at the points $x_1, x_2 \ldots x_n$, all of the same type and each one is in equilibrium, the equations

$$\sum_{\substack{i=1\\ i\neq j}}^{n}\frac{A}{x_j-x_i}+P(x_j)=0, \quad j=1, 2, \ldots n \qquad (2)$$

must hold. $P(x)$ is the appropriate component of the applied stress at the point x, *i. e.* the xy component for edge and the yz component for screw dislocations.

Equations (1) and (2) apply, more generally, to any set of parallel like dislocations lying in their common slip plane. If ψ is the angle between the line and Burgers vector of each dislocation, we then have

$$A=A_s\cos\psi+A_e\sin\psi,$$

$$P(x)=P_{yz}(x)\cos\psi+P_{xy}(x)\sin\psi,$$

since the screw and edge components interact independently with one another.

In all cases, therefore, the following condition must be satisfied at each dislocation: the component in the direction of **b** of the traction on the plane $y=0$ due to the other dislocations and to the applied stress must vanish.

It will be convenient to choose the unit of stress so that $A=1$. The condition of equilibrium is then

$$\sum_{i\neq j}\frac{1}{x_j-x_i}+P(x_j)=0, \quad j=1, 2 \ldots n. \qquad (3)$$

We may regard the x_i as the zeros of the polynomial

$$f=\prod_{i=1}^{n}(x-x_i). \qquad (4)$$

f has the convenient property that its logarithmic derivative is equal to the stress due to all the dislocations :

$$\frac{f'}{f}=\sum_{i=1}^{n}\frac{1}{x-x_i}.$$

The stress when the jth dislocation is missing is

$$\frac{f'}{f}-\frac{1}{x-x_j}. \qquad (5)$$

The value of this expression when $x=x_j$ is

$$\lim_{x\to x_j}\frac{(x-x_j)f'(x)-f(x)}{(x-x_j)f(x)}=\tfrac{1}{2}\frac{f''(x_j)}{f'(x_j)}, \qquad (6)$$

by double differentiation of numerator and denominator.

The conditions (2) can thus be written

$$\left.\begin{aligned} f(x_j)&=0\\ \tfrac{1}{2}\frac{f''(x_j)}{f'(x_j)}+\mathrm{P}(x_j)&=0\end{aligned}\right\}j=1,2\ldots n. \qquad (7)$$

Consider the differential equation

$$f''(x)+2\mathrm{P}(x)f'(x)+q(n,x)f(x)=0. \qquad (8)$$

Suppose that we can choose q so that this equation has a polynomial solution of the nth degree all of whose roots are real and distinct. Then if q does not have a pole at any of the roots the conditions (7) are satisfied and the problem is solved.

In the physical problem we may be interested in cases where certain of the dislocations are " locked " in fixed positions (Cottrell 1948). If a dislocation is locked at $x=x_\alpha$ we imply that it lies in a stress field $p(x)=\text{const}.\,\delta'(x-x_\alpha)$ where δ' is the derivative of Dirac's δ-function. The dislocation is in equilibrium at $x=x_\alpha$ where $p(x)=0$. Also, since $p'(x_\alpha)$ is negatively infinite, the equilibrium is stable and an added stress will shift the position of equilibrium only infinitesimally. The other dislocations are unaffected by $p(x)$. It is then convenient to omit $p(x)$ from the expression for the applied stress and to regard one of the dislocations as " locked " at the point x_α.

Suppose, then, that in addition to the n " free " dislocations there are locked dislocations at $x_{n+1}, x_{n+2} \ldots x_\nu$. The equilibrium positions of the free dislocations are found by regarding the stresses produced by the locked dislocations as forming part of the applied stress, so that (8) becomes

$$f''+2\left\{\mathrm{P}(x)+\sum_{\alpha=n+1}^{\nu}\frac{1}{x-x_\alpha}\right\}f'+q(n,x)f=0. \qquad (9)$$

In this case we expect q to have a pole at each x_α in order that f'/f shall not vanish there.

The solution will be of the form (4) and f'/f will give the stress due to the free dislocations alone. If we make the substitution

$$\mathrm{F}(x)=f(x) \prod_{\alpha=n+1}^{\nu} (x-x_\alpha)$$

(9) becomes

$$\mathrm{F}''+2\mathrm{PF}'+\frac{\mathrm{Q}(n, x)}{\prod_\alpha (x-x_\alpha)}\mathrm{F}=0, \quad \ldots \ldots (10)$$

where

$$\frac{\mathrm{Q}(n, x)}{\prod_\alpha (x-x_\alpha)}=q(n, x)-2\mathrm{P}(x)\sum_\alpha \frac{1}{x-x_\alpha}-\sum_\alpha \sum_{\substack{\beta \\ \alpha\neq\beta}} \frac{1}{(x-x_\alpha)(x-x_\beta)}$$

and α, β take the values $n+1$, $n+2 \ldots \nu$. At the zeros $x_1, x_2 \ldots x_n$ the conditions (7) are satisfied (free dislocations) whilst at the zeros x_{n+1}, $x_{n+2}, \ldots x_\nu$ the last term in (10) has a finite value owing to the vanishing of a factor in the denominator, and the second of the conditions (7) (with F replacing f) is not satisfied (locked dislocations). F′/F is the stress produced by locked and free dislocations.

If a particular problem has been solved in this way it remains to verify that the arrangement of dislocations is stable. The condition for this is that the x-derivative of the total stress at the position of a free dislocation (excluding its own stress) shall be negative, *i. e.* that

$$\lim_{x\to x_j} \frac{d}{dx}\left\{\frac{\mathrm{F}'(x)}{\mathrm{F}(x)}-\frac{1}{x-x_j}+\mathrm{P}(x)\right\}<0.$$

After some calculation this reduces to the condition that

$$\mathrm{I}(x)\equiv\mathrm{Q}-\mathrm{P}^2-\mathrm{P}'>0 \quad \ldots \ldots \ldots (11)$$

for $x=x_1, x_2 \ldots x_n$. I(x) is the " invariant " of equation (10) which can be reduced to the form

$$v''+\mathrm{I}(x)v=0 \quad \ldots \ldots \ldots (12)$$

by the substitution

$$v=\mathrm{F}\exp\left(\int \mathrm{P}dx\right). \quad \ldots \ldots \ldots (13)$$

v has the same zeros as F with possibly additional ones. A region in which $\mathrm{I}<0$ can usually be dealt with by the following argument. Suppose that v has a zero (which may be at infinity) in the region. Since v'' has the same sign as v the curve $v=v(x)$ is convex to the x-axis and there can be no other zero in the region. In particular, if $\mathrm{I}<0$ for $x>x'$ and $v\to 0$ as $x\to\infty$ then F can have no zero for $x>x'$.

It is easily seen that v'/v gives the total stress due to free and locked dislocations and the applied stress.

§3.

The general method will be illustrated by some cases of physical interest.

(i.) A row of n dislocations under zero applied stress, the outer two being locked at $x=\pm\mathrm{L}$. If $\mathrm{L}=1$, (9) becomes

$$f''+2\left\{\frac{1}{x-1}+\frac{1}{x+1}\right\}f'+q(n, x)f=0,$$

or if we take

$$q=\frac{n(n-1)-2}{1-x^2},$$

$$(1-x^2)f''-4xf'+\{n(n-1)-2\}f=0$$

which is satisfied by $f=P'_{n-1}(x)$, the first derivative of the $(n-1)$th Legendre polynomial. Hence

$$f=P'_{n-1}(x/L),\ F=(L^2-x^2)f=P'_{n-1}(x/L)\sqrt{(L^2-x^2)}.$$

In ordinary stress units the stress on one of the locked dislocations is

$$\pm AF''(1)/F'(1)=\pm\tfrac{1}{4}n(n-1)A/L.$$

The roots of f up to $n-1=8$ have been given by Kopal (1946).

(ii.) A row of n free dislocations with an applied stress proportional to x. Such a uniformly varying stress exists near the centre of a beam supported at the ends and uniformly loaded. We have $P=x(dP/dx)$ with dP/dx constant. The resulting force on a dislocation can be regarded as arising from a potential energy $V=-\frac{1}{2}bx^2(dP/dx)$. "Potential troughs" of this type may exist in real crystals owing to the presence of various defects (Mott and Nabarro 1948). If we choose $1/\sqrt{(dP/dx)}$ as the unit of length the equation for $F(=f)$ is

$$F''-2xF'+QF=0.$$

If $Q=2n$, (10) is satisfied by the nth Hermite polynomial, $H_n(x)$. The invariant $I=2n+1-x^2$ is positive if $|x|<\sqrt{(2n+1)}$ and negative otherwise. The reduced equation is satisfied by $v=H_n(x)\exp(-\frac{1}{2}x^2)$, so that by the argument following equation (13) all dislocations lie in the region $|x|<\sqrt{(2n+1)}$. (*cf.* also Appendix, § A2).

In ordinary units of length and stress

$$F=H_n\left\{x\sqrt{\left(\frac{dP/dx}{A}\right)}\right\}$$

and all dislocations lie in the region

$$|x|<\sqrt{\left\{\frac{(2n+1)A}{dP/dx}\right\}}.$$

(iii.) A set of $n-1$ free dislocations along the positive x-axis in equilibrium under the combined effect of a locked dislocation at the origin and a uniform stress tending to move them in the negative direction. This case is discussed in detail in § 4.

(iv.) Two locked dislocations as in (i.) with an applied stress as in (iii.). This leads to an equation occurring in the theory of the hydrogen molecular ion. There is an extensive literature (Wilson 1928, Teller 1930, Hylleraas 1937) but no general discussion of the polynomial solutions. As we should expect there are n polynomial solutions of order n, since we can put 0, 1, 2 . . . n dislocations in the interval $-L<x<L$ and the remainder in the interval $L<x$.

§ 4.

For case (iii.) above we have $P(x)$ constant and equal to say $-\tau_0$. Equation (8) becomes

$$f''+2(x^{-1}-\tau_0)f'+qf=0,$$

If we take $1/2\tau_0$ as the unit of length and put $q=(n-1)/x$ this becomes

$$xf''+(2-x)f'+(n-1)f=0 \quad . \quad . \quad . \quad . \quad . \quad . \quad (14)$$

which is satisfied by the first derivative of the nth Laguerre polynomial, $L_n'(x)$. Thus in ordinary units of length and stress

$$f=L_n'(2\tau_0 x/A), \quad F=(2\tau_0 x/A)\, L_n'\,(2\tau_0 x/A).$$

The spacing of the dislocations is the same as that of the radial nodes of a hydrogen atom in the ns state.

The invariant of (14), $I=nx^{-1}-\frac{1}{4}$ is positive if $x<4n$. The reduced equation is satisfied by $v=xL_n'(x)\exp(-\frac{1}{2}x)$ so that all zeros of f must occur for $x<4n$. A lower bound for the greatest root is obtained in the Appendix, § A.1. Since

$$L_n'(x)=-\sum_{k=0}^{n-1}\frac{n!\,(-x)^k}{k!\,(k+1)!\,(n-k-1)!} \quad . \quad . \quad . \quad . \quad (15)$$

there are no negative roots.

When x is small compared with n we may neglect $\frac{1}{4}$ in comparison with n/x and the reduced equation $v''+Iv=0$ becomes

$$v''+(n/x)v=0$$

with the solution

$$v=\sqrt{x}\,.\,J_1\{\sqrt{(4nx)}\} \quad . \quad . \quad . \quad . \quad . \quad . \quad . \quad . \quad . \quad (16)$$

so that near the origin the position of the ith dislocation is given by $x\cong j_i^2/4n$ where j_i is the ith zero of the Bessel function J_1. The position of the free dislocation nearest to the locked dislocation is $x_1\cong 3{\cdot}67/n$. Since $j_i\cong\pi i$ for large i the number, i, of dislocations between the origin and the point x is

$$i\cong(2/\pi)\sqrt{(nx)}, \qquad 1\leqslant i\leqslant n.$$

Returning to ordinary units of stress and length in which x becomes $2\tau_0 x/A$ we have the following approximate results for large n :

L, the length of the slip plane occupied by dislocations

$$=2nA/\tau_0 \quad . \quad . \quad . \quad . \quad . \quad . \quad . \quad . \quad . \quad . \quad (17)$$

the number of dislocations lying between 0 and x

$$=\frac{2}{\pi}\sqrt{\left(\frac{2n\tau_0 x}{A}\right)} \quad . \quad . \quad . \quad . \quad . \quad . \quad (18)$$

d, the distance between the locked and nearest free dislocation

$$=1{\cdot}84\ A/n\tau_0 \quad . \quad . \quad . \quad . \quad . \quad . \quad . \quad . \quad . \quad (19)$$

The difference between the L of equation (17) and the true distance between the locked and furthest free dislocation, L′ say, can be estimated from the inequality

$$1{\cdot}47n^{-2/3}-0{\cdot}55n^{-5/3}<\frac{\mathrm{L}-\mathrm{L}'}{\mathrm{L}}<2{\cdot}56n^{-2/3},$$

obtained by combining the result of § A.1 with an inequality given by Szegö (1939, equation 6.32.6). In what follows we shall refer to L (and not the accurate L′) as the length of the row of dislocations. Equation (18) is true in the sense that if $1\ll i\ll n$ the ratio of the right-hand side of (18) to the correct i approaches unity as n tends to infinity. Actually (18) is in error by a factor of only $4/\pi$ even if $i=n$ and in error by only one or two units for small i provided n is large. d must be larger than a few atomic distances if our results are to be valid, since the expression (1) is inaccurate very close to a dislocation. (Nabarro 1940).

If we move along the x-axis the discontinuity Δ in the displacement on crossing the slip plane changes by b each time a dislocation is passed. The curve $\Delta=\Delta(x)$ is thus stepped, but according to (17) and (18) it will approximate to the smooth curve

$$\Delta=\frac{2b\tau_0}{\pi\mathrm{A}}\sqrt{(\mathrm{L}x)}, \qquad \ldots\ldots\ldots \quad (20)$$

provided n is large and we do not go too far from the origin.

From the coefficients of x^{n-1} and x^{n-2} in (16) we find $(2\tau_0/\mathrm{A})\Sigma x_i=n(n-1)$ so that the centre of gravity of the free dislocations is at the point $x=n\mathrm{A}/2\tau_0=\frac{1}{4}\mathrm{L}$ whilst the centre of gravity of the n dislocations is at $x=(n-1)\mathrm{A}/2\tau_0$. The coefficients of x and x^0 give $\mathrm{A}\Sigma x_i^{-1}=(n-1)\tau_0$. According to (1) this is the stress at the locked dislocation produced by the free dislocations. The total stress on the locked dislocation is thus $n\tau_0$. This also follows from (14) which gives $f'(0)/f(0)=\frac{1}{2}(n-1)$ which must be multiplied by $2\tau_0$ to convert to ordinary units. Cottrell (1949) has already obtained this result by a virtual work argument.

The behaviour of the stress, τ, beyond the locked dislocation ($x<0$) is also of physical interest (Frank 1951). To discuss it, it is convenient to reverse the sign of x so that the dislocations lie on the negative x-axis. The most important region is that in which x is of the order of L, or, in the reduced units in which $2\tau_0/\mathrm{A}=1$, $x\sim n$. In these units the solution (16) with the sign of x changed is

$$v=\text{const.}\ y\,\mathrm{I}_1(y), \quad y=\sqrt{(4nx)}$$

and this is a good approximation when $x\ll 4n$. The total stress is given by v'/v, for which we find

$$\frac{v'}{v}=\sqrt{\left(\frac{n}{x}\right)}\cdot\frac{\mathrm{I}_0(y)}{\mathrm{I}_1(y)}, \quad x\ll 4n.$$

From known properties of Bessel functions $I_0/I_1 \to 1$ as $y \to \infty$ and $I_0/I_1 \to 2/y$ as $y \to 0$ so that $v'/v = \sqrt{(n/x)}$ if $1/4n \ll x \ll 4n$ and $v'/v = 1/x$ if $x \ll 1/4n$ or in ordinary units

$$\begin{aligned} \tau/\tau_0 &= \sqrt{(L/x)}, \quad \tfrac{1}{15}d \ll x \ll L \\ \tau &= A/x, \qquad \tfrac{1}{15}d \ll x. \end{aligned} \qquad (21)$$

Fig. 1.

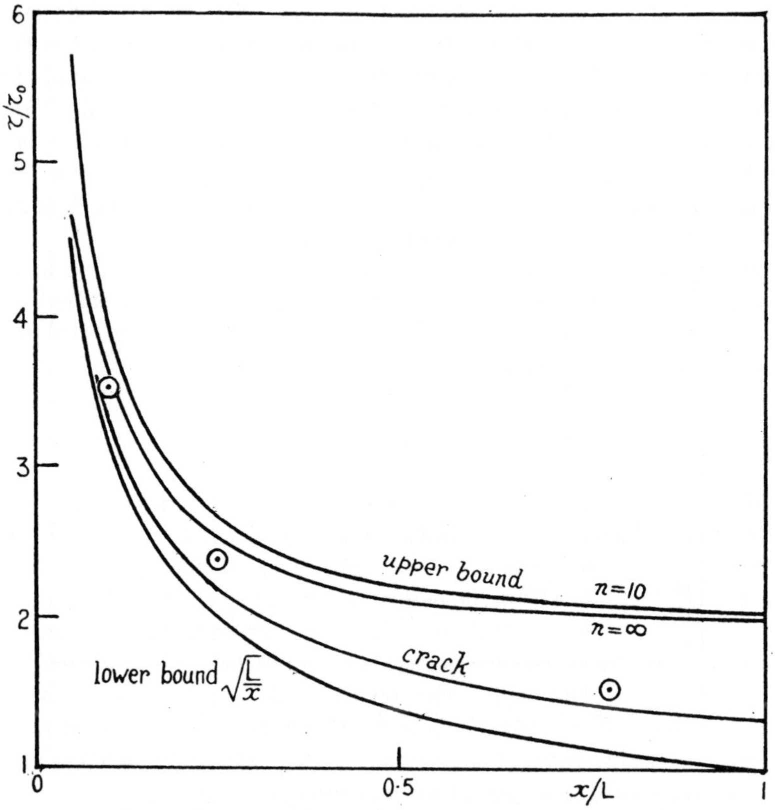

Shear stress in front of a row of held-up dislocations. Upper and lower bounds from equations (22). ⊙ exact values calculated from the polynomial (15) for n=10. For comparison the stress due to a crack of length 4L is given, according to equation (24).

The last equation implies that the effect of the locked dislocation is dominant at points much closer to it than is the nearest free dislocation. It is shown in the Appendix, (§ A3) that

$$\left.\begin{aligned} \frac{\tau}{\tau_0} &= \sqrt{\left(\frac{L}{x}\right)} \cdot \left[1+\theta\left\{\frac{x}{L}+\frac{1}{2n}\sqrt{\left(\frac{L}{x}\right)}\right\}\right], \quad 0<\theta<1 \\ \text{and also} &\qquad \tau > A/x, \quad \tau > \tau_0. \end{aligned}\right\} \qquad (22)$$

These relations enable us to set bounds to the departure of τ/τ_0 from the simple expression $\sqrt{(L/x)}$ for a given x; see figs. 1 and 2.

For large x we have

$$\tau = \tau_0 + nA/x \quad \text{or} \quad \tau/\tau_0 = 1 + \tfrac{1}{2}L/x, \; x \ll L, \qquad (23)$$

as can be seen from the inequality

$$\frac{nA}{x+L} < \frac{A}{x} + \sum_i \frac{A}{x+|x_i|} < \frac{nA}{x}$$

which follows from the facts that $0 < |x_i| < L$. Equation (23) shows that at large distances the stress is the sum of the applied stress and that due to all the dislocations regarded as congregated at the origin.

Fig. 2.

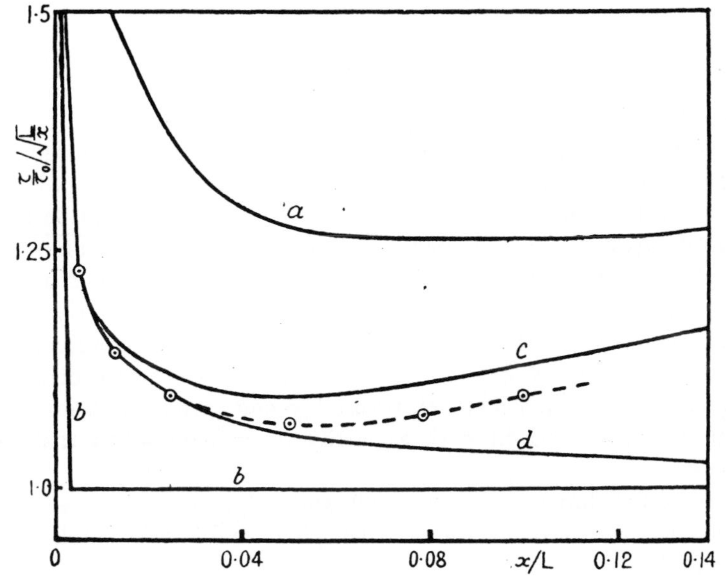

Shear stress in front of a row of ten held-up dislocations. a, b upper and lower bounds from equations (22). c, d upper and lower bounds from equation (a9) (Appendix, § A.3). ⊙, accurate values calculated from the polynomial (15). The stress falls off monotonically with x: the convexity arises from plotting $\tau/\tau_0\sqrt{(L/x)}$ instead of τ/τ_0.

Zener (1948) has compared a slip-band both to a crack and to an array of dislocations and it is interesting to compare the results of this section with those of Starr (1928) for a crack. He finds the following expressions for an infinitely narrow two-dimensional crack of length l extending along the x-axis from $x=0$ to $x=-l$ when the body containing it is subjected to a uniform applied shear stress $p_{xy}=\tau_0$. The stress in the plane $y=0$ beyond the tip of the crack is given by the exact formula

$$\frac{\tau}{\tau_0} = \frac{x+\frac{1}{2}l}{\sqrt{\{x(x+l)\}}} = \sqrt{\left(\frac{l}{4x}\right)} \cdot \left(1 + \frac{x}{l} - \frac{1}{2}\frac{x^2}{l^2} + \ldots\right). \qquad (24)$$

If the crack is not infinitely narrow but has a radius of curvature ρ at the end (24) is only correct if $x \gg \rho$. τ/τ_0 then reaches a maximum in the neighbourhood of $x=\rho$ and falls to zero at $x=0$. We have

$$\tau/\tau_0 = 1+\tfrac{1}{2}(l/2x)^2, \quad x \gg l \quad . \quad . \quad . \quad . \quad . \quad . \quad . \quad (25)$$

far from the crack.

The discontinuity in the displacement in crossing the x-axis, *i. e.* the relative displacement of the two faces of the crack, is given by the exact expression

$$\Delta = \frac{b\tau_0}{\pi \mathrm{A}} \sqrt{\{(l-|x|)|x|\}}, \quad -l<x<0$$

reducing to

$$\Delta = \frac{b\tau_0}{\pi \mathrm{A}} \sqrt{(l|x|)}, \quad x<0, \quad |x| \ll l \quad . \quad . \quad . \quad . \quad . \quad (26)$$

if we do not go too far from the tip of the crack.

The elastic constants have been expressed in terms of b and A (*cf.* equation (1)) to aid comparison, though of course the problem of the crack does not involve the Burgers vector. A is to be given its value for edge dislocations, $b\mu/2\pi(1-\sigma)$. It can be shown that the last four equations also apply to a crack in an anisotropic medium if we give A the value $b\mathrm{K}/2\pi$. They also cover the case of a crack in an isotropic body in a state of anti-plane strain with τ_0 and τ representing the yz stress component provided we put $\mathrm{A}=b\mu/2\pi$, the value appropriate to screw dislocations. Consequently these equations can be used to compare a crack under plane strain or anti-plane strain with an array of edge or screw dislocations respectively.

Comparison of (22) and (24) or of (20) and (26) shows that over a certain range of x the row of dislocations of length L simulates a crack of length 4L (*cf.* fig. 1).

It will be noted that near the origin the stress produced by the dislocations and the crack with a finite curvature at the tip diverge in opposite directions from the value $\tau_0\sqrt{(\mathrm{L}/x)}$.

From (23) and (25) it will be seen that for the dislocations $\tau-\tau_0$ falls off as x^{-1} at large distances, but for the crack as x^{-2}, that is, in the same way as for a group of positive dislocations near a group of an equal number of negative ones. It is also clear that the array of dislocations simulates only one end of the crack. These facts suggest that a better imitation of a crack would be given by the following sequence of dislocations : a locked positive dislocation, n free positive dislocations, n free negative dislocations, a locked negative dislocation. The free dislocations are prevented from coalescing by the applied stress and the locked dislocations prevent them from spreading and define the length of the equivalent crack (*cf.* Zener 1948, fig. 4). The method of the present paper is not capable of finding their equilibrium positions.

§ 5.

The polynomial F also has a physical interpretation for complex values of its argument. Let $Z = x + iy$. Then for a set of screw dislocations the displacement w and stresses at the point (x, y) are given in the isotropic case by

$$\phi + iw = \frac{b}{2\pi}\log F(Z),$$

$$p_{yz} + ip_{xz} = \frac{\mu b}{2\pi}\frac{F'(Z)}{F(Z)}.$$

ϕ is the electrostatic potential in the corresponding Stieltjes line-charge problem. These results follow from the fact that for a single screw dislocation at the origin $w = (b/2\pi)\tan^{-1}(y/x)$.

For an array of edge dislocations the results are less simple since problems in plane stress or strain cannot be solved in terms of a single complex variable. We find for the isotropic case

$$(p_{xx} + p_{yy}) + 2i\mu\varpi/(1-\sigma) = -2iG'$$

$$p_{xy} + \tfrac{1}{2}i(p_{xx} - p_{yy}) = G' + iyG''$$

where

$$G(Z) = A \log F(Z), \quad A = \mu b/2\pi(1-\sigma)$$

and ϖ is the rotation. The results follow from the value $\chi = -Ay \log r$ of the Airy function for an edge dislocation at the origin (Koehler 1941).

Acknowledgments.

Our thanks are especially due to Professor H. Heilbronn, who first indicated to us the connection between a set of algebraic equations of the form (2) and a differential equation of the form (8) upon which the whole analysis is based. He also obtained the important inequality of § A.1. We wish to thank Dr. C. S. Davis for much assistance and Mr. K. M. Baird for stimulating discussion.

APPENDIX.

§ A.1.

The reduced equation for the problem of § 4 is

$$v'' + \left(\frac{n}{x} - \frac{1}{4}\right)v = 0. \qquad \text{(a1)}$$

Let $X = x_{n-1}$ be the greatest root of v. We can arrange that $v > 0$ for $x > X$. Choose a constant $\xi > X$. We can write (a1) in the form

$$v'' + \lambda^2 v = g(x) \qquad \text{(a2)}$$

where

$$\lambda^2 = \left(\frac{n}{\xi} - \frac{1}{4}\right), \quad g(x) = n\left(\frac{1}{\xi} - \frac{1}{x}\right)v(x).$$

The solution of the inhomogeneous equation (a2) which satisfies $v(\mathrm{X})=0$ is

$$v(x)=\mathrm{A}\sin\lambda\,(x-\mathrm{X})+\lambda^{-1}\int_{\mathrm{X}}^{x}\sin\lambda\,(\mathrm{X}-x')\,g\,(x')\,dx'$$

where A is an arbitrary constant. Hence

$$v(\mathrm{X}+\pi/\lambda)=\lambda^{-1}\int_{\mathrm{X}}^{\mathrm{X}+\pi/\lambda}\sin\lambda\,(x'-\mathrm{X})\,g\,(x')\,dx'.$$

Since the left-hand side is positive $g(x')$ must be positive for at least part of the range of integration and so

$$\xi<\mathrm{X}+\pi/\lambda.$$

So far ξ is arbitrary. We now try to choose it to give as precise information as possible about X, bearing in mind that λ is a function of ξ. Put

$$\xi=4n/(1+\phi^2),\quad \lambda=\tfrac{1}{2}\phi.$$

Then

$$\mathrm{X}>\frac{4n}{1+\phi^2}-\frac{2\pi}{\phi} \qquad \text{(a3)}$$

and *a fortiori*

$$\mathrm{X}>4n(1-\phi^2-\pi/2n\phi). \qquad \text{(a4)}$$

The right-hand side of (a4) has a minimum when $\phi=(\pi/4n)^{1/3}$ so that

$$\mathrm{X}>4n[1-3(\pi/4)^{2/3}\,n^{-2/3}].$$

A slightly more precise lower limit could be found by minimizing the right-hand side of (a3).

§ A.2.

The same method applied to the Hermite polynomials occurring in § 3 (iii.) gives for the greatest root X

$$\mathrm{X}>\sqrt{(2n+1)}\,[1-\sqrt{(2\pi)}\,(2n+1)^{-3/4}].$$

§ A.3.

With the sign of x changed (a1) becomes

$$v''-\frac{n}{x}v=\tfrac{1}{4}v \qquad \text{(a5)}$$

which has the exact solution

$$v=(n-1)\,x\,\mathrm{L}_n'\,(-x)\exp\tfrac{1}{2}x \qquad \text{(a6)}$$

with a convenient choice of the arbitrary multiplying factor. If the right-hand side of (a5) were zero it would have the two independent solutions

$$v_1=y\,\mathrm{I}_1\,(y),\quad v_2=y\mathrm{K}_1\,(y),\quad \text{where } y=\sqrt{(4nx)}.$$

It is easily verified that v and v_1 agree in value and first derivative at the origin. Solving (a5) as an inhomogeneous equation in the usual way, making use of the relation

$$\mathrm{I}_0\,\mathrm{K}_1+\mathrm{K}_0\,\mathrm{I}_1=1/y, \qquad \text{(a7)}$$

we have

$$v=y\,\mathrm{I}_1(y)+y\{\mathrm{I}_1(y)\mathscr{K}-\mathrm{K}_1(y)\mathscr{I}\}/8n \qquad \text{(a8)}$$

with the abbreviations

$$\mathscr{I}=\int_0^x y'\,\mathrm{I}_1(y')\,v(x')\,dx', \quad \mathscr{K}=\int_0^x y'\mathrm{K}_1(y')\,v(x')\,dx', \quad y'=\sqrt{(4nx')}.$$

The term $y\{\ \}/8n$ vanishes with its first derivative at the origin so that solving the integral equation (a8) would give the exact solution (a6). Differentiating (a8) we have

$$v'=2n\,\mathrm{I}_0+\tfrac{1}{4}\,(\mathrm{I}_0\mathscr{K}+\mathrm{K}_0\mathscr{I}).$$

If we write (a8) in the form

$$v=y\mathrm{I}_1/[1-y\,(\mathrm{I}_1\mathscr{K}+\mathrm{K}_0\mathscr{I})/8nv]$$

we find, using (a7) again,

$$\frac{v'}{v}=\frac{2n\mathrm{I}_0}{y\mathrm{I}_1}\left(1+\frac{\mathscr{I}}{v}\,\frac{1}{8n\mathrm{I}_0}\right).$$

By finding upper and lower limits for $\mathscr{I}/v$ we can give upper and lower limits for v'/v. The simplest approximation is the following. Since $v(x)$ is a monotonically increasing function of x

$$0<\mathscr{I}/v<\int_0^x y'\,\mathrm{I}_1\,(y')\,\mathrm{d}x'=y^2\,\mathrm{I}_2(y)/2n$$

and so, using the relation $y\,\mathrm{I}_2=y\,\mathrm{I}_0-2\mathrm{I}_1$,

$$\frac{\mathrm{I}_0}{\mathrm{I}_1}<\frac{y}{2n}\,\frac{v'}{v}<\frac{\mathrm{I}_0}{\mathrm{I}_1}\left(1+\frac{y^2}{16n^2}\right)-\frac{y}{8n^2}. \qquad \text{(a9)}$$

This estimate could, of course, be improved. Using the recurrence relations for Bessel functions and the fact that, if $m<n$, $\mathrm{I}_m/\mathrm{I}_n$ decreases monotonically to unity as y increases we find the following inequalities:

$$\frac{\mathrm{I}_0}{\mathrm{I}_1}>1, \quad \frac{2}{y}<\frac{\mathrm{I}_0}{\mathrm{I}_1}<\frac{2}{y}+1.$$

With the help of these we can derive from (a9) this weaker relation free from Bessel functions :

$$\max\left(\frac{2}{y},1\right)<\frac{y}{2n}\,\frac{v'}{v}<1+\frac{2}{y}+\frac{y^2}{16n^2}.$$

The notation on the left implies that we are to choose the greater of $2/y$ and 1. Equation (22) in the text follows from the fact that

$$y=4n\sqrt{(x/\mathrm{L})}, \quad yv'/2nv=\tau\sqrt{x}/\tau_0\sqrt{\mathrm{L}}$$

in ordinary units.

References.

Burgers, J. M., 1939, *Proc. Kon. Ned. Akad. v. Wet.*, **42,** 293.

Cottrell, A. H., 1948, *Report of a Conference on the Strength of Solids* (London : Physical Society), p. 30 ; 1949, *Progress in Metal Physics* (London : Butterworths Scientific Publications), p. 105.

Eshelby, J. D., 1949, *Phil. Mag.*, **40,** [7], 903.

Frank, F. C., 1951, Symposium on Plastic Deformation of Crystalline Solids, May 1950, Carnegie Institute of Technology, Pittsburgh (in the press).

Hylleraas, E. A., 1937, *Ann. Inst. Henri Poincaré*, **7,** 121.

Koehler, J. S., 1941, *Phys. Rev.*, **60,** 397.

Kopal, Z., 1946, *Astrophys. J.*, **104,** 61.

Kulhmann, D., 1951, *Proc. Phys. Soc.*, **64,** 140.

Mott, N. F., and Nabarro, F. R. N., 1948, *Report of a Conference on the Strength of Solids* (London : Physical Society), p. 1.

Nabarro, F. R. N., 1940, *Proc. Phys. Soc.*, **52,** 34 ; 1951, *Some Recent Developments in Rheology* (London : United Trade Press Ltd., for the British Rheologists' Club), p. 38.

Starr, A. T., 1928, *Proc. Camb. Phil. Soc.*, **24,** 489.

Stieltjes, T. J., 1885, *Acta Math.*, **6,** 321.

Szegö, G., 1939, *Orthogonal Polynomials* (Amer. Math. Soc. Colloquium Publications, **23**), p. 136.

Teller, E., 1930, *Z. Phys.*, **61,** 458.

Wilson, A. H., 1928, *Proc. Roy. Soc.* A, **118,** 617, 635.

Zener, C., 1948, *Fracturing of Metals* (Symposium, American Society for Metals, Cleveland, Ohio), p. 3.

THE FORCE ON AN ELASTIC SINGULARITY

By J. D. ESHELBY
H. H. Wills Physical Laboratory, University of Bristol

(*Communicated by N. F. Mott, F.R.S.—Received* 18 *April* 1951)

CONTENTS

The parallel between the classical theory of elasticity and the modern physical theory of the solid state is incomplete; the former has nothing analogous to the concept of the force acting on an imperfection (dislocation, foreign atom, etc.) in a stressed crystal lattice. To remedy this a general theory of the forces on singularities in a Hookean elastic continuum is developed. The singularity is taken to be any state of internal stress satisfying the equilibrium equations but not the compatibility conditions. The force on a singularity can be given as an integral over a surface enclosing it. The integral contains the elastic field quantities which would surround the singularity in an infinite medium, multiplied by the difference between these quantities and those actually present. The expression for the force is thus of essentially the same form whether the force is due to applied surface tractions, other singularities or the presence of the free surface of the body ('image force'). A region of inhomogeneity in the elastic constants modifies the stress field; if it is mobile one can define and calculate the force on it. The total force on the singularities and inhomogeneities inside a surface can be expressed in terms of the integral of a 'Maxwell tensor of elasticity' taken over the surface. Possible extensions to the dynamical case are discussed.

1. Introduction

Modern theories of the solid state make use of the idea of the forces acting on imperfections in a crystal lattice, such as, for example, dislocations, foreign atoms, vacant lattice points, grain boundaries. The stress in the material arises from the presence of the imperfections and from any externally applied surface and body forces. If the applied forces are held constant, the total energy of the system (internal energy of the body plus potential energy of the sources of external force) is a function of the set of parameters necessary to specify the configuration of the imperfections. The negative gradient of the total energy with respect to the position of an imperfection may conveniently be called the force on it. This force, in a sense fictitious, is introduced to give a picturesque description of energy changes, and must not be confused with the ordinary surface and body forces acting on the material.

Even if there is only a single imperfection in the body and no externally produced stresses the force on it will not in general vanish. This 'image force' will depend on the shape of the boundary of the body and on the variation of the elastic constants from point to point in it.

The extra term which appears when there are also stresses due to applied forces and to other imperfections can be regarded as the force which these exert on the original imperfection.

In the usual theory of the elastic continuum the analogue of an imperfection (or rather of the stress-field associated with it) is some state of internal stress not produced by surface or body forces, for example, a nucleus of strain (Love 1927). The elastic analogue of an interstitial atom is a centre of dilatation supplemented by point singularities of higher order. The dislocations of physical theory have, of course, their analogues in certain of the elastic dislocations of Volterra (1907).

In many calculations it is possible to replace an imperfection by its elastic counterpart, allowing for the atomic structure and the departure from Hooke's law in an approximate fashion. The result may or may not be sensitive to the details of the approximation. The force on an interstitial atom or dislocation is of the insensitive type; its value when Hooke's law is assumed to be obeyed even for infinite strains does not differ significantly from the value with a reasonable approximation to the actual non-linear behaviour (Koehler 1941; Leibfried 1949; Bilby 1950). This suggests that one ought to be able to introduce the concept of the force on a singularity quite generally into the classical theory of elasticity, so completing the parallel between elastic continuum and crystal lattice. The object of this paper is to show how this can be done and to devise a simple way of calculating the force on a given elastic singularity.

We have first to allow that singularities can move through the medium, a possibility not envisaged in the classical theory but not inconsistent with it. To see the lines on which the theory must then be developed it is useful to make a comparison with electrostatics. The force on a point charge is usually taken as the starting point, and development of the theory leads to the concept of an energy density. The total energy of a point charge is found to be infinite, but this does not cause difficulty until the more sophisticated electrodynamic problems are reached. In the elastic case things are reversed—we know the energy density and must infer the force. The problem of the infinite self-energy of point and line singularities must be faced from the outset. Alternatively, Poisson's equation can be taken as the starting point for electrostatics; point charges are then to be regarded as limiting cases in which the distribution of charge has the form of a delta-function. Similarly, we can develop the elastic theory for states of internal stress with finite total strain energy and regard point and line singularities as limiting cases. From this point of view the elastic analogue of the electric field produced by a continuous charge distribution is the general state of self-strain in which the stresses satisfy the equilibrium equations but (unlike stresses arising from body and surface forces) not the compatibility conditions. The elastic displacements cannot be defined everywhere, but their place can be taken by the three stress functions of Maxwell or Morera. These provide the analogue of the electrostatic potential; the counterpart of the charge is the 'incompatibility tensor' formed of the six expressions which when equated to zero yield the compatibility conditions. We shall, however, work with the conventional representation in terms of stress, strain and displacement even though this leads to a certain amount of manœuvring to circumvent the lack of a displacement function in regions where the incompatibility tensor is not zero.

In what follows, the term 'singularity' is usually to be taken to mean an extended state of internal stress of the kind just discussed, rather than a singularity in the mathematical sense.

To make this idea more concrete we first describe (§3) a general type of surface singularity with finite though discontinuous stresses and show how various point and line singularities can be derived as limiting forms of it.

To be able to speak of the movement of a singularity one must decide precisely what is meant by saying that two states of internal stress in the same body represent the same singularity in different positions. This is discussed in §4 and the result used (§5) to define provisionally and calculate the force which surface tractions exert on a singularity. The image force and the force which one singularity exerts on another are dealt with in §7 by a simple argument which also justifies the provisional definition of §5. These results apply only to homogeneous (but anisotropic) bodies. In §8 the force on an inhomogeneity in a body free of internal stress is defined and calculated. Section 9 ties up certain loose threads in the previous argument and extends the results to bodies containing both internal stresses and inhomogeneities, and acted on by body forces as well as surface tractions. Section 10 discusses possible extensions, in particular to the dynamical behaviour of singularities.

Apart from recent work inspired by the needs of metal physics the present problem received some attention at the end of the last century, when elastic solids furnished models for the ether. Larmor (1897) discussed an elastic singularity formed by cutting a lens-shaped cavity in the material, giving a relative rotation to the faces, and cementing them together. He pointed out that a pair of these singularities would exert forces on each other. In a remarkable paper, Burton (1892) considered the equations of motion of 'strain figures', states of stress which are possible without applied forces in a medium with a non-linear stress-strain relation. Burton (as he himself realized) was inconsistent in assuming superposability of stresses in his non-Hookean solid; his results can, however, be interpreted in terms of internal stresses in a Hookean medium.

2. Notation and integral theorems

Rectangular Cartesian co-ordinates are denoted by x_i ($i = 1, 2, 3$). The elastic displacement vector has components u_i. Suffixes following a comma represent differentiation, so that, for example, $u_{i,j} = \partial u_i/\partial x_j$, $u_{i,jk} = \partial^2 u_i/\partial x_j \partial x_k$. A repeated suffix is to be summed over the values 1, 2, 3. The strain tensor e_{ij} and the stress tensor p_{ij} are defined by

$$\left.\begin{aligned} e_{ij} &= e_{ji} = \tfrac{1}{2}(u_{i,j}+u_{j,i}), \\ p_{ij} &= p_{ji} = c_{ijkl}e_{kl} = c_{ijkl}u_{k,l} \end{aligned}\right\}. \tag{1}$$

The elastic modulus tensor c_{ijkl} (which may be a function of position) is unaltered by interchanging i with j or k with l or the pair (ij) with the pair (kl). In an isotropic body

$$p_{ij} = \lambda e_{kk}\delta_{ij} + 2\mu e_{ij}. \tag{2}$$

In equilibrium with no body forces the divergence of the stress-tensor vanishes:

$$p_{ij,j} = 0. \tag{3}$$

The traction on an element of area dS with normal n_j is $p_{ij}n_j\,dS = p_{ij}\,dS_j$, where dS_j is an abbreviation for $n_j\,dS$. For the e_{ij} to be derivable from a displacement according to (1) the six compatibility conditions $S_{rs} = 0$ must be satisfied, where

$$S_{rs}(e_{mn}) \equiv e_{rs,ii} + e_{ii,rs} - e_{ri,si} - e_{si,ri} - (e_{ii,jj} - e_{ij,ij})\,\delta_{rs}. \tag{4}$$

12-2

Let (u_i, e_{ij}, p_{ij}) and (u'_i, e'_{ij}, p'_{ij}) be two sets of quantities each related by (1) and satisfying (3). We have

$$p_{ij}e'_{ij} = p'_{ij}e_{ij} = p_{ij}u'_{i,j} = p'_{ij}u_{i,j} \tag{5}$$

by the symmetry properties of the c's. If we form the vector $\mathbf{v}$ with components

$$v_j = p_{ij}u'_i - p'_{ij}u_i,$$

it easily follows from (3) and (5) that $\operatorname{div}\mathbf{v} = v_{i,i} = 0$. Hence, by Gauss's theorem,

$$\int_{\Sigma_1} v_j\,dS_j = \int_{\Sigma_2} v_j\,dS_j \quad \text{or} \quad \int_{\Sigma_1-\Sigma_2} v_j\,dS_j = 0, \tag{6}$$

where Σ_1 and Σ_2 are two closed surfaces which can be deformed into one another without passing through singularities of $\mathbf{v}$. If Σ is a closed surface containing no singularities of $\mathbf{v}$,

$$\int_{\Sigma} v_j\,dS_j = 0. \tag{7}$$

Equation (6) is essentially Betti's reciprocal theorem (Love 1927). We have assumed that $\mathbf{v}$ is single-valued. If it has to be made single-valued by a cut the two integrals in (6) differ by $\int v_j\,dS_j$ taken over both sides of the part of the cut surface intercepted between Σ_1 and Σ_2 (Colonnetti 1915).

If u'_i, p'_{ij} are corresponding displacement and stress, so are $u'_{i,l}, p'_{ij,l}$ as long as the body is homogeneous, and so by (6)

$$\int_{\Sigma_1-\Sigma_2} (p_{ij}u'_{i,l} - p'_{ij,l}u_i)\,dS_j = 0 \quad \text{if} \quad c_{ijkl,m} = 0. \tag{8}$$

Again, if we have a continuous series of elastic states specified by a parameter ξ then $\partial u'_i/\partial\xi$, $\partial p'_{ij}/\partial\xi$ are corresponding displacement and stress if u'_i, p'_{ij} are, provided the c's are independent of ξ; then (6) gives

$$\int_{\Sigma_1-\Sigma_2} (p_{ij}\,\partial u'_i/\partial\xi - u_i\,\partial p'_{ij}/\partial\xi)\,dS_j = 0 \quad \text{if} \quad \partial c_{ijkl}/\partial\xi = 0. \tag{9}$$

The following result will also be useful:

$$\int_{\Sigma} (w_{j,l} - w_{i,i}\delta_{jl})\,dS_j = -\int_{\sigma} \epsilon_{lij}w_i\,dx_j, \tag{10}$$

where Σ is a surface bounded by the curve σ, and w_i is single-valued and continuous. ϵ_{lij} is the completely antisymmetric tensor which is zero if two suffixes are equal and otherwise is $+1$ or -1 according as lij is an even or odd permutation of 123. Equation (10) is easily proved by applying Stokes's theorem to the tensor $\epsilon_{lij}w_i$ or by integrating $w_{i,i}$ over the volume generated by Σ during a small displacement parallel to the x_l-axis and then using Gauss's theorem. If w_i is multiple-valued on Σ but can be made single-valued by a cut C, we shall in general expect the integrand to have singularities at the ends of the cut. We can define the surface integral over Σ to be the limit of the integral over the region of Σ outside σ (figure 1*a*) as the latter shrinks on to C. Then (assuming Σ is closed apart from the cut)

$$\int_{\Sigma} (w_{j,l} - w_{i,i}\delta_{lj})\,dS_j = -\int_{C} \epsilon_{lij}\,\Delta w_i\,dx_j, \tag{11}$$

provided the circles A, B make no contribution in the limit. (We shall usually be concerned with the case where C is a closed curve, or two-dimensional problems where C becomes a point on a plane curve.) Δw_i is the jump in w_i on crossing C, the sign being chosen so that if

$$\Delta w_i = \lim_{a\to b}\{w_i(a) - w_i(b)\}$$

the vectors $\overline{ab}$, dx_i and the outward-drawn normal form a right-handed system.

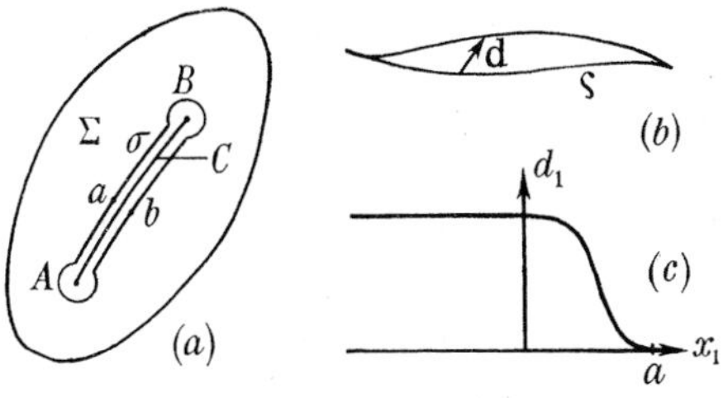

FIGURE 1

If we form a tensor $$t_{jl} = w_{j,l} - w_{i,i}\delta_{jl} \tag{12}$$

from any vector w_i, then for any closed surface Σ on which t_{jl} is single-valued

$$\int_\Sigma t_{jl}\, dS_j = 0,$$

even if there are singularities of t_{jl} inside Σ. This is to be contrasted with (7); in fact, $t_{jl,j}$ vanishes identically whilst $v_{j,j}$ does so only in virtue of the equation $p_{ij,j} = 0$.

3. Types of singularity

A large range of singularities can be regarded as particular cases of a type of surface singularity considered by Somigliana (1914) (cf. also Gebbia 1902; Mann 1949; Bogdanoff 1950). Following Neményi (1931) we shall call this singularity a Somigliana dislocation. It can be generated in the following way. Make a cut over a surface ς (open or closed) in an unstrained body and give the faces of the cut a small relative displacement, removing material where there would be interpenetration (figure 1 *b*). Fill in the remaining gaps and weld together. We are left with a system of internal strain which is completely characterized by giving as a function of position on ς the vector $\mathbf{d}$ which specifies the final separation of points originally adjacent on opposite sides of the cut. The stress $p_{ij}n_j$ (n_j is the normal to ς) is continuous across ς, but the individual components of p_{ij}, e_{ij} are in general discontinuous; these discontinuities can be calculated when $\mathbf{d}$ is known over ς. If $\mathbf{d}$ is a reasonably smooth function p_{ij} and e_{ij} will be finite everywhere except possibly near the edge of an open surface. (The condition for the displacement to be finite near the edge has been derived for the isotropic case by Gebbia.) Somigliana dislocations for which $\mathbf{d} = \mathbf{b} + \mathbf{r} \times \mathbf{c}$ with constant vectors $\mathbf{b}$ and $\mathbf{c}$ are the Weingarten dislocations discussed by Volterra. Dislocations for which $\mathbf{c} = 0$ are of particular interest; we shall call them physical dislocations. A number of physical dislocation loops lying in a surface (for example, near one of Frank & Read's (1950) dislocation sources) can be regarded as a general Somigliana dislocation for which $\mathbf{d}$ is a stepped function of position; at distances large compared with the spacing of the individual dislocations the stress is equivalent to that of a Somigliana dislocation with a continuously distributed **d**.

Near the boundary of a physical dislocation the stresses tend to infinity unless we remove the material in this region, as Volterra did. Alternatively, we can replace the physical dislocation by a Somigliana dislocation with **d** equal to **b** over most of ς but suitably tapered off near the boundary. Consider, for example, an infinite edge dislocation line along the x_3 axis with $\mathbf{d} = (b, 0, 0)$. For the tapered Somigliana dislocation we may, for example, take (figure 1c)

$$\begin{aligned} d_1 &= b(a^m - x_1^m)^n / a^{mn} \quad (x_1 > 0) \\ &= b \quad (x_1 < 0) \end{aligned}$$

with the half-plane $x_2 = 0, x_1 < a$ for ς. From a known relation between stress and the discontinuity in displacement (Nabarro 1947; Eshelby 1949b) the shear stress p_{12} in the plane $x_2 = 0$ can be shown to be proportional to $x^{m-1}(a^m - x^m)^{n-1} a^{-mn} \log |1 - a/x_1|$, plus a polynomial in x_1. Hence by choosing m and n large enough we can make p_{12} and as many of its derivatives as we like continuous on the plane $x_2 = 0$. Further calculation shows that the same is true for all components of p_{ij} both on and off this plane. It is clear that a physical dislocation bounded by an arbitrary curve can be similarly regarded as the limit as $a \to 0$ of a suitable Somigliana dislocation dependent on a parameter a specifying the taper near the edge.

To derive point and line singularities from the general Somigliana dislocation we take a sphere or tube of radius r for ς and a suitable distribution of **d**. We then let r decrease to zero, at the same time increasing **d** in such a way that the displacement at a fixed point of observation remains finite. For example, if ς is a sphere and **d** is normal to ς and of magnitude $d = \text{const.}\, r^{-2}$ we obtain a centre of dilatation.

Thus physical dislocations and point and line singularities with infinite self-energy can all be regarded as limiting cases of Somigliana dislocations of finite self-energy.

As already explained, we may also have a volume distribution of internal stress in which (3) is satisfied, but the e_{ij} derived from the p_{ij} with the aid of Hooke's law do not satisfy $S_{ij} = 0$ everywhere. The regions in which $S_{ij} \neq 0$ are to be regarded as the actual seat of the singularity. This is discussed more fully in §9.

4. The stress-field of a singularity in a finite body

We need first to decide what is meant by a particular type of singularity in a body of given shape and size. A point singularity will naturally be defined as a solution of the elastic equations with vanishing $p_{ij} n_j$ at the surface of the solid and becoming infinite in a prescribed way at a certain point. In the general case it will be more convenient to regard each type of singularity as defined by giving its stress field in an infinite body and then prescribing a process for finding the stress field of the same singularity in a given finite body.

Draw a closed surface Σ_0 in an infinite homogeneous elastic medium. We shall say that there is a singularity inside Σ_0 if the stresses in it could not be produced by body forces outside Σ_0. We shall suppose that, when the region within a certain closed surface Σ_S has been excluded, p_{ij} and $u_{i,j}$ are continuous and single-valued and u_i is continuous but not necessarily single-valued in the rest of the interior of Σ_0.

For a point or line-singularity Σ_S can be taken as a small sphere or narrow tube surrounding the point or line on which the stress and displacement become infinite. If u_i or p_{ij} become infinite on or discontinuous across a surface we have a surface singularity, and for Σ_S we may take a jacket closely enveloping the singular surface. The conditions imposed on u_i and $u_{i,j}$ have been chosen so that physical dislocations can be regarded as line singularities rather than surface singularities. (If the physical dislocation is replaced by a tapered Somigliana dislocation, as described in § 3, Σ_S must enclose the whole of the tapered part.) For a volume singularity Σ_S must enclose the region in which $S_{rs} \neq 0$. We shall assume that the surface tractions over any surface within Σ_O and surrounding Σ_S are in rigid-body equilibrium. A distribution of body force inside Σ_S with zero resultant and moment therefore qualifies as a singularity according to our definition, though it is not usually what we have in mind.

Let the displacement and stress of a certain singularity in the infinite medium be $u_i^\infty(\mathbf{x}), p_{ij}^\infty(\mathbf{x})$. Then we define $u_i^\infty(\mathbf{x}-\boldsymbol{\xi}), p_{ij}^\infty(\mathbf{x}-\boldsymbol{\xi})$ to be the corresponding quantities for the singularity after it has been moved distances ξ_1, ξ_2, ξ_3 parallel to the x_1-, x_2- and x_3-axes. For a point singularity it is natural to take for the ξ_i the co-ordinates of the mathematical singular point. In general, though nothing more is implied than that variation of $\boldsymbol{\xi}$ translates the stress-system rigidly, it is still convenient to speak of the point $\boldsymbol{\xi}$ as the position of the singularity.

Starting from the case of an infinite medium we define the same singularity in a finite body in the following way. Remove the material outside Σ_O without allowing the surface forces $p_{ij}^\infty n_j$ on Σ_O to relax, so that the displacements and strains remain unaltered within Σ_O. Next reduce these surface forces to zero. We are left with a singularity in a body whose surface Σ_O is stress-free. The displacement and stress are

$$\left.\begin{aligned} u_i^S &= u_i^\infty(\mathbf{x}-\boldsymbol{\xi}) + u_i^I(\mathbf{x},\boldsymbol{\xi}), \\ p_{ij}^S &= p_{ij}^\infty(\mathbf{x}-\boldsymbol{\xi}) + p_{ij}^I(\mathbf{x},\boldsymbol{\xi}), \end{aligned}\right\} \tag{13}$$

where u_i^I and p_{ij}^I (which we may call the image displacement and stress) have no singularities within Σ_O and are such that $p_{ij}^S n_j = 0$ on Σ_O, i.e.

$$p_{ij}^I n_j = -p_{ij}^\infty n_j \quad \text{on} \quad \Sigma_O. \tag{14}$$

We have

$$u_{i,l}^\infty = -\partial u_i^\infty/\partial\xi_l, \quad p_{ij,l}^\infty = -\partial p_{ij}^\infty/\partial\xi_l. \tag{15}$$

There are no corresponding relations for u_i^I, p_{ij}^I, since they depend on $\mathbf{x}, \boldsymbol{\xi}$ not merely through the differences $(\mathbf{x}-\boldsymbol{\xi})$. However, from (14) and (13)

$$(\partial p_{ij}^I/\partial\xi_l)\, n_j = -(\partial p_{ij}^\infty/\partial\xi_l)\, n_j = p_{ij,l}^\infty n_j \text{ on } \Sigma_O. \tag{16}$$

5. The force exerted on a singularity by surface tractions

Now apply surface tractions $p_{ij}^A n_j$ to Σ_O, producing displacement and stress u_i^A, p_{ij}^A in the body in addition to the u_i^S, p_{ij}^S already present. Let the singularity undergo a translation $\delta\xi$ parallel to the x_l-axis. The work done by the surface tractions is

$$\delta W = \delta\xi \int_{\Sigma_O} p_{ij}^A \frac{\partial u_i^S}{\partial \xi_l}\, dS_j + O(\delta\xi^2).$$

We tentatively define

$$F_l^A = \lim_{\delta\xi\to 0} \frac{\delta W}{\delta\xi} = \int_{\Sigma_O} p_{ij}^A \frac{\partial u_i^S}{\partial \xi_l}\, dS_j \tag{17}$$

as the force which the surface tractions exert on the singularity. Because $p^S_{ij}n_j = 0$ on Σ_O we can write

$$F^A_l = \int_\Sigma \left(p^A_{ij}\frac{\partial u^S_i}{\partial \xi_l} - u^A_i \frac{\partial p^S_{ij}}{\partial \xi_l}\right) dS_j$$

taken over the surface $\Sigma = \Sigma_O$. By (9) this integral can equally well be taken over any surface Σ in the body into which Σ_O can be deformed without entering Σ_S. By (13)

$$F^A_l = \int_\Sigma \left(p^A_{ij}\frac{\partial u^\infty_i}{\partial \xi_l} - u^A_i \frac{\partial p^\infty_{ij}}{\partial \xi_l}\right) dS_j + \int_\Sigma \left(p^A_{ij}\frac{\partial u^I_i}{\partial \xi_l} - u^A_i \frac{\partial p^I_{ij}}{\partial \xi_l}\right) dS_j.$$

The second term vanishes by (9) and (7). Using (15) we have simply

$$F^A_l = \int_\Sigma (u^A_i p^\infty_{ij,l} - p^A_{ij} u^\infty_{i,l})\, dS_j, \tag{17'}$$

a form in which we do not need to know u^I_i, p^I_{ij}. This can be further transformed by (11). Since the only multiple-valued quantity which we allow is the displacement of a physical dislocation for which $\Delta u^S_i = \Delta u^\infty_i = \text{const.}$ we have

$$F^A_l = \int_\Sigma (p^A_{ij,l} u^\infty_i - u^A_{i,l} p^\infty_{ij})\, dS_j + \Delta u^\infty_k \int_C \epsilon_{lij} p^A_{ik}\, dx_j. \tag{18}$$

The further form

$$F^A_l = \int_\Sigma \{(\tfrac{1}{2}p^\infty_{ik} u^A_{i,k}\delta_{jl} - p^\infty_{ij} u^A_{i,l}) + (\tfrac{1}{2}p^A_{ik} u^\infty_{i,k}\delta_{jl} - p^A_{ij} u^\infty_{i,l})\}\, dS_j \tag{19}$$

is easily verified; it holds whether u^∞_i is single-valued or not.

6. The centre of dilatation: dislocations

The simplest singularity is a centre of dilatation in an isotropic body (Love 1927). If it is at the origin

$$u^\infty_i = \delta x_i / r^3 \quad \text{with} \quad r^2 = x_1^2 + x_2^2 + x_3^2, \tag{20}$$

where δ is a constant. To find the force on it we use (18), taking for Σ a small sphere of radius r about the origin. Expanding the applied stress and displacement in a Taylor series we have

$$F^A_l = p^A_{ij,l}\int u^\infty_i\, dS_j + p^A_{ij,lm}\int x_m u^\infty_i\, dS_j - u^A_{i,l}\int p^\infty_{ij}\, dS_j - u^A_{i,lm}\int x_m p^\infty_{ij}\, dS_j + \dots,$$

where the A quantities are to be given their value at the origin. We have

$$\int u^\infty_i\, dS_j = \delta r^{-4}\int x_i x_j\, dS = \delta r^{-4}\delta_{ij}\int \tfrac{1}{3}(x_1^2+x_2^2+x_3^2)\, dS = \tfrac{4}{3}\pi\delta\delta_{ij}.$$

In a similar way we find, using (1) and (2),

$$\int x_m p^\infty_{ij}\, dS_j = -(16\pi/3)\,\mu\delta\delta_{mi}.$$

The term in $\int p^\infty_{ij}\, dS_j$ vanishes since the singularity is in equilibrium. The remaining terms are of order r. Since F^A_l is independent of the choice of r they can make no contribution whether we let r tend to zero or not. Thus

$$F^A_l = \tfrac{4}{3}\pi\delta(p^A_{ii,l} + 4\mu u^A_{i,il}) = 4\pi\delta\{(1-\sigma)/(1+\sigma)\}p^A_{ii,l},$$

where $\sigma = \lambda/2(\lambda+\mu)$ is Poisson's ratio. In general

$$\mathbf{F} = -12\pi\delta\{(1-\sigma)/(1+\sigma)\}\,\mathrm{grad}\,p^A, \tag{21}$$

where
$$p^A = -\tfrac{1}{3}(p^A_{11}+p^A_{22}+p^A_{33})$$

is the applied hydrostatic pressure at the position of the singularity.

The following model of an interstitial atom has been used by Bilby (1950). An elastic sphere of radius $(1+\alpha)r_0$ is forced into a spherical hole of radius r_0 in an infinite block of the same material. It can easily be shown (Mott & Nabarro 1940) that for $r>r_0$ the displacement is given by (20) with $\delta = \alpha r_0^3(1+\sigma)/3(1-\sigma)$, whilst for $r<r_0$ the material is uniformly compressed.

If we regard this as a volume singularity we may take for Σ any sphere of radius $r>r_0$. The force must then be given by (21), since this expression did not depend on the size of the surface over which we integrated. Alternatively, we may regard it as a Somigliana dislocation with a discontinuity αr_0 in the radial displacement across the sphere $r = r_0$. Σ must now be supplemented by a closed surface within the sphere $r = r_0$ in order that it may completely embrace the singular surface. Since this new surface can be contracted to a point it makes no contribution and the force is still given by (21). This is an exact result; F^A_l is proportional to the value of $\mathrm{grad}\,p^A$ at the centre of the sphere $r = r_0$, and there are no terms of order r_0 to be added.

For the two-dimensional problem of an infinite straight dislocation line along the x_3-axis with p^A_{ij}, u^A_i independent of x_3 we may use (18) to calculate F^A_l, the force per unit length, taking for Σ a cylinder of radius r and unit length with its axis along x_3. C is then unit length of a generator of this cylinder.

We could use explicit expressions for $u^\infty_i, p^\infty_{ij}$, but this would be tiresome in the anisotropic case. It is more interesting to see what is the least information we need about the dislocation in order to find F^A_l. The fundamental property is

$$\oint u_{i,j}\,dx_j = b_i \tag{22}$$

for any closed circuit embracing the dislocation line. This, however, does not completely define the singularity. Without altering (22) we could add single-valued line singularities (e.g. a line of dilatation) coincident with the dislocation line. They may be excluded by requiring that

$$\lim_{r\to 0} r u^\infty_i = 0, \tag{23}$$

where r is the distance of the point x_i from the singular line.

If we put $\Delta u^\infty_i = b_i$ and expand the A quantities about the origin (18) becomes

$$F^A_l = u^A_{i,l}\int_\Sigma p^\infty_{ij}\,dS_j + \epsilon_{lk3}p^A_{ik}b_i + O(r).$$

The first term vanishes since the dislocation is in equilibrium. Hence

$$F^A_1 = p^A_{2i}b_i, \quad F^A_2 = -p^A_{1i}b_i. \tag{24}$$

For a pure screw dislocation with $\mathbf{b} = (0, 0, b)$

$$F^A_1 = p^A_{23}b, \quad F^A_2 = -p^A_{13}b,$$

and for a pure edge dislocation with $\mathbf{b} = (b, 0, 0)$

$$F_1^A = p_{12}^A b, \quad F_2^A = -p_{11}^A b. \tag{24'}$$

In place of (24′) Koehler (1941) originally found in the isotropic case

$$F_1^A = \frac{2}{\pi}\frac{\lambda+\mu}{\lambda+2\mu} p_{12}^A b.$$

His method is equivalent to evaluating (17′) over a square surrounding the dislocation and omitting the first term in the integrand.

We can similarly show that for a loop of physical dislocation of arbitrary form

$$F_l^A = \lim_{r\to 0}\left\{\int_\Sigma u_{i,l}^A p_{ij}^\infty \, dS_j + O(rl)\right\} + b_i \epsilon_{klj} \int_\sigma p_{ik}^A \, dx_j, \tag{25}$$

where Σ is a narrow tube of radius r and total length l embracing the singular line σ. The first term in (25) will vanish if we require that not only is $\int p_{ij}^\infty \, dS_j$ zero when taken over Σ but also when it is taken over any part of Σ intercepted between two planes perpendicular to the dislocation line (cf. Burgers 1939). Actually there is no harm in including in the surface of integration the two cross-sections of the tube, since according to (23) p_{ij}^∞ ultimately behaves like r^{-1}. An equivalent condition for the vanishing of the surface integral in (25) is thus that any element of volume is in static equilibrium whether it is traversed by the singular line or not. If this is admitted,

$$F_l^A = b_1 \epsilon_{klj} \int_\sigma p_{ik}^A \, dx_j.$$

This is consistent with a force $\epsilon_{klj} b_i p_{ik}^A s_j$ per unit length on a dislocation line at a point where its unit tangent vector has components s_j (Peach & Koehler 1950; Nabarro 1951). We cannot prove this by the present method, which only considers translations of the loop without change of form.

For the general Somigliana dislocation

$$F_l^A = \int_\varsigma d_i p_{ij,l}^A \, dS_j,$$

with the ς and $\mathbf{d}$ of § 3.

7. The image force and the forces between singularities

Consider a body containing a singularity S whose energy-density $\mathscr{W} = \frac{1}{2} p_{ij}^S u_{i,j}^S$ is everywhere finite. The internal elastic energy of the body,

$$W_{\text{int.}} = \int \mathscr{W} \, dv,$$

is a function of the parameters ξ_i defining the position of the singularity. We may regard

$$F_l^I = -\partial W_{\text{int.}} / \partial \xi_l,$$

which measures the rate of decrease of $W_{\text{int.}}$ when the singularity is displaced in the x_l direction, as the force acting on it in the absence of applied forces or other singularities. Since F_l^I depends on the existence of the free surface it will be called the image force.

If there is a second singularity T whose energy-density is likewise finite everywhere we shall now have $\mathscr{W}=\frac{1}{2}(p_{ij}^S+p_{ij}^T)(u_{i,j}^S+u_{i,j}^T)$, and if S is moved whilst T remains fixed, $-\partial W_{\text{int.}}/\partial\xi_l$ will have a value differing from F_l^I. If

$$-\partial W_{\text{int.}}/\partial\xi_l = F_l^I+F_l^T, \tag{26}$$

we can regard F_l^T as the force which T exerts on S. Similarly, if there are in addition constant externally applied stresses we shall find, say,

$$-\partial W_{\text{int.}}/\partial\xi_l = F_l^I+F_l^T+F_l^{A'}.$$

If $W_{\text{ext.}}$ is the potential energy of the source of the applied stress, $-\partial W_{\text{ext.}}/\partial\xi_l$ is F_l^A as defined in §5. The total force on the singularity S can then be defined as

$$F_l = -\partial(W_{\text{int.}}+W_{\text{ext.}})/\partial\xi_l = F_l^I+F_l^T+F_l^A+F_l^{A'},$$

and $F_l\delta\xi$ is the decrease in energy of the whole system when S is displaced from ξ_i to $\xi_i+\delta_{il}\delta\xi$. Consequently our previous definition of F_l^A is reasonable only if we can show that $F_l^{A'}$ vanishes. To calculate F_l^I and F_l^T in a simple way for a homogeneous medium we use this known result:

An elastic body reacts to applied forces in the same way whether it is self-stressed or not. (27)

(See, for example, Southwell (1936) and the discussion and references in Neményi (1931). Gebbia (1902) stated (27) for the case of Somigliana dislocations.)

The following discussion is valid for singularities with finite self-energy. To apply the results to singularities with infinite self-energy we must make the following explicit assumption:

Singularities with infinite self-energy can be regarded as limiting cases of singularities with finite self-energy, and when we make the passage to the limit the expression for the force is still valid.

Alternatively we can simply lay down these two axioms:

(i) The statement (27) is true also for singularities with infinite self-energy.

(ii) The plausible subtraction of infinities implicit in equation (28) below is allowable.

As in §4 we take the singularity in an infinite medium, describe a surface Σ_0 around it and cut out the body bounded by Σ_0 without allowing the surface forces to relax. In this way we get a body with the singularity in the position specified by the parameters ξ_i. If we had translated Σ_0 a distance $-\delta\xi$ parallel to the x_l-axis before cutting it, we should have got a body equivalent to the first but with the singularity in the position specified by the parameters $\xi_i+\delta_{il}\delta\xi$. The difference of energy in these two cases before the surface tractions are relaxed is (figure 2*a*)

$$\frac{1}{2}\int_{\Sigma_0} p_{ik}^\infty e_{ik}^\infty\, dS_l\,\delta\xi + O(\delta\xi^2). \tag{28}$$

(It is here that any necessary 'subtraction of infinities' occurs; energy in the shaded region of figure 2*a* makes no contribution.) When the surface tractions are relaxed in the first case the elastic energy is reduced by an amount

$$-\frac{1}{2}\int_{\Sigma_0} p_{ij}^\infty u_i^I\, dS_j.$$

(Here we invoke (27).) The corresponding quantity in the second case is

$$-\left(1+\delta\xi\frac{\partial}{\partial\xi_l}\right)\frac{1}{2}\int_{\Sigma_0} p_{ij}^\infty u_i^I\, dS_j + O(\delta\xi^2).$$

13-2

Consequently, after relaxing the surface tractions the energy for the first case less that for the second is

$$\frac{1}{2}\int_{\Sigma_0}\{p_{ik}^{\infty}e_{ik}^{\infty}\delta_{lj}-\partial(p_{ij}^{\infty}u_i^I)/\partial\xi_l\}\,dS_j\,\delta\xi+O(\delta\xi^2).$$

This must be equal to $F_l^I\delta\xi+O(\delta\xi^2)$. Rearranging and using (11) and (15) we have

$$F_l^I=\int_{\Sigma_0}p_{ij}^I\frac{\partial u_i^S}{\partial\xi_l}dS_j+\frac{1}{2}\int_{\Sigma_0}\left(\frac{\partial p_{ij}^I}{\partial\xi_l}u_i^I-p_{ij}^I\frac{\partial u_i^I}{\partial\xi_l}\right)dS_j+\int_{\Sigma_0}(\tfrac{1}{2}p_{ik}^{\infty}u_{i,k}^{\infty}\delta_{jl}-p_{ij}^{\infty}u_{i,l}^{\infty})\,dS_j.$$

The second term vanishes by (7) and (9). According to (8) the last term is unchanged if Σ_0 is deformed in any way as long as it still encloses the singularity. Hence this term is a constant characteristic of the singularity in an infinite medium and entirely independent of the size and shape of the body containing it. When we let Σ_0 tend to infinity the term vanishes if u_i^{∞} behaves like r^{-1} or $\log r$ at large distances in a three-dimensional or two-dimensional problem. For isotropy the theory of biharmonic functions shows that these conditions are fulfilled if the stresses do in fact fall off at all with r, as they must do if there is not to be a singularity at infinity. It is fairly obvious that the same will be true for anisotropy. We return to this point in §9 and meanwhile assume that the three quantities

$$F_l^{\infty}=\int_{\Sigma}(\tfrac{1}{2}p_{ik}^{\infty}u_{i,k}^{\infty}\delta_{jl}-p_{ij}^{\infty}u_{i,l}^{\infty})\,dS_j \tag{29}$$

vanish for any surface Σ surrounding the singularity. Manipulating the surviving integral in F_l^I as we did (17), we have, for example,

$$F_l^I=\int_{\Sigma}(u_i^I p_{ij,l}^{\infty}-p_{ij}^I u_{i,l}^{\infty})\,dS_j.$$

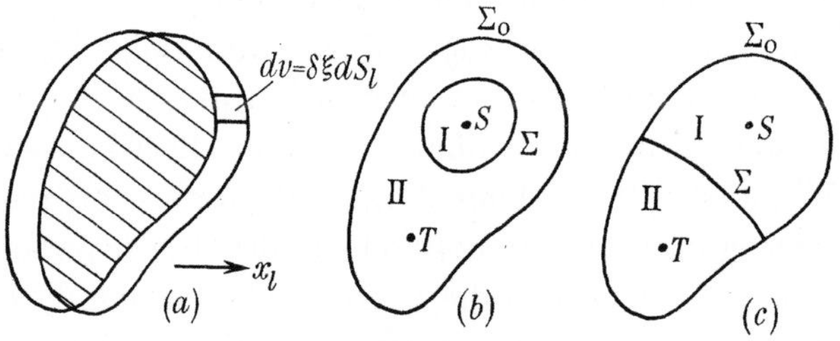

FIGURE 2

The case where there is a fixed singularity T can be treated by a slightly more elaborate set of imaginary operations. (We shall not trouble to insert terms of order $\delta\xi^2$.) Draw a surface Σ in Σ_0 surrounding S but not T, thus dividing the body into parts I and II (figure 2*b*). (The final result is the same with the scheme of figure 2*c*, where it is easier to imagine separating I from II, but the argument becomes more prolix if Σ and Σ_0 are not separate.) Let u_i be the total displacement and put $u_i'=u_i-u_i^{\infty}=u_i^I+u_i^T$ and similarly for p_{ij} and p_{ij}'. Carry out the following steps:

(i) Move S from ξ_i to $\xi_i+\delta_{il}\delta\xi$ and record the new surface tractions

$$\{p_{ij}+\delta\xi(\partial p_{ij}/\partial\xi_l)\}\,n_j \tag{30}$$

across Σ. Move S back to ξ_i.

(ii) Cut the surface Σ and remove I without allowing the tractions on the surface of I and the surface of the cavity in II to relax. Alter the surface tractions on I from $p_{ij}n_j$ to $p_{ij}^{\infty}n_j$. The energy entering I in this process is

$$\frac{1}{2}\int_{\Sigma}(p_{ij}+p_{ij}^{\infty})\,(u_i^{\infty}-u_i)\,dS_j = -\int_{\Sigma}(p_{ij}^{\infty}+\tfrac{1}{2}p_{ij}')\,u_i'\,dS_j.$$

(iii) Change the position of S from ξ_i to $\xi_i+\delta_{il}\delta\xi$ and adjust the surface tractions in I to $\{p_{ij}^{\infty}+\delta\xi(\partial p_{ij}^{\infty}/\partial\xi_l)\}n_j$. The increase of energy in I is obviously given by (28) with Σ in place of Σ_0.

(iv) Alter the surface traction on I to (30). The energy entering I is

$$(1+\delta\xi\,\partial/\partial\xi_l)\int_{\Sigma}(p_{ij}^{\infty}+\tfrac{1}{2}p_{ij}')\,u_i'\,dS_j.$$

(v) Alter the surface traction on the cavity in II to (30). The energy entering II is

$$-\delta\xi\int_{\Sigma}p_{ij}(\partial u_i/\partial\xi_l)\,dS_j.$$

The minus sign is correct if the normal is directed from the cavity into II.

(vi) The stress and displacement are now equal at corresponding points of the surface of I and the cavity in II. Replace I in the cavity and rejoin the surface Σ. There is no energy change in I or II.

Adding the different energy changes and using (29) we have

$$-\frac{\partial W}{\partial\xi_l} = \int_{\Sigma}\left\{(p_{ij}^{T}+p_{ij}^{I})\frac{\partial u_i^{\infty}}{\partial\xi_l}-\frac{\partial p_{ij}^{\infty}}{\partial\xi_l}(u_i^{T}+u_i^{I})\right\}dS_j+\frac{1}{2}\int_{\Sigma}\left(p_{ij}'\frac{\partial u_i'}{\partial\xi_l}-\frac{\partial p_{ij}'}{\partial\xi_l}u_i'\right)dS_j, \tag{31}$$

and this must equal $F_l^T+F_l^I$. The second integral vanishes by the usual argument, and from (15) and (26) we have

$$F_l^T = \int_{\Sigma}(p_{ij,l}^{\infty}u_i^{T}-p_{ij}^{T}u_{i,l}^{\infty})\,dS_j.$$

Finally, let us repeat the previous argument for the case where there are also applied surface tractions $p_{ij}^{A}n_j$ on Σ_0. The only differences are first that the primed quantities must be interpreted as

$$u_i' = u_i^I+u_i^T+u_i^A,\quad p_{ij}' = p_{ij}^I+p_{ij}^T+p_{ij}^A,$$

and secondly that in stage (v) the displacement on Σ_0 alters from u_i to $u_i+\delta\xi(\partial u_i/\partial\xi_l)$ and the surface forces do work. The following terms must thus be added to (31):

$$\int_{\Sigma}\left(p_{ij}^{A}\frac{\partial u_i^{\infty}}{\partial\xi_l}-\frac{\partial p_{ij}^{\infty}}{\partial\xi_l}u_i^{A}\right)dS_j-\int_{\Sigma_0}p_{ij}^{A}\frac{\partial u_i}{\partial\xi_l}dS_j.$$

But the first term is simply F_l^A as defined in §5, and the second is $-F_l^A$ since

$$\partial u_i/\partial\xi_l = \partial(u_i^{\infty}+u_i^{I})/\partial\xi_l = \partial u_i^{S}/\partial\xi_l,$$

u_i^T and u_i^A remaining constant. Consequently the presence of an externally-applied stress makes no difference at all to the change of internal energy when the singularity is displaced. Thus $F_l^{A'}$ vanishes as we required above.

Let $\phi_l(A,\infty)$ denote one of the expressions for F_l^A in §5 or the explicit expression for a particular singularity. Then clearly $F_l^I = \phi_l(I,\infty)$, with an obvious notation. Thus if surface tractions equal and opposite to $p_{ij}^I n_j$ are applied there is no total force on the singularity: $F_l^A + F_l^I = 0$. This can be described picturesquely by saying that since the total stress and displacement are p_{ij}^∞ and u_i^∞ the singularity thinks that it is in neutral equilibrium in an infinite medium with no applied stress. Similarly, $F_l^T = \phi_l(T, \infty)$, with the proviso that when $\phi_l(T, \infty)$ takes the form of a surface integral the surface of integration Σ shall separate S from T.

The expression for the total force is

$$F_l = F_l^A + F_l^I + F_l^T = \int_\Sigma P_{jl}\, dS_j, \tag{32}$$

where $$P_{jl} = -p_{ij}\, u_{i,l} + \tfrac{1}{2} p_{ik}\, e_{ik}\, \delta_{jl}.$$

This easily follows by taking ϕ_l in the form (19), adding F_l^∞ and noting that by (3) and (11)

$$\int_\Sigma (p'_{ij}\, u'_{i,l} - \tfrac{1}{2} p'_{ik}\, e'_{ik}\, \delta_{jl})\, dS_j = 0,$$

with $u'_i = u_i^A + u_i^I + u_i^T$, etc. Since we have reintroduced F_l^∞ the truth of (32) does not depend on its vanishing. S and T can of course each be a group of singularities. Consequently (32) gives the force on all the singularities within Σ exerted by

(i) their image stresses,

(ii) externally applied stresses,

(iii) all the singularities outside Σ.

By analogy with electrostatics we may call P_{jl} the Maxwell tensor of elasticity. Note that P_{jl} is not symmetric; l specifies the component of the force and j is used to form an inner product with the surface element. In an infinite medium free from applied stresses the force which one group of singularities exerts on a second group is equal and opposite to the force which the second exerts on the first. More generally action and reaction are equal and opposite in the following sense: the total force on the singularities in I (figure 2c) plus the total force on the singularities in II is equal to the total force on all the singularities within Σ_0, the integrals over the partition Σ cancelling.

The force which one infinite straight dislocation exerts on another parallel to it can be found from (24) with b_i equal to the Burgers vector of one dislocation and p_{ij}^A equal to the stress produced by the other. For a pair of edge dislocations the results agree with those of Leibfried (1949) and Bilby (1950) but not with the original calculations of Koehler (1941). If Koehler's calculation is repeated in bipolar co-ordinates the result agrees with Leibfried, Bilby and the present paper. The error appears to arise from the fact that Koehler had to split the total energy into self-energy and interaction terms which are difficult to evaluate over the same area. It is easy, for example, to evaluate the former over a circle and the latter over a rectangle. When we let these two areas of integration tend to infinity the result obtained depends on their relative size and shape. Instead of neglecting the energy within small cylinders surrounding the dislocation lines we may cut out these cylinders leaving stress-free holes at the centres of the dislocations, in the manner of Volterra. This problem can be solved rigorously by an extension of the analysis of Dean & Wilson (1947). The result differs from that obtained from (24) by terms which vanish with vanishing radius of the stress-free cylinders. The additional terms represent the image force on one dislocation due to the hole at the other.

It should be noted that singularities with single-valued displacements cannot have their centres excluded by stress-free surfaces in this way. For example, we can only annul the traction over a sphere surrounding a centre of dilatation by superimposing a hydrostatic pressure (giving the wrong behaviour at infinity) or an equal centre of compression at the same point, in which case the stress is zero everywhere.

The force exerted on a centre of dilatation by a dislocation, obtained by replacing p^A in (21) by the hydrostatic pressure of the dislocation, agrees with Bilby's result. It may be noted that Cottrell's (1949) process of creating one singularity in the stress field of another also gives the correct result. To create Bilby's model of an interstitial atom we have to make a spherical cut of radius r_0 and blow it up into a spherical annulus of thickness αr_0 (in the notation of § 6) against the applied stress. This leads at once to an interaction energy of which (21) is the gradient. We defer discussing Leibfried's result for this case to § 9 since it involves elastic inhomogeneities.

Even when F_l^A is known for a certain singularity the determination of F_l^I still requires the solution of a boundary-value problem to find p_{ij}^I. In some two-dimensional isotropic problems we can use existing solutions of related problems.

For a screw dislocation at the origin in a cylinder whose cross-section is bounded by the curve Σ_0 we must find a harmonic function u_3 for which $\partial u_3/\partial n$ vanishes on Σ_0 (no surface traction) and behaving like $\tan^{-1}(x_2/x_1)$ near the origin. The logarithmic potential, ϕ, of a charge at the origin inside a curve Σ_0 at constant potential is harmonic, satisfies $\partial\phi/\partial s = 0$ on Σ_0 and behaves like $\log r$ near the origin. Clearly u_3 is the harmonic function conjugate to ϕ. The conjugate function need not actually be constructed if only the stresses are required, since $p_{23}/\mu = u_{3,2} = \phi_{,1}$ and $p_{31}/\mu = u_{3,1} = -\phi_{,2}$. The case of a screw dislocation in a rectangular prism can be dealt with by the analysis of Courant & Hilbert (1931, p. 333; see also Leibfried & Dietze 1949). As a simpler case we see that the image stress for a screw dislocation distant r from the centre of a cylinder of radius R is given by an image dislocation of opposite sign distant R^2/r from the centre and on the same radius. The image force is $(\mu b^2/2\pi)\, r/(R^2-r^2)$, directed radially outwards. More generally the force on a screw dislocation is the same (in suitable units) as that on the line-charge in the associated electrostatic problem.

Again, the deflexion w of a plate and the Airy stress function χ of a plane-strain problem both satisfy the biharmonic equation with the boundary conditions $w = 0$, $\partial w/\partial n = 0$ and $\partial(\partial\chi/\partial x_1)/\partial s = 0$, $\partial(\partial\chi/\partial x_2)/\partial s = 0$ at a clamped edge and free surface respectively. The boundary conditions can be made to coincide if the plate is given a small rotation, or a linear expression in x_1, x_2 is added to χ, which does not alter the stresses. Near a concentrated load at $\boldsymbol{\xi}$, w behaves like $(\mathbf{x}-\boldsymbol{\xi})^2 \log(\mathbf{x}-\boldsymbol{\xi})^2$. For χ this singularity represents a Weingarten dislocation made by inserting a narrow wedge with its apex at $\boldsymbol{\xi}$. For an edge dislocation at the same point with Burgers's vector $\mathbf{b}$, χ behaves like $\mathbf{b}'.(\mathbf{x}-\boldsymbol{\xi})\log(\mathbf{x}-\boldsymbol{\xi})^2$ with $\mathbf{b}'$ perpendicular to $\mathbf{b}$. Hence if w is the deflexion for a plate clamped around the curve Σ_0 and loaded at (ξ_1, ξ_2), then $\chi_W = \text{const.}\, w$ and

$$\chi_E = \text{const.}\left(b_1\frac{\partial}{\partial\xi_2} - b_2\frac{\partial}{\partial\xi_1}\right) w$$

are the respective stress functions for a Weingarten dislocation or an edge dislocation with $\mathbf{b} = (b_1, b_2)$ at (ξ_1, ξ_2) in a cylinder with Σ_0 as stress-free boundary. The results of Michell for a disk (Love 1927, p. 491) can be used to discuss an edge dislocation in a circular cylinder. Koehler (1941) has treated the case where the slip plane is a diameter of the cylinder.

As a three-dimensional example consider the image force urging a centre of dilatation towards the surface of a semi-infinite solid. The shear stress over the plane $x_3 = 0$ due to a centre of dilatation at $(0, 0, Z)$ in an infinite medium can be removed by putting an equal centre of dilatation at $(0, 0, -Z)$. This does not contribute to the image force since it produces a pure shear. The effect of annulling the remaining normal pressure on the plane $x_3 = 0$ can be found by integration from the expression (Love 1927, p. 191) for the effect of a concentrated load on the surface of a semi-infinite body. We find

$$\left(\frac{\partial p^I_{ii}}{\partial x_3}\right)_{x_3=Z} = \frac{3\mu(1+\sigma)\,\delta}{2Z^4},$$

so that by (21)

$$F^I = 6\pi\mu(1-\sigma)\,\delta^2 Z^{-4}$$

directed towards the free surface. For the Bilby singularity of § 6

$$F^I = \tfrac{1}{3}\pi\alpha^2 E r_0^2 \frac{1+\sigma}{1-\sigma}\left(\frac{r_0}{Z}\right)^4,$$

where E is Young's modulus and Z is the distance of the centre of the sphere below the free surface. This force can be regarded as derived from a potential

$$U = -\tfrac{1}{9}\pi\alpha^2 (E r_0^3)\,\frac{1+\sigma}{1-\sigma}\left(\frac{r_0}{Z}\right)^3.$$

If we apply these results to an interstitial atom or lattice vacancy we see that in thermal equilibrium the concentration of these defects should be greater near the surface than in the body of the material. However, even with a fairly generous choice of the misfit constant α, U is of the order of kT at room temperature only when Z is of the order of r_0. At this point the approximation of an elastic continuum breaks down. When we reach a depth at which the continuum approximation is reasonable the concentration will have reached substantially its bulk value.

Two centres of dilatation in an infinite medium do not interact with one another (so long as linear isotropic elastic theory is valid), since each produces a pure shear stress. On the other hand, in the neighbourhood of a free surface each of them is acted on by its own and the other's image stress. It is clear that each one of a group of n centres of dilatation close together in comparison with their distance from the surface will be acted on by n times the force it would experience in isolation. However, if one centre is kept fixed at $(0, 0, Z)$ and the image force on it due to a second centre at (x_1, x_2, x_3) is averaged for all x_1, x_2 the result is zero, and so there is no long-range multiplicative effect.

8. The force on an inhomogeneity

Suppose that the elastic constants of an inhomogeneous body free from internal stress depend on three parameters ξ_i according to the relation

$$c_{ijkm} = c_{ijkm}(x_n - \xi_n). \tag{33}$$

The ξ_i might, for example, be the co-ordinates of a foreign inclusion in an otherwise uniform body. If fixed surface tractions are applied to the surface (so that $\partial(p_{ij}n_j)/\partial\xi_l = 0$ on Σ_0) and one of the parameters is changed, we have with the notation of § 7

$$-\frac{\partial W_{\text{ext.}}}{\partial \xi_l} = \int_{\Sigma_0} p_{ij}\frac{\partial u_i}{\partial \xi_l}\,dS_j = \frac{\partial}{\partial \xi_l}\int p_{ij}\,e_{ij}\,dv = 2\frac{\partial W_{\text{int.}}}{\partial \xi_l},$$

so that of the work done by the external forces in a small change of ξ_l half disappears and half remains as an increase of the internal energy of the body. Consequently we can define

$$F_l = +\partial W_{\text{int.}}/\partial \xi_l$$

as the force which the applied surface tractions exert on the inhomogeneity.

We have
$$\frac{\partial W_{\text{int.}}}{\partial \xi_l} = -\frac{1}{2}\int \frac{\partial c_{ijkm}}{\partial \xi_l}\,e_{ij}\,e_{km}\,dv. \tag{34}$$

This can be seen as follows. The difference of energy between two bodies of the same size and shape (distinguished by unprimed and primed quantities) with different inhomogeneous elastic constants and acted on by the same surface tractions is

$$\delta W = \frac{1}{2}\int (p'_{ij}\,e'_{ij} - p_{ij}\,e_{ij})\,dv.$$

But
$$\int (p_{ij} - p'_{ij})\,e_{ij}\,dv = \int_{\Sigma_0} (p_{ij} - p'_{ij})\,u_i\,dS_j = 0,$$

since $p_{ij}n_j = p'_{ij}n_j$ at the surface. Hence we can replace $p_{ij}\,e_{ij}$ by $p'_{ij}\,e_{ij}$ in δW and similarly $p'_{ij}\,e'_{ij}$ by $p_{ij}\,e'_{ij}$. Hence

$$\delta W = \frac{1}{2}\int (c_{ijkm} - c'_{ijkm})\,e'_{ij}\,e_{km}\,dv,$$

from which (34) follows. From (33) and (11)

$$\begin{aligned}
F_l &= \frac{1}{2}\int c_{ijkm,\,l}\,u_{i,\,j}\,u_{k,\,m}\,dv \\
&= \frac{1}{2}\int \{(c_{ijkm}\,u_{i,\,j}\,u_{k,\,m})_{,\,l} - 2c_{ijkm}\,u_{i,\,j}\,u_{k,\,lm}\}\,dv \\
&= \int_{\Sigma_0} (\tfrac{1}{2}p_{km}\,u_{k,\,m}\,\delta_{lj} - p_{jk}\,u_{k,\,l})\,dS_j,
\end{aligned}$$

so that F_l is given by the 'Maxwell tensor'. If the body is homogeneous except in a limited region, Σ_0 can be replaced by a surface Σ enclosing that region.

The stress and displacement can be written as

$$p_{ij} = p^A_{ij} + p^S_{ij}, \quad u_i = u^A_i + u^S_i,$$

where u^A_i is the displacement which the given surface tractions would produce in the body if it were homogeneous throughout, and u^S_i can be regarded as the disturbance produced by the inhomogeneity. The stresses p^A_{ij} could only occur in the actual body if there were body forces of density $f_i = -(c_{ijkm}\,u^A_{k,\,m})_{,\,j}$ present. Consequently, the displacement and stress are the same as would be produced by body forces $-f_i$ in the inhomogeneous body. f_i vanishes where the body is homogeneous and the volume integral of f_i is zero. We may say that the surface forces 'induce' in the inhomogeneity a singularity specified by u_i^S, p_{ij}^S and then exert a force on it. We can write $u_i^S = u_i^\infty + u_i^I$ as for a real singularity, and similarly for p_{ij}^S; u_i^∞ is the displacement which the body forces $-f_i$ would produce if the body were continued to infinity, the additional material being homogeneous. F_l^∞ (equation (29)) clearly vanishes, and so we can write

$$F_l = \phi_l(A, \infty) + \phi_l(I, \infty),$$

just as for a real singularity.

As a simple example consider a body with shear and bulk moduli μ and K containing a spherical inclusion of radius r_0 and bulk modulus K'. It is easy to show that, under a uniform hydrostatic pressure p, u_i^S is given by (20) with

$$\delta = p\frac{r_0^3}{4\mu}\frac{3(K'-K)}{4\mu+3K'}$$

if the body is infinite. If p is not constant F_l will be given approximately by (21) with this value of δ. This is inaccurate because the variation of p will induce higher order singularities and the result must be multiplied by $1+O(r_0\,|\operatorname{grad} p\,|/p)$. We have also neglected the image terms, which will be small if the inhomogeneity is far from the surface of the solid. For a spherical cavity ($K' = 0$) the approximate result is

$$\mathbf{F} = \tfrac{3}{2}\pi\frac{r_0^3}{\mu}\frac{\sigma}{1-\sigma}\operatorname{grad}(p^A)^2.$$

Such a cavity could migrate by evaporation and condensation or surface migration of molecules in the cavity or by diffusion of defects in the body of the solid. Similarly, for an incompressible inclusion ($K' = \infty$)

$$\mathbf{F} = -\tfrac{3}{2}\pi\frac{r_0^3}{\mu}\frac{1-\sigma}{1+\sigma}\operatorname{grad}(p^A)^2. \tag{35}$$

9. The general case

It would be tedious to treat the general case, where there are both singularities and inhomogeneities, by the methods of §7, 8. We shall therefore definitely adopt the point of view that singularities can be regarded as limiting cases of extended states of internal stress. We can then carry out volume integration without scruple; the only difficulty is the impossibility of defining a displacement function everywhere.

In order to make clear the nature of these internal stresses we consider a process (Timoshenko 1934) which would actually produce them. Reissner (1931) gives a more formal account.

Cut an unstrained body into infinitesimal cubes and give each one a permanent strain e_{ij}^*, for example by plastic deformation or by adding and removing material on its faces. Apply stresses $-c_{ijkl}e_{kl}^*$ to them; they become cubes again and can be welded together in the same relative position to give the original body. Since the surface stresses thus built in are not generally equal and opposite on the adjacent faces of neighbouring cubes, we are left with a distribution of body force of amount $(c_{ijkl}e_{kl}^*)_{,j}$ per unit volume. If we superimpose a distribution of body force

$$f_i = -(c_{ijkl}e_{kl}^*)_{,j}, \tag{36}$$

we are left with a body free of body forces but in a state of self-stress p^S_{ij} given by

$$p^S_{ij} = -p^*_{ij} + p^\dagger_{ij},$$

where $p^*_{ij} = c_{ijkl}\,e^*_{kl}$ and $p^\dagger_{ij}$ is a solution of $p^\dagger_{ij,j} + (c_{ijkl}\,e^*_{kl})_{,j} = 0$ chosen so that p^S_{ij} satisfies the prescribed boundary conditions. Such a state of self-stress satisfies the equilibrium equations $p^S_{ij,j} = 0$, but the strains derived from it by Hooke's law do not satisfy the compatibility conditions since according to (4) $S_{ij}(e^S_{kl}) = S_{ij}(e^*_{kl}) \neq 0$, the e^*_{ij} being arbitrary functions. In regions where $e^*_{ij} = 0$ and hence $p^*_{ij} = 0$ the state of the medium is indistinguishable from that produced by body forces (36) and a displacement function can be defined. Where $S_{ij} \neq 0$ the stresses cannot be imitated by body forces. Suppose that S_{ij} is known as a function of position. Then if we use Hooke's law to replace the e's by p's in (4) we get the Michell-Beltrami compatibility equation modified by the presence of the term S_{ij}. This inhomogeneous equation together with the equilibrium equations and the boundary conditions determines the stress uniquely, just as in the homogeneous case.

To the first order in the e^*_{ij}, $R_{prst} = \epsilon_{ipr}\,\epsilon_{jst}\,S_{ij}$ is the Riemann tensor derived from the metric $g_{ij} = \delta_{ij} + 2e^*_{ij}$. Formally, this means that our set of deformed cubes could be fitted together without stress in the non-Euclidean space defined by the g_{ij}. This is quite intelligible in the two-dimensional case. Let a thin lamina be cut into small squares each of which is given a permanent strain $e^*_{11}, e^*_{22}, e^*_{12}$. The pieces could be fitted together without stress as a mosaic on a surface with Gaussian curvature $K = 2\,e^*_{12,12} - e^*_{11,22} - e^*_{22,11}$ at each point. Conversely, a self-stressed lamina can relieve its stress by buckling into a suitable surface.

When e^*_{ij} is known the displacement corresponding to $p^\dagger_{ij}$ in a homogeneous medium is clearly

$$u^\dagger_m(\mathbf{x}') = c_{ijkl}\int u_{mi}(\mathbf{x}' - \mathbf{x})\, e^*_{kl,j}\,(\mathbf{x})\, dv,$$

where $u_{mi}(\mathbf{x})$ is the displacement at $\mathbf{x}$ produced by a unit concentrated force in the x_i direction at the origin. u_{mi} can be found from the six equilibrium equations $c_{ijkl}\,u_{kn,lj} = \delta_{in}\,\delta(\mathbf{x})$ by taking their Fourier transforms and solving the resulting algebraic equations. The result is

$$u_{kn}(\mathbf{x}) = -(2\pi)^{-3}\iiint_{-\infty}^{\infty} K_{nk}\exp(-ik_i x_i)\, dk_1\, dk_2\, dk_3,$$

where K_{ik} is the matrix reciprocal to $c_{ijkl}\,k_j\,k_l$. By changing to the dimensionless integration variables $l_i = k_i r$, where r is the distance from 0 to $\mathbf{x}$, it is easy to show that u_{kn} is the product of r^{-1} and a function of direction. Since $u^\dagger_i$ is in fact the actual displacement u^∞_i in regions where $S_{ij} = 0$, this verifies the vanishing of F^∞_l assumed in § 7.

For the type of internal stress we are considering the assumption (27) follows immediately. The energy of a body subject to external and self-stresses p^A_{ij} and p^S_{ij} differs from the sum of the energies it would have if p^A_{ij} or p^S_{ij} were present separately by the interaction term $\frac{1}{2}\int (p^S_{ij}\,e^A_{ij} + p^A_{ij}\,e^S_{ij})\, dv$. The first term can be written as $\frac{1}{2}\int (p^S_{ij}\,u^A_i)_{,j}\, dv = \frac{1}{2}\int p^S_{ij}\,u^A_i\, dS_j = 0$, since $p^S_{ij}\,n_j = 0$ on the surface. The second interaction term cannot be treated in the same way, since e^S_{ij} cannot be derived from a displacement everywhere. However, by (5) it is equal to the first term. Hence the interaction term is zero. The behaviour of the body towards applied forces is determined by the way its energy depends on $p^A_{ij}\,n_j$ and (27) follows.

Similarly, we can express the interaction energy between two singularities as an integral over a surface separating them. In figure 2*b*, u_i^S exists in II and u_i^T in I but not vice versa. We can, however, express the interaction energy in I in terms of u_i^T and in terms of u_i^S in II. By applying Gauss's theorem we find the interaction energy in the form

$$E_{ST} = \int_\Sigma (p_{ij}^S u_i^T - p_{ij}^T u_i^S)\, dS_j.$$

We are now in a position to deal with the general case where internal stresses and inhomogeneities exist in the same body. The aggregate of inhomogeneities and sources of internal stress within a surface Σ are to be regarded as a complex singularity within it. The state of internal stress (so far as it does not arise from sources outside Σ) is to be defined in terms of the stress, strain or (where it can be defined) displacement in an infinite body with the same inhomogeneous elastic constants c_{ijkl} within Σ but homogeneous (with $c_{ijkl} = c_{ijkl}^0$) outside Σ. In the actual body we shall have as before $p_{ij}^S = p_{ij}^\infty + p_{ij}^I$, etc. The generalized image quantities thus defined arise from the presence of the boundary and also from the inhomogeneities lying between Σ and the surface of the body Σ_0. They can be found in principle as follows. Impose the strains e_{ij}^∞ on the body. The equations of equilibrium can only be satisfied if body forces $-(c_{ijkl}e_{kl}^\infty)_{,j}$ are present. Hence the image quantities are those which would be produced by body forces

$$f_i = (c_{ijkl}e_{kl}^\infty)_{,j}, \tag{37}$$

and surface tractions $-p_{ij}^\infty n_j$ acting on the body. The f_i are zero within Σ since $p_{ij,j}^\infty = 0$ there, the singularity being assumed as before to be in equilibrium in the infinite medium.

The force on the singularity in region II due to sources of stress in I and surface tractions will be

$$F_l = -(W_{\text{int.}} + W_{\text{ext.}})/\partial\xi_l, \tag{38}$$

where the internal stress and elastic constants depend on $\boldsymbol{\xi}$ in the following way:

$$\begin{array}{ll} e_{ij}^\infty = e_{ij}^\infty(\mathbf{x}-\boldsymbol{\xi}) & \text{inside and outside } \Sigma, \\ c_{ijkm} = c_{ijkm}(\mathbf{x}-\boldsymbol{\xi}) & \text{inside } \Sigma, \\ c_{ijkm} \text{ is independent of } \boldsymbol{\xi} & \text{outside } \Sigma. \end{array}$$

For simplicity we shall assume that within a narrow shell containing Σ there are no sources of internal stress and $c_{ijkm} = c_{ijkm}^0$. This ensures that a displacement exists on Σ and that a small translation $\delta\xi_l$ does not engender discontinuities of c_{ijkm} across Σ.

In I we can write $e_{ij} = e'_{ij}(\mathbf{x}, \boldsymbol{\xi}^0) + e''_{ij}(\mathbf{x}, \boldsymbol{\xi})$ in terms of its value for some fixed value $\boldsymbol{\xi}^0$ of $\boldsymbol{\xi}$ and a variable part expressing the dependence on $\boldsymbol{\xi}$. Then although e'_{ij} cannot everywhere be derived from a displacement, e''_{ij} can. The contribution to (38) from the region I is

$$\begin{aligned} \frac{\partial W_{\text{int. II}}}{\partial\xi_l} &= \frac{1}{2}\int_{\text{I}} \frac{\partial}{\partial\xi_l}(p_{ij}e_{ij})\, dv = \int_{\text{I}} p_{ij}\frac{\partial e''_{ij}}{\partial\xi_l}\, dv \\ &= \int_{\text{I}} \left(p_{ij}\frac{\partial u''_i}{\partial\xi_l}\right)_{,j} dv = \int_{\Sigma_0-\Sigma} p_{ij}\frac{\partial u_i}{\partial\xi_l}\, dS_j, \end{aligned} \tag{39}$$

since $p_{ij,j} = 0$ and u'_i is independent of $\boldsymbol{\xi}$.

The contribution from the surface tractions is

$$\frac{\partial W_{\text{ext.}}}{\partial\xi_l} = -\int_{\Sigma_0} p_{ij}\frac{\partial u_i}{\partial\xi_l}\, dS_j. \tag{40}$$

In region II we put $e_{ij} = e^{\infty}_{ij} + e^{\backprime}_{ij}$. Then

$$\frac{\partial W_{\text{int. II}}}{\partial \xi_l} = \frac{1}{2}\frac{\partial}{\partial \xi_l}\int_{\text{II}} (p^{\backprime}_{ij}\, e^{\backprime}_{ij} + p^{\infty}_{ij}\, e^{\infty}_{ij} + p^{\infty}_{ij}\, e^{\backprime}_{ij} + p^{\backprime}_{ij}\, e^{\infty}_{ij})\, dv.$$

Since $e^{\backprime}_{ij}$ can be derived from a displacement the first and third terms can be converted into surface integrals by writing, for example, $p^{\backprime}_{ij}\, e^{\backprime}_{ij}$ as $(p^{\backprime}_{ij}\, u^{\backprime}_i)_{,j}$ and using Gauss's theorem. So can the second, since although e^{∞}_{ij} cannot be derived from a displacement we have

$$\partial(p^{\infty}_{ij}\, e^{\infty}_{ij})/\partial \xi_l = -(p^{\infty}_{ij}\, e^{\infty}_{ij})_{,l}.$$

The fourth term is equal to the third. In this way we find

$$\frac{\partial W_{\text{int. II}}}{\partial \xi_l} = \frac{1}{2}\int_{\Sigma} \left\{\frac{\partial}{\partial \xi_l}(p^{\backprime}_{ij}\, u^{\backprime}_i + p^{\infty}_{ij}\, u^{\backprime}_i - p^{\backprime}_{ij}\, u^{\infty}_i - p^{\infty}_{ij}\, u^{\infty}_i) - (p^{\infty}_{ik}\, e^{\infty}_{ik})\, \delta_{lj}\right\} dS_j. \tag{41}$$

Adding (39), (40) and (41) we find after rearranging and using (11) that F_l differs from the right-hand side of (32) by the terms

$$\frac{1}{2}\int_{\Sigma} (p^{\infty}_{ij}\, e^{\backprime}_{ij} - p^{\backprime}_{ij}\, e^{\infty}_{ij})\, dS_l \tag{42}$$

and

$$\frac{1}{2}\int_{\Sigma} \left\{p^{\backprime}_{ij}\left(\frac{\partial}{\partial x_l} + \frac{\partial}{\partial \xi_l}\right) u^{\backprime}_i - u^{\backprime}_i \left(\frac{\partial}{\partial x_l} + \frac{\partial}{\partial \xi_l}\right) p^{\backprime}_{ij}\right\} dS_j. \tag{42'}$$

The expression (42) vanishes since $p^{\infty}_{ij}\, e^{\backprime}_{ij} = p^{\backprime}_{ij}\, e^{\infty}_{ij}$ on Σ. The divergence of the integrand of (42′) can be shown to vanish by expressing it entirely in terms of displacements and noting that $\partial(c_{ijkm})/\partial \xi_l = c_{ijkm,l}$ inside Σ. Consequently (42′) is zero. (The conditions for us to apply (8) and (9) to the terms in $\partial/\partial x_l$ and $\partial/\partial \xi_l$ separately are not satisfied.)

A more careful analysis shows that the requirement that c_{ijkm} shall be constant on Σ and in its neighbourhood is unnecessary. The choice of the constant c^0_{ijkm} clearly makes no difference to the values of u^I_i, p^I_{ij} calculated from (37) nor to the values of u^{∞}_i and p^I_{ij} on Σ and hence cannot affect F_l. The modifications when body forces f_i are present in I are easily made. As externally applied forces they are to be treated on the same footing as the surface tractions. We thus have to add a term

$$-\int_{\text{I}} f_i \frac{\partial u''_i}{\partial \xi_l}\, dv$$

to (40). But since now $p_{ij} = -f_i$ an identical term must be subtracted from (41) and the rest of the calculation remains the same. If we wish to split F_l into F^A, F^I, etc., we have to show that (29) continues to hold even though the hypothetical infinite solid used to define p^{∞}_{ij} is inhomogeneous within Σ. It is easy to see that the stresses and displacements outside Σ are just those that would be produced by a system of body forces $-c^0_{ijkm}\, e_{km,j}$ within Σ if the interior of Σ were homogeneous, and (29) follows as before. Again, as noted in § 3, there is no reason why p^{∞}_{ij} should not be produced in whole or part by an equilibrating system of body forces $f^S_i = f^S_i(\mathbf{x}-\boldsymbol{\xi})$ within II. If they are present $p^{\infty}_{ij,j} = -f^S_i$ and a term

$$\frac{\partial}{\partial \xi_l}\int_{\text{II}} f^S_i\, u^{\backprime}_i\, dv$$

has to be added to the right-hand side of (41). F_l is then still given by the Maxwell tensor if $W_{\text{int. II}}$ is redefined as the sum of the elastic energy within II plus the potential energy of the sources of f^S_i.

14-3

It is thus clear that $F_l\delta\xi_l$ with

$$F_l=\int_\Sigma P_{jl}dS_j=\int_\Sigma(-p_{ij}u_{i,l}+\tfrac{1}{2}p_{ik}u_{i,k}\delta_{jl})\,dS_j$$

represents quite generally the decrease of the total potential energy of the system (namely, the elastic energy of the material and the potential energy of applied forces) when all sources of internal stress and inhomogeneity within Σ are given a small displacement $\delta\xi_l$. The energy $F_l\delta\xi_l$ is available for conversion into kinetic energy or dissipation by some process not considered in the elastic theory.

It will be seen from the foregoing that a singularity with a prescribed p_{ij}^∞ on and outside Σ can be regarded as arising from either (i) a suitable distribution of S_{ij} within Σ or (ii) a distribution of body forces f_i^S within Σ. The f_i^S are equal to the fictitious forces (36) associated with S_{ij}. F_l is the same for (i) and (ii). The representation (ii) fits into normal elastic theory where surface and body forces are regarded as the sources of the stress field. On the other hand (i) is a more natural representation of the singularities of physical theory, since they do not in fact arise from body forces.

The requirement of § 5 that surface tractions should be applied to a free surface of the body is rather unrealistic. It is certainly not satisfied, for example, when the body containing the singularity is a specimen in a tensile testing machine. We can, however, regard the specimen plus testing machine as a single body to which the results of this section may be applied. (Mechanical details can be accommodated to this picture, e.g. a bearing surface is a region where the shear modulus but not the bulk modulus vanishes.) Stresses are applied to the singularity by a weight rigidly connected to the machine (body forces), by tightening a screw (internal stress) or through a knife-edge (applied stresses according with the requirements of § 5).

It should be noted that there will be an image force (in the generalized sense) on a singularity even in a body of uniform composition if the orientation of c_{ijkm} varies with position, as it does on crossing a grain or twin boundary. There will thus be a force on a singularity near such a boundary quite apart from the effect of the array of dislocations into which the boundary can be decomposed.

Where there is a sudden discontinuity Δc_{ijkm} in the elastic constants across a surface Σ_d there will be a term $\Delta c_{ijkm}e_{kl}^\infty\,\partial H(\nu)/\partial x_j$ in the force (37) from which the image quantities are derived. Here ν is the distance measured from and normal to Σ_d and H is Heaviside's unit step function. Since $\partial H(\nu)/\partial x_j=\delta(\nu)\,\partial\nu/\partial x_j=\delta(\nu)\,n_j$, this term represents a surface distribution of force $\Delta c_{ijkm}e_{kl}^\infty n_j$ per unit area of Σ_d. When the c's vanish on one side of the surface we clearly get the surface traction $-p_{ij}^\infty n_j$ of § 4. In other words, we need not distinguish those image effects due to inhomogeneities from those due to the free surface; a finite body can be regarded as an infinite one for which the c's vanish outside Σ_0. Similarly, by letting the c's approach infinity on one side of Σ_d we can treat the image effect of a rigid wall, though the limiting process requires care.

Leibfried's (1949) expression for the interaction energy between a foreign atom and an edge dislocation does not agree with our result that the force between two singularities can be found by integrating P_{jl} over a surface separating them. Leibfried's model of the foreign atom is similar to Bilby's (cf. § 6), but the inner sphere is incompressible. According to our results the force on the foreign atom is given by the sum of (21) and (35) with p^A equal to the hydrostatic pressure produced by the dislocation, i.e. the force on a point centre of dilatation which produces the same stress field for $r > r_0$ plus a term quadratic in the dislocation stresses representing the force on the rigid sphere regarded as an inhomogeneity. With reversed sign the sum of these two terms is also the force which the foreign atom exerts on the dislocation, the inhomogeneity term becoming the force on the dislocation due to its image in the rigid sphere. If the interaction energy is to give the force correctly we must calculate it starting from an initial state independent of the relative position of the two singularities. A suitable initial state is furnished by the dislocation in an infinite homogeneous medium. First create a singularity of Bilby's type with its strength adjusted to equal Leibfried's. The work required for this is given by Leibfried's equation (18). Now cut out the centre of the Bilby singularity and insert the rigid sphere. The energy in the sphere cut out can be used to repay some of the energy expended. The interaction term in the energy of the sphere is found to be equal to Leibfried's (19*b*). Hence the required interaction energy is the difference between Leibfried's (18) and (19*b*), i.e. his equation (19*a*). As an alternative starting point we may take the rigid inclusion with its stress field in the absence of the dislocation. Then, provided the elastic medium is attached to the surface of the inclusion, the interaction energy necessary to create the dislocation is again given (apart from image terms) by Leibfried's equation (19*a*), in agreement with our results.

10. The dynamic case

The equations of motion of an elastic medium $p_{ij,j} - \rho\ddot{u}_i = 0$ (ρ is the density) are the Euler equations arising from a variational principle with the Lagrangian density function

$$\mathscr{L} = \tfrac{1}{2}\rho\dot{u}_i\dot{u}_i - \tfrac{1}{2}c_{ijkl}u_{i,j}u_{k,l}$$

(Love 1927, p. 166). Associated with such a variational problem there is a 'canonical energy-momentum tensor'

$$T_{jl} = \frac{\partial\mathscr{L}}{\partial u_{i,j}}u_{i,l} - \mathscr{L}\delta_{jl} \quad (j, l = 1, 2, 3, 4;\ x_4 = t)$$

(cf. Wenzel 1949). This gives

$$T_{jl} = P_{jl} - \tfrac{1}{2}\rho\dot{u}_i\dot{u}_i\delta_{jl} \quad (j, l = 1, 2, 3).$$

The components $T_{j4} = -p_{ij}\dot{u}_i$ form the vector giving the flux of energy (Love 1927, p. 177) and T_{44} is the energy density. The three components $T_{4l} = \rho\dot{u}_i u_{i,l}$ should represent a momentum density, but their interpretation is not immediate. It is natural to suggest that T_{jl} is the necessary generalization of P_{jl} for dynamical problems, though we can hardly hope to justify this by the simple kind of argument which served to find P_{jl}.

Linear elastic theory alone is insufficient to determine the motion of a singularity without some additional assumption. (Compare the problem of the equation of motion of a classical electron.) Consider, for example, a point singularity. We can write down a solution of the elastic equations which has a singular point moving in a prescribed manner. When the velocity of the singular point is not uniform there is a net radiation of energy from it. On the other hand, a singularity moving in an applied stress field absorbs energy from the latter in virtue of the force F_l^A. The physical counterpart of the elastic singularity has a finite energy density near the centre and cannot act as an unlimited source or sink of energy. A reasonable condition to impose on the elastic singularity is thus the following: the energy flux through a small sphere of radius r_0 about the singular point and moving with it is zero. This is sufficient to determine the equation of motion if the position of the singularity can be specified by a single parameter; in the general case we should have to try to interpret the momentum flux and apply a similar argument to it. For a centre of dilatation moving slowly in a straight line the motion is the same as that of a particle of mass $8\pi\rho\delta^2/r_0^3$ acted on by the force (21) (cf. also Burton 1892). It would, of course, be absurd to apply this result to the motion of an interstitial atom, whose motion is clearly not controlled merely by inertial forces. Frenkel & Kontorowa's (1938) caterpillar, in which it is the state of being interstitial rather than a particular atom which moves, is a closer analogue of a moving centre of dilatation, but even here dissipative effects are probably dominant.

A dislocation is a more hopeful example of an elastic singularity the motion of whose physical counterpart may perhaps be controlled chiefly by inertial forces. Consider a screw dislocation of infinite length parallel to the x_3-axis with its centre moving along the x_1-axis. The first step in finding its equation of motion is to calculate the displacement field surrounding it when it moves in a prescribed (supernatural) way. Suppose that the material is isotropic and that the dislocation is oscillating so that the position of its centre is $x_1 = \xi(t)$, $x_2 = 0$ with

$$\xi = l\,e^{i\omega t}. \tag{43}$$

The displacement at the point of observation (x_1, x_2) at time t is

$$u_3 = -\tfrac{1}{4}ibklH_1^{(2)}(kr)\,e^{i\omega t}\sin\theta \tag{43'}$$

plus a term independent of t (Eshelby 1949*a*), where

$$k = \omega/c, \quad c^2 = \mu/\rho, \quad r^2 = x_1^2 + x_2^2, \quad \tan\theta = x_2/x_1.$$

Apply to the left-hand sides of (43) and (43′) the operator (Lapwood 1949)

$$\frac{1}{2\pi i}\int_C (\ldots)\frac{d\omega}{\omega},$$

where C is the real ω-axis indented below the origin. Equation (43) becomes

$$\xi = lH(t), \tag{44}$$

representing a sudden jump of the dislocation through a distance l. The corresponding displacement, found by using the relations

$$\delta(t) = \frac{1}{2\pi}\int_C e^{i\omega t}\,d\omega, \quad H_1^{(2)}(z) = -\frac{2}{\pi}\int_0^\infty e^{-iz\cosh v}\cosh v\,dv$$

is

$$u_3 = -\frac{bl}{2\pi r}\sin\theta\frac{ct}{\surd(c^2t^2 - r^2)}H(ct - r).$$

A general motion $\xi = \xi(t)$ can be regarded as a succession of jumps of amount $l = \dot{\xi}(t)\,dt$ in time dt executed at the instantaneous position of the dislocation. Hence

$$u_3 = -\frac{b}{2\pi}x_2\int_{-\infty}^{\tau_0}\frac{c(t-\tau)}{[x_1 - \xi(\tau)]^2 + x_2^2}\frac{\dot{\xi}(\tau)\,d\tau}{\surd\{c^2(t-\tau)^2 - [x_1 - \xi(\tau)]^2 - x_2^2\}}, \tag{45}$$

where τ_0 is the value less than t for which the square root vanishes; it is unique if $\xi(\tau)$ is less than c for all τ less than t. Apart from its rather precarious derivation (45) has the required properties—it satisfies the elastic wave equation and has a discontinuity b across the line joining $(-\infty, 0)$ to the instantaneous position of the singular point.

It will be seen that the displacement and stress field depend on the whole previous motion. This is, of course, because following the jump (44) the effects arising from segments of the dislocation line with successively larger values of x_3 continue to arrive at a point in the plane $x_3 = 0$ ever afterwards. We may try to find the equation of motion by the energy argument used for the point singularity or by tapering the dislocation in the manner of figure 1*c* and requiring that the external and self-forces should balance. Owing to the way in which the dislocation is haunted by its past we obtain an integral equation of motion, or equivalently a differential equation of infinite order.

The problem is even more intractable in the case of an arbitrary dislocation loop. The integral now extends only over a finite time interval but has a more involved spatial dependence and is complicated by the existence (in the isotropic case) of two velocities of wave propagation. Moreover, the problem is not simply one of translation of the loop; we have to find its change of form as a function of time. It is perhaps not worth while pursuing this problem until it is decided whether the motion of a dislocation is in fact governed by inertia (Frank 1949) or by quasi-viscous forces (Leibfried 1950).

References

Bilby, B. A. 1950 *Proc. Phys. Soc.* A, **63**, 191.
Bogdanoff, J. L. 1950 *J. Appl. Phys.* **21**, 1258.
Burgers, J. M. 1939 *Proc. Acad. Sci. Amst.* **42**, 293.
Burton, C. V. 1892 *Phil. Mag.* [5], **33**, 191.
Colonnetti, G. 1915 *R.C. Accad. Lincei* [5], **24** (1), 404.
Cottrell, A. H. 1949 *Progress in metal physics*, **1** (ed. Chalmers). London: Butterworth.
Courant, R. & Hilbert, D. 1931 *Methoden der Math. Physik*, **1**. Berlin: Springer.
Dean, W. R. & Wilson, A. H. 1947 *Proc. Camb. Phil. Soc.* **43**, 205.
Eshelby, J. D. 1949*a* *Proc. Roy. Soc.*, A, **197**, 396.
Eshelby, J. D. 1949*b* *Phil. Mag.* [7], **40**, 903.
Frank, F. C. 1949 *Proc. Phys. Soc.* A, **62**, 131.
Frank, F. C. & Read, W. T. 1950 *Phys. Rev.* **79**, 722.
Frenkel, J. & Kontorowa, T. 1938 *Phys. Z. Sowjet.* **13**, 1.
Gebbia, M. 1902 *Ann. Mat. pura appl.* [3], **7**, 141.
Koehler, J. S. 1941 *Phys. Rev.* **60**, 397.
Lapwood, E. R. 1949 *Phil. Trans.* A, **242**, 63.
Larmor, J. 1897 *Phil. Trans.* A, **190**, 205.
Leibfried, G. 1949 *Z. Phys.* **126**, 781.
Leibfried, G. 1950 *Z. Phys.* **127**, 344.
Leibfried, G. & Dietze, H.-D. 1949 *Z. Phys.* **126**, 790.
Love, A. E. 1927 *Theory of elasticity*. Cambridge University Press.
Mann, E. H. 1949 *Proc. Phys. Soc.* A, **199**, 376.
Mott, N. F. & Nabarro, F. R. N. 1940 *Proc. Phys. Soc.* **52**, 86.
Nabarro, F. R. N. 1947 *Proc. Phys. Soc.* **59**, 256.
Nabarro, F. R. N. 1951 *Phil. Mag.* [7], **42**, 213.

Neményi, P. 1931 *Z. angew. Math. Mech.* **11**, 59.
Peach, M. & Koehler, J. S. 1950 *Phys. Rev.* [2], **80**, 436.
Reissner, H. 1931 *Z. angew. Math. Mech.* **11**, 1.
Somigliana, C. 1914 *R.C. Accad. Lincei* [5], **23** (1), 463.
Somigliana, C. 1915 *R.C. Accad. Lincei* [5], **24** (1), 655.
Southwell, R. V. 1936 *Theory of elasticity*. Oxford: Clarendon Press.
Timoshenko, S. 1934 *Theory of elasticity*. New York: McGraw-Hill.
Volterra, V. 1907 *Ann. Ec. Norm. Sup.* [3], **24**, 400.
Wenzel, G. 1949 *Quantum theory of fields*. New York: Interscience.

CXL. *Dislocations in Thin Plates.*

By J. D. Eshelby and A. N. Stroh,
H. H. Wills Physical Laboratory, University of Bristol*.

[Received August 15, 1951.]

Abstract.

The stress due to a screw dislocation passing normally through an infinite plate or a disc is largely confined to the neighbourhood of the dislocation line, in contrast to the case of a dislocation in an infinite medium. Two screw dislocations in a plate attract or repel one another with a short-range force in place of the inverse first power law for infinite parallel dislocations. The stress due to an edge dislocation is not essentially different in the plate and infinite body so long as the plate remains flat, but in some circumstances the stress may be largely relieved by buckling of the plate.

§ 1. Introduction.

Forty (1951) has studied the growth of tabular crystals containing one or more dislocations. Dawson and Vand (1950, 1951) have published photographs of thin crystals with growth spirals terminating on dislocations. It seems worth while, therefore, to contrast the properties of a dislocation in a thin plate and in an infinite body.

We discuss in detail the following configurations of a straight screw dislocation: (i.) meeting normally the surface of a semi-infinite body, (ii.) running normally through an infinite plate and (iii.) along the axis of a disc, with or without a stress-free hole excluding the core of the dislocation. Case (i.) may be of interest in connection with the detailed topography of the surface near the point of emergence of a dislocation (Frank 1951). Fig. 25 of Forty's (1951) paper corresponds almost exactly to case (iii.).

The corresponding problem for an edge dislocation is only briefly touched on.

It is interesting to note that the steps in Forty's growth spirals may be an arbitrary number of lattice spacings (of the order of 100), so that we have to deal with effectively the classical Volterra dislocation with arbitrary Burgers vector, in contrast with the theory of metals where the Burgers vector is assumed to be limited to one of a few simple lattice displacements.

---

* Communicated by Professor N. F. Mott, F.R.S.

§2. A Screw Dislocation Normal to the Surface of a Semi-infinite Body.

For a screw dislocation in an infinite body along the z-axis of cylindrical coordinates (r, θ, z) the non-vanishing components of displacement and stress are

$$u_z = \frac{b}{2\pi}\theta, \qquad \tau_{\theta z} = \frac{\mu b}{2\pi}\frac{1}{r}. \quad . \quad . \quad . \quad . \quad . \quad . \quad (1)$$

The elastic image field which annuls the traction due to the dislocation on the plane $z=0$ is

$$u_\theta = -\frac{b}{2\pi}\frac{r}{R+z}, \quad \tau_{\theta z} = -\frac{\mu b}{2\pi}\frac{r}{R(R+z)}, \quad \tau_{r\theta} = \frac{\mu b}{2\pi}\left(\frac{z^2}{R^3} + \frac{1}{R+z}\right), \quad (2)$$

with $R^2 = r^2 + z^2$ and the other components zero. It is produced by a distribution of couples along the negative z-axis, twisting about the z-axis and with density proportional to distance from the origin. This may be verified by integrating twice with respect to z the expression given by Love (1927, p. 187) for a point-couple. The sum of the states (1) and (2) thus gives the elastic field about a screw dislocation perpendicular to the free surface of the semi-infinite solid $z>0$.

§3. A Screw Dislocation in a Plate or Disc.

To solve the problem of a screw dislocation in an infinite plate we must annul the traction on the planes $z=\pm d$. It is easy to see that this can be done by introducing an infinite series of images of the type (2). A useful expression is, however, more easily found as follows. We take from the image representation the fact that only u_θ, $\tau_{\theta z}$ and $\tau_{r\theta}$ do not vanish. Then

$$\tau_{\theta z} = \mu \partial u_\theta/\partial z, \qquad \tau_{r\theta} = \mu(\partial u_\theta/\partial r - u_\theta/r)$$

and the equilibrium condition is

$$\frac{\partial^2 u_\theta}{\partial r^2} + \frac{1}{r}\frac{\partial u_\theta}{\partial r} - \frac{u_\theta}{r^2} + \frac{\partial^2 u_\theta}{\partial z^2} = 0.$$

A simple solution is $u_\theta = \exp(\pm kz)J_1(kr)$. Multiplication by a function of k and integration yields a more general solution. The solution of our problem is easily found to be

$$u_\theta = -\frac{b}{2\pi}\int_0^\infty \frac{\sinh kz}{\cosh kd} J_1(kr)\frac{dk}{k}, \quad . \quad . \quad . \quad . \quad . \quad (3)$$

for then

$$(\tau_{\theta z})_{z=\pm d} = -\frac{\mu b}{2\pi}\int_0^\infty J_1(kr)\, dk = -\frac{\mu b}{2\pi}\frac{1}{r}.$$

Since

$$\operatorname{sech} x = 2\sum_0^\infty (-1)^n \exp\{-(2n+1)x\}$$

and

$$\int_0^\infty e^{-kx} J_1(kr)\frac{dk}{k} = \frac{r}{x+\sqrt{(x^2+r^2)}}$$

(3) becomes, with $d_n=(2n+1)\,d$,

$$u_\theta=-\frac{b}{2\pi}\sum_{n=0}^{\infty}(-1)^n\left\{\frac{r}{d_n-z+\sqrt{[(d_n-z)^2+r^2]}}-\frac{r}{d_n+z+\sqrt{[(d_n+z)^2+r^2]}}\right\},$$

exhibiting u_θ as the sum of a set of images of the type (2). Returning to (3), u_θ can be expressed as a series of modified Bessel functions K_1 by contour integration or by following the analysis of Riemann (1855) in a similar problem. In this way the elastic field of a screw dislocation passing perpendicularly through an infinite plate of thickness $2d$ is found to be

$$\left.\begin{aligned}
&u_z=\frac{b}{2\pi}\theta,\quad u_\theta=-\frac{b}{2\pi}\frac{z}{r}+\frac{b}{\pi}\Sigma\,(\sin\tfrac{1}{2}n\pi)\,\frac{2}{n\pi}\mathrm{K}_1\left(\frac{n\pi r}{2d}\right)\sin\frac{n\pi z}{2d},\\
&\tau_{\theta z}=\frac{\mu b}{\pi d}\Sigma\,(\sin\tfrac{1}{2}n\pi)\,\mathrm{K}_1\left(\frac{n\pi r}{2d}\right)\cos\frac{n\pi z}{2d},\\
&\tau_{r\theta}=\frac{\mu b}{\pi}\frac{z}{r^2}-\frac{\mu b}{\pi d}\Sigma\,(\sin\tfrac{1}{2}n\pi)\,\mathrm{K}_2\left(\frac{n\pi r}{2d}\right)\sin\frac{n\pi z}{2d},\\
&u_r=0,\qquad \tau_{rr}=\tau_{\theta\theta}=\tau_{zz}=\tau_{rz}=0.
\end{aligned}\right\}\quad (4)$$

with summation over integral n.

The case of a screw dislocation in the annular disc bounded by the surfaces $z=\pm d$, $r=r_i$, r_0 $(r_i<r_0)$ can be treated by replacing each K_1 in the u_θ of (4) by a linear combination of K_1 and I_1 and adjusting the constants to annul the traction on the cylindrical surfaces. The result is

$$\left.\begin{aligned}
&u_z=\frac{b}{2\pi}\theta,\quad u_\theta=-\frac{b}{2\pi}\frac{z}{r}+\Sigma\left\{\mathrm{A}_n\mathrm{I}_1\left(\frac{n\pi r}{2d}\right)+\mathrm{B}_n\mathrm{K}_1\left(\frac{n\pi r}{2d}\right)\right\}\sin\frac{n\pi z}{2d},\\
&\tau_{\theta z}=\frac{\mu\pi}{2d}\Sigma n\left\{\mathrm{A}_n\mathrm{I}_1\left(\frac{n\pi r}{2d}\right)+\mathrm{B}_n\mathrm{K}_1\left(\frac{n\pi r}{2d}\right)\right\}\cos\frac{n\pi z}{2d},\\
&\tau_{r\theta}=\frac{\mu b}{\pi}\frac{z}{r^2}+\frac{\mu\pi}{2d}\Sigma n\left\{\mathrm{A}_n\mathrm{I}_2\left(\frac{n\pi r}{2d}\right)-\mathrm{B}_n\mathrm{K}_2\left(\frac{n\pi r}{2d}\right)\right\}\sin\frac{n\pi z}{2d},\\
&u_r=0,\quad \tau_{rr}=\tau_{\theta\theta}=\tau_{zz}=\tau_{rz}=0,
\end{aligned}\right\}\quad (5$$

where

$$\mathrm{A}_n=\frac{4b}{\pi^2}\frac{1}{n^3}\sin\tfrac{1}{2}n\pi\,\frac{(2d/n\pi r_i)^2\mathrm{K}_2(n\pi r_0/2d)-(2d/n\pi r_0)^2\mathrm{K}_2(n\pi r_i/2d)}{\mathrm{I}_2(n\pi r_0/2d)\mathrm{K}_2(n\pi r_i/2d)-\mathrm{I}_2(n\pi r_i/2d)\mathrm{K}_2(n\pi r_0/2d)}$$

and B_n is obtained from A_n by writing I_2 for K_2 in the numerator.

The energy required to form the dislocation in the annulus is

$$\mathrm{W}=\tfrac{1}{2}b\int_{-d}^{d}dz\int_{r_i}^{r_0}dr\,\tau_{\theta z}.\quad (6)$$

When the outer radius r_0 is infinite

$$\begin{aligned}
\mathrm{W}&=\frac{8\mu b^2 d}{\pi^3}\sum_{n\ \mathrm{odd}}\frac{\mathrm{K}_0(n\pi r_i/2d)}{n^2(n\pi r_i/2d)^2\mathrm{K}_2(n\pi r_i/2d)},\\
&=\frac{\mu b^2}{4\pi}\,.\,2d\,.\,\frac{2}{3}\left(\frac{d}{r_i}\right)^2,\qquad r_i\gg d,\\
&=\frac{\mu b^2}{4\pi}\,.\,2d\,.\,\ln\left(\frac{d}{2{\cdot}24 r_i}\right),\qquad r_i\ll d.
\end{aligned}\quad (7)$$

5 B 2

When $r_i=0$, $r_0=\infty$ the dislocation at the origin exerts a force

$$F=b'\int_{-d}^{d}\tau_{\theta z}\,dz=\frac{4bb'\mu}{\pi^2}\sum_{n\,\mathrm{odd}}n^{-1}K_1\left(\frac{n\pi r}{2d}\right) \quad . \quad . \quad . \quad . \quad (8)$$

on another screw dislocation with Burgers vector b' distant r from it.

Since when $x>1$ $I_m(x)$ and $K_m(x)$ respectively increase and decrease exponentially with x the elastic state is given closely by

$$u_z=\frac{b}{2\pi}\theta, \qquad u_\theta=-\frac{b}{2\pi}\frac{z}{r}, \qquad \tau_{r\theta}=\frac{\mu b}{\pi}\frac{z}{r^2}, \qquad \tau_{\theta z}=0, \quad . \quad (9)$$

if the point r is a few multiples of d away from both the inner and outer edges of the annulus. (9) represents exactly the combined effect of (1) together with (2) and its reflection in the plane $z=0$, with couples of opposite hand. In (4) the terms in K annul the effect of that part of the couple distribution which would lie inside the plate.

§ 4. Screw Dislocation : Discussion.

It will be seen that the elastic state of the plate (equations (4)) is quite different from that of a slab of width $2d$ marked out perpendicular to a screw dislocation in an infinite body (equations (1)). The shear stress $\tau_{\theta z}$ is confined to the neighbourhood of the dislocation. That this must be so is clear : $\tau_{\theta z}$ vanishes at the surface of the plate, and far from the dislocation this state of affairs must persist throughout the thickness of the plate. The energy of a screw dislocation in an infinite plate, with its core excluded by a stress-free hole, is finite, while for a dislocation in an infinite cylinder the integral (6), $(\mu b^2/4\pi)\,.\,2d.\,\ln\,(r_0/r_i)$, diverges as $r_0\to\infty$. Again, in an infinite body two parallel screw dislocations attract like electrostatic line-charges, with a force $(\mu bb'/2\pi)\,.\,2d/r$ per length $2d$, whilst each term in (8) is the force due to a line-charge made up of particles attracting with the Yukawa potential const. $r^{-1}\exp\{-n\pi r/2d\}$. Because $\tau_{r\theta}$ does not vanish the screw dislocation will also interact with an edge dislocation running parallel to the z-axis. Since $\tau_{r\theta}$ is an odd function of z there will be no net force between them, but only a couple. It is also clear that the image-force attracting the dislocation towards the edge of a lamina will be very small unless the dislocation is only a few multiples of d from the edge. In the simple case where the dislocation is distant D from the edge of a semi-infinite plate this force is given by (8) with $b'=b$, $r=2D$. For comparison the image force on a length $2d$ of a screw dislocation running at a distance D parallel to the free surface of a semi-infinite solid is $\mu b^2d/D$.

§ 5. Edge Dislocations.

The stresses τ_{rr}, $\tau_{\theta\theta}$, $\tau_{r\theta}$ produced by an edge dislocation along the z-axis in an infinite body are derived from the Airy function $\chi=\mathrm{const.}\;r\ln r\sin\theta$, while $\tau_{zz}=\nu(\tau_{rr}+\tau_{\theta\theta})$ and the remaining components are zero. To solve the problem of an edge dislocation traversing a plate we must find an image stress system giving $\tau_{rz}=0$ and $\tau_{zz}=\text{const.} \sin \theta/r$ on the planes $z=\pm d$. The necessary analysis would be similar to that of Sneddon (1946). We should have to evaluate integrals like those in his § 5, but with Bessel functions of higher order on account of the angular dependence. The results would not be simple, and the case of an edge dislocation in an annular disc would be quite intractable. The general nature of the result can, however, be made out quite simply. If in the solution (Love, *op. cit.* p. 225) for an infinite hollow cylinder with an edge dislocation we replace λ by $2\lambda\mu/(\lambda+2\mu)$ the resulting expressions give the stresses and displacements averaged across the thickness of a disc cut from the cylinder and having surfaces free of stress (generalized plane stress: Love, *op. cit.*, p. 207). The average stresses will fall off as $1/r$ and the elastic state will not be very different from what it was when the disc formed part of the infinite cylinder.

As long as the plate containing the edge dislocation remains flat there is thus no widespread relaxation of stress, as there is for the screw dislocation. However, the energy may be reduced by buckling of the plate. Take a sheet of paper with a hole in it and make a tuck of constant width b (small compared with the size of the hole) running from the hole to the edge of the sheet. The paper will form a surface given roughly by the equation $z=(b\sqrt{2} \sin \theta)/2\pi$. If the paper is flattened out to form a plane dislocated lamina a large amount of strain-energy will be introduced. On the other hand we should not expect buckling to occur in a thick dislocated disc.

To find out the relations between the Burgers vector and the thickness and inner and outer radii of the annular disc for which buckling is energetically favourable we should have to use the theory of plates with strain in the middle surface (v. Kármán (1910)), modified to take account of initial stress. These equations are non-linear and it is difficult to solve them with the necessary boundary conditions. It seems clear, however, that in certain circumstances an edge dislocation will be able to relieve most of its stress by slight buckling of the plate, except within a few multiples of d from its centre, leading to a state of affairs similar to that discussed for the screw dislocation.

References.

Dawson, I. M., and Vand, V., 1950, *Nature*, **165,** 295; 1951, *Proc. Roy. Soc.* **206**A 555.

Forty, A. J., 1951, *Phil. Mag.*, [7], **42,** 670.

Frank, F. C., 1951, *Acta Cryst.* In the press.

v. Kármán, Th., 1910, *Ency. der Math. Wiss. IV*, [4], 348, (Leipzig: Teubner).

Love, A. E. H., 1927, *Mathematical Theory of Elasticity*, (Cambridge: University Press).

Riemann, B., 1855, *Ann. der Phys.*, **95,** 130.

Sneddon, I. N., 1946, *Proc. Camb. Phil. Soc.*, **42,** 260.

ANISOTROPIC ELASTICITY WITH APPLICATIONS TO DISLOCATION THEORY*

J. D. ESHELBY,† W. T. READ,‡ and W. SHOCKLEY‡

The general solution of the elastic equations for an arbitrary homogeneous anisotropic solid is found for the case where the elastic state is independent of one (say x_3) of the three Cartesian coordinates x_1, x_2, x_3. Three complex variables $z_{(l)} = x_1 + p_{(l)} x_2$ ($l = 1, 2, 3$) are introduced, the $p_{(l)}$ being complex parameters determined by the elastic constants. The components of the displacement (u_1, u_2, u_3) can be expressed as linear combinations of three analytic functions, one of $z_{(1)}$, one of $z_{(2)}$, and one of $z_{(3)}$. The particular form of solution which gives a dislocation along the x_3-axis with arbitrary Burgers vector (a_1, a_2, a_3) is found. (The solution for a uniform distribution of body force along the x_3-axis appears as a by-product.) As is well known, for isotropy we have $u_3 = 0$ for an edge dislocation and $u_1 = 0$, $u_2 = 0$ for a screw dislocation. This is not true in the anisotropic case unless the $x_1 x_2$ plane is a plane of symmetry. Two cases are discussed in detail, a screw dislocation running perpendicular to a symmetry plane of an otherwise arbitrary crystal, and an edge dislocation running parallel to a fourfold axis of a cubic crystal.

L'ÉLASTICITÉ ANISOTROPE ET SON APPLICATION À LA THÉORIE DES DISLOCATIONS

La solution générale des équations de l'élasticité pour un solide anisotrope, homogène est trouvée dans le cas où l'état élastique est indépendant d'une (mettons x_3) des trois coordonnées cartésiennes x_1, x_2, x_3. Trois variables complexes $z_{(l)} = x_1 + p_{(l)}x_2$, ($l = 1, 2, 3$) sont introduites, les $p_{(l)}$ étant des paramètres complexes déterminés par les constantes d'élasticité. Les composantes du déplacement (u_1, u_2, u_3) peuvent être exprimées comme des combinaisons linéaires de trois fonctions analytiques, une de $z_{(1)}$, une de $z_{(2)}$, et une de $z_{(3)}$. Une forme particulière de solution est trouvée, elle donne une dislocation le long de l'axe x_3 avec un vecteur de Burgers arbitraire (a_1, a_2, a_3). (En même temps apparait, comme sous-produit, la solution dans le cas d'une distribution uniforme de la force interne le long de l'axe x_3). Dans le cas d'isotropie, $u_3 = 0$ pour une dislocation-coin et $u_1 = 0$, $u_2 = 0$ pour une dislocation-vis. Ceci n'est pas vrai dans le cas d'anisotropie, à moins que le plan $x_1 x_2$ soit un plan de symétrie. Deux cas sont discutés en détail, une dislocation-vis perpendiculaire à un plan de symétrie d'un cristal, qui est d'autre part quelconque, et une dislocation-coin parallèle à un axe quaternaire d'un cristal cubique.

ANISOTROPE ELASTIZITÄT MIT ANWENDUNGEN AUF DIE THEORIE DER VERSETZUNGEN

Die allgemeine Lösung der Elastizitätsgleichungen für einen willkürlichen anisotropen Festkörper wird für den Fall angegeben, in dem der Elastizitätszustand von einer (zB. x_3) der drei Cartesischen Koordinaten x_1, x_2, x_3 unabhängig ist. Es werden drei komplexe Veränderliche $z_{(l)} = x_1 + p_{(l)} x_2$ ($l = 1, 2, 3$) eingeführt, wobei die $p_{(l)}$ komplexe Parameter sind, die durch die Elastizitätskonstanten bestimmt sind. Die Verschiebungskomponenten (u_1, u_2, u_3) können als lineare Kombinationen von drei analytischen Funktionen, nämlich als eine von z_1, eine von z_2 und eine von z_3, ausgedrückt werden. Eine spezielle Form der Lösung wurde für eine Versetzung in der x_3-Achse mit willkürlichem Burgers-Vektor (a_1, a_2, a_3) gefunden. (Als Nebenresultat ergibt sich die Lösung für eine gleichförmige Verteilung der Kraft entlang der x_3-Achse.) Bekanntlich gilt im isotropen Fall für eine Stufenversetzung $u_3 = 0$ und für eine Schraubenversetzung $u_1 = 0$ und $u_2 = 0$. Im anisotropen Fall trifft das nicht zu, ausser wenn die $x_1 x_2$ Ebene die Symmetrieebene ist. Zwei Fälle werden eingehend diskutiert: (1) Eine Schraubenversetzung, die senkrecht zu einer Symmetrieebene eines sonst willkürlichen Kristalles verläuft, und (2) Eine Stufenversetzung, die parallel einer der vierfachen Achsen eines kubischen Kristalles verläuft.

1. Introduction

This paper develops the general theory of anisotropic elasticity for a three-dimensional state of stress in which the stress is independent of one Cartesian coordinate. The general theory is applied to the dislocation theory of metals, which has reached a state where the refinement of allowing for anisotropy is sometimes justified. However, many other practical applications of the general elasticity results are possible, particularly in engineering stress analysis where the anisotropy of the material is important. The application in the present paper, however, is restricted to dislocation theory.

The problem has already been considered for edge dislocations in a previous paper [1].§ Unfortunately the assumption there that the problem can always be treated as one of plane strain is not justified unless the dislocation line runs perpendicular to a symmetry plane. In general the solutions given in [1] represent an edge dislocation plus a concentrated force at the origin along the dislocation line.

*Received November 24, 1952.

†Department of Physics, University of Illinois, Urbana, Illinois, U.S.A.

‡Bell Telephone Laboratories, Murray Hill, New Jersey, U.S.A.

§We take the opportunity to correct some errors in [1]. On p. 904, line 5, *for* s'_{41} *read* s_{41}; p. 905, line 6 from below, *for* f_n *read* f'_n; p. 906, equation (8), *for* y^2 *read* $\lambda^2_1 y^2$, $\lambda^2_2 y^2$ respectively in the arguments of the first and second logarithms; p. 907, lines 4, 2 from below *for* (6) or (8) . . . (10) *read* (7) or (9) . . . (11); p. 910, line 8 from below, *for* case *read* ease.

ACTA METALLURGICA, VOL. 1, MAY 1953

In the following sections we determine successively:

(i) The most general three-dimensional state of stress in an elastically distorted region when the stress is independent of one Cartesian coordinate.

(ii) The most general analytic (continuous, continuously differentiable and single valued) distribution of stress in the region surrounding a cylinder parallel to the axis of constant stress. The excluded region may be a hole through the crystal or a line imperfection, i.e. a bad region in the sense used by Read and Shockley [2] and Frank [3].

(iii) The state of stress around a dislocation.

2. Notation and Basic Relations

Unless the coordinate axes are simply related to the axes of a crystal of high symmetry, the stress-strain relations will involve a large number of non-vanishing elastic constants. In the general case, which we shall treat, there are twenty-one elastic constants relating the six stress components and the six strain components. Clearly, unless some short-hand notation is employed, the formulas become hopelessly cumbersome and the meaning is obscured. We shall use the simple three-dimensional Cartesian tensor notation as a means of expressing lengthy relations in a concise and easily comprehended form.

Calling the rectangular coordinates x_1, x_2, and x_3 instead of x, y, z, we suppose that the elastic state is independent of x_3. Let u_i be the ith component of displacement where i can be 1, 2, or 3. *Italic subscripts* $i, j, k, \ldots$ will take the range 1, 2, 3 and *Greek subscripts* α, β, the range 1, 2. The component of stress acting in the ith direction on the plane normal to the jth axis is τ_{ij}. The strains e_{ij} are related to the partial derivatives of displacement by

$$e_{ij} = \tfrac{1}{2}\left[\frac{\partial u_i}{\partial x_j} + \frac{\partial u_j}{\partial x_i}\right]. \tag{2.1}$$

The components of the strain tensor (2.1) are to be distinguished from the usual engineering strains which are not tensor components; they are equal to e_{ij} for $i = j$, but have twice the value (2.1) for $i \neq j$.

We shall use the customary summation convention, according to which summation is understood to be carried out over repeated ("dummy") subscripts.

The most general linear relation between stress and strain involves a fourth-order tensor C_{ijkl} and is given by

$$\tau_{ij} = C_{ijkl}\, e_{kl}\,. \tag{2.2}$$

Considerations of symmetry and energy conservation require that

$$C_{ijkl} = C_{jikl} = C_{ijlk} = C_{klij}\,, \tag{2.3}$$

these relations being responsible for the fact that there are only twenty-one different elastic constants. The constants can be transformed from one set of axes x_i to another set x'_i by the relation

$$C'_{pqrs} = \cos(x_i, x'_p)\cos(x_j, x'_q)\cos(x_k, x'_r)\cos(x_l, x'_s)C_{ijkl}\,. \tag{2.4}$$

When the coordinate axes are along the axes of a cubic crystal all but three of the different elastic constants vanish.

3. Equilibrium Conditions

We shall first determine the most general solution for an equilibrium state of stress in a region of elastic distortion. In the present section the conditions of equilibrium will be investigated and formulas derived for the force on an internal surface. In the following section the equilibrium equations will be expressed in terms of displacement and the most general solution for both displacement and stress obtained. (By formulating the problem in terms of displacement we avoid having to deal with the compatibility conditions.) The expression for the force on an internal surface will be needed in section 5 in interpreting the general expression for an analytic distribution of stress.

We begin by considering the equilibrium of an element of material in the crystal. Since nothing varies with x_3 we take a cylindrical element with axis parallel to the x_3-axis. Let A be the cross section of the cylinder in the $x_1 x_2$ plane and let C be the closed curve bounding A. The force (per unit length) exerted across C by the material outside the cylinder is

$$F_i = \int_C \tau_{i\beta}\, n_\beta\, dc\,. \tag{3.1}$$

where n_1 and n_2 are components of the normal to C.

For the material inside C to be in equilibrium, it is necessary that an external force $-F_i$ should be exerted on the cylinder. We now distinguish two cases:

1. The curve C is a reducible circuit; that is, C can be shrunk to a point without passing outside the material or cutting through a singularity in the stress distribution.

2. The curve C is not a reducible circuit; instead it encloses either a hole through the material in which external forces could be applied, or a bad

region where the theory to be developed does not apply.

In case 1, clearly the resultant force F_i must vanish, since there is no physical mechanism for applying an external force in the interior of a crystal. Thus for any choice of the circuit C which does not enclose a hole or singularity, the line integral in (3.1) vanishes. This implies the existence of a vector having components ϕ_i defined throughout the good region by

$$\frac{\partial \phi_i}{\partial x_1} = \tau_{i2}, \quad \frac{\partial \phi_i}{\partial x_2} = -\tau_{i1}, \tag{3.2}$$

from which the equilibrium conditions for stress

$$\frac{\partial \tau_{i\beta}}{\partial x_\beta} = 0 \tag{3.3}$$

are readily obtained.

Conversely, if we begin with the equilibrium equations (3.3), we can prove that the functions ϕ_i exist and that $F_i = 0$ for any reducible circuit.

We now consider case 2 above, where C encloses a hole or singularity. Assume that C does not pass through any singular points, so that the equilibrium equations are satisfied and the ϕ_i exist at all points on C. Then (3.1) gives

$$F_i = \int_C d\phi_i = \Delta\phi_i \tag{3.4}$$

where $\Delta\phi_i$ is the change in $\Delta\phi_i$ in going once around the curve C. When $\Delta\phi_i \neq 0$, ϕ_i is multiple valued. Alternatively we may make a mathematical cut in the $x_1 x_2$ plane; ϕ_i is then single valued and has a discontinuity $\Delta\phi_i$ across the cut.

To summarize: every equilibrium state of stress can be represented by a vector with components ϕ_i, every set of three functions ϕ_i represents an equilibrium state of stress, and the discontinuities in the functions ϕ_i correspond to a resultant force on an internal surface.

Rather than derive the differential equations for the ϕ_i (which would involve the compatibility conditions), in the next section we shall formulate the problem in terms of displacement, which has a more direct physical meaning, and use the equations of equilibrium (3.1). The ϕ_i are then obtained by integration.

However, there are some special cases in which it is convenient mathematically to represent the state of stress in the $x_1 x_2$ plane by a function ϕ called the Airy stress function, which is related to ϕ_1 and ϕ_2 by

$$\frac{\partial \phi}{\partial x_2} = -\phi_1, \quad \frac{\partial \phi}{\partial x_1} = \phi_2. \tag{3.5}$$

The relations (3.5) and the existence of ϕ follow from (3.2) and the condition $\tau_{12} = \tau_{21}$.

Only in special cases can ϕ be treated independently of ϕ_3. In such cases it is sometimes more convenient to express the state of stress in the $x_1 x_2$ plane in terms of the single function ϕ rather than in terms of the two functions u_1 and u_2.

4. General Solution

In this section we shall determine the most general solution for the displacement and thence for the stress in a region of elastic distortion. The condition of elastic distortion necessarily excludes singularities in stress, so that at all points in the elastic region the equations of equilibrium (3.1) hold. By expressing stress in terms of displacement in the three equilibrium equations, we obtain three equations for the three components of displacement.

First substituting (2.1) in (2.2) we have

$$\tau_{ij} = \tfrac{1}{2} C_{ijkl} \frac{\partial u_k}{\partial x_l} + \tfrac{1}{2} C_{ijkl} \frac{\partial u_l}{\partial x_k}. \tag{4.1}$$

However, by (2.3) the third and fourth subscripts of the elastic constants can be interchanged so that

$$\tau_{ij} = \tfrac{1}{2} C_{ijkl} \frac{\partial u_k}{\partial x_l} + \tfrac{1}{2} C_{ijlk} \frac{\partial u_l}{\partial x_k}, \tag{4.2}$$

where the two terms on the right are seen to be equal since the dummy subscripts k and l go through the same set of values. Therefore, taking account of the fact that nothing varies with x_3, we have

$$\tau_{i\beta} = C_{i\beta k\alpha} \frac{\partial u_k}{\partial x_\alpha} \tag{4.3}$$

and equation (3.3) becomes

$$C_{i\beta k\alpha} \frac{\partial^2 u_k}{\partial x_\alpha \partial x_\beta} = 0. \tag{4.4}$$

This is a set of three second-order linear partial differential equations, the solution of which is given by an arbitrary function of a linear combination of the variables x_1 and x_2. Thus,

$$u_k = A_k f[p_1 x_1 + p_2 x_2] = A_k f[p_\alpha x_\alpha] \tag{4.5}$$

where the A's and p's are determined by substituting into (3.7) which gives

$$A_k p_\alpha p_\beta C_{i\beta k\alpha} = 0. \tag{4.6}$$

This equation has a solution for the vector A_1, A_2, A_3 only if the determinant of the coefficients vanishes. This determinant is a sixth-order polynomial in p_1 and p_2. It is obvious that in (4.5) we can always take $p_1 = 1$; (4.6) is then a sixth-order equation for p_2. It is shown in the appendix that since the energy density is always positive the roots are necessarily complex, and since the coefficients are real they occur in conjugate pairs. The vector $A_{(l)k}$ corresponding to a given root $p_{(l)}$ is in general complex. The condition that the displacements be real requires that the imaginary parts of corresponding pairs of solutions shall cancel. Thus we take only three roots $p_{(1)}$, $p_{(2)}$, and $p_{(3)}$, no two of which are complex conjugates, and set

$$(4.7) \qquad u_k = \sum_{l=1}^{3} A_{(l)k} f_{(l)}[z_{(l)}]$$

where

$$z_{(l)} = x_1 + p_{(l)} x_2 .$$

It will be understood throughout that where complex expressions are used only the real part is to be taken. The subscripts in parentheses distinguish the three individual solutions corresponding to the three different values $p_{(1)}$, $p_{(2)}$, $p_{(3)}$ and are not to be confused with the subscripts denoting coordinate axes.

From (4.3) and (4.7) we find that the most general equilibrium distribution of stress in the elastic region is

$$(4.8) \quad \tau_{ij} = \sum_{l=1}^{3} [C_{ijk1} + p_{(l)} C_{ijk2}] A_{(l)k} \frac{df_{(l)}}{dz_{(l)}} [z_{(l)}]$$

and the corresponding functions ϕ_i, obtained by substituting (4.8) in (3.2) and integrating, are

$$(4.9) \quad \phi_i = \sum_{l=1}^{3} [C_{i2k1} + p_{(l)} C_{i2k2}] A_{(l)k} f_{(l)}[z_{(l)}] .$$

Alternatively we may verify directly that (4.9) agrees with the first of equations (3.2); that it also satisfies the second of (3.2) follows from (4.6) which now takes the form

$$A_k[C_{i1k1} + p_{(l)} C_{i1k2} + p_{(l)} C_{i2k1} + p_{(l)}^2 C_{i2k2}] = 0.$$

We shall find below that particular solutions of (4.7) can be found which represent dislocations lying on the x_3-axis or localized forces applied to the x_3-axis. In general to express any such situation it is necessary to combine terms arising from all three values of l. Thus in the subsequent analysis, solutions will be represented as sums over the three roots.

5. Continuous Distribution of Stress Surrounding a Singularity

In the previous sections we obtained a general solution valid at all points in a region where the distortion is elastic, i.e. in a good region. No restrictions were placed on the shape of the good region, nor was it assumed that the stresses were continuous. Actually the equilibrium equations, which hold in a good region, require that the tractions be continuous across any surface but leave open the possibility of a discontinuity in the normal stress on a plane perpendicular to the surface of discontinuity.

In the present section we shall assume that the distortion is elastic everywhere in the crystal except in a cylindrical region along the x_3-axis. This excluded region may be either a hole through the crystal or a line imperfection, or bad region, of the type discussed in section 1. The good region is therefore bounded by two cylindrical surfaces, the outer one being the external surface of the crystal.

In this section we shall also introduce the requirement that the stress in the good region be continuous, single valued and have continuous derivatives. We then investigate the conditions imposed on the arbitrary functions $f_{(l)}$ $[z_{(l)}]$ of the last section by the above conditions ((ii) of section 1) on the stress in the good region, and the shape of the good region. We shall reserve for the appendix the discussion of the most general solution consistent with these requirements, and use expressions for $f_{(l)}$ sufficient for the needs of physical dislocation theory.

From (4.8) we see that if τ_{ij} is to be analytic and single valued in a region of the crystal, it is sufficient (though not necessary, see the appendix) that $df_{(l)}$ $[z_{(l)}]/dz_{(l)}$ be an analytic single-valued function in the corresponding region of the complex $z_{(l)}$ plane, which will be bounded by two closed curves corresponding to the inner and outer boundaries of the actual region in the $x_1 x_2$ plane. Therefore by Laurent's theorem $df_{(l)}[z_{(l)}]/dz_{(l)}$ can be expressed as a power series with both positive and negative powers of $z_{(l)}$. Hence

$$(5.1) \quad f_{(l)}[z_{(l)}] = \frac{D_{(l)}}{\pm 2\pi i} \ln z_{(l)} + \sum_{n=-\infty}^{\infty} C_{(l)n} z_{(l)}^n$$

where for convenience in connection with (5.2) below the sign of $2\pi i$ is taken to be same as the sign of the imaginary part of $p_{(l)}$.*

It is at once seen that $f_{(l)}[z_{(l)}]$ is not a single-valued function since it changes by $D_{(l)}$ per revolution about the x_3-axis, the circuit being taken in the positive direction. Thus if we make a cut joining the

*A positive circuit around the origin of the $x_1 x_2$ plane is a $\pm$ circuit in the $z_{(l)}$ plane depending on the sign of the imaginary part of $p_{(l)}$. Note that there is no summation over l. Such a summation will always be indicated explicitly.

inner and outer boundaries there will be a discontinuity

$$\Delta u_k = \sum_{l=1}^{3} A_{(l)k} D_{(l)} \tag{5.2}$$

in the displacement across it.

All the other terms in (5.1) give single-valued displacements.

It is also seen, from (3.4) and (4.9), that the logarithmic term in (5.1) is the only one which contributes to the resultant force on the internal boundary

$$F_i = \Delta\phi_i = \sum_{l=1}^{3} [C_{i2k1} + p_{(l)} C_{i2k2}] A_{(l)k} D_{(l)}. \tag{5.3}$$

From (4.8) the stress corresponding to (5.1) is found to be

$$\tau_{ij} = \sum_{l=1}^{3} [C_{ijk1} + p_{(l)} C_{ijk2}] A_{(l)k}$$

$$\left\{\frac{D_{(l)}}{\pm 2\pi i} \frac{1}{z_{(l)}} + \sum_{n=-\infty}^{\infty} n C_{(l)n} z_{(l)}^{n-1}\right\}. \tag{5.4}$$

From (5.4) it is seen that each value of l corresponds to two linearly independent stress distributions (two because the constants are complex) each giving a discontinuity in displacement and a force. In general all three l's are required to represent any state of stress, solutions for individual l's having no simple physical meaning. We shall later consider a simple case where two l's can represent the state of stress in the $x_1 x_2$ plane, the third l giving the stress components in the x_3 direction.

In the next section we shall discuss the terms in $D_{(l)}$ which represent dislocations and uniform distributions of force along the x_3-axis. The terms in $C_{(l)\pm n}$ enable us to satisfy arbitrary boundary conditions on the inner and outer cylindrical surfaces. The C's will not give a net force on a boundary (this is provided for by the D's) but in general the $C_{(l)-1}$ give a net couple on it.

For certain values of the elastic constants equation (4.6) may give multiple roots for p_2. Such cases are most easily dealt with by passing to the limit from a neighboring set of elastic constants for which this is not so (cf. [1]).

6. Dislocations

In this section we shall assume that the excluded cylindrical region is a line imperfection, or bad region. This requires the vanishing of the resultant force F_i exerted across the boundary between the good and bad regions since there is no mechanism whereby an external force could be applied to the bad region to maintain equilibrium.

In the previous sections we have found that for stress fields independent of x_3 the discontinuity in displacement is a constant. As is well known, in an actual crystal this constant must either vanish or be equal to a *lattice translation or slip vector* defined as *the shortest vector connecting two atoms in the crystal structure which have identical surroundings.* In a face-centered cubic metal, there are six possible slip vectors, one in each of the six [110] directions; in a body-centered cubic structure there are four possible slip vectors corresponding to the four [111] directions.

When the discontinuity in displacement vanishes, the bad region is either an extra row of atoms or a missing row. Since such a situation would be unlikely to occur unless two unlike dislocations on adjacent slip planes ran together we shall consider this as a special case of a dislocation array and proceed to investigate the stress field around a dislocation.

If the slip vector is a_k, the conditions for a dislocation are

$$\Delta u_k = a_k, \quad F_i = 0. \tag{6.1}$$

We should also see whether there is a net couple on the inner boundary. It may be shown to vanish however we choose the $D_{(l)}$ (cf. [1]).

From (5.2) and (5.3), conditions (6.1) become

$$a_i = \sum_{l=1}^{3} A_{(l)i} D_{(l)} \qquad (i = 1, 2, 3), \tag{6.2}$$

$$0 = \sum_{l=1}^{3} A_{(l)k} D_{(l)} [C_{i2k1} + p_{(l)} C_{i2k2}] \qquad (i = 1, 2, 3). \tag{6.3}$$

Equations (6.2) and (6.3) are a system of six linear equations for the determination of the real and imaginary parts of $D_{(1)}$, $D_{(2)}$, and $D_{(3)}$. Thus in general all three l's are required to represent any simple dislocation. Similarly all three l's are required in the general case to represent a concentrated internal force at the origin.

In [1] it was assumed that a solution could always be found by taking $u_3 = 0$, that is, $A_3 = 0$. This is allowable in the sense that it will certainly represent some state of stress. For an edge dislocation it is then possible to prescribe the conditions $\Delta u_1 = a_1$, $\Delta u_2 = a_2$, $F_1 = 0$, $F_2 = 0$. F_3 is then determined and in general does not vanish unless $C_{\alpha\beta k3} = 0$. Hence, except when the $x_1 x_2$ plane is a symmetry plane, the expressions in [1] represent the dislocations they purport to, plus a force along the x_3-axis necessary to maintain a state with $u_3 = 0$. The presence of F_3 was overlooked in the original paper.

In an infinite body with the requirement that the stresses shall vanish at infinity, the $C_{(l)n}$ with $n > 0$ are excluded and so also is the more general type of dislocation considered in the appendix. Then along any straight line through the origin the stresses can be expressed as a series of inverse powers of $r = (x_1^2 + x_1^2)^{\frac{1}{2}}$. The term in r^{-1} is fixed by the $D_{(l)}$ and the higher terms by the $C_{(l)n}$, $n < 0$, that is, by conditions at the inner boundary. Thus at large distances from the origin the state of stress is determined solely by the Burgers vector.

By adjusting the $C_{(l)\pm n}$ we can make the outer surface of the cylinder stress-free. If we cut out a cylinder of finite length there will be a distribution of stress on the ends. This will have zero resultant, but not in general zero moment [4]. When these end stresses are removed the cylinder will thus be twisted and the u_i will not be independent of x_3 even if end effects are neglected.

7. Dislocations in Special Cases

In this section we consider the case where the $x_1 x_2$ plane is a plane of symmetry. Then the problem is considerably simplified in that the components of stress in the $x_1 x_2$ plane can be expressed in terms of derivatives of u_1 and u_2 only, and the stress components in the x_3 direction in terms of u_3 only; that is, $C_{\alpha\beta k3} = 0$. This means that the matrix expression for A_k splits so that A_3 does not mix with A_1 and A_2, or in other terms, $A_{(3)1} = A_{(3)2} = A_{(1)3} = A_{(2)3} = 0$.

We shall treat u_3 first for the general case where $x_1 x_2$ is a plane of crystal symmetry and then deal with u_1 and u_2 together for the more specialized case where the x_3-axis is a crystal axis in a cubic crystal. The latter results are implicit in [1], but we here carry the calculations out explicitly and reach an expression for the shear stress on any plane parallel to the slip plane of the dislocation. A knowledge of this quantity is sufficient for many problems; for example, determination of the possible stable arrangements of two-dimensional arrays of dislocations or calculation of the energies of single dislocations [1] and grain boundaries [5].

(a) *Screw Dislocation in a Simple Case*

It is convenient here to adopt a conventional notation, replacing the fourth-order tensor C_{ijkl} by the elastic constants $c_{11}, c_{12}, c_{13}, \ldots, c_{66}$, where the pair of subscripts 11 is replaced by 1, 22 by 2, 33 by 3, 23 by 4, 13 by 5, and 12 by 6. Then in the present symmetrical case where $c_{14} = c_{15} = c_{16} = c_{24} = c_{25} = c_{26} = 0$, we have

$$(7.1) \qquad \tau_{23} = c_{44}\frac{\partial u_3}{\partial x_2} + c_{45}\frac{\partial u_3}{\partial x_1}, \qquad \tau_{13} = c_{54}\frac{\partial u_3}{\partial x_2} + c_{55}\frac{\partial u_3}{\partial x_1}.$$

Substituting these into the equilibrium equation in the x_3 direction

$$(7.2) \qquad \frac{\partial \tau_{13}}{\partial x_1} + \frac{\partial \tau_{23}}{\partial x_2} = 0$$

gives an equation corresponding to (3.9) of the general case:

$$(7.3) \qquad c_{55}\frac{\partial^2 u_3}{\partial x_1^2} + 2c_{45}\frac{\partial^2 u_3}{\partial x_1 \partial x_2} + c_{44}\frac{\partial^2 u_3}{\partial x_2^2} = 0,$$

the solution to which is an arbitrary function of $x_1 + px_2$ where p is complex and is a root of

$$(7.4) \qquad c_{44}p^2 + 2c_{45}p + c_{55} = 0.$$

The two roots of (7.4) are complex conjugates and the condition that u_3 be real requires that the two solutions also be complex conjugates. We therefore take either root and set

$$(7.5) \qquad u_3 = f(x_1 + px_2),$$

where it is understood that only the real part of the complex right-hand side is to be taken. The condition that the stress be analytic requires that the arbitrary function be of the form (4.2) where only the logarithmic term need be considered here. We therefore express u_3 in terms of the complex constant D,

$$(7.6) \qquad u_3 = \frac{D}{\pm 2\pi i}\ln(x_1 + px_2).$$

If a_3 is the component of the slip vector in the x_3 direction we have: real part of $D = a_3$. From (4.9) the stress function ϕ_3 is

$$\phi_3 = [C_{3231} + p\, C_{3232}]f(x_1 + px_2),$$

or in the present notation

$$(7.7) \qquad \phi_3 = [c_{45} + p\, c_{44}]f(x_1 + px_2).$$

Therefore the vanishing of the net force $F_3 = \Delta\phi_3$ in the x_3 direction gives

$$(7.8) \qquad D(c_{45} + p\, c_{44}) = 0$$

which is the third of equations (6.3) for the present case and determines the imaginary part of D.

The stresses are then readily determined from (7.1).

When the axis of the dislocation is parallel to a crystal axis in a cubic crystal, $c_{45} = 0$ and $c_{55} = c_{44}$, so that equation (7.3) reduces to the Laplace equation, and u_3 and the associated stresses are given by the same formulas as in the isotropic case, the constant c_{44} being used for the shear modulus [6]. This leads to $z = x_1 + ix_2$ so that $\ln z = \ln r + i\theta$ in cylindrical coordinates. The first term represents a concentrated force along the x_3-axis, the second a screw dislocation. Another simple example of practical interest is a pure screw dislocation parallel to a ⟨110⟩ direction in face centered cubic. Taking the x_1- and x_2-axes parallel to [110] and [011] respectively and using the elastic constants c_{11}, c_{12}, c_{44} referred to cube axes, we have $-p^2 = 2\, c_{44}/(c_{11} - c_{12}) = \beta^2$. It is convenient to express the two non-vanishing stress components in cylindrical coordinates:

$$(7.9)\qquad \tau_{\theta z} = \frac{c_{44}}{2\pi\beta}\frac{a_3}{r}, \qquad \tau_{rz} = \frac{c_{44}}{2\pi\beta}\frac{a_3}{r}\frac{(1-\beta^2)\sin\theta\cos\theta}{\cos^2\theta + \beta^2\sin^2\theta}.$$

(b) *Edge Dislocation in a Simple Case*

We now consider the state of stress in the x_1x_2 plane when the axis of the dislocation is along the axis of a cubic crystal. As we have seen in the last section, u_3 can be treated independently and is the same as in the isotropic case. We thus have only two equations for the determination of the displacement and stress in the x_1x_2 plane, and there are two independent terms in the general solution which is of the form (3.8). The constants A_1, A_2, and p are given by (3.9) which in this case becomes

$$(7.10a)\qquad \begin{aligned} (C_{1111} + p^2 C_{1212})A_1 + (C_{1122} + C_{1221})pA_2 &= 0,\\ (C_{2211} + C_{2121})p A_1 + (C_{2112} + p^2 C_{2222})A_2 &= 0, \end{aligned}$$

or in terms of the small c's,

$$(7.10b)\qquad \begin{aligned} (c_{11} + p^2 c_{44})A_1 + (c_{12} + c_{44})p A_2 &= 0,\\ (c_{12} + c_{44})p A_1 + (c_{44} + p^2 c_{11}) A_2 &= 0. \end{aligned}$$

The determinant vanishes for the four values of p given by

$$(7.11)\qquad p_{(1)} = e^{i\alpha},\ p^*_{(1)} = e^{-i\alpha},\ p_{(2)} = -e^{i\alpha},\ p^*_{(2)} = -e^{-i\alpha},$$

where α is given by

$$(7.12)\qquad \alpha = \tfrac{1}{2}\cos^{-1}\left[\frac{2c_{12}c_{44} + c_{12}^2 - c_{11}^2}{2c_{11}c_{44}}\right],$$

and is real for the common cubic metals (see the appendix). The corresponding eigenvectors A_k may be chosen with $A_1 = 1$ so that the values of $A_{(l)2} \equiv A_{(l)}$ are

$$(7.13)\qquad A_{(2)} = -A_{(1)} = \frac{c_{44}e^{i\alpha} + c_{11}e^{-i\alpha}}{c_{12} + c_{44}}.$$

The general solution for the displacement is therefore

$$(7.14)\qquad \begin{aligned} u_1 &= f_{(1)}[z_{(1)}] + f_{(2)}[z_{(2)}]\\ u_2 &= A_{(1)}f_{(1)}[z_{(1)}] + A_{(2)}f_{(2)}[z_{(2)}], \end{aligned}$$

where $z_{(1)} = x_1 + e^{i\alpha}x_2$ and $z_{(2)} = x_1 - e^{i\alpha}x_2$. For an analytic distribution of stress the f's, as we have seen, are of the form (4.2) where only the logarithmic term need be considered. Expressing these functions in the form (6.1), we have to determine the two complex constants $D_{(1)}$ and $D_{(2)}$. These are given by (6.3) for the slip vector and (6.4) for the vanishing of the net force. For the present case (6.3) becomes

$$(7.15)\qquad \begin{aligned} a_1 &= D_{(1)} + D_{(2)},\\ a_2 &= A_{(1)}D_{(1)} + A_{(2)}D_{(2)} = A_{(2)}[D_{(1)} - D_{(2)}]. \end{aligned}$$

(6.4) becomes

$$0 = D_{(1)}[C_{1221}A_{(1)} + C_{1212}\,p_{(1)}] + D_{(2)}[C_{1221}A_{(2)} + p_{(2)}C_{1212}]$$

$$0 = D_{(1)}[C_{2211} + p_{(1)}C_{2222}A_{(1)}] + D_{(2)}[C_{2211} + p_{(2)}C_{2222}A_{(2)}]$$

which may be expressed in terms of c_{11}, c_{12}, c_{44} and α as

$$(7.16)\qquad \begin{aligned} 0 &= [A_{(1)} + e^{i\alpha}][D_{(1)} - D_{(2)}]\\ 0 &= [c_{12} + c_{11}e^{i\alpha}A_{(1)}][D_{(1)} + D_{(2)}] \end{aligned}$$

where, as throughout, only the real part of complex expressions are taken. Equations (9.7) and (9.8) are four linear equations for the real and imaginary parts of the two complex constants from which the displacements and then the stresses are obtained.

When the slip vector is in the x_1 direction so that $a_2 = 0$, we have

$$(7.17)\qquad D_{(1)} = D_{(2)} = -\frac{a_1 i}{2\sin 2\alpha}\left\{\frac{e^{2i\alpha}c_{11} - c_{12}}{c_{11}}\right\}.$$

The most important quantity to be determined in applications of dislocation theory is the shear stress on the slip plane. In the present case this is

$$\tau_{12} = c_{44}\left\{\frac{\partial u_1}{\partial x_2} + \frac{\partial u_2}{\partial x_1}\right\} = a_1\frac{(c_{11}^2 - c_{12}^2)}{4\pi c_{11}\sin 2\alpha}\left[\frac{1}{x_1 + e^{i\alpha}x_2} + \frac{1}{x_1 - e^{i\alpha}x_2}\right]e^{i\alpha}$$

$$(7.18) \quad = a_1 \frac{(c_{11}+c_{12})}{2\pi}\sqrt{\frac{c_{44}\,c'_{44}}{c_{11}\,c'_{11}}} \cdot \left[\frac{x\,(x_1^2-x_2^2)}{(x_1^2+x_2^2)^2 - \frac{2(c_{11}+c_{12})}{c_{11}}\left\{1-\frac{c'_{44}}{c_{44}}\right\}x_1^2\,x_2^2}\right]$$

where c'_{11}, c'_{12}, c'_{44} are the elastic constants referred to axes rotated counter-clockwise through 45° about the x_3-axis and are given by

$$(7.20) \quad c'_{44} = \frac{c_{11}-c_{12}}{2}, \quad c'_{11} = \frac{c_{11}+c_{12}+2c_{44}}{2}, \quad c'_{11}+c'_{12} = c_{11}+c_{12}.$$

APPENDIX

(i) *The General Solution: Wedge Dislocations*

The most general type of multiple-valued displacement consistent with continuous and twice-differentiable strains is known from the classical work of Weingarten and Volterra (see, for example, the review article by Nabarro [7]). From their results it appears that by assuming in section 5 that $df_{(l)}/dz_{(l)}$ is analytic and single valued we have excluded Volterra's dislocation of order VI. (The first three orders are included in the analysis of the text, whilst for orders IV and V the elastic state would not be independent of x_3.) Weingarten's and Volterra's results are very general, being essentially geometrical, and depending neither on Hooke's law nor the equations of equilibrium. Nevertheless it is interesting to see how the most general dislocation state arises in our analysis. The stress is assumed to be analytic and single valued; this is equivalent to assuming that the strain is single valued and analytic. Since the strain is somewhat more convenient to work with, we use (2.1) to obtain the strain corresponding to the general equilibrium solution (4.7). This gives

$$(A.1) \quad e_{ij} = \sum_{l=1}^{3} A_{(l)i}\,p_{(l)j} + A_{(l)j}\,p_{(l)i}\,\frac{df_{(l)}}{dz_{(l)}}[z_{(l)}]$$

where $p_{(l)1}=1$, $p_{(l)2}=p_{(l)}$, $p_{(l)3}=0$ as in the text.

Equations (A.1) are a system of only five linear equations ($i=j=3$ being trivial) for the real and imaginary parts of the three f's. Thus (A.1) can be solved for five of these functions in terms of the sixth function and the strains. For example if we let $df_{(l)}/dz_{(l)} = U_{(l)} + iV_{(l)}$ we can solve (A.1) for $U_{(l)}$ in terms of $V_{(l)}$ and e_{ij}. Differentiating this solution with respect to the real and imaginary parts of $z_{(l)}$ gives two equations for the first derivatives of $U_{(l)}$ and $V_{(l)}$, the real and imaginary parts of $z_{(l)}$ being taken as the independent variables. These two equations added to the Cauchy-Riemann equations make four equations, which can be solved for the four first derivatives in terms of the strains, which are analytic single-valued functions. Thus $d^2f_{(l)}/d^2z_{(l)}$ is an analytic single-valued function in the good region and, by Laurent's theorem can be represented by a series of positive and negative powers of $z_{(l)}$. Therefore the most general form of $f_{(l)}$ is

$$(A.2) \quad f_{(l)}[z_{(l)}] = \frac{B_{(l)}}{\pm 2\pi i}\,z_{(l)}\ln z_{(l)} + \frac{D_{(l)}}{\pm 2\pi i}\ln z_{(l)} + \sum_{n=-\infty}^{\infty} C_{(l)n}\,z_{(l)}^n$$

the sign of $2\pi i$ being chosen as in the text.

(A.2) differs from the expression (5.1) only in the addition of the term in $B_{(l)}$. We now consider the physical meaning of the additional term.

Unlike the other constants in (A.2) the real and imaginary parts of the $B_{(l)}$ are not independent but are related by the five equations for the vanishing of the discontinuity in strain $\Delta e_{ij}=0$ which from (A.1) and (A.2) gives

$$(A.3) \quad \sum_{l=1}^{3} [A_{(l)i}\,p_{(l)j} + A_{(l)j}\,p_{(l)i}]\,B_{(l)} = 0.$$

Whatever new feature is introduced by admitting the terms in $B_{(l)}$ may thus be described by a single real constant. We shall now find the physical meaning of this constant.

First consider the force $F_i = \Delta\phi_i$ corresponding to the $B_{(l)}$. From (4.9) and the first term of (A.2) we have

$$(A.4) \quad F_i = -\sum_{l=1}^{3} C_{i2k\alpha}\,p_{(l)\alpha}\,A_{(l)k}\,B_{(l)}\,z_{(l)}.$$

Remembering that $p_{(l)1}=1$ and $p_{(l)3}=0$, we can write (A.4) as

$$(A.5) \quad F_i = -x_1\,C_{i2\beta\alpha}\sum_{l=1}^{3} p_{(l)\alpha}\,A_{(l)\beta}\,B_{(l)} - x_2\,C_{i2\beta\alpha}\sum_{l=1}^{3} p_{(l)\alpha}\,A_{(l)\beta}\,B_{(l)}\,p_{(l)\alpha}.$$

The first term vanishes because $C_{i2\beta\alpha}$ is symmetrical in α and β while by (A.3) the sum over l is unsymmetrical. This fact enables us to write the second term as

$$(A.6) \quad F_i = -x_2\,C_{i\beta k\alpha}\sum_{l=1}^{3} p_{(l)\alpha}\,p_{(l)\beta}\,A_{(l)k}\,B_{(l)},$$

which vanishes by the equilibrium conditions in the form (4.6). Thus the $B_{(l)}$ give no contribution to the force and equation (5.3) for F_i in terms of the $D_{(l)}$ is still valid.

Let us now consider the discontinuity in displacement corresponding to the $B_{(l)}$ term in (A.2). This is

$$\text{(A.7)} \qquad \Delta u_k = \sum_{l=1}^{3} A_{(l)k} B_{(l)} p_{(l)\alpha} x_\alpha .$$

If we define

$$\text{(A.8)} \qquad \omega = \sum_{l=1}^{3} A_{(l)1} B_{(l)} p_{(l)2}$$

and make use of (A.3), (A.7) can be written in the vector form

$$\text{(A.9)} \qquad \Delta \bar{u} = \bar{\omega} \times \bar{X}$$

where $\bar{X}$ has components x_1 and x_2 and $\bar{\omega}$ is parallel to the x_3-axis and of magnitude ω (A.8). Thus the discontinuity in displacement corresponding to the $B_{(l)}$ term in the general expression (A.2) represents a rigid body rotation of the adjoining surfaces about the x_3-axis. This would occur if a cylinder were cut and sprung open or a wedge-shaped piece were cut out and the adjoining surfaces cemented together. In a crystal there would clearly be a relative rotation ω of the lattices on opposite sides of the cut, so that the wedge solution is inadmissible in applications to a crystal lattice.

(ii) *Proof that the $p_{(l)}$ are Complex*

The condition that the energy density be positive for any state of strain is

$$\text{(A.10)} \qquad C_{ijkm}\, e_{ij}\, e_{km} = C_{ijkm} \frac{\partial u_i}{\partial x_j} \frac{\partial u_k}{\partial x_m} > 0$$

for any real set of e_{ij} or $\partial u_i / \partial x_j$ not all of which vanish.

Suppose that in (4.6), $p_1 = 1$ and p_2 is real. Then the ratios of the A_k found by solving the three equations (4.6) will be real. Hence the A_k may be chosen real and not all zero. Multiplying (4.6) by A_i and summing over i we have

$$A_i A_k p_\alpha p_\beta C_{i\beta k\alpha} = 0.$$

Accordingly the choice

$$\frac{\partial u_k}{\partial x_\alpha} = A_k p_\alpha \; (\alpha = 1, 2), \quad \frac{\partial u_k}{\partial x_3} = 0$$

will lead to a zero energy density for a non-zero strain. Hence the assumption that p_2 is real is inadmissible.

We can also give a direct proof from the determinantal equation itself in the special cases considered. It is convenient to write (A.10) in the form

$$\text{(A.11)} \qquad \sum_{i=1}^{6} \sum_{j=1}^{6} c_{ij}\, e_i\, e_j > 0$$

with $e_{11} = e_1, \ldots, 2e_{23} = e_4, \ldots$. Putting $e_4 = p$, $e_5 = 1$, and the other e_i zero we have $c_{44}\, p^2 + 2c_{45}\, p + c_{55} > 0$, showing that the p of equation (7.4) cannot be real.

The quantity α (equation 7.12) can be written in the more expressive form

$$\alpha = \cos^{-1} \sqrt{\left\{ \tfrac{1}{2} \frac{1}{1-\sigma} \left(1 - \frac{1}{A}\right) \right\}}$$

where $\sigma = c_{12}/(c_{11} + c_{12})$ is Poisson's ratio for extension in the $\langle 100 \rangle$ direction and $A = 2\, c_{44}/(c_{11} - c_{12})$ is the "anisotropy factor" for a cubic crystal [8]. (We are indebted to Dr. H. Brooks for pointing this out.) We have to show that α is never purely imaginary. In (A.11) put successively $e_1 = 1$; $e_4 = 1$; $e_1 = \pm e_2$; $e_1 = e_2 = e_3$, with the unspecified e_j zero in each case. We find that c_{11}, c_{44}, $c_{11} \pm c_{12}$, $c_{11} + 2c_{12}$ are all positive. (Note that $c_{11} = c_{22} = c_{33}$ and $c_{12} = c_{23} = c_{31}$ in a cubic crystal.) From these relations we easily show that

$$-1 < \sigma < \tfrac{1}{2} \text{ and } A > 0.$$

It follows that $\cos \alpha$ is purely imaginary if $0 < A < 1$ or purely real and less than unity if $1 < A < \infty$, and so α is complex or purely real. From Table 3 of [8] it is seen that the common cubic metals fall into the class with real α.

The quantity β^2 introduced at the end of section 7a is equal to A and so the associated p is purely imaginary.

References

1. ESHELBY, J. D. Phil. Mag., 7, **40** (1949) 903.
2. READ, W. T. and SHOCKLEY, W. In "Imperfections in Nearly Perfect Crystals" (New York, John Wiley and Sons, 1952), p. 77.
3. FRANK, F. C. Phil. Mag., 7, **42** (1951) 209.
4. ESHELBY, J. D. J. Appl. Phys., **24** (1953) 176.
5. READ, W. T. and SHOCKLEY, W. In "Imperfections in Nearly Perfect Crystals" (New York, John Wiley and Sons, 1952), p. 352.
6. BURGERS, J. M. Proc. Acad. Sci. Amst., **42** (1939) 378.
7. NABARRO, F. R. N. Advances in Physics, **1** (1952) 269.
8. ZENER, C. Elasticity and Anelasticity of Metals (Chicago University Press, 1948), p. 16.

JOURNAL OF APPLIED PHYSICS VOLUME 24, NUMBER 2 FEBRUARY, 1953

Screw Dislocations in Thin Rods*

J. D. ESHELBY
Department of Physics, University of Illinois, Urbana, Illinois
(Received June 13, 1952)

In connection with Galt and Herring's observation on thin whiskers of tin, the properties of a screw dislocation in a cylinder are worked out. When all boundary conditions are taken into account, the image force tends to keep the dislocation along the axis. Only when it is displaced about half-way to the surface does the image force tend to pull it out of the rod. Generators of the cylindrical rod become helices when the dislocation is introduced. The dislocation can be ejected from the rod by twisting or bending it suitably.

INTRODUCTION

GALT and Herring[1] have described the mechanical properties of thin whiskers of tin. Recent work on the theory of crystal growth[2] suggests that there may be a dislocation with a substantial screw component running parallel to the axis of the whisker. On the other hand, their enhanced strength might suggest that the whiskers are free from dislocations. It may therefore be useful to work out the properties of a rod containing a screw dislocation as a problem in isotropic elastic theory. The results from the theory of the torsion and bending of beams which we use may be found, for example, in Sokolnikoff's[3] book.

We first of all treat the problem as one in antiplane strain in which the displacement w is everywhere parallel to the z axis and is independent of z. The only nonvanishing stress components are τ_{zx} and τ_{zy}, which may be regarded as the x and y components of a vector $\boldsymbol{\tau}=\mu$ grad w, where μ is the shear modulus. The equilibrium condition is div $\boldsymbol{\tau}=\mu\nabla^2 w=0$.

The displacement due to a screw dislocation through the point $(x=\xi,\ y=0)$ when the surface of the infinite cylinder $x^2+y^2=R^2$ is stress-free is

$$w=\frac{b}{2\pi}\tan^{-1}\frac{y}{x-\xi}-\frac{b}{2\pi}\tan^{-1}\frac{y}{x-R^2/\xi}. \tag{1}$$

The stress is also zero on any other circle for which $(\xi,\ 0)$ and $(R^2/\xi,\ 0)$ are inverse points. Equation (1), therefore, equally well represents the displacement when the core of the dislocation is excluded by a small cylinder of radius r_0. If $r_0 \ll R$, the center of the small circle is practically at $(\xi,\ 0)$. The w of (1) is multiple-valued: we can make it single-valued by a cut joining the circles r_0 and R, most conveniently along the x axis (Fig. 1).

The elastic energy per unit length of the cylinder is

$$W=\tfrac{1}{2}\mu\int(\text{grad } w)^2 dxdy$$

$$=\tfrac{1}{2}\mu\int_C w\frac{\partial w}{\partial n}ds$$

taken round the circuit C in Fig. 1. If we let r_0 tend to zero and reject an infinite constant independent of ξ, the result is

$$W=\frac{\mu b^2}{4\pi}\ln(R^2-\xi^2).$$

Figure 2(a) shows that the dislocation is in unstable equilibrium when it is at $\xi=0$; at any other point there is an "image-force,"

$$F=-\frac{\partial W}{\partial \xi}=\frac{\mu b^2}{2\pi}\frac{\xi}{R^2-\xi^2},$$

tending to drive it out of the cylinder.

On any cross section of the cylinder there is a couple

$$M=\mu\int\left(x\frac{\partial w}{\partial y}-y\frac{\partial w}{\partial x}\right)dxdy$$

$$=\mu\int_C w(xm-yl)ds,$$

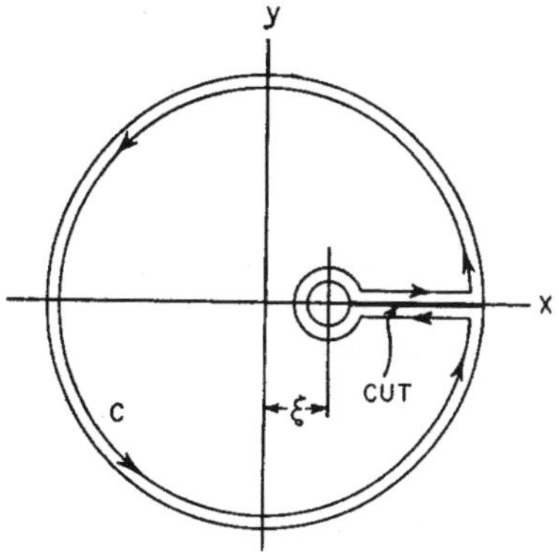

FIG. 1. Circuit used in determining couple and energy.

* Work supported by the U. S. Office of Naval Research.
[1] Conyers Herring and J. K. Galt, Phys. Rev. **85**, 1060 (1952).
[2] See the review article by F. C. Frank, Advances in Physics (Philosophical Magazine Supplement) **1**, 91 (1952).
[3] Ivan S. Sokolnikoff, *Mathematical Theory of Elasticity* (McGraw-Hill Book Company, Inc., New York, 1946).

where (l, m) is the normal to the curve C. When $r_0 \to 0$, this gives

$$M(\xi)=\tfrac{1}{2}\mu b(R^2-\xi^2).$$

We can regard the couple as being produced by the surface tractions on the ends of the cylinder at infinity which are necessary to maintain a state of antiplane strain. If we drop the antiplane strain requirement, we can get rid of M by applying a couple $-M$; this gives a twist in which two cross sections of the cylinder, unit distance apart, undergo a relative rotation

$$\alpha(\xi)=\frac{2M}{\mu\pi R^4}=\frac{b}{\pi R^2}(1-\xi^2/R^2). \qquad (2)$$

This reduces the energy by $\tfrac{1}{2}\alpha M$ per unit length, and W becomes

$$W=\frac{\mu b^2}{4\pi}\left[\ln(R^2-\xi^2)-\frac{(R^2-\xi^2)^2}{R^4}\right]. \qquad (3)$$

Figure 2(b) shows that the dislocation is now in stable (strictly, metastable) equilibrium at the center. Only if it is displaced further than

$$\xi_{\max}=(1-2^{-\frac{1}{2}})^{\frac{1}{2}}R=0.54R$$

from the center will the dislocation be pulled out of the cylinder by the image force.

If we now cut out a rod of finite length l from the infinite cylinder, there will be a distribution of surface traction on each end which has zero resultant and zero moment. According to St. Venant's principle, the effects of relaxing these surface tractions will penetrate only to a distance of order R from the ends, so that if $R/l \ll 1$ we may still take (3) as the energy per unit length of the finite cylinder. The effective force on the dislocation is now

$$F=-\frac{\partial W}{\partial \xi}=\frac{\mu b^2}{4\pi R^2}\xi\left[\frac{R^2}{R^2-\xi^2}-2\frac{R^2-\xi^2}{R^2}\right]. \qquad (4)$$

It has a maximum and a minimum

$$F=\mp 0.216\frac{\mu b^2}{4\pi R} \quad \text{at} \quad \xi=\pm 0.322R. \qquad (5)$$

The force can also be derived directly from the rule that the force on a screw dislocation running parallel to the z axis is

$$F_x=b\tau_{zy}, \quad F_y=-b\tau_{zx}, \qquad (6)$$

where the stresses are to be evaluated at the position of the dislocation after rejecting the stresses which the dislocation would produce if the medium were extended to infinity. In the present case τ_{zy} and τ_{zx} are the stresses arising from the second term in (1) and from the twist (2). The energy method we have used leads to less worry over signs.

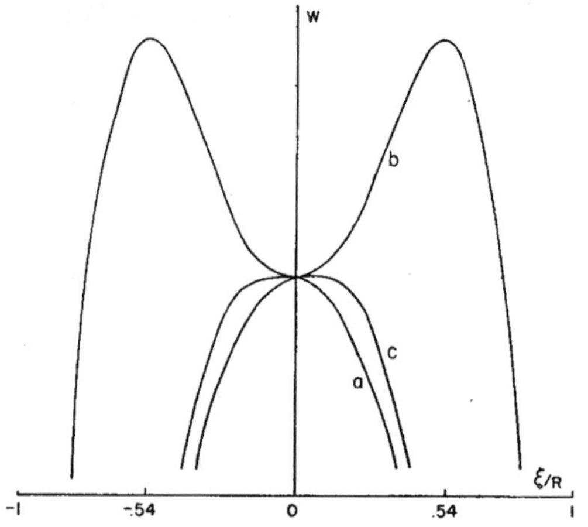

FIG. 2. Energy of a screw dislocation in a cylinder.

BEHAVIOR IN A TWISTED ROD

It is obvious that the point $\xi=0$ again becomes unstable if an external couple sufficient to undo the the twist (2) is applied. We shall show that actually a couple of only half this value is needed. Let a constant external couple M' be applied; to fix ideas, suppose that a weight is hung from a cord wound round one end of the rod. A change in the position of the dislocation will change the strain energy of the rod and raise or lower the weight. To find the equilibrium position we must find the value of ξ which minimizes the sum of the strain energy of the rod and the potential energy of the weight or other external device producing M'. This external energy is $l\alpha(\xi)M'$ in all, or

$$bM'\xi^2/\pi R^4 \qquad (7)$$

per unit length of the rod, plus a constant independent of ξ. The strain energy can be found as follows. Before applying M', it is given by Eq. (3). If Hooke's law is valid the body reacts to external forces in the same way whether it has internal stresses or not. Hence, the change in strain energy when M' is applied, being equal to the work done by M', is independent of ξ. In other words, the strain energy of the rod contains no cross term between the externally produced stress and the internal stress arising from the dislocation.[4] Thus, the total energy of the system, but for a constant, is simply the sum of (3) and (7). As M' increases, the maxima of Fig. 2(b) are lowered and shifted nearer the origin. When

$$M'=\tfrac{1}{4}\mu bR^2,$$

the maxima have moved in to the origin, swamping the minimum there and restoring instability [Fig. 2(c)].

The stress-strain curve of the rod in torsion will have the asymmetric form shown in Fig. 3, the horizontal part representing the disappearance of the twist

[4] F. R. N. Nabarro, Advances in Physics 1, Sec. II 3.5, 269 (1952).

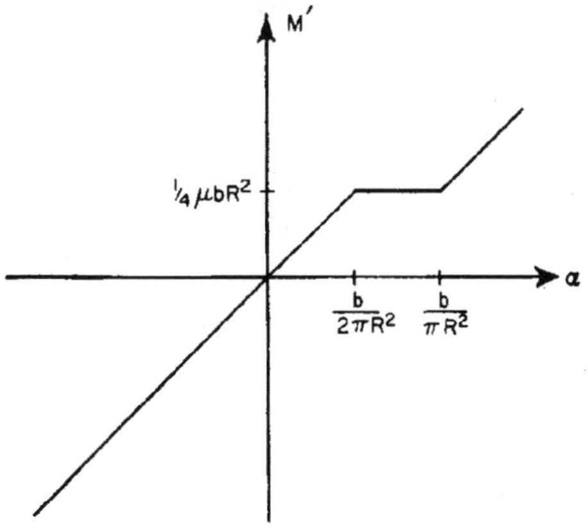

FIG. 3. Ideal torsional stress-strain curve of a whisker containing a dislocation.

when the dislocation leaves the rod. If the sign of b is changed, the curve must be inverted through the origin.

BEHAVIOR IN A BENT ROD

Because of its all-or-nothing effect we cannot use an applied couple to move the dislocation about in a controlled way. This can be done by clamping one end of the rod and deflecting the free end with a transverse load. The deflection of the free end d and the load W are connected by

$$d=\frac{4}{3\pi}\frac{Wl^3}{ER^4},$$

where E is Young's modulus. We shall work in terms of d, since in practice the end of the rod would be deflected by pushing with an adjustable probe rather than by hanging a weight on the end.

If the load acts parallel to the y axis, the force urging the dislocation along the x axis is given by (6) with τ_{zy} equal to the stress in the rod due to bending. The result is

$$F=\tfrac{3}{4}(3+2\nu)\mu b\frac{R^2d}{l^3}\left\{1-\frac{1-2\nu}{3+2\nu}\frac{\xi^2}{R^2}\right\}, \qquad (8)$$

where ν is Poisson's ratio. The term in ξ^2 vanishes when $\nu=\frac{1}{2}$, and never reaches $\frac{1}{10}$ of the constant term when $\nu=\frac{1}{3}$.

We may note in passing that for a beam of any section bent in flexure by a load along a principal axis the force on a dislocation is everywhere normal to the lines of shearing stress;[5] these lines are thus contours of an energy surface $W(x,y)$ for which $\mathbf{F}=\text{grad } W$.

To find the position of the dislocation for a given deflection, we have to solve for ξ the equation obtained by setting the sum of (4) and (8) equal to zero. As a sketch of the derivative of curve b of Fig. 2 shows, only the numerically smallest root gives a stable position. For small d

$$\xi=3(3+2\nu)\pi R^4d/bl^3.$$

If (8) exceeds the absolute value of the F of (5), there is no stable position and the dislocation leaves the rod as a result of the bending. The necessary deflection is

$$d=\frac{0.072b}{\pi(3+2\nu)}\left(\frac{l}{R}\right)^3, \qquad (9)$$

if we neglect the ξ^2 term in (8).

According to (2) the angle α varies as d (and hence ξ) is altered, so that there is a coupling between the deflection and twist of the rod.

When the dislocation leaves the rod, there will be an additional nonelastic deflection d' which remains when the load is removed. It can be found by equating Wd' to the force (8) integrated from 0 to R. This gives

$$d'=4bl/3\pi R.$$

The plastic bending giving rise to d' will occur at the clamped end of the rod (or at the root of a whisker growing on a metal surface) so that after the dislocation is expelled and the load removed the rod will be inclined at an angle

$$\theta=4b/3\pi R$$

to its original direction.

The type of bending we have just considered (flexion) must be clearly distinguished from, for example, the case in which the rod is bent into a circular arc of radius ρ. In this latter case τ_{zx} and τ_{zy} vanish, and the bending produces no force on the dislocation. However, if, following Mott and Nabarro,[6] we ascribe a line tension μb^2 to the dislocation, there will be an effective force $\mu b^2/\rho$ per unit length perpendicular to the dislocation line. The radius of curvature necessary for this to overcome the maximum image force (4) is

$$\rho=58R,$$

quite a modest degree of curvature for a thin rod. According to elementary beam theory, the maximum strain in a longitudinal filament is then $R/\rho=1.7$ percent. We may also apply this to the flexure problem: the maximum radius of curvature is at the clamped end and equal to $l^2/3d$. The deflection for the dislocation to leave the rod owing to its line tension is thus

$$d=0.0058l^2/R. \qquad (10)$$

Whether the dislocation leaves the rod because of this effect or because of the bending stresses [Eq. (9)] depends on the ratio between b/R and R/l.

Attributing a line tension to the dislocation is only a crude approximation to the solution of the difficult problem of calculating the energy of a dislocation line

[5] See, e.g., Ivan S. Sokolnikoff, reference 3, Fig. 43.

[6] Mott and Nabarro, Conference on Strength of Solids, Physical Society (London), 1948, p. 1.

as a function (or rather functional) of its shape and relation to the surface of the body containing it. Moreover, in a strongly curved lattice we must note that strictly the Burgers vector is not constant along the dislocation line, being everywhere parallel to a local lattice vector. Thus, the angle between the Burgers vector and direction of a straight dislocation running through a curved lattice will vary along its length, and so too will the energy per unit length, since it depends on this angle. As a result, the dislocation of lowest energy joining two given points may not be the straight line suggested by the line tension concept, since by deforming it into a curve the inclination of Burgers vector and tangent vector might be given a more favorable value at each point. Hence, it is not clear that the previous calculation is correct even in principle. If it is, the exact numerical results are still doubtful.

POSSIBLE EXPERIMENTS

Galt and Herring[1] found that when bent their "whiskers" behaved nonelastically at a strain of 2 or 3 percent. The conditions of bending probably did not correspond exactly with either of the cases we have discussed. The case of bending into a circular arc gives a maximum longitudinal strain of 1.7 percent. The type of deformation they found (a sharp bend through a large angle) does not agree with the deformation produced by the escape of one dislocation. The rough agreement of the theoretical and experimental critical strains may therefore be an accident, or it may indicate that the movement of the original dislocation triggers off more extensive plastic deformation. According to the calculation of Mackenzie,[7] slip of one atomic plane over another can occur without the aid of dislocations for strains of about 3 percent. Galt and Herring's results are therefore consistent either with there being no dislocation in the whisker or with the presence of a dislocation which plays no part in the deformation.

It is clear that other types of experiment would be more suitable for establishing the presence or absence of a dislocation in the whisker. When the dislocation is along the axis, the twist (2) amounts to about 50° per cm for $b=3\times10^{-8}$ cm and $R=10^{-4}$ cm. It should be possible to detect this by x-ray diffraction. However, it is not merely a question of a slow variation of lattice orientation along the length of the specimen. The relative rotation of the extreme cross sections illuminated by a beam of reasonable width will be quite large. In addition to the twist, there is the displacement $w=(b/2\pi)\tan^{-1}(y/x)$. The diffraction pattern caused by the latter alone has been calculated by Wilson.[8]

It might also be possible to measure the torsional stress-strain curve of a whisker, using a small superimposed tension to prevent the specimen raveling up. For a specimen 1 cm long our simple model suggests that the dislocation should be ejected after one end of the specimen has been twisted through about 25° relative to the other, giving rise to a further 25° of plastic twist (Fig. 3).

We have assumed that the dislocation can glide in any plane containing its Burgers vector. In practice, certain crystallographic planes will be preferred. Such a plane will be itself twisted so that the way in which the dislocation leaves the rod will be rather complicated. Moreover, the movement of the original dislocation may trigger off more extensive plastic deformation. Still, we might expect that nonelastic behavior will begin at an angle of twist of about 25° (or, in general, $bl/2\pi R^2$).

CONCLUSIONS

Contrary to what one might expect, the image force on a screw dislocation parallel to the axis of a long rod tends to drive it towards the axis so long as its distance from the axis is less than 0.54 of the radius of the rod. Under deformations of the rod below a certain limit it will therefore not tend to leave the rod. Hence, from the increased strength of metallic whiskers we cannot immediately conclude that they are free from dislocations. Suitable torsion and flexion experiments might reveal their presence.

The author would like to thank Professor J. S. Koehler for helpful discussions.

[7] J. K. Mackenzie, thesis, Bristol (1949).

[8] Wilson, Research 2, 541 (1949).

The Equation of Motion of a Dislocation*

J. D. ESHELBY
Department of Physics, University of Illinois, Urbana, Illinois

(Received November 28, 1952)

The elastic field surrounding an arbitrarily moving screw dislocation is found, and a useful analogy with two-dimensional electromagnetic fields is pointed out. These results are applied to a screw dislocation accelerating from rest and approaching the velocity of sound asymptotically. The applied stress needed to maintain this motion is found on the assumption that the Peierls condition is satisfied near the center of the dislocation. A general integral equation of motion is derived for a simplified dislocation model, and the kind of behavior it predicts is illustrated.

I. INTRODUCTION

AN extensive literature exists dealing with the properties of dislocations at rest (see the recent review article by Nabarro[1]). There has been much less discussion of their motion, perhaps because it is still uncertain whether it is governed by friction or inertia. Though at present most authors seem to suppose that dissipative processes are dominant, some still adopt the dynamic point of view initiated by Frank.[2] It therefore seems worth while to investigate the dynamical behavior of dislocations on the assumption that dissipative effects are negligible.

Ideally, we should discuss the change in shape with time of a dislocation loop in an applied stress field. Here we only consider the plane problem of an infinite straight dislocation. As a further simplification we shall suppose it is a pure screw dislocation. The problem is then one in antiplane strain, and only a single velocity of sound is involved instead of the two which would appear if there were an edge component. (We assume the medium is isotropic). Frank has shown[3] that uniformly moving screw dislocations behave in a manner reminiscent of particles in the special theory of relativity (the velocity of transverse elastic waves taking the place of that of light), and we shall sometimes find it convenient to use relativistic terminology.

In Secs. II and III we find the elastic field surrounding an arbitrarily moving screw dislocation and show how it is related to the electromagnetic field of a moving line-charge. This is illustrated (Sec. IV) by the case of motion from rest with constant proper acceleration, and the applied field necessary to maintain this particular motion is found by requiring the Peierls condition to be satisfied. In Sec. V a general equation of motion is derived for a simplified model of a dislocation, and the kind of behavior it predicts is exemplified in Sec. VI.

II. THE DISPLACEMENT ROUND A MOVING SCREW DISLOCATION[4,5]

The expression[6]

$$w=\tfrac{1}{4}blkH_1^{(2)}(kr)\sin\theta\, e^{i\omega t},\quad k=\omega/c, \tag{1}$$

is the time-dependent part of the displacement field around a screw dislocation oscillating along the x axis, the position of its center being

$$\xi=le^{i\omega t}. \tag{2}$$

As a solution of the wave equation, $c^2\nabla^2 w-\partial^2 w/\partial t^2=0$, Eq. (1) is characterized by the boundary conditions

$$\lim_{\epsilon\to 0} w(x;\, y=\pm\epsilon)=\pm\tfrac{1}{2}bl\delta(x)e^{i\omega t},$$

so that it represents a region of alternating slip across the x axis concentrated at the origin, the product of area and amplitude of slip being bl, in the limit when $l=\text{const}b^{-1}\to 0$.

Let us apply to both (1) and (2) the operator

$$\frac{1}{2\pi i}\int_C(\cdots)\frac{d\omega}{\omega}, \tag{3}$$

where C is the real ω-axis indented below the origin. From (2) we get

$$\xi=lH(t),\quad \text{where}\quad H(t)=\begin{matrix}1\\0\end{matrix}\quad\text{for}\quad t\gtrless 0.$$

To find the associated displacement we replace the Bessel function in (1) by the integral representation

$$H_1^{(2)}(kr)=-\frac{2}{\pi}\int_1^\infty \frac{e^{-ikrv}}{\sqrt{(v^2-1)}}vdv,$$

apply Eq. (3), and interchange the order of integration. Since

$$\frac{1}{2\pi}\int_C e^{i\omega z}d\omega=\delta(z),$$

* Part of this work was carried out at the H. H. Wills Physical Laboratory, University of Bristol, Bristol, England; the rest under U. S. Office of Naval Research contract.

[1] F. R. N. Nabarro, Advances in Phys. **1**, 271 (1952).

[2] F. C. Frank, *Report of a Conference on the Strength of Solids* (Physical Society, London, 1948), p. 46. [*Added in proof:* See also N. F. Mott, Phil. Mag. **43**, 1151 (1952); Fisher, Hart, and Pry, Phys. Rev. **87**, 958 (1952).]

[3] F. C. Frank, Proc. Phys. Soc. (London) **A62**, 131 (1949).

[4] F. R. N. Nabarro, Phil. Mag. **42**, 1224 (1951).

[5] J. D. Eshelby, Phil. Trans. Roy. Soc. **A244**, 87 (1951).

[6] J. D. Eshelby, Proc. Roy. Soc. (London) **A197**, 396 (1949).

we find

$$w=\frac{bl}{2\pi r}\sin\theta\int_1^\infty\frac{\delta(v-ct/r)}{\sqrt{(v^2-1)}}vdv$$

$$=\frac{bl}{2\pi r}\sin\theta\frac{ct}{\sqrt{(c^2t^2-r^2)}}H(ct-r).$$

If the jump occurs along the line $\theta=\theta_0$ instead of $\theta=0$, we have only to replace θ by $\theta-\theta_0$. A continuous motion in which the position of the center of the dislocation is given by

$$x=\xi(t),\quad y=\eta(t)$$

can be looked upon as a series of jumps from (ξ,η) to $(\xi+\dot{\xi}dt,\eta+\dot{\eta}dt)$ in successive intervals of time dt. Hence the displacement produced by the moving dislocation is

$$w=\frac{b}{2\pi}\int_{-\infty}^{\tau_0}\frac{c(t-\tau)\{(y-\eta)\dot{\xi}-(x-\xi)\dot{\eta}\}}{(x-\xi)^2+(y-\eta)^2}\frac{d\tau}{s}. \quad (4)$$

Here ξ and η are functions of τ, while

$$s^2=c^2(t-\tau)^2-(x-\xi)^2-(y-\eta)^2,$$

and τ_0 is that root of $s^2=0$ which is less than t. τ_0 is unique if the velocity of the dislocation is always less than c, which we assume is the case.

III. AN ELECTROMAGNETIC ANALOGY

Physically more important than w itself are its time-derivative $\dot{w}$ and the stresses $p_{zx}=\mu\partial w/\partial x$, $p_{zy}=\mu\partial w/\partial y$. ($\mu$ is the shear modulus.) They are rather hard to derive from (4) by differentiation, since this yields an infinite term from the variation of the upper limit plus a divergent integral. We may take an arbitrary value for τ_0 and follow the differentiation by an integration by parts. When τ_0 approaches its correct value the term from the variation of the upper limit cancels the integrated part, leaving a finite result.

For the time-derivative of (1) we may write

$$\dot{w}=-\frac{\partial}{\partial y}\{\tfrac{1}{4}ibl\omega H_0^{(2)}(kr)e^{i\omega t}\},$$

and repeating for the zero-order Bessel function the argument leading from (1) to (4), we find, for a dislocation moving along the x axis,

$$\dot{w}=-\frac{\partial I}{\partial y},\quad\text{where}\quad I=\frac{bc}{2\pi}\int_{-\infty}^{\tau_0}\dot{\xi}\frac{d\tau}{s}.$$

This suggests that we might be able to derive the velocity and stresses from a set of potentials involving simpler integrals than that for w. The substitution

$$r=c(t-\tau),\quad z^2=r^2-(x-\xi)^2-y^2$$

changes I into the form

$$I=\frac{b}{4\pi c}\int_{-\infty}^{\infty}\left[\frac{\dot{\xi}}{r-(x-\xi)\dot{\xi}/c}\right]_{\tau=t-r/c}dx.$$

If we identify the velocities of light and sound, this is seen to be the x component of the vector potential of an electrical line-charge moving in the same way as the dislocation.

There is in fact a close resemblance between antiplane strain elastic problems and electromagnetic fields in which all quantities are independent of the z coordinate. Consider the case where $E_z=H_x=H_y=0$, and make the identification

$$\dot{w}=H_z/\sqrt{\rho},\quad p_{zx}=-E_y\sqrt{\mu},\quad p_{zy}=E_x\sqrt{\mu}. \quad (5)$$

(We use Heaviside units.) Then all Maxwell's equations are satisfied identically except $\text{curl}\,\mathbf{E}+(1/c)(\partial\mathbf{H}/\partial t)=0$ which becomes

$$\partial p_{zx}/\partial x+\partial p_{zy}/\partial y=\rho\partial^2 w/\partial t^2,\quad\text{or}$$

$$c^2\nabla^2 w-\partial^2 w/\partial t^2=0, \quad (6)$$

the elastic equilibrium or wave equation. If there is a moving line-charge with linear density λ, Gauss's relation for the conservation of charge states that

$$\oint\mathbf{E}\cdot d\mathbf{s}=\oint(lE_x+mE_y)ds=\lambda\quad\text{or}\quad 0, \quad (7)$$

according as the circuit does or does not enclose the charge. Using (5) this becomes

$$\oint\left(l\frac{\partial w}{\partial y}-m\frac{\partial w}{\partial x}\right)ds=\oint\frac{\partial w}{\partial s}ds=\frac{\lambda}{\sqrt{\mu}}\quad\text{or}\quad 0, \quad (8)$$

showing that the elastic counterpart of the charge is a dislocation with a Burgers vector of magnitude $\lambda/\sqrt{\mu}$ directed along the z axis.

With the aid of this analogy we can complete the scheme for deriving the velocity and stresses from a set of potentials:

$$p_{zy}=-\mu^{\frac{1}{2}}\left(\frac{\partial\varphi}{\partial x}+\frac{1}{c}\frac{\partial A_x}{\partial t}\right),\quad p_{zx}=\mu^{\frac{1}{2}}\left(\frac{\partial\varphi}{\partial y}+\frac{1}{c}\frac{\partial A_y}{\partial t}\right),$$

$$\dot{w}=\rho^{-\frac{1}{2}}\left(\frac{\partial A_y}{\partial x}-\frac{\partial A_x}{\partial y}\right), \quad (9a)$$

where

$$\{\varphi,A_x,A_y\}=\frac{b\mu^{\frac{1}{2}}}{2\pi c}\int_{-\infty}^{\tau_0}\{c,\dot{\xi}(\tau),\dot{\eta}(\tau)\}\frac{d\tau}{s}. \quad (9b)$$

It is easy to verify that if we make the translation expressed by (5), the electromagnetic energy density and Poynting's vector become the elastic energy density and the elastic energy-flow vector:[7]

$$S_x=p_{zx}\dot{w},\quad S_y=p_{zy}\dot{w}. \quad (10)$$

The electromagnetic momentum density becomes not the ordinary elastic momentum density $\rho\dot{w}$ but the quasi momentum used, for example, in discussing the

[7] A. E. H. Love, *Mathematical Theory of Elasticity* (Cambridge University Press, Cambridge, 1927).

collisions of phonons with one another and with electrons. For a uniformly moving screw dislocation this quasi momentum agrees with the effective momentum introduced by Frank.[3] The Maxwell tensor becomes the "Maxwell tensor of elasticity."[5] The Lorentz force on the line-charge becomes

$$F_x = b(p_{zy} + \dot{w}v_y), \quad F_y = -b(p_{zx} + \dot{w}v_y),$$

where v is the velocity of the dislocation and $\dot{w}$, p_{zx}, p_{zy} refer to an applied stress-field. The terms independent of v are the orthodox expressions for the force on a stationary dislocation. Nabarro[8] has shown that the terms in v are physically significant. He has also pointed out that in the scattering of sound waves by a screw dislocation, the force and quasi momentum are related in the same way as ordinary force and momentum.

If there is a charge density σ in the electromagnetic problem, the equation $\mathrm{div}\mathbf{E} = \sigma$ becomes

$$\partial e_{zy}/\partial x - \partial e_{zx}/\partial y = \mu^{-\frac{1}{2}}\sigma, \tag{11}$$

in terms of the strains $e_{zx} = p_{zx}/\mu$, $e_{zy} = p_{zy}/\mu$. A continuous distribution of σ corresponds to a state of antiplane self-stress. A general state of self-stress can be specified[5] by a tensor S_{ij}; for antiplane strain the non-vanishing components are

$$S_{zx} = -\mu^{-\frac{1}{2}}\partial\sigma/\partial y, \quad S_{zy} = \mu^{\frac{1}{2}}\partial\sigma/\partial x.$$

A varying S_{ij} formally represents a state of plastic flow.

In place of (5) we might have taken an electromagnetic field in which $E_x = E_y = H_z = 0$ and made the identification,

$$\dot{w} = -E_z/\sqrt{\rho}, \quad p_{zx} = -H_y\sqrt{\mu}, \quad p_{zy} = H_x\sqrt{\mu}. \tag{12}$$

In place of (6) and (11) we should have

$$\partial p_{zx}/\partial x + \partial p_{zy}/\partial y + i_z/c = \rho\partial^2 w/\partial t^2, \quad \partial e_{zy}/\partial x - \partial e_{zx}/\partial y = 0,$$

where i_z, the current in the z direction, has to be equated to c times the body force (which must be parallel to the z axis to preserve the antiplane strain character). In contradistinction to (5) the correspondence (12) is adapted to problems with body force but no self-stress. We can, however, simulate the stress field of a dislocation with the help of a double layer of force along the x axis extending from the center of the dislocation to infinity. The electromagnetic field of the corresponding double current sheet can be derived from a vector potential $(0, 0, A_z)$. However, A_z is equal to w and the analogy is too good to be useful.

A possible third analogy is suggested by the static case. It is natural to identify the displacement of a stationary screw dislocation with the scalar potential of a current-carrying wire; (p_{zx}, p_{zy})—which transforms like a vector for rotation about the z axis—is then proportional to (H_x, H_y). However, this analogy cannot be extended to the time-dependent case since the equilibrium equation $\mathrm{div}(p_{zx}, p_{zy}) = \rho\partial^2 w/\partial t^2$ must come from one of Maxwell's curl equations, which alone contain time derivatives. Hence, the identification of the stress with **E** or **H** must always have the crosswise character of Eqs. (5) and (12).

IV. A DISLOCATION WITH CONSTANT PROPER ACCELERATION

As an example we consider a case of accelerated motion. Since a uniform acceleration would finally bring the velocity above c, we consider a dislocation at rest for negative t and thereafter moving with constant acceleration in its own rest system:

$$\begin{aligned} &\xi = x_0, \quad \eta = 0 && \text{for } t < 0; \\ &\xi^2 - c^2 t^2 = x_0^2, \quad \eta = 0 && \text{for } t > 0. \end{aligned} \tag{13}$$

For $t > 0$ there are the simple relations

$$\dot{\xi} = c^2 t/\xi, \quad \beta \equiv \sqrt{(1 - \dot{\xi}^2/c^2)} = x_0/\xi, \quad \partial^2\xi/\partial t^2 = \beta^3 c^3/x_0.$$

The dislocation starts with an acceleration c^2/x_0, conveniently specified by its initial x coordinate x_0, and approaches the velocity c asymptotically.

It is convenient to use the abbreviation

$$R^2 = x^2 - c^2 t^2,$$

where x, t refer to the point and time of observation. Then at any time points near the dislocation line are distinguished by small $x_0 - R$ and small y.

Choosing units in which $c = 1$, we have to evaluate (9b) with

$$\begin{aligned} &s^2 = \tau^2 - 2t\tau + s_0^2, \quad \dot{\xi} = \dot{\eta} = 0 && \text{for } \tau < 0; \\ &s^2 = 2x\xi - 2t\tau - R^2 - y^2 - x_0^2, \quad \dot{\xi} = \tau/\xi, \quad \dot{\eta} = 0 && \text{for } \tau > 0; \end{aligned}$$

where

$$s_0 = \sqrt{(2xx_0 - R^2 - y^2 - x_0^2)}$$

is the interval between $(x_0, 0, 0)$ and (x, y, t). The integral for $\tau < 0$ is elementary; we omit an infinite constant. The integrals for $\tau > 0$ yield to the substitution

$$\tan^2\psi = s^2/\{(R - x_0)^2 + y^2\},$$

giving

$$\begin{aligned} \varphi &= b\mu^{\frac{1}{2}}/2\pi(x_0/R)^{\frac{1}{2}}(2/kR)\{tk'\tan\psi - x[\Delta\tan\psi \\ &\qquad + \tfrac{1}{2}(1 + k'^2)F - E]\} + (b/2\pi)\ln(t + s_0), \\ A_x &= b\mu^{\frac{1}{2}}/2\pi(x_z/R)^{\frac{1}{2}}(2/kR)\{xk'\tan\psi - t[\Delta\tan\psi \\ &\qquad + \tfrac{1}{2}(1 + k'^2)F - E]\}, \quad A_y = 0, \end{aligned} \tag{14}$$

where with the usual notation for elliptic integrals

$$\Delta = \sqrt{(1 - k^2\sin^2\psi)}, \quad F(\psi, k) = \int_0^\psi \frac{d\psi}{\Delta}, \quad E(\psi, k) = \int_0^\psi \Delta d\psi$$

and

$$\tan\psi = s_0 k/k'\sqrt{(4x_0 R)}, \quad k^2 = 1 - k'^2 = 4x_0 R/\{(R + x_0)^2 + y^2\}.$$

The displacement can be expressed in terms of elliptic integrals of the first and third kinds. Equation (14) is only valid when

$$c^2 t^2 - (x - x_0)^2 - y^2 > 0, \quad t > 0, \quad x > ct,$$

but this is all we shall need.

[8] F. R. N. Nabarro, Proc. Roy. Soc. (London) **A209**, 278 (1951).

Near the center of the dislocation p_{zy} is given approximately by

$$p_{zy}=\frac{\mu b}{2\pi}\left\{\frac{x'}{\beta r'^2}-\frac{1}{2x_0}\ln\frac{8x_0}{r'}\left[\frac{2x_0}{s_0{}^2}(t-s_0)\right]-\frac{t-s_0}{x_0 s_0}-\frac{s_0}{2x_0(t+s_0)}\right\}, \quad (15)$$

where now

$$s_0=\surd\{t^2-[x-\xi(t)]^2\}$$

is the interval from $(x_0, 0, 0)$ to the center of the dislocation at time t and

$$x'=x-\xi(t), \quad r'^2=x'^2/\beta^2(t)+y^2.$$

r' is thus the radial distance from the center of the dislocation measured in its instantaneous rest-coordinates. Equation (14) is subject to the limitations

$$r'\ll x_0, \quad s_0\gg r'. \quad (16)$$

We shall be interested in values of r' of the order of a few lattice spacings. The first condition then means that the velocity acquired by the dislocation in the time that sound travels one lattice spacing shall be small compared with the velocity of sound. Otherwise expressed, the proper acceleration of the dislocation must be small compared with the acceleration (about 10^{18} cm/sec^2) of an atom oscillating with an amplitude of one lattice spacing at the frequency of a lattice vibration near the Debye limit. The second condition requires that the diameter of the disturbed region, $(x-x_0)^2+y^2\leqslant t^2$, which spreads from the starting position of the dislocation shall be large compared with the lattice spacing. Since $s_0\sim t$ for small t, this means that t must be large compared with the period of a lattice vibration.

To find the applied stress necessary to produce the motion (13), we have to introduce nonlinearity into the problem. Following Nabarro[8] we shall try to satisfy the Peierls law relating stress and displacement at the slip plane. The Peierls law appropriate to a screw dislocation with Burgers vector b and separation a between atomic planes parallel to the slip plane is

$$p_{zy}(y=\tfrac{1}{2}a)=-\mu/2\pi b/a\,\sin 4\pi/bw(y=\tfrac{1}{2}a). \quad (17)$$

This is satisfied by the elastic solution for a screw dislocation which has been moving uniformly for all time with velocity v:

$$w=\frac{b}{2\pi}\tan^{-1}\frac{y\surd(1-v^2/c^2)}{(x-vt)}.$$

Now, the first term in (15) is simply the stress produced by a uniformly moving dislocation which at time t happens to coincide in position and velocity with the accelerated dislocation. For the accelerated dislocation the condition (17) would be satisfied if we could impress on every point of the material a displacement equal to the difference between the displacements of the uniformly moving and accelerated dislocations. The nonlinear behavior is confined to the neighborhood of the center of the dislocation, and so we may hope that it will be enough if the impressed displacement is correct at least in this region. On the planes $y=\pm\frac{1}{2}a$ the last three terms of (15) are slowly varying functions of x' near $x'=0$, in view of (16). Hence the impressed displacement required can be produced by a uniform applied stress,

$$p_{zy}{}^A(t)=\frac{\mu b}{2\pi x_0}\left\{\frac{1}{2}\ln\left(\frac{32x_0{}^2}{as_0}\,\frac{t-s_0}{s_0}\right)+\frac{s_0}{2(t+s_0)}+\frac{t-s_0}{s_0}\right\}, \quad (18)$$

equal and opposite to the last three terms in (15) taken at the point $x'=0$, $y=\frac{1}{2}a$.

When (18) is multiplied through by b, it becomes a relation between the force on the dislocation and its acceleration. Formally, b and a are independent parameters corresponding to charge and diameter of charge in the associated electromagnetic problem. It is, therefore, natural to interpret the logarithmic term, which diverges when a approaches zero, as the effective mass of the dislocation. In ordinary units,

$$F_x=bp_{zy}{}^A=(1-\dot{\xi}^2/c^2)^{-\frac{3}{2}}(\rho b^2/4\pi)\{\ln f(t)\}\partial^2\xi/\partial t^2+g(t),$$

where $f(t)$ is the argument of the logarithm in (18) and $g(t)$ is b times the second and third terms. This has the form of the relativistic equation of motion of a particle with a slowly varying rest mass $(\rho b^2/4\pi)\ln f(t)$ and a radiation reaction term $g(t)$. It has already been suggested by Frank[3] that a rest mass $(\rho b^2/4\pi)\ln(r_1/a)$ can be ascribed to a screw dislocation, where r_1 is a not very well defined length. In the present case, $f(t)=8ct/a$ for small t and r_1 is, not unreasonably, of the order of the disturbed region surrounding the dislocation. As t increases, r_1 is a complicated function of the two lengths x_0 and ct. For large t, the dislocation takes up a position at a distance x_0 behind the leading edge of the disturbed region, and r_1 becomes $16x_0$. The second term in (18) comes from the part of the integral for φ with $\tau<0$, and may be associated with the discontinuity in $\partial^2\xi/\partial t^2$ at $t=0$. It approaches zero as t increases. The last term in (18) increases as $\surd t$; we may perhaps connect it with the fact that the accelerating dislocation is continually catching up the radiation it has already emitted.

It is clear that, if a is of the same order as b and x_0/b is reasonably large, the stress necessary to maintain the motion is given approximately by

$$p_{zy}{}^A/\mu=(b/4\pi x_0)\ln(x_0/b), \quad (19)$$

except at the beginning of the motion and in the extreme relativistic region. If $p_{zy}{}^A/\mu=10^{-4}$, (19) gives $x_0\sim 10^4 b$. In covering a distance $10^4 b$, the dislocation acquires 87 percent of the velocity of sound, in agreement with the calculations of Frank[9] and Leibfried and Dietze.[10]

V. THE GENERAL EQUATION OF MOTION

In principle the rectilinear motion of a dislocation in a given applied stress field could be found as follows. We generalize the Peierls-Nabarro equation to include time-dependent states. This will give $p_{zy}(x,t)$ on $y=\pm\frac{1}{2}a$ as an integral involving a general discontinuity $\delta w(x,t)$ in displacement across the slip plane. The stress so found at any point is to be equated to a prescribed function of δw at the same point. Finally, we have to find a solution of the resulting integral equation which has the character of a moving dislocation superimposed on the required applied stress field.

The generalized Peierls-Nabarro equation is easily set up. From the results of Sec. II it follows that, if $\delta w=H(x)H(t)$, then

$$p_{zy}=-\mu\surd(c^2t^2-x^2)H(ct-|x|)/2\pi ctx$$

at the slip plane. A general $\delta w(x,t)$ can be written as

$$\delta w(x,t)=\int_{-\infty}^{\infty}dx'\int_{-\infty}^{\infty}d\tau\frac{\partial^2\delta w(x',\tau)}{\partial x'\partial\tau}H(x-x')H(t-\tau),$$

and the corresponding stress will be

$$p_{zy}=-\frac{\mu}{2\pi}\int_{-\infty}^{\infty}dx'\int_{-\infty}^{\infty}d\tau\frac{\partial^2\delta w(x',\tau)}{\partial x'\partial\tau}\times\frac{\surd\{c^2(t-\tau)^2-(x-x')^2\}H\{c(t-\tau)-|x-x'|\}}{(x-x')c(t-\tau)}. \quad (20)$$

Integrating by parts with respect to τ, we have the generalized Peierls-Nabarro equation

$$f[\delta w(x,t)]=-\frac{\mu}{2\pi}\int_{-\infty}^{\infty}dx'\int_{-\infty}^{t-|x-x'|/c}d\tau\frac{\partial\delta w(x',\tau)}{\partial x'}\times\frac{x-x'}{c(t-\tau)^2\surd\{c^2(t-\tau)^2-(x-x')^2\}}, \quad (21)$$

where f is a function with period b reducing to $\mu\delta w/a$ for small δw.

The possibility of finding interesting solutions of (21) seems remote. Part of the difficulty lies in the fact that the solution would give us more than we need, the shape of the dislocation, specified by $\delta w(x,t)$, instead of merely its position, defined for example as the point where $\partial\delta w/\partial x$ has a maximum or minimum. As a first step we shall try to find an approximate equation of motion in the following way. The shape of the dislocation will be assumed to be independent of its motion, i.e.,

$$\delta w(x,t)=\delta w[x-\xi(t)].$$

We find the rate of flow $W(t)$ of energy into the slip plane. If W is known (for example if there are no dissipative processes and no accumulation of energy in the gap between the elastic half-planes of the Peierls model, so that $W=0$) this gives an integral equation connecting ξ with the applied stress. According to (10),

$$W=\int_{-\infty}^{\infty}(p_{zy}{}^A+p_{zy})\delta\dot{w}dx, \quad (22)$$

where $p_{zy}{}^A$ refers to the applied field and p_{zy} to the field of the dislocation. We assume that $p_{zy}{}^A$ depends on t but not on x. The first term in the integral is simply $bp_{zy}{}^A$, since we must have

$$\int_{-\infty}^{\infty}\frac{\partial\delta w}{\partial x}dx=-b$$

if the dislocation is to have strength b. In principle p_{zy} could be found from (20), but the following method makes it clearer what choice of δw will lead to a simple result. Returning to the electromagnetic analogy, suppose that there is a charge distribution,

$$\sigma=\eta\{x-\xi(t)\}\delta(y),$$

which is confined to the plane $y=0$ and which moves rigidly with velocity $\dot{\xi}(t)$. The current is $i_x=\eta\delta(y)\dot{\xi}$ and the potentials satisfy

$$\nabla^2\varphi-\ddot{\varphi}=-\eta\delta(y),\quad \nabla^2A_x-\partial^2A_x/\partial t^2=-\eta\delta(y)\dot{\xi}.$$

(We use units in which $c=1$.) The discontinuity in H_z across the x axis is equal to the strength of the current sheet $\eta\dot{\xi}$. Using Eq. (5) the elastic interpretation is that there is a discontinuity δw across the x axis for which

$$\delta\dot{w}=\eta\dot{\xi}/\surd\rho \quad\text{or}\quad \partial\delta w/\partial x=-\eta/\surd\rho.$$

The latter quantity does represent a "dislocation density," in the sense that a continuous distribution of infinitesimal dislocations along the x axis with total strength

$$db=-(\partial\delta w/\partial x)dx \quad (23)$$

between the points x and $x+dx$ would give the same elastic field. It is clear that the stresses will be unaltered if, to conform to the Peierls model, the two half-spaces $y\gtrless 0$ are separated by a gap of width a provided the same δw is maintained by the interatomic forces.

By (9a) the stress satisfies

$$\nabla^2p_{zy}-\ddot{p}_{zy}=\mu^{\frac{1}{2}}\{(1-\dot{\xi}^2)\partial\eta/\partial x+\eta d^2\xi/dt^2\}\delta(y). \quad (24)$$

In terms of the Fourier transforms,

$$\bar{p}(\mathbf{k})=\frac{1}{2\pi}\int_{\mathbf{r}}p_{zy}(\mathbf{r})e^{i\mathbf{k}\cdot\mathbf{r}}d\mathbf{r},$$

$$\bar{\eta}(\mathbf{k})=\bar{\eta}(k_x)=\frac{1}{2\pi}\int_{\mathbf{r}}\eta(x)\delta(y)e^{i\mathbf{k}\cdot\mathbf{r}}d\mathbf{r}=\frac{1}{2\pi}\int_{-\infty}^{\infty}\eta(x)e^{ik_xx}dx$$

[9] F. C. Frank, Pittsburgh Symposium on Plastic Deformation, p. 89 (U. S. Office of Naval Research, 1950).
[10] G. Leibfried and H.-D. Dietze, Z. Physik **126**, 781 (1949).

(so that $\bar{\eta}$ is the transform of $\eta\delta(y)$, not of η), Eq. (24) becomes the ordinary differential equation,

$$k^2\bar{p}+d^2\bar{p}/dt^2=-\mu^{\frac{1}{2}}\{d^2\xi/dt^2-ik_x(1-\dot{\xi}^2)\}\bar{\eta}(k_x,t),$$

with the solution:

$$\bar{p}=-\mu^{\frac{1}{2}}\int_{-\infty}^{t}\left\{\frac{d^2\xi(\tau)}{d\tau^2}-ik_x[1-\dot{\xi}^2(\tau)]\right\}\times\bar{\eta}(k_x,\tau)\frac{\sin k(t-\tau)}{k}d\tau. \quad (25)$$

It is easy to show that

$$\bar{\eta}(k_x,t)=\bar{\eta}(k_x,0)e^{ik_x\xi(t)},$$

assuming for convenience that $\xi(t)=0$ for $t=0$. According to Parseval's theorem,

$$\int_{\mathbf{r}}f_1(\mathbf{r})f_2(\mathbf{r})d\mathbf{r}=\int_{\mathbf{k}}\bar{f}_1(\mathbf{k})\bar{f}_2{}^*(\mathbf{k})d\mathbf{k},$$

if $\bar{f}_n$ is the Fourier transform of f_n. Applying this to (25) we have

$$\int_{-\infty}^{\infty}p_{zy}\delta\dot{w}dx=\rho^{\frac{1}{2}}\int_{\mathbf{r}}p_{zy}\eta\dot{\xi}\delta(y)d\mathbf{r}=\rho^{\frac{1}{2}}\dot{\xi}(t)\int_{\mathbf{k}}\bar{p}\bar{\eta}^*d\mathbf{k}$$

$$=\dot{\xi}(t)\int_{-\infty}^{t}d\tau\int_{\mathbf{k}}d\mathbf{k}\left\{\frac{d^2\xi(\tau)}{d\tau^2}-ik_x[1-\dot{\xi}^2(\tau)]\right\}|\bar{\eta}(k_x,0)|^2\times\exp\{ik_x[\xi(\tau)-\xi(t)]\}\frac{\sin k(t-\tau)}{k},$$

which gives the equation of motion,

$$bp_{zy}{}^A=\frac{W(t)}{\dot{\xi}(t)}+\pi\int_{-\infty}^{t}d\tau\int_{-\infty}^{\infty}dk_x\left\{\frac{d^2\xi(\tau)}{d\tau^2}-ik_x[1-\dot{\xi}^2(\tau)]\right\}\times\exp\{ik_x[\xi(\tau)-\xi(t)]\}|\bar{\eta}(k_x,0)|^2J_0\{k_x(t-\tau)\}. \quad (26)$$

We have now to choose an expression for δw. A physically reasonable form is

$$\delta w=-\frac{b}{\pi}\tan^{-1}\frac{\frac{1}{2}a}{x-\xi(t)},$$

since it satisfies the Peierls equation approximately for small velocities. Then we have

$$\eta(x,0)=\frac{b\rho^{\frac{1}{2}}}{2\pi}\frac{a}{x^2+\frac{1}{4}a^2},\quad \bar{\eta}(k_x,0)=\frac{b\rho^{\frac{1}{2}}}{2\pi}e^{-\frac{1}{2}a|k_x|}. \quad (27)$$

The expression (26) can then be evaluated by expanding exp{ } in a power series and using the relation

$$\int_0^{\infty}e^{-zv}J_0(\rho v)v^ndv=\frac{n!}{(\rho^2+z^2)^{\frac{1}{2}n+\frac{1}{2}}}P_n\left\{\frac{z}{(\rho^2+z^2)^{\frac{1}{2}}}\right\}.$$

The result with $n=0$ is well known; regarding it as the potential of a point charge in cylindrical coordinates we reach the general case by differentiating n times with respect to z and comparing with the potential of a linear multipole in polar coordinates,

$$\partial^n r^{-1}/\partial z^n=(-1)^n n!r^{-n-1}P_n(\cos\theta)\quad\text{with}\quad r^2=\rho^2+z^2,\ z=r\cos\theta.$$

Because our physical assumptions are invalid for large velocities, it would be pointless to carry the expansion beyond terms of order $\dot{\xi}/c$. In addition the higher terms give a nonlinear relation between the stress and ξ and its derivatives. Because of the dependence on past history we must be content to give the remainder in terms of $V(t)$, the greatest value of $|\dot{\xi}(\tau)|$ for $\tau\leqslant t$. The result is

$$bp_{zy}{}^A=W(t)/\dot{\xi}(t)+\frac{\rho b^2}{4\pi}\int_{-\infty}^{t}\frac{d^2\xi(\tau)/d\tau^2}{\{(t-\tau)^2+a^2/c^2\}^{\frac{1}{2}}}d\tau+\frac{\rho b^2}{8\pi}\int_{-\infty}^{t}\frac{a^2/c^2}{\{(t-\tau)^2+a^2/c^2\}^{\frac{3}{2}}}\frac{d}{d\tau}\frac{\xi(t)-\xi(\tau)}{t-\tau}d\tau+O[V^2(t)/c^2]. \quad (28)$$

In estimating the remainder we use the fact that $[\xi(t)-\xi(\tau)]/(t-\tau)$, the mean velocity between t and τ, must be numerically less than $V(t)$ and that $|P_n(x)|\leqslant 1$.

In the electromagnetic analogy, Eq. (22) divided through by $\dot{\xi}$ expresses the fact that the total force on a rigidly moving charge distribution balances a retarding force $-W/\dot{\xi}$. Thus our method is equivalent to that of Lorentz for the electron. We could use the same argument for the dislocation provided we admit that there is a force $(p_{zy}{}^A+p_{zy})db$ on each element of the distribution (23) according to the usual rule. The analogy with Lorentz' method suggests that we might apply to a point dislocation [$a\to 0$ in Eq. (27)] recent methods which give an equation of motion for a point electron. This is not so, however. First, these methods introduce advanced potentials either explicitly or surreptitiously, and whatever they may signify in electrodynamics we cannot very well allow advanced quantities in our elastic problem, particularly as in two dimensions they would involve integrals over the whole future motion of the dislocation. Secondly, they eliminate the electromagnetic mass, allowing an arbitrary inertial mass to be ascribed to the electron, whereas a dislocation is clearly the analog of a weightless charged rod with purely electromagnetic mass.

Equation (26) is the true equation of motion of a hypothetical "rigid" dislocation. It takes no account of the change of shape of the function $\delta w(x,t)$ which a rigorous solution of (21) would give. For small accelerations the main feature of the change in shape would presumably be a Lorentz contraction appropriate to the instantaneous velocity $\dot{\xi}$. Because of the lack of contraction in our model, there are none of the relativistic effects encountered in Sec. IV. Indeed, nothing untoward happens to Eq. (26) even if the velocity is supersonic. It is easy to show from (26) and (27) that a moving supersonic dislocation would experience a retarding force $bp_{zy}{}^A=(v^2/c^2-1)^{\frac{1}{2}}\mu b^2/2\pi a$ even in the absence of dissipative effects. Physically, this is because it is continually creating a greater disturbed region, or from another point of view, because the leading elements of the dislocation density distribution exert a force on the trailing elements, but not conversely, since each element produces a disturbance only in a wake behind it. In principle a "rigid" dislocation can reach supersonic velocities, but not one which contracts to zero width as the velocity of sound is approached. Admittedly a dislocation which obeys the Peierls law (or some generalization of it) contracts, but the effective width of the source can hardly be much less than one interatomic spacing, so that a supersonic dislocation is a formal possibility.

VI. EXAMPLES AND DISCUSSION

Because nothing is certainly known about dissipative effects, we shall give some examples of dislocation motion on the assumption that W is zero. The possibility of a contribution to W from the nonlinear region $-\frac{1}{2}a<y<\frac{1}{2}a$ of the Peierls model is taken up later.

(a) Constant acceleration f starting from rest at $t=0$. Here $d^2\xi/dt^2=fH(t)$ and for $t>0$

$$bp_{zy}{}^A=\frac{\rho b^2}{4\pi}f\sinh^{-1}\frac{ct}{a}+\frac{\rho b^2}{8\pi}f\frac{t}{t+(t^2+a^2/c^2)^{\frac{1}{2}}}.$$

This gives an effective mass $(\rho b^2/4\pi)\ln(2e^{\frac{1}{2}}ct/a)$ for large t, agreeing roughly with the results of Sec. IV.

(b) Impulsive change of velocity from 0 to v at $t=0$. Here $d^2\xi/dt^2=v\delta(t)$ and $bp_{zy}{}^A$ can be found from the previous result by differentiating and multiplying by v/f. Roughly speaking, we have to apply a constant stress for a time of order a/c followed by a stress falling off as $1/t$ ever afterwards. Since the acceleration is zero, except initially, we cannot calculate a mass directly. However, the time integral of the applied force times the velocity is the work done by the applied stress, and if this is equated to $\frac{1}{2}mv^2$ we find the same effective mass as before.

(c) Sinusoidal oscillation: $d^2\xi/dt^2=(d^2\xi/dt^2)_0\cos\omega t$. The solution can be given in terms of Bessel and Struve functions of imaginary argument. For $\omega a/c\ll 1$ it becomes

$$bp_{zy}{}^A=m(d^2\xi/dt^2)_0\cos(\omega t-\alpha),$$

with

$$m=\frac{\rho b^2}{4\pi}\left\{\left(\frac{\pi}{2}\right)^2+\left(\ln\frac{2c}{e^\gamma\omega a}\right)^2\right\}^{\frac{1}{2}},$$

$$\tan\alpha=\tfrac{1}{2}\pi/\ln\left(\frac{2c}{e^\gamma\omega a}\right),\quad e^\gamma=1.78\cdots,$$

in rough agreement with Nabarro.[8]

As it stands, our equation of motion gives the stress required to maintain a given acceleration. Of more interest is the inverse equation giving the acceleration under a given applied field. The exact solution of the last problem, which connects a sinusoidal stress with a sinusoidal acceleration, could be used to invert (28) by expressing the arbitrary applied stress as a Fourier integral. The result is not very helpful and it is more convenient to use Laplace transforms. Consider the case where $p_{zy}{}^A$ (written simply p in what follows) is zero for $t<0$. If we introduce the Laplace transform,

$$\mathfrak{L}\{f\}=\int_0^\infty e^{-st}f(t)dt,$$

and use the result[11]

$$\mathfrak{L}\left\{\int_0^t f_1(t-\tau)f_2(\tau)d\tau\right\}=\mathfrak{L}\{f_1\}\mathfrak{L}\{f_2\},\qquad(29)$$

we find

$$\mathfrak{L}\{d^2\xi/dt^2\}=(8\pi/\rho b)\mathfrak{L}\{p\}[2h(x)+h''(x)-x^{-1}h'(x)+x^{-1}-2x^{-2}]^{-1},\qquad(30)$$

with

$$h(x)=\tfrac{1}{2}\pi\{\mathrm{H}_0(x)-Y_0(x)\},\quad x=sa/c,$$

where Y_0 is the usual second solution of Bessel's equation and H_0 is Struve's function. (Their difference is monotonic though each is oscillatory.) Formally this solves the problem of finding $d^2\xi/dt^2$ in terms of p; we have only to find the function of which $[\;]^{-1}$ in (30) is the transform and apply (29). However, the necessary Mellin inversion integral is not simple since Y_0 is multiple-valued. But since

$$h(x)\sim x-\ln(\tfrac{1}{2}e^\gamma x),\quad x\ll 1;\quad h(x)\sim x^{-1},\quad x\gg 1,$$

we can find ξ for large and small t with the help of a theorem[11] which states that if

$$\mathfrak{L}\{f\}\sim s^{-\beta}L(s^{-1})\quad\text{for}\quad s\rightarrow 0(\infty),$$

where

$$L(ux)/L(x)\rightarrow 1\quad\text{for}\quad x\rightarrow 0(\infty),$$

then

$$\int_0^t f(\tau)d\tau\sim t^\beta L(t)/\Gamma(\beta+1)\quad\text{for}\quad t\rightarrow\infty\,(0).$$

(d) A constant stress p_0 is applied at $t=0$. Here $\mathfrak{L}\{p\}=p_0s^{-1}$, and applying the theorem to (30), we have

$$\xi(t)\sim\frac{4\pi}{\rho b}p_0\frac{t}{\ln(ct/e^\gamma a)},\qquad t\gg a/c;$$

$$\sim\frac{8\pi a}{3\rho bc}p_0,\qquad t\ll a/c.$$

[11] G. Doetsch, *Laplace-Transformation* (Julius Springer, Berlin, 1937).

For large t the dislocation gathers speed more slowly than a Newtonian particle; the effective mass is $(\rho b^2/4\pi)\ln(ct/e^\gamma a)$. As well as a free dislocation subjected to a suddenly applied force, we may consider (d) as describing a locked dislocation which breaks free when a gradually increasing stress reaches a certain value and also, though more schematically, a Frank-Read source which passes suddenly from a stable to an unstable state when a critical stress is reached.

(e) An arbitrarily varying stress is applied between $t=0$ and $t=t_1$, so that the total impulse is

$$P=\int_0^{t_1} bp(t)dt.$$

For $s\ll t_1$ $\mathfrak{L}\{p\}=P/b$ and the theorem gives

$$\dot{\xi}(t)\sim\frac{4\pi}{\rho b^2}P\frac{1}{\ln(ct/e^\gamma a)},\qquad t\gg a/c,\, t\gg t_1.$$

A dislocation started by an impulse and then allowed to run free gradually loses speed. Again a mass proportional to $\ln(ct/e^\gamma a)$ is appropriate since Newton's law, $d(m\dot{\xi})/dt=0$, is then satisfied during the free motion.

Like the example of Sec. IV the results (a), (b), (d), and (e) fit Frank's picture of a mass $(\rho b^2/4\pi)\ln(r_1/a)$, with r_1 of the order of the radius of the disturbed region surrounding the starting point. In all these cases $\dot{\xi}$ is a monotonic function of t, and presumably the elastic field in the disturbed region, logarithmically speaking, does not differ much from that round a uniformly moving dislocation. For a more complicated motion (e.g., oscillation) this is no longer true.

It can be shown that the results we have obtained are largely independent of the exact form of the density distribution η provided that, like the particular form (27), it is a bell-shaped curve with a width of order a and enclosing an area $\rho^{\frac{1}{2}}b$.

Even if we assume that there are no dissipative effects, W will not be zero if there is any variation in the energy stored in the gap at the slip plane. The calculation of this energy is difficult because of a certain ambiguity in extending the Peierls approximation to dynamic problems. If we take as our model two semi-infinite elastic solids separated by a gap a, the contribution to W is zero since the weightless interatomic forces have no kinetic energy and their total potential energy is independent of time, depending only on the shape of the prescribed function δw. On the other hand, if we apply the Peierls recipe of replacing the crystal lattice by a continuum only outside the gap, we are left with a number of half-atoms adhering to the faces of the gap, and their kinetic energy changes at a rate

$$2\times\tfrac{1}{2}\rho a\frac{\partial}{\partial t}\int_{-\infty}^{\infty}(\tfrac{1}{2}\delta\dot{w})^2dx=\frac{\rho b^2}{2\pi}\frac{d\xi}{dt}\frac{d^2\xi}{dt^2},$$

if we use the $\delta\dot{w}$ corresponding to (27). The W term in (28) becomes $\rho b^2(d^2\xi/dt^2)/2\pi$, an addition of "inertial" mass to the "electromagnetic" mass represented by the other terms. The modification to our examples is small.

If this or some similar account of the effect of the inertia of the matter in the slip plane is accepted, the results of Sec. IV and the generalized Peierls equation (21) need modification. In particular this spoils the elegant result that an antiplane elastic field satisfying the Peierls condition, or some generalization of it, continues to do so when moving uniformly with an appropriate Lorentz contraction.

It may be as well to contrast our calculation of the effect of the inertia of the material in the slip plane with Nabarro's.[8] In his problem the contribution to $\dot{w}$ at the slip plane from the applied field (a sound wave) is not zero, and the half-atoms above and below the slip plane move in the same direction. In our problem it is assumed that the external stress is applied in such a way that $\dot{w}$ is an odd function of y, and so the half-atoms move in opposite sense above and below the slip plane. The corresponding term in Nabarro's calculation is of an order higher than he retains.

We should also make a slight correction to allow for the fact that strictly not only the dislocation but also the applied elastic field makes a contribution to $\delta\dot{w}$ in (22). If we assume that this contribution is approximately $a\partial^2w^A/\partial y\partial t$, as it would be if there were no nonlinearity at the slip plane, it can be shown that a term $ab\dot{p}_{zy}{}^A/2c$ must be added to the left-hand side of (28). This will be partly offset by a contribution of the same order to the term in W; the exact value depends on the details of the nonlinear behavior in the slip plane. In any case the correction will be negligible if $d(\ln p_{zy}{}^A)/dt\ll c/a\sim10^{13}$.

In conclusion it should be emphasized that the foregoing is only a first step towards solving the much harder problem of the motion of a dislocation loop. It is not clear how much light the present calculations throw on the general problem. The curious behavior expressed by the integral equation (28) is, of course, due to the fact that a given element of the dislocation is acted on not only by the applied stress, but also by delayed disturbances from other parts of the dislocation line. The same effect will presumably be important in determining how a dislocation loop spreads under the influence of an applied stress.

I should like to thank Dr. G. Lee-Whiting for valuable help.

A Tentative Theory of Metallic Whisker Growth*

J. D. Eshelby
University of Illinois, Urbana, Illinois
(Received June 4, 1953)

PEACH'S[1] very pretty explanation of the formation of metallic whiskers[2] seems to be ruled out by the observation[3] that they grow at the root. The growth seems to be influenced[4] by the atmosphere over the surface. The energy required to form a fresh surface may be more than repaid if it is attacked avidly enough by the atmosphere; there is then an effective negative surface tension $-\gamma$. We might have perhaps $\gamma b^2 \sim 1$ *ev* (b = lattice constant). The ratio γ/μ (μ = shear modulus) would then be about 1A. The surface tension forces on a small hump on the surface [Fig. 1(a)] obviously have the right character (a central pull

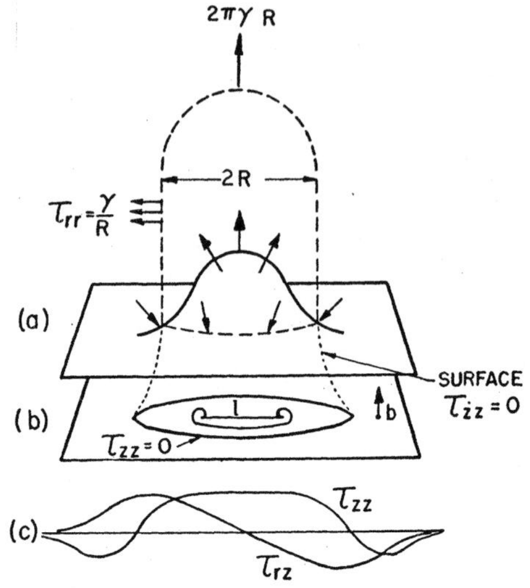

FIG. 1. Model for whisker growth.

surrounded by a restraining pressure) to "wire-draw" it into a whisker according to intuitive ideas of plastic flow. For the observed whisker size ($R \sim 10^{-4}$ cm) the stress of order γ/R in and just below the hump might exceed the actual yield stress, though not the theoretical yield stress ($\sim\mu/20$). However, on this small scale one must consider in detail how flow is catalyzed by dislocation.

The model of Fig. 1 lends itself to rough calculations. A Frank-Read source of length l and vertical Burgers vector b lies in a horizontal plane (b) at a depth of order l below the hump. The stress τ_{zz} [Fig. 1 (c)] makes the source emit a dislocation loop by "climb." The loop expands in the plane (b) until it reaches a radius where $\tau_{zz}=0$. Here the stress τ_{rz} assisted by image forces makes the loop glide vertically, so adding one atomic layer to the base of the hump. When by repetition of this a reasonable whisker has grown, τ_{zz} will be $2\gamma/R$ in the whisker and about γ/R at the source. To operate the source, τ_{zz} must be at least $\mu b/l$. With $\gamma/\mu \sim 1\text{A} \sim b$, this will be so if $R \sim l$. The stress at the source ultimately falls off both for $R \gg l$ and $R \ll l$ if the source depth stays constant. The whisker radius is thus tied to the length of a Frank-Read source, which is usually supposed to be about one micron. The surface $\tau_{zz}=0$ forms a "stress funnel" which guides each loop more or less unerringly to break surface at the base of the whisker, and so keeps its diameter constant. If the motion of the source is not to be stopped by the back-pressure of the vacancies it emits, there must be suitable sinks for them. It can be shown that surface tension changes the volume of a body of any shape with compressibility χ by $\frac{2}{3}\gamma\chi$ times its surface area. For a macroscopic specimen the corresponding mean pressure, $\frac{2}{3}\gamma\times$(surface/volume) would fall far short of the value required to make Frank-Read sources in the interior act as the necessary sinks. However, it should not be hard to find them at the surface. Frank[5] has shown that even with positive surface tension it may be energetically an advantage for a dislocation reaching the surface to develop a hollow core. Or again, if we had a depression instead of a hump in Fig. 1 the source would work in reverse, absorbing vacancies and deepening the depression.

We may use a calculation of Mott's[6] to find the rate of growth. He showed that, if a cube has normal stresses P on one pair of opposite faces and $-P$ on another pair, a volume $V \sim NbD(Pb^3/kT)$ of material is transferred from one pair of faces to the other in unit time if there are N points on dislocations which can absorb or emit vacancies and the coefficient of self-diffusion is D. Our case is analogous. P is γ/R times a factor κ depending on the detailed stress-distribution, including a possible stress concentration if successive loops help one another. N is about l/b, the number of lattice sites per loop, times the fraction β (perhaps $\ll 1$) of them which can emit or absorb vacancies times n, the number of loops in transit between source and surface at one time. The rate of change of the whisker length h will thus be

$$\dot{h} = V/\pi R^2 \sim \kappa\beta n D(b/l^2)(\gamma b^2/kT).$$

With the value of D for tin at room temperature,[7] we can get a growth rate of a millimeter or a centimeter per year with $\kappa\beta n \sim 100$ or 1000. The small number of accidental coincidences of sources and suitable surface irregularities may be enough to account for the number of whiskers per unit area. If not, we might suppose that the sources build their own humps by operating initially without stress as the result of a subsaturation of vacancies due to a change of temperature.

Many variations of this model are possible. The transfer of loops to the surface might occur by the formation and joining up of secondary loops in a vertical plane as in "prismatic punching,"[8] where the stress distribution is similar. The whiskers might then be prismatic. Professor Seitz (to whom the writer is indebted for helpful discussions) has suggested a mechanism involving a spiral prismatic dislocation,[9] which is the internal counterpart of spiral growth on the surface and which leaves a screw dislocation along the axis of the whisker.

* Work done under U. S. Office of Naval Research contract.
1 Milton O. Peach, J. Appl. Phys. **23**, 1401 (1952).
2 C. Herring and J. K. Galt, Phys. Rev. **85**, 1060 (1952).
3 S. Eloise Koonce and S. M. Arnold, J. Appl. Phys. **24**, 365 (1953).
4 Compton, Mendizza, and Arnold, Corrosion **7**, 327 (1951).
5 F. C. Frank, Acta Cryst. **4**, 497 (1951).
6 N. F. Mott, *IXe Conseil de Physique* (*Solvay*), *Rapports* (Stoops, Brussels, 1952), p. 515; Proc. Phys. Soc. (London) **B64**, 379 (1951), §7.
7 P. J. Fensham, Australian J. Sci. Research **A3**, 91 (1950).
8 Frederick Seitz, Phys. Rev. **79**, 723 (1950). Here *ABFE* corresponds to our plane (b) and *DCHG* is the root of the whiskers.
9 Frederick Seitz, Phys. Rev. **79**, 1003 (1950).

Geometrical and Apparent X-Ray Expansions of a Crystal Containing Lattice Defects*

J. D. ESHELBY
Department of Physics, University of Illinois, Urbana, Illinois
(Received June 24, 1953)

THE results reported by Miller and Russell[1] may be illustrated by a direct calculation for a spherical crystal. Using the notation of their previous paper[2] the displacement caused by a point imperfection is $\Delta\mathbf{r}=A(a/r)^3\mathbf{r}$, and the geometrical volume change it produces is $\Delta V_G=\int\Delta\mathbf{r}\cdot\mathbf{n}dS=4\pi a^3A$, being proportional to the solid angle subtended by the surface of the crystal at the imperfection. The relative expansion as determined by x-rays is equal to the relative contraction of the reciprocal lattice, $-[\Delta h_1(100)+\Delta h_2(010)+\Delta h_3(001)]$. This may be evaluated with the help of reference 2, Appendix 1, if we allow ourselves to replace sums like ΣL_1^2 and $\Sigma\Delta L_1L_1$ by $a^2\int x^2dv$ and $a^2\int(\Delta r)_xxdv$. The integrals are of the type occurring in potential theory. For an imperfection with position vector ξ relative to the center of a spherical crystal of radius R we easily find the following expression for the apparent volume change as determined by x-rays:

$$\Delta V_X=10\pi Aa^3(1-\xi^2/R^2). \tag{1}$$

When $\xi=0$, $\Delta V_X=2\cdot5\Delta V_G$ as in reference 2, but when $\xi>\sqrt{(3/5)}R$, $\Delta V_X<\Delta V_G$. If there are a large number of defects scattered randomly through the sphere, the mean ΔV_X per defect will be given by replacing ξ^2 in (1) by its mean value over the sphere, namely, $3R^2/5$. Thus the mean ΔV_X is equal to ΔV_G, and the total geometrical and x-ray expansions are equal.

We have neglected the fact that the displacement $\Delta\mathbf{r}=A(a/r)^3\mathbf{r}$ leads to nonvanishing stresses at the surface of the crystal, and so cannot be correct. It happens that for the problem above the necessary additional "image terms" affect ΔV_G and ΔV_X equally. The relation $\Delta V_G=\Delta V_X$ can in fact be established for a body of arbitrary shape uniformly filled with point defects, but only if the image terms are included. These terms make an important contribution. In fact,

$$\Delta V_G=\Delta V_X=4\pi a^3A\cdot\frac{3(1-\sigma)}{1+\sigma} \tag{2}$$

times the number of imperfections. The factor involving σ (Poisson's ratio) is about 1.5 for metals and 1.8 for alkali halides. For ΔV_G the result (2) has been given by Seitz.[3]

We hope to publish a more detailed account later.

* Work supported by the U. S. Office of Naval Research.
[1] P. H. Miller, Jr., and B. R. Russell, J. Appl. Phys. 24, 1248 (1953).
[2] P. H. Miller, Jr., and B. R. Russell, J. Appl. Phys. 23, 1163 (1952).
[3] Frederick Seitz, Revs. Modern Phys. 18, 384 (1946).

JOURNAL OF APPLIED PHYSICS VOLUME 25, NUMBER 2 FEBRUARY, 1954

Distortion of a Crystal by Point Imperfections*

J. D. ESHELBY
Department of Physics, University of Illinois, Urbana, Illinois

(Received August 17, 1953)

The expression $\mathbf{u}=c\mathbf{r}/r^3$, (where c is a constant) sometimes assumed for the displacement around a point imperfection (interstitial or substitutional impurity, lattice vacancy) gives a nonzero stress at the surface of the solid. The additional "image displacement" necessary to insure that this stress vanishes is usually neglected, but may be important. For example, it accounts for from 30 to 50 percent of the volume change produced by such defects. This and other effects of the image term are discussed. Miller and Russel have pointed out that for a point imperfection near the center of a sphere the apparent volume change deduced from measurements of the x-ray lattice constant is greater than the geometrical volume change. It is shown that the reverse is true when the defect is near the surface, and that for a large number of defects scattered uniformly through the sphere the geometrical and x-ray expansions are equal. It can be shown quite generally that a body of arbitrary shape is expanded uniformly by a statistically uniform distribution of point imperfections, and that the x-ray diffraction pattern is altered in the way to be expected for such an expansion. To establish this, however, it is essential to take the image terms into account.

I. INTRODUCTION

As a model for the distortion of a crystal lattice by a substitutional or interstitial atom or a vacant lattice site, a number of authors have taken a center of dilatation in an isotropic elastic continuum. In many cases calculations have been made using the elastic displacement appropriate to a center of dilatation in an infinite medium. This displacement cannot be correct since it would give a nonvanishing stress at the free surface of the body. It is convenient to regard the actual displacement as the sum of the displacement in an infinite medium and an "image" displacement due to the perturbing effect of the boundary.

It is the object of this paper to show that the image term cannot be neglected in certain applications. In particular, if it is omitted the volume change due to one or more point singularities will be underestimated by a factor of 1.5 for metals or 1.8 for alkali halides. Again, if a calculation is made in which the effects of the free boundary are ignored, it is found that any nonspherical body uniformly filled with point singularities would suffer a change of shape as well as a change of size. When the boundary effect is taken into account there is a change of volume without change of shape. Moreover, a calculation of the change of x-ray lattice constant gives a result consistent with the geometrical deformation only if the image displacements are included. Thus, unusually, the boundary effects simplify the problem instead of complicating it.

II. A POINT SINGULARITY IN AN INFINITE BODY

The elastic displacement caused by a center of dilatation in an infinite medium is[1]

$$\mathbf{u}^\infty = c\frac{\mathbf{r}}{r^3} = -c\,\mathrm{grad}(1/r), \tag{1}$$

where c is a constant, the "strength" of the singularity. (We use the affix ∞ to emphasize that (1) is only valid in an infinite medium.) The displacement has the same form as the field of an electrostatic point charge. The real justification for taking (1) as a rough description of the elastic field of a point imperfection in an infinite crystal which has been idealized as a homogeneous isotropic elastic continuum is that it is the only spherically symmetrical displacement which satisfies the equations of elasticity and does not increase with r. Still, it is convenient to have some sort of detailed elastic model. As a model of an interstitial or substitutional atom we might take an elastic sphere of radius $(1+\epsilon)r_0$ forced into a spherical hole of radius r_0 in an infinite block of the same material. It is easy to show that for $r>r_0$ the displacement is given by (1) with $c=\epsilon r_0^3(1+\sigma)/3(1-\sigma)$ and that for $r<r_0$ there is a uniform compression. (σ is Poisson's ratio.) The surface of the hole suffers an outward displacement c/r_0^2, increasing the volume within it by $4\pi c$.

This model must not be taken too literally. The lattice constant of gold is decreased by a little dissolved silver. The misfitting sphere model would suggest that therefore a little dissolved gold would increase the lattice constant of silver, and this is not true. All that we can hope for is that a particular type of singularity in a particular matrix will be characterized by a constant c which can be used consistently to describe various phenomena.

Again, a foreign atom in a lattice exerts forces on its neighbors differing from the "standard" forces they would experience in the perfect lattice. When we assimilate the lattice to a continuum, the standard forces are absorbed in the elastic properties of the medium, but the additional forces due to the imperfection are still outstanding. Hence as an alternative to the sphere-in-hole model we may take a cluster of point forces to represent a point lattice imperfection in the elastic approximation. For a sufficiency symmetrical relation of the interstitial or substitutional

* This work was supported by the U. S. Office of Naval Research.

[1] A. H. Love, *Mathematical Theory of Elasticity* (Cambridge University Press, Cambridge, England, 1924).

atom and the lattice, the forces will be equivalent to three equal "double forces without moment" at right angles. The displacement will then be of the form (1) which can be considered[1] as due to a distribution of body force

$$\mathbf{F}(\mathbf{r}) = -12\pi cK \,\mathrm{grad}\delta(\mathbf{r}). \tag{2}$$

(K is the bulk modulus.) In a less symmetrical case (e.g., carbon in iron) the cluster of forces will be equivalent to three unequal double forces without moment and the resulting displacement will be more complicated than (1). However, it is probably not justifiable to introduce this refinement without at the same time considering the anisotropy of the material, and we shall not consider it further.

The displacement (1) can be produced by heating a point of the elastic medium, assumed to be nonconducting. This can be seen by imagining that the misfit in the first model was produced by heating a sphere with originally no misfit, or in terms of the second model by noticing that the thermal stress[2] due to a temperature distribution $T(\mathbf{r})$ is the same as the stress produced by a density of body force proportional to grad T, which for the hot spot $T=\text{const}\ \delta(\mathbf{r})$ would be of the form (2). This analogy will be useful in discussing a body containing a large number of imperfections.

In general the stress associated with a displacement $\mathbf{u}$ whose Cartesian components are u_1, u_2, u_3 is

$$p_{ij} = \lambda\delta_{ij}\,\mathrm{div}\mathbf{u} + \mu\left(\frac{\partial u_i}{\partial x_j} + \frac{\partial u_j}{\partial x_i}\right), \tag{3}$$

where λ and μ are Lamé's constants. From the electrostatic interpretation it is at once clear that the divergence and curl of (1) vanish, so that the stress produced by the singularity when in an infinite body is simply

$$p_{ij}^{\infty} = 2\mu \frac{\partial u_i^{\infty}}{\partial x_j}. \tag{4}$$

III. A POINT SINGULARITY IN A FINITE BODY WITH A STRESS-FREE SURFACE

Consider now a center of dilatation in a finite body with a free surface. First mark out the surface S of the proposed body in an infinite block of material and introduce the singularity. The displacement is given correctly by (1). Across any surface element of S there is a stress† $p_{ij}n_j dS$, where $\mathbf{n}$ is the normal to S. Thus if we remove the material outside S the displacement will continue to be $\mathbf{u}^{\infty}$ only if we apply a distribution of surface traction $p_{ij}n_j$ to S. Removing this distribution to give a body with a stress-free surface is equivalent to applying an additional distribution $-p_{ij}n_j$. The elastic state will then be given by

$$\mathbf{u} = \mathbf{u}^{\infty} + \mathbf{u}^{I}, \quad p_{ij} = p_{ij}^{\infty} + p_{ij}^{I}, \tag{5}$$

where the image stress p_{ij}^I is the stress which surface tractions $-p_{ij}n_j$ would produce in the body and the image displacement $\mathbf{u}^I$ is related to p_{ij}^I by (3). p_{ij}^I is clearly free of singularities within S and satisfies

$$(p_{ij}^I + p_{ij}^{\infty})n_j = 0 \quad \text{on} \quad S. \tag{6}$$

Unlike $\mathbf{u}^{\infty}$, $\mathbf{u}^I$ has in general a nonvanishing divergence.

If the surface traction and surface displacement are prescribed (they are not, of course, independent) the elastic field inside the body can be found by integration. When only the surface traction is given, we need the appropriate elastic Green's function, known for only a few simple shapes. Since we know the image traction $p_{ij}^I n_j$ but not $\mathbf{u}^I$ on S we cannot in general calculate the image field in the body.

The change in volume of the solid can be divided into parts ΔV^{∞} and ΔV^I arising from the two terms in (5). We have at once

$$\Delta V^{\infty} = \int_S \mathbf{u}^{\infty}\cdot\mathbf{n}dS = 4\pi c, \tag{7}$$

the integral being, according to (1), c times the solid angle subtended by the surface at the singularity. This result can be seen at once for the sphere-in-hole model. When the sphere is inserted, the volume of the hole increases by $4\pi c$, and since $\mathrm{div}\mathbf{u}^{\infty}=0$ outside the hole, this increase is transmitted unchanged to S. Again for the ideal mathematical singularity for which (1) holds for all $\mathbf{r}$ we have strictly not $\mathrm{div}\mathbf{u}^{\infty}=0$ but rather

$$\mathrm{div}\mathbf{u}^{\infty} = -c\nabla^2(1/r) = 4\pi c\delta(\mathbf{r}) \tag{8}$$

with a delta-function of expansion at $r=0$. A formal volume integration gives (7).

Although we cannot find the image deformation in detail, we can find ΔV^I with the help of the rule that the volume change of a body subjected to a distribution of surface traction $\mathbf{T}$ per unit area is

$$\frac{1}{3K}\int_S \mathbf{r}\cdot\mathbf{T}dS = \frac{1}{3K}\int_S x_i p_{ij} n_j dS, \tag{9}$$

where p_{ij} is the stress produced by $\mathbf{T}$ and K is the bulk modulus. For

$$\int x_i p_{ij} n_j dS = \int \frac{\partial}{\partial x_j}(x_i p_{ij}) dv$$
$$= \int p_{jj} dv = 3K \int \mathrm{div}\mathbf{u} dv \tag{9'}$$

since $\partial x_i/\partial x_j = \delta_{ij}$ and in elastic equilibrium $\partial p_{ij}/\partial x_j = 0$.

To find ΔV^I we must put $p_{ij}n_j = p_{ij}^I n_j = -p_{ij}^{\infty} n_j$ with p_{ij}^{∞} from (4). But the operator $x_i(\partial/\partial x_i)$ applied to $u_k^{\infty} = cx_k/r^3$ merely multiplies it by -2. Hence the

[2] S. Timoshenko, *Theory of Elasticity* (McGraw-Hill Book Company, Inc., New York, New York, 1934).

† Throughout the paper we use the convention that a repeated suffix is to be summed over the values 1, 2, 3.

integral (9) is equal to the integral (7) times $4\mu/3K$ and so

$$\Delta V^I = 4\pi c \frac{2(1-2\sigma)}{1+\sigma}. \quad (10)$$

Thus the total volume change on introducing a center of dilatation of strength c into any homogeneous isotropic body with a stress-free surface is

$$\Delta V = \Delta V^\infty + \Delta V^I = \gamma \Delta V^\infty = 4\pi c\gamma, \quad (11)$$

where the constant

$$\gamma = 3(1-\sigma)/(1+\sigma) \quad (12)$$

is about 1.5 for metals ($\sigma \sim \frac{1}{3}$) and 1.8 for alkali halides ($\sigma \sim \frac{1}{4}$). This result has already been given by Seitz.[3] It is also implicit in the known result for the interaction energy of a center of dilatation and an external hydrostatic pressure.[4] This energy is the product of ΔV and the hydrostatic pressure.

If in (9′) we take p_{ij} to be any state of purely internal stress, so that the surface integral is zero we arrive at the known result that the volume of a body is unaffected by internal stresses provided Hooke's law is valid. The volume expansion we have calculated is not in disagreement with this. Consider the sphere-in-hole model of Sec. II, and suppose for clearness that ϵ is negative. Then the unstrained state is a body with a spherical hole of radius r_0 containing a sphere of radius $r_0(1-|\epsilon|)$ and an unoccupied volume $4\pi r_0^3|\epsilon| = 4\pi|c|\gamma$. In the strained state the surfaces of the hole and sphere have been drawn together and welded. Since the total volume of *material* is unchanged, the volume enclosed by the outer surface must have decreased by an amount equal to the volume eliminated between sphere and hole. This simple derivation does not supersede our more elaborate treatment since to relate the expansion to other effects arising from the stress field of the imperfections we need to know something about $\mathbf{u}^\infty$ and $\mathbf{u}^I$ separately. If there are departures from Hooke's law, the volume change of the material is not quite zero. This is discussed in Sec. VI.

IV. DEFORMATION PRODUCED BY A LARGE NUMBER OF SINGULARITIES

If there are a number of singularities in the body, we have

$$\mathbf{u}^\infty(\mathbf{r}) = \sum_n c_n \frac{\mathbf{r}-\mathbf{r}_n}{|\mathbf{r}-\mathbf{r}_n|^3}, \quad (13)$$

with summation over all imperfections. The image stress p_{ij}^I is that stress which has no singularities within S and on S annuls the surface traction calculated from (13). $\mathbf{u}^I$ is the associated displacement.

If the defects are all alike we have

$$\Delta V/V = 4\pi\gamma c f/\Omega, \quad (13')$$

where f is the atomic fraction of defects and Ω is the volume per atom of the matrix.

Of special interest is the case where there are a large number of identical centers spread through the body, with an approximately uniform density of n centers per unit volume. Here we can also say something about the change of shape if we allow ourselves to replace the actual distribution by a continuous distribution of infinitesimal centers with the same total strength per unit volume. (We shall try to justify this below.) If as before the body is embedded in an infinite matrix this averaged displacement is clearly

$$\langle \mathbf{u}^\infty(\mathbf{r})\rangle = cn \int_V \frac{\mathbf{r}-\mathbf{r}'}{|\mathbf{r}-\mathbf{r}'|^3} dv(\mathbf{r}') \quad (14)$$

at any rate outside S. (It is not at once clear that we can give any meaning to the average displacement in a medium riddled with singularities.)

The comparison already made with thermal expansion suggests that a body uniformly filled with infinitesimal centers of dilatation will, like a uniformly heated solid, expand uniformly. The following is a formal proof. We first show that the displacement given by the volume integral (14) can be dupilicated outside S by a distribution of body force over S of uniform magnitude

$$p = K\Delta V/V = 4\pi c n\gamma K$$

per unit area and directed along the outward normal at each point. A point force $\mathbf{F}$ in an infinite medium produces a displacement[1]

$$\mathbf{u} = A\mathbf{F}/r + B\mathbf{r}(\mathbf{r}\cdot\mathbf{F})/r^3,$$
$$B = 1/16\pi\mu(1-\sigma), \quad A = (3-4\sigma)B$$

at a vector distance $\mathbf{r}$ from it. Each element of the surface contributes a force $d\mathbf{F} = p\mathbf{n}dS$ and the net displacement is

$$u_i = p\int_S \left\{\frac{A}{r}\delta_{ij} + \frac{B}{r^3}x_i x_j\right\} n_j dS$$
$$= p\int \frac{\partial}{\partial x_j}\{\ \} dv = p(A-B)\int \frac{-x_i}{r^3} dv$$
$$= \langle u_i^\infty \rangle.$$

(We have taken the point of observation to be the origin.)

Suppose now that S is marked out in an infinite medium, that the uniform distribution if infinitesimal centers of dilatation is introduced and that a layer of body force equal and opposite to that just described is distributed over S. The combined effect of these last two steps gives zero displacement outside S and leaves S its original shape and size. If the unstrained matrix is now cut away, the elastic state of the body bounded by S is unchanged, but the layer of body force becomes

[3] Frederick Seitz, Revs. Modern Phys. 18, 384 (1946).
[4] See, for example, J. D. Eshelby, Trans. Roy. Soc. (London) A244, 87 (1951).

a hydrostatic pressure p acting on the surface. When this is removed S undergoes the uniform expansion

$$\langle u_i \rangle = (4/3)\pi n c \gamma x_i. \qquad (15)$$

Though $\langle \mathbf{u}^\infty \rangle + \mathbf{u}^I$ represents a uniform expansion, $\langle \mathbf{u}^\infty \rangle$ and $\mathbf{u}^I$ separately do not, except for a spherical body. Thus omission of the image terms in addition to giving a volume change lacking the factor γ (which could be absorbed in the usually unknown constant c) would also give a nonuniform expansion. Suppose, for example, that S is an ellipsoid with semi-axes $\mathfrak{a}$, $\mathfrak{b}$, $\mathfrak{c}$. Then $\langle \mathbf{u}^\infty \rangle = -cn\,\mathrm{grad}\varphi$ where

$$\varphi = \tfrac{1}{2}(D - Ax_1 - Bx_2 - Cx_3) \qquad (16)$$

with the usual notation for the potential of an ellipsoid.[5] Thus if image terms are neglected, the ellipsoid becomes another ellipsoid with semi-axes $\mathfrak{a}+\Delta\mathfrak{a}$, etc., where

$$\frac{\Delta\mathfrak{a}}{\mathfrak{a}} = cnA, \quad \frac{\Delta\mathfrak{b}}{\mathfrak{b}} = cnB, \quad \frac{\Delta\mathfrak{c}}{\mathfrak{c}} = cnC. \qquad (17)$$

Since $A+B+C=4\pi$ the volume change agrees with (7). As an example, if $\mathfrak{a}:\mathfrak{b}:\mathfrak{c} = 2:1:1$ then also $A:B:C = 2:1:1$ very nearly. Again, a cube would become barrel-shaped, with the increase in distance between opposite face centers equal to about three times the increase in distance between two adjacent corners. These departures from uniform expansion, if they existed, would be hard to detect by macroscopic measurements, but as we shall see, there is a similar effect on the lattice constants determined by x-ray diffraction, and this would be easily observable.

We must now try to justify replacing the sum (13) by the integral (14). Just this question arises when we replace the field of a set of gravitating or electrified particles by the field of a continuous body. This kind of problem can be attacked in various ways, none completely satisfying; the following seems a plausible line for the elastic case.

The observable macroscopic displacement at a point of a body in which the displacement fluctuates on a microscopic scale is essentially the displacement of the centroid of a region large compared with the scale of the fluctuations. It is therefore reasonable to define the gross (or macroscopic) displacement at a point as the average of the actual (microscopic) displacement taken over the volume of a sphere whose radius R is large enough to contain many fluctuations, but is small compared with the dimensions of the body.

Let us find the gross displacement $\langle \mathbf{u}^\infty \rangle$ corresponding to (14) for a point inside the body. The volume integral over the sphere R of the displacement caused by a singularity at $\mathbf{r}$ relative to its center is equal to the attraction on a charge c at $\mathbf{r}$ produced by a distribution of charge of unit density filling the sphere, i.e., $(4/3)\pi c\mathbf{r}$ if $r<R$ and $(4/3)\pi c R^3\mathbf{r}/r^3$ if $r>R$. Thus

$$\langle \mathbf{u}^\infty \rangle = \frac{c}{R^3}\sum_{r_n<R} \mathbf{r}_n + c\sum_{r_n>R}\frac{\mathbf{r}_n}{r_n^3}.$$

As R increases and becomes reasonably large compared with $n^{-\frac{1}{3}}$ the first term should be small, being proportional to the position vector of the centroid of a large number of points taken at random in a sphere. At the same time, as R increases it becomes more and more reasonable to replace the second term by an integral since the distance of all the singularities from the point at which their effects are summed is large compared with the mean distance between them. We thus reach (14) but with a sphere of radius R about the point of observation excluded from the integration. However, by a known result of potential theory (14) is unaltered by this omission provided the sphere lies wholly within S.

Having defined the gross displacement we can calculate a gross stress from it with the aid of (3). This should be the true macroscopic stress, i.e., it should give correctly the surface traction required to prevent relative motion of the faces of a macroscopic cut made in the material.

We can obviously generalize these results to a distribution of imperfections in which n is a function of position. The microscopic $\mathbf{u}^\infty$ is still given by (13) but the gross displacement is

$$\langle \mathbf{u}^\infty \rangle = c\int_V \frac{n(\mathbf{r}')(\mathbf{r}-\mathbf{r}')}{|\mathbf{r}-\mathbf{r}'|^3} dv(\mathbf{r}'), \qquad (18)$$

and in the averaging process we must have $n^{-\frac{1}{3}} \ll R \ll n^{-\frac{1}{3}} |\mathrm{grad}(n^{-\frac{1}{3}})|$.

The gross or microscopic $p_{ij}{}^I$ is the stress produced by surface tractions $-p_{ij}{}^\infty n_j$ calculated from (13) or (14,18). These two surface tractions will differ by rapidly fluctuating quantities with a "wavelength" of the order of $n^{-\frac{1}{3}}$ whose effect will be confined to a surface layer of the same order. Thus in the bulk of the material the image quantities are efficiently smoothed by St. Venant's principle whether we smooth the infinity quantities or not, and we need not distinguish between their gross and microscopic values.

V. EFFECT OF POINT SINGULARITIES ON THE X-RAY DIFFRACTION PATTERN

Miller and Russel[6] originally suggested that there is a difference between the geometrical volume change of a crystal containing imperfections and the volume change deduced from the change of x-ray lattice constant. They calculate that for a uniform distribution of point imperfections the latter should be about twice the former.[7] Huang[8] and Teltow[9] found that they were

[5] O. D. Kellog, *Foundations of Potential Theory* (Julius Springer, Berlin, 1929), p. 194.

[6] P. H. Miller, Jr., and B. R. Russel, J. Appl. Phys. **23**, 1163 (1952).

[7] They have since withdrawn this factor. P. H. Miller, Jr., and B. R. Russel, J. Appl. Phys. **24**, 1248 (1953).

[8] K. Huang, Proc. Roy. Soc. (London) **A190**, 102 (1947).

[9] J. Teltow, Ann. Physik **12**, 111 (1953).

equal. These authors omit image effects. Miller and Russel calculate the image effect for a single imperfection at the center of a sphere. They take a collection of such spheres with the interstices filled up to represent a body containing many imperfections and argue that the image effect is negligible because the surface traction across these spheres is by no means zero. As we have seen, the effect of the image term on the volume change can be taken into account quite generally, and we shall find that the same is true for its effect on the x-ray lattice constant.

For an undisturbed crystal the scattering power plotted in reciprocal space will be a set of patches with their maxima at the points of the reciprocal lattice. If the crystal is distorted by external forces or internal imperfections the patches will be displaced and deformed. We may regard the new maxima as defining a new reciprocal lattice and we can then speak of the deformation of the reciprocal lattice corresponding to a deformation of the crystal.

Let the position of a point of the crystal or reciprocal lattice before and after distortion be

$$\mathbf{r}=L_i\mathbf{a}_i, \quad \mathbf{r}+\Delta\mathbf{r}=(L_i+\Delta L_i)\mathbf{a}_i$$

and

$$\mathbf{B}=h_i\mathbf{b}_i, \quad \mathbf{B}+\Delta\mathbf{B}=(h_i+\Delta h_i)\mathbf{b}_i.$$

Miller and Russel derive the following relation valid within a region around the origin of reciprocal space which decreases as the degree of crystal distortion increases:

$$\Delta h_i A_{ij}=-h_i B_{ij} \quad (j=1,2,3), \tag{19}$$

where

$$A_{ij}=\sum L_iL_j, \quad B_{ij}=\sum \Delta L_iL_j,$$

$\sum$ implying summation over all points of the crystal lattice. We have simplified equations (1,2,3) in reference 6, appendix 1, by taking the origin at the lattice point nearest the center of gravity of the crystal: then the sums $\sum L_i$ will differ from zero by integers small compared with the number of atoms in the crystal and may be neglected. Equation (19) has the solution

$$\Delta h_i=-h_kC_{ki}, \quad C_{ki}=B_{kj}A_{ji}' \tag{20}$$

where A_{ij}' is the matrix reciprocal to A_{ij}. C_{ij} is not in general symmetric even if B_{ij} is. Thus (20) is an infinitesimal affine transformation in reciprocal space. In other words the reciprocal lattice points are shifted as if they were embedded in an imaginary elastic continuum which undergoes a deformation in which all the strain and rotation components

$$e_{ij}=\tfrac{1}{2}(\partial\Delta h_i/\partial h_j+\partial\Delta h_j/\partial h_i)=-\tfrac{1}{2}(C_{ij}+C_{ji}),$$
$$\omega_{ij}=\tfrac{1}{2}(\partial\Delta h_i/\partial h_j-\partial\Delta h_j/\partial h_i)=-\tfrac{1}{2}(C_{ij}-C_{ji})$$

are in general different from zero and constant for small h_i. This implies, for example, that from x-ray measurements of the position of low-order spots alone, a distorted cubic crystal could not be distinguished from an unstrained monoclinic crystal slightly misoriented.

Following Huang and Teltow we may replace the displacement expressed as a sum of the effects of the separate imperfections by an integral if the distribution is statistically uniform. This corresponds to the averaging process of the previous section, and we may put‡ $a\Delta L_i=\langle u_i\rangle$. The summation $\sum$ may also be replaced by an integral and we have

$$A_{ij}=\frac{1}{a^5}\int_V x_ix_jdv, \quad B_{ij}=\frac{1}{a^5}\int_V x_i\langle u_j\rangle dv. \tag{21}$$

But we know that when the image terms are included $\langle u_i\rangle$ has the form (15), so that $A_{ij}=(4/3)\pi ncB_{ij}$ and (20) has the solution

$$\frac{\Delta h_1}{h_1}=\frac{\Delta h_2}{h_2}=\frac{\Delta h_3}{h_3}=-(4/3)\pi nc\gamma.$$

(The A_{ij} are products of inertia of the crystal and $|A_{ij}|$ could only vanish if the moment of inertia of the crystal about some axis were zero.) Thus the reciprocal lattice contracts uniformly in the way we should expect for a crystal which had suffered the uniform expansion (15).

This simple result would be destroyed if the image terms were omitted. We should have

$$B_{ij}=-\frac{cn}{a^5}\int_V x_i\frac{\partial\varphi}{\partial x_j}dv,$$

where

$$\varphi=\int_V dv/r$$

is the potential of the specimen if filled with charge of unit density. For an ellipsoidal crystal φ is a quadratic function of the coordinates (Eq. (16)), and it follows that $-\Delta h_1/h_1$, $-\Delta h_2/h_2$, $-\Delta h_3/h_3$ are equal, respectively, to the three quantities (17). In this case the (incorrect) deformation of the reciprocal lattice agrees with the (incorrect) macroscopic deformation with image effects omitted. For a body of arbitrary shape this is no longer true, since there is no simple relation between A_{ij} and B_{ij}. X-ray measurements would indicate that the cubic unit cell had become monoclinic, and its change of volume would bear no simple relation to the change in volume of the crystals calculated with or without image effects. These complications are implicit in Huang's calculations. Each of the two terms in his expression (I) is similar to our (14). Thus Huang's (I) does not represent a uniform expansion except for a sphere, and even for the sphere the expansion lacks a factor γ. Actually Huang is concerned with the more general question of the distribution of scattering power between the reciprocal lattice points. It is easy to see that in his expressions (II) and (III) c must be replaced by $c\gamma$. Since c is eventually eliminated with the help of his

‡ For simplicity we now suppose the crystal is simple cubic with lattice spacing a.

Eq. (18) in which also c must be replaced by $c\gamma$ the final result is unaffected, as also is the important result that the sharpness of the diffraction peaks is unaltered by the uniform distribution of defects.

Miller and Russel found by calculating the $\sum L_i \Delta L_i$ numerically that for a single singularity at the center of a sphere the expansion deduced from x-rays was 2.5 times the geometrical expansion. They originally deduced from this that there should be a similar difference (with the factor reduced from 2.5 to about 2) when there are a large number of imperfections distributed at random.

The previous discussion indicates that this is incorrect, but does not show exactly how the discrepancy arises. This discrepancy exists whether image terms are introduced or not (though introducing them would reduce the factor from 2 to $1+1/\gamma$) and for simplicity we omit them in the following discussion.

Let us find the effect of a single imperfection at a point $\boldsymbol{\xi}$ relative to the center of a spherical crystal of radius R. We can save calculation by finding the dilatation of the reciprocal lattice and not its detailed distortion. Putting $\langle u_i \rangle = c(x_i - \xi_i)/|\mathbf{x}-\boldsymbol{\xi}|^3$ in (21) we find after some manipulation

$$\frac{\Delta V^\infty}{V} = -\frac{\partial \Delta h_i}{\partial h_i} = \frac{15}{4\pi}\frac{c}{R^5}\left(1+\xi_i\frac{\partial}{\partial \xi_i}\right)\int_V \frac{dv}{|\mathbf{x}-\boldsymbol{\xi}|}.$$

The integral is the gravitational potential at $\boldsymbol{\xi}$ in a solid sphere of unit density and has the value $\frac{2}{3}\pi(3R^2-\xi^2)$. Thus

$$\Delta V^\infty = 10\pi c(1-\xi^2/R^2). \tag{22}$$

If we exclude from the integration a sphere of radius r_0 about $\boldsymbol{\xi}$ a term $-r_0^2/R^2$ must be introduced inside the bracket. Since r_0 can presumably be taken of atomic dimensions we shall omit it.

The reason for the discrepancy now becomes clear. The expression (22) is greater or less than the geometrical ΔV^∞ according as ξ is greater or less than $\sqrt{(3/5)}R = 0.77R$. For $\xi=0$ it is 2.5 times the geometrical value, in agreement with Miller and Russel. If there are a large number of defects scattered at random throughout the sphere, the average contribution to ΔV^∞ per imperfection is given by replacing ξ^2 in (22) by its mean value for all points of the sphere,

$$\langle \xi^2 \rangle = 4\pi \int_0^R r^4 dr \Big/ 4\pi \int_0^R r^2 dr = \tfrac{3}{5}R^2.$$

This gives a mean x-ray ΔV^∞ of $4\pi c$, in agreement with the geometrical value (7). There seems to be nothing objectionable in the replacement of sums by integrals, and presumably Miller and Russel's summation procedure would give the geometrical value if carried out for a reasonable number of imperfections scattered throughout a sphere.

As a corollary we see that with a given number of defects the x-ray expansion depends strongly on their (nonuniform) distribution, while the geometrical expansion is, of course, independent of the distribution. Thus, for a sphere the ratio of the x-ray ΔV^∞ to the geometrical ΔV^∞ can be given any value between 2.5 and 0 by sweeping the defects towards the center or surface. When the two expansions are corrected for the image terms the ratio lies between $1+1.5/\gamma$ and $1-1/\gamma$.

VI. DISCUSSION

The equality of the x-ray and geometrical expansions of a crystal uniformly filled with point imperfections has a bearing on the question whether they are of Schottky or Frenkel type. Revised calculations by Miller and Russel[7] gave a difference of not more than about 10 or 20 percent between the two expansions in place of their original factor of two. The residual difference may be partly the result of the neglect of image terms.

Turning to the factor γ introduced into the expansion by the image terms it looks at first sight as if one could redefine the strength of an imperfection as $c\gamma$ and then forget about the image effects. This is not so. The contributions to the expansion from $\mathbf{u}^\infty$ and $\mathbf{u}^I$ are in the ratio $1:(\gamma-1)$. But the ratio of their contributions to any other phenomenon will in general be quite different, since $\mathbf{u}^\infty$ is a rapidly fluctuating function of position associated with a pure shear, whereas $\mathbf{u}^I$ is smooth except near the surface with an associated stress which is chiefly a hydrostatic pressure in a body of reasonable shape; c is thus the basic constant which we might hope to determine from measurements of one effect and apply in calculating the value of another. The following examples illustrate this.

Nabarro[10] has discussed the hardness of dilute alloys. The hardening is due almost entirely to p_{ij}^∞, the slowly varying p_{ij}^I playing scarcely any part. He eliminates c with the help of the observed change of lattice constant with composition but neglects the image effect. This can be put right by dividing his theoretical constant $\alpha=0.05$ by γ before comparing it with the values deduced from experiment. The spread of the latter is too great to say if this gives any improvement.

Dexter[11] has calculated the effects of $\mathbf{u}^\infty$ and $\mathbf{u}^I$ on the electrical conductivity of metals. The former gives a temperature independent and the latter a temperature dependent contribution. They are of the same order at room temperature.

Overhauser[12] has considered the modification of rates of diffusion by the strain due to point imperfections. The effects arising from $\mathbf{u}^\infty$ and $\mathbf{u}^I$ depend differently on temperature, and the image term may become important at high temperatures.

Zener[13] has calculated the change of shear modulus caused by the $\mathbf{u}^\infty$ field of point imperfections. Evidently

[10] F. R. N. Nabarro, Proc. Phys. Soc. (London) **58**, 669 (1946).
[11] D. L. Dexter, Phys. Rev. **87**, 768 (1952).
[12] A. W. Overhauser, Phys. Rev. **90**, 393 (1953).
[13] C. Zener, Acta Cryst. **2**, 163 (1949).

the dilatation due to the image field should give rise to an analogous change of bulk modulus. (As we have seen, $\mathbf{u}^\infty$ produces no dilatation of the material between the imperfections.) In fact we should have a change

$$\frac{\Delta K}{K}=\frac{d\log K}{d\log V}\frac{\Delta V^I}{V}=\frac{d\log K}{d\log V}3\frac{\gamma-1}{\gamma}\frac{dl}{l} \tag{23}$$

proportional to the relative change dl/l of lattice constant but of opposite sign, since $d\log K/d\log V$ is negative. According to a simple theory[14] verified quite well by experiment $-d\log K/d\log V$ is equal to $2\Gamma+4/3$, where Γ is Grüneisen's constant. Zener's $\Delta\mu/\mu$ on the other hand depends on the shear strain energy associated with an imperfection and so is proportional to $|dl/l|$.

By an atomic calculation Dienes[15] has found that one percent of interstitials increase the bulk modulus of copper by 6.8 percent and of sodium by 1.9 percent. His calculations deal in principle with what happens in the immediate neighborhood of the imperfection, so it is reasonable to suppose that they take account of $\mathbf{u}^\infty$ and that the effects of $\mathbf{u}^I$ should be added. The results of Dienes and of Huntington and Seitz[16] suggest that we may take $4\pi c/\Omega$ to be about 0.6 and 0.9, respectively, for an interstitial in copper or sodium. By (13′) and (23) the corresponding percentage changes for one percent interstitials are lattice constant 0.9 (Cu), 1.4 (Na), bulk modulus −4.6 (Cu), −6.9 (Na). Our calculation cannot claim any accuracy, but it suggests that the image term is important, perhaps even large enough to change the sign of $\Delta K/K$.

The change of shear modulus found by Zener's argument depends on nonlinear behavior in the region close to the imperfection (see below) and it is therefore in principle already included in a calculation such as Dienes' which considers the balance of interatomic forces near the imperfection.

It was pointed out at the end of Sec. III that for a purely elastic model the actual volume of material was unaltered. Zener[17] has shown how in such a case the departure from Hooke's law gives a volume change, and we ought to consider what effect this will have on our calculations. This change has the value

$$\Delta V' = -(1+d\log K/d\log V)W_d/K - (1+d\log\mu/d\log V)W_s/K,$$

where W_d and W_s are the total dilatational and shear strain energies associated with the internal stress. The first term can be considered as a small correction to ΔV^I corresponding to using in (9) the value of K appropriate to the lattice as expanded. The shear strain associated with the sphere-in-hole model of Sec. II is $8\pi\mu c^2/r_0^3$ (see reference 8). If we take the case of interstitials in a face-centered cubic material and put r_0 equal to the distance from the interstitial to the nearest face-centered atom we have

$$\Delta V'/V = -(1+d\log\mu/d\log V)16\pi(\mu/K)(c/\Omega)^2 f.$$

This is about one tenth of (13′) for interstitials in copper if as before we take $c/\Omega\sim 0.05$. However, even if the ratio were quite large our previous calculations would not be upset. Of the shear strain energy $8\pi\mu c/r_0^3$ associated with an imperfection nearly ninety percent is contained within a sphere of radius $2r_0$. Thus the correction comes entirely from an additional expansion confined to the immediate neighborhood of the imperfection. This extra volume change is transmitted to the surface by the displacement $\mathbf{u}^\infty$ which must have the form (1) as soon as the elastic region is reached. Thus the effect we are considering merely "renormalizes" the constant c in (1) and it is correct to use the modified value not only in (13′) but also in discussing any other phenomenon in which only strains in the elastic region are important.

Point imperfections may expand the lattice by mechanisms other than misfit. Take for example an alkali halide crystal with doubly charged foreign cations in some positions together with an equal number of vacant anion sites. The continuum model is a dielectric with positive and negative charges embedded in it. Round each there is an elastic stress field due to electrostriction. The dilatation is large near a charge and falls off rapidly with distance. The stress field is essentially that of a center of dilatation. A rough calculation suggests that its strength is of the same order as for a reasonable degree of misfit. Since it is proportional to the square of the charge, the effect should exist although the number of positive and negative charges is equal. Again, if atoms of different valency are introduced substitutionally into a metal, the resulting change in the number of electrons in the conduction band will produce a change of lattice constant quite apart from any misfit effects.

It is not the aim of this paper to advocate the elastic approach, but to show that when it is used neglect of the presence of the free surface may lead to qualitative untidinesses (such as nonuniform deformation by a uniform distribution of defects) and quantitative errors which may be important.

[14] J. C. Slater, Phys. Rev. **57**, 744 (1940).
[15] G. J. Dienes, Phys. Rev. **86**, 228 (1952).
[16] H. B. Huntington and F. Seitz, Phys. Rev. **61**, 315 (1942).
[17] C. Zener, Trans. Am. Inst. Mining Met. Engrs. **147**, 361 (1942).

THE ELASTIC INTERACTION OF POINT DEFECTS*

J. D. ESHELBY†

In an isotropic material it is known that two point defects regarded as centres of dilatation interact with one another only indirectly through the modification of their elastic fields by the surface of the body. In a cubic material there is an additional direct interaction energy equal to the product of the inverse cube of their separation and a function of direction whose average over all angles is zero. This function is evaluated approximately. If in a more refined elastic model we replace each defect by a centre of dilatation plus a small region where the elastic constants differ from those of the matrix there is an additional interaction proportional to the inverse sixth power of the distance between them.

INTERACTION ÉLASTIQUE DES DÉFAUTS PONCTUELS

Dans un milieu isotrope, il est connu que deux défauts ponctuels, considérés comme des centres de dilatation, ne présentent d'interaction qu'indirectement par la modification de leurs champs élastiques par la surface du corps. Dans un métal cubique, il y a directement une énergie d'interaction complémentaire égale au produit de l'inverse du cube de leur distance et d'une fonction de la direction dont la valeur moyenne est nulle; cette fonction a été estimée. Si dans un modèle plus précis, nous remplaçons chaque défaut par un centre de dilatations plus un petit domaine où les constantes élastiques sont différentes de celles de la matrice, on obtient alors une interaction complémentaire proportionnelle à l'inverse de la sixième puissance de la distance de ces défauts.

DIE ELASTISCHE WECHSELWIRKUNG VON PUNKTFÖRMIGEN FEHLSTELLEN

Es ist bekannt, dass in einem isotropen Material zwei punktförmige Fehlstellen, die schematisch als Dilatationskernen betrachtet werden, nur indirekt über die Veränderung ihrer elastischen Felder an der Oberfläche des Körpers eine Wechselwirkung aufeinander ausüben. In einem kubischen Material besteht eine zusätzliche direkte Wechselwirkung, die dem reziproken Wert der dritten Potenz ihres Abstandes entspricht und eine Funktion der Richtung, dessen Winkelmittelwert verschwindet. Diese Funktion wird angenähert bestimmt. Wenn wir in einem verfeinerten elastischen Modell jeden Fehlstelle durch ein Dilatationskern plus einer kleinen Region, in der die elastischen Konstanten von der des umgebenden Materials verschieden sind, ersetzen, besteht eine zusätzliche Wechselwirkung, die dem reziproken Wert der sechsten Potenz des Abstandes zwischen ihnen proportional ist.

1. INTRODUCTION

The idea of a purely elastic interaction between dislocations and point defects (interstitial and substitutional atoms, lattice vacancies) has proved very useful. Also, when other effects have been allowed for or are negligible, good results can be obtained from an elastic treatment of the interaction of one point defect with another. In the present paper we summarize the known results for the interaction between point defects when they are idealized as centres of dilatation in an isotropic elastic continuum and extend them to the case of cubic anisotropy. For the isotropic case alone we consider the additional interaction when the defects are represented as a superposition of a centre of dilatation and a small region whose elastic constants differ from those of the matrix.

2. THE ISOTROPIC CASE

The displacement around a centre of dilatation of strength c is[1]

$$\mathbf{u}=\mathbf{u}^{\infty}+\mathbf{u}^{I}, \tag{1}$$

where

$$\mathbf{u}^{\infty}=c\mathbf{r}/r^{3}$$

and $\mathbf{u}^{I}$ is the image displacement necessary to satisfy the boundary conditions. In a body with a free surface and Poisson's ratio σ the centre produces a change of volume

$$\Delta V=4\pi\gamma c, \quad \text{with} \quad \gamma=3(1-\sigma)/(1+\sigma)$$

made up of a contribution $4\pi c$ from $\mathbf{u}^{\infty}$, representing a simple radial pushing out of the surface without any dilatation of the medium, and a contribution $4\pi(\gamma-1)c$ from $\mathbf{u}^{I}$ which does involve dilatation.

If external forces or a state of internal stress produce a hydrostatic pressure p at a centre of dilatation there is an interaction energy

$$E_{\text{int}}=\Delta V p \tag{2}$$

independent of the shear stress[2,3,4] (Cottrell's original expression corresponds to using ΔV^{∞} instead of ΔV). If we neglect image terms, two centres of dilatation do not interact[5] since each produces a pure shear stress. The indirect interaction via the image terms cannot be calculated in general, but for a large number of defects scattered uniformly through the medium the image terms add up to give practically a uniform dilatation. Thus the interaction energy of one defect with all the others is

$$E_{\text{int}}=K\Delta V\Delta V^{I}C/\Omega, \tag{2'}$$

* Received March 17, 1955.

† Department of Physical Metallurgy, University of Birmingham, Birmingham, England.

where C is the atomic concentration of defects, Ω the volume per atom, K the bulk modulus. Combining this with the "self-energy" required to insert each defect in a perfect lattice Friedel[6] has given a good account of the heat and entropy of formation of AuNi alloys.

A familiar elastic model of a point defect is an elastic sphere forced into a hole too small for it. If sphere and matrix have the same elastic constants the results above still hold even if the sphere is not infinitesimal, provided it does not overlap another or intersect the surface of the body. However if its elastic constants (λ',μ') differ from those of the matrix (λ,μ) there is an additional interaction with a stress-field. This can be treated separately from the misfit effect, and so we consider a perfectly-fitting inclusion. If surface forces produce a strain e_{ij} in the homogeneous body, the increase of its elastic energy when the inhomogeneous sphere is introduced is given[4] by the volume-integral

$$\Delta E=\tfrac{1}{2}\int\{(\lambda-\lambda')ee'+2(\mu-\mu')e_{ij}e_{ij}'\}dv \qquad (3)$$

taken over the inclusion; e_{ij}' is the strain in the inclusion. (Repeated suffixes are supposed to be summed over the values 1, 2, 3. $e=e_{ii}$ is the dilatation.) The interaction energy is

$$E_{\text{int}}=-\Delta E$$

and this is also the interaction energy when e_{ij} is due to some system of internal stress instead of to surface forces.

We may find the relation between e_{ij}' and e_{ij} as follows. If a spherical cavity perturbs a uniform shear e_{12}^0 the surface of the cavity deforms as if it were the surface of a solid sphere which had undergone a shear αe_{12}^0, where[7]

$$\alpha=(15K+20\mu)/(9K+8\mu).$$

If we superimpose an additional shear e_{12}^x everywhere the shear at infinity is $e_{12}=e_{12}^0+e_{12}^x$. In the hole it is $e_{12}'=\alpha e_{12}^0+e_{12}^x$, but the surface tractions at its surface are equivalent to those on a sphere with uniform shear stress μe_{12}^x. This surface traction can evidently be provided by inserting a sphere with shear modulus μ' such that $\mu' e_{12}'=\mu e_{12}^x$. A little algebra gives

$$e_{12}'=\mu\alpha e_{12}/(\mu-\mu'+\alpha\mu').$$

In just the same way, starting from the fact that a spherical hole perturbing a uniform dilatation e undergoes a fractional volume change βe where

$$\beta=(3K+4\mu)/4\mu$$

we find

$$e'=K\alpha e/(K-K'+\beta K')$$

for a spherical inclusion of bulk modulus K' perturbing a uniform hydrostatic pressure. It follows by superposition that e_{ij}' is uniform in the inclusion and is an isotropic linear function of the e_{ij}, say

$$e_{ij}'=Ae\delta_{ij}+2Be_{ij},$$

where A and B can be determined from the two special cases. Equation (3) will evidently give

$$\Delta E=\tfrac{1}{2}\Omega\{\Lambda e^2+2Me_{ij}e_{ij}\},$$

where Λ and M are functions of λ, λ', μ, μ' and Ω is the volume per atom. The actual expressions for Λ and M are rather clumsy and not very significant. It is more reasonable to relate them to a macroscopic quantity, the change of an elastic constant with number of defects, just as the strength c could be related to the macroscopic change of lattice parameter. A reasonably dilute solution of n defects per unit volume will add an amount $n\Delta E$ to the normal energy density $\tfrac{1}{2}\lambda e^2+\mu e_{ij}e_{ij}$, so that the apparent elastic constants will be

$$\lambda_{\text{app}}=\lambda+C\Lambda, \quad \mu_{\text{app}}=\mu+CM$$

if C is the atomic concentration of defects. For a point defect of strength c we have $e_{ij}e_{ij}=6c^2/r^6$ and so the interaction energy between two defects distant r apart and producing volume-changes ΔV_1, ΔV_2 is

$$E_{\text{int}}=6\Omega(M_1\Delta V_2^2+M_2\Delta V_1^2). \qquad (4)$$

For a pair of interstitials in copper Tucker and Sampson[8] suggest that $\Delta V\sim 3\Omega$ and Dienes' work[9] gives $M/\mu\sim 7$. This gives

$$E_{\text{int}}\sim 10(a_0/r)^6\ \text{eV}, \qquad (5)$$

where a_0 is the lattice parameter. This may well be an over-estimate, but suggests that the interaction may not be negligible in regions where r is large enough for the elastic theory to be applicable. For a vacancy $|\Delta V|$ may, perhaps, be about $\tfrac{1}{5}$ of ΔV for an interstitial, and if $|M|$ is also smaller than it is for an interstitial this would give a (presumably attractive) interaction energy of a hundredth or a thousandth of (5) or, say, of the order of $kT(a_0/r)^6$ at room temperature. Stripp and Kirkwood[10] have found a contribution $0.1kT(a_0/r)^6$ to the free energy from a pair of vacancies, arising from their influence on lattice vibrations. Except at very high temperatures the term (4) may be more important.

Equation (4) implies that two defects with positive M_1, M_2 ("hard spots") will repel each other, the opposite of Crussard's[11] conclusion.

3. THE CUBIC CASE

The elastic field (1) may be considered as a solution of the elastic equation when the body forces reduce to three equal crossed "double forces without moment."[7] Formally, the force density is

$$\mathbf{f}=-G\,\text{grad}\,\delta(\mathbf{r}). \qquad (6)$$

G may be related to ΔV by the expression

$$\left\{\int \mathbf{T}\cdot\mathbf{r}dS+\int \mathbf{f}\cdot\mathbf{r}dv\right\}\Big/3K$$

for the volume change produced by surface tractions $\mathbf{T}$ and body forces $\mathbf{f}$. In the present case $\mathbf{T}=0$ (free surface) and

$$\Delta V=-\frac{G}{3K}\int x_i\frac{\partial}{\partial x_i}\delta(\mathbf{r})dv=\frac{G}{K}. \quad (7)$$

(Equation (2) of reference 1 is thus incorrect.)

We may likewise take the elastic field of a center of dilatation in a cubic material to be a solution of the appropriate equations with the body-force[(6)]. This leads to the following form for the dilatation:

$$e=D\delta(\mathbf{r})+F(\mathbf{l})/r^3+\mathrm{div}\mathbf{u}^I, \quad (8)$$

where $\mathbf{l}=\mathbf{r}/r$ and

$$\int_{4\pi} F(\mathbf{l})d\omega=0. \quad (9)$$

This follows from a previous discussion[4] of the displacement due to a point force. An explicit expression for $F(\mathbf{l})$ in finite terms is impossible. The volume change is still given by (7) but its splitting into ΔV^∞ and ΔV^I is more complicated. Evidently the first term in (8) contributes an amount D to ΔV^∞. The contribution from the second term is zero if the surface of the material is a sphere centred on the imperfection, but is no longer so if, for example, the sphere is indented throughout a region over which $F(\mathbf{l})$ does not change sign. The contribution is proportional to

$$\lim_{\epsilon\to 0}\int d\omega\int_0^{r(\mathbf{l})} F(\mathbf{l})\frac{dr}{r}=\int d\omega F(\mathbf{l})\log r \quad (10)$$

if the polar diagram of the surface about the defect is $r=r(\mathbf{l})$. If a line is drawn through any interior point P of an ellipsoid meeting its surface in A and B then $AP\cdot PB$ depends only on P and not on the direction of the line. Thus (10), which can be written

$$\tfrac{1}{2}\int d\omega F(\mathbf{l})\log\{r(\mathbf{l})r(-\mathbf{l})\}$$

vanishes in view of (9) for a point defect anywhere within an ellipsoid. Nevertheless, the F-term can make no contribution to the total volume change produced by a "uniform random" distribution of a large number of defects in a body of any shape. For, by an argument due to Zener[12] or by a generalization of the analytical argument given for the isotropic case[1] it follows that macroscopically there is a uniform shape-independent dilatation. A contribution from the F-terms would be shape-dependent, and since it vanishes for the ellipsoid it must do so always. It ought to be possible to show directly that (10) vanishes when averaged over all possible positions of the defect.

Equation (2) is still true in the cubic case. For[4] the interaction of a singularity S with an elastic field T is given by the integral

$$\int (p_{ij}{}^S u_i{}^T - p_{ij}{}^T u_i{}^S)dS_j$$

over a surface enclosing the singularity. If S can be considered to be due to a density of body-force $\mathbf{f}^S$ then $\partial p_{ij}/\partial x_j=-f_i{}^S$ and the integral becomes

$$-\int f_i{}^S u_i{}^T dv,$$

or, in our case,

$$G\int \mathbf{u}^T\cdot\mathrm{grad}\delta(\mathbf{r})dv=-G\int\delta(\mathbf{r})\,\mathrm{div}\mathbf{u}^T dv$$

which in view of (7) is just ΔV times the value of p due to T at the singularity.

We now try to get an estimate of D and $F(\mathbf{l})$. We have to solve

$$c_{44}\nabla^2 u_1+(c_{12}+c_{44})\frac{\partial e}{\partial x_1}+d\frac{\partial^2 u_1}{\partial x_i{}^2}=G\frac{\partial}{\partial x_i}\delta(\mathbf{r}) \quad (11)$$

and two similar equations, with

$$d=c_{11}-c_{12}-2c_{44}.$$

By taking the divergence we have

$$(c_{12}+2c_{44})\nabla^2 e+d\Sigma_i\frac{\partial^3 u_i}{\partial x_i{}^3}=G\nabla^2\delta(\mathbf{r}). \quad (12)$$

Following Leibfried[13] we put $c_{ij}=c_{ij}{}^0+c_{ij}{}'$ where the $c_{ij}{}^0$ are effective isotropic constants found by averaging the c_{ij} over all orientations; explicitly

$$5c_{11}{}^0=3c_{11}+2c_{12}+4c_{44}$$
$$5c_{12}{}^0=c_{11}+4c_{12}-2c_{44}$$
$$5c_{44}{}^0=c_{11}-c_{12}+3c_{44}.$$

If we write $\mathbf{u}=\mathbf{u}^0+\mathbf{u}'+\ldots$ and regard the quantities with affixed ^{0} and $'$ as of the zeroth and first order we may solve (11) by successive approximation. The zeroth order gives the usual isotropic solution

$$(c_{12}{}^0+2c_{44}{}^0)e^0=G\delta(\mathbf{r})$$
$$-4\pi(c_{12}{}^0+2c_{44}{}^0)\mathbf{u}^0=G\,\mathrm{grad}(1/r).$$

Then (12) gives for e'

$$\nabla^2\left\{(c_{12}{}^0+2c_{44}{}^0)e'+(c_{12}{}'+2c_{44}{}')e^0 - \frac{Gd}{8\pi(c_{12}{}^0+2c_{44}{}^0)}\Pi r\right\}=0,$$

noting that $\nabla^2 r=2/r$ and writing

$$\Pi\equiv\Sigma_i\partial^4/\partial x_i{}^4.$$

A solution is evidently given by deleting the ∇^2. We must be careful to avoid dropping a delta-function in operating with $\prod$ on r. If we add and subtract the mean value of $\prod$ over all angles,

$$\prod = \tfrac{3}{5}\nabla^4 + (\prod - \tfrac{3}{5}\nabla^4),$$

the first term gives a multiple of $\delta(\mathbf{r})$ whilst the second term, having zero mean value over angles, does not. We find

$$e = \frac{G}{c_{11}^0}\delta(\mathbf{r}) + \frac{15}{8\pi}\frac{Gd}{(c_{11}^0)^2}\frac{1}{r^7}(x_1^4 + x_2^4 + x_3 - \tfrac{3}{5}r^4). \quad (13)$$

The interaction energy between a pair of defects which produce volume changes ΔV_1, ΔV_2 is thus

$$E_{\text{int}} = -\frac{15}{32\pi\gamma^2}\frac{\Delta V_1}{\Omega}\frac{\Delta V_2}{\Omega}\Omega d\frac{a_0^3}{r^3}\Gamma,$$

where

$$\Gamma = l^4 + m^4 + n^4 - \tfrac{3}{5} = \tfrac{2}{5} - 2(l^2m^2 + m^2n^2 + n^2l^2), \quad a_0^3 = 4\Omega$$

if (l,m,n) are the direction cosines of the line joining the defects. Schmid and Boas[14] have given a stereographic plot of $l^2m^2 + m^2n^2 + n^2l^2$ from which we can see that Γ has a maximum value of .40 in the 100-direction, a minimum of $-.27$ in the 111-direction and a value $-.10$ in the 110-direction where there is a saddle-point. Thus for any sign of d, ΔV_1, ΔV_2 there are directions along which the interaction is attractive. These are the 111 directions for metals (d negative) and like sign of the ΔV. For a pair of interstitials in copper with $\Delta V \sim 3\Omega$ we find

$$E_{\text{int}} \sim 7.5(a_0/r)^3\Gamma \text{ eV}.$$

The image interaction (2′) still has the same form. ΔV^∞ is now the coefficient of $\delta(\mathbf{r})$ in (13) and thus $\Delta V^I/\Delta V = (\gamma-1)/\gamma$ as in the isotropic case, provided we form γ from the c_{ij}^0.

REFERENCES

1. J. D. Eshelby, J. Appl. Phys. **25**, 255 (1954).
2. A. H. Cottrell, *Report on Strength of Solids* (Physical Society London, 1948), p. 30.
3. B. A. Bilby, Proc. Phys. Soc. **63A**, 191 (1950).
4. J. D. Eshelby, Phil. Trans. Roy. Soc. **244**, 87 (1951).
5. F. Bitter, Phys. Rev. **37**, 1526 (1931).
6. J. Friedel, Advances in Physics **3**, 446 (1954).
7. A. E. H. Love, *Mathematical Theory of Elasticity* (Cambridge, 1952).
8. C. W. Tucker, Jr. and J. B. Sampson, Acta Met. **2**, 433 (1954).
9. G. J. Dienes, Phys. Rev. **86**, 228 (1952).
10. K. F. Stripp and J. G. Kirkwood, J. Chem. Phys. **22**, 1579 (1954).
11. C. Crussard, Métaux et Corrosion **25**, 203 (1950).
12. C. Zener, Phys. Rev. **74**, 639 (1948).
13. G. Leibfried, Z. Phys. **135**, 23 (1953).
14. E. Schmid and W. Boas, *Plasticity of Crystals* (F. A. Hughes & Co., 1950), p. 192.

Note on the Heating Effect of Moving Dislocations*

In a recent paper,[1] Freudenthal and Weiner have proposed that thermal stresses produced during fatigue are severe enough to create microcracks; these stresses are due to highly localized "thermal flashes," associated with repeated and reversed slip processes, which in the extreme case may raise the temperature near the slip plane by some 200°C. The heating effect of dislocations moving along slip planes is in fact of importance in many aspects of plastic deformation. In this note the heating effect is calculated in a different manner, and some of the assumptions of Freudenthal and Weiner are discussed.

We take the dislocation to be a moving line-source of heat of strength

$$q = b\tau V \text{ erg cm}^{-1}\text{ sec}^{-1}$$

where b is its Burgers vector, V its velocity, and τ is the applied shear stress. The temperature at (x,y) is[2]

$$T = \frac{q}{2\pi K} e^{x/\Lambda} K_0(r/\Lambda), \; r^2 = x^2 + y^2.$$

x,y are rectangular co-ordinates centered on the dislocation, $y = 0$ is the slip-plane, and x is positive behind the dislocation. K is the heat conductivity and κ is the thermal diffusivity (K divided by the specific heat per unit volume). The Bessel function $K_0(z)$ has the asymptotic forms

$$K_0(z) \sim -\ln z, \qquad z \gg 1;$$

$$K_0(z) \sim \left(\frac{\pi}{2z}\right)^{\frac{1}{2}} e^{-z}, \quad z \ll 1.$$

The length

$$\Lambda = 2\kappa/V$$

fixes the scale of the temperature pattern. It will usually be many atomic distances, so that near the dislocation

$$T = \frac{b\tau V}{2\pi K} \ln \frac{\Lambda}{r}$$

(compare Seitz[3]). Even for fairly extreme values of the constants, T is of the order of only a fraction of a degree a few atomic distances from the dislocation.

The question now is, do the temperature fields of a succession of dislocations add up to give an appreciable rise? We can make an estimate in two extreme cases. Consider a procession of n equally-spaced dislocations spread over a distance λ. Then if $\Lambda \gg \lambda$ the maximum temperature will be about

$$T = \frac{q}{2\pi K} \sum_{\nu=1}^{n} \ln \frac{\Lambda \nu}{\lambda} \sim \frac{q}{2\pi K} n \ln \frac{n\Lambda}{\lambda}$$

and the temperature will be n times what it would be at a distance λ/n (inter-dislocation spacing) for a single dislocation.

If $\Lambda \ll \lambda$, the maximum temperature will be about

$$T = \frac{q}{2\pi K} \int_{\lambda/n}^{\lambda} \left(\frac{\pi\Lambda}{2r}\right) \frac{dr}{\lambda/n} \sim \frac{q}{2\pi K} \left(\frac{\pi\Lambda}{2\lambda}\right)^{\frac{1}{2}} n$$

which is identical with Freudenthal and Weiner's equation (5):

$$T = \frac{\tau}{mK} \left(\frac{\kappa\lambda}{\pi}\right)^{\frac{1}{2}}$$

where m is the spacing between successive dislocations in units of b. As before, there will be a temperature-rise about n times that due to a single dislocation, provided the factor $(\Lambda/\lambda)^{\frac{1}{2}}$ is not too unfavorable.

To determine the actual temperature rise, we have to estimate plausible values for V, λ, n, and decide if one of the approximations $\Lambda \gg \lambda$, $\Lambda \ll \lambda$ is appropriate. V has a maximum value of some fraction of the velocity of sound, say $\frac{1}{3} V_0 \sim 10^5$ cm/sec. It is commonly assumed that all the dislocations on one slip-plane come from the same Frank-Read source; if the sweeping velocity of the source is about the same as the velocity of the moving dislocations, the spacing between successive dislocations will be about l, the length of the Frank-Read source. In annealed material, $l \sim 10^{-4}$ cm, and, with $n = 500$, λ becomes about 10^{-1} cm. Thus, $\Lambda \ll \lambda$, and the second form of our equation is the most appropriate. Freudenthal and Weiner achieve a temperature rise of about 8°C on a slip-plane in aluminum, with 500 dislocations moving at $\frac{1}{3} V_0$ in a procession 10^{-4} cm in length; whereas our model suggests that this procession has an effective length of 10^{-1} cm, giving a temperature-rise of about half a degree. Similarly, for the hard light alloy, 24S-T, we would suggest a temperature-rise of about two degrees rather than fifty degrees.

This discrepancy, of course, arises from the differing values, in the two models, of the length λ of the procession of dislocations. We have considered the case in which dislocations move freely along the slip-plane with a spacing determined by their rate of emission from the Frank-Read source, whereas Freudenthal and Weiner consider a procession of dislocations uniformly separated by about five interatomic distances. Bunching of the dislocations in the slip-plane may occur because of retarding forces, but it is unlikely that the average spacing could reach the value suggested by Freudenthal and Weiner. Their distribution would require very high stresses to keep the dislocations together, and the stress around the array would then exceed the theoretical strength of the material. If the bunching was so severe that the distribution was similar to that of a static pile-up of n dislocations, the distance between first and last, λ, would be about n^2 times the distance between the leading pair,[4] as against n times this distance for Freudenthal and Weiner's equally spaced procession. We believe then that the value of λ for 500 dislocations, about 10^{-4} cm from the simple picture of the Frank-Read source, might be reduced to 10^{-2} cm if bunching should occur, but that the figure of 10^{-4} cm is too small.

The effect of stacking individual slip processes into nearby parallel slip-planes increases considerably the possible rise in temperature. Freudenthal and Weiner's equation 7 is the same as that, obtained by different arguments, by Cottrell[5] and before that by Orowan.[6] For a given shear stress, the rise in temperature varies linearly with the amount of glide on each plane and inversely with the spacing between the slip-planes, so that a large shear on many closely-packed planes could give an appreciable rise in temperature. For this magnified heating effect it is necessary that the nearby slip processes occur simultaneously, and Freudenthal and Weiner suggest that this is not highly probable. The microscopical evidence presented by Brown[7] and Forsyth[8] suggests that, during fatigue, slip only occurs simultaneously on planes a micron or more apart. the "striations" widening and intensifying during the test by the addition of individual slip processes one after the other. Furthermore, if displacements of 1500 Å were to occur on many planes 200 Å apart, as Freudenthal and Weiner suggest, the material would suffer the equivalent of a homogeneous shear of seven atomic diameters on each atom plane; and this is clearly not achieved over large volumes of the material. The nearest approximation to this case is fine slip,[9] in which the displacement is about 50 Å on many slip-planes 200 Å apart. Here the maximum rise in temperature might be a few degrees if many of the slip processes occurred simultaneously.

It seems improbable then that there can be appreciable local heating near the glide-plane, except perhaps at very high rates of loading. The creation of cracks during normal reversed stressing may be explained by the pile-up of dislocations,[10] and the softening processes by the local generation of vacancies.[11]

J. D. ESHELBY
P. L. PRATT
Department of Physical Metallurgy
Birmingham University, England

References

1. A. M. FREUDENTHAL and J. H. WEINER *J. Appl. Phys.* **27,** 44 (1956).
2. H. S. CARSLAW and J. C. JAEGER *Conduction of Heat in Solids* p. 224, Oxford (1947).
3. F. SEITZ *Advances in Physics* **1,** 43 (1952).
4. J. D. ESHELBY, F. C. FRANK, and F. R. N. NABARRO *Phil. Mag.* **42,** 351 (1951).
5. A. H. COTTRELL *Dislocations and Plastic Flow in Crystals* p. 6 Oxford (1953).
6. E. OROWAN in *Principles of Rheological Measurement* p. 180 Nelson (1949).
7. A. F. BROWN *Advances in Physics* **1,** 427 (1952).
8. P. J. E. FORSYTH *J. Inst. Metals* **80,** 181 (1952/53); **82,** 449 (1953/54); P. J. E. FORSYTH and C. A. STUBBINGTON *R.A.E. Report Met.* 76 (1953), *Met.* 78 (1954).
9. D. KUHLMANN-WILSDORF and H. WILSDORF *Acta Met.* **4,** 394 (1953).
10. N. F. MOTT *J. Phys. Soc. Japan* 10 (8), 650 (1955).
11. T. BROOM and J. H. MOLINEUX *J. Inst. Metals* **83,** 528 (1954/55).

* Received March 14, 1956.

Supersonic Dislocations and Dislocations in Dispersive Media

By J. D. ESHELBY

Department of Metallurgy, University of Birmingham

MS. received 23rd *March* 1956

Abstract. The conditions governing the motion of a screw dislocation at velocities above and below the velocity of shear waves are contrasted. For supersonic motion the Peierls–Nabarro equation becomes a differential equation, in contrast to the subsonic integral equation. It has a solution representing a moving restoration of fit along a slip plane across which there was originally complete misfit.

In a dispersive medium where the sound velocity decreases with decreasing wavelength a dislocation moving between the maximum and minimum sound velocities suffers a retarding force, which may be calculated if the shape of the dislocation is supposed to be known.

§ 1. Introduction

It is well known that the energy of a moving screw dislocation becomes infinite in 'relativistic' fashion as its velocity approaches the velocity c of transverse elastic waves (Frank 1949, Leibfried and Dietze 1949). There are, nevertheless, dislocation-like solutions of the elastic equations which move with velocities greater than c and which satisfy the Peierls condition or a prescribed generalization of it. A radiation of elastic energy away from the slip plane is always involved. For such supersonic velocities the Peierls–Nabarro equation becomes a differential equation in contrast to the integral equation of the subsonic case. The simplest solutions describe a crystal in which the atoms immediately above and below the slip plane start in a state of complete misfit; the passage of a supersonic ' dislocation ' then restores proper fit. These solutions may have some relation to the propagation of diffusionless transformations by dislocations.

In an actual crystal there is the complication of dispersion, the fact that the velocity of an elastic wave depends on its wavelength (and also on its direction and polarization). The sound velocity decreases from a maximum for infinite wavelengths to a minimum for the shortest wavelengths which the atomic structure allows. If it has a velocity below this minimum the dislocation, within the limitations of the elastic or Peierls–Nabarro model, moves freely. Above this minimum it is partly supersonic and so suffers a retarding force. The applied stress required to maintain its motion can be estimated if the width of the dislocation is known: disappointingly, it depends exponentially on the width. A complete investigation would involve setting up and solving the Peierls–Nabarro equation for the dispersive case. This is not attempted.

§ 2. Subsonic and Supersonic Dislocations

The elastic displacement w of a screw dislocation moving uniformly with velocity $v<c$ can be built up as a Fourier integral from elementary solutions of the wave equation

$$\frac{\partial^2 w}{\partial x^2}+\frac{\partial^2 w}{\partial y^2}-\frac{1}{c^2}\frac{\partial^2 w}{\partial t^2}=0 \qquad \ldots\ldots(1)$$

of the type

$$w=\exp\{i(x-vt)\mp\beta y\}, \quad \beta=(1-v^2/c^2)^{1/2} \qquad \ldots\ldots(2)$$

above and below the slip plane, that is, from plane waves moving parallel to the x axis and attenuated perpendicular to the direction of propagation. Stress and displacement on a plane $y=$constant are related by

$$p_{zy}(x,y)=\mu\frac{\partial w}{\partial y}=\frac{\mu\beta}{\pi}\int_{-\infty}^{\infty}\frac{1}{\xi-x}\frac{dw(\xi,y)}{d\xi}\,d\xi.$$

In the model of Peierls and Nabarro the dislocation is held together by a certain law of force connecting stress and displacement on the atomic planes $y=\pm\frac{1}{2}a$ adjacent to the slip plane, say

$$p_{zy}(x,\tfrac{1}{2}a)=-f[w(x,\tfrac{1}{2}a)]. \qquad \ldots\ldots(3)$$

The simple law

$$p_{zy}(x,\tfrac{1}{2}a)=-\frac{\mu b}{2\pi a}\sin\frac{4\pi w(x,\frac{1}{2}a)}{b} \qquad \ldots\ldots(4)$$

is satisfied by the displacement

$$w=\frac{b}{2\pi}\tan^{-1}\frac{\beta y}{x-vt}. \qquad \ldots\ldots(5)$$

The stress p_{zy} and velocity dw/dt of the elementary waves (2) are in quadrature, so that in any solution built up from them there is no net flux of energy perpendicular to the slip plane and the interatomic forces which give rise to the law (3) take care of the local energy transfers involved.

On the other hand if we look for a state of anti-plane strain moving uniformly with a velocity greater than c the wave equation takes the hyperbolic form

$$\frac{\partial^2 w}{\partial y^2}-\gamma^2\frac{\partial^2 w}{\partial(x-vt)^2}=0, \quad \gamma=i\beta=(v^2/c^2-1)^{1/2}$$

with the general solution

$$w=w_1(x-vt+\gamma y)+w_2(x-vt-\gamma y) \qquad \ldots\ldots(6)$$

i.e. plane waves of velocity c inclined to the x axis at such an angle that along a line $y=$constant the profile of w propagates without change of form at a velocity v greater than c.

The solution (6) in the upper half-plane combined with the complementary solution

$$w=-w_1(x-vt-\gamma y)-w_2(x-vt+\gamma y)$$

in the lower half-plane will give a moving system of self-stress which could be maintained by a suitable law of force (3) across the slip plane. Physically, however, w_1 and w_2 are not on the same footing. They represent, respectively, plane waves moving obliquely out of and into the slip plane. Thus whereas w_1 can reasonably be said to be 'due to' the moving dislocation, the ingoing w_2 can only be maintained by elaborate arrangements at the outer boundary of the medium, and is better considered to be part of the externally applied elastic field. (In the electromagnetic analogy (Nabarro 1951, Eshelby 1953) w_1 and w_2 are associated with the retarded and advanced potentials of the ribbon of charge which replaces the displacement discontinuity across the slip plane.)

In particular for an infinite medium with a constant applied stress p^A,

$$w = \pm w_1(x - vt \pm \gamma y) + p^A y.$$

Since now the stress due to the dislocation is

$$p_{zy} = \mu \frac{\partial w_1}{\partial y} = \gamma\mu \frac{\partial w_1}{\partial x} = \gamma\mu \frac{\partial w}{\partial x}$$

the Peierls integral equation is replaced by the *differential* equation

$$\gamma\mu \frac{dw}{dx} + p^A = -f(w) \qquad \text{......(7)}$$

in the supersonic case.

Suppose first that $p^A = 0$. Since p_{zy} and dw/dt are proportional there is everywhere a flux of energy out of the slip plane, and we might imagine that there would be no steady solution. But taking, for example, the law (4) there is the solution

$$w = \frac{b}{2\pi} \tan^{-1} \exp\{-2(x - vt)/\gamma a\}. \qquad \text{......(8)}$$

w is zero before the passage of the disturbance, $\frac{1}{4}b$ afterwards. Comparison of (5) with figure 1(a) shows that $w = \frac{1}{4}b$ means perfect fit across the slip plane and $w = 0$ perfect misfit. (For clarity figures 1 (a) and (b) are drawn as if for an edge dislocation.) Let the upper half of the perfect crystal be shifted by $\frac{1}{2}b$ relative to the lower; then (8) evidently describes a restoration of fit progressing from left to right (figure 1(b)).

Things are much the same for an arbitrary law of force. We have

$$x - vt = -\gamma\mu \int_0^w \frac{dw}{f(w)}$$

and figure 1(c) shows that, as x goes from ∞ to $-\infty$, w swings from an unstable to a stable zero of f. The driving force to maintain the motion comes from the

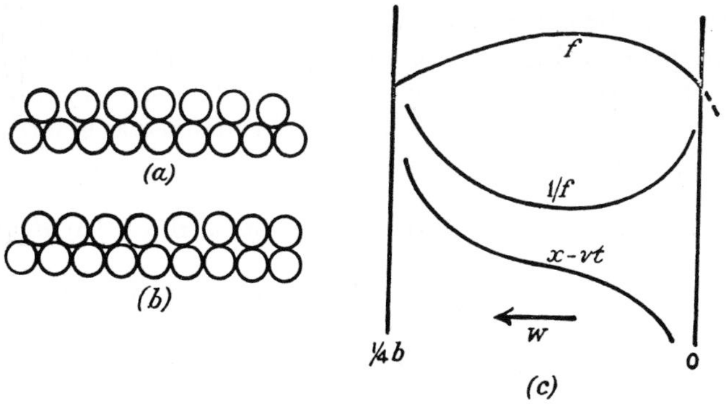

Figure 1

energy released as the region of misfit disappears. If there is a moderate applied stress we may introduce an effective force $f_1 = f(w) + p^A$. Then if p^A does not exceed the theoretical yield stress (the magnitude of the most negative value of f)

X–2

f_1 has zeros and figure 1(c) still applies, reading f_1 for f. For p^A greater than the theoretical yield stress f_1 is always positive and w is a linear function of $x-vt$ with a superimposed periodic ripple.

§ 3. The Effect of Dispersion

Consider now a dispersive medium. The relation between frequency $\omega/2\pi$ and wave number k will have the general shape shown in figure 2. (We suppose that ω is independent of the direction of k.) $c(k)=\omega(k)/k$ will be a

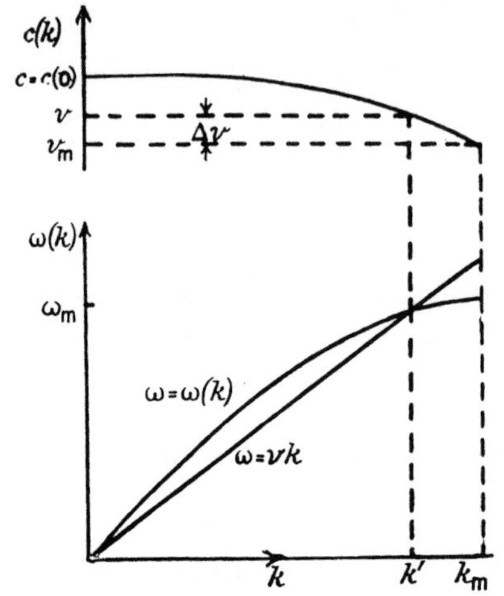

Figure 2.

decreasing function of k. Let k_m be the largest value of k permitted by the atomic structure. Then a dislocation with velocity v will be subsonic with respect to all plane waves if a line of slope v on the diagram fails to cut the (ω, k) curve for k less than k_m. If, however, it intersects the curve (say at k'), it is partly subsonic, partly supersonic.

If the displacement at the slip plane is

$$w(x-vt, \tfrac{1}{2}a) = \int_{-\infty}^{\infty} u(k) \exp ik(x-vt)\, dk$$

a typical element in the corresponding integral for the displacement $w(x-vt, y)$ above the slip plane will have the form $\exp ik\{x-vt+i\beta(y-\tfrac{1}{2}a)\}$ (compare equation (2)) if $0<|k|<k'$ and the form $\exp ik\{x-vt+\gamma(y-\tfrac{1}{2}a)\}$ (compare equation (6) with $w_2=0$) if $k'<|k|<k_m$. β and γ, defined as before, now depend on k through $c(k)$. We may thus write

$$w(x-vt, y) = \int_{-\infty}^{\infty} u(k) \exp ik\{x-vt+i\beta(y-\tfrac{1}{2}a)\}\, dk$$

with the following convention for the ambiguous sign inherent in the square roots:

$$\left.\begin{aligned} \beta(k) &= -\beta(-k) > 0, \qquad k^2<k'^2 \\ \gamma(k) &= i\beta(k) > 0, \qquad k^2>k'^2. \end{aligned}\right\} \qquad \ldots\ldots(9)$$

The first condition ensures that the subsonic components fall off with increasing distance from the slip plane and the second that the supersonic components are outgoing waves.

The stress at the slip plane will be

$$p_{zy}(x, \tfrac{1}{2}a) = \int p(k) \exp(ikx)\, dk$$

with

$$p(k) = -\mu(k)\, k\, \beta(k)\, u(k)$$

where $\mu(k)$ is an effective elastic constant which we may take to be proportional to $c^2(k)$ for the supersonic components. Its value for the subsonic components will not be needed. (For brevity we have put $t=0$ and here and below we drop the limits of integration $\pm\infty$.) Similarly

$$\frac{dw}{dt}(x, \tfrac{1}{2}a) = \int q(k) \exp ikx\, dk$$

with $q(k) = -vik\,u(k)$. The flux of energy out of the slip plane is

$$E = -2\int p_{zy} \frac{dw}{dt}\, dx$$

which by Parseval's theorem (Titchmarsh 1937) may be written

$$E = -4\pi \int p(k)\, q(-k)\, dk$$

or with the help of (9)

$$E = 8\pi v \int_{k'}^{k_m} \mu(k)\, k^2 \gamma(k)\, u(k)\, u(-k)\, dk.$$

As we should expect, the subsonic components make no contribution. If v exceeds v_m only slightly, γ alone will vary appreciably over the range of integration; the other factors may be replaced by their value for $k=k_m$. Moreover, since $\partial\omega/\partial k = 0$ at $k = k_m$ we may put $c(k) = \omega(k)/k \simeq \omega_m/k$ and

$$\gamma = [v/c(k)+1]^{1/2}\,[v/c(k)-1]^{1/2} \simeq 2^{1/2}(k-k')^{1/2}/k_m^{1/2}.$$

Hence

$$E = (16\pi 2^{1/2}/3)\, v\mu(k_m)\, k_m\, u(k_m)\, u(-k_m)\, (k_m-k')^{3/2}/k_m^{3/2}.$$

The final factor is $(\Delta v/v_m)^{3/2}$ where Δv is the excess of v over v_m. Then if, for definiteness, we give k_m and $\mu(k_m)$ the values π/b, $4\mu/\pi^2$ appropriate to the 100 direction in a simple cubic lattice of spacing b we have

$$E = (64\,.\,2^{1/2}/3b)\, v\mu\, u(k_m)\, u(-k_m) (\Delta v/v_m)^{3/2}.$$

To find the form of $w(x)$ and hence $u(k)$ we should need to have a solution of the Peierls–Nabarro equation generalized for a dispersive medium. Failing this we suppose that w has the familiar form

$$w = (b/2\pi) \tan^{-1}(x/\zeta) \qquad \ldots\ldots(10)$$

with some value of ζ. Then

$$u = (b/2\pi) \exp(-\zeta\,|k|). \qquad \ldots\ldots(11)$$

E must be equated to bvp^A, the rate of working of the applied stress maintaining the motion. Rounding off the numerical coefficient we have finally

$$p^A/\mu \simeq (\Delta v/v_m)^{3/2} \exp\{-2\pi\zeta/b\}. \qquad \ldots\ldots(12)$$

§ 4. Discussion

The present calculations were originally undertaken in the hope that dispersion might provide a retarding mechanism which though unlikely to be the dominant effect, would at least provide an upper limit to the velocity of dislocations sufficiently unambiguous to answer some questions about their behaviour. (For a discussion of other dissipation mechanisms see Seeger (1955).) Unfortunately, the presence of the exponential factor in (12) prevents us from saying much more than that a non-zero retarding force sets in above v_m. This factor appears because the effect depends on the strength of those Fourier components (11) of the displacement (10) which have a wavelength of order b. Even if we could solve the appropriate dispersive Peierls–Nabarro equation precisely the result would be extremely sensitive to the law of force $f(w)$ assumed. There must evidently be a similar sensitive factor associated with any property of a dislocation in which the short wavelength components play the dominant role (e.g. the Peierls anchoring force). When the velocity of the dislocation is only slightly greater than v_m (so that $\Delta v/v_m \ll 1$) the proportion of supersonic components in the Fourier spectrum of its elastic field is small. We might perhaps assume that the dislocation is, so to speak, mainly held together by the subsonic components. The width ζ is then fixed by the Peierls–Nabarro integral equation. With the law of force (4) this gives $p^A/\mu \sim 10^{-3}$ for $\Delta v/v_m \sim 1/10$. The exponential in (12) is the square root of that in Nabarro's (1951) expression for the Peierls force, and is equal to the exponential in the modification proposed by Huntington (1955).

The effect we have discussed is entirely analogous to the retardation of an electron by Čerenkov radiation when it moves through a dispersive dielectric, but with an important difference of detail. In the Čerenkov effect the electron may be treated as a point charge. Analogously we could formally consider a 'point dislocation', putting $\zeta = 0$ in (12) to obtain the simple result $p^A/\mu \sim (\Delta v/v_m)^{3/2}$. But in fact the atomic structure determines not only the dispersive properties of the medium, but also the width of the dislocation, and so introduces the exponential factor.

Returning to the case of a non-dispersive medium we have seen in § 2 that equation (7) purports to have as solutions all physically significant states of internal stress which can be supported by the law (4) and which move uniformly with velocity greater than c. However, Seeger (1953) has discussed certain 'oszillatorische Eigenbewegungen' which satisfy the wave equation (1) in the elastic region and the law (4) across the slip plane, but which are not solutions of (7). The displacement is just of the form (6) with $w_1(x) = w_2(x)$, so that there is an equal mixture of ingoing and outgoing waves. There is no net transfer of energy into or out of the slip plane. In general this state of affairs could only be maintained by an elaborate mechanism at the outer surface of the medium. In the special case of an infinite plate of suitable thickness the ingoing wave becomes the reflection of the outgoing. The surfaces would, however, have to be 'acoustically flat' even for the shortest wavelengths involved, that is, to within a fraction of an atomic spacing. Moreover, the possibility of maintaining the motion depends on the plate being infinitely long, for the disturbance can be loosely described as a procession of 'dislocations' each of which depends for the maintenance of its motion on the wake of all its predecessors multiply reflected by the surfaces of the plate. Of course solutions of the type (8) are open to the reverse objection: strictly they require a complete absence of ingoing waves. Nevertheless the motion might be expected to persist even in face of the largely unorganized ingoing waves resulting from internal scattering and reflection at the microscopically rough outer boundary, particularly as we now have in the misfit energy a definite driving force located at the slip plane.

Figure 1 (*b*) is reminiscent of a spreading stacking fault. If we take the movement of a particular atom from an unstable to stable position relative to its neighbours as representing a more complex energy-releasing rearrangement of atoms we may take our solution as symbolically representing the motion of a transformation dislocation moving in the boundary between two phases. (Compare for example, Bilby and Christian 1956.)

References

BILBY, B. A., and CHRISTIAN, J. W., 1956, *Institute of Metals Monograph and Report Series*, No. 18, 121.

ESHELBY, J. D., 1953, *Phys. Rev.*, **90,** 248.

FRANK, F. C., 1949, *Proc. Phys. Soc.* A, **62,** 131.

HUNTINGTON, H. B., 1955, *Proc. Phys. Soc.* B, **68,** 1043.

LEIBFRIED, G., and DIETZE, H. D., 1949, *Z. Phys.*, **126,** 790.

NABARRO, F. R. N., 1951, *Proc. Roy. Soc.* A, **209,** 278.

SEEGER, A., 1953, *Z. Naturf.*, **8a,** 47; 1955, *Handbuch der Physik*, VII/1, p. 622 (Berlin : Springer).

TITCHMARSH, B. C., 1937, *Fourier Integrals* (Oxford : Clarendon Press).

The Continuum Theory of Lattice Defects

J. D. ESHELBY
University of Birmingham, Birmingham, England

I. Introduction

1. Relation between Lattice Defects and Continuum Theory

Among the imperfections to which a crystal is subject,[1] some (interstitial and impurity atoms, vacant lattice sites, dislocations . . .) are relatively permanent. The introduction of one of them generally alters the

[1] F. Seitz, *in* "Imperfections in Nearly Perfect Crystals" (W. Shockley, ed.), Chapter 1. Wiley, New York, 1952.

position of every lattice point. Obviously in calculation we cannot take every lattice point into account explicitly in a crystal of any size, and must be content to treat the greater part of the crystal as a continuum. In favorable cases the exact behavior in the regions where the continuum approximation is inappropriate is unimportant and can be taken into account by giving suitable values to certain parameters appearing in the continuum solution.

The continuum analog of a crystal containing imperfections is an elastic body in a state of stress not produced by surface and body forces. The appropriate tool for handling the "continuum theory of lattice defects" is thus the usual theory of elasticity modified to include internal stress. Unlike the residual stresses encountered in engineering practice, these internal stresses have to be considered as capable of moving about in the medium. Such mobile "strain figures" were discussed by Burton[2] and Larmor[3] when elastic models of the ether were in vogue. Recent interest in solid state physics has stimulated further development. It is the object of the present review to emphasize some of the background principles and to illustrate them by specific examples chosen to bring out the peculiar features involved. Naturally the continuum theory can hardly be expected to answer questions of current interest about the more intimate behavior of lattice defects (e.g., the binding energy of two adjacent point defects). On the other hand, the theory perhaps suffers from the disadvantage that its limitations are more immediately obvious than are those of other approximate methods which have to be used in dealing with the solid state, for it sometimes gives good results even in what appear to be extreme cases.

2. Basic Ideas and Survey of Topics

Of the properties of lattice defects, only some can be expected to survive and still be describable in the continuum idealization. The theory of elasticity is concerned with the relation between the deformation of a body and the energy content of itself and its surroundings. Thus we are effectively limited to a discussion of the deformations and energy changes associated with the presence of defects.

The first problem is to find a way of transcribing defects into their continuum analog. This can usually be done in a plausible *ad hoc* way for particular types of defect (Section a). It is also possible to develop a general theory based on an internal stress "source function" bearing the same relation to the internal strain as charge does to electric field in electrostatics (Section 4b). Closely related to this is the description of

[2] C. V. Burton, *Phil. Mag.* [5] **33,** 191 (1892).
[3] J. Larmor, *Phil. Trans. Roy. Soc.* **A190,** 205 (1897).

internal stress in terms of a continuous distribution of dislocations (Section 9d).

Figure 1 shows a body containing a number of defects S, T, and interacting with its surroundings, typified by a weight W and a spring P. If S moves about, the deformation and elastic energy of the body change. At the same time the changes of shape of its outer surface communicate themselves to W and P and alter their potential energy.

The shape of the body is related to the position of a defect in a rather complicated way. When the defect is moved its elastic field is not simply transported with it bodily, since this would usually violate whatever boundary conditions may have been imposed at its surface. It is often convenient to divide the elastic field into a part which *is* transported bodily with the defect ("field in an infinite medium") and a remainder ("image field") which adjusts itself so that the boundary conditions are

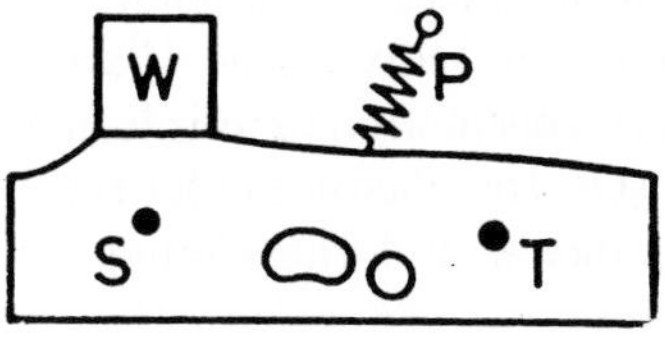

FIG. 1. To illustrate Section 2.

satisfied. We shall see that the image field often plays an unexpectedly important role. There is an analogy with electrostatic problems involving charges in a finite medium whose dielectric constant is large enough to confine the field effectively to its interior, as opposed to the case of charges in free space, where inconvenient surface integrals can be relegated to infinity.

Generally it will not be enough to describe a defect by its position alone; for example, a dislocation loop may change its shape. Let $\alpha, \beta \ldots$ be a (possibly infinite) set of parameters sufficient to characterize the configuration of the defects. Both the elastic energy of the body E_{el} and the potential energy E_{ext} of any external mechanism connected with it will depend on the parameters. Rather than E_{el} and E_{ext} individually, the quantity of physical interest is their sum, the total energy[4,5,6]

$$E_{tot} = E_{el}(\alpha,\beta \ldots) + E_{ext}(\alpha,\beta \ldots).$$

If the parameters are able to vary (subject to certain constraints), it is E_{tot} and not E_{el} or E_{ext} which must be minimized with respect to them to

[4] B. A. Bilby, *Proc. Phys. Soc.* (*London*) **A63,** 3 (1950).
[5] M. O. Peach, *J. Appl. Phys.* **22,** 1359 (1951).
[6] J. D. Eshelby, *Phil. Trans. Roy. Soc.* **A244,** 87 (1951).

find the equilibrium state. In fact the distinction between internal and external energy is artificial, though convenient. Consider, for example, a dislocation in a specimen strained in a tensile testing machine by tightening a screw. We may regard this as a case of a defect in a body (the specimen) acted on by external forces, or as a defect in a complicated self-stressed body (specimen plus machine).

From a thermodynamic point of view E_{tot} is likewise the important quantity. The properties of a nonisolated system may be derived from a knowledge of its enthalpy or Gibbs free energy under adiabatic or isothermal conditions. Although we shall usually regard E_{el} as "purely mechanical" it is strictly the body's internal energy in the adiabatic case, or its Helmholtz free energy in the isothermal case.[7,8] It follows that E_{tot} is its enthalpy or Gibbs free energy, for these quantities are introduced precisely to give an account of the internal energy or Helmholtz free energy of the body plus the energy of its environment under the guise of considering a property of the body alone. If we take this wider point of view we may also derive thermodynamic information from the temperature variation of E_{tot}. (On the elastic model this variation will be determined by thermal expansion and the change of elastic constants with temperature.)

In the infinitesimal theory of elasticity two or more elastic fields may be superimposed. The expression for E_{tot} will then be made up of "self-energy" terms quadratic in the individual fields together with interaction terms involving products of pairs of fields. It is often convenient to deal with the interaction energies rather than with E_{tot}, particularly when the self-energy terms are formally infinite. Even when there are such infinite terms it is possible to "subtract them out" and find a simple expression for the interaction terms (Section 6).

In accordance with usage in analytical mechanics and thermodynamics we may call

$$F(\alpha) = -\partial E_{tot}/\partial \alpha$$

the generalized force associated with the parameter α. Equilibrium is determined by equating to zero the generalized forces corresponding to those parameters which are supposed to be freely variable. In nonequilibrium problems the generalized forces, being derivatives of the free energy, are the driving forces which provide the raw material for a kinetic calculation of the rate of approach to equilibrium by arguments outside the scope of a continuum theory.

[7] I. S. Sokolnikoff, "Mathematical Theory of Elasticity." McGraw-Hill, New York, 1946.

[8] A. E. Green and W. Zerner, "Theoretical Elasticity," p. 72. Oxford Univ. Press, London and New York, 1954.

If it is sufficient to give the Cartesian coordinates x, y, z specifying the position of a defect, we may call

$$\mathbf{F} = -(\partial/\partial x,\ \partial/\partial y,\ \partial/\partial z)E_{\text{tot}} \tag{2.1}$$

the force on the defect in the narrower sense. It is often convenient to subdivide $\mathbf{F}$ along the following lines. Consider the force on the defect S in Fig. 1.

(i) If S is the only defect in the body and W and P are absent, $E_{\text{tot}} = E_{\text{el}}$ will vary with the position of the defect. Because in a homogeneous body the existence of $\mathbf{F}$ is evidently related to the presence of the surface, we may speak of it as an image force $\mathbf{F}^I$, in analogy with the nomenclature in electrostatics. The surface of an internal cavity O will make a contribution to the image force. The cavity will still make a contribution even if it is filled with material, provided its elastic constants differ from those of the remainder of the body. More generally, inhomogeneities of the medium will contribute to $\mathbf{F}^I$. In fact we may simply say that $\mathbf{F}^I$ is due to inhomogeneities if we regard the body as part of an infinite body whose elastic constants are zero outside a certain region.

(ii) If the defect T is introduced, the force on S will have a different value; say $\mathbf{F} = \mathbf{F}^I + \mathbf{F}^T$, and we may call $\mathbf{F}^T$ the force which T exerts on S.

(iii) If surface tractions are next applied (as by W and P), $\mathbf{F}$ becomes, say, $\mathbf{F}^I + \mathbf{F}^T + \mathbf{F}^E$. Then $\mathbf{F}^E$ can be regarded as the force exerted on S by the surface tractions, or the external mechanism responsible for them.

(iv) If x, y, z now refer to the position of the cavity or region of elastic inhomogeneity O, E_{tot} will depend on x, y, z and we may speak of $\mathbf{F}$ as the force on the inhomogeneity.

The results of Section 5a lead at once to simple expressions for $\mathbf{F}^E$ and $\mathbf{F}^T$ (Section 5b), while in Section 6 we find an expression for the force on an inhomogeneity. In Section 7 we develop a general expression for the force on a singularity or inhomogeneity which embraces the foregoing results but which is not limited to infinitesimal deformations. In the infinitesimal case it completes the results of Section 5b by giving a formula for $\mathbf{F}^I$ analogous to the expressions for $\mathbf{F}^E$ and $\mathbf{F}^T$.

As a first illustration we take the familiar misfitting sphere model for a point defect. Image effects play an important part here. They make a large contribution to the volume change produced by the defect, and their retention is essential if we are to reach formally the intuitively obvious result that a uniform density of defects produces a uniform macroscopic dilatation of the body containing them (Section 8a). In Section 8b we take up the effect of point defects on the x-ray diffraction pattern of a crystal in relation to its macroscopic deformation. Image terms again make themselves felt in the elastic theory of the energy of solid solutions

(Section 8c). In Section 8d we refine the model of a point defect by including the effect of anisotropy, and in Section 8e we consider a defect as a lattice inhomogeneity.

In Section 9 we consider some particular topics in the theory of dislocations. (There are several excellent accounts of the general theory.[9,10,11]) Section 9a gives a formal derivation of the interaction energy of a dislocation loop with a stress-field. Section 9b deals with image effects, in particular the problem of a screw dislocation in a rod. This presents unexpected features and is of some interest in connection with the properties of metallic "whiskers." Dislocations in motion (Section 9c) present some intriguing theoretical problems, but at present they do not appear to be of much practical significance. Finally in Section 10 we gather together a few points relating to the behavior of lattice inhomogeneities on a large scale.

II. Formal Theory

3. Elements of Elasticity

The clearest approach to the usual infinitesimal theory, with which we shall be chiefly concerned, is by way of the general theory of finite deformation. Moreover, since some of our results hold in the general case, we give first a simple formulation of the theory of finite strain in a medium with an arbitrary stress-strain relation.

Throughout we use the convention that a repeated suffix is to be summed over the values 1, 2, 3 and that suffixes following a comma denote differentiation:

$$e_{ii} = e_{11} + e_{22} + e_{33}, \qquad u_{i,j} = \partial u_i/\partial x_j, \qquad u_{i,jk} = \partial^2 u_i/\partial x_j \partial x_k,$$
$$p_{ij,j} = p_{i1,1} + p_{i2,2} + p_{i3,3}.$$

The symbol δ_{ij} has the value 1 or 0 according to whether i and j are or are not equal. The symbol ϵ_{ijk} has the value 1 if ijk is an even permutation of 123, -1 if it is an odd permutation and is zero otherwise.

The state of finite strain produced in a medium (conveniently visualized as a transparent jelly) by body and surface forces may be described most vividly thus. Imagine space partitioned into small cubes by the network of a rectangular coordinate system x_i. Within the body we embed a network of threads coinciding with the coordinate net. When the medium is strained, the embedded net becomes a curvilinear coordinate system $\bar{x}_i$ (Fig. 2a), and the shape and size of any small mesh (cubic before

[9] F. R. N. Nabarro, *Advances in Phys.* 1, 269 (1952).
[10] W. T. Read, Jr., "Dislocations in Crystals." McGraw-Hill, New York, 1953.
[11] A. Seeger, "Handbuch der Physik," 3rd ed., p. 383. Springer, Berlin, 1955.

deformation) gives an immediate picture of the deformation in its neighborhood.

The vector **u** joining a point x_i of the undeformed coordinate net to a point $\tilde{x}_i$ of the deformed net with $\tilde{x}_1 = x_1$, $\tilde{x}_2 = x_2$, $\tilde{x}_3 = x_3$ is evidently the finite displacement undergone by the particle of material originally

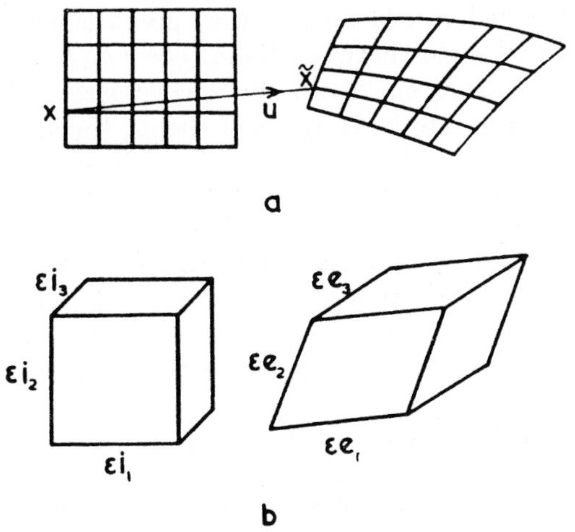

FIG. 2. Finite deformation.

at x_i. As a vector field, the displacement may be considered to be a function of the rectilinear x_i or the curvilinear $\tilde{x}_i$. Let $\mathbf{u}(x_i)$ denote the vector arrow whose tail is at x_i, $\mathbf{u}(\tilde{x}_i)$ the arrow whose head is at $\tilde{x}_i$. The relations

$$\mathbf{u}(x_i) = \mathbf{u}(\tilde{x}_i), \qquad \partial\mathbf{u}(x_i)/\partial x_j = \partial\mathbf{u}(\tilde{x}_i)/\partial\tilde{x}_j$$

if

$$x_i = \tilde{x}_i$$

merely expresses the fact that every **u**-arrow joins the points x_i, $\tilde{x}_i$ with identical coordinate numbers. Thus, mathematically, we need not distinguish the x_i (Lagrangian coordinates) from the $\tilde{x}_i$ (embedded coordinates). Let u_i be the components of **u** along the unit vectors $\mathbf{i}_1$, $\mathbf{i}_2$, $\mathbf{i}_3$ of the undeformed coordinate system:

$$\mathbf{u} = u_m\mathbf{i}_m = u_1\mathbf{i}_1 + u_2\mathbf{i}_2 + u_3\mathbf{i}_3.$$

A small cube with edges $\epsilon\mathbf{i}_1$, $\epsilon\mathbf{i}_2$, $\epsilon\mathbf{i}_3$ (Fig. 2b) before deformation becomes a parallelepipedal mesh of the deformed coordinate system with edges $\epsilon\mathbf{e}_1$, $\epsilon\mathbf{e}_2$, $\epsilon\mathbf{e}_3$ where

$$\mathbf{e}_i = \frac{\partial}{\partial x_i}(x_m\mathbf{i}_m + \mathbf{u}) = (\delta_{mi} + u_{m,i})\mathbf{i}_m. \tag{3.1}$$

For the purposes of the theory of elasticity we need to disentangle from the $\mathbf{e}_i$ (or the $u_{m,i}$) a measure of deformation free from any reference to the *orientation* of the elementary mesh. The six scalar products

$$g_{ij} = \mathbf{e}_i \cdot \mathbf{e}_j = u_{i,j} + u_{j,i} + u_{m,i}u_{m,j} + \delta_{ij}$$

evidently provide such a measure, for they give the lengths of the edges ($\epsilon g_{11}^{\frac{1}{2}}$ . . .) and the angles between them ($\cos^{-1} g_{12}/g_{11}^{\frac{1}{2}}g_{22}^{\frac{1}{2}}$ . . .) and so enable us to reconstruct the mesh the correct size and shape without telling us how to orientate it. In place of the g_{ij}, the strain components

$$e_{ij} = \tfrac{1}{2}(g_{ij} - \delta_{ij}) \tag{3.2}$$

are generally used.

Let the material all round the elementary mesh be cut away and let such forces be applied to its free surfaces that it retains its shape, size, and orientation. Let $\epsilon^2\mathbf{p}_j$ be the force on the face which, before deformation, had the positive x_j axis normal to it. Then the p_{ij} defined by resolving $\mathbf{p}_j$ along $\mathbf{i}_1$, $\mathbf{i}_2$, $\mathbf{i}_3$,

$$\mathbf{p}_j = p_{ij}\mathbf{i}_i = p_{1j}\mathbf{i}_1 + p_{2j}\mathbf{i}_2 + p_{3j}\mathbf{i}_3,$$

are the (unsymmetrical) Boussinesq[12] stress components. The equation of equilibrium of the mesh is easily shown to be

$$\frac{\partial p_{ij}}{\partial x_j} + f_i = 0 \tag{3.3}$$

where

$$\mathbf{f} = f_m\mathbf{i}_m$$

is the body force per unit mesh of the deformed (or equally well, the undeformed) coordinate net. Consideration of the work done in a small additional deformation of the body shows that

$$p_{ij} = \frac{\partial W}{\partial u_{i,j}} \tag{3.4}$$

where W is the density of elastic energy per unit mesh.

In the infinitesimal linear theory,[7] which we shall use unless otherwise stated, second-order terms in the strain tensor are neglected,

$$e_{ij} = \tfrac{1}{2}(u_{i,j} + u_{j,i}) \tag{3.5}$$

and W is taken to be a general quadratic expression in the e_{ij}:

$$W = \tfrac{1}{2}c_{ijkl}e_{ij}e_{kl}. \tag{3.6}$$

[12] L. Brillouin, "Les Tenseurs en Méchanique et en Élasticité," p. 246. Masson, Paris, 1949.

The suffixes of the elastic constants c_{ijkl} have the same symmetry as those of $e_{ij}e_{kl}$, that is, i and j or k and l or (ij) and (kl) may be interchanged without altering c_{ijkl}. The stress tensor is now symmetrical,

$$p_{ij} = p_{ji} = c_{ijkl}e_{kl} = c_{ijkl}u_{k,l}. \tag{3.7}$$

Moreover, in an isotropic medium

$$p_{ij} = \lambda e_{mm}\delta_{ij} + 2\mu e_{ij}. \tag{3.8}$$

The equilibrium equation is

$$p_{ij,j} = 0 \tag{3.9}$$

in the absence of body forces and

$$p_{ij,j} + f_i = 0 \tag{3.10}$$

in their presence. In the isotropic case (3.10) may be written in terms of the displacement:

$$\mu\nabla^2\mathbf{u} + (\lambda + \mu) \text{ grad div } \mathbf{u} + \mathbf{f} = 0. \tag{3.11}$$

For finite strain, the problem of finding a rotation which when combined with the deformation e_{ij} will send the cube with edges $\epsilon\mathbf{i}_1$, $\epsilon\mathbf{i}_2$, $\epsilon\mathbf{i}_3$ into the parallelepiped with edges $\epsilon\mathbf{e}_1$, $\epsilon\mathbf{e}_2$, $\epsilon\mathbf{e}_3$ is rather complex. In the linear theory, however, we may define the rotation to be half the curl of the displacement,

$$\tilde{\omega}_i = -\tfrac{1}{2}\epsilon_{ijk}u_{j,k}$$

or more conveniently as the antisymmetric tensor

$$\tilde{\omega}_{ij} = \tfrac{1}{2}(u_{i,j} - u_{j,i}). \tag{3.12}$$

We have

$$\tilde{\omega}_{ij} = -\epsilon_{ijk}\tilde{\omega}_k, \qquad \tilde{\omega}_k = -\tfrac{1}{2}\epsilon_{kij}\tilde{\omega}_{ij}.$$

With an eye to later application, it is convenient to have a definition of the rotation directly in terms of the $\mathbf{e}_i$ and $\mathbf{i}_i$ without reference to the $u_{i,j}$. If $\mathbf{e}_i$ and $\mathbf{i}_i$ differ only infinitesimally, the magnitude of the vector $\mathbf{i}_1 \times \mathbf{e}_1$ is the angle through which $\mathbf{i}_1$ must be rotated to coincide with $\mathbf{e}_1$. Its direction is the axis about which the rotation must be performed. Then $\tfrac{1}{2}(\mathbf{i}_1 \times \mathbf{e}_1 + \mathbf{i}_2 \times \mathbf{e}_2) \cdot \mathbf{i}_3$ is the x_3 component of the averages of the rotations of the edges $\epsilon\mathbf{i}_1$, $\epsilon\mathbf{i}_2$ of that face of the elementary cube of Fig. 2b which has $\mathbf{i}_3$ for normal. In fact we may put

$$\tilde{\boldsymbol{\omega}} = \tfrac{1}{2}(\mathbf{i}_1 \times \mathbf{e}_1 + \mathbf{i}_2 \times \mathbf{e}_2 + \mathbf{i}_3 \times \mathbf{e}_3) \tag{3.13}$$

for then

$$\tilde{\omega}_3 = -\tilde{\omega}_{12} = \tfrac{1}{2}(\mathbf{i}_1 \times \mathbf{e}_1 + \mathbf{i}_2 \times \mathbf{e}_2) \cdot \mathbf{i}_3 = \tfrac{1}{2}(u_{2,1} - u_{1,2})$$

by (3.1), in agreement with (3.12).

From (3.5) and (3.12), we have

$$\tilde{\omega}_{ij,k} = e_{ki,j} - e_{kj,i} \quad \text{or} \quad \tilde{\omega}_{l,k} = -\epsilon_{lij}e_{ki,j}. \tag{3.14}$$

The two tensor fields e_{ij}, $\tilde{\omega}_{ij}$ cannot be strain and rotation in an elastic field with a displacement function unless they satisfy (3.14). But this is not enough to ensure the existence of a displacement, for the line integral

$$\tilde{\omega}_l(Q) - \tilde{\omega}_l(P) = -\int_P^Q \epsilon_{lij}e_{ki,j}dx_k \tag{3.15}$$

giving the difference in the rotation at points P, Q must be independent of the path joining them. The curl of the integrand must be zero, that is we must have

$$S_{ij} = 0 \tag{3.16}$$

where

$$S_{ij}(e_{pq}) = -\epsilon_{ikm}\epsilon_{jln}e_{kl,mn}. \tag{3.17}$$

It is shown directly in works on the theory of elasticity that the vanishing of S_{ij} in a region is the necessary and sufficient condition for the existence of a displacement there.

Consider a narrow tube in the unstrained material with its axis parallel to the x_3 axis. In the strained state $\tilde{\omega}_{3,3}dx_3$ is the relative rotation of two of its cross sections separated by dx_3, while $\tilde{\omega}_{3,1}$, $\tilde{\omega}_{3,2}$ are its curvatures about the x_1 and x_2 axes. We may call $\tilde{\omega}_{i,j}$ the curvature tensor.

It will be convenient to summarize here some elementary theorems and manipulations which will be needed later.

If (u_i, e_{ij}, p_{ij}) and (u_i', e_{ij}', p_{ij}') are two sets of quantities, each related by (3.5), (3.7), and satisfying (3.9), we obviously have

$$p_{ij}e_{ij}' = p_{ij}'e_{ij} = p_{ij}u_{r,j}' = p_{ij}'u_{i,j} = (p_{ij}u_i')_{,j} = (p_{ij}'u_i)_{,j}. \tag{3.18}$$

Thus the vector

$$v_j = p_{ij}u_i' - p_{ij}'u_i \tag{3.19}$$

has zero divergence, and so by Gauss's theorem

$$\int_{\Sigma_1} v_j dS_j = \int_{\Sigma_2} v_j dS_j \tag{3.20}$$

for any two surfaces Σ_1, Σ_2 which can be deformed into one another without encountering singularities of v_j. (We use dS_j as an abbreviation for $n_j dS$, where n_j is the normal to the surface and dS is the surface element.) In particular if Σ contains no singularities of v_j,

$$\int_\Sigma v_j dS_j = 0. \tag{3.21}$$

If the material is homogeneous ($c_{ijkl,m} = 0$), the difference between the elastic field u_i', p_{ij}' and the same field advanced bodily a short distance along the x_l axis satisfies the elastic equations. Thus in (3.19), (3.20), (3.21) we may replace u_i', p_{ij}' by $u_{i,l}'$, $p_{ij,l}'$.

If p_{ij}, p_{ij}' satisfy (3.10) instead of (3.9), we have

$$\int_\Sigma (p_{ij}u_i' - p_{ij}'u_i)dS_j = \int (f_i'u_i - f_iu_i')dv$$

(Betti's reciprocal theorem[13,14]).

We also have

$$\int p_{ij}e_{ij}'dv = \int p_{ij}u_{i,j}'dv = \int \{(p_{ij}u_i')_{,j} + f_iu_i'\}dv = \int p_{ij}u_i'dS_j + \int f_iu_i'dv$$

provided only that

$$p_{ij,j} + f_i = 0 \qquad \text{and} \qquad e_{ij}' = \tfrac{1}{2}(u_{i,j}' + u_{j,i}'). \tag{3.22}$$

Here p_{ij} and e_{ij}' need satisfy no other conditions and need not be related in any way. In particular they need not be possible stress and strain tensors in the same material. With $u_i' = x_i$, (3.22) gives

$$\int p_{ii}dv = \int_\Sigma p_{ij}x_idS_j + \int f_ix_idv. \tag{3.23}$$

In a homogeneous isotropic or cubic medium, $p_{ii} = 3Ke_{ii}$, where K is the bulk modulus. We have then the expression

$$\Delta V = \frac{1}{3K}\int \mathbf{r}\cdot\mathbf{T}dS + \frac{1}{3K}\int \mathbf{r}\cdot\mathbf{f}dv \tag{3.24}$$

for the volume change produced by a body force density $\mathbf{f}$ and surface tractions $\mathbf{T}$. If $\mathbf{f}$ and $\mathbf{T}$ are zero, $\Delta V = 0$ even if the body is in a state of self-stress in which p_{ij} does not vanish throughout the interior.

We shall also need Stokes's theorem in the relatively unfamiliar form

$$\int_\Sigma w_{\ldots j,l}dS_j = \int_\Sigma w_{\ldots m,m}dS_l \tag{3.25}$$

for a closed surface. This follows by applying Stokes's theorem in its usual form to the quantity $\epsilon_{lij}w_{\ldots i}$, or by applying Gauss's theorem to the body generated by giving the surface a small displacement parallel to the x_l axis (see Fig. 7).

4. Specification of Internal Stress

a. Somigliana Dislocations

To pass from the crystal lattice with defects to its elastic analog, we must be able to associate with each type of defect a suitable state of

[13] A. E. H. Love, "Mathematical Theory of Elasticity." Cambridge U. P., London, 1952.

[14] S. Timoshenko and J. N. Goodier, "Theory of Elasticity." McGraw-Hill, New York, 1951.

internal stress in the continuum. For brevity we shall refer to these states of internal stress as "singularities."

Most of the singularities of physical interest are particular cases of a general type of dislocation described by Somigliana.[15] To construct a Somigliana dislocation, mark out in the elastic body a surface C bounded by a curve c and make a cut coinciding with C. Give each pair of points adjacent to one another on opposite sides of the cut a relative displacement $\mathbf{d}$ (Fig. 3), scraping away material where there would be interpenetration. Fill in the remaining gaps with additional material and cement together. This evidently leaves the material in a state of internal stress. The stress $p_{ij}n_j$ (where n_j is the normal to C) is continuous across

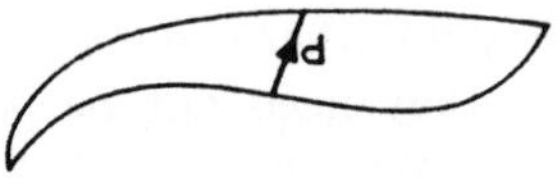

FIG. 3. A Somigliana dislocation.

the surface of discontinuity, but the various components of stress and strain p_{ij}, e_{ij} in general are not. It is physically obvious, and can be proved mathematically[15,16] that a knowledge of $\mathbf{d}$ as a function of position over C, together with the boundary conditions at the surface of the body, completely determines the resulting state of internal stress. If $\mathbf{d}$ is a reasonably smooth function, stress and strain will be finite everywhere except possibly at c.

If $\mathbf{d}$ has a constant value, we have the usual dislocations of solid state theory, the dislocations of types 1, 2, 3, of Volterra.[17] If $\mathbf{d} = \mathbf{r} \times \boldsymbol{\omega}$, where $\mathbf{r}$ is the position vector and $\boldsymbol{\omega}$ a constant, we have Volterra's dislocations of types 4, 5, 6. Physically we may take them to represent twist and tilt boundaries made up of an array of dislocations which, in the spirit of the continuum approximation, have been replaced by a continuous distribution of infinitesimal dislocations.

To make a model of a point defect, we take for C a small sphere with a suitable distribution of $\mathbf{d}$ over the surface. If we let the radius of the sphere tend to zero and, at the same time, increase $\mathbf{d}$ in such a way that the displacement at a fixed distance from the sphere remains finite, we obtain a point singularity in the mathematical sense. For many purposes it is an adequate representation of a physical point defect.

As a simple example we might take $\mathbf{d}$ constant in magnitude and directed radially. Somigliana's recipe is then equivalent to the following.

[15] C. Somigliana, *Atti accad. nazl. Lincei Rend. Classe sci. fis. mat. e nat.* **23**(1) 463 (1914); **24**(1) 655 (1915).

[16] M. Gebbia, *Ann. Mat. Pura Appl.* **7**, 141 (1902).

[17] V. Volterra, *Ann. Éc. Norm. Sup.* **24**, 400 (1907).

Cut a sphere out of the matrix, alter its radius by adding or removing material, and reinsert in the matrix. This is just the familiar misfitting-sphere model for a substitutional or interstitial atom.

It is convenient to divide the elastic field of the singularity into two parts, u_i^∞ and p_{ij}^∞, the value it would have in an infinite medium and an "image field" u_i^I, p_{ij}^I chosen so that $u_i^\infty + u_i^I$, $p_{ij}^\infty + p_{ij}^I$ satisfy the conditions imposed at the surface of the actual finite body containing the singularity. Then if we give Somigliana's surface of discontinuity a displacement $\boldsymbol{\xi}$, the elastic field will change from

$$u_i = u_i^\infty(x_k) + u_i^I(x_k), \qquad p_{ij} = p_{ij}^\infty(x_k) + p_{ij}^I(x_k) \tag{4.1}$$

to

$$u_i = u_i^\infty(x_k - \xi_k) + u_i^I(x_k,\xi_k), \qquad p_{ij} = p_{ij}^\infty(x_k - \xi_k) + p_{ij}^I(x_k,\xi_k) \tag{4.2}$$

The ∞ field undergoes a rigid displacement, but the image field changes in a more complicated way which can only be found by solving a boundary-value problem. The exact form of u_i^∞, p_{ij}^∞ can be fixed by requiring that p_{ij}^∞ shall approach zero at large distances, at least as r^{-2} in three-dimensional problems and at least as r^{-1} in two dimensions.

b. *The Incompatibility Tensor*

The way in which we introduced the stress field associated with a given defect is analogous to a development of electrostatics which begins by postulating that the field of a point-charge is $e\mathbf{r}/r^3$. Electrostatics may alternately be developed starting from the concept of a charge density which is the "source" of the field and determines it by way of Poisson's equation. The field of a point charge is then found by specializing the density to have the form of a delta function. Something analogous can be done in the elastic case. We start with a body in a state of internal stress and find a "source function of internal stress" which, when prescribed, determines the internal stress if suitable boundary conditions are given.

In engineering practice, the state of internal stress of a body is investigated by cutting a piece off and seeing how it or the remainder deforms. We may idealize this process as follows.[14,6] Let a small cubical element be marked out in the body and then cut out of it. Its shape and size will alter; in other words, it will spontaneously undergo a certain strain, say e_{ij}. By repeating this process for every point we obtain a field $e_{ij}(x_k)$ which serves to specify the state of internal strain. (Evidently this e_{ij} is minus the strain derived from the internal stress using Hooke's law.) Unlike a strain field arising from external forces, e_{ij} will not, in general, satisfy the compatibility conditions (3.16). We can see this most clearly by reversing the foregoing argument. Cut a stress-free body into elementary cubes and give each one a permanent strain e_{ij}^* so that the field $e_{ij}^*(x_k)$ has continuous first and second derivatives but is otherwise arbitrary. Then, in general, $S_{ij}(e_{mn}^*) \neq 0$. Now pull the elements back to their original cubical form and size by suitable forces applied to their surfaces and cement them together. Then remove the distribution of body forces resulting from the building-in of these surface forces. This induces in the body an additional strain e_{ij}' for which $S_{ij}(e_{mn}') = 0$. The internal stresses are now those derived from $e_{ij}^* + e_{ij}'$ by Hooke's law. If now the body is redissected, each element will undergo a spontaneous strain $e_{ij} = e_{ij}^* + e_{ij}'$ for which $S_{ij}(e_{mn}) = S_{ij}(e_{mn}^*) \neq 0$.

Evidently if to the internal strain there is added a strain produced by external forces, and therefore derivable from a displacement, the value of S_{ij} is unaltered. Thus in a sense the "incompatibility tensor" S_{ij} separates the internal from the external strain. It is in fact a suitable source function for internal stress. In other words, given $S_{ij}(\mathbf{r})$ as a function of position, we can in principle solve the relation

$$-\epsilon_{ikm}\epsilon_{jln}e_{kl,mn} = S_{ij}(\mathbf{r}) \tag{4.3}$$

for the e_{kl}. A solution is

$$e_{ij}(\mathbf{r}) = \frac{1}{4\pi}\int \frac{S_{ij}(\mathbf{r}') - S_{mm}(\mathbf{r}')\delta_{ij}}{|\mathbf{r} - \mathbf{r}'|}\,dv$$

if S_{ij}, or less restrictively its normal component $S_{ij}n_j$, vanishes at the boundary of the body. (This follows from the solution of a similar problem in the general theory of relativity.[18]) If this boundary condition is not satisfied, less elegant solutions are still possible.

To any such solution we can add the general solution of $S_{ij} = 0$, namely $e_{ij}^0 = \frac{1}{2}(u_{i,j}^0 + u_{j,i}^0)$ with arbitrary u_i^0. The complete determination of the state of internal stress when $S_{ij}(\mathbf{r})$ is prescribed thus goes as follows. Find any solution e_{ij} of (4.3). With the aid of Hooke's law, find the stresses p_{ij} and hence the body-forces $f_i = -p_{ij,j}$ and surface tractions $p_{ij}n_j$ necessary to maintain them. By standard elastic theory determine the compatible strain e_{ij}^0 arising from equal and opposite forces. Then the stress in the body is that derived from $e_{ij} + e_{ij}^0$ by Hooke's law.

Such calculations are simplified by introducing a stress function $\chi_{ij} = \chi_{ji}$ related to p_{ij} in the same way as e_{ij} is to S_{ij}:

$$p_{ij} = -\epsilon_{ikm}\epsilon_{jln}\chi_{kl,mn}. \tag{4.4}$$

Clearly for any χ_{kl} we have $p_{ij,j} = 0$. Southwell[19] and Kuzmin[20] have

[18] A. S. Eddington, "Mathematical Theory of Relativity," p. 128. Cambridge U. P., New York, 1923.

[19] R. V. Southwell, *Phil. Mag.* [7] **30**, 253 (1940).

[20] R. O. Kuzmin, *Compt. rend. acad. sci. U.R.S.S.* **49**, 326 (1945).

shown directly that any symmetric tensor with vanishing divergence can, conversely, be represented in the form (4.4), and indeed with one or other of the restrictions $\chi_{12} = \chi_{23} = \chi_{31} = 0$ (Maxwell's form) or

$$\chi_{11} = \chi_{22} = \chi_{33} = 0$$

(Morera's form).[13] Kröner[21] has reduced the problem to manageable form for the isotropic case and obtains the direct relation

$$\nabla^4 \chi_{ij} = 2\mu \left(S_{ij} + \frac{\sigma}{1-\sigma} S_{mm}\delta_{ij} \right) \qquad (\sigma = \text{Poisson's ratio})$$

between incompatibility tensor and stress function. He has also discussed the anisotropic case.[22] The elastic energy of a self-stressed body can be expressed in the form

$$\tfrac{1}{2}\int \chi_{ij} S_{ij} dv$$

plus certain surface terms which vanish if $S_{ij}n_j$ vanishes at its surface.[23,21]

From a given state of incompatible strain, e_{ij}, we can construct a tensor

$$g_{ij} = \delta_{ij} + 2e_{ij} \tag{4.5}$$

on the pattern of (3.2). If we take the g_{ij} as a metric tensor associated with our ordinary Euclidean coordinate system, we thereby define a geometry which is in general not Euclidean but Riemannian. The test for this is whether the Riemann tensor formed from the g_{ij} vanishes or not. In three dimensions, where the four-suffixed Riemann tensor R_{prst} has only six independent components, we may equally well use the two-component tensor $S_{ij} = \epsilon_{ipr}\epsilon_{jst}R_{prst}$[24] which, with (4.5), can be shown to be identical with the S_{ij} of (3.17). Eckart[25] has shown that this Riemannian geometry has a simple physical meaning. The non-Euclidean arc length

$$s = \int_c (g_{ij}dx_i dx_j)^{\frac{1}{2}} \tag{4.6}$$

along any curve c drawn in the body is the actual length of a thin curved rod with c as axis when it has been cut out and allowed to relax its internal stresses. If we take a rectangular closed path $ABCD$ (Fig. 4a) and calculate (4.6) along each of the four sides we shall find in general that $s_{AB} \neq s_{CD}$, $s_{BC} \neq s_{DA}$. Thus when a filamentary loop enclosing $ABCD$ is cut out,

[21] E. Kröner, *Z. Physik* **139**, 175 (1954).
[22] E. Kröner, *Z. Physik* **141**, 386 (1955).
[23] R. V. Southwell, *Proc. Roy. Soc.* **A154**, 4 (1936).
[24] A. J. McConnell, "Applications of the Absolute Differential Calculus," p. 154. Blackie, London, 1936.
[25] C. Eckart, *Phys. Rev.* **73**, 373 (1948).

it must be cut through (say at A) in order to relax its stresses completely. Moreover, the cut will define the two ends of a vector AA' (Fig. 4b). We may relate this to Frank's[26] discussion of the Burgers circuit in a dislocated crystal. We traverse a circuit in a region of "good" crystal surrounding "bad" crystal. For each interatomic step we make in the real crystal, we make a corresponding step in a perfect "comparison" crystal. When we have come back to the starting point in the real crystal, we are still a certain vector distance (closure failure) from the starting point in the comparison crystal. If we desire, we may dispense with a separate comparison crystal and have the work of traversing the comparison circuit done for us automatically. Dissect out a thin loop enclosing the circuit in the real crystal and cut through the loop. We are left with a perfect crystal, admittedly with an odd shape, which can serve as a

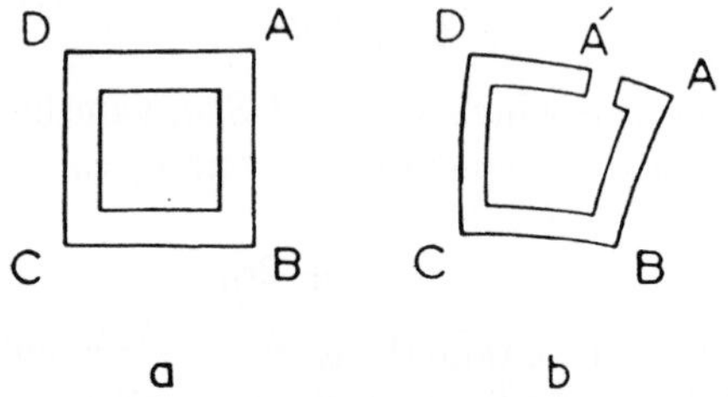

FIG. 4. To illustrate Section 4b.

comparison crystal, in which the Burgers circuit is already marked out and the closure failure is directly indicated by the gap AA'.

The physical significance of S_{ij} can be seen as follows. Equation (3.15) gives the difference of the rotation at the ends of a path drawn in a region where $S_{ij} = 0$. Consider a closed path c embracing a region where $S_{ij} \neq 0$. The integral will not in general vanish. In fact Stokes's theorem and (3.17) give for its value

$$\Delta\bar{\omega}_l = -\int_C S_{lm} dS_m$$

where C is any cap bounded by c. Consider first a state of plane strain where S_{33} is the only nonvanishing component of S_{ij}, and let S_{33} vanish everywhere except in a small patch near the origin, so that we may write $S_{33} = \omega\delta(x_1)\delta(x_2)$. Then

$$\Delta\bar{\omega}_3 = -\int_C S_{33} dS_3 = -\omega.$$

This describes the state of internal strain resulting from cutting out a wedge of material of angle ω and cementing the faces of the cut together

[26] F. C. Frank, *Phil. Mag.* [7] **42,** 809 (1951).

(Fig. 5a). In physical terms, this represents a tilt grain boundary of angle ω terminating at the origin. Thus S_{33} is a measure of the number of terminations of tilt boundaries per unit area. More generally, S_{ij} measures the flux of Volterra dislocations of types 4, 5, 6 or, in other words, the number of tilt and twist boundaries which terminate in unit area. The exact relation can be developed in detail, but it is evident that S_{ij} is not an adequate measure of the density of physical dislocations which give rise to a discontinuity of displacement, not of rotation. The Bianchi identity $S_{ij,j} = 0$ expresses the fact that a Volterra dislocation of general type cannot end in the medium.

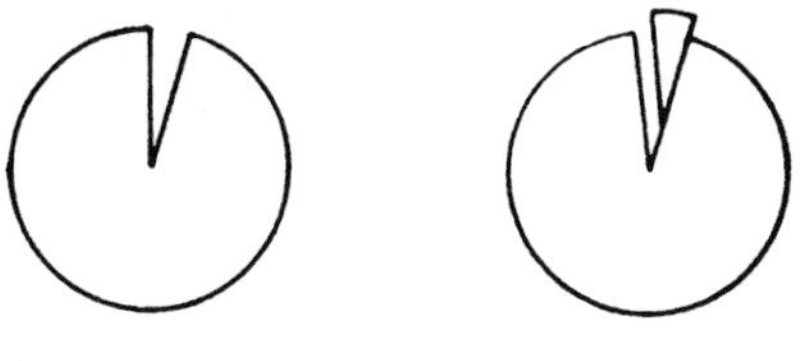

a b

FIG. 5. Relation between edge and "wedge" dislocations.

An edge dislocation can be made by cutting out a parallel-sided fissure and closing the gap. This can be done by removing a wedge and inserting a wedge of the same angle at an adjacent point (Fig. 5b). Thus to describe an edge dislocation at the origin we should have to take $S_{33} = \text{const}\ \partial\{\delta(x_1)\delta(x_2)\}/\partial x_2$. More generally, if we take Volterra tilt and twist dislocations to be analogous to current-carrying wires, edge and screw dislocations are analogous to closely-spaced wires carrying opposite currents.[27] We shall treat the description of internal stress in terms of dislocations in Section 9d.

5. Elastic Interaction Energies

a. Interaction Energies between Stress Systems

Suppose that in the body whose surface is Σ_0 we have one system of internal stress S whose sources lie entirely within the surface Σ (Fig. 6) and another system T whose sources lie entirely outside Σ. If E_S and E_T are the values of the total elastic energy when S or T alone exists in the body, we may write the total energy when they coexist in the form $E_S + E_T + E_{\text{int}}(S,T)$. Here

$$E_{\text{int}}(S,T) = \tfrac{1}{2}\int (p_{ij}{}^S e_{ij}{}^T + p_{ij}{}^T e_{ij}{}^S)\,dv$$

is, by definition, the interaction energy between S and T. According to

[27] E. Kröner, *Proc. Phys. Soc.* (*London*) **A421,** 55 (1955).

(3.18) the two terms in the integral are equal. The volume integral can be reduced to a surface integral if we note that one or other of the strains can be written in terms of displacements inside and outside Σ. In fact $e_{ij}{}^S$ can

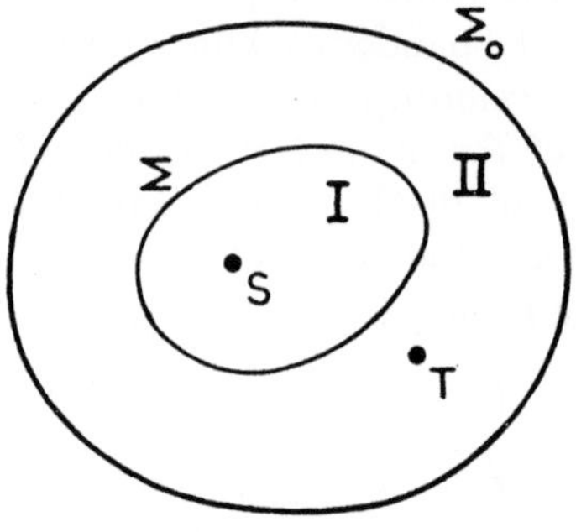

FIG. 6. To illustrate Section 5.

be written as $\frac{1}{2}(u_{i,j}{}^S + u_{j,i}{}^S)$ in region II and $e_{ij}{}^T$ as $\frac{1}{2}(u_{i,j}{}^T + u_{j,i}{}^T)$ in region I, but not conversely. Hence we have

$$E_{\text{int}}(S,T) = \int_{\text{I}} p_{ij}{}^S u_{i,j}{}^T dv + \int_{\text{II}} p_{ij}{}^T u_{i,j}{}^S dv.$$

Because of the equilibrium equations (3.9) $p_{ij}{}^S u_{i,j}{}^T = (p_{ij}{}^S u_i{}^T)_{,j}$. Gauss's theorem converts the first term into

$$\int_{\Sigma} p_{ij}{}^S u_i{}^T dS_j.$$

Similarly the second term becomes

$$\int_{\Sigma_0} p_{ij}{}^T u_i{}^S dS_j - \int_{\Sigma} p_{ij}{}^T u_i{}^S dS_j.$$

The minus sign is correct if in $dS_j = n_j dS$, n_j is supposed to be the outward normal to Σ. The integral over Σ_0 vanishes since $p_{ij}{}^T n_j = 0$ on Σ_0. Thus we have an expression[6]

$$E_{\text{int}}(S,T) = \int_{\Sigma} (p_{ij}{}^S u_i{}^T - p_{ij}{}^T u_i{}^S) dS_j \tag{5.1}$$

for the interaction energy between S and T in the form of an integral over a surface separating them. From the derivation it is clear that the choice of Σ is arbitrary so long as it lies in a region where both $u_i{}^S$ and $u_i{}^T$ exist; analytically, the divergence of the integrand vanishes in the region between two such surfaces.

Now let $p_{ij}{}^T$ and $u_i{}^T$ be the stress and displacement produced by surface tractions $p_{ij}{}^T n_j$ instead of by a source of internal stress; $u_i{}^T$ exists throughout the body. Thus for the interaction term in the elastic energy we have

$$E_{\text{int}}^*(S,T) = \int p_{ij}{}^S u_{i,j}{}^T dv = \int (p_{ij}{}^S u_i{}^T)_{,j} dv = \int_{\Sigma_0} p_{ij}{}^S u_i{}^T dS_j.$$

This vanishes however, since $p_{ij}{}^S n_j = 0$ at the surface of the body. Thus

the interaction term in the elastic energy between a system of internal stress and a system of external stress is zero. (5.2)

The response of a body to external forces can be derived from its elastic energy by Castigliano's and related theorems.[14] Hence (5.2) says physically that the response of a body to external forces is the same whether it is self-stressed or not.[28]

This does not, of course, mean that there is no interaction energy between the internal and external stresses, since we must include the potential energy of the external mechanism giving rise to the latter. We can in fact show that (5.1) is still a good measure of the interaction energy when T refers to an external stress. The requirement that Σ separate S from T evidently means now that Σ shall lie within Σ_0 but outside the sources of S. In particular we may put $\Sigma = \Sigma_0$. Then

$$E_{\text{int}}(S,T) = -\int_{\Sigma_0} p_{ij}{}^T u_i{}^S dS_j.$$

To be a sensible interaction energy $E_{\text{int}}(S,T)$ must have the following property: $E_{\text{int}}(S'',T) - E_{\text{int}}(S',T)$ is the difference of the energy of the whole system for two different states of internal stress S'' and S' and the same external stress T, insofar as it depends on cross terms between S'' and T or S' and T. The energy of the system is made up of the elastic energy of the body and the potential energy of the mechanism producing the surface traction. We have just seen that the former makes no contribution to the interaction energy. The change of the potential energy is the negative of the work done by the external forces in passing from S' to S'', that is

$$-\int_{\Sigma} p_{ij}{}^T (u_i{}^{S''} - u_i{}^{S'}) dS_j.$$

This is just $E_{\text{int}}(S'',T) - E_{\text{int}}(S',T)$ as calculated from (5.1). Hence, quite generally, (5.1) gives the interaction energy between S and an elastic field produced by internal or external stress, or, by an easy generalization, a combination of both. In place of (5.1) we may write

$$E_{\text{int}}(S,T) = \int_{\Sigma} \{(p_{ij}{}^S + p_{ij}{}^U) u_i{}^T - p_{ij}{}^T (u_i{}^S + u_i{}^U)\} dS_j$$

[28] R. V. Southwell, "Theory of Elasticity." Oxford Univ. Press, London and New York, 1936.

where u_i^U, p_{ij}^U is any elastic field free of singularities within Σ, for, by (3.21), the additional terms give no contribution. In other words, in place of u_i^S we can use any "wrong" elastic field which has the same singularities inside Σ. In particular we may put $u_i^U = -u_i^I$ and obtain the result

$$E_{\text{int}}(S,T) = \int_\Sigma (p_{ij}^\infty u_i^T - p_{ij}^T u_i^\infty) dS_j \tag{5.3}$$

which is the most generally useful. Here u_i^∞ and u_i^I are the displacement in an infinite medium and the image displacement discussed in Section 4.

It may happen that a fictitious distribution of body force f_i^S inside Σ can be found which produces the same stress on and outside Σ as does the actual source of internal stress within Σ. Then Gauss's theorem reduces (5.1) to the volume integral

$$E_{\text{int}}(S,T) = -\int \mathbf{f}^S \cdot \mathbf{u}^T dv \tag{5.4}$$

taken over the interior of Σ.

These results are closely connected with the Green's function for the boundary-value problems of elasticity.[13] Suppose that for some point singularity at P we can evaluate the interaction energy explicitly in the form

$$E_{\text{int}}(S,T) = \varphi(u_i^T, p_{ij}^T \text{ at } P). \tag{5.5}$$

Combining this with (5.1) we have a formula for evaluating φ at P from the applied surface tractions. Evidently u_i^S is the appropriate Green's function. Similarly, combining (5.5) with (5.3) we see that merely with the help of the "Green's function for an infinite medium," u_i^∞, we can find φ only if we know both surface traction and displacement. Thus, for example, the result (8.9) gives McDougall's[13] formula for dilatation in terms of surface traction, while the interaction energy for an infinitesimal dislocation loop (see Section 9a) gives effectively Lauricella's[13] relation for determining shear stress in terms of surface data.

b. The Force on a Singularity

From the foregoing we can easily find an expression for the force on the singularity S due to another stress system T, in the sense explained in Section 2. Evidently the force in the x_l direction is

$$F_l = \lim_{\epsilon \to 0} \epsilon^{-1} \{ E_{\text{int}}(S',T) - E_{\text{int}}(S,T) \}$$

where S' stands for the singularity S after it has been advanced a distance ϵ along the x_l axis. To find the elastic field of S', we may shift the field of S bodily and make an adjustment to ensure that the boundary conditions are still satisfied. Thus

$$u_i^{S'} = u_i^S - \epsilon u_{i,l}^S + u_i'$$
$$p_{ij}^{S'} = p_{ij}^S - \epsilon p_{ij,l}^S + p_{ij}'.$$

The field u_i', p_{ij}' is free of singularities within Σ, and, by (3.21), makes no contribution to $E_{\text{int}}(S,T)$. Thus from (5.1) we have at once

$$F_l = \int_\Sigma (p_{ij,l}^S u_i^T - p_{ij}^T u_{i,l}^S) dS_j \tag{5.6}$$

or, splitting this into ∞ field and image field terms and applying (3.21) to the latter, we obtain

$$F_l = \int_\Sigma (p_{ij,l}^\infty u_i^T - p_{ij}^T u_{i,l}^S) dS_j. \tag{5.7}$$

These results are still true if the affixes S, T or ∞, T are interchanged:

$$F_l = \int_\Sigma (p_{ij,l}^T u_i^S - p_{ij}^S u_{i,l}^T) dS_j \tag{5.8}$$
$$= \int_\Sigma (p_{ij,l}^T u_i^\infty - p_{ij}^\infty u_{i,l}^T) dS_j \tag{5.9}$$

for the difference between (5.6) and (5.8) is

$$\int_\Sigma (p_{ij}^T u_i^S - p_{ij}^S u_i^T)_{,l} dS_j = \int_\Sigma (p_{ij}^T u_i^S - p_{ij}^S u_i^T)_{,j} dS_l$$
$$= \int_\Sigma (p_{ij}^T u_{i,j}^S - p_{ij}^S u_{i,j}^T) dS_l = 0$$

by (3.25) and (3.18). (We have assumed that u_i^S is single-valued on Σ; the case where this is not so is discussed in reference 6.) Equation (5.9) follows from (5.8) on rejecting the image field as before. Any of these expressions for F_l give the image force if u_i^T, p_{ij}^T are replaced by the image field u_i^I, p_{ij}^I. This follows by a rather tedious extension of the present argument[6] or more simply from the results of Section 7.

6. Interaction Energies between Stresses and Inhomogeneities

Suppose that a body is subject to prescribed surface tractions over its surface Σ_0 and that the elastic constants c_{ijkl} are functions of position. Let the elastic constants change to some other function of position c_{ijkl}' and let the new values of the elastic quantities be distinguished by primes, the prescribed surface tractions remaining unaltered. The increase of elastic energy is

$$\delta E_{\text{el}} = \tfrac{1}{2}\int (p_{ij}' e_{ij}' - p_{ij} e_{ij}) dv \tag{6.1}$$
$$= \tfrac{1}{2} \int_{\Sigma_0} p_{ij}(u_i' - u_i) dS_j. \tag{6.2}$$

The work done by the external forces during the alteration, $-\delta E_{\text{ext}}$, is clearly just twice (6.2). Thus

$$\delta(E_{\text{el}} + E_{\text{ext}}) = -\delta E_{\text{el}} = \tfrac{1}{2}\delta E_{\text{ext}}. \tag{6.3}$$

Of the work done by the external forces, half disappears and half goes to increase the internal energy of the body.

Equation (6.1) can be written in the form

$$\delta E_{\text{el}} = \tfrac{1}{2}\int (p_{ij}e_{ij}' - p_{ij}'e_{ij})dv = \tfrac{1}{2}\int (c_{ijkm}' - c_{ijkm})e_{ij}'e_{km}dv \tag{6.4}$$

for the difference between (6.1) and (6.4) can be transformed by (3.18) into

$$\tfrac{1}{2}\int_{\Sigma_0} (p_{ij}' - p_{ij})(u_i' - u_i)dS_j$$

which vanishes in view of the boundary condition $p_{ij}n_j = p_{ij}'n_j$. We may also write

$$\delta E_{\text{el}} = \tfrac{1}{2}\int (s_{ijkm}' - s_{ijkm})p_{ij}'p_{km}dv \tag{6.5}$$

where the coefficients are those giving e_{ij} in terms of p_{ij}, $e_{ij} = s_{ijkm}p_{km}$.

The x_1 component of the effective force on the elastic inhomogeneity is evidently given by taking $c_{ijkm}'(x_1,x_2,x_3) = c_{ijkm}(x_1 - \epsilon, x_2,x_3)$ calculating (6.4), dividing by $-\epsilon$, and letting ϵ tend to zero. Thus

$$F_l = \tfrac{1}{2}\int c_{ijkm,l}e_{ij}e_{km}dv = \tfrac{1}{2}\int \{(c_{ijkm}e_{ij}e_{km})_{,l} - 2c_{ijkm}e_{ij,l}e_{km}\}dv$$
$$= \int (W_{,l} - p_{ij}e_{ij,l})dv$$

where W is the elastic energy-density. The second term may be written as $-p_{ij}u_{i,jl} = -(p_{ij}u_{i,l})_{,j}$ since $p_{ij,j} = 0$. Gauss's theorem then gives[6]

$$F_l = \int_{\Sigma_0} (W\delta_{jl} - p_{ij}u_{i,l})dS_j. \tag{6.6}$$

It follows from the discussion in Section 7 that (6.8) also gives the force on an inhomogeneity due to a system of internal stress provided Σ_0 is taken to be a surface separating the inhomogeneity from the source of internal stress.

According to the discussion in Section 2, Eq. (6.3) states that in an adiabatic change the changes of enthalpy and internal energy are equal and opposite, or that in an isothermal change the changes of the Gibbs and Helmholtz free energies are equal and opposite. For a thermodynamic system in which the deformation is described sufficiently by giving the specific volume V, there is the adiabatic relation

$$H = E - V(\partial E/\partial V)_S$$

between enthalpy H and internal energy E and the isothermal relation

$$G = F - V(\partial F/\partial V)_T$$

between Gibbs free energy G and Helmholtz free energy F. If E or F is a quadratic function of V we have $H = -E$, $G = -F$. Equation (6.3) is just the generalization of this to the more complex elastic case. It depends on the energy density being quadratic in the strains.

7. The Energy-Momentum Tensor of the Elastic Field

It is possible to develop a general expression[6] for the force on an elastic singularity or inhomogeneity which embraces the foregoing results and which, moreover, is valid for finite strain and an arbitrary stress-strain relation. If the latter is to be true, we must use the total displacement and stress throughout, since, in a nonlinear system, the division of elastic quantities into parts due to image terms, internal and external stress systems and the like no longer has any meaning.

We begin with the simple case of a body whose free surface Σ_0 is subject to surface tractions and which contains some singularity S, that is, a source of internal stress or an elastic inhomogeneity. For the moment, we suppose that the displacements are infinitesimal.

We first find the change in the elastic energy of the body E_{el} on moving S a small distance ϵ in the direction of the positive x_1 axis. We can do this in two stages: (i) at each point (x_1,x_2,x_3) we replace the value $\varphi(x_1,x_2,x_3)$ of any quantity associated with the elastic field by $\varphi(x_1 - \epsilon, x_2,x_3)$, and (ii) we adjust the surface values of φ, as thus changed, so that they again conform with the boundary conditions. In stage (i) the change of elastic energy evidently is

$$\begin{aligned} \delta E_{el}^{(i)} &= -\epsilon \int \frac{\partial W}{\partial x_1} dv + O(\epsilon^2) \\ &= -\epsilon \int_{\Sigma_0} W dS_1 + O(\epsilon^2) \end{aligned} \tag{7.1}$$

where W is the energy density. If we suppose that for some reason W or its derivative cannot be defined throughout the interior of Σ_0, we may evaluate $\delta E_{el}^{(i)}$ by keeping the φ fixed and shifting Σ_0 by ϵ in the direction of the negative x_1 axis. Evidently $\delta E_{el}^{(i)}$ is the volume integral of W over the unshaded area of Fig. 7 with due regard for sign; this again gives (7.1). The figure also makes it clear that the shaded area contributes nothing. Thus, in the case of singularities for which W becomes formally infinite, we have managed to "subtract out the infinities."

If the displacement at the surface is u_i before stage (i), it will be $u_i - \epsilon u_{i,1} + O(\epsilon^2)$ at its conclusion. Let its value at the conclusion of stage

(ii) be u_i^{final}. Similarly, if the surface traction is $p_{ij}n_j$ initially, it will be $(p_{ij} - \epsilon p_{ij,1})n_j + O(\epsilon^2)$ at the end of stage (i). During the course of stage (ii), it will be $(p_{ij} - \epsilon p_{ij,1})n_j + p_{ij}'n_j$, where p_{ij}' will vary during the adjustment in a way depending on the degree of "hardness" of the external mechanism. In any case it will be of order ϵ. Thus the energy entering the body during stage (ii) is

$$\delta E_{\text{el}}^{(ii)} = \int_{\Sigma_0} p_{ij}(u_i^{\text{final}} - u_i + \epsilon u_{i,1})dS_j + O(\epsilon^2). \tag{7.2}$$

Consider next the change of E_{ext}. The surface traction has changed from

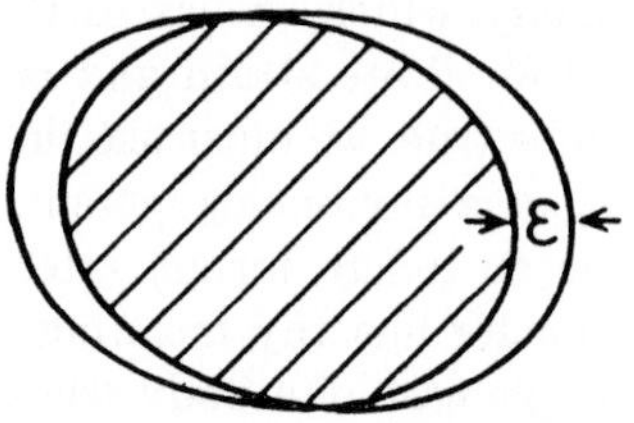

FIG. 7. To illustrate Section 7.

$p_{ij}n_j$ to $p_{ij}n_j + O(\epsilon)$ and the surface has moved through a distance $u_i^{\text{final}} - u_i$ (of order ϵ) at each point. Thus

$$\delta E_{\text{ext}} = - \int_{\Sigma_0} p_{ij}(u_i^{\text{final}} - u_i)dS_j + O(\epsilon^2) \tag{7.3}$$

and so

$$\delta(E_{\text{el}} + E_{\text{ext}}) = \epsilon \int_{\Sigma_0} (p_{ij}u_{i,1} - W\delta_{1j})dS_j + O(\epsilon^2).$$

Fortunately the expression $u_i^{\text{final}} - u_i$ which is generally incalculable and which would contribute non-negligible terms of order ϵ to (7.2) and (7.3), disappears from their sum. For the force in the x_1 direction we thus have

$$F_1 = - \lim_{\Sigma \to 0} \epsilon^{-1}\delta(E_{\text{el}} + E_{\text{ext}}) = \int_{\Sigma_0} (W\delta_{1j} - p_{ij}u_{i,1})dS_j.$$

We could, of course, equally well have displaced the singularity parallel to the x_2 or x_3 axis. Thus

$$F_l = \int_{\Sigma} (W\delta_{jl} - p_{ij}u_{i,l})dS_j \tag{7.4}$$

taking for Σ the surface Σ_0 of the body. By expressing W and p_{ij} in terms of c_{ijkl} and $u_{k,l}$, it is easy to show that the divergence of the integrand, $W_{,l} - (p_{ij}u_{i,l})_{,j}$ vanishes wherever $c_{ijkl,m} = 0$, that is wherever the material is homogeneous. Thus the integral (7.4) can be taken over any surface Σ into which Σ_0 can be deformed without entering a region in which u_i

cannot be defined or where the elastic constants vary with position. Thus Σ can be any surface embracing, but not cutting, inhomogeneities or sources of internal stress.

The extension to the case where there is a force on S arising from sources of internal stress which are outside Σ (Fig. 6), in addition to or instead of those arising from surface tractions, is immediate. No difference is made in $E_{el} + E_{ext}$ if we redefine E_{el} to be the elastic energy within Σ and re-define E_{ext} to be the sum of the elastic energy between Σ and Σ_0 and the energy of the external mechanism. We can then repeat the previous calculation, now taking Σ as the surface of the "body" and regarding everything outside Σ as the mechanism producing surface tractions on Σ. This is quite legitimate since we were careful, in the original derivation of (7.4), to impose no limitation on the external mechanism, beyond the continuity requirement that a change of order ϵ in any one of the quantities $p_{ij}n_j$, u_i, E_{ext} involve a change of the same order in the other two. Evidently we arrive at (7.4) again, with Σ being any surface separating the singularity S from the surface tractions and the sources of internal stress which we regard as exerting a force on it.

When there is no source of internal stress outside Σ and E_{ext} vanishes, (7.4) gives the image force on S arising from the imposed boundary conditions and any inhomogeneities in II (Fig. 6), since nothing in the argument excludes the latter. Boundary conditions for which $E_{ext} = 0$ are: surface traction zero over some regions of Σ_0, displacement constant over the remainder. When the boundary condition is $p_{ij}n_j = 0$ all over Σ_0, (7.4) reduces to

$$F_l = \int_{\Sigma_0} W dS_l$$

and we may say that the energy change in a small displacement simply arises from the movement of the peripheral elastic field of S into and out of Σ_0 in stage (i) (Fig. 7). The adjustment in stage (ii) makes no further contribution.

In stage (i) the whole elastic field was supposed to be shifted. This means, in particular, that the elastic contents underwent the change $c_{ijkl}(x_1,x_2,x_3) \rightarrow c_{ijkl}(x_1 - \epsilon, x_2,x_3)$. Thus even if there are no sources of internal stress within Σ, F_l will not be zero if there are elastic inhomogeneities within Σ_0 Eq. (7.4) then gives the force on them in agreement with (6.6). When there are both sources of internal stress and inhomogeneities within Σ, F_l gives the combined force on them. There is no way of separating the two contributions.

Equation (7.4) is also valid for finite strain and for an arbitrary stress-strain relation if the x_i are interpreted as the Lagrangian x_i of Section 3 and u_i, p_{ij} are the finite displacement and Boussinesq stress components there defined. dS_j is to be taken as the surface element before deformation. The proof is the same word for word except that the vanishing of the divergence of the integrand of (7.4) now follows from the relation

$$\frac{\partial W(u_{m,n})}{\partial x_l} = \frac{\partial W}{\partial u_{i,j}} u_{i,jl}$$

which is valid wherever W depends only on the $u_{i,j}$ and not explicitly on the x_i.

To sum up, the integral

$$F_l = \int_\Sigma P_{jl} dS_j \tag{7.5}$$

with

$$P_{jl} = W\delta_{jl} - p_{ij}u_{i,l}$$

gives the force on all sources of stress and elastic inhomogeneities in I (Fig. 6) arising from surface stresses on Σ_0, from sources of internal stress in II and from the image effects associated with the boundary conditions and elastic inhomogeneities in II.

In the linear infinitesimal case, we may split the elastic field into terms relating to S in an infinite medium (∞), the corresponding image term (I), the field arising from other sources of internal stress (T) and from external forces (E):

$$u_i = u_i^\infty + u_i^I + u_i^T + u_i^E, \qquad p_{ij} = p_{ij}^\infty + p_{ij}^I + p_{ij}^T + p_{ij}^E.$$

Then

$$F_l = F_l^I + F_l^T + F_l^E$$

where

$$F_l^X = \int_\Sigma (u_i^X p_{ij,l}^\infty - p_{ij}^X u_{i,l}^\infty) dS_j, \qquad X = I,\, E,\, T.$$

To see this one need only verify with the help of (3.18) and (3.25) that a typical cross term in (7.5), say

$$(X,Y) = \int_\Sigma (\tfrac{1}{2}p_{mn}{}^X e_{mn}{}^Y \delta_{jl} - p_{ij}{}^X u_{i,l}{}^Y) dS_j, \qquad X,\, Y = I,\, E,\, T,\, \infty$$

is equal to (Y,X) and to

$$\tfrac{1}{2}\int_\Sigma (u_i^X p_{ij,l}{}^Y - p_{ij}{}^X u_{i,l}{}^Y) dS_j.$$

This expression vanishes by (3.21) unless one or other of the quantities X, Y stands for ∞. The term (∞, ∞) vanishes in view of the limitations imposed on u_i^∞ in Section 4. We have thus recovered the results of Section 5 with the extension to the case of the image force, as promised there.

The spatial part of the canonical energy-momentum tensor of the time-independent elastic field is P_{jl}. It is interesting to set up the complete four-dimensional tensor for the general time-dependent field. The equation of motion in the embedded coordinates of Fig. 2a is found by replacing the force f_i by $f_i - \rho \ddot{u}_i$ in (3.3). With (3.4) this gives

$$-\frac{\partial}{\partial x_j}\frac{\partial W}{\partial u_{i,j}} + \frac{\partial}{\partial t}\rho \dot{u}_i = f_i. \tag{7.6}$$

This is the equation of motion derived from the Lagrangian density

$$L = \tfrac{1}{2}\rho \dot{\mathbf{u}}^2 - W(u_{i,j})$$

for the free elastic field together with an external force density f_i not taken account of in the Lagrangian. The methods of field theory[29] enable us to derive an energy-momentum tensor

$$T_{\eta\lambda} = (\partial L/\partial u_{i,\eta})u_{i,\lambda} - L\delta_{\eta\lambda}$$
$$(\eta,\lambda = 1, 2, 3, 4;\ x_4 = t;\ u_4 = 0).$$

Its components are

$$T_{jl} = P_{jl} - \tfrac{1}{2}\rho\dot{\mathbf{u}}^2\delta_{jl}, \qquad T_{44} = W + \tfrac{1}{2}\rho\dot{\mathbf{u}}^2 \tag{7.7}$$
$$s_j = T_{j4} = -p_{ij}\dot{u}_i, \qquad g_l = T_{4l} = \rho\dot{u}_i u_{i,l}.$$

If the medium is homogeneous, there is the conservation law

$$\partial T_{jl}/\partial x_j + \partial g_l/\partial t = f_i u_{i,l}. \tag{7.8}$$

Here T_{44} is the energy density and s_j the energy flux vector.[13] The "field momentum" density g_l differs from the true momentum density $G_l = \rho\dot{u}_l$. We may give the following formal interpretation. Consider an imaginary particle able to move through the medium, and take for its generalized coordinates $x_i(t)$ the values of x_1, x_2, x_3 associated with the point of the embedded coordinate net of Fig. 2a with which it coincides at time t. (The shape of the coordinate net changes, of course, with time.) Its equation of motion will be

$$\frac{d}{dt}\frac{\partial T}{\partial \dot{x}_l} - \frac{\partial T}{\partial x_l} = Q_l \tag{7.9}$$

where $T(x_i,\dot{x}_i)$ is its kinetic energy and Q_l is the generalized force acting on it. In particular, we can identify the particle with a small element of the elastic medium, say the elementary mesh of Fig. 2b. Throughout its motion $\dot{x}_l = 0$. This does not mean, however, that its generalized momentum $\partial T(x_i,\dot{x}_i)/\partial\dot{x}_l$ vanishes. In fact the momentum is easily shown to be $\epsilon^3\rho(\dot{u}_l + \dot{u}_i u_{i,l})$, that is, $\epsilon^3(G_l + g_l)$. Thus the field momentum density

[29] G. Wenzel, "Quantum Theory of Fields." Interscience, New York, 1949.

is the difference between the true momentum and the generalized momentum per unit mesh when the motion of the medium is referred to the coordinate system deforming with it. Equation (7.8) becomes, term for term

$$\epsilon^3\partial(G_i + g_i)/\partial t - \epsilon^3\rho\dot{u}_n u_{n,j} = \epsilon^3\{f_n + p_{nj,j}\}[\delta_{ni} + u_{n,i}]. \qquad (7.10)$$

The $\epsilon^3\{\ \}$ are the Cartesian components of the force on the element, made up of the applied force and the force exerted on it by its neighbors. The factor [ ] converts this to the generalized force. Equation (7.10) can be transformed into

$$\frac{\partial}{\partial t}(G_i + g_i) - \frac{\partial}{\partial x_j}(p_{ij} - T_{ji}) = f_n(\delta_{ni} + u_{n,i})$$

which is just the result of adding (7.6) and (7.8). If there is a region v outside which the disturbance is zero, integration gives

$$\frac{d}{dt}\int_v (G_i + g_i)dv = \int_v f_n(\delta_{ni} + u_{n,i})dv \qquad (7.11)$$

that is, the rate of change of true plus field momentum is equal to the sum of the generalized external forces acting on all the elementary meshes. This takes a more interesting form if f_n is derivable from a potential depending only on the absolute position of the element (and on time), so that $f_n = \partial V/\partial(x_i + u_i)$. The integrand on the right of (7.10) is simply $\partial V/\partial x_i$ and hence the integral vanishes if V vanishes outside v. Thus if the elastic field is varying as a result of an interaction with, say, electrified particles which move within it, changes in momentum may be calculated correctly by assuming a fictitious momentum density $-g_i$ in place of the true G_i.[30]

A number of points in the theory of fields receive a simple interpretation in the case of the elastic field when it is realized that the x_i are embedded (Lagrangian) coordinates. In other words, the u_i have the dual role of field variables and components of an actual displacement of the material. Thus, for example, the fact that certain "spin" terms have to be introduced to obtain conservation of angular momentum is closely related to the circumstance that $x_i + u_i$ and not x_i is the appropriate lever arm for taking moments.

These results refer to a homogeneous medium where L does not depend explicitly on x_i. In a medium with internal stress and elastic inhomogeneities, we have

$$P_{jl,j} = \partial W(x_k,u_{i,k})/\partial x_l \qquad (7.12)$$

[30] W. Brenig, *Z. Physik* **143**, 168 (1955).

for the static case in the absence of body forces. The right-hand side denotes the explicit dependence on x_l when W is regarded as a function of the independent variables x_k and the $u_{i,k}$. Equation (7.11) is closely related to our derivation of (7.4). However, there are certain difficulties connected with the direct use of (7.12) which our method by-passes. In the simplest cases it is possible to extend the methods of the present section to dynamical problems (see Section 9c).

III. Applications

8. Point Defects

a. Distortion of Crystals

As the simplest elastic model of a substitutional or interstitial atom we take a sphere ("inclusion") forced into a spherical hole of slightly different size in an infinite block ("matrix") of elastic material.

It is clear that $\mathbf{u}^\infty$ must be spherically symmetric and must not increase with distance outside the inclusion. In fact

$$\mathbf{u}^\infty = c\mathbf{r}/r^3 = -c \operatorname{grad} (1/r) \tag{8.1}$$

where the constant c is a measure of the "strength" of the defect. Equation (8.1) is of the same form as the field around a charged particle. Thus $\operatorname{div} \mathbf{u}^\infty = 0$, $\nabla^2\mathbf{u}^\infty = 0$ and (3.11) is obviously satisfied with $\mathbf{f} = 0$. A second solution $\mathbf{u} = \text{const} \cdot \mathbf{r}$ also satisfies (3.11) since $\operatorname{div} \mathbf{u}$ is constant and $\nabla^2\mathbf{u} = 0$. The sum of these solutions is the general solution of the second-order equation in r to which (3.11) reduces for spherical symmetry. Hence (8.1) is the only solution which satisfies our conditions. The corresponding stress is simply

$$p_{ij}^\infty = 2\mu u_{i,j}^\infty. \tag{8.2}$$

This follows from (3.8) and the fact that both the divergence and curl of the displacement vanish. When a defect is introduced at any point within a closed surface Σ_0 in the infinite matrix, a surface element dS with normal $\mathbf{n}$ moves and sweeps out a volume $\mathbf{u}^\infty \cdot \mathbf{n}dS$. The volume enclosed by Σ_0 increases by

$$\Delta V^\infty = \int_{\Sigma_0} \mathbf{u}^\infty \cdot \mathbf{n}dS = c \int_{\Sigma_0} \frac{\mathbf{r} \cdot \mathbf{n}}{r^3}\, dS = 4\pi c \tag{8.3}$$

the integral being simply the total solid angle subtended by Σ_0 at the defect. We note that there is a volume change even though $\operatorname{div} \mathbf{u}^\infty$ is zero in the matrix.

Consider (8.1) for a moment as a solution of the elastic equations valid for all $\mathbf{r}$, even though in our application it does not hold inside the inclusion. Formally we have

$$\operatorname{div} \mathbf{u}^{\infty} = -c\nabla^2(1/r) = 4\pi c\delta(\mathbf{r})$$

and

$$\nabla^2\mathbf{u}^{\infty} = -c \operatorname{grad} \nabla^2(1/r) = 4\pi c \operatorname{grad} \delta(\mathbf{r}).$$

Then (3.11) shows that $\mathbf{u}^{\infty}$ can be produced by a density of body force

$$\mathbf{f} = -4\pi c(\lambda + 2\mu) \operatorname{grad} \delta(\mathbf{r}). \tag{8.4}$$

In Cartesian coordinates, grad $\delta(\mathbf{r})$ has components

$$\left(\frac{\partial}{\partial x_1}, \frac{\partial}{\partial x_2}, \frac{\partial}{\partial x_3}\right) \delta(\mathbf{r})$$

and (8.4) formally represents three equal "double forces without moment"[13] at right angles (Fig. 8a).

If we wish Σ_0 to be a free surface, we must add to $\mathbf{u}^{\infty}$ the image displacement $\mathbf{u}^I$ produced by surface tractions $-p_{ij}^{\infty}n_j$ distributed over Σ_0. A complete solution is possible only in the simplest cases, but we can always find the volume change ΔV^I due to $\mathbf{u}^I$. According to (3.24) and (8.2)

$$\Delta V^I = -\frac{2\mu}{3K}\int_{\Sigma_0} x_i \frac{\partial}{\partial x_i} u_k^{\infty} dS_k. \tag{8.5}$$

Since u_k^{∞} is homogeneous of degree -2, the integral in (8.5) is -2 times the integral in (8.3) and

$$\Delta V^I = 4\pi c \frac{2(1 - 2\sigma)}{1 + \sigma}$$

(σ is Poisson's ratio). The total volume change is

$$\Delta V = \Delta V^{\infty} + \Delta V^I = 4\pi c\gamma \tag{8.6}$$

where

$$\gamma = 3\frac{1 - \sigma}{1 + \sigma} = \frac{3K + 4\mu}{3K}. \tag{8.7}$$

We may also find ΔV directly from (3.24), inserting the body forces (8.4) and zero surface tractions and using the result

$$\int \mathbf{r} \cdot \{\operatorname{grad} \delta(\mathbf{r})\} dv = \int x_i \frac{\partial}{\partial x_i} \delta(\mathbf{r}) dv = -\int \delta(\mathbf{r}) \frac{\partial x_i}{\partial x_i} dv = -3.$$

Here ΔV^I is quite a substantial fraction of ΔV^{∞}, being one-half if σ is $\frac{1}{3}$, $\frac{4}{5}$ if σ is $\frac{1}{4}$. Unlike ΔV^{∞}, it arises from an actual dilatation of the matrix, although we cannot, in general, calculate how this dilatation is distributed.

The interaction energy of the point defect with another system of internal or external stress T may be found by using any of the results in Section 5. For example, noting that (8.4) may be written

$$\mathbf{f} = -\Delta VK \text{ grad } \delta(\mathbf{r}) \tag{8.8}$$

(5.2) gives

$$\begin{aligned} E_{\text{int}} &= \Delta VK \int \mathbf{u}^T \cdot \text{grad } \delta(\mathbf{r}) dv = -\Delta VK \int \delta(\mathbf{r}) \text{ div } \mathbf{u}^T dv \\ &= \Delta V p^T \end{aligned} \tag{8.9}$$

where p^T is the hydrostatic pressure produced at the defect by the field T. In particular two point defects of the type considered here interact only through their image fields, since div $\mathbf{u}^\infty = 0$.[31]

If there are N defects in the body, its volume change will be $4\pi\gamma cN$. We can also say something about the change of *shape* of a body containing a large number of defects if we are prepared to admit a lack of rigor of the kind involved in the transition from the theory of a set of point charges to the electrostatics of a continuous charge distribution.[32] Let the defects be uniformly scattered throughout the body with a mean density of n defects per unit volume.

Consider first a sphere. The following results are almost obvious from the foregoing discussion and considerations of symmetry. If the sphere forms part of an infinite medium, introduction of the defects increases its volume by a fraction $4\pi cn$ and leaves its surface a sphere, apart from small ripples whose scale is set by the mean distance between defects, namely about $n^{-\frac{1}{3}}$. The dilatation is zero between the defects. When the sphere is cut out of its matrix it undergoes an additional fractional change of volume $4\pi cn(\gamma - 1)$, associated this time with a uniform dilatation of the material. Its surface remains a sphere, again apart from ripples. We may summarize these results thus:

(i) $\mathbf{u}^\infty$ alone or $\mathbf{u}^I$ alone produces a change of size without change of shape. (8.10)

(ii) $\mathbf{u}^\infty$ and $\mathbf{u}^I$ together produce a change of size without change of shape. The fractional change in volume is $4\pi c\gamma n$. (8.11)

(iii) Between the defects there is a uniform dilatation $4\pi c(\gamma - 1)n$ which is less by a factor $(\gamma - 1)/\gamma$ than that suggested by the change in the volume enclosed by the surface of the body. (8.12)

We now try to show that (ii) and (iii) remain valid for a body of arbitrary shape, but that (i) does not. If the body whose surface is Σ_0 is

[31] F. Bitter, *Phys. Rev.* **37**, 1526 (1931).

[32] J. D. Eshelby, *J. Appl. Phys.* **25**, 255 (1954).

embedded in an infinite matrix we have

$$\mathbf{u}^{\infty}(\mathbf{r}) = c\sum_m \frac{\mathbf{r}-\mathbf{r}_m}{|\mathbf{r}-\mathbf{r}_m|^3}. \tag{8.13}$$

At a point outside Σ_0, and sufficiently far from Σ_0 for the distance to the nearest defect to be large compared with the mean distance between defects, the displacement is approximately

$$\bar{\mathbf{u}}^{\infty}(\mathbf{r}) = cn\int \frac{\mathbf{r}-\mathbf{r}'}{|\mathbf{r}-\mathbf{r}'|^3}\,dv. \tag{8.14}$$

In accordance with the spirit of the usual discrete-to-continuous transition, we shall suppose that this is valid right up to Σ_0. Equation (8.14) has the same form as the electric field due to a uniform change density $4\pi cn$ filling Σ_0, so that the deformation of Σ_0 on introducing the defects is certainly not a uniform expansion. A direct calculation of the image traction $-p_{ij}^{\infty}n_j$ at each point of Σ_0, followed by a calculation of the field it would produce in the body with Σ_0 as free surface evidently is impossible. Thus we shall use an indirect approach. According to (8.4), (8.1) may be written

$$u_i^{\infty} = -4\pi\gamma K\frac{\partial}{\partial x_j}\sum_m U_{ij}(\mathbf{r}-\mathbf{r}_m)$$

where $U_{ij}(\mathbf{r})$ is the value of $u_i(\mathbf{r})$ when a unit point force acts at the origin parallel to the x_j axis. Thus

$$\begin{aligned}\bar{u}_i^{\infty} &= 4\pi c\gamma nK\int_V \frac{\partial}{\partial x_j'}U_{ij}(\mathbf{r}-\mathbf{r}')dv\\ &= 4\pi c\gamma nK\int_{\Sigma_0} U_{ij}(\mathbf{r}-\mathbf{r}')n_j dS.\end{aligned}$$

This shows that outside Σ_0 $\bar{\mathbf{u}}^{\infty}$ can be considered to be caused by a layer of body force on each element dS of Σ_0 of amount $4\pi\gamma cnKdS$ and directed along its normal.

We now carry out the following sequence of operations:

(i) Mark out the surface Σ_0 of the proposed body in an infinite medium.

(ii) Introduce the distribution of defects inside Σ_0. A change of size and shape is undergone by Σ_0.

(iii) Apply a body force $-4\pi\gamma cnK\mathbf{n}dS$ to each element of Σ_0. Now Σ_0 is restored to its size and shape in stage (i) and the displacement in the matrix is everywhere zero.

(iv) Cut away this unstrained matrix, scraping right up to the layer of body force, but not removing it. Nothing is altered within Σ_0. Now Σ_0 is the actual surface of the body, but is subject to a hydrostatic pressure $4\pi\gamma cnK$, since the layer of body force has now become a surface traction; Σ_0 still has the shape and size it had in stage (i).
(v) Remove the hydrostatic pressure. The body undergoes a uniform dilatation $4\pi\gamma cn$.

The displacement due to the layer of body force is clearly given by $-\bar{\mathbf{u}}^\infty$ inside Σ_0 as well as in the matrix. The displacement after stage (iii) or (iv) is thus $\mathbf{u}^\infty - \bar{\mathbf{u}}^\infty$ and is $\mathbf{u}^\infty + \mathbf{u}^I$ after stage (v). Since these differ by a uniform expansion $\mathbf{u} = \frac{4}{3}\pi\gamma cn\mathbf{r}$, we have

$$\mathbf{u}^I = \tfrac{4}{3}\pi\gamma cn\mathbf{r} - \bar{\mathbf{u}}^\infty \tag{8.15}$$

showing that $\mathbf{u}^I$ is not uniform. Its dilatation, however, is constant, for $\bar{\mathbf{u}}^\infty$ is $-$ grad φ, where φ is the potential of a uniform charge density cn filling Σ_0. Thus div $\bar{\mathbf{u}}^\infty = -\nabla^2\varphi = 4\pi cn$ and div $\mathbf{u}^I = 4\pi cn(\gamma - 1)$. The expression (8.14) also provides a reasonable value for the "macroscopic" displacement (excluding the image term) for points within the body. We define the macroscopic displacement at a point as the actual microscopic displacement averaged over a sphere of radius R large compared with the distance between defects. From the fact that (8.14) is a potential function, it is easy to show[32] that the macroscopic displacement so defined is

$$\mathbf{u}^\infty = \frac{c}{R^3}\sum_{\mathbf{r}_m<R}\mathbf{r}_m + c\sum_{\mathbf{r}_m>R}\frac{\mathbf{r}_m}{\mathbf{r}_m{}^3}$$

where $\mathbf{r}_m$ is the position vector from the point we are interested in to the defect. When R is large enough, the first term, being proportional to the position vector of the center of gravity of a large number of points taken at random in a sphere, should approach zero. The second term may be replaced by an integral, since the distance between the defects is small compared with any of the r_m. This integral is just (8.14) with the sphere R omitted from the volume of integration; however, the omitted part of the integral is proportional to the gravitational attraction at the center of a homogeneous sphere, that is, zero. Thus for points within the body, (8.14) gives the macroscopic displacement omitting image terms.

Since the image displacement defined by (8.15) is derived from the smoothed $\bar{\mathbf{u}}^\infty$, it needs no averaging to give its macroscopic value. Indeed we can go further. If we were to deduce the image traction $-p_{ij}{}^\infty n_j$ from the exact (8.13) instead of the smoothed (8.14), the results would differ only by terms fluctuating on the scale of the interdefect distance. According to

St. Venant's principle this difference would make itself felt only to a depth of the same order; the smoothed and unsmoothed $\mathbf{u}^I$ would agree in the bulk of the material.

Adding the macroscopic image and ∞ displacements and using (8.15), we have finally, for the total macroscopic displacement, the uniform expansion

$$\bar{\mathbf{u}} = \bar{\mathbf{u}}^{\infty} + \mathbf{u}^I = \tfrac{4}{3}\pi\gamma cn\mathbf{r}. \tag{8.16}$$

Thus we have verified that for shapes other than spherical (8.11) and (8.12) remain true, although (8.10) does not.

The more general case in which the density of defects is a non-uniform function of position $n(\mathbf{r})$ is now easily disposed of. Let the body be dissected into elementary cubes in each of which n is nearly constant. Each will undergo a uniform dilatation $4\pi\gamma cn(\mathbf{r})$. In the undissected body these expansions are inhibited and lead to distortion and internal stress. The problem is thus identical with the determination of the elastic state of a nonuniformly heated body if we identify the temperature T with n and the linear coefficient of thermal expansion α with one-third of the volume change produced by one defect. Thus

$$T(\mathbf{r}) = n(\mathbf{r}), \qquad \alpha = \tfrac{4}{3}\pi\gamma c.$$

For particular problems, we can draw on the methods already developed for calculating thermal stress.[14] Here we consider a simple problem of some physical interest.[33] Suppose that a thin surface layer of a massive body has been filled with defects (for example by irradiation), so that n is a function of depth which has fallen effectively to zero in a distance small compared with the dimensions of the body. Let n have the value n_s at the surface. The expansion of an element at the surface is unhindered perpendicular to the surface, but cannot take place parallel to it. Thus the free expansion $e_{11} = e_{22} = e_{33} = 4\pi\gamma cn_s$ must be supplemented by an additional deformation e_{ij}' in which $e_{11}' = e_{22}' = -e_{33}$ and $p_{33}' = 0$. (We take the x_3 axis along the normal to the surface.) An easy calculation gives a total expansion perpendicular to the surface

$$e_{33} + e_{33}' = e_{33}(1 + \sigma)/(1 - \sigma)$$

and a stress (compressive if $c > 0$) of magnitude $4\pi\gamma cn_sE/3(1 - \sigma)$ across any plane perpendicular to the surface (E is Young's modulus). Thus an x-ray determination of the spacing of lattice planes parallel to the surface would give $(1 + \sigma)/(1 - \sigma) \sim 2$ times the change of lattice constant that would be observed for a body uniformly filled with n_s defects per unit volume. If the x-ray beam penetrates to a depth where

[33] D. Binder and W. J. Sturm, *Phys. Rev.* **96,** 1519 (1954).

$n(\mathbf{r})$ departs appreciably from n_s, or if the defects are distributed down to a depth which is not small compared with the dimensions of the body, a detailed calculation using the theories of thermal stress and x-ray diffraction is necessary. For a cylinder in which n depends only on distance from the axis or for a plate in which n depends only on depth and is symmetric about the midplane, we have

$$\frac{e_s}{\bar{e}} = \frac{1+\sigma}{1-\sigma}\frac{n_s}{\bar{n}} - \frac{2\sigma}{1-\sigma}. \tag{8.17}$$

Here e_s is the expansion normal to the surface at the surface, $\bar{e}$ is the fractional change of radius or thickness, n_s is the concentration of defects at the surface, and $\bar{n}$ is the average concentration of defects. The ratio (8.17) has the value deduced above for $n_s \gg \bar{n}$ and approaches unity as n_s approaches $\bar{n}$.

b. Effect on X-Ray Diffraction

In Section 8a we took it as obvious that the change of x-ray lattice parameter in a crystal uniformly expanded by lattice defects would be just what one would infer from its change of macroscopic dimensions. Doubt was thrown on this[34] at one time but apparently the intuitive result is true.[35,36,32] This is confirmed by Huang's[37] results. He considered a spherical crystal containing a uniform random distribution of defects and took the expression (8.13) for the displacement of the lattice points. He found a change in the positions of x-ray reflections consistent with a volume change equal to the ΔV^∞ of (8.3) per defect. Huang omitted image effects, but his results may be taken to apply to a sphere subjected to a uniform hydrostatic pressure just sufficient to annul the image terms, which for a sphere are equivalent to a uniform hydrostatic tension. Removal of this pressure evidently would affect the change of x-ray lattice constant and the geometrical dimensions in the same way; both would be multiplied by a factor γ. For a shape of crystal other than spherical, the conditionally convergent sums of the type (8.13) involved in Huang's calculation are dependent on the shape of the crystal, and omission of the image terms would involve more complex errors than the mere omission of a factor γ. His method cannot be extended easily to the general case, since it rests on the fact that the displacement of the atom at $\mathbf{r}_m$ arising from the defect at $\mathbf{r}_n$ depends only on $\mathbf{r}_m - \mathbf{r}_n$. This is no longer true when

[34] P. H. Miller, Jr. and B. R. Russell, *J. Appl. Phys.* **23,** 1163 (1952).
[35] P. H. Miller, Jr. and B. R. Russell, *J. Appl. Phys.* **24,** 1248 (1953).
[36] J. Teltow, *Ann. Physik* **12,** 111 (1953).
[37] K. Huang, *Proc. Roy. Soc.* **A190,** 102 (1947).

image terms are included. We shall give an alternative argument based on a result of Miller and Russell.[34]

Suppose that the base vectors of the perfect crystal lattice are $\mathbf{a}_1$, $\mathbf{a}_2$, $\mathbf{a}_3$ and that those of the corresponding reciprocal lattice are $\mathbf{b}_1$, $\mathbf{b}_2$, $\mathbf{b}_3$. The atoms of the crystal are taken to be situated at the points $\mathbf{r} = L_i\mathbf{a}_i$ with integral L_i, and the maxima of scattering power in reciprocal space are at the points $\mathbf{k} = h_i\mathbf{b}_i$ with integral h_i. If the crystal is distorted, the lattice points move to the neighboring points $(L_i + \Delta L_i)\mathbf{a}_i$, while the maxima of scattering power move to $(h_i + \Delta h_i)\mathbf{b}_i$. Miller and Russell derive the following relation between Δh_i and ΔL_i,

$$\Delta h_i \Sigma L_i L_j + h_i \Sigma \Delta L_i L_j = 0 \tag{8.18}$$

valid for small integral h_i. The summations are over all points of the crystal lattice and the origin of coordinates is at the center of gravity of the crystal. If we assume that we may take the macroscopic displacement (8.16) for the displacement of the lattice points, we have

$$\Delta L_i = \tfrac{4}{3}\pi\gamma c n L_i.$$

Equation (8.18) shows at once that the reciprocal lattice undergoes a uniform contraction equal and opposite to the uniform dilatation of the crystal lattice; in other words the fractional change of lattice constant is in fact equal to the fractional change in the linear dimensions of the crystal. The replacement of sums by integrals involved in using (8.14) in place of (8.13) seems justified in this particular application. It would be inadequate in calculating the influence of the defects on line profiles or the scattering power between the points of the reciprocal lattice. Omission of the image terms would lead to a nonuniform deformation of the reciprocal lattice. There would not be a simple relation between the change of x-ray lattice constant and macroscopic deformation. Their retention, here as elsewhere, in addition to being physically correct, makes the solution much simpler.

If a spherical crystal of radius R contains one defect at a distance ξ from its center, it can be shown[32] that

$$\Delta V_X^\infty = \tfrac{5}{2}(1 - \xi^2/R^2)\Delta V_G^\infty$$

where ΔV_G^∞ is the geometrical change of volume and ΔV_X^∞ is the change that would be inferred from x-ray measurements. To avoid an awkward elastic calculation, image effects are neglected. When $\xi = 0$,

$$\Delta V_X^\infty = 2{\cdot}5\Delta V_G^\infty.$$

Miller and Russell based their original argument on this result. However, if the defect is more than about three-quarters of the way from the center to the surface of the sphere, ΔV_X^∞ is less than ΔV_G^∞. In fact if ξ^2 is given its mean value over the sphere, $\frac{3}{5}R^2$, we have $\Delta V_X^\infty = \Delta V_G^\infty$. Hence a uniform density of defects in the sphere would give equal x-ray and geometrical expansions. Clearly the equality will not be affected by reintroducing the image terms.

c. Solid Solutions

So far we have been able to treat a point defect simply as a center of dilatation. To discuss solid solutions we must relate the strength c of the center of dilatation to the details of the sphere-in-hole model. This is simple when the sphere and matrix have the same elastic constants. Let $V_{\text{mis}} = V_i - V_h$ be the excess of the volume of the sphere V_i over the volume of the hole V_h before the one is inserted in the other. It is easier to visualize the state of affairs if V_{mis} is negative, for then we may start by putting the sphere loosely into the hole. We can take this configuration as an unstrained body to which (3.24) may be applied. When we draw the surfaces of sphere and hole together and cement them we have the same body in a state of internal stress. According to (3.24), however, the volume of material is unchanged. Hence the empty volume eliminated, V_{mis}, must be balanced by an equal decrease in the volume enclosed by the bounding surface of the matrix.[13,38] Thus when inclusion and matrix have the same elastic constants we have simply

$$\Delta V = V_{\text{mis}}, \qquad c = V_{\text{mis}}/4\pi\gamma.$$

When the matrix and inclusion are of different materials, (3.23) tells us only that the volume integral of the hydrostatic pressure is zero in the assembled system, and a more explicit calculation is necessary. We could solve the general elastic equations, matching surface traction and displacement at the boundary between matrix and inclusion. The following method is less tedious, however, and also gives directly the information we shall need below concerning the elastic energy of a defect.

When the inclusion is put into the hole, their common boundary will evidently take up some intermediate position. Let ΔV_h, ΔV_i be the changes of volume of hole and inclusion. The inclusion will evidently be subject to a uniform hydrostatic pressure; its elastic energy will be

$$E_i = \tfrac{1}{2}K_i(\Delta V_i)^2/V_i. \tag{8.19}$$

In the infinite matrix the displacement will be given by (8.1). A direct calculation using (8.2), (3.8), (3.6) shows that the energy density at distance r is

$$\mu e_{ij}e_{ij} = 6\mu c^2/r^6.$$

[38] F. Seitz, *Revs. Mod. Phys.* **18,** 384 (1946).

Hence, by integration, the energy exterior to a sphere of volume V_h is $E_m = 32\pi^2c^2/3V_h$. But ΔV_h is just the ΔV^∞ associated with c by Eq. (8.3). Hence

$$E_m = \tfrac{2}{3}\mu_m(\Delta V_h)^2/V_h. \tag{8.20}$$

We may imagine the hole to have been blown up by an internal pressure. Comparison of (8.19) and (8.20) shows that the change of volume of the hole and the internal pressure are related just as are the change of volume and external pressure for a solid sphere, provided the "effective bulk modulus for expanding a hole" is taken to be $\tfrac{4}{3}\mu_m$.[39] Since the internal pressure in the hole must equal the external pressure on the inclusion, we have

$$\tfrac{4}{3}\mu_m\Delta V_h/V_h = -K_i\Delta V_i/V_i$$

or, to the first order, simply

$$\tfrac{4}{3}\mu_m\Delta V_h + K_i\Delta V_i = 0.$$

However, we also have the relation $\Delta V_h - \Delta V_i = V_{\text{mis}}$, whence

$$\Delta V_h = \Delta V^\infty = 4\pi c = V_{\text{mis}}/\gamma'. \tag{8.21}$$

Here

$$\gamma' = \frac{3K_i + 4\mu_m}{3K_i}$$

is formed on the pattern of (8.7) but from the bulk modulus of the inclusion and the shear modulus of the matrix. We shall suppose that the relation (8.6) is not affected appreciably by the presence of the inhomogeneous inclusion. Then (8.21) gives

$$\Delta V = \gamma V_{\text{mis}}/\gamma'.$$

From (8.19) and (8.20) we find for the total energy of the defect

$$E_s = E_i + E_m = \tfrac{2}{3}\mu_m\gamma'(4\pi c)^2/V_h. \tag{8.22}$$

We shall refer to E_s as the "self-energy" of the defect to distinguish it from its interaction energy with any other stress system, in particular that arising from other defects.

Consider a dilute substitutional solid solution of a metal M_1 in a metal M_2. Let Ω_1, Ω_2 be their atomic volumes. If we take a crystal containing N M_2 atoms and replace CN of them by M_1 atoms, the volume of the crystal becomes $N\Omega_2 + CN\Delta V$ where ΔV is the volume expansion due to one M_1 atom. Thus the mean volume per atom at a (small) atomic con-

[39] C. Zener, *Phys. Rev.* **74**, 639 (1948).

centration C of solute atoms is, on our elastic model,

$$\Omega(C) = \Omega_2 + C(V_i - V_h)$$

provided $\gamma = \gamma'$ (identical elastic constants for solvent and solute). If M_1 and M_2 have the same crystal structure, it seems reasonable to put

$$\Omega_1 = kV_i, \qquad \Omega_2 = kV_h. \tag{8.23}$$

Then

$$\Omega(C) = C\Omega_1 + (1 - C)\Omega_2 + (k - 1)(\Omega_1 - \Omega_2)C. \tag{8.24}$$

The question of determining k really lies outside the scope of the continuum theory. The choice $k = 1$ leads to the law of additivity of atomic volumes

$$\Omega(C) = C\Omega_1 + (1 - C)\Omega_2$$

or, on the linear approximation, equally well to the law of additivity of atomic radii (Vegard's law)

$$r(C) = Cr_1 + (1 - C)r_2. \tag{8.25}$$

To the same approximation, the fractional rate of change of lattice constant $(dr/dC)/r$ has the constant value

$$\epsilon = (r_1 - r_2)/r_1 \simeq (r_1 - r_2)/r_2$$

right across the composition diagram.

There seems to be no compelling reason to take $k = 1$. We might be tempted, for example, to take the radius of the hole or inclusion equal to the nearest neighbor distance in the appropriate metal. For face-centered cubic this would give $k = (3\sqrt{2})/\pi$ and upset the agreement with (8.25). We shall take the approximate validity of Vegard's law as a justification for putting $k = 1$ in the discussion of the energy of alloys.

When the solute and solvent have different elastic constants, we have by (8.21)

$$\Omega(C) = \Omega_2 + \gamma C(V_i - V_h)/\gamma'.$$

With $k = 1$ this gives

$$r(C) = Cr_1 + (1 - C)r_2 + \beta C$$

where

$$\beta = \frac{4\mu_2}{3\gamma'}\left(\frac{1}{K_2} - \frac{1}{K_1}\right)(r_1 - r_2).$$

Thus the actual value of $r(C)$ should lie above or below the value predicted by Vegard's law according to whether β is positive or negative.[40,41]

[40] B. J. Pines, *J. Phys. U.S.S.R.* **3**, 308 (1940).
[41] J. Friedel, *Phil. Mag.* [7] **46**, 514 (1955).

Friedel[41] has shown that this is qualitatively correct and has discussed the quantitative agreement.

Consider next the elastic energy of the alloy. Each successive solute atom added contributes E_s (8.22) and an interaction energy with the image field of all its predecessors. The image hydrostatic pressure is $-K\Delta V^I$ times the number of defects per unit volume. Since the image pressure builds up linearly with composition, the mean image interaction per solute atom is $\frac{1}{2}K\Delta V\Delta V^I$. If we put $k = 1$ in (8.23), the energy per atom of alloy is easily found to be

$$E(C) = E_s C \left\{1 - \frac{\gamma(\gamma - 1)}{\gamma'(\gamma' - 1)} C\right\}. \tag{8.26}$$

The free energy per atom is

$$F(C) = E(C) - TS_{\text{mix}}$$

where S_{mix} is the configurational entropy of mixing. The entropy $-\partial F/\partial T$ is made up of S_{mix} and an additional term

$$\Delta S = -\partial E(C)/\partial T$$

where ΔS may be estimated by assuming that E depends on T only through the variation with temperature of the elastic constants in (8.26).[42,42a] Friedel[42] finds good agreement between the theoretical and experimental values of E and ΔS for AuNi alloys.

If the solvent and solute are nearly alike elastically, we may put $\gamma = \gamma'$ in (8.26). Moreover, E_s will be the same for the insertion of an M_1 atom into an M_2 matrix or conversely. Then the formula

$$E(C) = \frac{6\mu\Omega}{\gamma} \epsilon^2 C(1 - C) \tag{8.27}$$

is valid for both $C \ll 1$ and $(1 - C) \ll 1$, and we may perhaps take it as a reasonable interpolation for intermediate compositions. Equation (8.27) has the simple parabolic dependence on composition predicted by the chemical theory of alloys. The constant involved depends only on the elastic constants, the atomic volume and the misfit constant ϵ, equal to the fractional rate of change of lattice parameter with composition.

From (8.27) we can give a formal derivation of Hume-Rothery's rule that if $|\epsilon|$ exceeds 15%, solubility is severely limited. According to the chemical theory of alloys,[43] there is a dome-shaped two-phase region on the temperature-composition diagram with a maximum for $C = \frac{1}{2}$ at

[42] J. Friedel, *Advances in Phys.* **3**, 446 (1954).

[42a] E. S. Machlin, *Trans. Am. Inst. Mining Met. Engrs.* **200**, 592 (1954).

[43] A. H. Cottrell, "Theoretical Structural Metallurgy." Edward Arnold, London, 1954.

a temperature T such that kT is half the coefficient of $C(1 - C)$ in (8.27) (k is Boltzmann's constant). If there is to be no miscibility gap, T must be less than the melting-point T_m of the alloy. This gives

$$|\epsilon| < \left(\frac{kT_m}{\mu\Omega}\frac{\gamma}{3}\right)^{\frac{1}{2}}.$$

Reasonable values for the constants give about 15% for the limiting misfit.[42] More elegantly we may use Leibfried's[44] theory of melting; it gives directly $kT_m/\mu\Omega = 0.042$. Or again, this quantity may be written as $RT_m/\mu V_M$ where R is the gas constant and V_M the volume per mole. In this form we may relate Hume-Rothery's rule to two other empirical rules. Richard's rule[45] states that the entropy of melting is nearly R, so that $kT_m/\mu\Omega$ may be equated to the latent heat of melting per unit volume divided by the shear modulus. Bragg[46] has noted that this is nearly 0.034 for many metals. With $\gamma = 1.5$ these two values for $kT_m/\mu\Omega$ give 14.5 and 13% for $|\epsilon|$, respectively.

d. Point Defects in Anisotropic Media

Two of the point defects discussed in Section 8a do not interact with one another except indirectly via their image fields. This behavior depends

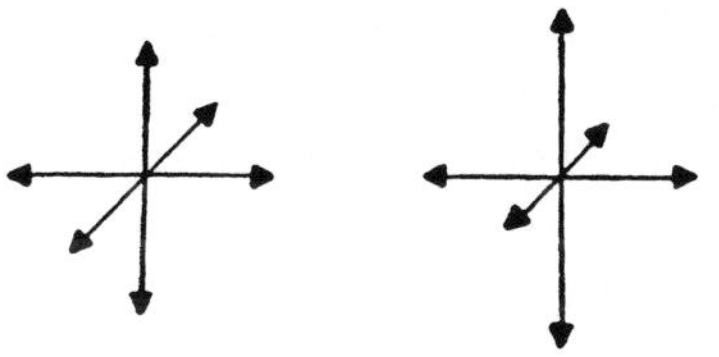

a b

FIG. 8. Crossed double forces.

on the satisfaction of two rather special conditions: (i) the interaction energy is proportional to the dilatation produced at one defect by the other; (ii) the dilatation produced by either defect (omitting image terms) is zero.

Condition (i) can be upset by choosing a less symmetrical defect. In the sphere-in-hole model we may replace the misfitting sphere by an ellipsoid, or, more manageably, we can replace the equal double forces of

[44] G. Leibfried, *Z. Physik* **127**, 344 (1949).

[45] L. S. Darken and R. W. Gurry, "Physical Chemistry of Metals." McGraw-Hill, New York, 1953.

[46] W. L. Bragg, "Symposium on Internal Stresses," p. 221. Institute of Metals, London, 1947.

Fig. 8a by unequal double forces (Fig. 8b). With the horizontal doublets of equal magnitude, Fig. 8b is a model for an interstitial carbon atom in iron.[39] For this case we may write in place of (8.4)

$$f_i = -a_{ij}\frac{\partial}{\partial x_j}\delta(\mathbf{r}).$$

A repetition of the argument leading from (8.8) to (8.9) now gives

$$E_{\text{int}} = -a_{ij}u_{i,j}{}^T = -a_{ij}e_{ij}{}^T. \tag{8.28}$$

The last step follows from the fact that the force density f_i must produce no twisting moment. This requires that a_{ij} be symmetric. The interaction no longer depends on the dilatation but on a more general linear combination of the strain or stress components of the field T with which the defect is interacting.

Again, if we drop the limitation to an isotropic medium, condition (ii) is no longer satisfied even if we take the symmetrical force system of Fig. 8a.

We consider in more detail a cubic material containing a point defect with a cubically symmetric elastic field. The equilibrium equations are

$$c_{44}\nabla^2 u_1 + (c_{12}+c_{44})\frac{\partial e}{\partial x_1} + d\frac{\partial^2 u_1}{\partial x_1{}^2} + f_1 = 0 \tag{8.29}$$

and two similar equations. The c_{ij} are the elastic constants c_{ijkl} in the usual abbreviated notation.[7] The quantity

$$d = c_{11} - c_{12} - 2c_{44}$$

vanishes for isotropy. To define the field of the defect we may require the displacement to fall off with distance and to have cubic symmetry, or equivalently we may solve (8.29) with f_i given by (8.4). If we take the latter point of view, we see at once that the interaction energy with a stress system T is given by (8.9), the transition from (8.8) to (8.9) being equally valid for the cubic case. That the coefficient ΔV in (8.8) is still the total volume change produced by the defect follows from (3.24), which is also true in the cubic case. (The bulk modulus is $K = c_{11} + 2c_{12}$.)

We could find the field of the defect merely by differentiation if we knew the displacement arising from a point force in a cubic medium. Unfortunately the elastic field caused by a point-force in any medium other than an isotropic or hexagonal one cannot be given explicitly,[47,48] a fact which considerably hinders the solution of any but trivial three-

[47] I. M. Lifshitz and L. N. Rosenzweig, *Zhur. Eksptl. i Teort. Fiz.* **17**, 783 (1947).
[48] E. Kröner, *Z. Physik* **136**, 402 (1953).

dimensional elastic problems in an anisotropic medium. We have to be content, therefore, with an approximate solution.

Write $c_{ij} = c_{ij}^0 + c_{ij}'$ where c_{ij}^0 satisfies the condition for isotropy, $c_{11}^0 - c_{12}^0 - 2c_{44}^0 = 0$. If we treat the c_{ij}' as small, we may solve (8.29), (8.8) by successive approximation. To the second order the dilatation is found to be[49]

$$e^\infty(\mathbf{r}) = \frac{\Delta VK}{c_{11}^0}\left\{\delta(\mathbf{r}) + \frac{15}{8\pi}\frac{d}{c_{11}^0}\frac{x_1^4 + x_2^4 + x_3^4 - \frac{3}{5}r^4}{r^7}\right\} \tag{8.30}$$

The value of c_{11}^0 depends on how we split off an isotropic component from the c_{ij}. Lifshitz and Rosenzweig[47] limit themselves to "weak anisotropy" and in effect take $c_{11}^0 = c_{11}$. Leibfried's[50] method of averaging gives

$$c_{11}^0 = \tfrac{3}{5}(c_{11} + 2c_{12} + 4c_{44}).$$

For many materials there seems to be no way of arranging that the c_{ij}' shall be convincingly smaller than the c_{ij}^0. It should be a fair approximation to replace c_{11}^0 in (8.30) by $\lambda + 2\mu$, where λ, μ are the Lamé constants for the bulk material in polycrystalline form.

The value of ΔV^∞ is found by integrating (8.30) over all space: it is thus the coefficient of $\delta(\mathbf{r})$. (The second term is zero when averaged over all directions.) The relation between ΔV, ΔV^∞, and ΔV^I is easily shown to be the same as in (8.6) with $\gamma = K/c_{11}^0$, that is, the value calculated from the averaged isotropic constants.

The interaction energy between two such defects for which $\Delta V = \Delta V_1$, $\Delta V = \Delta V_2$ is from (8.30) and (8.9)

$$E_{\text{int}} = -\frac{15d}{8\pi\gamma^2}\Delta V_1 \Delta V_2 \frac{\Gamma}{r^3}$$

with

$$\Gamma = l^4 + m^4 + n^4 - \tfrac{3}{5}$$

where r is the distance between them and (l,m,n) are the direction cosines of the line joining them. As a function of angle, Γ has a maximum in the 100 direction, a minimum in the 111 direction, and a saddle-point in the 110 direction. Thus whatever the signs of ΔV_1, ΔV_2, d may be, there is a direction for which the interaction is attractive.

By an extension of the argument of Section 8a it may be shown that a uniform distribution of these defects gives a uniform macroscopic dilatation. It will no longer be exactly true that there is a uniform dilatation between the defects, because of the second term in (8.30). Since this term averages to zero over angles, however, the results obtained for the

[49] J. D. Eshelby, *Acta Metallurgica* **3**, 487 (1955).
[50] G. Leibfried, *Z. Physik* **135**, 23 (1953).

energy of an alloy should not be affected unless the solute atoms take up ordered positions relative to one another.

e. Point Defects as Inhomogeneities

When calculating the "strength" of a point defect in Section 8c in terms of the sphere-in-hole model, we considered the general case in which the sphere had different elastic constants from its surroundings. On this model the defect is both a source of internal stress and an elastic inhomogeneity in the sense of Section 6. So far we have neglected its interaction as an inhomogeneity. On the linear theory this can be treated separately from its effect as a source of stress, so that we consider a perfectly fitting sphere with elastic constants λ', μ' embedded in a medium with constants λ, μ. Let surface forces produce a uniform strain e_{ij}^T in the homogeneous medium. When the sphere is introduced, the change of total energy is

$$E_{\text{int}} = \tfrac{1}{2}\int \{(\lambda' - \lambda)e^T e' + 2(\mu' - \mu')e_{ij}^T e_{ij}'\} dv$$

according to (6.4). The integral is taken only over the sphere (since the elastic constants do not change outside it), and e_{ij}' is the strain in the inclusion. It can be shown[49] that e_{ij}' is uniform. It is a linear function of the e_{ij}^T, and by symmetry it must be an isotropic function, say

$$e_{ij}' = Ae^T\delta_{ij} + 2Be_{ij}^T \tag{8.31}$$

so that

$$E_{\text{int}} = -\tfrac{1}{2}\Omega\{\Lambda(e^T)^2 + 2\mathrm{M}e_{ij}^T e_{ij}^T\} \tag{8.32}$$

where Ω is a volume which, in applications, may conveniently be the volume per atom; Λ and M have the dimensions of elastic constants and can be calculated. Their ratio is a definite function of λ, μ, λ', μ', but it would be taking the model too seriously to suppose that this relation will be satisfied if, for example, we apply (8.32) to the interaction of a vacant lattice site with a stress field. It is better to regard them as independent constants which may, in principle, be found from the macroscopic elastic constants of a material containing a large number of defects. In fact, according to (6.3), with constant external loading, the normal elastic energy density $\tfrac{1}{2}\lambda(e^T)^2 + \mu e_{ij}^T e_{ij}^T$ of the medium is increased by $-nE_{\text{int}}$ when n defects per unit volume are introduced. The apparent elastic constants are thus

$$\lambda_{\text{app}} = \lambda + C\Lambda, \qquad \mu_{\text{app}} = \mu + C\mathrm{M}$$

if C is the atomic concentration of defects.

We have supposed that e_{ij}^T is uniform, but we may take (8.32) to apply also to a nonuniform field, provided it varies little over a distance of the order of the size of the inclusion. Thus the force exerted on the inhomogeneity by a stress field T is

$$F_l = \Omega\{\Lambda e^T\delta_{ij} + 2Me_{ij}{}^T\}e_{ij,l}{}^T. \tag{8.33}$$

Although we have derived this for the particular case where $e_{ij}{}^T$ is produced by externally applied forces, it must hold also when $e_{ij}{}^T$ is an internal stress, for the expression (6.6), from which in effect we derived (8.33), is the same as the general expression (7.4) which covers all cases.

Equation (8.32) is just the same as (8.28) with a_{ij} a linear function of $e_{ij}{}^T$. Calculation shows that, whereas the perturbation $e_{ij}' - e_{ij}{}^T$ produced by the presence of the inhomogeneous sphere has a uniform value given by (8.31) inside the sphere, it has the form of the stress field produced by the forces of Fig. 8b outside the sphere. Thus we may say that the applied field "induces" a complex point defect in the inhomogeneity and then exerts a force on it.

For two point defects 1 and 2 a distance r apart, we have from (8.32), (8.2), (8.6)

$$E_{\text{int}} = -6\Omega(M_1\Delta V_2{}^2 + M_2\Delta V_1{}^2)/r^6$$

in an obvious notation, and, by differentiation, a radially directed force

$$F = -36\Omega(M_1\Delta V_2{}^2 + M_2\Delta V_1{}^2)/r^7. \tag{8.34}$$

Had we found the force by evaluating (6.6) over a surface surrounding defect 1, we might have expected to get only the first term in (8.33). However, a detailed calculation shows that the second term appears as the image force on 1 due to the inhomogeneous sphere 2, so that (8.34) is correct.

9. Dislocations

a. *Interaction Energy*

To carry out a formal calculation of the interaction energy of the stress field S of a dislocation loop with another stress field T without using explicit expressions for the field of the dislocation, it seems to be necessary to make the three following assumptions:

(i) The displacement changes by a constant vector $\mathbf{b}$ on traversing any circuit c embracing the dislocation line:

$$\int_c u_{i,j}{}^S dx_j = b_i.$$

(ii) If $\mathbf{r}$ is the position vector from any fixed point on the dislocation line then

$$\lim_{r\to 0} ru_i{}^S(\mathbf{r}) = 0.$$

(iii) The integral

$$\int p_{ij}{}^S dS_j$$

vanishes even when taken over the surface bounding a volume which is traversed by the dislocation line.

The essential character of the dislocation is expressed by (i). Assumption (ii) excludes other line singularities, e.g. a line of dilatation[13] coinciding with the dislocation line, whereas (iii) excludes the possibility of distributions of body force along the dislocation line. Assumption (ii) ensures that the integral in (iii) will converge even though the stresses become infinite where the dislocation cuts the surface.

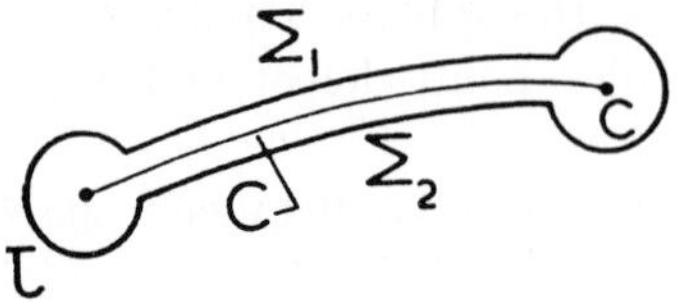

FIG. 9. To illustrate Section 9a.

Let C be any cap bounded by the dislocation line. We use (5.1) and take for Σ a surface, closely enveloping C, made up of surfaces Σ_1 and Σ_2 parallel to C and joined by a tube τ whose axis is the dislocation line (Fig. 9). As the radius of τ approaches zero, the second term in (5.1) vanishes because of (ii), since $p_{ij}{}^T$ is supposed to be continuous in the neighborhood of the tube. If we divide the tube into many small segments, we see that the first term in (5.1) also vanishes as the tube contracts, in virtue of (iii) and the continuity of $u_i{}^T$. We are left with the contributions of Σ_1 and Σ_2. The contributions from the first term of (5.1) cancel and the remaining term gives

$$E_{\text{int}}(S,T) = \int_{\Sigma_1+\Sigma_2} (-p_{ij}{}^T u_i{}^S) dS_j = b_i \int p_{ij}{}^T dS_j \tag{9.1}$$

since $u_i{}^S$ has a discontinuity b_i across C.

Let the shape of the loop be altered by giving a short segment of it having length l and direction $\mathbf{s}$ a small displacement $\boldsymbol{\xi}$. This adds to C a new surface element and $l\mathbf{s} \times \boldsymbol{\xi}$ is the product of its area and normal vector. The change in (9.1) is

$$\delta E_{\text{int}}(S,T) = l b_i p_{ij}{}^T \epsilon_{jkl} s_k \xi_l.$$

We may thus regard

$$F_l = \epsilon_{kjl} b_i p_{ij}{}^T s_k \tag{9.2}$$

as the force per unit length on the dislocation.[51,52]

[51] M. O. Peach and J. S. Koehler, *Phys. Rev.* **80**, 436 (1950).
[52] F. R. N. Nabarro, *Phil. Mag.* [7] **42**, 213 (1951).

For an infinite edge or screw dislocation along the x_3 axis, (9.2) gives the well-known results

$$F_1 = bp_{12}{}^T, \qquad F_2 = -bp_{11}{}^T \text{ (edge)} \tag{9.3}$$
$$F_1 = bp_{23}{}^T, \qquad F_2 = -bp_{13}{}^T \text{ (screw)}. \tag{9.4}$$

Koehler's[53] pioneer calculation gave an incorrect numerical factor for F_1 in (9.3) which, of course, causes no trouble if we are only interested in equilibrium with $F_1 = 0$. His method was equivalent to evaluating (7.5) with the term $p_{ij}u_{i,l}$ omitted. The result then depends on the shape of Σ. Read and Shockley's[54] method is equivalent to evaluating the same expression with the term $W\delta_{ij}$ omitted. The result again depends on the choice of Σ, but for their choice (a pair of parallel planes above and below the slip plane) gives the correct result. Leibfried[55] first clearly stated that the interaction term in the internal energy between an internal and an external stress system is zero. His results appeared to show that (5.1) gave correctly the force exerted on a dislocation by surface tractions or another dislocation, but not by a point defect. This difficulty has been resolved.[5,6]

Detailed discussion of the interaction between various configurations of dislocations may be found in references 9 and 10. Blin[56] has given an expression for the interaction energy between two dislocation loops in the form of a line integral.

Nabarro[57] constructed a solution of the Peierls-Nabarro equation representing two edge dislocations and a uniform external stress and verified that the latter was just what was required by (9.3) to give zero total force on either dislocation. Since this is one of the few cases where direct contact can be made between the elastic theory and an approximate atomic theory it seems worthwhile to sketch the solution of the corresponding problem for screw dislocations, where the analysis is quite simple.

The displacement around a screw dislocation in an infinite isotropic medium is everywhere parallel to the dislocation line (which we choose as z axis) and of magnitude

$$w = \frac{b}{2\pi}\tan^{-1}\frac{y}{x} = \frac{b\theta}{2\pi} \tag{9.5}$$

in Cartesian or polar coordinates. The Peierls-Nabarro condition requires that the stress and displacement at the atom plane adjacent to the slip

[53] J. S. Koehler, *Phys. Rev.* **60**, 397 (1941).
[54] W. T. Read and W. Shockley, *Phys. Rev.* **78**, 275 (1950).
[55] G. Leibfried, *Z. Physik* **126**, 781 (1949).
[56] J. Blin, *Acta Metallurgica* **3**, 199 (1955).
[57] F. R. N. Nabarro, *Proc. Phys. Soc. (London)* **59**, 256 (1947).

plane shall satisfy the relation

$$p_{zy} = -\frac{\mu b}{2\pi a}\sin\frac{4\pi w}{b}, \tag{9.6}$$

where a is the spacing of atom planes parallel to the slip plane.

It is well known that the purely elastic solution (9.5) itself satisfies (9.6). In fact we have

$$p_{zy} = \mu\frac{\partial w}{\partial y} = -\frac{\mu b}{2\pi}\frac{\cos\theta}{r} = -\frac{\mu b}{4\pi y}\sin 2\theta$$

which satisfies (9.6) with $y = \frac{1}{2}a$.

According to (9.4), two screw dislocations separated by a distance $2l$ exert a force $\mu b^2/4\pi l$ on one another. They should be kept apart, therefore, by an applied stress $-\mu b/4\pi l$. Let the dislocation be situated at the vertices $A(l,0)$, $B(-l,0)$ of the triangle ABC, $C(x,y)$ being any arbitrary point. From the properties of the triangle, the displacement and stress at C arising from the two dislocations and the applied stress are easily seen to be

$$w(x,y) = \frac{b}{2\pi}C - \frac{b}{4\pi l}y + w_0 \tag{9.7}$$

$$p_{zy}(x,y) = \frac{\mu b}{4\pi y}(\sin 2A + \sin 2B) - \frac{\mu b}{4\pi l} \tag{9.8}$$

where w_0 is an arbitrary constant. We shall show that this purely elastic solution satisfies (9.6) with only trivial modifications. To do this we evidently need a relation between the sines of $2A$, $2B$, $2C$ for a fixed value of y. Join the vertices of ABC to the center of its circumscribed circle to form three isosceles triangles whose equal sides are, in each case, radii $R = l \csc C$ of the circle and which embrace angles $2A$, $2B$, $2C$. Their areas must add up to ly, the area of ABC. This gives at once

$$\sin 2A + \sin 2B + \sin 2C = 2yl/R^2 = (y/l)(1 - \cos 2C)$$

or

$$y^{-1}(\sin 2A + \sin 2B) - l^{-1} = -[(y^{-2} + l^{-2})^{\frac{1}{2}}] \sin 2\{C + \tan^{-1}(y/l)\}.$$

When $y = \frac{1}{2}a/\{1 - (a/2l)^2\}^{\frac{1}{2}} \equiv y_0$, the factor [ ] is $2/a$, so that (9.3) and (9.4) satisfy (9.6) on the plane $y = y_0$ (instead of on the required plane $y = \frac{1}{2}a$), provided we give to w_0 the value $-G\{\tan^{-1}(y_0/l) + y_0/l\}$ in the upper half-plane and, to preserve the antisymmetry, an equal and opposite value in the lower half-plane. Thus the Peierls solution is derived from the elastic solution merely by removing the slab between $y = \pm y_0$, narrowing the gap to a, and giving a certain constant shift to the upper and lower half-planes. Thus, within the limits of the Peierls-Nabarro approximation, we have found a state of affairs in which each atom is in equilibrium under the action of the same applied stress as is required by the elastic theory to maintain the dislocations in equilibrium on the continuum approximation.

b. Image Effects

The discussion of the interaction of dislocations with free surfaces usually involves rather lengthy calculations. We take up first the problem of a screw dislocation in a cylinder. The calculation is simple, but the result is rather unexpected. It turns out, in fact, that the image force need

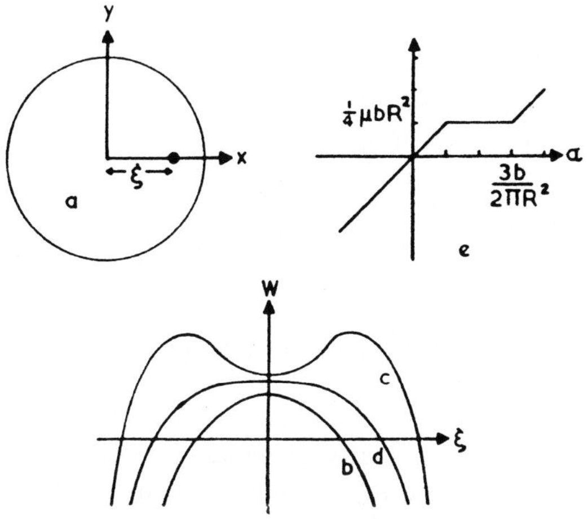

FIG. 10. Behavior of a screw dislocation in a cylinder.

not always tend to make a source of internal stress move towards the surface.

As may easily be verified, the displacement about a screw dislocation at the point $x = \xi$, $y = 0$ (Fig. 10a) in an isotropic infinite cylinder whose surface $x^2 + y^2 = R^2$ is free of stress has the form

$$w = \frac{b}{2\pi}\tan^{-1}\frac{y}{x-\xi} - \frac{b}{2\pi}\tan^{-1}\frac{y}{x - R^2/\xi} \tag{9.9}$$

if we stipulate a state of antiplane strain. Equation (9.9) is made up of an expression like (9.5), centered at $(\xi,0)$, and a similar expression of opposite sign centered at $(R^2/\xi,0)$. The "image displacement" in this case is just that produced by an image dislocation at the image point in the sense of electrostatics. The image force is directed radially outwards and is inversely proportional to the dislocation-image distance. It is convenient to write

$$F_x = -\partial W/\partial \xi \tag{9.10}$$

with

$$W = (\mu b^2/4\pi) \ln (R^2 - \xi^2). \tag{9.11}$$

Equation (9.11) is sketched in Fig. 10b; evidently the dislocation is in unstable equilibrium when $\xi = 0$ and will tend to leave the cylinder if disturbed.

A calculation like this, which assumes an infinite rod and a state of antiplane strain, is, however, unrealistic. On any cross section of the cylinder there are tractions which have zero resultant but give a twisting moment about the axis of the cylinder whose magnitude can easily be shown to be[58]

$$M = \tfrac{1}{2}\mu b(R^2 - \xi^2). \tag{9.12}$$

Thus for a finite rod cut from the infinite cylinder, the displacement will only retain the form (9.9) if suitable tractions are distributed over the ends,[59] and (9.10) is really the sum of the image force and the force due to these tractions. To find the true image force in a cylinder free of all surface tractions we must get rid of these end couples. Their removal gives a twist per unit length

$$\alpha(\xi) = M/\tfrac{1}{2}\mu\pi R^4 = (b/\pi R^2)(1 - \xi^2/R^2) \tag{9.13}$$

to the rod, together with certain end corrections which we may neglect if the length of the cylinder is many times its diameter. The production of twist by a screw dislocation in a cylinder (or rather the converse) may easily be verified by slitting a length of thick-walled rubber tubing along a radial plane. Twisting the tube produces an obvious screw dislocation, which can be made permanent by coating the cut with rubber solution.

The total elastic field is found by superimposing (9.9) and the field arising from the twist (9.13). From it the true image force may be calculated. It is easily found that (9.11) must be replaced by

$$W = (\mu b^2/4\pi)[\ln (R^2 - \xi^2) - (R^2 - \xi^2)^2/R^4]. \tag{9.14}$$

As ξ increases, the elastic energy for the case in which twist is prevented decreases and the part of this energy which is released by allowing twist to take place also decreases, initially rather rapidly. The upshot is that W first increases and then decreases (Fig. 10c). The dislocation is now bound to the center of the rod by the image forces. Only if it is somehow displaced about half (more precisely 0.54) the distance to the surface, do the image forces tend to pull it out of the rod.

[58] J. D. Eshelby, *J. Appl. Phys.* **24,** 176 (1953).
[59] E. H. Mann, *Proc. Roy. Soc.* **A199,** 376 (1949).

These results may have some application in the study of metallic "whiskers." Present theories[60,61,62] suggest that whiskers which grow at the tip may depend for their growth on an axial screw dislocation, while those which grow at the root may be free of dislocations. Figure 10c assures us that an axial screw dislocation, if present, will be stable against quite large displacements from the axis. Equation (9.13) suggests that there should be an easily observable rotation of the crystal lattice as we move along the whisker. For example with $\xi = 0$, $b = 3.10^{-8}$ cm, $R = 10^{-4}$ cm, the twist is about 50 degrees per centimeter.

It is a simple matter to find out how much the whisker must be bent or twisted in order to dislodge the dislocation. For example, an external couple M' produces a shear stress proportional to the distance from the axis, and hence a force on the dislocation proportional to ξ. This can be taken into account by adding a term $M'b\xi^2/\pi R^4$ to (9.14). For small M' this blunts the maxima in Fig. 10c and moves them nearer the center. For

$$M' = \tfrac{1}{2}\mu b R^2 \tag{9.15}$$

the maxima coalesce at the origin (Fig. 10d). For greater values of M' the center becomes an unstable position and the dislocation should be ejected from the rod. What has just been said applies when the sign of M' is such as to produce a twist tending to undo the twist arising from the dislocation itself. (We see from (9.12) and (9.15) that M' need annul only half the twist caused by the dislocation.) A couple of the opposite sign would merely deepen the well in Fig. 10c and bind the dislocation more tightly to the axis. The torsional stress-strain curve for a whisker containing a screw dislocation should thus ideally have the form of Fig. 10e. The horizontal portion represents the disappearance of the twist (9.13) when ξ suddenly changes from 0 to R as the dislocation leaves the whisker. The behavior of the dislocation under other types of external loading may also be discussed.*

The problem of a screw dislocation along the axis of a finite cylinder, with end effects included, has been solved.[63] For a long cylinder (rod) the unimportance of the end effects is confirmed, whilst for a very short cylinder (disk) the image field is, so to speak, all end effect. The stress

* The writer is indebted to Professor F. X. Eder for pointing out an error in the version of Fig. 10e in reference 58. He also points out that a factor 2 is missing from the right-hand side of Eq. (4) of the same reference: this leads to errors of 2 or $\frac{1}{2}$ in some subsequent formulas.

[60] F. C. Frank, *Phil. Mag.* [7] **44**, 854 (1953).

[61] J. D. Eshelby, *Phys. Rev.* **91**, 755 (1953).

[62] G. W. Sears, *Acta Metallurgica* **3**, 361 (1955).

[63] J. D. Eshelby and A. N. Stroh, *Phil. Mag.* [7] **42**, 1401 (1951).

components p_{zy} and p_{zx} become nearly zero at distances from the dislocation greater than the thickness of the disk. Consequently two screw dislocations in a plate interact with a short-range force, in contrast to the inverse first power law interaction in an infinite medium.

Koehler[53] has considered the problem of an edge dislocation parallel to the axis of a circular cylinder. Let Fig. 10a now refer to an edge dislocation with its Burgers vector along the x axis. His results show that

$$F_x = \frac{\mu b^2}{2\pi(1-\sigma)} \frac{1}{R^2/\xi - \xi} \tag{9.16}$$

This is just the force arising from an edge dislocation at the image point, although the image field is in fact more complex.

Koehler actually solved the problem of a dislocation in an infinite cylinder in a state of plane strain. As in the case of the screw dislocation, there will be nonzero tractions on the ends of a finite rod cut from the cylinder, and we might suppose that their removal would modify F_x. They have, however, no resultant or moment, and so, according to St. Venant's principle, removing them gives rise only to small end effects. Apart from these, the elastic state of an edge dislocation in a long cylinder coincides with the state derived on the assumption of plane strain, and (9.16) gives correctly the image force per unit length. There is nothing analogous to the odd behavior of the screw dislocation; an edge dislocation in a rod will always tend to leave it. Evidently a mixed dislocation will be attracted to or repelled from the axis of a whisker according to the relative strength of its screw and edge components.

Head[64,65] has given an extensive discussion of the interaction of dislocations with plane boundaries. The boundary may be a free or clamped surface, a surface where tangential but not normal displacement is allowed (slipping surface), or the boundary between regions of differing elastic constants.

c. Dislocations in Motion

At one time it appeared that the dynamical behavior of dislocations might play an important role in the theory of plasticity. Roughly we may say that a moving dislocation exhibits dynamical behavior when the kinetic energy of the disturbance caused by its passage is comparable with its elastic strain energy. It now seems likely that frictional forces on a dislocation prevent this condition being realized in practice.

[64] A. K. Head, *Phil. Mag.* [7] **44,** 92 (1953).
[65] A. K. Head, *Proc. Phys. Soc.* (*London*) **B66,** 793 (1953).

A number of writers have discussed dislocations in uniform motion.[66–71] The case of a screw dislocation in an isotropic medium is particularly simple. The displacement w in a state of antiplane strain satisfies

$$\frac{\partial^2 w}{\partial x^2} + \frac{\partial^2 w}{\partial y^2} = 0 \tag{9.17}$$

when w is independent of time and

$$\frac{\partial^2 w}{\partial x^2} + \frac{\partial^2 w}{\partial y^2} - \frac{1}{c^2}\frac{\partial^2 w}{\partial t^2} = 0, \qquad c = (\mu/\rho)^{\frac{1}{2}}, \qquad \rho = \text{density}$$

when it is not. For an elastic field moving uniformly parallel to the x axis with velocity v, the displacement $w = \varphi(x - vt, y)$ satisfies

$$\frac{\partial^2 w}{\partial x^2} + \frac{1}{\beta^2}\frac{\partial^2 w}{\partial y^2} = 0 \qquad \beta = (1 - v^2/c^2)^{\frac{1}{2}}. \tag{9.18}$$

Thus if

$$w = w(x,y) \tag{9.19}$$

satisfies (9.17), then

$$w = w\left(\frac{x - vt}{\beta}, y\right) \tag{9.20}$$

satisfies (9.18). If we take for (9.19) the expression (9.5), the corresponding relation (9.20) still has the property characteristic of a screw dislocation, namely that w increases by b on encircling the point $x = vt$. The field of the moving dislocation is derived from the field of the static dislocation by giving it a "Lorentz contraction." The contracted field continues to satisfy the Peierls condition (9.6). Leibfried and Dietze[69] considered a screw dislocation moving in the midplane of a plate with free surfaces. They established the "relativistic" relation

$$E^v = E_{\text{kin}}{}^v + E_{\text{el}}{}^v + E_{\text{pot}}{}^v = E^0/\beta \tag{9.21}$$

for the total energy at velocity v, made up of contributions from the kinetic and elastic energy of the continuum and the potential energy of the atomic forces maintaining the law (9.6) at the slip plane.

These results may be generalized. If a static dislocation satisfies any generalization of the law (9.6), say $p_{zy}(x, \pm\frac{1}{2}a) = f[w(x,\frac{1}{2}a) - w(x,-\frac{1}{2}a)]$, its contracted moving version satisfies the same relation. Equation (9.21)

[66] F. C. Frank, *Rept. Conf. on Strength of Solids Univ. Bristol* p. 48, 1947.
[67] F. C. Frank, *Proc. Phys. Soc. (London)* **A62,** 131 (1949).
[68] J. D. Eshelby, *Proc. Phys. Soc. (London)* **A62,** 307 (1949).
[69] G. Leibfried and H. D. Dietze, *Z. Physik* **126,** 790 (1949).
[70] R. Bullough and B. A. Bilby, *Proc. Phys. Soc. (London)* **B67,** 615 (1954).
[71] A. W. Sáenz, *J. Rat. Mech. Analysis* **2,** 83 (1953).

continues to hold for this more general law, even if the slip plane does not lie at the center of the plate, and if the surfaces of the plate are either free ($\partial w/\partial y = 0$) or clamped ($w = 0$). The proofs are simple and do not require a knowledge of the explicit form of $w(x,y)$. Let the point ($x = X,y$) in the static solution $w = w(x,y)$ and the point ($x = \beta X,y$) in the moving solution $w = w(x/\beta,y)$ be called corresponding. (We take t to be zero.) Then at corresponding points the following pairs of quantities are evidently equal:

	(i)	(ii)	(iii)	(iv)	(v)	(vi)
static	w	$\dfrac{\partial w}{\partial y}$	$\dfrac{\partial w}{\partial x}$	$(\text{grad } w)^2$	dy	dx
moving	w	$\dfrac{\partial w}{\partial y}$	$\dfrac{1}{\beta}\dfrac{\partial w}{\partial x}$	$(\text{grad } w)^2 - \dfrac{\dot{w}^2}{c^2}$	dy	$\dfrac{dx}{\beta}$.

The line elements in (v), (vi) are supposed to be bounded by corresponding points. Column (iv) expresses the "relativistic" invariance of the Lagrangian density, and follows from (ii), (iii), since $\dot{w} = -v\partial w/\partial x$.

Columns (i), (ii) show that the two solutions satisfy either of the boundary conditions $w = 0$ or $\partial w/\partial y = 0$ together at the surfaces of the plate and that along the slip plane p_{zy} is the same function of w in the two cases; thus, if a certain law of force holds the static field together, it will hold its Lorentz-contracted version together. The potential energy per unit length of the slip plane depends only on the difference in w across the slip plane and so from (i) and (vi) $E_{\text{pot}}{}^{v} = \beta E_{\text{pot}}{}^{0}$. The total energy is thus

$$E^v = \tfrac{1}{2}\textstyle\int\{\mu(\text{grad } w)^2 + \rho\dot{w}^2\}dxdy + \beta E_{\text{pot}}{}^0$$

where the integral extends over the elastic region. The integral can be rearranged with an eye to making use of column (iv), and we easily find

$$E^v = \frac{E^0}{\beta} + \frac{v^2}{c^2\beta}\left[E_{\text{pot}}{}^0 + \frac{1}{2}\int\left\{\left(\frac{\partial w}{\partial x}\right)^2 - \left(\frac{\partial w}{\partial y}\right)^2\right\}dxdy\right]$$

where the integral is evaluated for the static case. By integration by parts and use of (9.17), the integral can be converted to

$$\int_{-\infty}^{\infty} xp_{zy}\frac{dw}{dx}\,dx\,\Bigg|_{y=-\frac{1}{2}a}^{y=\frac{1}{2}a}.$$

But by an argument of Foreman and Nabarro (reference 9, p. 360: put $\sigma = 0$ for antiplane strain), this is just $-E_{\text{pot}}{}^0$, which establishes (9.21). If a approaches zero, $E_{\text{pot}}{}^0$ becomes negligible compared with $E_{\text{el}}{}^0$ and we recover a formal result of Frank[67] for the purely elastic case.

The behavior of an edge dislocation in an isotropic medium or of any dislocation in an anisotropic medium is "relativistic with complications,"[67] owing to the existence of several velocities of sound.

We turn now to the properties of dislocations in nonuniform motion. Nabarro[72] has given a general method for calculating the elastic field of a dislocation loop which is changing its shape in an arbitrary way. The result takes a relatively simple form for the two-dimensional problem of a screw dislocation whose center moves in an arbitrary manner.[6,73] If the position of its center at time τ is $x = \xi(\tau)$, $y = \eta(\tau)$, the displacement is

$$w(x,y,t) = \frac{b}{2\pi} \int_{-\infty}^{\tau_0} \frac{c(t-\tau)\{(y-\eta)\dot{\xi} - (x-\xi)\dot{\eta}\}}{(x-\xi)^2 + (y-\eta)^2} \frac{d\tau}{s}$$

where $s^2 = c^2(t-\tau)^2 - (x-\xi)^2 - (y-\eta)^2$ and τ_0 is the root of $s^2 = 0$ which is less than t. Such an arbitrary motion will require an external stress system varying suitably in space and time to maintain it. For simple cases this field can be calculated by imposing suitable conditions at the center of the dislocation. For example, we may require that the Peierls-Nabarro condition be satisfied there,[74,73] or we may, in effect, require that the force which the dislocation exerts on itself balance the force due to the external field.[73] At any moment the field at a point on the dislocation is made up of the applied field and contributions from all other points of the dislocation. Because of the finite time of propagation of elastic disturbances, the present motion depends on the previous history of the motion extending back over a time of the order of the maximum dimension of the dislocation loop divided by the velocity of sound. As a result, the equation of motion of the dislocation takes the form of an integral equation giving the external field required to maintain the prescribed motion. The more interesting problems of finding the motion in a prescribed applied elastic field requires the inversion of this integral equation. It can be carried out approximately in simple cases.[73] Roughly speaking, the rectilinear motion of an infinite screw dislocation is the same as that of a Newtonian particle of mass $(\rho b^2/4\pi) \ln (R/a)$ acted on by a force $F = bp_{zy}$, where p_{zy} is the applied stress; R is a distance of the order of the dimensions of the disturbed region surrounding the dislocation. For motion which starts from rest at $t = 0$ and in which the distance of the dislocation from the starting point increases monotonically, $R \sim ct$. Thus, for example, a dislocation which is started by an impulsive force and then runs freely will slow down as its effective mass increases. For oscillatory motion with frequency ω, $R \sim c/\omega$.

In discussing a screw dislocation, we can make use of an electro-

[72] F. R. N. Nabarro, *Phil. Mag.* [7] **42,** 1224 (1951).
[73] J. D. Eshelby, *Phys. Rev.* **90,** 248 (1953).
[74] F. R. N. Nabarro, *Proc. Roy. Soc.* **A209,** 278 (1951).

magnetic analogy.[74,73] If the dislocation is parallel to the z axis, the only nonzero elastic quantities are the z component w of the displacement and the stresses p_{zx}, p_{zy}. Consider an electromagnetic field in which

$$E_z = H_x = H_y = 0$$

and all quantities are independent of z. Make the identification

$$\partial w/\partial t = H_z/\rho^{\frac{1}{2}}, \qquad p_{zx} = -E_y/\mu^{\frac{1}{2}}, \qquad p_{zy} = E_x/\mu^{\frac{1}{2}}.$$

Here μ and ρ are the shear modulus and density. We identify the velocity of shear waves $c = (\mu/\rho)^{\frac{1}{2}}$ with the velocity of light and use Heaviside units for the electromagnetic field. The electromagnetic energy and Poynting vector translate into the sum of the elastic and kinetic energy densities and the elastic energy-flux vector. The analog of a dislocation

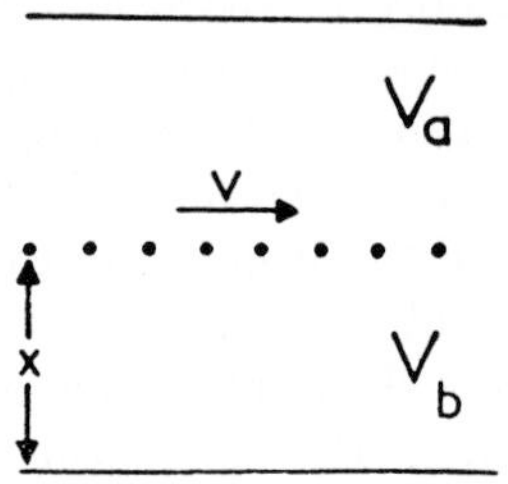

FIG. 11. Lorentz force on a procession of dislocations.

with Burgers vector b is a line charge of $b/\mu^{\frac{1}{2}}$ units per unit length. The force on a stationary charge translates correctly into the force on a stationary dislocation. The Lorentz force on a moving charge translates into a force perpendicular to the direction of motion of the dislocation of magnitude

$$F = \rho b v V \tag{9.22}$$

where v is the speed of the dislocation and V is the velocity of the medium along the z axis "at the center of the dislocation" (to avoid the singularity at the dislocation, we may take V to be the average of $\dot{w}$ over a small circle centered on the dislocation). Nabarro[74] has elucidated the physical meaning of the Lorentz force. The following is a crude illustration.

Let a closely spaced procession of n screw dislocations per unit length move with velocity v in a plate of unit thickness (Fig. 11). Their motion makes the blocks above and below the slip plane slide over each other in a direction perpendicular to the paper with relative velocity

$$V_a - V_b = bnv. \tag{9.23}$$

The effective velocity of the medium at the slip plane is $V = \frac{1}{2}(V_a + V_b)$;

we can give this any value, while still maintaining the relation (9.23). The kinetic energy per unit length and depth of the plate is

$$T = \tfrac{1}{2}\rho V_a^2(1 - x) + \tfrac{1}{2}\rho V_b^2 x = \tfrac{1}{2}\rho(V - \tfrac{1}{2}bnv)^2 + \rho bnvVx.$$

If we raise the height of the procession by δx, maintaining v and V constant, some external source must do work $\delta T = \rho bvnV\delta x$ and we may say that there is a force $F = \rho bvVn$ on the n dislocations (or a force (9.22) on each) resisting vertical motion.

Replace the screw dislocations by a procession of edge dislocations. The blocks now move with velocities V_a, V_b parallel to the slip direction: otherwise the argument is word for word the same as before and we obtain a Lorentz force (9.22) on an *edge* dislocation. Now V is the velocity of the medium in the slip-direction and F is still perpendicular to the slip plane. This Lorentz force on a procession of edge or screw dislocations agrees with what one gets by integrating the energy-momentum tensor (7.7) over a loop embracing unit length of the procession.

An aerofoil moving relative to a fluid generates a dislocation in it. To see this, suppose that the aerofoil is at rest with the fluid streaming past it. Consider a particle upstream straddling the critical streamline which divides at the nose of the aerofoil. The particle is split into two parts which traverse the upper and lower surfaces of the aerofoil and leave the trailing edge at different times, so that downstream they both lie on the critical streamline, but are separated by a certain distance, the "Burgers vector." This Burgers vector can be shown to be equal to the ratio of circulation and stream velocity, and so the lift is equal to the Lorentz force (9.22). The lift is a "real" force: if we remove the aerofoil, leaving a free vortex with the same circulation, the force takes on the "fictitious" character of the force on a dislocation.

By a very general argument,[75] it may be shown that an electron moving with velocity v through an isotropic flux of electromagnetic waves experiences a retarding force

$$F = -\alpha\sigma Wv/c \qquad (9.24)$$

where W is the energy density of the electromagnetic field, σ is the Thompson scattering cross section, and $\alpha = \frac{4}{3}$. If we take σ to be a prescribed constant, we can apply the same argument to a line charge moving in a flux of electromagnetic waves isotropic in the xy plane, and, hence, to the related dislocation problem. We should obtain (9.24) again, where now F is the retarding force per unit length of line charge or dislocation, σ is the scattering area of unit length, and α is changed from $\frac{4}{3}$ to $\frac{3}{2}$ by the

[75] L. Landau and E. Lifshitz, "The Classical Theory of Fields." Addison-Wesley, 1951.

transition from three to two dimensions. The same expression should be valid for a screw dislocation in a three-dimensional flux of sound waves if we reduce the value of α to allow for the fact that the "antiplane" sound waves contribute only a part of the total energy density W. Leibfried[44] has considered the corresponding problem for an edge dislocation. He obtains (9.24) with $\alpha = \frac{1}{10}$. By taking σ to be of the order of b and identifying W with the energy of thermal vibrations, he finds a retarding force which is sufficient to keep v a small fraction of c for any reasonable applied stress.

Nabarro[74] has given a critical discussion of Leibfried's theory. The scattering cross section of a screw dislocation for an elastic wave need not be assumed, but can be found from the electromagnetic analogy. Unlike the cross section for an electron it is not independent of the wave length of the scattered wave, but very nearly proportional to it. The arguments leading to (9.24) thus need modification. The result is found to be that F vanishes at least to order v/c. The electromagnetic analogy assumes that Hooke's law is valid even for infinite strains and neglects the atomic structure. In fact the disturbed region at the center of the dislocation may well act as a scatterer with a cross section of the order of b for waves near the top of the Debye spectrum, but smaller for long waves. This would lead to a force of the form (9.24) with an α presumably less than Leibfried's value. According to Granato and Lücke,[76] damping experiments enable the limits $0.015 < \alpha < 0.12$ to be set for dislocations in germanium.

d. Continuous Distributions of Dislocations

Having replaced the dislocations of a crystal lattice by their continuum analogs, it may be convenient to go a stage further and regard a body containing a large number of dislocations as being filled with a continuous distribution of dislocations. Following Nye[77] we may define a tensor α_{ij} whose ij element gives the sum of the x_i components of the Burgers vectors of all the dislocations threading unit area perpendicular to the x_j axis. The total Burgers vector of the dislocations threading a circuit c will be

$$b_i = \int_C \alpha_{ij} dS_j \tag{9.25}$$

where C is a cap bounded by c. So far α_{ij} merely provides a convenient description of the distribution of dislocations; we need to relate it to the deformation of the lattice and the state of internal stress. Since the

[76] A. Granato and K. Lücke, Technical Report, Contract No: DA-36-039 SC-52623 Part II. Brown University, 1955.

[77] J. F. Nye, *Acta Metallurgica* 1, 153 (1953).

internal stress is, in principle, adequately described by the incompatibility tensor, we should be able to express S_{ij} in terms of α_{ij}. Kröner[78] has made the necessary connection. On the other hand, the Riemannian geometry associated with S_{ij} is not adequate for description of a continuous distribution of dislocations. Kondo[79] and Bilby and co-workers[80] have given an elegant interpretation in terms of non-Riemannian geometry. We give a rather over-simplified sketch of this work, basing it directly on the idea of the Burgers circuit.

Let the comparison crystal be simple cubic with a unit cell defined by the vectors $\epsilon\mathbf{i}_1$, $\epsilon\mathbf{i}_2$, $\epsilon\mathbf{i}_3$ in the notation of Section 3 (ϵ need not be infinitesimal). The possibility of carrying out associated circuits in the real and comparison crystals implies that for every vector step $\epsilon\mathbf{i}_i$ we take in the comparison crystal we can pick out a step, say $\epsilon\mathbf{e}_i(P)$, at the corresponding point P in the real crystal. We may express $\mathbf{e}_i(P)$ in terms of the $\mathbf{i}_i$:

$$\mathbf{e}_i(P) = D_{ij}(P)\mathbf{i}_j \tag{9.26}$$

and conversely

$$\mathbf{i}_i = E_{ij}(P)\mathbf{e}_j(P) \qquad \text{with} \qquad D_{ij}E_{jk} = \delta_{ik}. \tag{9.27}$$

When there is a descrete distribution of dislocations, the Burgers circuit is drawn in "good" crystal which by definition is a region where the identification (9.26) can be made by inspection. If we pass to the case of a continuous distribution of dislocations, there is no "good" crystal, nor indeed any crystal lattice. Then (9.26) becomes a direct specification of the particular vector triplet $\mathbf{e}_j$ at a point of one continuum (representing the real crystal) which corresponds with the triplet $\mathbf{i}_j$ at the corresponding point of a second continuum (representing the comparison crystal). For ease of description, we shall adopt the hybrid point of view that the $\epsilon\mathbf{e}_j$ are lattice vectors of the real crystal, but that ϵ is small enough in comparison with the dimensions of the Burgers circuit for sums of lattice steps to be replaced by integrals.

Draw a closed circuit c in the real crystal. The element of path joining x_i to $x_i + dx_i$ is the vector $\mathbf{i}_i dx_i$ and the sum

$$\int_c \mathbf{i}_i dx_i$$

is zero. We can exhibit this explicitly as a sum of lattice steps in the real crystal by expressing $\mathbf{i}_i$ in terms of $\mathbf{e}_i$:

$$\int_c E_{ij}(P)\mathbf{e}_j(P)dx_i.$$

[78] E. Kröner, *Z. Physik* **142,** 463 (1955).
[79] K. Kondo, *Proc. 2nd Japan Natl. Congr. Appl. Mech. 1952,* p. 41 (1953).
[80] B. A. Bilby, R. Bullough, and E. Smith, *Proc. Roy. Soc.* **A231,** 263 (1955).

The sum of the corresponding steps in the comparison lattice is found by replacing $\mathbf{e}_j$ by its associated vector $\mathbf{i}_j$ in the comparison lattice. Thus the Burgers vector of the circuit is

$$\mathbf{b} = \int_c E_{ij}(P)\mathbf{i}_j dx_i.$$

By Stokes's theorem, this becomes the surface integral

$$\mathbf{b} = \mathbf{i}_j \int_C \epsilon_{ikl} E_{ij,l} dS_k$$

taken over a cap C bounded by c. Comparison with (9.25) gives

$$\alpha_{jk} = \epsilon_{ikl} E_{ij,l}.$$

The divergence $\alpha_{jk,k}$ vanishes, so that the choice of the cap C is arbitrary. This is the continuum analog of the rule that a dislocation cannot end in the material. The Burgers vector associated with a surface element dS_k is

$$d\mathbf{b} = \alpha_{jk}\mathbf{i}_j dS_k.$$

The *local* Burgers vector is, by definition, the vector in the real lattice which corresponds to the Burgers vector in the comparison lattice. It is found by replacing $\mathbf{i}_j$ by $\mathbf{e}_j = D_{jp}\mathbf{i}_p$:

$$d\mathbf{b}' = \alpha_{jk} D_{jl}\mathbf{i}_l dS_k.$$

If we choose to take the surface element in the form of an antisymmetric tensor dS_{mn}, where $dS_k = -\frac{1}{2}\epsilon_{kmn} dS_{mn}$ the components of $d\mathbf{b}'$ are easily found to be

$$db_p' = T_{mn}{}^p dS_{mn}$$

where

$$T_{mn}{}^p = \tfrac{1}{2} D_{jp}\{E_{mj,n} - E_{nj,m}\}. \qquad (9.28)$$

For simplicity we now suppose that $\mathbf{e}_i$ and $\mathbf{i}_i$ differ only infinitesimally from one another and write

$$D_{ij} = \delta_{ji} + U_{ji}$$

in analogy with (3.1). Then $E_{ij} = \delta_{ji} - U_{ji}$ and[80a]

$$\alpha_{ij} = -\epsilon_{mjp} U_{im,p}. \qquad (9.29)$$

We may still define a rotation in terms of the $\mathbf{e}_i$ and $\mathbf{i}_i$ by means of (3.13). This gives

$$\tilde{\omega}_{ij} = \tfrac{1}{2}(U_{ij} - U_{ji}) \qquad \text{or} \qquad \tilde{\omega}_k = -\epsilon_{kij} U_{ij}.$$

If we regard $\epsilon\mathbf{e}_1$, $\epsilon\mathbf{e}_2$, $\epsilon\mathbf{e}_3$ as the edges of a distorted unit cell in the real

[80a] B. A. Bilby, *Rept. Conf. Defects Crystalline Solids, Univ. Bristol*, p. 124, 1955.

crystal, it is reasonable to call

$$e_{ij} = \tfrac{1}{2}(\mathbf{e}_i \cdot \mathbf{e}_j - \delta_{ij}) = \tfrac{1}{2}(U_{ij} + U_{ji}) \qquad (9.30)$$

the strain components. This is the point of view of Kröner,[78] who supposes there exists in a crystal with a continuous distribution of dislocations a "nonsysmmetric state of strain" given by U_{ij} (replacing the $u_{i,j}$ of the compatible case), from which one can derive a symmetrical strain and a rotation which are no longer connected by (3.14). We have, rather,

$$\kappa_{ji} \equiv \tilde{\omega}_{j,i} = \alpha_{ij} - \tfrac{1}{2}\alpha_{mm}\delta_{ij} - \epsilon_{jmp}e_{im,p}. \qquad (9.31)$$

(Write $U_{im} = e_{im} + \tilde{\omega}_{im}$ in (9.29), express $\tilde{\omega}_{im}$ as a vector, and note that $\alpha_{mm} = -2\tilde{\omega}_{mm}$.) The incompatibility tensor corresponding to the strain (9.30) is easily found by operating on (9.31) with $\epsilon_{ikq}\partial/\partial x_q$; we find

$$S_{kj} = \epsilon_{ikq}(\alpha_{ij} - \tfrac{1}{2}\alpha_{mm}\delta_{ij})_{,q}.$$

Kröner[78] gave the explicitly symmetric form

$$S_{kj} = \tfrac{1}{2}\epsilon_{ikq}\alpha_{ij,q} + \tfrac{1}{2}\epsilon_{ijq}\alpha_{ik,q}.$$

The two expressions can be shown to be identical because of the relation $\alpha_{ik,k} = 0$. When α_{ij} is known, S_{ij} can be found, and hence the stresses are determined, if we assume with Kröner that they are derived from (9.30) by Hooke's law. Evidently if there is a uniform distribution of dislocations ($\alpha_{ij,k} = 0$), there is no stress.

The quantity (9.31) gives the rate of rotation of the triplet $\mathbf{e}_1$, $\mathbf{e}_2$, $\mathbf{e}_3$ as we follow it about the lattice, just as in the compatible case. It is Nye's[77] curvature tensor. Nye gave (9.31) for cases where the term in e_{mi} could be neglected. Kröner[78] gave the complete form, but with a slight difference in interpretation. When $\alpha_{ij} = 0$, and consequently $S_{ij} = 0$, (9.31) reduces to the ordinary compatible relation (3.14) giving the gradient of the rotation. Thus, in general, (9.31) gives the lattice rotation arising from dislocations and any compatible strain caused by body and surface forces. Kröner assumes, on the other hand, that the strain may be divided into a compatible and an incompatible part, and that the last term in (9.31) refers to only the latter. Thus κ_{ji} vanishes in his formulation if $\alpha_{ij} = 0$ and is a measure of the rotations due to the dislocations alone.

It is clear that $\alpha_{ij} - \tfrac{1}{2}\alpha_{mm}\delta_{ij}$ is a measure of the failure to satisfy the condition (3.14) for compatibility between strain and rotation, just as S_{ij} is a measure of the failure to satisfy the more rigorous strain compatibility conditions. We may give the following physical interpretation. Mark out a thin straight rod in the material and cut it out. It becomes a rod with curvature and torsion specified by $\kappa_{ji}' = \epsilon_{mjp}e_{mi,p}$ owing to relaxation of internal stress. If we annihilate the dislocations contained in it, it undergoes a further curvature and torsion specified by

$$\kappa_{ji}'' = -(\alpha_{ij} - \tfrac{1}{2}\alpha_{mm}\delta_{ij}).$$

The connection with non-Riemannian geometry comes about as follows. (We no longer assume that D_{ij} differs only slightly from δ_{ij}.) In an obvious sense, the vectors $\mathbf{e}_1(P)$ and $\mathbf{e}_1(Q)$ at points P, Q are "equivalent." More generally, we can say that $\mathbf{A}(P) = a_i(P)\mathbf{e}_i(P)$ and

$$\mathbf{A}(Q) = a_i(Q)\mathbf{e}_i(Q)$$

are equivalent if $a_1(P) = a_1(Q)$, $a_2(P) = a_2(Q)$, $a_3(P) = a_3(Q)$. Evidently $\mathbf{A}(P)$ and $\mathbf{A}(Q)$ are generated by the transformation (9.26) from equal vectors in the comparison crystal. Let $\mathbf{A}$ have rectangular coordinates A_i; then

$$\mathbf{A} = A_i\mathbf{i}_i = a_i\mathbf{e}_i = a_iD_{ij}\mathbf{i}_j$$

and $A_k = a_iD_{ik}$ or, multiplying by E_{kj} and using (9.27), $a_i = A_kE_{kj}$. The condition $a_i(P) = a_i(Q)$ is thus

$$A_k(P)E_{kj}(P) = A_k(Q)E_{kj}(Q).$$

If P, Q are neighboring points x_i, $x_i + dx_i$ and we put

$$A_i(Q) - A_i(P) = dA_i$$

this gives

$$dA_kE_{kj} = -A_kE_{kj,l}dx_l$$

or by (9.27)

$$dA_m = -L_{kl}{}^mA_kdx_l \tag{9.32}$$

where

$$L_{kl}{}^m = D_{jm}E_{kj,l}.$$

In the language of differential geometry, a relation like (9.32) prescribing which vectors at neighboring points of a coordinate network (manifold) are to be considered equivalent is known as a *linear connection* with coefficients $L_{kl}{}^m$. In a Riemannian geometry, $L_{kl}{}^m$ is symmetric in k and l. (In the geometry of Section 4, $L_{kl}{}^m = e_{lm,k} + e_{mk,l} - e_{kl,m}$ to the first order.) The geometry associated with the dislocated lattice is more complex since $L_{kl}{}^m$ is not symmetric. In fact, its antisymmetric part (torsion tensor) $\tfrac{1}{2}(L_{kl}{}^m - L_{lk}{}^m)$ is just the local Burgers vector density in the form (9.28).

10. Surface and Volume Defects

We saw in Section 6 that when the elastic constants of a body are changed from one function of position to another under constant external loading, half the work done by the external forces goes to increase the internal elastic energy. Since the calculation rests on a comparison of the equilibrium states before and after the change, it cannot tell us what happens to the missing half of the energy. It may be dissipated or it may reappear, for example, as kinetic or surface energy. Apart from its application to point defects (Section 8e), this result has a bearing on the behavior of grosser inhomogeneities of the lattice, in particular of cracks and of boundaries across which the orientation of the crystal changes.

A crack may be regarded as a narrow zone where the elastic constants are zero; the extension of a crack qualifies as a change in the distribution of elastic constants of the type we are considering. In the Griffith criterion for the spread of a crack, the energy made available in a small extension of the crack must be equal to the resulting increase of surface energy. According to (6.3), this means that at constant load the increase of surface energy must be equal to the *increase* of elastic energy. On the other hand, suppose that the body is strained by giving parts of its surface fixed displacements and leaving the rest free of traction. Then δE_{ext} is zero in any change and the criterion for spread of the crack is that the increase of surface energy shall be equal to the *decrease* of elastic energy.[81,82]

There is a force on a grain boundary or twin boundary in virtue of the fact that it represents an array of dislocations.[10] There is also a less obvious contribution arising from the fact that it is effectively a junction between regions with different elastic constants, even though the material is homogeneous. As we cross the boundary, the orientation of the crystal axes changes and so does the array of elastic coefficients c_{ijkl}. In this wider sense, the material is elastically inhomogeneous.

The force which a stress-field exerts on an element of the boundary, regarded as an array of dislocations, can be found by applying (9.2) to the dislocations it contains. In simple cases it can be found directly. Figure 12a illustrates schematically the experiment of Parker and Washburn[83] on the movement of small-angle grain boundaries. A beam is loaded at one end and contains a tilt boundary AB of angle ω. If the boundary moves a distance dx to the left, the load descends a distance ωdx and loses potential energy $W\omega dx$. Thus the force on the boundary has magnitude $W\omega$ and is directed to the left. If the load were upward, the force would be directed to the right.

Contrast Fig. 12a with the rather artificial situation shown in Fig. 12b. Here AB marks not a grain boundary but the junction between regions

[81] A. A. Griffith, *Phil. Trans. Roy. Soc.* **A221,** 163 (1920).

[82] E. Orowan, *Welding J.* **34,** 157s, 1955.

[83] E. R. Parker and J. Washburn, *Trans Am. Inst. Mining Met. Engrs.* **194,** 1076 (1952).

with different elastic constants. Suppose that the Young's modulus E' on the right is less than the Young's modulus E on the left. Per unit length, the beam is more flexible on the right than on the left. If the junction is moved to the left the beam as a whole becomes more flexible and the load descends, losing potential energy. If the load were directed upward instead of downward, moving the boundary to the left would make the tip of the beam move further upward, so that the potential energy of the load would again decrease. Thus there is a force on the

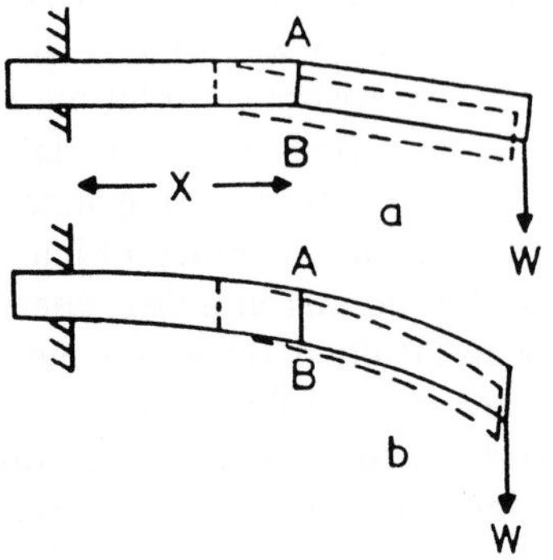

FIG. 12. To illustrate Section 10.

boundary in Fig. 12b which (unlike the force on the boundary in Fig. 12a) does not reverse when the load is reversed. According to (6.3), its magnitude is

$$F = -\frac{d}{dx}(E_{\mathrm{el}} + E_{\mathrm{ext}}) = -\frac{1}{2}\frac{d}{dx}E_{\mathrm{ext}}. \tag{10.1}$$

Elementary beam theory gives the deflection of the load as a function of x and leads to

$$F = \frac{M^2}{2I}\left(\frac{1}{E} - \frac{1}{E'}\right)$$

where M is the bending moment at x, and I is the moment of inertia of the cross section. (Strictly the boundary conditions necessary for (6.3) to hold are not exactly satisfied at the clamped end, but this makes no difference within the limits of elementary beam theory.)

Again, suppose that instead of the load W there is an external couple M twisting the beam about its axis. The twist per unit length in the two sections is $M/D\mu$ or $M/D\mu'$, where D is the torsional rigidity when $\mu = 1$. The total twist at the load end is thus

$$\theta = \frac{M}{D}\left(\frac{x}{\mu} + \frac{L - x}{\mu'}\right).$$

Here E_{ext} is $-M\theta +$ const. Thus (10.1) gives for the effective force on the junction

$$F = \frac{M}{2D}\left(\frac{1}{\mu} - \frac{1}{\mu'}\right). \tag{10.2}$$

These results may be applied to the more realistic case where the array of elastic coefficients c_{ijkl} differs on opposite sides of a grain boundary by taking E, E', μ, μ' to be effective moduli for the bending or twisting of anisotropic beams cut with slightly different orientations from the same crystal.

Generally, then, an element of a grain or twin boundary will experience two forces, namely F^D because it is an array of dislocations (or equivalently, because its movement alters the form of the crystal) and F^R because it is the junction of two regions in which the crystal axes are rotated with respect to one another. Apart from questions of relative magnitude, there is the qualitative distinction that when the applied stresses are reversed F^D reverses but F^R does not.

Usually we may expect F^R to be swamped by F^D. However, in Thomas and Wooster's[84] phenomenon of *piezocrescence* F^R comes into its own. If a quartz crystal transforms into its Dauphiné (or electrical) twin, its external form is unaltered, but its internal crystal structure is rotated 180° about the pseudo-hexad axis. Thus, if a region inside a crystal changes to its Dauphiné twin, it will constitute a region of elastic inhomogeneity unaccompanied by internal stress.

Thomas and Wooster carried out what in effect is the analog for F^R of the Washburn-Parker experiment for F^D. They found that a Dauphiné boundary in a quartz rod twisted by an applied couple moved (at high enough temperatures) in a direction determined by the sign of $(\mu^{-1} - \mu'^{-1})$ in (10.2) but independent of the sign of the applied couple. $(\mu^{-1} - \mu'^{-1})$ may be positive or negative according to the orientation of the crystal axes in the rod.

They have also given what in effect is a calculation of the force per unit area on an inhomogeneity boundary in an arbitrary stress field. Following them, we neglect both the possible discontinuity in the stress components across the boundary and the fact that moving the boundary will itself upset the applied stress distribution. Then (6.5) may be rewritten as

$$\delta E_{\text{el}} = \tfrac{1}{2}\int (s_{ijkm}' - s_{ijkm}) p_{ij} p_{km} dv$$

omitting the prime on p_{km}. If a surface element of the boundary is displaced a distance $\delta\xi$ in the x_l direction, it sweeps out a volume $dv = dS\delta\xi n_l$

[84] L. A. Thomas and W. A. Wooster, *Proc. Roy. Soc.* **A208** (1951).

where n_l is the normal to dS. Then δE_{el} is $dv \Delta s_{ijkm} p_{ij} p_{km}$, where Δs_{ijkm} is the value of s_{ijkm} on the side of the positive normal minus its value on the negative side. The force on the boundary per unit area is thus a normal pressure

$$F_l^R = \frac{\delta E_{\mathrm{el}}}{dS\,\delta\xi} = \Delta s_{ijkm} p_{ij} p_{km} n_l. \tag{10.3}$$

The coefficient of n_l in (10.3) has the form of an energy density derived from fictitious elastic constants Δs_{ijkm}. Unlike a true energy density, it may be positive at some points and negative at others. In equilibrium the twin boundary must coincide with the surface separating positive from negative regions, for there $F_l^R = 0$.

Thomas and Wooster verified this experimentally for a quartz plate in a complex state of stress. In their theory they correctly maximized the elastic energy, invoking Le Chatelier's principle. Stepanov[85] in similar work minimized the energy. This gives the shape of the boundary correctly but interchanges the twinned and untwinned regions.

It is interesting to compare the magnitudes of F^D and F^R for a grain or twin boundary in a metal. Let ω be the misorientation, p a typical stress component, and s a typical component of s_{ijkl}. The fractional change of s on crossing the boundary is of order ω. Then $F^D \sim \omega p$ and from (10.3) $F^R \sim \omega p^2 s \sim \omega p^2/\mu$ where μ is some elastic modulus. Hence F^R/F^D is of the order of the usually small quantity p/μ.

[85] A. V. Stepanov, *Zhur. Eksptl. i Teort. Fiz.* **20**, 438 (1950).

The determination of the elastic field of an ellipsoidal inclusion, and related problems

By J. D. Eshelby

Department of Physical Metallurgy, University of Birmingham

(*Communicated by R. E. Peierls, F.R.S.—Received* 1 *March* 1957)

It is supposed that a region within an isotropic elastic solid undergoes a spontaneous change of form which, if the surrounding material were absent, would be some prescribed homogeneous deformation. Because of the presence of the surrounding material stresses will be present both inside and outside the region. The resulting elastic field may be found very simply with the help of a sequence of imaginary cutting, straining and welding operations. In particular, if the region is an ellipsoid the strain inside it is uniform and may be expressed in terms of tabulated elliptic integrals. In this case a further problem may be solved. An ellipsoidal region in an infinite medium has elastic constants different from those of the rest of the material; how does the presence of this inhomogeneity disturb an applied stress-field uniform at large distances? It is shown that to answer several questions of physical or engineering interest it is necessary to know only the relatively simple elastic field inside the ellipsoid.

1. Introduction

In the physics of solids a number of problems present themselves in which the uniformity of an elastic medium is disturbed by a region within it which has changed its form or which has elastic constants differing from those of the remainder. Some of these problems may be solved for a region of arbitrary shape. Others are intractable unless the region is some form of ellipsoid. Fortunately, the general ellipsoid is versatile enough to cover a wide variety of particular cases. It is the object of this paper to develop a simple method of solving these problems.

When a twin forms inside a crystal the material is left in a state of internal stress, since the natural change of shape of the twinned region is restrained by its surroundings. A similar state of strain arises if a region within the crystal alters its unconstrained form because of thermal expansion, martensitic transformation, precipitation of a new phase with a different unit cell, or for some other reason. These examples suggest the following general problem in the theory of elasticity.

The transformation problem

A region (the 'inclusion') in an infinite homogeneous isotropic elastic medium undergoes a change of shape and size which, but for the constraint imposed by its surroundings (the 'matrix'), would be an arbitrary homogeneous strain. What is the elastic state of inclusion and matrix?

We shall solve this problem with the help of a simple set of imaginary cutting, straining and welding operations. Cut round the region which is to transform and remove it from the matrix. Allow the unconstrained transformation to take place. Apply surface tractions chosen so as to restore the region to its original form, put it back in the hole in the matrix and rejoin the material across the cut. The stress is now zero in the matrix and has a known constant value in the inclusion. The applied surface tractions have become built in as a layer of body force spread over the interface between matrix and inclusion. To complete the solution this unwanted layer is removed by applying an equal and opposite layer of body force; the additional elastic field thus introduced is found by integration from the expression for the elastic field of a point force.

So far nothing has been assumed about the shape of the inclusion. However, we shall find that if it is an ellipsoid the stress within the inclusion is uniform. This fact enables us to use the solution of the transformation problem as a convenient stepping-stone in solving a second set of elastic problems. Superimpose on the whole solid a uniform stress which just annuls the stress in the inclusion. The removal of the unstressed inclusion to leave a hole with a stress-free surface is then a mere formality, and we have solved the problem of the perturbation of a uniform stress field by an ellipsoidal cavity. More generally, suppose that the uniform applied stress does not annul the stress in the inclusion. Then the stress and strain in the inclusion are not related by the Hooke law of the material, since part of the strain arises from a non-elastic twinning or other transformation with which no stress is associated. The stress and strain are, however, related by the Hooke law of some hypothetical material, and the transformed ellipsoid may be replaced by an ellipsoid of the hypothetical material which has suffered the same total strain, but purely elastically. We have thus solved the following problem.

The inhomogeneity problem

An ellipsoidal region in a solid has elastic constants differing from those of the remainder (if, in particular, the constants are zero within the ellipsoid we have the case of a cavity). How is an applied stress, uniform at large distances, disturbed by this inhomogeneity?

The strain in the inclusion or inhomogeneity may be found explicitly in terms of tabulated elliptic integrals. The elastic field at large distances is also easy to determine. The field at intermediate points is more complex, but for many purposes we do not need to know it. In fact, knowing only the uniform strain inside the ellipsoid we can find the following items of physical or engineering interest:

(i) The elastic field far from an inclusion.

(ii) All the stress and strain components at a point immediately outside the inclusion.

(iii) The total strain energy in matrix and inclusion.

(iv) The interaction energy of the elastic field of the inclusion with another elastic field.

(v) The elastic field far from an inhomogeneity.

(vi) All the stress and strain components at a point immediately outside the inhomogeneity. (This solves the problem of stress concentration.)

(vii) The interaction energy of the inhomogeneity with an elastic field.

(viii) The change in the gross elastic constants of a material when a dilute dispersion of ellipsoidal inhomogeneities is introduced into it.

Problems (i) to (iv) can also be solved for an inclusion of arbitrary shape, (i) and (iv) trivially, (ii) and (iii) if we can evaluate the necessary integrals. They differ, of course, from the problems considered by Nabarro (1940) and Kröner (1954) in which the inclusion breaks away from the matrix. Problems (v) to (viii) can only be solved for the ellipsoid. They each have an analogue in the theory of slow viscous flow.

Many particular cases of these problems have been discussed. Robinson (1951) gives references to earlier work; see also Shapiro (1947), Sternberg, Eubanks & Sadowsky (1951). Apart from some increase in generality (we consider shear transformations and the disturbance of an arbitrary shear stress by an ellipsoidal inhomogeneity) our treatment is, perhaps, rather simpler and more direct than the orthodox method. Nowhere do we have to introduce ellipsoidal co-ordinates, search for suitable stress functions or match stress and displacement at an interface. Indeed, we do not even use the equations of elastic equilibrium explicitly except in certain of the applications (i) to (viii).

2. The general inclusion

We employ the usual suffix notation. A repeated suffix is summed over the values 1, 2, 3 and suffixes preceded by a comma denote differentiation:

$$u_{i,j} = \partial u_i/\partial x_j, \quad \phi_{,ik} = \partial^2\phi/\partial x_i \partial x_k.$$

The elastic displacement u_i, strain e_{ij} and stress p_{ij} are related by

$$e_{ij} = \tfrac{1}{2}(u_{i,j}+u_{j,i}), \tag{2·1}$$

$$p_{ij} = \lambda e_{mm}\delta_{ij}+2\mu e_{ij} \tag{2·2}$$

in an isotropic medium with Lamé constants λ, μ. When a particular set of elastic functions are distinguished by an affix (e.g. u_i^C, e_{ij}^C, p_{ij}^C), it is to be understood that they are related by (2·1) and (2·2). It is often convenient to split a second-order tensor f_{ij} into its scalar and so-called deviatoric parts:

$$f_{ij} = {}'f_{ij}+\tfrac{1}{3}f\delta_{ij},$$

where $$f = f_{mm} \quad \text{and} \quad {}'f_{ij} = f_{ij}-\tfrac{1}{3}f\delta_{ij}.$$

Thus, for example, (2·2) may be written

$$p = 3\kappa e, \quad {}'p_{ij} = 2\mu\, {}'e_{ij} \quad (\kappa = \lambda+\tfrac{2}{3}\mu), \tag{2·3}$$

and the inversion

$$e = p/3\kappa, \quad {}'e_{ij} = {}'p_{ij}/2\mu$$

is immediate, whereas to find e_{ij} in terms of p_{ij} from (2·2) is more difficult. For two tensors f_{ij}, g_{ij} we also have the convenient relation $f_{ij}g_{ij} = \tfrac{1}{3}fg+{}'f_{ij}\,{}'g_{ij}$; there are no cross-terms between the scalar and deviatoric parts. The elastic energy density is thus

$$\tfrac{1}{2}p_{ij}e_{ij} = \tfrac{1}{2}(\kappa e^2+2\mu\,{}'e_{ij}\,{}'e_{ij}) = \frac{1}{2}\left(\frac{1}{9\kappa}p^2+\frac{1}{2\mu}{}'p_{ij}\,{}'p_{ij}\right). \tag{2·4}$$

Following Robinson (1951) we shall give the name 'stress-free strain' to the uniform transformation strain e_{ij}^T which the inclusion would undergo in the absence of the matrix. The main problem is to find the 'constrained strain' e_{ij}^C in the inclusion when it transforms while it is embedded in the matrix and also the strain set up in the matrix, which we shall also call e_{ij}^C. Let S be the surface separating matrix and inclusion, n_i its outward normal and $\mathrm{d}S_i = n_i\,\mathrm{d}S$ the product of the normal and an element of S. We now carry out the steps outlined in the Introduction.

I. Remove the inclusion and allow it to undergo the stress-free strain e^T_{ij} without altering its elastic constants. Let

$$p^T_{ij} = \lambda e^T \delta_{ij} + 2\mu e^T_{ij}$$

be the stress derived from e^T_{ij} by Hooke's law. At this stage the stress in the inclusion and matrix is zero.

II. Apply surface tractions $-p^T_{ij} n_j$ to the inclusion. This brings it back to the shape and size it had before transformation. Put it back in the matrix and reweld across S. The surface forces have now become a layer of body force spread over S.

III. Let these body forces relax, or, what comes to the same thing, apply a further distribution $+p^T_{ij} n_j$ over S. The body is now free of external force but in a state of self-stress because of the transformation of the inclusion.

Since the displacement at $\mathbf{r}$ due to a point-force F_i at $\mathbf{r}'$ is (Love 1927)

$$U_j(\mathbf{r}-\mathbf{r}') = \frac{1}{4\pi\mu}\frac{F_j}{|\mathbf{r}-\mathbf{r}'|} - \frac{1}{16\pi\mu(1-\sigma)} F_l \frac{\partial^2}{\partial x_l \partial x_j} |\mathbf{r}-\mathbf{r}'|, \tag{2·5}$$

the displacement impressed on the material in stage III is

$$u^C_i(\mathbf{r}) = \int_S \mathrm{d}S_k p^T_{jk} U_j(\mathbf{r}-\mathbf{r}'), \tag{2·6}$$

where σ is Poisson's ratio. It will be convenient to take the state of the material at the conclusion of stage II as a state of zero displacement. This is reasonable, since the stress and strain in the matrix are then zero and the inclusion, though not stress-free, has just the geometrical form which it had before the transformation occurred. u^C_i is then the actual displacement in the matrix and inclusion. The strain in matrix or inclusion is

$$e^C_{ij} = \tfrac{1}{2}(u^C_{i,j} + u^C_{j,i}).$$

The stress in the matrix is derived from e^C_{ij} by Hooke's law:

$$p^C_{ij} = \lambda e^C \delta_{ij} + 2\mu e^C_{ij}.$$

On the other hand, the inclusion had a stress $-p^T_{ij}$ even before stage III, so that the stress in it is

$$p^I_{ij} = p^C_{ij} - p^T_{ij} = \lambda(e^C - e^T)\,\delta_{ij} + 2\mu(e^C_{ij} - e^T_{ij}), \tag{2·7}$$

where, according to our general convention, p^C_{ij} is the stress derived by Hooke's law from the strain e^C_{ij} in the inclusion.

By using Gauss's theorem and the equivalence of $\partial/\partial x_i$ and $-\partial/\partial x'_i$ when acting on $|\mathbf{r}-\mathbf{r}'|$, (2·7) may be made to read

$$u^C_i = \frac{1}{16\pi\mu(1-\sigma)} p^T_{jk} \psi_{,ijk} - \frac{1}{4\pi\mu} p^T_{ik} \phi_{,k}, \tag{2·8}$$

where

$$\phi = \int_V \frac{\mathrm{d}v}{|\mathbf{r}-\mathbf{r}'|} \quad \text{and} \quad \psi = \int_V |\mathbf{r}-\mathbf{r}'|\,\mathrm{d}v$$

are the ordinary Newtonian potential and the biharmonic potential of attracting matter of unit density filling the volume V bounded by S. Evidently

$$\nabla^2 \psi = 2\phi \tag{2·9}$$

and

$$\nabla^4 \psi = 2\nabla^2 \phi = \begin{cases} -8\pi \text{ inside } S, \\ 0 \text{ outside } S. \end{cases}$$

Generally we must know both ψ and ϕ. However, if we are interested only in the dilatation in the material, it is enough to know ϕ:

$$e^C = -\frac{1-2\sigma}{8\pi\mu(1-\sigma)} p^T_{ik}\phi_{,ik}. \tag{2·10}$$

Again, if e^T_{ij} is a pure dilatation $\frac{1}{3}e^T\delta_{ij}$, then

$$e^C_{il} = -\frac{1}{4\pi}\frac{1+\sigma}{3(1-\sigma)}e^T\phi_{,il},$$

a result due to Crum (Nabarro 1940). In this case the dilatation is $e^T(1+\sigma)/3(1-\sigma)$ in the inclusion and zero in the matrix. Thus, for example, with $\sigma = \frac{1}{3}$, the constraint of the matrix reduces the free expansion of the inclusion by a factor $\frac{2}{3}$.

The second derivatives of a potential function satisfying $\nabla^2 U = -4\pi\rho$ undergo a jump $\Delta U_{,ij} = -4\pi\Delta\rho n_i n_j$ on crossing a surface (with normal n_i) across which the density jumps by $\Delta\rho$ (Poincaré 1899). (This is perhaps more familiar in the form: the jump in attraction across a double layer is equal to its moment. In our problem $-\phi_{,i}$ is the potential of a double layer over S with unit moment directed along the x_i axis and $\phi_{,ij}$ is the corresponding force.) This gives for ϕ the expression

$$\phi_{,ij}(\text{out}) - \phi_{,ij}(\text{in}) = 4\pi n_i n_j \tag{2·11}$$

for the difference at adjacent points just inside and outside S. Applying the same argument in turn to $\psi_{,ij}$, which is the potential derived from the density $-2\phi_{,ij}/4\pi$, we have

$$\psi_{,ijkl}(\text{out}) - \psi_{,ijkl}(\text{in}) = 8\pi n_i n_j n_k n_l. \tag{2·12}$$

From (2·11), (2·12) and (2·8) we can find the stresses and strains just outside the inclusion from their values at an adjacent point just inside without having to solve the exterior problem at all. We easily find that

$$e^C(\text{out}) = e^C(\text{in}) - \frac{1}{3}\frac{1+\sigma}{1-\sigma}e^T - \frac{1-2\sigma}{1-\sigma}{'e^T_{ij}}n_i n_j$$

and

$$'e^C_{il}(\text{out}) = {'e^C_{il}}(\text{in}) + \frac{1}{1-\sigma}{'e^T_{jk}}n_j n_k n_i n_l - {'e^T_{ik}}n_k n_l - {'e^T_{lk}}n_k n_i$$

$$+\frac{1-2\sigma}{3(1-\sigma)}{'e^T_{jk}}n_j n_k \delta_{il} - \frac{1}{3}\frac{1+\sigma}{1-\sigma}e^T(n_i n_l - \tfrac{1}{3}\delta_{il}). \tag{2·13}$$

The C quantities are related by (2·2) and so are the T quantities. Thus either or both sides of these equations may be expressed in terms of stress without trouble. This solves problem (ii).

We can find a convenient alternative form for u^C_i by noting that (2·5) may be written as

$$16\pi\mu(1-\sigma)U_i = \frac{F_j}{|\mathbf{r}-\mathbf{r}'|}\left[(3-4\sigma)\delta_{ij} + \frac{(x_i-x'_i)(x_j-x'_j)}{|\mathbf{r}-\mathbf{r}'|^2}\right]. \tag{2·14}$$

Inserting this in (2·6) and using Gauss's theorem to convert to a volume integral we find

$$u^C_i(\mathbf{x}) = \frac{p_{jk}}{16\pi\mu(1-\sigma)}\int_V \frac{dv}{r^2}f_{ijk}(\mathbf{l}) = \frac{e^T_{jk}}{8\pi(1-\sigma)}\int_V \frac{dv}{r^2}g_{ijk}(\mathbf{l}), \tag{2·15}$$

where r and $\mathbf{l} = (l_1, l_2, l_3)$ are the length and direction of the line drawn from the volume element dv to the point of observation $\mathbf{x}$ and

$$f_{ijk} = (1-2\sigma)(\delta_{ij}l_k + \delta_{ik}l_j) - \delta_{jk}l_i + 3l_il_jl_k, \tag{2·16}$$

$$g_{ijk} = (1-2\sigma)(\delta_{ij}l_k + \delta_{ik}l_j - \delta_{jk}l_i) + 3l_il_jl_k. \tag{2·17}$$

For points $\mathbf{x}$ remote from the inclusion we may take everything except dv outside the sign of integration to obtain

$$u_i^C(\mathbf{x}) = Vp_{jk}^T f_{ijk}/16\pi\mu(1-\sigma)r^2 = Ve_{jk}^T g_{ijk}/8\pi(1-\sigma)r^2, \tag{2·18}$$

where r and $\mathbf{l}$ are now the distance and direction of $\mathbf{x}$ from the inclusion. This solves problem (i).

The strain energy density in the inclusion is $\frac{1}{2}p_{ij}^I e_{ij}^I$, where e_{ij}^I is the strain derived from p_{ij}^I by Hooke's law. By (2·7) the elastic energy in the inclusion is thus

$$\frac{1}{2}\int_V p_{ij}^I(e_{ij}^C - e_{ij}^T)\,dv. \tag{2·19}$$

The elastic energy in the matrix is

$$-\frac{1}{2}\int_S p_{ij}^C u_i^C\,dS_j = -\frac{1}{2}\int_S p_{ij}^I u_i^C\,dS_j = -\frac{1}{2}\int_V p_{ij}^I e_{ij}^C\,dv. \tag{2·20}$$

The first member exhibits it as the work done in setting up the elastic field by applying suitable forces to the surface S; the sign is correct if the normal points from inclusion to matrix. The second follows because displacement and normal traction are continuous across S. The third follows from Gauss's theorem, the equilibrium equation $p_{ij,j}^I = 0$ and the symmetry condition $p_{ij}^I = p_{ji}^I$. The total strain energy in matrix and inclusion is thus

$$E_{\text{el.}} = -\frac{1}{2}\int_V p_{ij}^I e_{ij}^T\,dv. \tag{2·21}$$

In the special case where e_{ij}^T is a uniform expansion we have at once from (2·10) and (2·21) $E_{\text{el.}} = 2\mu(e^T)^2V(1+\sigma)/9(1-\sigma)$, whatever be the shape of the cavity, as pointed out by Crum (Nabarro 1940).

The interaction energy of the elastic field u_i^C with another field u_i^A is (Eshelby 1951, 1956)

$$E_{\text{int.}} = \int_\Sigma (p_{ij}^C u_i^A - p_{ij}^A u_i^C)\,dS_j \tag{2·22}$$

taken over any surface Σ enclosing the inclusion. Let us take Σ to be a surface just outside S. Once again, since u_i^C and the normal stress are continuous across S, (2·22) can be converted into an integral over a surface just inside S, and hence into a volume integral over the inclusion:

$$E_{\text{int.}} = \int_V (p_{ij}^I e_{ij}^A - p_{ij}^A e_{ij}^C)\,dv.$$

The second term in the integrand is equal to $-p_{ij}^C e_{ij}^A$, so that

$$E_{\text{int.}} = -\int_V p_{ij}^T e_{ij}^A\,dv = -\int_V p_{ij}^A e_{ij}^T\,dv = -\int_\Sigma p_{ij}^A u_i^T\,dS_j. \tag{2·23}$$

25-2

This solves problem (iv). The same result is reached by evaluating (2·22) over a large sphere using the remote field (2·18). It is fortunate that we need only e_{ij}^T and not e_{ij}^C. In fact the last member of (2·23) has formally the appearance of being the work done against the external field in 'blowing up' the inclusion (regarded as rigid) to a final shape specified by e_{ij}^T. It is perhaps not obvious that this should be so, since the inclusion is not rigid and its final shape is described by a displacement u_i^C which may be quite complicated (e.g. it produces a barrel or pincushion distortion of the cubical inclusion which Cochardt, Schoek & Wiedersich (1955) consider). If we regard the inclusion as capable of moving through the matrix, as in the elastic model of a substitutional atom,

$$F_l = -\partial E_{\text{int.}}/\partial \xi_l \tag{2·24}$$

is the 'force' on the inclusion, where ξ_l is a vector specifying its position.

Let $E_{\text{trans.}}$ be the change of internal energy when the inclusion transforms in the absence of the matrix. Consider the sum

$$E = E_{\text{trans.}} + E_{\text{el.}} + E_{\text{int.}}.$$

Give their adiabatic values to λ, μ, κ and suppose that the constrained transformation occurs without any heat flow. Then E can be interpreted indifferently as the enthalpy change of the inclusion, the enthalpy change of the body (inclusion plus matrix) or the change of internal energy of the body and loading mechanism regarded as a single thermodynamic system. There is a similar interpretation for an isothermal process if we read 'Helmholtz free energy' for 'internal energy' and 'Gibbs free energy' for 'enthalpy' and give λ, μ, κ their isothermal values.

Since problems (v) to (viii) can only be solved for an ellipsoid their discussion is deferred to § 4.

As a simple example of the use of (2·18), suppose that we need the field at large distances from a dislocation loop of area A in the x_1x_2 plane with its Burgers vector along the positive x_3 axis. We have to insert a sheet of material of area A and thickness b. One way to do this is to cut out a disk of area A and height h, give it a permanent strain $e_{33}^T = b/h$ to increase its height by b and then force it back into the cavity. In (2·18) we have to put $V = Ah$, $e_{33}^T = b/h$ and the other e_{ij}^T equal to zero. Thus

$$u_i = bA\, g_{i33}/8\pi(1-\sigma)\, r^2.$$

Suppose next that the Burgers vector lies in the plane of the loop and, say, along the x_1 axis. We now give the disk a permanent shear $e_{13}^T = \frac{1}{2}b/h$, which gives its upper and lower surfaces a relative offset b, and re-insert it in the matrix. In the limit $h \to 0$ we have a displacement discontinuity b across the loop. Putting $V = Ah$, $e_{13}^T = e_{31}^T = \frac{1}{2}b/h$ and the other e_{ij}^T zero in (2·18) we get

$$u_i = bA g_{i13}/4\pi(1-\sigma)\, r^2, \tag{2·25}$$

reproducing a result of Nabarro's (1951). It is perhaps not quite clear that the restraint of the matrix will not reduce the offset to something less than b. Actually this is not true in the limit $h \to 0$. But we can see that (2·25) is correct by inserting $Ve_{13}^T = \frac{1}{2}Ab$ in (2·23); this gives $E_{\text{int.}} = -bAp_{13}^A$, which is the correct interaction energy for such a loop (Nabarro 1952). Indeed by the same argument we can find the remote field of an arbitrary loop of area A, normal n_i and Burgers vector b_i. The interaction energy is $-b_i p^A_{ij} n_j A$ for any p^A_{ij}. Equation (2·23) then shows that $Ve^T_{ij} = \frac{1}{2}(b_i n_j + b_j n_i)$ and (2·18) gives

$$u_i = Ab_i n_k g_{ijk}/8\pi(1-\sigma)\,r^2.$$

There is, in fact, a more general connexion with dislocation theory. The stress-free strain in the inclusion may always be imagined to be (or may actually be) the result of plastic deformation. A set of dislocation loops (with equal Burgers vectors) expanding from zero size on a close set of equally-spaced planes will give a shear if their Burgers vectors lie in the planes, or an extension perpendicular to the planes if their Burgers vectors are at right angles to the planes. (In the latter case their movement is non-conservative.) If the deformation occurs in the absence of the matrix these loops will, so to speak, disappear into free space. But if the inclusion is embedded in the matrix the dislocations will lodge in the surface S separating matrix from inclusion. S then becomes the discontinuity surface of a Somigliana (1914, 1915) dislocation. In this generalized type of dislocation there is a variable discontinuity d_i of displacement across S. In our model, d_i makes itself felt through the gaps and interpenetrations of matter which we should find if we tried to re-insert the transformed inclusion into the hole in the matrix without pulling it back to its original shape. It is easy to see that d_i has the value $\frac{1}{2}e^T_{ij}x'_j$ at a point x'_j of S and hence to verify, after some manipulation, that our expression for u^C_i agrees with Somigliana's.

3. The ellipsoidal inclusion

In discussing the elastic field inside an inclusion it is convenient to redefine the l_i in (2·16), (2·17) to be the direction cosines of a line drawn *from* the point of observation $\mathbf{x} = (x_1, x_2, x_3) = (x, y, z)$ *to* the volume element dv. This involves changing the sign of the integrals in (2·15). Let us first integrate over an elementary cone d$\omega(\mathbf{l})$ centred on the direction $\mathbf{l} = (l_1, l_2, l_3) = (l, m, n)$ with its vertex at $\mathbf{x}$. It gives a contribution $r(\mathbf{l})\,\mathrm{d}\omega$ to $\int \mathrm{d}v/r^2$. Thus

$$8\pi\mu(1-\sigma)\,u_i(\mathbf{x}) = -e^T_{jk}\int_{4\pi} r(\mathbf{l})\,d\omega(\mathbf{l})\,g_{ijk}(\mathbf{l}), \tag{3·1}$$

which gives the displacement at $\mathbf{x}$ in terms of an angular integration over the polar diagram $r = r(l, m, n)$ of the surface S as viewed from $\mathbf{x}$.

More briefly we could go directly from (2·6) to (3·1) by writing $r^{-1} = \frac{1}{2}\nabla^2 r$ in (2·14) applying Stokes's theorem in the form

$$\int_S w_{\dots j,l}\,\mathrm{d}S_j = \int_S w_{\dots i,i}\,\mathrm{d}S_l$$

and noting that $\mathrm{d}\omega = n_i r_i\,\mathrm{d}S/r^3$.

For the ellipsoid

$$X^2/a^2 + Y^2/b^2 + Z^2/c^2 = 1$$

$r(\mathbf{l})$ is the positive root of

$$(x+rl)^2/a^2 + (y+rm)^2/b^2 + (z+rn)^2/c^2 = 1,$$

that is, $$r(\mathbf{l}) = -f/g+(f^2/g^2+e/g)^{\frac{1}{2}}, \tag{3·2}$$

where $$g = l^2/a^2+m^2/b^2+n^2/c^2 \tag{3·3}$$

and $$f = lx/a^2+my/b^2+nz/c^2, \quad e = 1-x^2/a^2-y^2/b^2-z^2/c^2.$$

The sign of the square root is evidently correct, since e is positive if $\mathbf{x}$ is within the ellipsoid. In any case, we may omit this term when (3·2) is inserted in (3·1) since it is even in $\mathbf{l}$, whilst g_{ijk} is odd. To retain the advantages of suffix notation we introduce the 'vector' $$\lambda_1 = l/a^2, \quad \lambda_2 = m/b^2, \quad \lambda_3 = n/c^2.$$

Then $$u_i^C(\mathbf{x}) = \frac{x_m e_{jk}^T}{8\pi(1-\sigma)}\int_{4\pi}\frac{\lambda_m g_{ijk}}{g}\,d\omega \tag{3·4}$$

and the strains $$e_{il}^C(\mathbf{x}) = \frac{e_{jk}^T}{16\pi(1-\sigma)}\int_{4\pi}\frac{\lambda_i g_{ljk}+\lambda_l g_{ijk}}{g}\,d\omega \tag{3·5}$$

are uniform and depend only on the shape of the ellipsoid. The same is also true for an anisotropic medium. (This verifies a hypothesis of Frank's (private communication).) For it can easily be shown (see, for example, Eshelby 1951) that (2·5) has then to be replaced by $$U_j(\mathbf{r}) = F_i D_{ij}(\mathbf{l})/r,$$

where the functions of direction D_{ij} cannot generally be found in finite form. A repetition of the argument will evidently lead to an expression like (3·4), but with the g_{ijk} no longer given by (2·17).

It is convenient to write the relation between the constrained and stress-free strains in the inclusion in the form $$e_{il}^C = S_{ilmn}e_{mn}^T. \tag{3·6}$$

From the symmetry of the problem it is clear that the S_{ijkl} have some of the properties of the elastic coefficients of an orthorhombic crystal with its axes parallel to the axes of the ellipsoid, though relations of the form $S_{1122} = S_{2211}$ are not valid. Coefficients coupling an extension and a shear ($S_{1112}, S_{1123}, S_{2311}\ldots$) or one shear to another ($S_{1223}\ldots$) are zero. In fact, (3·5) vanishes if any one of l, m, n appears raised to an odd power in the integrand. The reduction of surface integrals of the type $\int l^{2i}m^{2j}n^{2k}\,d\omega/g$ to simple integrals has been given by Routh (1892). We find

$$\left.\begin{aligned} S_{1111} &= Qa^2I_{aa} + RI_a,\\ S_{1122} &= Qb^2I_{ab} - RI_a,\\ S_{1212} &= Q\tfrac{1}{2}(a^2+b^2)\,I_{ab}+R\tfrac{1}{2}(I_a+I_b),\end{aligned}\right\} \tag{3·7}$$

$$\left.\begin{aligned} &\text{where} \quad Q = \frac{3}{8\pi(1-\sigma)}, \quad R = \frac{1-2\sigma}{8\pi(1-\sigma)}, \quad \tfrac{1}{3}Q+R = \frac{1}{4\pi},\\ &\text{and} \quad I_a = \int\frac{l^2}{a^2}\frac{d\omega}{g} = 2\pi abc\int_0^\infty\frac{du}{(a^2+u)\,\Delta},\\ &I_{aa} = \int\frac{l^4}{a^4}\frac{d\omega}{g} = 2\pi abc\int_0^\infty\frac{du}{(a^2+u)^2\,\Delta},\\ &I_{ab} = \int\frac{l^2}{a^2}\frac{m^2}{b^2}\frac{d\omega}{g} = \tfrac{2}{3}\pi abc\int_0^\infty\frac{du}{(a^2+u)\,(b^2+u)\,\Delta},\end{aligned}\right\} \tag{3·8}$$

with $$\Delta = (a^2+u)^{\frac{1}{2}}\,(b^2+u)^{\frac{1}{2}}\,(c^2+u)^{\frac{1}{2}}.$$

The remaining coefficients are found by simultaneous cyclic interchange of $(1, 2, 3)$, (a, b, c), (l, m, n). I_a, I_b, I_c occur as coefficients in the expression

$$\phi = \tfrac{1}{2}(a^2-x^2)\,I_a + \tfrac{1}{2}(b^2-y^2)\,I_b + \tfrac{1}{2}(c^2-z^2)\,I_c$$

for the Newtonian potential within an ellipsoid of unit density. We have† (Kellogg 1929)

$$I_a = \frac{4\pi abc}{(a^2-b^2)\,(a^2-c^2)^{\frac{1}{2}}}(F-E),$$

$$I_c = \frac{4\pi abc}{(b^2-c^2)\,(a^2-c^2)^{\frac{1}{2}}}\left\{\frac{b(a^2-c^2)^{\frac{1}{2}}}{ac}-E\right\}, \tag{3·9}$$

where $F = F(\theta, k)$ and $E = E(\theta, k)$ are elliptic integrals of the first and second kinds of amplitude and modulus

$$\theta = \sin^{-1}(1-c^2/a^2)^{\frac{1}{2}}, \quad k = (a^2-b^2)^{\frac{1}{2}}/(a^2-c^2)^{\frac{1}{2}}$$

and it is assumed that $\quad a > b > c.$

The relations

$$I_a + I_b + I_c = 4\pi, \tag{3·10}$$

$$I_{aa} + I_{ab} + I_{ac} = 4\pi/3a^2, \tag{3·11}$$

$$a^2I_{aa} + b^2I_{ab} + c^2I_{ac} = I_a \tag{3·12}$$

follow from the ω integrals when we use the definition (3·3) of g and the relation $l^2+m^2+n^2 = 1$. Again, if we split the factor $(a^2+u)^{-1}(b^2+u)^{-1}$ in the u integral for I_{ab} into partial fractions we have $3(a^2-b^2)\,I_{ab} = I_b - I_a$. Thus when I_a, I_c have been calculated from (3·9) we have for I_b

$$I_b = 4\pi - I_a - I_c$$

and the remaining quantities are found from

$$I_{ab} = (I_b - I_a)/3(a^2-b^2), \tag{3·13}$$

$$I_{aa} = 4\pi/3a^2 - I_{ab} - I_{ac} \tag{3·14}$$

and their cyclic counterparts.

For the oblate spheroid $(a = b > c)$ with

$$I_a = I_b = \frac{2\pi a^2 c}{(a^2-c^2)^{\frac{3}{2}}}\left\{\cos^{-1}\frac{c}{a} - \frac{c}{a}\left(1-\frac{c^2}{a^2}\right)^{\frac{1}{2}}\right\} \tag{3·15}$$

the relation (3·13) fails, though not its analogues for I_{bc} or I_{ac}. But from the u integrals for I_{aa} and I_{ab} it is clear that $I_{aa} = 3I_{ab}$ and we may use (3·14). For the prolate spheroid $(b = c < a)$ we have

$$I_b = I_c = \frac{2\pi ac^2}{(a^2-c^2)^{\frac{3}{2}}}\left\{\frac{a}{c}\left(\frac{a^2}{c^2}-1\right)^{\frac{1}{2}} - \cosh^{-1}\frac{a}{c}\right\}, \tag{3·16}$$

and the remaining quantities may be determined similarly.

† Osborn (1945) has given curves for these quantities as functions of b/a and c/a. In his notation $I_a = L$, $I_b = M$, $I_c = N$.

For the elliptic cylinder $x^2/a^2+y^2/b^2 = 1$, $c\to\infty$ we have the simple results

$$I_a = 4\pi b/(a+b), \quad I_b = 4\pi a/(a+b), \quad I_c = 0,$$
$$I_{ab} = 4\pi/3(a+b)^2, \quad I_{aa} = 4\pi/3a^2 - I_{ab}, \quad I_{bb} = 4\pi/3b^2 - I_{ab}. \tag{3·17}$$

I_{ac}, I_{bc} and I_{cc} are zero. However, it is clear from (3·7) that what we really need is the limit of their products with c^2. In this sense (3·12) and (3·13) give

$$c^2 I_{ac} = \tfrac{1}{3}I_a, \quad c^2 I_{bc} = \tfrac{1}{3}I_b, \quad c^2 I_{cc} = 0. \tag{3·18}$$

The uniform rotation $\varpi^C_{il} = \frac{1}{2}(u^C_{i,l} - u^C_{l,i})$ in the inclusion may be written in a form

$$\varpi^C_{il} = \Pi_{iljk} e^T_{jk} \tag{3·19}$$

analogous to (3·6). The only non-zero components are Π_{1212}, Π_{2323}, Π_{3131} where, for example,

$$\Pi_{3131} = -\Pi_{1331} = (I_a - I_c)/8\pi. \tag{3·20}$$

These results are only valid in a co-ordinate system whose axes are parallel to the principal axes of the ellipsoid. For any other system there are still relations of the form (3·6), (3·19) and the new coefficients S_{ijkl}, Π_{ijkl} must be found by the usual law for transforming tensors.

Problems (i) to (iv) of § 1 are solved as for the inclusion of arbitrary shape (§ 2). The only simplification is that, since e^C_{ij} is uniform within the inclusion, (2·21) becomes

$$E_{\text{el.}} = -\tfrac{1}{2}Vp^I_{ij}e^T_{ij}, \quad V = \tfrac{4}{3}\pi abc, \tag{3·21}$$

and similarly (2·23) becomes

$$E_{\text{int.}} = -Vp^A_{ij}e^T_{ij} \tag{3·22}$$

if the applied field p^A_{ij} is also uniform. The field immediately outside the inclusion is found from (2·13) using the expressions

$$n_1 = x/a^2h, \quad n_2 = y/b^2h, \quad n_3 = z/c^2h, \quad h^2 = x^2/a^4+y^2/b^4+z^2/c^4 \tag{3·23}$$

for the components of the normal to an ellipsoid at the point x, y, z.

We have seen that usually it is sufficient to know only the elastic field within, just outside and far from the inclusion. The field at any point outside the inclusion can, of course, be found from (2·8) if we know the potentials ϕ and ψ. The expression for ϕ is well known (Kellogg 1929). Dirichlet (1839) calculated the exterior potential of an ellipsoid when the law of attraction is the inverse pth power of the distance. His derivation is only valid for $2 \leqslant p < 3$ and so does not cover the biharmonic case $p = 0$. However, his calculation of the force $-\operatorname{grad}\psi$ is valid for $p = 0$, and the derivatives are all we need to know. His result is

$$\frac{\partial\psi}{\partial x} = x\pi abc\int_\lambda^\infty \frac{U\,\mathrm{d}u}{(a^2+u)\,\Delta}, \quad \frac{\partial\psi}{\partial y} = \ldots,$$

where
$$U = 1 - \frac{x^2}{a^2+u} - \frac{y^2}{b^2+u} - \frac{z^2}{c^2+u}$$

and λ is the positive root of $U(u) = 0$. The integral can be reduced to elliptic integrals by the same substitutions as serve for ϕ. (For the details see, for example, Byrd & Friedman (1954).) For an external point ϕ and ψ are respectively first and second degree polynomials in x^2, y^2, z^2 whose coefficients are integrals of the type (3·8) with the lower limit changed from 0 to λ. These coefficients can be made to depend on the integrals (3·9) with argument

$$\theta = \sin^{-1}(a^2-c^2)^{\frac{1}{2}}/(a^2+\lambda)^{\frac{1}{2}}$$

and modulus

$$k = (a^2-b^2)^{\frac{1}{2}}/(a^2-c^2)^{\frac{1}{2}}.$$

They depend on x, y, z through λ.

4. The ellipsoidal inhomogeneity

The inhomogeneity problem for the ellipsoid is solved in the way outlined in § 1. On the elastic field e^C_{ij} due to an ellipsoidal inclusion with arbitrary e^T_{ij} we superimpose a uniform strain e^A_{ij}. The deformation of the surface of the inclusion is specified by the strain $e^C_{ij}+e^A_{ij}$. Because a part e^T_{ij} of this strain is not associated with any stress (compare equation (2·7)) the uniform stress in the inclusion is given by applying Hooke's law not to $e^C_{ij}+e^A_{ij}$, but rather to $e^C_{ij}+e^A_{ij}-e^T_{ij}$. In the notation of (2·3) the strain in the inclusion is

$$e = e^C+e^A, \quad 'e_{ij} = 'e^C_{ij}+'e^A_{ij}, \tag{4·1}$$

but the stress in it is

$$p = 3\kappa(e^C+e^A-e^T), \quad 'p_{ij} = 2\mu('e^C_{ij}+'e^A_{ij}-'e^T_{ij}). \tag{4·2}$$

Take an ellipsoid the same shape and size as the untransformed inclusion and made of an isotropic material with elastic constants λ_1, μ_1, $\kappa_1 = \lambda_1+\frac{2}{3}\mu_1$ different from those of the matrix and inclusion. Subject this ellipsoid to the strain (4·1). If this treatment develops the stress (4·2) in it, it may be used to replace the inclusion with continuity of displacement and surface traction across the interface. We can always ensure that the correct stress is developed by choosing λ_1, μ_1 suitably. It is only necessary that they should satisfy the relations

$$\kappa_1(e^C+e^A) = \kappa(e^C+e^A-e^T) \tag{4·3}$$

and

$$\mu_1('e^C_{ij}+'e^A_{ij}) = \mu('e^C_{ij}+'e^A_{ij}-'e^T_{ij}). \tag{4·4}$$

Actually, it is the uniform applied field e^A_{ij} and the elastic constants of the inhomogeneity which are prescribed, and (4·3), (4·4) are equations which have to be solved for e^T_{ij} in terms of e^A_{ij}, λ_1, μ_1 after eliminating e^C_{ij} with the help of the relation $e^C_{ij} = S_{ijkl}e^T_{kl}$. Equations (4·3) and (4·4) are not as simple as they appear, since each of e^C, $'e^C_{ij}$ depends on both e^T and $'e^T_{ij}$. However, for the shear components the solution is immediate, since the S_{ijkl} do not couple different shears:

$$e^T_{13} = \frac{\mu-\mu_1}{2(\mu_1-\mu)\,S_{1313}+\mu}\,e^A_{13}, \ldots. \tag{4·5}$$

On the other hand, for the components e^T_{11}, e^T_{22}, e^T_{33} we have to solve the three simultaneous equations

$$(\lambda_1-\lambda)\,S_{mmpq}\,e^T_{pq}+2(\mu_1-\mu)\,S_{ijpq}\,e^T_{pq}+\lambda e^T+2\mu e^T_{ij}$$
$$= (\lambda-\lambda_1)\,e^A+2(\mu-\mu_1)\,e^A_{ij}, \quad ij = 11, 22, 33.$$

(Only e^T_{11}, e^T_{22}, e^T_{33} appear in the pq summations since, e.g. $S_{1112} = 0$.)

From our derivation it is clear that e^T_{ij} found in this way is the stress-free strain of a certain transformed inclusion which, with the given applied field e^A_{ij}, could replace the inhomogeneity without altering the stress or displacement anywhere. We shall call this imaginary transformed inclusion the 'equivalent inclusion'. Outside the inclusion the elastic field u_i, e_{ij}, p_{ij} is the sum of the applied field u^A_i, e^A_{ij}, p^A_{ij} and the field u^C_i, e^C_{ij}, p^C_{ij} of the equivalent inclusion. It is convenient to regard the latter as the field 'of' or 'due to' the inhomogeneity. It measures the perturbation of the applied field by the inhomogeneity and may be found from e^T_{ij} by the methods of §§ 2, 3. In particular the remote field of the inhomogeneity follows at once from (2·18). Inside the inhomogeneity the total strain is

$$e = \frac{\kappa}{\kappa - \kappa_1} e^T, \quad 'e_{ij} = \frac{\mu}{\mu - \mu_1} 'e^T_{ij}, \tag{4·6}$$

by (4·3) and (4·4). The field immediately outside the inhomogeneity is found from (2·13) and (3·23).

The question of the interaction of a stress field and an elastic inhomogeneity has been discussed elsewhere (Eshelby 1951, 1956). We shall need the following result. If the initial elastic constants κ, μ of a body loaded by surface tractions are changed to arbitrary functions of position $\kappa^*(\mathbf{x})$, $\mu^*(\mathbf{x})$ the surface tractions being held constant, then the total energy of the system increases by

$$E_{\text{int.}} = -\frac{1}{2}\int \{(\kappa - \kappa^*)\, ee^* + 2(\mu - \mu^*)\, 'e_{ij}\, 'e^*_{ij}\}\, \mathrm{d}v, \tag{4·7}$$

where e_{ij}, e^*_{ij} are the strains before and after the change. By total energy we mean the sum of the elastic energy of the body and the potential energy of the loading mechanism. $E_{\text{int.}}$ is by definition the interaction energy of the applied stress and the inhomogeneity described by the non-uniform elastic constants κ^*, μ^*. The increase of elastic energy arising from the change is

$$\Delta E = -E_{\text{int.}}. \tag{4·8}$$

Equation (4·7) is valid also if e_{ij} is the strain arising from sources of internal stress or because the material is strained by rigid grips, but in this case (4·8) is replaced by

$$\Delta E = +E_{\text{int.}}. \tag{4·9}$$

In the present case the change of elastic constants is confined to the interior of the ellipsoid and e_{ij}, e^*_{ij} are uniform there. Thus

$$\begin{aligned} E_{\text{int.}} &= -\tfrac{1}{2}V\{(\kappa - \kappa_1)\, e^A(e^A + e^C) + 2(\mu - \mu_1)\, 'e^A_{ij}('e^A_{ij} + 'e^C_{ij})\} \\ &= -\tfrac{1}{2}V(\kappa e^A e^T + 2\mu\, 'e^A_{ij}\, 'e^T_{ij}) = -\tfrac{1}{2}V p^A_{ij} e^T_{ij} \end{aligned} \tag{4·10}$$

by (4·8) and (2·3) (V is the volume of the ellipsoid). This solves problem (vii). It will be noticed that (4·10) is just half the expression (3·22) for the equivalent inclusion. The physical interpretation is as follows. Equation (4·10) is still approximately correct if p^A_{ij} is a slowly varying function of position. The 'force' on the inhomogeneity regarded as an elastic singularity is again given by (2·24). F_l depends only on the remote field of the singularity and not at all on whether its field is permanent (internal stress) or merely 'induced' in an inhomogeneity by an elastic field. The factor $\frac{1}{2}$ ensures that this is so, for clearly (4·10) varies with position twice as fast when e_{ij}^T is a linear function of e_{ij}^A as it does when e_{ij}^T is constant.

The effective bulk elastic constants for a material containing a uniform dispersion of ellipsoidal inhomogeneities (not necessarily all of the same form or orientation) may be calculated as follows. Consider a specimen of unit volume. Let E_0 be its elastic energy when it is free of inhomogeneities and certain surface tractions produce a uniform stress p_{mn}^A in it. If we introduce the inhomogeneities, keeping the surface tractions constant, the elastic energy is augmented by $-\Sigma E_{\text{int.}}(p_{mn}^A)$, the sum of the interaction energies of all the inhomogeneities with the particular stress p_{mn}^A (compare equation (4·8)). Since E_0 and $E_{\text{int.}}$ are both quadratic functions of p_{mn}^A we may write

$$\tfrac{1}{2}s_{ijkl}p_{ij}^A p_{kl}^A = E_0 - \Sigma E_{\text{int.}}(p_{mn}^A). \tag{4·11}$$

The s_{ijkl} are then the elastic compliance constants which would be inferred from the work done in applying the given type of loading. To find the individual components of s_{ijkl} we put, say, $p_{11}^A = 1, p_{22}^A = 1, p_{ij}^A = 0$ otherwise, and obtain s_{1122}, and so forth. Unless the ellipsoids are spheres, or have their orientations distributed at random, the effective constants s_{ijkl} will be anisotropic. The effective elastic moduli c_{ijkl} may be found from the relation $c_{ijkl}s_{klmn} = \delta_{im}\delta_{jn}$. It would not do to find them directly by equating the right-hand side of (4·11) to $\frac{1}{2}c_{ijkl}e_{ij}^A e_{kl}^A$, since that equation was derived by considering a process at constant load. Rather, we must consider a unit volume given a uniform macroscopic strain by a rigid framework which keeps the surface displacements fixed when the inhomogeneities are introduced. This leads to

$$\tfrac{1}{2}c_{ijkl}e_{ij}^A e_{kl}^A = E_0 + \Sigma E_{\text{int.}}(e_{mn}^A). \tag{4·12}$$

(For the difference in sign compare (4·8) and (4·9).)†

5. Discussion

It has been shown how the problems listed in § 1 may be solved for the ellipsoid. Explicit solutions for the general case would be very clumsy and in this section only a few special cases are considered.

For a sphere of radius a, we have at once from (3·10), (3·11) and the symmetry of the problem, $I_a = I_b = I_c = 4\pi/3$ and $I_{aa} = I_{bb} = 3I_{ab} = \ldots = 4\pi/5a^2$. We find the following expressions for the constrained strain e_{ij}^C inside the transformed sphere in terms of the stress-free strain e_{ij}^T:

$$e^C = \alpha e^T, \quad 'e_{ij}^C = \beta' e_{ij}^T,$$

where
$$\alpha = \frac{1}{3}\frac{1+\sigma}{1-\sigma}, \quad \beta = \frac{2}{15}\frac{4-5\sigma}{1-\sigma}.$$

For a spherical inhomogeneity with elastic constants κ_1, μ_1 in an applied field e_{ij}^A the equivalent e_{ij}^T are given by

$$e^T = Ae^A, \quad 'e_{ij}^T = B'e_{ij}^A,$$

† Failure to appreciate this point led to an error in a previous paper (Eshelby 1955). It may be rectified by changing the sign of the right-hand side of the equation for ΔE on p. 488, col. 2, line 5. In addition a factor r^{-6} is missing from the right-hand side of equation (4).

where $$A = \frac{\kappa_1 - \kappa}{(\kappa - \kappa_1)\alpha - \kappa} = \frac{\kappa - \kappa_1}{\kappa}\frac{4\mu + 3\kappa}{4\mu + 3\kappa_1}$$

and $$B = \frac{\mu_1 - \mu}{(\mu - \mu_1)\beta - \mu}.$$

The interaction energy is†

$$E_{\text{int.}} = \tfrac{1}{2}V\left\{\frac{A}{9\kappa}p^A p^A + \frac{B}{2\mu}{}'p^A_{ij}{}'p^A_{ij}\right\}.$$

By (2·13) the stress just outside the inhomogeneous sphere is easily found to be

$$p = p^A - \frac{1+\sigma}{1-\sigma}B'p^A_{ij}n_i n_j,$$

$$'p_{il} = (1+\beta B)\,'p^A_{il} - B('p^A_{ik}n_k n_l + 'p^A_{lk}n_k n_i) + \frac{B}{1-\sigma}\,'p^A_{jk}n_j n_k n_i n_l$$

$$+\frac{1-2\sigma}{3(1-\sigma)}B'p^A_{jk}n_j n_k \delta_{il} - \frac{1-2\sigma}{3(1-\sigma)}Ap^A(n_i n_l - \tfrac{1}{3}\delta_{il}).$$

In particular, for the stress at the surface of a spherical cavity ($\kappa_1 = 0$, $\mu_1 = 0$) perturbing a uniform stress-field p^A_{ij} we find

$$p_{il} = \frac{15}{7-5\sigma}\left\{(1-\sigma)(p^A_{il} - p^A_{ik}n_k n_l - p^A_{lk}n_k n_i) + p^A_{jk}n_j n_k n_i n_l \right.$$
$$\left. - \sigma p^A_{jk}n_j n_k \delta_{il} + \frac{1-5\sigma}{10}p^A(n_i n_l - \delta_{il})\right\}.$$

The expression given by Landau & Lifshitz (1954) is clearly incorrect, since the surface traction $p_{il}n_l$ formed from it does not vanish.

According to the discussion leading to (4·7), the energy density of a body containing a volume fraction v of inhomogeneous spheres is

$$\frac{1}{2}\left\{\frac{1}{9\kappa}(1+Av)p^A p^A + \frac{1}{2\mu}(1+Bv)'p^A_{ij}{}'p^A_{ij}\right\},$$

and so the effective bulk elastic constants are

$$\kappa_{\text{eff.}} = \kappa/(1+Av), \quad \mu_{\text{eff.}} = \mu/(1+Bv).$$

Since we have neglected the interaction between the spheres these expressions will only be valid for small v and we may equally well write

$$\kappa_{\text{eff.}} = \kappa(1-Av), \quad \mu_{\text{eff.}} = \mu(1-Bv).$$

When we let κ_1, μ_1 approach zero or infinity we recover known results for a material containing empty spherical cavities (Mackenzie 1950) or rigid and incompressible spherical inclusions (Hashin 1955). For arbitrary κ_1, μ_1 the expression for $\kappa_{\text{eff.}}$ agrees with Bruggemann's (1937). Bruggemann's expression for $\mu_{\text{eff.}}$ is independent of the Poisson's ratio of the matrix and can hardly be correct. It is derived by considering the perturbation of the non-uniform elastic field in a sphere twisted in pure torsion when a spherical inclusion is introduced at the centre. This is obviously not typical of

* There is an error in a previous paper (Eshelby 1951); the expressions for δ, $\mathbf{F}(K'=0)$, $\mathbf{F}(K'=\infty)$ on page 104 should all be multiplied by $4\mu/3K$.

the effect of an inhomogeneity at an arbitrary point in the sphere. A well-known analogy (Goodier 1936) between problems in elasticity and viscosity enables us to interpret $\mu_{\text{eff.}}$ as the effective viscosity of a suspension of rigid spheres in a liquid of viscosity μ provided we put $\mu_1 = \infty$, $\sigma = \frac{1}{2}$. This gives Einstein's expression $\mu_{\text{eff.}} = (1+2{\cdot}5v)\mu$. Kynch (1956) has discussed the value of v above which the mutual interaction of the spheres makes this expression inaccurate. Evidently $\kappa_{\text{eff.}}, \mu_{\text{eff.}}$ will be subject to a similar limitation.

The problem of an ellipsoidal inclusion which has undergone a simple shear is of interest in connexion with twinning and martensitic or other diffusionless transformations. Suppose, then, that an ellipsoidal region of volume V undergoes a pure shear transformation in which $e^T_{13} = e^T_{31}$ are the only non-zero components of e^T_{ij}. Then (3·21) and (2·7) give

$$E_{\text{el.}} = 2\gamma\mu V(e^T_{13})^2 \tag{5·1}$$

with

$$\gamma = 1 - 2S_{1313} \tag{5·2}$$

for the total elastic energy in matrix and inclusion. $2\mu V(e^T_{13})^2$ is the energy necessary to pull the inclusion back to its original shape in the absence of the matrix. Alternatively, it is the energy we should find if the inclusion transformed whilst embedded in an imaginary rigid matrix. Thus we may regard γ as a measure of the extent to which the matrix is able to accommodate the transformation. It also describes the partition of the total strain energy between matrix and inclusion, for from (2·19) and (2·20) we have

$$\frac{\text{energy in matrix}}{\text{energy in inclusion}} = \frac{1-\gamma}{\gamma},$$

so that for good accommodation (small γ) most of such energy as remains is in the matrix. We also have from (2·7) the value

$$p^I_{13}/(-p^T_{13}) = \gamma$$

for the ratio between the actual stress in the inclusion to the value it would have if the transformation occurred in a rigid matrix.

For an oblate spheroid we have

$$\gamma = \frac{2-\sigma}{1-\sigma}\frac{I_a}{4\pi} - \tfrac{2}{3}Q\frac{c^2}{a^2-c^2}(4\pi - 3I_a)$$

with I_a given by (3·15). For a sphere $\gamma = (7-5\sigma)/15(1-\sigma)$ and for a needle $(b = c \ll a)$, $\gamma = \frac{1}{2}$. The values for sphere and needle are about the same if Poisson's ratio is in the neighbourhood of $\frac{1}{3}$.

If the inclusion is a thin plate in the form of an ellipsoid with its c axis much less than its a and b axes we have

$$\gamma = \eta c/b,$$

where η has the following values:

$$\left.\begin{aligned} \eta &= E(k) + \frac{\sigma}{1-\sigma}\frac{K(k)-E(k)}{k^2}; && a > b,\ k = (1-b^2/a^2)^{\frac{1}{2}},\ k' = b/a; \\ &= \frac{E(k)}{k'} + \frac{\sigma}{1-\sigma}\frac{1}{k'}\frac{E(k)-k'^2K(k)}{k^2}; && b > a,\ k = (1-a^2/b^2)^{\frac{1}{2}},\ k' = a/b; \\ &= \pi(2-\sigma)/4(1-\sigma); && a = b. \end{aligned}\right\} \tag{5·3}$$

Here $E(k)$ and $K(k)$ are complete elliptic integrals. It follows that for a very thin plate γ approaches zero and there is complete accommodation. If the operative shear is e^T_{12} instead of e^T_{13}, tending to deform the plate in its own plane, the corresponding accommodation factor $1-2S_{1212}$ approaches unity as the thickness of the plate decreases and there is no accommodation. We may compare these results with the case where e^T_{ij} is a pure dilatation. Then, as we saw in § 2,

$$E_{\text{el.}} = 2\mu V(e^T)^2(1+\sigma)/9(1-\sigma)$$

whatever is the shape of the inclusion. In a rigid matrix the energy would be $\frac{1}{2}\kappa V(e^T)^2$ and so the accommodation factor is always $2(1-2\sigma)/3(1-\sigma)$ or $\frac{1}{3}$ for $\sigma = \frac{1}{3}$.

As a rough illustration of how these results might be used, consider the formation of martensite in iron. Zener (1946) has shown that the thermodynamics of the process suggest that a strain energy of 290 cal is associated with each mole transformed. Suppose that the transformation involves a 5 % volume expansion and a 10° shear, so that $e^T = 0{\cdot}05$, $e^T_{13} = 0{\cdot}009$. It is easily seen that the total strain energy is the sum of the values it would have if the dilatation or shear acted alone. With $\mu = 8\times10^{11}\,\text{dyn/cm}^2$, $\sigma = \frac{1}{4}$, and V one molar volume, the shape-independent dilatational contribution to the energy is 25 cal. This leaves 265 cal for the shear contribution. The quantity $2\mu V(e^T_{13})^2$ has the value 1900 cal. If the transformed region is supposed to be an ellipsoid the accommodation factor is thus $\gamma = 265/1900 = 0{\cdot}14$, and this tells us something about its shape. For example, if it is assumed to be a circular disk, (5·3) shows that its thickness/diameter ratio must be 0·08. In the presence of an applied stress the free energy change associated with the transformation becomes $E_{\text{el.}}+E_{\text{int.}}$ instead of $E_{\text{el.}}$ For the case we have been considering equation (3·22) gives $E_{\text{int.}} \sim 8\cdot10^{-9}(-\frac{1}{3}p^A) - 3\cdot10^{-8}p^A_{13}$ cal/mole if the applied field p^A_{ij} is measured in dynes/cm².

For a cavity, the equations (4·3), (4·4) for the ellipsoidal inhomogeneity simplify to

$$e^T_{ij} - e^C_{ij} = e^T_{ij} - S_{ijkl}e^T_{kl} = e^A_{ij}; \tag{5·4}$$

we shall only consider this case.

Suppose that an ellipsoidal cavity is perturbing a simple shear $e^A_{13} = S/2\mu$. We have at once for the equivalent stress-free strain, putting $\mu_1 = 0$ in (4·5),

$$e^T_{13} = e^A_{13}/\gamma, \tag{5·5}$$

with the notation of (5·2). The interaction energy is

$$E_{\text{int.}} = -\tfrac{1}{2}Vp^A_{ij}e^T_{ij} = -VS^2/2\mu\gamma. \tag{5·6}$$

If we let the c axis of the ellipsoid become very small we have an elliptical crack. From (5·3) it is clear that $V\gamma$ remains finite as c approaches zero and the interaction energy of the crack with the applied shear stress S is

$$E_{\text{int.}} = -2\pi ab^2S^2/3\mu\eta.$$

Consider next the displacement of the faces of the crack. If the c axis is still finite, the displacement of a point x_i at the surface of the cavity is

$$u^C_i = (e^C_{ij} + \varpi^C_{ij})x_j$$

plus the displacement u_i^A due to the applied field. We suppose that $u_i^A = 0$ in the plane of the crack. If we evaluate u_i^C from (3·6) and (3·19) and pass the limit $c \to 0$ and use (5·5) we find that in the plane of the crack

$$\pm u_1^C = (bS/\mu\eta)(1 - x_1^2/a^2 - x_2^2/b^2)^{\frac{1}{2}} \equiv \tfrac{1}{2}\Delta u_1, \qquad (5\cdot7)$$
$$u_2^C = 0, \quad u_3^C = \alpha x_1, \quad \alpha = \pi(1-2\sigma)\, b e_{13}^A / 4(1-\sigma)\, a\eta;$$

the + and − signs refer to the upper and lower faces of the crack. Thus the plane of the crack is tilted through an angle α, but it remains a plane. The relative displacement Δu_1 of the faces is everywhere parallel to the x_1 axis and has an ellipsoidal distribution.

There is a problem in dislocation theory closely related to the theory of the sheared crack. Under the influence of a stress $p_{13}^A = S$ dislocation loops expand in the $x_1 x_2$ plane from a source at the origin and pile up against an elliptical barrier until their back-stresses annul p_{13}^A at the source. What is their distribution if each loop is in equilibrium under the combined action of the other loops and the applied stress? In the limit of a large number of loops with very small Burgers vectors the crack and dislocation problems coincide (cf. Leibfried 1954). If the source has its Burgers vector parallel to the x_1 axis the dislocations are of pure edge type where they cross the x_1 axis and the number of them crossing a length dx_1 is $dx_1\, \partial \Delta u_1(x_1, x_1 = 0)/\lambda\, \partial x_1$. Where they cross the x_2 axis they are of pure screw type and their density is $\partial \Delta u_1(x_1 = 0, x_2)/\lambda\, \partial x_2$. The interaction energy of the loops is given by (5·6). In diagrams the tip of an array of piled-up dislocations is often drawn curling up or down. The remarks following (5·7) do not support this.

As a further example consider a spheroidal cavity ($b = c$) in a material subject to a simple tension stress T. If the a axis coincides with the direction of T we need only know e_{11}^T to find the interaction energy. The non-zero components of e_{ij}^A are e_{11}^A, $e_{22}^A = e_{33}^A = -\sigma e_{11}^A$. From (5·4) we have

$$e_{11}^T = \epsilon e_{11}^A, \quad \text{where} \quad \epsilon = \frac{(1 - S_{33} - S_{23}) - 2\sigma S_{13}}{(1 - S_{33} - S_{23})(1 - S_{11}) - 2S_{13}S_{31}}, \qquad (5\cdot8)$$

with the abbreviated notation $S_{11} = S_{1111}$, $S_{13} = S_{1133}$ ....

On the other hand, if the a axis is at right angles to the direction of T we only need to know e_{33}^T. The non-vanishing applied strains are e_{33}^A, $e_{11}^A = e_{22}^A = -\sigma e_{33}^A$ and we find

$$e_{33}^T - e_{22}^T = \frac{1+\sigma}{1 + S_{23} - S_{33}} e_{33}^A, \quad e_{33}^T + e_{22}^T = \frac{(1-\sigma)(1 - S_{11}) - 2\sigma S_{31}}{(1 - S_{33} - S_{22})(1 - S_{11}) - 2S_{13}S_{31}} e_{33}^A \qquad (5\cdot9)$$

or, say,

$$e_{33}^T = \zeta e_{33}^A.$$

The interaction energies are respectively

$$E_{\text{int.}}(\|) = -\tfrac{1}{2} V \epsilon T^2 / E \qquad (5\cdot10)$$

and

$$E_{\text{int.}}(\perp) = -\tfrac{1}{2} V \zeta T^2 / E, \qquad (5\cdot11)$$

where E is Young's modulus.

If the direction of T remains unaltered and the cavity changes from the parallel to the perpendicular orientation, the interaction energy changes from (5·10) to (5·11).

The parallel or transverse orientation is energetically favourable according as the spheroid is oblate or prolate. We may take the case of the prolate spheroid as illustrating the orienting effect of an applied stress on a di-vacancy in a metal (A. Seeger, private communication). For $a/c = 2$, $\sigma = \frac{1}{3}$ we find $\epsilon = 2{\cdot}24$, $\zeta = 5{\cdot}88$.

When a approaches zero the numerator of ϵ becomes $1-2\sigma$ and the denominator approaches zero as $a\pi(1-2\sigma)/4c(1-\sigma^2)$. The product ϵV in (5·10) remains finite and we reproduce Sack's (1946) value $-8c^3(1-\sigma^2)T^2/2E$ for the interaction energy of a penny-shaped crack in tension.

The total strain at the surface of the cavity $e^C_{ij}+e^A_{ij}$ is, according to (5·4), given by the right-hand sides of (2·13) with e^C, $'e^C_{ij}$ replaced by e^T, $'e^T_{ij}$. The normal is given by (3·23) and the stress concentration can be found from (2·3). For the spheroid in tension we must supplement (5·8) by

$$e^T_{22} = e^T_{33} = e^A_{11}(\epsilon - \epsilon S_{11} - 1)/2S_{13}$$

and (5·9) by

$$(1-S_{11})\,e^T_{11} = S_{12}(e^T_{22}+e^T_{33}) - \sigma e^A_{33}.$$

These results actually apply to a quite general state of triaxial stress symmetrical about the polar axis of the spheroid, for in the applied strain

$$e^A_{22} = e^A_{33} = -\sigma e^A_{11}, \quad e^A_{12} = e^A_{23} = e^A_{31} = 0,$$

we may take σ to be any number unconnected with Poisson's ratio. (The σ implicit in the S_{ij} must, of course, be put equal to Poisson's ratio.) The stress concentration about an ellipsoid in shear is found similarly from (5·5).

Two-dimensional problems involving an infinite elliptic cylinder can be dealt with similarly, using (3·17) and (3·18). The interaction energy per unit length is

$$E_{\text{int.}} = -\tfrac{1}{2}Ap^A_{ij}e^T_{ij},$$

where A is the cross-sectional area of the cylinder. The passage to the limit $b\to 0$ or $a \to 0$ is in this case very easy and we can derive well-known results for cracks in plane strain tension or shear. Another simple case is that of a crack joining the points $x = \pm a$, $y = 0$ perturbing a uniform stress $p^A_{23} = S$. Both before and after the introduction of the crack there is a state of anti-plane strain in which the displacement is everywhere perpendicular to the xy plane. The interaction energy and relative shift of the faces of the crack are

$$E_{\text{int.}} = -\pi a^2 S^2/2\mu \quad \text{and} \quad \Delta u_3 = (2S/\mu)\,(a^2-x^2)^{\frac{1}{2}}. \tag{5·12}$$

Several writers have derived approximate expressions for the reduction of energy by a crack (or an array of dislocations simulating a crack) by supposing that the applied stress is effectively relaxed to zero in a region about the crack whose dimensions are of the order of the width of the crack. (In the same sense we might say that in (5·10) or (5·11) the 'energy drainage volume' of the cavity was ϵ or ζ times its geometrical volume.) This method gives correct results, but the logic behind it is not clear. If the applied stress is maintained by constant external loads, the elastic energy is *increased* by a certain amount when the crack is introduced (compare equation (4·8)). At the same time, the loading mechanism has to expend twice this amount of work. Thus the decrease of total energy ($-E_{\text{int.}}$) is numerically equal to the increase of elastic energy. On the other hand, if the applied stress is due to sources of internal strain, or if the body is strained by rigid clamps, the elastic energy (now the total energy) clearly *decreases* (compare equation (4·9)). But even here the decrease is in no simple sense located near the crack. This may easily be shown explicitly for the case of the crack in anti-plane strain (equation (5·12)). In terms of the elliptic co-ordinates ξ, η defined by

$$x = a\cosh\xi\cos\eta, \quad y = a\sinh\xi\sin\eta,$$

the displacement is

$$u_3 = (aS/\mu)\cosh\xi\sin\eta. \tag{5·13}$$

For, when ξ is large enough for hyperbolic sine and cosine to be indistinguishable it reduces to the uniform state of shear

$$u_3 = Sy/\mu = (aS/\mu)\sinh\xi\sin\eta, \tag{5·14}$$

whilst the traction on a curve $\xi = \text{const.}$ is proportional to $\partial u_3/\partial\xi$ and so vanishes on the limiting ellipse $\xi = 0$ defining the crack. The energy in the small rectangle $\mathrm{d}\xi\,\mathrm{d}\eta$ at ξ, η is $\frac{1}{2}\mu\{(\partial u_3/\partial\xi)^2 + (\partial u_3/\partial\eta)^2\}\,\mathrm{d}\xi\,\mathrm{d}\eta$. If we evaluate this using the displacement without ((5·14)) and with ((5·13)) the crack, we find that the change of energy density ΔE at any point due to the introduction of the crack is given by

$$\Delta E\,\mathrm{d}\xi\,\mathrm{d}\eta = \tfrac{1}{2}(S^2/\mu)\cos 2\eta\,\mathrm{d}\xi\,\mathrm{d}\eta. \tag{5·15}$$

We may say that there is stress relaxation between the hyperbolas $\eta = \frac{1}{4}\pi, \frac{3}{4}\pi$ and stress concentration outside them. The integral of (5·15) over any ellipse with the ends of the crack for foci is precisely zero. By judiciously deforming the ellipse we can find a curve within which the energy 'relaxation' is positive or negative. Attempts to evaluate interaction energies in this way lead not only to errors of sign (which may be corrected by common sense), but also to incorrect numerical factors.

The problem of a rigid and incompressible ellipsoidal inhomogeneity is also relatively simple, since (4·3) and (4·4) reduce to

$$S_{ijkl}e^T_{kl} = -e^A_{ij}.$$

From the solution Goodier's (1936) analogy enables us to find the perturbation of the slow motion of a viscous fluid when a solid ellipsoid is immersed in it. We have only to put $\sigma = \frac{1}{2}$ in the matrix and interpret μ, u_i, e_{ij} and p_{ij} as viscosity, velocity, rate of strain and stress. The energy density becomes half the rate of dissipation of energy per unit volume. Equation (4·11) or (4·12) enables us to find the viscosity of a dilute suspension of ellipsoids. $E_{\text{int.}}$ is positive for a rigid inclusion and so the viscosity is increased. Equation (4·11) now states that a viscometer working at constant load will produce a lower rate of deformation and so will dissipate less energy, whilst equation (4·12) states that a viscometer working at constant speed will have to work harder to maintain a prescribed rate of strain.

For a single immersed ellipsoid the increase in the rate of energy dissipation is clearly twice $E_{\text{int.}}$ for the related elastic problem. The calculation is much simplified by the fact that for $\sigma = \frac{1}{2}$, $R = 0$ in (3·7), whilst the dilatations e^A, e^T, e^C are all zero. We can easily verify, for example, Jeffery's (1922) expression for the energy dissipation by a prolate spheroid, as amended by Eisenschitz (1933). To find the viscosity of a dispersion of spheroids it is necessary to decide what orientation they will take up. The elastic analogy suggests (though it does not prove) that they will ultimately orient themselves so as to minimize the energy dissipated. This agress with Jeffery's hypothesis, verified experimentally by Taylor (1923).

References

Bruggemann, D. A. G. 1937 *Ann. Phys., Lpz.*, **29**, 160.

Byrd, P. F. & Friedman, M. D. 1954 *Handbook of elliptic integrals*, p. 4. Berlin: Springer-Verlag.

Cochardt, A. W., Schoek, G. & Wiedersich, H. 1955 *Acta Met.* **3**, 533.

Dirichlet, G. L. 1839 *Verh. K. Preuss. Akad. Wiss.* p. 18 (*Werke*, **1**, 383).

Eisenschitz, R. 1933 *Z. phys. Chem.* A, **163**, 133.

Eshelby, J. D. 1951 *Phil. Trans.* A, **244**, 87.

Eshelby, J. D. 1955 *Acta Met.* **3**, 487.

Eshelby, J. D. 1956 In *Progress in solid state physics* (ed. Seitz & Turnbull), **3**, 79. New York and London: Academic Press.

Goodier, J. N. 1936 *Phil. Mag.* [7], **22**, 678.

Hashin, Z. 1955 *Bull. Res. Coun. Israel*, C, **5**, 46.

Jeffery, G. B. 1922 *Proc. Roy. Soc.* A, **102**, 161.

Kellogg, O. D. 1929 *Potential theory*. Berlin: Springer-Verlag.

Kröner, E. 1954 *Acta Met.* **2**, 301.

Kynch, G. J. 1956 *Proc. Roy. Soc.* A, **237**, 90.

Landau, L. D. & Lifshitz, E. M. 1954 *Mekhanika Sploshnykh Sred*, p. 666. Moscow: Gostekhizdat.

Leibfried, G. 1954 *Z. angew. Phys.* **6**, 251.

Love, A. E. H. 1927 *Theory of elasticity*. Cambridge University Press.

Mackenzie, J. K. 1950 *Proc. Phys. Soc.* B, **63**, 2.

Nabarro, F. R. N. 1940 *Proc. Roy. Soc.* A, **175**, 519.

Nabarro, F. R. N. 1951 *Phil. Mag.* [7], **42**, 1224.

Nabarro, F. R. N. 1952 *Advanc. Phys.* **1**, 269.

Osborn, J. A. 1945 *Phys. Rev.* **67**, 351.

Poincaré, H. 1899 *Théorie du Potentiel Newtonien*, p. 118. Paris: Carré et Naud.

Robinson, K. 1951 *J. Appl. Phys.* **22**, 1045.

Routh, D. J. 1892 *Analytical statics*, **2**, 121. Cambridge University Press.

Sack, R. A. 1946 *Proc. Phys. Soc.* **58**, 729.

Shapiro, G. S. 1947 *Dokl. Akad. Nauk SSSR*, **58**, 1309.

Somigliana, C. 1914 *R.C. Accad. Lincei*, [5], **23** (1), 463.

Somigliana, C. 1915 *R.C. Accad. Lincei*, [5], **24** (1), 655.

Sternberg, E., Eubanks, R. A. & Sadowsky, M. A. 1951 *J. Appl. Phys.* **22**, 1121.

Taylor, G. I. 1923 *Proc. Roy. Soc.* A, **103**, 58.

Zener, C. 1946 *Trans. Amer. Inst. Min. (Metall.) Engrs*, **167**, 513.

DISCUSSION

Eshelby (partly written): The relative rotation $d\theta$ of the ends of a whisker containing a screw dislocation when it is stretched by an amount dl can be calculated by modifying the analysis of Green and Shield (1951) of the simultaneous finite extension and infinitesimal torsion of a cylinder of arbitrary cross section, made of isotropic material with an arbitrary stress-strain relation. When the extension is small the result is

$$d\theta = -\alpha_0\, 2(1+\sigma)\, dl \left(\frac{I}{S} - 1\right) \tag{1}$$

in which α_0 is the twist per unit length produced by the dislocation in the unstretched whisker, σ is Poisson's ratio, I is the polar moment of inertia of the cross section, and S is the torsional rigidity with the shear modulus set equal to unity. Equation (1) differs from the estimate of Gomer (1958) by the factor (I/S -1). The effect thus depends on the shape of the cross section but not on the position of the dislocation except for its effect on α_0. For a circle (I/S - 1) = 0. For any other cross section this term is greater than zero (Pólya and Szegö, 1951), and there is untwisting. For small departures from a circle this term is rather small, e.g., 0.045 for a regular hexagon or $1/4\,(ab^{-1} + ba^{-1})^2$ for an ellipse of semi-axes a, b.

To demonstrate Eq. (1) one first needs an expression for the twist produced by a screw dislocation in a cylinder whose cross section is an arbitrary region R. It can be found as follows. First find the antiplane strain displacement, $(0,\, 0,\, W^D)$ say, which represents the dislocation in a cylinder with its side surface (but not its ends) stress-free. Then the twist α when the ends are stress-free as well is found from the relation

$$\int_R \left(x\frac{\partial W^D}{\partial y} - y\frac{\partial W^D}{\partial x}\right) dx\, dy + \alpha S = 0 \tag{2}$$

with

$$S = \int_R \left(x^2 + y^2 + x\frac{\partial \phi^T}{\partial y} - y\frac{\partial \phi^T}{\partial x}\right) dx\, dy$$

which ensures that there is no net twisting moment. ϕ^T is the usual torsion function, and S is the geometrical torsional rigidity, that is, the ordinary torsional rigidity divided by the shear modulus. The twist can be written in the form (Eshelby, 1958)

$$\alpha = K\, b/A \tag{3}$$

where b is the Burgers vector, A the area of R and K is a numerical factor depending only on the shape (but not size) of R and on the relative position of the dislocation in the cross section.

Green and Shield have solved the problem of a cylinder of arbitrary cross section subject to simultaneous finite extension by forces on the ends and small twist by couples on the ends. The notation used here follows Green and Zerna (1954). Let the stretching multiply the length of the cylinder by λ and its transverse dimensions by μ, and let its cross section in the stretched state happen to coincide with the cross section R considered above. On the finite extension superimpose an additional small displacement

$$U = -\epsilon yz, \qquad V = \epsilon xz, \qquad W = \epsilon\phi(x, y) \tag{4}$$

It is found that the equations of finite deformation are satisfied to order ϵ^2 if

$$\nabla^2 \phi = 0 \text{ in } R \tag{5}$$

and

$$\left(\frac{\partial\phi}{\partial x} - y\right)\frac{\partial F}{\partial x} + \left(\frac{\partial\phi}{\partial y} + x\right)\frac{\partial F}{\partial y} = 0 \tag{6}$$

on the curve $F(x, y) = 0$ bounding R.

For the couple necessary to maintain this displacement Green and Shield find the expression

$$M \propto \epsilon(\lambda^2 - \mu^2)\, I + \int_R \epsilon\mu^2 \left[x\left(\frac{\partial\phi}{\partial y} + x\right) - y\left(\frac{\partial\phi}{\partial x} - y\right)\right] dx\, dy \tag{7}$$

where

$$I = \int_R (x^2 + y^2)\, dx\, dy \quad .$$

For our application we shall not need to know the constant of proportionality in Eq. (7).

In Green and Shield's application ϕ is the ordinary torsion function ϕ^T for R according to the infinitesimal theory. The difference from the infinitesimal theory makes itself felt in Eq. (7). The analysis remains valid with a more general form for ϕ ; in particular we may take

$$W = \epsilon\phi = \epsilon\phi^T + W^D$$

which satisfies Eqs. (5) and (6). Equation (4) now represents the displacement due to the combined effect of a dislocation with Burgers vector b and the applied couple given by Eq. (7), superimposed, of course, on the extension represented by λ, μ. To find the twist α due to the dislocation alone we equate Eq. (7) to zero and eliminate WD with the help of Eq. (2):

$$\epsilon = \frac{\alpha \mu^2 S}{(\lambda^2 - \mu^2) I + \mu^2 S} .$$

Finally we have to relate α to the twist of the unstretched whisker. Let a suffixed O refer to the unstretched state. Then evidently in (3) $b = \lambda b_0$, $A = \mu^2 A_0$, so that $\alpha = \lambda \alpha_0/\mu^2$. The ratio I/S is independent of λ and μ. The relative rotation of the end of the cylinder is thus

$$\theta = \epsilon \lambda l_0 = \frac{\alpha l_0 \lambda^2}{\mu^2 + (\lambda^2 - \mu^2) I/S} \qquad (8)$$

For small extensions ($\lambda \sim 1$, $\mu \sim 1$) this reduces to (1) if we identify the contraction ratio $\sigma = (1 - \mu)/(\lambda - 1)$ with Poisson's ratio. For a circle I = S and according to Eq. (8), $d\theta = 0$ even for a large extension.

References

Eshelby, J.D., Phil. Mag., 3, 440 (1958).

Green, A.E. and Shield, R.T., Phil. Trans. (London), A 244, 47 (1951).

Green, A.E. and Zerna, S., "Theoretical Elasticity," Oxford Univ. Press (1954).

Gomer, R., J. Chem. Phys., 28, 457 (1958).

Polya, G. and Szego, G., "Isoperimetric Inequalities in Mathematical Physics," Princeton Univ. Press (1951).

Hillig: Can you make any estimates of the surface diffusion coefficients required for the growth of mercury whiskers? Do they show any indications of an anomalously high growth rate, such as apparently occurs for the potassium whiskers?

Gomer: We do make an estimate of the surface diffusion coefficient, at -78°C, from th terminal length of the whiskers. I don't remember the numerical value, but it was fairly normal yielding an activation energy of 1.1 kcal.

Charged Dislocations and the Strength of Ionic Crystals†

By J. D. Eshelby, C. W. A. Newey and P. L. Pratt
Department of Physical Metallurgy, University of Birmingham

and A. B. Lidiard
Department of Physics, University of Reading

[Received November 13, 1957]

Abstract

If the energies required to form positive and negative ion vacancies in an ionic crystal are unequal, then in thermal equilibrium dislocations in the crystal will be electrically charged and surrounded by a Debye–Hückel cloud of vacancies. If the vacancy cloud is immobile a finite force is required to separate the dislocation from the cloud, and so the crystal will possess a yield stress below which plastic flow will not occur. The presence of divalent impurities modifies the magnitude of the charge on a dislocation, and may even reverse it. If precipitation of the impurity or association of impurity atoms and vacancies can occur, the concentration of impurities may be a complicated function of temperature. The yield stress of the crystal may then exhibit maxima and minima when plotted as a function of temperature. Experimental results showing this behaviour are presented and tentatively compared with the theory.

§ 1. Introduction

The number of Schottky defects (vacant anion and cation sites) in an ionic crystal is normally calculated by minimizing the free energy of the system subject to the condition that the crystal be electrically neutral throughout. However, as Lehovec (1953) has pointed out, the actual state of affairs is more complicated when we suppose that vacancies can only be formed at the surface of the crystal. (Compare also Frenkel 1946, Grimley and Mott 1947 and Grimley 1950.) For definiteness we consider the case of sodium chloride. Here sodium-ion vacancies are more easily formed than are chlorine-ion vacancies. Hence when the temperature is raised from the absolute zero an excess of sodium-ion vacancies is emitted from the surface into the crystal, leaving a net positive charge on the surface. The resulting space-charge discourages the emission of further sodium vacancies and encourages the emission of chlorine vacancies. Calculation shows that in equilibrium the bulk of the crystal is electrically neutral, but that there is a positive charge on the surface, balanced by an equal and opposite negative charge cloud penetrating some distance into the crystal. Because of the presence of

† Communicated by the Authors.

this double layer there is a potential difference between the surface and interior of the crystal.

It is generally admitted that vacancies can be formed at dislocations as well as at the surface of a crystal. Thus we should expect (Pratt 1957) dislocations in an ionic crystal to be charged (by having an excess of jogs of one sign), and to be surrounded by a sheath of vacancies of predominantly the opposite sign. Clearly a grain boundary or mosaic boundary should also be charged and surrounded by a balancing layer of charge, whether we regard the boundary as essentially a junction between two separate blocks of crystal, or as an array of dislocations.

We have pictured the charges on surfaces and dislocations as arising from the emission of an excess of vacancies of one sign. This is not necessary: any initial distribution of defects would re-arrange itself to give the same final equilibrium configuration. There should be similar

Fig. 1

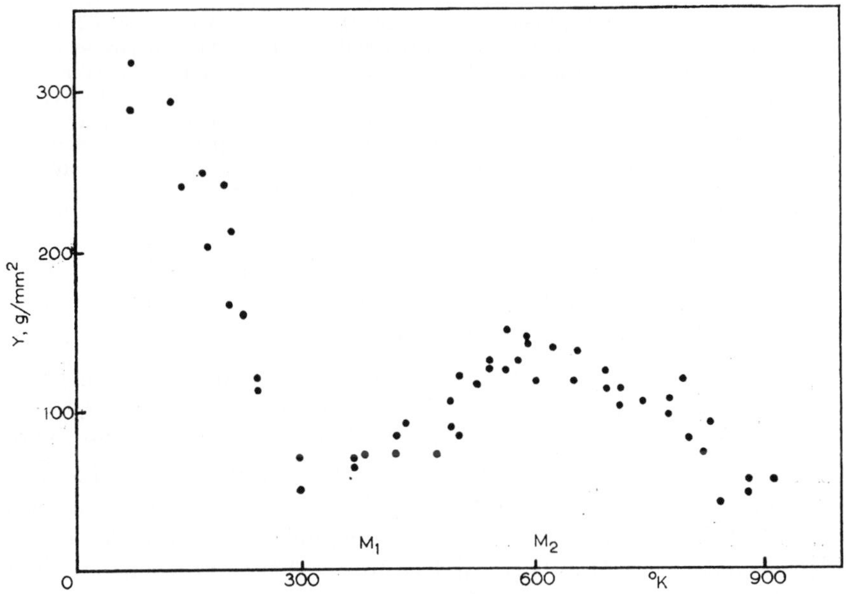

Yield stresses of rock salt specimens at various temperatures.

effects where Frenkel (interstitial plus vacancy) defects predominate, even though there is no difficulty in imagining their formation and disappearance in the body of the lattice as well as at surfaces and dislocations.

Suppose that the temperature is low enough for the charge cloud around a dislocation to be immobile. The electrostatic field of the vacancies is such as to pull the charged dislocation back to the axis of the cloud if it is displaced, and so a certain minimum mechanical stress will be required to pull the dislocations away from their clouds and cause plastic flow.

Thus we have a mechanism which may in part determine the yield point of an ionic crystal.

If a sodium chloride crystal contains divalent impurities mobile enough to play their part in forming the ion atmospheres the above account must be modified. At high temperatures there is qualitatively no modification, but at low temperatures the impurities cause the sign of the charge on a dislocation (or surface) to reverse. At one particular intermediate temperature the dislocations are uncharged, and at this temperature the yield strength of the crystal should drop to zero if the present mechanism were the only one determining it. Again, at low temperatures some of the impurity may precipitate out in a new phase, and thus reduce the concentration of impurity which take part in ion-cloud formation. If this is so the yield stress can have a succession of maxima and minima when plotted as a function of temperature. The measured points in fig. 1 do indeed show this sort of behaviour.

The strength of ionic crystals will not, of course, be determined entirely by the mechanism discussed here. The factors commonly considered in connection with the strength of metals will also be operative, for example the elastic interactions between point defects and dislocations. We have thought it best, however, to present a preliminary account of the theory of charged dislocations ignoring these complications. Bassani and Thomson (1956) have dealt with the same topic, though from a different point of view. They showed that in NaCl a positive-ion vacancy is attracted into that part of the core of a (110) dislocation where the lattice is compressed, with a binding energy of about 0·4 ev relative to normal lattice positions. Divalent impurity ions were found to be little attracted into dislocation cores. From a knowledge of the binding energy of a point defect and a dislocation it is not possible to calculate a yield stress without introducing additional assumptions. Our approach, on the other hand, emphasizes the role of the charge cloud around a dislocation, and so we are able to obtain a yield stress by a purely electrostatic calculation. We may note that Bassani and Thomson's model of a set of sites along a dislocation at which vacancies may be trapped with a certain binding energy is formally included in our ‘vacancy source’ model so long as the predicted charge on the dislocation does not require these sites to be completely emptied. Evidently the connection between the two models needs further examination.

§ 2. The Charge on a Crystal Surface

We give first a brief sketch of the generalization of Lehovec's results to the case where mobile divalent impurities are present. (A more detailed account will appear elsewhere.)

Let N be the number of cation (or anion) sites and cN the mean number of impurity atoms per unit volume. Let n_+, n_-, n_i be the numbers per unit volume of cation vacancies, anion vacancies and impurity atoms at a point where the electrostatic potential is v. (We suppose that $v=0$ at the surface of the crystal.) In what follows we assume that n_+, n_-, n_i, cN are all small compared with N. We use an affixed ∞ to denote the value of any quantity far from the surface of the crystal. T, e, k are the absolute temperature, the electronic charge and Boltzmann's constant.

If g_+, g_- are the free energies of formation of anion and cation vacancies the vacancy concentrations at a given point are

$$n_+=N \exp \{-(g_+-ev)/kT\}$$
$$n_-=N \exp \{-(g_-+ev)/kT\}.$$

The concentration of impurities must, according to Boltzmann's barometric formula, depend on v through a factor $\exp(-ev/kT)$. The constant corresponding to $g_\pm$, however, is indeterminate, since for the present we do not allow impurities to enter or leave the system. But at large distances from the surface n_i must approach the bulk value cN, and so

$$n_i=cN \exp \{-e(v-v_\infty)/kT\}.$$

Similarly n_+^∞ and n_-^∞ must agree with the values deduced neglecting the Lehovec effect. Thus

$$\left.\begin{aligned} n_+^\infty&=N \exp \{-(g_+-ev_\infty)/kT\}=N\alpha \\ n_-^\infty&=N \exp \{-(g_-+ev_\infty)/kT\} \\ &=n_+^\infty-n_i^\infty=N(\alpha-c) \end{aligned}\right\} \quad . \quad . \quad . \quad . \quad (1)$$

where

$$\alpha=\tfrac{1}{2}c+\{(\tfrac{1}{2}c)^2+\exp[-(g_++g_-)/kT]\}^{1/2}. \quad . \quad . \quad . \quad (2)$$

A derivation of (2) has been given by Lidiard (1957). For our purpose it is enough to note that $\alpha=\exp[-\tfrac{1}{2}(g_++g_-)/kT]$ in the region of intrinsic conductivity but that at low enough temperatures the number of positive-ion vacancies equals the number of divalent impurities, so that $\alpha=c$.

v_∞ is determined by the first equation of (1) which may be put in the form

$$ev_\infty/kT=\ln \alpha+g_+/kT. \quad . \quad . \quad . \quad . \quad . \quad . \quad . \quad (3)$$

Thus

$$n_+=N\alpha \exp p$$
$$n_-=N(\alpha-c) \exp(-p)$$
$$n_i=Nc \exp(-p)$$

where

$$p=e(v-v_\infty)/kT.$$

The charge per unit volume,

$$\rho=-e(n_+-n_--n_i)=2N\alpha \sinh p$$

is related to the potential by Poisson's equation $\nabla^2 v=-4\pi\rho/\epsilon$ where ϵ is the static dielectric constant of the material, and so

$$\nabla^2 p=\kappa^2 \sinh p \quad . \quad . \quad . \quad . \quad . \quad . \quad . \quad . \quad (4)$$

where

$$\kappa^2 = 8\pi e^2 N\alpha/\epsilon kT. \qquad (5)$$

The solution of (4) for the plane boundary has been given by Lehovec (1953). (It occurs also in the theory of colloids; see for example Verwey and Overbeek 1948.) The potential has effectively risen from 0 to v_∞ at a depth κ^{-1} below the surface. (Our κ and Lehovec's characteristic length λ are related by $\kappa\lambda = \sqrt{2}$.) The charge Q on the surface and the field E just inside the crystal are given by

$$Q = -\frac{\epsilon}{4\pi} E = -\frac{2\epsilon\kappa kT}{4\pi e} \sinh \frac{ev_\infty}{2kT}. \qquad (6)$$

The solution for an internal surface is clearly obtained by putting two surface solutions back-to-back. The charge on the internal surface is

$$Q_{\text{int}} = 2Q. \qquad (7)$$

§ 3. The Force Required to Move a Charged Dislocation

We consider next the field about a dislocation. We should have to solve (4) in cylindrical coordinates. Unfortunately it is not possible to solve this non-linear equation exactly. If p is small enough for us to replace $\sinh p$ by p the solution is

$$p = AK_0(\kappa r) + BI_0(\kappa r) \qquad (8)$$

where K_0, I_0 are the modified Bessel functions normally so denoted (Watson 1945); the equation may also be solved if p is large enough for $\sinh p$ to be replaced by $\frac{1}{2}\exp p$. In any case it is not quite clear what boundary conditions we should impose at the dislocation line, and so we shall use the following argument to find the charge on a dislocation.

Consider a mosaic boundary made up of edge dislocations spaced a distance $h \ll \kappa^{-1}$ apart. Macroscopically the charge per unit area is $2Q$. Microscopically, however, the charge is not uniformly spread, but resides on the dislocations. Thus the charge per unit length of dislocation is

$$\sigma = 2Qh. \qquad (9)$$

Let h increase. At first σ increases linearly with h. But when h passes through the value κ^{-1} the charge clouds of the dislocations will no longer overlap and the dislocations are effectively independent. Thus it seems reasonable to suppose that the charge on an isolated dislocation is given by (9) with h set equal to κ^{-1} times a numerical constant near to unity. If we take this constant to be unity the charge on an isolated dislocation is

$$\sigma = 2Q/\kappa. \qquad (10)$$

For lack of the precise solution of (3) we use (8) to find the field surrounding the dislocation. If the field is to fall off with distance the term in I_0 must be rejected. The charge on the dislocation is then

$$\sigma = -\lim_{r\to 0} \frac{\epsilon}{4\pi} \cdot 2\pi r \frac{dv}{dr} = \frac{\epsilon AkT}{2e}$$

using the relations

$$K_0'(x) = -K_1(x), \quad \lim_{x\to 0} xK_1(x) = 1. \quad . \quad . \quad . \quad . \quad . \quad . \quad (11)$$

Comparing with (8) we have

$$v - v_\infty = (2\sigma/\epsilon)K_0(\kappa r).$$

The potential due to the line-charge on the dislocation is

$$-(2\sigma/\epsilon)\ln r + \text{const}.$$

and so that due to the charge cloud surrounding it is

$$v_{\text{cloud}} = (2\sigma/\epsilon)[K_0(\kappa r) + \ln(\kappa r)] + \text{const}.$$

Fig. 2

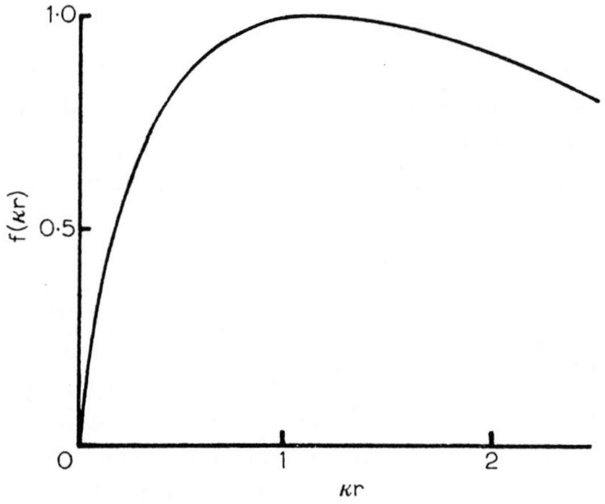

The restoring force on a charged dislocation.

If the dislocation moves a distance r from the centre of the cloud, the cloud remaining fixed, it is subject to a force

$$F = -\sigma \frac{d}{dr} v_{\text{cloud}}$$
$$= (2\sigma^2\kappa/\epsilon)[K_1(\kappa r) - (\kappa r)^{-1}]$$

by (11). If the movement is the result of the force $b\tau$ due to the resolved shear stress τ acting on a dislocation with Burgers vector b we must have $F + b\tau = 0$ in equilibrium, or

$$\tau = 0{\cdot}80(\sigma^2\kappa/\epsilon b)f(\kappa r)$$

where $f(x)$ (fig. 2) is the function $x^{-1} - K_1(x)$ divided by its maximum value, 0·40, which occurs when $x = 1{\cdot}02$. The shear stress required to pull the dislocation away from its cloud is thus

$$Y = \tau_{\max} = 0{\cdot}80\sigma^2\kappa/\epsilon b. \quad . \quad . \quad . \quad . \quad . \quad . \quad (12)$$

We shall suppose that g_+ and g_- are represented adequately by the linear functions of temperature

$$\left.\begin{aligned} g_+(T)&=g_+{}^0-kT\ln A_+ \\ g_-(T)&=g_-{}^0-kT\ln A_- \end{aligned}\right\} \quad . \quad . \quad . \quad . \quad . \quad . \quad . \quad (13)$$

so that in the intrinsic conductivity range

$$\alpha=(A_+A_-)^{1/2}\exp\{-\tfrac{1}{2}(g_+{}^0+g_-{}^0)/kT\}.$$

It is convenient to introduce in addition the quantity

$$\alpha'=\exp(-g_+/kT)=A_+\exp(-g_+{}^0/kT). \quad . \quad . \quad . \quad (14)$$

Evidently α' is the concentration of positive-ion vacancies which we should infer if we neglected the presence of impurities and also the requirement of bulk neutrality. We may call it the ' naive ' concentration.

Figure 3 (*a*) illustrates how v_∞ varies with temperature. As a function of $1/kT$ the curve $\ln \alpha'$ is of uniform slope $-g_+{}^0$. The curve $\ln \alpha$ has the constant value $\ln c$ in the impurity conduction range, and changes over to a line of slope $-\frac{1}{2}(g_+{}^0+g_-{}^0)$ in the intrinsic range. According to (3) ev_∞/kT is numerically equal to the vertical distance between the $\ln \alpha'$ and $\ln \alpha$ curves, and is positive where $\ln \alpha'$ is uppermost. At a certain temperature T_a the two curves intersect and v_∞ is zero. Following the usage in colloid theory we shall call such a temperature an isoelectric temperature. At T_a the charge on a surface or dislocation (eqns. (6) and (10)) is zero and, according to (12), the yield stress Y vanishes as shown schematically in the lower part of fig. 3 (*a*). In fact at an isoelectric point the naive and true vacancy concentrations coincide and the neglect of electrostatic effects is justified. If T_a lies in the impurity conduction range its value is

$$T_a=g_+/k\ln(c^{-1}).$$

Even in the absence of vacancies the two curves may intersect at high temperatures provided $g_+{}^0<g_-{}^0$ and $A_+<A_-$. At such an ' intrinsic ' isoelectric point $g_+=g_-$ (eqn. (13)), that is, the naive concentration of negative vacancies has just caught up with the naive concentration of positive vacancies. The position of the intrinsic isolectric point will be altered by the presence of impurities since α approaches

$$\tfrac{1}{2}c+\{(\tfrac{1}{2}c)^2+A_+A_-\}^{1/2}$$

rather than $(A_+A_-)^{1/2}$ at infinite temperature.

We have supposed above that the impurity concentration c remains constant throughout the impurity range. In fact, if the heat of solution of the impurity is positive it will precipitate out in a new phase at low temperatures (provided equilibrium is attained) and we shall have

$$c=A\exp(-B/kT) \quad . \quad . \quad . \quad . \quad . \quad . \quad (15)$$

with nearly constant A and B. As the temperature rises (15) will reach

P.M. F

a value $c=c_0$ at which all the precipitate is re-dissolved and at higher temperatures c will remain at the value c_0. Thus in the impurity range, at low temperatures, the $\ln \alpha$ curve will have a sharp knee and will intersect or fail to intersect the $\ln \alpha'$ curve in one of the ways indicated in fig. 3 (*b*) according to the particular values of g_+, B, A_+, A. Two of

Fig. 3

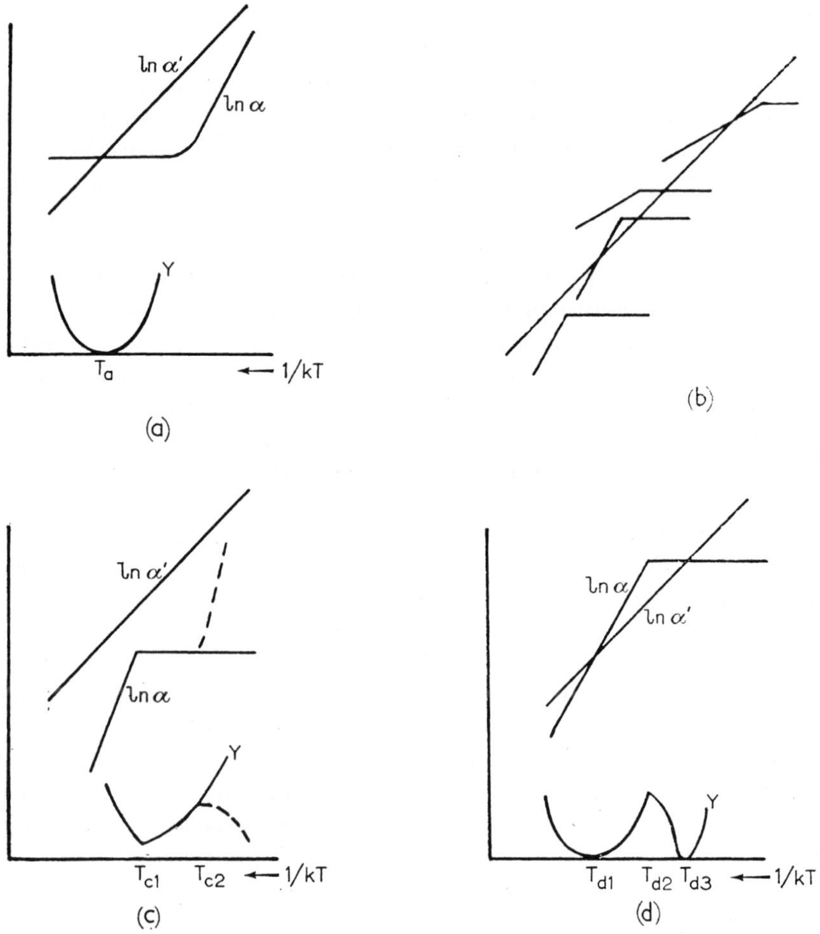

these possibilities are shown in greater detail in fig. 3 (*c*), (*d*) together with qualitative sketches of the behaviour of the yield stress. In fig. 3 (*d*) there are two isoelectric points T_{d1} and T_{d3} and a maximum in Y at T_{d2}. (Since (12) is dominated by the behaviour of the hyperbolic sine the maxima of $|\,ev_\infty/kT\,|$ and Y will very nearly coincide.) In fig. 3 (*c*) there is a minimum in the yield stress at T_{c1}. (We ignore the dotted curves for the present.) The values

$$T_{d1}=(B-g_+)/k \ln A$$

$$T_{d3}=g_+/k \ln (c_0^{-1})$$

$$T_{d2}=T_{c1}=B/k \ln (A/c_0)$$

are easily verified: the position of the maximum or minimum is independent of g_+. If there were a number of impurities with different values of A, B, c_0 the $\ln \alpha$ curve could in principle have a complicated zigzag shape and give rise to a succession of isoelectric points and maxima and minima.

There is also the possibility that vacancies and impurities may associate at low temperatures (Lidiard 1954). As in the case of precipitation this will lead to a fall in the $\ln \alpha$ curve below a certain temperature, but the knee at T_{c1} or T_{d2} will be smooth instead of abrupt, and the change of slope small.

§ 4. The Values of g_+ and g_-

In most problems relating to Schottky defects in ionic crystals (conductivity, diffusion) it is sufficient to know only the sum g_++g_- of the free energies of formation, since it is this sum which determines the bulk concentration of defects. However, in order to predict yield stresses from eqn. (12) it is necessary to know the values of g_+ and g_- individually. Moreover, we have implicitly assumed that the values of g_+ and g_- for the formation of vacancies at a surface are still appropriate when the vacancies are formed at a dislocation, and this assumption evidently needs examination.

It is clear that if dislocations are supposed to be in thermal equilibrium in a crystal g_++g_- for a dislocation must be equal to g_++g_- for a surface, since otherwise Schottky defects would distil over from one to the other. We can establish the equality more directly with the help of an imaginary experiment. Consider a crystal containing, say, a single edge dislocation. We have to show that no work is required to transfer an Na–Cl pair from a jog on the dislocation to a kink in a step on the external surface. Cleave the crystal in two along a plane containing the dislocation line. This will require the expenditure of a certain amount of work. If the plane is suitably chosen one of the new surfaces formed will be flat, and the other will be flat except for a single step representing the termination of the extra half-plane of Na–Cl pairs associated with the dislocation. Transfer an Na–Cl pair from a kink in this step (formerly a jog on the dislocation) to a kink in a step on the external surface. This requires no work. If the halves of the crystal are now joined together again the work expended in cleaving is recovered.

This argument cannot be applied to the transfer of a single ion, since the transfer is not a 'repeatable step' which leaves the physical state of the crystal unchanged. Thus we cannot infer that g_+ for the dislocation

F 2

is the same as g_+ for the surface. Strictly speaking even the transfer of a neutral pair shifts the position of a jog and alters the form of the dislocation and its relation to the free surface, so that the energies of cleaving and re-assembling do not exactly balance. The difference, negligible in the present context, provides a driving force which tends to make the dislocation climb out of the crystal.

Etzel and Maurer's (1950) experiments give the value

$$\alpha = 5{\cdot}2 \exp\{-1{\cdot}01 \text{ ev}/kT\} \quad . \quad . \quad . \quad . \quad (16)$$

for the concentration of Schottky defects in the intrinsic range. Thus (cf. eqn. (13))

$$g_+{}^0 + g_-{}^0 = 2{\cdot}02 \text{ ev}$$

and $A_+A_- = (5{\cdot}2)^2$. It is reasonable to suppose that the frequency of vibration of an ion bordering a vacancy is reduced whether the vacancy is positive or negative. In this case both A_+ and A_- are greater than unity and

$$1 < A_+ = 5{\cdot}2/A_- < 5{\cdot}2. \quad . \quad . \quad . \quad . \quad . \quad (17)$$

This means that $g_\pm$ at room temperature can differ from $g_\pm = g_\pm{}^0$ at 0°K by a tenth of an electron-volt or so (compare (13)).

The calculations of Mott and Littleton (1938) and their later modifications are mainly aimed at finding the sum $g_+{}^0 + g_-{}^0$ which determines the bulk concentration of defects. The separate energies, g_+', g_-' say, required to remove a positive or a negative ion from the bulk of the crystal are found during the course of the calculation. When a positive–negative pair is replaced on the surface of the crystal or a dislocation an amount of energy L, the lattice energy per ion pair, is recovered, and so $g_+{}^0 + g_-{}^0 = g_+' + g_-' - L$.

In order to find $g_+{}^0$ and $g_-{}^0$ individually we have to decide what fraction of L is recovered when a positive ion alone or a negative ion alone is replaced on a surface kink or dislocation jog. At first glance we might think that this fraction would be the same for a positive ion as for a negative ion if only Coulomb forces and nearest-neighbour anion–cation repulsive forces were operative. This would give $g_+{}^0 = g_+' - \frac{1}{2}L$, $g_-{}^0 = g_-' - \frac{1}{2}L$. This is Lehovec's assumption. In this case it seems more reasonable to use the results of Mott and Littleton or of Brauer (1952), who neglect next-nearest neighbour interactions, rather than those of Bassani and Fumi (1954) who do not. Mott and Littleton's figures give $g_+{}^0 = 0{\cdot}65$ ev, $g_-{}^0 = 1{\cdot}21$ ev, Brauer's $g_+{}^0 = 0{\cdot}70$ ev, $g_-{}^0 = 1{\cdot}41$ ev. However, the assumption that L is split equally in this way ignores the fact that on adding, say, a positive ion the ionic displacements in the vicinity of the kink or jog change from those appropriate to a negative kink or jog to those for a positive kink or jog : and conversely on addition of a negative ion. This means that the energy gained on adding a positive ion from infinity will be different from that gained on adding a negative ion, in the same way that the different displacements around positive and negative vacancies make g_+' and g_-' different. Indeed it is possible that this difference is comparable with or even equal to the difference between g_+' and g_-'. In the extreme case g_+^0 and g_-^0 will each be equal to one half the energy of formation of a Schottky pair. Lehovec's assumption corresponds to the opposite extreme where the displacements round a kink or jog are primarily determined by other factors and are not sensitive to the sign of the kink or jog. Evidently without detailed calculations all we can say is that probably g_+^0 lies between 0·6 and 1·0 ev and that the values for surface and dislocation need not be identical.

§ 5. Discussion

As it stands the theory is unlikely to give good numerical results even with the correct values of g_+ and g_-. The replacement of sinh p by p in (4) is not justified unless

$$|ev_\infty/kT| < 1 \qquad \qquad (18)$$

in the case of the plane solution. A similar or more severe restriction will apply to the cylindrical solution. With any reasonable values of the constants (18) is only obeyed within a few tens of degrees on either side of an isoelectric point. Even if the mathematical approximation were adequate outside these regions the high defect concentration which it predicts near a dislocation would require the physical assumptions to be refined. Moreover, in just those regions where the theory is at its best (near an isoelectric point) other hardening mechanisms, masked elsewhere, will prevent the yield stress falling to zero. However, we may hope that the experimental maxima and minima in the yield stress will coincide with the maxima and minima of $|ev_\infty/kT|$. In the present paper we therefore limit ourselves to a tentative comparison of the critical temperatures in figs. 1 and 3.

The points in fig. 1 show the yield stresses of a series of rock-salt specimens measured at various temperatures. The experimental details will be given elsewhere. Briefly, all specimens were cleaved from one large melt-grown single crystal, annealed for half an hour at 650°C, furnace-cooled to room temperature, and finally re-heated and held at the testing temperature for fifteen minutes before measuring the stress–strain curve. The divalent impurity content of the material is unknown, but it is estimated to be one part in 10^6. For brevity we give the labels $\mathbf{M_2}$, $\mathbf{M_1}$ **to the maximum and minimum in fig. 1.**

If g_+ is less than g_- than ln α' must be above ln α at high temperatures. The transition from intrinsic to impurity controlled conduction must then be associated with a maximum in $|ev_\infty/kT|$ (compare the dotted curves in fig. 3 (*c*)). This could perhaps be identified with M_2 if allowance is made for experimental scatter and the uncertainty in the impurity content. Identification of the minimum M_1 with any of the possibilities of fig. 3 enables one to write

$$g_+ = kT_{M_1} \ln (1/c),$$

c being the free impurity concentration at M_1. We then find that g_+ is less than 0·6 ev unless (locating M_1 at 320°K) c is less than about 10^{-9}. Even if some of the impurity is precipitated or associated with vacancies such a value would seem unreasonably low. (Compare Haven's (1955) measurements of the solubilities of Ca, Cd and Mn in NaCl.) This line of interpretation therefore leads us to a value of g_+ considerably less than g_-, and which is probably about 0·4 ev. It is interesting to note the agreement with Bassani and Thomson's (1956) value for the binding energy between a dislocation and a positive ion vacancy.

An interpretation based on the alternative starting point that g_+ is greater than g_- cannot be sustained. For we should then have to suppose that the maximum M_2 was associated with the onset of precipitation and that the minimum M_1 represented a crossing of the $\ln \alpha$ and $\ln \alpha'$ lines. For this to occur the heat of solution of the impurities would have to be greater than g_+, i.e. greater than 1 ev, which would be quite inconsistent with the conductivity results.

Thus we may say that comparison of our theory with experimental results on NaCl suggests that g_+ is considerably smaller than g_-, perhaps only a quarter of it. It suggests that the maximum M_2 is associated with the change from impurity-controlled conduction to intrinsic conduction, and that the minimum M_1 is associated with the onset of precipitation of the principal impurities. It would obviously be very interesting to have experimental results for crystals containing known amounts of deliberately added impurities. We should expect a reduction of strength at high temperatures (associated with the increased $\ln \alpha$) with the removal of the maximum to higher temperatures (associated with a higher temperature for the ' knee ' in the $\ln \alpha$ curve) and possibly the appearance of a new minimum at intermediate temperatures.

It is possible, however, that the fall in yield stress above M_2 has another explanation, namely that the charge cloud surrounding a dislocation is mobile enough to move with it at high temperatures. Electrical phenomena on deforming rock salt at various temperatures gives some support to this idea.

If a dislocation moves away from its charge cloud an electric polarization is produced, proportional to the product of charge and distance moved. The movement of charged dislocations should thus give rise to a potential between opposite faces of a crystal, or to a flow of charge between electrodes connected by a low-resistance circuit (Pratt 1957). Plastic deformation is proportional to the distance moved by the dislocations, and so we might at first sight expect proportionality of voltage (or charge) and plastic deformation. However, though the dislocations are all of the same sign electrically, they can have opposite signs mechanically, in the sense that for a dislocation with a given Burgers vector we can always find another with an equal and opposite Burgers vector. Uniform deformation will involve the movement of nearly equal numbers of dislocations with opposite Burgers vectors in opposite directions, and there will be no large net electrical polarization. In inhomogeneous deformation, on the other hand, a net flow of dislocations of one mechanical sign in the same direction can occur. Marked electrical effects of this type were indeed observed by Fischbach and Nowick (1955) under such conditions.

Caffyn and Goodfellow (1955) observed smaller electrical effects in homogeneously deformed rock salt, and they found, moreover, that the effects decreased with temperature, disappearing altogether above 600°K. This suggests that at high temperatures the cloud follows the dislocation, so that no separation of charge is possible. In a temperature range where defects impeding a dislocation can only just keep up with its motion we may expect to observe a serrated stress–strain curve (compare Cottrell 1953). It is interesting to note that Classen–Nekludowa (1929) observed jerky flow in rock salt at 500°K and above.

We conclude by mentioning some phenomena which might be associated with charged dislocations. Though we have been considering Schottky defects there should be similar effects when Frenkel defects predominate.

By plastic bending it is possible to introduce an excess of dislocations of one mechanical sign into a crystal. A crystal treated in this way should give large charge displacements or voltages when deformed homogeneously. In such a crystal it would also in principle be possible to observe the converse effect, plastic deformation by the 'electrophoretic' movement of dislocations in an applied electric field.

The movement of a dislocation under stress modifies the elastic constants of a material and, in an alternating stress field, there is a contribution to the mechanical damping. A charged dislocation should make analogous contributions to the dielectric constant and dielectric loss when it moves under the influence of an applied electric field. However, the restoring force which binds a dislocation to the centre of its charge cloud is proportional to $r \ln (1/\kappa r)$ rather than to r when it is given a small displacement r (compare fig. 2). Thus for a charged dislocation these effects, if observable, should be markedly non-linear.

A dislocation is repelled by the charge on a crystal surface or internal mosaic boundary made up of dislocations with spacing less than κ^{-1}. According to (6) and (10) the repulsive force near the wall is $\pi/0{\cdot}80$ or about four times the force necessary to tear a dislocation from its charge cloud. Thus when a dislocation has been pulled away from its charge cloud it may be held up by internal boundaries or the surface of the crystal until the stress is raised somewhat. This only applies when the dislocations in the boundary are closely spaced in comparison with κ^{-1}. If κ^{-1} is small compared with the spacing each dislocation is surrounded by an individual ion cloud which completely screens its charge and electrostatically the boundary offers almost no hindrance to the passage of a dislocation through it. In a crystal with divalent impurities the screening length κ^{-1} drops sharply as we pass from the impurity to the intrinsic conductivity range. According to this model a mosaic boundary behaves like a Venetian blind, closed to dislocations at low temperatures, but open at high temperatures.

Again, two charged dislocations screened by their charge clouds exert a short-range repulsion on one another. If the dislocations are, say, a distance $r=\frac{1}{2}\kappa^{-1}$ apart the electrostatic force between them is 80% of its unscreened value $2\sigma^2/\epsilon r$. Their elastic fields give rise to a force of the order of $\mu b^2/2\pi r$, where μ is a suitable elastic constant and b is the Burgers vector. These two forces are approximately equal if σ is of the order of one electronic charge per atom plane. Consider, for example, a regular array of edge dislocations forming a vertical tilt boundary. If one dislocation, together with its ion cloud, is displaced out of the plane of the boundary the elastic fields of its neighbours tend to pull it back, but their electric fields tend to repel it further. In some circumstances an energetically reasonable compromise might be for successive dislocations to lie alternately to left and right of the mean plane of the boundary. Such an arrangement has been observed in ionic crystals (Mitchell 1956).

Lehovec (1953) has shown that in a pure NaCl crystal there should be an extra conductivity proportional to the surface area, since in the ion cloud near the surface the concentration of cation vacancies (which carry most of the electrolytic current) is greater than it is in the body of the crystal. His analysis is easily extended to the case where there are mobile divalent impurities. Evidently the concentration of positive-ion vacancies near a surface exceeds or falls short of its bulk value according as v_∞ is negative or positive. Thus a surface increases the conductivity for temperatures where $\ln \alpha'$ is above $\ln \alpha$ in fig. 3, but decreases it where $\ln \alpha$ is above $\ln \alpha'$. According to the argument leading to (7) and (9), one square centimetre of an internal boundary made up of dislocations spaced more closely than κ^{-1} should affect the conductivity in the same way as two square centimetres of external surface, while an isolated dislocation should be equivalent to a strip of external surface of width $2\kappa^{-1}$.

A number of crystallographic points need further discussion. Charge cannot form on pure screw dislocations. It may be energetically advantageous for a screw dislocation to turn towards edge orientation so that it may build up a charge with its balancing cloud. Again, if a charged segment of dislocation forming a Frank–Read source of length l expands into a large loop its original charge will not be spread uniformly over the loop. A simple argument suggests that only a segment of length l will be charged, and that the line joining its centre to the centre of the unexpended Frank–Read source will be parallel to the Burgers vector. Subsequent dislocations from this source will be initially uncharged.

References

Bassani, F., and Fumi, F. G., 1954, *Nuovo Cim.*, **11**, 274.
Bassani, F., and Thomson, R., 1956, *Phys. Rev.*, **102**, 1264.
Brauer, P., 1952, *Z. Naturforsch.*, **7a**, 372.
Caffyn, J. E., and Goodfellow, T. L., 1955, *Nature, Lond.*, **176**, 878.
Classen-Nekludowa, M., 1929, *Z. Phys.*, **55**, 555.

COTTRELL, A. H., 1953, *Phil. Mag.*, **44,** 829.
ETZEL, H. W., and MAURER, R. J., 1950, *J. chem. Phys.*, **18,** 1003.
FISCHBACH, D. B., and NOWICK, A. S., 1955, *Phys. Rev.*, **98,** 1543.
FRENKEL, J., 1946, *Kinetic Theory of Liquids* (Oxford : Clarendon Press), p. 37.
GRIMLEY, T., 1950, *Proc. roy. Soc.* A, **201,** 40.
GRIMLEY, T., and MOTT, N. F., 1947, *Disc. Faraday Soc.*, **1,** 3.
HAVEN, Y., 1955, *Rep. Conf. on Defects in Crystalline Solids* (London : The Physical Society), p. 261.
LEHOVEC, K., 1953, *J. chem. Phys.*, **21,** 1123.
LIDIARD, A. B., 1954, *Phys. Rev.*, **94,** 29.
LIDIARD, A. B., 1957, *Handbuch der Physik,* XX (Berlin : Springer-Verlag), p. 245.
MITCHELL, J. E., 1956, *Lake Placid Conference*, to appear.
MOTT, N. F., and LITTLETON, M. J., 1938, *Trans. Faraday Soc.*, **34,** 485.
PRATT, P. L., 1957, *Inst. Metals Monograph and Report Series*, **23,** 99.
VERWEY, E. J. W., and OVERBEEK, J. TH. Q., 1948, *The Theory of Lyophobic Colloids* (Amsterdam : Elsevier).
WATSON, G. N., 1945, *Bessel Functions* (Cambridge : University Press), p. 77.

The Twist in a Crystal Whisker Containing a Dislocation†

By J. D. Eshelby
Department of Metallurgy, University of Birmingham

[Received February 10, 1958]

Abstract

The twist due to a screw dislocation parallel to the axis of an isotropic cylinder of arbitrary cross section can be found from the solution of the ordinary torsion problem for the same cylinder. Some particular cases are worked out. The results are also valid for certain kinds of anisotropy.

§ 1. Introduction

It has been pointed out (Mann 1949, Eshelby 1953) that when a screw dislocation is introduced along the axis of a circular cylinder the cylinder suffers a uniform twist equal (in radians per unit length) to the Burgers vector of the dislocation divided by the area of the cross section of the cylinder. Recently twists have been observed and accurately measured in crystal ' whiskers ' whose cross sections are not circles but, for example, regular hexagons or rectangles (Webb *et al.* 1957, Webb and Forgeng 1958, Dragsdorf and Webb 1958).

In § 2 we give a general method for finding the twist due to a screw dislocation in a cylinder of general cross section. The dislocation line is parallel to the cylinder axis, but its position in the cross section is arbitrary. The problem reduces to the solution of the ordinary torsion problem for the cylinder. Some particular cases are discussed in § 3. These results refer to an elastically isotropic material, but in § 4 it is shown that they can also be applied with little or no modification to anisotropic crystals in a number of particular situations which are likely to occur in actual whiskers.

§ 2. Solution of the Twist Problem

Let R denote the whisker cross section and C the curve bounding it, and let $\alpha(x, y)$ be the twist produced by a screw dislocation parallel to the whisker axis and passing through the point (x, y). In order to calculate α we give the whisker an arbitrary additional twist α_1 by applying couples to its ends.

To obtain the stresses associated with α_1 we have to find a function ψ satisfying

$$\nabla^2\psi = 0 \text{ in } R, \qquad \psi = \tfrac{1}{2}(x^2 + y^2) \text{ on } C. \quad . \quad . \quad . \quad . \quad (1)$$

† Communicated by the Author.

(See any book on the theory of elasticity: our notation agrees with that of Sokolnikoff 1946.) In terms of the auxiliary function

$$\Psi = \psi - \tfrac{1}{2}(x^2 + y^2) \quad . \quad . \quad . \quad . \quad . \quad . \quad . \quad . \quad . \quad (2)$$

the non-vanishing stresses are

$$\tau_{xz} = \mu\alpha_1 \,\partial\Psi/\partial y, \qquad \tau_{yz} = -\mu\alpha_1 \,\partial\Psi/\partial x \quad . \quad . \quad . \quad . \quad (3)$$

where μ is the shear modulus.

If the dislocation has a Burgers vector b these stresses exert a force

$$F_x = b\tau_{yz} = -\mu b\alpha_1 \,\partial\Psi/\partial x$$
$$F_y = -b\tau_{xz} = -\mu b\alpha_1 \,\partial\Psi/\partial y$$

on unit length of it. Thus Ψ is a potential whose gradient determines the force on the dislocation. (Compare the analogous role of the flexure function in a bent whisker (Eshelby 1953).)

The work done by the applied couples (per unit length of whisker) when the dislocation at (x, y) moves right out of the whisker along any path s is

$$\int (F_x\,dx + F_y\,dy) = -b\alpha_1\mu \int \frac{\partial\Psi}{\partial s}\,ds = -b\alpha_1\mu\,\Psi(x, y)$$

since $\Psi = 0$ on C. But this work may be calculated directly: if D is the torsional rigidity of the whisker it is simply $-\alpha_1\alpha(x, y)D$, the work done (per unit length of whisker) by the couple $\alpha_1 D$ as the twist due to the dislocation falls from its original value $\alpha(x, y)$ to the value zero when the dislocation reaches C. Thus $b\alpha_1\Psi = \alpha_1\alpha D$ and we have our basic result

$$\alpha(x, y) = \Psi(x, y)\mu b/D. \quad . \quad . \quad . \quad . \quad . \quad . \quad . \quad . \quad (4)$$

Since

$$D = 2\mu \int_R \Psi\,dx\,dy$$

we have

$$\alpha(x, y) = \tfrac{1}{2}b\Psi(x, y) \Big/ \int_R \Psi\,dx\,dy. \quad . \quad . \quad . \quad . \quad . \quad . \quad (5)$$

The integral of α over R is $\tfrac{1}{2}b$. This implies that a uniform distribution of small dislocations of total strength b per unit area spread over the cross section produces a total twist $\tfrac{1}{2}b$, in agreement with a result in the continuum theory of dislocations (see, for example, Nye 1952, Bilby *et al.* 1958).

This physical argument (which is equivalent to an application of Colonnetti's theorem), is, perhaps, formally unsatisfactory in that we have in effect subtracted the infinite self-energy of the dislocation from the two sides of the energy balance $-\alpha_1\alpha D = -b\alpha_1\Psi$, and so we sketch an analytical proof which also suggests a way to determine α when the Ψ-function is not available. It is adapted from the method used in a previous paper (Eshelby 1953). We imagine that we have obtained a solution of the elastic equations representing a dislocation in a whisker whose side-surfaces are stress-free, and in which the displacement w is required to be everywhere parallel to the whisker axis. In this state of anti-plane strain the whisker is not twisted but the stresses on its end-surfaces have a certain moment M about the whisker axis. If the necessary end-couples are not supplied the whisker will evidently have a twist $-M/D$.

Suppose that the dislocation is at $\mathbf{r}' = (x', y')$. Then the displacement is harmonic, behaves like $(b/2\pi)\tan^{-1}[(x-x')/(y-y')]$ near $\mathbf{r}'$, and, since the sides of the whisker are stress-free, its normal derivative $\partial w/\partial n$ vanishes on C (compare (6) below). It is most convenient to use the conjugate function χ related to w by

$$\frac{\partial w}{\partial x} = \frac{\partial \chi}{\partial y} = \tau_{xz}/\mu, \qquad \frac{\partial w}{\partial y} = -\frac{\partial \chi}{\partial x} = \tau_{yz}/\mu. \quad . \quad . \quad . \quad . \quad (6)$$

If s denotes arc-length along C, (6) shows that $\partial\chi/\partial s = -\partial w/\partial n$, and so on C χ has a constant value which we may take to be zero. Further, χ is harmonic and behaves like $(b/2\pi)\ln|\mathbf{r}-\mathbf{r}'|$ near $\mathbf{r}'$. More compactly, χ is defined by the requirements:

$$\nabla^2\chi = b\delta(\mathbf{r}-\mathbf{r}') \quad \text{in } R, \quad \chi = 0 \quad \text{on } C. \quad . \quad . \quad . \quad . \quad (7)$$

This implies that $-2\pi\chi/b$ is the Green's function which solves Dirichlet's problem for the contour C. That is to say, if $f(\mathbf{r})$ is some function defined only for values of $\mathbf{r}$ on C then the expression

$$F(\mathbf{r}') = \frac{1}{b}\int_C f(\mathbf{r})\frac{\partial}{\partial n}\chi(\mathbf{r},\mathbf{r}')\,ds(\mathbf{r}) \quad . \quad . \quad . \quad . \quad . \quad (8)$$

regarded as a function of $\mathbf{r}'$ is harmonic and takes the value $f(\mathbf{r}')$ on C.

The end couple required to maintain anti-plane strain is given by

$$M = \int_R (x\tau_{yz} - y\tau_{xz})\,dx\,dy = \mu\int_R \left(x\frac{\partial w}{\partial y} - y\frac{\partial w}{\partial x}\right)dx\,dy \quad . \quad . \quad (9)$$

or, in terms of χ,

$$M = -\tfrac{1}{2}\mu\int_R (\operatorname{grad} r^2)\,.\,(\operatorname{grad}\chi)\,dx\,dy$$

$$= \tfrac{1}{2}\mu\int_R r^2\nabla^2\chi\,dx\,dy - \tfrac{1}{2}\mu\int_C r^2\frac{\partial\chi}{\partial n}\,ds \quad . \quad . \quad . \quad . \quad (10)$$

by Green's theorem. According to (7) the first term is $\tfrac{1}{2}\mu b r'^2$. Equation (8) implies that the second term is harmonic in R and reduces to $-\tfrac{1}{2}\mu b r'^2$ on C; it is therefore identical with $-\mu b\psi(x', y')$(eqn. (1)). Thus, dropping the primes, $M(x, y)$ is identical with $-\mu b\Psi(x, y)$. To find the twist when the end couples are removed we divide $-M$ by the torsional rigidity D and so recover (4).

The sign convention we have used for the Burgers vector gives positive twist for positive b. The actual relation between Burgers vector and direction of twist is most simply stated thus. Suppose that there is a set of crystal planes perpendicular to the axis of a cylindrical whisker. When a screw dislocation is introduced they join up to form a spiral staircase. A creature ascending this staircase would follow a helical path with a certain sense. The helix formed by a generator of the undislocated cylinder has the opposite sense, and, naturally, a much coarser pitch.

When Ψ is not easily available it may be practicable to evaluate M by mapping the whisker cross section conformally on to the unit circle. Suppose that the dislocation is at the origin and that the relation

$$z \equiv x + iy = f(\zeta) \quad . \quad . \quad . \quad . \quad . \quad . \quad . \quad . \quad (11)$$

maps the whisker contour C on to the circumference γ of the unit circle $|\zeta| \leqslant 1$ in the complex ζ-plane and in such a way that $z=0$ maps into $\zeta=0$. (The complex variable (11) is, of course, not the same as the z which appears as a suffix in (3).) We shall use a suffixed ζ to distinguish quantities taken in the ζ-plane from those in the z-plane. Let us associate with each point ζ the value of χ at the corresponding point in the z-plane. We obtain a function $\chi_\zeta(\zeta) = \chi[z(\zeta)]$ which is easily seen to be the χ-function for a dislocation at the centre of the unit circle γ, that is,

$$\chi_\zeta(\zeta) = (b/2\pi)\ln|\zeta|.$$

It follows from the theory of conformal mapping that

$$\frac{\partial \chi}{\partial n} ds = \frac{\partial \chi_\zeta}{\partial n_\zeta} ds_\zeta$$

if ds, ds_ζ are corresponding elements of C and γ. Since $\partial\chi_\zeta/\partial n_\zeta = \mu b/2\pi$ eqn. (10) with $\boldsymbol{r}'=0$ may be written as

$$|M| = \frac{\mu b}{4\pi}\int_\gamma |z|^2 d\theta.$$

If the transformation (11) can be expressed in the form

$$f(\zeta) = \sum_{n=1}^{\infty} a_n \zeta^n$$

then

$$|M| = \frac{\mu b}{4\pi}\int_\gamma \sum_m \sum_n a_m a_n^* \exp\{i(m-n)\theta\}\, d\theta$$

since $\zeta = \exp i\theta$ on γ. Thus we have the simple result

$$|M| = \tfrac{1}{2}\mu b \sum_1^\infty |a_n|^2 \quad . \quad . \quad . \quad . \quad . \quad . \quad . \quad . \quad (12)$$

and $\alpha = |M|/D$.

The point $z=0$ can be chosen at will within C, and so our assumption that $z=f(\zeta)$ maps $z=0$ into $\zeta=0$ does not in principle limit the position of the dislocation in the whisker cross section. However, there will usually be one position of the origin for which $f(\zeta)$ takes on a manageable form, and this may not coincide with the position of the dislocation. We should then be faced with the general problem of determining ψ or Ψ at an arbitrary point within C using, for example, the analysis of Morris (1940) (see Sokolnikoff 1946) of which the above is a simple special case.

2 H 2

§ 3. Some Examples

It is convenient to express the twist in the form

$$\text{twist } \alpha = \kappa \times \frac{\text{Burgers vector } b}{\text{area of whisker cross section } A}.$$

We have, for example, $\kappa = 1$ for any ellipse, $\kappa = 10/9$ for an equilateral triangle, $\kappa = 1{\cdot}048$ for a square, $\kappa = 1{\cdot}015$ for a regular hexagon. In each case the dislocation is supposed to be at the centre of the cross section. The Ψ-functions for ellipse and equilateral triangle are simple expressions, (Sokolnikoff 1946, § 36, § 37) and for these and some other cross sections there is no difficulty in finding κ for a dislocation at any point in the cross section. The value for the square follows from the discussion of the rectangle below. In the case of the hexagon κ was calculated from (12) using the transformation

$$z(\zeta) = \int_0^\zeta \frac{dt}{(1-t^6)^{1/3}} = \zeta + \sum_{n=1}^{\infty} A_n \zeta^{6n+1}$$

with

$$A_n = \frac{2.8.14\ldots\{2+6(n-1)\}}{n!\,6^n(6n+1)}$$

which transforms a regular hexagon of side $a = z(1) = 2^{-5/2}3^{-1/2}\pi^{-1}[\Gamma(\tfrac{1}{3})]^2$ into the unit circle (Kober 1957). For the torsional rigidity the value $D = 1{\cdot}0359\,\mu a^4$ quoted by Pólya and Szegő (1951) was used.

In the case of a rectangle there are rapidly-converging series for $\Psi(x, y)$ and D. (see, for example, Sokolnikoff 1946, § 38). The following values refer to a dislocation at the centre of a rectangle of sides l and $h \leqslant l$.

l/h	1	1·5	2	3	5	10	∞
κ	1·048	1·026	0·996	0·948	0·858	0·800	$\frac{3}{4}$

If we put

$$\Psi(0,0) = \tfrac{1}{4} - (8/\pi^3)\,\mathrm{sech}\,(\pi l/2h),$$

$$D/\mu l h^3 = \tfrac{1}{3} - (64/\pi^5)\{\tanh(\pi l/2h) + 3^{-5}\tanh(3\pi l/2h)\}$$

the error in

$$\kappa = \frac{\Psi(0,0)/h^2}{D/\mu l h^3}$$

will not exceed $\frac{1}{5}$%. If $l \geqslant 3h$ we have

$$\kappa = \tfrac{3}{4}/(1 - 0{\cdot}630h/l)$$

to better than $\frac{1}{5}$%.

The twist due to a dislocation not at the centre of the cross section of the rectangle is easily found from the series for $\Psi(x, y)$. As an example, suppose that in a rectangle of sides 1 unit by 2 units the dislocation is displaced from the centre to a point equidistant from three sides. Then κ is reduced from 0·996 to 0·850, or by about 12%.

For any whisker cross section C a well known analogy enables us to visualize roughly how the twist varies with the position of the dislocation. A hole having C for boundary is cut in a plate and a membrane is stretched across it. If a uniform pressure is applied to one side the deflection of the membrane at (x, y) is proportional to $\Psi(x, y)$ and hence to α or κ. There is also a membrane analogy for the function χ of § 2. Let the uniform pressure be removed and instead let the membrane be deflected by a single point-force applied at a point corresponding to the position of the dislocation. Then the deflection at (x, y) is proportional to $\chi(x, y)$. The result following eqn. (5) says, in effect, that deflection by a multitude of point-forces and deflection by a pressure mean the same thing.

The expression for the twist needs modification if the dislocation has a hollow core. Suppose that this takes the form of a circular cylinder of radius r_0 small compared with the dimensions of the cross section. In the state of anti-plane strain described by χ (§ 2), there is no stress across a curve $\chi = \text{const.}$ Around the dislocation these curves will be very nearly circles, and so the elastic field will not be upset by the removal of the core, but $|M|$ will be reduced by $\frac{1}{2}\mu b r_0^2$. Thus (4) must be replaced by

$$\alpha(x, y) = \mu b\{\Psi(x, y) - \tfrac{1}{2}r_0^2\}/D.$$

Strictly, D as well as $|M|$ is decreased by the presence of the hole, but it is not so easy to calculate the correction. If the dislocation happens to be at a centre of high symmetry of the boundary curve C the curves $\Psi = \text{const.}$ around the dislocation will be nearly circular, and D is reduced by $\frac{1}{4}\pi\mu r_0^4$. Whether it has this simple form or not the correction to D will be negligible compared with the correction to $|M|$.

§ 4. Anisotropic Whiskers

If the whisker axis coincides with a four-fold or six-fold axis of the crystal the equations governing anti-plane strain and torsion are identical with those for an isotropic medium with $\mu = c_{55}$ (Voigt 1929, Eshelby *et al.* 1953) and our previous expressions for the twist can be used without modification.

Again, suppose that the whisker axis coincides with the z-axis of the crystal and that the elastic coefficients $c_{14}, c_{15}, c_{16}, c_{24}, c_{25}, c_{26}$ and c_{45} vanish. Then it is possible to have a state of anti-plane strain with the displacement parallel to the whisker axis. The governing equation is

$$\frac{\partial \tau_{xz}}{dx} + \frac{\partial \tau_{yz}}{\partial y} = 0 \quad \text{with } \tau_{xz} = c_{55}\frac{\partial w}{\partial x}, \quad \tau_{yz} = c_{44}\frac{\partial w}{\partial y}.$$

We need to find a solution representing a screw dislocation and from it we have to calculate the end-couple required to maintain anti-plane strain. Suppose that the dislocation is at (x', y') and that the boundary of the whisker cross section is the curve

$$C : x = F(y);$$

for example, $x = \pm(a^2 - y^2)^{1/2}$ when C is a circle.

Introduce the variables

$$\xi = (c_{44}/c_{55})^{1/2}x, \quad \eta = y. \quad \dots\dots\dots (13)$$

When C is described in the (x, y) plane the curve

$$\Gamma : \xi = (c_{44}/c_{55})^{1/2} F(\eta)$$

is described in the (ξ, η) plane. It is easy to show that the substitution (13) reduces the anisotropic problem to the corresponding isotropic problem in the sense that we have to find a function w to satisfy $\partial^2 w/\partial\xi^2 + \partial^2 w/\partial\eta^2 = 0$ with the boundary conditions $w \sim (b/2\pi)\tan^{-1}[(\xi-\xi')/(\eta-\eta')]$ near (ξ', η') and $\partial w/\partial\nu = 0$ on Γ, regarding ξ, η as ordinary Cartesian coordinates. (ν is the normal to Γ.) The integral (9) translates into

$$M = c_{55}\int\left(\eta\frac{\partial w}{\partial\xi} - \xi\frac{\partial w}{\partial\eta}\right) d\xi\, d\eta$$

taken over the interior of Γ. This is the same as M for a fictitious isotropic whisker bounded by Γ and having a shear modulus $\mu = c_{55}$. On the other hand, it can be shown (Sokolnikoff 1946, § 51) that the torsional rigidity of the anisotropic whisker is $(c_{55}/c_{44})^{1/2}$ times the torsional rigidity of the fictitious whisker. Thus the twist in the real whisker is equal to the twist in the fictitious whisker mutiplied by the factor $(c_{44}/c_{55})^{1/2}$. This factor can be eliminated from the expression for the coefficient κ of § 3 (the fictitious whisker has $(c_{44}/c_{55})^{1/2}$ times the area of the real whisker) and we are left with the following simple result:

> Let the whisker axis coincide with the z-axis of a crystal in which the elastic constants c_{14}, c_{15}, c_{16}, c_{24}, c_{25}, c_{26} and c_{45} vanish. Let the dislocation be at the point $x = X$, $y = Y$ within the whisker contour $x = F(y)$. Then the coefficient κ is the same as κ for a dislocation at $x = (c_{44}/c_{55})^{1/2}X$, $y = Y$ in an isotropic whisker bounded by the curve $x = (c_{44}/c_{55})^{1/2}F(y)$.

In addition to the more obvious cases this also applies when the whisker axis is in the basal plane of a hexagonal crystal. (We use ' hexagonal ' in the narrower sense, to denote a crystal which has only five independent elastic constants.) The direction of the axis in the basal plane is immaterial. What we have called c_{44} and c_{55} would in this case be the quantities normally tabulated as c_{44} and c_{66} respectively.

We may use our previous results for the isotropic case when the transformation (13) alters the shape of the whisker cross section but not its type. For example, an ellipse becomes an ellipse, and the rectangle $x = \pm\frac{1}{2}h$, $y = \pm\frac{1}{2}l$ becomes the rectangle $x = \pm\frac{1}{2}h\,(c_{44}/c_{55})^{1/2}$, $y = \pm\frac{1}{2}l$. On the other hand a rectangle whose sides are not parallel to the coordinate axes becomes a parallelogram, a figure we have not treated.

REFERENCES

BILBY, B. A., GARDNER, L. R. T., and SMITH, E., 1958, *Acta Met.*, **6,** 29.
DRAGSDORF, R. D., and WEBB, W. W., 1958 (to be published).
ESHELBY, J. D., 1953, *J. appl. Phys.*, **24,** 176.
ESHELBY, J. D., READ, W. T., and SHOCKLEY, W., 1953, *Acta Met.*, **1,** 252.
KOBER, H., 1957, *Dictionary of Conformal Representations* (New York : Dover Publications).
MANN, ELIZABETH H., 1949, *Proc. roy. Soc.* A, **199,** 376.
MORRIS, ROSA M., 1940, *Proc. London Math. Soc.*, **46,** 81.
NYE, J. F., 1952, *Acta Met.*, **1,** 153.
PÓLYA, G., and SZEGŐ, G., 1951, *Isoperimetric Inequalities in Mathematical Physics* (Princeton : University Press).
SOKOLNIKOFF, I. S., 1946, *Mathematical Theory of Elasticity* (New York : McGraw-Hill Book Company.)
VOIGT, W., 1929, *Lehrbuch der Kristallphysik* (Leipzig : Teubner).
WEBB, W. W., DRAGSDORF, R. D., and FORGENG, W. D., 1957, *Phys. Rev.*, **108,** 498.
WEBB, W. W., and FORGENG, W. D., 1958, *J. appl. Phys.*, **28,** 1449.

The Elastic Model of Lattice Defects

By J. D. Eshelby

With 1 Figure

Abstract

It is shown that between rigid misfitting spheres in an elastic continuum there is a repulsive interaction if the spheres are bonded to the medium. The effect of other boundary conditions is discussed, and a simplified method is given for calculating the interaction between various types of defect.

Recently Teltow[1]) has discussed the energy of interaction between a pair of impurity atoms in a crystal. The impurity atoms are represented by two rigid and incompressible spheres K_1, K_2 of radii ϱ_1, ϱ_2 forced into two spherical holes K_1^0, K_2^0 (each of radius ϱ_0) in an elastic matrix. Teltow finds that the total strain energy of the medium decreases as the spheres approach one another, that is, there is an attractive force between them. This is the reverse of the writer's conclusion[2]). (Compare also Crussard[3]).)

We shall consider the special case in which $\varrho_2 = \varrho_0$. K_2 then fits the hole K_2^0 exactly and so produces no strain in the medium. This greatly simplifies the discussion, and at the same time nothing essential is lost.

When K_2 and the hole K_2^0 are absent the energy density due to the strain set up by K_1 is a certain function ε of position, and the total elastic energy is the volume integral of ε over all the medium exterior to K_1^0. Teltow supposes that the presence of K_2 and K_2^0 does not affect the value of ε at any point not within K_2^0. Consequently when K_2 and K_2^0 are introduced the total elastic energy falls by an amount equal to the integral of ε taken over the space now occupied by K_2^0.

The question of the boundary conditions to be imposed at the junction of hole and sphere is obviously open to discussion. Teltow's condition, that the presence of the defect does not disturb a pre-existing stress field, is appropriate when the sphere is made of the same material as the matrix, but it does not seem very plausible when the sphere is rigid. Consider, for example, two elastic bodies A and B of the same form. B contains a large number of

---

1) J. Teltow, Ann. Physik **19**, 169 (1956).

2) J. D. Eshelby, Acta Met. **3**, 487 (1955). A factor r^{-6} must be inserted on the right-hand side of eqn. (4). There is also an inconsistency of sign which may be corrected as follows. Change the sign of the expression for ΔE on p. 488, col. 2, l. 5. The expressions for λ_{app}, μ_{app} are correct, for eqn. (3) refers to a change of elastic constants at constant external load, and a decrease of elastic energy at constant load implies an increase of apparent elastic constants.

3) C. Crussard, Acta Met. **4**, 555 (1956).

defects made by inserting rigid spheres into holes of the same size, while A is made of the same material as B but is free of defects. A and B are deformed by imposing identical surface displacements on them. The elastic energy in A is the volume integral of the energy density. On Teltow's assumption the elastic energy in B is given by the same integral with the interiors of the spheres omitted from the domain of integration. Thus for equal deformations B contains less elastic energy than A, and so its macroscopic elastic constants are less than A's. It seems unlikely that with any physically acceptable boundary conditions a material containing rigid inclusions could be, for example, more compressible than one which does not.

The simplest condition we can impose is to require that the surfaces of hole and sphere be firmly attached to each other. For brevity we shall refer to this boundary condition as bonding. We may define it operationally as follows: a spherical hole is excavated in a perfect infinite stress-free continuum; the sphere, coated with a suitable cement, is forced into the hole and the cement is allowed to set. Every point of the surface of the hole is opposite a definite point on the sphere, and this relation is undisturbed during subsequent deformation of the medium. Moreover, when the sphere is inserted every point on the surface of the hole undergoes a purely radial displacement. When we depart from this ideal case certain extra precautions must be taken. If the medium is in a state of stress we must cut not a spherical hole, but rather one which would be spherical if the stress were absent. Moreover, in this case, and also if the hole is made near some non-uniformity in the elastic constants (a free surface, another embedded rigid sphere) insertion of the sphere will not usually give to the surface of the hole a displacement which is purely radial; to ensure that it is purely radial we must apply suitable forces until the cement has set. Some such detailed prescription is necessary if we are to avoid inconsistencies. We may perhaps justify the bonding condition as follows. Our model of the perfect crystal is the perfect elastic continuum. If the impurity atom is replaced by a normal lattice atom the misfitting rigid sphere must clearly be replaced by a perfectly-fitting sphere with the same elastic constants as the matrix. Unless we bond sphere and hole together we shall not reproduce the elastic properties of the perfect continuum, since the surfaces of sphere and hole may slide over one another or separate when a stress is applied. Thus bonding is implicit in our model when the „impurity" atom is really a normal atom, and so it seems reasonable to adopt it universally. Again, if we are dealing with an impurity atom which produces no stress-field (perfectly fitting sphere) it is natural to define the defect simply as a region of the continuum whose elastic constants differ from those of the remainder. This approach, too, effectively implies bonding.

We shall now try to demonstrate that with this boundary condition K_1 and K_2 repel one another. It is clear that whatever interaction there may be will decrease as the distance between K_1 and K_2 increases. Also there is no stress-field associated with K_2 since we have assumed that $\varrho_2 = \varrho_0$. Thus we have to show that the energy of the elastic field produced by K_1 is greater when K_2 is present than when it is not.

Let us mark out spherical surfaces K_1^0, K_2^0 in the continuum, remove the interior of K_1^0 (but not the interior of K_2^0) and insert K_1. The elastic energy is $E_1 = 8\,\pi\,S\,\varrho_0\,(\varrho_1 - \varrho_0)^2$. K_2^0 is no longer a sphere. We pull the surface K_2^0

back to the form of a sphere of radius ϱ_0 with the help of a suitable layer of body-force spread over it. The work done by these forces, ω say, goes to increase the elastic energy of the medium. The interior of is now stress-free: we remove it, leaving the layer of body-force still applied to the surface of the cavity, and cement in the rigid sphere K_2. When the cement has set we may remove the forces. This does not alter the energy of the system since points on K_2^0 now cannot move. Thus we have shown that the energy of the medium is E_1 when K_2 is absent and $E_1 + \omega$ when it is present, and that ω is positive.

Again, the elastic energy of the medium is equal to the work required to force K_1 into K_1^0. It is perhaps almost obvious that this work will be greater when a rigid inclusion is present than when it is not. Hashin[4]) has shown that a material of shear modulus S and Poisson's ratio μ which contains a uniform distribution of small rigid spheres bonded into it behaves macroscopically like a material with shear modulus

$$S' = S\,(1 + \alpha\,\varphi) \tag{1}$$

where

$$\alpha = \frac{15\,(1-\mu)}{2\,(4-5\,\mu)} \tag{2}$$

and φ is the volume fraction occupied by the spheres. The energy required to insert K_1 into such a medium is $E_1' = 8\,\pi\,S'\,\varrho_0\,(\varrho_1 - \varrho_0)^2$. Since $\alpha > 0$, $E_1' > E_1$. Thus it is harder to insert K_1 into a medium if it contains many small embedded spheres, and we may reasonably suppose that the same will be true when there is one large embedded sphere.

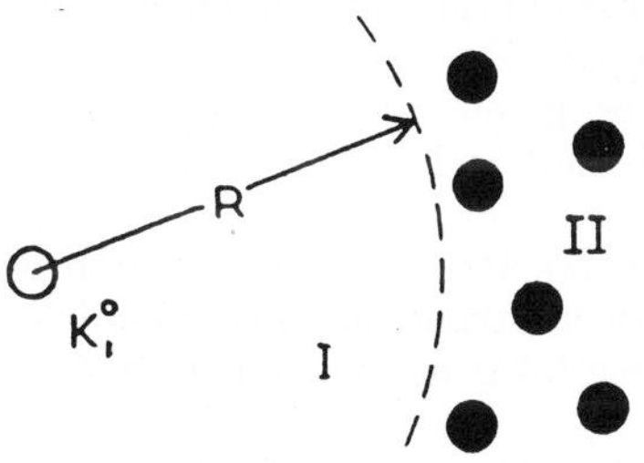

Fig. 1. Model for calculating interaction energy

In one of these two arguments we supposed that K_1 was inserted before K_2, in the other K_2 before K_1. Each shows that the interaction is repulsive. The second of them can be made quantitative in the following way.

Figure 1 shows the hole K_1^0 situated in an elastic medium divided into regions I, II by a spherical surface of radius R. Throughout the infinite region II defects are uniformly scattered at the rate of n per unit volume. Each defect is made by cutting a spherical hole of radius ϱ_0 and cementing into it a rigid incompressible sphere also of radius ϱ_0. Region I is free from defects. With the help of an internal pressure the hole K_1^0 is now inflated to a radius ϱ_1, and the rigid sphere K_1 is inserted.

The work W done by the internal pressure is stored as the elastic energy in I and II. To calculate it we treat II as a continuous medium with an effective shear modulus S'. This is reasonable if R is much greater than both ϱ_0 and the mean distance $n^{-\frac{1}{3}}$ between the defects in II. According to (1)

$$\frac{S'}{S} - 1 = \alpha \cdot V_2 \cdot n \tag{3}$$

where V_2 is the volume of K_2^0.

[4]) Z. Hashin, Bull. Res. Council of Israel 5 C, 46 (1955).

The elastic field may be found by fitting together two solutions to the problem of a thick spherical shell under internal and external pressure[5]). We find the value

$$p = \frac{4\,S}{\varrho_0}(\varrho_1 - \varrho_0)\left[1 + \left(\frac{S'}{S} - 1\right)\left\{\frac{3\,k + 4\,S}{3\,k + 4\,S'}\right\}\left(\frac{\varrho_0}{R}\right)^3\right]$$

for the pressure which must be developed in K_1^0 to increase its radius from ϱ_0 to ϱ_1. (k is the bulk modulus of the matrix). We may assume that $S' - S$ is small, and replace the factor { } by unity. The work done in expanding K_1^0 is half the product of p and the change in volume. Thus

$$W = 8\,\pi\,S\,\varrho_0\,(\varrho_1 - \varrho_0)^2\left[1 + \left(\frac{S'}{S} - 1\right)\left(\frac{\varrho_0}{R}\right)^3\right] \tag{4}$$

is the elastic energy of the material when K_1 has been forced into K_1^0. Let us decrease R by ΔR. W increases by $\Delta W = -\Delta R \cdot \partial W/\partial R$. This increase results from introducing $4\,\pi\,R^2\,\Delta R\,n$ new rigid spheres at a distance R from K_1^0. If we had introduced only one sphere at this distance, the increase would have been $\omega = \Delta W/4\,\pi\,R^2\,\Delta R\,n$. Using (3) we find

$$\omega = 6\,\alpha\,V_2\,S\left(\frac{\varrho_1 - \varrho_0}{\varrho_0}\right)^2\left(\frac{\varrho_0}{R}\right)^6. \tag{5}$$

Evidently ω is the elastic energy of the medium when K_2 is distant R from K_1, minus the elastic energy of the medium when K_2 and its hole K_2^0 are absent. As μ increases from 0 to $^1/_2$, α changes almost linearly from 1.875 to 2.5. Hence ω is always positive, and the defects repel one another with a force proportional to the inverse seventh power of their separation.

A slightly more vivid derivation could be given by assuming that the embedded spheres all lie within a spherical shell and noting how the work required to insert K_1 in K_1^0 varies with the radius of the shell. The result is the same. Hashins's calculation of S' takes no account of interaction between the embedded spheres, so our method of inferring the effect of one from the effect of a large number is valid. Of course (5) is no langer true if R is of the order of ϱ_0.

Another boundary condition which lends itself to calculation is what one may call slipping. The surfaces of sphere and matrix are free to slide over one another, but cannot separate. In terms of our model we may suppose the interface to be lubricated with a grease tenacicos enough to prevent the surfaces parting. If we limit curselves to the case of an incompressible medium we may make use of a hydrodynamical analogy.

The solution of an elastic problem in an incompressible medium gives the solution of a related problem in slow viscous flow if we interpret stress, strain and shear modulus as stress, rate of strain and coefficient of viscosity respectively. With this re-interpretation (1) is an expression for the effective viscosity S' of a suspension of rigid spheres in a liquid of viscosity S which adheres to the surface of each sphere. Indeed if we put $\mu = {}^1/_2$ in (2) we recover Einstein's[6]) value $\alpha = 2.5$.

If the liquid slips freely over the spheres instead of adhering to them the viscosity is given by (1) with $\alpha = 1$[7]) and, reversing the analogy, so also is

---

5) A. E. H. Love, Mathematical Theory of Elasticity, Cambridge 1927, p. 142.
6) A. Einstein, Ann. Physik [4] 19, 289 (1906); 34, 591 (1911).
7) R. Eisenschitz, Physik. Z. 34, 411 (1933).

the shear modulus of a medium containing slipping rigid embedded spheres. Consequently the interaction energy of K_1 with one freely slipping sphere is given by (5) with $\alpha = 1$.

Thus when $\mu = {}^1/_2$ the interaction energy decreases by a factor 2.5 when we change from bonding to slipping, but it does not reverse its sign. No doubt there is a similar relation for other values of μ. The object of the foregoing calculation is not to advocate the slipping condition, but rather to show that the interaction remains repulsive when the boundary conditions are drastically altered. Nevertheless, it is known that interstitial atoms in body-centred metals are extremely mobile and it might be that a misfitting slipping sphere would be the most appropriate elastic model for them.

We may also apply the argument based on Fig. 1 to discuss the interaction of K_1 with other kinds of defect. If the defects in II are empty spherical cavities the effective shear modulus of II is given[8]) by (1) with $\alpha = -15(1-\mu)/(7-5\mu)$ and the interaction energy is given by (5) with this value of α. The interaction is now attractive. We have here a rough model of the interaction between an impurity atom and a lattice vacancy, supposing that the vacancy itself gives rise to no stress-field. More generally if the defects in II are elastic spheres of shear modulus S_1 bonded into holes of the same size in the matrix the shear modulus of II is given[9]) by (1) with

$$\alpha = \left(\frac{2}{15}\,\frac{4-5\mu}{1-\mu} + \frac{S}{S_1 - S}\right)^{-1} \tag{2'}$$

and with this value of α, (5) gives the ineraction of K_1 with K_2 when K_2 is a perfectly fitting sphere with shear modulus S_1 and arbitrary compressibility. The interaction is attractive or repulsive according as S_1 is less than or greater than S. The interaction energy of this defect with a general stress field is[9]).

$$\omega = \frac{1}{2}\,V_2\left(\frac{\alpha'}{k}\,p^2 + \frac{\alpha}{2S}\,I\right) \tag{6}$$

where p_{ij} ist the stress at the position to be occupied by K_2 and K_2^0 but before they are introduced,

$$\alpha' = \left(\frac{1}{3}\,\frac{1+\mu}{1-\mu} + \frac{k}{k_1 - k}\right)^{-1}$$

$$p = \frac{1}{3}\,(p_{11} + p_{22} + p_{33})$$

$$3\,I = (p_{11} - p_{22})^2 + (p_{22} - p_{33})^2 + (p_{33} - p_{11})^2 + 6\,(p_{12}^2 + p_{23}^2 + p_{31}^2)$$

and α is given by (2'). k and k_1 are the bulk moduli of matrix and sphere[10]). If p_{ij} is produced by some source of internal stress (for example, K_1) (6) is simply the increase of elastic energy when the defect is introduced. When p_{ij} is wholly or partly produced by external forces (6) also includes the increase in the potential energy of the loading mechanism. In deriving (6) it is assumed

---

[8]) J. K. Mackenzie, Proc. Phys. Soc. (London) (B) **63**, 361 (1950).

[9]) J. D. Eshelby, Proc. Roy. Soc. London (A) **241**, 376 (1957).

[10]) The expression given previously (J. D. Eshelby, Philos. Trans. Roy. Soc. London (A) **877**, 87 (1951)) for the gradient of the term in p^2 is incorrect. The expressions for δ, $F\,(K = 0)$, $F\,(K = \infty)$ on page 104 should be multiplied by $4\,\mu/3\,K$ in the notation of that paper. The expression given by Friedel (Les Dislocations, Gauthier-Villars 1956, p. 237 equation (13.14)) for ω when $I = 0$ must be multiplied by ${}^1/_2$.

that p_{ij} varies little within K_2^0. For the interaction of K_1 and K_2 this is equivalent to the condition $R \gg \varrho_0$.

We have supposed throughout that one defect produces no stress field. In the general case we have[2], with an obvious notation

$$\omega = \omega\,(\varrho_2 = \varrho_0, \varrho_1 \neq \varrho_0) + \omega\,(\varrho_1 = \varrho_0, \varrho_2 \neq \varrho_0) \tag{7}$$

with an error of the same order as the error already introduced by the limitation $R \gg \varrho_0$. The first term in (7) depends on the stress field at large distances from K_1, and not on the boundary conditions at the interface between K_1 and K_1^0 or on whether K_1 is rigid or elastic. It must be calculated from (5) (or (6)) with the value of α appropriate to the character of K_2 (rigid bonded or slipping sphere, cavity, elastic sphere). The second term is calculated in the same way, reversing the roles of K_1 and K_2.

Birmingham (England), Department of Metallurgy, University of Birmingham.

Bei der Redaktion eingegangen am 6. Mai 1957.

STRESS INDUCED ORDERING AND STRAIN-AGEING IN LOW CARBON STEELS*

D. V. WILSON† and B. RUSSELL† *with an Appendix by* J. D. ESHELBY‡

Part of the increase in yield stress during the strain-ageing of a low-carbon steel develops too rapidly to be explained by long-range diffusion. The rate at which the initial rapid rise in yield stress develops and the dependence of its magnitude on the dissolved solute content are shown to be those expected from a contribution due to stress-induced ordering of the interstitial solute atoms in the stress fields of dislocations. A simple theoretical treatment gives values in reasonable agreement with experiment.

LA RÉORGANISATION ATOMIQUE DUE AUX TENSIONS ET LE VIEILLISSEMENT DE DÉFORMATION DANS LES ACIERS A FAIBLE TENEUR EN CARBONE

L'augmentation de la limite élastique pendant le vieillissement de déformation d'un acier à faible teneur en carbone se fait trop rapidement pour être expliquée par une diffusion à grande distance. Les auteurs montrent que la vitesse avec laquelle s'accroit, au début, la limite élastique et la dépendance de son amplitude avec la teneur en éléments dissous sont celles qu'on s'attendrait à trouver en tenant compte de la réorganisation atomique provoquée par des tensions des atomes interstitiels situés dans le champ de tension des dislocations. Un raisonnement théorique simple donne des valeurs en bon accord avec l'expérience.

SPANNUNGSINDUZIERTE ORDNUNG UND RECKALTERUNG BEI KOHLENSTOFFARMEN STÄHLEN

Ein Teil der Fliessspannungserhöhung bei der Reckalterung eines kohlenstoffarmen Stahls entwickelt sich zu rasch, um durch Diffusion über grössere Entfernung hinweg erklärbar zu sein. Die Geschwindigkeit, mit der sich die anfänglich rasche Erhöhung der Fliessspannung entwickelt und die Abhängigkeit ihrer Grösse vom Gehalt an gelöstem Legierungszusatz entsprechen den Erwartungen in Bezug auf einen Beitrag aufgrund spannungsinduzierter Ordnungseinstellung der auf Zwischengitterplätzen gelösten Atome in den Spannungsfeldern von Versetzungen. Eine einfache theoretische Behandlung ergibt Werte, die ausreichend mit dem Experiment übereinstimmen.

1. INTRODUCTION

Nabarro[1], in 1948, pointed out that the diffusion of carbon in iron can affect mechanical properties by three mechanisms. These are the stress-induced ordering of carbon atoms among the possible sets of interstitial sites (Snoek[2]), the segregation of carbon to form dislocation atmospheres (Cottrell[3]) and the precipitation of iron carbide particles.

Each of these kinds of solute redistribution can occur during the strain-ageing of steel. The Cottrell mechanism is of first importance in causing the return of the sharp yield point, while an increase in the steel's ability to work-harden and reduction in its ductility in the later stages of ageing are probably associated with precipitation of carbides or nitrides.[4]

The possible contribution to the rise in yield stress made by ordering of solute atoms in the stress fields of dislocations has generally been ignored. Recently, however, Schoeck[5] has discussed the dragging force on moving dislocations due to this effect. Since only atomic jumps between neighbouring lattice sites are involved, ordering will be more rapid by orders of magnitude than is segregation to form atmospheres. In the case of carbon and nitrogen in iron any contribution to strain-ageing made by this effect should be completed within a few seconds at room temperatures.

It is expected that the magnitude of the effects due to such ordering will be proportional to the amount of dissolved interstitial solute. The size of the yield point which develops as a result of atmosphere locking, on the other hand, can be shown to be essentially independent of the solute content provided this exceeds the small quantity, (generally a few thousandths of an atom per cent), required to complete atmosphere formation.[6]

A treatment due to Eshelby which is given in the appendix to this paper, provides a method for calculating the force required to move a dislocation when ordering has taken place. This suggests that the effect on yield stress should be appreciable in quenched steels.

In the experiments described here evidence of the effects of stress-induced ordering of the interstitial solutes, in prestrained low-carbon steel, has been sought by measuring the changes in yield stress during the early stages of strain-ageing.

* Received October 9, 1958.
† Department of Industrial Metallurgy, The University, Birmingham 15, England.
‡ Department of Physical Metallurgy, The University, Birmingham 15, England.

2. EXPERIMENTAL METHOD

A steel of the following composition in weight per cent was used in the majority of the experiments: C 0.039; N 0.0044; Mn 0.40; S 0.017; Si and P 0.008; Ni and Cu 0.05.

Specimens of 0.050×0.60 in. cross-section were prestrained through the initial yield, aged and then tested in a hard beam machine. The observation of small yield points was simplified because ageing could be carried out without removing the specimen from the machine for all the shorter times.

The amount of solute in solid solution was varied by heating specimens *in vacuo* at suitable temperatures in the range 200° to 700°C, for times sufficient to reach equilibrium, and then quenching in water. An extremely low solute content was obtained by heating in wet hydrogen at 800°C for 48 h.

3. RESULTS

Fig. 1 summarizes results given by a series of specimens all quenched from 600°C, prestrained 4 per cent, unloaded and then aged at 20°C for times up to 15 h before testing. Initially these specimens would have, in weight per cent, about 0.01 carbon and 0.0044 nitrogen in solution.

The increase in yield stress, ΔY, is taken as the difference between the upper yield stress, after strain-ageing for time t, and the flow stress observed at the end of prestraining (Fig. 1). The maximum rise in yield stress due to strain ageing, ΔY_{max}, was obtained only after very prolonged ageing ($>10^4$ min at 60°C). If we assume, for atmosphere formation, that ΔY is proportional to the amount of solute collected by the dislocations then, according to the Cottrell–Bilby equation,[7] ΔY should increase proportionately to $t^{2/3}$ during the early stages of ageing. In Fig. 1 the portion of the curve BCD shows this now familiar feature of strain ageing due to atmosphere formation. If ΔY were proportional to the amount of solute segregating throughout the ageing process the dislocation density required to give the rate of ageing indicated by the linear portion BC would be about 2×10^{11} lines per cm^2. This is reasonable (but probably high because the solute becomes less effective in raising the yield stress in the later stages of ageing). Fig. 1 shows there was an initial rise in ΔY which occurred much too rapidly to be accounted for by atmosphere formation.

In low-carbon steel specimens the extent of this initial rapid ageing was found to be strongly dependent on the quenching temperature but the effect was absent in specimens from which the carbon and nitrogen had been removed by annealing in wet hydrogen. The effect was insensitive to grain size in the range 50 to 1800 grains per mm^2 and appeared to be only weakly dependent on the amount of prestrain.

Fig. 2 shows the extent of the rapid ageing effect observed in specimens of differing dissolved solute contents, all prestrained 4 per cent and aged for 2 min at 20°C. The increase in yield stress was evidently close to being proportional to the amount of dissolved solute, but the increase was greater if ageing was carried out without removing the prestraining load. The total interstitial solute content of the hydrogen-treated specimens was probably less than 0.0001 weight per cent. These were quenched from 700°C but gave no evidence of an increase in yield stress if

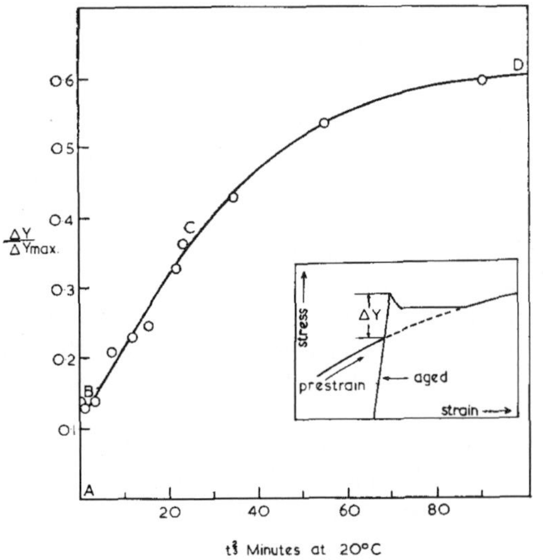

FIG. 1. The fractional return of the yield point as a function of $t^{2/3}$ for specimens quenched from 600°C and prestrained 4 per cent.

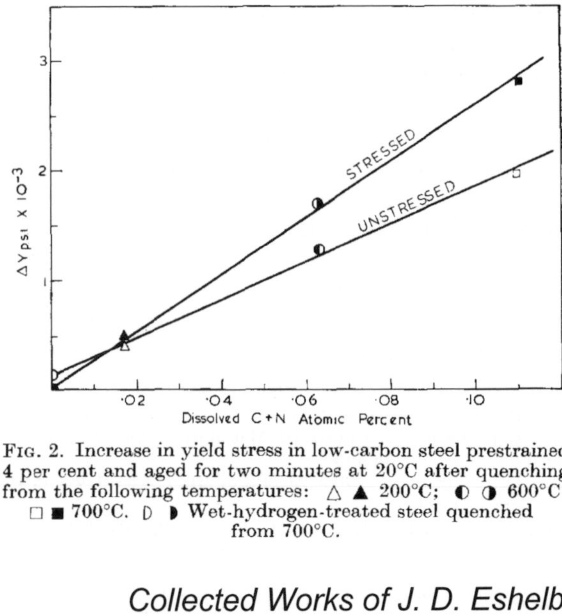

FIG. 2. Increase in yield stress in low-carbon steel prestrained 4 per cent and aged for two minutes at 20°C after quenching from the following temperatures: △ ▲ 200°C; ◐ ◑ 600°C; □ ■ 700°C. D ◗ Wet-hydrogen-treated steel quenched from 700°C.

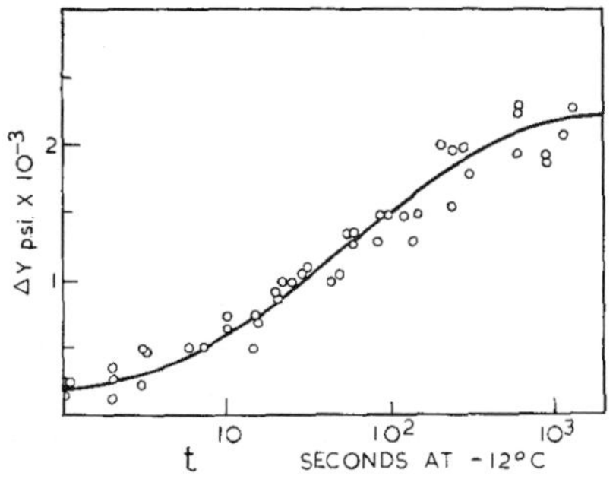

FIG. 3. Strain-ageing at −12°C in specimens quenched from 700°C and prestrained 4 per cent.

the load were maintained during ageing. Unloaded specimens showed a small effect which is probably not due to strain-ageing but may be related to dislocation rearrangement on unloading, as discussed by Haasen and Kelly.[(8)]

When ageing was carried out without removing the prestraining load it is likely that the apparent ageing rate was influenced by creep (amounting to about 0.08 per cent extension in 2 min). At 20°C the testing machine could not be unloaded sufficiently rapidly to obtain any estimate of the rate of rapid ageing in the unloaded condition, but experiments in which the specimen were unloaded rapidly to about 90 per cent of the prestrain load and then reloaded immediately, suggested that about two-thirds of the full rapid ageing effect developed within the first 2 to 4 sec at room temperature. Tests at reduced temperatures showed that about two-thirds of the effect developed in 2 min at −10°C but that there was negligible ageing in 2 min at −60°C. Fig. 3 shows results obtained with a series of specimens all quenched from 700°C and then tested and aged at −12°C. Ageing was in this case carried out with a reduced applied load which was generally between 80 and 90 per cent of the load at the end of prestrain. The results show that the initial rapid rise in yield stress was about 63 per cent complete in 100 sec at −12°C. This is in reasonable agreement with relaxation times observed in the case of the elastic after-effect due to the ordering of carbon in iron.[(9)]

Test pieces freshly quenched from temperatures above about 550°C showed irregular flow at room temperature during continuous straining at low strain rates. Fig. 4 illustrates stress–strain curves recorded at 20°C, 85°C and 130°C using specimens quenched from 700°C and strained at a rate of 10^{-4} sec^{-1}. The irregularities observed at room temperature may be described as repeated yielding rather than jerky flow, which is observed at high temperatures. At 85°C and higher temperatures precipitation of an appreciable

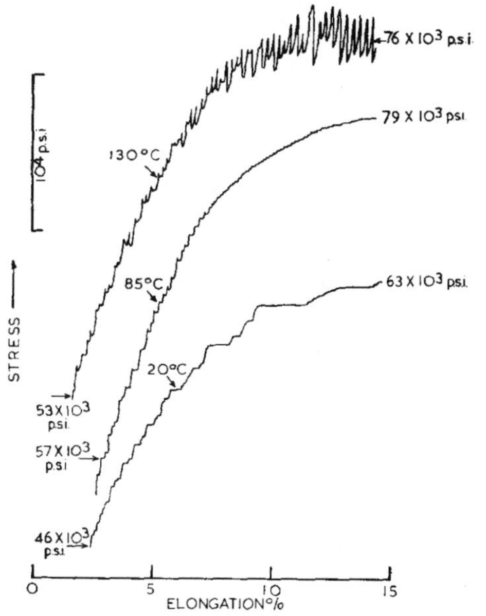

FIG. 4. Effect of temperature on the flow behaviour of low-carbon steel quenched from 700°C and strained at 10^{-4} sec^{-1}.

proportion of the dissolved carbon would occur once the specimen was deformed. This may account for the unusual character of the behaviour at 85°C.

4. CONCLUSIONS

Part of the increase in yield stress during the strain-ageing of a low-carbon steel develops too rapidly to be explained by long range diffusion of carbon and nitrogen to dislocations.

This initial rapid ageing shows characteristics to be expected from a contribution due to local ordering of the interstitial atoms in the stress fields of dislocations. Unlike the effects due to segregation, its magnitude is evidently proportional to the amount of dissolved solute. Thus the effect is small in slowly cooled steels (and, presumably, its contribution will always be small in the later stages of strain ageing, when the solute content of the matrix has been sufficiently reduced).

Experiments at −12°C, made with a reduced applied load, have shown that the rate is consistent with a process involving jumps of a single atomic spacing by the carbon atoms. Observed values of the upper yield points are in reasonable agreement with predictions based on Eshelby's treatment, particularly for specimens aged with only a small relaxation of the prestraining load.

Low-carbon steel quenched from temperatures above about 550°C shows repeated yielding at room temperature during continuous straining at low strain rates. This may also be due to stress-induced ordering

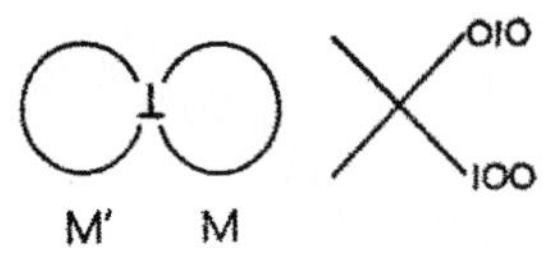

FIG. 5.

of carbon. The room temperature effect differs from high temperature jerky flow (associated with the migration of solute atoms with dislocations[1,10]) in that the former is only appreciable when the dissolved solute content exceeds several hundredths of an atom per cent.

APPENDIX

Suppose (Fig. 5) that a dislocation is introduced into a crystal of ferrite. Initially the carbon and nitrogen atoms are distributed randomly among the mid-points of 100, 010 and 001 cube edges. However, it may be energetically advantageous for them all to move into a particular one of these positions if the resulting tetragonality tends to annul the stress due to the dislocation. At large distances from the dislocation thermal agitation will over-ride this tendency. According to an estimate of Zener[11] a tensile strain of 0.005 induces substantial ordering at room temperature, and such a strain is to be found within about 20 atomic spacings of a dislocation.

With the configuration of Fig. 5 (an edge dislocation* with 110 slip plane) the interstitial atoms to the right of the dislocation will enter 010 positions. On the left the dislocation stress is reversed and they will enter 100 positions. We may say that the dislocation has trapped itself between two regions of martensite M and M'. In these regions there will be a shear stress $\pm\sigma$ superimposed on the stress field of the dislocation, and of such a sign as to oppose its motion. Any applied shear stress must exceed σ before the dislocation can move, provided there is no time for the interstitials to re-arrange themselves.

We tentatively identify σ with the resolved shear stress corresponding to ΔY. In doing so we assume that σ is additive to the other hardening mechanisms operating. The stress σ is proportional to the maximum tetragonality which the interstitials can produce, that is, to their concentration. Hence ΔY is proportional to the concentration, as observed. On this model ΔY should be substantially independent of temperature.

To estimate σ we note that, if the region M were not embedded in the ferrite matrix, ordering to form martensite would cause a fractional extension $0.86c$ in the 010 direction and a contraction $-0.08c$ in the 100 direction, where c is the atomic fraction of either carbon or nitrogen atoms[12] or, we shall assume, a mixture of them. This is equivalent to a shear strain

$$e = \tfrac{1}{2}(0.86 + 0.08)c \simeq \tfrac{1}{2}c$$

in the slip-plane. Complete inhibition of this free strain by a rigid matrix would evidently set up a stress $2\,\mu e$, where μ is the shear modulus. In fact, the matrix only partly inhibits the strain, giving rise to a stress $\sigma = 2\,\gamma\mu e$, where γ is an "accommodation factor".[13] In general, γ is a function of position, but it is constant if M is some form of ellipsoid. We shall assume that M is a long circular cylinder parallel to the dislocation axis. Then $\gamma = \tfrac{1}{2}$, taking Poisson's ratio to be $\tfrac{1}{4}$. (In the notation of [13] p. 391, $\gamma = 1 - 2S_{1212}$ where S_{1212} is to be found from (3.7) and (3.17) with $a = b$). In the most favourable case the applied shear stress on the slip-plane is half the applied tensile stress, and so

$$\Delta Y = \tfrac{2}{3}\mu c$$
$$= 6.6 \times 10^6 c \text{ lb/in}^2$$

if we take $\mu = 1.1 \times 10^7$ lb/in².

This is about twice the observed value. The geometrical arrangement of Fig. 5 was chosen to make the situation as clear as possible. As it happens, any other arrangement gives a somewhat lower value for ΔY.

* Since slip takes place in the 111 direction the dislocation must have a screw component perpendicular to the plane of the figure. It may be neglected provided (as we shall assume) the applied shear stress is in the plane of the figure.

REFERENCES

1. F. R. N. NABARRO, *Report on Strength of Solids*, p. 38. Physical Society, London (1948).
2. J. L. SNOEK, *Physica* **8**, 711, 734 (1941).
3. A. H. COTTRELL, *Report on Strength of Solids*, p. 30. Physical Society, London (1948).
4. B. B. HUNDY, *Metallurgia* **53**, 203 (1956).
5. G. SCHOECK, *Phys. Rev.* **102**, 1458 (1956).
6. D. V. WILSON and B. RUSSELL, *Acta. Mta.* To be published.
7. A. H. COTTRELL and B. A. BILBY, *Proc. Phys. Soc.* A **62**, 49 (1949).
8. P. HAASEN and A. KELLY, *Acta Met.* **5**, 192 (1957).
9. G. RICHTER, *Ann. Phys.* **32**, 683 (1938).
10. A. H. COTTRELL, *Phil. Mag.* **44**, 829 (1953).
11. C. ZENER, *Elasticity and Anelasticity of Metals*, p. 122. University of Chicago Press (1948).
12. K. H. JACK, *Proc. Roy. Soc.* A **208**, 200 (1951).
13. J. D. ESHELBY, *Proc. Roy. Soc.* A **241**, 376 (1957).

Scope and Limitations of the Continuum Approach

J. D. ESHELBY

University of Birmingham, Birmingham, England

ABSTRACT

In calculations relating to defects in a crystal lattice the solid state physicist draws largely on continuum concepts familiar to the engineer. Some topics relating to the two points of view are considered. By way of the theory of continuous distributions of dislocations, microscopic internal stresses may be traced to the lattice defects which are their source. A simple method for calculating the stresses in and around precipitations and transformed regions is outlined; in some cases it may be useful in dealing with the interactions between stresses on a microscopic scale. Generally, however, such interactions have to be discussed in terms of energy changes.

INTRODUCTION

It has been suggested that I should say something about the difference between the engineer's conception of stresses in an elastic continuum and the solid state physicist's idea of stresses in a crystal lattice. This is rather an embarrassing assignment, for, as anyone who opens a book on, say, the theory on dislocations will find, the physicist clings to the continuum point of view as long as he can.

I am sure that many of you will recall Lord RUTHERFORD's dictum that "all science is either mathematics or stamp-collecting". Had he been acquainted with some aspects of modern solid state physics he might, perhaps, have been tempted to amend it to "mathematics, stamp-collecting or playing with constructional toys". The solid-state theoretician has a kit of basic constructional elements with more or less well established properties, and these he fits together in an attempt to reproduce the observed behaviour of solids.

If he is a specialist in the mechanical properties of metals, his kit will be mainly made up of lattice defects, in particular dislocations and point imperfections. Sometimes he uses them in the ready-assembled form of Frank-Read sources, dislocation pile-ups, Friedel cages and so forth.

The theory of these defects has been largely developed by regarding the crystal as an elastic continuum. Of course at various points one is forced to come to terms with the discontinuous lattice structure of the crystal, but I am afraid that the

References p. 58.

various *ad hoc* ways in which this has been done do not add up to anything like a coherent philosophy of the difference between stresses in a continuum and stresses in a crystal lattice, and so I shall consider one or two particular topics which link the two points of view.

There is one theoretical approach which directly links the main topics of this symposium, large-scale internal stresses and atomic imperfections. There is the recently-developed theory of continuous distributions of dislocations, and it will perhaps be best to begin with a short account of it.

DISLOCATIONS AND INTERNAL STRESS

Suppose that we have a body containing internal stresses. If we limit ourselves to mechanical means there are two ways of investigating them. One is to make cuts in the body and see how this affects its shape. It is this method which connects up with the theory of continuous distributions of dislocations. There is, however, in principle a second, non-destructive method. This is the direct detection of internal stresses by their influence on the way in which the body behaves when it is deformed by externally-applied forces.

According to the normal infinitesimal theory the presence of internal stress in a body makes no difference to its response to applied forces, but this is no longer true on a more precise theory. A piano, for example, can be regarded as a rather complex self-stressed body, and the stress in one of its strings can be found by observing its sideways deflection by a concentrated force.

There is, in fact, a respectable theory, going back to CAUCHY, for dealing with the elastic behaviour of self-stressed bodies. The second-order effects which it predicts are, in a sense, partly geometrical and partly the result of departures from Hooke's law. One cannot make a precise separation, since the degree of departure from Hooke's law will depend on what measure of finite strain we decide to employ, a purely geometrical question. For example, in order to calculate the change of volume when a dislocation is introduced into a body we must know the rate of change of the elastic constants with stress. On the other hand, in some cases calculation can be carried one stage beyond the infinitesimal approximation knowing only the ordinary elastic constants. The following is an example.

Perhaps the simplest self-stressed body one can imagine is a body containing a single dislocation. GOMER[1] has grown mercury whiskers which probably contain a single axial screw dislocation. Such a dislocated whisker has a certain twist per unit length, α/l say. Second-order theory shows that when the length of the whisker is changed from l to $l + dl$ by a longitudinal tension the degree of twist also alters, and so the ends of the whiskers suffer a relative rotation

$$d\theta = l d\alpha + \alpha dl = -2(1 + \sigma)\left(\frac{GI}{D} - 1\right)\alpha dl.$$

References p. 58.

G is the shear modulus and σ Poisson's ratio, I the polar moment of inertia about the centre of gravity of the cross-section, and D is the torsional rigidity. Thus, even if the twist is not directly observable, we can still infer that there are internal stresses by observing the rotation $d\theta$. GOMER did indeed observe rotations of the right order when some of his whiskers were put under tension by electrostatic forces. If these rotations were not due to some other cause (and there are several possible ones) we have a rather pretty demonstration of the effect of internal stresses in modifying the reaction of a body to applied forces.

It seems unlikely that this method could be developed into a practical tool. The method of dissection, on the other hand, has been highly developed as a way of investigating internal stresses. In an idealized version it is closely connected with the geometrical theories of internal strain of MORIGUTI[2] and ECKART[3], and, at one remove, with the theory of continuous distributions of dislocations due to KONDO, BILBY, KRÖNER[4] and others. Imagine that we divide a self-stressed body into a large number of small cubes, (Fig. 1a) and cut the body up along the planes defining the cube faces. There results a collection of small deformed cubes which we may imagine to be loosely assembled so that it is apparent which faces were originally adjacent (Fig. 1b). Each deformed cube can be derived from the unde-

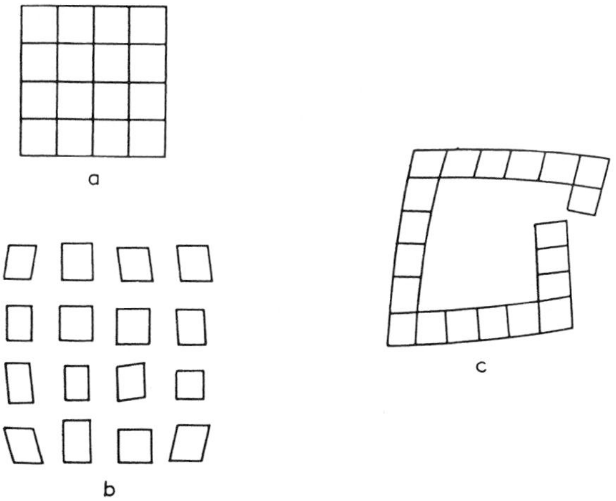

Fig. 1. Dissection of a self-stressed body. (*a*) Dissection into cubes. (*b*) Deformed cubes after dissection. (*c*) Frame built up of deformed cubes showing misfit (gap).

formed cube by a certain strain e_{ij}. Obviously the stress derived by Hooke's law from the strain—e_{ij} for any cube is the stress which existed in the undissected body at the point from which this particular cube came. If our subdivision has been fine enough we may regard the e_{ij} as continuous functions of position.

We might equally well have applied the imaginary dissection process to a body

References p. 58.

free from internal stress but deformed by applied forces. We should have obtained a collection of deformed cubes qualitatively indistinguishable from those in Fig. 1b. A distinction becomes apparent if we try to build up a picture frame (Fig. 1c). If the body was free of internal stress we shall succeed. If it contained internal stresses we shall be left with a misfit in the form of a gap (Fig. 1c) or interference.

Instead of wholesale dissection we might have followed the method of MORIGUTI AND ECKART. Draw any curve joining two points A and B in the self-stressed body. The length of the curve is the integral

$$s_0 = \int_A^B (dx^2 + dy^2 + dz^2)^{1/2}$$

taken along the curve. Now mark out a thin rod having this curve for axis and remove the surrounding material. The strains in the rod are relaxed, and its length is

$$s = \int_A^B (g_{xx} dx^2 + \ldots + 2g_{xy} dx dy + \ldots)^{1/2}$$

where

$$g_{xx} = 1 + 2\, e_{xx},\ g_{xy} = 2e_{xy} \ldots$$

and the e_{ij} are the strains we should have obtained by our original wholesale dissection process.

Suppose that we choose to regard s rather than s_0 as the "length" of a curve in the self-stressed body. We thereby define a Riemannian geometry with metric tensor g_{ij} in which lengths are, so to speak, not what one would expect. This gives a physically visualizable interpretation to Riemannian geometry; conversely, and this is more to the point, one can use some of the apparatus of higher differential geometry in discussing internal stress. The four-suffix curvature tensor R_{ijkl} plays an important role in Riemannian geometry. In three dimensions, fortunately, we may use instead the two-suffix symmetrical tensor S_{ij} related to it by

$$S_{xx} = R_{yzyz}, \ldots, \ S_{xv} = R_{yzzx}$$

In terms of the e_{ij} we have

$$S_{xx} = 2\,\partial^2 e_{yz}/\partial y \partial z - \partial^2 e_{yy}/\partial z^2 - \partial^2 e_{zz}/\partial y^2 \ldots$$

$$S_{xy} = \partial^2 e_{xy}/\partial z^2 + \partial^2 e_{zz}/\partial x \partial y - \partial^2 e_{xz}/\partial y \partial z - \partial^2 e_{yz}/\partial x \partial z \ldots \quad (1)$$

The relations $S_{ij} = 0$ are the "compatibility conditions" which must be satisfied by ordinary strains induced by external forces, and so S_{ij} may appropriately be called "incompatibility tensor".

The S_{ij} have a fairly simple physical interpretation. If we cut from the self-

References p. 58.

stressed solid a rod in the form of a closed loop (Fig. 2a) its internal stresses will not be completely relaxed unless we cut through it. The two sides of the cut will spring apart and define a vector $\Delta\mathbf{u}$ (AA' in Fig. 2b). We have, of course, reached the same state of affairs as in Fig. 1c, but by a slightly different route.

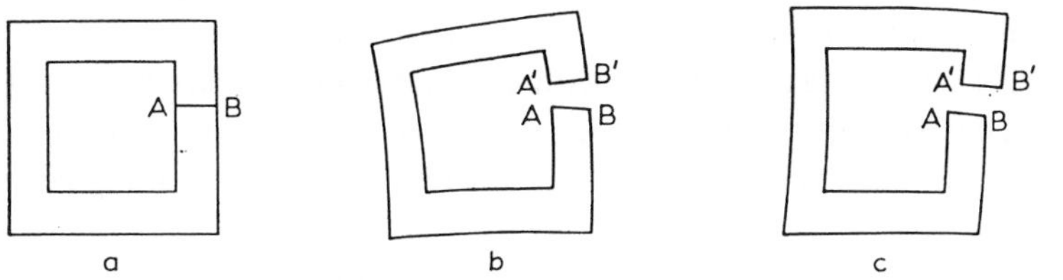

Fig. 2. An alternative way of dissecting. (a) Closed loop as cut from self-stressed body. (b) Shape after cutting through to remove long-range stresses. (c) Shape after removal of dislocations.

There is a connection between Δu and the S_{ij} tensor, but it is rather complicated, principally because the value of Δu depends on where we cut through the loop. However, in addition to flying apart, the two sides of the cut suffer a relative rotation; the face $A'B'$ can be derived from the face AB by the rigid body displacement $\Delta\mathbf{u} + \mathbf{r} \times \Delta\boldsymbol{\omega}$ where $\Delta\boldsymbol{\omega}$ is a vector representing a small rotation. It can be shown that though $\Delta\mathbf{u}$ varies with the position of the cut, $\Delta\boldsymbol{\omega}$ does not. In fact, if the loop lies in the yz-plane, with sides dy, dz then

$$\Delta\omega_x = -S_{xx}\mathrm{d}y\mathrm{d}z,\ \Delta\omega_y = -S_{yz}\mathrm{d}y\mathrm{d}z,\ \Delta\omega_z = -S_{zx}\mathrm{d}y\mathrm{d}z\,. \tag{2}$$

(The derivation just consists in expressing in mathematical terms the stacking up of elementary deformed cubes illustrated in Fig. 1c).

When the S_{ij} are known, the internal stresses can in principle be calculated. If we express the strains in eqn. (1) in terms of stresses and use the equilibrium conditions we obtain a set of equations

$$\nabla^2\sigma_{xx} + \frac{1}{1+\sigma}\frac{\partial^2\Theta}{\partial x^2} = 2GS_{xx} - \frac{2G}{1-\sigma}S$$

$$\nabla^2\sigma_{xy} + \frac{1}{1+\sigma}\frac{\partial^2\Theta}{\partial x\partial y} = 2GS_{xy} \tag{3}$$

where $\Theta = \sigma_{xx} + \sigma_{yy} + \sigma_{zz},\ S = S_{xx} + S_{yy} + S_{zz}.$

When there is no internal stress, $S_{ij} = 0$ and eqn. (3) are the Mitchell–Beltrami equations which must be satisfied by any stress produced by applied surface forces. In conjunction with the equilibrium equations and the boundary conditions, they determine the stresses.

References p. 58.

The same is true when the S_{ij} are not zero, only now the Mitchell–Beltrami equations are inhomogeneous because of the terms in S_{ij} on the right. We may compare eqn. (3) with Poisson's equation $\nabla^2 V = -4\pi\varrho$ connecting electric potential V and charge ϱ. This equation enables us to determine the field V when we know its "source" ϱ. In a similar manner eqn. (3) lets us calculate the internal stresses from their "source" S_{ij}.

We have not yet made contact with the theory of dislocations. If dislocations traversed the arms of the loop in Fig. 2a when it formed part of the original self-stressed body, they will continue to do so when it is cut out and freed from macroscopic stress.

In electrostatics one can usually neglect the discrete nature of the electric charges and treat the charge density as if it were continuous. NYE[5] introduced a similar treatment for dislocations. Since two vectors are associated with a dislocation (its Burgers vector and the direction of the dislocation line) a rather more elaborate specification of the "dislocation density" is necessary. One must introduce a tensor α_{ij} whose ij term gives the sum of the x_i -components of the Burgers vectors of all the dislocations threading unit area perpendicular to the x_j-axis. Evidently terms like α_{zz} with equal suffixes refer to the flux of screw dislocations, and terms like α_{yz} with unequal suffixes refer to the flux of edge dislocations. NYE was able to relate α_{ij} to the curvature and twist of the crystal lattice. Let K_{yx} be the angle through which the crystal lattice rotates about the y-axis per unit length travelled along the x-axis, K_{zz} the angle the lattice twists through about the z-axis per unit distance travelled along the z-axis, and so on. NYE established these relations between K_{ji} and α_{ij}:

$$K_{yx} = \alpha_{xy}, \ldots \tag{4}$$

$$K_{zz} = \tfrac{1}{2}\alpha_{zz} - \tfrac{1}{2}(\alpha_{xx} + \alpha_{yy}), \ldots \tag{5}$$

The interpretation of eqn. (4) is straightforward (the difference in the order of the suffixes is conventional). A vertical wall of edge dislocations with Burgers vectors b and spacing h is equivalent to a tilt boundary of angle b/h; a sequence of such walls set up with spacing d produces a curvature $(b/h)/d$. This is just equivalent to eqn. (4) since hd is the area monopolized by each dislocation.

The interpretation of eqn. (5) requires somewhat more intense geometrical thinking. It can also be derived from elastic theory. It can be shown that a single screw dislocation running along the axis of a circular cylinder gives it a twist equal to b/A, the Burgers vector divided by the cross-section of the cylinder. On the other hand, if the dislocation runs close to the surface of the cylinder the twist is almost zero. Thus if n dislocations are distributed uniformly over the cross-sec-

References p. 58.

tion we should expect to find a twist equal to nb/A times some factor between 0 and 1. It turns out that this factor is $\frac{1}{2}$ whether the cross-section is circular or not. This accounts for the first term in eqn. (5). Again, if a rubber cube is twisted so that lines marked on opposite faces undergo a relative rotation θ, it will be found that lines marked on either of the remaining pairs of faces undergo a relative rotation $-\theta$. This accounts for the second term in eqn. (5).

The relations (4 and 5) apply to a crystal which is free of macroscopic stress. If it is not, we must add to K_{ji} the curvatures associated with the stress.

We can now see the type of imaginary dissection process which must be carried out in order to determine the distribution of dislocations in a crystal. Cut out a number of thin rods at a point in the specimen. When they are freed from the surrounding material the relaxation of internal stress will give them certain curvatures and twists. Ignore these. Give the rods some treatment which gets rid of the dislocations, and measure the change in curvature and twist which this induces. From eqns. (4) and (5) calculate α_{ij}, the distribution of dislocations.

If we apply this process to the loop of Fig. 2 we can get a relation between the incompatibility tensor S_{ij} and the dislocation density α_{ij}. The loop of Fig. 2b has a certain gap $\Delta\mathbf{u}$ and rotation $\Delta\boldsymbol{\omega}$ due to removal of long-range stresses.

Annihilate the dislocations. $\Delta\mathbf{u}$ will change by an amount equal to the sum of the Burgers vectors of the dislocations embraced by the loop (Fig. 2c). There will be no relative rotation of the faces of the cut if the dislocation density is uniform across the loop, since opposite arms will rotate and twist by the same amount. But if α_{ij} is not uniform across the loop there will be a rotation, which is easily calculated to be

$$\Delta\omega_x' = \left(\frac{\partial K_{xz}}{\partial y} - \frac{\partial K_{xy}}{\partial z}\right) \mathrm{d}y\mathrm{d}z,\ \Delta\omega_y' = \left(\frac{\partial K_{yz}}{\partial y} - \frac{\partial K_{yy}}{\partial z}\right) \mathrm{d}y\mathrm{d}z \ldots \tag{6}$$

Now suppose that instead of cutting the loop and then removing the dislocations, we remove the dislocations from the uncut loop (Fig. 2c), arranging that they all pass out along the path AB. As each one leaves the loop it displaces AB relative to $A'B'$ by an amount equal to its Burgers vector, but it does not tilt or twist AB relative to $A'B'$. Thus in the final stage (Fig. 2c) AB and $A'B'$ are displaced, but not tilted or twisted relative to each other. This means that the rotations $\Delta\boldsymbol{\omega}$ (eqn. 2) and $\Delta\boldsymbol{\omega}'$ (eqn. 6) must exactly cancel. Thus, using eqns. (5) and (6) to turn the K's into α's we have

$$S_{xx} = \frac{\partial\alpha_{zx}}{\partial y} - \frac{\partial\alpha_{yx}}{\partial z},\ S_{xy} = \frac{\partial\alpha_{zy}}{\partial y} - \frac{\partial\alpha_{yy}}{\partial z} + \frac{1}{2}\frac{\partial}{\partial z}(\alpha_{xx} + \alpha_{yy} + \alpha_{zz}), \ldots \tag{7}$$

Though the method is rather untidy we have reached the basic relation (7) be-

References p. 58.

tween dislocation density and incompatibility tensor, and hence via eqn. (3) the connection between dislocation density and internal stress.

We have seen that the stress in the material is governed by the distribution of dislocations in it. On the other hand, the distribution of the dislocations is presumably determined by the forces acting on them, and hence by the stresses in the material. From the interlocking of these two facts we might hope to develop a dislocation theory of macroscopic plasticity. So far no one has done this in detail. The main trouble is that to relate dislocation distribution and stress is not enough. In addition one must introduce a yield criterion. An indication of the ideas that would be involved can be obtained from READ's[6] account of the elastic–plastic bending of a beam.

Suppose the beam is bent to a curvature K and that initially the deformation is everywhere elastic. Elementary beam theory gives the relation

$$\frac{\mathrm{d}\sigma_{xx}}{\mathrm{d}y} = EK \ (E = \text{Young's modulus})$$

between the curvature and the variation of the fibre-stress across the beam (y denotes distance measured perpendicular to the length of the beam). If plastic flow has occurred and part of the beam is filled with dislocations we must subtract from K the non-elastic curvature they produce. If we are concerned only with the edge dislocations running perpendicular to the plane of bending, the appropriate curvature is α_{xy} according to eqn. (4). Thus we arrive at READ's relation

$$\frac{\mathrm{d}\sigma_{xx}}{\mathrm{d}y} = E\,(K - \alpha_{xy}) \tag{8}$$

READ considers first the case where the stress σ_c required to generate dislocations is less than the stress σ_0 required to keep them moving. As K is increased the deformation is elastic until σ_0 is exceeded at the outer fibres. Then dislocations will move in towards the neutral axis until they come to rest at a position where $\sigma_{xx} = \sigma_0$. The presence of these dislocations modifies the stress-field experienced by subsequent dislocations moving in from the surfaces.

In the equilibrium there is a region around the neutral axis which no dislocations have reached ($\alpha_{xy} = 0$) flanked on each side by regions where the stress has the constant value σ_0. According to eqn. (8) σ_{xx} is a linear function of y in the inner region, and α_{xy} has the constant value K in the outer regions. This is, of course, in accordance with the elementary theory of the elastic–plastic deformation of beams. READ also considers the more interesting case where the stress σ_c required to create a dislocation is greater than the stress required to keep it moving. When the stress reaches σ_c at the surface, dislocations are generated there. They cannot remain near the surface, but seek a position where $\sigma_{xx} = \sigma_0$. Again there is a dislo-

References p. 58.

cation-free region about the neutral axis flanked by regions containing a constant density of dislocations, but they in turn are flanked by regions extending to the surface which are free from dislocations, but through which dislocations have passed on their way to the inner regions. Eqn. (8) enables the problem to be solved quantitatively. READ obtains the interesting result that if σ_0 is large enough in comparison with σ_c the load (bending moment) *versus* deformation (curvature) curve exhibits a "yield point".

CALCULATIONS OF INTERNAL STRESS

We have seen how the internal stresses due to a known distribution of dislocations can, in principle, be found from eqns. (3) and (7). A more practical problem is the determination of the internal stresses left in a body which has been plastically deformed by given loads, and this involves solving the much more difficult problem of determining the dislocation distribution itself. There is, however, a class of internal stress problems of metallurgical interest in which the dislocation distribution is directly prescribed.

Internal stresses are set up whenever a region of a crystal tries to change its form but is partly prevented from doing so by the surrounding material. The change of form may, for example, be a volume change due to the formation of a precipitate with a lattice constant different from that of the parent lattice, or it may be a combined shear and volume change due to a martensitic or other diffusionless transformation. Any homogeneous change of shape of this type can be produced by the passage of a suitable set of (real or fictitious) dislocations through the transformed region. If the transformation occurs while the region is embedded in an untransformed matrix these dislocations are left at the interface between the transformed and untransformed regions. Thus the calculation of the stresses set up by the transformation is reduced to the problem of determining the stress due to a distribution of dislocations over a certain surface. This particular problem can, however, be solved in a more direct manner. Fig. 3 illustrates a simple routine for doing this[7].

A region R undergoes a transformation, which in the absence of the surrounding matrix would be a uniform shear and dilatation changing its shape from A to B. Because of the restraint imposed by the matrix both the transformed region and the matrix are left in a state of stress. Cut R out of the matrix, (Fig. 3a), let the transformation from A to B take place, (Fig. 3b), apply surface forces to change its shape back from B to A, (Fig. 3c), put it back in the hole in the matrix and weld the cut surfaces together. At this stage (Fig. 3d) there is a calculable uniform stress in R and no stress in the matrix, but the surface forces of Fig. 3c have got built in as a layer of body force. To get rid of this layer and complete the calculation we

References p. 58.

superimpose an equal and opposite layer of body force. Provided the transformed region has the same elastic constants as the matrix, the effect of removing the layer can be calculated by integration starting from the known expression for the elastic field produced by a point force. In this way we arrive (Fig. 3c) at the required state of internal stress.

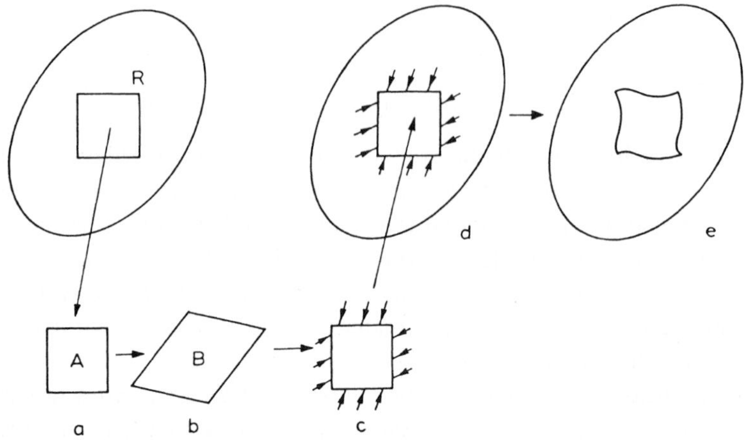

Fig. 3. Routine for calculation of stresses set up by transformation of constrained region. (Schematic)

In the figure the transformed region has been drawn as a rectangle in order to make the changes of shape clear. The calculations for this shape are not easy, nor is it of special interest physically. The shapes of metallographic interest (spheres, lenses, needles, plates) approximate more or less closely to some form of ellipsoid. The calculation for the general ellipsoid with three unequal axes is quite straightforward. It is found that after transformation the stresses inside the ellipsoid are uniform. Because of this fortunate fact (which is not true for any other shape) the method of Fig. 3 can be extended in several ways. One can drop the requirement that the transformed material must have the same elastic constants as the untransformed. Or one can solve the problem of an ellipsoidal precipitate which is not coherent with the matrix; here we are not given the transformation strain $A \rightarrow B$ directly, but only the volume misfit and the additional fact that for minimum strain energy there must be a hydrostatic pressure within the ellipsoid after the transformation is completed[8].

If we let a lump of misfitting precipitate dwindle until only one atom is left we reach the "misfitting sphere" model of an impurity atom. It is clear that the continuum approximation ought to be quite inadequate here. Nevertheless, this model, and refinements of it, have been used successfully in metal physics. The

References p. 58.

reason is that in many cases what is important is the elastic field at large distances from the defect, and, in principle at least, the continuum theory can give an adequate description of it. If we decide to use this model the first problem is to settle what size we are to take for the hole and sphere. It has become traditional to equate the volumes of hole and sphere to the volume-per-atom of solvent and impurity. There does not seem to be any obvious reason for doing this. We might, for example, be tempted to fit the impurity into a cage bounded by its neighbours. However, the atomic volume approach predicts a variation of mean lattice parameter with composition which agrees with Vegard's law, and this gives it some kind of *a posteriori* justification. Far away from a defect the stresses can certainly be treated by ordinary elastic theory. The general form of the stress field is dictated by symmetry, and it would seem that we have only to determine one constant, the "strength" of the imperfection, from atomic considerations. There are, however, a number of practical difficulties. In the first place there is the unfortunate mathematical fact that even the purely elastic problem cannot be solved explicitly except for an isotropic or hexagonal material. Again, in the usual type of lattice calculation one finds the atomic arrangement around a defect by adjusting the positions of the atoms so as to minimise the energy. But according to the isotropic elastic solution nearly all the strain energy resides within a region extending two or three inter-atomic distances from the defect, and we may expect that something of the sort will be true also for the crystal lattice. This means that a small misplacement of some of the more remote atoms would scarcely affect the energy but could be fatal to the fitting together of the continuum and lattice solutions.

Despite these difficulties quite a lot of work has been based on the misfitting sphere model. It has also been extended to interstitial atoms and lattice vacancies. BROOKS[9], for example, has treated vacancies by regarding them as spherical cavities pulled inwards by surface tension.

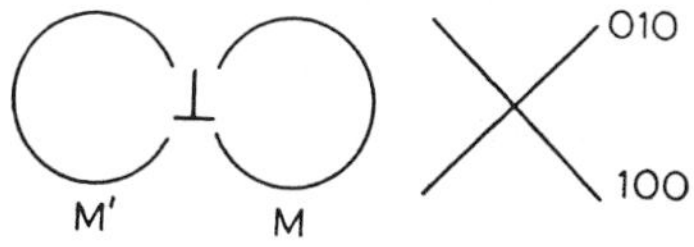

Fig. 4. To illustrate the interpretation of the increase of yield stress in iron after straining due to local rearrangement of carbon atoms.

When we turn to dislocations, relations between continuum and lattice are much happier. The "strength" of a dislocation (its Burgers vector b) is determined by the lattice geometry. We can, if we like, regard a dislocation loop as a plate one atom thick which has undergone a transformation which offsets its faces by b, and use the method of Fig. 4 to calculate its stress-field. In whatever way the calculation is done we find that at a distance r from the dislocation the strains are of order

References p. 58.

$b/2\pi r$ so that at quite a modest distance they have fallen to a value within the scope of the theory of elasticity. In the case of the dislocation the principle tool for making a more detailed connection between continuum and crystal lattice is the Peierls–Nabarro equation. The slip plane of the dislocation is regarded as a gap between two blocks of elastic continuum, across which interatomic forces act. Suppose that the faces of these blocks are given a relative displacement which varies arbitrarily along the slip-plane. It is possible to write down an expression (in the form of an integral) for the forces across the slip-plane which are necessary to maintain this displacement. When some law relating force and relative displacement across the slip-plane is prescribed we have an integral equation for the displacement. Among its solutions there is one which has the general character of a dislocation. For the sort of force laws which have been investigated this solution does not differ greatly from the elastic solution, except in the immediate neighbourhood of the centre, where the stress remains finite.

THE INTERACTION OF INTERNAL STRESSES

In the theory of the solid state one is not merely interested in the internal stress field as a static entity, but also in how it changes, and in how different systems of internal stress interact with one another. There, too, continuum ideas can carry us a long way.

The engineer is familiar with the fact that internal stresses can be removed or redistributed by annealing. On the crystal lattice level one would say that dislocations and point defects had moved through the lattice carrying their stress-fields along with them. In addition to this migration of stress-fields there are more subtle effects which can alter internal stresses.

To take an example, suppose that a dislocation is suddenly introduced into a crystal. Then, if there is time for diffusion, too-small impurities can migrate to the compressed side of an edge dislocation, and too-large atoms to the dilated side. This is an example of the movement of pre-existing stresses to offset the stress-field of the dislocation. There are, however, a number of other physical changes which can reduce the strain due to the dislocation. If the crystal is made of an alloy which can undergo an order–disorder transformation, and in which the lattice parameter varies with the degree of order, a suitable variation of order from point to point can annul the dilatation due to the dislocation, and the shear stress as well, if directional ordering is possible. In body-centred iron, local rearrangement of carbon or nitrogen atoms can produce tetragonality which offsets the shear stress of the dislocation. These effects involve no long-range diffusion of atoms. Suppose for the moment that one or more of these effects succeed in completely annulling the stress-field of the dislocation. Then if the dislocation is suddenly

References p. 58.

removed there is left behind a "ghost dislocation" in the shape of a stress-field which is an exact replica of that of the dislocation, but with reversed sign. Ultimately the ghost will be dissipated by thermal agitation, but while it exists it attracts the original dislocation just as if it were a real dislocation. Thus a certain force is necessary to dislodge the dislocation from its original position.

In fact, the ghost will be imperfect. First, far from the dislocation, where its stress-field is weak, thermal agitation will prevent the necessary atomic rearrangements. Secondly, the high strains near the centre of the dislocation will be only partly annulled, since the effects we are considering can produce only a limited amount of strain. (The alloy, for example, cannot do any better than order or disorder completely.)

The ghost field at a particular point is not determined solely by the atomic rearrangement near that point. If a single volume-element undergoes rearrangement it creates a stress inside itself and also produces stresses in the surrounding material (compare Fig. 3). The stress at a particular point thus depends on what has happened in all the volume-elements of the material. A precise calculation, taking into account this non-local property and also the effect of thermal agitation would be quite complicated. In certain cases, however, a rough calculation using only simple macroscopic ideas is possible.

As an illustration I should like to consider the interpretation of some measurements of strain-ageing in iron carried out recently in Birmingham. WILSON AND RUSSELL strained specimens of low-carbon steel, held them at constant load for varying lengths of time, and then measured the increase of yield stress on further loading. The curve showing this increase as a function of time can be separated into two components. The first follows the Cottrell–Bilby $t^{\frac{2}{3}}$ law and is evidently due to the migration of carbon atoms to dislocations. The second component reaches a constant value, say ΔY, in a few seconds at room temperature. The magnitude of ΔY is proportional to the carbon content. These two facts suggest that ΔY is due to local rearrangement of carbon atoms without long-range diffusion. Fig. 4 indicates a possible interpretation. Initially the carbon atoms are distributed randomly among the mid-points of the [100], [010] and [001] cube edges of the ferrite lattice. When the dislocation is introduced the carbon atoms will tend to enter the [010] positions on the right and the [100] positions on the left, since the resulting tetragonality will partly annul the strain-field of the dislocation. This effect will be negligible above and below the dislocation where the shear stress is small and will fade out at large distances because of thermal agitation. Thus the ghost dislocation in effect reduces to patches of martensite M and M' on either side of the dislocation. The stress in M can be estimated if it is assumed that inside M the migration of carbon atoms to [010] positions is complete. If M transformed to martensite in the absence of the surrounding material it would undergo a tetrag-

References p. 58.

onal strain which is equivalent to a shear strain in the slip-plane; the magnitude of the strain is proportional to c, the atomic fraction of carbon. The observed variation of the lattice constants of martensite with composition show that the constant of proportionality is very nearly $\frac{1}{2}$. Thus if the transformation took place in a perfectly rigid matrix it would set up a shear stress $\sigma = Gc$ where G is the shear modulus. In fact M is only partly restrained by its elastic surroundings and the shear stress σ is only γGc where γ is an "accomodation factor" which can be calculated if the shape of M is known. As a reasonable guess we may take M to be a circular cylinder parallel to the dislocation. Then $\gamma = 1/3$. If a certain stress σ_0 is necessary to move the dislocation to the right when M is absent, then a stress $\sigma_0 + \sigma$ will be necessary when it is present. In the most favourable case the applied shear stress in the slip-plane is half the applied tensile load. Thus $\Delta Y = 2/3\, Gc$. This is about twice the observed value. The particular geometrical arrangement of Fig. 4 was chosen to make the situation as clear as possible. As it happens, any other arrangement gives a somewhat lower value for ΔY, and we can say that the simple "macroscopic" treatment agrees quite well with the experiments.

It is not often that the interaction of defects can be treated in this direct way. More usually it is necessary to work in terms of energy. Suppose that we have a body containing defects which produce internal stresses and loaded by external forces which produce additional stresses. It is perhaps worth emphasizing that the important quantity is the total energy E_{tot} of the system, that is, the sum of the internal energy of the body E_{int} and the potential energy E_{ext} of the external devices (springs, weights, etc.) responsible for the applied forces.

If, for example, we want to find out whether a particular defect is in equilibrium we must test whether E_{tot} (and not E_{int}) is stationary for a small displacement. In thermodynamic language, we must minim'ze the enthalpy (in an adiabatic process) or the Gibbs free energy (in an isothermal process) rather than the internal energy or He'mholtz free energy. For a system approaching equilibrium the rate of decrease of E_{tot} is the appropriate driving force to use in a kinetic calculation. The contrast between E_{tot} and E_{int} is well illustrated by the problem of a stressed body containing a crack (OROWAN[10]). Suppose that a sheet of rubber is stretched and held by immovable grips. Then $E_{\text{tot}} = E_{\text{int}}$. If we make a cut in the sheet the stress is partially relaxed and E_{int} decreases. Suppose next that the sheet is stretched by a weight. If we plunge a knife into the sheet to create a crack, the spring-constant of the sheet is reduced and the weight descends. After oscillating up and down it finally comes to rest. Evidently the weight has lost potential energy, that is, E_{ext} has decreased. It can be shown that of this potential energy half has been dissipated and half has gone to *increase* the strain energy. That is, $-\delta E_{\text{tot}} = \delta E_{\text{int}} = -\frac{1}{2}\delta E_{\text{ext}}$. In this case confusion between internal and total energy does no harm, provided the sign is determined by common sense, but in other cases it

References p. 58.

may lead to an incorrect numerical factor as well as an error in sign. A direct continuum calculation of the total energy would be very precarious, since the major contribution to the internal energy comes from the immediate vicinity of the defects and it is just in these regions that the continuum approach is least adequate. However, we are usually not interested in the actual value of the total energy, but only in how it changes as defects move about. It turns out that such energy changes can be calculated with fair accuracy without knowing the details of the elastic fields near the defects. The situation is quite analogous to that in electrostatics where one can treat the energies of interaction between charges without knowing their (possibly infinite) self-energies.

In many cases the energy changes can be expressed most conveniently and vividly in terms of the "force on a lattice defect". Let A stand for the state of a body with a defect (we need not specify exactly what it is) at the point (x, y, z), and let B refer to a state where the defect is at $(x + \varepsilon, y, z)$. Then we may define

$$F_x = - \lim_{\varepsilon \to 0} \frac{1}{\varepsilon} \left\{ E_{\text{tot}}(B) - E_{\text{tot}}(A) \right\}$$

to be the x-component of the "force" on the defect. The remaining components F_y, F_z of the "force" are defined similarly. Evidently when the defect suffers a displacement $\mathrm{d}x$, $\mathrm{d}y$, $\mathrm{d}z$, the decrease in E_{tot} is $F_x \mathrm{d}x + F_y \mathrm{d}y + F_z \mathrm{d}z$.

In order to calculate F_x we have first to define what we mean by saying that states A and B are due to the same source of internal stress (defect), but that in B it has been shifted a distance ε parallel to the x-axis. To do this for the case of finite deformation and an arbitrary stress–strain relation is not easy, but we can give a reasonable operational recipe for constructing state B from state A. It falls into two stages. Suppose that $f(x, y, z)$ stands for any one of the quantities (stress, energy density, elastic constants and so forth) associated with the elastic field. Then the first stage of the recipe is

(i) Replace $f(x, y, z)$ at each point (x, y, z) by its value at the neighbouring point $(x - \varepsilon, y, z)$ on the left:

$$f(x, y, z) \to f(x - \varepsilon, y, z) = f - \varepsilon \frac{\partial f}{\partial x} + O(\varepsilon^2) \ldots \qquad (9)$$

In this way we get a new state which is elastically permissible, and in which the source of internal stress has been displaced in the way required. It is not, however, the state we want, since it does not satisfy whatever boundary conditions we may have imposed. The surface tractions, for example, will not match the prescribed surface loading. To correct this we carry out stage (ii).

(ii) Adjust the surface tractions and displacements to conform with the boundary conditions.

References p. 58.

It is now easy to calculate the force on the defect. The result can be expressed as the integral

$$F_x = \int_S \left(W n_x - \frac{\partial \mathbf{u}}{\partial x} \cdot \mathbf{T} \right) \mathrm{d}S \tag{10}$$

taken over the surface S of the body. W is the energy density and $\mathbf{u}$ the displacement. $\mathbf{T}$ is the surface traction per unit area and (n_x, n_y, n_z) are the components of the normal to the surface. To derive this expression we have only to calculate the energy changes associated with steps (i) and (ii)[11].

In stage (i) the change in elastic energy is just

$$\delta E_{\mathrm{el}}\,(\mathrm{i}) = \int_V W\,(x - \varepsilon, y, z)\,\mathrm{d}x\mathrm{d}y\mathrm{d}z - \int_V W\,(x, y, z)\,\mathrm{d}x\mathrm{d}y\mathrm{d}z$$

with the volume integrals taken over the volume V inside S. But the first integral is equal to the integral of $W(x, y, z)$ taken over the volume V' obtained by shifting V an amount ε to the left. Thus δE_{el} (i) is the volume integral of W taken over the unshaded shell of Fig. 5 with due allowance for sign. In this region the volume element can be written as $-\varepsilon n_x \mathrm{d}S$ and so

$$\delta E_{\mathrm{el}}\,(\mathrm{i}) = -\,\varepsilon \int W n_x\,\mathrm{d}S. \tag{11}$$

The shaded region makes no contribution and we only need to know the form of the energy density near the surface S. If, for example, W is formally infinite somewhere inside S the infinity is "subtracted out" in forming (11).

Next we have to find the change in elastic energy in stage (ii). At the beginning of stage (ii) the surface displacement is, according to eqn. (9)

$$\mathbf{u} - \varepsilon \frac{\partial \mathbf{u}}{\partial x}$$

and at the end of stage (ii) it has some value $\mathbf{u}'$, say, which could only be found by detailed calculation. Throughout stages (i) and (ii) the surface traction $\mathbf{T}$ only varies by a quantity of order ε. Thus to order ε^2

$$\delta E_{\mathrm{el}}\,(\mathrm{ii}) = \int \left[\mathbf{u} - \varepsilon \frac{\partial \mathbf{u}}{\partial x} - \mathbf{u}' \right] \cdot \mathbf{T}\,\mathrm{d}S$$

is the increase in elastic energy due to the work done on S during stage (ii).

The work done by the external loading mechanism in the change from state A to state B is

$$-\,\delta E_{\mathrm{ext}} = \int [\mathbf{u} - \mathbf{u}'] \cdot \mathbf{T}\,\mathrm{d}S.$$

References p. 58.

On adding ∂E_{el} (i), δE_{el} (ii) and δE_{ext} and dividing by $-\varepsilon$ we arrive at eqn. (10).

So far eqn. (10) applies only to the case of a single defect in a body free from other defects but stressed by external loads. We know, however, that the degree of hardness or softness of the loading mechanism (that is, the extent to which $\mathbf{T}$ varies as the surface displacement changes from $\mathbf{u}$ to $\mathbf{u}'$, affects the energy changes only to order ε^2, and hence has no effect on F_x. This simple remark enables us to make a rather wide generalisation. Suppose that in Fig. 5 S is now any closed surface drawn *inside* the body and enclosing some particular defect. Inside the body

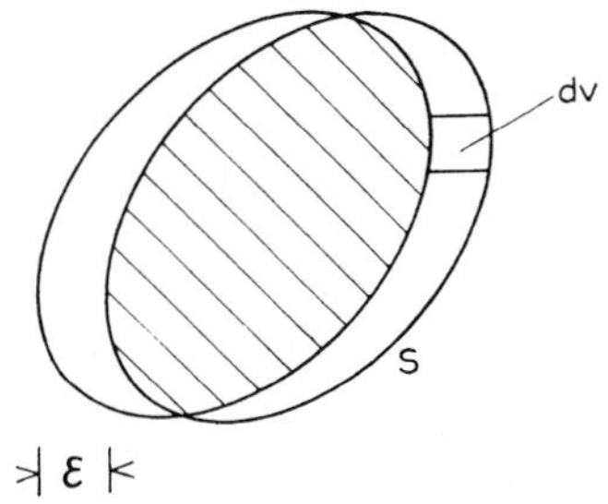

Fig. 5. Surface of integration used in calculating the force on a defect.

but outside S let there be other defects giving rise to internal stresses. Let the body also be stressed by loads applied to its surface. We may calculate the force on the defect inside S by repeating the previous argument, regarding the self-stressed material outside S together with the mechanism applying the surface loads as a single rather complicated loading machine. Thus to find the force on a given defect due to other defects and to surface loads we have only to enclose it in a surface S which isolates it from the other defects and from the surface of the body, and evaluate the integral (10) over S.

If we confine ourselves to the usual linear theory the expression for F_x can be transformed to a manageable form which gives well known results for the forces between defects, and the forces exerted on defects by applied stresses. Its real value, however, lies in the fact that it gives some idea of how far these results are valid. The derivation of eqn. (10) does not require the deformation to be infinitesimal, nor, since nothing specific is assumed about the form of the energy-density, need Hooke's law be obeyed. In fact, all we require is that the defect can be surrounded by a surface S which isolates it from other defects, and that an energy-density can be defined in the neighbourhood of S. What goes on inside the shaded region of Fig. 5 does not matter. Suppose, for example, that we have a cubic crystal containing a point defect with a cubically symmetrical stress field, and that we can draw S far enough away from it for the equations of linear elasticity to be valid on S. The surface integral in eqn. (10) can be transformed to give the familiar

References p. 58.

result $F_x = \Delta V \partial p / \partial x$, where ΔV is the amount by which the volume of the body changes when the defect is introduced, and p is the hydrostatic pressure which would be produced at the position of the defect by other sources of stress if the defect itself were absent. The fact that the stress fields due to the defect and to other sources of stress cannot, in fact, be simply superimposed near the defect is irrelevant.

It would be nice to be able, in this and other cases, to decide just when the continuum treatment becomes inadequate. It is not hard to fix a point where one should begin to feel uncomfortable about using continuum methods; the trouble is that if one were to act on this feeling a number of practically successful calculations would have to be rejected. This must be my excuse for having emphasised the applicability of the continuum approach rather than its limitations.

REFERENCES

1 R. Gomer, *J. Chem. Phys.*, *28* (1958) 457.
2 S. Moriguti, *Oyo Sugaki Rikigaku*, *1* (1947) 29, 87.
3 C. Eckart, *Phys. Rev.*, *73* (1948) 373.
4 For general reviews see E. Kröner, *Ergebnisse der angewandten Mathematik. 5, Kontinuumstheorie der Versetzungen und Eigenspannungen*, Springer Verlag, Berlin, 1958; B. Bilby, *Progess in Solid Mechanics*, Vol. 1, North Holland Publishing Co., 1959.
5 J. F. Nye, *Acta Met.*, *1* (1953) 153.
6 W. Thornton Read, *Acta Met.*, *5* (1957) 83.
7 J. D. Eshelby, *Proc. Roy. Soc. (London)*, *A 241* (1957) 376.
8 F. R. N. Nabarro, *Proc. Roy. Soc. (London)*, *A 175* (1940) 519.
9 H. Brooks, *Impurities and Imperfections*, Am. Soc. Metals, 1955.
10 E. Orowan, *Welding J.*, *34* (1955) 157 s.
11 J. D. Eshelby, *Solid State Physics*, Vol. 3, edited by F. Seitz and D. Turnbull, Academic Press Inc., New York, 1956, p. 79.

The elastic field outside an ellipsoidal inclusion

By J. D. Eshelby
Department of Metallurgy, University of Birmingham

(*Communicated by R. E. Peierls, F.R.S.—Received* 5 *June* 1959)

The results of an earlier paper are extended. The elastic field outside an inclusion or inhomogeneity is treated in greater detail. For a general inclusion the harmonic potential of a certain surface distribution may be used in place of the biharmonic potential used previously. The elastic field outside an ellipsoidal inclusion or inhomogeneity may be expressed entirely in terms of the harmonic potential of a solid ellipsoid. The solution gives incidentally the velocity field about an ellipsoid which is deforming homogeneously in a viscous fluid. An expression given previously for the strain energy of an ellipsoidal region which has undergone a shear transformation is generalized to the case where the region has elastic constants different from those of its surroundings. The Appendix outlines a general method of calculating biharmonic potentials.

1. Introduction

In a previous paper (Eshelby 1957; to be referred to as I) a method was given for finding the stresses set up in an elastic solid when a region within it (the 'inclusion') undergoes a change of form which would be a uniform homogeneous deformation if the surrounding material were absent. It was also shown that the results for the inclusion can be used to find how a uniform stress is disturbed by the presence of an ellipsoidal cavity, or more generally an ellipsoidal region whose elastic constants differ from those of the remaining material. It was emphasized in I that the elastic field inside the inclusion can be calculated without having to find the field outside the inclusion, and that a good deal of information can be derived from a knowledge of the internal field alone. Consequently the question of determining the field outside the inclusion or inhomogeneity was only briefly touched on. In the present paper the external problem is considered in more detail.

We first show (§ 2) that the biharmonic potential introduced in I in discussing the general inclusion may be replaced by the harmonic potential of a certain surface distribution. In § 3 the displacement due to an ellipsoidal inclusion is given in a form which involves only the harmonic potential of a solid ellipsoid, and it is reduced to a form suitable for numerical calculation of the stress. The Appendix describes a general method for calculating biharmonic potentials; an explicit expression is obtained for the ellipsoid.

It will be convenient to note here some errors in I. A shortcoming in the notation on p. 379 is corrected in § 2 of the present paper. On p. 380, last line, *for* p_{jk} *read* p^T_{jk}. The arithmetic (though not the principle) is incorrect in the second paragraph of p. 392. On the same page, sixth line from the bottom, *for* $V\gamma$ *read* V/γ.

2. The general inclusion

In I the following problem was solved. An inclusion bounded by a surface S in an infinite homogeneous and isotropic elastic medium undergoes a change of form which in the absence of the surrounding material would be an arbitrary homogeneous strain e^T_{ij}: to find the elastic state inside and outside S.

It was shown that the resulting displacement u^C_i is the same as that produced by a layer of body-force spread over S of amount $p^T_{ij}n_j\,\mathrm{d}S$ on each surface element $\mathrm{d}S$, where p^T_{ij} is the stress derived from e^T_{ij} by Hooke's law. The displacement produced at $\mathbf{r}$ by a point-force F_i at $\mathbf{r}'$ is

$$U_i = U_{ij}F_j,$$

where
$$U_{ij} = \frac{1}{4\pi\mu}\frac{\delta_{ij}}{|\mathbf{r}-\mathbf{r}'|} - \frac{1}{16\pi\mu(1-\sigma)}\frac{\partial^2}{\partial x_i \partial x_j}|\mathbf{r}-\mathbf{r}'|. \tag{2·1}$$

Hence
$$u^C_i(\mathbf{r}) = \int_S \mathrm{d}S\, p^T_{jk} n_k U_{ij}(\mathbf{r}-\mathbf{r}'). \tag{2·2}$$

(This corrects a deficiency of notation in (I, 2·5) and (I, 2·6).) On substituting (2·1) in (2·2) and converting the surface integrals to volume integrals we have

$$u^C_i = \frac{1}{16\pi\mu(1-\sigma)} p^T_{jk}\psi_{,ijk} - \frac{1}{4\pi\mu} p^T_{ik}\phi_{,k} \tag{2·3}$$

with
$$\phi = \int_V \frac{\mathrm{d}v}{|\mathbf{r}-\mathbf{r}'|} \quad \text{and} \quad \psi = \int_V |\mathbf{r}-\mathbf{r}'|\,\mathrm{d}v, \tag{2·4}$$

where the integrals are taken over the volume V enclosed by S.

The harmonic potential ϕ and the biharmonic potential ψ have the following properties:

$$\nabla^2\psi = 2\phi; \quad \nabla^4\psi = 2\nabla^2\phi = \begin{cases} -8\pi & \text{inside } S, \\ 0 & \text{outside } S. \end{cases} \tag{2·5}$$

The quantities
$$\phi,\ \phi_{,i},\ \psi,\ \psi_{,i},\ \psi_{,ij},\ \psi_{,ijk} \tag{2·6}$$

are continuous across S, while $\phi_{,ij}$ and $\psi_{,ijkl}$ have the discontinuities

$$\left.\begin{aligned} \phi_{,ij}(\text{out}) - \phi_{,ij}(\text{in}) &= 4\pi n_i n_j, \\ \psi_{,ijkl}(\text{out}) - \psi_{,ijkl}(\text{in}) &= 8\pi n_i n_j n_k n_l, \end{aligned}\right\} \tag{2·7}$$

where n_i is the outward normal to S.

The Appendix gives one method of reducing the determination of ψ to a problem in ordinary potential theory. Alternatively, u^C_i can be expressed in terms of ϕ and an additional harmonic function in the following way.

Equation (2·3) can be rewritten in terms of a harmonic vector B_i and a harmonic scalar β (Papkovich–Neuber):

$$u^C_i = B_i - \frac{1}{4(1-\sigma)}(x_m B_m + \beta)_{,i} \tag{2·8}$$

with and
$$\left.\begin{aligned} 4\pi\mu B_i &= -p^T_{ik}\phi_{,k} \\ 4\pi\mu\beta &= p^T_{jk}(x_j\phi_{,k} - \psi_{,jk}). \end{aligned}\right\} \tag{2·9}$$

In verifying that β is harmonic, and elsewhere, the relation

$$\nabla^2(pq) = p\nabla^2 q + q\nabla^2 p + 2p_{,m}q_{,m}$$

is useful.

The determination of β can be reduced to a standard problem in the theory of the ordinary harmonic potential. The normal derivative of β in the neighbourhood of the interface S is

$$\partial\beta/\partial n = \beta_{,m} n_m = (\delta_{jm}\phi_{,k} + x_j\phi_{,km} - \psi_{,jkm})\, n_m p^T_{jk}/4\pi\mu.$$

On crossing S, $\phi_{,km}$ increases abruptly by $4\pi n_k n_m$; $\phi_{,k}$ and $\psi_{,jkm}$ on the other hand are both continuous. Thus the corresponding increase in $\partial\beta/\partial n$ is

$$(\partial\beta/\partial n)\,(\text{out}) - (\partial\beta/\partial n)\,(\text{in}) = p^T_{jk} x_j n_k/\mu. \tag{2·10}$$

Since β falls to zero far from the inclusion (2·9) implies that β is the potential of a layer of attracting matter distributed with surface density $-p^T_{jk} x_j n_k/4\pi\mu$ over the interface.

3. The ellipsoidal inclusion

When the inclusion is bounded by the ellipsoid

$$x^2/a^2 + y^2/b^2 + z^2/c^2 = 1$$

the harmonic potential is

$$\phi = \pi abc \int_\lambda^\infty \frac{U\,du}{\Delta}, \tag{3·1}$$

where
$$U(u) = 1 - x^2/(a^2+u) - y^2/(b^2+u) - z^2/(c^2+u) \tag{3·2}$$

and
$$\Delta = (a^2+u)^{\frac{1}{2}}\,(b^2+u)^{\frac{1}{2}}\,(c^2+u)^{\frac{1}{2}}. \tag{3·3}$$

For an external point λ in the greatest root of $U(u) = 0$ and for an internal point $\lambda = 0$.

A similar expression for ψ is derived in the Appendix (equation (A 1)). This completes the formal solution of the inclusion problem. However, to reach a point where numerical calculation is possible it is more convenient to find the derivatives of ψ directly in terms of ϕ. Consider the function

$$f = x_1\phi_{,2} - \psi_{,12}$$

which figures as the coefficient of p^T_{12} in (2·9). It is harmonic inside and outside the inclusion, falls to zero at infinity and according to (2·7) there is a discontinuity

$$(\partial f/\partial n)\,(\text{out}) - (\partial f/\partial n)\,(\text{in}) = 4\pi x_1 n_2$$

in its normal derivative across the interface. The function

$$g = x_1\phi_{,2} - x_2\phi_{,1},$$

familiar in the hydrodynamic theory of the rotating ellipsoid, is harmonic and falls to zero at infinity. The discontinuity in its normal derivative across S is

$$(\partial g/\partial n)\,(\text{out}) - (\partial g/\partial n)\,(\text{in}) = 4\pi(x_1 n_2 - x_2 n_1) = 4\pi x_1 n_2(1 - b^2/a^2)$$

since the normal to the ellipsoid at x_i has components (I, 3·23)

$$n_1 = x_1/a^2h, \quad n_2 = x_2/b^2h, \quad n_3 = x_3/c^2h$$

and so $b^2x_1n_2 = a^2x_2n_1$. Hence the harmonic functions f and $ga^2/(a^2-b^2)$ are identical, since they behave similarly at infinity and have the same discontinuity in their normal derivative. Thus

$$f = \frac{a^2}{a^2-b^2}(x_1\phi_{,2}-x_2\phi_{,1})$$

and so
$$\psi_{,12} = \frac{a^2}{a^2-b^2}x_2\phi_{,1}+\frac{b^2}{b^2-a^2}x_1\phi_{,2}.$$

Similarly
$$\psi_{,23} = \frac{b^2}{b^2-c^2}x_3\phi_{,2}+\frac{c^2}{c^2-b^2}x_2\phi_{,3},$$

$$\psi_{,31} = \frac{c^2}{c^2-a^2}x_1\phi_{,3}+\frac{a^2}{a^2-c^2}x_3\phi_{,1}.$$

The requirement that $(\psi_{,ij})_{,k}$ be equal to $(\psi_{,ik})_{,j}$ leads to the identity

$$(b^2-c^2)\,x_1\phi_{,23}+(c^2-a^2)\,x_2\phi_{,31}+(a^2-b^2)\,x_3\phi_{,12} = 0$$

which may be verified independently.

If we were to try to find $\psi_{,11}$ by treating the function $x_1\phi_1-\psi_{,11}$ in the same way as f we should have to construct a harmonic function whose normal derivative has a discontinuity $4\pi x_1n_1$, and this is rather laborious (cf. Dyson 1891). But in fact there is no need to determine $\psi_{,11}$, $\psi_{,22}$ and $\psi_{,33}$ since the third derivatives which appear in (2·3) can all be made to depend on the four quantities ϕ, $\psi_{,12}$, $\psi_{,23}$, $\psi_{,31}$. If i, j, k are not all equal we have, trivially, $\psi_{,112} = (\psi_{,12})_{,1}$ and so forth, while if i, j, k are all equal we can use

$$\psi_{,111} = 2\phi_{,1}-(\psi_{,12})_{,2}-(\psi_{,13})_{,3}$$

and the two similar relations which are obtained by differentiating $\nabla^2\psi = 2\phi$.

In this way we find from (2·8)

$$8\pi(1-\sigma)\,u_1^C = \frac{e_{22}^T-e_{11}^T}{a^2-b^2}\frac{\partial}{\partial x_2}(a^2x_2\phi_{,1}-b^2x_1\phi_{,2})+\frac{e_{33}^T-e_{11}^T}{c^2-a^2}\frac{\partial}{\partial x_3}(c^2x_1\phi_{,3}-a^2x_3\phi_{,1})$$
$$-2\{(1-\sigma)\,e_{11}^T+\sigma(e_{22}^T+e_{33}^T)\}\,\phi_{,1}-4(1-\sigma)\,(e_{12}^T\phi_{,2}+e_{13}^T\phi_{,3})+\frac{\partial}{\partial x_1}\bar{\beta}, \qquad (3·4)$$

where
$$\bar{\beta} = \frac{2e_{12}^T}{a^2-b^2}(a^2x_2\phi_{,1}-b^2x_1\phi_{,2})+\frac{2e_{23}^T}{b^2-c^2}(b^2x_3\phi_{,2}-c^2x_2\phi_{,3})$$
$$+\frac{2e_{31}^T}{c^2-a^2}(c^2x_1\phi_{,3}-a^2x_3\phi_{,1}).$$

The expressions for u_2^C and u_3^C follow by cyclic permutation of $(1, 2, 3)$ and (a, b, c).

The harmonic potential can be written in the form

$$\phi = \frac{2\pi abc}{l^3}\left\{\left[l^2-\frac{x^2}{k^2}+\frac{y^2}{k^2}\right]F(\theta,k)+\left[\frac{x^2}{k^2}-\frac{y^2}{k^2k'^2}+\frac{z^2}{k'^2}\right]E(\theta,k)+\frac{l}{k'^2}\left[\frac{C}{AB}y^2-\frac{B}{AC}z^2\right]\right\},$$

where
$$A = (a^2+\lambda)^{\frac{1}{2}},\quad B = (b^2+\lambda)^{\frac{1}{2}},\quad C = (c^2+\lambda)^{\frac{1}{2}},\quad a > b > c,$$
$$l = (a^2-c^2)^{\frac{1}{2}},\quad k = (a^2-b^2)^{\frac{1}{2}}/(a^2-c^2)^{\frac{1}{2}},\quad k' = (b^2-c^2)^{\frac{1}{2}}/(a^2-c^2)^{\frac{1}{2}},$$

and F, E are elliptic integrals of modulus k and argument $\theta = \sin^{-1}(l/A)$. The differentiations required to find u_i^C, e_{ij}^C or p_{ij}^C can be carried out with the help of the relations

$$\partial F/\partial\lambda = -\tfrac{1}{2}l/ABC, \quad \partial E/\partial\lambda = -\tfrac{1}{2}lB/A^3C,$$

$$\partial\lambda/\partial x = 2x/Ah, \quad \partial\lambda/\partial y = 2y/Bh, \quad \partial\lambda/\partial z = 2z/Ch,$$

where

$$h^2 = x^2/A^4 + y^2/B^4 + z^2/C^4.$$

It is evident from (3·1) that λ may be treated as a constant in forming the *first* derivatives of ϕ. Finally λ has to be put equal to the greatest (in fact positive) root of

$$\lambda^3 - L\lambda^2 + M\lambda - N = 0, \tag{3·5}$$

where

$$L = r^2 - R^2,$$

$$M = a^2x^2 + b^2y^2 + c^2z^2 - a^2b^2 - b^2c^2 - c^2a^2 + r^2R^2,$$

$$N = a^2b^2c^2\left(\frac{x^2}{a^2} + \frac{y^2}{b^2} + \frac{z^2}{c^2} - 1\right)$$

with

$$R^2 = a^2 + b^2 + c^2, \quad r^2 = x^2 + y^2 + z^2.$$

4. Discussion

The field outside a homogeneous transformed inclusion is found by substituting the appropriate e_{ij}^T in (3·4). The external perturbing field of an inhomogeneous ellipsoid is found by inserting the e_{ij}^T of the 'equivalent inclusion', calculated from the unperturbed field e_{ij}^A by solving (I, 4·3, 4·4). It is possible to treat in the same way the case where the inhomogeneous ellipsoid has elastic constants, c_{ijkl}^*, say, which are anisotropic. It is only necessary to replace (I, 4·3, 4·4) by the six equations

$$c_{ijkl}^*(e_{ij}^C + e_{ij}^A) = \lambda(e^C + e^A - e^T)\,\delta_{ij} + 2\mu(e_{ij}^C + e_{ij}^A - e_{ij}^T) \tag{4·1}$$

and solve them for e_{ij}^T. It may seem pointless to solve the problem for an anisotropic inhomogeneity when we cannot deal with an anisotropic matrix. However, the result has been used by Kröner (1958) in discussing the elastic constants of anisotropic aggregates (cf. also Hershey 1954).

There is another problem which can be solved in terms of an equivalent inclusion. This is the case of an ellipsoidal region which undergoes a transformation strain, e_{ij}^{T*} say, and which in addition has elastic constants (which need not be isotropic) different from those of the matrix. From the present point of view it is irrelevant whether the difference in elastic constants existed originally or developed during or after the transformation. For brevity we shall refer to this inclusion as E^* and use E to denote our standard inclusion which has the same elastic constants as the matrix and has undergone a transformation strain e_{ij}^T. We suppose that before transformation E and E^* are of identical form. After transformation E can be replaced by E^*, with continuity of displacement and stress, provided that E^* when constrained to have the same final form as E develops the same stresses as E. This requires that the conditions

$$p_{ij}^I \equiv \lambda(e^C - e^T)\,\delta_{ij} + 2\mu(e_{ij}^C - e_{ij}^T) = c_{ijkl}^*(e_{kl}^C - e_{kl}^{T*}) \tag{4·2}$$

be satisfied, or, if the inclusion is isotropic,

$$p^I \equiv \kappa(e^C - e^T) = \kappa^*(e^C - e^{T*}), \quad 'p^I_{ij} \equiv 2\mu('e^C_{ij} - 'e^T_{ij}) = 2\mu^*('e^C_{ij} - 'e^{T*}_{ij}) \tag{4·3}$$

with the notation of I, § 2. Here e^C_{ij} specifies the final form of E or E^*, and e^T_{ij} or e^{T*}_{ij} is the part of e^C_{ij} which produces no stress. If we replace e^C_{kl} by $S_{klmn}e^T_{mn}$ (cf. I, 3·6) (4·1) becomes a set of equations from which e^T_{ij} can be determined when e^{T*}_{ij} and c^*_{ijkl} are prescribed. The elastic field outside the inclusion is given by (3·4) with these values of e^T_{ij}; the stress inside is, of course, given by (4·2) or (4·3) directly. The elastic energy in the matrix is given by (I, 2·20), while the energy in the inclusion is given by (I, 2·19) with e^T_{ij} replaced by e^{T*}_{ij}. Consequently the total elastic energy is

$$E_{\text{el.}} = -\tfrac{1}{2}Vp^I_{ij}e^{T*}_{ij}.$$

When only the non-diagonal components of e^{T*}_{ij} are involved the solution of (4·3) for e^C_{ij} or e^T_{ij} is simple. We may, for example, generalize a result in I (p. 391) which has been used in discussing the formation of martensite (Christian 1958, 1959; Kaufman 1959). Let an ellipsoidal region with elastic constants μ^*, κ^* in a matrix with constants μ, κ undergo a shear transformation in which e^{T*}_{13} is the only non-vanishing component. Then we have

$$E_{\text{el.}} = 2\gamma^*\mu V(e^{T*}_{13})^2 \tag{4·4}$$

and

$$\frac{\text{energy in matrix}}{\text{energy in inclusion}} = \frac{\mu^*}{\mu}\frac{1-\gamma}{\gamma}, \tag{4·5}$$

where

$$\gamma^* = \frac{\gamma\mu^*}{\gamma\mu + (1-\gamma)\mu^*}$$

and γ is the accommodation coefficient defined by (I, 5·2). For good accommodation (small γ) $E_{\text{el.}}$ is fairly insensitive to the relative values of μ and μ^*, though (4·5) is not. In fact if μ^*/μ increases from $\frac{1}{2}$ to 2 the ratio (4·5) increases by a factor 4, but (4·4) only by a factor $(1+\gamma)/(1-\frac{1}{2}\gamma)$. The value of σ to be used in calculating γ is that for the matrix. The expressions (4·4) and (4·5) are independent of κ^*, as they must be, since the inclusion is in pure shear. To deal with the case where e^{T*}_{11}, e^{T*}_{22}, e^{T*}_{33} are not zero we should have to solve a set of simultaneous equations. Robinson (1951) has given an extensive discussion of the case where e^{T*}_{ij} is a pure dilatation.

If the inclusion is entirely rigid ($c^*_{ijkl} \to \infty$) we must have, according to (4·1),

$$e^C_{ij} = S_{ijkl}e^T_{kl} = e^{T*}_{ij}; \tag{4·6}$$

that is, the constrained strain e^C_{ij} is equal to the transformation strain e^T_{ij}, or, in engineering language, there is no spring-back. In other words, the problem becomes that of finding the elastic field outside an ellipsoidal surface on which the displacement is required to take the value

$$u_i = (e^C_{ij} + \varpi^C_{ij})x_j. \tag{4·7}$$

This is not a completely arbitrary linear function of the x_i, since if e^C_{ij} is prescribed, e^T_{ij} may be found from (4·6) and ϖ^C_{ij} is then fixed by (I, 3·19). The origin of this connexion lies in the fact that there is no net couple acting on the rigid inclusion E^*, since there was none on the inclusion E which it replaced. To obtain an arbitrary ϖ_{ij}^C we should have somehow to apply a suitable external couple to the rigid inclusion. Edwardes (1893) has discussed the elastic field about an embedded ellipsoid which is given an arbitrary small rotation in this way without change of form. Daniele (1911) has solved the combined problem of determining the elastic field outside an ellipsoidal surface which is subject to the linear displacement (4·7) with independent e_{ij}^C and ϖ_{ij}^C.

It was pointed out in I (p. 387) that the elastic solution of the problem of a rigid ellipsoidal inclusion perturbing a uniform stress could be adapted to give the solution of a related problem in slow viscous motion. We can use the results for a rigid transformed inclusion in the same way to solve the following problem. An ellipsoid in a fluid of viscosity μ is undergoing a steady homogeneous change of form specified by the rate-of-strain tensor e_{ij}^{T*}, that is, its semi-axes are changing length at the rate $\dot{a} = e_{11}^{T*}$, ... and the angles between them are altering at the rate $\dot{\theta}_{ab} = 2e_{12}^{T*}, \ldots$; to find the velocity of the fluid. Solve (4·6) for e_{ij}^T, putting $\sigma = \frac{1}{2}$ in the relations (I, 3·7) which define the S_{ijkl}. Insert these values of e_{ij}^T in (3·4) and put $\sigma = \frac{1}{2}$. Then u_i^C is the velocity required.

Several authors (see Robinson (1951) and the references he gives) have solved problems concerning ellipsoidal inclusions and inhomogeneities by using ellipsoidal co-ordinates. It is perhaps worth comparing this approach with the method outlined in I and the present paper. It is clear that in order to set up the problem and solve it formally there is no need to introduce ellipsoidal co-ordinates. If our object is merely to find the stresses immediately outside the ellipsoid we calculate the S_{ijkl} from b/a and c/a by consulting a table of elliptic integrals or Osborn's (1945) curves. The e_{ij}^T are either given or have to be calculated by solving the set of simultaneous equations (4·1) or (4·2) when e_{ij}^A or e_{ij}^{T*} are the initial data. The e_{ij}^C for an internal point are then calculated from the e_{ij}^T and S_{ijkl} and the stress or strain immediately outside the ellipsoid follows from (I, 2·13). The elastic energy, interaction energy and the remote field also follow from (I, 2·21, 3·22, 4·10, 2·18) without trouble. If the calculation is to stop at this point the use of ellipsoidal co-ordinates seems to offer no advantages.

In order to find the elastic field at an arbitrary external point we have to insert the appropriate e_{ij}^T in (3·4) and carry out several tedious but straightforward differentiations. Finally (3·5) must be solved for λ. This equation reduces to a quadratic if the point of interest lies on one of the co-ordinate planes. (It is also a quadratic for arbitrary x, y, z if the ellipsoid is a spheroid.) Again, if we are only interested in points on, say, the x-axis we have simply $\lambda = x^2 - a^2$. If the field at a general point is really required we may use the tables given by Emde (1940) or by Jahnke & Emde (1943) where, in effect, λ/L is tabulated as a function of M/L^2 and N/L^3. Alternatively, if it is merely required to survey the stresses around the ellipsoid it may be sufficient to pick a triplet of values for x, y, λ, calculate the corresponding z from $U(\lambda) = 0$ (equation (3·2)) and repeat the process for other choices of x, y, λ.

It has to be admitted that, except in the simplest cases, a calculation of the external field is laborious. The real value of having a solution for the ellipsoid with three unequal axes lies rather in the fact that it furnishes a number of useful quantities (for example, elastic and interaction energies, the interior and remote fields, the stresses at the inner surface of the matrix) without much effort even in the general case, and that from it can be obtained the solutions of a number of more specialized problems.

The use of ellipsoidal co-ordinates does not really simplify the calculation of the external field. If we have already available a three-dimensional ellipsoidal co-ordinate network (constructed for the appropriate value of k) we may locate points relative to the ellipsoid without having to solve a cubic equation. If not, it is necessary to carry out the equivalent of finding all three roots of (3·5) in order to obtain the ellipsoidal co-ordinates of a point whose Cartesian co-ordinates are specified. The stresses are somewhat complicated expressions involving elliptic functions of the ellipsoidal co-ordinates. They appear in the analysis referred to local rectangular axes parallel to the curvilinear co-ordinate lines at the point under consideration. There does not seem to be any special advantage in this orientation of axes except perhaps at the surface of the ellipsoid, and there, as we have seen, a much simpler treatment is adequate.

Appendix. A method of calculating biharmonic potentials

If into the integrand of the expression for ψ (equation (2·4)) we introduce a factor unity in the form

$$1 = \frac{(x_i - x'_i)(x_i - x'_i)}{|\mathbf{r}-\mathbf{r}'|^2} = \frac{r^2 - 2x_i x'_i + r'^2}{|\mathbf{r}-\mathbf{r}'|^2}$$

we obtain

$$\psi = r^2\phi - 2x_i \int_V \frac{x'_i \mathrm{d}v}{|\mathbf{r}-\mathbf{r}'|} + \int_V \frac{r'^2 \mathrm{d}v}{|\mathbf{r}-\mathbf{r}'|}.$$

The integrals are the harmonic potentials of solids bounded by S and having variable densities $\rho(\mathbf{r}) = x_i$, $\rho(\mathbf{r}) = r^2$.

Ferrers (1877) and Dyson (1891) have shown how to calculate the harmonic potential of a solid ellipsoid whose density is proportional to a polynominal in x, y, z. (There is a factor 2 missing from Ferrers's expression (p. 10) for the potential of a density distribution proportional to x). From their results we obtain

$$\psi = \pi abc \int_\lambda^\infty \left\{ \tfrac{1}{2} \frac{\mathrm{d}}{\mathrm{d}u} \left(\frac{U^2 u^2}{\Delta} \right) - \tfrac{1}{4} \frac{U^2 u}{\Delta} \right\} \mathrm{d}u. \tag{A 1}$$

At first sight it seems as if the first term could be removed by a trivial integration. Unfortunately, although the contribution from the lower limit is zero, that from the upper limit is infinite. In fact the two terms in the integrand of (A 1) behave for large u like $\frac{1}{4}u^{-\frac{1}{2}}$ and $-\frac{1}{4}u^{-\frac{1}{2}}$ respectively, so that they must be taken together to secure convergence. We can, however, treat the integral arising from the first term as an infinite constant independent of x, y, z in the following sense. Let us put formally

$$\psi + \text{const.} = -\tfrac{1}{4}\pi abc \int_\lambda^\infty \frac{U^2 u}{\Delta} \mathrm{d}u \tag{A 2}$$

and differentiate with respect to x_i, ignoring the undefined contribution from the variation of the upper limit. (The lower limit gives no contribution, since $U = 0$ there.) The result is the correct expression for $\psi_{,i}$. Dirichlet (1839) has given an expression for the potential of a solid ellipsoid when the law of attraction is the p^{-1}th power, valid only for $2 \leqslant p < 3$. If we put $p = 0$ (the biharmonic case) we obtain the divergent integral (A 2). Hadamard (1923) has given a method for extracting a 'finite part' from a certain type of divergent integral. Applied to the integral on the right of (A 2) it gives precisely (A 1).

References

Christian, J. W. 1958 *Acta Metall.* **6**, 379.
Christian, J. W. 1959 *Acta Metall.* **7**, 218.
Daniele, E. 1911 *Nuovo Cim.* [6], **1**, 211.
Dirichlet, G. L. 1839 *Verh. K. Preuss Akad. Wiss.* p. 18 (*Werke*, **1**, 383).
Dyson, F. W. 1891 *Quart. J. Pure Appl. Math.* **25**, 259.
Edwardes, D. 1893 *Quart. J. Pure Appl. Math.* **26**, 270.
Emde, F. 1940 *Tables of elementary functions.* Leipzig: Teubner.
Eshelby, J. D. 1957 *Proc. Roy. Soc.* A, **241**, 376.
Ferrers, N. M. 1877 *Quart. J. Pure Appl. Math.* **14**, 1.
Hadamard, J. 1923 *Lectures on Cauchy's problem*, p. 133. New Haven: Yale University Press.
Hershey, A. V. 1954 *J. Appl. Mech.* **21**, 1.
Jahnke, E. & Emde, F. 1943 *Tables of functions.* New York: Dover Publications.
Kaufman, L. 1959 *Acta Metall.* **7**, 216.
Kröner, E. 1958 *Z. Phys.* **151**, 504.
Osborn, J. A. 1945 *Phys. Rev.* **67**, 351.
Robinson, K. 1951 *J. Appl. Phys.* **22**, 1045.

CHAPTER III

ELASTIC INCLUSIONS AND INHOMOGENEITIES

BY

J. D. ESHELBY

Department of Physical Metallurgy,
University of Birmingham, England

CONTENTS

§ 1. Introduction

This review is concerned with the two following problems in the infinitesimal theory of elasticity, and with their inter-relation and generalization.

(i) The transformation problem.

A region (the 'inclusion') in a homogeneous elastic medium undergoes a permanent change of form which, in the absence of the constraint imposed by its surroundings (the 'matrix'), would be a prescribed uniform strain. To find the elastic field in matrix and inclusion.

(ii) The inhomogeneity problem.

A region in an otherwise homogeneous elastic medium has elastic constants differing from those of the remainder. To find how an applied stress, uniform at large distances, is disturbed by the inhomogeneity.

We shall not consider two-dimensional problems, where complex variable methods can be used, and the number of special cases which may be solved is unlimited. The three-dimensional inhomogeneity problem has been discussed extensively, particularly the special case of a cavity, i.e. an 'inhomogeneity' with vanishing elastic constants. The inclusion problem has received less attention, but is encountered in the discussion of various phenomena in solid-state physics, for example martensitic transformations and the formation of precipitates. STERNBERG [1958] has given an excellent annotated bibliography of the three-dimensional inhomogeneity (and inclusion) problems which have been solved.

We shall be largely concerned with the special case where the inclusion or inhomogeneity takes the form of the general ellipsoid with three unequal axes. There are two reasons for this. First, it appears to be the most general case whose solution can be given in a manageable form. Secondly, in this particular case there is a close connexion between the transformation and inhomogeneity problems. It stems from the fact that, as we shall see, the stress is constant throughout an ellipsoidal inclusion which has undergone a uniform transformation. As an illustration of this connexion suppose that it is required to solve the inhomogeneity problem for an ellipsoidal cavity. On the stress-field due to the inclusion superimpose everywhere a uniform stress equal and opposite to the uniform stress in the inclusion. The inclusion is then free of stress and may be removed. We are left with a stress-field which becomes uniform at infinity and which gives zero traction on the surface of the ellipsoid, as required. The general problem of an ellipsoidal inhomogeneity can be handled by an extension of this argument.

Among closed surfaces the ellipsoid alone has this convenient property. It shares it with other second-degree surfaces, and the analysis of §§ 3,4 can in fact be applied with trivial modifications to hyperboloids and paraboloids. However, the properties of such infinite 'inclusions' are not of much interest, and we shall not consider them. The corresponding in homogeneity problem is essentially the problem of stress-concentration by hyperboloidal notches. The cases which are of practical use have been discussed by NEUBER [1958].

We begin (§ 2) by finding a solution to problem (i) for an inclusion of arbitrary shape. The argument used is somewhat intuitive, but it is verified that the solution which is obtained does in fact satisfy the conditions of the problem. In § 3 the special case of an ellipsoidal inclusion is worked out to a point where numerical calculation is possible. Section 4 begins with a discussion of the inhomogeneity problem for an inhomogeneous region of arbitrary shape (though not much progress can be made) and the solution for the ellipsoidal inclusion is obtained from the results of § 2 in the way already indicated. By taking advantage of the connexion between inclusions and Somigliana dislocations (§ 5) it is possible to solve the problem of an ellipsoidal inhomogeneity perturbing a non-uniform stress-field. In § 6 we present some selected physical applications of the theory.

We shall use the familiar suffix notation. Repeated suffixes are to be summed over the values 1, 2, 3 and suffixes following a comma will denote differentiation with respect to the Cartesian coordinates x_1, x_2, x_3, e.g. $u_{i,j} = \partial u_i/\partial x_j$, $\psi_{,ijk} = \partial^3\psi/\partial x_i \partial x_j \partial x_k$.

Displacement u_i and strain e_{ij} are related by

$$e_{ij} = \tfrac{1}{2}(u_{i,j} + u_{j,i})\,. \tag{1.1}$$

Stress p_{ij} and strain are related by

$$p_{ij} = \lambda e_{kk}\delta_{ij} + 2\mu e_{ij} \tag{1.2}$$

in an isotropic medium with Lamé constants λ, μ. If a second pair of quantities p'_{ij}, e_{ij} satisfy

$$p'_{ij} = \lambda e'_{kk}\delta_{ij} + 2\mu e'_{ij}$$

then

$$p_{ij}e'_{ij} = p'_{ij}e_{ij}\,. \tag{1.3}$$

A set of quantities bearing a common affix, e.g. $\overset{\mathrm{C}}{u_i}$, $\overset{\mathrm{C}}{e_{ij}}$, $\overset{\mathrm{C}}{p_{ij}}$ will be supposed to be related by (1.1) and (1.2) unless otherwise stated. It will sometimes be convenient to use the notation

$$f = f_{kk}\,, \qquad 'f_{ij} = f_{ij} - \tfrac{1}{3}f_{kk}\delta_{ij} \tag{1.4}$$

to denote the scalar and deviatoric parts of a second-order tensor.

We shall often make use of the formula

$$\nabla^2(pq) = p\nabla^2 q + q\nabla^2 p + 2p_{,k}q_{,k} \tag{1.5}$$

for calculating the Laplacian of a product, and of Gauss's theorem in the form

$$\int_S A_{\ldots}\mathrm{dS}_k = \int_V A_{\ldots,k}\mathrm{d}v \tag{1.6}$$

where S is a closed surface enclosing the volume V. Here and elsewhere dS_k is an abbreviation for $n_k\mathrm{dS}$, where dS is an element of surface and n_k is its normal.

§ 2. The General Transformed Inclusion

2.1. The elastic field

In this section we give a formal solution of the following problem: A region bounded by a closed surface S in a homogeneous isotropic elastic medium undergoes a change of form which but for the constraint imposed by the surrounding material would be an arbitrary homogeneous strain $\overset{\mathrm{T}}{e_{ij}}$. To find the resulting elastic field inside and outside S.

It will be convenient to refer to the material inside S as the 'inclusion' and to the material outside S as the 'matrix'. The strain will be called the 'transformation strain', or, following Robinson [1951], the 'stress-free strain'. We assume that adjacent points immediately inside and outside S suffer no relative displacement; in other words the inclusion is 'bonded' to the matrix before, during and after the transformation. The problem can also be formulated as follows. To find a state of self-stress in an infinite body, with the following property: on making a cut over a prescribed closed surface S we are left with a stress-free cavity and a stress-free solid bounded by surfaces S_1, S_2 such that S_2 is transformed into S_1 by the homogeneous strain $-\overset{\mathrm{T}}{e}_{ij}$.

TIMOSHENKO and GOODIER [1951] have given a method for finding the elastic field in a material in which each volume element alters its unconstrained shape. Our basic inclusion problem is merely the special case in which the change of shape is identical for all the volume elements inside a certain surface S and is zero for all elements outside S. The method of calculation we shall use (ESHELBY [1957]) is essentially equivalent to theirs.

It is first necessary to decide how to define the displacement. Let us fix our attention on some marked point $\boldsymbol{r}$ in the material, with coordinates $x_i(\boldsymbol{r})$ and suppose that, as we watch, the transformation takes place gradually, by some physical mechanism unspecified. Every point of the medium moves, and when the transformation is over the marked point will have different coordinates, $x_i(\boldsymbol{r}) + \overset{\mathrm{C}}{u}_i(\boldsymbol{r})$ say. We take $\overset{\mathrm{C}}{u}_i$ as our displacement function. The state of zero displacement is the state of the material before the transformation has occurred.

We may contrast this definition of the displacement with another, perhaps equally natural, one. Suppose that after the transformation has occurred we make a cut over the surface S. Points on either side of the cut will move relatively as the stresses relax. For simplicity suppose that the two faces of the cut shrink away from each other everywhere, leaving a gap. During the relaxation every point of the matrix or inclusion suffers a certain displacement, $-\overset{\mathrm{D}}{u}_i$ say. In place of $\overset{\mathrm{C}}{u}_i$ we might take $\overset{\mathrm{D}}{u}_i$ as our displacement function. In the matrix both $\overset{\mathrm{C}}{u}_i$ and $\overset{\mathrm{D}}{u}_i$ are measured from a state in which the matrix is free of stress. Consequently in the matrix $\overset{\mathrm{C}}{u}_i = \overset{\mathrm{D}}{u}_i$. In the inclusion $\overset{\mathrm{C}}{u}_i$ is measured from a state in which the inclusion is untransformed and unstressed, whereas $\overset{\mathrm{D}}{u}_i$ is measured from a state in which the inclusion has transformed but is free of stress because the constraint due to the matrix has been removed. Because of the gap which appears when matrix and inclusion are cut apart $\overset{\mathrm{D}}{u}_i$ is discontinuous across S, whereas $\overset{\mathrm{C}}{u}_i$ is continuous. It is generally more convenient to use $\overset{\mathrm{C}}{u}_i$, but we shall refer to $\overset{\mathrm{D}}{u}_i$ again in § 5.

The displacement $\overset{\mathrm{C}}{u}_i$ will be calculated with the help of a sequence of imaginary cutting, straining and welding operations. This approach is somewhat alien to the usual methods of applied mathematics; the argument can, of course, be considered to be a purely heuristic one which points to a result (eq. (2.8)) whose validity has to be tested.

We shall suppose, to begin with, that the matrix extends to infinity. In the unstrained medium mark out the boundary S of the proposed transformed inclusion. Make a cut over S and remove the inclusion. Allow the transformation to occur. After it has suffered the uniform transformation $\overset{\mathrm{T}}{e}_{ij}$ the inclusion can no longer be fitted without strain into the cavity from which it was taken. Apply surface tractions $-\overset{\mathrm{T}}{p}_{ij}n_j$ to the surface of the inclusion, where

$$\overset{\mathrm{T}}{p}_{ij} = \lambda \overset{\mathrm{T}}{e}_{mm}\delta_{ij} + 2\mu \overset{\mathrm{T}}{e}_{ij} . \tag{2.1}$$

This produces a strain $-\overset{\mathrm{T}}{e}_{ij}$ in the inclusion and restores it to the form it had before transformation. Put the inclusion back in the cavity, still maintaining the surface tractions. Weld the material together across S. The surface tractions thus become an embedded layer of body force of amount

$$\mathrm{d}F_i = -\overset{\mathrm{T}}{p}_{ij}n_j\mathrm{dS} \tag{2.2}$$

on each element dS of S. The matrix is now unstressed and there is a uniform stress $-\overset{\mathrm{T}}{p}_{ij}$ in the inclusion. Further, every point in matrix or inclusion has the same position as it had before the transformation. That is, as regards displacement, the material is in the initial state from which we have agreed to measure $\overset{\mathrm{C}}{u}_i$. This state only differs from the required final state by the presence of the layer of body force (2.2). To get rid of this unwanted layer we apply an equal and opposite layer of body force

$$\mathrm{d}F_i = +\overset{\mathrm{T}}{p}_{ij}n_j\mathrm{dS} = \overset{\mathrm{T}}{p}_{ij}\mathrm{dS}_j \tag{2.3}$$

over S. The displacement induced by this operation is the displacement which we are trying to calculate.

A point force F_i at $\boldsymbol{r}'$ produces a displacement

$$u_i(\boldsymbol{r}) = U_{ij}F_j$$

at $\boldsymbol{r}$, where† (LOVE [1954])

$$U_{ij} = \frac{1}{4\pi\mu}\frac{\delta_{ij}}{|\boldsymbol{r}-\boldsymbol{r}'|} - \frac{1}{16\pi\mu(1-\sigma)}\frac{\partial^2}{\partial x_i \partial x_j}|\boldsymbol{r}-\boldsymbol{r}'|$$

$$= \frac{1}{4\pi\mu}\left(\tfrac{1}{2}\nabla^2 - \frac{1}{4(1-\sigma)}\frac{\partial^2}{\partial x_i \partial x_j}\right)|\boldsymbol{r}-\boldsymbol{r}'| \tag{2.4}$$

since

$$|\boldsymbol{r}-\boldsymbol{r}'|^{-1} = \tfrac{1}{2}\nabla^2|\boldsymbol{r}-\boldsymbol{r}'| . \tag{2.5}$$

† σ is Poisson's ratio.

Hence

$$u_i^{\mathrm{C}}(\boldsymbol{r}) = \int_{\mathrm{S}} \mathrm{dS}_k p_{jk}^{\mathrm{T}} U_{ij}(|\boldsymbol{r}-\boldsymbol{r}'|)\,. \tag{2.6}$$

But by Gauss's theorem

$$\int |\boldsymbol{r}-\boldsymbol{r}'|\,\mathrm{dS}_k = \int \frac{\partial}{\partial x_k'} |\boldsymbol{r}-\boldsymbol{r}'|\,\mathrm{d}v = -\frac{\partial}{\partial x_k}\int |\boldsymbol{r}-\boldsymbol{r}'|\mathrm{d}v$$

and so

$$u_i^{\mathrm{C}} = \frac{1}{16\pi\mu(1-\sigma)} p_{jk}^{\mathrm{T}}\,\psi_{,ijk} - \frac{1}{4\pi\mu} p_{ik}^{\mathrm{T}}\,\varphi_{,k} \tag{2.7}$$

or

$$u_i^{\mathrm{C}} = \frac{1}{8\pi(1-\sigma)} e_{jk}^{\mathrm{T}}\,\psi_{,ijk} - \frac{1}{2\pi} e_{ik}^{\mathrm{T}}\,\varphi_{,k} - \frac{\sigma}{4\pi(1-\sigma)} e^{\mathrm{T}}\,\varphi_{,i} \tag{2.8}$$

where

$$\varphi(\boldsymbol{r}) = \int_V \frac{\mathrm{d}v}{|\boldsymbol{r}-\boldsymbol{r}'|} \quad \text{and} \quad \psi(\boldsymbol{r}) = \int_V |\boldsymbol{r}-\boldsymbol{r}'|\,\mathrm{d}v\,. \tag{2.9}$$

φ is the ordinary (harmonic or Newtonian) potential of attracting matter of unit density filling the volume V bounded by S. ψ is the corresponding biharmonic potential. Geometrically, ψ/V is the mean distance of the point $\boldsymbol{r}$ from all the points inside S.

From (2.5)

$$\nabla^2\psi = 2\varphi\,. \tag{2.10}$$

The following results follow from the theory of attraction (POINCARÉ [1899], MACMILLAN [1958]).

$$\nabla^4\psi = 2\nabla^2\varphi = \begin{cases} -8\pi \text{ outside S} \\ 0 \text{ inside S} \end{cases} \tag{2.11}$$

$$\varphi,\ \varphi_{,i} \text{ are continuous across S} \tag{2.12}$$

$$\varphi_{,ij}\,(\text{out}) - \varphi_{,ij}\,(\text{in}) = 4\pi n_i n_j\,. \tag{2.13}$$

The last equation gives the difference in the second derivative at two adjacent points immediately inside and outside S at a point where the normal to S is n_i. We shall use a similar notation for other quantities which are discontinuous across S. Eq. (2.13) is a re-statement of the result that the discontinuity in attraction across a double layer is 4π times its moment. More generally, a function satisfying

$$\nabla^2 U = -4\pi\varrho \tag{2.14}$$

suffers a discontinuity in its second derivatives

$$U_{,kl}\,(\text{out}) - U_{,kl}\,(\text{in}) = -4\pi\{\varrho(\text{out}) - \varrho(\text{in})\}n_k n_l \tag{2.15}$$

on crossing a surface across which ϱ is discontinuous. But $\psi_{,ij}$ satisfies (2.14) with $\varrho = -2\varphi_{,ij}/4\pi$ and so from (2.15) we obtain the relation

$$\psi_{,ijkl}\,(\text{out}) - \psi_{,ijkl}\,(\text{in}) = 8\pi n_i n_j n_k n_l\,. \tag{2.16}$$

By similar arguments one finds that

$$\psi\,,\;\psi_{,i}\,,\;\psi_{,ijk}\ \text{are continuous across S.} \tag{2.17}$$

The stress in the matrix is

$$\overset{\mathrm{C}}{p}_{ij} = \lambda \overset{\mathrm{C}}{u}_{m,m}\,\delta_{ij} + \mu(\overset{\mathrm{C}}{u}_{i,j} + \overset{\mathrm{C}}{u}_{j,i})\,. \tag{2.18}$$

Since the inclusion was already subject to a uniform stress $-\overset{\mathrm{T}}{p}_{ij}$ before the body force (2.3) was applied, the stress in the inclusion is

$$\overset{\mathrm{I}}{p}_{ij} = \overset{\mathrm{C}}{p}_{ij} - \overset{\mathrm{T}}{p}_{ij} \tag{2.19}$$

with $\overset{\mathrm{C}}{p}_{ij}$ derived from $\overset{\mathrm{C}}{u}_{ij}$ as in (2.18).

We outline a method of verifying formally that the proposed solution (2.8), (2.18), (2.19) does in fact solve the inclusion problem. From (2.7) it follows that $\overset{\mathrm{C}}{u}_i$ satisfies the equilibrium equation

$$\mu\nabla^2 u_i + (\lambda + \mu)u_{m,mi} = 0$$

and from (2.12), (2.17) that it is continuous across S. If the stress is defined by (2.18) and (2.19) the relations (2.13) and (2.16) show that $p_{ij}n_j$ is continuous across S. Let an additional displacement $-\overset{\mathrm{C}}{u}_i(\boldsymbol{r}')$ be imposed on all points $\boldsymbol{r}'$ or the inner boundary of the matrix. By the uniqueness theorem of the theory of elasticity (LOVE [1954]) it will produce an additional displacement $-\overset{\mathrm{C}}{u}_i(\boldsymbol{r})$ throughout the matrix, and so leave it stress-free. Likewise an additional displacement $-\overset{\mathrm{C}}{u}_i(\boldsymbol{r}')$ imposed on all points $\boldsymbol{r}'$ of the surface of the inclusion induces a displacement $-\overset{\mathrm{C}}{u}_i(\boldsymbol{r})$ throughout its interior and so by (2.19) leaves it in a state of uniform stress $-\overset{\mathrm{T}}{p}_{ij}$. At this point the inner surface of the matrix and the outer surface of the inclusion still fit perfectly, because of the continuity of $\overset{\mathrm{C}}{u}_i$. If the tractions $-\overset{\mathrm{T}}{p}_{ij}n_j$ on the inclusion are reduced to zero the inclusion suffers a uniform strain $\overset{\mathrm{T}}{e}_{ij}$ and we are left with an unstressed matrix and an unstressed inclusion between whose surfaces there is the required misfit.

Equation (2.7) can be written in the Boussinesq-Papkovich-Neuber form

$$\overset{\mathrm{C}}{u}_i = B_i - \frac{1}{4\pi(1-\sigma)}\,(x_m B_m + \beta)_{,i} \tag{2.20}$$

with harmonic B_i and β:

$$4\pi\mu B_i = -\overset{\mathrm{T}}{p}_{ik}\,\varphi_{,k}\,, \qquad 4\pi\mu\beta = \overset{\mathrm{T}}{p}_{jk}\,f_{jk} \tag{2.21}$$

where

$$f_{ij} = x_i\varphi_{,j} - \psi_{,ij}\,. \tag{2.22}$$

That f_{ij} (and hence β) is harmonic inside and outside the inclusion follows from (1.5) and (2.10). Further f_{ij} behaves like r^{-1} for large r, while its normal derivative

$$\partial f_{ij}/\partial n = \varphi_{,j}n_i + x_i\varphi_{,jk}n_k - \psi_{,ijk}n_k$$

suffers a discontinuity $4\pi x_i n_i n_k n_k = 4\pi x_i n_i$ on passing through S, by (2.13) and (2.16). Hence f_{ij} is the harmonic potential of a layer of attracting matter distributed over S with surface density $x_i n_j$. In this way the biharmonic potential ψ can be replaced by the harmonic potential β of a certain surface layer.

It is interesting to see how much information can be obtained when our knowledge of φ and ψ is incomplete. We know in any case that φ and ψ behave as V/r and Vr for large r, and hence, from (2.7) that the field at large distance from the inclusion is given by

$$\overset{\mathrm{C}}{u}_i(\boldsymbol{r}) = \frac{V\overset{\mathrm{T}}{e}_{jk}\,g_{ijk}}{8\pi(1-\sigma)r^2} \tag{2.23}$$

where

$$g_{ijk} = (1-2\sigma)(\delta_{ij}l_k + \delta_{ik}l_j - \delta_{jk}l_i) + 3l_il_jl_k$$

and l_i is a unit vector joining the origin to the point of observation $\boldsymbol{r}$. If only φ is known we can find the dilatation and rotation:

$$e^{\mathrm{C}} = -\frac{1-2\sigma}{8\pi\mu(1-\sigma)}\overset{\mathrm{T}}{p}_{ik}\,\varphi_{,ik} \tag{2.24}$$

$$4\pi\overset{\mathrm{C}}{\omega}_{il} = 2\pi(\overset{\mathrm{C}}{u}_{i,l} - \overset{\mathrm{C}}{u}_{l,i}) = \overset{\mathrm{T}}{e}_{lk}\,\varphi_{,ki} - \overset{\mathrm{T}}{e}_{ik}\,\varphi_{,kl}\,. \tag{2.25}$$

If $\overset{\mathrm{T}}{e}_{ij}$ happens to be a pure dilatation we can find the complete field in terms of φ:

$$\overset{\mathrm{C}}{u}_i = -\frac{1+\sigma}{12\pi(1-\sigma)}\overset{\mathrm{T}}{e}\,\varphi_{,i} \qquad (\overset{\mathrm{T}}{e}_{ij} = \tfrac{1}{3}\overset{\mathrm{T}}{e}\,\delta_{ij}) \tag{2.26}$$

(Goodier [1937]). In this special case the dilatation has the constant value $e^{\mathrm{T}}(1+\sigma)/3(1-\sigma)$ inside the inclusion. In the matrix the dilatation is zero. Consequently the value of the bulk modulus $\varkappa^*$ of the matrix is irrelevant and (2.26) applies also to the case of an inclusion of elastic constants $\mu,\varkappa$ in a matrix with constants $\mu,\varkappa^*$ if we give to σ the value appropriate to the inclusion (CRUM, quoted by NABARRO [1940]; ROBINSON [1951]).

Again, it may happen that it is easier to calculate φ and ψ for points within the inclusion than for points outside it. (We shall see that this is so for an ellipsoidal inclusion.) Since the strains involve $\varphi_{,ij}$ and $\psi_{,ijkl}$ we can use (2.13) and (2.16) to find the strain $\overset{\mathrm{C}}{e_{ij}}(\text{out})$ (and hence the stress) at a point immediately outside the inclusion from the values $\overset{\mathrm{C}}{e_{ij}}(\text{in})$ at the adjacent point immediately inside the inclusion. Expressed in terms of the dilatational and deviatoric parts of $\overset{\mathrm{C}}{e_{ij}}$ the result is

$$\overset{\mathrm{C}}{e}(\text{out}) = \overset{\mathrm{C}}{e}(\text{in}) - \frac{1}{3}\frac{1+\sigma}{1-\sigma}\overset{\mathrm{T}}{e} - \frac{1-2\sigma}{1-\sigma}\,{}'\overset{\mathrm{T}}{e_{ij}}\, n_i n_j$$

$${}'\overset{\mathrm{C}}{e_{il}}(\text{out}) = {}'\overset{\mathrm{C}}{e_{il}}(\text{in}) + \frac{1}{1-\sigma}\,{}'\overset{\mathrm{T}}{e_{jk}}\, n_j n_k n_i n_l - {}'\overset{\mathrm{T}}{e_{ik}}\, n_k n_l - {}'\overset{\mathrm{T}}{e_{lk}}\, n_k n_i \quad (2.27)$$

$$+ \frac{1-2\sigma}{3(1-\sigma)}\,{}'\overset{\mathrm{T}}{e_{jk}}\, n_j n_k\, \delta_{il} - \frac{1}{3}\frac{1+\sigma}{1-\sigma}\overset{\mathrm{T}}{e}(n_i n_l - \tfrac{1}{3}\delta_{il}) \,.$$

The foregoing results all refer to a transformed inclusion in an infinite matrix. If the matrix has a finite boundary the displacements will be of the form

$$\overset{\mathrm{F}}{u_i} = \overset{\mathrm{C}}{u_i} + \overset{\mathrm{im}}{u_i}$$

and the stress will be

$$\overset{\mathrm{F}}{p_{ij}} = \overset{\mathrm{C}}{p_{ij}} + \overset{\mathrm{im}}{p_{ij}}$$

in the matrix and

$$\overset{\mathrm{F}}{p_{ij}} - \overset{\mathrm{T}}{p_{ij}}$$

in the inclusion. The 'image field' $\overset{\mathrm{im}}{u_i}$, $\overset{\mathrm{im}}{p_{ij}}$ is free of singularities in the medium and is determined by the requirement that the sum of the C-field and the image-field shall satisfy whatever boundary conditions are imposed on the outer surface of the matrix. If the outer boundary S_0 is stress-free we must have

$$\overset{\mathrm{im}}{p_{ij}}\, n_j = -\overset{\mathrm{C}}{p_{ij}}\, n_j \quad \text{on} \quad S_0$$

that is, the image-field is the field produced by surface tractions $-\overset{\mathrm{C}}{p_{ij}}n_j$ acting on the outer surface. If the outer surface is held immovable we must have

$$\overset{\mathrm{im}}{u_i} = -\overset{\mathrm{C}}{u_i} \quad \text{on} \quad S_0 \,. \quad (2.28)$$

Thus when the C-quantities are known the determination of the image quantities reduces to a standard boundary-value problem.

It is sometimes convenient to have a formal expression analogous to (2.6) exhibiting $\overset{\mathrm{F}}{u}_i$ as the displacement induced by the layer of body force (2.3). When the boundary condition is (2.28) (rigidly held boundary) we may evidently write

$$\overset{\mathrm{F}}{u}_i(\boldsymbol{r}) = \int_{\mathrm{S}} \mathrm{dS}_k \, \overset{\mathrm{T}}{p}_{jk} \, U_{ij}(\boldsymbol{r}, \boldsymbol{r}') \tag{2.29}$$

where $F_j U_{ij}(\boldsymbol{r},\boldsymbol{r}')$ is the displacement at $\boldsymbol{r}$ due to a point-force $\boldsymbol{F} = (F_1, F_2, F_3)$ acting at $\boldsymbol{r}'$ in a body at whose surface the displacement is required to be zero.

With a suitable alternative definition of $U_{ij}(\boldsymbol{r}, \boldsymbol{r}')$ (2.29) also applies to the case where the outer boundary is stress-free. However, we cannot simply say that $F_j U_{ij}$ is the displacement at $\boldsymbol{r}$ due to a point-force at $\boldsymbol{r}'$ in a body with a stress-free surface. No such solution of the elastic equations exists, since the integral of the surface traction over S_0 must be equal to $-\boldsymbol{F}$. Instead we define $U_{ij}(\boldsymbol{r}, \boldsymbol{r}')$ as the displacement in the body with stress-free surfaces due to a point-force $\boldsymbol{F}$ at $\boldsymbol{r}'$, a point-force $-\boldsymbol{F}$ at the origin and a double force (LOVE [1954]) at the origin of moment $-\boldsymbol{F} \times \boldsymbol{r}'$. Since the resultant and moment of this set of forces are both zero the condition of zero traction over S_0 can be satisfied. It can easily be shown that when the elementary forces (2.3) are summed over S the corresponding auxiliary forces and moments at the origin add up to zero. Consequently, with this definition of U_{ij} the expression (2.29) actually gives the displacement due to the layer of body force (2.3) alone.

2.2. ENERGY RELATIONS

The elastic energy associated with the inclusion (i.e. the energy in the inclusion plus the energy in the matrix) can be calculated very simply by following the energy changes which occur during the imaginary operations leading to (2.6). Suppose first that the matrix is infinite. When the inclusion has been welded back into the matrix but is still held in its untransformed shape by the layer of body force (2.2) the energy in the matrix is zero and the energy in the inclusion is

$$\tfrac{1}{2} \int_V \overset{\mathrm{T}}{p}_{ij} \, \overset{\mathrm{T}}{e}_{ij} \, \mathrm{d}v \, . \tag{2.30}$$

When the layer of body force is relaxed each element of S moves through a distance $\overset{\mathrm{C}}{u_i}$ as the force on it falls from $\mathrm{d}F_i$ to zero. The amount of energy extracted from the elastic solid during the relaxation is thus

$$-\tfrac{1}{2}\int \overset{\mathrm{C}}{u_i}\,\mathrm{d}F_i = \tfrac{1}{2}\int_S \overset{\mathrm{T}}{p_{ij}}\,\overset{\mathrm{C}}{u_i}\,\mathrm{dS}_j = \tfrac{1}{2}\int \overset{\mathrm{T}}{p_{ij}}\,\overset{\mathrm{C}}{e_{ij}}\,\mathrm{d}v\,. \tag{2.31}$$

The energy remaining in the medium is found by subtracting (2.31) from (2.30):

$$E_\infty = -\tfrac{1}{2}\int_V \overset{\mathrm{T}}{p_{ij}}\,(\overset{\mathrm{C}}{e_{ij}} - \overset{\mathrm{T}}{e_{ij}})\,\mathrm{d}v = -\tfrac{1}{2}\int_V \overset{\mathrm{I}}{p_{ij}}\,\overset{\mathrm{T}}{e_{ij}}\,\mathrm{d}v\,. \tag{2.32}$$

The suffix ∞ emphasizes that this is the energy for an inclusion in an infinite matrix. Evidently we only need to know the elastic field inside the inclusion. If the stress-free strain is a pure dilatation $\overset{\mathrm{T}}{e_{ij}} = \tfrac{1}{3}e^{\mathrm{T}}\delta_{ij}$ we have by (2.26)

$$E_\infty = \frac{2}{9}\mu V\,\frac{1+\sigma}{1-\sigma}\,(e^{\mathrm{T}})^2 \tag{2.33}$$

for an inclusion of any shape. According to the argument following (2.26) the expression is still correct if the bulk moduli of matrix and inclusion differ (NABARRO [1940], ROBINSON [1951]).

Suppose next that the inclusion is situated in a finite matrix bounded by the surface S_0. No difference is made to (2.30) if we cut away the part of the unstrained matrix exterior to S_0. The energy removed by relaxing the layer of body force is given by (2.31) with $\overset{\mathrm{C}}{u_i}$ replaced by $\overset{\mathrm{C}}{u_i} + \overset{\mathrm{im}}{u_i}$. The elastic energy due to the stress-field of the inclusion is thus

$$E_{\mathrm{inc}} = E_\infty + E_{\mathrm{im}}$$

where the 'image term'

$$E_{\mathrm{im}} = -\tfrac{1}{2}\int_V \overset{\mathrm{T}}{p_{ij}}\,\overset{\mathrm{im}}{e_{ij}}\,\mathrm{d}v = -\tfrac{1}{2}\int_V \overset{\mathrm{im}}{p_{ij}}\,\overset{\mathrm{T}}{e_{ij}}\,\mathrm{d}v$$

varies with the position of the inclusion. As an example we consider the case of a sphere of volume $V(1+e^{\mathrm{T}})$ forced into a spherical cavity of volume V in a semi-infinite solid. In the infinite solid $\overset{\mathrm{C}}{u_i}$ is given by (2.26) with $\varphi = V/r$. E_∞ is given by (2.33). To find E_{im} we have to calculate $\overset{\mathrm{im}}{e_{kk}}$. It is the dilatation produced by surface tractions $-\overset{\mathrm{C}}{p_{ij}}n_j$ acting on the free surface of the semi-infinite solid, and can be found by well-known methods (LOVE [1954]). As the dilatation is harmonic we need only find its value at the centre of the inclusion, since the mean value of harmonic function of a sphere is equal to its value at the centre. The result is

$$E_{\mathrm{inc}} = E_\infty\left\{1 - \frac{1}{4}(1+\sigma)\,\frac{a^3}{h^3}\right\} \tag{2.34}$$

where a is the radius of the sphere and h is the distance of its centre from the free surface. If $\sigma = \frac{1}{3}$ the elastic energy is reduced to $\frac{2}{3}E_\infty$ if the inclusion just reaches the surface ($h = a$).

It is worth recalling at this point that, in thermodynamic terms, 'elastic energy' represents internal energy under adiabatic conditions and Helmholtz free energy under isothermal conditions (SOKOLNIKOFF [1946]). A calculation such as the foregoing covers both cases; the distinction only appears when we decide to insert either the adiabatic or isothermal values of the elastic constants.

In some applications it is necessary to consider the changes of energy which occur when an inclusion is formed in a body which is already stressed by externally applied loads. For simplicity we shall suppose that the body is stressed by surface tractions which do not vary when the outer surface S_0 of the body is slightly deformed by the introduction of the inclusion; in engineering language this is the case of 'dead loading'. The stress due to the inclusion must satisfy $\overset{F}{p}_{ij}n_j = 0$ on S_0.

Let the external loads produce stress and strain $\overset{A}{p}_{ij}$, $\overset{A}{e}_{ij}$, not necessarily uniform. Before the inclusion has transformed the elastic energy in the material is

$$E_A = \tfrac{1}{2}\int \overset{A}{p}_{ij}\,\overset{A}{e}_{ij}\,\mathrm{d}v \tag{2.35}$$

with the integral extending over the whole volume of the material. Suppose next that the body is subject to the combined action of the stresses due to the load and the internal stresses due to the inclusion. We might expect that the elastic energy would be the sum of (2.35), (2.34) and a cross term, representing an 'interaction energy'. But in fact the cross term is zero. To see this, suppose that the transformation occurs first, in the absence of external loads. The elastic energy is E_{inc}. Now let the load be applied. Within the limits of the usual infinitesimal theory of elasticity the body responds to external forces just as it would if it were not self-stressed by the transformed inclusion. The work done on the body by the load is thus

$$\tfrac{1}{2}\int_{S_0} \overset{A}{p}_{ij}\,\overset{A}{u}_i\,\mathrm{d}S_j = \tfrac{1}{2}\int \overset{A}{p}_{ij}\,\overset{A}{e}_{ij}\,\mathrm{d}v = E_A$$

and so the total elastic energy is simply $E_{\text{inc}} + E_A$. The same conclusion can be reached analytically. If the volume integral of the energy density is converted into a surface integral over S and the outer surface of the matrix it will be found that the cross term between the A- and F-terms vanishes because $\overset{F}{p}_{ij}n_j = 0$ on S_0.

Despite the lack of a cross term in the elastic energy it is possible to define a physically meaningful interaction energy. (For a general discussion in the context of solid state theory cf. PEACH [1951], ESHELBY [1951, 1956].) To introduce the concept of interaction energy we begin by enquiring whether in the presence of the stresses arising from the external load it is energetically possible for the inclusion to form spontaneously. At first sight it appears as if the answer is no, since the elastic energy increases by the necessarily positive quantity E_{inc} when the inclusion is introduced. However, we have to consider not simply the elastic energy of the material, but rather the total energy, E_{tot} say, of the closed system made up of the body and the loading mechanism. When the transformation occurs in the presence of the external load the increase in the potential energy of the loading mechanism is equal to the work which the surface tractions $\overset{\text{A}}{p}_{ij}n_j$ do on the body as the surface displacements change from $\overset{\text{A}}{u}_i$ to $\overset{\text{A}}{u}_i + \overset{\text{F}}{u}_i$. Thus the increase in the energy of the whole system (inclusion plus matrix plus loading mechanism) is

$$\Delta E_{\text{tot}} = E_{\text{inc}} + E_{\text{int}} \tag{2.36}$$

where

$$E_{\text{int}} = -\int_{S_0} \overset{\text{A}}{p}_{ij} \overset{\text{F}}{u}_i \, \mathrm{d}S_j \, . \tag{2.37}$$

We may write

$$E_{\text{tot}} = E_0 + E_{\text{A}} + E_{\text{inc}} + E_{\text{int}} \tag{2.38}$$

where E_0 is the potential energy of the loading mechanism in the absence of the inclusion. In (2.38) the first two terms depend only on the elastic field due to the load and the third only on the field due to the inclusion. The last term, which depends on both these fields, may be appropriately called the interaction energy.

We can now answer the question posed in the preceding paragraph. If the applied stress is chosen so that $\Delta E_{\text{tot}} < 0$ (and this can always be done) it is energetically possible for the transformation to take place spontaneously. Generally ΔE_{tot}, whether it is positive or negative, may be called the energy of formation of the inclusion in the applied stress-field. This concept is familiar in thermodynamics. In fact, if the transformation leading to the formation of the inclusion occurs under adiabatic conditions ΔE_{tot} represents, in thermodynamic language, the increase in the enthalpy of the body, while if the transformation takes place under isothermal conditions ΔE_{tot} is the increase in its Gibbs free energy. This follows at once from the definition of these quantities. The increase in enthalpy associated with some change in the state of a thermally isolated system is equal to the increase in its internal energy plus the work which it does on its environment during the course of the change. For a change at constant temperature the Gibbs free energy is similarly defined, replacing 'internal energy' by 'Helmoltz free energy'. That ΔE_{tot} is the change in enthalpy or Gibbs free energy follows from the fact that, as we have seen, the 'elastic energy' has to be identified with the internal energy in adiabatic processes and with Helmholtz free energy in isothermal processes. Thus ΔE_{tot} is the enthalpy or Gibbs free energy of formation of the inclusion.

The expression (2.39) for the interaction energy can be put into a more useful form by the following artifice. We re-write (2.37) as

$$E_{\text{int}} = -\int_{S_0} (\overset{A}{p}_{ij} \overset{F}{u}_i - \overset{F}{p}_{ij} \overset{A}{u}_i)\, dS_j\,. \tag{2.39}$$

Because $\overset{F}{p}_{ij} \overset{A}{u}_i = 0$ on S_0 the added term is zero. Consider the divergence of the integrand. It can be split into two terms

$$(\overset{\text{im}}{p}_{ij} \overset{A}{u}_i - \overset{A}{p}_{ij} \overset{\text{im}}{u}_i)_{,j} = \overset{\text{im}}{p}_{ij} \overset{A}{e}_{ij} - \overset{A}{p}_{ij} \overset{\text{im}}{e}_{ij} \tag{2.40}$$

and

$$(\overset{C}{p}_{ij} \overset{A}{u}_i - \overset{A}{p}_{ij} \overset{C}{u}_i)_{,j} = \overset{C}{p}_{ij} \overset{A}{e}_{ij} - \overset{A}{p}_{ij} \overset{C}{e}_{ij}\,. \tag{2.41}$$

By (1.3) the expression (2.41) is zero both in the matrix and the inclusion and so the image terms make no contribution. On the other hand, (2.40) vanishes in the matrix but not in the inclusion, since $\overset{C}{p}_{ij}$ is discontinuous across S. Hence in (2.39) we may replace $\overset{F}{p}_{ij}$, $\overset{F}{u}_i$ by $\overset{C}{p}_{ij}$, $\overset{C}{u}_i$ and carry out the integration over the boundary of the inclusion instead of over the outer surface of the body; that is

$$E_{\text{int}} = \int_{S} (\overset{C}{p}_{ij} \overset{A}{u}_i - \overset{A}{p}_{ij} \overset{C}{u}_i)\, dS_j\,. \tag{2.42}$$

In (2.43) we may replace $\overset{C}{p}_{ij}$ by $\overset{I}{p}_{ij}$ since $\overset{C}{p}_{ij} u_j = \overset{I}{p}_{ij} n_j$ on S. The integral can then be converted to a volume integral over the inclusion:

$$E_{\text{int}} = \int_{V} (\overset{I}{p}_{ij} \overset{A}{e}_{ij} - \overset{A}{p}_{ij} \overset{C}{e}_{ij})\, dS_j\,.$$

By (1.3) the integrand is equal to $(\overset{I}{p}_{ij} - \overset{C}{p}_{ij})\, \overset{A}{e}_{ij}$ and so by (2.19)

$$E_{\text{int}} = -\int_{V} \overset{T}{p}_{ij} \overset{A}{e}_{ij}\, dv = -\int_{V} \overset{A}{p}_{ij} \overset{T}{e}_{ij}\, dv\,. \tag{2.43}$$

Evidently to find the interaction energy we need only know the stress-free strain e_{ij}^{T}; it is unnecessary to solve the elastic problems associated with the determination of u_i^{C} or u_i^{im}.

§ 3. The Ellipsoidal Inclusion

3.1. The elastic field

When the inclusion is bounded by the ellipsoid

$$x^2/a^2 + y^2/b^2 + z^2/c^2 = 1$$

the elastic field may be found explicitly. The form of the harmonic potential φ is well-known (Kellogg [1929]). There is an analogous expression for the biharmonic potential ψ (Eshelby [1959b]), but in fact all the derivatives of ψ which enter (2.7) can be found in terms of derivatives of φ. Let us compare f_{12} (eq. (2.7)) with the function $g = a^2(x_1\varphi_{,2} - x_2\varphi_{,1})/(a^2 - b^2)$ which plays a role in the hydrodynamic theory of rotating ellipsoids. Each of these functions is harmonic inside and outside the ellipsoid, is continuous across its surface and falls to zero at infinity. Across the surface there is a discontinuity $4\pi x_1 n_2$ in the normal derivative of f_{12} (eq. (2.22)). The corresponding discontinuity for g is $4\pi a^2(x_1n_2 - x_2n_1)/(a^2 - b^2)$. This is simply $4\pi x_1 n_2$ in view of the relation $b^2x_1n_2 = a^2x_2n_1$ which follows from the expressions

$$n_1 = x_1/a^2h\,,$$
$$n_2 = x_2/b^2h\,, \quad (h^2 = x^2/a^4 + y^2/b^4 + z^2/c^4)$$
$$n_3 = x_3/c^2h\,,$$

for the components of the normal to an ellipsoid. Hence f_{12} and g are identical, being harmonic potentials of the same surface distribution. We therefore have

$$f_{12} = x_1\varphi_{,2} - \psi_{,12} = \frac{a^2}{a^2-b^2}(x_1\varphi_{,2} - x_2\varphi_{,1})$$

and similarly for the other f_{ij} with $i \neq j$. Thus

$$\psi_{,12} = \frac{a^2}{a^2-b^2}x_2\varphi_{,1} + \frac{b^2}{b^2-a^2}x_1\varphi_{,2}$$
$$\psi_{,23} = \frac{b^2}{b^2-c^2}x_3\varphi_{,2} + \frac{c^2}{c^2-b^2}x_2\varphi_{,3}$$
$$\psi_{,31} = \frac{c^2}{c^2-a^2}x_1\varphi_{,3} + \frac{a^2}{a^2-c^2}x_3\varphi_{,1}\,.$$

It is more difficult to derive expressions for f_{11}, f_{22}, f_{33} and hence for $\psi_{,11}$, $\psi_{,22}$, $\psi_{,33}$. However, there is no need to obtain them, since all the third derivatives $\psi_{,ijk}$ appearing in (2.8) can be made to depend on φ, $\psi_{,12}$, $\psi_{,23}$, $\psi_{,31}$. We may write, for example,

$$\psi_{,112} = (\psi_{,12})_{,1} \qquad \psi_{,111} = 2\varphi_{,1} - (\psi_{,12})_{,2} - (\psi_{,13})_{,3} \,.$$

The first relation is trivial; the second is obtained by differentiating $\nabla^2\psi = 2\varphi$ with respect to x_1.

Substitution in (2.8) gives

$$\begin{aligned} 8\pi(1-\sigma)u_1^{\mathrm{C}} = {} & \frac{e_{22}^{\mathrm{T}} - e_{11}^{\mathrm{T}}}{a^2 - b^2}\frac{\partial}{\partial x_2}(a^2x_2\varphi_{,1} - b^2x_1\varphi_{,2}) + \\ & \frac{e_{33}^{\mathrm{T}} - e_{11}^{\mathrm{T}}}{c^2 - a^2}\frac{\partial}{\partial x_3}(c^2x_1\varphi_{,3} - a^2x_3\varphi_{,1}) - \\ & 2\{(1-\sigma)e_{11}^{\mathrm{T}} + \sigma(e_{22}^{\mathrm{T}} + e_{33}^{\mathrm{T}})\}\varphi_{,1} - \\ & 4(1-\sigma)(e_{12}^{\mathrm{T}}\varphi_{,2} + e_{13}^{\mathrm{T}}\varphi_{,3}) + \frac{\partial}{\partial x_1}\bar{\beta} \end{aligned} \tag{3.1}$$

where

$$\begin{aligned} \bar{\beta} = {} & \frac{2e_{12}^{\mathrm{T}}}{a^2 - b^2}(a^2x_2\varphi_{,1} - b^2x_1\varphi_{,2}) + \\ & \frac{2e_{23}^{\mathrm{T}}}{b^2 - c^2}(b^2x_3\varphi_{,2} - c^2x_2\varphi_{,3}) + \\ & \frac{2e_{31}^{\mathrm{T}}}{c^2 - a^2}(c^2x_1\varphi_{,3} - a^2x_3\varphi_{,1}) \,. \end{aligned}$$

u_2^{C} and u_3^{C} are found by cyclic permutation of (1, 2, 3), (a, b, c).

At an internal point

$$\varphi = \tfrac{1}{2}(a^2 - x^2)I_a + \tfrac{1}{2}(b^2 - y^2)I_b + \tfrac{1}{2}(c^2 - z^2)I_c \tag{3.2}$$

where I_a, I_b, I_c are constants depending only on the axial ratios of the ellipsoid. Consequently u_i^{C} is a linear function of the x_i and the strain and stress are uniform within the inclusion, as stated in the Introduction. The constant strains e_{ij}^{C} are linear functions of the e_{ij}^{T} and we may write

$$e_{ij}^{\mathrm{C}} = S_{ijkl}\, e_{kl}^{\mathrm{T}} \,. \tag{3.3}$$

The S_{ijkl} are symmetric in ij and in kl, but in general S_{ijkl} is different from S_{klij}. It is easy to verify that these coefficients vanish unless they are of the form S_{iiii}, S_{iijj} or S_{ijij} ($i \neq j$; no summation). That is, unlike shears are not coupled, and shears are not coupled to extensions. From (3.1) we obtain

$$
\begin{aligned}
8\pi(1-\sigma)S_{1111} &= \frac{a^2 I_a - b^2 I_b}{a^2-b^2} + \frac{a^2 I_a - c^2 I_c}{a^2-c^2} - (1-\sigma)I_a \\
8\pi(1-\sigma)S_{1122} &= -b^2\frac{I_a - I_b}{a^2-b^2} - (1-2\sigma)I_a \\
8\pi(1-\sigma)S_{1212} &= -\frac{1}{2}\frac{a^2+b^2}{a^2-b^2}(I_a - I_b) + \tfrac{1}{2}(1-2\sigma)(I_a+I_b)\,;
\end{aligned}
\tag{3.4}
$$

the remaining coefficients may be obtained by cyclic interchange. (For an alternative way of obtaining these coefficients see ESHELBY [1957].) The rotation inside the inclusion is also constant. We have at once from (2.25) and (3.2)

$$
\begin{aligned}
4\pi\overset{\mathrm{C}}{\omega}_{12} &= (I_b - I_a)\,\overset{\mathrm{T}}{e}_{12}, \\
4\pi\overset{\mathrm{C}}{\omega}_{23} &= (I_c - I_b)\,\overset{\mathrm{T}}{e}_{23}\,, \\
4\pi\overset{\mathrm{C}}{\omega}_{31} &= (I_a - I_c)\,\overset{\mathrm{T}}{e}_{31}\,.
\end{aligned}
\tag{3.5}
$$

In their role of demagnetising factors I_a, I_b, I_c have been plotted as functions of b/a and c/a by OSBORN [1945] with the notation $I_a = L$, $I_b = M$, $I_c = N$. They may also be found from tables of elliptic integrals $F(\theta,k)$, $E(\theta,k)$ using the relations

$$
\begin{aligned}
I_a &= \frac{4\pi abc}{(a^2-b^2)(a^2-c^2)^{\frac{1}{2}}}(F-E), \\
I_b &= 4\pi - I_a - I_c, \\
I_c &= \frac{4\pi abc}{(b^2-c^2)(a^2-c^2)^{\frac{1}{2}}}\left\{\frac{b(b^2-c^2)^{\frac{1}{2}}}{ac} - E\right\} \\
k^2 &= \frac{a^2-b^2}{a^2-c^2}, \qquad \theta = \sin^{-1}\left(1-\frac{c^2}{a^2}\right)^{\frac{1}{2}}.
\end{aligned}
\tag{3.6}
$$

(Compare (3.7) below.)

For a point outside the ellipsoid the potential takes the form (KELLOGG [1929], MACMILLAN [1958])

$$
\varphi = \frac{2\pi abc}{l^3}\left\{\left[l^2 - \frac{x^2}{k^2} + \frac{y^2}{k^2}\right]F(\theta,k) + \left[\frac{x^2}{k^2} - \frac{y^2}{k^2k'^2} + \frac{z^2}{k'^2}\right]E(\theta,k) + \frac{l}{k'^2}\left[\frac{C}{AB}y^2 - \frac{B}{AC}z^2\right]\right\}
\tag{3.7}
$$

where

$$
\left.
\begin{aligned}
A &= (a^2+\lambda)^{\frac{1}{2}}, \quad B = (b^2+\lambda)^{\frac{1}{2}}, \quad C = (c^2+\lambda)^{\frac{1}{2}} \\
l &= (a^2-c^2)^{\frac{1}{2}}, \quad k^2 = 1-k'^2 = \frac{a^2-b^2}{a^2-c^2}
\end{aligned}
\right\},
\tag{3.8}
$$

$$a^2 > b^2 > c^2 \,, \tag{3.9}$$

and F, E are elliptic integrals of modulus k and argument

$$\theta = \sin^{-1}(l/A) \,. \tag{3.10}$$

λ is the greatest (and in fact the only positive) root of

$$x^2/A^2 + y^2/B^2 + z^2/C^2 = 1 \,. \tag{3.11}$$

Equation (3.7) also gives the potential at an internal point if we put $\lambda = 0$; this gives (3.6).

To carry out the differentiations necessary to find the displacement or stress outside the inclusion one can make repeated use of

$$\partial F/\partial\lambda = -\tfrac{1}{2}l/ABC, \qquad \partial E/\partial\lambda = -\tfrac{1}{2}lB/A^3C$$

$$\partial\lambda/\partial x = 2x/Ah, \ldots, \qquad h^2 = x^2/A^4 + y^2/B^4 + z^2/C^4 \,.$$

In forming the first (but not the higher) derivatives of φ, λ may be treated as a constant. The condition (3.9) is not really necessary. It ensures that $0 < k^2 < 1$ and $0 < \theta < \frac{1}{2}\pi$. If it is violated $F(\theta,k), E(\theta,k)$ can be made to depend on $F(\theta_1,k_1)$, $E(\theta_1,k_1)$ with $0 < k_1^2 < 1$, $0 < \theta_1 < \frac{1}{2}\pi$ with the help of known transformations. (See, for example, BYRD and FRIEDMAN [1954].) This is useful if it becomes convenient to ignore (3.9) at a late stage in a calculation.

The results we have obtained can only be applied to the sphere after a tedious passage to the limit. However, we may use the following expressions for the potentials of a sphere of radius a:

$$\left.\begin{aligned} \varphi &= \frac{4}{3}\pi a^2\left(\frac{3}{2} - \frac{1}{2}\frac{r^2}{a^2}\right), & r < a \\ &= \frac{4}{3}\pi a^2 \frac{a}{r}, & r > a \end{aligned}\right\} \tag{3.12}$$

$$\left.\begin{aligned} \psi &= \frac{4}{3}\pi a^4\left(\frac{3}{4} + \frac{1}{2}\frac{r^2}{a^2} - \frac{1}{20}\frac{r^4}{a^4}\right), & r < a \\ &= \frac{4}{3}\pi a^4\left(\frac{1}{5}\frac{a}{r} + \frac{r}{a}\right), & r > a \end{aligned}\right\}. \tag{3.13}$$

The expression for φ is well-known. ψ may be found by integrating

$$\frac{1}{r^2}\frac{\mathrm{d}}{\mathrm{d}r}\left(r^2\frac{\mathrm{d}\psi}{\mathrm{d}r}\right) = 2\varphi$$

and calculating $\psi(0)$ by direct integration:

$$\psi(0) = 4\pi\int_0^a r \cdot r^2\,\mathrm{d}r = \pi a^4 \,.$$

For the sphere (3.3) reduces to

$$e^{\mathrm{C}} = \alpha e^{\mathrm{T}}, \qquad {}'e^{\mathrm{C}}_{ij} = \beta\, {}'e^{\mathrm{T}}_{ij}$$

with

$$\alpha = \frac{1}{3}\frac{1+\sigma}{1-\sigma}, \qquad \beta = \frac{2}{15}\frac{4-5\sigma}{1-\sigma}. \tag{3.14}$$

KRÖNER [1958b] has given the S_{ijkl} (with the notation $\overset{-1}{w}_{ijkl}$) for prolate and oblate spheroids.

For some purposes it may be unnecessary to deal with the relatively complex field outside the inclusion. A knowledge of the e^{T}_{ij} alone is enough to give the field far from the inclusion (2.23) or the interaction energy with an applied field (2.43). When the numerical coefficients S_{ijkl} have been computed we can find the elastic field inside the inclusion, and also, with the help of (2.27), the field at points in the matrix immediately outside the inclusion.

The displacement (3.1) at points external to the ellipsoid may be regarded as the solution to one or other of the following boundary-value problems:

(i) To find the elastic field falling to zero at infinity and having the displacement

$$u_i = (e^{\mathrm{C}}_{ij} + \omega^{\mathrm{C}}_{ij})x_j \qquad (e^{\mathrm{C}}_{ij}, \omega^{\mathrm{C}}_{ij} \text{ constant})$$

over the surface of an ellipsoid.

(ii) To find the elastic field falling to zero at infinity and giving surface tractions

$$T_i = p^{\mathrm{I}}_{ij} n_j \qquad (p^{\mathrm{I}}_{ij} \text{ constant})$$

on the surface of an ellipsoid.

The solution (3.1) is designed to give the constrained elastic field directly when the stress-free strain e^{T}_{ij} is known. Thus in using it to solve (i) or (ii) the first step would have to be the solution of (3.3), or (2.19) and (3.3) to find the e^{T}_{ij} appropriate to the prescribed u^{C}_i or p^{I}_{ij}.

Regarded as a solution of (i) the solution (3.1) is closely related to one already given by DANIELE [1911]. Again, if a constant stress-field $p^{\mathrm{A}}_{ij} = -p^{\mathrm{I}}_{ij}$ is superimposed on the solution of (ii) the ellipsoidal surface is free of stress and we are left with a solution representing an ellipsoidal cavity perturbing a uniform stress p^{A}_{ij}. The solution has been given by SADOWSKY and STERNBERG [1949] for the case where $p^{\mathrm{A}}_{12} = p^{\mathrm{A}}_{23} = p^{\mathrm{A}}_{31} = 0$.

Daniele determined the displacements in an infinite medium outside an ellipsoidal surface over which the displacement is required to be

$$u_i = \xi_{ij} x_j \tag{3.15}$$

with constant coefficients ξ_{ij}. He assumed that there was a solution of the form

$$\left.\begin{aligned} u_i &= \alpha_{ij}\varphi_{,j} + \alpha_{ijkl} x_j \varphi_{,kl} \\ e &= \lambda_{ij}\varphi_{,ij} \end{aligned}\right. \quad (\alpha_{ij},\ \alpha_{ijkl},\ \lambda_{ij}\ \text{constant})\ .$$

The α_{ij} and α_{ijkl} were determined by substituting in the equilibrium equations, equating the coefficients of $\varphi_{,ij}$ and $\varphi_{,ijk}$ to zero (subject to the conditions $\nabla^2\varphi_{,ij} = 0$, $\nabla^2\varphi_{,ijk} = 0$) and applying the boundary condition (3.15). Actually Daniele's solution is more general than (3.1) since we cannot prescribe $\overset{C}{e}_{ij}$ and $\overset{C}{\omega}_{ij}$ independently. If the $\overset{C}{e}_{ij}$ are given the $\overset{T}{e}_{ij}$ follow from (3.3) and the $\overset{C}{\omega}_{ij}$ are then fixed by (3.5). This connection between the $\overset{C}{e}_{ij}$ and $\overset{C}{\omega}_{ij}$ is due to the fact that there is no external couple acting on the inclusion. Thus, in physical terms, Daniele's solution gives the field about a rigid embedded inclusion which suffers a homogeneous deformation and is then rotated from its equilibrium position by an external couple. (The case where the ellipsoid is rotated but not deformed was discussed earlier by EDWARDES [1893].) When the requirement of zero couple is imposed Daniele's solution agrees with (3.1).

Sadowsky and Sternberg presented their solution in ellipsoidal coordinates. The problem they had in mind was that of an ellipsoidal cavity perturbing a stress field which at infinity is uniform and has its principal axes parallel to those of the ellipsoid; ROBINSON [1951] and NIESEL [1953] have pointed out that it can be applied to the inclusion problem. Robinson treats the case where $\overset{T}{e}_{ij}$ is a pure dilatation; Niesel considers the more general case where $\overset{T}{e}_{11}$, $\overset{T}{e}_{22}$, $\overset{T}{e}_{33}$ are unequal and $\overset{T}{e}_{12}$, $\overset{T}{e}_{23}$, $\overset{T}{e}_{31}$ are zero.

Sadowsky and Sternberg's solution is expressed in the form (2.20) with

$$\left.\begin{aligned} B_1 = \mathscr{C}A_1X, \quad B_2 = \mathscr{C}A_2Y, \quad B_3 = \mathscr{C}A_3Z, \\ \beta = \mathscr{C}A_4F_1 + \mathscr{C}A_5F_2 \end{aligned}\right\}. \tag{3.16}$$

(We have introduced the constant $\mathscr{C} = -2(1-\sigma)/\mu$ to make our notation agree with Sadowsky and Sternberg's.) The A_n are numerical coefficients and X, Y, Z, F_1, F_2 are harmonic functions chosen so that the boundary conditions can be satisfied. They are related to the potential φ (eq. (3.7)) as follows:

$$X = -vk\,\varphi_{,1}, \qquad Y = +vkk'^2\,\varphi_{,2}, \qquad Z = -v\,\frac{k'^2}{k}\,\varphi_{,3}$$

$$F_1 = -vk^3 f_{ii} = vk^3(2\varphi - x_i\varphi_{,i})$$

$$F_2 = \gamma_{bc} f_{11} + \gamma_{ca} f_{22} + \gamma_{ab} f_{33}$$

with

$$\gamma_{qr} = \frac{2(p-1)}{3\pi abc}(d^2 - q^2)(d^2 - r^2) \qquad (q,r = a,b,c)$$

$$v = \frac{l^3}{4\pi abc}, \qquad d^2 = a^2 - 2\,\frac{a^2 - b^2}{p(1+k^2)}$$

$$p = 2 + \frac{2(k'^2 + k^4)^{\frac{1}{2}}}{1 + k^2}.$$

For completeness we have given the values of the numerical coefficients connecting φ and X, Y, Z, F_1, F_2. We note that f_{11}, f_{22}, f_{33} (eq. (2.22)) only enter the solution through the two particular linear combinations F_1 and F_2. This is a result of Sadowsky and Sternberg's requirement that F_1 and F_2 (and also X, Y, Z) shall take the form

$$f(\alpha_1)\,g(\alpha_2)\,h(\alpha_3) \tag{3.17}$$

when expressed in terms of their ellipsoidal coordinates α_1, α_2, α_3. Thus the A_n cannot be found by direct comparison of (3.17) and (2.21). For example, if we choose A_4 and A_5 so that the coefficients of f_{11} and f_{22} agree with (2.7), the coefficient of f_{33} is fixed and so $\overset{\mathrm{T}}{p}_{33}$ cannot be prescribed at will. This could be remedied by adding to β a further harmonic function, $\mathscr{C}A_5'F_2'$ say, where F_2' is derived from F_2 by changing the sign of the radical in the equation defining p. (Cf. SADOWSKY and STERNBERG [1949] eq. (41).) There is, however, no need to do this. In a representation such as (2.20) it is possible to modify B_i and β simultaneously and leave $\overset{\mathrm{C}}{u}_i$ unchanged. Sadowsky and Sternberg give a matrix relation (their eq. (50)) which enables the A_n to be determined in such a way that the resulting stress annuls the surface tractions on the ellipsoidal surface due to uniform stresses $p_{11} = \sigma_1$, $p_{22} = \sigma_2$, $p_{33} = \sigma_3$.

According to NIESEL [1953] it is convenient to include the solution F_2' in β if we wish to pass to the limiting case of a spheroid. To extend Sadowsky and Sternberg's solution to the case where the principal axes of the stress at infinity are not parallel to the axes of the cavity (or to apply it to the inclusion problem with non-vanishing $\overset{\mathrm{T}}{e}_{12}$, $\overset{\mathrm{T}}{e}_{23}$, $\overset{\mathrm{T}}{e}_{31}$) it would be necessary to add to β terms $\mathscr{C}A_6(f_{12} + f_{21})$, $\mathscr{C}A_7(f_{23} + f_{32})$, $\mathscr{C}A_8(f_{31} + f_{13})$. These terms have the required form (3.17) when transcribed into ellipsoidal coordinates.

The use of ellipsoidal coordinates does not seem to offer much advantage. As we have seen, the problem may be set up and solved formally in Cartesian coordinates. To find the elastic field at a given point (x, y, z) outside the ellipsoid we have to solve the cubic (3.11) for λ. We appear to be spared this when ellipsoidal coordinates are used. But, in fact, we must have already prepared an ellipsoidal coordinate network in order to be able to locate points relative to the ellipsoid, and, further, a different network is required for each value of the axial ratio b/a.

LURIE [1952] has given a solution to Sadowsky and Sternberg's problem in the disconcertingly simple form

$$B_1 = (M/a^2)\varphi_{,1}\,, \qquad B_2 = (M/b^2 - N)\varphi_{,2}\,, \qquad B_3 = (M/c^2 - N)\varphi_{,3}$$
$$\beta = N(x_i\varphi_{,i} - 2\varphi) + P\varphi$$

where M, N, P are disposable constants. (Our M, N, P differ from Lurie's by a common multiplicative factor.) Unfortunately the contributions to $\overset{\mathrm{C}}{u_i}$ from the terms in N and P are both of the form const. $\varphi_{,i}$. (This illustrates the fact mentioned above that different choices of B_i,β can give the same elastic field.) Consequently there are really only two disposable constants, and so only two of $\overset{\mathrm{T}}{p}_{11}$, $\overset{\mathrm{T}}{p}_{22}$, $\overset{\mathrm{T}}{p}_{33}$ (or, in the cavity problem, σ_1, σ_2, σ_3) can be prescribed independently.

3.2. THE INHOMOGENEOUS INCLUSION

So far we have been concerned with a homogeneous inclusion, that is, one which has the same elastic constants as the matrix. We consider next the case where the elastic constants of matrix and inclusion are different. As in § 2 we imagine that the inclusion undergoes a transformation specified by a uniform stress-free strain whilst constrained by the matrix, and try to calculate the resulting elastic field. This must be distinguished from the problem considered in § 4 below. There we have, in effect, a *perfectly fitting* inhomogeneous insertion cemented into a cavity in the matrix, and consequently the material is everywhere stress-free in the absence of applied forces.

We have seen that (2.26) still applies when the bulk moduli of matrix and inclusion differ. This seems to be the only simple statement one can make for an inclusion of arbitrary shape. For the ellipsoid, on the other hand, we may solve the general problem very simply by taking advantage of the fact that the stress is constant inside a homogeneous ellipsoidal inclusion (ROBINSON [1951], NIESEL [1953], ESHELBY [1957]).

For brevity let E denote the ellipsoidal inclusion we have been considering hitherto. It has the same elastic constants λ, μ as the matrix, and it has suffered a permanent change of shape characterized by the stress-free strain $\overset{\mathrm{T}}{e}_{ij}$ while embedded in the matrix. The strain $\overset{\mathrm{C}}{e}_{ij}$ relates its final form to its form before transformation. Take a second ellipsoid E^* which to begin with has the same form as E had before its transformation and which has elastic constants λ^*, μ^*. Let E^* undergo a stress-free strain $\overset{\mathrm{T}*}{e}_{ij}$. To E^* apply surface tractions chosen so as to produce a uniform elastic strain $\overset{\mathrm{C}}{e}_{ij} - \overset{\mathrm{C}*}{e}_{ij}$ in it. It then has precisely the same form as the embedded inclusion E. If this treatment should happen also to produce in E^* stresses identical with those in E we can replace E by E^* without upsetting the continuity of displacement and surface traction across the interface. The condition for this to be possible is (cf. eq. (2.19))

$$\overset{\mathrm{I}}{p}_{ij} = \lambda(\overset{\mathrm{C}}{e} - \overset{\mathrm{T}}{e})\delta_{ij} + 2\mu(\overset{\mathrm{C}}{e}_{ij} - \overset{\mathrm{T}}{e}_{ij})$$
$$= \lambda^*(\overset{\mathrm{C}}{e} - \overset{\mathrm{T}*}{e})\delta_{ij} + 2\mu^*(\overset{\mathrm{C}}{e}_{ij} - \overset{\mathrm{T}*}{e}_{ij}). \qquad (3.18)$$

When the values of λ^*, μ^* and $\overset{\mathrm{T}*}{e}_{ij}$ for the inhomogeneous inclusion are given we can solve (3.18) for the $\overset{\mathrm{T}}{e}_{ij}$. The elastic field inside and outside the inhomogeneous inclusion is then identical with that of a homogeneous inclusion with the stress-free strain $\overset{\mathrm{T}}{e}_{ij}$.

To find $\overset{\mathrm{T}}{e}_{ij}$ for the equivalent homogeneous inclusion we use (3.3) to express $\overset{\mathrm{C}}{e}_{ij}$ in terms of $\overset{\mathrm{T}}{e}_{ij}$. For the non-diagonal components we have simply

$$\overset{\mathrm{T}}{e}_{12} = \frac{\mu^*}{2(\mu^* - \mu)S_{1212} + \mu}\overset{\mathrm{T}*}{e}_{12}, \qquad \overset{\mathrm{T}}{e}_{23} = \ldots .$$

To find $\overset{\mathrm{T}}{e}_{11}$, $\overset{\mathrm{T}}{e}_{22}$, $\overset{\mathrm{T}}{e}_{33}$ we have to solve the set of three simultaneous equations

$$(\lambda^* - \lambda)\overset{\mathrm{C}}{e} + \lambda\overset{\mathrm{T}}{e} + 2(\mu^* - \mu)S_{ijkl}\overset{\mathrm{T}}{e}_{kl} = \lambda^*\overset{\mathrm{T}*}{e} + 2\mu^*\overset{\mathrm{T}*}{e}_{ij}$$
$$(ij = 11, 22, 33)\ .$$

Only $\overset{\mathrm{T}}{e}_{11}$, $\overset{\mathrm{T}}{e}_{22}$, $\overset{\mathrm{T}}{e}_{33}$ enter the kl-summation, and in the first term we have

$$\overset{\mathrm{C}}{e} = \frac{1-2\sigma}{4\pi(1-\sigma)}(I_a\overset{\mathrm{T}}{e}_{11} + I_b\overset{\mathrm{T}}{e}_{22} + I_c\overset{\mathrm{T}}{e}_{33}) + \frac{\sigma}{1-\sigma}\overset{\mathrm{T}}{e} \qquad (3.19)$$

by (2.24) and (3.2).

It is a simple matter to calculate the elastic energy E^*_∞ associated with an inhomogeneous (ellipsoidal) inclusion in an infinite matrix. For the equivalent homogeneous inclusion we have from (2.32)

$$E_\infty = -\tfrac{1}{2}V\overset{\mathrm{I}}{p}_{ij}\overset{\mathrm{T}}{e}_{ij} \qquad (V = \tfrac{4}{3}\pi abc)\ . \qquad (3.20)$$

The energy in the matrix is the same for E and E^*. The internal stress is $\overset{\mathrm{I}}{p}_{ij}$ for both, but the effective strain is $\overset{\mathrm{C}}{e}_{ij} - \overset{\mathrm{T}}{e}_{ij}$ for E and $\overset{\mathrm{C}}{e}_{ij} - \overset{\mathrm{T*}}{e}_{ij}$ for E^*. Thus we have to add $\frac{1}{2}V\overset{\mathrm{I}}{p}_{ij}(\overset{\mathrm{C}}{e}_{ij} - \overset{\mathrm{T*}}{e}_{ij}) - \frac{1}{2}V\overset{\mathrm{I}}{p}_{ij}(\overset{\mathrm{C}}{e}_{ij} - \overset{\mathrm{T}}{e}_{ij})$ to (3.20), which gives the simple result (cf. ROBINSON [1951])

$$E^*_\infty = -\tfrac{1}{2}V\overset{\mathrm{I}}{p}_{ij}\,\overset{\mathrm{T*}}{e}_{ij}\,. \tag{3.21}$$

The results of this section may be adapted to the case of an ellipsoidal cavity containing a fluid under pressure. If the pressure P of the fluid is prescribed it is only necessary to put $\overset{\mathrm{I}}{p}_{ij} = -P\delta_{ij}$ in (3.18) and solve for the $\overset{\mathrm{T}}{e}_{ij}$ or $\overset{\mathrm{T*}}{e}_{ij}$. If instead we are given the excess volume v of fluid introduced (measured at zero pressure) we have to put $e^{\mathrm{T}} = v$ and require $\overset{\mathrm{I}}{p}_{ij}$ to have the form $\frac{1}{3}p^{\mathrm{I}}\delta_{ij}$. The solution of (3.18) then gives p^{I}, $'\overset{\mathrm{T}}{e}_{ij}$ and $\overset{\mathrm{T*}}{e}_{ij}$.

§ 4. The Ellipsoidal Inhomogeneity

4.1. THE ELASTIC FIELD

In this section we shall take up the second problem mentioned in the introduction. It may be formulated as follows:

An infinite solid has elastic constants λ^*, μ^* inside a region bounded by a closed surface S (the 'inhomogeneity') and elastic constants λ, μ in the region outside S (the 'matrix'). To find the elastic field everywhere when the strain is required to reduce to the constant value $\overset{\mathrm{A}}{e}_{ij}$ far from S.

Although the problem can only be solved in detail for an ellipsoidal inhomogeneity it is convenient to start from the case where the form of the inhomogeneity is arbitrary. The problem can be reduced to the determination of the elastic field produced by a certain layer of body-force distributed over S. To see this, suppose that the strain $\overset{\mathrm{A}}{e}_{ij}$ is impressed throughout the medium. The displacement is then

$$\overset{\mathrm{A}}{u}_i = \overset{\mathrm{A}}{e}_{ij}x_j$$

plus an inessential rigid-body displacement. The equilibrium equations are satisfied inside and outside S. The traction on the inner boundary of the matrix is $(\lambda e^{\mathrm{A}}\delta_{ij} + 2\mu\overset{\mathrm{A}}{e}_{ij})n_j$, but the traction on the outer surface of the inhomogeneity is $(\lambda^* e^{\mathrm{A}}\delta_{ij} + 2\mu^*\overset{\mathrm{A}}{e}_{ij})n_j$. Consequently the required state of strain can only be maintained if there is a layer of body-force of surface density $\{(\lambda^* - \lambda)e^{\mathrm{A}}\delta_{ij} + 2(\mu^* - \mu)\overset{\mathrm{A}}{e}_{ij}\}n_j$ spread over S. To find the actual elastic field we apply an equal and opposite layer of body force of surface density

$$T_i = \{(\lambda - \lambda^*)e^{\mathrm{A}}\delta_{ij} + 2(\mu - \mu^*)\overset{\mathrm{A}}{e_{ij}}\}n_j \tag{4.1}$$

and calculate the displacement $\overset{\mathrm{C}}{u_i}$ which it induces in the medium. The final displacement is then

$$u_i = \overset{\mathrm{A}}{e_{ij}}\,x_j + \overset{\mathrm{C}}{u_i}\,.$$

$\overset{\mathrm{C}}{u_i}$ is evidently given by the expression

$$\overset{\mathrm{C}}{u_i}(\boldsymbol{r}) = \{(\lambda - \lambda^*)e^{\mathrm{A}}\delta_{kj} + 2(\mu - \mu^*)\overset{\mathrm{A}}{e_{kj}}\}\int_{\mathrm{S}} U_{ik}(\boldsymbol{r}, \boldsymbol{r}')\,\mathrm{dS}_j\,, \tag{4.2}$$

where $U_{ik}(\boldsymbol{r}, \boldsymbol{r}')$ is the i-component of the displacement at $\boldsymbol{r}$ when a unit point-force is applied at $\boldsymbol{r}'$ parallel to the x_k-axis. Because the medium is inhomogeneous U_{ik} depends on $\boldsymbol{r}$ and $\boldsymbol{r}'$ separately and not simply on $|\boldsymbol{r} - \boldsymbol{r}'|$ as does the corresponding quantity in (2.4). It is not possible to determine U_{ik} for an arbitrary form of S, and hence a transformation of (9.2) corresponding to the step from (2.6) to (2.7) cannot be made. However, we shall find that the formulation in terms of a layer of force is useful in deriving certain energy relations.

When the inhomogeneity has the form of an ellipsoid the solution can be found from the solution for the ellipsoidal inclusion by making use of the fact that the stress in the inclusion is uniform. For the special case of a cavity ($\lambda^* = 0$, $\mu^* = 0$) the method has already been outlined in § 1. The general ellipsoidal inhomogeneity is handled similarly. On the elastic field of a homogeneous inclusion with stress-free strain $\overset{\mathrm{T}}{e_{ij}}$ superimpose a uniform strain $\overset{\mathrm{A}}{e_{ij}}$. Let

$$\overset{\mathrm{A}}{p_{ij}} = \lambda e^{\mathrm{A}}\delta_{ij} + 2\mu\overset{\mathrm{A}}{e_{ij}}$$

be the corresponding stress. The stress in the inclusion is now

$$\overset{\mathrm{inc}}{p_{ij}} = \overset{\mathrm{I}}{p_{ij}} + \overset{\mathrm{A}}{p_{ij}} = \overset{\mathrm{C}}{p_{ij}} - \overset{\mathrm{T}}{p_{ij}} + \overset{\mathrm{A}}{p_{ij}}\,,$$

and the strain in the inclusion is

$$\overset{\mathrm{inc}}{e_{ij}} = \overset{\mathrm{C}}{e_{ij}} + \overset{\mathrm{A}}{e_{ij}}\,.$$

On account of the term $-\overset{\mathrm{T}}{p_{ij}}$ in (2.19) (which appears because there is no stress associated with the stress-free transformation strain $\overset{\mathrm{T}}{e_{ij}}$) $\overset{\mathrm{inc}}{p_{ij}}$ and $\overset{\mathrm{inc}}{e_{ij}}$ are not related by Hooke's law for material with elastic constants λ, μ. They are, however, related by Hooke's law for a material with constants λ^*, μ^* provided these satisfy

$$\overset{\mathrm{inc}}{p_{ij}} = \lambda^* e^{\mathrm{inc}}\,\delta_{ij} + 2\mu^*\overset{\mathrm{inc}}{e_{ij}}\,,$$

that is, if

$$(\lambda-\lambda^*)(\overset{\mathrm{C}}{e}+\overset{\mathrm{A}}{e})\delta_{ij}+2(\mu-\mu^*)(\overset{\mathrm{C}}{e}_{ij}+\overset{\mathrm{A}}{e}_{ij}) = (\lambda\overset{\mathrm{T}}{e}\,\delta_{ij}+2\mu\overset{\mathrm{T}}{e}_{ij})\equiv\overset{\mathrm{T}}{p}_{ij}\,. \quad (4.3)$$

An ellipsoid made of material with these constants can be used to replace the transformed inclusion with continuity of stress and displacement across the interface provided that in its unstressed state this ellipsoid coincides in shape and size with the untransformed inclusion. This replacement does not alter the stresses inside or outside the ellipsoid; they remain the same as the stresses due to a homogeneous transformed inclusion with stress-free strain $\overset{\mathrm{T}}{e}_{ij}$, together with the original applied stress $\overset{\mathrm{A}}{p}_{ij}$.

The argument has been presented as if $\overset{\mathrm{A}}{e}_{ij}$ and $\overset{\mathrm{T}}{e}_{ij}$ were given and λ^*, μ^* were to be found. In the actual inhomogeneity problem λ^*, μ^* and $\overset{\mathrm{A}}{e}_{ij}$ (or $\overset{\mathrm{A}}{p}_{ij}$) are given and we have to determine $\overset{\mathrm{T}}{e}_{ij}$. To do this we express $\overset{\mathrm{C}}{p}_{ij}$ and $\overset{\mathrm{C}}{e}_{ij}$ in terms of S_{ijkl} and $\overset{\mathrm{T}}{e}_{kl}$ and substitute in (4.3). This gives

$$(\lambda^*-\lambda)S_{mmkl}\,\overset{\mathrm{T}}{e}_{kl}\,\delta_{ij}+2(\mu^*-\mu)S_{ijkl}\,\overset{\mathrm{T}}{e}_{kl}+\lambda\overset{\mathrm{T}}{e}\,\delta_{ij}+2\mu\overset{\mathrm{T}}{e}_{ij}$$
$$=(\lambda-\lambda^*)\overset{\mathrm{A}}{e}\,\delta_{ij}+2(\mu-\mu^*)\overset{\mathrm{A}}{e}_{ij}\,. \quad (4.4)$$

As in the case of (3.18) the solution for the non-diagonal $\overset{\mathrm{T}}{e}_{ij}$ is immediate,

$$\overset{\mathrm{T}}{e}_{12}=\frac{\mu-\mu^*}{2(\mu^*-\mu)S_{1212}+\mu}\overset{\mathrm{A}}{e}_{12}, \qquad \overset{\mathrm{T}}{e}_{23}=\ldots,$$

while for $\overset{\mathrm{T}}{e}_{11}$, $\overset{\mathrm{T}}{e}_{22}$, $\overset{\mathrm{T}}{e}_{33}$ we have the three simultaneous equations

$$(\lambda^*-\lambda)\overset{\mathrm{C}}{e}+2(\mu^*-\mu)S_{ijkl}\,\overset{\mathrm{T}}{e}_{kl}+\lambda\overset{\mathrm{T}}{e}+2\mu\overset{\mathrm{T}}{e}_{ij}=(\lambda-\lambda^*)\overset{\mathrm{A}}{e}+2(\mu-\mu^*)\overset{\mathrm{A}}{e}_{ij} \quad (4.5)$$

(ij = 11, 22, 33) with the value (3.19) for e^{C}.

For a sphere (4.4) reduces to

$$\overset{\mathrm{T}}{e}=A\overset{\mathrm{A}}{e}, \qquad '\overset{\mathrm{T}}{e}_{ij}=B\,'\overset{\mathrm{A}}{e}_{ij} \quad (4.6)$$

where

$$A=A\{\varkappa,\varkappa^*\}=\frac{\varkappa^*-\varkappa}{(\varkappa-\varkappa^*)\alpha-\varkappa}\,,$$
$$B=B\{\mu,\mu^*\}=\frac{\mu^*-\mu}{(\mu-\mu^*)\beta-\mu} \quad (4.7)$$

with the values (3.14) for α, β.

The $\overset{\mathrm{T}}{e}_{ij}$ found in this way is the stress-free strain of an 'equivalent homogeneous inclusion' from which the elastic field can be calculated. Note that $\overset{\mathrm{T}}{e}_{ij}$ goes to zero with $\overset{\mathrm{A}}{e}_{ij}$. Consequently the solution does correspond to a perfectly fitting inhomogeneous ellipsoid in a body which is stress-free when not acted on by external forces.

The displacement inside and outside the ellipsoid is

$$u_i = \overset{\mathrm{A}}{u}_i + \overset{\mathrm{C}}{u}_i \,. \tag{4.8}$$

The term $\overset{\mathrm{A}}{u}_i$ represents the unperturbed displacement; it is equal to $\overset{\mathrm{A}}{e}_{ij}x_j$ plus an arbitrary rigid-body displacement. The term $\overset{\mathrm{C}}{u}_i$ represents the perturbation due to the presence of the ellipsoid; it is calculated from (3.1) with the $\overset{\mathrm{T}}{e}_{ij}$ of the equivalent inclusion. The stress is

$$p_{ij} = \overset{\mathrm{A}}{p}_{ij} + \overset{\mathrm{C}}{p}_{ij}$$

outside the ellipsoid and

$$p_{ij} = \overset{\mathrm{A}}{p}_{ij} + \overset{\mathrm{I}}{p}_{ij}$$

inside it. The form

$$p_{ij} = \overset{\mathrm{A}}{p}_{ij} + \lambda \overset{\mathrm{C}}{e} \delta_{ij} + 2\mu \overset{\mathrm{C}}{e}_{ij}$$

is valid inside and outside the ellipsoid; in calculating the interior field we use the 'wrong' elastic constants λ, μ in place of λ^*, μ^* and this compensates for the change from $\overset{\mathrm{C}}{p}_{ij}$ to $\overset{\mathrm{I}}{p}_{ij}$.

It is perhaps worth mentioning that the results of this section and of § 3.2 for the inhomogeneous inclusion can be generalized to the case where the matrix is isotropic but the interior of the ellipsoid is anisotropic and has elastic constants $\overset{*}{c}_{ijkl}$, say. For the transformation problem it is only necessary to replace (3.18) by

$$\lambda(\overset{\mathrm{C}}{e} - \overset{\mathrm{T}}{e})\delta_{ij} + 2\mu(\overset{\mathrm{C}}{e}_{ij} - \overset{\mathrm{T}}{e}_{ij}) = \overset{*}{c}_{ijkl}\,(\overset{\mathrm{C}}{e}_{ij} - \overset{\mathrm{T*}}{e}_{ij}) \,. \tag{4.9}$$

For the inhomogeneity problem (4.4) becomes

$$\lambda(S_{mmkl}\,\overset{\mathrm{T}}{e}_{kl} - \overset{\mathrm{T}}{e} + \overset{\mathrm{A}}{e})\delta_{ij} + 2\mu(S_{ijkl}\,\overset{\mathrm{T}}{e}_{kl} - \overset{\mathrm{T}}{e}_{ij} + \overset{\mathrm{A}}{e}_{ij})$$
$$= \overset{*}{c}_{ijpq}\,(S_{pqkl}\,\overset{\mathrm{T}}{e}_{kl} + \overset{\mathrm{A}}{e}_{pq}) \,. \tag{4.10}$$

By solving (4.9) or (4.10) the $\overset{\mathrm{T}}{e}_{ij}$ of the equivalent inclusion can be found.

The case of an anisotropic ellipsoid in an isotropic medium may seem rather artificial. It finds an application, however, in the theory of aggregates of anisotropic crystals (cf. § 6).

The argument leading to the expression (2.6) for the displacement u_i^C due to a transformed inclusion of any shape applies equally to an anisotropic material. To obtain an explicit expression for u_i^C analogous to (2.7) we should have to know the form of the displacement due to a point force in an anisotropic material. Unfortunately an expression explicit enough for our purpose cannot be obtained (FREDHOLM [1900], LIFSHITZ and ROSENZWEIG [1947], KRÖNER [1953b]). It is, however, possible to carry the analysis far enough to show that the stress is uniform inside a transformed anisotropic ellipsoid in an anisotropic matrix (ESHELBY [1957]).

We may give a picturesque interpretation to (4.8) by saying that the applied field 'induces' an inclusion in the inhomogeneity, with a stress-free strain e_{ij}^T proportional to the applied stress (cf. KRÖNER [1958a]).

If the inhomogeneity is in a finite body we may write the perturbing field of the inclusion in the form

$$u_i^F = u_i^C + u_i^{im}, \qquad p_{ij}^F = p_{ij}^C + p_{ij}^{im}$$

as in § 2.1, with the image terms chosen so that the boundary conditions on the outer surface S_0 of the body are satisfied. They are not, of course, quite identical with the image terms for the equivalent inclusion in a homogeneous medium.

4.2. Energy relations

It is sometimes necessary to compare the elastic behaviour of a body containing an inhomogeneity with the behaviour of a similar body which is homogeneous. In such cases it is convenient to imagine that we have a single body and that the inhomogeneity may be introduced or removed at will.

When an inhomogeneity is introduced into a body already stressed by some external mechanism there will be a change in its elastic energy. At the same time there may be a change in the potential energy of the loading mechanism. As in § 2.2 we can define a change in total energy (or in enthalpy or Gibbs free energy). We shall calculate this quantity for an ellipsoidal inhomogeneity in a body subjected to two types of loading, rigidly imposed surface displacements and constant surface tractions.

Consider first the case where a constant displacement is imposed on the outer boundary by a perfectly rigid external mechanism, producing uniform stress and strain p_{ij}^A, e_{ij}^A in the absence of the inhomogeneity.

The perturbing field when the inhomogeneity is introduced must thus satisfy the condition

$$\overset{\mathrm{F}}{u_i} = \overset{\mathrm{C}}{u_i} + \overset{\mathrm{im}}{u_i} = 0 \quad \text{on} \quad \mathrm{S}_0 \,. \tag{4.11}$$

The elastic energy E_{A} in the medium when the inclusion is absent is found by integrating the (constant) energy density over the whole volume of material. Let the inhomogeneity be introduced and, as in § 4.1, let the material be held in a state of uniform strain $\overset{\mathrm{A}}{e_{ij}}$ by a layer of body force $-T_i$ (eq. (4.9)). The elastic energy is found by replacing λ, μ by λ^*, μ^* in that part of the volume integral for E_{A} which refers to the interior of S. Thus at this stage the elastic energy is $E_{\mathrm{A}} + W_1$ where

$$W_1 = \tfrac{1}{2}V[(\lambda^* - \lambda)\overset{\mathrm{A}}{e}\delta_{ij} + 2(\mu^* - \mu)\overset{\mathrm{A}}{e_{ij}}]\overset{\mathrm{A}}{e_{ij}} \,.$$

If the layer of force $-T_i$ is relaxed to zero each element dS of S suffers a displacement $\overset{F}{u_i}$ and an amount of energy

$$W_2 = \tfrac{1}{2}\int_{\mathrm{S}} T_i \overset{\mathrm{F}}{u_i} \mathrm{dS}$$

is withdrawn from the medium. By Gauss's theorem and (4.11) this can be put in the form

$$W_2 = \tfrac{1}{2}V[(\lambda - \lambda^*)\overset{\mathrm{A}}{e}\delta_{ij} + 2(\mu - \mu^*)\overset{\mathrm{A}}{e_{ij}}]\overset{\mathrm{C}}{e_{ij}} - E_{\mathrm{im}}$$

where

$$E_{\mathrm{im}} = \tfrac{1}{2}[(\lambda - \lambda^*)\overset{\mathrm{A}}{e}\delta_{ij} + 2(\mu - \mu^*)\overset{\mathrm{A}}{e_{ij}}]\int_V \overset{\mathrm{im}}{e_{ij}}\,\mathrm{d}v \,.$$

The increase in the elastic energy when the inhomogeneity is introduced is thus

$$\Delta E_{\mathrm{el}} = W_1 - W_2 \,.$$

This is also the increase in the total energy, ΔE_{tot}, since the rigid loading mechanism does no work. With the help of (4.3) we have

$$W_1 - W_2 = -\tfrac{1}{2}V\overset{\mathrm{A}}{p_{ij}}\overset{\mathrm{T}}{e_{ij}} + E_{\mathrm{im}} \,.$$

We consider next the case where the body is loaded by constant surface tractions $\overset{\mathrm{A}}{p_{ij}}n_j$. The perturbing field due to the introduction of the inhomogeneity must now satisfy

$$\overset{\mathrm{F}}{p_{ij}}n_j = (\overset{\mathrm{C}}{p_{ij}} + \overset{\mathrm{im}}{p_{ij}})n_j = 0 \quad \text{on} \quad \mathrm{S}_0 \,.$$

The elastic energy when the inhomogeneity is introduced and the layer of force $-T_i$ is present is again given by $E_{\mathrm{A}} + W_1$, and W_2 still represents the energy removed on relaxing the layer of force. But in addition there will now be a movement of the outer boundary in which the surface tractions do an amount of work

$$W_3 = \int_{S_0} \overset{A}{p}_{ij} \overset{F}{u}_i \, dS_j$$

on the body. Thus

$$\Delta E_{el} = W_1 - W_2 + W_3 .$$

If we are interested in the change in total energy we do not need to know the value of W_3; evidently W_3 also represents the decrease in the energy of the loading mechanism and so

$$\Delta E_{tot} = \Delta E_{el} - W_3 = W_1 - W_2 .$$

It is in fact not difficult to establish the relation

$$W_3 = 2W_2 - W_1 .$$

We make use of the same device as in (2.39) and write

$$W_3 = \int_{S_0} (\overset{A}{p}_{ij} \overset{F}{u}_i - \overset{F}{p}_{ij} \overset{A}{u}_i) dS_j , \tag{4.12}$$

change the surface of integration from S_0 to S and convert to a volume integral over the inhomogeneity. The relation (4.12) then follows if we note that in the inhomogeneity

$$\overset{A}{p}_{ij} = \lambda \overset{A}{e} \delta_{ij} + 2\mu \overset{A}{e}_{ij}$$

by definition, but that, contrary to the general rule laid down in § 1

$$\overset{im}{p}_{ij} = \lambda^* \overset{im}{e} \delta_{ij} + 2\mu^* \overset{im}{e}_{ij}$$

since $\overset{im}{p}_{ij}$, $\overset{im}{e}_{ij}$ represent the actual image field in the inhomogeneity.

If we introduce the notation

$$E_{int} = -\tfrac{1}{2} V \overset{A}{p}_{ij} \overset{T}{e}_{ij} + E_{im} \tag{4.13}$$

these results may be summarised as follows:

(i) for a rigidly-held boundary

$$\Delta E_{tot} = E_{int}, \qquad \Delta E_{el} = \Delta E_{tot} , \tag{4.14}$$

(ii) for a boundary subject to constant loads

$$\Delta E_{tot} = E_{int}, \qquad \Delta E_{el} = -\Delta E_{tot} . \tag{4.15}$$

The difference in sign between (4.14) and (4.15) may be illustrated by considering the case where the inhomogeneity takes the form of a crack, that is, a narrow region in which the elastic constants are zero. Introduction of the crack into a body strained by rigidly-imposed surface displacements will obviously reduce the elastic energy. On the other hand the presence of the crack makes the body more 'flexible'. Consequently given applied loads will deform it more, do more work on it and so increase its elastic energy.

The first term in (4.13) is precisely half the interaction energy (2.43) for the equivalent homogeneous inclusion. The image term E_{im} will have a different value in (i) and (ii) since e_{ij}^{im} is derived from different boundary conditions. If the inhomogeneity is far from the boundaries it will be small compared with $-\frac{1}{2}Vp_{ij}^{\mathrm{A}}e_{ij}^{\mathrm{T}}$, and we may say that provided the initial stress p_{ij}^{A} is the same in each case, the total energy changes in cases (i) and (ii) are the same, but that the changes in elastic energy are equal and opposite. It is possible to extend the analysis to the mixed case where a constant displacement is imposed on part of S_0 and a constant load on the remainder. As one might expect, ΔE_{tot} is still given by (4.14) (with the appropriate e_{ij}^{im} inserted in E_{im}), but no general statement can be made about the relation between ΔE_{el} and ΔE_{tot}. This is no drawback since it is ΔE_{tot} which is the physically important quantity.

§ 5. Relation to the Theory of Dislocations

There is a close relation between the inclusion problem and the theory of dislocations, more particularly with the general type of dislocation introduced by SOMIGLIANA [1914, 1915]). It will be convenient, however, to begin with the more familiar type of dislocation which plays a part in the physical theory of plasticity (NABARRO [1952], SEEGER [1955]). For want of a better name we may refer to these as VOLTERRA [1907] dislocations, though in fact they only correspond to the first three of his six classes. For our purposes a Volterra dislocation may be defined as a state of self-stress in which the displacement has discontinuity b_i, the Burgers vector, across a surface bounded by an open or closed curve, the dislocation line. If the dislocation line forms a closed loop and lies in a plane, the dislocation is characterised by giving the form and orientation of the loop and the value of the Burgers vector.

The displacement at large distances from a Volterra dislocation loop of area A, Burgers vector b_i and normal n_i situated at the origin is

$$u_i = \frac{A b_j n_k g_{ijk}}{8\pi(1-\sigma)r^2} \tag{5.1}$$

with the notation of (2.23) (BURGERS [1939], NABARRO [1951]). This can be put in the alternative form

$$u_i = \frac{Ab_j n_k}{8\pi(1-\sigma)} \vartheta_{ijk}\, r \tag{5.2}$$

where

$$\vartheta_{ijk} = \frac{\partial^3}{\partial x_i \partial x_j \partial x_k} - \left\{\sigma\delta_{jk}\frac{\partial}{\partial x_i} + (1-\sigma)\delta_{ij}\frac{\partial}{\partial x_k} + (1-\sigma)\delta_{ik}\frac{\partial}{\partial x_j}\right\}\nabla^2 . \tag{5.3}$$

The limiting process

$$A \to 0, \quad b_j \to \infty, \quad Ab_j \to s_j \tag{5.4}$$

gives an elementary infinitesimal dislocation loop of strength s_j and normal n_k. The displacement at any distance from it is given by (5.1) or (5.2) if we identify Ab_j with s_j.

Comparison of (5.1) with (2.23) shows that the remote field of a (finite) dislocation loop is the same as the remote field of an inclusion of arbitrary shape whose volume and stress-free strain satisfy

$$Ve_{ij}^{\mathrm{T}} = \tfrac{1}{2}(b_i n_j + b_j n_i) . \tag{5.5}$$

To an elementary dislocation loop there corresponds an elementary inclusion. It is natural to imagine it as a platelet coinciding with the loop. Its thickness and stress-free strain may be so chosen that in the limit (5.5) agrees with (5.4). A finite plane dislocation loop may be formed by spreading elementary loops over a plane surface. When each loop is replaced by an equivalent elementary inclusion we obtain an inclusion in the form of a thin disc, and the elastic field which it produces is the same as that of the dislocation. It is, in fact, possible to build up directly an expression for the elastic field of a finite plane dislocation loop starting from an inclusion in the form of a disc of small (and ultimately vanishing) thickness. The quantity B_i (2.7) is then the potential of a plane lamina, and the f_{ij} (2.22) are the potentials of certain double layers. The discontinuities in potential and attraction on crossing them can be found by elementary electrostatics, and the e_{ij}^{T} may then be so adjusted that there is a discontinuity b_i in the displacement u_i^{C} between points on opposite faces of the disc. This procedure, however, does not completely determine the e_{ij}^{T}; it is necessary to impose further conditions to ensure, for example, that there is not a line of dilatation running round the rim of the disc.

We turn now to the more fruitful connection between inclusions and the general Somigliana dislocation. A Somigliana dislocation can be constructed as follows. Make a cut over a surface S (open or closed) and give the two faces of the cut an arbitrary small relative displacement, removing material where there would be interpenetration. Fill in any gaps and weld the material together again. Let $b_i(\boldsymbol{r})$ be the relative displacement at the point $\boldsymbol{r}$ on the cut. There are several ways of calculating the resulting state of self-stress. The simplest is to regard the Somigliana dislocation as equivalent to a distribution of elementary dislocations over S, the strength of the loop associated with the element of area $dS(\boldsymbol{r})$ being $b_j(\boldsymbol{r})dS$. By (5.2) the displacement is

$$u_i^{\mathrm{D}} = \frac{1}{8\pi(1-\sigma)} \vartheta_{ijk} I_{jk} \tag{5.6}$$

where

$$I_{jk} = \int_S b_j(\boldsymbol{r}')\,|\boldsymbol{r}-\boldsymbol{r}'|\,dS_k .$$

This can be shown to be equivalent to the expression given by SOMIGLIANA [1915].

To see the connection with the inclusion problem suppose that a cut is made over the surface S of the Somigliana dislocation. The faces spring apart leaving a gap $b_i(\boldsymbol{r})$ (in some places the 'gap' may in fact be an interpenetration of material). As we saw in § 2.1 the inclusion behaves in just this fashion. If we make a cut over the interface a gap

$$b_j(\boldsymbol{r}) = e_{jl}^{\mathrm{T}} x_l \tag{5.7}$$

appears. (To (5.7) we might possibly add a rigid body displacement; for the moment we ignore it.) Consequently the inclusion is equivalent to a Somigliana dislocation in which S is a closed surface coinciding with the interface and having the discontinuity (5.7). We should expect the displacement calculated from (5.7) and (5.6) to be not u_i^{C} but rather the quantity which, in anticipation, we called u_i^{D} in § 2.1. That is, it should coincide with u_i^{C} outside the inclusion and differ from u_i^{C} by the displacement $e_{il}^{\mathrm{T}} x_l$ inside the inclusion. This may be verified analytically. With (5.7) we have, using Gauss's theorem,

$$I_{jk} = e_{jl}^{\mathrm{T}} \int_S x_l' \,|\boldsymbol{r}-\boldsymbol{r}'|\,dS_k = e_{jl}^{\mathrm{T}} \int_V \frac{\partial}{\partial x_k'} \{x_l' |\boldsymbol{r}-\boldsymbol{r}'|\}\,dv$$

$$= e_{jk}^{\mathrm{T}} \psi + e_{jl}^{\mathrm{T}} X_{lk}$$

where ψ is the biharmonic potential (2.9) and

$$X_{lk} = \int_V x_l' \frac{\partial}{\partial x_k'} |\boldsymbol{r}-\boldsymbol{r}'|\,dv .$$

When inserted in (5.6) the term in ψ gives $\overset{\mathrm{C}}{u_i}$ in the form (2.8). It can be verified that

$$\vartheta_{ijk} X_{lk} = 2(1-\sigma)\delta_{ij} \nabla^2 \int_V \frac{x_l'}{|\boldsymbol{r}-\boldsymbol{r}'|} dv . \qquad (5.8)$$

The integral is the harmonic potential of matter of density x_l filling S, and so its Laplacian is $-4\pi x_k$ inside S and zero outside. Consequently

$$\begin{aligned} \vartheta_{ijk} X_{lk} &= -8\pi(1-\sigma)\delta_{ij} x_l \quad \text{inside S} \\ &= 0 \quad \text{outside S} \end{aligned} \qquad (5.9)$$

and so

$$\begin{aligned} \overset{\mathrm{D}}{u_i} &= \overset{\mathrm{C}}{u_i} - \overset{\mathrm{T}}{e_{il}} x_l \quad \text{inside S} \\ &= \overset{\mathrm{C}}{u_i} \quad \text{outside S} \end{aligned} \qquad (5.10)$$

as expected. If we had included a rigid-body rotation $\overset{\mathrm{T}}{\omega_{jl}} x_l$ in (5.7) I_{jk} would have contained the additional terms $\overset{\mathrm{T}}{\omega_{jk}}\psi + \overset{\mathrm{T}}{\omega_{jl}} X_{lk}$. The first contributes nothing to $\overset{\mathrm{D}}{u_i}$, since ϑ_{ijk} is symmetric in jk. According to (5.8) the second term leaves $\overset{\mathrm{D}}{u_i}$ unchanged outside S. Inside S it gives a rigid-body rotation which in a structureless medium produces no observable effect.

It is clear that treatment in terms of a Somigliana dislocation offers no computational advantages when applied to an inclusion which has undergone a homogeneous stress-free strain. However, we may use the formula (5.6) to extend our results to the case where the inclusion suffers any small change of shape, not necessarily a homogeneous deformation.

Suppose that the inclusion undergoes a permanent change of form which, if the matrix were absent, would not be associated with stress. To specify this 'stress-free change of form' it is only necessary to give the displacement of each point of its surface,

$$\overset{\mathrm{T}}{u_i} = \overset{\mathrm{T}}{u_i}(\boldsymbol{r}), \qquad \boldsymbol{r} \text{ on S} . \qquad (5.11)$$

Alternatively, to maintain the analogy with § 2.1, we could introduce a variable stress-free strain $\overset{\mathrm{T}}{e_{ij}}(\boldsymbol{r})$ defined throughout the interior of the inclusion and giving a displacement agreeing with (5.11) on S. (We recall (SOKOLNIKOFF [1946]) that specification of a (compatible) strain throughout a region fixes the displacement in it to within a rigid body displacement.) Since $\overset{\mathrm{T}}{e_{ij}}$ would be largely arbitrary it seems better to work with $\overset{\mathrm{T}}{u_i}$ in the general case.

If we try to put the transformed inclusion into the matrix there will be a misfit $\overset{\mathrm{T}}{u}_i$ at each point of S. When this misfit is removed by suitably straining the material and welding together corresponding points on either side of S we are left with a Somigliana dislocation for which $b_i = \overset{\mathrm{T}}{u}_i$. Thus to find the elastic field when the inclusion is constrained by the matrix we have only to put $b_j(\mathbf{r}') = \overset{\mathrm{T}}{u}_j(\mathbf{r}')$ in (5.6). The resulting $\overset{\mathrm{D}}{u}_i$ is the displacement measured from an initial state in which the matrix is unstressed and the inclusion is transformed and stress-free. Consequently the stress is given by

$$\overset{\mathrm{D}}{p}_{ij} = \lambda \overset{\mathrm{D}}{u}_{m,m}\, \delta_{ij} + \mu(\overset{\mathrm{D}}{u}_{i,j} + \overset{\mathrm{D}}{u}_{j,i})$$

both inside and outside the inclusion.

This problem can also be solved by the method of § 2.1. Let

$$\overset{\mathrm{T}}{u}_i = \overset{\mathrm{T}}{u}_i(\mathbf{r}), \quad \mathbf{r} \text{ inside or on S} \quad ,$$

be any convenient continuous displacement which coincides with (5.11) on S, and put

$$\overset{\mathrm{T}}{p}_{ij} = \lambda \overset{\mathrm{T}}{u}_{m,m}\, \delta_{ij} + \mu(\overset{\mathrm{T}}{u}_{i,j} + \overset{\mathrm{T}}{u}_{j,i}) \quad .$$

Remove the transformed inclusion from the matrix and apply surface tractions $-\overset{\mathrm{T}}{p}_{jk} n_j$ and a distribution of body force of amount $+\overset{\mathrm{T}}{p}_{jk,k}$ per unit volume. This produces a displacement $-\overset{\mathrm{T}}{u}_i$ and so restores the inclusion to its untransformed shape. Cement it back in the matrix and relax the unwanted forces. This gives the displacement (cf. (2.6))

$$\overset{\mathrm{C}}{u}_i = \int_{\mathrm{S}} \mathrm{dS}_k\, \overset{\mathrm{T}}{p}_{jk}(\mathbf{r}')\, U_{ij} - \int_V \mathrm{d}v\, \frac{\partial p_{jk}}{\partial x'_k}(\mathbf{r}')\, U_{ij}\, .$$

It can be shown by the same sort of analysis as led to (5.10) that $\overset{\mathrm{C}}{u}_i = \overset{\mathrm{D}}{u}_i$ in the matrix and that $\overset{\mathrm{C}}{u}_i = \overset{\mathrm{D}}{u}_i + \overset{\mathrm{T}}{u}_i$ in the inclusion. Since $\overset{\mathrm{T}}{u}_i$ is by definition a stress-free displacement the stresses calculated from $\overset{\mathrm{C}}{u}_i$ and $\overset{\mathrm{D}}{u}_i$ are identical.

The solution for the elastic field due to an inclusion which has suffered a non-uniform transformation does not seem to have any obvious applications. However, for an ellipsoidal inclusion the solution has a property which enables it to be used to find how an ellipsoidal cavity or inhomogeneity perturbs a non-uniform stress-field of fairly general type. This property generalizes the result that a uniform transformation strain leads to a uniform strain in the constrained ellipsoid. It may be stated thus: if the stress-free transformation displacement is a polynomial in x_1, x_2, x_3 of degree N, then the displacement is also a polynomial of degree N inside the constrained ellipsoid. We shall only outline its derivation and application. The details of the calculation can easily be filled in.

The analysis involves a number of polynomials in x_i or x'_i with constant coefficients. We shall denote them by script capitals and indicate only the argument and degree. Thus, for example, $\mathscr{P}(M, x)$ stands for a polynomial in x_1, x_2, x_3 whose highest term is of degree M. Similarly $\mathscr{G}_{ijk}(M, x'), i,j = 1, 2, 3$ denotes a set of twenty-seven polynomials in x'_1, x'_2, x' all of the same degree, and so forth. We shall also use R as an abbreviation for $|\boldsymbol{r} - \boldsymbol{r}'|$.

Let the ellipsoidal inclusion undergo the transformation specified by

$$u_i^{\mathrm{T}}(\boldsymbol{r}) = \mathscr{T}_i\,(N,x) \qquad (\boldsymbol{r} \text{ on S})\,.$$

If we put $b_i = u_i^{\mathrm{T}}$ in (5.6) and use Gauss's theorem we find that I_{jk} has the form

$$I_{jk} = \int_V \mathscr{E}_{jk}(N-1, x')R\,\mathrm{d}v + \int_V \mathscr{T}_i(N,x')\,\frac{\partial R}{\partial x'_k}\,\mathrm{d}v\,. \tag{5.12}$$

In the first integrand introduce the factor

$$1 = \frac{(x_m - x'_m)(x_m - x'_m)}{|\boldsymbol{r} - \boldsymbol{r}'|^2}$$

and in the second write $\partial R/\partial x'_k$ as $(x'_k - x_k)/R$. After a little re-arrangement (5.12) takes the form

$$I_{jk} = x_m x_m \int_V \mathscr{F}_{jk}(N-1, x')\,\frac{\mathrm{d}v}{R} + x_p \int_V \mathscr{G}_{pjk}(N, x')\,\frac{\mathrm{d}v}{R}$$
$$+ \int_V \mathscr{H}_{jk}(N+1, x')\,\frac{\mathrm{d}v}{R}\,. \tag{5.13}$$

The integrals are the harmonic potentials of solid ellipsoids whose densities are polynomial functions of the co-ordinates. FERRERS [1877] and DYSON [1891] have discussed the potentials of inhomogeneous ellipoids of this type. Their results show that when the density is of the form $\mathscr{P}(M,x)$ the potential is of the form $\mathscr{Q}(M+2, x)$ inside the ellipsoid. The coefficients of the polynomial $\mathscr{Q}$ can be calculated from the coefficients of $\mathscr{P}$ and the three quantities I_a, I_b, I_c (eq. (3.2)). Outside the ellipsoid the potential is a similar polynomial in x_1, x_2, x_3 but with coefficients which are themselves functions of position. (Compare the relation between (3.2) and (3.7).) These variable coefficients can be expressed in terms of the coefficients of $\mathscr{P}$ and the quantities

A, B, C, F, E of (3.8), (3.10). Consequently, by (5.13) and (5.6) the constrained displacement is a polynomial of the form

$$u_i^{\mathrm{D}}(\boldsymbol{r}) = \mathscr{D}_i(N,x)$$

inside the ellipsoid. Thus the constrained and unconstrained displacements of the inclusion are similar polynomials with calculable relations between their coefficients.

The solution of the cavity problem follows the lines of § 4.1. Superimpose a displacement

$$u_i^{\mathrm{A}}(\boldsymbol{r}) = -\mathscr{D}_i(N,x) \tag{5.14}$$

everywhere. The total displacement $u_i^{\mathrm{A}} + u_i^{\mathrm{D}}$ is zero inside the ellipsoid. Hence the inclusion is unstressed, and it can be removed without disturbing the matrix. The coefficients in $\mathscr{T}_i$ can be chosen so as to make (5.14) any required polynomial of degree N. From these coefficients the coefficients of $\mathscr{F}$, $\mathscr{G}$, $\mathscr{H}$ (eq. (5.13)) can be found. The field $u_i^{\mathrm{A}} + u_i^{\mathrm{D}}$ in the matrix then follows from (5.13) and Ferrers' and Dyson's results. The extension to the ellipsoidal inhomogeneity follows as in § 4.1.

The calculations are not too unwieldy for small N. With $N = 2$ the displacement

$$u_i^{\mathrm{A}} = \alpha_{ij}^{\mathrm{A}} x_j + \beta_{ijk}^{\mathrm{A}} x_j x_k$$

covers the case of an applied stress which is a combination of simple torsion, bending and flexure. Solutions, based on other methods, have been given for a number of such problems involving spheroids and spheres (NEUBER [1958], SEN [1933], DAS [1953]).

§ 6. Applications

The application of the results reviewed here to the actual calculation of stress in and about inclusions and inhomogeneities calls for no special comment. The engineering application of the theory of cavities to the calculation of stress concentrations has been thoroughly treated by NEUBER [1958]. In this section we indicate some of the applications to physical problems.

Expressions for the elastic energy and interaction energy of an inclusion find a use in the theory of martensitic transformations. (For a general review see KAUFMAN and COHEN [1958].) Suppose that a metal can exhibit two crystal structures, γ and α, the former stable at high temperatures and the latter at low temperatures. Ideally we might expect that on cooling the whole of a single crystal of the metal would undergo a uniform stress-free strain in the sense of § 2, the e_{ij}^{T} being the deformation which carries the γ-lattice into the α-lattice. In a martensitic transformation, however, the low-temperature phase first appears in the form of inclusions of α embedded in the γ matrix. Thus, in considering the thermodynamics of the transformation, one must take into account the elastic energy of the misfitting inclusions of α and also, if there is an externally applied stress, their interaction energy with the latter.

The Gibbs free energy change associated with the formation of a martensitic inclusion may be written in the form

$$\Delta G = \Delta G_{\mathrm{chem}} + \Delta G_{\mathrm{surf}} + E_{\mathrm{inc}} + E_{\mathrm{int}}\,. \tag{6.1}$$

ΔG_{chem} is the 'chemical' contribution, the free energy change which would occur if the inclusion underwent its stress-free strain in the absence of the matrix. The theory of elastic inclusions can tell us nothing about it. ΔG_{surf} is a contribution due to the interface between matrix and inclusion. It may be estimated by using the theory of dislocations. E_{inc} is the elastic energy associated with the inclusion, and E_{int} is the interaction energy with any externally applied stress which may be present. According to the discussion in § 2.2 the terms E_{inc} and E_{int} taken together make up the elastic contribution to the change of Gibbs free energy.

It will usually be accurate enough to identify E_{int} with E_{∞}, the value in an infinite medium, and to ignore the difference between the elastic constants of matrix and inclusion. If the inclusion can be considered to be some form of ellipsoid E_{∞} can be calculated from (3.3) and (2.32) when the transformation strain e_{ij}^{T} is known. Suppose, for example, that the inclusion takes the form of a plate in the x_1x_2-plane and that the transformation is made up of a shear parallel to this plane through an angle s, a uniform dilatation Δ and an extension ξ perpendicular to the plane of the plate. Then

$$e_{13}^{T} = e_{31}^{T} = \tfrac{1}{2}s, \qquad e_{11}^{T} = e_{22}^{T} = \tfrac{1}{3}\Delta\,, \qquad e_{33}^{T} = \xi + \tfrac{1}{3}\Delta \tag{6.2}$$

and the remaining components are zero. If the plate has the form of a thin oblate spheroid with $c \ll a = b$ we have

$$\frac{1-\sigma}{\mu V} E_{\infty} = \frac{1}{8}\pi(2-\sigma)\frac{c}{a}s^2 + \frac{2}{9}(1+\sigma)\,\Delta^2 + \frac{1}{4}\pi\frac{c}{a}\xi^2 + \frac{1}{3}\pi(1+\sigma)\frac{c}{a}\xi\Delta \tag{6.3}$$

(CHRISTIAN [1958]). There is also a fairly simple expression for E_∞ when the inclusion is a flat ellipsoid with $a > b \gg c$ (ESHELBY [1957]). The expression (6.3) has been used by CHRISTIAN [1958, 1959] and KAUFMAN [1959] to discuss the nucleation of martensite.

The interaction energy may be found from (2.43). In an actual experiment the applied stress will usually be a uniaxial tension; p_{ij}^{A} then takes the form $\tau n_i n_j$, where τ is the magnitude of the tension and n_i is its direction. If the transformation strain is given by (6.2) we have

$$E_{\text{int}} = -\tau C V \tag{6.4}$$

with

$$C = s \cos\beta \sin\beta + \xi \cos^2\beta + \tfrac{1}{3}\Delta$$

where β is the angle between the x_3-axis and the direction of the tension, and the latter is assumed to lie in the x_1x_3-plane. A result equivalent to (6.4) was first obtained (with $\Delta = 0$) by PATEL and COHEN [1953] (cf. also MACHLIN and WEINIG [1953]). FISHER and TURNBULL [1953] have also discussed the effect of an applied stress on the free energy change associated with martensite formation. In effect, they take the interaction energy to be the cross-term in the elastic energy between the applied field and the field due to the inclusion. As we have seen (§ 2.2), this quantity should be zero. Since they only integrate the energy density over the immediate neighbourhood of the inclusion they obtain a finite result which, however, is not quite correct numerically and which, in addition, has the wrong sign. To correct this they have to make an arbitrary reversal of sign in the relation between applied stress and applied strain.

When a crystal having the high-temperature γ-structure is cooled, ΔG decreases. According to elementary thermodynamics a martensitic inclusion of the α-phase can form when a temperature is reached at which ΔG is zero. In fact a nucleation barrier has to be overcome and the transformation will occur when ΔG reaches some finite negative value, say ΔG_{nuc}. Let T_s be the temperature at which $\Delta G = \Delta G_{\text{nuc}}$ in the absence of external stress, i.e. with $E_{\text{int}} = 0$ in (6.1). When there is an applied stress the term E_{int} will alter the temperature at which $\Delta G_{\text{nuc}} = \Delta G$ to, say, $T_s + \Delta T_s$. If ΔG_{nuc} is independent of temperature and ΔG_{chem} is the only strongly temperature-dependent term in (6.1) we must have

$$\frac{\mathrm{d}\Delta G_{\text{chem}}}{\mathrm{d}T} \Delta T_s + E_{\text{int}} = 0\,,$$

and so from (6.4) we obtain the expression

$$\frac{\mathrm{d}T_s}{\mathrm{d}\tau} = C \Big/ \frac{1}{V} \frac{\mathrm{d}\Delta G_{\text{chem}}}{\mathrm{d}T} \tag{6.5}$$

for the rate of change of transformation temperature with applied tensile stress. C can be found from crystallographic measurements, and the rate of change of chemical free energy with temperature may be obtained from thermodynamic data. Patel and Cohen found excellent agreement between theory and experiment for certain iron-nickel and iron-nickel-carbon alloys. They also showed how (6.5) should be modified when ΔG_{nuc} varies with temperature.

For martensitic inclusions the transformation strain is essentially a pure shear. For particles of precipitate formed by diffusion it is essentially a dilatation. For an ellipsoidal precipitate the strain energy can be found from (3.21) with $e_{ij}^{\mathrm{T}*} = \frac{1}{3}e^{\mathrm{T}}\delta_{ij}$. ROBINSON [1951] has given a detailed treatment. In these calculations it is assumed that the inclusion is coherent with the matrix. NABARRO [1940] and KRÖNER [1954] have treated the problem of finding the minimum strain energy due to an inclusion which has broken away from the matrix. The displacement need not be continuous across the interface, but the matrix and inclusion are supposed to be in contact everywhere. The volume misfit is prescribed, but the stress-free shape of the inclusion has to be determined so as to minimize the energy. According to KRÖNER [1953a, 1954] the energy is a minimum for the state in which the stress in the inclusion is purely hydrostatic. (Matrix and inclusion may be made of different anisotropic materials.)

The theory of cracks, i.e. empty cavities one of whose dimensions is evanescent, has played a part in Griffiths' treatment of rupture (see, for example, SNEDDON [1951]) and, more recently, in the theory of the brittle and ductile fracture of metals. The crack is supposed to have associated with it a surface energy γA, where A is its surface area and γ is a constant representing true surface energy or an effective surface energy associated with plastic deformation. According to the discussion in § 4.2 the total energy of the system made up of the body containing the crack and the external loading mechanism is

$$E_{\text{int}} + \gamma A + \text{const.}$$

By studying the way in which E_{int} and A vary when the form of the crack is altered slightly one can find whether it is energetically favourable for the crack to spread or contract.

We may find the necessary properties of an elliptical crack from the results of §§ 4.1, 4.2 by putting $\lambda^* = \mu^* = 0$ and letting the c-axis tend to zero. The equivalent stress-free strain $\overset{\mathrm{T}}{e}_{ij}$ approaches infinity for a fixed applied stress $\overset{\mathrm{A}}{p}_{ij}$, but the products $V\overset{\mathrm{T}}{e}_{ij}$ (and hence also the potentials φ, ψ) remain finite.

There exist a number of calculations for the interaction energy of a circular or two-dimensional crack (INGLIS [1913], STARR [1928], SACK [1946], SEGEDIN [1951]). They may be deduced from the following results for the elliptical crack

$$\frac{x_1^2}{a^2} + \frac{x_2^2}{b^2} = 1 \, .$$

(i) The applied stress is a pure tension $\overset{\mathrm{A}}{p}_{33}$ normal to the plane of the crack (tensions $\overset{\mathrm{A}}{p}_{11}$, $\overset{\mathrm{A}}{p}_{22}$ evidently have no effect). Then

$$E_{\text{int}} = -\frac{2\pi(1-\sigma)ab^2}{3E(k)} \frac{(\overset{\mathrm{A}}{p}_{33})^2}{\mu} \, . \tag{6.6}$$

This is easily deduced from the results of GREEN and SNEDDON [1949].
(ii) The applied stress is a pure shear $\overset{\mathrm{A}}{p}_{13}$ in the plane of the crack. Then (ESHELBY [1957])

$$E_{\text{int}} = -\frac{2\pi ab^2}{3\eta} \frac{(\overset{\mathrm{A}}{p}_{13})^2}{\mu}$$

where

$$\eta = E(k) + \frac{\sigma}{1-\sigma} \frac{K(k) - E(k)}{k^2} \, . \tag{6.7}$$

In (6.6) and (6.7) E, K are complete elliptic integrals of modulus $k = (1 - b^2/a^2)^{\frac{1}{2}}$. Eq. (6.7) is valid for both $a > b$ and $a < b$. In the latter case the elliptic integrals may be reduced to real form with the help of the relation

$$K(k_1) = (1-k^2)^{\frac{1}{2}} K(k), \qquad E(k_1) = E(k)/(1-k^2)^{\frac{1}{2}},$$
$$\text{where} \quad k_1 = ik/(1-k^2)^{\frac{1}{2}} \, .$$

(Compare the remark following (3.6).) For the circular crack $a = b$, $E = \frac{1}{2}\pi$, $\eta = \pi(2-\sigma)/4(1-\sigma)$. STROH [1958] has calculated the interaction energy for two-dimensional cracks in anisotropic materials.

The concept of the interaction energy of an inhomogeneity with an applied stress simplifies the calculation of the bulk elastic constants of elastically inhomogeneous aggregates. Suppose, for example, that we wish to calculate the effective elastic constants of a material of bulk modulus $\varkappa$ and shear modulus μ containing a dispersion of spherical inhomogeneities with elastic constants $\varkappa^*$, μ^* (ESHELBY [1957]). In the absence of the inhomogeneities unit volume of the material has elastic energy

$$E_0 = \frac{1}{2}\left(\frac{1}{9\varkappa} \overset{A}{p} \overset{A}{p} + \frac{1}{2\mu} {}'\overset{A}{p}_{ij} {}'\overset{A}{p}_{ij}\right) \tag{6.8}$$

if it is subjected to the uniform stress

$$\overset{A}{p}_{ij} = \tfrac{1}{3}\overset{A}{p}\delta_{ij} + {}'\overset{A}{p}_{ij}$$

(for the notation see § 1). If the spheres are introduced the elastic energy becomes

$$E = E_1 - \Sigma E_{\text{int}} \tag{6.9}$$

by (4.15), where Σ denotes summation over all the spheres. If we ignore the image term in (4.13) and also the interaction between the spheres (so that we limit ourselves to a dilute dispersion) (6.9) may be written as

$$E = \frac{1}{2}\left\{\frac{1}{9\varkappa}(1 + Av)\overset{A}{p}\overset{A}{p} + \frac{1}{2\mu}(1 + Bv)\, {}'\overset{A}{p}_{ij} {}'\overset{A}{p}_{ij}\right\}$$

where v is the fraction of the volume of material occupied by the inhomogeneous spheres and A, B are given by (4.7). Comparing with (6.8) we see that the effective elastic constants are

$$\varkappa_{\text{eff}} = \varkappa/(1 + Av), \qquad \mu_{\text{eff}} = \mu/(1 + Bv)$$

or, since the analysis is only valid for $v \ll 1$,

$$\varkappa_{\text{eff}} = \varkappa(1 - Av), \qquad \mu_{\text{eff}} = \mu(1 - Bv)\,. \tag{6.10}$$

These expressions have been obtained in another manner by HASHIN (REINER [1958]).

Much attention has been given to a related problem: namely, to determine the macroscopic elastic constants, $\overset{0}{c}_{ijkl}$, $\overset{0}{s}_{ijkl}$ say, of a polycrystalline aggregate whose actual constants c_{ijkl}, s_{ijkl} vary from grain to grain, not because the material is inhomogeneous, but because the crystal orientation varies. The estimates $\overset{0}{c}_{ijkl} = \bar{c}_{ijkl}$ and $\overset{0}{s}_{ijkl} = \bar{s}_{ijkl}$ are associated with the names of Voigt and Reuss. Here the bar denotes an average over all crystal orientations, weighted if necessary according to the relative frequency of each orientation. HILL [1952] has shown that

$$\bar{s}'_{ijkl} \leqslant \overset{0}{s}_{ijkl} \leqslant \bar{s}_{ijkl}$$

$$\bar{c}'_{ijkl} \leqslant \overset{0}{c}_{ijkl} \leqslant \bar{c}_{ijkl}$$

where $\bar{c}'_{ijkl}$ and $\bar{s}'_{ijkl}$ are, respectively, the tensors inverse to $\bar{s}_{ijkl}$ and $\bar{c}_{ijkl}$. For an aggregate of cubic crystals the upper and lower estimates of the bulk modulus coincide and we have exactly

$$\varkappa = \tfrac{1}{3}(c_{11} + 2c_{12}) \,. \tag{6.11}$$

Hill's relation seems to mark the limit of what can be proved precisely. To go further it is necessary to make assumptions which may be physically plausible, but cannot be proved conclusively. Most of the attempts in this direction are rather outside the scope of this article. HERSHEY'S [1954] and KRÖNER'S [1958b] calculations, however, make use of the concept of an anisotropic inhomogeneity in an isotropic matrix. We shall not give their physical arguments, but instead present a simplified treatment which involves essentially the same mathematics.

We confine ourselves to an aggregate of cubic crystals which is macroscopically isotropic. Let the effective elastic constants be $\varkappa$, μ. Imagine that the material of each grain is replaced by isotropic material with constants $\varkappa$, μ, but that the boundaries between grains can still be distinguished. We thus obtain, trivially, an 'equivalent isotropic aggregate' with the same overall elastic constants as the polycrystalline aggregate. It seems clear that this equivalent aggregate must pass the following test: if any representative sample of its grains have their original anisotropic constants restored the bulk elastic constants are unaltered. In applying the test we may take the collection of re-transformed grains to be so far apart that the effect of the perturbing field of one grain on another may be neglected. Of course the isotropic aggregate must also pass the more severe test that its bulk elastic constants are unchanged when *every* grain is restored to its original anisotropic state. However, we may hope that if the weaker test gives definite values for $\varkappa$ and μ, then these will be the same as those which would result from applying the severer test.

We shall suppose that the set of test grains have their crystal axes oriented at random and that they may be treated as spheres. This second assumption is hard to justify rigorously, but without it not much progress can be made. In the presence of a uniform applied strain $\overset{\mathrm{A}}{e}_{ij}$ the $\overset{\mathrm{T}}{e}_{ij}$ of the equivalent inclusion for one of the grains can be written in the form

$$\overset{\mathrm{T}}{e}_{ij} = D_{ijkl}\, \overset{\mathrm{A}}{e}_{kl} \;;$$

the D_{ijkl} depend on the orientation of the grain. When the anisotropic test grains are introduced into the equivalent isotropic material its elastic constants are changed by an amount proportional to the sum of their interaction energies with the applied field, by the same argument as led to (6.9). Thus the test requires that

$$\Sigma\,(\overset{\mathrm{A}}{p}_{ij}\,\overset{\mathrm{T}}{e}_{ij}) = \Sigma\,(\overset{\mathrm{A}}{p}_{ij}\,\overset{\mathrm{A}}{e}_{kl}\,D_{ijkl}) = 0 \tag{6.12}$$

where Σ implies summation over all the re-transformed grains. If we choose fixed coordinate axes, $\overset{\mathrm{A}}{p}_{ij}\overset{\mathrm{A}}{e}_{kl}$ is the same for each term of the summation but D_{ijkl} varies from term to term. It is therefore more convenient to choose for each term axes parallel to the crystal axes of the grain in question. Then D_{ijkl} is the same for each term, but $\overset{\mathrm{A}}{p}_{ij}\overset{\mathrm{A}}{e}_{kl}$ is not. The calculation of the sum $\Sigma\overset{\mathrm{A}}{p}_{ij}\overset{\mathrm{A}}{e}_{kl}$ could be carried out as follows. Assign fixed values to the principal strains of $\overset{\mathrm{A}}{e}_{ij}$, form $\overset{\mathrm{A}}{p}_{ij}\,\overset{\mathrm{A}}{e}_{kl}$, average the result over all orientations of the principal axes of the strain and multiply by the number of terms. The resulting expression will have the form of the most general isotropic tensor which has the symmetry of the suffixes in $\overset{\mathrm{A}}{p}_{ij}\overset{\mathrm{A}}{e}_{kl} = \overset{\mathrm{A}}{p}_{kl}\overset{\mathrm{A}}{e}_{ij}$. Consequently it must have the same form as the elastic constant tensor c_{ijkl} of an isotropic medium, that is

$$\Sigma\overset{\mathrm{A}}{p}_{ij}\,\overset{\mathrm{A}}{e}_{kl} = c_1\delta_{ij}\delta_{kl} + c_2(\delta_{ik}\delta_{jl} + \delta_{il}\delta_{jk})$$

with arbitrary c_1, c_2. Inserting this in (6.12) we get $D_{iijj} = 0$ and $D_{ijij} = 0$, or in view of the symmetry (spherical grains, cubic crystal referred to its principal axes)

$$D_{1111} + 2D_{1122} = 0 \tag{6.13}$$

and

$$D_{1111} + 2D_{1212} = 0\,. \tag{6.14}$$

The values of the D_{ijkl} are easily calculated from (4.6) and (4.7) if we bear in mind that when extended along its axes a cubic crystal behaves like an isotropic material with $\varkappa^* = \frac{1}{3}(c_{11} + 2c_{12})$, $\mu^* = \frac{1}{2}(c_{11} - c_{22})$ and that for a shear of the type e_{12} it behaves like an isotropic material with $\mu^* = c_{44}$. Eq. (6.14) gives $A\{\varkappa, \frac{1}{3}(c_{11} + 2c_{12})\} = 0$ or $\varkappa = \frac{1}{3}(c_{11} + 2c_{12})$ in agreement with the already known result (6.11). The expression for the shear modulus is more interesting. Eq. (6.14) gives

$$\tfrac{2}{3}B\{\mu, \tfrac{1}{2}(c_{11} - c_{12})\} - B\{\mu, c_{44}\} = 0\,.$$

On inserting the value of β (eq. (3.14)) and clearing of fractions we are left with a cubic equation:

$$16\mu^3 + 2(5c_{11} + 4c_{12})\mu^2 - 2c_{44}(7c_{11} - 4c_{12})\mu \\ - c_{44}(c_{11} - c_{12})(c_{11} + 2c_{12}) = 0 \qquad (6.15)$$

for determining μ. It has only one real positive root.

HERSHEY [1954] obtained a quartic equation for μ which in fact is what one obtains on multiplying (6.15) by $8\mu + 9\varkappa$. Evidently this introduces no new positive root. KRÖNER [1958b] obtained both the quartic and cubic and gave an argument to show that their respective positive roots are upper and lower bounds for μ, and that since these roots coincide the value so obtained is exact apart from statistical uncertainties due to the fact that the aggregate contains only a finite number of grains.

GOODIER [1936] has emphasized that there is a useful analogy between problems in the slow motion of viscous liquids and elastic problems for incompressible solids. In general special care has to be taken in extrapolating elastic solutions to the incompressible case $\sigma = \frac{1}{2}$, since we have, in effect, to make the simultaneous transition $u_{k,k} \to 0$, $\lambda \to \infty$. However, in the case of the solution (2.8) to the general inclusion problem there is no trouble. With $\sigma = \frac{1}{2}$ it satisfies

$$\mu\nabla^2 \boldsymbol{u}^{\mathrm{C}} = 0, \qquad \operatorname{div} \boldsymbol{u}^{\mathrm{C}} = 0$$

outside the inclusion. These are just the Stokes equations governing the velocity $\boldsymbol{u}^{\mathrm{C}}$ in an incompressible fluid of viscosity μ when inertial effects may be neglected and the hydrostatic pressure, p_0, say, is independent of position. Consequently (2.8) with $\sigma = \frac{1}{2}$ and $\overset{\mathrm{C}}{u_i}$ interpreted as a velocity represents a certain state of viscous flow. The associated stress is

$$p_{ij} = \mu(\overset{\mathrm{C}}{u}_{i,j} + \overset{\mathrm{C}}{u}_{j,i}) - p_0\delta_{ij}\,.$$

Eq. (2.8) describes the flow about a solid which is deforming in such a way that each point $\boldsymbol{r}$ of its surface S has a velocity $\overset{\mathrm{C}}{u_i}(\boldsymbol{r})$. Since we cannot prescribe $\overset{\mathrm{C}}{u_i}(\boldsymbol{r})$, but only the constants $\overset{\mathrm{T}}{e_{ij}}$ the analogy is not of much use in the general case. Eq. (3.1), however, (with $\sigma = \frac{1}{2}$) describes the viscous flow around an ellipsoid which is undergoing a change of shape specified by the constant rate-of-strain tensor $\overset{\mathrm{C}}{e_{ij}}$, and the $\overset{\mathrm{T}}{e_{ij}}$ appropriate to prescribed $\overset{\mathrm{C}}{e_{ij}}$ can be found be solving (3.3).

The solution to the problem of an ellipsoidal inhomogeneity perturbing a uniform strain $\overset{\mathrm{A}}{e_{ij}}$ has a simple viscous interpretation when the ellipsoid is perfectly rigid and incompressible ($\lambda^* \to \infty$, $\mu^* \to \infty$). Eq. (4.3) gives $\overset{\mathrm{C}}{e_{ij}} + \overset{\mathrm{A}}{e_{ij}} = 0$ on S or equivalently

$$\overset{\mathrm{C}}{u_i} + \overset{\mathrm{A}}{u_i} = 0 \quad \text{on} \quad \mathrm{S}\,. \qquad (6.16)$$

Translated into terms of velocity this is the condition that the liquid shall adhere to the surface of the ellipsoid, and so the field $u_i = \overset{\mathrm{C}}{u_i} + \overset{\mathrm{A}}{u_i}$ gives the velocity about a solid ellipsoid immersed in the uniform flow specified by the rate-of-strain tensor $\overset{\mathrm{A}}{e_{ij}}$. The elastic energy density translates into half the rate of dissipation of energy per unit volume. (In the one case we are concerned with the familiar half stress times strain, in the other with stress times rate of deformation.) Consequently $2E_{\mathrm{int}}$ (eq. (4.13)) is equal to the additional rate of dissipation of energy, $\dot{E}_{\mathrm{diss}}$ say, when a solid is introduced into a viscous liquid at whose boundary constant velocities are maintained.

If the image term may be neglected in (4.13) the interaction energy for an inhomogeneity is half the interaction energy for the equivalent transformed inclusion, and it follows that $\dot{E}_{\mathrm{diss}}$ is given by the right hand side of (2.42), or, in view of (6.16), by

$$\dot{E}_{\mathrm{diss}} = -\int_S (\overset{\mathrm{C}}{p_{ij}} + \overset{\mathrm{A}}{p_{ij}})\overset{\mathrm{C}}{u_i}\,\mathrm{d}S_j$$

$$= \int_S \boldsymbol{p}_n \cdot \boldsymbol{u}^{\mathrm{C}}\,\mathrm{d}S$$

where $\boldsymbol{p}_n$ is the actual surface traction on the solid (with the sign convention usual in hydrodynamics) and $\boldsymbol{u}^{\mathrm{C}}$ is the perturbation of velocity due to the presence of the solid. This is identical with a result of BRENNER's [1958]. The expression (6.10) for the effective shear modulus of a dispersion of spheres gives the Einstein viscosity formula $\mu_{\mathrm{eff}} = (1 + \frac{5}{2}v)\mu$ when we put $\mu^* = \infty$, $\sigma = \frac{1}{2}$. The difference in sign between (4.14) and (4.15) has the following interpretation in terms of viscosity. The viscosity is always increased by the introduction of solid particles. Consequently a viscometer working at constant load produces a lower rate of deformation and so less energy dissipation, while a viscometer working at constant speed has to work harder to maintain the prescribed rate of strain.

The applications considered above have all been essentially macroscopic. Elastic inclusions and inhomogeneities have also been found useful as models of lattice defects in crystals. (For a general account see, for example, FRIEDEL [1956] or ESHELBY [1956].)

Suppose that one atom of a crystal lattice is replaced by a foreign atom. The foreign atom will generally have a different size from the host atoms. As a simple elastic model we may take a spherical hole in an isotropic continuum into which a misfitting elastic sphere is inserted (BILBY [1950]). Since the elastic constants of the foreign atom (in so far as one can speak of them for a single atom) will differ from those of the host atoms the misfitting elastic sphere will in general have to be assigned elastic constants differing from those of the matrix. It is useful to consider two limiting cases. (i) A pure inclusion; the sphere and inclusion have the same elastic constants, and the misfit gives rise to a permanent elastic field. (ii) A pure inhomogeneity; there is no misfit, but the elastic constants of sphere and matrix differ. There is no permanent elastic field, but a field can be 'induced' by an applied field. The general case, intermediate between (i) and (ii), corresponds to the inhomogeneous inclusion of § 3.2.

It is not at all obvious that such a crude model will be of any use in discussing the behaviour of lattice defects. We shall discuss some of the consequences of taking the model seriously and then try to indicate the reasons for its success.

The elastic field of a pure inclusion is given by (2.26) with the value (3.12) for φ and e^{T} equal to the fractional volume misfit between hole and sphere. The stress falls off as r^{-3} and its hydrostatic component is zero. The field of a pure inhomogeneity in an applied field $\overset{\mathrm{A}}{e_{ij}}$ is found by calculating the equivalent $\overset{\mathrm{T}}{e_{ij}}$ from (4.6) and inserting them in (2.8) with the values (3.12), (3.13) for φ and ψ.

It may be shown that the expressions (2.43) and (4.13) for the interaction energies of inclusions and inhomogeneities remain valid when $\overset{\mathrm{A}}{p_{ij}}$, $\overset{\mathrm{A}}{e_{ij}}$ refer to the field produced by some source of internal stress, another inclusion or a dislocation for example. Thus the interaction energy between two pure inclusions is proportional to $\overset{\mathrm{A}}{p_{ij}}\overset{\mathrm{T}}{e_{ij}}$, where $\overset{\mathrm{A}}{p_{ij}}$ is the stress produced by one inclusion and $\overset{\mathrm{T}}{e_{ij}}$ is the stress-free strain associated with the other. Since $\overset{\mathrm{T}}{e_{ij}} = \frac{1}{3}e^{\mathrm{T}}\delta_{ij}$ the interaction energy is proportional to p^{A}. But neither defect produces a hydrostatic pressure, and so the interaction energy is zero (BITTER [1931]). Again, let $\overset{\mathrm{A}}{p_{ij}}$ refer to the field of a pure inclusion and let $\overset{\mathrm{T}}{e_{ij}}$ be the equivalent stress-free strain which it induces in a pure inhomogeneity at a distance r from it. Since the $\overset{\mathrm{A}}{p_{ij}}$ are proportional to r^{-3} and the $\overset{\mathrm{T}}{e_{ij}}$ are proportional to the $\overset{\mathrm{A}}{p_{ij}}$ it follows that the interaction energy between a pure inclusion and a pure inhomogeneity is proportional to the inverse sixth power of the distance between them. The interaction energy between two defects each of which is represented by a misfitting and inhomogeneous sphere is also proportional to r^{-6}. Its numerical value may be found by a detailed calculation (ESHELBY [1958]).

The stress-field at a distance r from a dislocation line is proportional to r^{-1}. It follows by the same kind of argument as before that the dislocation has an interaction energy proportional to r^{-1} with a pure inclusion but an interaction energy proportional to r^{-2} with a pure inhomogeneity. This difference allows one to distinguish experimentally between the two types of defect. Since the interaction energy is a function of position there is an effective force

$$\boldsymbol{F} = -\operatorname{grad} E_{\text{int}}(\boldsymbol{r}) \tag{6.17}$$

on the defect. If the defect is capable of diffusing through the lattice, a drift velocity

$$\boldsymbol{v} = D\boldsymbol{F}/kT$$

is superimposed on its random motion, where D is the diffusion coefficient, k is Boltzmann's constant and T is the absolute temperature. Thus, if dislocations are introduced into a crystal containing defects, the latter will be attracted to the dislocations. The resulting depletion of defects in the bulk of the material can be detected by measuring suitable physical properties (electrical resistance, internal friction). Calculation shows that for an r^{-1} interaction the number of defects drawn into the dislocations is proportional to $t^{\frac{2}{3}}$, where t is the time since the interaction between defects and dislocations began. If, on the other hand, there is an r^{-2} interaction the expression $t^{\frac{2}{3}}$ is replaced by $t^{\frac{1}{2}}$ (FRIEDEL [1959]).

For a defect which can be represented by an inserted sphere which is both misfitting and inhomogeneous the interaction energy with a dislocation will evidently have the form $Ar^{-1} + Br^{-2}$. Even if the coefficient B is relatively large the A-term will dominate at large distances and we should expect the $t^{\frac{2}{3}}$ law to be most nearly obeyed. The $t^{\frac{2}{3}}$ law has, in fact, been verified for carbon and nitrogen diffusing in iron. We might, perhaps, regard a vacant lattice site as a pure inhomogeneity ($A = 0$). However, such calculations as have been made (e.g. TEWORDT's [1958] for copper) indicate that there is an appreciable displacement of the atoms bordering the vacant site. On the elastic model this means that there is a stress-field associated with the vacancy even in the absence of an applied field, and so $A \neq 0$.

Rather surprisingly WINTENBERGER's [1957] measurements on vacancies in aluminium follow the $t^{\frac{1}{2}}$ law, which indicates that, in aluminium at least, lattice vacancies behave as pure inhomogeneities.

We shall now try to indicate why the simple misfitting sphere model has been relatively successful. In the first place it is reasonable to suppose that sufficiently far from a lattice defect the crystal can be treated as an elastic continuum. The displacement representing the disturbance due to the defect can be expanded in ascending inverse powers of r (the distance from the defect) each provided with a suitable angular factor. If we treat the material as isotropic the leading term of the expansion has precisely the form (2.23) with arbitrary symmetric $\overset{\mathrm{T}}{e}_{ij}$. Thus, if the material is isotropic, (2.23) gives the elastic field at large distances from the most general type of point defect. In many of the common metals the elastic field of the defect will have cubic symmetry. This physical condition, combined with the artificial limitation to isotropy, requires that $\overset{\mathrm{T}}{e}_{ij}$ shall be of the form $\frac{1}{3}e^{\mathrm{T}}\delta_{ij}$. According to (2.26) the remote field is independent of the shape of the inclusion and we may take it to be a sphere. In this way we recover the misfitting sphere model. In some cases (in particular, interstitial carbon and nitrogen in iron) the assumption that $\overset{\mathrm{T}}{e}_{ij}$ is a pure dilatation is inadequate, but the displacement may still be taken to have the form (2.23) with suitably chosen values for the $\overset{\mathrm{T}}{e}_{ij}$.

The interaction energy (2.42) can be re-written in the form

$$E_{\mathrm{int}} = \int_{\mathrm{S}'} (\overset{\mathrm{C}}{p}_{ij}\, \overset{\mathrm{A}}{u}_i - \overset{\mathrm{A}}{p}_{ij}\, \overset{\mathrm{C}}{u}_i)\mathrm{dS}_j \qquad (6.18)$$

where S′ is any surface drawn in the material and enclosing the inclusion. (The expressions (2.42) and (6.18) are identical because the divergence of the integrand is zero between S and S′.) The expression for the interaction energy was derived on the assumption that the infinitesimal theory could be applied everywhere. This is obviously not true near an inclusion representing an atomic defect. However, it may be shown that for (6.18) to be correct it is only necessary that the infinitesimal theory be valid in the neighbourhood of the surface S′, a much less severe requirement. The following is an outline of the argument (cf. ESHELBY [1959a]). We may form an expression for the x_l-component of the effective force (6.17) by subtracting from (6.18) the corresponding expression with $\overset{\mathrm{C}}{p}_{ij}$, $\overset{\mathrm{C}}{u}_i$ replaced by $\overset{\mathrm{C}}{p}_{ij} + \overset{\mathrm{C}}{p}_{ij,l}\,\varepsilon$, $\overset{\mathrm{C}}{u}_i + \overset{\mathrm{C}}{u}_{i,l}\,\varepsilon$, dividing by ε and letting ε approach zero. The result is

$$F_l = \int_{\mathrm{S}'} (\overset{\mathrm{C}}{p}_{ij,l}\, \overset{\mathrm{A}}{u}_i - \overset{\mathrm{A}}{p}_{ij}\, \overset{\mathrm{C}}{u}_{i,l})\mathrm{dS}_j\,. \qquad (6.19)$$

It is possible (ESHELBY [1956]) to derive a general expression for F_l which is valid for an arbitrary non-linear stress-strain relation and for finite deformation. Like (6.19) this expression takes the form of an integral over a surface surrounding the defect on which the force is to be calculated. If it is permissible to apply the infinitesimal theory on this surface the general expression reduces to (6.19). Hence (6.19), (6.18) or the simple formula (2.43) can be used whenever it is possible to draw a surface S′ enclosing the defect and no other source of internal stress and far enough from it for the strains on S′ to be reasonably small. Applied to the interaction of two point defects, for example, this means that (2.42) can be used if the linear theory is approximately obeyed halfway between them.

The treatment of lattice defects by continuum methods sometimes gives useful results even in cases too extreme for the above considerations to apply. FRIEDEL [1954] has shown, for example, that in some cases it is possible to get a reasonable estimate for the energy of solution of atoms of a metal X in a metal Y by associating with each dissolved X-atom a strain energy (3.21), taking for the volumes of hole and inclusion the atomic volumes of X and Y and for (λ, μ), (λ^*, μ^*) the ordinary elastic constants of the metals Y and X. An amusing example is provided by JACOBS' [1954] calculation of the effect of hydrostatic pressure on the frequency of the absorption band of an F-centre in an alkali halide crystal. An F-centre is an electron trapped at a negative-ion vacancy. An empirical rule of MOLLWO's [1933] states that in passing from one alkali halide crystal with the sodium chloride structure to another the product νa^2 remains constant; ν is the frequency of the maximum of the F-absorbtion band and a is the lattice parameter. It is reasonable to suppose that the same relation will govern the behaviour of a given alkali halide when its lattice parameter is changed by compression. We should then have the relation $-(\mathrm{d}\nu/\nu)/(\mathrm{d}a/a) = 2$. In place of 2, Jacobs' experiments gave values between 4.4 and 3.4. This discrepancy can be reconciled if we admit that the characteristics of an F-centre are determined by the positions of the atoms bordering the vacancy. Then Mollwo's rule must be interpreted as $\nu R^2 = \text{const.}$, where R is, say, the distance of a neighbouring atom from the centre of the vacancy, so that

$$-\frac{\mathrm{d}\nu/\nu}{\mathrm{d}R/R} = 2\,. \tag{6.20}$$

On passing from one type of crystal to another $(\mathrm{d}R/R)/(\mathrm{d}a/a)$ is unity, since R/a depends only on the lattice geometry. However, when a given crystal is compressed, $\mathrm{d}a/a$ and $\mathrm{d}R/R$ are not identical. In fact, if we idealize the vacancy as a spherical hole in an isotropic continuum we have

$$\frac{\mathrm{d}R/R}{\mathrm{d}a/a} = \frac{e^{\mathrm{A}} + e^{\mathrm{C}}}{e^{\mathrm{A}}} \tag{6.21}$$

in the notation of (4.6), with $\varkappa^* = 0$ in A. For the alkali halides we may put $\sigma = \frac{1}{4}$; the ratio (6.21) is then 2.25, and from (6.20) and (6.21) we have

$$-\frac{\mathrm{d}\nu/\nu}{\mathrm{d}a/a} = 4.5$$

in much better agreement with experiment. The above is admittedly rather a travesty of Jacobs' argument, but in his more rigorous calculation he also found it necessary to introduce the magnification factor (6.21).

References

BILBY, B. A., 1950, Proc. Phys. Soc. A **63** 191.
BITTER, F., 1931, Phys. Rev. **37** 1526.
BRENNER, H., 1958, The Physics of Fluids **1** 388.
BURGERS, J. M., 1939, Proc. Acad. Sci. Amsterdam **42** 293.
BYRD, P. F. and M. D. FRIEDMAN, 1954, Handbook of Elliptic Integrals (Springer-Verlag, Berlin 1954).
CHRISTIAN, J. W., 1958, Acta Met. **6** 377.
CHRISTIAN, J. W., 1959, Acta Met. **7** 218.
DANIELE, E., 1911, Nuovo Cimento [6] **1** 211.
DAS, S. C., 1953, Bull. Calcutta Math. Soc. **45** 55.
DYSON, F. W., 1891, Q.J. Pure Appl. Math. **25** 259.
EDWARDES, D., 1893, Q.J. Pure Appl. Math. **26** 70.
ESHELBY, J. D., 1951, Phil. Trans. Roy. Soc. A **244** 87.
ESHELBY, J. D., 1956, in Solid State Physics (ed. Seitz and Turnbull) **3** 79.
ESHELBY, J. D., 1957, Proc. Roy. Soc. A **241** 376.
ESHELBY, J. D., 1958, Ann. Physik [7] **1** 116.
ESHELBY, J. D., 1959a, in Internal Stresses and Fatigue in Metals (ed. Rassweiler and Grube; Elsevier, Amsterdam, 1959) 41.
ESHELBY, J. D., 1959b, Proc. Roy. Soc. A **252** 561.
FERRERS, N. M., 1877, Q.J. Pure Appl. Math. **14** 1.
FISHER, J. C. and D. TURNBULL, 1953, Acta Met. **1** 310.
FREDHOLM, I., 1900, Acta Math. **23** 1.
FRIEDEL, J., 1954, Advances in Physics **3** 446.
FRIEDEL, J., 1956, Les Dislocations (Gauthier-Villars, Paris, 1956).
FRIEDEL, J., 1959, in Internal Stresses and Fatigue in Metals (ed. Rassweiler and Grube; Elsevier, Amsterdam, 1959) 220.
GOODIER, J. N., 1936, Phil. Mag. **22** 678.
GOODIER, J. N., 1937, Phil. Mag. **23** 1017.
GREEN, A. E. and I. N. SNEDDON, 1949, Proc. Camb. Phil. Soc. **46** 159.
HERSHEY, A. V., 1954, J. Appl. Mech. **21** 236.
HILL, R., 1952, Proc. Phys. Soc. A **65** 349.
INGLIS, C. E., 1913, Trans. Inst. Naval Arch. **55** 219.
JACOBS, I. S., 1954, Phys. Rev. **93** 993.
KAUFMAN, L., 1959, Acta Met. **7** 216.
KAUFMAN, L. and M. COHEN, 1958, Progress in Metal Physics **7** 165.
KELLOGG, O. D., 1929, Potential Theory (Springer-Verlag, Berlin, 1929).
KRÖNER, E., 1953a, Diplomarbeit (Stuttgart, 1953).

KRÖNER, E., 1953b, Z. Phys. **136** 404.
KRÖNER, E., 1954, Acta Met. **2** 302.
KRÖNER, E., 1958a, Kontinuumstheorie der Versetzungen und Eigenspannungen (Springer-Verlag, Berlin, 1958).
KRÖNER, E., 1958b, Z. Phys. **151** 504.
LIFSHITZ, I. M. and N. ROSENZWEIG, 1947, J. Eksp. Teor. Fiz. **17** 783.
LOVE, A. E. H., 1954, Mathematical Theory of Elasticity (Cambridge University Press, 1954).
LURIE, A. I., 1952, Doklady Akad. Nauk SSSR **87** 709.
MACHLIN, E. S. and S. WEINIG, 1953, Acta Met. **1** 480.
MACMILLAN, W. D., 1958, The Theory of the Potential (Dover Publications, New York, 1958) 175.
MOLLWO, E., 1933, Z. Phys. **85** 56.
NABARRO, F. R. N., 1940, Proc. Roy. Soc. A **175** 519.
NABARRO, F. R. N., 1951, Phil. Mag. **42** 1224.
NABARRO, F. R. N., 1952, Advances in Physics **1** 269.
NEUBER, H., 1958, Kerbspannungslehre (Springer-Verlag, Berlin, 1958).
NIESEL, W., 1953, Inauguraldissertation, Karlsruhe.
OSBORN, J. A., 1945, Phys. Rev. **67** 351.
PATEL, J. R. and M. COHEN, 1953, Acta Met. **1** 531.
PEACH, M. O., 1951, J. Appl. Phys. **22** 1359.
POINCARÉ, H., 1899, Théorie du Potentiel Newtonien (Carré et Naud, Paris, 1899) 118.
REINER, M., 1958, Encyclopedia of Physics VI (Springer-Verlag, Berlin, 1958) 528.
ROBINSON, K., 1951, J. Appl. Phys. **22** 1045.
SACK, R. A., 1946, Proc. Phys. Soc. **58** 729.
SADOWSKY, M. A. and E. STERNBERG, 1949, J. Appl. Mech. **16** 149.
SEEGER, A., 1955, Encyclopedia of Physics VII (1) 383.
SEGEDIN, C. M., 1951, Proc. Camb. Phil. Soc. **47** 396.
SEN, B., 1933, Bull. Calcutta Math. Soc. **25** 107.
SNEDDON, I. N., 1951, Fourier Transforms (McGraw-Hill Book Company, New York, 1951).
SOKOLNIKOFF, I. S., 1946, Mathematical Theory of Elasticity (McGraw-Hill Book Company, New York, 1946).
SOMIGLIANA, C., 1914, R. C. Accad. Lincei [5] **23** (1) 463.
SOMIGLIANA, C., 1915, R. C. Accad. Lincei [5] **24** (1) 655.
STARR, A. T., 1928, Proc. Camb. Phil. Soc. **24** 489.
STERNBERG, E., 1958, Appl. Mech. Rev. **11** 1.
STROH, A. N., 1958, Phil. Mag. **30** 623.
TEWORDT, L., 1958, Phys. Rev. **109** 61.
TIMOSHENKO, S. and J. N. GOODIER, 1951, Theory of Elasticity (McGraw-Hill Book Company, New York, 1951) 425.
VOLTERRA, V., 1907, Ann. École Norm. Super. [3] **24** 400.
WINTENBERGER, M., 1957, C. R. Acad. Sci. Paris **244** 2800.

Dislocations in Visco-elastic Materials

By J. D. Eshelby
Department of Physical Metallurgy, University of Birmingham

[Received March 23, 1961]

Abstract

An expression is found for the force required to move a dislocation steadily through a visco-elastic medium, in particular Zener's standard linear solid. It is applied to the case of a screw dislocation in iron exhibiting the Snoek effect. Schoeck and Seeger's atomic treatment of the same problem contains an error. When this is corrected the two methods give solutions of the same form.

§ 1. Introduction

Zener (1948) has given the name 'standard linear solid' to a simple type of visco-elastic material in which stress σ, strain e and their rates of change are connected by the equation

$$\sigma+\tau_1\dot{\sigma}=M_{\mathrm{R}}(e+\tau_2\dot{e}). \qquad (1)$$

If unit strain is suddenly imposed the stress is $M_{\mathrm{U}}=\tau_2 M_{\mathrm{R}}/\tau_1$ to begin with (unrelaxed modulus). It then falls to a lower value M_{R} (relaxed modulus) with a time of relaxation τ_1. If unit stress is suddenly applied the strain rises from $1/M_{\mathrm{U}}$ to $1/M_{\mathrm{R}}$ with a relaxation time $\tau_2>\tau_1$. The quantity

$$\Delta=\frac{M_{\mathrm{U}}-M_{\mathrm{R}}}{M_{\mathrm{R}}}=\frac{\tau_2-\tau_1}{\tau_1}$$

measures the 'strength' of the relaxation process. If σ and e both depend on time through a factor exp $(i\omega t)$ the relation (1) is equivalent to

$$\sigma=\mathscr{M}(\omega)e \qquad (2)$$

where

$$\mathscr{M}(\omega)=M_{\mathrm{R}}\frac{1+i\omega\tau_2}{1+i\omega\tau_1} \qquad (3)$$

is a complex elastic modulus. If $\mathscr{M}(\omega)$ has a more general form (subject to the requirement that it always leads to positive strain-energy and positive rate of dissipation) (2) describes the behaviour of a general linear visco-elastic solid (Hunter 1960).

It may seem rather academic to consider the movement of dislocations in such materials. However, a number of solid-state phenomena can be described in terms of relaxed and unrelaxed elastic moduli and a relaxation time. The Snoek effect (re-orientation of carbon atoms in alpha-iron) is the classic example, and there are others of the same kind. Certain effects involving phonon or electron relaxation can be fitted into the same pictur(In § 2 we find an expression for the force required to keep a dislocation moving uniformly through the visco-elastic medium. Section 3 treats the particular case of a screw dislocation moving through iron showing the Snoek effect. We find it necessary to treat the non-linear region near the dislocation in a rather unsatisfactory way. Schoeck and Seeger's (1959) atomic treatment of the same problem in principle handles this region satisfactorily, but it contains an error. When this is corrected (§ 4) the result agrees in form with the visco-elastic solution, though the numerical value is different, and probably more reliable.

§ 2. Retarding Force on a Dislocation Moving in a Visco-elastic Medium

Consider a linear visco-elastic solid in which the displacement is everywhere parallel to the z-axis and of the form $w=W(x, y)\exp(i\omega t)$. The only non-vanishing stress components are

$$\sigma_{zx}=\mathcal{M}(\omega)\partial w/\partial x, \quad \sigma_{zy}=\mathcal{M}(\omega)\partial w/\partial y$$

where $\mathcal{M}(\omega)$ is the complex shear modulus. If the motion is slow enough for the inertial forces to be neglected (we assume this throughout) the equilibrium equation is

$$\mathcal{M}(\omega)\nabla^2 w=0. \quad \ldots \ldots \quad (3\,a)$$

Let us impress on the face $y=0$ of the semi-infinite block $y\geqslant 0$ a sinusoidal corrugation $w=\exp\{-i(kx-\omega t)\}$ moving along the x-axis with constant velocity $v=\omega/k$. Experience with this kind of problem suggests that the displacement will penetrate into the block according to the law

$$w=\exp(-ly-ikx') \quad \ldots \ldots \quad (4)$$

where

$$x'=x-vt, \quad v=\omega/k.$$

Equation (3 a) is in fact satisfied if we set l equal to k. On $y=0$ the velocity of the material is $\dot{w}=i\omega\exp(-ikx')$ and all the stress components vanish except for σ_{zy} which has the value $\sigma=-k\mathcal{M}(\omega)\exp(-ikx')$. Cancelling a factor $i\omega$ and taking the real part we find that to produce a velocity-distribution

$$\dot{w}=\cos kx' \quad \ldots \ldots \quad (5)$$

on the surface of the block we must apply a stress

$$\sigma=v^{-1}\,\mathrm{re}\,\{\mathcal{M}(kv)\exp(ikx')\} \quad \ldots \ldots \quad (6)$$

to it. If we build up a more general surface velocity in the form of the Fourier integral

$$\dot{w}=\int_0^\infty f(k)\cos kx'\,dk \quad \ldots \ldots \quad (7)$$

the stress necessary to maintain it is evidently

$$\sigma=v^{-1}\int_0^\infty f(k)\{\mathrm{re}\,\mathcal{M}(kv)\sin kx'-\mathrm{im}\,\mathcal{M}(kv)\cos kx'\}\,dk. \quad (8)$$

(We note in passing that by (4) $\dot{w}(x', y)$ is given by (7) with a factor $\exp(-k|y|)$ inserted in the integrand. Since this is independent of $\mathscr{M}(\omega)$, $w(x', y)$ is the same as it would be for a disturbance moving in a non-relaxing medium.) The work done by the surface forces in unit time is

$$\dot{E} = -2\int_0^\infty \sigma(x')\dot{w}(x')\,dx'. \qquad (9)$$

Now take the block $y<0$ and impress a velocity distribution equal and opposite to (7) on its surface. The stresses match those on the surface of the block $y \geqslant 0$ and the two may be stuck together. If w is chosen suitably we shall have a moving screw dislocation. The total rate of dissipation of energy in the two blocks is $2\dot{E}$ and the retarding force on the dislocation is $F=2\dot{E}/v$. According to Parseval's theorem (Titchmarsh 1937) if

$$g_m(x') = \int_0^\infty G_m(k)\cos kx'\,dk, \quad m=1, 2$$

then

$$\int_0^\infty g_1(x')g_2(x')\,dx' = \tfrac{1}{2}\pi\int_0^\infty G_1(k)G_2(k)\,dk.$$

Thus from (7), (8) and (9)

$$F(v) = -\frac{2\pi}{v^2}\int_0^\infty f^2(k)\,\mathrm{im}\,\mathscr{M}(kv)\,dk. \qquad (10)$$

Equation (10) applies to a general visco-elastic medium. If $\mathscr{M}(\omega)$ has the simple form (3) appropriate to a standard linear solid with $M_R=\mu_R$ it becomes

$$F(v) = \frac{2\pi}{v}\mu_R(\tau_2-\tau_1)\int_0^\infty \frac{kf^2(k)}{1+k^2v^2\tau_1^2}\,dk.$$

If we take the usual expression

$$w = \frac{b}{2\pi}\tan^{-1}\frac{x'}{\zeta}$$

for the displacement at the slip plane of a dislocation of width ζ then

$$\dot{w} = -\frac{vb}{2\pi}\frac{\zeta}{x'^2+\zeta^2} = -\frac{vb}{2\pi}\int_0^\infty \exp(-k\zeta)\cos kx'\,dk$$

and

$$F(v) = \frac{\mu_R(\tau_2-\tau_1)b^2}{2\pi v\tau_1^2}\int_0^\infty \frac{\exp(-2k\zeta)k\,dk}{k^2+(v\tau_1)^{-2}}$$
$$= \frac{\mu_R b^2}{4\pi}\frac{\Delta}{\zeta}\mathscr{F}\left(\frac{v\tau_1}{2\zeta}\right) \qquad (11)$$

where (see figure)

$$\mathscr{F}(z) = z^{-1}\{-\cos(z^{-1})\mathrm{Ci}(z^{-1}) - \sin(z^{-1})\mathrm{si}(z^{-1})\} \qquad (12)$$

3 P 2

in the notation of Jahnke and Emde (1945). We have

$$\mathscr{F}(z)=z, \qquad z \ll 1$$

$$=z^{-1}\ln(z/\gamma), \quad z \gg 1 \qquad (\ln\gamma=0\cdot5772\ldots),$$

$$\mathscr{F}_{\max}=0\cdot35 \text{ at } z=1\cdot3,$$

and so for low velocities the force is proportional to v,

$$F=\frac{\mu_R}{8\pi}\frac{b^2}{\zeta^2}\tau_1\Delta v, \quad v\ll 2\zeta/\tau_1 \quad . \quad . \quad . \quad (13)$$

while beyond a màximum

$$F_{\max}=\frac{0\cdot35}{4\pi}\frac{\mu_R b^2\Delta}{\zeta} \quad \text{at } v_{\max}=1\cdot3(2\zeta/\tau_1)$$

it falls off with increasing v.

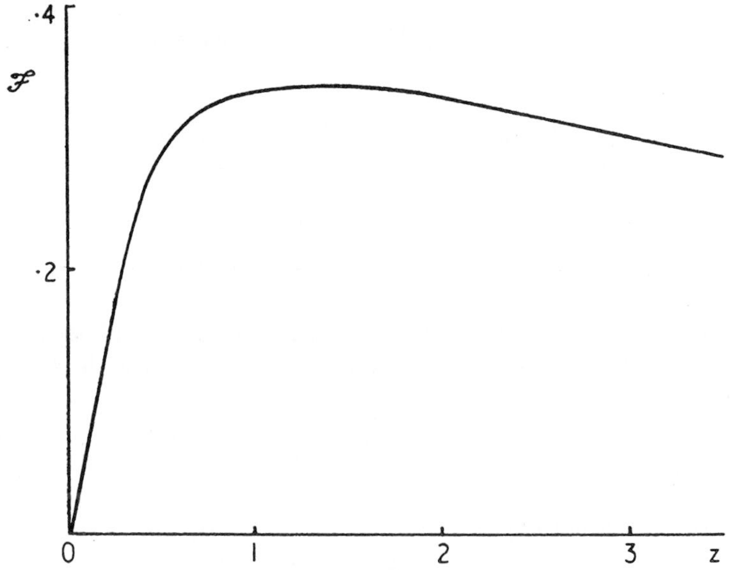

Following Mason (1960) we may derive the low velocity limit (13) directly. Far from the moving dislocation the Fourier components of its field will have frequencies small enough for us to write

$$\mathscr{M}(\omega)=\mu_R[1+i\omega(\tau_2-\tau_1)]$$

or equivalently

$$\sigma=\mu_R e+\eta\dot{e} \quad . \quad . \quad . \quad . \quad . \quad . \quad . \quad . \quad (14)$$

where

$$\eta=\mu_R\tau_1\Delta$$

is a small viscosity in parallel with the stiffness μ_R. The rate of flow of energy into unit volume is $\sigma\dot{e}=\mu_R e\dot{e}+\eta\dot{e}^2$. The second term represents a dissipation at the rate $\eta\,(\operatorname{grad}\dot{w})^2$ per unit volume, whereas the first term corresponds to stored energy which is recovered if e goes through a cycle and returns to its initial value. According to the remark following (8) we may take $w=(b/2\pi)\tan^{-1}(x'/y)$ far from the dislocation. The total rate of dissipation outside a circle of radius r_0 is thus

$$\dot{E}=\frac{\eta v^2}{4\pi}\frac{b^2}{r_0^{\,2}} \qquad (15)$$

if (14) is a good approximation for $r>r_0$. Suppose that physical considerations (not directly connected with the visco-elastic behaviour of the medium) dictate an inner cut-off radius r_0. Then for small enough velocities (14) will be a good approximation for all $r\geqslant r_0$ and (15) is the whole dissipation. The retarding force is equal to (13) with $r_0=2\zeta$.

The expression (11) is related to the reduction in the energy of a stationary dislocation due to relaxation. If the dislocation is suddenly introduced its energy is $(\mu_U b^2/4\pi)\ln(r_1/r_2)$ to begin with and $(\mu_R b^2/4\pi)\ln(r_1/r_2)$ when relaxation is complete. The energy loss is

$$E_{\text{rel}}=\tfrac{1}{2}W_0\ln(r_1/r_2) \qquad (16)$$

where

$$W_0=(\mu_U-\mu_R)b^2/4\pi$$

and we can write

$$F(v)=\frac{W_0}{2\zeta}\mathscr{F}\left(\frac{v\tau_1}{2\zeta}\right). \qquad (17)$$

Equations (16) and (17) can be generalized to cover the case of any straight dislocation (screw, edge or mixed) in an anisotropic medium with relaxed and unrelaxed elastic constants $c_{ij}^{\,R}$, $c_{ij}^{\,U}$, subject to two limitations. These are (i) all the c_{ij} have the same relaxation time τ_1, though the individual relaxation strengths $\Delta_{ij}=(c_{ij}^{\,U}-c_{ij}^{\,R})/c_{ij}^{\,R}$ may be different, and (ii) the width parameter ζ has the same value for the edge and screw components of the dislocation. The elastic energy of the dislocation can be written in the form

$$E=\frac{Kb^2}{4\pi}\ln\frac{r_1}{r_2}$$

where K is a function of the direction of the Burgers vector and of the c_{ij}. In some cases K can be found immediately from the c_{ij}, in others it may be necessary to solve a cubic equation (Foreman 1955). The energy relaxation and force are given by (16) and (17) with

$$W_0=(K_U-K_R)b^2/2\pi$$

where K_U, K_R are the values of K calculated from the $c_{ij}^{\,U}$ and the $c_{ij}^{\,R}$ respectively. The proof consists in writing out the relations analogous to (7) and (8) which connect displacement and stress on the face of a semi-infinite block, calculating the rate of dissipation as before and comparing the result with Foreman's expression for the energy as an integral over the slip-plane.

If both relaxation strengths and relaxation times are different for the various elastic constants things are more complicated. Consider, for example, an edge dislocation in an isotropic medium. The relation between displacement and stress at the surface of a semi-infinite block is now of the form (7, 8) with

$$\mathscr{M}(\omega)=\frac{\mu(\omega)}{1-\nu(\omega)}=2\mu(\omega)\frac{3\kappa(\omega)+\mu(\omega)}{3\kappa(\omega)+4\mu(\omega)} \quad . \quad . \quad . \quad (18)$$

where $\mu(\omega)$, $\kappa(\omega)$, $\nu(\omega)$ are the complex shear modulus, bulk modulus and Poisson's ratio. This is just a generalization to the visco-elastic case of a result given by Nabarro (1947). So long as we are concerned with a general visco-elastic material the expression (10) for the force is in principle no more complicated than it would be for a screw dislocation with $\mathscr{M}(\omega)=\mu(\omega)$. We might, however, be interested in a material which behaves as a standard linear solid under either shear or compression, but with a different relaxation time and strength for each of these types of deformation. Then $\mathscr{M}(\omega)$ (eqn. (18)) no longer has the simple form (3) and $F(v)$ depends on v through a more complicated function than the $\mathscr{F}$ of eqn. (11). It can easily be evaluated. If Δ and τ_1 happen to be the same for shear and compression, Poisson's ratio is independent of frequency and $F(v)$ is equal to (11) divided by $1-\nu$.

§ 3. Snoek Drag on a Screw Dislocation

Carbon atoms in a crystal of alpha-iron can occupy positions at the mid-points of 100, 010 or 001 cube edges. If a stress is applied in, say, the 100 direction there is a certain initial extension. Subsequently the crystal extends further as carbon atoms jump from 010 and 001 positions into the 100 positions. If the stress had been applied in the 111 direction there would have been no relaxation. Thus the material behaves as an anisotropic standard linear solid (Snoek effect). Polder (1945) gives the value

$$\tau_1=4{\cdot}4\times 10^{-14}\exp(9000/T)\text{ sec}$$

for the relaxation time. This is about one second at room temperature. A carbon atom need only jump one interatomic spacing to change from, say, an 010 to a 100 position, and no long-range diffusion is involved. Indeed it is often convenient to ignore the change of position altogether and think of the carbon atom as a ' dipole ' which can take up one of three orientations. By analogy with paramagnetism, Kröner (1958) has given the name ' paraelastic ' to materials which behave as visco-elastic solids because of the re-orientation of atomic dipoles of some kind. Zener (1948) has suggested that in some circumstances the interaction between the carbon atoms may be enough to make them all take up the same orientation spontaneously and so form martensite. If this is in fact so, iron containing carbon may reasonably be called a ferroelastic material.

Nabarro (1948) pointed out that (apart from forming carbide particles on them) carbon atoms in iron can interact with dislocations in two ways. They may re-orient to form a 'Snoek atmosphere'. This is a rapid process. Given time they may migrate over large distances and form a Cottrell atmosphere. The difference between the effects of the Snoek and Cottrell atmospheres is nicely demonstrated by Wilson and Russell's (1959) experiments on strain-ageing in low carbon steels.

We shall apply the results of § 2 to find the retarding effect of the Snoek atmosphere about a uniformly moving screw dislocation.

According to Polder (1945), the relaxed and unrelaxed moduli for alpha-iron containing an atomic fraction c of carbon are related by

$$c_{11}{}^{\mathrm{R}}=c_{11}{}^{\mathrm{U}}-2a,\ c_{12}{}^{\mathrm{R}}=c_{12}{}^{\mathrm{U}}+a,\ c_{44}{}^{\mathrm{R}}=c_{44}{}^{\mathrm{U}} \quad . \quad . \quad (19)$$

with

$$a=\tfrac{1}{6}(c_{11}{}^{\mathrm{U}}-c_{12}{}^{\mathrm{U}})\beta^2a_0{}^3c/kT$$

where a_0 is the lattice parameter, $\beta=1{\cdot}00$ is the rate of change of the axial ratio of martensite with carbon content and k, T are Boltzmann's constant and the absolute temperature. (As we ignore interaction between the carbon atoms we put Polder's constant χ equal to zero.)

Following Schoeck and Seeger (1959) we consider a screw dislocation in the 111 direction. This is not one of the cases for which K_{R} and K_{U} can be found easily. Since in any case we shall not be able to determine the length ζ very accurately we shall simply average the relaxed and unrelaxed anisotropic moduli and treat the problem as an isotropic one. If, with Voigt, we average the c_{ij} rather than the s_{ij} we obtain

$$\mu_{\mathrm{U}}-\mu_{\mathrm{R}}=\tfrac{3}{5}a.$$

With Polder's figures this gives

$$\Delta=1{\cdot}9\times10^4\,c/T. \quad . \quad . \quad . \quad . \quad . \quad . \quad . \quad (20)$$

Zener (1948) gives $\Delta=2{\cdot}5\times10^4\,c/T$ for extension in the 100 direction, for which the effect is strongest.

We have next to choose the length ζ. For a given ζ the maximum shear strain due to the dislocation is $b/4\pi\zeta$. In an ordinary elastic problem we should choose ζ so that the maximum strain does not exceed the value for which Hooke's law becomes invalid. This would be wrong in the present calculation. If we apply a slowly increasing stress to the material the shear modulus is initially μ_{R}. Above a certain stress, however, nearly all the carbon atoms will have entered the favoured sites, and thereafter the rate of increase of stress with strain is nearly equal to μ_{U}, and anelastic effects are greatly reduced. Thus we must choose ζ in such a way that the maximum strain $b/4\pi\zeta$ does not exceed the value at which, so to speak, Hooke's law breaks down for the difference-modulus $\mu_{\mathrm{R}}\Delta=\mu_{\mathrm{U}}-\mu_{\mathrm{R}}$. At normal temperatures this is much less than the strain at which Hooke's law breaks down for μ_{U} or μ_{R}. Zener (1948) gives a curve (p. 122) which shows how the number of carbon atoms in preferred positions varies with strain and temperature. The curve becomes markedly non-linear at a strain of about 0·005 at room temperature, or a strain 0·005 $T/300$ at absolute temperature T. We shall therefore require ζ to satisfy

$$b/4\pi\zeta = 2 \times 10^{-5}\, T. \qquad (21)$$

From (11), (20) and (21) we find that the stress required to maintain a velocity v is

$$\sigma = F(v)/b = 3 \times 10^{11}\, c\mathscr{F}(v\tau_1/2\zeta) \text{ dynes/cm}^2 \qquad (22)$$

or very nearly $10^{11}\, c$ dynes/cm² at $v_{\max}$.

§ 4. The Method of Schoeck and Seeger

In the previous section our treatment of the anelastically non-linear region near the dislocation was very crude. Schoeck and Seeger (1959) have already discussed the same problem. Their treatment, which is atomistic, runs into no special difficulties near the dislocation. However, it contains an error. To show this we first give a brief account of their method. The error is easily corrected, and the result is then substantially the same as ours.

Schoeck and Seeger begin by calculating the interaction energy between a stationary dislocation and the equilibrium Snoek atmosphere surrounding it. The result is

$$U_0 = W_0 \ln (3L/R)$$

where

$$W_0 = \frac{2\pi c}{a_0^3} \frac{A^2}{kT},$$

$$R = \frac{A}{kT}$$

and

$$A = 1{\cdot}84 \times 10^{-20} \text{ dyne-cm}^2.$$

The length L is a cut-off radius which is taken to be 10^{-4} cm. The factor 3 in the argument of the logarithm stands for exp (1·1). U_0 is approximately twice our expression (16) because it is an interaction energy, not a self-energy. They then develop an ingenious method designed to avoid having to calculate the precise form of the Snoek atmosphere round the moving dislocation. For this purpose they define a function $V(t)$ which is " the energy a fictitious dislocation at the origin would acquire in the field of the impurity atoms ordered by a dislocation line at x " and another function $V_0(x)$ which is the energy the fictitious dislocation would acquire if the atmosphere about the real dislocation at x were in complete equilibrium (as it would be, for example, if the real dislocation had been at rest at x for a long time). They show that $V(t)$ and $V_0(x)$ are related by the equation

$$\frac{\partial V(t)}{\partial t} = -\bar{\nu}\,[V(t) - V_0(x)]$$

where $\bar{\nu}$ is an atomic jump frequency. For the force on a uniformly moving dislocation they find

$$F(v) = \left[\frac{1}{v}\frac{\partial V}{\partial t}\right]_{t=0} = \frac{\bar{\nu}}{v}\left[V_0(0) - \int_0^\infty V_0(vt)\exp(-\bar{\nu}t)\,dt\right]. \qquad (23)$$

Since the actual form of $V_0(x)$ is not known Schoeck and Seeger assume that it will be adequate to insert in (23) a trial function which satisfies two conditions: (i) $V_0(0) = U_0$, (ii) $V_0(x)$ falls rapidly to zero outside the range $|x| \sim R$. They actually take

$$V_0(x) = U_0 \exp(-x^2/R^2). \qquad (24)$$

This gives a force which can be written in the form

$$F(v) = \frac{W_0}{R}\ln\left(\frac{3L}{R}\right)\mathscr{G}\left(\frac{2v}{\bar{\nu}R}\right) \qquad (25)$$

where $\mathscr{G}(z)$ is a function similar to our $\mathscr{F}(z)$ with the following properties:

$$\mathscr{G}(z) = z,\ \ z \ll 1; \ \ \mathscr{G}(z) = 2z^{-1},\ \ z \gg 1;$$

$$\mathscr{G}_{\max} = \tfrac{1}{2} \text{ at } z = \tfrac{4}{3}.$$

If we suppose that $R \sim \zeta$ and $\bar{\nu} \sim \tau_1^{-1}$ (25) gives roughly the same velocity dependence as (11). Numerically (25) is about six times greater than our expression (11). There is nothing unreasonable in this. However, the disagreement is more than numerical. Equation (25) contains a factor $\ln(3L/R)$, estimated by Schoeck and Seeger to be $41/2\pi = 6{\cdot}5$. Our expression contains no such logarithmic cut-off factor. Indeed the discussion in § 2 leading to (15) shows that far from the dislocation the rate of dissipation per unit volume falls off as r^{-4}, and so it is difficult to see why an outer cut-off should be necessary.

In fact the logarithmic factor only appears because (24) (or any other function of short range) is a physically inadmissible trial function. Under moderate stresses the material behaves as an anisotropic standard linear solid with relaxed and unrelaxed moduli c_{ij}^{R}, c_{ij}^{U} related by (19). Thus at large distances from the dislocation the difference between its elastic fields in the relaxed and unrelaxed condition must be the same as the field of a dislocation in an ordinary elastic medium with constants $c_{ij}^{\mathrm{U}} - c_{ij}^{\mathrm{R}}$. The interaction energy $V_0(x)$ of this difference-field with a fictitious dislocation at a distance x from it must thus behave like $\ln x$ for large x. Only a detailed calculation could give the value of $V_0(x)$ for small x, but by definition we must have $V_0(0)=U_0$. A trial function which meets these requirements is

$$V_0(x)=\tfrac{1}{2}W_0\ln\frac{L^2}{x^2+r_0^2} \qquad (26)$$

with

$$r_0=\tfrac{1}{3}R=A/3kT.$$

If this is inserted in (23) we get

$$F(v)=\frac{W_0}{r_0}\mathscr{F}(v/\bar{\nu}r_0)$$

or

$$\sigma=F/b=6{\cdot}0\times 10^{11}c\mathscr{F}(v/\bar{\nu}r_0). \qquad (27)$$

With $\bar{\nu}=\tau_1^{-1}$ and $r_0=2\zeta$ the stress is twice as large as our (22), but the variation with velocity is the same. (Actually r_0 is about four times the value of 2ζ given by (21).)

Schoeck and Seeger compare the result of their calculation with a value

$$\sigma=4500c \text{ kg/mm}^2$$

which they deduce from experiment. (c is the atomic fraction of carbon.) Their original expression (25) gives $\sigma=F_{max}/b=6300c$ kg/mm^2, but the revised expression (27) only gives $2000c$ kg/mm^2. Our estimate (22) gives $1000c$ kg/mm^2. We can get better agreement if we identify the experimental value not with the force needed to attain the velocity v_{max}, but rather with the force, F^* say, required to pull the dislocation away from its Snoek atmosphere suddenly, so that there is no time for the carbon atoms to move. Schoeck and Seeger find that $F^*=F_{max}$. But it is easy to see that F^* is simply the maximum value of $\partial V_0(x)/\partial x$. If $V_0(x)$ has the form (26) then

$$F^*=W_0/r_0=2{\cdot}9F_{max}$$

which gives $\sigma=6000c$ kg/mm^2. A rough calculation for the force to pull an edge dislocation suddenly from its Snoek atmosphere gives $\sigma=2500c$ kg/mm^2 (Eshelby 1959: the quantity ΔY must be divided by 2 to convert to a shear stress. On p. 631, col. 2, l. 20, $\gamma=\frac{1}{2}$ should read $\gamma=\frac{1}{3}$).

These values cannot be taken very seriously, since they are sensitive to the detailed form of $V_0(x)$ near the origin. For example, it is easy to show from (23) that

$$F(v)=-\bar{\nu}^{-1}V_0''(0)v+O(v^2)$$

for small v. Obviously a knowledge of the value of $V_0(x)$ at $x=0$ and of its behaviour at infinity tells us nothing about its curvature at the origin. F_{max} and F^* are somewhat less sensitive, but it is clear that really reliable values could only be obtained by calculating $V_0(x)$ in the same way as Schoeck and Seeger calculated $V_0(0)$.

References

Eshelby, J. D., 1959, Appendix to Wilson, D. V., and Russell, B., *Acta Met.*, **7**, 628.

Foreman, A. J. E., 1955, *Acta Met.*, **3**, 322.

Hunter, S. C., 1960, *Progress in Solid Mechanics*, **1**, 1.

Jahnke, E., and Emde, F., 1945, *Tables of Functions* (New York: Dover Publications).

Kröner, E., 1958, *Kontinuumstheorie der Versetzungen und Eigenspannungen.* (Berlin: Springer–Verlag).

Mason, W. P., 1960, *J. acoust. Soc. Amer.*, **32**, 458.

Nabarro, F. R. N., 1947, *Proc. phys. Soc., Lond.*, **59**, 256; 1948, *Report on Strength of Solids* (London: Physical Society), p. 38.

Polder, D., 1945, *Philips Res. Rep.*, **1**, 5.

Schoeck, G., and Seeger, A., 1959, *Acta Met.*, **7**, 469.

Titchmarsh, B. C., 1937, *Fourier Integrals* (Oxford: Clarendon Press).

Wilson, D. V., and Russell, B., 1959, *Acta Met.*, **7**, 628.

Zener, C., 1948, *Elasticity and Anelasticity of Metals* (Chicago: University Press).

The interaction of kinks and elastic waves

By J. D. Eshelby

Department of Physical Metallurgy, University of Birmingham

(*Communicated by G. V. Raynor, F.R.S.—Received 3 August* 1961)

A kink on a dislocation in an isotropic elastic medium is treated as a 'point defect' with a certain mass, constrained to move along a line and subject to a radiation reaction. A value for the mass is obtained from the well known stretched-string model, and the radiation reaction is found by calculating the rate at which an oscillating kink radiates energy into the medium. It is found that the kink has a scattering cross-section for elastic waves which is proportional to the square of its width. For long waves the cross-section is independent of frequency, in contrast to the case of ordinary point defects. A kink moving through an isotropic flux of elastic waves experiences a retarding force proportional to the product of its velocity and the energy density of the waves. In connexion with a similar result for the retarding force on a dislocation moving rigidly it has been suggested that the expression for the energy density should include the zero-point energy. A formal quantum-mechanical calculation shows that this is not so in the case of a kink.

1. Introduction

Several authors (Leibfried 1949; Nabarro 1951*b*; Lothe 1960) have discussed the retarding force which is experienced by a dislocation when it moves through a flux of lattice vibrations. Some of the difficulties of the problem arise from the fact that the dislocation is a line singularity, so that, for example, although it may be said to have a mass, the numerical value of the mass is a function (or rather functional) of its present and past state of motion. These difficulties become less acute if one tries to calculate the retarding force, not on a dislocation moving as a whole, but on a kink moving along the dislocation. The word 'kink' refers to the configuration in the neighbourhood of the point where an otherwise straight dislocation crosses over from one minimum of the Peierls potential to the next (figure 1 (*a*)).

Recently, Lothe (1962) has given a detailed physical discussion of the various effects which may contribute to the frictional force on a moving dislocation or kink. Here we concentrate on one aspect, the problem of describing the interaction between a kink and the elastic waves in an isotropic medium. It is not suggested that the results are particularly realistic, but there are some interesting points of principle. An attempt is made to avoid 'relativistic' arguments of the kind which are useful in discussing screw dislocations (Nabarro 1951*b*; Lothe 1960). Even if we admitted them we should still be faced with what Frank (1949) has called 'relativity with complications' because of the existence of two limiting velocities. The following is an outline of the paper.

A well-known model (§2) suggests that a kink can be treated as a particle constrained to move along a line, and having a certain mass. In §3 we find the rate at which an oscillating kink radiates energy into the elastic medium in which the dislocation is embedded, and from it is deduced the radiation reaction term in the equation of motion of the kink. From the results of §§2, 3 it is possible (§4) to calculate the cross-section for the scattering of elastic waves by a kink. Section 5 is concerned with the main problem: to find the retarding force on a kink which is sliding along a dislocation and interacting with the thermal vibrations of the medium. There are a number of pitfalls, and our first attempt at a solution leads to a 'retarding' force with the wrong sign. The nature of the error can be made out by using the same method to calculate (incorrectly) the analogous retarding force on a charged particle moving through a flux of electromagnetic radiation. Comparison with a valid method of solution shows that the error arises from the neglect of higher-order terms in the radiation reaction. When the analogous terms have been calculated for the kink and inserted in its equation of motion a reasonable value is obtained for the retarding force. This procedure does not, of course, involve any appeal to an electromagnetic analogy; the electromagnetic problem is used merely as a touchstone to reveal the shortcomings of a general method of solution.

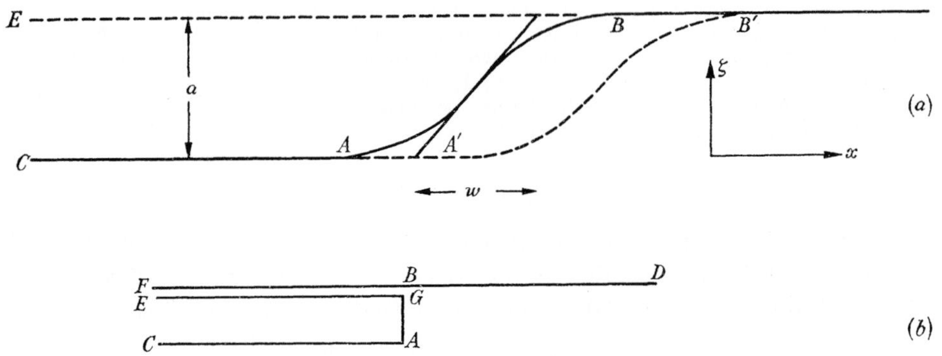

FIGURE 1

In this way it is found that there is a retarding force on the kink which is a numerical multiple of the product of the scattering cross-section, the thermal energy density in the material, and the ratio of the velocity of the kink to the velocity of sound. This is similar to Leibfried's original result for the retarding force on one atomic length of a dislocation. Leibfried suggested that the expression for the energy density should include the zero-point energy as well as the thermal energy. The classical calculation of § 5 cannot tell us whether the same is true or not for a kink. However, a rather crude quantum-mechanical calculation (§ 6) shows that the zero-point contribution is to be excluded. In addition, it confirms the results of § 5. It also provides a convenient method for correcting the previous results to allow for the finite width of the kink.

Section 6 is concerned with some properties of straight (unkinked) dislocations which follow easily from the calculations for a kink.

2. The string model

In this section we recall some properties of a well known model which in effect treats the dislocation as a heavy string under tension lying on a piece of corrugated iron. If the dislocation has a mass ρ per unit length, a line tension S, and is subject to a Peierls potential $\mathscr{V}(\zeta)$ which depends on its transverse displacement ζ, the equation of motion is

$$\rho\,\partial^2\zeta/\partial t^2 - S\,\partial^2\zeta/\partial x^2 + \partial\mathscr{V}(\zeta)/\partial\zeta = 0. \tag{2·1}$$

The properties of this equation (mostly with a sinusoidal potential) have been discussed by a number of writers, though not necessarily in connexion with the present problem. (For references see Seeger 1956.) Here we indicate some properties of the solution which are independent of the exact form of $\mathscr{V}$. We merely suppose that it is periodic with period a, that it has simple minima, and, for convenience, that $\mathscr{V}(0) = 0$.

For the static case ($\partial\zeta/\partial t = 0$) (2·1) has the first integral

$$\tfrac{1}{2}S\zeta'^2 - \mathscr{V}(\zeta) = C, \quad \zeta' = \mathrm{d}\zeta/\mathrm{d}x,$$

where C is a constant which determines the character of the solution

$$x = (\tfrac{1}{2}S)^{\frac{1}{2}} \int^{\zeta} \{\mathscr{V}(\zeta) + C\}^{-\frac{1}{2}}\,\mathrm{d}\zeta.$$

The case $C > 0$ corresponds to a succession of kinks of the same sign, and $C < 0$ to an (unstable) succession of kinks of alternating sign. If $C = 0$, x varies once from $-\infty$ to $+\infty$ as ζ moves from one zero of $\mathscr{V}$ to the next, and there is a single kink. We shall only consider this case. Then

$$\tfrac{1}{2}S\zeta'^2 - \mathscr{V}(\zeta) = 0. \tag{2·2}$$

Following Frank & van der Merwe (1949) we define the width w of the kink to be the length of the projection on the x axis of that part of the steepest tangent which is intercepted between the lines $\zeta = 0$, $\zeta = a$ (cf. figure 1 (a)). By (2·2)

$$w = a/|\zeta'|_{\text{max.}} = a(S/2\mathscr{V}_{\text{max.}})^{\frac{1}{2}}$$

and the width depends only on the maximum value of $\mathscr{V}$, and not, for example, on the Peierls stress

$$\tau_P = b^{-1}|\partial\mathscr{V}/\partial\zeta|_{\text{max.}}$$

(b is the Burgers vector of the dislocation.) The energy of the kink is

$$W_{\text{k}} = \int_{-\infty}^{\infty} (\tfrac{1}{2}S\zeta'^2 + \mathscr{V})\,\mathrm{d}x.$$

According to (2·2) the two terms make equal contributions:

$$W_{\text{k}} = S\int_{-\infty}^{\infty} \zeta'^2\,\mathrm{d}x = (2S)^{\frac{1}{2}} \int_0^a \mathscr{V}^{\frac{1}{2}}(\zeta)\,\mathrm{d}\zeta. \tag{2·3}$$

Figure 1 (a) suggests that in (2·3) we may with fair accuracy take ζ' to be $\zeta'_{\text{max.}}$ between $x = \frac{1}{2}w$ and $x = -\frac{1}{2}w$, and zero elsewhere. Thus we should expect to have generally

$$W_{\text{k}} = 2\eta w\mathscr{V}_{\text{max.}} = \eta a^2 S/w, \tag{2·4}$$

with η close to unity.

If $\mathscr{V}(\zeta)$ is required to be sinusoidal and to give a prescribed Peierls stress it must have the form

$$\mathscr{V}(\zeta) = \frac{\tau_P ab}{\pi}\sin^2\frac{\pi\zeta}{a}, \tag{2·5}$$

which gives

$$\zeta = (2a/\pi)\tan^{-1}\exp(\pi x/w),$$
$$w = (\pi a S/2\tau_P b)^{\frac{1}{2}},\ W_{\text{k}} = (2/\pi)(a^2S/w).$$

Since for a single kink ζ only covers the range $0 \leqslant \zeta \leqslant a$, $\mathscr{V}$ need not be explicitly periodic. For example, the polynomial

$$\mathscr{V}(\zeta) = 8S\zeta^2(a-\zeta)^2/w^2a^2 \tag{2·6}$$

gives

$$\zeta = \tfrac{1}{2}a + \tfrac{1}{2}a \tanh(2x/w),$$

$$W_k = \tfrac{2}{3}a^2S/w, \quad \tau_P = (8\sqrt{3}/9)(Sa/w^2).$$

It is easy to verify that if $\zeta = f(x)$ is a static solution of (2.1) then the uniformly moving 'Lorentz-contracted' solution

$$\zeta = f(x'), \quad x' = (x-vt)/\beta$$

with

$$\beta = (1-v^2/c_k^2)^{\frac{1}{2}}, \quad c_k = (S/\rho)^{\frac{1}{2}} \tag{2·7}$$

satisfies the time-dependent equation (2·1). The total energy is

$$E(v) = \int_{-\infty}^{\infty} \{\tfrac{1}{2}S(\partial\zeta/\partial x)^2 + \tfrac{1}{2}\rho(\partial\zeta/\partial t)^2 + \mathscr{V}(\zeta)\}\,dx$$

which can be transformed to read

$$E(v) = E(0)/\beta + (\beta^{-1}-\beta)\int_{-\infty}^{\infty} \{\tfrac{1}{2}Sf'^2(x) - \mathscr{V}\}\,dx.$$

But for a single kink the integral vanishes by (2·2), and so there is the 'relativistic' relation

$$E(v) = W_k/\beta = W_k + \tfrac{1}{2}m_k v^2 + \ldots, \tag{2·8}$$

where

$$m_k = W_k/c_k^2$$

is the effective mass of the kink. The velocity c_k also appears in the dispersion law

$$\omega^2 = c_k^2(k^2+k_0^2), \quad k_0^2 = S^{-1}(\partial^2\mathscr{V}/\partial\zeta^2)_{\zeta=0}, \tag{2·9}$$

which governs the propagation of small-amplitude disturbances along the dislocation; it is, in fact, their maximum group velocity.

If there are several kinks on the same dislocation line it is necessary to distinguish between positive kinks with, say, $\zeta'(x) > 0$ and negative kinks with $\zeta'(x) < 0$. At low temperatures a dislocation with its ends pinned in the same valley of the Peierls potential will be free from kinks. At higher temperatures kinks may be formed in pairs by thermal activation. The problem of finding their equilibrium number is essentially the same as the problem of finding the number of electrons and positrons in a hot vacuum (Landau & Lifshitz 1958), or the number of electrons and holes in a non-degenerate semi-conductor with band gap $2W_k$. The only difference is that the problem is one-dimensional. The result is

$$n_+ = n_- = (2\pi m_k kT/h^2)^{\frac{1}{2}} \exp(-W_k/kT), \tag{2·10}$$

where n_+, n_- are the number of positive and negative kinks per unit length of dislocations. If higher terms in (2·8) are important (2·10) must be multiplied by $(2z/\pi)^{\frac{1}{2}} \exp zK_1(z)$, where $z = W_k/kT$ and K_1 is the Bessel function usually so denoted. Equation (2·10) differs from Lothe & Hirth's (1959) result by a factor $kT/\hbar\omega_0$, where $\omega_0 = k_0 c_k$ with the notation of (2·9).

15-2

A dislocation anchored at the points $x = 0$, $\zeta = 0$ and $x = X$, $\zeta = Na$ necessarily contains N 'intrinsic' kinks at zero temperature. According to the law of mass action the number of kinks at temperature T is given by

$$n_+ = n_- + n_0 = \tfrac{1}{2}n_0 + (\tfrac{1}{4}n_0^2 + A^2)^{\frac{1}{2}},$$

where A is the right-hand side of (2·10) and $n_0 = N/X$ is the linear density of (positive) intrinsic kinks. (Cf. Landau & Lifshitz 1958, or the corresponding relation for intrinsic and thermal lattice defects (Lidiard 1957).)

If a pair of kinks of opposite sign run together from infinite separation an amount of energy $2W_k$ may be recovered. Most of this energy will be released when the separation of the kinks is w or less. There is thus a short-range attraction between them represented by a potential well of depth $2W_k$ and width of order w. There is, however, also a long-range attraction with potential

$$U(x) = -\frac{\mu a^2 b^2}{8\pi(1-\nu)}(1+\nu-3\nu\sin^2\theta)\frac{1}{|x|}, \tag{2·11}$$

where x is the distance between the kinks, μ, ν are the shear modulus and Poisson's ratio, b is the magnitude of the Burgers vector of the dislocation, and θ is the angle between the Burgers vector and the x axis. Equation (2·11) can be established by calculating the interaction energy between two hairpin-shaped dislocations applied to a straight dislocation so as to form a pair of kinks (cf. figure 1 (*b*).) Equation (2·1) does not predict the existence of U because the static coupling between the dislocation and the surrounding elastic medium is only rather crudely represented by the constant S. Extrapolated to $x = w$, (2·11) gives $U(\omega) \sim \frac{1}{4}W_k$ if, with Seeger, Donth & Pfaff (1957) we put $S = \frac{1}{2}\mu b^2$. It may perhaps be important in problems concerning the diffusion and interaction of kinks. Brailsford (1961) has presented a model in which a curved dislocation is resolved into a succession of sharp kinks, and finds that it then has no line tension. Since the interaction between kinks is ignored this is not surprising. One can show, in fact, that re-introducing the long-range interaction is, macroscopically speaking, equivalent to reinstating the line tension.

If we choose to say that $U(x)$ is due to the absorption and emission by the kinks of virtual phonons of the Debye spectrum we may also say that the short-range interaction is due to 'heavy phonons' propagated along the dislocation. Since these obey (2·9) their mass is $\hbar k_0/c_k$ and so they should produce a force of range k_0^{-1}, which for a potential of the type (2·5) or (2·6) is indeed of order w.

In what follows we shall make the assumption that under the influence of an applied stress the kink behaves like a particle of mass m_k constrained to move along a line. This is admittedly a large extrapolation from the results of this section. One can go some way towards justifying it by a perturbation treatment of (2·1), but the analysis becomes clumsy and will not be reproduced here. Even if the assumption is granted it is still necessary to know something more about the interaction between kink and medium before the action of elastic waves on a kink can be discussed. As the first step we shall calculate the rate of radiation of energy from an oscillating kink.

3. The radiation from a kink

To calculate the rate at which energy is radiated from an oscillating kink it is necessary to find the oscillating elastic field around it. We follow a method due to Nabarro (1951*a*). If the kink is moved a small distance ξ along the dislocation the change in the elastic field at a remote point is equal to the elastic field associated with a small dislocation loop of Burgers vector **b** and area $A = a\xi$. In figure 1*a*, if the kink moves from AB to $A'B'$ the loop in question is $AA'B'B$. If the kink oscillates with frequency $\omega/2\pi$ and amplitude ξ_0 the area is a function of time:

$$A(t) = a\xi_0 \cos\omega t. \tag{3·1}$$

Negative A corresponds to the case where $A'B'$ is to the left of AB. To conform with Nabarro's notation we temporarily take the xz plane as slip-plane and the z axis as the direction of **b**. The time-dependent part of the field is given by

$$u_i = \mu ab\xi_0(U_{iy,z}+U_{iz,y}), \tag{3·2}$$

where U_{ij} is the displacement in the i-direction due to a point force at the origin of magnitude $\cos\omega t$ directed along the j axis. In other words, u_i is the same as the displacement produced by a system of oscillating force-doublets represented by the singular force density

$$f_x = 0, \quad f_y = F_0 \cos\omega t\, \partial\delta(\mathbf{r})/\partial z, \quad f_z = F_0 \cos\omega t\, \partial\delta(\mathbf{r})/\partial y, \tag{3·3}$$

where $$F_0 = \mu ab\xi_0, \quad \delta(\mathbf{r}) = \delta(x)\,\delta(y)\,(z).$$

We have (Rayleigh 1896)

$$U_{ij} = \mathscr{R}(\chi_{,ij}+w\delta_{ij}),$$

with $$w = (4\pi\mu r)^{-1} \exp i\omega(t-r/c_t) \tag{3·4}$$

and $$\chi = \psi_t - \psi_l, \tag{3·5}$$

where $$\psi_t = (c_t^2/\omega^2)\, w, \tag{3·6}$$

and ψ_l is the same expression with c_t replaced by c_l throughout; c_t and c_l are the velocities of transverse and longitudinal waves. (We use this rather clumsy notation with an eye to a later generalization.) We shall only need the displacement near the kink. If we put

$$u_i = u_i' \cos\omega t + u_i'' \sin\omega t,$$

the amplitude of the out-of-phase component is

$$\{u_x'', u_y'', u_z''\} = -(20\pi\mu c_r^3)^{-1} F_0\omega^3\{0, z, y\} + O(r^2\omega^2/c_t^2), \tag{3·7}$$

where the velocity c_r defined by

$$\frac{1}{c_r^3} = \frac{1}{c_t^3}\left[1+\frac{2}{3}\frac{c_t^5}{c_l^5}\right]$$

usually does not differ from c_t by more than a few parts per cent. The in-phase component u_i' (which we shall not need) agrees with Nabarro's expression for a static loop of area $a\xi_0$.

The obvious way to find the rate of radiation from the kink is to integrate the energy flux over a large sphere surrounding it. However, we can save some calculation by noting that the rate of radiation must be equal to the rate at which the fictitious forces (3·3) do work on the medium. If a distribution of body force of amount f_i per unit volume produces a displacement u_i the rate at which energy is fed into the medium is

$$\int f_i \dot{u}_i \,\mathrm{d}v. \tag{3·8}$$

Only u_i'' contributes to the mean rate, $\mathscr{S}$ say. Thus

$$\mathscr{S} = -\frac{1}{2}\frac{F_0^2\omega^4}{20\pi\mu c_\mathrm{r}^3}\int\left(z\frac{\partial\delta(\mathbf{r})}{\partial z}+y\frac{\partial\delta(\mathbf{r})}{\partial y}\right)\mathrm{d}v.$$

$$= \frac{\mu a^2b^2}{20\pi}\frac{\xi_0^2\omega^4}{c_\mathrm{r}^3}. \tag{3·9}$$

For the particular motion $\xi = \xi_0 \cos\omega t$ this may be written

$$\mathscr{S} = \gamma m_\mathrm{k}\overline{\dot{v}^2}, \quad \gamma = \mu a^2/10\pi m_\mathrm{k} c_\mathrm{r}^3, \tag{3·10}$$

which may be compared with the expression $\mathscr{S} = \frac{2}{3}e^2\overline{\dot{\mathbf{v}}^2}/c^3$ for the rate of radiation from an electron (Heitler 1954). Just as in the electromagnetic case it can be shown that (3·10) is equivalent to a radiation reaction self-force

$$K_\mathrm{S} = \gamma m_\mathrm{k}\ddot{\mathbf{v}} \tag{3·11}$$

on the kink. One can also find K_S directly. According to a basic result of dislocation theory the force acting on the abrupt kink of figure 1 (*b*) is

$$f_x = -ab\tau, \tag{3·12}$$

where
$$\tau = n_i\beta_j p_{ij} = \mu n_i \beta_i(u_{i,j}+u_{j,i})$$

is the applied shear stress at the kink resolved on the slip-plane (normal **n**) and in the direction of the Burgers vector (direction $\boldsymbol{\beta}$). The minus sign follows from the convention we have implicitly adopted for the sign of the Burgers vector. It is easy to show that (3·12) is also true for the total force on the smooth kink of figure 1 (*b*) provided τ is uniform in the neighbourhood of the kink. Consider the values of τ at a point **R** near the kink due to the kink itself. Its time-dependent part is, from (3·10), (3·7)

$$\tau = -m_\mathrm{k}\gamma\ddot{v} + M(\mathbf{R})\dot{v} + O(\mathbf{R}), \tag{3·13}$$

where $\ddot{v} = \omega^3\xi_0 \sin\omega t$, $\dot{v} = -\omega^2\xi_0 \omega s\,\omega t$. A Fourier resolution shows that (3·13) also holds for arbitrary small motion. The first term is finite as $\mathbf{R}\to 0$ and represents the radiation reaction (3·11). The second term, proportional to the acceleration $\dot{v}$, diverges as $\mathbf{R}\to 0$. We suppose that it represents (or can be absorbed into) the inertial reaction $-m_\mathrm{k}\dot{v}$, so that the kink obeys the equation of motion

$$\dot{v} - \gamma\ddot{v} = f_x/m_\mathrm{k}, \tag{3·14}$$

where f_x is the applied force.

4. The scattering cross-section of a kink

Consider a plane longitudinal or transverse wave

$$u_i = As_i \cos(k_m x_m - \omega t) \tag{4·1}$$

propagating in the direction $\mathbf{l} = \mathbf{k}/k$ with polarization $\mathbf{s}$ ($|\mathbf{s}| = 1$) parallel or perpendicular to $\mathbf{l}$. Its mean energy density is

$$\epsilon = \tfrac{1}{2}\rho A^2 \omega^2,$$

where ρ is the density of the medium. On a kink at the origin it exerts a force (cf. (3·13))

$$f_x = f_0 \sin \omega t, \tag{4·2}$$

with
$$f_0 = ab_i n_j p_{ij\max.} = \mu ab A k \Phi, \tag{4·3}$$

where
$$\Phi(\mathbf{l}, \mathbf{s}) = (\mathbf{s}\,.\,\mathbf{n})(\mathbf{l}\,.\,\boldsymbol{\beta}) + (\mathbf{l}\,.\,\mathbf{n})(\mathbf{s}\,.\,\boldsymbol{\beta})$$

($\mathbf{n}$ is the normal to the slip-plane, $\boldsymbol{\beta}$ a unit vector in the direction of the Burgers vector). We note that

$$f_0^2 = 2\mu a^2 b^2 (c_t/c)^2 \epsilon \Phi^2,$$

where $c = c_l$ or $c = c_t$ according as the wave is longitudinal or transverse.

Under the influence of the force (4·2) the kink executes an oscillation

$$\xi = \xi_0 \sin(\omega t - \alpha), \tag{4·4}$$

where, according to (3·14),

$$\xi_0 = f_0/m_k \omega^2, \quad \sin\alpha = \gamma\omega,$$

so long as
$$\gamma\omega \ll 1. \tag{4·5}$$

The total cross-section σ for scattering of the wave (4·1) by the kink is found by equating the product of σ and the incident energy flux ϵc to the rate of radiation (3·9). This gives

$$\sigma = \sigma_0 (c_t/c)^3 \Phi^2(\mathbf{l}, \mathbf{s}), \tag{4·6}$$

with
$$\sigma_0 = \frac{1}{10\pi} \frac{(\mu a^2 b^2)^2}{m_k^2 c_r^3 c_t},$$

and where
$$c = c_l, \quad \mathbf{s} = \mathbf{l}$$

for a longitudinal wave and
$$c = c_t, \quad \mathbf{s}\,.\,\mathbf{l} = 0$$

for a transverse wave. If (4·5) is not satisfied (4·6) must be multiplied by $(1 + \gamma^2\omega^2)^{-1}$. The influence of the finite kink width on σ is discussed in § 6.

The details of the scattered wave could be found by examining the behaviour of (3·10) for large r. However, for an incompressible medium ($c_l \to \infty$) the longitudinal waves may be ignored and the differential cross-section can then be found by an argument familiar in aerial theory. Consider a kink situated in a flux of transverse lattice vibration in thermal equilibrium. We may imagine it to be absorbing and re-emitting acoustic radiation. If equilibrium is to be preserved its transmitting and receiving polar diagrams must be the same, polarization for polarization. Thus the cross-section for scattering a wave $(\mathbf{l}, \mathbf{s})$ into angle $d\Omega'$ centred on $\mathbf{l}'$, with polarization $\mathbf{s}'$, must be of the form $d\sigma = \text{const.} \times \Phi^2(\mathbf{l},\mathbf{s})\,\Phi^2(\mathbf{l}', \mathbf{s}')\,d\Omega'$. The constant is fixed by the requirement that $\mathrm{d}\sigma$ shall yield (4·6) when summed over angles and polarizations. According to equation (A1) in the appendix, the sum of Φ^2 over polarizations and angles is $8\pi/5$ and so

$$\frac{\mathrm{d}\sigma}{\mathrm{d}\Omega'} = \frac{5\sigma_0}{8\pi}\Phi^2(\mathbf{l},\mathbf{s})\,\Phi^2(\mathbf{l}',\mathbf{s}'). \tag{4·7}$$

To a good approximation we may put $c_t = c_r = c_k$. Then

$$\sigma_0 = \frac{1}{10\pi\eta^2}\frac{\mu b^2}{S}w^2$$

by (2·4) and (2·8). The rather different potentials (2·5) and (2·6) both give $\eta \simeq 0{\cdot}7$ and we may perhaps take this value to be typical of any reasonable potential. If we also put $S = \frac{1}{2}\mu b^2$ we have

$$\sigma_0 \simeq \tfrac{1}{7}w^2, \tag{4·8}$$

and the cross-section is proportional to the square of the kink width. It is instructive to make a comparison with the electron. For a point charge the Thomson cross-section defines an effective radius r_e through $\sigma_{Th} \sim r_e^2$. The samedistance appears in the expression for the electromagnetic mass, $m \sim e^2/r_e c^2$. If the charge has a finite physical extension of order r_{ph} then $m \sim e^2/r_{ph}c^2$, but we still have $\sigma_{Th} \sim r_e^2$, at least for low frequencies. For a point charge the equation of motion has solutions representing self-accelerated motion, but for a charge of finite extent they disappear if, roughly, r_{ph} is not much less than r_e (cf. Bellomo 1955). For a kink the place of e is taken by $\mu^{\frac{1}{2}}ab$, which determines its coupling to the elastic field, and hence also the interaction (2·10) between two kinks analogous to $-e^2/r$. Equations (2·8) and (4·8) show that the kink width w plays the role of both r_e and r_{ph}, and so we may hope not to be troubled with self-acceleration.

5. The retarding force on a kink

Before we calculate the force on a kink moving through a flux of elastic waves it will be helpful to examine a related phenomenon, the force experienced by a charged particle moving through a flux of electromagnetic radiation. An electron of charge e and mass m moving through an isotropic flux of radiation is subject to a retarding force

$$\mathbf{F} = -\tfrac{4}{3}\sigma_{Th}\mathscr{E}\mathbf{v}_0/c \tag{5·1}$$

(Landau & Liftshitz 1951). Here $\mathscr{E}$ is the electromagnetic energy density, $\mathbf{v}_0$ is the velocity of the charge (supposed to be much less than c, the velocity of light) and

$$\sigma_{Th} = \tfrac{8}{3}\pi\,(e^2/mc^2)^2$$

is the Thomson scattering cross-section. However, as a kink can only move along a line its true analogue is not a free charge but a charged bead sliding along a smooth wire. We therefore give a simple derivation of (5·1) and then apply the same method to the rather artificial (but perfectly proper) problem of a charge drifting through a flux of radiation but constrained to move along a line.

An electromagnetic wave

$$\mathbf{E} = \mathbf{E}^0\cos\omega(t-\mathbf{l}.\mathbf{r}/c), \quad \mathbf{H} = \mathbf{l}\times\mathbf{E}, \tag{5·2}$$

with energy density $\epsilon = (\mathbf{E}^0)^2/8\pi$ exerts a mean force $\bar{\mathbf{f}} = \sigma_{\mathrm{Th}}\epsilon\mathbf{l}$ with x component

$$f_x = \sigma_{\mathrm{Th}}\epsilon\cos\theta, \quad l_x = \cos\theta \tag{5·3}$$

on a charge at rest (Landau & Lifshitz 1951). (In what follows 'at rest' means 'oscillating but with no mean velocity'.) It is helpful to have a physical picture of how the force (5·3) originates (cf. Nabarro 1951*b*). The equation of motion of the charge is

$$m\dot{\mathbf{v}} - \mathbf{K}_{\mathrm{S}} = e\mathbf{E} + (e/c)\,\mathbf{v}\times\mathbf{H},$$

where

$$\mathbf{K}_{\mathrm{S}} = \tfrac{2}{3}e^2\ddot{\mathbf{v}} \tag{5·4}$$

is the radiation reaction. Let us first ignore $\mathbf{K}_{\mathrm{S}}$. $\mathbf{E}$ makes the charge oscillate perpendicular to $\mathbf{l}$, but exerts no mean force on it. The resulting velocity interacts with $\mathbf{H}$ to give an oscillating Lorentz force which, however, has a zero mean value since $\mathbf{v}$ and $\mathbf{H}$ are precisely out of phase. When we restore the radiation reaction $\mathbf{K}_{\mathrm{S}}$, $\mathbf{v}$ and $\mathbf{H}$ are no longer exactly out of phase and the Lorentz force has a mean value (5·3). Suppose now that the charge is moving with velocity v_0 along the x axis. To find the mean force on it due to the wave (5·2) we first make a Lorentz transformation to the rest system of the charge. In this system the expression analogous to (5·3) is

$$\bar{f}'_x = \sigma_{\mathrm{Th}}\epsilon'\cos\theta',$$

where $\epsilon', \cos\theta'$ are the transformed quantities for the wave (5·2). To order v_0/c they are

$$\epsilon' = \epsilon\{1 - 2(v_0/c)\cos\theta\},$$
$$\cos\theta' = \cos\theta - (v_0/c)\sin^2\theta.$$

When we transform back to the original system the force is unchanged to order v_0^2/c^2, or indeed to any order if it is suitably defined. Thus for the moving charge the mean force is

$$\bar{f}_x = \sigma_{\mathrm{Th}}\epsilon\{\cos\theta - (v_0/c)(1+\cos^2\theta)\}.$$

To find the force in an isotropic flux of radiation we average over all directions of $\mathbf{l}$ ($\cos^2\theta \to \frac{1}{3}$) and replace ϵ by the total energy density $\mathscr{E}$. This gives (5·1).

In the case of the charged bead threaded on a wire along the x axis the y and z components of the force are balanced by the reaction of the wire, and its motion under the influence of the wave (5·2) is governed by

$$\dot{v} - \gamma\ddot{v} = (e/m)\,E_x^0\cos\omega t, \quad (\gamma = 2e^2/3mc^3).$$

If $\gamma\omega \ll 1$ the displacement of the bead is thus

$$\xi = -(eE_x^0/m\omega^2)(\cos\omega t - \gamma\omega\sin\omega t).$$

There now seems to be no mean force on it. However we note (cf. Leibfried 1949) that the bead does not experience the electric field precisely at $x = 0$ but rather at $x = \xi$. Thus to the second order the force on it is

$$f_x = eE_x^0\cos\omega(t - l_x\xi/c),$$

which for small ξ has the mean value

$$\bar{f}_x = \frac{1}{2}\frac{(eE_x^0)^2}{mc}\gamma\cos\theta. \tag{5·5}$$

To find the corresponding expression for a bead drifting along the wire with velocity v_0 we transform into and back from its rest system as before. The result is

$$\bar{f}_x = \sigma_{\mathrm{Th}} \epsilon \cos^2 \phi \{ \cos \theta - (v_0/c) \sin^2 \theta \}, \tag{5·6}$$

where ϕ is the angle between $\mathbf{E}$ and the x axis. The effect of an isotropic flux of radiation is found by averaging over polarization ($\cos^2 \phi \to \frac{1}{2} \sin^2 \theta$) and over direction ($\sin^4 \theta \to 8/15$) and replacing ϵ by $\mathscr{E}$. The result is

$$F_x = -\tfrac{4}{15} \sigma_{\mathrm{Th}} \mathscr{E} v_0/c, \tag{5·7}$$

that is, one-fifth of the value for a free charge (5·1).

Both (5·1) and (5·7) depend on the phase shift induced by the radiation reaction (5·4). It is not particularly significant that one arises from a second-order term in the velocity, the other from a second-order term in the displacement. If we had chosen to carry out the calculation of (5·7) in a co-ordinate system moving obliquely to the wire part of the effect would have come from the Lorentz force.

The mean force which an elastic wave exerts on a kink at rest can be found by the argument which led to (5·5). To the first order the wave (4·1) excites the motion (4·4). We now note with Leibfried that the kink actually samples the stress field of the wave at the point $x = \xi$ rather than at $x = 0$. The mean force on it is thus the time average of

$$f_x = f_0 \sin \{ \omega t - k_x \xi(t) \}, \tag{5·8}$$

with ξ given by (4·4). If $k_x \xi$ is small this gives

$$\bar{f}_x = \tfrac{1}{2} f_0 k_x \xi_0 \sin \alpha,$$

which can also be written as

$$\bar{f}_x = \sigma l_x \epsilon. \tag{5·9}$$

We may give the following formal interpretation to (5·9) (cf. Nabarro 1951*b*). The incident wave has quasi-momentum $\mathbf{l}\epsilon/c$ per unit volume. In unit time energy $\sigma \epsilon c$ and hence quasi-momentum $\sigma \mathbf{l} \epsilon$ is removed from the wave. By symmetry the x component of the scattered momentum is zero, and so the force is given by (5·9).

The extension of (5·9) to the case where the kink is moving is not so easy as it was in the electromagnetic problem since we cannot now use transformation arguments of the kind which led from (5·5) to (5·6). To bring out the need for a rather elaborate treatment we first present a simple argument which leads to an absurd conclusion, and then show how it is to be corrected.

Suppose that under the influence of the wave (4·1) the kink carries out the motion

$$\xi = v_0 t + \xi_0' \sin (\omega' t - \alpha'), \tag{5·10}$$

where v_0 is its constant drift velocity. To the first approximation the wave exerts a force

$$f_x = f_0 \sin (\omega t - k_x v_0 t)$$

on it, and so the wave can actually maintain the proposed motion (to first order) if

$$\omega' = \omega (1 - l_x v_0/c). \tag{5·11}$$

Then in place of (4·4) we have

$$\xi_0' = f_0 m \omega'^2, \quad \alpha' = \gamma \omega'. \tag{5·12}$$

Note that there is no question of replacing $k = \omega/c$ by ω'/c in (4·3); the maximum shear stress produced by the wave is unaffected by the motion of the kink. To the next approximation there is a mean force due to the wave, found by substituting (5·10) in (5·8) and taking the time-average. The result is

$$\bar{f}_x = \tfrac{1}{2} f_0 k_x \xi_0 (1 - l_x v_0/c)^{-1}$$

or, to terms of order v_0^2/c^2,

$$\bar{f}_x = \sigma_0 \Phi^2 \epsilon (c_t/c)^3 (l_x + l_x^2 v_0/c). \tag{5·13}$$

To find the mean force on a kink moving through an isotropic flux of waves we have to sum (5·13) over the Debye spectrum. For the longitudinal contribution we replace the angular factors by their average over all directions **l**, put $c = c_l$ and replace ϵ by $\mathscr{E}_l$, the energy density of the longitudinal waves. The transverse contribution is found similarly, but in addition we have to sum over the two polarizations associated with each direction **l**. The necessary sums and averages are evaluated in the appendix. The result is

$$\bar{f}_x = f_l + f_t, \tag{5·14}$$

where

$$f_l = \tfrac{16}{105} \sigma_0 \frac{c_t^4}{c_l^4} \Psi_l \mathscr{E}_l \frac{v_0}{c_t}, \tag{5·15}$$

$$f_t = \tfrac{2}{35} \sigma_0 \Psi_t \mathscr{E}_t \frac{v_0}{c_t} \tag{5·16}$$

and

$$\Psi_l = \tfrac{3}{4}\beta_x^2 + \tfrac{1}{4}\beta_y^2, \quad \Psi_t = \tfrac{4}{3}\beta_x^2 + \tfrac{5}{6}\beta_y^2. \tag{5·17}$$

In these expressions $\beta_x = \cos\theta$, $\beta_y = \sin\theta$, where θ is the angle between the dislocation and its Burgers vector, and $\mathscr{E}_t$ includes the energy of both polarizations.

The expression (5·14) appears reasonable until we notice that its positive sign implies that a kink moving through an isotropic flux experiences a force accelerating it in the direction in which it is going. This is obviously wrong.

To see where the error lies we return to the corresponding problem for an electron and imagine that we have to solve it without using relativistic transformation properties. If we follow the argument used for the kink we obtain for the free charge half the expression (5·1). The error might pass unnoticed if we did not know the correct value. For the constrained charge, on the other hand, we obtain (5·7) with the factor $-\frac{4}{15}$ replaced by $+\frac{1}{15}$, and so, as for the kink, there is an obvious error of sign.

Evidently in each case (kink, free or constrained charge) there is some error which reduces or even reverses the correct value. For a charge it can be isolated by comparing the relativistic and non-relativistic treatments. One finds that for a charge which is not on the average at rest the expression (5·4) for $\mathbf{K}_s$ is not sufficiently accurate. It is necessary to take account of the term of next higher order. Then (Heitler 1954)

$$\mathbf{K}_s = \frac{2e^2}{3c^3}\ddot{\mathbf{v}} + \frac{2e^2}{3c^5}(3\mathbf{v}\dot{\mathbf{v}}^2 + 2\mathbf{v}^2\ddot{\mathbf{v}}), \tag{5·18}$$

$$= \mathbf{K}_i + \mathbf{K}_{ii}$$

say. For a motion compounded of oscillation and drift with velocity $\mathbf{v}_0$, $\mathbf{K}_s$ has a non-vanishing mean value

$$\overline{\mathbf{K}}_s = \overline{\mathbf{K}}_{ii} = -\frac{2e^2}{3c^5}\mathbf{v}_0\overline{\dot{\mathbf{v}}^2}. \tag{5·19}$$

To find the retarding force we now argue as follows. If we ignore $\mathbf{K}_{ii}$ we find, as before, that the wave provides the force necessary to maintain the oscillatory motion, and in addition a certain mean force $\overline{\mathbf{f}}$. When we introduce K_{ii} the oscillatory component of the motion is not disturbed to the first order, but the radiation reaction exerts a mean force (5·19). Thus to maintain the motion we must supply a constant force $-(\overline{\mathbf{f}}+\overline{\mathbf{K}}_s)$ over and above the oscillatory force provided by the wave. Hence the mean force exerted by the wave is $\overline{\mathbf{f}}+\overline{\mathbf{K}}_s$ rather than merely $\overline{\mathbf{f}}$. When averaged over an isotropic flux $\overline{\mathbf{f}}+\overline{\mathbf{K}}_s$ reproduces (5·1) and (5·7) correctly.

To complete the calculation of the retarding force on a kink we evidently have to find the analogue of (5·18) for a kink. To do this it is necessary to find the elastic field near a kink which is moving along the x axis according to the arbitrary law

$$x = \xi(t). \tag{5·20}$$

We note first that the kinked dislocation $CABD$ (figure 1b) can be constructed by superimposing a straight dislocation FD and a hairpin-shaped dislocation $CAGE$; the segments EG and FB cancel if $CAGE$ and FD have the same Burgers vector when traversed in the directions $C \to E$, $F \to D$ respectively. We thus have to find the field due to a hairpin whose cross-piece is moving along the x axis according to (5·20). The motion can be considered to be the result of the continual addition or removal of infinitesimal loops at the tip (Nabarro 1951a). Instead of using Nabarro's results directly it will be more convenient to extend the treatment of §3 to the case where the area of the loop is an arbitrary function of time $A(t)$ and apply the result to the case where $A(t)$ is a step function. We cannot now be sure that the higher terms in the radiation reaction will, like (3·11), be independent of the angle between the dislocation and its Burgers vector (in fact they are not), and so we must drop the limitation that the Burgers vector shall lie along the z axis. We easily show that the displacement due to the loop is

$$u_i = \mu b\{2n_j\beta_k\chi_{,ijk} + (n_k\beta_i + n_i\beta_k)w_{,k}\},$$

where w, χ have the values (3·4), (3·5) if the area of the loop is given by $A = \exp(\mathrm{i}\omega t)$ rather than by (3·1). For a loop whose area is an arbitrary function of time we could find the appropriate w, ψ_t, ψ_l by a Fourier resolution. However, it is obvious from the linearity of the equations that if $A = f(t)$ the new w will be $(4\pi\mu r)^{-1}f(t-r/c_t)$ and that the new ψ_t will be related to the new w by the equation

$$\nabla^2\psi_t + w = 0, \tag{5·21}$$

which connects (3·4) and (3·6). For the creation of a loop of area $A_0(t')$ at time t' in the position $x = \xi(t')$, $y = 0$, $z = 0$ we have

$$A(t) = A_0(t')\,H(t-t'),$$

where H is the unit step function. The corresponding w is

$$w = (4\pi\mu R')^{-1} H(t-t'-R'/c_t),$$

where
$$R'^2 = \{x-\xi(t')\}^2+y^2+z^2.$$

The progress of the hairpin along the x axis can be regarded as a succession of jumps of amount $\xi dr = v\,dt$ in successive time intervals dt. Each jump is equivalent to the creation of an infinitesimal loop of area $av\,dt$ at the instantaneous position of the kink. The appropriate w-function is thus

$$w = \frac{a}{4\pi\mu}\int_{-\infty}^{\infty} \frac{H(t-t'-R'/c_t)}{R'} v(t')\,dt'. \tag{5·22}$$

The reduction of (5·22) to an expression involving instantaneous rather than retarded quantities can be carried out rapidly by a device due to Ivanenko & Sokolov (1950). Expand the H-function as a formal Taylor series:

$$H(t-t'-R'/c_t) = H(t-t') + \sum_{n=1}^{\infty} \frac{1}{n!}\left(-\frac{R'}{c_t}\right)^n \delta^{(n-1)}(t-t').$$

This gives at once

$$w = w_0 - \frac{a}{4\pi\mu}\sum_{n=1}^{\infty} \frac{1}{(n+1)!\,c_t}\left(\frac{\partial}{c_t\partial t}\right)^n (vR^n),$$

where
$$R^2 = \{x-\xi(t)\}^2+y^2+z^2$$

and
$$(4\pi\mu/a)\,w_0 = \ln\{x-\xi(t)+R\}+v(t)/c_t+\text{const.}$$

From (5·21)
$$\psi_t = \psi_{t0} + \frac{a}{4\pi\mu}\sum_{n=1}^{\infty} \frac{1}{(n+3)!\,c_t}\left(\frac{\partial}{c_t\partial t}\right)^n (vR^{n+2}),$$

where ψ_{t0} is the contribution from w_0. ψ_l is the same expression with c_l replacing c_t.

For the resolved shear stress at a point near the kink (or rather, near the head of the hairpin) we find, after some calculation, the expression

$$\tau = \tau_0(\mathbf{R}) - K_s/ab + M(\mathbf{R})\,\dot v + O(\mathbf{R}) \tag{5·23}$$

in place of (3·13). Here

$$K_s = \frac{\mu a^2 b^2}{10\pi c_T^3}\ddot v + \frac{\mu a^2 b^2}{7\pi}\left[\frac{\Psi_t}{c_t^5} + \frac{4\Psi_l c_t^2}{c_l^7}\right](3v\dot v^2 + 2v^2\ddot v), \tag{5·24}$$

with the notation of (5·17), and τ_0 is the contribution from $w_0, \psi_{t0}, \psi_{l0}$. We give (5·23) the same interpretation as (3·13), so that $M\dot v$ is an inertial term and K_s is the radiation reaction term we are looking for. We may suppose that τ_0 includes the contribution from the straight dislocation which must be laid alongside the hairpin to convert it into a kinked dislocation. Since τ_0 is independent of $\dot v, \ddot v, \ldots$, it represents the field of a kink moving uniformly. We have assumed that such a motion is possible in the absence of applied force, and so we must suppose that the forces corresponding to τ_0 are self-equilibrating and contribute nothing to the equation of motion.

For the particular motion (5·10) with the values (5·11), (5·12) for ω', ξ_0', α' the mean value of (5·24) can be written in the form

$$\bar K_s = -\tfrac{10}{7}\sigma_0 \Phi^2 \epsilon \frac{c_T^3}{c c_t}\left(\Psi_t + \frac{4c_t^7}{3c_l^7}\Psi_l\right)\frac{v_0}{c_t}. \tag{5·25}$$

To complete the calculation we argue as follows. First omit the second term in (5·24). As before we find that a wave which can provide the necessary oscillatory force on the kink also exerts a steady mean force (5·14) on it. When the second term of K_s is restored the oscillatory motion is undisturbed to the first order, but the radiation reaction exerts an additional steady mean force (5·25). Thus to maintain the proposed motion we must supply not only the wave (4·1) but also a steady force equal and opposite to

$$F = \bar{f}_x + \bar{K}_s. \tag{5·26}$$

If the necessary force $-F$ is not supplied the mean motion $v_0 t$ will accelerate as if the kinks were acted on by the force (5·26). Thus (5·26) is the true expression for the mean force which the wave (4·1) exerts on the kink; the division into a part $\bar{f}_x$ 'really' due to the wave and a remainder due to a neglected term in K_s, is, of course, quite arbitrary.

To find the retarding force in an isotropic flux we have to sum (5·26) over the Debye spectrum. Equations (5·15), (5·16) give the contribution from f_x, and the contributions from K_s are found similarly. They are

$$K_l = -\tfrac{2}{21}\sigma_0 \frac{c_r^3}{c_l^2 c_t}\left[\Psi_t + \frac{4c_t^7}{3c_l^7}\Psi_l\right]\mathscr{E}_l \frac{v_0}{c_t}, \tag{5·27}$$

$$K_t = -\tfrac{2}{7}\sigma_0 \frac{c_r^3}{c_t^3}\left[\Psi_t + \frac{4c_t^7}{3c_l^7}\Psi_l\right]\mathscr{E}_t \frac{v_0}{c_t}, \tag{5·28}$$

and so the total retarding force is

$$F = f_l + f_t + K_l + K_t,$$

where the individual terms are given by (5·15), (5·16), (5·27) and (5·28). F is negative for positive v_0, as required.

We record two results for the special case of an incompressible medium, $c_l \to \infty$. The force (5·26) due to a single wave is

$$F = \sigma_0 \Phi^2 \epsilon \left[l_x\left(1 + l_x \frac{v_0}{c_t}\right) - \tfrac{10}{7}\Psi_t \frac{v_0}{c_t}\right]. \tag{5·29}$$

The retarding force in an isotropic flux of transverse waves is

$$F = -\tfrac{8}{35}\sigma_0 \Psi_t \mathscr{E}_t \frac{v_0}{c_t}. \tag{5·30}$$

6. An alternative treatment

Equation (5·30) is similar to Leibfried's (1949) expression for the retarding force on one atomic length of dislocation. He found

$$F = -\tfrac{1}{10}\sigma \mathscr{E} v_0 / c_t, \tag{6·1}$$

where σ is of the order of the square of the interatomic distance and $\mathscr{E} = \mathscr{E}_l + \mathscr{E}_t$. In effect Leibfried assumed that small elements of the dislocation could move and scatter separately, and so his calculation really concerns the behaviour of a small isolated segment of dislocation, AB say (figure 1 (*b*)). But according to the rules of dislocation theory AB must form part of a continuous dislocation line. We might, for example, suppose that it trailed behind it the segments AC, BF, but then it would be pulled to the left by their line tension. On the other hand, if we extend the segment AB to the left from A and to the right from B the line tensions balance and the segment AB can move freely. But it is then precisely a kink in the dislocation CD. Thus in a sense Leibfried's calculation refers to a kink rather than to one atomic length of dislocation.

We shall make a numerical comparison in § 8, but here we are concerned with a point of principle raised by Leibfried. He suggested that the zero-point energy should contribute to the retarding force on the same footing as the thermal component (cf. also Leibfried 1956). It would be disastrous if the same were true for an electron moving through a flux of radiation. In the first place the zero-point contribution to (5·1) or (5·7) would be infinite. However, suppose for the sake of argument that the contribution is actually finite, perhaps as a result of the decrease of scattering cross-section at high frequencies. An electron moving through a hot furnace experiences a retarding force proportional to its velocity relative to the furnace because the radiation flux is isotropic only in the rest system of the furnace. If zero-point terms were included the electron could detect its own motion even in a cold furnace with no flux of radiation, and this is absurd.

We cannot immediately apply this argument to a kink or dislocation moving through a crystal lattice, since even if we treat the lattice as a continuum we must still introduce an upper cut-off frequency. The zero-point contribution to the energy density is finite and, further, an elastic singularity moving in the medium might be able to detect an apparent change in the cut-off frequency and so infer its velocity.

It is possible to give a rather naïve quantum-mechanical treatment of the kink problem which shows formally that the zero-point contribution is to be excluded. Since the calculation also serves to verify the rather precarious results of the previous section it is perhaps worth presenting. For simplicity we shall consider only an incompressible medium ($c_l = \infty$), so that longitudinal waves can be ignored. Also we shall suppose that the velocity c_k of (2·7) is equal to c_t, the velocity of transverse waves, and write $c_t = c_k = c$ simply.

The quantization of the acoustic field is a standard problem. In the context of the theory of elasticity it may be accomplished rapidly by using a result of Larmor's (quoted by Love 1952, p. 167) according to which the elastic Hamiltonian

$$H = \int_V \{\mu(u_{i,j}+u_{j,i})^2 + \tfrac{1}{2}\rho \dot{u}_i^2\}\, \mathrm{d}V$$

can be rewritten as

$$H = \int_V \tfrac{1}{2}\rho\{c^2(\operatorname{curl}\mathbf{u})^2 + \dot{\mathbf{u}}^2\}\,\mathrm{d}V,$$

plus certain surface integrals which vanish if we impose periodic boundary conditions at the surface of the volume V.

If we identify the velocities of sound and light and put

$$\mathbf{u} = (4\pi\rho c^2)^{-\frac{1}{2}}\mathbf{A},$$

the modified expression for H becomes the Hamiltonian for an electromagnetic field (with vector potential $\mathbf{A}$) enclosed in the same volume. Comparison with the electromagnetic case shows that the displacement operator is

$$\mathbf{u} = \Sigma_{\mathbf{k}}\mathbf{s}_{\mathbf{k}}\{\exp(\mathrm{i}\mathbf{k}\,.\,\mathbf{r})\,q_{\mathbf{k}} + \exp(-\mathrm{i}\mathbf{k}\,.\,\mathbf{r})\,q_{\mathbf{k}}^{\dagger}\}, \tag{6·2}$$

where $\mathbf{s}_{\mathbf{k}}$ is the polarization vector for mode $\mathbf{k}$ and $q_{\mathbf{k}}$, $q_{\mathbf{k}}^{\dagger}$ are annihilation and creation operators for phonons of wave vector $\mathbf{k}$. The elastic field may be described by a state vector $|n_{\mathbf{k}}, n_{\mathbf{k}'}, \ldots\rangle$ whose argument is the set of numbers which specifies the number of phonons in the modes $\mathbf{k}, \mathbf{k}', \ldots$. The operators $q_{\mathbf{k}}$ and $q_{\mathbf{k}}^{\dagger}$ have matrix elements

$$\langle n_{\mathbf{k}}-1|q_{\mathbf{k}}|n_{\mathbf{k}}\rangle = \langle n_{\mathbf{k}}|q_{\mathbf{k}}^{\dagger}|n_{\mathbf{k}}-1\rangle = (\hbar/2k\rho c V)^{\frac{1}{2}}\,n_{\mathbf{k}}^{\frac{1}{2}},$$

in which one $n_{\mathbf{k}}$ changes by unity, the remaining (unwritten) $n_{\mathbf{k}}$ being equal, each to each, in the initial and final states. All other matrix elements are zero.

We shall treat the kink as a (non-relativistic) particle of mass $m = m_{\mathrm{k}}$ constrained to slide along the x axis, or, equivalently, as a particle confined to a tube of ultimately vanishing cross-section. Its wave function will have the form

$$\psi(x,y,z) = \phi(x)\,\Psi(y,z).$$

All we need to know about Ψ is that it is zero except very close to the x axis, and that it is normalizable, so that, say,

$$\Psi\Psi^{*} = \delta(y)\,\delta(z).$$

Then, if the kink is confined to a normalization length L, but is otherwise free,

$$\phi(x) = L^{-\frac{1}{2}}\exp(\mathrm{i}Kx).$$

For the interaction energy between the kink and the elastic field we take the classical expression

$$H_{\mathrm{int.}} = \int^{x} ab\beta_i n_j p_{ij}(x,0,0)\,\mathrm{d}x \tag{6·3}$$

representing the work required to bring the kink from an arbitrary point on the x axis to the point x in the presence of an applied stress p_{ij}. With (6·2) this takes the form

$$H_{\mathrm{int.}} = \mu ab\Sigma_{\mathbf{k}}\,\Phi_{\mathbf{k}}(\mathbf{l}_{\mathbf{k}})_x^{-1}\{\exp(\mathrm{i}\mathbf{k}\,.\,\mathbf{r})\,q_{\mathbf{k}} + \exp(-\mathrm{i}\mathbf{k}\,.\,\mathbf{r})\,q_{\mathbf{k}}^{\dagger}\}, \tag{6·4}$$

where

$$\Phi_{\mathbf{k}} = (\mathbf{s}_{\mathbf{k}}\,.\,\mathbf{n})(\mathbf{l}_{\mathbf{k}}\,.\,\boldsymbol{\beta}) + (\mathbf{l}_{\mathbf{k}}\,.\,\mathbf{n})(\mathbf{s}_{\mathbf{k}}\,.\,\boldsymbol{\beta}) \quad (\mathbf{l}_{\mathbf{k}} = \mathbf{k}/k).$$

The matrix element for a transition in which the kink momentum changes from K to K' and $n_{\mathbf{k}}$ changes to $n_{\mathbf{k}} \pm 1$ is

$$\begin{aligned}
&\langle K', n_{\mathbf{k}} \pm 1|\,H_{\mathrm{int.}}|K, n_{\mathbf{k}}\rangle \\
&= \int \psi^{*}(K')\,\langle n_{\mathbf{k}} \pm 1|\,H_{\mathrm{int.}}|\,n_{\mathbf{k}}\rangle\,\psi(K)\,\mathrm{d}V \\
&= \left.\begin{matrix}(n_{\mathbf{k}}+1)^{\frac{1}{2}} \\ n_{\mathbf{k}}^{\frac{1}{2}}\end{matrix}\right\}(\hbar/2k\rho c V)^{\frac{1}{2}}\,\mu ab\Phi_{\mathbf{k}}(\mathbf{l}_{\mathbf{k}})_x^{-1} \\
&\quad \times L^{-1}\int \exp \mathrm{i}(-K'x \mp \mathbf{k}\,.\,\mathbf{r} + Kx)\,\delta(y)\,\delta(z)\,\mathrm{d}x\,\mathrm{d}y\,\mathrm{d}z \\
&= \mu ab\Phi_{\mathbf{k}}(\mathbf{l}_{\mathbf{k}})_x^{-1}\,(\hbar/2k\rho c V)^{\frac{1}{2}}\left\{\begin{matrix}(n_{\mathbf{k}}+1)^{\frac{1}{2}} \\ n_{\mathbf{k}}^{\frac{1}{2}}\end{matrix}\right.
\end{aligned}$$

if

$$K = K' \pm k_x \tag{6·5}$$

and zero otherwise. It is easy to show that unless the kink is supersonic (a possibility we do not consider) (6·5) is incompatible with the equation of energy conservation

$$\hbar^2K^2/2m = \hbar^2K'^2/2m \pm \hbar ck,$$

and so we must consider second-order transitions.

A transition in which $K, n_{\mathbf{k}}, n_{\mathbf{k}'}$ become $K', n_{\mathbf{k}}-1, n_{\mathbf{k}'}+1$ can take place by way of two intermediate states: I, the kink absorbs $\mathbf{k}$ and then emits $\mathbf{k}'$, and II, the kink emits $\mathbf{k}'$ and then absorbs $\mathbf{k}$. The matrix element is easily found to be

$$M(K', n_{\mathbf{k}}-1, n_{\mathbf{k}'}+1; K, n_{\mathbf{k}}, n_{\mathbf{k}'}) = \frac{(\mu ab)^2\hbar}{2\rho cV}\frac{\Phi_{\mathbf{k}}\Phi_{\mathbf{k}'}}{ll'}\frac{n_{\mathbf{k}}^{\frac{1}{2}}(n_{\mathbf{k}'}+1)^{\frac{1}{2}}}{k^{\frac{1}{2}}k'^{\frac{1}{2}}}\Delta, \tag{6·6}$$

where $$\Delta = (E_{\mathrm{O}}-E_{\mathrm{I}})^{-1}+(E_{\mathrm{O}}-E_{\mathrm{II}})^{-1}$$

and $$l = (\mathbf{1}_{\mathbf{k}})_x,\; l' = (\mathbf{1}_{\mathbf{k}'})_x$$

are the projections of $\mathbf{k}/k$, $\mathbf{k}'/k'$ on the x axis.

The rate at which M causes transitions in which $\mathbf{k}'$ is scattered into the solid angle $\mathrm{d}\Omega'$ is

$$R = \frac{2\pi}{\hbar}|M|^2\rho_{\mathrm{F}},$$

where $$\rho_{\mathrm{F}} = \frac{k'^2}{\pi hc}\frac{V\mathrm{d}\Omega'}{4\pi}\Big/\left[\frac{\partial E_{\mathrm{F}}}{\partial(\hbar ck')}\right]_{l'=\mathrm{const.},\, E_{\mathrm{F}}=E_{\mathrm{O}}}$$

(cf. Heitler 1954 p. 214). As in §5 we shall only retain terms of order v/c, where v is the kink velocity. We have

$$v/c = K/k_{\mathrm{C}},$$

where $$k_{\mathrm{C}} = 2\pi/\lambda_{\mathrm{C}} = mc/\hbar,$$

and λ_{C} is the appropriate 'Compton wavelength'. We shall also ignore quantities of higher order than k/k_{C}.

To this order,

$$\left.\begin{aligned}\Delta &= \frac{ll'}{mc^2}\{1+(v/c)\,(l+l')\},\\ \frac{k'}{k} &= 1+(v/c)(l'-l),\\ \rho_{\mathrm{F}} &= \frac{k^2}{\pi hc}\frac{V\mathrm{d}\Omega'}{4\pi}\{1+(v/c)\,(3l'-2l)\},\end{aligned}\right\} \tag{6·7}$$

and so $$R = \tfrac{5}{2}\sigma_0 c\frac{n_{\mathbf{k}}(n_{\mathbf{k}'}+1)}{V}\Phi_{\mathbf{k}}^2\Phi_{\mathbf{k}'}^2\frac{\mathrm{d}\Omega'}{4\pi}\{1+(v/c)\,(4l'+l)\}. \tag{6·8}$$

To find the scattering cross-section of the kink we put $n_{\mathbf{k}'} = 0$ and equate the scattered energy $R\hbar ck$ to the product of $\mathrm{d}\sigma$ and the incident energy flux $(n_{\mathbf{k}}\hbar ck)c/V$. This gives

$$\mathrm{d}\sigma = \tfrac{5}{2}\sigma_0\Phi_{\mathbf{k}}^2\Phi_{\mathbf{k}'}^2\{1+(v/c)\,(4l'+l)\}\frac{\mathrm{d}\Omega'}{4\pi}, \tag{6·9}$$

$$\sigma = \sigma_0\Phi_{\mathbf{k}}^2(1+lv/c), \tag{6·10}$$

which for $v = 0$ agrees with (4·7).

According to (6·7) the transition (6·6) results in a change

$$\hbar\Delta K = \hbar(k_x - k'_x) = \hbar k(l' - l)(1 + vl'/c) \tag{6·11}$$

in the kink momentum. Thus if only the wave $\mathbf{k}$ is present ($n_{\mathbf{k}'} = 0$) the force on the kink is

$$F = \sum_{\mathbf{k}'} R\hbar\Delta K = -\tfrac{5}{2}\sigma_0 \frac{cn_{\mathbf{k}}}{V}\hbar k \sum_{\mathbf{s}'} \int \frac{X\,\mathrm{d}\Omega'}{4\pi}, \tag{6·12}$$

where $$X = \Phi_{\mathbf{k}}^2 \Phi_{\mathbf{k}'}^2 (l' - l)\{1 + (v/c)(5l' + l)\}.$$

When the summations over $\mathbf{s}'$ and $\mathrm{d}\Omega'$ are carried out (see the appendix), (6·12) is found to be the same as the classical result (5·29) with the value

$$\epsilon = n_{\mathbf{k}}\hbar kc/V$$

for the energy density of the wave.

To calculate the retarding force in an isotropic flux we take the momentum change (6·11) for each possible transition, multiply by the appropriate rate (6·8) and sum over all transitions. At the same time we replace the $n_{\mathbf{k}}$ by their average values $\bar{n}_{\mathbf{k}}$ It is convenient to take together the transition (6·6) and the inverse transition in which $n_{\mathbf{k}} \to n_{\mathbf{k}} + 1$, $n_{\mathbf{k}'} \to n_{\mathbf{k}'} - 1$; as they involve equal and opposite momentum changes they give terms of opposite sign. Thus

$$F = -\tfrac{5}{2}\sigma_0 c \sum_{\mathbf{s}} \sum_{\mathbf{k}} \tfrac{1}{2} N\hbar k \int \frac{X\,\mathrm{d}\Omega'}{4\pi}, \tag{6·13}$$

where $$N = \bar{n}_{\mathbf{k}}(\bar{n}_{\mathbf{k}'} + 1) - (\bar{n}_{\mathbf{k}} + 1)\bar{n}_{\mathbf{k}'}.$$

The factor $\tfrac{1}{2}$ prevents the same transition being counted twice. We assume that $\bar{n}_{\mathbf{k}}$ depends only on $|\mathbf{k}|$, so that we have

$$N = \bar{n}_{\mathbf{k}} - \bar{n}_{\mathbf{k}'} = -k(v/c)(l' - l)\,\mathrm{d}\bar{n}_{\mathbf{k}}/\mathrm{d}k$$

by (6·7). Since N itself is of order v/c the term of order v/c in X may be neglected. We may put

$$\sum_{\mathbf{k}} (\ldots) = \frac{V}{(2\pi)^3} \int (\ldots)\, k^2\,\mathrm{d}k\,\mathrm{d}\Omega$$

and so the energy density is

$$\mathscr{E}_{\mathrm{t}} = \sum_{\mathbf{s}} \sum_{\mathbf{k}} \bar{n}_{\mathbf{k}}\hbar ck/V = \frac{1}{\pi^2} \int \hbar c\bar{n}_{\mathbf{k}} k^3\,\mathrm{d}k \tag{6·14}$$

if we exclude the zero-point terms. Using the results in the appendix we obtain

$$F = \tfrac{2}{35}\sigma_0 \Psi_{\mathrm{t}} \frac{v}{c} \int \frac{\hbar c}{\pi^2} \frac{\mathrm{d}\bar{n}_{\mathbf{k}}}{\mathrm{d}k} k^4\,\mathrm{d}k.$$

An integration by parts converts this to

$$F = -\tfrac{8}{35}\sigma_0 \Psi_{\mathrm{t}} \mathscr{E}_{\mathrm{t}} v/c,$$

with the value (6·14) for $\mathscr{E}_{\mathrm{t}}$. If the distribution $\bar{n}_{\mathbf{k}} = \bar{n}_{\mathbf{k}}(k)$ has a sharp cut-off at the Debye limit we appear to have also a contribution from the integrated part. However, it is easy to see that this is an artefact arising from the inclusion of transitions for which $\mathbf{k}$ or $\mathbf{k}'$ would lie outside the Debye sphere. We have thus recovered (5·30) with the additional information that $\mathscr{E}_t$ is not to include the zero point energy.

The formalism of this section also provides a convenient way of correcting the previous results to allow for the finite width of the kink. An equivalent calculation could, of course, be carried out classically. In writing the interaction energy in the form (6·3) we in effect assumed that the kink had the rectangular form of figure 1 (*b*) and that the difference in the stress at points such as *C*, *E* which are separated by the kink 'height' *a* could be neglected. In fact the interaction energy is the product of *b* and the resolved shear stress integrated over the tapered loop *CABDE* of figure 1 (*a*). We shall continue to ignore the variation across the height of the kink; this is reasonable except close to the Debye limit. On the other hand, if *w* is a fairly large multiple of *a* we cannot ignore the variation of stress across the width of the kink. Thus a better approximation to the interaction energy is

$$H_{\text{int.}} = \int_P^Q b\beta_i n_j p_{ij}(x', 0, 0)\, \zeta(x')\, \mathrm{d}x', \tag{6·15}$$

where *P*, *Q* are the points where the dislocation line cuts the boundary of the normalization volume *V* and $\zeta = \zeta(x)$ is the function which describes the shape of the kink (§2). The position of the kink has now to be specified by giving the co-ordinate *x* of some identifiable point on it, say the point $(x, \zeta = \frac{1}{2}a)$ where it crosses the maximum of the potential $\mathscr{V}$. For the sharp kink of figure 1 (*b*), $\zeta(x') = aH(x-x')$ and (6·15) reduces to (6·3). It is easy to show that for a kink of finite width we have simply to replace $\Phi_{\mathbf{k}}$ in (6·4) by $\Phi_{\mathbf{k}}\mathscr{F}(k_x)$, where

$$\mathscr{F}(k_x) = \int_{-\infty}^{\infty} a^{-1}\zeta'(x') \exp(\mathrm{i}k_x x')\, \mathrm{d}x'.$$

For a sharp kink $a^{-1}\zeta'(x)$ is a delta function and $\mathscr{F} = 1$. Generally $a^{-1}\zeta'(x)$ is a bell-shaped curve enclosing unit area, and we can regard it as a distribution function showing how the smooth kink is built up from a succession of infinitesimal sharp kinks.

For the potentials (2·5) and (2·6) we have

$$\mathscr{F}(k_x) = \operatorname{sech}(\tfrac{1}{2}wk_x),$$

and

$$\mathscr{F}(k_x) = \tfrac{1}{4}\pi wk_x \operatorname{cosech}(\tfrac{1}{4}\pi wk_x),$$

respectively. Either of these forms can be represented roughly by

$$\left.\begin{aligned}\mathscr{F}(k_x) &= 1\ (|k_x| < k_{\mathrm{m}})\\ &= 0\ (|k_x| > k_{\mathrm{m}})\end{aligned}\right\} \tag{6·16}$$

with, say,

$$k_{\mathrm{m}} \simeq 3/w,$$

though in fact the cut-off is not particularly sharp.

To correct the differential cross-section we have only to multiply (6·9) by the factor

$$\mathscr{F}^2(kl)\,\mathscr{F}^2(kl'), \tag{6·17}$$

but a difficult integration would be needed to find the correction to the corresponding total cross-section (6·10). Similarly, the factor (6·17) must be attached to *X* in the

16-2

expression (6·13) for the total retarding force. We shall only consider the special case where (i) the material is incompressible, (ii) the dislocation is a pure screw, (iii) the temperature is so high that all phonon oscillators are equally excited, and (iv) $\mathscr{F}$ has the form (6·16). The total retarding force can then be found by summing (5·29) over all waves which have $|k_x| < k_m$. A straightforward calculation shows that (5·30) must be multiplied by the factor

$$\mathscr{C}(\kappa) = \tfrac{15}{128}[20\kappa + 47\kappa^3(\ln\kappa + \tfrac{1}{6}) - \tfrac{223}{10}\kappa^5 + 3\kappa^7], \tag{6·18}$$

where

$$\kappa = k_m/k_D \tag{6·19}$$

and $ck_D/2\pi$ is the Debye frequency. When κ is small (that is, when the kink is many atomic spacings wide) we have $\mathscr{C}(\kappa) = \frac{75}{32}\kappa$. In this case the only part of the Debye sphere which contributes is a diametral slab of thickness $2k_m$ and so we might have expected to have $\mathscr{C}(\kappa) = 2\pi\kappa/(\frac{4}{3}\pi) = \frac{3}{2}\kappa$. The numerical coefficient is larger because waves incident nearly perpendicular to the dislocation ($l_x \sim 0$) are particularly effective since in (5·29) the term in Ψ_t is not offset by the term in l_x^2.

From what is known about the physical parameters of real kinks it seems likely that the coupling between kinks and lattice vibrations is too strong for elementary perturbation theory to be strictly applicable. However, this does not affect our use of it to confirm and extend the classical results of §5.

7. Dislocations

In this section we consider some points concerning dislocations which are closely related to the foregoing calculations. The results of §2 can easily be adapted to give the radiation from a dislocation oscillating as a rigid whole. In the expression (3·2) for the displacement due to the elementary oscillating loop put $a = dx$ and integrate with respect to x from $-\infty$ to $+\infty$. The result is the displacement due to a loop in the form of an infinite strip lying in the xz plane and parallel to the x axis. It has a variable width $\xi = \xi_0 \cos\omega t$ and its Burgers vector is directed along the z axis. Its elastic field is clearly the same as the variable part of the field of an infinite edge dislocation oscillating with amplitude ξ_0.

The field of the oscillating dislocation can be found from (3·2) with $F_0 = \mu b\xi_0$ provided that in the equations defining w, χ we replace $r^{-1}\exp(-ikr)$ by

$$\int_{-\infty}^{\infty} \frac{\exp\{-ik(r^2+x^2)^{\frac{1}{2}}\}}{(r^2+x^2)^{\frac{1}{2}}}\,dx = -iH_0^{(2)}(kr).$$

There is now a state of plane strain in the yz plane, and r stands for $(y^2+z^2)^{\frac{1}{2}}$. In place of (3·7) we have

$$\{u''_y, u''_z\} = -\tfrac{1}{16}b\xi_0\omega^2c_t^2(c_t^{-4}+c_l^{-4})\{z, y\}.$$

Equation (3·3) is still valid if we interpret it two-dimensionally, that is if $F_0 = \mu b\xi_0$ and $\delta(\mathbf{r}) = \delta(y)\,\delta(z)$. The rate of radiation is found as before from the integral (3·8), taken now over a slab of unit thickness. The rate of radiation per unit length is

$$\mathscr{S} = \tfrac{1}{16}\mu b^2\xi_0^2 c_t^2\left(\frac{1}{c_t^4}+\frac{1}{c_l^4}\right)\omega^3. \tag{7·1a}$$

An oscillating screw dislocation can be treated in the same way, starting once more from the elementary loop and this time integrating with respect to z. In place of (7·1 a) we get

$$\mathscr{S} = \tfrac{1}{16}\mu b^2 \xi_0^2 c_{\mathrm{t}}^{-2} \omega^3. \tag{7·1b}$$

A value given previously (Eshelby 1949) was too small by a factor 2. Equation (7·1 b) agrees with Donth's (1957) value and with Nabarro's (1951 b) equation

$$F = \{-m(\omega) + \mathrm{i}R(\omega)\}\,\omega^2 \xi, \tag{7·2}$$

which connects applied force F and displacement ξ when both vary as $\exp(\mathrm{i}\omega t)$. If ω is not too near the Debye frequency R has the constant value

$$R = \tfrac{1}{8}\rho b^2 = \tfrac{1}{8}\mu b^2/c_{\mathrm{t}}^2, \tag{7·3}$$

and the rate at which the force does work on the dislocation is precisely (7·1 b).

The real and imaginary parts of the coefficient of $\omega^2\xi$ in (7·2) are connected by the dispersion relation

$$m(\omega) = m(\infty) + \frac{2}{\pi}\int_0^\infty \frac{\omega' R(\omega') - \omega R(\omega)}{\omega'^2 - \omega^2}\, \mathrm{d}\omega'.$$

$m(\infty)$ may be taken to be the value of m when the frequency is so high that the disturbance is confined to the core of the dislocation. If we take R to be equal to (7·3) up to a cut-off frequency $\omega_{\mathrm{max.}}$ and zero beyond it we have

$$m(\omega) = m(\infty) + \frac{1}{4\pi}\rho b^2 \ln(1 + \omega_{\mathrm{max.}}/\omega)$$

or for not too high frequencies simply

$$m(\omega) = \frac{\rho b^2}{4\pi} \ln \frac{\omega_{\mathrm{max.}}}{\omega},$$

which gives the well-known logarithmic dependence with the correct coefficient.

This tells us nothing new. But we can apply the same argument to the edge dislocation where the resistive term is known but the calculation of the mass is a considerable undertaking (Weertman 1961). The only difference is in the coefficient R; we evidently have

$$\frac{m_{\mathrm{edge}}}{m_{\mathrm{screw}}} = \frac{\mathscr{S}(\omega)_{\mathrm{edge}}}{\mathscr{S}(\omega)_{\mathrm{screw}}} = 1 + \frac{c_{\mathrm{t}}^4}{c_{\mathrm{l}}^4},$$

if the magnitude of the Burgers vector and the type of motion is the same in each case. This agrees with Weertman's calculation for uniformly moving dislocations.

We saw in § 5 that omission of higher-order terms in the radiation reaction reduced the expression for the retarding force by 50 % for a free charge and reversed its sign for a constrained charge. For a free screw dislocation a corresponding omission results in a reduction of precisely a hundred percent. (By a free dislocation is meant one on which the hypothetical Lorentz force can act.) This explains why Nabarro (1951 b) found (incorrectly) that there was no retarding force in this case. One can also look at the matter in the light of (6·9) and (6·10). Generally, motion of the kink alters both the total and differential cross-sections. For a wave incident normal to the dislocation ($l = 0$) the total cross-section is unaffected by the motion, but according to (6·9) the kink scatters energy, and hence also momentum, more strongly forwards than backwards, and the resulting recoil contributes to the retarding force. Things are similar for a dislocation scattering as a whole, and an argument such as Nabarro's, which is based on energy balance and the total cross-section, may not give the correct result.

8. Discussion

According to §§5 and 6 a kink has a scattering cross-section for long waves proportional to the square of its width. It is not, however, in any direct sense an obstacle with an area of order w^2; the w^2 dependence arises from the fact that the mass of a kink is proportional to w^{-1}. In a sense a kink is a point defect, but whereas the scattering cross-section is proportional to the fourth power of the frequency for orthodox point defects, it is independent of frequency for a kink, at least for long waves. This is because a kink scatters by 'fluttering' (Ziman 1960), whereas ordinary point defects are in effect inhomogeneities in the density and elastic constants. Thus at low enough temperatures kinks should contribute to the T^{-3}-law thermal resistance in non-metals and superconductors. Their contribution is likely to be small in comparison with the contributions from the surface of the specimen and grain boundaries.

Lothe & Hirth (1959) have used the value

$$F = \tfrac{1}{10} ab\mathscr{E}\frac{v}{c_t} \tag{8·1}$$

for the retarding force on a kink, that is, Leibfried's expression (6·1) with $\sigma = ab$. It is interesting to compare (8·1) with our result.

We shall suppose that the dislocation is pure screw and that σ_0 has the value (4·8). As the contribution from the longitudinal waves is small we may use (5·30) with $\mathscr{E}_t = \frac{2}{3}\mathscr{E}$. It is also necessary to introduce the correction factor (6·18). For a face-centred cubic lattice the choice (6·19) for the cut-off wave number is nearly equivalent to putting $\kappa = a/w$. Thus

$$F \simeq \tfrac{1}{40} ab\mathscr{E}\frac{v}{c_t} f\left(\frac{w}{a}\right)$$

where

$$f\left(\frac{w}{a}\right) = \frac{w^2}{a^2}\mathscr{C}\left(\frac{a}{w}\right)$$

For large enough w/a, F increases linearly with the width of the kink, since the scattering cross-section is proportional to w^2 and $\mathscr{C}$ is proportional to w^{-1}. If we had followed Lothe (1962) in cutting off all waves with wavelengths less than w we should have had $\mathscr{C} \sim w^{-3}$ and the retarding force would have been proportional to w^{-1}. In fact the effective wave length as seen by the kink is $\lambda/\cos\theta$, where θ is the angle between the direction of the wave and the dislocation line. Thus it seems more reasonable to omit only those waves for which $\lambda/\cos\theta < w$ and this is nearly equivalent to using the value (6·16) for $\mathscr{F}$. From (6·18) we have the following approximate values for f:

w/a	1	2	3	5	8	$x > 8$
$f(w/a)$	1	3	4	10	15	$2{\cdot}3x$

Thus if the width is two or three atomic spacings our result agrees with (8·1), but if with Lothe (1962) we suppose that $w \sim 7b \sim 8a$ it somewhat exceeds it.

Expressions similar to (8·1) have been given for the force on a dislocation moving as a whole (Lothe 1962). The conclusion of § 6 that the zero point energy is to be ignored should be valid in this case also. A similar point arises in Donth's (1957) treatment of the rate of creation of kink pairs by an applied stress. At one stage in the analysis the zero-point and thermal energies enter on the same footing, but thereafter a high-temperature approximation is used. Niblett (1961) observed a Bordoni peak in copper at about one-fifth of the Debye temperature. The zero-point energy is then about ten times the thermal energy, and the role of the zero point term would have to be clarified before Donth's analysis could be applied in such a case.

The writer would like to thank the Director of RIAS, Baltimore, for hospitality and Dr Jens Lothe for helpful comment.

Appendix

For transverse waves we require the sum

$$J = \Phi^2(\mathbf{l}, \mathbf{s}_1) + \Phi^2(\mathbf{l}, \mathbf{s}_2),$$

where
$$\Phi(\mathbf{l}, \mathbf{s}) = (\mathbf{s}\,.\,\mathbf{n})\,(\mathbf{l}\,.\,\boldsymbol{\beta}) + (\mathbf{l}\,.\,\mathbf{n})\,(\mathbf{s}\,.\,\boldsymbol{\beta}),$$

and $\mathbf{s}_1$, $\mathbf{s}_2$ are unit polarization vectors at right angles to one another and to the unit propagation vector $\mathbf{l}$. Put

$$\Phi^2 = \boldsymbol{\phi}\,.\,\mathbf{ss}\,.\,\boldsymbol{\phi}, \quad \text{where} \quad \boldsymbol{\phi} = (\mathbf{l}\,.\,\boldsymbol{\beta})\,\mathbf{n} + (\mathbf{l}\,.\,\mathbf{n})\,\boldsymbol{\beta}.$$

Then
$$J = \boldsymbol{\phi}\,.(\mathbf{s}_1\mathbf{s}_1 + \mathbf{s}_2\mathbf{s}_2)\,.\,\boldsymbol{\phi} = \boldsymbol{\phi}\,.\,(\mathbf{I} - \mathbf{ll})\,.\,\boldsymbol{\phi} = \boldsymbol{\phi}^2 - (\boldsymbol{\phi}\,.\mathbf{l})^2$$

since
$$\mathbf{s}_1\mathbf{s}_1 + \mathbf{s}_2\mathbf{s}_2 + \mathbf{ll} = \mathbf{I},$$

where $\mathbf{I}$ is the identity dyadic. With $\mathbf{n}^2 = \boldsymbol{\beta}^2 = 1$, $\mathbf{n}\,.\,\boldsymbol{\beta} = 0$ we have

$$J = (\mathbf{l}\,.\,\boldsymbol{\beta})^2 + (\mathbf{l}\,.\,\mathbf{n})^2 - 4(\mathbf{l}\,.\,\boldsymbol{\beta})^2\,(\mathbf{l}\,.\,\mathbf{n})^2.$$

We also need certain angular averages over all directions of $\mathbf{l}$, denoted by $\langle \ldots \rangle$. They are most easily found by expanding in Cartesian components and using the relations
$$\langle l_x^2 \rangle = \tfrac{1}{3}, \quad \langle l_x^4 \rangle = 3\langle l_x^2 l_y^2 \rangle = \tfrac{1}{5}, \quad \langle l_x^2 l_z^4 \rangle = 3\langle l_x^2 l_y^2 l_z^2 \rangle = \tfrac{1}{35}.$$

For transverse waves
$$\langle J \rangle = \tfrac{2}{5}, \quad \langle l_x^2 J \rangle = \tfrac{16}{105}\beta_x^2 + \tfrac{2}{21}\beta_y^2. \tag{A. 1}$$

For longitudinal waves $\mathbf{l}$ and $\mathbf{s}$ coincide and we have similarly

$$\langle \Phi^2 \rangle = \tfrac{1}{15}, \quad \langle l_x^2 \Phi^2 \rangle = \tfrac{4}{35}\beta_x^2 + \tfrac{4}{105}\beta_y^2. \tag{A. 2}$$

REFERENCES

Bellomo, E. 1955 *Nuov. Cim.* **2**, 456.
Brailsford, A. D. 1961 *Phys. Rev.* **122**, 778.
Donth, H. 1957 *Z. Phys.* **149**, 111.
Eshelby, J. D. 1949 *Proc. Roy. Soc.* A, **197**, 396.
Frank, F. C. 1949 *Proc. Phys. Soc.* A, **62**, 131.
Frank, F. C. & van der Merwe, J. H. 1949 *Proc. Roy. Soc.* A, **198**, 205.
Heitler, W. 1954 *The quantum theory of radiation.* Oxford: Clarendon Press.
Ivanenko, D. & Sokolov, A. 1950 *Klassicheskaya teoriya polya*, p. 194. Moscow-Leningrad: Gostekhizdat.
Landau L. D. & Lifshitz, E. M. 1958 *Statistical physics*, p. 325. London: Pergamon Press.
Landau L. D. & Lifshitz, E. M. 1951 *The classical theory of fields.* Cambridge, Mass.: Addison-Wesley.
Leibfried, G. 1949 *Z. Phys.* **127**, 344.
Leibfried, G. 1956 In *Deformation and flow of solids* (ed. R. Grammel), p. 25. Berlin: Springer-Verlag.
Lidiard, A. B. 1957 *Encyclopedia of physics*, **20**, 285. Berlin: Springer-Verlag.
Lothe, J. 1960 *Phys. Rev.* **117**, 704.
Lothe, J. 1962 To be published.
Lothe, J. & Hirth, J. P. 1959 *Phys. Rev.* **115**, 543.
Love, A. E. H. 1952 *Theory of elasticity*, p. 167. Cambridge University Press.
Nabarro, F. R. N. 1951*a* *Phil. Mag.* **42**, 1224.
Nabarro, F. R. N. 1951*b* *Proc. Roy. Soc.* A, **209**, 278.
Niblett, D. H. 1961 *J. Appl. Phys.* **32**, 895
Rayleigh, Lord 1896 *Theory of sound*, §378. London: Macmillan.
Seeger, A. 1956 *Phil. Mag.* **1**, 651.
Seeger, A., Donth, H. & Pfaff, F. 1957 *Disc. Faraday Soc.* **23**, 19.
Weertman, H. 1961 In *Response of metals to high velocity deformation* (ed. P. G. Shewmon & V. F. Zackay), p. 205. New York: Interscience.
Ziman, J. M. 1960 *Electrons and phonons*, p. 233. Oxford: Clarendon Press.

J. Mech. Phys. Solids, 1962, Vol. 10, pp. 27 to 34. Pergamon Press Ltd. Printed in Great Britain

THE ENERGY AND LINE TENSION OF A DISLOCATION IN A HEXAGONAL CRYSTAL

By Y. T. CHOU*
Department of Metallurgy, University of Cambridge

and J. D. ESHELBY
Department of Physical Metallurgy, University of Birmingham

(*Received* 10*th July*, 1961)

SUMMARY

CERTAIN results of the theory of dislocations in isotropic materials are extended to the case of a dislocation lying in the basal plane of a hexagonal crystal. Expressions are found for the energy of a circular loop and for the line tension of a dislocation. Numerical results are presented for graphite.

1. INTRODUCTION

NABARRO (1952), KRÖNER (1958) and KROUPA (1960) have discussed the energy of a circular dislocation loop in an isotropic material. It is not possible to give an explicit solution of the same problem for a general anisotropic material. However, in the case of a hexagonal crystal the elastic constants are rotationally invariant with respect to the hexagonal axis and an analytic solution can be given (Section 2).

In Section 3 we obtain an expression for the line tension of a dislocation line lying in the basal plane of a hexagonal crystal. It may be of use in determining the stacking fault energy in such highly anisotropic materials as graphite (Section 4).

2. THE ENERGY OF A CIRCULAR LOOP

We consider a circular dislocation of radius a lying in the basal plane of a hexagonal crystal. Take Cartesian and cylindrical coordinates (x, y, z), (r, θ, z) with origin at the centre of the loop and with z directed along the hexagonal axis.

Because of the presence of the loop the material is in a state of self-stress in which there is a discontinuity in displacement across the plane $z = 0$, distributed according to the law

$$\left.\begin{aligned} \delta\mathbf{u} &= \mathbf{b}, \qquad r < a \\ &= 0, \qquad r > a \end{aligned}\right\} \tag{1}$$

where $\mathbf{b}$ is the Burgers vector of the loop. The total elastic energy is

$$W = \tfrac{1}{2}\int_0^{2\pi} d\theta \int_0^{a-r_0} r dr\, \mathbf{b}\cdot\boldsymbol{\sigma}\cdot\mathbf{n} \tag{2}$$

*On leave from Edgar C. Bain Laboratory for Fundamental Research, United States Steel Corporation.

where σ is the stress tensor at the plane $z = 0$, $\mathbf{n}$ is a unit vector normal to the loop, and r_0 is a small cut-off length corresponding to the core radius of the loop.

Thus we have to find the stress σ corresponding to the displacement discontinuity (1). To do this we shall, in effect, resolve (1) into its Fourier components, find the stress associated with each component and then make a Fourier synthesis to find the total stress.

For the present we shall assume that the Burgers vector is parallel to the plane of the loop and denote it by $\mathbf{b}_{\parallel}$. For this case the tensor σ can be regarded as a vector in the xy plane with components σ_{zx}, σ_{zy}.

For a state of plane strain in which $\delta\mathbf{u}$ has the particular form

$$\delta u_x = \sin mx, \qquad \delta u_y = 0 \tag{3}$$

the corresponding stress on the plane $z = 0$ is

$$\sigma_{zx} = -\tfrac{1}{2} K_e m \sin mx, \qquad \sigma_{zy} = 0 \tag{4}$$

where

$$K_e = (\bar{c}_{13} + c_{13}) \left\{ \frac{c_{44} (\bar{c}_{13} - c_{13})}{c_{33} (\bar{c}_{13} + c_{13} + 2 c_{44})} \right\}^{\frac{1}{2}}, \qquad \bar{c}_{13} = (c_{11} c_{33})^{\frac{1}{2}} \tag{5}$$

(see the Appendix).

It can be shown similarly that for the state of anti-plane strain with

$$\delta u_x = 0, \qquad \delta u_y = \sin mx \tag{6}$$

the stress on $z = 0$ is

$$\sigma_{zx} = 0, \quad \sigma_{zy} = -\tfrac{1}{2} K_s m \sin mx \tag{7}$$

where

$$K_s = \{\tfrac{1}{2} c_{44} (c_{11} - c_{12})\}^{\frac{1}{2}}. \tag{8}$$

Because the material is isotropic with respect to rotation about the z-axis the direction of the x-axis in (3) or (6) is arbitrary. Hence more generally, if

$$\delta\mathbf{u} = (\mathbf{d}_{\parallel} + \mathbf{d}_{\perp})\, e^{i\mathbf{m}\cdot\mathbf{r}} \tag{9}$$

where $\mathbf{d}_{\parallel}$ and $\mathbf{d}_{\perp}$ are parallel and perpendicular to $\mathbf{m}$, the corresponding stress is

$$\sigma = -\tfrac{1}{2} (K_e \mathbf{d}_{\parallel} + K_s \mathbf{d}_{\perp})\, m\, e^{i\mathbf{m}\cdot\mathbf{r}}. \tag{10}$$

For the circular dislocation loop the displacement discontinuity (1) can be expressed by Weber's discontinuous integral (Watson 1922) in the form

$$\begin{aligned} \delta\mathbf{u} &= \mathbf{b}_{\parallel} \int_0^\infty a J_0(rm) J_1(am)\, dm \\ &= \frac{\mathbf{b}_{\parallel}}{2\pi} \int \frac{a}{m} J_1(am)\, e^{i\mathbf{m}\cdot\mathbf{r}}\, dS \end{aligned} \tag{11}$$

where J_0 and J_1 are zeroth- and first-order Bessel functions, $dS = m dm\, d\theta$ and θ is the angle between $\mathbf{m}$ and $\mathbf{r}$. By resolving $\mathbf{b}_{\parallel}$ parallel and perpendicular to $\mathbf{m}$ and applying (9) to each element of the integrand of (11) we can find σ as a function of position $\mathbf{r}$. Actually all we need to evaluate (2) is $\sigma \cdot \mathbf{b}_{\parallel}$. A little geometry gives

$$\sigma \cdot \mathbf{b}_{\parallel} = \frac{ab^2_{\parallel}}{4\pi} \int J_1(am) \{K_e \cos^2\theta' + K_s \sin^2\theta'\}\, e^{i\mathbf{m}\cdot\mathbf{r}}\, dS \tag{12}$$

where θ' is the angle between $\mathbf{m}$ and $\mathbf{b}_{\parallel}$. With the help of the relations

$$J_n(t) = \frac{1}{\pi i^n} \int_0^\pi e^{it\cos\phi} \cos n\phi \, d\phi \tag{13}$$

and

$$\int_0^{2\pi} e^{it\cos\phi} \sin n\phi \, d\phi = 0 \tag{14}$$

we find

$$\boldsymbol{\sigma} \cdot \mathbf{b}_{\parallel} = \frac{ab^2_{\parallel}}{4} \int_0^\infty m J_1(am) \{(K_e + K_s) J_0(rm) - (K_e - K_s) J_2(rm) \cos 2\theta''\} \, dm \tag{15}$$

where θ'' is the angle between $\mathbf{b}_{\parallel}$ and $\mathbf{r}$. Using Gubler's relation (WATSON 1922)

$$\int_0^\infty J_\mu(am) J_\nu(rm) \, dm = \frac{r^\nu \Gamma\left(\frac{\mu+\nu+1}{2}\right)}{a^{\nu+1} \Gamma(\nu+1) \Gamma\left(\frac{\mu-\nu+1}{2}\right)} \times F\left(\frac{\mu+\nu+1}{2}, \frac{\nu-\mu+1}{2}; \nu+1; \frac{r^2}{a^2}\right), \quad r < a \tag{16}$$

and

$$K(k) = \frac{\pi}{2} F(\tfrac{1}{2}, \tfrac{1}{2}; 1; k^2)$$

$$E(k) = \frac{\pi}{2} F(-\tfrac{1}{2}, \tfrac{1}{2}; 1; k^2)$$

this becomes

$$\boldsymbol{\sigma} \cdot \mathbf{b}_{\parallel} = (K_e + K_s) \frac{b^2_{\parallel}}{2\pi a} \frac{d}{dk} [k K(k)] - (K_e - K_s) \frac{2b^2_{\parallel}}{\pi a} \cos 2\theta'' \, k^2 \frac{d^2}{d(k^2)^2} E(k) \tag{17}$$

where Γ and F are respectively the gamma and hypergeometric functions and K and E are the complete elliptic integrals with modulus $k = r/a$. The second term in (17) contributes nothing to the integral (2). The other term gives

$$W_{\parallel} = (K_e + K_s) b^2_{\parallel} a [K(k_0) - E(k_0)] \tag{18}$$

where $k_0 = (a - r_0)/a$. The only physically significant case is that in which $r_0 \ll a$. Then

$$W_{\parallel} = \tfrac{1}{4} (K_e + K_s) b^2_{\parallel} a \left(\ln \frac{8a}{r_0} - 2\right). \tag{19}$$

In the case of isotropic crystals, (19) reduces to

$$W_{\parallel} = \tfrac{1}{4} \left(\frac{2-\nu}{1-\nu}\right) \mu b^2_{\parallel} a \left(\ln \frac{8a}{r_0} - 2\right) \tag{20}$$

where μ and ν are the shear modulus and Poisson's ratio.

A circular loop of dislocation in which the Burgers vector $\mathbf{b}_{\perp}$ is normal to the plane of the loop (produced for example by the collapse of a penny-shaped vacancy disc on the basal plane) can be treated similarly using the fact that if on the plane $z = 0$

$$\delta u_z = \sin mx \tag{21}$$

then

$$\sigma_{zz} = -\tfrac{1}{2} K_n m \sin mx \tag{22}$$

where

$$K_n = (\bar{c}_{13} + c_{13}) \left\{ \frac{c_{44}(\bar{c}_{13} - c_{13})}{c_{11}(\bar{c}_{13} + c_{13} + 2c_{44})} \right\}^{\frac{1}{2}} = \left(\frac{c_{33}}{c_{11}}\right)^{\frac{1}{2}} K_e. \tag{23}$$

The distribution of $\sigma_{zz} \cdot \mathbf{b}_\perp$ over the circle is given by the first term of (17) with $(K_e + K_s)$ replaced by $2K_n$. The energy is thus

$$W_\perp = \tfrac{1}{2} K_n b^2_\perp a \left(\ln \frac{8a}{r_0} - 2\right). \tag{24}$$

In isotropic crystals (24) reduces to

$$W_\perp = \tfrac{1}{2}\left(\frac{1}{1-\nu}\right) \mu b^2_\perp a \left(\ln \frac{8a}{r_0} - 2\right). \tag{25}$$

For the general case in which $\mathbf{b} = (b_x, b_y, b_z)$ the energy may be found by putting $b^2_\parallel = b_x^2 + b_y^2$ in (19) and $b^2_\perp = b_z^2$ in (24) and adding the results, since there is no interaction between $(\delta u_x, \delta u_y)$ and σ_{zz}, nor between δu_z and $(\sigma_{zx}, \sigma_{zy})$.

The integrals (12) and (15) exist only in the sense of the theory of generalized functions. However, the validity of our results may be verified by inserting a convergence factor $\exp(-lm)$ in the integrand of (11) and repeating the calculation. On letting l approach zero we recover (17). Insertion of the convergence factor in (11) is actually equivalent to rounding off the sharp corners of the distribution (1) so as to simulate the elastic field of a dislocation with a finite width of order l.

3. The Line Tension

The line energy of a straight dislocation lying in the basal plane of a hexagonal crystal, with its Burgers vector in the plane, is

$$E(\theta) = \frac{b^2_\parallel}{4\pi} (K_e \sin^2\theta + K_s \cos^2\theta) \ln \frac{R}{r_0} \tag{26}$$

per unit length of dislocation, where θ is the angle between the dislocation line and the Burgers vector. Equation (26) follows from the calculations of Foreman (1955). It can also be derived by the method of Section 2, taking the displacement discontinuity to be $\delta\mathbf{u} = \mathbf{b}_\parallel, x > 0$; $\delta\mathbf{u} = 0, x < 0$ in place of (1). R is a somewhat ill-defined length depending on the surroundings of the dislocation.

One can also introduce the concept of the line tension of a dislocation, T say, defined as the product of the radius of curvature of a segment of dislocation and the normal force (per unit length) acting on it. In particular, if the force is due to a stacking fault bounded by the dislocation we have the much-used relation

$$\gamma = T/\rho b_\parallel \tag{27}$$

where γ is the stacking fault energy and ρ is the radius of curvature of the dislocation.

The line energy $E(\theta)$ and the line tension $T(\theta)$ are not equal. On the contrary they are related by

$$T(\theta) = E(\theta) + d^2 E(\theta)/d\theta^2 \tag{28}$$

(see DE WIT and KOEHLER 1959). The same relation connects the edge energy and effective edge tension of a two-dimensional crystal nucleus (BURTON, CABRERA and FRANK 1951) or the surface energy and surface tension of a (cylindrical) crystal surface or grain boundary (HERRING 1951, 1953). HERRING's analysis of the equilibrium of grain boundaries suggests that it may be useful to introduce yet another quantity, the 'vector line tension' $\mathbf{T}$ which must be applied to the (imaginary) free end of a dislocation to hold it in equilibrium. HERRING shows that where several grain boundaries meet the equilibrium condition

$$\Sigma \mathbf{T} = 0$$

must be satisfied, where

$$\mathbf{T} = \mathbf{t}\, E(\theta) + \mathbf{n}\, dE(\theta)/d\theta \tag{29}$$

and $\mathbf{t}$, $\mathbf{n}$ are unit vectors tangential and normal to the grain boundary. The same relation governs the equilibrium of a set of dislocations in the same plane where they meet at a node. It may be established by the argument which HERRING used for grain boundaries, or by varying the end-points in the variational treatment which leads to (28).

In a sense (29) is the basic relation. It states that to keep a unit length of dislocation in equilibrium we must apply a longitudinal tension $E(\theta)$ to stop it contracting and a couple $dE(\theta)/d\theta$ to stop it rotating into an orientation of lower energy. If the segment has a radius of curvature ρ there will be a force $d\mathbf{T}/ds$ on it with normal and tangential components

$$\mathbf{n} \cdot d\mathbf{T}/ds = \rho \left[E(\theta) + d^2 E(\theta)/d\theta^2\right], \qquad \mathbf{t} \cdot d\mathbf{T}/ds = 0 \tag{30}$$

as may be verified by using the equations

$$ds/\rho = d\theta = \mathbf{n} \cdot d\mathbf{t} = -\,\mathbf{t} \cdot d\mathbf{n}$$

which relate the arc-length s of a curve to θ, ρ, $\mathbf{n}$, $\mathbf{t}$. Evidently (30) is equivalent to (28).

From (26) and (28) we have

$$T(\theta) = \frac{b^2_{\parallel}}{4\pi} \{K_e \sin^2\theta + K_s \cos^2\theta + 2(K_e - K_s)\cos 2\theta\} \ln \frac{R}{r_0}. \tag{31}$$

Thus in general $T(\theta)$ and $E(\theta)$ are different. However, for a pure screw or pure edge dislocation they coincide,

$$T_e = E_e = \frac{K_e}{4\pi} b^2_{\parallel} \ln \frac{R}{r_0} \tag{32}$$

$$T_s = E_s = \frac{K_s}{4\pi} b^2_{\parallel} \ln \frac{R}{r_0}. \tag{33}$$

For a dislocation with its Burgers vector perpendicular to the basal plane E is independent of θ and

$$T_n = E_n = \frac{K_n}{4\pi} b^2_{\perp} \ln \frac{R}{r_0}. \tag{34}$$

If the dislocation has an arbitrary Burgers vector the line energy or line tension is the sum of (26) and (34) or (31) and (34).

The expressions (26) and (31) may be applied to a small element of a dislocation if the value of R is chosen suitably. If the element forms part of a loop we expect R to be of the order of the linear dimensions of the loop. STROH (1954) has given a careful discussion of this point.

It is interesting to note that for the loop of Section 2 (with $\mathbf{b}_\perp = 0$) the mean energy per unit length is the arithmetic mean of (32) and (33) (cf. NABARRO 1952).

4. A NUMERICAL EXAMPLE: GRAPHITE

As an illustrative example we consider the case of a single crystal of graphite. According to the theoretical calculations of BOWMAN and KRUMHANSL (1958) the elastic constants are

$$c_{11} = 4c_{12} = 1{\cdot}13 \times 10^{13}, \qquad c_{44} = 2{\cdot}3 \times 10^{10}, \qquad c_{33} \geqslant 1{\cdot}8 \times 10^{11}$$

dyn/cm². K_s can be calculated from these values, but the value of c_{13} must also be known to calculate K_e and K_n. Recently, BAKER, CHOU and KELLY (unpublished work) have estimated K_e to be $4{\cdot}24 \times 10^{11}$ dyn/cm², based on the measurements of stacking fault widths. Using this value of K_e and taking $c_{33} = 1{\cdot}8 \times 10^{11}$* they deduced the missing constant c_{13} to be $7{\cdot}7 \times 10^{11}$ dyn/cm², in reasonable agreement with the value $6{\cdot}4 \times 10^{11}$ originally reported by RILEY (1945) and later corrected to $9{\cdot}4 \times 10^{11}$ by DAVIDSON and LOSTY (1958).

Values for $W_{\|}$, $W_\perp$, T_e, T_s and T_n based on these figures are listed in the second column of Table 1. The values in the remaining columns were calculated from the isotropic formulas using values for the shear modulus μ and Poisson's ratio ν derived in two different ways. The values

$$\mu = 2{\cdot}1 \times 10^{12}\ \text{dyn/cm}^2, \qquad \nu = 0{\cdot}25$$

TABLE 1

	Anisotropic	*Isotropic*		
		Voigt	*Reuss*	
$W_{\|} \Big/ \frac{b^2_{\|} a}{4}\left(\ln\frac{8a}{r_0} - 2\right)$	0·74	4·8	0·12	$\times 10^{12}$ ergs/cm³
$W_\perp \Big/ \frac{b^2_\perp a}{4}\left(\ln\frac{8a}{r_0} - 2\right)$	0·11	5·5	0·14	$\times 10^{12}$ ergs/cm³
$T_e \Big/ \frac{b^2_{\|}}{4\pi}\ln\frac{R}{r_0}$	0·42	2·8	0·071	$\times 10^{12}$ dyn/cm²
$T_s \Big/ \frac{b^2_{\|}}{4\pi}\ln\frac{R}{r_0}$	0·31	2·1	0·048	$\times 10^{12}$ dyn/cm²
$T_n \Big/ \frac{b^2_\perp}{4\pi}\ln\frac{R}{r_0}$	0·054	2·8	0·071	$\times 10^{12}$ dyn/cm²

*The possible error introduced by the choice of the minimum value has relatively little effect on our results.

used in the column headed 'Voigt' were found by averaging the elastic constants c_{ij}. The figures in the column headed 'Reuss' were calculated from the values

$$\mu = 4{\cdot}8 \times 10^{10}\,\mathrm{dyn/cm^2}, \qquad \nu = 0{\cdot}32$$

obtained by averaging the compliance moduli s_{ij}.

The table shows clearly that, for graphite at least, the use of anisotropic formulas is essential.

Appendix

The following calculation is based on NABARRO's treatment (1947) of the isotropic case. For a state of plane strain in the xz plane the stresses may be written in the form

$$\sigma_{xx} = \frac{\partial^2 \chi}{\partial z^2}, \qquad \sigma_{zz} = \frac{\partial^2 \chi}{\partial x^2}, \qquad \sigma_{zx} = -\frac{\partial^2 \chi}{\partial z\, \partial x} \tag{A1}$$

where χ is a suitable stress function. Stress and strain are related (see ESHELBY 1949) by

$$\begin{aligned} e_{xx} &= \bar{s}_{11}\sigma_{xx} + \bar{s}_{13}\sigma_{zz} + \bar{s}_{15}\sigma_{zx} \\ e_{zz} &= \bar{s}_{31}\sigma_{xx} + \bar{s}_{33}\sigma_{zz} + \bar{s}_{35}\sigma_{zx} \\ e_{zx} &= \bar{s}_{51}\sigma_{xx} + \bar{s}_{53}\sigma_{zz} + \bar{s}_{55}\sigma_{zx} \end{aligned} \tag{A2}$$

where, for a hexagonal material,

$$\bar{s}_{11} = \frac{s^2_{11} - s^2_{12}}{s_{11}}, \qquad \bar{s}_{13} = \bar{s}_{31} = \frac{(s_{11} - s_{12})\,s_{13}}{s_{11}}, \qquad \bar{s}_{33} = \frac{s_{11}s_{33} - s^2_{13}}{s_{11}}$$

$$\bar{s}_{55} = s_{44}, \qquad \bar{s}_{15} = \bar{s}_{51} = \bar{s}_{35} = \bar{s}_{53} = 0. \tag{A3}$$

Inserting equations (A1) and (A2) into the compatibility equation

$$\frac{\partial^2 e_{zz}}{\partial x^2} + \frac{\partial^2 e_{xx}}{\partial z^2} = \frac{\partial^2 e_{zx}}{\partial z\,\partial x} \tag{A4}$$

we have

$$\left(\frac{\partial^2}{\partial x^2} + \alpha_1 \frac{\partial^2}{\partial z^2}\right)\left(\frac{\partial^2}{\partial x^2} + \alpha_2 \frac{\partial^2}{\partial z^2}\right)\chi = 0 \tag{A5}$$

where

$$\alpha_1\alpha_2 = \bar{s}_{11}/\bar{s}_{33}, \qquad \alpha_1 + \alpha_2 = (2\bar{s}_{13} + \bar{s}_{55})/\bar{s}_{33}.$$

A solution satisfying (A5) and the boundary conditions

$$\sigma_{zz}(z = 0) = 0 \quad \text{and} \quad \sigma_{ij}(z = \infty) = 0$$

is

$$\chi = C\left(e^{-m\alpha_1^{-\frac{1}{2}}z} - e^{-m\alpha_2^{-\frac{1}{2}}z}\right)\cos mx. \tag{A6}$$

The corresponding stresses are

$$\begin{aligned} \sigma_{xx} &= C\left(\alpha_1^{-1}e^{-m\alpha_1^{-\frac{1}{2}}z} - \alpha_2^{-1}e^{-m\alpha_2^{-\frac{1}{2}}z}\right)m^2\cos mx \\ \sigma_{zz} &= -C\left(e^{-m\alpha_1^{-\frac{1}{2}}z} - e^{-m\alpha_2^{-\frac{1}{2}}z}\right)m^2\cos mx \\ \sigma_{zx} &= -C\left(\alpha_1^{-\frac{1}{2}}e^{-m\alpha_1^{-\frac{1}{2}}z} - \alpha_2^{-\frac{1}{2}}e^{-m\alpha_2^{-\frac{1}{2}}z}\right)m^2\sin mx \end{aligned} \tag{A7}$$

On the boundary plane $z = 0$ they reduce to

$$\begin{aligned} \sigma_{xx}(z=0) &= C(\alpha_1^{-1} - \alpha_2^{-1})\, m^2 \cos mx \\ \sigma_{zz}(z=0) &= 0 \\ \sigma_{zx}(z=0) &= -\,C(\alpha_1^{-\frac{1}{2}} - \alpha_2^{-\frac{1}{2}})\, m^2 \sin mx \end{aligned} \tag{A8}$$

For $z = 0$ we have $\partial u_x/\partial x = \bar{s}_{11}\,\sigma_{xx}$ simply and so the corresponding displacement u_x on $z = 0$ is

$$u_x(z=0) = C\,\bar{s}_{11}(\alpha_1^{-1} - \alpha_2^{-1})\, m \sin mx. \tag{A9}$$

Choosing $C = \frac{1}{2}\left[\bar{s}_{11}(\alpha_1^{-1} - \alpha_2^{-1})\, m\right]^{-1}$ we see that to the displacement

$$u_x(z=0) = \tfrac{1}{2}\sin mx \tag{A10}$$

there corresponds the stress

$$\sigma_{zx}(z=0) = -\,\tfrac{1}{2} K_e\, m \sin mx \tag{A11}$$

where

$$\begin{aligned} K_e &= (\alpha_1^{-\frac{1}{2}} - \alpha_2^{-\frac{1}{2}})\left[\bar{s}_{11}(\alpha_1^{-1} - \alpha_2^{-1})\right]^{-1} \\ &= \left[\bar{s}_{11}\left\{(2\bar{s}_{13} + \bar{s}_{55}) + 2(\bar{s}_{11}\bar{s}_{33})^{\frac{1}{2}}\right\}\right]^{-\frac{1}{2}} \\ &= (\bar{c}_{13} + c_{13})\left\{\frac{c_{44}(\bar{c}_{13} - c_{13})}{c_{33}(\bar{c}_{13} + c_{13} + 2c_{44})}\right\}^{\frac{1}{2}} \end{aligned} \tag{A12}$$

and $\bar{c}_{13} = (c_{11}c_{33})^{\frac{1}{2}}$.

If in (A7) we replace z by $|z|$ we obtain a solution in which the stress and displacement fall to zero at both $z = +\infty$ and $z = -\infty$. There is now a discontinuity in u_x across $z = 0$ equal to twice the expression (A10), but the shear stress on $z = 0$ is still given by (A11). This establishes equations (3), (4) of the text. The relations (6), (7) and (21), (22) may be derived similarly.

ACKNOWLEDGMENTS

One of us (Y.T.C.) wishes to thank Professor A. H. COTTRELL, F.R.S., for his interest and encouragement and the Edgar C. Bain Laboratory for Fundamental Research of United States Steel Corporation for financial support.

REFERENCES

BOWMAN, J. C. and KRUMHANSL, J. A. 1958 *J. Phys. Chem. Solids* **6**, 367.
BURTON, W. K., CABRERA, N. and FRANK, F. C. 1951 *Phil. Trans.* A **243**, 299.
DAVIDSON, H. W. and LOSTY, H. H. W. 1958 In *Mechanical Properties of Non-Metallic Brittle Materials*, ed. by W. H. WALTON (Butterworth, London).
DE WIT, G. and KOEHLER, J. S. 1959 *Phys. Rev.* **116**, 1113.
ESHELBY, J. D. 1949 *Phil. Mag.* **40**, 903.
FOREMAN, A. E. J. 1955 *Acta Met.* **3**, 322.
HERRING, C. 1951 In *The Physics of Powder Metallurgy*, ed. by W. E. KINGSTON (McGraw-Hill, New York).
1953 In *Structure and Properties of Surfaces*, ed. by R. GOMER and C. S. SMITH (Chicago University Press).
KRÖNER, E. 1958 *Kontinuumstheorie der Versetzungen und Eigenspannungen* (Springer, Berlin).
KROUPA, F. 1960 *Czech. J. Phys.* B **10**, 284.
NABARRO, F. R. N. 1947 *Proc. Phys. Soc. Lond.* **59**, 256.
1952 *Advanc. Phys.* **1**, 269.
RILEY, D. P. 1945 *Proc. Phys. Soc. Lond.* **57**, 486.
STROH, A. N. 1954 *Proc. Phys. Soc. Lond.* B **67**, 427.
WATSON, G. N. 1922 *Theory of Bessel Functions* (Cambridge University Press).

phys. stat. sol. 2, 1021 (1962)

Department of Physical Metallurgy, University of Birmingham

The Distortion and Electrification of Plates and Rods by Dislocations

By

J. D. ESHELBY

It is shown that some problems concerning the distortion of plates and rods by dislocations can be solved very simply with the help of Colonetti's theorem. The relation between the plastic deformation and the electrical potentials produced by the movement of charged dislocations is also examined.

Es wird gezeigt, daß einige Probleme im Zusammenhang mit der Verzerrung von Platten und Stäben durch Versetzungen auf sehr einfache Weise mit Hilfe des Colonettischen Satzes gelöst werden können. Der Zusammenhang zwischen der plastischen Verformung und den elektrischen Potentialen, verursacht durch die Bewegung der geladenen Versetzungen, wird untersucht.

1. Introduction

KROUPA [1] and SIEMS, DELAVIGNETTE and AMELINCKX [2] have shown that an edge dislocation, with Burgers vector b, running at a height y above the middle surface of a plate of thickness $2c$ causes it to bend through an angle

$$\beta = \frac{3}{4} b (c^2 - y^2)/c^3 \tag{1}$$

in the way indicated in Fig. 1.

Equation (1) was obtained by a detailed elastic calculation. Problems of this kind can also often be solved very simply by a combination of Saint-Venant's principle and the Betti-Rayleigh reciprocal theorem [3], in particular the special case known as Colonetti's theorem [4].

In Fig. 1 make a physical cut AB from the dislocation to the surface and at the same time apply suitable tractions to the surfaces of the cut so that the stresses in the plate are unaltered. This converts the problem into one concerning a notched plate loaded along the surfaces of the notch. Saint-Venant's principle assures us that the elastic strain will be inappreciable a few multiples of c from A. Thus the overall deformation will be, at most, a bending through some angle β about the dislocation as axis[1]).

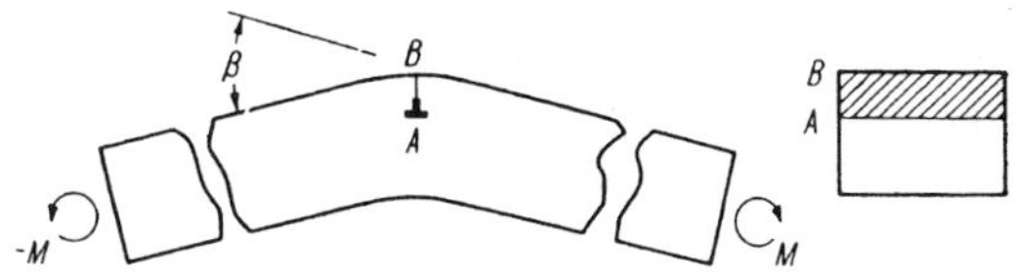

Fig. 1. Bending of a plate by an edge dislocation

[1]) The term 'buckling' used by SIEMS et al. is perhaps misleading, since it suggests the existence of two equilibrium states, one stable, the other unstable. MITCHELL and HEAD [5] have treated a case of buckling by a dislocation.

67*

To find the magnitude of β we start again with the uncut and undislocated plate and apply couples $\pm M$ to a pair of edges as shown in the figure. Create the dislocation just below the surface at B and allow it to move by climb to its final position A. During the movement the stresses in the plate do a certain amount of work E_1 on the dislocation. E_1 is equal to b times the fibre stress in the plate integrated over the part of the cross-section shown shaded in the figure, that is, over the area swept through by the dislocation. Elementary beam theory gives $E_1 = \frac{3}{4} b\, M\, (c^2 - y^2)/c^3$. On the other hand, as the dislocation is introduced the edges of the plate undergo a relative rotation β and the couples do an amount of work $E_2 = \beta\, M$. But the work done on the dislocation must be equal to the work done by the couples. (This is the physical content of Colonetti's theorem). Thus $E_1 = E_2$, or $\beta = E_1/M$, in agreement with (1).

In the next section we treat a rather more general case by the same method. Section 3 discusses the relation between deformation produced by dislocations and the potentials which are developed if they carry electrical charges.

2. Bending and Twisting of Plates and Rods by a Dislocation

Fig. 2a represents part of a uniform rod (or plate) directed along the z axis and having its principal axes coincident with the x and y axes. Let it be required to find the bending produced by a dislocation ABC which lies wholly in the plane $z =$ const and has Burgers vector (b_x, b_y, b_z).

Apply couples $\pm M$ about the x axis to the free ends of the rod. This induces a stress distribution whose only non-zero components is

$$\tau_{zz} = M\, y/I_x \tag{2}$$

where I_x is the geometrical moment of inertia of the cross-section about the x axis. The work required to introduce the dislocation against this stress is the integral of $b_z\, \tau_{zz}$ taken over the area ABCDA.

The work done by the external forces as the dislocation is introduced is $\beta_x\, M$, where β_x is the relative rotation of the ends of the rod about the x axis due to the presence of the dislocation. These two quantities must be equal, and so

$$\beta_x = (b_z/I_x) \int_{\Sigma} y\, dx\, dy \tag{3}$$

where

$$I_x = \int_{\Sigma_0} y^2\, dx\, dy$$

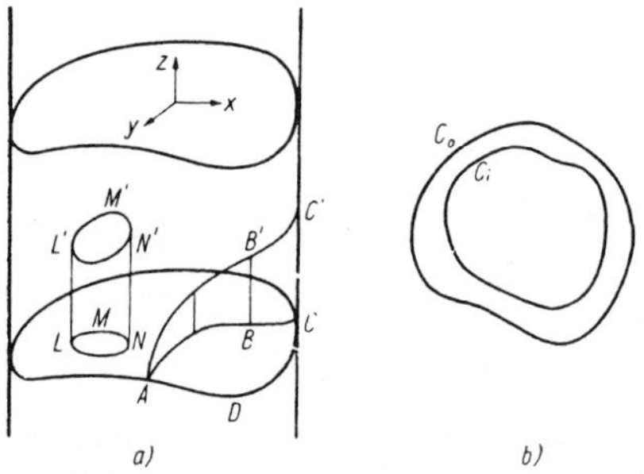

Fig. 2a. Rod or plate containing an arbitary dislocation
Fig. 2b. A hollow whisker

and Σ_0 represents the whole cross-section of the rod.

For the configuration of Fig. 1, (3) gives (1) at once.

Generally there will also be a rotation about the y axis equal to

$$\beta_y = (b_z/I_y) \int_{\Sigma} x\, dx\, dy \tag{4}$$

with

$$I_y = \int_{\Sigma_0} x^2\, dx\, dy$$

and so in all a rotation through $\beta = (\beta_x^2 + \beta_y^2)^{1/2}$ about an axis with direction cosines $(\beta_x/\beta, \beta_y/\beta, 0)$.

If the dislocation forms a closed loop LMN which does not meet the surface there is a rotation given by (3) and (4) where now Σ denotes the area enclosed by LMN.

These results are easily extended to the case where the dislocation does not lie wholly in a plane $z =$ const. Consider an infinitesimal loop L′M′N′ of area dS and normal (n_x, n_y, n_z). To create it an amount of work

$$b_i \, n_j \, \tau_{ij} \, dS \tag{5}$$

must be done against the applied stress [4]. For the stress (2) this reduces to $b_z \, \tau_{zz} \, n_z \, dS$. This is the same as the work required to create the loop LMN, the projection of L′M′N′ on the plane $z =$ const, since τ_{zz} is the same for the two loops and their areas are in the ratio 1 to n_z. The same equality holds if L′M′N′ is a finite loop, as may be seen by dividing it into a collection of elementary loops. Thus the bending due to the loop L′M′N′ is given by (3), (4) with $\Sigma =$ LMN. Similarly the bending due to the dislocation AB′C′ is given by these equations with $\Sigma =$ ABCDA where ABC is the projection of AB′C′ on the plane $z =$ const.

As a simple example consider a dislocation with Burgers vector $(0, 0, b_z)$ along some diameter of a circular rod of radius R. Equation (3) gives $\beta = 8\,b_z/3\,\pi\,R$. Suppose now that half of the dislocation line glides parallel to the axis so that the dislocation becomes a pure screw along the axis, terminated by edge segments running to the surface along oppositely-directed[2]) radii. Since the projection is unaltered β is unchanged, but the bending will obviously now take place in two steps of $\frac{1}{2}\,\beta$ at the points of exit. Thus if an axial screw dislocation leaves a whisker along a radius the whisker will be kinked through an angle $4\,b/3\,\pi\,R$. This agrees with a result already obtained otherwise ([6], second equation below (9)).

In addition to bending the rod the dislocation AB′C′ of Fig. 1b will in general produce a relative rotation of its ends through some angle β_z about the z axis. This angle can be found in the same way as β_x, β_y.

Apply couples $\pm M$ about the z axis to the ends of the rod. They induce a stress whose only non-zero components are

$$\tau_{xz} = \mu\,\alpha_1\,\partial\Psi/\partial y, \quad \tau_{yz} = -\,\mu\,\alpha_1\,\partial\Psi/\partial x \tag{6}$$

where Ψ is a torsion function satisfying

$$\nabla^2\,\Psi + 2 = 0 \tag{7}$$

over the cross-section and vanishing on its boundary [3]. In (6) μ is the shear modulus and

$$\alpha_1 = M/\mu\,S$$

is the twist (rotation per unit length) produced by M. The quantity

$$S = 2 \int_{\Sigma_0} \Psi(x, y)\,dx\,dy$$

is the geometrical torsional rigidity (the physical torsional rigidity with $\mu = 1$).

[2]) Strictly speaking, radii which are oppositely directed in the undislocated rod, or in the final state when the rotation due to the screw segment (cf. section 3) is removed by suitable couples.

When the dislocation is introduced the ends of the rod suffer a relative rotation β_z and the couples do an amount of work $\beta_z M = \beta_z \alpha_1 \mu S$. To find β_z we have to equate $\beta_z M$ to the work done on the dislocation by the stresses (6). This is the integral of (5), that is of

$$\mu \alpha_1 (b_z n_x + b_x n_z) \frac{\partial \Psi}{\partial y} dS - \mu \alpha_1 (b_z n_y + b_y n_z) \frac{\partial \Psi}{\partial x} dS ,$$

taken over any surface spanning the closed curve made up of AB′C′ and any line lying in the surface of the rod and joining C′ to A. The surface integral may be transformed into a line integral round AB′C′A by Stokes's theorem. In this way we obtain

$$\beta_z = \beta_z' + \beta_z''$$

where

$$\beta_z' = (b_z/S) \int \Psi \, dz \tag{8}$$

and

$$\beta_z'' = -(1/S) \int \Psi (b_x dx + b_y dy) . \tag{9}$$

The line-integrals (8) and (9) are to be taken along AB′C′; the closing line from C′ back to A gives no contribution since $\Psi = 0$ at the surface of the rod. For the closed loop L′M′N′ the integrals must be taken round L′M′N′. The second integral (9) may equally well be taken along ABC or LMN, the projections of AB′C′ or L′M′N′ on the plane $z =$ const (Note that $\partial \Psi / \partial z = 0$).

If the dislocation lies wholly in the plane of the cross-section only (9) contributes to β_z. For example, for a circular rod $\Psi = \frac{1}{2} (R^2 - r^2)$, $S = \pi R^4$ and so a dislocation running along a diameters produces a rotation of magnitude $2 b \cos \theta / 3 \pi R$ where θ is the angle between the Burgers vector and the dislocation line.

Consider next a straight segment of screw dislocation of length l running parallel to the axis of the rod and joined to the surface by short segments at the ends. If the straight portion passes through (x_0, y_0) its contribution β_z'' to β_z is $l \alpha$, where

$$\alpha = b_z \Psi (x_0, y_0)/S . \tag{10}$$

If l is large enough the contribution of the end segments may be neglected and the length l suffers a uniform twist (10). This case has been treated elsewhere [7].

Suppose that the screw dislocation now winds up into the form of a uniform helix of radius r_0 with the original dislocation line as axis. According to (8) the effect of this is to replace $\Psi(x_0, y_0)$ in (10) by the average of $\Psi(x, y)$ round a circle of radius r_0 centered on (x_0, y_0). Equation (7) shows that we may write $\Psi(x, y) = \varphi(x, y) - \frac{1}{2} (x^2 + y^2)$ where φ is harmonic. But by a well-known theorem the mean value of a harmonic function over a circle is equal to its value at the centre. Also, it is easy to show that the mean value of $x^2 + y^2$ round the circle is $x_0^2 + y_0^2 + r_0^2$. Thus the twist becomes

$$\alpha = b_z \left\{ \Psi(x_0, y_0) - \frac{1}{2} r_0^2 \right\} \Big/ S . \tag{11}$$

This is identical with the twist due to a straight screw dislocation with a hollow core having a radius r_0 small compared with the dimensions of the cross-section of the rod [7]. (When (11) refers to the helix, however, there is no limitation on the magnitude of r_0).

The helical dislocation also bends the rod. If it contains N turns its projection on z = const is a circle described N times and by (3), (4) each turn gives a contribution $\beta_{x0} = \pi\, b_z\, r_0^2\, y_0/I_x$, $\beta_{y0} = \pi\, b_z\, r_0^2\, x_0/I_y$. If r_0 and the pitch d of the helix are small compared with the dimensions of the cross-section the total bending will be evenly distributed and the rod will have a curvature $\varrho^{-1} = (\beta_{x0}^2 + \beta_{y0}^2)/d$. Under the combined action of this curvature and the twist (10) the centre line of the rod will take the form of a helix of pitch $2\,\pi\,\alpha/(\varrho^{-2} + \alpha^2)$ and radius $\varrho^{-1}/(\varrho^{-2} + \alpha^2)$.

In the above example the original screw dislocation develops an edge component when it winds into a helix, and it is this edge component (with its Burgers vector parallel to the axis) which is responsible for the curvature. Consider now a straight edge dislocation parallel to the axis, and thus with its Burgers vector in a radial direction. It has been suggested (by analogy with the fact that an axial screw dislocation produces twist), that such an edge dislocation would produce curvature in the rod. We can show that this is incorrect. If it existed the curvature could be calculated as follows. Apply bending couples $\pm M$ to the ends of the rod. The work required to introduce the dislocation is equal to M times the relative rotation of the ends, that is, to M times the product of the curvature and the length of the rod. But the stress (2) exerts no force on the dislocation, the couples have to do no work, and so the dislocation produces no curvature.

As a final example we calculate the twist in a thin-walled tube (of arbitrary cross-section and variable thickness) containing a dislocation or, what comes to the same thing, in a whisker containing a dislocation with a hollow core so large as almost to fill the cross-section. Let the inner and outer boundaries C_o, C_i (Fig. 2b) enclose areas A_o, A_i respectively. The stresses are given by (6). Ψ still satisfies (7) and takes the value 0 on C_o. On C_i it has some other constant value k say. It is easy to show that the twist is given by (10) with $\Psi(x_0, y_0) = k$. It is shown in books on the theory of elasticity [3] that a good approximation to the torsional rigidity is

$$\mu\, S = \mu\, k\,(A_o + A_i)\,.$$

If we actually needed to find the torsional rigidity we should have to estimate k. However, it cancels from the expression (10) to give the simple result

$$\alpha = b_z/(A_o + A_i)\,. \tag{12}$$

Since (12) is only valid for a thin tube it is not clear that it really any better than the simpler

$$\alpha = b_z/2\,A_o\,.$$

However, if, for example, C_o and C_i happen to be similar ellipses, concentric and similarly situated, (12) is exact over the whole range $0 \leq A_i/A_o < 1$. For a dislocation near the centre of a solid rod whose cross-section is reasonably equiaxed we have $\alpha \simeq b_z/A_o$ [7] and so the twist is roughly halved if the rod is hollowed out to form a tube.

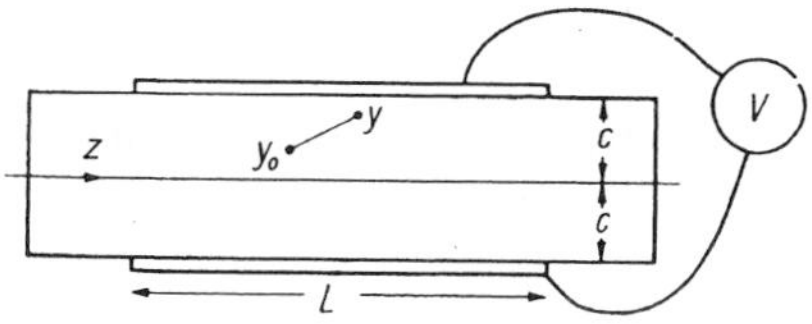

Fig. 3. Dislocation charge displacement in a dielectric plate

The results of this section are not limited to isotropic media. The expressions (3), (4) for β_x, β_y are valid whenever terminal couples produce the simple stress-dislocation (2). Similarly the results for β_z and α depend only on the stresses in torsion being given correctly by (6). They thus remain true in an anisotropic material provided there is fourfold or sixfold symmetry about the axis of the rod [7]. They may be adapted to some other cases in the way described in reference [7], section 4.

3. Electric Polarization by Dislocation Motion

It appears that dislocations in ionic crystals carry an electric charge. Thus their motion gives rise to electric polarization and hence to differences in potential between different parts of the crystal. (For recent work and earlier references cf. [8, 9]). In this section we examine the connection, unfortunately not very direct, between plastic deformation and potential difference for the case of a uniformly bent plate.

Fig. 3 represents a crystal plate (with dielectric constant ε) fitted with electrodes of length L on the upper and lower surfaces. If a dislocation originally at (y_0, z_0) moves to (y, z) carrying with it a fixed charge σ per unit length and leaving a compensating charge cloud $-\sigma$ behind at (y_0, z_0) a voltage

$$V = (4\pi\sigma/L\varepsilon)(y - y_0) \tag{13}$$

is set up between the electrodes. Equation (13) may be derived by a detailed electrostatic calculation or by using Green's reciprocation theorem [10]. For the present problem the theorem states that the difference of potential between the electrodes (top minus bottom) due to a charge σ at (y, z) is equal to the potential at (y, z) when charges $+\sigma, -\sigma$ are simultaneously given to the upper and lower electrodes respectively. The latter quantity is $4\pi\sigma y/L\varepsilon$ and (13) follows.

If the plate is bent plastically so that many dislocations move the resulting voltage will be

$$V = (4\sigma\pi/\varepsilon) K_1 \tag{14}$$

where

$$K_1 = L^{-1} \Sigma (y - y_0) \tag{15}$$

and Σ represents summation over all dislocations between the electrodes.

Suppose, alternatively that the dislocations are initially uncharged but pick up a charge proportional to the distance they move. If all slip planes are equally inclined to the z axis we may conveniently write the charge picked up as $\gamma |y - y_0|$. It is easy to show that (13) is still valid, where now σ is the charge picked up by the time the dislocation is halfway between (y_0, z_0) and (y, z). Thus for a plastically bent beam we have in place of (14)

$$V = (4\pi\gamma/\varepsilon) K_2 \tag{16}$$

where

$$K_2 = L^{-1} \Sigma \frac{1}{2} (y - y_0) |y - y_0| . \tag{17}$$

Another possibility is that the dislocations pick up charge until some maximum σ_{max} is reached, and then move without further change of charge. If the distance required to pick up σ_{max} is small compared with the average total distance moved the potential difference will be nearly equal to (14) with $\sigma = \sigma_{max}$.

If the width of the plate perpendicular to the figure is not large compared with its thickness the value of V will be modified by fringe effects. The alteration will be much reduced, however, if the dielectric constant ε is large compared with unity.

In addition to the voltage the movement of the dislocations will produce a plastic curvature. According to (3) each dislocation contributes a deflection $\frac{3}{4} b_z (y^2 - c^2)/c^3 - \frac{3}{4} b_z (y_0^2 - c^2)/c^3$ so that the curvature will be

$$\varkappa = \frac{3}{4} |b_z| K_3$$

where

$$K_3 = L^{-1} \Sigma (b_z/|b_z|)(y^2 - y_0^2)/c^3 . \tag{18}$$

The possibility that fresh dislocations are formed during plastic deformation can easily be allowed for. In a two-dimensional model the generation of loops from a source would be described as the separation of two initially coincident dislocations with opposite Burgers vectors. Thus we only have to suppose that the summations in K_1, K_2, K_3 contain a suitable number of pairs of terms corresponding to dislocations with opposite values of b_z, the same initial y_0 and different final y.

In reality deformation will take place by the movement of dislocation segments anchored at the ends and by the spreading of loops. However, the relation between V and $\varkappa$ should be much the same as the two-dimensional model predicts. For the sake of simplicity suppose that deformation occurs by the spreading of rectangular loops with their edges in and perpendicular to the plane of the figure. Edges in the plane of the figure contribute nothing to V or $\varkappa$ when they move perpendicular to the plane of the figure, and the remaining edges add up to give what is in effect a two-dimensional distribution. Simultaneous measurement of V and $\varkappa$ would give some information about the charge σ or the pick-up constant γ if we could estimate the ratios K_1/K_3 or K_2/K_3, but there are several difficulties. In the first place we should expect that in simple uniform bending K_1 or K_2 would be nearly zero because contributions from dislocations above and below the neutral axis would cancel. However, this difficulty can be overcome by embedding one electrode in the neutral surface [9]. V is still given by (14) or (16) with the summation in K_1 or K_2 confined to the dislocations which lie between the electrodes. (K_3, of course, must still be taken over all dislocations.) However, there is still no unique relation between V and $\varkappa$. The value of one of the terms in (18) is unaltered if we interchange the initial and final positions of the dislocation and at the same time reverse its Burgers vector, but the corresponding term in (15) or (17) has its sign changed. Thus two schemes of dislocation motion which give the same $\varkappa$ may give different V.

Read [11] has given an elegant dislocation model of the plastic deformation of a bent beam for the case where the stress required to create a dislocation exceeds the stress required to move it. He gives expressions which show how the distributions of dislocation density and stress vary as deformation proceeds. The stress is always largest at the upper and lower surfaces. If we admit that dislocations are always created where the stress is greatest we could thus evaluate K_1, K_2, K_3 by summation over Read's distribution, taking $y_0 = c$ for all dislocations above the neutral axis and $y_0 = -c$ for those below it. However, it is not certain that deformation in, say, a bent rock-salt crystal conforms to this model. It is possible that dislocations move in from the surface and trigger off the formation of fresh dislocations which then move both inwards and outwards. In this case it is not possible to be certain even of the sign of the voltage developed.

References

[1] F. Kroupa, Czech. J. Phys. **9**, 488 (1959).
[2] R. Siems, P. Delavignette, and S. Amelinckx, phys. stat. sol. **2**, 421 (1962).
[3] I. S. Sokolnikoff, Mathematical Theory of Elasticity, McGraw Hill Book Company 1956.
[4] F. R. N. Nabarro, Adv. Phys. **1**, 269 (1952).
[5] L. H. Mitchell and A. K. Head, J. Mech. Phys. Solids **9**, 131 (1961).
[6] J. D. Eshelby, J. appl. Phys. **24**, 176 (1953).
[7] J. D. Eshelby, Phil. Mag. **3**, 440 (1958).
[8] G. Remaut, phys. stat. sol. **2**, 576 (1962).
[9] J. E. Caffyn and T. L. Goodfellow, Proc. Phys. Soc. **79**, 1285 (1962).
[10] J. H. Jeans, Electricity and Magnetism, University Press, Cambridge 1915 (p. 92).
[11] W. T. Read jr., Acta metall. **5**, 83 (1957).

(Received June 8, 1962)

phys. stat. sol. 3, 2057 (1963)

Max-Planck-Institut für Metallforschung, Stuttgart, und Institut für theoretische und angewandte Physik der Technischen Hochschule Stuttgart

The Distribution of Dislocations in an Elliptical Glide Zone

By

J. D. Eshelby[1])

The following problem, of interest in the theory of work-hardening, is considered. Under the influence of an applied stress dislocation loops are emitted from a source and pile up against an elliptical barrier. To determine their equilibrium number and distribution.

Es wird ein Problem betrachtet, das für die Theorie der Verfestigung von Bedeutung ist: Unter dem Einfluß einer angelegten Spannung werden Versetzungsschleifen von einer Quelle emittiert und stauen sich an einem elliptischen Hindernis auf. Die Anzahl der Versetzungen im Gleichgewicht und ihre Verteilung sollen bestimmt werden.

Introduction

The following problem arises in the theory of work-hardening [1]. Under the influence of an applied stress τ a dislocation source S (Fig. 1) emits dislocation loops which pile up against a more or less elliptical barrier. It is required to find the maximum number n of piled-up dislocations, that is, the number which produce a back-stress $-\tau$ at S and so inhibit the emission of further loops.

Leibfried [2] solved the problem for a circular barrier in a medium with Poisson's ratio $\nu = 0$. He found

$$n = \alpha \frac{2 \tau L}{\pi b G}, \tag{1}$$

where L is the diameter of the circular barrier, b the common Burgers vector of the dislocations and G is the shear modulus. The numerical factor α is unity. It is our object to find the value of α when Poisson's ratio is not zero and the barrier is elliptical rather than circular.

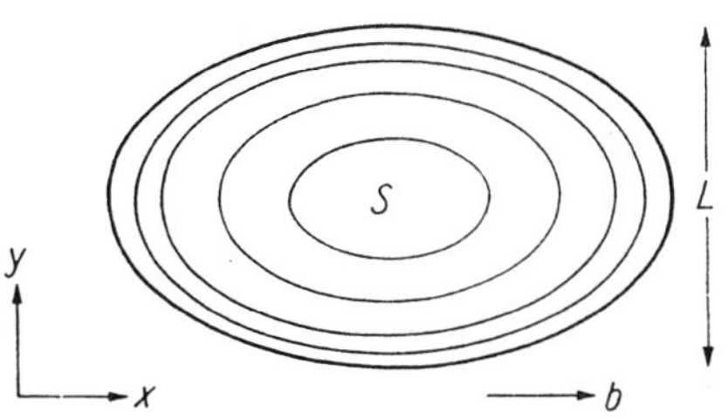

Fig. 1. An elliptical glide zone

1) Permanent address: Department of Physical Metallurgy, University of Birmingham.

The Symmetrical Case

In this section we treat the case where the applied shear stress and the Burgers vector of the dislocation loops are both parallel to one of the principal axes of the elliptical barrier. Consider the relative displacement Δu of adjacent points immediately above and below the xy-plane (the common slip plane of the dislocations). As we move in from the barrier towards the source Δu increases by an amount b (directed parallel to the Burgers vector) each time we pass a dislocation. The surface $\Delta u = \Delta u(x, y)$ is thus the same as the upper surface of a stack of discs each of thickness b and bounded by one of the dislocation loops. In the limit of a large number of dislocations this stepped surface may be approximated to by a smooth dome-shaped surface. The number of dislocations n is the maximum height of the stepped surface or dome, divided by b.

For the dislocations to be in equilibrium the shear stress (resolved parallel to the Burgers vector) must be zero at every point of the xy-plane traversed by a dislocation line. In the limit of a large number of dislocations it is plausible to require that the resolved shear stress shall vanish at *every* point of the xy-plane inside the barrier. The problem then becomes that of finding the relative displacement of the two faces of a freely slipping crack in a body subject to an applied shear stress τ. This problem has been treated by Segedin [3] for a circular crack in a medium of arbitrary Poisson's ratio. The formal equivalence with Leibfried's problem can be seen as follows. Leibfried reduces his problem to the solution of an integral equation relating to the distribution of electric charge on a circular disc, while Segedin uses results due to Green [4] which are based on what is essentially the same integral equation, with the same electrostatic interpretation [5].

From Segedin's results we may infer that for a circular barrier in a material with arbitrary Poisson's ratio the number of dislocations n is given by (1) with

$$\alpha = \frac{2\,(1-\nu)}{2-\nu}\,.$$

We shall assume that results for an elliptical barrier may be derived in the same way from the theory of an elliptical crack under shear [6]. Suppose that the barrier in Fig. 1 is the ellipse

$$\frac{x^2}{\left(\frac{1}{2}L'\right)^2} + \frac{y^2}{\left(\frac{1}{2}L\right)^2} = 1$$

and that the applied shear stress τ acts parallel to one of the principal axes, say along the x-axis. (This corresponds to the situation in the theory of [1].) The relative displacement of the faces of the crack, and hence also the common Burgers vector of the loops, is parallel to the x-axis. Its magnitude is

$$\Delta u = \frac{L\,\tau}{G\,\eta}\left\{1 - \frac{x^2}{\left(\frac{1}{2}L'\right)^2} - \frac{y^2}{\left(\frac{1}{2}L\right)^2}\right\}^{1/2}, \tag{2}$$

where

$$\eta = E(k) + \frac{\nu}{1-\nu}\frac{k'^2}{k^2}[K(k) - E(k)],$$

$$k = (1 - L^2/L'^2)^{1/2},\ k' = L/L',\ L' > L,$$

$$\eta = \frac{E(k)}{k'} + \frac{\nu}{1-\nu}\frac{1}{k^2 k'}[E(k) - k'^2 K(k)],$$

$$k = (1 - L'^2/L^2)^{1/2},\ k' = L'/L,\ L' < L,$$

$$\eta = \pi(2-\nu)/4(1-\nu),\ L' = L.$$

(In [6] the factor k'^2 was accidentally omitted from the first of these equations.) $E(k)$ and $K(k)$ are complete elliptic integrals of modulus k.

As in the circular case the number of dislocations is equal to the maximum value of Δu, divided by b, namely $L\,\tau/G\,\eta\,b$. This n is given by (1) with

$$\alpha = \pi/2\,\eta.$$

Fig. 2 shows how α varies with the axial ratio L'/L for two values of Poisson's ratio, 0 and $\frac{1}{3}$. In the application to work-hardening $\nu \sim \frac{1}{3}$ and $L' \sim 1.5\,L$ [1].

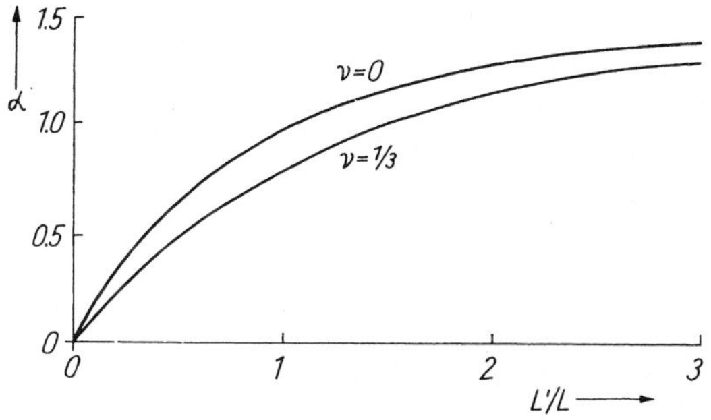

Fig. 2. Correction factor α to LEIBFRIED's formula

The constant α is then very nearly unity and LEIBFRIED's original formula, equation (1) with $\alpha = 1$, may be used. (Note that L is the diameter *transverse* to the Burgers vector.)

The individual dislocation loops coincide with the curves $\Delta u =$ const. They are thus ellipses similar to the elliptical barrier. N such elliptical pile-ups per unit volume give a plastic strain

$$\varepsilon = N\int \Delta u\,dx\,dy = \frac{\pi N L^2 L'}{6\,\eta}\frac{\tau}{G} = \frac{2}{3}N\,A\,n\,b$$

so that the average area of a loop is $\frac{2}{3}$ that of the barrier, $A = \frac{1}{4}\pi\,L\,L'$.

The General Case

So far we have supposed that the directions of the applied shear stress and of the Burgers vector both coincide with one of the principal axes of the ellipse. This is the important case. For completeness we outline a treatment of the case where these two directions are arbitrary and in general different.

For a general shear stress τ_{zx}, τ_{zy} the solution of the crack problem is a superposition of the solutions for the cases $\tau_{zx} \neq 0$, $\tau_{zy} = 0$ and $\tau_{zy} \neq 0$, $\tau_{zx} = 0$. The former we have already solved, the latter is dealt with similarly, interchanging the roles of L, L'. The displacement discontinuity has an ellipsoidal distribution similar to (2) and is everywhere parallel to a fixed direction which is a rather complicated function of the ratio τ_{zx}/τ_{zy}. If this direction happens to coincide with that of the required Burgers vector the dislocation loop problem can be solved as before. However, in general it will not. To deal with this case we recall that a dislocation is not affected by a component of the shear stress perpendicular to its Burgers vector. This fact can be introduced into the crack model by supposing that the dividing surface of the crack is finely corrugated parallel to the Burgers vector, so that the crack can support stresses transverse to the Burgers vector, but not parallel to it. Write the applied stress in the form

$$\tau_{zx} = \tau'_{zx} + \tau''_{zy}\,, \qquad \tau_{zy} = \tau'_{zy} + \tau''_{zy}\,, \tag{3}$$

where the stress vector $(\tau''_{zx}, \tau''_{zy})$ is perpendicular to the proposed Burgers vector. Solve the crack problem for (τ'_{zx}, τ'_{zy}), ignoring $(\tau''_{zx}, \tau''_{zy})$ and choose the ratio τ'_{zx}/τ'_{zy} so that the direction of Δu coincides with that of the prescribed Burgers vector. This condition together with (3) and the requirement that τ''_{zx}, τ''_{zy} be transverse fixes τ'_{zx}, τ'_{zy} uniquely. Finally interpret the crack solution as a dislocation distribution in the way already indicated.

Strictly speaking the line tension of the dislocation loops will modify their shape and distribution. However, it is not difficult to show that in the limit of a large number of loops this effect is negligible.

References

[1] S. Mader, A. Seeger, and C. Leitz, J. appl. Phys. (1963), in press.
[2] G. Leibfried, Z. angew. Phys. **6**, 251 (1954).
[3] C. M. Segedin, Proc. Camb. Phil. Soc. **47**, 396 (1951).
[4] A. E. Green, Proc. Camb. Phil. Soc. **45**, 251 (1949).
[5] E. T. Copson, Proc. Edin. Math. Soc. **8**, 14 (1947).
[6] J. D. Eshelby, Proc. Roy. Soc. **A 241**, 376 (1957).

(Received August 16, 1963)

APPENDIX I

ON THE ELASTIC INTERACTION BETWEEN INCLUSIONS

J. D. ESHELBY

Consider two misfitting inclusions A and B bounded by surfaces S_A, S_B embedded in a matrix having the same elastic constants as the inclusions. The elastic energy inside S_A is

$$\frac{1}{2}\int_{S_A+S_B} u_i^I p_{ij} n_j \, \mathrm{d}S, \tag{I1}$$

where u_i^I is the displacement of a point on the surface of A or B relative to its unstrained state, p_{ij} is the stress and n_j is the outward unit normal vector of the surface element dS. The energy of the matrix is, similarly,

$$-\frac{1}{2}\int_{S_A+S_B} u_i^M p_{ij} n_j \, dS, \tag{I2}$$

where the minus sign is correct if n_j is the inward normal of the matrix and thus the outward normal to A or B. There is no contribution from the outer surface of the matrix since we suppose it free of stress. Adding (I1) and (I2) and noting that $p_{ij}n_j$ is continuous across S_A or S_B we have the expression

$$E = \frac{1}{2}\int_{S_A+S_B} (u_i^I - u_i^M) p_{ij} n_j \, dS \tag{I3}$$

for the total elastic energy of the system.

Suppose now that the elastic constants of A and B change to values differing from those of the matrix, and not necessarily the same for A and B. The stress and displacement will change everywhere to, say, p_{ij}', u_i', and by the same argument as before the total energy will be

$$E' = \frac{1}{2}\int_{S_A+S_B} (u_i^{I'} - u_i^{M'}) p_{ij}' n_j \, dS \tag{I4}$$

(there is nothing in the argument leading to (I3) which requires the elastic constants of the matrix and inclusion to be the same).

We are interested in comparing E with E' in the case where the elastic constants are different but the misfits of the inclusions are the same, so that

$$u_i^I - u_i^M = u_i^{I'} - u_i^{M'} \tag{I5}$$

on S_A and S_B. It proves convenient to use (I5) to rewrite (I3) and (I4) as

$$E = \frac{1}{2}\int_{S_A+S_B} (u_i^{I'} - u_i^{M'}) p_{ij} n_j \, dS \tag{I6}$$

and

$$E' = \frac{1}{2}\int_{S_A+S_B} (u_i^I - u_i^M) p_{ij}' n_j \, dS \tag{I7}$$

The first term in (I6) may be converted as follows into a volume integral over the volumes V_A, V_B enclosed by S_A, S_B:

$$\frac{1}{2}\int_{S_A+S_B} u_i^{I'} p_{ij} n_j \, dS$$
$$= \frac{1}{2}\int_{V_A+V_B} \frac{\partial (u_i^{I'} p_{ij})}{\partial x_j} \, dV$$
$$= \frac{1}{2}\int_{V_A+V_B} p_{ij} \frac{\partial u_i^{I'}}{\partial x_j} \, dV$$
$$= \frac{1}{2}\int_{V_A+V_B} p_{ij} e_{ij}' \, dV. \qquad \text{(I8)}$$

The first step is an application of Gauss' theorem, the second of the equilibrium condition $\partial p_{ij}/\partial x_j = 0$ and the third follows from the definition of strain, $e_{ij}' = \frac{1}{2}(\partial u_i'/\partial x_j + \partial u_j'/\partial x_i)$ and the fact that $p_{ij} = p_{ji}$. The affix I on u_i' (or e_{ij}') may be dropped without ambiguity; only on the surfaces S_A, S_B where u_i' is discontinuous need we distinguish between $u_i^{I'}$ and $u_i^{M'}$.

The second term in (I6) may be likewise transformed into a volume integral similar to (I8) but extended over the matrix. The minus sign disappears for the same reason as it appeared, namely that on S_A, S_B the outward normals of the matrix and inclusions are oppositely directed. Thus

$$E = \frac{1}{2}\int_V p_{ij} e_{ij}' \, dV$$

and similarly

$$E' = \frac{1}{2}\int_V p_{ij}' e_{ij} \, dV,$$

where V is the whole volume of material. Hence the energy change on altering the elastic constants is

$$E' - E = \frac{1}{2}\int_V (p_{ij}' e_{ij} - p_{ij} e_{ij}') \, dV.$$

If the Lamé elastic constants of the material are λ, μ before the change and λ', μ' after it (where λ', μ' may be different in V_A, V_B and the matrix) we have

$$p_{ij} = \lambda e \delta_{ij} + 2\mu e_{ij}$$
$$p_{ij}' = \lambda' e' \delta_{ij} + 2\mu' e_{ij}'$$

and so

$$E' - E = \frac{1}{2}\int_V \{(\lambda' - \lambda) e e' + 2(\mu' - \mu) e_{ij} e_{ij}'\} \, dV,$$

or if we write $\lambda' - \lambda = \delta\lambda$, $\mu' - \mu = \delta\mu$, $e_{ij}' = e_{ij} + \delta e_{ij}$,

$$E' - E = \frac{1}{2}\int_V (\delta\lambda e^2 + 2\delta\mu e_{ij} e_{ij}) \, dV + \varepsilon, \qquad \text{(I9)}$$

where

$$\varepsilon = \frac{1}{2}\int_V (\delta\lambda \delta e e + 2\delta\mu \delta e_{ij} e_{ij}) \, dV.$$

Generally we should have to solve a difficult boundary value problem in order to find the changes in strain δe_{ij}. However, if $\delta\lambda$, $d\mu$ are small of the first order the δe_{ij} are also small of the first order and so ε is of the second order and the energy change is given correct to the first order by (I9) with $\varepsilon = 0$, an expression which only involves finding the e_{ij} for the simple misfitting but homogeneous case.

Let e_{ij}^A, e_{ij}^B be the strains produced by A alone or B alone. Then since $\delta\lambda = 0$, $\delta\mu = 0$ in the matrix we have to the first order

$$E' - E = \frac{1}{2}\int_{V_A+V_B} \delta\lambda (e^A e^A + e^B e^B + 2e^A e^B) \, dV$$
$$+ \int_{V_A+V_B} \delta\mu (e_{ij}^A e_{ij}^A + e_{ij}^B e_{ij}^B + 2e_{ij}^A e_{ij}^B) \, dV. \qquad \text{(I10)}$$

Outside an inclusion of any shape for which the transformation strain specifying the misfit is a pure dilatation the strain is a pure shear, provided the perturbing effect of the outer free surface of the matrix may be neglected. Thus $e^A = 0$ in B and $e^B = 0$ in A, so that the term in $e^A e^B$ is zero. For the same reason $e^A e^A$ contributes nothing to the integral over V_B and $e^B e^B$ nothing to the integral over V_A. If we extract those terms in (I10) which depend upon the relative positions of A and B we have the interaction energy that we seek, namely

$$E_{\text{int}} = \int_{V_A} \delta\mu_A (e_{ij}^B e_{ij}^B + 2e_{ij}^A e_{ij}^B) \, dV$$
$$+ \int_{V_B} \delta\mu_B (e_{ij}^A e_{ij}^A + 2e_{ij}^A e_{ij}^B) \, dV. \qquad \text{(I11)}$$

The remaining terms combine to give us the change in the "self-energy" of the system. since these terms are independent of the relative positions of A and B. Hence,

$$\delta E_{\text{self}} = \int_{V_A} (\delta\mu_A e_{ij}^A e_{ij}^A + \tfrac{1}{2}\,\delta\lambda_A e^A e^A) \, dV$$
$$+ \int_{V_B} (\delta\mu_B e_{ij}^B e_{ij}^B + \tfrac{1}{2}\,\delta\lambda_B e^B e^B) \, dV. \qquad \text{(I12)}$$

Finally, if the inclusions are spherical the e_{ij} are pure dilatations within them. Thus the terms in $e_{ij}^A e_{ij}^B$ in (I11) vanish because inside A $e_{ij}^A = \frac{1}{3} e^A \delta_{ij}$, so that $e_{ij}^A e_{ij}^B = \frac{1}{3} e^A e^B$ which vanishes because $e^B = 0$ in A, and similarily for B. Therefore, for spherical inclusions we are left with

$$E_{\text{int}} = \int_{V_A} \delta\mu_A e_{ij}^B e_{ij}^B \, dV + \int_{V_B} \delta\mu_B e_{ij}^A e_{ij}^A \, dV. \qquad \text{(I13)}$$

Furthermore, the term $e_{ij}{}^A e_{ij}{}^A$ in A may be simplified to $\frac{1}{3}e^A e^A$ and similarly for the term $e_{ij}{}^B e_{ij}{}^B$ in B. The change in self-energy for spherical inclusions thus becomes

$$\delta E_{\text{self}} = \int_{V_A} \tfrac{1}{2}\,\delta K_A e^A e^A\, dV + \int_{V_B} \tfrac{1}{2}\,\delta K_B e^B e^B\, dV, \quad \text{(I14)}$$

where $\delta K = \delta\lambda + \frac{2}{3}\,\delta\mu$ is the change in the bulk modulus. If A and B are made of the same material δE_{self} is independent of how a given amount of inclusion material is distributed between A and B.

APPENDIX II

AN APPROXIMATE EVALUATION OF THE INTERACTION FOR γ' IN Ni–Al

We wish to evaluate the integrals in the equation

$$E_{\text{int}} = \delta\mu \int_{V_A} e_{ij}{}^B e_{ij}{}^B\, dV + \delta\mu \int_{V_B} e_{ij}{}^A e_{ij}{}^A\, dV \quad \text{(II1)}$$

for the spherical particles A and B in Fig. II1. We will follow the terminology used by Eshelby,[19] in which summation over repeated indices is understood. We are concerned, specifically, with the interaction between precipitates whose transformation strains are pure dilatations,

$$e_{ij}{}^T = \tfrac{1}{3}\, e^T \delta_{ij}, \quad \text{(II2)}$$

where e^T is the dilatation of A and B, and δ_{ij} is the Kronecker delta.

Consider the integral on the left in equation (II1), E_A. The displacements in the matrix due to B are given by

$$u_i{}^B = \alpha \frac{\partial \varphi_B}{\partial x_i}; \quad \alpha = \frac{-(1+\sigma)}{12\pi(1-\sigma)}\, e^T \quad \text{(II3)}$$

where σ is Poisson's ratio and φ_B is given by

$$\varphi_B = \frac{4\pi a^3}{3r} = \frac{V_B}{r}; \quad r > a. \quad \text{(II4)}$$

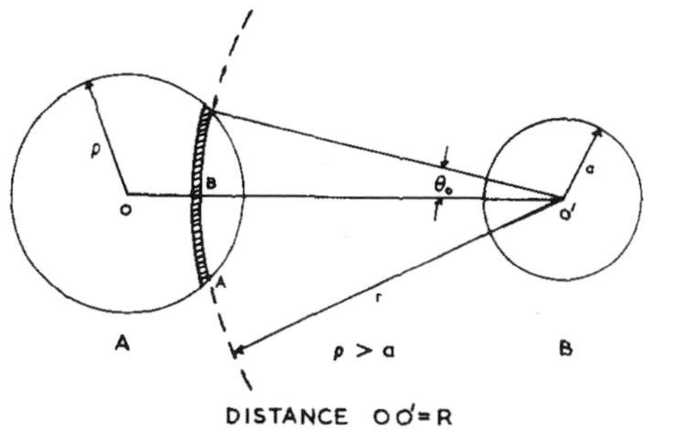

Fig. 11.1. Diagram showing the definition of the parameters used in Appendix II.

The $e_{ij}{}^B$ are given by

$$e_{ij}{}^B = \frac{1}{2}\left(\frac{\partial u_i{}^B}{\partial x_j} + \frac{\partial u_j{}^B}{\partial x_i}\right) = \alpha \frac{\partial^2 \varphi_B}{\partial x_i \partial x_j}. \quad \text{(II5)}$$

Performing the differentiation of (II4) indicated in (II5) and the summation in (II1), we get

$$E_A = 6\alpha^2 V_B{}^2 \delta\mu \int_{V_A} \frac{dV}{r^6}. \quad \text{(II6)}$$

Using the geometry due to Hamaker[25] in Fig. II1, we may evaluate the integral in (II6). Area A_{ABC} is given by

$$A_{ABC} = \int_0^{\theta_0} 2\pi r^2 \sin\theta\, d\theta, \quad \text{(II7)}$$

with θ_0 defined by

$$\rho^2 = R^2 + r^2 - 2rR\cos\theta_0. \quad \text{(II8)}$$

Thus

$$dV = \frac{\pi r}{R}\{\rho^2 - (R-r)^2\}\, dr, \quad \text{(II9)}$$

and

$$E_A = 6\pi\alpha^2 V_B{}^2 \delta\mu \int_{R-\rho}^{R+\rho} \{\rho^2 - (R-r)^2\}\frac{dr}{r^5}. \quad \text{(II10)}$$

The integral (II10) is readily evaluated, and after some algebra we obtain

$$E = \frac{8\pi}{81}\left(\frac{1+\sigma}{1-\sigma}\right)^2 (e^T)^2 a^6 \delta\mu \frac{\rho^3}{(R^2-\rho^2)^3}. \quad \text{(II11)}$$

To find E_B we need only to interchange the parameters of A and B in (II11).

Thus

$$E_{\text{int}} = \beta\left\{\frac{a^6\rho^3}{(R^2-\rho^2)^3} + \frac{\rho^6 a^3}{(R^2-a^2)^3}\right\}, \quad \text{(II12)}$$

with

$$\beta = \frac{8\pi}{81}\left(\frac{1+\sigma}{1-\sigma}\right)^2 (e^T)^2 \delta\mu. \quad \text{(II13)}$$

We can see immediately from (II12) that for $R \gg \rho$ (hence $R \gg a$ (see Fig. II1)), $E_{\text{int}} \propto 1/R^6$, which is the standard result for the interaction between two inhomogeneous point sources of dilatation[19] as was mentioned in the text.

We note, as a general observation, that the strength of the interaction increases as $(e^T)^2$ [see equation II13)] and also with increasing particle size. Thus, if $\delta\mu < 0$, as it is for γ' in a Ni matrix, we would expect that the attraction between particles will increase as they grow, which is what we observe. Also, for those systems in which there is a small misfit, we would expect that the degree of alignment would be considerably reduced, or absent. This accounts for the observation that there is no alignment of γ' particles when they are spherical (Section 3.3).

Let us consider the two cases of greatest interest to this investigation.

Case 1. *Two particles of equal size*

We will estimate the strength of the attraction between two γ' particles by substituting in equation (II13) the appropriate parameters. Taking 700°C as a representative ageing temperature, and assuming $\sigma = 1/3$ for both γ' and Ni, we get, from Fig. 10 (see text), $\delta\mu = (3/8)\ \delta E = -8.63 \times 10^{10}$ dyn/cm². We may calculate e^T for the coherent γ' precipitate from the lattice parameters, $a_{\gamma'} = 3.56$ Å; $a_M = 3.54$ Å; $e^T = 3\,\Delta a/a_M = 1.7 \times 10^{-2}$. This gives $\beta = -3.09 \times 10^7$ dyn/cm². Typically then, for $\rho < 40$ Å, $|E_{\text{int}}| < kT$ ($T = 1000$°K) for all $R > 2\rho$. We can thus assume that the interaction between nuclei, for example, is negligible. For $\rho > 40$ Å the interaction increases rapidly. Thus when $\rho = 100$ Å, $|E_{\text{int}}| = kT$ for $R \simeq 3\rho$ or when the minimum separation between the particles is equal to a radius. When $\rho = 500$ Å, the same criterion gives $R \simeq 6.5\ \rho$. For our purposes we might want the interaction to be somewhat stronger than this for small particle sizes. Nevertheless, in view of the approximate nature of the calculations, it would appear that the interaction is strong enough to exert its influence, even though we have ignored E_s by assuming the γ' particles to be spherical.

Case 2. *One particle is much greater than the other*

This will approximate the interaction between a large precipitate and a solute atom in the matrix. Although $\delta\mu$ and e^T will usually be different for the inclusion and atom, nothing essential will be lost if they are taken as equal. In this case (II12) will apply. Taking the radius of the dissolved atom, a, as 1 Å, and using the parameters appropriate to γ', we can see that $|E_{\text{int}}| < kT$ for all $R > \rho + a$ when $\rho < 10^4$ Å. We may conclude that the interaction is negligible for all of the particle sizes we are likely to encounter. We can thus expect that the "direct" interaction will not affect the solute diffusion rate. It is difficult to estimate the "shape-effect" interaction between a solute atom and a cube-shaped precipitate, although it is probably small. We note, however, that it is necessary to carry out the integration in equation (4) of the text only over the volume of the precipitate. Insofar as the solute atom can be considered a spherical inclusion, the integral over its volume vanishes.

REFERENCES

1. A. Guinier, *Solid State Physics* **9,** 293. Academic Press, New York (1959).
2. A. Kelly and R. B. Nicholson, *Progress in Materials Science* **10,** 149. Pergamon Press, Oxford (1963).
3. J. Manenc, *J. Phys. Radium Paris* **23,** 830 (1962).
4. J. Manenc, *Rev. Met.* **54,** 867 (1957).
5. C. Bückle, B. Genty and J. Manenc, *Revue Métall, Paris* **56,** 247 (1959).
6. Y. A. Bagaryatskii and Y. D. Tiapkin, *Soviet Phys. Crystallogr.* **2,** 414 (1957).
7. R. O. Williams, *Trans. Am. Inst. Min. metall. Engrs* **215,** 1026 (1959).
8. J. W. Cahn, *Acta Met.* **9,** 795 (1961); **10,** 907 (1962).
9. D. H. Ben Israel and M. E. Fine, *Acta Met.* **11,** 1051 (1965).
10. A. Taylor and R. W. Floyd, *J. Inst. Metals* **81,** 25 (1952).
11. I. M. Lifshitz and V. V. Slyozov, *J. Phys. Chem. Solids* **19,** 35 (1961).
12. C. Wagner, *Z. Elektrochem.* **65,** 581 (1961).
13. R. A. Swalin and A. Martin, *Trans. Am. Inst. Min. metall. Engrs* **206,** 567 (1956).
14. M. Hillert, *Acta Met.* **9,** 525 (1961).
15. R. A. Oriani, *Acta Met.* **12,** 1399 (1964).
16. C. A. Johnson, to be published.
17. M. J. Blackburn, Ph.D. Dissertation, University of Cambridge (1961).
18. J. Manenc, *Acta Met.* **7,** 124 (1959).
19. J. D. Eshelby, *Progress in Solid Mechanics* **2,** 89. North-Holland, Amsterdam (1961).
20. J. D. Eshelby, *Solid State Physics* **3,** 79. Academic Press, New York (1956).
21. R. G. Davies and N. S. Stoloff, *Trans. Am. Inst. Min. metall. Engrs* **233,** 714 (1965).
22. P. E. Armstrong and H. L. Brown, *Trans. Am. Inst. Min. metall. Engrs* **230,** 962 (1964)
23. F. R. N. Nabarro, *Proc. phys. Soc. Lond.* **52,** 90 (1940).
24. J. H. Westbrook, *Z. Kristallogr.* **110,** 21 (1958).
25. H. C. Hamaker, *Physica* **4,** 1058 (1937).

BRIT. J. APPL. PHYS., 1966, VOL. 17

A simple derivation of the elastic field of an edge dislocation

J. D. ESHELBY

Department of Physics, Cavendish Laboratory, University of Cambridge

MS. received 29*th March* 1966, *in revised form* 20*th June* 1966

Abstract. The elastic field of an edge dislocation is found in a simple manner by making use of the relation between an edge dislocation and a 'wedge' dislocation made by inserting or removing a narrow wedge of material.

1. Introduction

The rather forbidding expression

$$u = \frac{b}{2\pi}\left(\tan^{-1}\frac{y}{x} + \frac{1}{2(1-\sigma)}\frac{xy}{r^2}\right)$$
$$v = \frac{b}{2\pi}\left(\frac{1-2\sigma}{2(1-\sigma)}\ln\frac{1}{r} + \frac{1}{2(1-\sigma)}\frac{y^2}{r^2}\right) \qquad (1)$$

(Love 1920 §156A, Burgers 1939, Koehler 1941) for the displacement around an edge dislocation in an infinite isotropic material plays an important part in elementary dislocation theory. The following simple derivation may therefore be of use. In it an attempt has been made to use as little as possible of the apparatus of elasticity theory and to present the argument in the form of a series of steps which can be seen to lead more or less inevitably to the displacement field of an edge dislocation. This is achieved by first relating the edge dislocation to what may be called a 'wedge' dislocation, a somewhat simpler state of internal stress on which geometrical and physical intuition can be more easily exercised (Eshelby 1951 §7).

2. The relation between wedge and edge dislocations

Throughout we shall be concerned with a plane problem in an infinite medium, but for

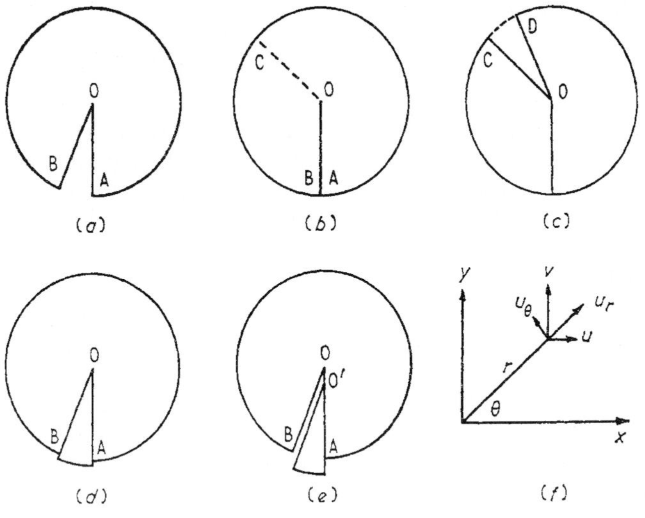

Wedge and edge dislocations.

clarity we draw the material as a circular cylinder (see figure). We assume that the component of the elastic displacement perpendicular to the figure is everywhere zero.

To make a positive wedge dislocation ((*a*)) we remove a wedge AOB of material of small angle Ω, force the faces AO, BO into contact and cement them together ((*b*)). To make a negative wedge dislocation we make a radial cut along AO ((*d*)) and force in a wedge of small angle Ω, cementing together along AO, BO.

In (*b*) make a radial cut OC. The material will spring open to form the gap COD (of angle Ω) in (*c*), and the stresses will disappear; all that has happened is that the sector BOC has been rotated rigidly from the position it had in (*a*). If COD (*c*) is now plugged with a wedge of angle Ω nothing is changed, but formally we are left with a body which contains a positive wedge dislocation along OA and a negative one along OC, and yet is free of stress. This means that a negative wedge dislocation made along any radius cancels the stress field of a positive one made along any other radius. Hence (i) their stress fields are equal and opposite and (ii) the stress field of a wedge dislocation is cylindrically symmetric.

Statement (i) implies that the stresses derived from the two displacements $\mathbf{u}_{w^+}$ and $-\mathbf{u}_{w^-}$ (with an obvious notation) are equal. Hence these displacements can differ only by a rigid-body displacement. If the two wedge dislocations are formed as in (*a*), (*d*) with a common point O and a common undeviated direction OA we have simply

$$\mathbf{u}_{w^-} = -\mathbf{u}_{w^+}. \tag{2}$$

It is indeed almost obvious that if we remove and insert a wedge at the same place we have in effect done nothing.

Let us now make a positive wedge dislocation as before and then a negative one with its apex a small distance $d = OO'$ lower down ((*e*)). One way of doing this is as follows.

(i) Cut along AO, BO.

(ii) Slide the wedge down so that its apex is at O′, and cement along AO′. We are now left with an unstressed body containing a parallel-sided radial fissure of width $b = \Omega d$.

(iii) Close up the fissure and cement its faces together.

The result of closing a parallel-sided fissure of width *b*, however, is to produce an edge dislocation with a Burgers vector of magnitude *b*. In atomic terms we have removed a half-plane of atoms. (The argument here is closely related to Thompson and Millard's (1952) demonstration that a step in a twin boundary is equivalent to a dislocation.) Hence

$$\begin{aligned}\mathbf{u}_e(x, y) &= \mathbf{u}_{w^+}(x, y) + \mathbf{u}_{w^-}(x, y + d)\\ &= \mathbf{u}_{w^+}(x, y) - \mathbf{u}_{w^+}(x, y + d)\end{aligned}$$

by (2). Hence finally, since $b = \Omega d$,

$$\mathbf{u}_e = -\frac{b}{\Omega}\frac{\partial}{\partial y}\mathbf{u}_{w^+} \tag{3}$$

at points distant a few multiples of *d* from the centre, and so the elastic field of an edge dislocation can be obtained from that of a wedge dislocation simply by differentiation. Evidently in (3) we may replace the displacement **u** by any quantity linearly related to it, for example a stress or strain component in Cartesian coordinates, or the Airy stress function.

3. The elastic fields

We shall use rectangular or polar coordinates and resolve the elastic displacement **u** into either rectangular components u, v or radial and tangential components u_r, u_θ as may be convenient ((*f*)).

In §2 we saw that the stress field of a wedge dislocation has cylindrical symmetry. The stress field is also obviously unchanged when reflected across the radius AO of (*b*) and hence across any radius. Thus two of the principal stresses are everywhere radial and transverse to a radius, and their magnitudes are independent of θ. The same statement must be true of the principal strains. Consequently a polar grid scribed on (*a*), made up of equally spaced radii and equally spaced circles becomes in (*b*) a similar grid with the radii still equally spaced (though by a slightly greater amount) but with the spacing of the circles no longer necessarily equal. Hence u_θ is proportional to r and linear in θ, while u_r is independent of θ.

The geometry of (*a*), (*b*) shows that in fact

$$u_\theta = \frac{\Omega}{2\pi}(\theta + \tfrac{1}{2}\pi)\, r. \tag{4}$$

It is convenient to write u_r in the form

$$u_r = \frac{\Omega}{2\pi} rf(r). \tag{5}$$

We cannot expect u_r to be zero, since when a hoop of material is stretched by closing the gap it will tend to contract radially.

The rectangular components u, v of the displacement are

$$\begin{aligned} u &= \frac{\Omega}{2\pi}\{fx - (\theta + \tfrac{1}{2}\pi)y\} \\ v &= \frac{\Omega}{2\pi}\{(\theta + \tfrac{1}{2}\pi)\, x + fy\} \end{aligned} \tag{6}$$

for the wedge dislocation, and by (3)

$$\begin{aligned} u &= \frac{b}{2\pi}\left(\theta + (1 - rf')\frac{xy}{r^2}\right) + \frac{b}{4} \\ v &= \frac{b}{2\pi}\left(-f + (1 - rf')\frac{y^2}{r^2}\right) - \frac{b}{2\pi} \end{aligned} \tag{7}$$

for the edge dislocation.

In order to find f we must require that the displacements satisfy the equilibrium equations. The following method, though possibly not the most direct, appears to require the least effort. If we introduce the dilatation and rotation

$$e = \frac{\partial u}{\partial x} + \frac{\partial v}{\partial y}, \quad \omega = \frac{1}{2}\left(\frac{\partial v}{\partial x} - \frac{\partial u}{\partial y}\right) \tag{8}$$

the equilibrium equations (Love 1920)

$$(1 - 2\sigma)\, \nabla^2 u + \frac{\partial e}{\partial x} = 0, \quad (1 - 2\sigma)\, \nabla^2 v + \frac{\partial e}{\partial y} = 0 \tag{9}$$

may, as is well known, be put into the simpler looking forms

$$\frac{\partial(\alpha e)}{\partial r} = \frac{\partial \omega}{\partial y}, \quad \frac{\partial(\alpha e)}{\partial y} = -\frac{\partial \omega}{\partial x} \tag{10}$$

respectively, where $\alpha = (1 - \sigma)/(1 - 2\sigma)$ and σ is Poisson's ratio. Equations (10) hold for any orientation of the x and y axes. If we take them to be in the directions of r increasing and θ increasing at (r, θ) we get

$$\frac{\partial(\alpha e)}{\partial r} = \frac{1}{r}\frac{\partial \omega}{\partial \theta}, \quad \frac{1}{r}\frac{\partial(\alpha e)}{\partial \theta} = -\frac{\partial \omega}{\partial r}. \tag{11}$$

From the geometry of (*a*), (*b*) it is clear that for a wedge dislocation the rotation is

$$\omega = \frac{\Omega}{2\pi}(\theta + \tfrac{1}{2}\pi). \tag{12}$$

We have assumed that, as its name implies, ω actually represents the rotation of the material, but (12) also follows from (4), (5) and (8); the unknown f contributes nothing. With (12) one of the pair (11) confirms that e is independent of θ and the other gives

$$\alpha e = \frac{\Omega}{2\pi}(\ln r + \text{const.}). \tag{13}$$

To relate u_r or f to e we note that when the wedge dislocation is introduced the volume of a disk of unit thickness and radius r changes by an amount which can be written either as the product of $u_r(r)$ and the circumference or as the volume integral of the dilatation e over its interior, so that

$$2\pi r u_r(r) = \int_0^r 2\pi e r dr$$

which gives

$$rf(r) = \frac{2\pi}{\Omega} u_r = \frac{r}{2a} \ln r \tag{14}$$

together with a term const. r. The latter represents an arbitrary uniform expansion superimposed on the wedge dislocation. We shall omit it.

The field of the wedge dislocation, found by substituting $f = (1/2a) \ln r$ in (6), is of some interest on its own account. It represents the field of a terminating tilt boundary made up of edge dislocations with Burgers vectors b and spacing $h = b/\Omega$, provided we keep a few multiples of h away from the boundary itself.

Substitution of $f = (1/2a) \ln r$ in (7) gives the displacement due to an edge dislocation. It is identical with (1) if we omit the final terms in (7); they only represent an unimportant rigid-body translation.

The symmetry argument preceding (4) shows that for a wedge dislocation the traction across any circle $r =$ const. is purely radial and independent of θ, so that the total force and couple on the circuit are both zero. Consequently there are no unwanted forces or couples lurking at the singularity at $r = 0$, nor are any induced by differentiating to obtain the field of an edge dislocation.

The stresses may be found by differentiating the displacements. It is interesting to note that in order to find the hydrostatic pressure P due to an edge dislocation we only need the purely geometrical equation (12), the second of (10) and the basic relation (3) between edge and wedge dislocations:

$$P = -Ke_{\mathrm{e}} = \frac{bK}{\Omega}\frac{\partial}{\partial y} e_{\mathrm{w}} = -\frac{bK}{\Omega a}\frac{\partial}{\partial x}\omega_{\mathrm{w}} = \frac{\mu b}{2\pi}\frac{2(1+\sigma)}{3(1-\sigma)}\frac{\sin\theta}{r}$$

where $K = 2\mu(1+\sigma)/3(1-2\sigma)$ is the bulk modulus and μ is the shear modulus.

Alternatively it is not hard to construct the Airy stress function χ for a wedge or edge dislocation, in terms of which the stresses are

$$\sigma_{xx} = \frac{\partial^2\chi}{\partial y^2}, \quad \sigma_{yy} = \frac{\partial^2\chi}{\partial x^2}, \quad \sigma_{xy} = -\frac{\partial^2\chi}{\partial x \partial y}$$

$$\sigma_{zz} = \sigma(\sigma_{xx} + \sigma_{yy}), \quad \sigma_{xz} = 0, \quad \sigma_{yz} = 0. \tag{15}$$

From (15) and the relation $Ke = \frac{1}{3}(\sigma_{xx} + \sigma_{yy} + \sigma_{zz})$ between the dilatation and the hydrostatic part of the stress we have

$$(1+\sigma)\nabla^2\chi = 3Ke. \tag{16}$$

For the wedge dislocation χ is evidently independent of θ, so that with (13) we have

$$\frac{1}{r}\frac{d}{dr}\left(r\frac{d\chi}{dr}\right) = \frac{2\mu}{1-\sigma}\frac{\Omega}{2\pi}(\ln r + \text{const.})$$

which on repeated integration gives (Yoffe 1958)

$$\chi = \frac{\mu}{1-\sigma}\frac{\Omega}{2\pi}\left(\tfrac{1}{2}r^2 \ln r - \tfrac{1}{4}r^2\right) \tag{17}$$

plus terms $Ar^2 + B \ln r$ with arbitrary A, B. The first of these merely represents an arbitrary superimposed two-dimensional uniform expansion, and we may put $A = 0$. When substituted back into the left-hand side of (16) the term in $\ln r$ makes a contribution proportional to $\nabla^2(\ln r) = 2\pi\,\delta(x)\,\delta(y)$. There is no such singular concentration of dilatation at the origin in (13), so we must put $B = 0$. Otherwise there would be a line of dilatation (Love 1920) superimposed on the wedge dislocation.

To find χ for the edge dislocation we have, according to (3), to differentiate (17) with respect to y and multiply by $-b/\Omega$. This gives the well-known result (Koehler 1941)

$$\chi = -\frac{\mu}{1-\sigma}\frac{b}{2\pi}y \ln r.$$

Acknowledgments

The writer would like to thank the referees for help in removing errors and obscurities.

References

BURGERS, J. M., 1939, *Proc. K. Ned. Akad. Wet.*, **42**, 293–325.
ESHELBY, J. D., 1951, *Phil. Trans.* A, **244**, 87–112.
KOEHLER, J., 1941, *Phys. Rev.*, **60**, 397–410.
LOVE, A. E. H., 1920, *Mathematical Theory of Elasticity* (Cambridge: University Press).
THOMPSON, N., and MILLARD, D. J., 1952, *Phil. Mag.*, **43**, 422–40.
YOFFE, E., 1958, *Phil. Mag.*, **3**, 8–18.

The Velocity of a Wave along a Dislocation

By T. LAUB and J. D. ESHELBY†
Cavendish Laboratory, Cambridge, England

[Received 13 August 1966]

ABSTRACT

The behaviour of a small-amplitude infinite standing or travelling sine wave on a dislocation is examined. For a certain range of frequencies and wavelengths no elastic energy is radiated, but in general a suitable applied stress is required to maintain the motion. When, however, a certain relation between frequency and wavelength is satisfied no applied forces are necessary and the motion is self-maintaining. This relation may be expressed analytically in the limit of very long wavelengths, but in general it must be evaluated numerically. Curves are given for edge, screw and 60° dislocations in materials with various Poisson ratios. In the course of the calculation a value for the line tension of a sinusoidally deformed stationary dislocation is also obtained.

§ 1. INTRODUCTION

THE expression for the line tension of a stationary dislocation in an isotropic material is well known (cf. § 3). Expressions have also been given for the mass per unit length of a rigidly-moving dislocation (Weertman 1961, Eshelby 1962). It is tempting to apply these results to the string model of a dislocation, and say that the velocity of a wave travelling along it is the square root of the ratio of line tension to mass per unit length. Both the line tension and the mass contain logarithmic factors which depend, respectively, on the shape of the dislocation and the details of its motion. If we assume that these logarithmic factors cancel we find that the velocity of a wave along the dislocation may exceed the velocity, c_t, of transverse elastic waves. For example, if Poisson's ratio, σ, is one-third, the velocity is $c_t\sqrt{2}$ for a pure screw dislocation. This implausible result is not altered by any reasonable adjustment of the logarithmic factors. However, there is no real paradox, since the string model is only an approximation. In the present paper we attempt a direct calculation of the velocity of a sinusoidal wave of small amplitude travelling along a dislocation, or, what amounts to the same thing, of the frequency versus wave-number relation for a travelling or stationary wave.

The results of a similar calculation have been reported by Pegel (1966). So far as they overlap, our results agree with his.

† Now at University of Sheffield.

In § 2 we lay the groundwork for our calculation. In § 3 we use the results obtained to derive an expression for the line tension of a sinusoidally deformed dislocation. Finally, in § 4, we find the characteristic velocity of a sine wave travelling along an infinitely long dislocation.

Throughout, we find it convenient to treat first the particular cases of pure edge and screw dislocations ; the results for a dislocation of mixed character then follow without difficulty.

§ 2. Strain Field ; Radiation

We consider a dislocation with Burgers vector b lying in the x-direction and oscillating in the xz-plane in the form of a standing sine wave :

$$\xi=\xi_0 \cos \kappa x \cos \omega t. \qquad (1)$$

Nabarro (1951) introduced a method for synthesizing the elastic field of a static or deforming loop of dislocation by summing the fields of a suitable distribution of infinitesimal dislocation loops. In the present case the dislocation whose line follows the curve (1) may be derived from the straight dislocation $\xi=0$ by applying a loop of area $\xi\, dx$ to each element dx of the straight dislocation. The field of an infinitesimal oscillating loop is known† (Eshelby 1962), and from it the field of the oscillating dislocation may be found by integration.

For an edge dislocation the displacement is found to be:

$$\left.\begin{aligned}
u_x^{\mathrm{e}} &= u_x^{\mathrm{e}^0}+\operatorname{Re}\frac{b\xi_0}{2\pi}(-\kappa \sin \kappa x)\exp(i\omega t)\\
&\quad\times 2\left(\frac{c_{\mathrm{t}}^2}{c_{\mathrm{l}}^2}\right)\frac{yz}{r^2}\{P_{\mathrm{t}}^2K_2(P_{\mathrm{t}}r)-P_{\mathrm{l}}^2K_2(P_{\mathrm{l}}r)\},\\
u_i^{\mathrm{e}} &= u_i^{\mathrm{e}^0}+\operatorname{Re}\frac{b\xi_0}{2\pi}\cos \kappa x \exp(i\omega t)\\
&\quad\times\left[2\left(\frac{c_{\mathrm{t}}^2}{c_{\mathrm{l}}^2}\right)\left[\frac{x_i}{r^2}\{P_{\mathrm{t}}^2K_2(P_{\mathrm{t}}r)-P_{\mathrm{l}}^2K_2(P_{\mathrm{l}}r)\}\right.\right.\\
&\qquad\left.-\frac{yzx_i}{r^2}\{P_{\mathrm{t}}^3K_3(P_{\mathrm{t}}r)-P_{\mathrm{l}}^3K_3(P_{\mathrm{l}}r)\}\right]\\
&\qquad\left.-P_{\mathrm{t}}K_1(P_{\mathrm{t}}r)\left(\frac{z}{r}\delta_{iy}+\frac{y}{r}\delta_{iz}\right)\right],
\end{aligned}\right\} \qquad (2)$$

where the subscript i takes the values y or z and x_i is to be replaced by y or z. $\mathbf{u}^{\mathrm{e}^0}$ is the displacement due to the undisturbed dislocation,

$$r^2=y^2+z^2, \quad P_{\mathrm{t}}^2=\kappa^2-\frac{\omega^2}{c_{\mathrm{t}}^2}, \quad P_{\mathrm{l}}^2=\kappa^2-\frac{\omega^2}{c_{\mathrm{l}}^2},$$

where c_{t}, c_{l} are the velocities of transverse and longitudinal sound waves.

† In this reference the definition of the quantity ψ_{l} (below (3.6) on p. 227, and preceding (5.23) on p. 235) is incorrect. It should read "ψ_{l} is $c_{\mathrm{t}}^2/c_{\mathrm{l}}^2$ times the same expression with c_{t} replaced by c_{l} throughout".

$K_n(x)$ is the nth order modified Bessel function of the second kind and $H_n^{(2)}(x)$ is a Bessel function of the third kind ; for real x they are connected by the relation $(i)^n K_n(ix) = (-i\pi/2)H_n^{(2)}(x)$.

It may be verified that these displacements satisfy the equations of motion, and that close to the dislocation they agree with the field of a stationary dislocation displaced by the amount (1).

The rate of radiation per unit length is given by :

$$\mathscr{S} = \frac{1}{32}\mu b^2\xi_0{}^2 c_t{}^2\omega^3\left(\frac{1}{c_t{}^4}+\frac{1}{c_l{}^4}-\frac{2\kappa^2}{\omega^2 c_l{}^2}\right), \quad 0<\kappa<\omega/c_l$$

$$= \frac{1}{32}\mu b^2\xi_0{}^2 c_t{}^2\omega^3\left(\frac{1}{c_t{}^4}-\frac{K^4}{\omega^4}\right), \quad \omega/c_l<\kappa<\omega/c_t$$

$$=0, \quad \omega/c_t<\kappa, \quad . \quad . \quad . \quad . \quad (3)$$

where μ is the shear modulus. This reduces to a known result (Eshelby 1962) when κ approaches zero.

Later we shall be interested in the free motion of the oscillating dislocation. These solutions must obviously occur, if at all, in the non-radiating region.

The strain may be obtained by differentiating (2). For $\kappa r \ll 1$ the shear strain in a plane parallel to the plane of motion of the dislocation is given by:

$$e_{yz}{}^{e} = \left\{ e_{yz}{}^{e^0} + \frac{b\xi_0}{2\pi}\cos\kappa x\cos\omega t\left(\frac{c_t{}^2}{\omega^2}\right)\frac{1}{r^2}(P_t{}^2-P_l{}^2)\left(1-\frac{8y^2z^2}{r^4}\right)\right\}$$

$$+\frac{b\xi_0}{4\pi}\cos\kappa x\cos\omega t\left(\frac{c_t{}^2}{\omega^2}\right)\left[\tfrac{1}{2}P_l{}^4\ln(\tfrac{1}{2}P_l r)-\tfrac{1}{2}P_t{}^4\ln(\tfrac{1}{2}P_t r)\right.$$

$$\left.-\left(\frac{\omega^2}{c_t{}^2}\right)P_t{}^2\ln(\tfrac{1}{2}P_t r)-(P_t{}^4-P_l{}^4)\left(\tfrac{1}{2}\gamma+\tfrac{1}{8}-\frac{y^2z^2}{r^4}\right)-\left(\frac{\omega^2}{c_t{}^2}\right)P_t{}^2\gamma\right]$$

$$-\frac{b\xi_0}{16}\cos\kappa x\sin\omega t\left(\frac{c_t{}^2}{\omega^2}\right)\left[P_l{}^4 H(\omega/c_l-\kappa)\right.$$

$$\left.-\left(P_t{}^4-2\frac{c_t{}^2}{\omega^2}P_t{}^2\right)H(\omega/c_t-\kappa)\right], \quad . \quad . \quad . \quad . \quad . \quad . \quad . \quad . \quad . \quad (4)$$

where $H(x)$ is the unit step function defined by :

$$H(x)=0, \quad x\leqslant 0, \quad H(x)=1, \quad x>0.$$

We shall not need the other $e_{ij}{}^{e}$.

Similarly for a screw dislocation lying in the x-direction, and oscillating in the form of the standing sine wave (1), we have :

$$
\left.\begin{aligned}
u_x^{\mathrm{s}} &= u_x^{\mathrm{s}^\circ} + \operatorname{Re}\frac{b\xi_0}{2\pi}\cos\kappa x \exp(i\omega t)\frac{y}{r}\Bigg[2\left(\frac{c_{\mathrm{t}}^2}{\omega^2}\right)\kappa^2\{P_{\mathrm{t}}K_1(P_{\mathrm{t}}r) \\
&\qquad - P_1K_1(P_1r)\} - P_{\mathrm{t}}K_1(P_{\mathrm{t}}r)\Bigg], \\
u_y^{\mathrm{s}} &= u_y^{\mathrm{s}^\circ} + \operatorname{Re}\frac{b\xi_0}{2\pi}(-\kappa\sin\kappa x)\exp(i\omega t)\Bigg[2\left(\frac{c_{\mathrm{t}}^2}{\omega^2}\right)\Bigg\{\frac{y^2}{r^2}(P_{\mathrm{t}}^2K_2(P_{\mathrm{t}}r) \\
&\qquad - P_1^2K_2(P_1r)) - \frac{1}{r}(P_{\mathrm{t}}K_1(P_{\mathrm{t}}r) - P_1K_1(P_1r)\Bigg\} + K_0(P_{\mathrm{t}}r)\Bigg], \\
u_z^{\mathrm{s}} &= u_z^{\mathrm{s}^\circ} + \operatorname{Re}\frac{b\xi_0}{2\pi}(-\kappa\sin\kappa x)\exp(i\omega t)\,2\left(\frac{c_{\mathrm{t}}^2}{\omega^2}\right)\frac{zy}{r^2}\{P_{\mathrm{t}}^2K_2(P_{\mathrm{t}}r) \\
&\qquad - P_1^2K_2(P_1r)\}.
\end{aligned}\right\} \quad (5)
$$

The rate of radiation per unit length is given by :

$$
\begin{aligned}
\mathscr{S} &= \frac{1}{32}\mu b^2\xi_0^2c_{\mathrm{t}}^2\omega^3\left(\frac{1}{c_{\mathrm{t}}^4} - \frac{3\kappa^2}{\omega^2c_{\mathrm{t}}^2} + \frac{4\kappa^2}{\omega^2c_1^2}\right), \quad 0<\kappa<\omega/c_1 \\
&= \frac{1}{32}\mu b^2\xi_0^2c_{\mathrm{t}}^2\omega^3\left(\frac{1}{c_{\mathrm{t}}^4} - \frac{3\kappa^2}{\omega^2c_{\mathrm{t}}^2} + \frac{4\kappa^4}{\omega^4}\right), \quad \omega/c_1<\kappa<\omega/c_{\mathrm{t}} \\
&= 0, \qquad \omega/c_{\mathrm{t}}<\kappa.
\end{aligned} \quad (6)
$$

For $\kappa r \ll 1$ one of the strain components is :

$$
\begin{aligned}
e_{xy}^{\mathrm{s}} &= \left\{e_{xy}^{\mathrm{s}^\circ} + \frac{b\xi_0}{4\pi}\cos\kappa x\cos\omega t\frac{1}{r^2}\frac{y^2-z^2}{r^2}\right\} \\
&+ \frac{b\xi_0}{4\pi}\cos\kappa x\cos\omega t\Bigg[2\kappa^2\left(\frac{c_{\mathrm{t}}^2}{\omega^2}\right)\{P_{\mathrm{t}}^{2\mathrm{L}}\ln(\tfrac{1}{2}P_{\mathrm{t}}r) - P_1^2\ln(\tfrac{1}{2}P_1r)\} \\
&+ (\kappa^2 - \tfrac{1}{2}P_{\mathrm{t}}^2)\ln(\tfrac{1}{2}P_{\mathrm{t}}r) + \left\{2\kappa^2\left(\frac{c_{\mathrm{t}}^2}{\omega^2}\right)(P_{\mathrm{t}}^2 - P_1^2) - \tfrac{1}{2}P_{\mathrm{t}}^2\right\}\left(\gamma - \tfrac{1}{2} - \frac{y^2}{r^2}\right) \\
&+ \kappa^2\gamma\Bigg] + \frac{b\xi_0}{8}\cos\kappa x\sin\omega t\Bigg[2K^2\left(\frac{c_{\mathrm{t}}^2}{\omega^2}\right)P_1^2H(\omega/c_1-\kappa) \\
&+ \left\{-2\kappa^2\left(\frac{c_{\mathrm{t}}^2}{\omega^2}\right)P_{\mathrm{t}}^2 + \tfrac{1}{2}P_{\mathrm{t}}^2 - \kappa^2\right\}H(\omega/c_{\mathrm{t}}-\kappa)\Bigg]. \quad \ldots\ldots \quad (7)
\end{aligned}
$$

We shall not need the others.

§ 3. Line Tension

We first use the results of § 2 to calculate the line tension of a stationary dislocation. This serves to explain the method we shall use in § 4, and is of interest for its own sake.

Consider first a pure edge dislocation. With $\omega = 0$, (4) becomes :

$$
e_{yz}^{\mathrm{e}} = \left\{e_{yz}^{\mathrm{e}^\circ} - \frac{b\xi_0}{2\pi}\cos\kappa x\frac{1}{r^2}\left(1 - \frac{c_{\mathrm{t}}^2}{c_1^2}\right)\left(1 - \frac{8y^2z^2}{r^4}\right)\right\}
$$

$$
+ \frac{b\xi_0}{4\pi}\cos\kappa x\left[-\frac{c_{\mathrm{t}}^2}{c_1^2}\kappa^2\left\{\ln(\tfrac{1}{2}\kappa r) + \gamma + \tfrac{1}{2} - \frac{y^2z^2}{r^4}\right\} + \kappa^2\left(\tfrac{1}{2} - \frac{y^2z^2}{r^4}\right)\right]. \quad (8)
$$

Here the first term (enclosed in curly brackets) represents the strain due to a straight dislocation coinciding with the actual dislocation at $(x,z = \xi)$, that is, it is simply equal to $[1-\xi(\partial/\partial z)]e_{yz}^{e\circ}$. We may, if we wish, go further and say that the straight dislocation is tangential to the curved one at this point ; to terms of order ξ^2 this makes no difference.

To find the applied stress which is necessary to maintain the shape (1) we now argue as follows. The straight dislocation is in elastic equilibrium. Following Peierls and Nabarro we imagine that the material between a pair of planes, $y = \pm\frac{1}{2}a$ say, is removed and that the elastic field outside the gap is maintained by interatomic forces acting across it. We may then say that the straight dislocation is in equilibrium because each atom is in equilibrium. For the curved dislocation this equilibrium is disturbed by the presence of the second term in (8), but it may be restored by applying an equal and opposite strain. The applied strain found in this way may be converted into a stress by multiplying by 2μ, and into a force F per unit length by a further multiplication by b.

If we satisfy this condition exactly at the position $(x, y = \pm\frac{1}{2}a, z = \xi)$ immediately above and below the dislocation, it will be approximately satisfied in a fairly large region surrounding the dislocation because of the logarithmic dependence on r. The terms in (y^2z^2) with $z = \xi$ give a term of order ξ^2 and must thus be discarded in our linear approximation. The remaining constant terms may be absorbed into the logarithm by suitably adjusting the value r_0 of r. We further note that, to first order, $(-\xi_0\kappa^2\cos\kappa x)^{-1}$ is the radius of curvature ρ, say, of the dislocation at x and so we obtain a conventional line-tension formula of the form :

$$F = T/\rho, \qquad (9)$$

where, for the edge dislocation, the line tension has the value :

$$T_e = \frac{\mu b^2}{4\pi}\frac{1-2\sigma}{1-\sigma}\ln\left(\frac{2}{\kappa r_0}\right). \qquad (10)$$

The same analysis for a screw dislocation gives the line tension :

$$T_s = \frac{\mu b^2}{4\pi}\frac{1+\sigma}{1-\sigma}\ln\left(\frac{2}{\kappa r_0}\right). \qquad (11)$$

A mixed dislocation with an angle θ between its Burgers vector and the direction of its line may be regarded as the superposition of edge and screw dislocations with respective Burgers vectors $b\sin\theta$, $b\cos\theta$. The strain field near the dislocation again takes the form of that due to a coincident straight dislocation together with additional terms proportional to the curvature. The straight dislocation would be in equilibrium if the applied shear strain, e_{ij}^{A} say, resolved on its slip-plane and in the direction of its Burgers vector, namely:

$$\cos\theta e_{xy}^{A} + \sin\theta e_{yz}^{A} \qquad (12)$$

were zero. It is unaffected by the component resolved perpendicular to the Burgers vector. The expression analogous to (12) but formed from the complete field of the mixed dislocation can be written as :

$$\cos^2\theta e_{xy}{}^{\mathrm{s}} + \sin^2\theta e_{yz}{}^{\mathrm{e}} + \cos\theta\sin\theta(e_{xy}{}^{\mathrm{e}} + e_{yz}{}^{\mathrm{s}}), \quad . \quad . \quad . \quad (13)$$

where $e_{xy}{}^{\mathrm{s}}$, $e_{yz}{}^{\mathrm{e}}$ are given by (7), (4) respectively, and $e_{yz}{}^{\mathrm{s}}$, $e_{xy}{}^{\mathrm{e}}$ could be calculated similarly. As before, we imagine that if the curved form of the dislocation is to be maintained the terms in (13) which do not represent the field of the straight dislocation must be cancelled by the applied strain. It is found that the cross-term in $\cos\theta\sin\theta$ only gives terms of order ξ^2, and so must be omitted. If the applied strain is, as before, converted into a force F per unit length we easily find that (9) must be satisfied with

$$T = T_{\mathrm{s}}\cos^2\theta + T_{\mathrm{e}}\sin^2\theta$$
$$= \frac{\mu b^2}{4\pi}\left(\frac{1+\sigma-3\sigma\sin^2\theta}{1+\sigma}\right)\ln(R/r_0), \quad . \quad . \quad . \quad . \quad (14)$$

with $R = 2/\kappa$. This agrees in form with accepted values of the line tension. In general the length R depends on the precise shape of the deformed dislocation. We see that for a sinusoidal deformation we must take R to be the wavelength divided by π.

§ 4. Dispersion Law ; Wave Velocity

By a similar argument (cf. Eshelby 1953) it is possible to calculate the applied stress required to maintain the oscillatory motion (1) of our sinusoidally deformed dislocation. Here we only consider the case of free motion, that is, one which requires no applied force to maintain it, and therefore we require the impressed stress to be zero. In this way we shall obtain a dispersion law $\omega = \omega(\kappa)$ relating the frequency ω to the wave-number κ, which may conveniently be written :

$$\omega(\kappa) = \eta(\kappa)\kappa c_{\mathrm{t}}. \quad . \quad . \quad . \quad . \quad . \quad . \quad . \quad . \quad . \quad (15)$$

On our linear approximation we may also superpose on a dislocation two suitably phased standing waves having the same κ and ω so as to obtain a travelling sine wave, whose characteristic velocity would be given by :

$$c_{\mathrm{d}} = \eta(\kappa)c_{\mathrm{t}}. \quad . \quad . \quad . \quad . \quad . \quad . \quad . \quad . \quad . \quad (16)$$

We consider first an edge dislocation. The term in curly brackets in (4) represents, in the neighbourhood of x, the strain due to a dislocation displaced statically by (1) and coinciding locally with the moving dislocation. Equally well it represents the field of a straight dislocation coincident with and moving uniformly with the same instantaneous velocity as the actual dislocation, since the field of a uniformly moving dislocation differs from that of a stationary one only by terms of order $v^2/c^2 \sim \xi_0{}^2\omega^2$, and we are only retaining terms linear in ξ_0. The motion of the atoms near the centre of a uniformly moving dislocation is a dynamically possible one. Consequently the motion of the curved dislocation will also be possible if we apply an external field which annuls the second and third terms in (4), and we could use this argument to find their magnitude.

However, we are at present only interested in the case where no external force is needed, and so the second and third terms must both vanish. If $\omega/c_t \leqslant \kappa$ the third term does not appear and we have the condition for free motion :

$$\begin{aligned}-E \equiv \ln(\tfrac{1}{2}\kappa r_0)\{2\alpha-\eta^2(1+\alpha^2)\}+\ln(1-\eta^2)\,.\,\tfrac{1}{2}(\eta^{-2}-\eta^2)\\ -\ln(1-\alpha\eta^2)\,.\,\tfrac{1}{2}(\eta^{-2}-2\alpha+\alpha^2\eta^2)\\ +\gamma\{2\alpha-\eta^2(1+\alpha^2)\}+\tfrac{1}{4}\{2\alpha-2+\eta^2(1+\alpha^2)\}=0, \quad . \quad . \quad (17)\end{aligned}$$

where we have introduced the abbreviations :

$$\eta=\omega/\kappa c_t, \quad \alpha=c_t^2/c_l^2.$$

For later use we note the relations :

$$\sigma=\frac{1-2\alpha}{2(1-\alpha)}, \quad 4\sigma(1-\sigma)=\frac{2\alpha}{2\alpha-1-\alpha^2}.$$

It is easy to verify that in the range $\omega/c_t>\kappa$, where the third term in (4) does not vanish automatically, there is no solution, as is to be expected from the fact that the motion is then radiative (§ 2).

For a screw dislocation (7) leads, similarly, to the following condition for free motion :

$$\begin{aligned}S\equiv\ln(\tfrac{1}{2}\kappa r_0)(2\alpha-\tfrac{3}{2}+\tfrac{1}{2}\eta^2)+\ln(1-\eta^2)(\eta^{-2}-\tfrac{3}{4}+\tfrac{1}{4}\eta^2)\\ -\ln(1-\alpha\eta^2)(\eta^{-2}-\alpha)+(\gamma-\tfrac{1}{2})(2\alpha-\tfrac{3}{2}+\tfrac{1}{2}\eta^2)+\tfrac{1}{2}=0, \qquad (18)\end{aligned}$$

if $\omega/c_t\leqslant\kappa$, and again there is no solution for $\omega/c_t>\kappa$.

For a mixed dislocation we must apply the same argument to the combination of strains as that which appears in § 3. As in the static case the cross-term in $\cos\theta\sin\theta$ is found to be of a higher order in ξ than we are retaining. Thus the condition for free motion is, for $\omega/c_t\leqslant\kappa$:

$$S\cos^2\theta+E\sin^2\theta=0,$$

i.e.

$$\begin{aligned}\ln(\tfrac{1}{2}\kappa r_0)[\cos^2\theta\{2\alpha-\tfrac{3}{2}+\tfrac{1}{2}\eta^2\}-\sin^2\theta\{2\alpha-\eta^2(1+\alpha^2)\}]\\ +\ln(1-\eta^2)[\cos^2\theta\{\eta^{-2}-\tfrac{3}{4}+\tfrac{1}{4}\eta^2\}-\sin^2\theta\,.\,\tfrac{1}{2}\{\eta^{-2}-\eta^2\}]\\ -\ln(1-\alpha\eta^2)[\cos^2\theta\{\eta^{-2}-\alpha\}-\sin^2\theta\,.\,\tfrac{1}{2}\{\eta^{-2}-2\alpha+\alpha^2\eta^2\}]\\ +\cos^2\theta\{(2\alpha-\tfrac{3}{2}+\tfrac{1}{2}\eta^2)(\gamma-\tfrac{1}{2})+\tfrac{1}{2}\}-\sin^2\theta\{\gamma(2\alpha-\eta^2(1+\alpha^2))\\ +\tfrac{1}{4}(2\alpha-2+\eta^2(1+\alpha^2))\}=0. \quad . \quad . \quad . \quad . \quad (19)\end{aligned}$$

Again, there is no solution for $\omega/c_t>\kappa$.

For arbitrary κ the solutions of (19) can only be obtained numerically, but it is possible to obtain analytical expressions for the asymptotic values of η in the limit of long wavelengths ($\kappa\to0$). Examination of (19) shows that two cases must be distinguished. If

$$\tan^2\theta<4\sigma(1-\sigma), \quad . \quad . \quad . \quad . \quad (20)$$

then

$$\eta=1 \quad (\kappa\to0),$$

and the corresponding limiting velocity is c_t. On the other hand, if

$$\tan^2\theta>4\sigma(1-\sigma),$$

402

then

$$\eta^2 = \frac{(\frac{3}{2} - 2\alpha)\cos^2\theta + 2\alpha\sin^2\theta}{\frac{1}{2}\cos^2\theta + (1+\alpha^2)\sin^2\theta} \quad (\kappa \to 0), \quad . \quad . \quad . \quad . \quad (21)$$

and the corresponding limiting velocity is ηc_t.

The figure gives some typical results of numerical solutions of (19).

§ 5. Discussion

The results of § 4 show that a dislocation behaves as a weakly dispersive one-dimensional medium for the propagation of infinite wave trains. For very long waves on an edge dislocation the velocity agrees with that calculated from the string model on the assumption that the logarithmic factors in the line tension and mass cancel. In all other cases the velocities are less than those given by the string model, and they never exceed the velocity of transverse elastic waves.

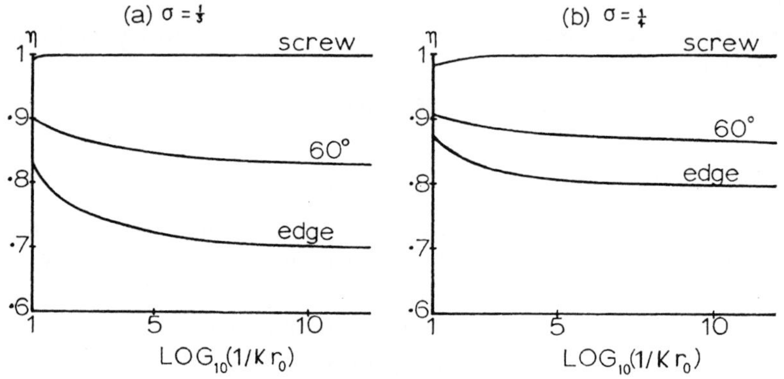

Velocity ηc_t versus wave-number κ curves for edge, screw and 60° dislocations in materials with Poisson's ratio, σ, having the values (*a*) $\frac{1}{3}$ and (*b*) $\frac{1}{4}$.

If $\omega/\kappa > c_t$ a travelling or standing wave radiates (eqns. (3), (6)) and a suitable applied stress is required to maintain it. There are, of course, also possible motions of the form (1) in the non-radiative region which do not obey the dispersion law $\omega = \omega(\kappa)$ which we have derived. They also require a suitable applied stress to maintain them, but it does no net work and the dislocation behaves as a pure reactance.

Our results, which refer to an infinite train of progressive or standing waves, cannot be directly applied to, for example, the oscillation of a dislocation loop held at two pinning points. A comparison with electromagnetic theory will make this clear. An infinite wire carrying a current distribution proportional to $\cos \kappa x \cos \omega t$ does not radiate provided ω/κ is less than the velocity of electromagnetic waves. However, if we suppress all but one half wave-length of the current distribution we are left, approximately, with the case of a half-wave aerial. The oscillation is then damped, and the current distribution is not quite sinusoidal. It is to be expected, similarly, that a finite section of dislocation pinned at the ends will have damped proper vibrations in which (1) is replaced by $\xi = f(x) \cos \omega t \exp(-\lambda t)$, where $f(x)$ is nearly, though not exactly, sinusoidal. It is hoped to discuss these points elsewhere.

Acknowledgments

One of us (T.L.) would like to thank the Council for Scientific and Industrial Research (South Africa) for financial support during the course of this work.

References

Eshelby, J. D., 1953, *Phys. Rev.*, **90,** 248 ; 1962, *Proc. roy. Soc.* A, **266,** 222.
Nabarro, F. R. N., 1951, *Phil. Mag.*, **42,** 1224.
Pegel, B., 1966, *Phys. Stat. Sol.*, **14,** K165.
Weertman, H., 1961, *Response of Metals to High Velocity Deformation*, edited by P. G. Shewmon and V. F. Zackay (New York : Interscience), p. 205.

THE INTERPRETATION OF TERMINATING DISLOCATIONS[1]

J. D. ESHELBY AND T. LAUB
Cavendish Laboratory, Cambridge, England

A physical meaning is suggested for Li's terminating dislocation, based on Heaviside's interpretation of the magnetic field, which the formula of Biot and Savart gives when applied to an incomplete circuit.

1. INTRODUCTION. TERMINATING DISLOCATIONS

When one has to calculate the stresses due to a dislocation configuration made up of straight lines (a hexagon or a tetrahedron, for example), it is useful to know the stress field of an angular dislocation, that is to say a dislocation made up of two semi-infinite straight segments meeting at an angle (Fig. 1(*a*)).

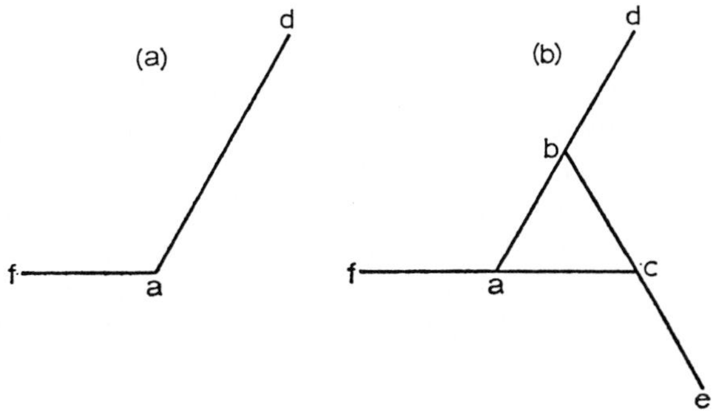

FIG. 1. Angular dislocations.

Yoffe (1960) introduced the angular dislocation, calculated its elastic field, and showed how it could be used to find the field of a dislocation in the form of a polygon. Thus, in Fig. 1(*b*) the triangular dislocation *abc* can be regarded as being made up of the angular dislocations *fad*, *dbe*, *ecf*. If *ab*, *bc*, *cd* are arranged to have the same Burgers vector, the two coincident dislocations *fa* automatically have opposite Burgers vectors and cancel, and similarly for *db*, *ec*.

Li (1964) went a step further and noticed that the expression for the stress due to an angular dislocation could be split into two parts, each of which represented, in some sense, the stress due to one arm of the angle. In this way, we reach the concept of a semi-infinite dislocation which terminates at one end. (Brown (1964) also found it convenient to introduce the idea of a terminating dislocation.) One can, of course, then go on to consider a dislocation which terminates at both ends, so that a loop like *abc* (Fig. 1(*b*)) can be regarded as being made up, not of angular dislocations, but of singly terminated

[1]Presented at an international conference on the Deformation of Crystalline Solids, held in Ottawa, August 22–26, 1966.

Canadian Journal of Physics. Volume 45 (1967)

dislocations *ad*, *be*, *cf*, together with singly terminated dislocations *bd*, *ce*, *af* of opposite sign, or, more simply, as being composed of three doubly terminated dislocations *ab*, *bc*, *ca*. The stress found by adding together the stresses of the appropriate terminating dislocations is the same as that obtained by integrating Peach and Koehler's (1950) line integral round the loop *abc*. Consequently if we only use them to build up the stress fields of closed loops, we do not have to ask what terminated dislocations mean, if anything. Still, it is interesting, and perhaps useful, to know the answer to this question.

2. INTERPRETATION WITH THE HELP OF THE MAGNETIC ANALOGY

The clue to the physical meaning of terminated dislocations is provided by the well-known analogy between dislocations and current-carrying wires, Burgers vector and stress, corresponding, respectively, to current and magnetic field. This analogy is only qualitative, and is also in some respects misleading because like dislocations repel but wires carrying like currents attract. Consequently a pair of dislocation loops with the same Burgers vector interact roughly like two wire loops carrying opposite currents, but there is no magnetic analogue for three loops. For the same reason, a dislocation has a positive line tension, and tends to straighten out, but a flexible current-carrying filament has a negative line tension (Thompson 1964) and tends to ravel up.

In the 1840's Grassmann (1845) introduced the magnetic analogue of an angular dislocation, the angular current (Winkelstrom) as an aid in discussing the fields of current loops; Fig. 1(*b*) is, in fact, copied from his paper. Later Heaviside (1888) considered the analogue of a terminating dislocation. Biot and Savart's law states that the magnetic field $\boldsymbol{H}$ produced by a current I flowing in a closed loop of wire l is given, in suitable units, by the line integral

$$\boldsymbol{H}(\boldsymbol{r}) = I \int_l \frac{\mathrm{d}\boldsymbol{l} \times \boldsymbol{R}}{R^3}, \tag{1}$$

where $\boldsymbol{R} = \boldsymbol{r} - \boldsymbol{r}'$ is the vector joining the point of observation $\boldsymbol{r}$ to the position $\boldsymbol{r}'$ of the line element $\mathrm{d}\boldsymbol{l}$.

There is nothing to stop us working out the integral (1) over a finite unclosed length of wire with apparent violation of current conservation at the ends. What does the result represent?

In effect, Heaviside gave the following answer. Imagine that the wire is insulated except at the tips, and that it is immersed in a conducting liquid. At one end the current flows radially out into the liquid and at the other end it flows radially in (Fig. 2(*a*)), so that the current is, in fact, continuous (Fig. 2(*b*)). Then (1) integrated along the wire gives the magnetic field of the whole current distribution, namely the current in the wire plus the backflow in the liquid. If several wire segments each carrying the same current are joined up to make a closed loop, the inflows and outflows at the junctions cancel and the presence of the liquid is irrelevant.

This obviously suggests the following interpretation for Li's semi-infinite terminating dislocation: it is a bundle of infinitesimal dislocations of total Burgers vector $\boldsymbol{b}$ which splay out in all directions at the "cut" end (Fig. 2(*c*)). We thus have a naturally occurring case of a continuous distribution of dis-

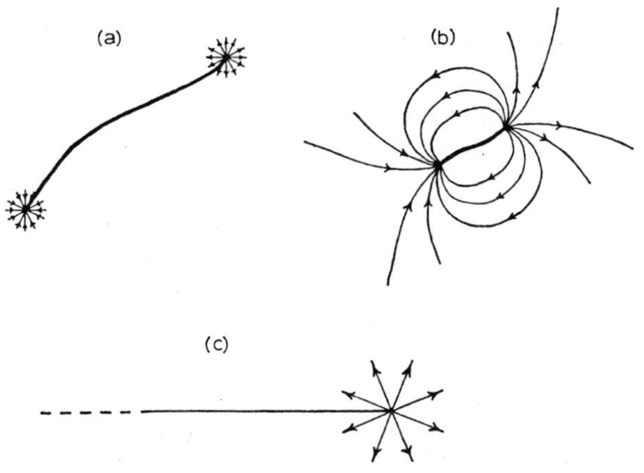

FIG. 2. Terminating currents and dislocations.

locations (Kröner 1958; Bilby 1960). Such a distribution may be specified by a tensor α_{ij} defined so that $db_i = \alpha_{ij}n_j\, dS$ is the total Burgers vector of all the infinitesimal dislocations which thread a surface element of area dS and normal n_j. It is a reasonable guess that, like the return current from the end of a semi-infinite wire, the infinitesimal dislocations spread out uniformly in all directions from the cut end. If this is so, we must have

(2) $$\alpha_{ij} = b_i x_j / 4\pi r^2,$$

where r, x_j are the distance and Cartesian coordinates of the point of observation relative to the cut end, for then the flux of dislocations is the same through all cross sections of any cone with its vertex at the cut end, and also the total flux is b_i.

We can verify that this is a legitimate interpretation in the following way. In the first place, the strains e_{ij} derived from Li's stresses with the help of Hooke's law are, as is to be expected, incompatible, that is to say that, when substituted into the relation

(3) $$\eta_{ij} = -\epsilon_{ikm}\epsilon_{jln}\partial_m\partial_n e_{kl},$$

they give a nonzero value for the incompatibility tensor η_{ij} (also denoted by S_{ij}). (We use ∂_m as an abbreviation of $\partial/\partial x_m$, and ϵ_{ikm} is the usual symbol which is completely antisymmetric in i, k, m and has $\epsilon_{123} = 1$.) On the other hand η_{ij} is related to the dislocation density tensor α_{ij} by

(4) $$\eta_{ij} = \epsilon_{ptq}\partial_q(\alpha_{pj} - \tfrac{1}{2}\alpha_{mm}\delta_{pj}).$$

In principle, a knowledge of η_{ij} completely determines the stress field, provided the boundary conditions and any body forces are also prescribed.

A direct calculation shows that the result of substituting Li's stresses (converted to strains) into (3) is the same as the result of substituting (2) into (4). Hence the dislocation distribution (2) does indeed produce these stresses. To (2) we must add a singular distribution of α_{ij} to represent the dislocation line itself. If the dislocation is along the negative x_3 axis, the appropriate distribution is

(5) $$\alpha_{ij} = b_i\delta_{j3}\delta(x_1)\delta(x_2)H(-x_3),$$

where $\delta(x)$ is a delta function and $H(x) = 1$, $x > 0$, $H(x) = 0$, $x < 0$, so that formally $H'(x) = \delta(x)$. The divergences of (2) and (5) are $\partial_j\alpha_{ij} = b_i\delta(\boldsymbol{r})$ and $\partial_j\alpha_{ij} = -b_i\delta(\boldsymbol{r})$ respectively, so that the total α_{ij} is divergence-free, as it must be.

3. THE RATIONAL DISLOCATION ELEMENT

If in Fig. 2(*b*) we let the vector joining the ends of the wire become the infinitesimal vector dl_i, we get what Heaviside calls a "rational current element". The current density is

(6) $$dj_i = Idl_i\delta(\boldsymbol{r}) + Idl_m\partial_m\partial_i(1/4\pi r),$$

where $I\,dl_i$ is, of course, to be interpreted in the sense of a doublet, that is $dl_i \to 0$, $I \to \infty$, with $I\,dl_i$ remaining constant. The first term represents the current in the wire, in the form of a blob of current density of total amount $I\,dl_i$, and the second is the current in the liquid, the flow associated with an equal source and sink separated by dl_i. We can also give the second term another interpretation. The current element is equivalent to a charge q moving with velocity v_i, provided $qv_i = I\,dl_i$. The electric field of the charge is $E_i = -\partial_i(qr^{-1})$. (We suppose that it is moving slowly: since only the product qv_i is specified, this can always be arranged.)

The displacement current is thus

$$\frac{1}{4\pi}\frac{\partial E_i}{\partial t} = -\frac{1}{4\pi}v_m\partial_m E_i = qv_m\partial_m\partial_i(1/4\pi r),$$

so that (6) may be regarded as the sum of the conduction and displacement currents due to the moving charge. This interpretation seems to have no application to dislocations, since, although they behave somewhat like currents, what they are currents of is not clear. With the help of the formal relation $\nabla^2(1/r) = -4\pi\delta(\boldsymbol{r})$ eq. (6) may be written as

(7) $$dj_j = I\,dl_j(\partial_m\partial_j - \nabla^2\delta_{mj})(1/4\pi r).$$

In this form it is obvious that the divergence of dj_i is zero. It may be verified directly by calculating $(1/4\pi)$ curl $d\boldsymbol{H}$, where $d\boldsymbol{H}$ is the contribution of $d\boldsymbol{l}$ to (1), and using a suitable vector identity.

We may similarly introduce a "rational dislocation element", a doubly terminated dislocation whose ends are separated by the small vector dl_i. Its dislocation density tensor is obtained from (6) or (7) simply by changing the scalar I into the vector b_i:

(8) $$d\alpha_{ij} = b_i dl_m(\partial_m\partial_j - \nabla^2\delta_{mj})(1/4\pi r).$$

If we lay a set of such elements end to end along the negative x_3 axis, we obtain, as expected, the density (2) plus (5) appropriate to the semi-infinite dislocation.

The stress associated with the distribution (8) can be found by a method due to Kröner (1958). We first calculate the incompatibility tensor corresponding to (8), substitute it into Kröner's differential equation for the stress function χ_{ij}, solve it, and calculate the stresses from χ_{ij}. The arbitrary complementary solution may be fixed by requiring that the stress fall off with increasing distance from the element. The result is

$$(9)\quad 8\pi \mathrm{d}\sigma_{ij}/\mu = [b_n \mathrm{d}l_s \partial_m(\epsilon_{jmn}\delta_{si} + \epsilon_{imn}\delta_{sj})(1/r) + (1-\sigma)^{-1} b_n \mathrm{d}l_s \partial_m \epsilon_{smn}(\partial_i\partial_j - \nabla^2\delta_{ij})r] + \{-\tfrac{1}{2}\mathrm{d}l_s\partial_s b_n \partial_m(\epsilon_{jmn}\partial_i + \epsilon_{imn}\partial_j)r\},$$

where μ is the shear modulus and σ is Poisson's ratio.

A finite (terminated) dislocation can be made up by joining rational dislocation elements (with the same b_i) end to end. The stress is evidently

$$(10)\qquad \sigma_{ij} = \int \frac{\mathrm{d}\sigma_{ij}}{\mathrm{d}l}\,\mathrm{d}l.$$

If the integral is carried round a closed loop, the term in { } in (9) contributes nothing to (10) because of the operator $\mathrm{d}l_s\partial_s$. What is left is Peach and Koehler's (1950) expression for the stress due to a loop; see, for example, de Wit (1960), eq. (7.4). The extra term { } ensures that (9) satisfies the equilibrium equations. There is a certain distribution of body force associated with the main term [ ] and an equal and opposite distribution associated with { }. As it is compatible, the term { } does not interfere with the dislocation density. When a set of elements is joined together to make a closed loop, the body forces associated with [ ] cancel out along with the fluxes of infinitesimal dislocations at the joints, and it does not matter whether the term { } is retained or not. Li's expression for the stresses of a semi-infinite dislocation are the result of (9) with { } omitted.

Evidently there will be an energy of interaction between two rational current or dislocation elements. For a pair of his current elements (i, d$\boldsymbol{l}$), (i', d$\boldsymbol{l}'$) separated by the vector $\boldsymbol{R}$ Heaviside obtained

$$(11)\qquad E_{\mathrm{int}} = -\tfrac{1}{2}ii'\left\{\frac{\mathrm{d}\boldsymbol{l}\cdot\mathrm{d}\boldsymbol{l}'}{R} + \frac{(\mathrm{d}\boldsymbol{l}\cdot\boldsymbol{R})(\mathrm{d}\boldsymbol{l}'\cdot\boldsymbol{R})}{R^3}\right\},$$

"if I have correctly calculated it". One can verify that he had done so by way of the interpretation of (6) in terms of moving charges. Equation (11) should then be, and in fact is, equal to the appropriate term in Darwin's approximate Lagrangian for a pair of moving charges (Landau and Lifshitz 1962).

The corresponding expression for a pair of rational dislocation elements ($\boldsymbol{b}$, d$\boldsymbol{l}$), ($\boldsymbol{b}'$, d$\boldsymbol{l}'$) can be found by calculating the total elastic energy of their combined field, retaining only cross terms. The calculation can be simplified by an intelligent use of Green's theorem, but the result is considerably more involved than (11). It can be written

$$E_{\mathrm{int}} = \frac{\mu}{96\pi}(\boldsymbol{b}\times\nabla)\cdot(\boldsymbol{b}'\times\nabla)(\mathrm{d}\boldsymbol{l}\times\nabla)\cdot(\mathrm{d}\boldsymbol{l}'\times\nabla)R^3 + \frac{\mu}{16\pi}[\mathrm{d}\boldsymbol{l}\boldsymbol{b}\nabla][\mathrm{d}\boldsymbol{l}'\boldsymbol{b}'\nabla]R + \frac{\mu}{8\pi}\frac{\sigma}{1-\sigma}[\mathrm{d}\boldsymbol{l}\boldsymbol{b}'\nabla][\mathrm{d}\boldsymbol{l}'\boldsymbol{b}\nabla]R,$$

where $[\mathrm{d}\boldsymbol{l}\boldsymbol{b}\nabla] = (\mathrm{d}\boldsymbol{l}\times\boldsymbol{b})\cdot\nabla$, or in suffix notation

$$(12)\quad E_{\mathrm{int}} = \frac{\mu}{8\pi} b_n \mathrm{d}l_s \partial_m b'_\nu \mathrm{d}l'_\sigma \partial'_\mu \left[\epsilon_{kmn}\epsilon_{k\mu\nu}\delta_{s\sigma} + \frac{1+\sigma}{1-\sigma}\epsilon_{smn}\epsilon_{\sigma\mu\nu}\right]R - \left\{\frac{\mu}{96\pi} b_n \partial_m{}' b_\nu{}' \partial_\mu \mathrm{d}l_s \partial_s \mathrm{d}l_\sigma{}' \partial_\sigma \epsilon_{kmn}\epsilon_{k\mu\nu} R^3\right\}.$$

The energy of interaction of two closed loops may be found by integrating with respect to $\mathrm{d}l$ and $\mathrm{d}l'$. Because of the operators $\mathrm{d}l_s\partial_s$, $\mathrm{d}l_\sigma\partial_\sigma$, the term in { } makes no net contribution, and what is left is identical with an expression given by Kröner for the interaction energy.

So long as we only join rational dislocation elements together to form closed loops, the extra terms in (9) and (12) serve no useful purpose. However, we may obviously go on to consider volume distributions of such elements, and then these terms are essential.

It is sometimes convenient to regard a state of self-stress as caused by the distribution of some type of source of internal stress. Possible sources are a distribution of infinitesimal dislocation loops of density γ_{ij} (Kroupa 1963), a distribution of initial or transformation strain e_{ij}^{T} (Timoshenko and Goodier 1951; Eshelby 1961), a distribution of infinitesimal dislocations specified by α_{ij} or of infinitesimal disclinations (Nabarro 1965) specified by η_{ij}. The distributions $\gamma_{ij}(\boldsymbol{r})$ or $e_{ij}^{\mathrm{T}}(\boldsymbol{r})$ may be prescribed at will, but $\alpha_{ij}(\boldsymbol{r})$ and $\eta_{ij}(\boldsymbol{r})$ must be functions whose divergence vanishes.

We may now add to these a new type of source, a distribution of rational dislocation elements specified by a tensor, $m_{ij}(\boldsymbol{r})$ say, equal to the sum of the moments $b_i\mathrm{d}l_j$ of all the elements in unit volume. This tensor, like γ_{ij} and e_{ij}^{T} may be made an arbitrary function of position. The relation between m_{ij} and α_{ij} is similar to that between magnetization $\boldsymbol{M}$ and magnetic induction $\boldsymbol{B}$; when $\boldsymbol{M}$ or m_{ij} (with nonzero divergence) has been prescribed, $\boldsymbol{B}$ or α_{ij} may be calculated from it and automatically has zero divergence. Possibly m_{ij} may prove to be a useful addition to the list of source functions.

ACKNOWLEDGMENT

One of us (T. Laub) would like to thank the Council for Scientific and Industrial Research (South Africa) for financial support during the course of this work.

REFERENCES

BILBY, B. 1960. Progress in solid mechanics, Vol. 1 (North-Holland, Amsterdam), p. 331.
BROWN, L. M. 1964. Phil. Mag. **10**, 441.
DE WIT, R. 1960. Solid state physics, Vol. 10, *edited by* Seitz (Academic Press, New York), p. 249.
ESHELBY, J. D. 1961. Progress in solid mechanics, Vol. 2 (North-Holland, Amsterdam), p. 89.
GRASSMANN, H. G. 1845. Ann. Physik, **64**, 1.
HEAVISIDE, O. 1888. The Electrician, Vol. 18, p. 229 (Electrical Papers, **2**, 500, Macmillan, London, 1892).
KRÖNER, E. 1958. Kontinuumstheorie der Versetzungen und Eigenspannungen (Springer-Verlag, Berlin).
KROUPA, F. 1963. Czechoslov. J. Phys. **A13**, 301.
LANDAU, L. D. and LIFSHITZ, E. M. 1962. Classical theory of fields (Addison-Wesley, Reading, Mass.), p. 193.
LI, J. C. M. 1964. Phil. Mag. **10**, 1097.
NABARRO, F. R. N. 1965. *In* Intern. Conf. Electron Diffraction and Crystal Defects, *edited by* R. C. Gifkin (Melbourne), p. II L-1.
PEACH, M. and KOEHLER, J. S. 1950. Phys. Rev. **80**, 436.
THOMPSON, W. B. 1964. Introduction to plasma physics (Pergamon Press, Oxford), p. 107.
TIMOSHENKO, S. and GOODIER, J. N. 1951. Theory of elasticity (McGraw-Hill, New York), p. 425.
YOFFE, E. H. 1960. Phil. Mag. **5**, 161.

CHAPTER 2 Stress analysis: theory of elasticity

This chapter summarizes the elements of the theory of elasticity, illustrated by some problems basic in fracture mechanics. It is not absolutely necessary to absorb the whole chapter to appreciate the succeeding one. However, it is hoped that readers not already familiar with the topic may find that the application of complex variable theory to elasticity is not in fact very difficult, particularly if we start with the simple case of anti-plane strain (Mode III deformation).

DISPLACEMENT AND STRAIN

When a body is deformed a small particle of the material at (x, y, z) will move to

$$x' = x + u(x, y, z)$$

$$y' = y + v(x, y, z)$$

$$z' = z + w(x, y, z)$$

where u, v, w are the components of the *displacement vector.*

A sufficiently small cube with edges parallel to the coordinate axes and of length ϵ (Fig. 1) will be deformed into a parallelepiped. Knowledge of the new lengths of three concurrent edges and the angles between them defines the deformation, though not the orientation, of the parallelopiped (Fig. 2). For small (infinitesimal) deformation the *strain components* e_{xx}, e_{xy}, etc. are defined by new edge lengths $= \epsilon\,(1 + e_{xx}), \epsilon(1 + e_{yy}), \epsilon(1 + e_{zz})$, angles between them $= \tfrac{1}{2}\pi - 2e_{xy}, \tfrac{1}{2}\pi - 2e_{yz}, \tfrac{1}{2}\pi - 2e_{zx}$. A little geometry shows that

$$e_{xx} = \partial u/\partial x, \ldots, e_{xy} = \tfrac{1}{2}\,(\partial u/\partial y + \partial v/\partial x) \qquad (1)$$

STRESS

Suppose a hole is excavated in the material with one face flat and of area A and normal **n** (Fig. 3). If the displacements and strains in the rest of the material are not to alter (no relaxation), a distribution of forces must be applied to the freshly formed surfaces. If **F** is the total force applied to the flat face, then

$$\sigma = \lim_{A \to 0} \mathbf{F}/A$$

is the *stress* on a surface with normal **n**. There will be a different σ for each choice of **n**: $\sigma = \sigma(\mathbf{n})$. However, if σ is known for three different **n**'s it can be found for any **n**. Let σ become $\sigma_x, \sigma_y, \sigma_z$ (vectors) when $\mathbf{n} = (1, 0, 0)$, $(0, 1, 0)$, $(0, 0, 1)$, respectively. The set of their components,

$$\begin{array}{llll} \sigma_x: & \sigma_{xx}, & \sigma_{xy}, & \sigma_{xz} \\ \sigma_y: & \sigma_{yx}, & \sigma_{yy}, & \sigma_{yz} \\ \sigma_z: & \sigma_{zx}, & \sigma_{zy}, & \sigma_{zz} \end{array}$$

make up the *stress tensor*, usually displayed in a picture like Fig. 4.

If we consider the equilibrium of a penny-shaped piece of material (Fig. 5) we get

$$\sigma(\mathbf{n}) = -\sigma(-\mathbf{n}).$$

Next consider the equilibrium of a tetrahedron (Fig. 6) with front face of area A and normal $\mathbf{n} = (l, m, n)$. The other faces have their edges parallel to the coordinate axes. Their areas are Al, Am, An, and their normals are $(-1, 0, 0)$, $(0, -1, 0)$, $(0, 0, -1)$. So if the material inside the tetrahedron is in equilibrium, we must have

$$A\sigma(\mathbf{n}) - Al\sigma_x - Am\sigma_y - An\sigma_z = 0$$

(The minuses come from $\sigma(\mathbf{n}) = -\sigma(-\mathbf{n})$).

Consequently

$$\sigma(\mathbf{n}) = \sigma_x l + \sigma_y m + \sigma_z n$$

so that the force per unit area on the plane with normal **n** has

$$\begin{array}{ll} \text{x-component} & \sigma_{xx}l + \sigma_{xy}m + \sigma_{xz}n, \\ \text{y-component} & \sigma_{yx}l + \sigma_{yy}m + \sigma_{yz}n, \\ \text{and z-component} & \sigma_{zx}l + \sigma_{zy}m + \sigma_{zz}n. \end{array} \tag{2}$$

Also, by taking moments we get

$$\sigma_{xy} = \sigma_{yx}, \quad \sigma_{yz} = \sigma_{zy}, \quad \sigma_{zx} = \sigma_{xz}.$$

EQUILIBRIUM EQUATIONS

The x-components of the forces on the six faces of the small cube of Fig. 1 are approximately

$$\epsilon^2\left(\pm\sigma_{xx} + \tfrac{1}{2}\,\epsilon\frac{\partial\sigma_{xx}}{\partial x}\right), \quad \epsilon^2\left(\pm\sigma_{xy} + \tfrac{1}{2}\,\epsilon\frac{\partial\sigma_{xy}}{\partial y}\right), \quad \epsilon^2\left(\pm\sigma_{xz} + \tfrac{1}{2}\,\epsilon\frac{\partial\sigma_{xz}}{\partial z}\right)$$

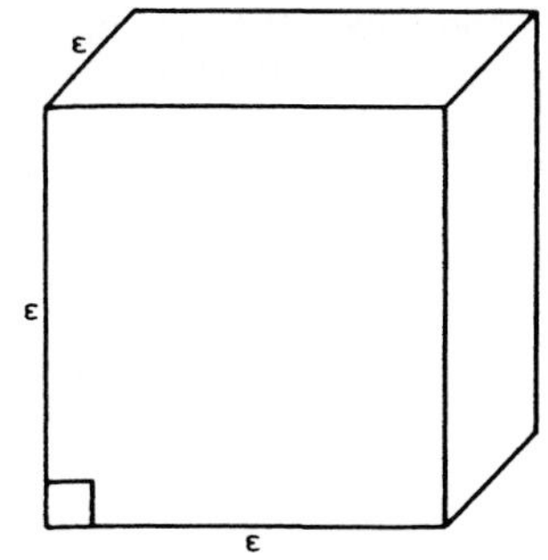

Fig. 1

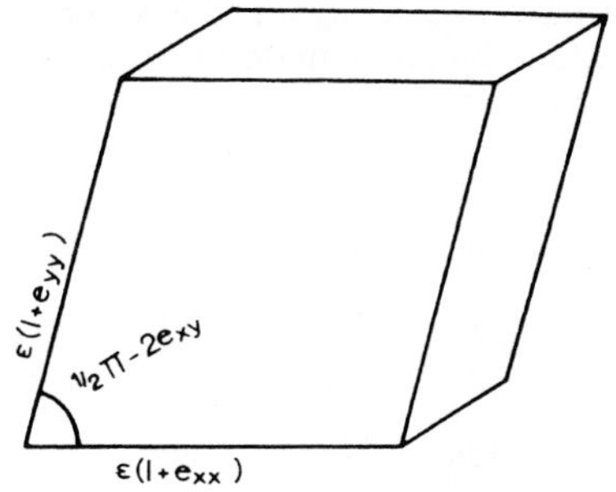

Fig. 2

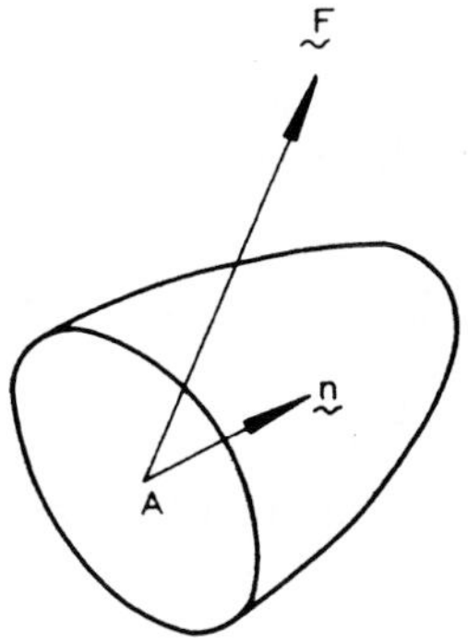

Fig. 3

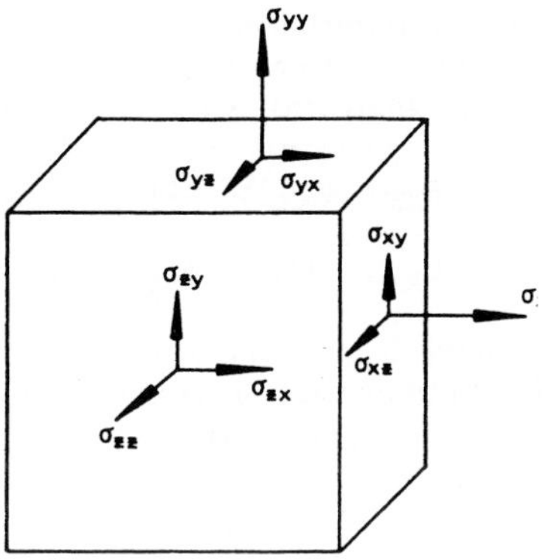

Fig. 4

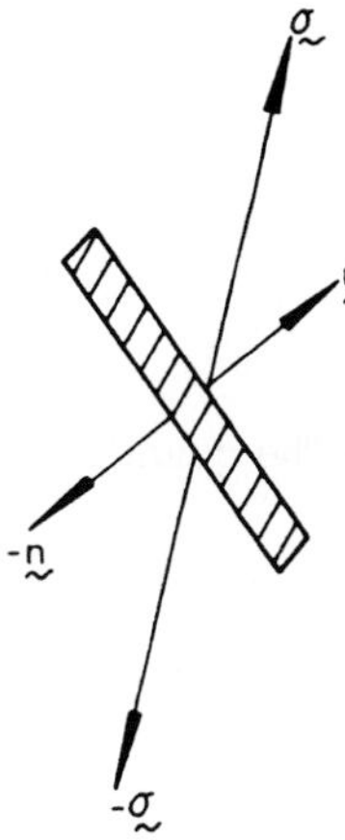

Fig. 5

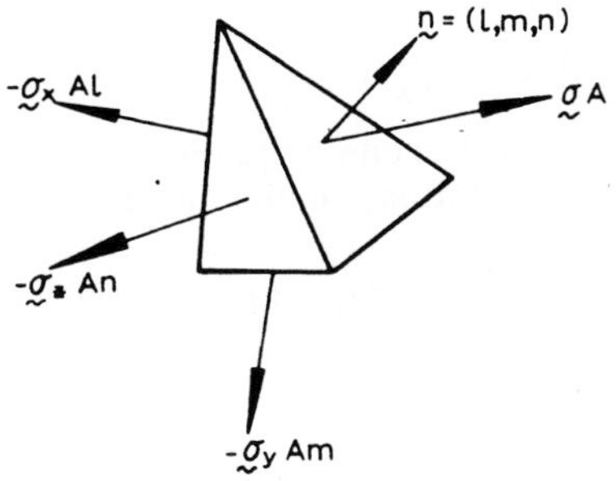

Fig. 6

where σ_{xx}, $\partial\sigma_{xx}/\partial x$, and so on refer to the centre of the cube. If these are added up and equated to zero we get the first of the three equilibrium equations

$$\frac{\partial\sigma_{xx}}{\partial x}+\frac{\partial\sigma_{xy}}{\partial y}+\frac{\partial\sigma_{xz}}{\partial z}=0$$

$$\frac{\partial\sigma_{yx}}{\partial x}+\frac{\partial\sigma_{yy}}{\partial y}+\frac{\partial\sigma_{yz}}{\partial z}=0 \qquad (3)$$

$$\frac{\partial\sigma_{zx}}{\partial x}+\frac{\partial\sigma_{zy}}{\partial y}+\frac{\partial\sigma_{zz}}{\partial z}=0.$$

HOOKE'S LAW

In the infinitesimal theory it is assumed that each strain component is a linear function of all the stress components, and that the converse is true. If the material is *isotropic* the relations are

$$e_{xx}=\sigma_{xx}/E-\nu(\sigma_{yy}+\sigma_{zz})/E,\ \ldots,\ e_{xy}=\sigma_{xy}/2\mu \qquad (4)$$

$$\sigma_{xx}=\lambda(e_{xx}+e_{yy}+e_{zz})+2\mu e_{xx},\ \ldots,\ \sigma_{xy}=2\mu e_{xy}.$$

where E is Young's modulus, ν Poisson's ratio, λ and μ the Lamé constants, μ (often called G) being the shear modulus. There are only two independent constants, for example,

$$E=2(1+\nu)\mu, \qquad \lambda=E\nu/(1+\nu)(1-2\nu).$$

If equation (1) is substituted into equation (4) and equation (4) into equation (3), we get

$$(1-2\nu)\,\nabla^2 u+\partial e/\partial x=0 \qquad (5)$$

$$(1-2\nu)\,\nabla^2 v+\partial e/\partial y=0 \qquad (6)$$

$$(1-2\nu)\,\nabla^2 w+\partial e/\partial z=0 \qquad (7)$$

with $e=e_{xx}+e_{yy}+e_{zz}$,

the equilibrium equations in terms of the displacement. (The quantity e is the dilatation, i.e. the change of volume per unit volume).

TWO-DIMENSIONAL ELASTIC PROBLEMS

Here

$$\frac{\partial}{\partial z}\,(\text{anything})=0.$$

If the displacement is in the x, y plane, so that $u \neq 0, v \neq 0, w = 0$, a state of *plane strain* exists. If it is perpendicular to the xy plane, so that $w \neq 0$, $u = 0, v = 0$, there is a state of *anti-plane strain.*

Figure 7a shows a body with a slit crack. In Fig. 7b the crack is opened by a tensile force; in Fig. 7c its faces are slid over one another by a shear in the plane of the figure; and in Fig. 7d the stresses and corresponding displacements of the crack faces are into and out of the paper, as indicated by the conventional signs for arrow heads and tails.

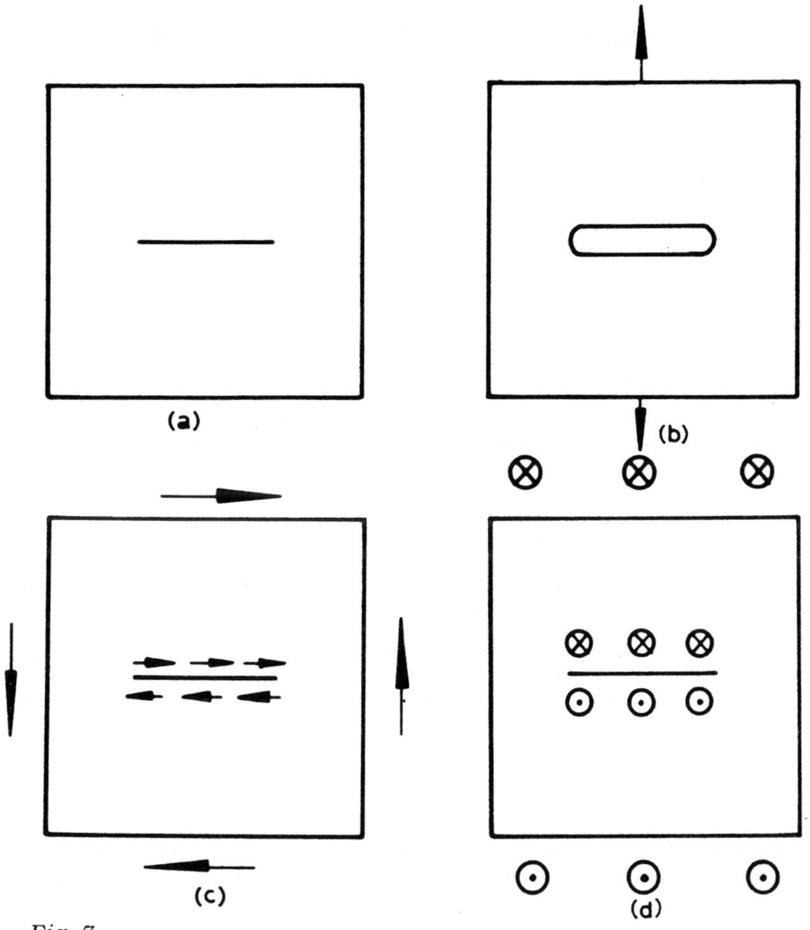

Fig. 7

In Figs. 7b and 7c there is a state of plane strain, and in Fig. 7d anti-plane strain. In fracture mechanics, b, c, d are referred to as Mode I, II, III deformations, respectively.

ANTI-PLANE STRAIN (Mode III deformation)

Anti-plane strain is much simpler than plane strain, and it also provides an easy introduction to the use of functions of a complex variable in the theory of elasticity.

Since $u = v = 0$ and $\partial w/\partial z = 0$ the only non-vanishing stress components are

$$\sigma_{zx} = \mu \partial w/\partial x, \qquad \sigma_{zy} = \mu \partial w/\partial y \tag{8}$$

and, since $e = 0$, the equilibrium equations reduce to the single one

$$\nabla^2 w = 0$$

which says that the displacement w is a harmonic function.

Two-dimensional harmonic functions can be handled nicely with the help of the theory of (analytic) *functions of a complex variable*. To begin with, a point x, y is represented in the usual way by the complex variable

$$z = x + iy.$$

(It is nothing to do with the third Cartesian coordinate, which plays no role in what follows, apart from figuring as a suffix in equation (8)).

For our purposes it is enough to define a function of a complex variable as what one obtains by taking a reasonable function f(x) of x and replacing x by $z = x + iy$.

For example, from $f(x) = x^2$, $f(x) = \sin x$, we get

$$f(z) = (x + iy)^2 = x^2 - y^2 + 2ixy \tag{8a}$$

$$\begin{aligned} f(z) = \sin(x + iy) &= \sin x \cos iy + \cos x \sin iy \\ &= \sin x \cosh y + i \cos x \sinh y. \end{aligned}$$

As these examples indicate we can always separate the real and imaginary parts and write

$$f(z) = \phi\,(x, y) + i\,\psi(x, y).$$

However, if we just insert arbitrary functions of x, y we shall not usually get a function of the complex variable z, in the sense in which we are using the term. Take, for example, $\phi = x^2 - y^2$, $\psi = -2xy$. Then we get equation (8a) with the sign of the second term changed, so that $f = (x - iy)^2$, which is a function of the complex variable $\bar{z} = x - iy$, but not of z.

Differentiating by the chain rule, we get

$$\frac{\partial f}{\partial x} = f'(z)\frac{\partial z}{\partial x} = f'(z), \tag{10}$$

$$\frac{\partial f}{\partial y} = f'(z)\frac{\partial z}{\partial y} = if'(z) \tag{11}$$

where

$$f'(z) = \frac{df(z)}{dz}$$

Equations (10) and (11) give the basic relation

$$\frac{\partial f}{\partial x} = \frac{\partial f}{\partial (iy)} = \frac{df}{dz} \tag{12}$$

of which the first equality is an obvious test of whether or not f depends merely on the *sum* of x and iy. If equation (12) is applied to equation (9) and the real and imaginary parts are separated, we get

$$\frac{\partial \phi}{\partial x} = \frac{\partial \psi}{\partial y}, \quad \frac{\partial \phi}{\partial y} = -\frac{\partial \psi}{\partial x}, \tag{13}$$

the *Cauchy-Riemann relations*. Next, if one of equations (13) is differentiated with respect to x and the other with respect to y and the cross-term is eliminated, we get

$$\nabla^2 \phi = 0, \nabla^2 \psi = 0$$

so that any analytic function provides us with two real harmonic functions either of which can be identified with the anti-plane strain displacement. For definiteness we shall usually put $w = \phi$, so that

$$f(z) = w + i\psi \tag{14}$$

or, equivalently,

$$w = \text{real part of } f.$$

In equation (14), ψ is mere ballast to make up a complex function, but when we generalize to plane strain problems, ψ as well as ϕ becomes useful.

The simplest choice for f is some power of z:

$$f(z) = Az^n = Ar^n e^{in\theta} = Ar^n (\cos n\theta + i \sin n\theta).$$

If A is either purely real or purely imaginary, then

$$w \propto r^n \cos n\theta \text{ or } w \propto r^n \sin n\theta.$$

If n is integral, these are not very interesting. But if $n = \frac{1}{2}$ we have just what we want to describe the tip of a crack. Figure 8b is a perspective drawing of what goes on near the right-hand end of the crack of Fig. 7d. With the polar coordinates indicated there is a discontinuity in $w = w(r, \theta)$ across the plane of the crack, so that $w(r, \pi) \neq w(r, -\pi)$. If

$$f(z) = -\,i\,B\,z^{1/2} \quad (B \text{ real})$$

then

$$w = B\,r^{1/2} \sin \tfrac{1}{2}\theta \tag{15}$$

so that ahead of the crack $w(r, 0) = 0$, but behind it $w(r, \pi) = +Br^{1/2}$ on the top surface, and $w(r, -\pi) = -Br^{1/2}$ on the bottom surface, giving a discontinuity as required. However, we also require that the faces of the crack are stress-free, i.e. $\sigma_{zy} = 0$ for $\theta = \pm\pi$. σ_{zy} can be obtained by differentiating equation (15) with respect to y, but it is easier, here, and especially in more complicated cases, to go back to the complete f(z). From equations (12), (13), and (7) we get

$$\mu\, f'(z) = \sigma_{zx} - i\sigma_{zy} \tag{16}$$

so that we only need to differentiate with respect to z to get the stresses directly. With equation (15), (16) gives

$$\mu\, f'(z) = -\tfrac{1}{2}\, i\, \mu\, B\, z^{-1/2} = -\tfrac{1}{2}\, i\mu B r^{-1/2}(\cos \tfrac{1}{2}\theta - i \sin \tfrac{1}{2}\theta)$$

so that

$$\sigma_{zx} = -\tfrac{1}{2}\, \mu B r^{-1/2} \sin \tfrac{1}{2}\theta, \quad \sigma_{zy} = \tfrac{1}{2}\, \mu B r^{-1/2} \cos \tfrac{1}{2}\theta \tag{17}$$

and $\sigma_{zy} = 0$ on $\theta = \pm\pi$.

Similarly, near the other end of the crack (Fig. 8a), using a new origin for (r, θ),

$$f(z) = Cz^{1/2}, w = Cr^{1/2} \cos \tfrac{1}{2}\theta, \quad \sigma_{zy} = \tfrac{1}{2}\, Cr^{-1/2} \sin \tfrac{1}{2}\theta$$

so that beyond the crack ($\theta = \pi$) $w = 0$, whereas moving from $\theta = \pi$ round to $\theta = 0$ gives $w = Cr^{1/2}$, and moving from $\theta = \pi$ to $\theta = 2\pi$ gives $w = -Cr^{1/2}$, while $\sigma_{zy} = 0$ on the faces $\theta = 0, \theta = 2\pi$.

In both Figs. 8a and 8b we get into trouble if we cross the crack, because $\sin \tfrac{1}{2}\theta$ or $\cos \tfrac{1}{2}\theta$ do not reproduce their original values when θ increases by 2π, but only when it increases by 4π. Of course, this is just what enables these functions to represent a crack. To avoid this trouble we make a mathematical *cut* coinciding with the physical crack, and forbid ourselves to cross it.

These simple 'one-ended crack' solutions are not very realistic inasmuch as the surface tractions required to maintain them are somewhat artificial. For example, to maintain (15) in a circular cylinder centred on the crack

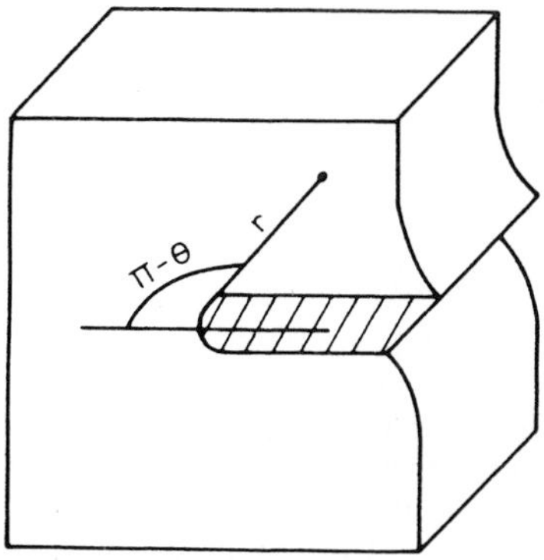

Fig. 8a

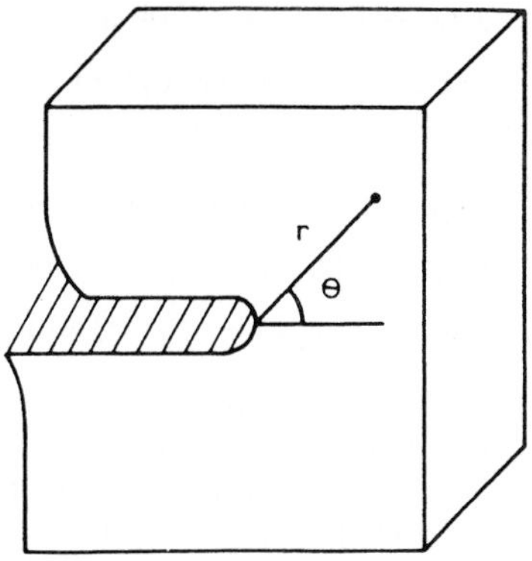

Fig. 8b

tip, surface forces proportional to $\sin\frac{1}{2}\theta$ would have to be applied round its circumference.

However, they will help us to solve the next problem, that of a crack extending from $x = -a$ to $x = +a$, perturbing a uniform shear stress $\sigma_{zy} = \tau, \sigma_{zx} = 0$, (Fig. 7d). The displacement can be expected to look like Fig. 8a at one end and like Fig. 8b at the other, i.e. with the notation of Fig. 9,

$$w \sim r_2^{1/2} \sin \tfrac{1}{2}\theta_2 \text{ near } -a, \quad w \sim r_1^{1/2} \sin \tfrac{1}{2}\theta_1 \text{ near } a.$$

The corresponding complex functions are

$$f_2 = \text{const } r_2^{1/2} e^{i1/2\theta_2}, \quad f_1 = \text{const } r_1^{1/2} e^{i1/2\theta_1} \tag{18}$$

or

$$f_2 = \text{const } (z + a)^{1/2}, \quad f_1 = \text{const } (z - a)^{1/2}. \tag{19}$$

It is easy to see that it is no good adding f_1 and f_2 to obtain the complete solution. However, and this is the vital trick, if two complex functions are multiplied together another is obtained which can be interpreted as the solution of some elastic problem or other. In the present case f_2 is nearly constant near a, and f_1 is nearly constant near $-a$, so that f_1f_2 reproduces the behaviour expressed by equations (18) and (19) near the two tips. We therefore try

$$f(z) = -i\,(\tau/\mu)(z^2 - a^2)^{1/2} = -i\,(\tau/\mu)(r_1r_2)^{1/2}\exp\tfrac{1}{2}\, i(\theta_1 + \theta_2)$$

where the arbitrary constant has been chosen with an eye on the final result. The stresses are given by equation (16) with

$$f'(z) = -i(\tau/\mu)\, z/(z^2 - a^2)^{1/2}$$
$$= -i(\tau/\mu)(r/r_1^{1/2} r_2^{1/2}) \exp i\,(\theta - \tfrac{1}{2}\theta_1 - \tfrac{1}{2}\theta_2)$$

so

$$w = (\tau/\mu)(r_1 r_2)^{1/2} \sin(\tfrac{1}{2}\theta_1 + \tfrac{1}{2}\theta_2) \tag{20}$$
$$\sigma_{zy} = \tau\,(r/r_1^{1/2} r_2^{1/2}) \cos(\theta - \tfrac{1}{2}\theta_1 - \tfrac{1}{2}\theta_2)$$
$$\sigma_{zx} = \tau\,(r/r_1^{1/2} r_2^{1/2}) \sin(\theta - \tfrac{1}{2}\theta_1 - \tfrac{1}{2}\theta_2)$$

Figure 9 shows that for large r, $\theta_1 = \theta_2 = \theta$, and $r_1 = r_2 = r$, so that $\sigma_{zy} = \tau$, $\sigma_{zx} = 0$ at large distances. Also, on the upper face of the crack $\theta = \pm\pi$, $\theta_1 = \pi$, $\theta_2 = 0$, so that $\sigma_{zy} = 0$ there, and similarly $\sigma_{zy} = 0$ on the lower face. To keep things single-valued a mathematical cut is made from $x = -a$ to $x = +a$. On the x-axis inside and outside the crack (Fig. 10) we have

$$w = \pm(\tau/\mu)(a^2 - x^2)^{1/2}\ \begin{matrix}\text{above}\\ \text{below}\end{matrix}\ \text{inside},\ w = 0\ \text{outside} \tag{21}$$

$$\sigma_{zx} = \pm\tau x/(a^2 - x^2)^{1/2}\ \begin{matrix}\text{above}\\ \text{below}\end{matrix}\ \text{inside},\ \sigma_{zx} = 0\ \text{outside} \tag{22}$$

$$\sigma_{zy} = \tau\,|x|/(a^2 - x^2)^{1/2}\ \text{outside},\ \sigma_{zy} = 0\ \text{inside}. \tag{23}$$

These are obtained by giving $\theta, \theta_1, \theta_2$ the appropriate values $0, \pm\pi$, which takes care of factors ± 1, i, and then putting $r = |x|$, $r_1 = |x - a|$, $r_2 = |x + a|$.

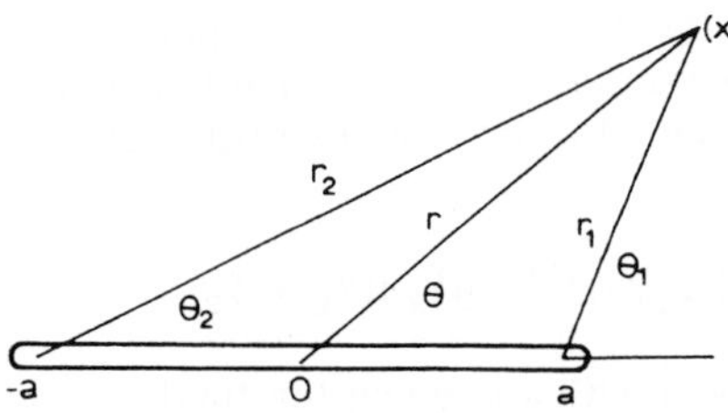

Fig. 9

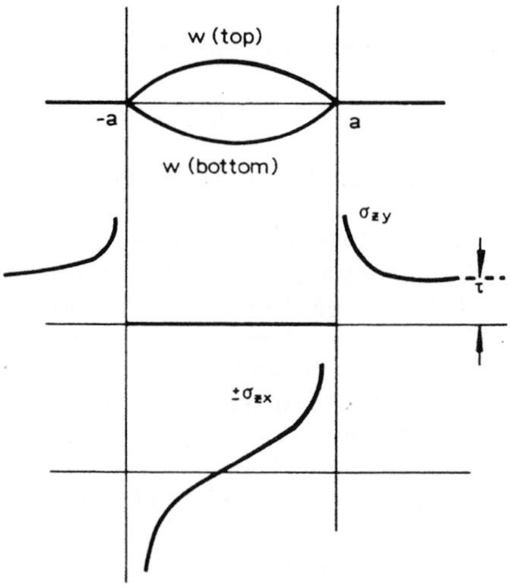

Fig. 10

PLANE STRAIN

Here $w = 0$ and u, v satisfy equations (5) and (6). They are not harmonic, but biharmonic, i.e. $\nabla^2\nabla^2 u = 0, \nabla^2\nabla^2 v = 0$. (Proof: apply $\partial/\partial x$ to equation (5), $\partial/\partial y$ to equation (6), and add to get $\nabla^2 e = 0$; substitute into equations (5) and (6) operated on with ∇^2). Furthermore, u, v are not just any pair of biharmonic functions, since they are coupled together by the second terms in equations (5) and (6). However, u, v can be expressed in terms of *two* complex functions, or equivalently two pairs of conjugate functions, say

$$f(z) = \phi + i\psi, F(z) = \Phi + i\Psi.$$

There are several ways of presenting the results, all of which are equivalent.

A general plane solution of equations (5) and (6) is

$$u = \phi - \frac{1}{2(1-\nu)} \frac{\partial}{\partial x} (y\psi) + \Psi \tag{24}$$

$$v = \psi - \frac{1}{2(1-\nu)} \frac{\partial}{\partial y} (y\psi) + \Phi \tag{25}$$

(Note that the final terms Ψ, Φ are 'the wrong way round' compared with the initial ϕ, ψ).

The solution is general in the sense that (i) any $(\phi, \psi), (\Phi, \Psi)$ give a solution of the equilibrium equations; (ii) given a displacement which satisfies the equilibrium equations, suitable (ϕ, ψ) (Φ, Ψ) can be found which, when substituted in equations (24) and (25), reproduce it. Point (i) can be checked by substitution, but (ii) is harder to establish; one way is outlined in the Appendix.

Equations (24) and (25) can also be written as

$$u = \phi - \frac{1}{2\mu} \frac{\partial \chi}{\partial x} \tag{26}$$

$$v = \psi - \frac{1}{2\mu} \frac{\partial \chi}{\partial y} \tag{27}$$

with

$$\chi = \frac{\mu}{1-\nu} (y\,\psi + \Omega)$$

where

$$\partial\Omega/\partial x = -2(1-\nu)\,\Psi, \quad \partial\Omega/\partial y = -2(1-\nu)\,\Phi.$$

The single harmonic function Ω which replaces Φ, Ψ can be guessed or calculated as

$$\Omega = -2(1-\nu) \text{ imaginary part of } \int^{z} F(z)\, dz.$$

The quantity χ is the *Airy stress function;* it can be verified that it satisfies

$$\nabla^2\nabla^2\chi = 0. \tag{28}$$

In terms of this function the stresses are very simple:

$$\sigma_{xx} = \partial^2\chi/\partial y^2, \sigma_{yy} = \partial^2\chi/\partial x^2, \sigma_{xy} = -\partial^2\chi/\partial x\partial y. \tag{29}$$

Actually, as is easily checked, any χ on substitution into equation (29) gives stresses which satisfy the equilibrium equations, but the strains calculated from them by Hooke's law cannot be derived from a displacement unless equation (28) is satisfied.

Although χ gives the stresses, the corresponding ϕ and ψ must be found (see equations (26) and (27)) in order to obtain the displacements, and this is not very easy. So in fracture mechanics, where the interest is often in crack openings and so forth, analysis which works entirely in terms of χ is not particularly advantageous. Although equation (28) guarantees the existence of a displacement, it does not guarantee that it will be single-valued, so that a proposed χ must be checked for unwanted dislocations, and also the unwanted point forces and couples which may lurk at singularities. All this is nearly as tedious as working directly in terms of displacements.

To get *Muskhelishvili's* form of the general solution (actually invented by Filon) equations (24) and (25) are tidied up so that they are more symmetrical in x and y:

$$u = \phi - \frac{1}{4(1-\nu)} \frac{\partial}{\partial x} (y\,\psi + x\,\phi) + \Psi_1 \tag{30}$$

$$v = \psi - \frac{1}{4(1-\nu)} \frac{\partial}{\partial y} (y\,\psi + x\,\phi) + \Phi_1 \tag{31}$$

The values of the new Φ_1, Ψ_1 can be inferred by comparison with equations (24) and (25), and it can be checked that they are a conjugate pair, arbitrary since Φ, Ψ are arbitrary.

Multiply equation (31) by i and add it to equation (30). By freely applying the Cauchy-Riemann relations and equation (12) the result can be thrown into Muskhelishvili's form

$$u + iv = (3-4\nu)\, F_1(z) - z\,\overline{F_1'(z)} - \overline{F_2(z)}$$

where $F_1(z)$, $F_2(z)$ are two arbitrary complex functions, equal, though here it does not particularly matter, to $(\phi + i\psi)/4(1-\nu)$ and $i(\Phi_1 + i\Psi_1)$, respectively. The bar over a symbol denotes its complex conjugate, i.e. what is obtained by replacing i by −i everywhere, not just in x + iy; this must not be confused with the idea of conjugate (real) functions.

The stresses are given by

$$\sigma_{xx} + \sigma_{yy} = \text{real part of } 8\mu\ F_1'(z)$$

$$\tfrac{1}{2}(\sigma_{yy} - \sigma_{xx}) + i\,\sigma_{xy} = 2\mu\ [\bar{z}\ F_1''(z) + F_2'(z)].$$

It will be convenient to describe *Westergaard's* method after illustrating the use of the general solution by solving a couple of plane strain problems which are basic in fracture mechanics.

CRACK PERTURBING UNIFORM SHEAR (Fig. 6c)

A little thought shows that there is no tensile stress across the x-axis. This can be secured by putting $\Phi = \Psi = 0$ (or $\Omega = 0$) in equations (24) and (25). Then σ_{yy} vanishes on y = 0 because of the y in χ. Suppose a pair ϕ, ψ is chosen for the plane strain problem, and at the same time ϕ is used as the displacement w in an anti-plane problem, then, among other analogies, there are the following:

anti-plane	plane
$w = \phi(x, 0)$ on x-axis	$u = \phi(x, 0)$ on x-axis
$\sigma_{zy} = \mu\partial\phi/\partial y$ on x-axis	$\sigma_{xy} = [\mu/(1-\nu)]\ \partial\phi/\partial y$ on x-axis

That is, on the plane y = 0 the displacements parallel to the plane are the same, and the shear stresses across it are the same apart from a factor $(1-\nu)$. In particular, if the shear stress is zero on some part of the x-axis in one problem, it is zero on the same part of the x-axis in the other.

This suggests that for ϕ, apart from a multiplying constant, the w found for the anti-plane sheared crack (equation (20)), should be taken. This fixes the conjugate ψ and so a tentative solution is given by equations (24) and (25) with

$$\phi = [(1-\nu)/\mu] S\ (r_1 r_2)^{1/2} \sin \tfrac{1}{2}(\theta_1 + \theta_2)$$

$$\psi = -[(1-\nu)/\mu]\ S\ (r_1 r_2)^{1/2} \cos \tfrac{1}{2}(\theta_1 + \theta_2) \qquad (32)$$

$$\chi = -S\ r(r_1 r_2)^{1/2} \cos \tfrac{1}{2}(\theta_1 + \theta_2) \sin\theta$$

$$\Phi = 0, \Psi = 0, \Omega = 0$$

and the notation of Fig. 8. The form of the multiplicative constant is chosen in anticipation of the final result. When r is large the r's and θ's become equal and $\chi = -Sxy$, so that $\sigma_{xy} = S, \sigma_{xx} = 0, \sigma_{yy} = 0$. Thus, equation (32) actually describes a shear crack disturbing a uniform shear S.

CRACK PERTURBING A UNIFORM TENSION (Fig. 6b)

Here it can be seen that $\sigma_{xy} = 0$ on $y = 0$. This can be secured, as is easily checked, by putting

$$\Phi = \psi/2(1-\nu), \quad \Psi = -\phi/2(1-\nu)$$

so that equations (24) and (25) become

$$u = \frac{1-2\nu}{2(1-\nu)}\phi - \frac{1}{2(1-\nu)} y \frac{\partial\psi}{\partial x}$$

$$v = \psi - \frac{1}{2(1-\nu)} y \frac{\partial\psi}{\partial y} \tag{33}$$

As before ϕ, ψ are chosen, but now ψ is identified with the w of an associated anti-plane problem, and the following analogies are obtained:

anti-plane	plane
$w = \psi(x, 0)$ on $y = 0$	$v = \psi(x, 0)$ on $y = 0$
$\sigma_{zy} = \mu\, \partial\psi/\partial y$ on $y = 0$	$\sigma_{yy} = [\mu/(1-\nu)]\, \partial\psi/\partial y$ on $y = 0$

So wherever there is no shear across $y = 0$ in the anti-plane case there is no tension in the plane case. This suggests identifying ψ with a multiple of w for the anti-plane shear crack, equation (20).

This gives

$$\psi = [(1-\nu)/\mu]\, T\, (r_1 r_2)^{1/2} \sin\tfrac{1}{2}(\theta_1 + \theta_2)$$

$$\phi = [(1-\nu)/\mu]\, T\, (r_1 r_2)^{1/2} \cos\tfrac{1}{2}(\theta_1 + \theta_2) \tag{34}$$

which are to be fed into equation (33). The stress function χ is too complicated to be of much use in this case. For large r, $\psi \to [(1-\nu)/\mu]\, T\, y$, $\phi \to [(1-\nu)/\mu]\, T\, x$ which gives $\sigma_{xx} = \sigma_{yy} = T, \sigma_{xy} = 0$ at infinity, so that equation (34) actually solves the problem of a crack perturbing a uniform two-dimensional hydrostatic pressure. To get a uniform uniaxial stress

$$\sigma_{yy} = T, \quad \sigma_{xx} = \sigma_{xy} = 0 \text{ at infinity}$$

it is necessary to superimpose everywhere the uniform stress

$$\sigma_{xx} = -T, \sigma_{yy} = \sigma_{xy} = 0 \qquad (35)$$

Imposing the stress (35) does not upset the boundary conditions at the crack, because it produces no stress on the crack surfaces, but it does produce an extra displacement

$$u = -\frac{1-\nu}{2\mu} Tx, \quad v = \frac{\nu}{2\mu} Ty$$

which must be added to equation (33).

Westergaard has given two types of solution in terms of a complex function Z (z). The first gives stresses

$$\sigma_{xx} = \text{Re } Z - y \text{ Im } Z'$$
$$\sigma_{yy} = \text{Re } Z + y \text{ Im } Z' \qquad (36)$$
$$\sigma_{xy} = -y \text{ Re } Z$$

It gives $\sigma_{xy} = 0$ on the x-axis and is adapted to type I deformation. As usual Re and Im stand for 'real part of' and 'imaginary part of'. Putting

$$Z(z) = \mu f'(z)/(1-\nu), \quad f(z) = \phi + i\psi \qquad (37)$$

the corresponding displacements are given by equation (33), which was used for the tensile crack. Conversely, of course, equation (36) gives the stresses for the displacement of equation (33).

The second solution gives stresses

$$\sigma_{xx} = 2 \text{ Im } Z + y \text{ Re } Z$$
$$\sigma_{yy} = -y \text{ Re } Z \qquad (38)$$
$$\sigma_{xy} = \text{Re } Z - y \text{ Im } Z'$$

This gives $\sigma_{yy} = 0$ on the axis and is adapted to type II deformation. The displacements are given by equations (24) and (25) with equation (37) for ϕ and ψ and $\Phi = \Psi = 0$, i.e., the special form of the general solution which was used for the sheared crack.

APPENDIX: DERIVATION OF EQUATIONS (24) AND (25)

If in addition to the dilatation $e = \partial u/\partial x + \partial v/\partial y$ the *rotation* $\omega = \frac{1}{2}(\partial v/\partial x - \partial u/\partial y)$ is introduced, the equilibrium equations (5) and (6) may be manipulated to read

$$\frac{1-\nu}{1-2\nu}\frac{\partial e}{\partial y} = \frac{\partial \omega}{\partial y}, \quad \frac{1-\nu}{1-2\nu}\frac{\partial e}{\partial y} = -\frac{\partial \omega}{\partial x}$$

so that $(1-\nu)\,e/(1-2\nu)$ and ω are conjugate functions and

$$g(z) = \frac{1-\nu}{1-2\nu}\, e + i\omega \tag{a}$$

is an analytic function of z. Obviously (u, v) must involve quantities which when suitably differentiated yield e and ω, and so we start off by integrating (a) and write

$$\int^{z} g(z)\, dz = f(z) = \phi + i\psi \tag{b}$$

As a first trial we take the displacement

$$u_1 = \phi, \quad v_1 = \psi .$$

With the help of equations (12) and (13) it is found that (u_1, v_1) gives the correct ω, but that the dilatation

$$\frac{\partial u_1}{\partial x} + \frac{\partial v_1}{\partial y} = \frac{\partial \phi}{\partial x} + \frac{\partial \psi}{\partial y} = 2\frac{\partial \phi}{\partial x} = 2\frac{\partial \psi}{\partial y} = \frac{2(1-\nu)}{1-2\nu}\, e$$

is wrong by a numerical factor. Also, (u_1, v_1) does not satisfy the equilibrium equations.

The dilatation of (u_1, v_1) can be altered without affecting its rotation if a gradient is added to it to form the second trial displacement

$$u_2 = \phi - \frac{\partial F}{\partial x}, \quad v_2 = \psi - \frac{\partial F}{\partial y} \tag{c}$$

Its divergence,

$$\frac{\partial u_2}{\partial x} + \frac{\partial v_2}{\partial y} = 2\frac{\partial \psi}{\partial y} - \nabla^2 F = \frac{2(1-\nu)}{1-2\nu}\, e - \nabla^2 F,$$

will actually be equal to e if

$$\nabla^2 F = \frac{1}{1-2\nu}\, e = \frac{1}{1-\nu}\frac{\partial \psi}{\partial y} . \tag{d}$$

But $\nabla^2(y\psi) = y\nabla^2\psi + 2\,\partial\psi/\partial y = 2\,\partial\psi/\partial y$, since ψ is harmonic. So equation (d) can be written as

$$\nabla^2 \left\{ F - \frac{1}{2(1-\nu)}\, y\psi \right\} = 0 .$$

An obvious solution is obtained by striking out the ∇^2 to get

$$F = \frac{1}{2(1-\nu)} y\psi. \tag{e}$$

With this F, equation (c) gives the correct e, ω, and, as it happens, (u_2, v_2) satisfies the equilibrium equations. However, (u_2, v_2) is not necessarily the final answer. Still, we can certainly write

$$u = \phi - \frac{\partial F}{\partial x} + \Psi, \quad v = \psi - \frac{\partial F}{\partial y} + \Phi \tag{f}$$

where (Ψ, Φ) is just the difference between the right answer and (u_2, v_2). Since by construction (u, v) and (u_2, v_2) have the same dilatation and rotation, the dilatation and rotation of the displacement (Ψ, Φ) must be zero, i.e.

$$\frac{\partial \Psi}{\partial x} + \frac{\partial \Phi}{\partial y} = 0, \quad \frac{\partial \Psi}{\partial y} - \frac{\partial \Phi}{\partial x} = 0,$$

which are the Cauchy-Riemann equations which say that Φ and Ψ are a conjugate pair, as the notation has anticipated.

In equation (f) we have reproduced equations (24) and (25) and in addition we know how in principle ϕ, ψ, Φ, Ψ could be derived from a given (u, v): from u, v calculate e, ω and by integrating get ϕ, ψ (equation (b)), and finally insert u, v, ϕ, ψ into equation (f) to get Φ, Ψ.

In equation (d) $\partial\psi/\partial y$ could have been replaced by the equal quantity $\partial\phi/\partial x$, and then equation (e) would have read

$$F = \frac{1}{2(1-\nu)} x\,\phi \tag{g}$$

which gives a general alternative solution to equations (24) and (25), favouring the y-axis rather than the x-axis. If F is taken to be the average of equations (e) and (g), then equations (30) and (31) of the text can be derived directly. Of course, whenever F is changed Φ, Ψ must be altered as well, if (u, v) are to stay the same.

In equation (b) (and in the text when finding the function Ω which appears in χ) a complex function has been integrated. There is no need to know anything about the theory of complex integration; equation (b) merely means that an f(z) must be found which when differentiated with respect to z gives g. For example, if $g = z^{-1/2}$, then $f = 2\,z^{1/2}$, as for a real variable.

CHAPTER 3 Stress analysis: fracture mechanics

In fracture mechanics the interest lies in the elastic state near the crack tip. This state is characterized by three quantities K_I, K_{II}, K_{III}, the stress intensity factors. If these quantities are the same for two crack tips, even though the crack geometries and the types of loading are different, then conditions are identical near the two tips. Closely related to the stress intensity factor is the crack extension force G which gives the amount of energy released from the system cracked-specimen-plus-loading-mechanism for unit advance of the crack front. This chapter discusses K and G and then briefly outlines the physical interpretation of fracture criteria and the plasticity around crack tips.

STRESS INTENSITY FACTOR

The stresses near the right-hand tips of the three cracks in Fig. 7b, c, d can be found from equations (34), (32), and (20) of the previous chapter, by putting $r_2 \simeq 2a, \theta_2 \simeq 0$. The results can be written for Mode I:

$$\left.\begin{aligned}
\sigma_{xx} &= K_I(2\pi r_1)^{-1/2}.\ \tfrac{1}{4}(3\cos \tfrac{1}{2}\theta_1 + \cos \tfrac{5}{2}\theta_1) \\
\sigma_{yy} &= K_I(2\pi r_1)^{-1/2}.\ \tfrac{1}{4}(5\cos \tfrac{1}{2}\theta_1 - \cos \tfrac{5}{2}\theta_1) \\
\sigma_{xy} &= K_I(2\pi r_1)^{-1/2}.\ \tfrac{1}{4}(-\sin \tfrac{1}{2}\theta_1 + \sin \tfrac{5}{2}\theta_1) \\
\Delta v &= K_I(2\pi)^{-1/2}\, r_1^{1/2}.\ 4(1-\nu)/\mu
\end{aligned}\right\} \qquad (39)$$

for Mode II:

$$\left.\begin{aligned}
\sigma_{xx} &= K_{II}(2\pi r_1)^{-1/2}.\ \tfrac{1}{4}(-7\sin \tfrac{1}{2}\theta_1 - \sin \tfrac{5}{2}\theta_1) \\
\sigma_{yy} &= K_{II}(2\pi r_1)^{-1/2}.\ \tfrac{1}{4}(-\sin \tfrac{1}{2}\theta_1 + \sin \tfrac{5}{2}\theta_1) \\
\sigma_{xy} &= K_{II}(2\pi r_1)^{-1/2}.\ \tfrac{1}{4}(3\cos \tfrac{1}{2}\theta_1 + \cos \tfrac{5}{2}\theta_1) \\
\Delta u &= K_{II}(2\pi)^{-1/2}\, r_1^{1/2}.\ 4(1-\nu)/\mu
\end{aligned}\right\} \qquad (40)$$

and for Mode III:

$$\left.\begin{aligned}\sigma_{zx} &= K_{III}(2\pi r_1)^{-1/2}.\ (-\sin \tfrac{1}{2}\theta)\\ \sigma_{zy} &= K_{III}(2\pi r_1)^{-1/2}.\ (\cos \tfrac{1}{2}\theta)\\ \Delta w &= K_{III}(2\pi)^{-1/2}\, r_1^{1/2}.\ 4/\mu\end{aligned}\right\} \qquad (41)$$

Here

$$K_I = (\pi a)^{1/2}T, K_{II} = (\pi a)^{1/2}S, K_{III} = (\pi a)^{1/2}\tau \qquad (42)$$

where $\sigma_{yy} = T, \sigma_{xy} = S, \sigma_{zy} = \tau$ are the uniform stresses which the cracks of Fig. 7 are perturbing. The unwritten stress components are zero, except that in equations (39) and (40), as always in plane strain, $\sigma_{zz} = \nu(\sigma_{xx} + \sigma_{yy})$. The complete displacements have not been written out, but only the displacement discontinuity at a distance r_1 behind the tip, i.e., in equation (39) $\Delta v = v(r_1, \theta_1 = \pi) - v(r_1, \theta_1 = -\pi)$ and similarly for Δu and Δw.

The constants K are *stress intensity factors*, defined by the condition that the appropriate stress (σ_{yy} for I, σ_{xy} for II, σ_{zy} for III) ahead of the crack (i.e. for $\theta_1 = 0$) is $K(2\pi r_1)^{-1/2}$. The factor $(2\pi)^{-1/2}$ is conventional and tends to cancel out nicely in many calculations.

So long as the calculations are confined to the plane of the crack the cases I, II, III can be treated together. The only difference is the factor $(1-\nu)$ which distinguishes Δv and Δu from Δw, and this can be taken care of by introducing an elastic modulus M which takes the value μ or $\mu/(1-\nu)$ as appropriate. Then

$$\sigma(s) = s^{-1/2}\, K/(2\pi)^{1/2} \qquad (43)$$

$$\Delta U(-s) = (-s)^{1/2}\, (K/M)(8/\pi)^{1/2} \qquad (44)$$

where

$$K = (\pi a)^{1/2}\, \sigma^{app} \qquad (45)$$

and σ, ΔU, K, M are to be interpreted according to the following table:

Mode		σ	ΔU	K	M
plane strain	I	σ_{yy}	Δv	K_I	$\mu/(1-\nu) = \frac{1}{2}E(1-\nu^2)$
	II	σ_{xy}	Δu	K_{II}	$\mu/(1-\nu) = \frac{1}{2}E(1-\nu^2)$
	III	σ_{zy}	Δw	K_{III}	μ
plane stress	I	$\bar{\sigma}_{yy}$	$\overline{\Delta v}$	$\bar{K}_I$	$(1+\nu)\mu = \frac{1}{2}E$
	II	$\bar{\sigma}_{xy}$	$\overline{\Delta u}$	$\bar{K}_{II}$	$(1+\nu)\mu = \frac{1}{2}E$

(46)

(For the moment the part below the double line will be ignored). The symbol σ^{app} stands for the applied stress which the crack is perturbing; it is to be supplied with the same suffixes as σ, the stress in the plane of the crack a distance s ahead of the crack. In ΔU, $-s$ is the (positive) distance behind the tip.

The above results actually apply quite generally near the border of any crack, in a body not necessarily of infinite extent and perturbing an applied stress not necessarily uniform. For a crack which is not two-dimensional, (for example, a penny-shaped or elliptical one), the origin of coordinates is supposed to be taken at the point of interest with z tangential to the crack border, y perpendicular to the crack plane, and x (or s) measured outward from the border of the crack (Fig. 11).

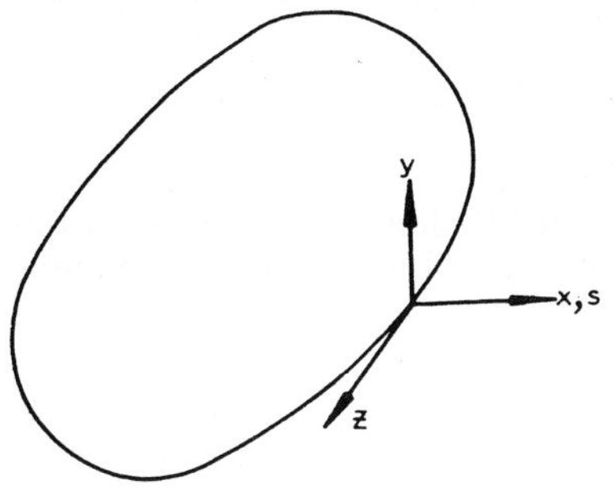

Fig. 11

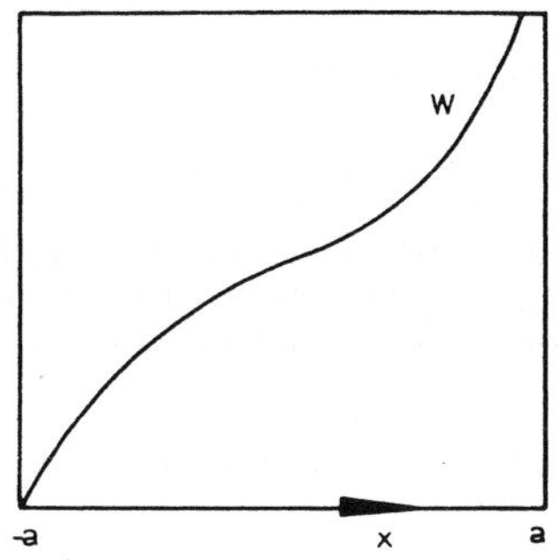

Fig. 12

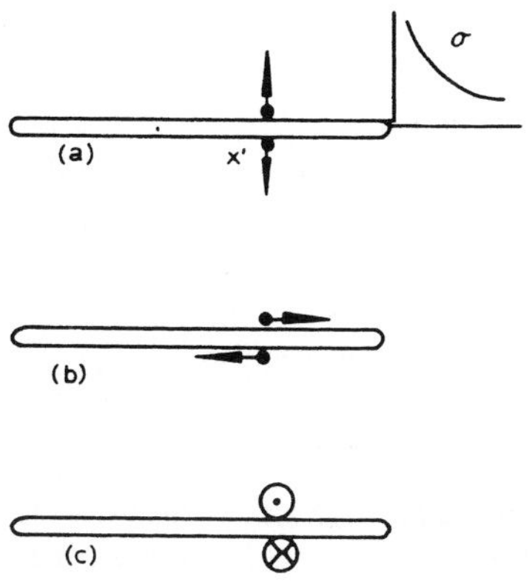

Fig. 13

Of course, K is no longer given by equation (45) which is meaningless in the context of a more general geometry and a varying σ^{app}. A lot of effort has gone into calculating K_I, K_{II}, K_{III} for various loadings and geometries and some of the results will now be reviewed.

PLANE CRACK UNDER NON-UNIFORM LOADING

Suppose first that there is still a crack from (—a, 0) to (a, 0) in an infinite body, but that it is perturbing a non-uniform stress. Let $\sigma^{app}(x, y)$ be the applied stress, i.e. the stress the prescribed surface loads or displacements would maintain in the material in the absence of the crack. For example, σ^{app}_{yy} might vary linearly with x, as in the middle of a bent beam.

For non-uniform σ^{app}, equation (45) remains valid provided the replacement

$$\sigma^{app} \rightarrow \sigma^{app}_{av}$$

is made, where σ^{app}_{av} is some suitable average.

This average only depends on the value of σ^{app} along the segment of the x-axis occupied by the crack. However, it is not a uniform average, but rather the weighted average

$$\sigma^{app}_{av} = \int_{-a}^{a} W(x')\sigma^{app}(x', y = 0) = dx' \tag{47}$$

where the weight function has the form

$$W(x') = (a + x')^{1/2}/\pi(a - x')^{1/2} \text{ with } \int_{-\infty}^{\infty} W(x')dx' = 1$$

which is sketched in Fig. 12.

Consequently,

$$K = (\pi a)^{-1/2} \int_{-a}^{a} (a + x')^{1/2}(a - x')^{-1/2} \sigma^{app}(x')dx'. \tag{48}$$

If $\sigma^{app}(x) = \sigma^{app}(-x)$, an equivalent expression is

$$K = (a/\pi)^{1/2} \int_{-a}^{a} (a^2 - x'^2)^{-1/2} \sigma^{app}(x')dx'. \tag{49}$$

Of course, σ^{app} is evaluated on y = 0, and its suppressed suffixes are the same as those on σ, in accordance with equation (46). If σ^{app} has non-zero components σ^{app}_{yy}, σ^{app}_{xy}, σ^{app}_{zy} on the x-axis, these three quantities have to be inserted successively into equation (48) to obtain K_I, K_{II}, and K_{III}.

Because of its importance one derivation of equation (48) will be outlined.

Suppose that the loading is such that only σ_{yy}^{app} is non-zero along the line of the crack, though σ_{xx}^{app} and σ_{xy}^{app} will then usually not be zero elsewhere. In the absence of the crack, the stresses at an arbitrary point will be, say

$$\sigma_{ij}(x, y) = \sigma_{ij}^{app}(x, y) \tag{50}$$

(σ_{ij} is just a shorthand way of writing 'σ_{xx}, σ_{xy}, etc.'). When the crack is introduced they will be different, say,

$$\sigma_{ij}(x, y) = \sigma_{ij}^{app}(x, y) + \sigma_{ij}^{crack}(x, y) \tag{51}$$

where σ^{crack} is the perturbation introduced by the crack.

Before the crack is cut there are forces $-\sigma_{yy}^{app}(x, 0)$ per unit length on the upper face of the proposed crack and $+\sigma_{yy}^{app}(x, 0)$ on the lower face, by equation (2), and the fact that the normals to the upper and lower faces are, respectively, (0, −1, 0) and (0, 1, 0). So if the crack is made and then the above-mentioned forces are applied to its surfaces, the crack will fail to open and the stresses in the medium will be given by equation (50) instead of equation (51). In other words the surface forces produce stresses which everywhere cancel σ^{crack}, i.e. produce stresses $-\sigma^{crack}$ in the medium. Consequently, forces $+\sigma^{app}$ per unit length of the upper face together with forces $-\sigma^{app}$ on the lower face produce a stress $\sigma^{crack}(x, y)$ in the material. Consequently,

$$\sigma^{crack}(x, y) = \int_{-a}^{a} \sigma(x'; x, y)\sigma^{app}(x')dx'$$

where $\sigma(x'; x, y)$ is the stress produced at x, y when a unit positive force is applied to the upper face at x' together with a unit negative force at x' on the lower face, Fig. 13a. Actually only the stress just beyond the right-hand tip is of interest, so we can write more simply,

$$\sigma(s) = \int_{-a}^{a} \sigma(x'; s)\sigma^{app}(x')dx' \tag{52}$$

where $\sigma(s)$ is the stress a short distance ahead of the crack in the crack plane and $\sigma(x'; s) = \sigma(x'; x = a + s, y = 0)$. According to equation (51), σ^{app} should be added to equation (52) to obtain the total stress, but since σ^{crack} is normally singular at the tip and σ^{app} is not, the omission does not usually matter.

In the appendix it is shown that

$$\sigma(x'; s) = (2a)^{-1/2}\pi^{-1}(a + x')^{1/2}(a - x')^{-1/2}\, s^{-1/2} \tag{53}$$

which when inserted in equation (52) and compared with equation (43) gives equation (48).

If instead of σ_{yy}^{app} either σ_{xy}^{app} or σ_{zy}^{app} is the only non-zero component, the argument goes through as before, applying stresses to prevent the crack surfaces sliding rather than opening. Equation (53) is still correct, but $\sigma(x'; s)$ stands for σ_{xy} or σ_{zy} and refers to point forces applied as in Fig. 13b or Fig. 13c. If all three components of σ^{app} are non-zero, K_I, K_{II}, K_{III} are calculated by simply inserting σ_{xx}^{app}, σ_{xy}^{app}, σ_{zy}^{app} successively into equation (48), as already explained; there are no cross-terms.

The above derivation makes it clear that equation (48) can be used to obtain the crack tip stresses when the crack is loaded *internally*. For example, if there is a gas pressure P in the crack (hydrogen in steel), it is only necessary to put $\sigma^{app} = P$ in equation (48). The result is then, of course, just equation (45) with $\sigma^{app} = P$; but non-uniform internal loading comes in, for example, in the theories of Barenblatt, of Dugdale, and of Bilby, Cottrell and Swinden (see later).

CRACKS IN FINITE BODIES

Figure 14a represents a crack of length a, reaching the boundary of a semi-infinite body. Suppose it is subject to uniform Mode III loading. The problem of a crack of length 2a in an infinite medium (Fig. 14b) has already been solved. The displacement w is even in x, so $\sigma_{zx} = \mu\partial w/\partial x$ is zero on the symmetry line AB, so that if we saw along AB and the material on the left is thrown away, the stresses in the remainder are unaltered, leaving the solution for Fig. 14a; the value of K_{III} is the same for (a) as for (b), namely, $\sigma(\pi a)^{1/2}$.

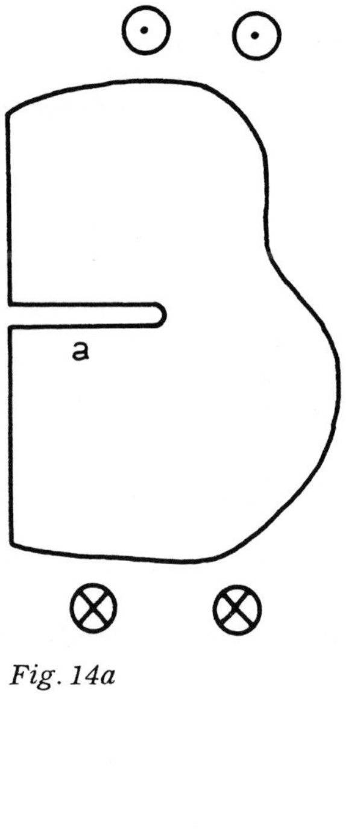

Fig. 14a

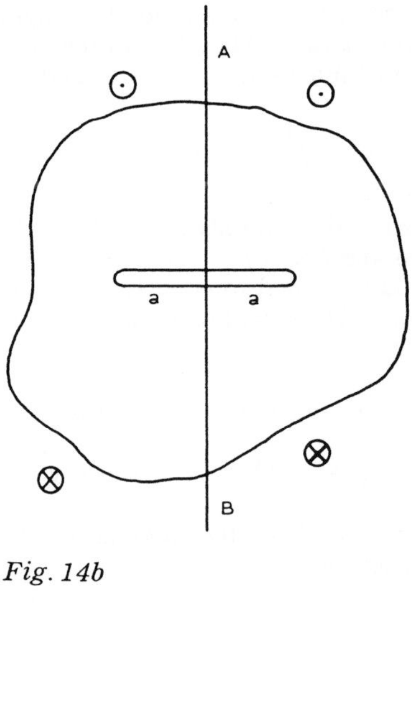

Fig. 14b

The same trick cannot be used for Mode I loading by a tension T. There is no shear stress across AB, but σ_{xx} is not zero across it. If the material to the left is discarded, the situation of Fig. 14a is arrived at, but with an unwanted distribution of pressure over the vertical surface, and the effect of its removal has to be calculated. The result of a complicated calculation is to show that K_I changes from the $(\pi a)^{1/2}T$ of Fig. 14b to $(1.12\ \ldots)$ $(\pi a)^{1/2}T$ for Fig. 14a. The idea of starting with an approximate solution which does not quite satisfy the boundary conditions and then adjusting it until it does is, of course, a commonplace of applied mathematics. In the Science of Fracture Mechanics, however, it has been dignified with the name of *boundary collocation*.

Figure 15 shows a set of three cracks of length 2a whose centres are separated by 2b. The K value can be determined at any of the six tips. In general, it depends on an integral similar to that of equation (48), only much worse, extending over all three cracks. However, if it is supposed that Fig. 15 represents the middle three of an infinite set of equal and equally spaced cracks, things simplify, at any rate if the applied stress is constant

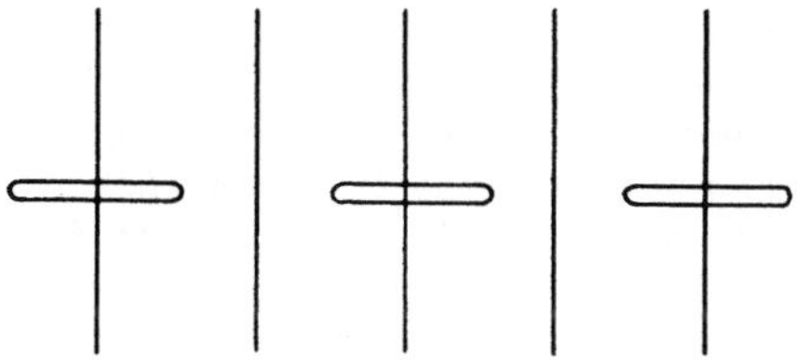

Fig. 15

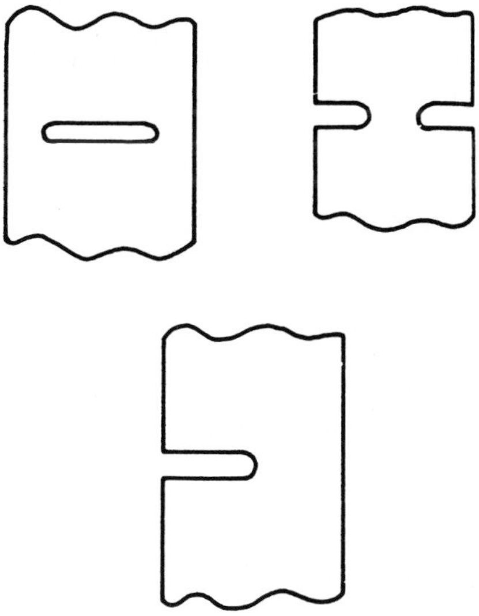

Fig. 16

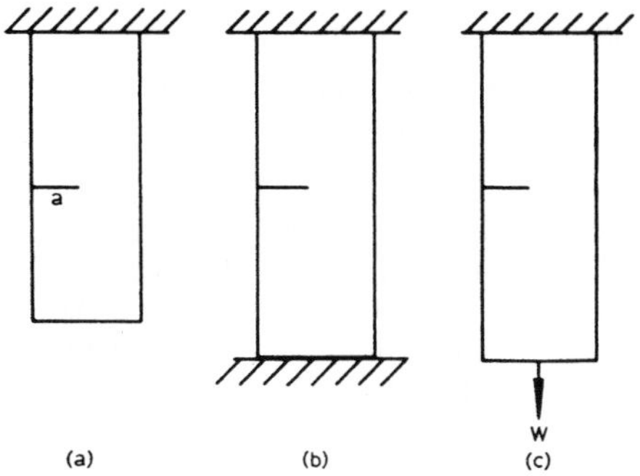

Fig. 17

or has the same periodicity as the cracks. For a constant Mode III load τ the appropriate complex stress function is

$$\sigma_{zx} - i\,\sigma_{zy} = \tau W$$

with

$$W = \sin(\pi z/2b)/[\sin^2(\pi z/2b) - \sin^2(\pi a/2b)]^{1/2}$$

and for Modes I, II with loads T, S, the appropriate Westergaard stress functions for insertion into equations (36) or (37) are TW and SW, respectively.

Nobody is particularly interested in an infinite array of cracks for its own sake, but in the case of anti-plane strain something more interesting can be derived from it. There is no stress across any of the vertical lines in Fig. 15, so that the solution W also applies to the configurations of Fig. 16. As before, this method cannot be applied directly to Modes I and II.

PLANE STRESS

Suppose a plate is cut out of one of the infinite cracked solids of Figs. 7b or 7c. If the plane strain elastic solution which has already been derived is to remain correct in the plate, forces must be applied to the surfaces of the plate equal to the stresses which existed across those surfaces when the plate formed part of the original infinite solid, namely, $+\sigma_{zz}(x, y)$ on the front face and $-\sigma_{zz}(x, y)$ on the back face. The actual elastic field in the plate with stress-free surfaces is thus the original plane strain solution plus the elastic field set up when forces $-\sigma_{zz}(x, y)$ and $+\sigma_{zz}(x, y)$ are applied, respectively, to the front and back surfaces, $\sigma_{zz}(x, y)$ being the stress predicted by the plane strain solution. The plate is then in a complicated three-dimensional state of stress in which, generally speaking, all of the three displacement and six stress components are non-zero.

Two approximations are conventional in plate theory:

(i) *plane stress.* Here it is assumed that $\sigma_{zz}, \sigma_{zx}, \sigma_{zy}$ vanish everywhere

(ii) in *generalized plane stress* it is assumed, less restrictively, that σ_{zz} vanishes everywhere, but that σ_{zy} and σ_{zx} only vanish at the surfaces of the plate.

If (i) or (ii) is assumed, it is not difficult to derive the following results by integrating the equations of elasticity across the thickness of the plate. Let $\bar{u}, \bar{\sigma}_{xx}, \bar{\sigma}_{xy}$, etc., denote averages across the plate thickness. Then the barred quantities are related as the unbarred ones would be in a corresponding state of plane strain in a material with the same shear modulus μ and a fictitious Poisson's ratio

$$\nu^* = \nu/(1 + \nu). \tag{54}$$

If it is assumed that the conditions for plane or generalized plane stress are reasonably satisfied for a crack, equation (54) can be used to fill equation (46) in below the double line. From the average $\bar{\sigma}$ the average of K over the width of the plate can be inferred. Evidently the relation between the plate case and the corresponding infinite plane strain case is just

$$K_{I,II}\,(\text{plane stress}) = K_{I,II}\,(\text{plane strain})$$

for the stress intensity factors, though the crack displacements change slightly since they depend on Poisson's ratio. Actually the stresses near the tip do not satisfy the assumption (i) or (ii) very well. It is no good saying, as it can be said for, say, a plate with a circular hole in it, that things will be practically plane stress if the plate is thin enough, because near the tip there is no relevant length for the thickness to be small in comparison with.

OTHER CRACK GEOMETRIES

Problems relating to the K values for penny-shaped and elliptical cracks under uniform loading have been solved (the three K's vary round the periphery). If there should ever be any call for it, a method exists for finding the stresses produced by an elliptical crack perturbing any applied stress which can be expanded in a power series in x, y, z about the centre of the crack.

The shapes of fracture toughness testing specimen now in use pretty well defy analysis. However, as will be seen, K values, or rather the closely related G values, can be found experimentally.

ENERGY RELEASE RATE

Figure 17a shows a stress-free body containing a crack; Fig. 17b shows the same body stressed by imposing an extension z and clamping the ends. In Fig. 17c the same body is in the same state of extension but now maintained by a weight W. Suppose now that the length of the crack increases by δa. In Fig. 17b the total elastic energy E_{el} of the material obviously decreases. In Fig. 17c when the crack extends by δa the body becomes more flexible, the weight descends and does work on the body, and E_{el} increases. At the same time the potential energy E_{pot} of the weight (or other loading mechanism) decreases, so that, as is to be expected, the sum $E_{el} + E_{pot}$ suffers a decrease. This decrease represents the work which could be derived from the system when the crack tip moves by δa. Expressed per unit length of crack edge and per unit increase δa it is the *energy release rate*, or *crack extension force* G.

In order to find G it is simplest not to bother about E_{el} and E_{pot} separately, but to calculate directly the energy which could be extracted by suitable operations at the tip.

Suppose first that it is a plane problem, i.e. the specimen extends to infinity above and below the plane of the paper, and that the crack is, say, subject to Mode I deformation. Allow the right-hand tip to advance from $x = a$ to $x = a + \delta a$ and apply forces to the freshly formed surfaces so as to prevent them separating. So far the elastic energy and the energy of the loading mechanism are unchanged. Next allow the applied forces to relax to zero (Fig. 18). On each unit length of the top fresh surface the stress falls from $\sigma(s)$ to 0, where $\sigma(s)$ is the stress at $s = x - a$, and the displacement changes from 0 to $\frac{1}{2}\Delta U(-s')$, where $-s' = a + \delta a - x$ distance measured from the tip of the crack at its new position $a + \delta a$, and similarly for the bottom surface. The work recoverable during the relaxation is thus

$$\delta E = \tfrac{1}{2} \int_a^{a+\delta a} \sigma(s)\Delta u(-s')dx. \tag{55}$$

σ can be taken directly from equation (43). $\Delta U(-s')$ is obtained from equation (44) by replacing $-s$ by $-s'$ and modifying the formula to allow for the fact that the crack now extends from $-a$ to $a + \delta a$ instead of from $-a$ to a. It is easy to convince oneself that this second correction only changes things to order δa^2, so it will be ignored. Hence equation (55) becomes

$$\delta E = (K^2/\pi M) \int_a^{a+\delta a} (a + \delta a - x)^{1/2}(x - a)^{-1/2}dx$$

The integral has the value $\frac{1}{2}\pi\delta a$, so that

$$\delta E = \tfrac{1}{2} \frac{K^2}{M} \delta a.$$

By definition the ratio of δE to δa is the *crack extension force*

$$G = K^2/2M. \tag{56}$$

As it represents the energy released per unit length of crack when the tip advances by unit distance, it has the dimensions of force per unit length, lb/in, say, though in fracture mechanics it is normally quoted as so many in lb/in^2, an energy per unit area.

For Type I deformation $M = \mu/(1 - \nu)$. Equation (56) obviously also holds for Type II or III deformation with the appropriate M from equation (46). The argument is just the same, except that the newly formed faces do not separate but slide over one another.

Figure 19 represents the more realistic case of a plane cracked area which increases by the small area shown shaded and having $l >> \delta a$. The energy obtainable during this change is, by the same argument,

$$\tfrac{1}{2} \int_A \sigma\Delta U \, dA \tag{57}$$

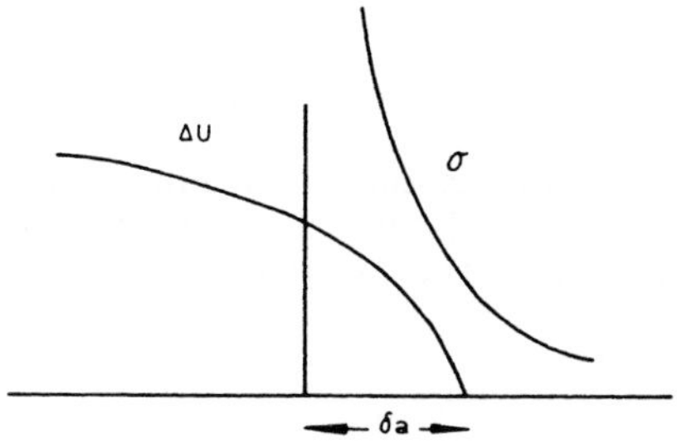

Fig. 18

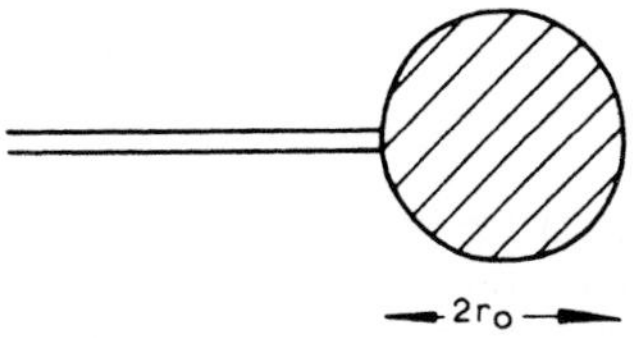

Fig. 21

Fig. 22

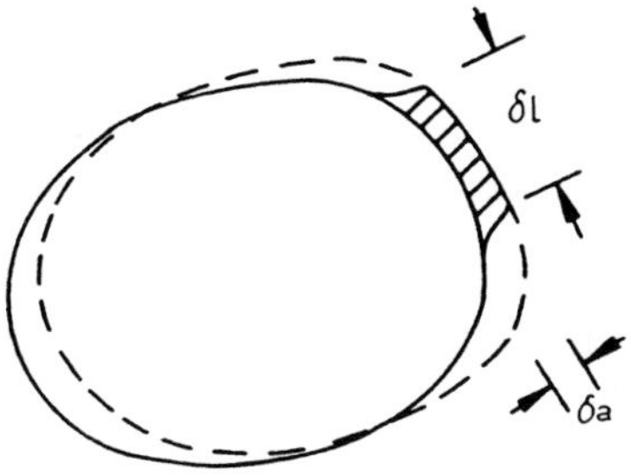

Fig. 19

Fig. 20

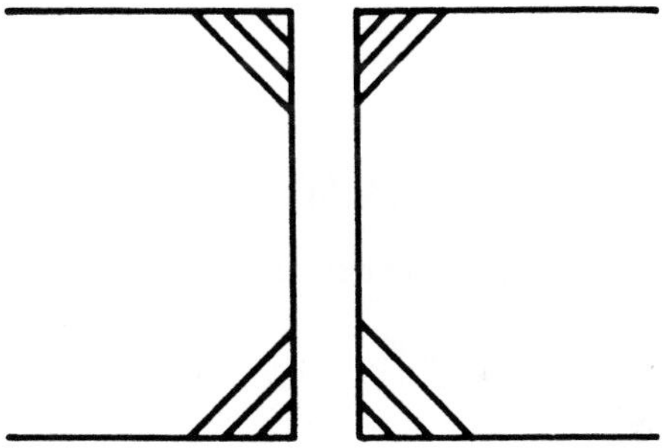

Fig. 23

where A is the shaded area, σ is the stress due to the original crack, and ΔU is the displacement due to the extended crack. σ can be calculated from equations (39), (40), or (41) with suitable local axes, but ΔU is more troublesome. It is usually assumed that it is the same as ΔU for the original crack advanced bodily as shown by the dotted line. This is hard to justify rigorously, but is almost certainly correct except near the ends of the shaded area. If this is granted, equation (57) is the same as equation (55) multiplied by l and the crack extension force

$$G = \lim_{\substack{l \to 0 \\ \delta a \to 0}} \frac{\text{energy extracted}}{l\delta a}$$

is still given by equation (56). It has been assumed that the loading is pure Mode I, II, or III. If it is a mixture of them, it is clear that Δv, Δu, Δw have work done on them only by σ_{yy}, σ_{xy}, σ_{yz}, respectively, so that

$$G = \frac{(1-\nu)K_I^2}{2\mu} + \frac{(1-\nu)\,K_{II}^2}{2\mu} + \frac{K_{III}^2}{2\mu}$$

where the individual terms may be called G_I, G_{II}, G_{III}.

One can make out where the energy $G\,\delta a$ realeased comes from by studying Fig. 17c.

$$E_{el} = \tfrac{1}{2}\, W\, z, \tag{58}$$

and z is proportional to W, say

$$z = C\,(a)\,W$$

where the *compliance* C(a) obviously depends on the length a of the crack. So equation (58) may be rewritten either as

$$E_{el} = \tfrac{1}{2}\, C\,(a)\, W^2 \tag{59}$$

or as

$$E_{el} = \tfrac{1}{2}\, z^2/C\,(a). \tag{60}$$

The potential energy of the weight, taken to be zero when $z = 0$, is

$$E_{pot} = -z\,W$$

or

$$E_{pot} = -C\,(a)\,W^2.$$

Suppose the crack lengthens by δa. To find the corresponding δE_{el} equation (59) must be used rather than equation (60) because we are working at constant W. Then

$$\delta E_{el} = \tfrac{1}{2}\,[\partial C(a)/\partial a]\; W^2 \delta a \tag{61}$$

and

$$\delta E_{pot} = -[\partial C(a)/\partial a]\; W^2 \delta a$$

so that

$$\delta E_{tot} = \delta E_{el} + \delta E_{pot} = \tfrac{1}{2}\,\delta E_{pot} = -\delta E_{el} \tag{62}$$

and of the work done by W, half goes to increase E_{el}, and the other half is released as $Gl\delta a$. Here l is the length of crack edge which moves; normally it will be the width of a plate containing the crack. From the above equation we easily get

$$G = W\delta z/\delta a$$

so that G can be found by observing how far the load moves for a given small increase in crack length, and hence K can be inferred from equation (56) if it is required.

In Fig. 17b the load has stretched the specimen by the same amount z as before and then the ends of the specimen have been clamped. Since now E_{pot} is constant, for a small change in crack length

$$\begin{aligned}\delta E_{tot} = \delta E_{el} &= \tfrac{1}{2}\, z^2\,[\partial C(a)/\partial a]\,\delta a/c^2(a)\\ &= -\tfrac{1}{2}\,[\partial C(a)/\partial a]\,W^2\delta a\end{aligned}$$

so that from equations (61) and (62) the energy released is the same as it was in Fig. 17c, but now it has all come from a *decrease* in E_{el} which is numerically equal to the amount by which E_{el} *increased* in Fig. 17c.

Of course, with the arrangement of Fig. 17b the change in compliance cannot be measured and so G cannot be computed, but our argument shows that G, defined in terms of the total energy change, is the same for dead soft and dead hard loading. As it might be expected, it is also the same for types of loading intermediate between these extremes.

FRACTURE CRITERIA

Although the stress intensity factors K_I, K_{II}, K_{III} are defined in terms of the stress ahead of the crack, a knowledge of these (or equivalently of G_I, G_{II}, G_{III}) completely determines the elastic field near the tip (see equations (39), (40), and (41)). This has the following consequences. Suppose there are two specimens A and B made of the same material and each containing a crack, the specimen shapes and the geometries of the cracks being different, and let each be loaded in a different arbitrary manner. Then we can say that, despite all these differences, the physical state of the material at the tip of the A crack will be the same as the physical state of the material at the tip of the B crack if it happens to be true that K_I, K_{II}, and K_{III} for the A tip are equal, respectively, to K_I, K_{II}, and K_{III} for the B tip. Consequently, the physical behaviour of the two tips would be expected to be the same. Suppose, for example, that $K_{II} = K_{III} = 0$ in both cases. Then if it is found that tip A starts to propagate (or changes from slow to rapid propagation) when K_I reaches some critical value K_{IC}, it would be expected that when the B crack has been loaded in such a way that at its tip $K_I = K_{IC}$, it also will start to propagate (or change from slow to rapid propagation).

This conclusion is not upset if there are departures from the results predicted by the elementary linear elastic theory which is being used, provided these departures are confined to a small enough region around the tip. Imagine a material whose non-linear behaviour can be switched on or off at will. With the non-linearity switched off, load up the A and B cracks to the same K values. The non-linearity is now switched on. The non-linear forces will be identical at the two tips, and they will act on the same environment (the initial linear elastic stress fields) and so will produce identical effects. Similarly if we have an imaginary material in which the ability to flow plastically can be switched on or off. Actually more care is necessary here in that the loading schedules of the A and B specimens must be such as to produce identical histories of the K values at the A and B tips.

From the above point of view, a criterion in which a crack starts to extend when a value K_C (or of G_C which is a definite function of K_C) characteristic for the material is reached, just says that the same thing happens at the tip of any crack when the physical state of the material round the tip reaches the same physical state. This is very reasonable, and in a sense no more needs to be said. However, one or two of the ways in which the criterion can be given some physical background will be outlined.

Evidently, the above argument covers theories in which crack propagation starts when the stress at a prescribed distance from the tip, or the total energy in a certain volume about the tip, reaches a definite value.

The theory of *Griffith* for brittle solids when transcribed into the language of fracture mechanics says that for a crack tip to be able to extend, the energy $G\delta a$ released when it advances by δa must be equal to or exceed the energy $2\gamma\delta a$ of the freshly formed surfaces, where γ is the surface energy per unit area of one of the two newly formed surfaces.

The theory of *Orowan* and *Irwin* suggests that when crack extension is accompanied by plastic deformation, the work done in plastic deformation per unit crack extension can be regarded as a fictitious surface energy γ_p which may be used in place of γ in the Griffith theory. More precisely γ should be replaced by $\gamma + \gamma_p$, but the first term is usually negligible.

Barenblatt supposes that near the ends of a crack its surfaces are acted on by 'cohesive forces' $\sigma^{coh}(x)$, indicated in Fig. 20 by the vertical lines. He requires that the displacement discontinuity ΔU shall not behave abruptly like $(-s)^{1/2}$ near the tip, but shall join smoothly onto the x-axis like $(-s)^{3/2}$. The condition for this is $K = 0$ (see the appendix) which is equivalent to the condition that there shall be no stress singularity beyond the tip. K results from the combined action of the applied stresses and of the cohesive forces on the crack faces, each of which can be calculated from the same formula, equation (48) or its extension to more complicated geometries. (Recall that equation (48) applies both to internal and external loads). So Barenblatt's criterion becomes

$$K\ (\text{applied}) + K\ (\text{cohesive}) = 0. \tag{63}$$

Barenblatt assumes that as the load increases the cohesive forces adjust themselves so that equation (63) is satisfied. Finally, however, a point comes when they are no longer able to do so, a stress singularity develops and the crack starts to extend. He assumes that when the crack is just on the point of moving (so that equation (63) is still satisfied) the shape of the crack tip is, for a particular material, a definite function $\Delta U = \Delta U(-s)$ of distance $-s$ behind the tip. (This is equivalent to the argument above that a fixed K(applied) produces a unique physical state near the tip even in the presence of non-linear effects.) If it is further supposed that σ^{coh} is a definite function of ΔU at each point, it follows that σ^{coh} is also a unique function of x when the crack is just on the point of moving, and so the second term of equation (63), found by inserting $\sigma^{coh}(x)$ into, say, equation (48) is a definite constant $-K_c$ characteristic of the material, negative since the cohesive forces tend to close the crack. Consequently, equation (63) gives a fracture criterion K(applied) = K_c of the usual type. Barenblatt actually uses a 'modulus of cohesion' M equal to $(\frac{1}{2}\pi)^{1/2}\, K_c$.

PLASTIC FLOW AT CRACK TIPS

Near the tip of a crack the intense stresses would be expected to produce plastic deformation (Fig. 21). As a rough guess it might be expected that the plastic zone would cover the area where the elastic solution gives stresses exceeding the yield stress σ_Y, so that $r_0^{-1/2}\, K/(2\pi)^{1/2} \sim \sigma_y$ or equivalently

$$r_0 \sim \frac{K^2}{2\pi\sigma_Y{}^2} \tag{64}$$

If the crack tip advances, the plastic region will look something like Fig. 22. If the crack is in a plate, an end-on view will look like Fig. 23. Since the faces of the plate are stress-free, the stresses near the edges of the plate will be roughly like those in a tensile specimen, and slip will occur on 45° planes, and thus extend in from the faces to a depth of order r_0. If the plate thickness is much larger than r_0, the crack will propagate with only small plastic lips, but if the thickness is comparable with r_0; the plastic zones will extend over most of the cross-section.

For Mode III deformation the form of the plastic zone at the tip can be found. If the zone is short compared with the crack, it is, as suggested in the figure, circular, with radius given exactly by equation (64), where σ_Y is the shear yield stress, and K is the value of K_{III} in the absence of plastic flow. Outside the plastic region the elastic field is the same as that of a crack with the same K_{III}, but with its centre at the centre of the plastic zone.

Several people, in particular Dugdale and Bilby, Cottrell and Swinden have used the following one-dimensional model of the formation of plastic zones at the ends of cracks. Figure 24 shows a crack of length 2c under uniform

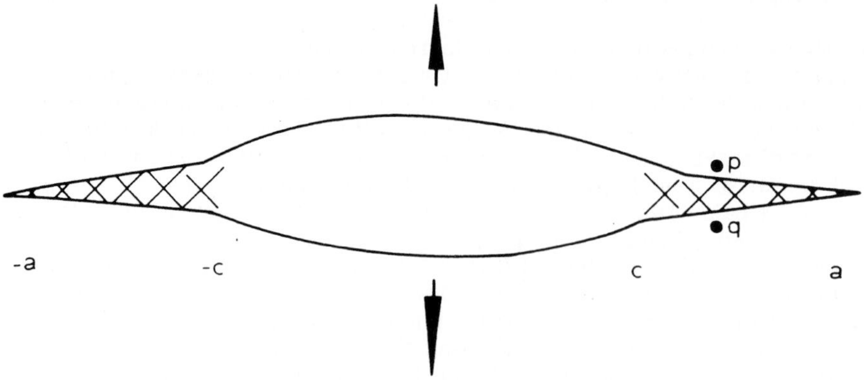

Fig. 24

Mode I loading and having narrow plastic zones of length a − c at each end, represented by hatching. In the hatched regions the tensile stress is equal to σ_Y. As a result of the plastic deformation points p, q just above and below the plastic zone which were originally close together have moved apart by ΔU. In fact, the problem can be treated as one about a fictitious crack of length 2a acted on by the external load and also by internal forces applied to the crack surfaces between −a and −c, and c and a, just as in Barenblatt's theory, except that now the 'cohesive force' is a constant and not a function of ΔU. The length of the plastic zone is fixed by the postulate that there shall be no stress singularity at a. The condition for this is contained in equation (63), where the first term is calculated from equation (49) with $\sigma^{app} = \sigma =$ const. from −a to +a, and the second term is found from equation (49) with $\sigma^{app} = -\sigma_Y$ between −a and −c and between c and a, and zero between −c and c. When this is worked out the following relation between a and c is obtained:

$$c/a = \cos(\pi\sigma/2\sigma_Y) \tag{65}$$

If the plastic zone has length l short compared with c, equation (65) gives

$$l = a - c = \frac{\pi^2}{8}\frac{\sigma^2}{\sigma_Y{}^2}c$$

which is comparable with equation (64), since $K = (\pi a)^{1/2}\sigma$. Admittedly, the assumed narrow shape of the plastic zone is not in exact agreement with the theoretical picture given in Fig. 21 for Mode III, but for Mode I, (which has not been done analytically), plastic flow is in some cases confined to a fairly narrow zone and the relationship expressed in equation (65) is confirmed by experiment.

APPENDIX: DERIVATION OF EQUATION (53)

We start with anti-plane strain, and have to find the elastic field due to the two point forces of Fig. 13c. The displacement due to a unit point force at the origin in an infinite medium must depend only on r and satisfy $\nabla^2 w = 0$. So it must be of the form C ln r. By equation (2) the total force acting on a cylinder of unit length about the z axis is

$$\int (p_{zx} l + p_{zy} m) r d\theta = \mu \int \frac{\partial w}{\partial r} r d\theta = 2\pi\mu C \tag{66}$$

and since this plus the point force must give zero for equilibrium, $C = -1/2\pi\mu$. Also since $\ln z = \ln(r \exp i\theta) = \ln r + i\theta$

$$\begin{aligned} f(z) &= -\ln z/(2\pi\mu) \\ \mu f' &= \sigma_{zx} - i\sigma_{zy} = -1/2\pi\mu z \end{aligned} \tag{67}$$

Except at $r = 0$, (67) gives no stress on the plane $y = 0$, so we can cut along it and discard the lower block. The force along the z axis is now half what it was before, i.e. the integral (66) taken over a semicircle, so that for $y > 0$ (67) also represents a force ½ acting at the boundary of the semi-infinite solid. Hence

$$\mu f' = -1/\pi\mu(z - x') \tag{68}$$

represents the effect of a force +1 at $(x = x', y = 0)$ in the half-space $y > 0$, and, similarly,

$$\mu f' = +1/\pi\mu(z - x') \tag{69}$$

represents the effect of a force −1 at $(x = x', y = 0)$ in the half-space $y < 0$. Evidently, $\mu f'$ for Fig. 13c must behave like equation (68) near p and equation (69) near q. At large distances it must behave like a doublet of (67), $f' \sim 1/z^2$. Also it may be expected that the characteristic crack factor $(a^2 - z^2)^{1/2}$ will appear. If equation (68) is multiplied by

$$g = (a^2 - x'^2)^{1/2}/(a^2 - z^2)^{1/2}$$

then a new $\mu f'$ is obtained which behaves as required at infinity. Further g approaches 1 as z approaches x' on the upper surface and −1 as z approaches x' on the lower side, so all conditions are met and

$$\mu f' = -(a^2 - x'^2)^{1/2}/\mu(a^2 - z^2)^{1/2}(z - x'). \tag{70}$$

With $z = a + s$ and $s > 0$, equation (70) is real and equal to what was called $\sigma(x'; s)$ in the text. The expression for it can be put in a more convenient shape by using the identity

$$\frac{(a^2 - x'^2)^{1/2}}{(x^2 - a^2)^{1/2}} + \frac{(x^2 - a^2)^{1/2}}{(a^2 - x'^2)^{1/2}} = \frac{(x - x')(x + x')}{(x^2 - a^2)^{1/2}(a^2 - x'^2)^{1/2}}.$$

Substituting $\sigma(x'; s)$ into equation (52) gives

$$\sigma(s) = (2\pi s)^{-1/2} K - \sigma'(s)$$

where

$$\sigma'(s) = (2as)^{1/2} \pi^{-1} \int_{-a}^{a} \frac{\sigma^{app}(x')dx}{(a^2 - x'^2)^{1/2}(x - x')}$$

which is what is required provided we can show that σ' is not singular at the tip. To see that it is not, note that an upper bound to the integral can be obtained by giving σ^{app} its largest value in the range of integration, since the rest of the integrand does not change sign. A standard integral is left which is equal to $\pi(x^2 - a^2)^{-1/2} \simeq \pi(2as)^{-1/2}$, and the singular $s^{-1/2}$ is cancelled by the $s^{1/2}$ in front of the integral.

Similarly, when $s < 0$ (70) is purely imaginary and gives $\sigma_{zx} = \mu\partial w/\partial x$, whence an expression can be found for Δw just behind the tip, namely, $(-s)^{1/2}(K/\mu)(8/\pi)^{1/2}$. It is not hard to show that if $K = 0$, then ΔU starts off like $(-s)^{3/2}$.

As explained earlier, a change can be made directly to Mode I or II. For σ no change is necessary except in the suppressed suffixes, and to go from Δw to Δu or Δv, μ is changed into M (see equation (46)).

BIBLIOGRAPHY

There seems to be no good book dealing with stress analysis as it applies to fracture mechanics. This should be remedied by the appearance of Vols. 1 and 2 of 'Fracture', ed. M. Liebowitz, 1968 Academic Press.

A treatise on the mathematics of cracks is promised shortly:

I. N. Sneddon and M. Lowengrub: 'Crack Problems in the Theory in Elasticity' 1968 Wiley.

Probably the most sensible introduction to the theory of elasticity is

S. Timoshenko and J. N. Goodier: 'The Theory of Elasticity'.

It has a good account of the use of the complex variable (Muskhelishvili, conformal transformation) and does the elliptical crack perturbing a uniform stress.

Another useful introduction to the theory of elasticity is

M. Filonenko-Borodich: 'The Theory of Elasticity'.

The careful reader may like to know that it can be obtained in the following three editions, with identical (English) text: Foreign Language Publishing House, Moscow, hardback, 13/6, Dover paperback, 16/-, Noordhoff hardback, 55/-. It does not deal with cracks, but has an account of Muskhelishvili's method. Oddly enough Filonenko-Borodich seems to be the only Russian who, in effect, admits that Filon did it first.

J. C. Jaeger: 'Elasticity, Fracture and Flow', Methuen's Monographs, contains a leisurely development of the elements of elasticity. Fracture is treated from the point of view of rock and soil mechanics. The second edition contains an account of Muskhelishvili's method, illustrated by the elliptical crack. The account of the Griffith criterion cannot be recommended.

The second Russian edition of
L. D. Landau and E. M. Lifshitz: 'Theory of Elasticity'
contains (and presumably the second English edition (Pergamon) will contain) a short account of cracks in the manner of Barenblatt, using dislocation theory for the mathematics.

G. I. Barenblatt: 'The Mathematical Theory of Equilibrium Cracks in Brittle Fracture', Advances in Applied Mechanics, **7**, 55-129; 1962 Academic Press
contains an account of Russian work on cracks.

'Fracture Toughness Testing and its Application', ASTM Special Technical Publication No. 381
contains a compendium of K-values by P. C. Paris and C. M. Sih, and other useful articles.

THE FLOW OF ENERGY INTO THE TIP OF A MOVING CRACK

C. Atkinson* and J. D. Eshelby*

ABSTRACT

A modification of Craggs' method for calculating the flow of energy into the tip of a moving crack is proposed. For a plane crack extending uniformly at both ends the flow falls to zero at the Rayleigh velocity, contrary to Craggs' result, but in agreement with that of Broberg.

INTRODUCTION

Craggs[1] has calculated the rate at which elastic and kinetic energy disappears into the tip of a moving crack (the energy release rate) by integrating the flow of energy across a small circle centered on the tip of the crack. For a plane crack whose ends are extending with equal and opposite uniform velocities under the influence of a tensile stress[1, 2] he finds that the energy release rate falls to zero at about 0.7 times the velocity of shear waves. For other, somewhat simpler solutions representing moving cracks various investigators have found that the energy release rate reaches zero precisely at the Rayleigh velocity, i.e. about 0.9 times the shear wave velocity. It is the object of this paper to find out whether there is a definite difference between the case considered by Craggs and that considered by the others, or whether there is some error in Craggs' analysis.

ANALYSIS

The quasi–static case

We consider the plane problem of a straight crack in the plane $x_2 = 0$ in a cylinder of arbitrary cross section whose surface is subject to any type of loading independent of x_3 (Fig. 1). The me-

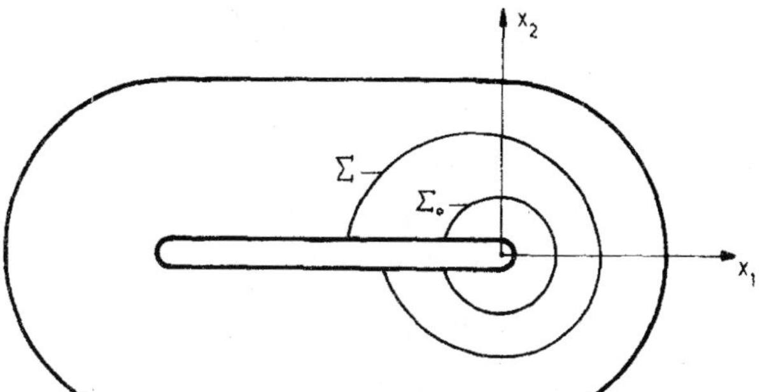

Fig. 1.

dium is linear, but not necessarily isotropic, with displacement u_i and stress p_{ij}. We use the expression dS_j as an abbreviation for $n_j dS$, where n_j and dS are the normal to and an element of a contour such as Σ or Σ_0. The elastic field is a mixture of plane strain and anti–plane strain, so that in (1) below, j is to be summed over 1, 2, but i over 1, 2, 3.

---

* Department of the Theory of Materials, University of Sheffield, Sheffield, England.

Craggs' expression for the energy release rate may be written in the form

$$vG = \int_{\Sigma_0} p_{ij}\dot{u}_i dS_j \tag{1}$$

where (Fig. 1) Σ_0 is a small circle whose center is the instantaneous position of the crack tip. The expression (1) is the total flux of energy (elastic plus kinetic) through Σ_0[3].

We first examine (1) in the limit of very small v. The kinetic energy may then be neglected and G should reduce to the crack opening force of fracture theory, or, in the language of dislocation theory, the force on the leading dislocation of the equivalent pile–up[4]. In this quasi–static case we may write

$$\dot{u}_i = v\frac{\partial u_i}{\partial \xi} \tag{2}$$

where ξ is the x_1–coordinate of the crack tip, and $u_i = u_i(x_1, x_2, \xi)$ is regarded as depending on ξ as a parameter. Then (1) becomes

$$G = \int_{\Sigma_0} p_{ij}\frac{\partial u_i}{\partial \xi} dS_j. \tag{3}$$

When applied to special cases for which G is known (3) gives incorrect results, and this suggests that (1) is likely to be incorrect in the dynamic case also.

To find the reason for the error, and correct it, we compare (3) with expressions which are known to give correct results for G, for example

$$G = \int_{\Sigma}\left(W\delta_{1j} - p_{ij}\frac{\partial u_i}{\partial x_1}\right) dS_j, \tag{4}$$

where $W = {}^1/_2\, p_{mn}\, \partial u_m/\partial x_n$ is the elastic energy density,

$$G = -{}^1/_2\int_{\Sigma}\left(p_{ij}\frac{\partial u_i}{\partial x_1} - u_i\frac{\partial p_{ij}}{\partial x_1}\right) dS_j \tag{5}$$

or

$$G = {}^1/_2\int_{\Sigma}\left(p_{ij}\frac{\partial u_i}{\partial \xi} - u_i\frac{\partial p_{ij}}{\partial \xi}\right) dS_j \tag{6}$$

where Σ is an arbitrary contour (in particular it can be Σ_0) embracing the tip of the crack (Fig. 1). Equation (4) follows from the general theory of the elastic energy–momentum tensor[5], and (5) may be obtained from it by an integration by parts along Σ. Equation (6) follows from the general theory of the force on an elastic inhomogeneity[5]; it is the same as Sanders' result[6], except that Σ only embraces the end of the crack in which we are interested. In deriving (4) or (6) from the general body the crack has to be regarded as an infinitely narrow region with zero elastic constants, and a certain amount of goodwill is necessary in making the passage to the limit, but it is also possible to devise satisfactory ad hoc proofs specifically for the crack. It is clear that (3) will not usually give the same value as one of (4), (5) or (6) taken over Σ_0. Apart from this fact, (3) has one rather unsatisfactory qualitative feature. For each of (4), (5) and (6) the divergence of the integrand is zero, so that Σ may be deformed arbitrarily, as long as it does not embrace the other tip, without altering G. The divergence of the integrand in (3) does not vanish, and so if Σ_0 were taken to be, say, a small square instead of a circle the value of G would be different.

It is possible to modify Craggs' expression for the energy flow so that it gives the correct result in the quasi–static case in the following way. Intuitively one expects that when the crack tip moves it will, to a good approximation, carry its elastic field with it, that is, that $\partial u_i/\partial x_1 \approx -\partial u_i/\partial \xi$ at small enough distances r from the tip. With the sole assumption that the displacement near the tip behaves like $r^{1/2}$ one can in fact show that

$$\frac{\partial u_i}{\partial \xi} = -\frac{\partial u_i}{\partial x_1} + O(r^{1/2}) \tag{7}$$

so that, although both $\partial u_i/\partial \xi$ and $-\partial u_i/\partial x_1$ behave singularly like $r^{-1/2}$ their difference is finite. Equation (7) holds for a straight crack in a cylindrical body of arbitrary cross–section, subject to arbitrary surface loads independent of x_3. Some minimum assumption such as $u_i \sim r^{1/2}$ is obviously necessary in order to let the analysis know that we are dealing with a crack and not some other singularity.

If we combine (2) and (7) the expression (4) may be rewritten as

$$vG = \int_{\Sigma} (p_{ij}\dot{u}_i + vW\delta_{1j})dS_j \tag{8}$$

with a fractional error of order r, where r is, say, the greatest diameter of Σ, so that in the limit as Σ closes up onto the crack tip (8) will give the same result as (4).

The integral (8) may be given the following interpretation. Imagine that the contour Σ (Fig. 1) is moving with velocity v parallel to the x_1–axis, so that it keeps up with the crack tip. Then (8) is the rate at which energy flows into the moving region bounded by Σ. The first term, as in (1), is the rate at which the material outside Σ does work on the material inside Σ and the second term gives the energy which is carried in with the new material which enters Σ on the right and is carried out in the material which leaves on the left.

Thus in the quasi–static case we may use Craggs' basic idea of calculating the energy flow through an infinitesimal circuit about the crack tip provided we allow the circuit to move with the tip and take care to include the convective energy flow.

The dynamic case

In the quasi–static case (8), valid only for small Σ, has no particular advantage, apart from intuitiveness, over (4) which is valid for finite Σ. However, in the general dynamic case no exact analogue of (4) is known, and probably none exists. (See below). We shall therefore consider the obvious generalization of (8) in which the elastic energy density W is replaced by the total energy density

$$E = W + {}^1/_2\, \rho\dot{u}_i\dot{u}_i$$

made up of W and the kinetic energy density (ρ is the density of the medium).

The integral

$$I(\Sigma) = \int_{\Sigma} (p_{ij}\dot{u}_i + vE\delta_{1j})dS_j \tag{9}$$

gives the total flow of energy through the contour Σ, which is regarded as a surface embracing the tip of the crack and moving with it with the instantaneous velocity of the tip v; the second term now accounts for the convective transfer of both elastic and kinetic energy.

The total energy inside Σ is given by

$$\epsilon(\Sigma) = \int E \, dx_1 dx_2$$

where the integral extends over the interior of Σ . ϵ varies with time as the result of the inflow (9) and the outflow vG at the tip. Hence

$$I(\Sigma) = vG + \dot{\epsilon}(\Sigma). \tag{10}$$

Generally we cannot tell how much of I represents vG and how much $\dot{\epsilon}$. However, in what one may call 'steady' motion, with

$$\dot{u}_i(x_1, x_2, t) = u_i(x_1 - vt, x_2) \tag{11}$$

$\dot{\epsilon} = 0$, and I gives vG correctly. The type of motion given by (11) is exemplified by Yoffe's solution(7) for a uniformly moving crack of constant length. Several solutions satisfying (11) have been given for semi–infinite cracks. In some of them the motion is partly maintained by tractions applied to the surfaces of the crack. Our analysis covers these cases provided the tip and a finite region behind it are free of tractions. The surface Σ must then embrace only that part of the crack.

In more complicated cases (11) will not usually be satisfied even if the crack tip is moving uniformly. In particular it is not satisfied for the crack extending at both ends studied by Broberg(2) and by Craggs(1). Even here, however, the error in assuming eq. (11) valid becomes less and less the closer we are to the tip. This suggests that (9) might give vG correctly provided Σ is small enough.

Any state of non–steady motion can, of course, be written in the form

$$u_i(x_1, x_2, t) = u_i^{\circ}(x_1 - vt, x_2) + u_i'(x_1, x_2, t)$$

in an infinity of ways, but in favorable cases we may hope to be able to choose u_i°, u_i' so that close enough to the tip u_i' is much smaller than u_i°. In $\dot{\epsilon}$ the terms quadratic in u_i° make no contribution, and the remainder, containing terms quadratic in u_i' and cross–terms between u_i° and u_i' will be small, and we may expect $\dot{\epsilon}(\Sigma)$ to become negligible in comparison with $I(\Sigma)$ for a small enough circuit. Thus if we can show, by inspection or detailed calculation, that

$$\lim_{\Sigma \to 0} \dot{\epsilon}(\Sigma)/I(\Sigma) = 0 \tag{12}$$

then

$$vG = \lim_{\Sigma \to 0} \int_{\Sigma} (p_{ij}\dot{u}_i + vE\delta_{1j})\, dS_j. \tag{13}$$

If this is to be a sensible result we must require that (13) is ultimately independent of the shape of Σ as well as of its size. If terms in u_i' ultimately make no contribution this requirement will be automatically fulfilled, since when u_i is replaced by u_i° in the integrand its divergence vanishes. We may then, of course, use u_i° in place of u_i in evaluating (13).

It is not possible to give an estimate of u_i' comparable with the relation (7) of the static case, and hence show that (12) is always true. Indeed it is quite likely that for, say, a sharply accelerating crack it is not. The following rather artificial counter–example shows that no general expression exists which gives vG in the form of an integral over Σ with divergence–free integrand and containing derivatives of u_i up to a finite order. Suppose that we are given an expression $I'(\Sigma)$ which purports to do this. We apply it to a state of steady motion. It must then agree with (9), so that $I'(\Sigma) = I'(\Sigma_0) = vG$. At a point outside Σ let us create a concentrated pulse of elastic energy directed so as ultimately to intercept the crack tip. When the pulse is outside Σ the integrals $I'(\Sigma)$ and $I'(\Sigma_0)$ are correct. When the pulse intersects one or other of the surface Σ, Σ_0 the corresponding integral is temporarily incorrect. When it reaches the tip and distorts the elastic field there, both integrals are incorrect. Hence I' cannot always give the correct result.

Application to the Broberg–Craggs crack.

To conclude we return to the question raised at the end of the Introduction and apply (13) to the case studied by Broberg(2) and Craggs(1). The crack extends uniformly and symmetrically under the influence of a constant tensile stress, the tips moving with velocities $+v$ and $-v$. It may be verified that the requirement (12) is met.

Some of the quantities required in evaluating (13) are given explicitly by Broberg(2) and the remainder may be derived from his analysis. Near the right–hand tip we have

$$\dot{u}_1 = -(Q/2\rho c_d \beta_0^3)(\ell/r)^{1/2}\left\{(2k^2-\beta_0^2)F_1(\beta_0) - 2k^2(1-\beta_0^2/k^2)^{1/2}F_1(\beta_0/k)\right\}$$

$$\dot{u}_2 = -(Q/2\rho c_d \beta_0^3)(\ell/r)^{1/2}\left\{(2k^2-\beta_0^2)F_2(\beta_0) - 2k^2F_2(\beta_0/k)\right\}$$

$$p_{11} + p_{22} = (Q/\beta_0^2)(\ell/r)^{1/2}(1-k^2)(2k^2-\beta_0^2)(1-\beta_0^2)^{-1/2}F_1(\beta_0)$$

$$p_{22} = (Q/\beta_0^4)(\ell/r)^{1/2}\left\{-{}^1/_2(2k^2-\beta_0^2)^2(1-\beta_0^2)^{-1/2}F_1(\beta_0) + \right.$$
$$\left. + 2k^3(k^2-\beta_0^2)^{1/2}F_1(\beta_0/k)\right\}$$

$$p_{12} = (Q/\beta_0^4)(\ell/r)^{1/2}k^2(2k^2-\beta_0^2)\left\{F_2(\beta_0) - F_2(\beta_0/k)\right\}$$

where

$$F_1(\beta_0) = \left[\frac{(1+\beta_0^2\sin^2\theta)^{1/2} + \cos\theta}{1-\beta_0^2\sin^2\theta}\right]^{1/2},$$

$$F_2(\beta_0) = \left[\frac{(1+\beta_0^2\sin^2\theta)^{1/2} - \cos\theta}{1-\beta_0^2\sin^2\theta}\right]^{1/2},$$

and similarly for $F_1(\beta_0/k)$, $F_2(\beta_0/k)$.

The notation is similar to Broberg's; r, θ are polar coordinates centered at the instantaneous position of the right hand tip, with $\theta = 0$ along the positive x_1–axis, and $\ell = vt$ is half the length of the crack at time t, while $\beta_0 = v/c_d$ and $k = c_r/c_d$ where c_r and c_d are the velocities of transverse and longitudinal waves respectively. The quantity Q is an expression involving complete elliptic integrals of the first and second kinds; the reader is referred to Broberg's paper for its exact form.

When the above expressions for the stresses and velocities are substituted into (13), terms in $F_m(\beta_0)\,F_n(B_0/k)$, which would be troublesome to evaluate, cancel, leaving only terms of the form $F_1(\beta_0)\,F_2(\beta_0)$, $F_1(\beta_0/k)\,F_2(\beta_0/k)$, $F_1^2(\beta_0) - F_2^2(\beta_0)$ and $F_1^2(\beta_0/k) - F_2^2(\beta_0/k)$.

The final result is

$$vG = \frac{\pi Q^2 t\,[4k^3(k^2-\beta_0^2)^{1/2}(1-\beta_0^2)^{1/2} - (2k^2-\beta_0^2)^2]}{2\rho\beta_0^4(1-\beta_0^2)^{1/2}}.$$

This differs from Craggs' result[1] but agrees with Broberg's[8]. His $\dot{W}$ is twice our vG since he considers the outflow at both tips. Equating the expression in square brackets to zero gives the usual equation for determining the Rayleigh velocity. Hence G vanishes at the Rayleigh velocity, and one can verify that it vanishes at no lower velocity.

Received September 8, 1967.

REFERENCES

1. Craggs, J.W. *Fracture of Solids*, Interscience, New York, 1963, p. 51.

2. Broberg, K.B. Arkiv för Fysik, *18*, 159 (1960).

3. Love, A.E.H. *The Mathematical Theory of Elasticity*, Cambridge University Press, 1927, p. 177.

4. Bilby, B.A. and Eshelby, J.D. *A Treatise on Fracture*, ed. H. Liebowitz, Wiley, New York, 1968.

5. Eshelby, J.D. Solid State Physics, *3*, 79 (ed. Seitz and Turnbull, Academic Press, New York), 1956; also, Phil. Trans., *A 244*, 87 (1951).

6. Sanders, J.L. J. Appl. Mech., *27*, 352 (1960).

7. Yoffe, E.H. Phil. Mag. *42*, 739 (1951).

8. Broberg, K.B. J. Appl. Mech., *31*, 546.

RÉSUMÉ – On propose une modification de la méthode de Craggs pour calculer le flux d'énergie qui entre dans l'extrémité d'une fissure en mouvement. Pour une fissure à deux dimensions dont les extrémités se propagent uniformément en directions contraires le flux se réduit à zéro à la vitesse d'ondes de Rayleigh. Ce résultat est en accord avec celui de Broberg, mais non avec celui de Craggs.

ZUSAMMENFASSUNG – Eine Modifikation des Craggs'schen Verfahrens zur Berechnung des Energieflusses in die Spitze eines sich bewegenden Risses wird vorgeschlagen. Im Fall eines zweidimensionalen Risses, dessen beide Enden gleichmässig auswärts laufen, findet man, dass der Fluss bei der Geschwindigkeit der Rayleighwellen Null wird. Dieses Resultat stimmt mit der Behauptung von Broberg überein, aber mit dem Schluss von Craggs nicht.

CHAPTER 2

DISLOCATIONS AND THE THEORY OF FRACTURE

B. A. Bilby   *J. D. Eshelby*

Abstract: Dislocations enter the theory of fracture in two ways. First, as crystal dislocations, they play a role in the physics of fracture. Secondly, they can serve as convenient basic elements in the macroscopic treatment of fracture. This is a result of the fact that a crack is equivalent to a continuous array of dislocations. This equivalence is made use of in developing the mathematical theory of dislocation arrays and cracks. The problem of determining the equilibrium position of dislocations in a linear array is discussed for the cases of discrete dislocations and continuous distributions of dislocations. The basic problem in the theory of cracks is the determination of the way in which a crack modifies an applied stress field. The problem is first treated as one in antiplane strain (type III deformation), and it is then shown that the results obtained may easily be modified so as to apply to plane deformation (type I and type II deformation). The methods for handling energy changes which have been developed in fracture mechanics and dislocation theory are discussed and compared. These theoretical results are applied to a number of problems in fracture mechanics and to dislocation models of crack initiation. Among the topics reviewed are crack theories in which some attempt is made to take account of cohesive forces, as developed by Barenblatt, or of plastic relaxation at the crack tips, as developed by Bilby, Cottrell, and Swinden.

I. Introduction

In this chapter, we survey the contribution that dislocation theory has made to our understanding of the process of fracture in solids and particularly in crystalline solids. As one of the most important types of defect in crystalline solids and one connected intimately with their mechanical behavior, it would not be surprising if the crystal dislocation played an important role in our understanding of the fracture process of crystalline materials. Indeed, it does, but the dislocation concept is also of considerable value, in a wider sense, in understanding the fracture mechanics of all solids. This is because the dislocation in a continuum, as distinct from the physical crystal dislocation, plays an important role in the general theory of internal stresses. We must thus distinguish two quite distinct roles of dislocation theory in fracture mechanics and the fracture process. First, the presence of the crystal dislocations themselves in crystalline material enables us to understand how fracture may be initiated and how all the plastic relaxation phenomena associated with the presence of cracks and microscopic inhomogeneities take place. Secondly, we have also to examine the use of the dislocation concept in macroscopic fracture mechanics. Here it provides not only a description of dislocations associated with the plastic deformation of the continua considered, but also a representation of the cracks forming in them. By its use, therefore, a unified treatment embracing both the cracks themselves and the plastic deformation associated with them can be given in terms of dislocation theory.

In this chapter, we shall not attempt a comprehensive review either of the role of the crystal dislocation in the initiation of crystalline fracture or of the general principles of macroscopic fracture mechanics. Rather, we shall concentrate on the use of the methods of dislocation theory in the quantitative development of both these aspects.

We begin by specifying precisely the meanings we attach to dislocations in a continuum, to continuous distributions of infinitesimal dislocations, and to crystal dislocations physically observable in real crystalline matter. We note, in passing, that the scale of application of continuous distributions of dislocations may be vastly different. They may represent the relative displacement of two atomic planes in a crystal, or of macroscopic regions of a solid, or of rock strata in an earthquake. Physical parameters differ in these situations, but the general mathematical methods are common to all.

II. Dislocations

If, for the moment, we exclude the situation where there is a continuous distribution of dislocations throughout some three-dimensional region, then the idea of a dislocation requires a medium which is not completely rigid, but which offers some resistance to deformation. Dislocations are the type of defect produced when cuts are made in the medium, the two sides of the cut are displaced relative to each other, material being added or removed as necessary, and the cuts then rewelded.

The most general kind of singularity to which we shall need to refer is the Somigliana dislocation (Somigliana, 1914, 1915; see also Neményi, 1931). In Fig. 1, let S be a surface (open or closed) in a linear elastic solid, and let n_i be its unit normal (taken outwards if S is closed). Make a cut over S, displace the positive side of the cut (to which $\mathbf{n}$ points) with respect to the negative by a small vector $\mathbf{d}$, removing material to avoid interpenetration, if need be; fill in any gaps and reweld. Then we have a Somigliana dislocation over S, specified by the displacement jump $\mathbf{d}$. In the state of internal stress thus produced, the tractions

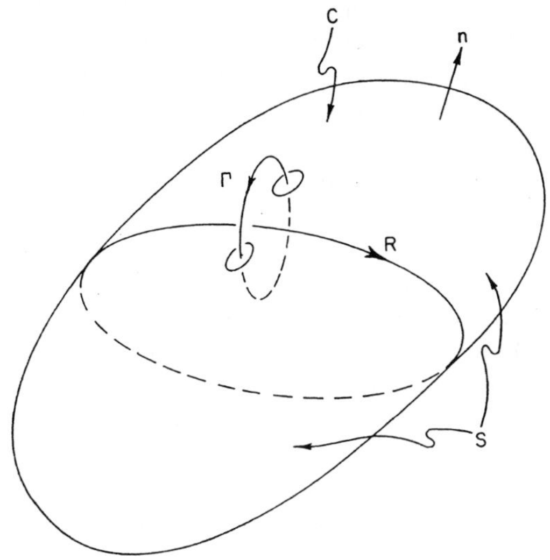

FIG. 1. A Somigliana dislocation.

$\sigma_{ij}n_i$ are continuous across S, but, in general, the stresses are not. However, in a linear elastic material, if the displacement $\mathbf{d}$ corresponds to a small relative rigid body motion of the two sides of the cut, then the stresses are continuous over S and we have a Volterra dislocation (Volterra, 1907). If $\mathbf{d}$ is constant over S, we have a dislocation of translation, or a physical dislocation. Now let S be open so that it becomes a cap C with a boundary R. Then we may speak of a Somigliana, Volterra, or physical dislocation ring, although the first still has a singular surface S where the stresses may be discontinuous. Along R, we must remove material for a Volterra or physical dislocation, unless we taper the relative displacement suitably in this region (thus making the whole singularity a Somigliana dislocation once more).

Assign a sense to R by requiring that a circuit Γ executed in right-hand screw sense about the sense of R shall pass through the cap in the same sense as the cap normal. Then, for a physical dislocation, the displacement $\mathbf{d}$ or its negative $-\mathbf{d}$ is called the Burgers vector of the dislocation ring R. If we take $\mathbf{b} = -\mathbf{d}$ we have the *RH/FS* convention (Bilby, Bullough, and Smith 1955), which we shall use in this paper. With this convention, we have

$$b_i = \int_\Gamma \frac{\partial u_i}{\partial x_j}\, dx_j \tag{1}$$

where u_i is the elastic displacement, and the integral is taken in the sense assigned to Γ. The ring R is completely specified by its Burgers vector b_i and an arrow indicating its line sense. Bilby (1950, 1952, 1967) has discussed the sign conventions and rules for determining the displacements associated with the motion of dislocation lines. These definitions are operational (Bilby, 1967) in the sense that they describe operations needed to introduce the dislocations into the solid. But, given a crystal, we can ask if there is a prescription (other than actually seeing the singularities themselves) for determining whether the crystal contains dislocations or not. We distinguish in the real crystal regions which are good and regions which are bad (Frank, 1951). A region is good if there exists everywhere in it a one-one correspondence of lattice directions with those of a perfect reference crystal. We execute a closed sequence of lattice steps in good crystal; this we call a Burgers circuit. We repeat the same sequence of steps in the reference crystal; this sequence we call the associated path. The associated path begins at some point S and ends at F which will not, in general, coincide with S. We call FS (which is a vector of the perfect reference lattice) the resultant Burgers vector associated with the Burgers circuit in the real crystal. Frank (1951) discusses the complications which arise if the Burgers circuit must pass through any bad crystal—as it must if it encloses an imperfect dislocation (that is, one whose Burgers vector is not a vector of the crystal translation group). The definition is not complete until we have assigned a sense to the Burgers circuit so that it is in right-hand screw relation to any (sensed) dislocation lines threading it. Then, if the associated path is executed in the same sense as the Burgers circuit, our choice *FS* gives the same sense of Burgers vector as the choice $\mathbf{b} = -\mathbf{d}$ in the operational definition; hence the name of the convention.

We shall make extensive use in the sequel of the idea of continuous distributions of dislocations with infinitesimal Burgers vectors.

For example, to represent the taper of the relative displacement of a straight dislocation, we can use the idea of a continuous distribution of parallel straight physical dislocations with infinitesimal Burgers vectors (Eshelby, 1949). Let the straight dislocation lie along Oz, and let $\phi(x)$ be the relative displacement when $\epsilon \to 0$ of the material on $y = +\epsilon$ with respect to that on $y = -\epsilon$. Then if $\phi(x)$ is represented by a distribution of infinitesimal dislocations, $\mathscr{D}(x)\, dx$ is the number of dislocations with Burgers vector b in dx and $d\phi = -b\mathscr{D}(x)$. In the Peierls–Nabarro model of a crystal dislocation along Oz (Peierls, 1940; Nabarro, 1947), the tangential stresses produced by such a distribution are made compatible with a sinusoidal law of force between two linear elastic blocks $y > 0$ and $y < 0$, and $\mathscr{D}(x)$ extends on the surface $y = 0$ over $-\infty < x < +\infty$.

More generally, we can consider a state of internal stress to be a state in which the compatibility equations are not satisfied throughout a volume V. Then no displacement function u_i exists in V, but we can still define a Pfaffian form giving the increment of displacement on moving a small distance dx_i

$$du_i = dx_j\, \beta_{ji} \tag{2}$$

However, the line integral of Eq. (2) taken round any infinitesimal circuit does not vanish. We then have a continuous distribution of dislocations specified by a dislocation tensor α_{ji} (Nye, 1953; Bilby, 1955; Bilby, Bullough, and Smith, 1955; Kröner, 1955; and for reviews see Eshelby, 1956; Kröner, 1958; Bilby, 1960; Kondo, 1955, 1958, 1962). The tensor α_{ji} specifies the component in the i direction of the resultant Burgers vector of all dislocation lines threading unit area of a circuit normal to the j direction.

If b_i is the resultant Burgers vector associated with any closed circuit Γ bounding a surface S, we have

$$b_i = \int_\Gamma dx_j\, \beta_{ji} \tag{3}$$

and so, using the theorem of Stokes,

$$b_i = \int_S dS_j \, \alpha_{ji} \tag{4}$$

where

$$\alpha_{ji} = \epsilon_{jmn} \, \partial_m \beta_{ni} \tag{5}$$

and $dS_j = \nu_j \, dS$ is a surface element of area dS and normal ν_j. In linear elasticity, if e_{ij} and ϖ_{ij} are the elastic strain and rotation, we have $\beta_{ij} = e_{ij} + \varpi_{ij}$ and

$$\alpha_{ji} = \epsilon_{jmn} \, \partial_m(e_{ni} + \varpi_{ni}) \tag{6}$$

If Γ is infinitesimal and bounds dS_j, we can write for the Burgers vector associated with dS_j

$$db_i = dS_j \, \alpha_{ji} \tag{7}$$

A distribution of discrete dislocation lines may be related to α_{ji} in the limit, as follows. Let there be $n^{(k)}$ dislocation lines of unit tangent vector $t_j^{(k)}$ threading a unit area normal to $t_j^{(k)}$ and having Burgers vectors $b_i^{(k)}$. Then the Burgers vector associated with a surface element $dS_j = \nu_j \, dS$ is

$$db_i = \sum_k n^{(k)} t_j^{(k)} \nu_j b_i^{(k)} \, dS \tag{8}$$

so that in the limit when $n^{(k)} \to \infty$ and $b_i^{(k)} \to 0$,

$$\alpha_{ji} = \sum_k n^{(k)} t_j^{(k)} b_i^{(k)} \tag{9}$$

A Somigliana dislocation is a surface distribution of α_{ji} or a surface dislocation (Bilby, 1955; Bilby, Gardner, and Smith, 1958; Bilby, Bullough, and Grinberg, 1964). Physical examples include tilt boundaries (Frank, 1950; Nye, 1953; Amelinckx, 1957) and interfaces in phase transformations (Bullough and Bilby, 1956).

In the sequel, we shall consider frequently a special surface distribution in which all dislocation lines are straight and parallel to the x_3 axis, and are concentrated along the plane $x_2 = 0$, and we shall write

$$\alpha_{3i}(x_1, x_2) = b\mathscr{D}_i(x_1) \, \delta(x_2)$$

The resultant Burgers vector associated with dx_1 is

$$\begin{aligned} db_i &= \alpha_{3i} \, dS_3 \\ &= \int_{x_2=-\infty}^{x_2=+\infty} \alpha_{3i}(x_1, x_2) \, dx_1 \, dx_2 \\ &= b\mathscr{D}_i(x_1) \, dx_1 \end{aligned} \tag{10}$$

Thus, $\mathscr{D}_i(x_1)\,dx_1$ is the number of dislocations with Burgers vectors of magnitude $b > 0$ in the i direction. When $\mathscr{D}_1$ and $\mathscr{D}_3$ are positive we have, with the *RH/FS* convention, distributions of positive edge and right-hand screw dislocations, respectively.

The general Somigliana dislocation is a continuous distribution of dislocation rings with infinitesimal Burgers vectors. A crack whose faces have undergone a relative displacement as a result of an applied stress may be regarded as an "induced" Somigliana dislocation which disappears as the applied stress is removed. In Sect. IV,B we discuss in detail how two-dimensional plane cracks may be represented by distributions of dislocations, and conversely. Leibfried (1954) and Eshelby (1963) have discussed circular and elliptical plane dislocation pileups in a similar manner. Before discussing this representation however, it is convenient to establish some important results on arrays of discrete dislocations, since the theory of such arrays is intimately related to that of the continuous distributions, and enables the limiting properties available in the continuous case to be more clearly understood.

III. Arrays of Discrete and Continuously Distributed Dislocations

The problem of finding the equilibrium form of a linear array of dislocations may be formulated as follows (Eshelby, Frank, and Nabarro, 1951):

A set of identical dislocations lies in the same plane (the plane $y = 0$). What position will they take up under the combined action of their mutual repulsions and the force exerted on them by a given applied stress, in general a function of position along the plane?

If the Burgers vectors of the dislocation lie in the plane $y = 0$ the latter is their common slip plane, and they can reach their equilibrium positions by glide. This is the physically important case. If their Burgers vectors are (or have a component) perpendicular to the plane $y = 0$, the latter is their climb plane, and they can only move by a diffusive process. Consequently, this case is not of much physical interest. However, a crack opened under tension may be simulated by a distribution of imaginary "crack" dislocations (cf. Sect. IV,B) in climb equilibrium, so that the problem has a formal interest. The details of the calculation are the same for either case.

As discussed in App. B, a dislocation in an applied stress field experiences a force which is the product of its Burgers vector and an appropriate component P of the stress. For a gliding edge dislocation, the appropriate component is σ_{xy} and, for a gliding screw dislocation, σ_{zy}.

For an edge dislocation climbing along the x axis, it is σ_{yy}. (This follows from the second equation of (B7) on interchanging x and y.) The stress $P(x)$ on the dislocation at x is made up of a contribution from the applied stress and a contribution from each of the other dislocations. It may easily be calculated from Eq. (75) with $a = 0$, and Eq. (79). The contribution from a dislocation at x' is $A/(x - x')$ where $A = Gb/2\pi(1 - \nu)$ for an edge dislocation in glide or climb and $A = Gb/2\pi$ for a screw dislocation. We shall be interested in cases where certain of the dislocations are locked at fixed positions x_α, say. They may be taken into account by writing $\sigma(x)$ in the form

$$\sigma(x) = \tau(x) + \sum_\alpha \frac{A}{x - x_\alpha} \tag{11}$$

where $\tau(x)$ is the externally applied stress.

If the unlocked dislocations are situated at $x = x_j$, the condition that each of them shall be acted on by zero total force is evidently

$$\sum_{i\neq j} \frac{A}{x_j - x_i} + \sigma(x_j) = 0, \qquad j = i, 2, \ldots, n \tag{12}$$

or, if, for convenience, we take A as our unit of stress

$$\sum_{i\neq j} \frac{1}{x_j - x_i} + P(x_i) = 0, \qquad j = 1, 2, \ldots, n \tag{13}$$

To solve Eq. (13), we follow Stieltjes (1885) (cf. also Szegö, 1939; Whittaker and Watson, 1927) and construct a polynomial

$$f = \prod_{i=1}^{n} (x - x_i) \tag{14}$$

whose zeros x_i are the (unknown) positions of the n free dislocations. Its logarithmic derivative

$$\frac{f'}{f} = \sum_{i=1}^{n} \frac{1}{x - x_i} \tag{15}$$

is equal to the stress at x due to all the free dislocations. The stress produced by all the free dislocations except the one at x_j is

$$\frac{f'}{f} - \frac{1}{x - x_j} = \frac{(x - x_j)f'(x) - f(x)}{(x - x_j)f(x)} \tag{16}$$

and the stress acting on the jth dislocation is the value of Eq. (16) at

$x = x_j$. The limit of the indeterminate fraction Eq. (16) can be found by differentiating its numerator and denominator twice and putting $x = x_j$ (l'Hôpital's rule). The result is $\frac{1}{2}f''(x_j)/f'(x_j)$. The polynomial f must thus satisfy

$$f(x_j) = 0, \qquad j = 1, 2, \ldots, n \qquad (17)$$
$$\frac{1}{2}\frac{f''(x_j)}{f'(x_j)} + P(x_j) = 0$$

The second of these equations expresses the fact that the total force on the jth dislocation is zero. Both these conditions will be met if $f(x)$ satisfies the differential equation

$$f''(x) + 2P(x)f'(x) + q(n, x)f(x) = 0 \qquad (18)$$

provided the roots of f are distinct and q does not have a pole at a zero of f.

Suppose that, in Eq. (11), $\tau(x)$ is a polynomial in x. Then, after clearing of fractions, Eq. (18) will take the form

$$K(x)f''(x) + 2L(x)f'(x) + M(n, x)f(x) = 0 \qquad (19)$$

where K and L are polynomials. Since f is a polynomial, M (the product of q and a polynomial) must also be a polynomial. M and f may, in principle, be found by writing K, L, M, and f as power series (the coefficients in K and L being known) and equating coefficients (Head and Thomson, 1962). Generally speaking, the resulting equations are nonlinear, and may be no easier to solve than Eq. (17). However, in a number of interesting cases, the result of equating the highest and next highest powers of x to zero is a pair of equations which give the form of M (and hence q) directly, and Eq. (18) takes the form of the differential equation satisfied by one of the classical orthogonal polynomials. The fact that q can be found in these cases without actually solving for f is helpful in connection with the continuum approximation. In these cases, one may alternatively merely consult a list of the differential equations satisfied by the orthogonal polynomials until one is found for which the terms in f'' and f' agree with Eq. (18). The appropriate f and q are then obvious. The following are some examples:

(1) A row of $n - 2$ free dislocations under no external stress, but prevented from spreading by locked dislocations at $x = \pm a$. With a as the unit of length Eq. (18) takes the form

$$(1 - x^2)f'' - 4xf' + (1 - x^2)\,q(n, x)f = 0 \qquad (20)$$

The first derivative $P'_{n-1}(x)$ of the $(n - 1)$th Legendre polynomial satisfies

$$(1 - x^2)f'' - 4xf' + \{n(n - 1) - 2\}f = 0 \tag{21}$$

so we must take

$$f = P'_{n-1}(x), \qquad q = \frac{n(n - 1) - 2}{1 - x^2} \tag{22}$$

The stress on the locked dislocation at $x = 1$ due to the free dislocations is f'/f evaluated at $x = 1$. By Eq. (21), this is $\frac{1}{4}\{n(n - 1) - 2\}$. If we add the force $\frac{1}{2}$ due to the locked dislocation at $x = -1$, the total is $\frac{1}{4}n(n - 1)$. In ordinary units of length and stress

$$q = \frac{n(n - 1) - 2}{a^2 - x^2} \tag{23}$$

the free dislocations lie at the roots of

$$f(x) = P'_{n-1}(x/a) \tag{24}$$

and the force on the locked dislocations at $x = \pm a$ is

$$F(\pm a) = \pm \tfrac{1}{4}n(n - 1)\, bA/a \tag{25}$$

(2) A row of n free dislocations in a uniformly varying stress $\tau = -\alpha x$. There are no locked dislocations. The dislocations climb up the sides of the potential well $V = \frac{1}{2}\alpha x^2$ as a result of their mutual repulsion. If we choose the unit of length so that $\alpha = 1$, Eq. (18) becomes

$$f'' - 2xf' + q(n, x)f = 0 \tag{26}$$

The nth Hermite polynomial, $H_n(x)$, satisfies

$$f'' - 2xf' + 2nf = 0 \tag{27}$$

so that we must take $q = 2n$, $f = H_n(x)$ or in ordinary units of length and stress

$$f = H_n(\alpha^{1/2}x/A^{1/2}), \qquad q = 2n\alpha \tag{28}$$

(3) A set of $n - 1$ free dislocations along the positive x axis, piled up against a locked dislocation at $x = 0$ by a uniform external stress $-\tau_0$. If we take $2\tau_0$ to be the unit of length Eq. (18) becomes

$$xf'' + (2 - x)f' + xqf = 0$$

The first derivative of the nth Laguerre polynomial, $L_n'(x)$, satisfies

$$xf'' + (2 - x)f' + (n - 1)f = 0 \tag{29}$$

so that $q = (n - 1)/x$, $f = L_n'(x)$. The stress on the locked dislocation is from Eq. (29), $f'(0)/f(0) = \frac{1}{2}(n - 1)$ due to the free dislocations and $-\frac{1}{2}$ due to the external stress, or in all $\frac{1}{2}n$. In ordinary units

$$f = L_n'(2\tau_0 x/A), \qquad q = (n - 1)\, A/2\tau_0 x \tag{30}$$

and the force on the locked dislocation is

$$F = -n\tau_0 \tag{31}$$

that is, the locked dislocation has to support the force τ_0 exerted by the external stress on itself and the $n - 1$ free dislocations.

The substitution

$$v = f(x) \exp\left(\int P(x)\, dx\right)$$

that is, by Eq. (11)

$$v = f(x) \prod_\alpha (x - x_\alpha) \exp\left(\int \tau(x)\, dx\right) \tag{32}$$

converts Eq. (18) into

$$v'' + I(x)\, v = 0 \tag{33}$$

with

$$I = q - P^2 - P'$$

The function v has the same zeros as f plus the additional ones x_α corresponding to the locked dislocation. It is easy to see that v'/v represents the total stress due to free and locked dislocations and the external stress τ.

According to Eq. (33), the curve $v = v(x)$ is concave or convex to the x axis according as $I > 0$ or $I < 0$. In the former case, the curve is oscillating with a succession of zeros corresponding to dislocations and, in the latter, it is monotonically increasing or decreasing. Thus, we expect to find the dislocations in the regions where $I(x) > 0$.

In the limit of a quasicontinuous distribution of dislocations, $I(x)$ will only vary by a small fraction of itself when x increases by an amount equal to the spacing between dislocations.

In the neighborhood of a particular point x, the solution of Eq. (33) will then look like the solution const. sin kx of $v'' + k^2 v = 0$ where the constant k is the local value of $I(x)$. More precisely, we can always write, for $I > 0$,

$$v = A(x) \sin\{[I(x)]^{1/2} x + \phi(x)\} \tag{34}$$

with suitable $A(x)$, $\phi(x)$. In the circumstances mentioned, A and ϕ will be slowly varying functions of x. If we neglect the variation of ϕ with x, the distance between two zeros of Eq. (34) is $d = \pi/[I(x)]^{1/2}$.

The dislocation density is thus

$$\mathscr{D}(x) = [I(x)]^{1/2}/\pi \tag{35}$$

which can be found from the value of P and q without considering the properties of the polynomial f.

For the examples given in the previous section, we have: (1) Locked dislocations at $x = \pm a$, no external stress. Here, $I = n(n-1)/(a^2 - x^2)$, and so, ignoring unity in comparison with n

$$\mathscr{D}(x) = n/\pi(a^2 - x^2)^{1/2} \tag{36}$$

(2) Dislocations restrained by a force $\tau = -\alpha x$, no locked dislocations. In ordinary units, $I = [(2n + 1)\,\alpha/A] - (\alpha^2 x^2/A^2)$ and

$$\mathscr{D} = (\alpha/\pi A)(a^2 - x^2)^{1/2} \tag{37}$$

where

$$a = (2nA/\alpha)^{1/2}$$

is half the distance between the outermost dislocations. (3) Dislocations piled up against a locked dislocation at $x = 0$ by a constant stress $-\tau_0$. Here, $I = 2n\tau_0/Ax - \tau_0{}^2/A^2$ in ordinary units and $\mathscr{D} = (\tau_0/\pi A)\{(2a - x)/x\}^{1/2}$ where $2a = 2nA/\tau_0$ is the distance between the locked and the last free dislocations. It will be convenient to shift the origin of x to a point halfway between them so as to make comparison with Eqs. (36) and (37) easier. Then

$$\mathscr{D} = \frac{\tau_0}{\pi A}\left(\frac{a - x}{a + x}\right)^{1/2} \tag{38}$$

with

$$a = nA/\tau_0$$

The distributions Eqs. (36)–(38) are plotted in Fig. 2. They may be compared with the results of the polynomial solution.

For large n, the roots of $P'_{n-1}(x/a)$ pile up at $x = \pm a$ and, correspondingly, in Fig. 2.I the distribution $\mathscr{D}$ becomes infinite like $(a - x)^{1/2}$ and $(a + x)^{1/2}$ at its extremities. On the other hand, in case (2) for large n, the roots of $H_n(\alpha^{1/2}x/A^{1/2}) = 0$, reading from the center, become successively more widely spaced and eventually stop. We may say that, instead of a pileup, there is a tailoff at either end. Correspondingly, $\mathscr{D}$ falls to zero at the ends of its range.

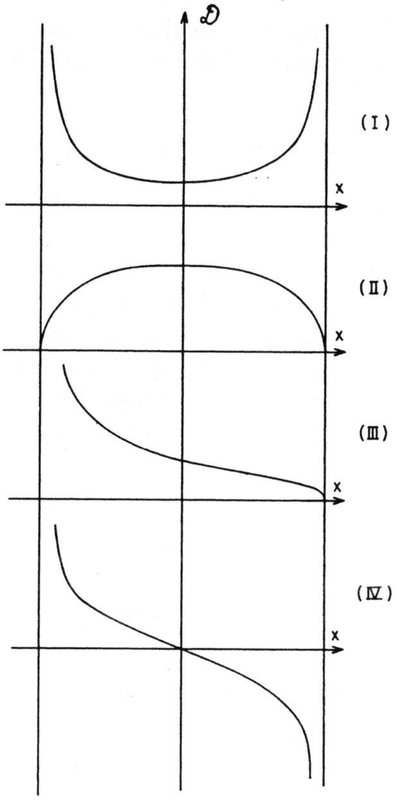

FIG. 2. Dislocation distributions.

In case (3), the polynomial solution Eq. (3) gives a pileup at one end of the row of dislocations and a tailoff at the other. This is reflected in the fact that the $\mathscr{D}$ curve of Fig. 2.III behaves like (1) at one end and (2) at the other.

The stress on the x axis at points outside the distribution may be found as follows. We have seen that the total stress is v'/v. We have the identity

$$\left(\frac{v'}{v}\right)^2 = \frac{v''}{v} - \left(\frac{v'}{v}\right)' = I(x) - \left(\frac{v'}{v}\right)'$$

9

For slowly varying $I(x)$ it is possible (Jefferys and Jefferys, 1962) to justify the omission of the term $(v'/v)'$. Then

$$\frac{v'}{v} = \pm[I(x)]^{1/2}$$

The sign is fixed by the requirement that v'/v must approach $P + n/x$ for large $|x|$. For a distribution of the type Eqs. (36), (37), and (38) we have in ordinary units

$$\tau(x) = -A\pi i\mathscr{D}(x)(|x|/x), \qquad |x| > a$$

that is, we extrapolate $\mathscr{D}$ to values of x for which it becomes imaginary and correct for this by the factor $i = (-1)^{1/2}$. If the square roots occurring in $i\mathscr{D}$ are always given the positive sign, the sign of τ changes on going from $x > a$ to $x < a$, as indicated by the factor $|x|/x$.

The force on a locked dislocation at x_l due to the free dislocation is $f'(x_l)/f(x_l)$. If we multiply Eq. (18) by $x - x_l$ and pass to the limit $x \to x_l$, the first term vanishes, the second becomes $2f'(x_l)$, because only the term $(x - x_l)^{-1}$ in P survives and we have

$$\frac{f'(x_l)}{f(x_l)} = -\frac{1}{2}\lim_{x\to x_l}(x - x_l)\, q(x)$$

If we add to both sides $P_1 = P - (x - x_l)^{-1}$, the external stress plus the stresses due to the locked dislocations except the one at x_l, we get

$$\frac{f'}{f} + P_1 = -\frac{1}{2}\lim_{x\to x_l}(x - x_l)[q - P^2 - P']$$

so that, by Eq. (35), in ordinary units the total force on the locked dislocation at x_l is

$$F(x_l) = -\tfrac{1}{2}bA\pi^2 \lim_{x\to x_l}(x - x_l)\mathscr{D}^2(x) \tag{39}$$

Applied to Eqs. (36) and (38), Eq. (39) reproduces Eqs. (25) and (31) correctly.

The stress due to a quasicontinuous distribution of dislocations can also be calculated as follows. The total Burgers vector of the dislocations between x' and $x' + dx'$ is $b\mathscr{D}(x')\,dx'$ and the stress which they produce at x is $A\mathscr{D}(x')\,dx'/(x - x')$. Hence, the total stress at x is

$$P(x) = A\int \frac{\mathscr{D}(x')}{x - x'}\,dx' \tag{40}$$

where the integral is taken over the region where $\mathscr{D}(x') \neq 0$, and x is supposed to lie outside that region.

Leibfried (1951; see also Head and Louat, 1955) has extended Eq. (40) to points x which lie inside the dislocation distribution, i.e., where $\mathscr{D}(x) \neq 0$. Consider the Eq. (12) governing the equilibrium of the dislocation at x_j under the combined stresses of other free dislocations at x_i and the applied stress $\sigma(x)$. Choose two points $x_j + \epsilon$, $x_j - \epsilon$ on either side of x_j with ϵ large enough to contain a large number of dislocations. Equation (12) can be written

$$\sum_{x_i < x_j - \epsilon} \frac{A}{x_j - x_i} + \sum_{x_i > x_j + \epsilon} \frac{A}{x_j - x_i} + \sum_{x_j - \epsilon < x_i < x_j + \epsilon} \frac{A}{x_j - x_i} = -P(x_j) \quad (41)$$

Since the interval 2ϵ is supposed to be large compared with the average spacing of the dislocations, the first two terms of Eq. (41) may with good accuracy be replaced by the integrals

$$A \int_{-a}^{x-\epsilon} \frac{\mathscr{D}(x')}{x - x'} dx' + A \int_{x+\epsilon}^{a} \frac{\mathscr{D}(x')}{x - x'} dx', \qquad x = x_j$$

(For convenience we suppose that the distribution of dislocations extends from $-a$ to $+a$.) In the region of the x axis covered by the third term in Eq. (41), the dislocations will, in the limit of a quasicontinuous distribution, be nearly equally spaced, and the dislocations on either side of the jth will push it equally from opposite sides, so that the third term is nearly zero. In the continuum limit, we may let ϵ approach zero, and Eq. (41) takes the form

$$A \int_{-a}^{+a} \frac{\mathscr{D}(x')}{x - x'} dx' = -P(x) \quad (42)$$

where the Cauchy principal value of the integral is to be taken.

It is shown in App. C that the general solution of the integral equation (42) is

$$\mathscr{D}(x) = \frac{1}{A\pi^2} \int_{-a}^{a} \left(\frac{a^2 - x'^2}{a^2 - x^2} \right)^{1/2} \frac{P(x')\, dx'}{x - x'} + \frac{C}{\pi(a^2 - x^2)^{1/2}} \quad (43)$$

The term in C is the solution of the homogeneous equation, Eq. (42) with $P = 0$. It is easy to verify that

$$\int_{-a}^{a} \mathscr{D}(x)\, dx = C,$$

the integral in Eq. (43) making no contribution.

The distributions, Eqs. (36)–(38), may be reproduced by inserting the appropriate P in Eq. (43) and equating C with n, the total number of dislocations. However, with a given P and arbitrary C, Eq. (43), will also give solutions in which $\mathscr{D}$ is positive in some regions and negative in others. This is the great merit of Leibfried's treatment; it can deal with mixtures of dislocations of both signs which cannot be handled by the polynomial method. In Leibfried's method, the constant C is chosen so as to give the correct total number of dislocations, or so as to lead to a pileup or tailoff at one end or the other, as required.

Since Eq. (43) is linear, a linear combination of two solutions is also a solution. For example, if we take a suitable combination of Eqs. (36) and (38) (Figs. 2.I and 2.II) we get the solution

$$\mathscr{D}(x) = -\frac{\tau_0}{\pi A}\frac{x}{(a^2 - x^2)^{1/2}} \tag{44}$$

of Fig. 2.IV. It evidently represents

$$\int_{-a}^{0} \mathscr{D}(x)\, dx = \frac{\tau_0 a}{\pi A}$$

positive dislocations piled up at $-a$ and an equal number of negative ones piled up at $+a$ under the influence of a uniform applied stress τ_0.

Even though it is not linear in $\mathscr{D}$, one can show that Eq. (39) is still true when $\mathscr{D}$ is not everywhere of one sign. Equation (39) is obviously also true if $\mathscr{D}$ is everywhere negative. Reversing the sign of $\mathscr{D}$ does not affect F, and this is correct, because, for example, if all the dislocations in Fig. 2.I are reversed in sign, the outermost ones are still pushed outwards. It is then tempting to say that as Eq. (39) only depends on the behavior of $\mathscr{D}$ near x_l it does not matter whether $\mathscr{D}$ changes sign elsewhere. However, one can do a little better, and this is perhaps worthwhile, since Eq. (39) provides a link between the concept of the force on a single dislocation and the crack opening force of fracture mechanics (see Sect. V).

For definiteness, consider the situation of Fig. 2.IV. The problem may be reduced to one which, in principle (though not in practice), is accessible to the polynomial method. Solve the problem of a set of positive dislocations between $-a$ and 0 piled up against a locked positive dislocation at $-a$ by the combined action of an external stress and a set of locked negative dislocations at general positions between 0 and a. Then adjust the positions of the locked dislocations until there is zero stress on each of them. Passing to the continuum limit, we get the appropriate positive $\mathscr{D}$ in the region $-a < x < 0$ and from the positions of the negative dislocations an (irrelevant) negative $\mathscr{D}$ between $x = 0$ and $x = a$. The force on the locked positive dislocation at $-a$ can then be calculated from Eq. (39). From this point of view, the negative dislocations contribute to the applied stress σ rather than to $\mathscr{D}$, but this does not matter, since Eq. (39) gives the sum of the forces due to the free dislocations and to the applied stress, whatever the latter may be. This type of argument is obviously not limited to the situation of Fig. 2.IV, and Eq. (39) holds generally.

There are, naturally, some features of the distribution which a knowledge of $\mathscr{D}$ alone cannot provide, even in the case of large n. In particular, one may wish to know the distance x_1, say, between the locked and the nearest free dislocation in case (3). In the units used in connection with Eq. (29)

$$I = n/x - \tfrac{1}{4}, \qquad v = xL_n'(x)\exp(-x/2)$$

In these units, n is approximately the distance between the first and last dislocations (cf. Eq. (38)), and so, in finding x_1 we may ignore $\frac{1}{4}$ in comparison with nx. Equation (29) then becomes $v'' + (n/x)v = 0$ which has the solution $v = Ax^{1/2}J_1\{(4nx)^{1/2}\} + Bx^{1/2}Y_1\{(4nx)^{1/2}\}$ where J_1 and Y_1 are the Bessel functions usually so denoted. The term in Y_1 must be rejected; otherwise, v would not vanish, for $x = 0$, as required by Eq. (32). Consequently, $x_1 = j_1{}^2/4n$ where $j_1 = 3.832$ is the lowest root (excluding $y = 0$) of the equation $J_1(y) = 0$. In ordinary units, this reads

$$x_1 = 1.84\, A/n\tau_0$$

We have now set out the analysis needed to determine equilibrium arrays of discrete and continuously distributed dislocations. Next, we consider the relation between dislocation arrays and cracks, prefacing our remarks by some results on stress fields round simple cracks, which will be of general application in the sequel.

IV. Cracks and Dislocations

A. Stress Fields Due to Cracks

In fracture mechanics, it has become customary to distinguish the three cases shown in Fig. 3.

In type I deformation, a load is applied which gives $\sigma_{xy} = 0$, $\sigma_{xx} = 0$, $\sigma_{yy} \neq 0$ on the plane $y = 0$. When a crack is made in this plane, its faces open up as shown. In type II deformation, the load gives $\sigma_{xx} = 0$, $\sigma_{yy} = 0$, $\sigma_{xy} \neq 0$, on $y = 0$ and the faces of the crack slide past one another, as indicated by the arrows. In type III deformation (antiplane strain), the applied stress gives $\sigma_{zx} = 0$, $\sigma_{zy} \neq 0$ on $y = 0$, and the faces of the crack move into and out of the figure, as indicated by the conventional arrowheads and tails. In each case, one needs to be

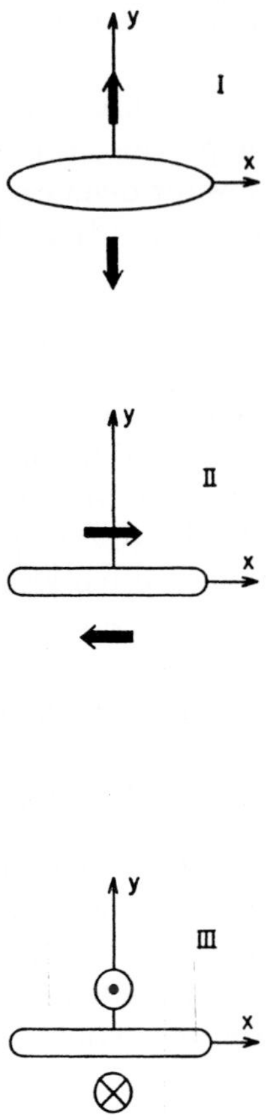

FIG. 3. Modes of deformation in fracture.

able to calculate the elastic field, when the crack is present, in terms of the applied field in the absence of the crack. For many purposes, it is sufficient to know this relation for points in the plane of the crack.

The solution of this problem is particularly simple for type III deformation, and, from it, solutions for related type I and II deformations are easily obtained. We therefore give a fairly detailed discussion of type III (antiplane strain) and at the end show how it may be adapted to give expressions appropriate to type I or II. In antiplane strain, the displacements u, v are zero and $\partial w/\partial z = 0$, so that the only nonvanishing stress components are

$$\sigma_{zy} = G\,\partial w/\partial y, \qquad \sigma_{zx} = G\,\partial w/\partial x \tag{45}$$

and the equilibrium condition is

$$\partial\sigma_{zx}/\partial x + \partial\sigma_{zy}/\partial y = G\,\nabla^2 w = 0 \tag{46}$$

The displacement is thus harmonic. We can generate suitable harmonic functions by taking any analytic function of the complex variable $z = x + iy$ and separating its real and imaginary parts:

$$f(z) = f(x + iy) = \varphi(x, y) + i\psi(x, y) \tag{47}$$

Since

$$f'(z) = df/dz = \partial f/\partial x = -i\,\partial f/\partial y$$

we have, on separating real and imaginary parts, the Cauchy–Riemann relations

$$\partial\varphi/\partial x = \partial\psi/\partial y, \qquad \partial\varphi/\partial y = -\partial\psi/\partial x \tag{48}$$

from which follow

$$\nabla^2\varphi = 0, \qquad \nabla^2\psi = 0$$

We may identify w with either φ or ψ. In what follows we shall put $w = \varphi$; ψ will play an independent role when we extend our results to type I and II deformation. We shall need certain special cases in what follows. The displacement due to a screw dislocation is

$$w = b/2\pi \tan^{-1}(y/x)$$

and the appropriate function f is evidently

$$f(z) = -i(b/2\pi)\ln z = (b/2\pi)(\theta - i\ln r) \tag{49}$$

The displacement due to a concentrated distribution of force along the z axis directed parallel to the z axis and of magnitude unity per unit length of the z axis is

$$w = -(1/2\pi)\ln r$$

To verify this, we calculate the traction across a surface element perpendicular to a radius. It is parallel to the z axis and of magnitude

$$\sigma_{zx}\frac{\partial x}{\partial r} + \sigma_{zy}\frac{\partial y}{\partial r} = G\frac{\partial w}{\partial r} = -\frac{G}{2\pi}\frac{1}{r}$$

The total traction across a circle around the origin is thus -1, and, since the system is in equilibrium, there must be a force $+1$ at the origin, as required. The appropriate f is

$$f(z) = -(1/2\pi) \ln z \tag{50}$$

Except at the origin, the elastic field Eq. (50) gives $\sigma_{zy} = 0$ on $y = 0$, so that we can remove either of the half spaces $y > 0$ or $y < 0$ leaving a point force acting on the surface of a semi-infinite solid. However, its magnitude is $\frac{1}{2}$, not 1, so that, for the elastic field due to a unit point force at (0, 0) on the boundary of the semi-infinite solid $y > 0$ or $y < 0$, we must put

$$f(z) = -(1/\pi) \ln z \tag{51}$$

The simplest cracklike solution is obtained by taking $f(z)$ to be a multiple of $z^{1/2}$. In fact, if

$$f = -iz^{1/2} = -ire^{i\theta/2}, \qquad f' = -\tfrac{1}{2}iz^{-1/2} = -\tfrac{1}{2}ir^{-1/2}e^{-i\theta/2} \tag{52}$$

then

$$w = r^{1/2} \sin \tfrac{1}{2}\theta, \qquad \sigma_{zx} = -G\tfrac{1}{2}r^{-1/2} \sin \tfrac{1}{2}\theta, \qquad \sigma_{zy} = G\tfrac{1}{2}r^{-1/2} \cos \tfrac{1}{2}\theta$$

If we make a mathematical cut along the x axis from $x = 0$ to $x = -\infty$ then r, $\theta = \pi$ represents a point on the negative x axis approached from above, and r, $\theta = -\pi$ represents the same point approached from below. According to Eq. (51), $\sigma_{zy} = 0$, $w = \pm r^{1/2}$ for $\theta = \pm\pi$, so that there is no stress across the negative x axis, but there is a discontinuity $2 \mid x \mid^{1/2}$ in w across it. Thus, the mathematical cut represents a physical stressfree crack extending along the negative x axis. On the positive x axis ahead of the crack, there is a stress $\sigma_{zy} = \frac{1}{2}Gx^{1/2}$. The traction across a circle $r =$ constant is $G\, \partial w/\partial r = \frac{1}{2}r^{1/2} \sin \frac{1}{2}\theta$, so that the elastic state Eq. (52) could be maintained in a cylinder slit along $\theta = \pi$ by applying longitudinal forces proportional to $\sin \frac{1}{2}\theta$ around its circumference.

In the same way, if we make a mathematical cut from $x = 0$ to $x = \infty$ the function

$$f = z^{1/2}, \qquad f' = \tfrac{1}{2}z^{-1/2} \tag{53}$$

with

$$w = r^{1/2} \cos \tfrac{1}{2}\theta, \qquad \sigma_{zx} = \tfrac{1}{2}Gr^{-1/2} \cos \tfrac{1}{2}\theta, \qquad \sigma_{zy} = \tfrac{1}{2}Gr^{-1/2} \sin \tfrac{1}{2}\theta$$

we obtain the elastic field associated with a crack extending along the positive x axis.

We now look for the elastic field associated with a uniform applied stress $\sigma_{zy} = \tau$, $\sigma_{zx} = 0$ perturbed by a crack extending along the x axis from $x = -a$ to $x = a$. For large z, we must have $Gf' = -i\tau$. Near $x = -a$ and $x = a$, we expect the solution to behave like Eqs. (52) and (53) with the origin suitably shifted, i.e.,

$$f \approx (z - a)^{1/2} \qquad \text{near} \quad z = a \tag{54}$$

$$f \approx i(z + a)^{1/2} \qquad \text{near} \quad z = -a \tag{55}$$

It is easily verified that adding Eqs. (54) and (55) does not give the solution we want. However, antiplane solutions in terms of a complex function f have the useful property that the product of two solutions is also the solution of some problem, so we consider a multiple of the product of Eqs. (54) and (55),

$$f = -(i\tau/G)(z^2 - a^2)^{1/2}, \qquad Gf' = -i\tau z/(z^2 - a^2)^{1/2} \tag{56}$$

where the "arbitrary" constant has been chosen with an eye on the final result. With the notation of Fig. 4

$$z = re^{i\theta}, \qquad z - a = r_1 e^{i\theta_1}, \qquad z + a = r_2 e^{i\theta_2}$$

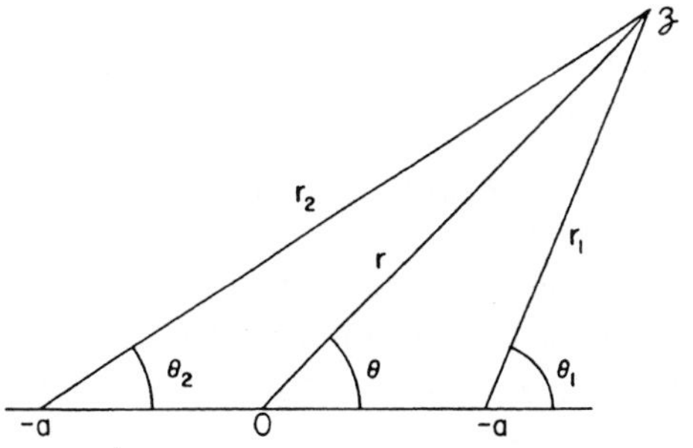

FIG. 4. Coordinates for discussing a crack.

and Eq. (56) gives

$$\begin{aligned} f &= -(i\tau/G)(r_1 r_2)^{1/2} \exp \tfrac{1}{2}i(\theta_1 + \theta_2) \\ f' &= -i\tau r(r_1 r_2)^{-1/2} \exp i[\theta - \tfrac{1}{2}(\theta_1 + \theta_2)] \end{aligned} \tag{57}$$

so that

$$\begin{aligned} w &= (\tau/G)(r_1 r_2)^{1/2} \sin \tfrac{1}{2}(\theta_1 + \theta_2) \\ \sigma_{zx} &= \tau r(r_1 r_2)^{1/2} \sin[\theta - \tfrac{1}{2}(\theta_1 + \theta_2)] \\ \sigma_{zy} &= \tau r(r_1 r_2)^{-1/2} \cos[\theta - \tfrac{1}{2}(\theta_1 + \theta_2)] \end{aligned} \tag{58}$$

At large distances from the crack, the figure shows that $r = r_1 = r_2$, $\theta = \theta_1 = \theta_2$ and the stress reduces to $\sigma_{zx} = 0$, $\sigma_{zy} = \tau$, as required. If we make a mathematical cut from $-a$ to $+a$, the upper face of the cut corresponds to $\theta = 0$ or π, $\theta_1 = \pi$, $\theta_2 = 0$ and the lower to $\theta = 0$ or $-\pi$, $\theta_1 = -\pi$, $\theta_2 = 0$ so that, by Eq. (54), the faces of the cut are stressfree and Eq. (53) satisfies all the conditions of the problem. The displacement discontinuity across the crack is, on inserting these values of θ, θ_1, θ_2 in Eq. (54),

$$\begin{aligned} \Delta w &= \lim_{\epsilon \to 0} [w(x, \epsilon) - w(x, -\epsilon)], \qquad \epsilon > 0 \\ &= (2\tau/G)(r_1 r_2)^{1/2} = (2\tau/G)(a^2 - x^2)^{1/2}, \qquad |x| < a \end{aligned}$$

The function $-i(z^2 - a^2)^{1/2} = (a^2 - z^2)^{1/2}$ which appears in Eq. (56) also plays an important role in more complicated crack problems. In what follows, Fig. 5 will often be useful. It shows the values assumed by

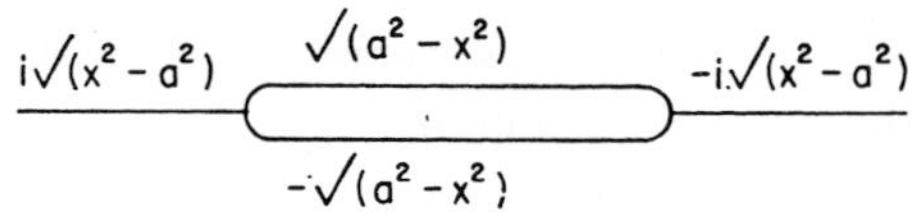

FIG. 5. Behavior of the function $(a^2 - z^2)^{1/2}$.

$(a^2 - z^2)^{1/2}$ on the x axis outside the crack and on the upper and lower faces of the crack. It may easily be verified by considering how θ_1 and θ_2 vary along a line parallel to the x axis and slightly above or below it. It is understood that $(a^2 - x^2)^{1/2}$ and $(x^2 - a^2)^{1/2}$ stand for $+\sqrt{(a^2 - x^2)}$ and $+\sqrt{(x^2 - a^2)}$, respectively.

We now consider the more general problem of finding how the presence of a crack perturbs an arbitrary applied stress σ_{zx}^A, σ_{zy}^A which is not necessarily uniform. Apply the stress σ_{zx}^A, σ_{zy}^A to the crackfree solid and mark out the site of the proposed crack, $|x| < a$, $y = 0$, but do not actually make a cut. At this stage, the traction on an element of the x axis with normal $(0, -1, 0)$, i.e., an element of the upper face of the proposed crack, is $-\sigma_{zy}^A(x, 0)$ and the traction on an element with normal $(0, 1, 0)$, an element of the lower face of the proposed crack is $+\sigma_{zy}^A$. Hence, if we now make the cut, the material will not relax if we apply tractions $-\sigma_{zy}^A$ to the upper face and $+\sigma_{zy}^A$ to the lower face, and the stress is still σ_{zx}^A, σ_{zy}^A. If we remove the tractions on the faces of the crack, the stress in the material becomes $\sigma_{zx}^A + \sigma_{zx}^C$, $\sigma_{zy}^A + \sigma_{zy}^C$. But removing the tractions $-\sigma_{zy}^A$, $+\sigma_{zy}^A$ is the same as applying additional tractions $+\sigma_{zy}^A$ and $-\sigma_{zy}^A$ to the upper and lower faces, respectively. Hence, finally

The elastic field of a crack perturbing an applied stress σ_{zx}^A, σ_{zy}^A is the same as the elastic field produced by applying traction σ_{zy}^A to the upper and $-\sigma_{zy}^A$ to the lower face of a crack in an otherwise stressfree body. (59)

The stresses applied to the faces can be regarded as the result of applying concentrated forces $\pm\sigma_{zy}^A\, dx$ to each element of the x axis. Consequently,

$$\sigma_{zx}^C(x, y) - i\sigma_{zy}^C(x, y) = G \int_{-a}^{a} f_0'(z, x')\, \sigma_{zy}^A(x')\, dx' \tag{60}$$

where $f_0'(z, x')$ is the complex stress due to a point force $+1$ at x' on the upper face, together with a point force -1 at x' on the lower face (Fig. 6).

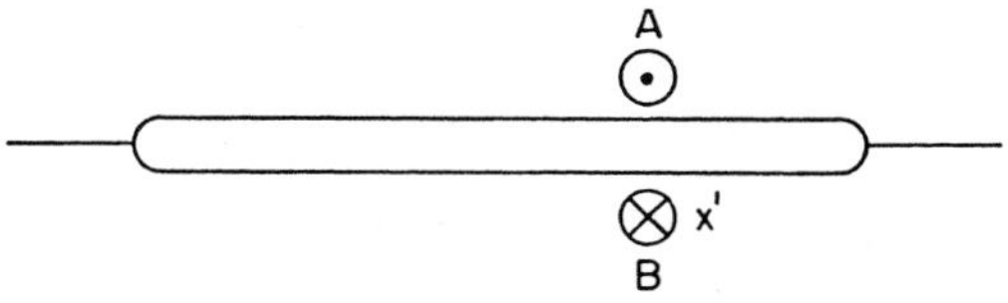

FIG. 6. Unit forces applied to crack surfaces.

It is clear that, near A (Fig. 6), f_0' must behave like the solution for a point force on the lower boundary of the semi-infinite solid $y > 0$, namely,

$$f' = -[G\pi(z - x')]^{-1} \tag{61}$$

(cf. Eq. (50)) and like the same expression with reversed sign near B. We further expect that it will reduce to the field constant z^{-2} of a force doublet at large distances from the crack and that, so as to reproduce Eq. (56) when σ_{zy}^A = constant, it will involve the expression $(a^2 - z^2)^{1/2}$. As in deriving Eq. (56), we use the fact that Eq. (61) remains a solution of the elastic equations when it is multiplied by another analytic function of z. A little thought suggests that $(a^2 - x'^2)^{1/2}/(a^2 - z^2)^{1/2}$ is a suitable multiplier, since it reduces to z^{-1} for large z, to $+1$ at A and to -1 at B (cf. Fig. 5). We therefore put

$$f_0'(z) = -\frac{1}{\pi G} \frac{(a^2 - x'^2)^{1/2}}{(a^2 - z^2)^{1/2}} \frac{1}{z - x'} \tag{62}$$

Since, on the x axis between $-a$ and $+a$, $f_0'(z)$ is purely real, it gives zero stress at the crack and so meets all requirements.

On the x axis outside the crack, Eqs. (60) and (62) give, on paying attention to Fig. 5,

$$\sigma_{zy}^{C} = \frac{\operatorname{sgn} x}{\pi(x^2 - a^2)^{1/2}} \int_{-a}^{a} \frac{(a^2 - x'^2)^{1/2}}{x - x'} \sigma_{zy}^{A}(x')\, dx', \; |x| > a \tag{63}$$

The identity

$$\frac{(a^2 - x'^2)^{1/2}}{(x^2 - a^2)^{1/2}} + \frac{(x^2 - a^2)^{1/2}}{(a^2 - x'^2)^{1/2}} = \frac{(x - x')(x + x')}{(x^2 - a^2)^{1/2}(a^2 - x'^2)^{1/2}} \tag{64}$$

converts Eq. (63) into

$$\sigma_{zy}^{C} = \frac{|x|}{(x^2 - a^2)^{1/2}} I_1 + \frac{a \operatorname{sgn} x}{(x^2 - a^2)^{1/2}} I_2 - (x^2 - a^2)^{1/2} \operatorname{sgn} x\, I_3 \tag{65}$$

where

$$\begin{aligned} I_1 &= \frac{1}{\pi} \int_{-a}^{a} \frac{\sigma_{zy}^{A}(x')\, dx'}{(a^2 - x'^2)^{1/2}} \\ I_2 &= \frac{1}{\pi a} \int_{-a}^{a} \frac{x' \sigma_{zy}^{A}(x')\, dx'}{(a^2 - x'^2)^{1/2}} \\ I_3 &= \frac{1}{\pi} \int_{-a}^{a} \frac{\sigma_{zy}^{A}(x')\, dx'}{(a^2 - x'^2)^{1/2}(x - x')} \end{aligned} \tag{66}$$

(In Eq. (63) and elsewhere $\operatorname{sgn} x = |x|/x$.)

One can show that, unlike the terms in I_1, I_2, the term in I_3 is not singular at the crack tips. Suppose, for definiteness, that $x > a$. The coefficient of $\sigma_{zy}^{A}(x')$ in the integrand does not change sign over the region of integration, and so

$$\begin{aligned} (x^2 - a^2)^{1/2} |I_3| &< (x^2 - a^2)^{1/2} \max |\sigma_{zy}^{A}| \int_{-a}^{a} \frac{dx'}{(a^2 - x'^2)^{1/2}(x - x')} \\ &= \max |\sigma_{zy}^{A}| \end{aligned} \tag{67}$$

by the first of the following standard results:

$$\begin{aligned} \int_{-a}^{a} \frac{dx'}{(a^2 - x'^2)^{1/2}(x - x')} &= \frac{\pi \operatorname{sgn} x}{(x^2 - a^2)^{1/2}}, \qquad |x| > a \\ &= 0, \qquad |x| < a \\ \int_{-a}^{a} \frac{dx'}{(a^2 - x'^2)^{1/2}} &= \pi \end{aligned} \tag{68}$$

Hence, for $-a > x > a$, the third term in Eq. (65) nowhere exceeds $\max |\sigma_{zy}^{A}|$, the numerically greatest value of σ_{zy}^{A} in the interval $|x| < a$.

We can get a better estimate for I_3 if we write it in the form

$$I_3 = \sigma_{zy}^A(x)\frac{1}{\pi}\int_{-a}^{a}\frac{1}{(a^2-x'^2)^{1/2}}\frac{dx'}{x-x'}$$

$$-\frac{1}{\pi}\int_{-a}^{a}\frac{\sigma_{zy}^A(x)-\sigma_{zy}^A(x')}{x-x'}\frac{dx'}{(a^2-x'^2)^{1/2}} \qquad (69)$$

The first term is $(\operatorname{sgn} x)(x^2-a^2)^{1/2}\,\sigma_{zy}^A(x)$, by Eq. (68). In the second term, the expression $[\sigma_{zy}^A(x)-\sigma_{zy}^A(x')]/(x-x')$ is, by Rolle's theorem, equal to $d\sigma_{zy}^A(x'')/dx''$ for some x'' lying between x and x'.

Hence, if

$$M(x) = \max \mid d\sigma_{zy}^A/dx' \mid$$

is the maximum absolute value of $d\sigma_{zy}^A(x')/dx'$ in the range $-a \leqslant x' \leqslant x$ for $x > a$ or in the range $x \leqslant x' \leqslant a$ for $x < a$, the second term is less in absolute value than

$$M\frac{1}{\pi}\int_{-a}^{a}\frac{dx'}{(a^2-x'^2)^{1/2}} = M$$

The stress due to the crack is thus

$$\sigma_{zy}^C(x) = \frac{\mid x \mid}{(x^2-a^2)^{1/2}}I_1 + \frac{a \operatorname{sgn} x}{(x^2-a^2)^{1/2}}I_2 - \sigma_{zy}^A(x)$$

$$+ (x^2-a^2)^{1/2}\,\vartheta(x)\,M(x) \qquad (70)$$

where

$$-1 < \vartheta(x) < 1$$

The total stress $\sigma_{zy} = \sigma_{zy}^C + \sigma_{zy}^A$ is obtained by omitting the term $-\sigma_{zy}^A(x)$ in Eq. (70). Thus, just beyond the end $\pm a$ of the crack, the total stress behaves like $\mid x \mp a \mid^{-1/2}$ unless the terms in I_1, I_2 each vanish or cancel one another, in which case it behaves like $\mid x \mp a \mid^{1/2}$. Evidently, if the stress is to be finite at the end a, we must have $I_1 + I_2 = 0$, if it is to be finite at $-a$, we must have $I_1 - I_2 = 0$, and, if it is to be finite at both ends, we must have $I_1 = 0$, $I_2 = 0$.

On the x axis inside the crack $\sigma_{zy}^C = -\sigma_{zy}^A$ by construction, but Eq. (60) now gives σ_{zx}^C.

We have

$$\sigma_{zx}^C = -\frac{1}{\pi}\int_{-a}^{a}\frac{(a^2-x'^2)^{1/2}}{(a^2-x^2)^{1/2}}\frac{\sigma_{zy}^A(x')\,dx'}{x-x'}, \qquad \mid x \mid < a \qquad (71)$$

on the upper face and an equal and opposite value on the lower face (cf. Fig. 6). Since $\sigma_{zx} = G\,\partial w/\partial x$, Eq. (71) may be rewritten as

$$\frac{\partial\,\Delta w}{\partial x} = -\frac{2}{G\pi}\int_{-a}^{a}\frac{(a^2-x'^2)^{1/2}}{(a^2-x^2)^{1/2}}\frac{\sigma_{zy}^{A}(x')\,dx'}{x-x'}, \qquad |\,x\,| < a \tag{72}$$

where $\Delta w(x)$ is the difference in displacement (upper minus lower) across the faces of the crack at x.

The identity Eq. (64) converts Eq. (71) into

$$\sigma_{zx}^{C} = -\frac{x}{(a^2-x^2)^{1/2}}I_1 - \frac{a}{(a^2-x^2)^{1/2}}I_2 - (a^2-x^2)^{1/2}I_3$$

with the notation Eq. (66). Since x is now within the range of integration, I_3 must be interpreted as a Cauchy principal value.

The first term in I_3 vanishes by Eq. (68). The second term is numerically less than $M(a)$, the maximum of $|\,d\sigma_{zy}^{A}\,dx'\,|$ over the range $|\,x'\,| < a$ so that

$$\sigma_{zx}^{C}(x) = -\frac{x}{(a^2-x^2)^{1/2}}I_1 - \frac{a}{(a^2-x^2)^{1/2}}I_2 + (a^2-x^2)^{1/2}\,\vartheta(x)\,M(a) \tag{73}$$

The condition that σ_{zx}^{C} shall be finite just inside one tip or the other is the same as the condition that σ_{zy}^{C} shall be finite just outside that tip.

In what follows, we shall need the stress field of what may be termed a dislocation with a cracked core. Introduce a dislocation at, say, the origin and then make a cut from $-a$ to a passing through the center of the dislocation. The appropriate complex stress Gf' must obviously have the following properties: f' behaves like Eq. (49) for large $|\,z\,|$, $\sigma_{zy} = 0$ on the faces of the crack, the integral of $\partial w/\partial x$ from $-a$ to a along the lower face of the crack and back from a to $-a$ along the upper face must be b. To find f', we again use the trick of multiplying the solution Eq. (49) by a complex function which leaves it unaltered at infinity and introduces the characteristic crack function $(z^2 - a^2)^{1/2}$. A suitable candidate for the multiplier is evidently $z/(z^2 - a^2)^{1/2}$ which gives

$$f' = -\frac{ib}{2\pi}\frac{1}{(z^2-a^2)^{1/2}} = -\frac{ib}{2\pi}\frac{1}{(r_1r_2)^{1/2}}e^{-\frac{1}{2}i(\theta_1+\theta_2)} \tag{74}$$

an expression which meets all requirements. On the x axis, the stresses are

$$\sigma_{zx} = 0, \qquad \sigma_{zy} = \frac{Gb}{2\pi}\frac{\operatorname{sgn} x}{(x^2-a^2)^{1/2}}, \qquad |\,x\,| > a$$

$$G\frac{\partial\,\Delta w}{\partial x} = \sigma_{zx} = -\frac{Gb}{2\pi}\frac{1}{(a^2-x^2)^{1/2}}, \qquad \sigma_{zy} = 0, \qquad |\,x\,| < a \tag{75}$$

We now consider how the foregoing results can be used to obtain analogous relations for type I or type II deformations.

The displacements associated with a general state of plane strain may be written in the form

$$\begin{aligned} u &= (1-\nu)\,\varphi - \frac{1}{2}\frac{\partial}{\partial x}(y\psi) + \Psi \\ v &= (1-\nu)\,\psi - \frac{1}{2}\frac{\partial}{\partial y}(y\psi) + \Phi \end{aligned} \tag{76}$$

(ν is Poisson's ratio). Here, φ, ψ are a pair of conjugate harmonic functions and so are Φ, Ψ, so that

$$\begin{aligned} \varphi + i\psi &= f(z) \\ \Phi + i\Psi &= F(z) \end{aligned} \tag{77}$$

where f and F are analytic functions (Love, 1952).

To deal with a state of type II deformation, we need to have $\sigma_{yy} = 0$ on the x axis. This can be secured by taking $\Phi = 0$, $\Psi = 0$. Then, on the x axis

$$u = (1-\nu)\,\varphi, \qquad v = \tfrac{1}{2}(1-2\nu)\,\psi, \qquad \sigma_{yy} = 0, \qquad \sigma_{xy} = G\,\partial\varphi/\partial y$$

If we use the same function $f(z)$ to construct simultaneously a type III antiplane solution (denoted by an affixed III) and a type II plane strain solution (denoted by an affixed II), there are the following relations between the stresses and displacements on the x axis:

$$\sigma_{zy}^{\text{III}} = \sigma_{xy}^{\text{II}}, \qquad \sigma_{zx}^{\text{III}} = G\frac{\partial w^{\text{III}}}{\partial x} = \tfrac{1}{2}\sigma_{xx}^{\text{II}} = \frac{G}{1-\nu}\frac{\partial u^{\text{II}}}{\partial x}$$

Consequently, the relations obtained above connecting the applied stress σ_{zy}^{A} with the stress and displacement σ_{zy}^{C} and w^{C} due to the crack can be used to find the stress σ_{xy}^{C} and displacement u^{C} induced in the same crack by an applied stress system which produces stresses σ_{xy}^{A}, $\sigma_{yy}^{A} = 0$ on the x axis. It is only necessary to make the replacements

$$\sigma_{zy}^{A} \to \sigma_{xy}^{A}, \qquad \sigma_{zy}^{C} \to \sigma_{xy}^{C}, \qquad \frac{\partial\,\Delta w}{\partial x} \to \frac{1}{(1-\nu)}\frac{\partial\,\Delta u}{\partial x} \tag{78}$$

in Eqs. (63), (65), (66), (69), (70), (72), and (73).

This analogy is not limited to the x axis. For example, if we put $\Phi = 0$, $\Psi = 0$ in Eq. (76) and take for f the function Eq. (56) appropriate to a crack perturbing a uniform applied stress $\sigma_{zy}^{A} = \tau$ we get the elastic field of the same crack perturbing a uniform stress $\sigma_{xy}^A = \tau$, or, if we take for f the function Eq. (49) for a screw dislocation, we get the displacements due to an edge dislocation with Burgers vector $b_x = (1 - \nu)\, b$, $b_y = 0$, $b_z = 0$. When the unwanted factor $(1 - \nu)$ is deleted, the result is the well-known expression

$$u = \frac{b}{2\pi} \tan^{-1}\left(\frac{y}{x}\right) + \frac{b}{2\pi} \frac{1}{2(1-\nu)} \frac{xy}{r^2}$$

$$v = \frac{b}{2\pi} \frac{1-2\nu}{2(1-\nu)} \ln \frac{1}{r} + \frac{b}{2\pi} \frac{1}{2(1-\nu)} \frac{y^2}{r^2} \tag{79}$$

To discuss type I deformation, we need to have $\sigma_{xy} = 0$ on the x axis in place of $\sigma_{yy} = 0$. This can be secured by restoring Φ, Ψ in Eq. (76) and choosing

$$\Phi = \tfrac{1}{2}\psi, \qquad \Psi = -\tfrac{1}{2}\varphi, \qquad \text{i.e.,} \qquad F(z) = -\tfrac{1}{2} i f(z)$$

Finally, to reinforce the analogy with type II and type III deformations, it is convenient to replace the conjugate pair φ, ψ by the (equally general) pair $-\psi$, φ. For the resulting displacement

$$u = -\frac{1}{2}(1-2\nu)\,\psi - \frac{1}{2}\frac{\partial}{\partial x}(y\varphi)$$

$$v = \frac{1}{2}(3-2\nu)\,\varphi - \frac{1}{2}\frac{\partial}{\partial y}(y\varphi) \tag{80}$$

we have, on the x axis

$$u = -\tfrac{1}{2}(1-2\nu)\,\psi, \qquad v = (1-\nu)\,\varphi, \qquad \sigma_{xy} = 0, \qquad \sigma_{yy} = G\,\partial\varphi/\partial y$$

and, as in the case of Eq. (76) we can convert a relation between σ_{zy}^A, σ_{zy}^C, $\partial\,\Delta w/\partial x$ along the x axis for type III deformations into one for type I deformations by making the replacements

$$\sigma_{zy}^A \to \sigma_{yy}^A, \qquad \sigma_{zy}^C \to \sigma_{yy}^C, \qquad \frac{\partial\,\Delta w}{\partial x} \to \frac{1}{1-\nu}\frac{\partial\,\Delta v}{\partial x} \tag{81}$$

B. Representation of Cracks by Dislocations

The elastic stress fields around loaded cracks, slip bands, Z-type kink bands, twins, and martensite plates all have qualitatively similar features when viewed on a suitable scale. This is because they all represent the same type of incompatibility in the medium, namely, that caused by the Somigliana dislocation (1914, 1915). Under appropriate conditions, therefore, we must expect any one of these singularities to be a source of any of the others (Frank and Stroh, 1952; Bilby and Bullough, 1954; Bilby and Entwisle, 1954). The similarity of the piled up group of slip dislocations to the freely slipping crack was recognized very early in the development of the theory of crystal dislocations (Zener 1948; Eshelby, Frank, and Nabarro 1951). The description of a crack with a normal displacement discontinuity by a continuous distribution of freely climbing "dislocations de clivage" is due to Friedel (1956, 1959, 1964) who has likened them to dislocations representing boundaries (Frank, 1950; Nye, 1953; Bilby, Gardner, and Smith, 1958). The representation of such a discontinuity by dislocations depends on the definition of the type of dislocations involved, and there is some confusion in the literature. To discuss the crack field and its interaction with other inclusions and inhomogeneities, it seems most satisfactory to adopt the following point of view and treat a crack as an inclusion at whose boundary there is zero surface traction, or a narrow zone where the elastic constants are zero (Eshelby, 1956, 1960).

Consider first a linear elastic medium containing a continuous distribution $\mathscr{D}_y(x)$ of infinitesimal dislocations parallel to z with line senses along [001] and lying in the region R: $-a < x < +a, y = 0$, with Burgers vectors in the y direction. The total Burgers vector between x and $x + dx$ is $b\mathscr{D}_y(x)\,dx$, and, with the RH/FS convention, the relative displacement $d\phi_y$ of the region $y > 0$ with respect to the region $y < 0$ is $d\phi_y = -b\mathscr{D}_y(x)\,dx$. If we take $b > 0$, then $\mathscr{D}_y(x)$ is positive in regions where the amount of material inserted decreases with x, or the amount of material removed increases with x; a symmetrical situation involving only inserted material and with $\mathscr{D}_y(x)$ positive for $x > 0$ and negative for $x < 0$ is depicted in Fig. 7. Along $y = 0$, the distribution produces everywhere a stress σ_{yy}^D, where

$$\sigma_{yy}^D = A \int_{-a}^{a} \frac{\mathscr{D}_y(x')\,dx'}{x - x'} \tag{82}$$

and $A = Gb/2\pi(1 - \nu)$. Now let there be an external stress σ_{yy}^A along $y = 0$ due to externally applied loads or to other sources of internal

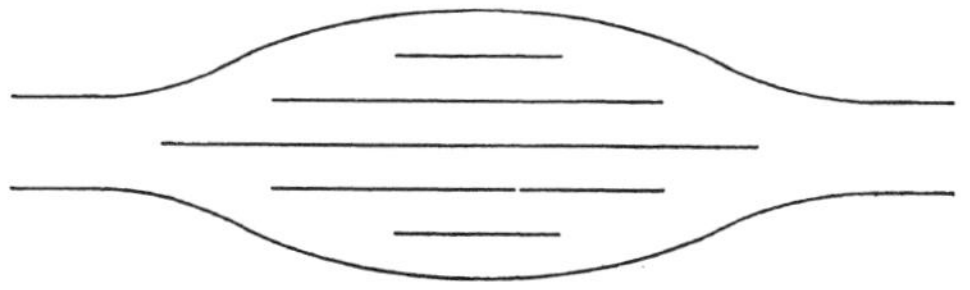

FIG. 7. Interpretation of crack dislocations: material initially inserted.

10

stress in the solid. If we now adjust $\mathscr{D}_y(x)$ so that $\sigma^A_{yy} + \sigma^D_{yy} = 0$ over R, then this region is free from traction. Consequently, the medium can be cut over R and all the inserted material can be removed without disturbing the surrounding field. The distribution of dislocations $\mathscr{D}_y(x)$ thus represents the discontinuity in normal displacement across a region $-a < x < +a, y = 0$ free from traction in an applied field σ^A_{ij}, and so describes the shape of a thin crack or cavity in this region. Conversely, let us regard $\mathscr{D}_y(x)$ as given and adjust σ^A_{yy} so that $\sigma^A_{yy} + \sigma^D_{yy} = 0$ over R. If $\mathscr{D}_y(x)$ was originally produced entirely by inserting material within the region R (as in Fig. 7), then if the external stress is relaxed after all this material has been removed, we have a solid which, when unstressed, just closes up so that $\mathscr{D}_y(x) = 0$ over R. The same distribution $\mathscr{D}_y(x)$ reappears on reloading to the same stress, and we may say that the external stress induces the dislocations along a slit over R. Further loading in tension leads to the induction of a distribution $|\mathscr{D}''_y(x)| > |\mathscr{D}_y(x)|$, and reloading to a reduced stress, to a distribution $|\mathscr{D}''_y(x)| < |\mathscr{D}_y(x)|$, which would correspond to greater and lesser amounts of originally inserted material respectively. The variation of $\mathscr{D}_y(x)$ under load, or the extension of a crack beyond $-a < x < +a$, can be regarded as due to the generation and propagation by formal outward climb from the origin of edge dislocations with empty half planes.

If $\mathscr{D}_y(x)$ was originally produced entirely by removing material within the region $-a < x < +a$, and rewelding over this region (Fig. 8a); then, after adjusting σ^A_{yy} so that $\sigma^A_{yy} + \sigma^D_{yy} = 0$ over R, we may again cut along R without disturbing the stress field. On now relaxing σ^A_{yy}, a

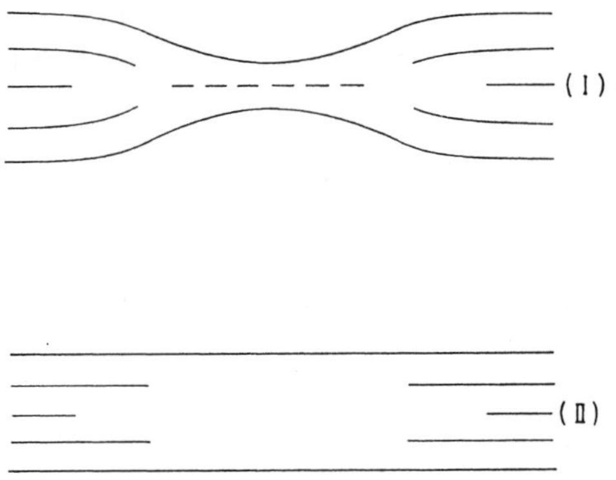

FIG. 8. Interpretation of crack dislocations: (a) Material removed and stresses applied to make tractions vanish over dotted region—crack dislocations present; (b) Unstressed cavity, but no crack dislocations.

cavity corresponding to $\mathscr{D}_y(x)$ over R appears when the body is unstressed (Fig. 8b). We can partially close the cavity by applying a reduced compression σ_{yy}^A, but we must remove more material if we wish to make a $\mathscr{D}_y'''(x)$ such that $|\mathscr{D}_y'''(x)| > | \mathscr{D}_y(x)$ everywhere in this case. We can do this, of course, by a formel generation and climb outwards from the origin of edge dislocations with empty half planes. The process is similar in sign to that in Fig. 7, since, in both cases, we are increasing the size of the cavity, although, in the first case, the cavity appears only when the solid is stressed, and in the second only when it is unstressed.

With this definition of $\mathscr{D}_y(x)$, the stresses produced by $\mathscr{D}_y(x)$ are those of an "ordinary" dislocation distribution. The distribution for given σ_{yy}^A is determined by the integral equation

$$\sigma_{yy}^A + A \int_{-a}^{a} \frac{\mathscr{D}_y(x')\,dx'}{x - x'} = 0 \tag{83}$$

and, when $\mathscr{D}_y(x)$ has been found, the total stress at any point is $\sigma_{ij}^A + \sigma_{ij}^D$ where

$$\sigma_{ij}^D(\mathbf{r}) = b \int_{-a}^{a} \mathscr{D}_y(x')\,\sigma_{ij}^y(x - x', y, z)\,dx' \tag{84}$$

and $\sigma_{ij}^y(\mathbf{r})$ is the stress field at $\mathbf{r}$ produced by a dislocation at the origin with Burgers vector [0, 1, 0]. The stress σ_{ij}^D due to the dislocations describing a crack which is not loaded on its internal surfaces is thus identical with the perturbation stress σ_{ij}^C produced in the applied field σ_{ij}^A by the presence of the crack and discussed in Sect. IV, A.

It should be noted that the crack dislocation defined in this way is a special type of dislocation, in a sense reciprocal to an ordinary dislocation. For its Burgers vector is related to its stress field in the same way as is that of an ordinary dislocation, but this vector has a reciprocal geometrical meaning because it is generated in a slit body by a field external to the dislocation itself. Usually, we define dislocations by discontinuities forced to occur across cuts in an unstressed body, the cuts being subsequently rewelded; crack dislocations are defined by discontinuities occuring naturally across unwelded cuts when a body is stressed. Perhaps the simplest way to regard a crack dislocation is as a type of imperfect dislocation whose associated sheet of bad crystal is a missing plane of atoms; in calculating its Burgers vector, the thickness of the missing plane must be included in the circuit by taking into account the appropriate normal displacement, as with a stacking fault involving such a displacement (Frank, 1951).

In Fig. 9, we show, for simplicity, a pair of ordinary (A, B) and a pair of crack (A', B') dislocations, both with line senses up from the paper, as indicated by the circles. Using the operational definition (*RH/FS* convention),

$$b_y = \oint \frac{\partial u_y}{\partial x_i} dx_i \tag{85}$$

the Burgers vectors of B and B' are both equal to $S'F'$, although the displacements round B are produced by forcing in the plane P, while those round B' arise because a cut along P' enables an external stress (here nonuniform) to produce a displacement discontinuity. Other definitions of crack dislocations would be possible, but this is the most convenient, since it makes them identical in all properties involving their stress fields with ordinary dislocations. For example, in Fig. 9, an applied tensile stress $\sigma_{yy}^A > 0$ forces both AB and $A'B'$ apart.

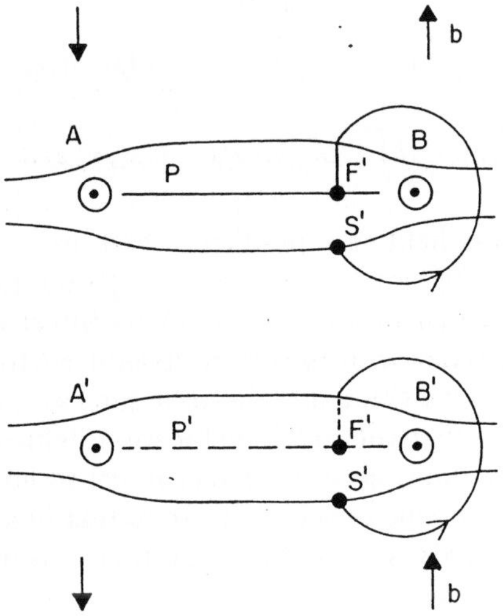

FIG. 9. Burgers vectors of ordinary and crack dislocations.

The situation may also be discussed in terms of the forces acting on the dislocations. Due to the external stress, the dislocations $b\mathscr{D}_y(x)\,dx$ at x experience a force $(\sigma_{yy}^A + \sigma_{yy}^D)\,b\mathscr{D}_y(x)\,dx$ in the positive x direction tending to promote their formal climb (by taking the Cauchy principal value of the integral we exclude the force exerted on a dislocation by its own stress). The distribution $\mathscr{D}_y(x)$ will be an equilibrium one if this force is zero everywhere in $-a < x < +a$, and we obtain again the condition $\sigma_{yy}^A + \sigma_{yy}^D = 0$ for this to occur. We discuss in Sect. V the rather delicate question of the force on the ends of the distribution at

$x = \pm a$. We shall see that the model implies that $\mathscr{D}_y(x)$ at an end point be either infinite or zero. Under stress, the crack shape at the tip is thus either rounded, with a vertical slope, or flat, closing smoothly usually as at a cusp. In the latter case, the force is zero, and, in the former, we must suppose that infinitely great short-range forces exist at $x = \pm a$, such as those due to a stress $\delta(x \pm a)$, where δ is the Dirac delta function. These short-range forces are called barriers, or blocks. Whether they are required or not depends on the nature of the solutions of Eq. (83) which we consider. The equation automatically limits $\mathscr{D}_y(x)$ to the region R, and, according to the form of $\sigma_{yy}^A(x)$, blocks may or may not be required, as we shall discuss.

Since, at least in isotropic linear elasticity, parallel dislocations with perpendicular Burgers vectors do not interact, a discussion in terms of forces on dislocations shows that a long crack parallel to z under combined plane strain (or stress) and antiplane strain may be discussed in terms of three independent distributions of dislocations $\mathscr{D}_x(x)$, $\mathscr{D}_y(x)$, and $\mathscr{D}_z(x)$ (cf. Eq. (10)). $\mathscr{D}_x$ obeys the same integral equation as $\mathscr{D}_y$ but with σ_{xy}^A replacing σ_{yy}^A; in the equation for $\mathscr{D}_z$, σ_{yz}^A replaces σ_{yy}^A and the constant A becomes $Gb/2\pi$. The same result is obtained by considering the stresses. $\mathscr{D}_x$, $\mathscr{D}_y$ and $\mathscr{D}_z$ correspond to cracks under modes II, I, and III, respectively, in the language used by engineers (Sect. IV,A); $\mathscr{D}_x$ and $\mathscr{D}_z$ represent freely slipping craks, with discontinuities in tangential displacement only. The crack dislocations in the distributions $\mathscr{D}_x$ and $\mathscr{D}_z$ do not require any special discussion. Though they form at unwelded slits in a stressed body rather than at slits displaced and subsequently rewelded in a body originally unstressed, no special interpretation of their Burgers vectors is required. Because no material need be removed or inserted along the crack during their formation and motion, they are thus essentially like ordinary dislocations which have no resistance to their motion.

We can also allow for the effect on the dislocation distribution representing the crack of tractions applied internally to its faces. This is readily done if the tractions applied at opposite faces of the crack are equal and opposite, so that they can be described by an additional stress along $y = 0$; this is the case, for example, if the crack is under internal pressure due to a fluid, or if its opposite faces rub together and experience a force of friction, or if normal or tangential forces of cohesion act across it. If the tractions applied to the opposite faces of the crack are not equal and opposite, then the discontinuity of displacement may still be described by crack dislocations, but a contribution to the stress due to a surface distribution of body forces must also be included; we shall not consider this complication here, although an expression for the unit force of this kind in antiplane strain can be readily obtained from the analysis in Sect. IV,A.

In terms of forces on the dislocations, an additional stress acting in opposition to the applied stress merely replaces the terms σ^A_{yx}, σ^A_{yy}, and σ^A_{yz} in the integral equations by $\sigma^{A'}_{yx} = \sigma^A_{yx}(x) - G_{yx}(x)$, $\sigma^{A'}_{yy} = \sigma^A_{yy}(x) - G_{yy}(x)$ and $\sigma^{A'}_{yz} = \sigma^A_{yz}(x) - G_{yz}(x)$, and the analysis proceeds as before, the total stress everywhere being $\sigma^A_{ij} + \sigma^{D'}_{ij}$ where $\mathscr{D}'$ is calculated from the modified integral equation. We use the minus sign, so that, for example, in a crack under external tension or shear, G_{ij} is positive if it represents friction or normal adhesion which oppose the relative displacement of the crack faces; for internal pressure, G_{yy} is negative. Alternatively, to find the total stress, we may use the arguments of Sect. III.

To apply Eq. (59) (modified, if necessary, according to Eqs. (78) and (81)), we use an applied stress $\sigma^{A'}_{ij} = \sigma^A_{ij} - G_{ij}$ and add to the resulting σ^C_{ij} the unperturbed field σ^A_{ij}. We emphasize again, however, that for a crack not loaded internally, we find $\mathscr{D}$ and σ^D_{ij} using an integral equation $\sigma^D_{ij} = -\sigma^A_{ij}$ in R and then $\sigma^A_{ij} + \sigma^D_{ij}$ is the total stress everywhere; for a crack loaded internally and in the same external field, we find $\mathscr{D}'$ and $\sigma^{D'}_{ij}$ using an integral equation $\sigma^{D'}_{ij} = -\sigma^{A'}_{ij} = -(\sigma^A_{ij} - G_{ij})$ in R, but now the total stress everywhere is $\sigma^A_{ij} + \sigma^{D'}_{ij}$, *not* $\sigma^{A'}_{ij} + \sigma^{D'}_{ij}$. If $\sigma^A_{ij} = 0$, we have for the crack not loaded internally no total stress, while for the crack internally loaded we have everywhere a total stress $\sigma^{D'}_{ij}$ (equal to G_{ij} on R) and *not* $\sigma^{D'}_{ij} - G_{ij}$, which would give zero stress on R. We can thus apply the analysis of Sect. IV,A to find expressions for the stresses and displacements around any array of crack or ordinary dislocations subject both to external forces and forces applied along the array itself. For type III deformation, the stress σ_{zy} on the x axis a short distance $s = x - a$ beyond the tip a of the crack is given by the first two terms of Eq. (70) with $(x^2 - a^2)^{1/2} = (2a)^{1/2} s^{1/2}$ and $|x|$ replaced by a. Similarly, σ_{zx} inside the crack is given by the first two terms of Eq. (73) with $(a^2 - x^2)^{1/2} = (2a)^{1/2}(-s)^{1/2}$ and x replaced by a in the numerator of the coefficient of I_1 (note that inside the crack $-s = a - x$ is positive). According to Eqs. (71) and (72), the resulting expression, when multiplied by $2/G$, is the gradient of the relative displacement of the faces of the crack, $\partial\,\Delta w/\partial x$ or $\partial\phi/\partial x$ and this, in turn, is $-b\mathscr{D}$, where $\mathscr{D}$ is the density of dislocations (Sect. II).

The above applies to the case where the total dislocation content of the crack is zero. If it is not, we must add the stresses Eq. (75) with b equal to the sum of the Burgers vectors of all the dislocations in the crack.

To adapt the results to type I or type II deformation, we have only to make the changes indicated in Eqs. (81) and (78). For the relation between σ^A_{zy} and σ_{zy}, this amounts to a mere change of suffixes, while, in the relation between σ_{zy}^A and $\mathscr{D}$, we make an appropriate change of suffixes in σ^A and introduce a factor $1/(1-\nu)$. This factor, and the factor b required to convert ϕ into $\mathscr{D}$, may be conveniently combined into the factor A of Sect. III. In this way, we obtain the following expression for the stress σ in the plane of the crack, the dislocation density $\mathscr{D}$, and the relative displacement ϕ of the faces of the crack in the neighborhood of the crack tip at a:

$$\sigma = \frac{K}{(2\pi)^{1/2}} s^{-1/2}, \quad \mathscr{D} = \frac{K}{\pi A(2\pi)^{1/2}}(-s)^{-1/2}, \quad \phi = \frac{2bK}{\pi A(2\pi)^{1/2}}(-s)^{1/2} \tag{86}$$

with

$$s = x - a \ll a, \qquad K = (\pi a)^{1/2}(I_1 + I_2 + I_d) \tag{87}$$

where

$$I_1 = \frac{1}{\pi}\int_{-a}^{a} \frac{\sigma^A(x)\,dx}{(a^2 - x^2)^{1/2}}, \qquad I_2 = \frac{1}{\pi a}\int_{-a}^{a} \frac{x\sigma^A(x)}{(a^2 - x^2)^{1/2}}\,dx \tag{88}$$

or, equivalently

$$I_1 + I_2 = \frac{1}{\pi a}\int_{-a}^{a} \frac{(a + x)^{1/2}}{(a - x)^{1/2}}\,\sigma^A(x)\,dx \tag{89}$$

and

$$I_d = \frac{A}{a}\int_{-a}^{a} \mathscr{D}(x)\,dx \tag{90}$$

is A/a times the total dislocation content of the crack, if any. The expression for ϕ is obtained by integrating $\mathscr{D}$, bearing in mind that $\phi = 0$ at $s = 0$. The symbols have the following meanings:

For type I deformation (plane strain tension)

$$\sigma = \sigma_{yy}, \qquad \sigma^A = \sigma_{yy}^A, \qquad A = Gb/2\pi(1 - \nu)$$

For type II deformation (plane strain shear)

$$\sigma = \sigma_{xy}, \qquad \sigma^A = \sigma_{xy}^A, \qquad A = Gb/2\pi(1 - \nu)$$

For type III deformation (antiplane strain)

$$\sigma = \sigma_{zy}, \qquad \sigma^A = \sigma_{zy}^A, \qquad A = Gb/2\pi$$

If the crack is loaded internally or acted on by cohesive forces in addition to the externally applied forces, σ^A must be replaced by $\sigma^{A\prime}$, as already explained. To apply Eqs. (86) and (87) near the other end $x = -a$ of the crack interpret s as

$$s = a - x \ll a$$

and reverse the signs of I_2 and I_d (but not I_1) or, alternatively, reverse the sign of I_d and replace Eq. (89) by the new definition

$$I_1 + I_2 = \frac{1}{\pi a} \int_{-a}^{a} \frac{(a-x)^{1/2}}{(a+x)^{1/2}} \sigma^A(x)\, dx$$

The above discussion readily extends to anisotropic materials, three independent distributions $\mathscr{D}_x$, $\mathscr{D}_y$, and $\mathscr{D}_z$ remaining as long as the elastic equations separate. The constant A is, however, replaced by constants K_x, K_y, and K_z calculated from the elastic moduli (Eshelby, 1949; Eshelby, Read, and Shockley, 1953; Stroh, 1958a; Atkinson, 1967; Atkinson and Heald, 1967).

We now consider a number of simple examples of linear dislocation arrays representing cracks in an isotropic medium, illustrating them by diagrams showing normal discontinuities due to $\mathscr{D}_y$; the discussion of the tangential components $\mathscr{D}_x$ and $\mathscr{D}_z$ is entirely similar. The general solution of the integral Eq. (82) is Eq. (43) where the second term or its negative is the solution of the homogeneous equation $\sigma^D_{yy}(x, 0) = 0$, corresponding to a crack loaded neither internally nor externally. This term appears only when the total dislocation content of the crack is not zero; from Eq. (C5) of App. C, $C = n$, where $n = \int_{-a}^{a} \mathscr{D}_{yy}(x')\, dx'$ is the total number of dislocations. The relative displacement corresponding to

$$\mathscr{D}_{yy} = \pm(n/\pi)(a^2 - x^2)^{-1/2} \tag{91}$$

is

$$\phi_y = \pm(nb/2)[(\pi/2) - \sin^{-1}(x/a)]$$

where we have made $\phi_y = 0$ at $x = +a$. With the RH/FS convention, the positive sign gives crack dislocations with their (empty) extra half planes pointing in the negative x direction (Fig. 10.I). The solution represents the wedge crack which forms if a cut over $-a < x < +a$ is made on the tension side of an ordinary edge dislocation at $x = -a$ with Burgers vector $b_y = +nb$, and so having real extra half planes over $x < -a, y = 0$ (Stroh, 1954). With the negative sign, there is interpenetration of amount nb for $x < -a$, tapering to zero over $-a < x < +a$. If we remove this interpenetrating material and reweld where $x < -a$, we obtain an ordinary edge dislocation at $x = -a$ of total Burgers vector $b_y = nb$ (with extra half planes where $x > +a$), with a stressfree slit on its compression side over $-a < x < +a$.

These wedge solutions correspond to the discrete arrays with dislocations at the zeros of Legendre polynomials (Sect. III). In the continuous distributions, $\mathscr{D}_y(x)$ is unbounded both at $x = +a$ and at $x = -a$.

Next, suppose that σ^A_{yy} is constant and equal to P, and that there is no resultant dislocation content. From Eq. (43), the solution is

$$\mathscr{D}_y(x) = (Px/\pi A)(a^2 - x^2)^{-1/2} \tag{92}$$

$$\phi_y(x) = (Pb/\pi A)(a^2 - x^2)^{1/2} \tag{93}$$

with $A = Gb/2\pi(1 - \nu)$.

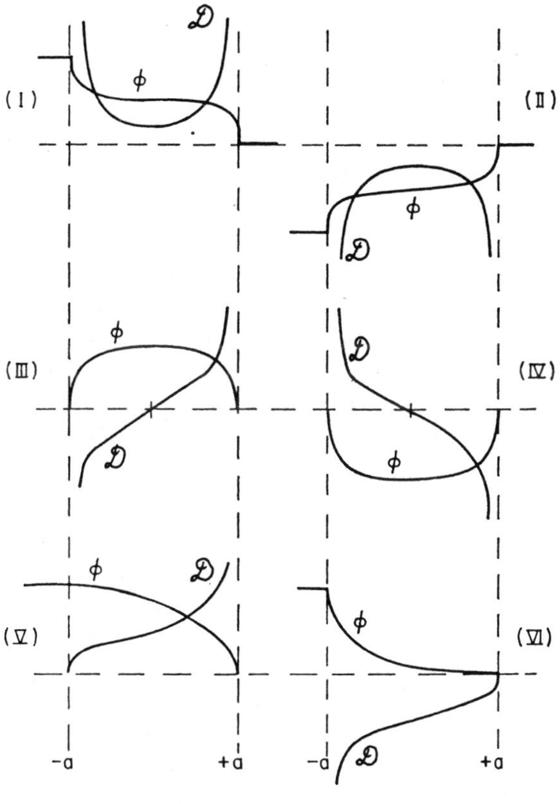

FIG. 10. Distribution of crack dislocations: ϕ = relative displacement; $\mathscr{D}$ = dislocation density. (I) No applied stress, $\mathscr{D} \to +\infty$ at $x = \pm a$. (II) No applied stress, $\mathscr{D} \to -\infty$ at $x = \pm a$. (III) Zero resultant $\mathscr{D}$. External stress $P > 0$, $\mathscr{D} \to \infty$ at $\pm a$. (IV) Zero resultant $\mathscr{D}$. External stress $P < 0$, $\mathscr{D} \to \infty$ at $\pm a$. (V) Suitable combination of (I) and (III). $P > 0$, $\mathscr{D} = 0$ at $x = -a$, $\mathscr{D} \to +\infty$ at $x = +a$. (VI) Suitable combination of (I) and (IV). $P < 0$, $\mathscr{D} = 0$ at $x = +a$, $\mathscr{D} \to -\infty$ at $x = -a$.

For $P > 0$, we obtain a slit crack over R under uniform tension; its form is an ellipse (Fig. 10.III). (For σ_{xy}^A and ϕ_x, see Leibfried (1951), Bilby and Bullough (1954), Head and Louat (1955); for ϕ_y, see Priestner and Louat (1963)). For $P < 0$, a physical interpretation is that over R an ellipse of material has been removed so that, under uniform compression, the two faces of the hole just close and are stressfree (Fig. 10d).

Again, $\mathscr{D}_y$ is unbounded at $x = \pm a$, and the ϕ_y curve is vertical at these end points. This is the general situation unless special conditions are imposed. If we combine Eqs. (91) and (92) in suitable proportions, we can study the effect of applied tension and compression on wedge cracks or on edge dislocations with cuts along the compression side of their extra half planes. In particular, by choosing

$$Pa = \pm nA = \pm nbG/2\pi(1 - \nu) \tag{94}$$

we can always make $\mathscr{D}_y = 0$ at either $x = +a$ or $x = -a$, but not both. The most interesting cases are as follows; they correspond to the Laguerre polynomial solutions for discrete arrays:

(1) $P > 0$, $\mathscr{D}_y > 0$, $\mathscr{D}_y(-a) = 0$. As P increases from zero, a wedge crack (Fig. 10.I) over R opens until, when $\mathscr{D}_y(-a) = 0$, ϕ_y is just flat at $x = -a$ (Fig. 10.III); at this stage n, a, and P are related by Eq. (94). It may be shown either by considering the total free energy of the crack in the applied field (Stroh, 1957b; Bullough, 1964) or, equivalently, by using the criterion (143) (Smith, 1965c), that, when the crack becomes unstable, Eq. (94) is always just satisfied. For given n, the critical stress P_{crit} is large and the critical crack length $2a_{\text{crit}}$ is small if the surface energy γ is large, and conversely; but $P_{\text{crit}}a_{\text{crit}}$ is independent of the surface energy. We discuss this result in Sect. VI, noting here only its simple geometrical interpretation (Fig. 10.V).

(2) $P < 0$, $\mathscr{D}_y > 0$, $\mathscr{D}_y(+a) = 0$. A wedge crack (Fig. 10.I) is compressed until it becomes just flat at $+a$ (Fig. 10.VI), when again Eq. (94) is satisfied. The external stress is just compensating the wedging effect of the dislocation and the stress is zero at the crack tip $x = +a$. The external stress is playing a role analogous to that of the cohesive forces (Barenblatt, 1959).

An example of a distribution where $\mathscr{D}_y$ is bounded at both $x = \pm a$ may be obtained by taking $\sigma_{yy}^A = -\alpha x$ (Leibfried, 1951). The general solution gives, for a total dislocation content of n,

$$\mathscr{D}_y(x) = \frac{\alpha}{A\pi}\left((a^2 - x^2)^{1/2} - \frac{a^2}{2(a^2 - x^2)^{1/2}}\right) + \frac{n}{\pi(a^2 - x^2)^{1/2}}$$

By choosing $n = \alpha a^2/2A$, we get

$$\mathscr{D}_y(x) = \frac{2n}{\pi a^2}(a^2 - x^2)^{1/2} \tag{95}$$

and, if $\phi_y(+a) = 0$,

$$\phi_y(x) = \frac{nb}{\pi}\left\{a^2\left(\frac{\pi}{2} - \sin^{-1}\frac{x}{a}\right) - \frac{x(a^2 - x^2)^{1/2}}{a^2}\right\} \tag{96}$$

Near $x = +a$, if $a - x = s$, $\phi_y(s) \sim \text{const}\ s^{3/2}$. The solution represents a wedge crack in a uniformly varying stress; for example, at the center of a bent beam. The end +a is in compression, and the end $-a$ in tension, $\phi_y(x)$ being flat at both ends (Fig. 11). In the discrete case, the dislocations are at the zeros of the Hermite polynomials (Sect. III). In this example, it may be readily verified that

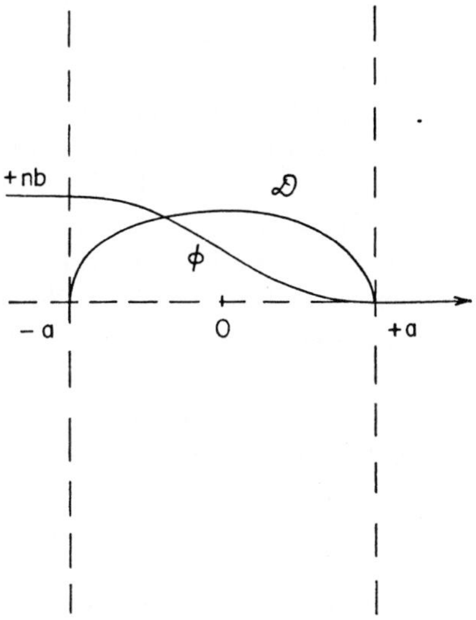

FIG. 11. Distribution of crack dislocations under applied stress $P = -\alpha x$.

$$\int_{-a}^{a} \frac{\sigma_{yy}^A(x')\,dx'}{(a^2 - x'^2)^{1/2}} = 0 \tag{97}$$

In fact, this is the condition which $\sigma_{yy}^A(x)$ must satisfy if a solution with $\mathscr{D}_y(\pm a) = 0$ is to exist. Muskhelishvili has given explicitly the formulas for $\mathscr{D}$ and the conditions which σ must satisfy if $\mathscr{D}$ is to be bounded or unbounded at any of the $2N$ endpoints of a distribution over N intervals of this type (Muskhelishvili, 1953; Head and Louat, 1955).

There are no problems of principle in extending these methods to arrays of collinear cracks, and an analysis for an infinite array of equally spaced identical cracks is given in App. C. There has been considerable recent work on the representation of plasticity at cracks by collinear dislocation arrays; these can be brought within the same framework by regarding the plastic regions as extensions of the crack loaded internally by stress representing the yield stress; both the plasticity and the cracking are given one common description in terms of dislocations, and fracture and relaxation become determined by a balance between the two types. Cracks and dislocation arrays which are not collinear, but are even curved and which are near free surfaces, notches, holes, or other perturbations in stress can still be treated, but with increasing complication. There is much current work in this field, which we will review briefly later. We can also consider, by these methods, the effect of repeated and reversed loading (Weertman, 1964, 1966). Another development arising when plasticity is represented by frictional loading of part of the crack is to use a yield criterion which may then couple the different dislocation distributions (Blackburn, 1966a, 1966b; Keer and Mura, 1965); see also the last authors for a discussion of the relation of the dislocation methods to those of Collins (1962), England (1963), and England and Green (1963). We can also make the friction a function of loading, displacement or $\mathscr{D}$, and so represent work hardening (Bilby and Swinden, 1965).

There is a rapid increase in difficulty when we consider problems where there is another degree of freedom and the dislocations are not straight, so that the crack, though plane, has a curved or polygonal contour. All cracks of this type represent particular examples of Somigliana dislocations, and so, too, of equilibrium distributions of translation dislocations, but only cracks in the form of circular and elliptical disks and under uniform loading have been discussed explicitly in terms of dislocation theory (Leibfried, 1954; Eshelby, 1963).

Hitherto, we have considered only slit cracks having one dimension infinitesimal. There has been some work on the use of dislocation theory to discuss cavities of finite thickness, inclusions, and inhomogeneities; the applications are not only to cavities but also to twins, kink bands, and martensite plates, all of which are relevant to cracking and fracture (Frank and Stroh, 1952; Eshelby, 1957, 1960). As with a slit, on loading there is a small perturbation of the boundary of the cavity. This we can describe by a Somigliana dislocation, or by a distribution of simpler dislocations. Their stress field σ_{ij}^D gives the perturbation produced in the applied stress by the cavity. Our approach is, again, to insert a Somigliana dislocation, apply an external field, and find a closed surface which is stressfree. We can then remove the material within it.

Let V be the region inside a closed surface S in a stress free elastic medium, and V' the medium outside S. We shall consider that V undergoes a transformation to change its size—for example, by the introduction or removal of material—and so introduces internal stress. We define three displacements u_i^T, u_i^C, and u_i^D; we speak of a contraction of the material V within S, for definiteness. Suppose the contraction occurs slowly, the material remaining joined continuously across S; then we call the displacement everywhere $u_i^C(\mathbf{r})$, both within and outside S. The zero of displacement is the untransformed material everywhere. Now, with u_i^C present, cut the stressed body along S, and let every point of the medium suffer a displacement $-u_i^D$ during the relaxation. In the matrix outside S, $u_i^D = u_i^C$, but inside S, $u_i^D = u_i^T - u_i^C$, where u_i^T is the displacement V would have undergone had it transformed while unconstrained by V'.

If we cut over S, allow V to transform and reweld, we obtain a Somigliana dislocation with $b_i(\mathbf{r}') = u_i^T(\mathbf{r}')$ over S. We can find u_i^D by integrating the expression for the field of the infinitesimal dislocation loops $b_j(\mathbf{r}')\, dS_k$ over S. A loop $b_j\, dS_k$ at the origin has a displacement at $\mathbf{r}$ (Burgers, 1939; Nabarro, 1951):

$$u_i(\mathbf{r}) = \frac{b_j\, dS_k}{8\pi(1-\nu)}\, \vartheta_{ijk}(r) \tag{98}$$

where

$$\vartheta_{ijk} = \frac{\partial^3}{\partial x_i\, \partial x_j\, \partial x_k} - \left[r\, \delta_{jk} \frac{\partial}{\partial x_i} + (1-\nu)\, \delta_{ij} \frac{\partial}{\partial x_k} + (1-\nu)\, \delta_{ik} \frac{\partial}{\partial x_j}\right] \nabla^2 \tag{99}$$

and so

$$u_i^D = \frac{1}{8\pi(1-\nu)}\, \vartheta_{ijk} \int_S b_j(\mathbf{r}')|\, \mathbf{r} - \mathbf{r}'\, |\, dS_k \tag{100}$$

Both inside and outside the inclusion the stress is

$$\sigma_{ij}^D = \lambda \frac{\partial u_m^D}{\partial x_m}\, \delta_{ij} + G\left(\frac{\partial u_i^D}{\partial x_j} + \frac{\partial u_j^D}{\partial x_i}\right)$$

We have now to superpose an applied field u_i^A and adjust u_i^D to $u_i^{D'}$ so that $(\sigma_{ij}^A + \sigma_{ij}^{D'})\, dS_j = 0$ on the surface S: $\sigma_{ij}^A + \sigma_{ij}^{D'}$ is then the total field of a cavity within S in an applied field σ_{ij}^A.

If S is an ellipsoidal surface, this procedure can be carried out in a simple way for an applied field of fairly general type. For such a surface, it can be shown from the properties of harmonic potentials of inhomogenous ellipsoids that, if on S, $u_i^T = T_i(N, x, y, z)$ where $T_i(N, x, y, z)$

is a polynomial in x, y, z of degree N, then within S, $u_i^D = D_i(N, x, y, z)$ where D_i is also a polynomial of degree N, and the relation between the coefficients of T_i and D_i can be calculated (Eshelby, 1960). Outside S, u_i^D can be expressed by Eq. (100) in terms of other polynomials whose coefficients can be calculated and which depend on x, y, and z. The case $N = 2$ covers an applied stress which is a combination of simple torsion, bending, and flexure. When one axis of the ellipsoid approaches zero, we have a crack, in the form of an elliptical disk, perturbing a general stress field.

V. Energy Considerations

In both dislocation theory and the theory of cracks, the energy of the system and its rate of variation with configuration play an important role. In this section, we indicate how the concepts current in dislocation theory and in fracture mechanics are related.

In Sect. III, we calculated in two special cases the force F on the locked dislocation at the head of a pileup and also obtained a general expression for it, (Eq. (39)) in the limit of a large number of dislocations. If the locked dislocation is allowed to move a distance δa, an amount of work $F\,\delta a$ is done on it at the expense of the energy of interaction between the dislocations (in other words, the elastic energy of the medium) and the potential energy of the mechanism providing the externally applied stress. At the head of a pileup, the dislocation density is given by Eq. (86), so that, from Eq. (39), with $x - a = -s$, we have

$$F = bK^2/4\pi A \tag{101}$$

Since a crack under shear or tension is equivalent to a certain distribution of dislocations, it is reasonable to suppose that $F\,\delta a$ calculated from the equivalent distribution $\mathscr{D}$ should represent the energy released when one tip of the crack moves a distance δa, or, in other words, that F is equal to the crack opening force $\mathscr{G}$ (Irwin, 1948, 1957). This can easily be shown.

Suppose that the crack is subjected to simple tension (mode I). Allow the right-hand tip of the crack to advance from $x = a$ to $x = a + \delta a$, but apply forces to the freshly formed surfaces so as to hold them closed. The elastic energy and the energy of the loading mechanism are unaltered. Now allow these forces to relax to zero so that the crack opens up from a to $a + \delta a$. In doing so, they extract an amount of energy

$$-\delta E_{\text{tot}} = \frac{1}{2}\int_a^{a+\delta a} \sigma(s)\,\Delta u(-s')\,dx \tag{102}$$

where σ is the stress ahead of the shorter crack and Δu is the difference in displacement across the lengthened crack, just behind its tip, so that

$$s = a - x, \qquad -s' = a + \delta a - x.$$

The stress σ may be found at once from the first of Eq. (86). The opening Δu may be found from the second of Eq. (86) modified to allow for the fact that the crack extends from $-a$ to $a + \delta a$ instead of from $-a$ to a. It is easy to show that ignoring this difference only leads to an error of order $(\delta a)^2$, provided $o^A(x)$ is not actually discontinuous near $x = a$, so we may use Eq. (102) as it stands, reading s' for s, and hence, with $t = x - a - \frac{1}{2}\delta a$, we have

$$-\delta E_{\mathrm{tot}} = \frac{1}{2}\frac{2bK^2}{2\pi^2 A}\int_{-\frac{1}{2}\delta a}^{\frac{1}{2}\delta a}\left(\frac{t + \frac{1}{2}\delta a}{\frac{1}{2}\delta a - t}\right)^{1/2} dt = \frac{bK^2}{4\pi A}\delta a$$

By definition, the decrease of energy of the system, divided by the crack extension δa, is the crack opening force $\mathscr{G}$ per unit edge length of the crack, and so

$$\mathscr{G} = bK^2/4\pi A \tag{103}$$

which is indeed equal to the force F (Eq. (101)) on the leading dislocation of the equivalent pileup. If the crack is subjected to pure shear (mode II or III) the argument is the same, except that, after allowing the extension δa, we must apply forces to the freshly formed surfaces to prevent relative sliding, and then relax. The result is evidently again Eq. (103) with the appropriate K and A.

For mode III deformation

$$\mathscr{G}_{\mathrm{III}} = \frac{K_{\mathrm{III}}^2}{2G} = \frac{1+\nu}{E}K_{\mathrm{III}}^2 \tag{104}$$

while for modes I and II,

$$\mathscr{G}_{\mathrm{I}} = \mathscr{G}_{\mathrm{II}} = \frac{1-\nu^2}{E}K_{\mathrm{I}}^2. \tag{105}$$

Figure 12 shows a body to whose outer surface Σ_0 tractions are applied by some loading mechanism. It contains a crack which can alter its length from BA to BA'. To begin with, we shall assume that the loading mechanism is perfectly soft, that is, as the body deforms the load on a particular element of its surface remains constant.

Let σ_{ij}, u_i be the stress and displacement in the body when the crack is AB. The elastic energy of the body is

$$E = \frac{1}{2}\int_{\Sigma_0} \sigma_{ij}u_i\, dS_j\,;$$

the integral should extend over the whole surface of the body, i.e., over Σ_0 and the surface of the crack as well, but, since the latter is stressfree, it makes no contribution.

Similarly, when the crack becomes $A'B$ the energy of the body is

$$E' = \tfrac{1}{2} \int_{\Sigma_0} \sigma'_{ij} u'_i \, dS_j$$

where σ'_{ij}, u'_i, different from σ_{ij}, u_i, are the new stress and displacement. Hence, if the crack tip grows from A to A' the elastic energy of the body increases by

$$\delta E_{\text{el}} = \tfrac{1}{2} \int (\sigma'_{ij} u'_i - \sigma_{ij} u_i) \, dS_j$$

$$= \tfrac{1}{2} \int \sigma_{ij}(u'_i - u_i) \, dS_j \, ; \tag{106}$$

the second form is the same as the first because for each surface element dS_j we have

$$\sigma'_{ij} \, dS_j = \sigma_{ij} \, dS_j \tag{107}$$

because of the constancy of the loading. When the crack tip moves from A to A', the surface displacements alter from u_i to u'_i and the loading mechanism does an amount of work

$$-\delta E_{\text{ext}} = \int \sigma_{ij}(u'_i - u_i) \, dS_j \tag{108}$$

which, as the notation indicates, is the decrease in the potential energy of the loading mechanism. From Eqs. (106) and (108), we see that

$$\delta E_{\text{ext}} = -2 \, \delta E_{\text{el}}$$

so that the change in the total energy is

$$\delta E_{\text{tot}} = \delta E_{\text{el}} + \delta E_{\text{ext}} = -\delta E_{\text{el}} = \tfrac{1}{2} \, \delta E_{\text{ext}} \tag{109}$$

When the crack extends, the body becomes more flexible and so, under fixed loading, its elastic energy δE_{el} increases. At the same time, the potential energy of the loading mechanism decreases by twice δE_{el}, so that there is an overall decrease in the total energy. It is therefore wrong to say, as is often done, that introducing a crack into a body or lengthening an existing one decreases the elastic energy, though, in view of Eq. (109), this can hardly lead to anything worse than an error in sign, which is usually silently corrected by common sense. Likewise, the argument that a penny-shaped or ribbon-shaped crack relaxes the elastic energy to zero inside a sphere or cylinder whose diameter is about the same as the crack width gives the wrong sign for δE_{el}, but is a useful mnemonic, since it gives about the right magnitude for δE_{tot}. (For the detailed examination of a simple case see Eshelby, 1957, p. 394.)

From Eqs. (106) and (108)

$$\delta E_{\text{tot}} = -\tfrac{1}{2} \int \sigma_{ij}(u_i' - u_i)\, dS_j \tag{110}$$

Because of Eq. (107), the integrand could equally well be written as $\sigma_{ij}'(u_i' - u_i)$ or even $\sigma_{ij}u_i' - \sigma_{ij}'u_i$. As in other such problems, there is a great advantage in using the last form. Equation (110) then becomes

$$\delta E_{\text{tot}} = -\tfrac{1}{2} \int_{\Sigma} (\sigma_{ij}u_i' - \sigma_{ij}'u_i)\, dS_j \tag{111}$$

(with $\Sigma = \Sigma_0$). For then

$$\begin{aligned}(\sigma_{ij}u_i' - \sigma_{ij}'u_i)_{,j} &= \sigma_{ij,j}u_i' - \sigma_{ij,j}'u_i + \sigma_{ij}u_{i,j}' - \sigma_{ij}'u_{i,j} \\ &= 0;\end{aligned}$$

the first two terms vanish by the equilibrium equations and the last two, which may be written as $\sigma_{ij}e_{ij}' - \sigma_{ij}'e_{ij}$, cancel when they are written out entirely in terms of strains with the help of Hooke's law. Consequently, the divergence of the integrand in Eq. (110) vanishes, and so, by Gauss' theorem, we may carry out the integration over any contour Σ into which Σ_0 may be deformed without intersecting the crack. In particular, we may take Σ to be the curve $\Sigma_1 = abc$ surrounding the crack plus the dotted curve *cda*. The straight parts of the dotted curve contribute nothing to Eq. (111), since, on the surface of the crack, $\sigma_{ij}\, dS_j = 0$ and $\sigma_{ij}'\, dS_j = 0$. The dotted circle contributes nothing when it is made small enough, if we assume that at a distance ρ from a crack tip $u_i \approx \rho^{1/2}$, $\sigma_{ij} \approx \rho^{-1/2}$.

We have thus succeeded in expressing δE_{tot} as an integral over a contour Σ_1 embracing only that end of the crack which is supposed to move. The terminations of Σ_1 on the two faces of the crack, *a*, *c*, need not be exactly opposite one another, though it is usually convenient to make them so.

Σ_1 may be further deformed into the keyhole-shaped loop about AA'. The circles ultimately contribute nothing. On AA', $\sigma_{ij}'\, dS_j = 0$. Hence, Eq. (111) becomes

$$\delta E_{\text{tot}} = -\tfrac{1}{2} \int_a^{a+\delta a} \sigma_{ij}(x)\, \Delta u_i'(x)\, dx \tag{112}$$

II

where we have taken the x axis along the crack with A at $(a, 0)$ and A' at $(a + \delta a, 0)$ and $\Delta u_i'$ is the value of u_i on the upper face of the crack minus its value on the lower face. Equation (112) is identical, apart from notation, with Eq. (102) and the crack-opening force is, as before

$$\mathscr{G} = -\lim_{\delta a \to 0} \frac{\delta E_{\text{tot}}}{\delta a}$$

We can also obtain a more general expression for $\mathscr{G}$ by going back to the finite contour Σ_1 about the crack tip. Let ξ denote the x coordinate of the crack tip. If we regard the displacement and stress as functions of the parameter ξ giving the configuration of the crack, we can write for points on Σ_1

$$u_i' = u_i + \frac{\partial u_i}{\partial \xi} \delta a + O(\delta a)^2$$

$$\sigma_{ij}' = \sigma_{ij} + \frac{\partial \sigma_{ij}}{\partial \xi} \delta a + O(\delta a)^2$$

and so, for Eq. (111)

$$\mathscr{G} = \frac{1}{2} \int_{\Sigma_1} \left(\sigma_{ij} \frac{\partial u_i}{\partial \xi} - u_i \frac{\partial \sigma_{ij}}{\partial \xi} \right) dS_j \tag{113}$$

The divergence of the integrand is zero. Hence, when we have taken the limit $\delta a \to 0$ and A and A' coincide we can, if we wish, deform Σ_1 in Eq. (113) into a small loop around the crack tip. It is intuitively plausible that, near the crack tip, shifting the tip by δa simply shifts the elastic field rigidly, that is, $\partial u_i/\partial \xi$ and $-\partial u_i/\partial x$ differ by an amount small compared with either of them, and similarly for $\partial \sigma_{ij}/\partial \xi$ and $-\partial \sigma_{ij}/\partial x$. This can easily be verified in specific cases, and then Eq. (113) may be rewritten as

$$\mathscr{G} = \frac{1}{2} \int_{\Sigma_1} \left(u_i \frac{\partial \sigma_{ij}}{\partial x} - \sigma_{ij} \frac{\partial u_i}{\partial x} \right) dS_j \tag{114}$$

where Σ_1 is an infinitesimal circuit about the tip. However, it is easy to verify that the divergence of the integrand is still zero, so Σ_1 may equally well be the finite circuit *abc* of Fig. 12. In contrast to Eq. (113), the expression Eq. (114) for $\mathscr{G}$ depends only on the elastic field of the crack before extension. Equation (114) is, in fact, generally valid, though a direct proof is rather involved. In connection with Eq. (116) below, we give a proof in terms of the equivalent array of dislocations.

If, in Fig. 12, A coincides with B, we have the case of the expansion of a crack from zero length to a finite length. Equation (111) then gives the change of total energy when a crack is introduced into a crackfree body. The initial $u_i = u_i^A$ say, and $\sigma_{ij} = \sigma_{ij}^A$ are due to the external forces acting on the crackfree body. The final stress and displacement may be written (cf. Sect. IV,A) as $\sigma'_{ij} = \sigma_{ij}^A + \sigma_{ij}^C$, $u'_i = u_i^A + u_i^C$ where σ_{ij}^C, u_i^C is the elastic field "due to" the crack.

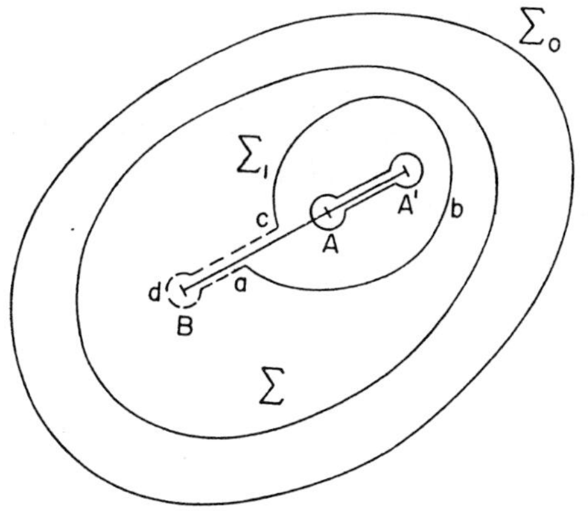

FIG. 12. To illustrate the calculation of the crack opening force.

Equation (111) becomes

$$\delta E_{\text{tot}} = -\tfrac{1}{2}\int_{\Sigma} (\sigma_{ij}^A u_i^C - \sigma_{ij}^C u_i^A)\, dS_j$$

where Σ is the surface of the body or any surface enclosing the crack. On the crack itself, $(\sigma_{ij}^A + \sigma_{ij}^C)\, dS_j = 0$, and σ_{ij}^A, u_i^A are continuous. Hence, letting Σ reduce to a loop tightly embracing the crack, we get

$$\delta E_{\text{tot}} = -\tfrac{1}{2}\int_{-a}^{a} \sigma_{ij}^A(x)\, \Delta u_i^C(x)\, dx \tag{115}$$

where the crack is supposed to run from $-a$ to a along the x axis. Thus, the change in energy is formally equal to "the work done in opening the crack against the applied stress." Equation (115) is important in Griffith's theory of rupture.

So far, we have supposed that the loading mechanism is perfectly soft. If it is not, the load on a surface element dS_j of Σ_0 will vary by an amount of order δa as the crack extends by δa. Since $u'_i - u_i$ is of order δa, Eqs. (106) and (108), and consequently Eq. (111) will be in error by a quantity of order $(\delta a)^2$. Consequently, the expressions for $\mathscr{G}$, which depend on terms of order δa only, are unaffected and remain valid for an arbitrarily hard or soft loading mechanism. We can use this fact to extend our results further to the case where the crack is subjected to stresses which arise partly from sources of internal stress, dislocations, or inclusions, for example. In Fig. 12, suppose there are such sources of internal stress between the surfaces Σ_0 and Σ. Then we can regard Σ as the surface of a body subject to loading by a complex loading mechanism made up of the loading mechanism properly speaking and the stressed material between Σ_0 and Σ. The same argument as before establishes Eq. (111) and, hence, the expressions we have derived from it are valid also for a combination of internal and externally produced stresses.

The foregoing results for the effective force on the head of a dislocation pileup or the tip of a crack are closely related to the theory of the elastic energy-momentum tensor (Eshelby, 1951, 1956). It can be shown that the force on an elastic singularity is given by

$$F_l = \int_\Sigma P_{jl}\, dS_j \tag{116}$$

where in

$$P_{jl} = W\,\delta_{jl} - \sigma_{ij}\frac{\partial u_i}{\partial x_l}$$

(the energy-momentum tensor) $W = \frac{1}{2}\sigma_{mn}\,\partial u_m/\partial x_n$ is the elastic energy density and Σ is a surface enclosing the singularity. If Σ embraces several singularities, the integral gives the sum of the forces on them. The force on the locked dislocation in the pileup in Fig. 13 is given by Eq. (116), where Σ is a loop embracing the locked dislocation only. But, if we replace Σ by Σ_1, which also embraces any number of the free dislocations in the pileup, the integral is unaltered, because there is no force on the unlocked dislocations. Thus, in the limit of a continuous distribution of dislocations, the force on the leading one may be found by

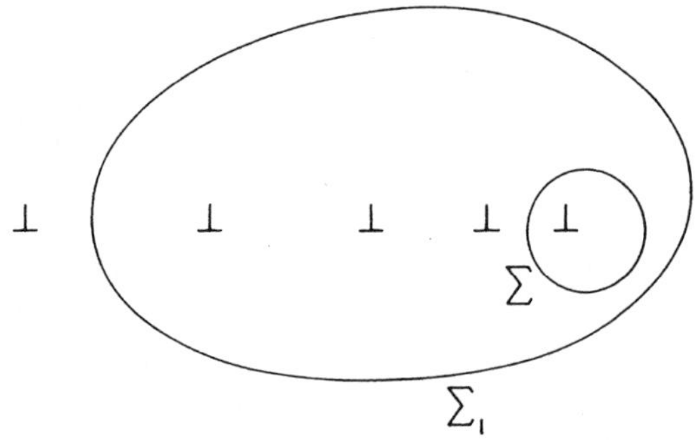

FIG. 13. Force on dislocation at crack tip.

integrating Eq. (116) over any loop embracing the head of the pileup. Consequently, if the crack of Fig. 12 is regarded as an equivalent array of dislocations, the crack-opening force $\mathcal{G}$ is equal to F_x calculated from Eq. (116). We can, in fact, establish the equality of Eqs. (114) and (116) directly. After a little rearrangement, we have

$$F_x - \mathcal{G} = \int_{\Sigma_1} \left[dS_x \frac{\partial}{\partial y} - dS_y \frac{\partial}{\partial x}\right] (\sigma_{iy} u_i) \tag{117}$$

The operator in square brackets is simply d/ds, where s denotes arc length along the curve *abc* of Fig. 12. Since the traction on the crack surfaces is zero, $\sigma_{iy} = 0$ at *a* and *c* and so Eq. (117) vanishes.

VI. Applications to Fracture Mechanics and to Dislocation Models of Crack Initiation

A complete discussion of the fracture of any "continuous" solid must consider the cohesive forces between its ultimate structural units as they undergo relative displacements which become large compared with their normal separation. It is, however, often reasonable in describing fracture to represent most of the solid by a continuum with suitable properties and to introduce the cohesive forces between the structural units as part of the boundary conditions. Such a model can be made to describe the extension and contraction of cracks without further assumptions. In a still simpler approach, we may treat the energetics of the extension of cracks heuristically by imagining that new surfaces form and by introducing the concept of a surface energy accompanying further cracking. There is then, however, no built-in mechanism in the model by which existing cracks or slits can extend or contract. However, the method is of wide application, since, by modifying suitably the definition of the surface energy, a large variety of types of crack extension can be treated. In this way, for instance, an unstable discontinuous ductile fracture may be treated in the same formal manner as a failure by brittle cleavage, the ductile "bridges" merely replacing "atomic bonds" (Cottrell, 1963a, 1965a).

There has been some debate about whether notches and cracks should be represented in a continuum model by slits or by boundaries with finite radii of curvature, such as ellipsoids of large eccentricity. In fact, the argument is largely irrelevant, despite the formal infinities in stress which accompany the use of slits in loaded continua. This is because, when the radius of curvature of a boundary becomes of the same order as the dimensions of the ultimate structural units considered in the discussion, the stresses obtained at points of physical interest will not differ significantly from those obtained by using a formal slit or sharp corner to represent the boundary in this region. Accordingly, the choice of model becomes largely a matter of mathematical convenience. It remains true, nevertheless, that the ultimate fracture process cannot be satisfactorily treated without considering the structural units as discrete. Moreover, the smaller the crack, the more essential does it become to take into account this discrete structure in order to obtain an adequate discussion of its properties. For this reason, we shall treat crack initiation only rather briefly in this section, mainly concentrating our attention on problems involving the extension and stability of existing cracks.

The simplest reasonably faithful model of a crack or slit able to extend is obtained as follows: for simplicity we take a crack extending in a plane in an infinite solid. We cut the solid along $y = 0$ (Fig. 14) and

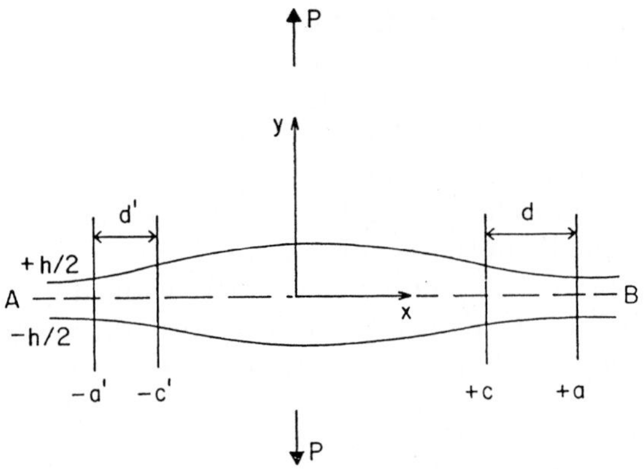

FIG. 14. Model of loaded crack.

separate the resulting semi-infinite blocks $y > 0$ and $y < 0$ so that their surfaces are at $y = h/2$ and $y = -h/2$, respectively. Now let an external stress be applied to produce a stress $\sigma_{yy} = P(x)$ on $y = \pm h/2$, and, at the same time, allow forces of cohesion to act between the faces of the blocks to produce a stress $\sigma_{yy} = Q$ on the surfaces $y = \pm h/2$. If we assume that $\sigma_{xy} = 0$ on $y = \pm h/2$ and that, on these surfaces, there arise elastic displacements in the y direction $v(x, h/2) = -v(x, -h/2)$, then the equation expressing the equilibrium of the blocks is

$$-A \int_{-\infty}^{+\infty} \frac{d\phi}{dx'} \frac{dx'}{x - x'} + P(x) - Q[\phi(x)] = 0 \tag{118}$$

where $\phi = v(x, h/2) - v(x, -h/2)$ is the relative displacement of the upper block with respect to the lower. Equation (118) follows from Eq. (83) by writing $-b\mathscr{D}(x') = d\phi/dx'$ and replacing σ^A_{yy} by $P(x) - Q[\phi(x)]$. This is a nonlinear singular integral equation of the same type as that appearing in the Peierls–Nabarro model of a dislocation (Peierls, 1940; Nabarro, 1947). However, the function $Q[\phi]$ is not now periodic in ϕ but has the general form shown in Fig. 15, being zero at

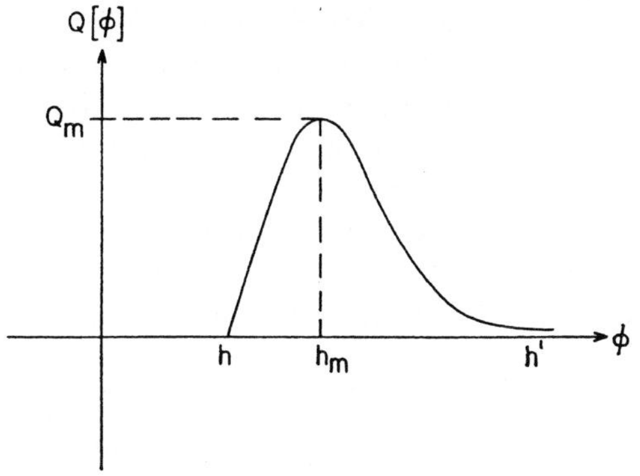

FIG. 15. Form of cohesive forces.

the equilibrium spacing h for the cohesive forces between the blocks, rising to a maximum Q_m when $h = h_m$ and again falling to zero as $\phi \to \infty$. This will be true whether $Q[\phi]$ represents the forces due to atomic bonds between separating crystal planes, or, for example, the forces due to a series of internal necks forming when a sequence of small cracks arising ahead of a large crack successively link up with it. The values of the parameters Q_m, h and h' (defined by the requirement that $Q[h']$ is already again very small) will, however, be very different in these two situations.

For any system of external loading leading to a "cracklike" solution of the Eq. (118), we expect $Q[\phi]$ to be nearly zero in two regions: one representing the crack itself where the separation of the surfaces $y = \pm h/2$ is much greater than h', and another more extensive region when $x \ll -a'$ and $x \gg a$ where $d\phi/dx \approx 0$ and $Q[\phi]$ is a linear function of ϕ corresponding to small linear elastic deformation of the solid. Between these regions are transition zone of width $d = a - c$, $d' = a' - c'$ where ϕ changes over from $\phi \approx h$ to $\phi \gg h'$. These regions may be called the crack fronts, and their width, or more strictly, the ratios $2d/(a + c)$ and $2d'(a' + c')$, depend on the shape of the $Q[\phi]$ curve and on $P(x)$. As we discuss later, it is instructive to compare the concept of the width of a crack front with the width of a dislocation in the Peierls–Nabarro model; the crack front, indeed, can be regarded as a new type of "dislocation" separating a region of $y = 0$ which has cracked, from one which has not.

Mathematical difficulties make the complete discussion of Eq. (118) a formidable problem. Elliot (1947) has used the stresses and displacements produced in a linear elastic medium along $y = \pm h/2$ by a crack $-c < x < +c$ under constant stress $\sigma_{yy} = P$ at infinity, to calculate the force law "$Q[\phi]$" required to maintain these stresses and displacements in the material $|y| > h/2$ when the material lying where $|y| < h/2$ is removed. He also discusses the penny-shaped crack in a similar way. In general, in an "inverse" approach of this type, the resulting "$Q[\phi]$" will depend not only on ϕ but on the other parameters of the problem such as P and c. Indeed, if it did not, it would follow that the stresses and displacements assumed on $y = \pm h/2$ corresponded to an exact solution of Eq. (118).

Goodier and Kanninen (1966) have treated the problem numerically, regarding $Q[\phi]$ as produced by a dense array of nonlinear springs with a gap $-c < x < +c$ to represent the crack. Numerical results are given when $Q[\phi]$ is represented by two linear branches, and by sine, exponential and inverse power laws.

Other treatments of the crack problem may be obtained by making further assumptions. In the outer regions, far from the crack where $d\phi/dx \approx 0$, $Q[\phi]$ may be represented approximately by filling in the region between the blocks once again with linear elastic material. For very small wedge cracks (cracked dislocations), Leonov and Panasyuk (1959) and Panasyuk (1960) have used this assumption, together with a linear cutoff for $Q[\phi]$ when $h > h_m$ (Fig. 15), to obtain a linear integral equation for ϕ. This they solve approximately, and use the solutions to discuss the reduction in theoretical strength to be expected from such a defect.

Barenblatt (1959) has developed a theory of cracks whose relation to Eq. (118) may be understood from the following considerations. As stated above, we can represent the cohesive forces $Q[\phi]$ by elastic material where $d\phi/dx \approx 0$, while on the other hand, when $c' < x < +c$, $\phi \gg h'$ and $Q \approx 0$ (Fig. 14). Further, the range of action of the cohesive forces is often much less than the length of the crack, so that $2d \ll (a + c)$ and $2d' \ll (a' + c')$. Thus, $Q[\phi]$ changes rapidly from $Q \approx 0$ to a value on the linear portion of the $Q[\phi]$ curve in small transition zones at the crack fronts. The shape of the crack over most of its length will thus not differ greatly from that given by the solution of the equation

$$-A \int_{-a'}^{a} \frac{d\phi}{dx'} \frac{dx'}{x - x'} + P(x) = 0 \tag{119}$$

This is the familiar linear integral equation arising in the theory of slit cracks or of continuous distributions of infinitesimal dislocations in a linear medium. In view of its relative simplicity, we are thus led to make the sweeping assumption that we can treat $Q[\phi]$ at the crack front as a function not of ϕ but of x, and so obtain, as some kind of approximation to the model correctly described by Eq. (118), the linear integral equation

$$-A\int_{-a'}^{a}\frac{d\phi}{dx'}\frac{dx'}{x-x'}+P(x)-G(x)=0 \tag{120}$$

Here, we now regard $G(x)$ as an unknown function of x, and treat it as independent of $P(x)$. The $\phi(x)$ obtained from Eq. (120) must approximate as far as possible to the solution of Eq. (118), and one essential requirement is that the stresses at $x = \pm a$ must be finite. From the theory of integral equations of this type (Sect. IV), this requires that $d\phi/dx$ falls smoothly to zero at the crack fronts, vanishing when, $x = a$ and $x = -a'$. As we discuss in Sect. IV, this gives the conditions

$$\int_{-a'}^{a}[P(x')-G(x')]\left(\frac{a-x'}{x'+a'}\right)^{1/2}dx'=0 \tag{121}$$

for smoothness at a' and

$$\int_{-a'}^{a}[P(x')-G(x')]\left(\frac{x'+a'}{a-x'}\right)^{1/2}dx'=0 \tag{122}$$

for smoothness at a. If $a = a'$ and $P(x) - G(x)$ is symmetrical about $x = 0$, Eqs. (121) and (122) are equivalent to the more familiar

$$\int_{-a}^{a}[a^2-x'^2]^{-1/2}[P(x')-G(x')]\,dx'=0 \tag{123}$$

Equations (121) and (122) determine a' and a, and so a crack length with smooth closure at both ends, for given $P(x)$ and $G(x)$. We have stated the conditions Eqs. (121) and (122) separately because if the crack is blocked by some obstacle at one (or both) of its ends, then a or a' (or both) are fixed a priori, and the conditions at the crack front are modified. In Eq. (118), this would be represented by a modification of $Q[\phi]$ at a or a' (or both); to represent an impassable barrier to the crack front, $Q[\phi]$ must increase with ϕ without limit when $x > a$ or $x < -a'$ (or both)(cf. Jaswon and Foreman, 1952). By suitable use of the conditions Eqs. (121) and (122), the approximate model can nevertheless represent all three possibilities.

We emphasize again the rather sweeping nature of the approximations leading to Eq. (120); the restriction of the integral to a finite range alters completely the mathematical nature of the solutions to be expected (introducing algebraic expressions involving $(a - x)^{1/2}$ and $(x + a')^{1/2}$), while the replacement of $Q[\phi]$ by $G(x)$ removes the real nonlinear character of the problem. Nevertheless, this approach has been used extensively by Barenblatt to develop a useful theory of the behavior of a solid containing cracks which is likely to be a reasonable approximation to the actual situation as long as the crack front is small compared to the crack itself. Barenblatt's theory is also closely related, on the one hand, to those of Griffith, Orowan, and Irwin (for a recent comparison see Willis, 1967), and, on the other, to recent work on simplified representations of plastic zones near cracks (Dugdale, 1960; Cottrell 1961; Bilby, Cottrell, and Swinden, 1963). Accordingly, we now present a more detailed account of Barenblatt's work, combining it with a discussion of that of these other authors.

We begin with some definitions and hypotheses. Cracks in the solid are represented by slits in a linear elastic body. On stressing the body, their faces undergo a relative displacement. As discussed in Sect. IV,B, this may be described by distributions of crack dislocations, induced along them, and, in conformity with our general development, this is the point of view we shall adopt in presenting Barenblatt's ideas.

A primary motivation for his approach is the recognition of a fundamental distinction between the theory of cracks and that of notches or holes, even in a linear solid. On loading either, the surfaces suffer a relative displacement, which will necessarily be small in a linear theory. Under this loading, however, while notches and holes are essentially static, cracks may sometimes extend or contract. As a problem in the theory of elasticity, therefore, a crack requires a model in which the boundary of the elastic region is a function of the applied load and must be determined; it is not sufficient to satisfy given boundary conditions on the initial surfaces of the solid. This introduces an essential nonlinearity into the theory of cracks, but, by suitable hypotheses, Barenblatt is able to state the problem of cracks as a boundary value problem in linear elasticity.

Cracks may be reversible or irreversible; a reversible crack is one which can extend or contract, the opposing faces rewelding on contraction. Real cracks are generally irreversible. By the contour of a crack we mean the boundary curve separating an area in the solid that has cracked from one which has not. As we have emphasized, this boundary region will have a finite width, just as does a slip dislocation. Let A (Fig. 16) be a curve representing the contour of a crack. For simplicity in developing the mathematical argument, the crack contour is assumed for the moment to be a plane curve.

A crack whose contour neither extends nor contracts under the applied load (which may be due to other sources of internal stress) is called an equilibrium crack. If we include all the constraints acting on the crack in the applied forces, then, for a crack to be an equilibrium crack, there must be no change in the total free energy for any small variation of the crack contour A. By the total free energy is meant the elastic energy plus the potential energy of the applied loads plus the

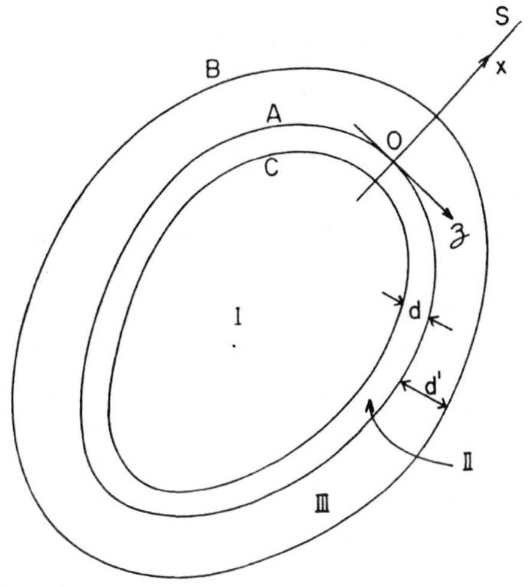

FIG. 16. Crack contours. Region I—cohesive forces zero. Region II—cohesive forces important. Region III—zone of plastic deformation.

work stored in cohesive forces on the crack surface. We shall later, by the usual arguments (Orowan, 1949; Irwin, 1948), include plastic work at the crack contour in the work done by cohesive forces; for the moment, we discuss truly brittle fracture and assume that this work is stored reversibly. The equilibrium of the crack may be stable, unstable, or neutral, according to the distribution of applied loads.

Barenblatt's theory of equilibrium cracks (Barenblatt 1959, 1960, 1961, 1962, 1964; Barenblatt and Cherepanov, 1961a, 1961b) is based on the following hypotheses. The crack is divided into an inner region I (Fig. 16) with contour C and a region II of width d between the contours C and A. The cohesive forces across the faces of the loaded crack are assumed to be important only over the boundary region II, and are taken to be zero throughout the inner region I. At any point of the crack contour, we consider a normal section S, intersecting the contour A at O, and take right-hand axes x, y, and z with origin O; x and z define the element of crack surface at O and y its normal. To develop the general theory, we assume that we can find for any such point O on A a small region $V: d \ll |\mathbf{r}| < s$ about it such that s is small compared with the crack surface and its radii of curvature, the radius of curvature of the crack contour, and the distances to the nearest other singularity in the elastic solid and to the outer boundary Σ_0. Then, within $|\mathbf{r}| = s$, the field of the loaded crack approximates to that of a slit $-2a < x < 0, y = 0$ with $a \gg s$ under a combination of plane and antiplane strain produced by the external loads.

The dislocation distribution representing the crack may thus be described, as far as it affects the region V, by three sets of lines parallel to Oz; $b\mathscr{D}_x = -d\phi_x/dx$, $b\mathscr{D}_y = -d\phi_y/dx$, and $b\mathscr{D}_z = -d\phi_z/dx$, where $\phi_i = (\phi_x, \phi_y, \phi_z)$ is the relative displacement across the crack surface.

We define the loading of the crack as follows. Let σ_{yx}^A, σ_{yy}^A, and σ_{yz}^A be the stresses which would be produced along the crack surface by applied loads and all other singularities of stress in the solid if the crack were not present. Now make the crack and apply a system of tractions to the cut surfaces corresponding to a stress σ_{yj}^A there so that the elastic field is everywhere unchanged (this means tractions $-\sigma_{yj}^A$ are applied to the (0, −1, 0) face, and tractions $+\sigma_{yj}^A$ to the (0, 1, 0) face (cf. Sect. IV,A).

Now suppose that, in its final state, the crack surface with normal (0, −1, 0) is to be subject to tractions $-\sigma_{yj}^I$ due to internally applied stress and tractions $-G_{yj}$ due to cohesive forces, the (0, 1, 0) face being subject to equal and opposite tractions; that is, along the crack there are finally to be tractions corresponding to a stress $+\sigma_{yj}^I + G_{yj}$. If σ_{yj}^I is negative, the crack is subject to internal pressure. A positive value of G_{yy} corresponds to the effect on the elastic medium of attractive cohesive forces in simple normal parting, while a positive value of G_{yx} (or G_{yz}) corresponds to a tangential restoring force (because of cohesion) opposing the relative displacements across the crack caused by a positive externally applied stress σ_{yx}^A (or σ_{yz}^A). Components of this latter type are relevant to some types of fracture, for example, quasibrittle fracture, when, as discussed below, the cohesive forces are caused mainly by plasticity.

We achieve such a final state by allowing a distribution of crack dislocations to form which produces a stress $\sigma_{ij}^{D'}$ everywhere, the distribution being chosen so that, on the crack, the stress $\sigma_{yj}^{D'}$ is $-\sigma_{yj}^A$ plus the final values $\sigma_{yj}^I + G_{yj}$ which arises from the internally applied stress and the cohesive forces. The final total stress σ_{ij}^T everywhere is then $\sigma_{ij}^T = \sigma_{ij}^A + \sigma_{ij}^{D'}$ and so has the value $\sigma_{yj}^I + G_{yj}$ along the crack, as required. The dislocation distribution is determined by the integral equation

$$\sigma_{yj}^{D'} = -\sigma_{yj}^{A} + \sigma_{yj}^{I} + G_{yj} \tag{124}$$

It may alternatively be regarded as due to the application of tractions to the crack faces which cancel the effect of the stress σ_{yj}^{A} and add the tractions finally required, namely, those which correspond to a stress $\sigma_{yj}^{I} + G_{yj}$ at the crack. Thus, $\sigma_{yj}^{D'}$ is due to the tractions $+\sigma_{yj}^{A} - \sigma_{yj}^{I} - G_{yj}$ applied to the crack face (0, −1, 0) and the opposite tractions $-\sigma_{yj}^{A} + \sigma_{yj}^{I} + G_{yj}$ applied to the crack face (0, 1, 0). Accordingly, for the crack loaded internally by the stresses $\sigma_{yj}^{I} + G_{yj}$ and externally by σ_{ij}^{A}, we can use Eq. (63) (suitably modified according to the deformation mode), with an effective applied stress at the crack of $\sigma_{yj}^{A'} = \sigma_{yj}^{A} - \sigma_{yj}^{I} - G_{yj}$, as discussed in Sect. IV,B.

We now make the following hypotheses (Barenblatt, 1959; Barenblatt and Cherepanov, 1961a, 1961b; Barenblatt, 1964):

(1) The width d of the region II is small, compared with the size of the whole crack.

(2) The shape of the normal section S of the crack between A and C (and hence the local distribution *with* x of the cohesive forces over the region II) does not depend on the applied loads, and depends only on the material and physical parameters (such as temperature, composition, and atmospheric pressure) controlling its behavior.

(3) The opposite faces of the crack are smoothly joined at the contour A so that $\mathscr{D}_i(x) = -b\, d\phi_i/dx = 0$ along A for $i = x, y, z$. From the theory of stressed slits, this is equivalent to the statement that the stresses at the end of the slit are finite.

The third hypothesis was motivated by earlier work of Khristianovich (Zheltov and Khristianovich, 1955) on the hydraulic fracture of oil bearing rock strata. A graphic description of a crack whose contour behaves according to the second hypothesis is that it is like a "zipper" (Sergaziev, quoted in Barenblatt, 1962). This second hypothesis does, however, require some qualification. It is assumed to hold only at points of the crack contour where, at the contour A, the cohesive forces have reached their maximum possible value. This is taken to be the case for all cracks which are about to extend at that point, and for all truly reversible equilibrium cracks. In a real solid, not only are cracks generally irreversible, but similar discontinuities may be introduced by cutting operations. At the contour of a cut or a crack remaining after the load has been reduced, it is supposed that the cohesive forces have fallen below their maximum value. A crack having the cohesive forces at their maximum value at at least one point on its A contour, and which therefore extends under infinitesimal increase of the load, is called by Barenblatt a mobile equilibrium crack in contrast to one which does not have this property, which is called an immobile equilibrium crack. As the load increases, an immobile equilibrium crack changes to a mobile equilibrium crack, the tip widening in the region II as the load grows until the crack begins to move; thereafter, the shape of the tip remains constant. For Barenblatt, however, all true cracks are equilibrium cracks in the sense that they obey hypothesis 3, closing smoothly at their outer contours.

The mathematical consequences of these hypotheses can be developed either by using the requirement of smooth closure, or that of finite stress. We first consider the latter. On the axis outside the crack contour we have in V for the stresses $\sigma_{yj}^{D'}$ from Eqs. (63), (78), and (87):

$$\sigma_{yj}^{D'} = \pi^{-1}(x^2 - a^2)^{-1/2} \int_{-a}^{a} \frac{(a^2 - x'^2)^{1/2}}{x - x'} \sigma_{yj}^{A'}(x')\, dx' \tag{125}$$

where

$$\sigma_{yj}^{A}(x) = \sigma_{yj}^{A}(x, 0) - \sigma_{yj}^{I}(x, 0) - G_{yj}(x, 0) \tag{126}$$

The total stress is $\sigma_{yj}^{I} = \sigma_{yj}^{D'} + \sigma_{yj}^{A}$. With $s = x - a \ll a$, we have from Eqs. (70), (86), and (87) of Sect. IV the following relations in the region V. When s is positive, that is, just outside the contour A, the total stress σ_{yj}^{T} is

$$\sigma_{yj}^{T}(s) = \frac{K_j}{(2\pi)^{1/2}} s^{-1/2} + \sigma_{yj}^{I}(0) + G_{yj}(0) + O(s^{1/2}) \tag{127}$$

while, when s is negative, that is, just inside A, the relative displacements ϕ_j across the crack are

$$\phi_j(s) = \frac{2bK_j}{\pi A(2\pi)^{1/2}} (-s)^{1/2} \tag{128}$$

In Eq. (128), the constant A has the value $Gb/2\pi(1 - \nu)$ for $j = 1, 2$ and $Gb/2\pi$ for $j = 3$. The constants K_j are given by

$$K_j = (\pi a)^{1/2} (I_{1j} + I_{2j} + I_{dj}) \tag{129}$$

where the integrals I_{1j}, I_{2j}, and I_{dj} are obtained by substituting in Eqs. (88)–(90) for $\sigma^{A}(x)$ the expressions $\sigma_{yj}^{A'}(x, 0)$ given by Eq. (126).

The energy relations at crack contours are examined in some detail in Sect. V. From the discussion given there, it is clear that we can use the following argument to assess the small change in total free energy accompanying a small increase in the area of the crack in the region V.

In V, the contour is approximately the line $x = 0, y = 0$; let it undergo a small perturbation so that when $-t \leqslant z \leqslant +t$ it becomes $x = h(z) > 0$, where $h, t \ll s$. The change in total free energy is $-\delta F$ (cf. Eq. (102)) (Barenblatt and Cherepanov, 1961a, 1961b; Barenblatt, 1959; Irwin, 1957)

$$\delta F = -\tfrac{1}{2}\int_{-t}^{t}\int_{0}^{h} \sigma_{yj}^{T}\phi_j' \, dx \, dz$$

where σ_{yj}^{T} is the total stress due to the unperturbed crack, and ϕ_j' is the displacement discontinuity across the perturbed crack. From Eqs. (127) and (128), in V where $h \ll s$, $\sigma_{yj}^{T} \propto x^{-1/2}$ and $\phi_j' \propto (h - x)^{1/2}$. Thus

$$\begin{aligned} \delta F &= -(1/G\pi)\{(1-\nu)(K_1^2 + K_2^2) + K_3^2\} \int_{-t}^{t}\int_{0}^{h} \left(\frac{h-x}{x}\right)^{1/2} dx \, dz \\ &= -(1/G\pi)\{(1-\nu)(K_1^2 + K_2^2) + K_3^2\} \, \delta S \end{aligned} \tag{130}$$

where δS is the area of the small perturbation. If the crack is to be an equilibrium crack, we must have $\delta F = 0$, and it follows at once that $K_1 = K_2 = K_3 = 0$. Thus, from Eqs. (127) and (128), the stresses at the crack contour are finite, and the crack closes smoothly. The third hypothesis is therefore really a consequence of the requirement that the crack be an equilibrium one.

We cannot evaluate the constants K_i without a knowledge of the distribution of stresses $\sigma_{yj}^{A'}$ over the whole crack surface. To proceed further without this knowledge, we write

$$K_i = K_i^0 + K_i^c \tag{131}$$

where K_i^c arises from the cohesive forces G_{yi} alone and K_i^0 from all other terms. We have

$$K_i^c = -(\pi a)^{1/2} \int_0^{2a} \left(\frac{2a - t}{t}\right)^{1/2} G_{yi}(a - t) \, dt \tag{132}$$

By the first hypothesis, G_{yi} is almost zero, except in the small regions $0 < t < d$ and $2a - d < t < 2a$ near the crack contour. Putting $q = 2a - t$, we can thus write

$$K_i^c = I_i + J_i \tag{133}$$

where

$$\begin{aligned} I_i &= -(2/\pi)^{1/2} \int_0^d t^{-1/2} G_{yi}(a - t) \, dt \\ J_i &= -(2\pi a^2)^{-1/2} \int_0^d q^{1/2} G_{yi}(q - a) \, dq \end{aligned} \tag{134}$$

We expect G_{yi} to increase monotonically from zero to its maximum value at $x = \pm a$ over the distance d, and it is clear that it is reasonable to assume that $|J_i| \approx (d/a)|I_i| \ll |I_i|$, so that we may neglect J_i in extimating K_i^c for the region V. Thus, finally, we can write

$$m_i = -(\pi/2)^{1/2} K_i^c = \int_0^d t^{-1/2} F_{yi}(t)\, dt \tag{135}$$

where $G_{yi}(a - t) = F_{yi}(t)$. By the second hypothesis, the function $F_{yi}(t)$ is independent of the applied load at all points on an equilibrium crack contour where the cohesive forces attain their maximum value. The m_i values for these points (where the crack contour is about to extend) are thus constants of the material, which we write M_i. Barenblatt calls the M_i the moduli of cohesion. The units of M_i are dyne-cm$^{-3/2}$ or (dyne cm^{-2})(cm$^{1/2}$). If the cohesive forces have not reached their maximum value, then $m_i < M_i$.

Since $K_i = 0$ always if the cracks are to be equilibrium cracks, we must have at points of a crack contour where the equilibrium is mobile:

$$K_i^0 = -K_i^c = (2/\pi)^{1/2} M_i \tag{136}$$

and at points of the contour where the equilibrium is immobile:

$$K_i^0 = (2/\pi)^{1/2} m_i < (2/\pi)^{1/2} M_i \tag{137}$$

Now let a body containing equilibrium cracks bounded by given initial contours Γ_0 be loaded, the loads increasing monotonically with a parameter λ. The contours Γ of the cracks existing at any λ will be determined by finding a solution of the elastic equations giving K_i^0 values on the contours Γ satisfying the following conditions:

(1) Where Γ is outside Γ_0, condition Eq. (136) is satisfied.

(2) Where Γ still coincides with Γ_0

$$K_i^0 \leqslant (2/\pi)^{1/2} M_i$$

If Γ lies wholly outside Γ_0, or indeed if the initial cracks are reversible, condition Eq. (136) must be satisfied everywhere on Γ. Then, the configuration Γ_0 of initial cracks is of no importance, except that it must be compatible with expansion to Γ under increasing load. If there are no solutions satisfying the above conditions, it may mean either that the given load causes the body to fracture or that mobile equilibrium cracks do not form from the initial cracks under the given load.

The moduli of cohesion M_i may be related to the surface energy γ.

The condition for an equilibrium crack to be mobile then becomes identical with that for crack extension using the Griffith theory, or its equivalent statement in terms of the crack extension force or strain energy release rate $\mathscr{G}$ (Irwin 1948, 1957).

In fact, if we replace K_i by K_i^0 in Eq. (130), δF simply becomes δE_{tot} in the notation of Sect. V, where δE_{tot} is the total free-energy change accompanying a small displacement of the crack excluding contributions from the forces of cohesion (that is, from the surface energy). The crack contour lies along the z axis, and, taking a length δz of the crack, we have $\delta S = \delta z\, \delta a$ for a displacement from a to $a + \delta a$. Per unit length of the crack front, we therefore have

$$\mathscr{G} = -\lim_{\delta a \to 0} \frac{\delta E_{\text{tot}}}{\delta a} = \mathscr{G}_{\text{I}} + \mathscr{G}_{\text{II}} + \mathscr{G}_{\text{III}} \tag{138}$$

where $\mathscr{G}_{\text{I}}$, $\mathscr{G}_{\text{II}}$, and $\mathscr{G}_{\text{III}}$ are given by Eqs. (104) and (105). Here we have used the relation $E = 2G(1 + \nu)$, and the fact that our K_i^0 are related to the stress-intensity factors K_{I}, K_{II}, and K_{III} by

$$K_1^0 = K_{\text{II}}, \qquad K_2^0 = K_{\text{I}}, \qquad K_3^0 = K_{\text{III}} \tag{139}$$

For the crack to be about to extend, that is, to be in mobile equilibrium, we must have, on the one hand, $\mathscr{G}$ equal to its critical value $\mathscr{G}_c = 2\gamma$ where γ is the effective surface energy of fracture, or, on the other, the relations Eqs. (136) and (139) between the moduli of cohesion M_i and the stress-intensity factors. Thus, we get

$$2\gamma = \mathscr{G}_c = \frac{1 + \nu}{E}\{(1 - \nu)(K_{\text{II}c}^2 + K_{\text{I}c}^2) + K_{\text{III}c}^2\} \tag{140}$$

$$= \frac{2(1 + \nu)}{\pi E}\{(1 - \nu)(M_1^2 + M_2^2) + M_3^2\} \tag{141}$$

We obtain separate relations for each mode of fracture occurring separately or if we assume that we can write $\gamma = \gamma_1 + \gamma_2 + \gamma_3$ each γ_i making an independent contribution to the appropriate mode. These relations between the moduli of cohesion and the surface energy at once bring Barenblatt's theory into complete correspondence with those of Griffith, Orowan, and Irwin.

The criteria Eqs. (140) and (141) may also be written in terms of the density of crack dislocations. We note from Eq. (86) (compare also Eq. (39)) that the stress-intensity factors K_i may be written

$$K_i^2 = 2\pi^3 A^2 \lim_{x \to a} (a - x)\, \mathscr{D}_i^2(x) \tag{142}$$

12

Thus, if we consider the crack dislocations $\mathscr{D}_i{}^0(x)$ induced by all the applied loads, but excluding the cohesive forces, we have, correspondingly, the intensity factors $K_i{}^0$ and the criterion Eq. (140) for instability may be written

$$\frac{8\gamma}{Gb^2\pi} = \lim_{x\to a}(a-x)\left\{\frac{[\mathscr{D}_1{}^0(x)]^2 + [\mathscr{D}_2{}^0(x)]^2}{(1-\nu)} + [\mathscr{D}_3{}^0(x)]^2\right\} \tag{143}$$

Smith (1965c) has given the criterion in this form. It has been used particularly to discuss the stability of wedge cracks, both by Smith (1965c) in generalising earlier results (Bilby and Hewitt, 1962; Bullough, 1964), and in new models of crack initiation (Smith, 1966a, 1966b, 1966c, 1966d; Atkinson and Heald, 1967). Murrell (1964a, 1964b) has compared Eq. (140) with alternative criteria proposed for the failure of rocks under combined stresses. The relations Eqs. (140), (141), and (143) are for isotropy and a combination of plane strain and antiplane strain. For a combination of plane stress and antiplane strain (that is, for "tunnel" cracks in thin plates) we discard the factors $1 - \nu^2$ in the definitions of $\mathscr{G}_{\mathrm{I}}$ and $\mathscr{G}_{\mathrm{II}}$ in Eq. (105). The effect of anisotropy on the modulus of cohesion has been discussed by Barenblatt and Cherepanov (1961b), while Sih, Paris, and Irwin (1965) have given modifications of the criterion Eq. (140) applicable to the anisotropic case. Paris and Sih (1965) have assembled a useful collection of results on stress-intensity factors for different states of loading and crack geometries. For a straight crack in simple tension, $\sigma_{yy} = P$, $K_{\mathrm{I}} = P(\pi a)^{1/2}$ and the criterion Eq. (140) becomes the Griffith formula in plane strain:

$$P = \left(\frac{2E\gamma}{(1-\nu^2)\,\pi a}\right)^{1/2}$$

It is a limitation of all theories using a linear elastic continuum to represent a solid containing singularities such as cracks that, near such singularities, the stresses and strains become too large for the linear theory to be reasonable. For example, if true brittle cleavage is being studied, the stress just outside the contour in Barenblatt's model (with no internal loading σ_{ij}^I) will be equal to the value of $G_{yi}(0)$, the maximum value reached by the cohesive stress in region II just within this contour. As this will be, say $E/10$, the linear theory must be inadequate. In fact, however, in a large class of fractures of practical importance, the onset of plasticity at stress singularities intervenes to introduce a different type of nonlinearity, which is much more important than the actual deviations from nonlinearity of interatomic bonds in tension. It is, however, possible to make considerable progress in understanding by viewing the crack front on a different scale and regarding the "cohesive energy" as arising largely from the localized plastic work associated with the fracture front (Orowan, 1949; Irwin, 1948). In this way, a large class of unstable semiductile and ductile fractures can be discussed. The onset of plasticity at crack fronts, which distorts the gross features of the elastic stress fields, is a principal limitation of current linear elastic fracture mechanics (Paris and Sih, 1965), and much effort is being devoted to the development of an analysis to embrace these effects using the theory of macroplasticity (for a recent account, see McClintock and Irwin, 1965). We shall not discuss this work here, but shall confine our attention to a simple description of these effects using a dislocation model which provides general understanding by making possible the analytical correlation of some of the important parameters, rather than by assessing them by numerical methods.

First, we note that, to include quasibrittle fracture in Barenblatt's model, a further contour B and a region III of width d' outside A (Fig. 16) is introduced.

Figure 17 shows the normal section S at $\dot{O}$. The linear elastic medium is now outside the surface B', instead of outside A', and the effect of the plastic work in the shaded region is incorporated by defining modified forces of cohesion causing stresses $G_{yj}(x)$ over the regions II and III: $G_{yj}(x)$ is equal to the true cohesive stress $G^c_{yj}(x)$ in region II and to the stresses $G^P_{yj}(x)$ needed for separation of the surfaces B' by plastic work in region III. The work done by G^P_{yj} is usually much greater than that done by G^c_{yj}, and we can generally disregard the latter. The cohesive stress in the region $d + d'$ is thus roughly constant and equal to a yield stress Y, and we obtain for example in mode I a modulus of cohesion M_2 (Barenblatt, 1962):

$$M_2 = \int_0^{d_1} t^{-1/2} Y\, dt = 2Y d_1^{1/2} \tag{144}$$

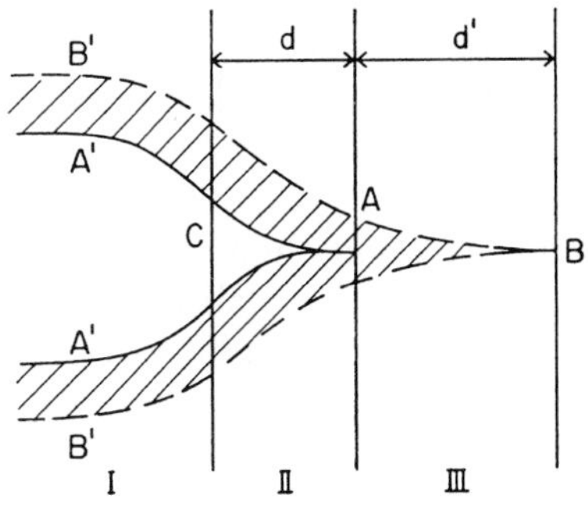

FIG. 17. Section of crack tip (after Barenblatt).

A model of this type, in which there are zones of constant stress at the ends of linear cracks, has been used independently in various applications by a number of workers, and we shall now discuss it more fully.

A freely slipping crack extending from $-c < x < +c$ (in mode II or III) is represented by crack dislocations, and zones beyond it $-a < x < -c$ and $c < x < a$ by slip dislocations whose motion is opposed by a friction stress representing the yield stress (Bilby, Cottrell, and Swinden, 1963; first reported by Cottrell, 1961). From the condition that $\mathscr{D}(\pm a) = 0$, we obtain

$$c/a = \cos(\pi R/2\sigma_1) \tag{145}$$

where R is the applied shear at infinity. The relative displacement at the tip of the crack $\phi(\pm c)$ is given by

$$\pi^2 A\phi(c) = 2cb\sigma_1 \ln(a/c) \tag{146}$$

There is no stress on $x = 0$ in antiplane strain, so that the model then also represents a surface slit notch of length c in the half space $x > 0$. The constant A then has the value $Gb/2\pi$, and the dislocations are pure screw. Hult and McClintock (1957) have discussed a sharp notch of angle θ and depth c in a nonhardening, elastic-plastic, semi-infinite solid in antiplane strain. The value of a/c given by the dislocation model Eq. (145) agrees very well with that obtained by Hult and McClintock for $\theta = 0$ and with $\sigma_1 = k$ where k is the yield stress in the criterion $\sigma_{yz}^2 + \sigma_{xz}^2 = k^2$, despite the great difference in shape between the plastic zones in the two models. Using a critical displacement criterion $\phi(c) = \phi_{\text{crit}}$ for fracture, the results have been used to discuss the range of dangerous notch lengths over which fracture below general yield may occur in a structure where regions of size a reach a stress R. (For recent discussions of the displacement criterion, see Cottrell, 1965b; Wells, 1963, 1965; Burdekin, Stone, and Wells, 1965; McClintock and Irwin, 1965). The formulas (145) and (146) have also been correlated (Cottrell, 1963a, 1965; Hahn and Rosenfield, 1965, 1967) with data of Tetelman (1964), Hahn and Rosenfield (1965), Allen (1961), Wundt (1959), and Lubahn and Yukawa (1958); the last two include results on 9-inch sections of turbine steel and show the size effect strikingly. For small stresses when the plasticity is localized at the crack tip $\sigma_1 \ll R$, $a - c = r \ll c$, and so Eqs. (145) and (146) give

$$2r/c = (\pi R/2\sigma_1)^2 \tag{147}$$

and

$$\phi(c) = 2\sigma_1 br/\pi^2 A \tag{148}$$

With $\gamma \approx \sigma_1 \phi(c)$, these give the Griffith–Orowan–Irwin relation (Bilby, Cottrell, and Swinden, 1963). Cottrell (1965a, 1965b) has also used the model as an approximate solution to the integral Eq. (118) with a pulse function law, $\sigma = \sigma_1$, $0 \leqslant \phi \leqslant \phi_c$; $\sigma = 0$, $\phi > \phi_c$, for the cohesive forces, using Eqs. (147) and (148) in plane strain to obtain

$$R = \{2E\gamma/\pi(1 - \nu^2)\, c\}^{1/2} \tag{149}$$

with $\gamma = \frac{1}{2}\phi(c)\,\sigma_1$, σ_1 now being interpreted as the theoretical fracture stress. Leonov and Onyshko (1961) obtained the same result, in an investigation of very fine brittle cracks with this model, and report that the limiting stress obtained by combining Eqs. (145) and (146) was given by Vitvitskiĭ and Leonov (1960). Leonov and Onyshko (1961) use a displacement criterion because, in their model, the cohesive forces are sharply cut off at a critical relative displacement. They also treat a smoothly closing brittle wedge crack on the tensile side of a single dislocation under tension; Burdekin, Stone, and Wells (1965) have used exactly the same model to correlate Eqs. (145) and (146) with data on crack opening displacements in fractures with plasticity at the crack tip. Dugdale (1960) considered the model under tension in plane stress or strain and obtained the relation Eq. (145) by requiring the stress to be finite at $x = \pm a$. This he found to correlate well with the extent of plastic zones round slits in thin steel sheets of yield stress σ_1 under tension R, including slits of length c cut from one edge.

In applying the model used by Bilby, Cottrell, and Swinden to a wide range of types of fracture Cottrell (1962, 1965) has distinguished between cumulative and noncumulative modes. In a cumulative mode, the same distribution of dislocations describing the crack front propagates essentially unchanged through the specimen during fracture; the contribution to fracture accumulates continuously with the distance traveled by the dislocation. In a noncumulative mode, each incremental growth of the crack front requires the injection of more dislocations and a/c increases rapidly, producing an inherently stable situation. Most types of ductile fracture (e.g., shear, fibrous, tensile cup-and-cone, and plastic tearing) occur not by cracking but by sliding off and are noncumulative. Examples of cumulative fracture are brittle cleavage, mode I plane stress necking in thin sheets by formal edge dislocations, mode III tearing in thin sheets by formal screw dislocations, and various types of ductile and semiductile discontinuous fractures involving internal necking (Cottrell; 1963a, 1963b).

By comparing the relations between R/σ_1 and $c\phi(c)/\epsilon$ where $\epsilon = \sigma_1/G$, given by Eqs. (147) and (148) and the Griffith-type relation between them given by Eq. (149), Cottrell (1963b) has estimated that

$$c \approx \sigma_1 \phi(c)/G \tag{150}$$

is a criterion for notch sensitivity. When c is much less than this, fracture occurs when $R \approx \sigma_1$, when c is much greater, a Griffith-type relation, Eq. (149), holds and the material is notch sensitive, fracturing when $R \ll \sigma_1$, that is, before general yield. In the low stress limit, given by Eq. (148), we have

$$r = \frac{\pi^2 A\phi(c)}{2\sigma_1 b} = \frac{E\gamma}{\sigma_1{}^2} \frac{\pi}{4(1-\nu^2)} \tag{151}$$

$$\approx \frac{E\gamma}{\sigma_1{}^2} \tag{152}$$

with $\gamma = \frac{1}{2}\sigma_1\phi(c)$. Thus r is a constant of the material, and Cottrell has compared this width of the crack front with the dislocation width in the Peierls–Nabarro approximation. In fact, the modulus of cohesion in this limiting case is by Eq. (144), $M_2 = 2\sigma_1 r^{1/2}$, and the relation Eq. (151) is just that Eq. (141) between M_2 and the surface energy. This special model thus shows in the limit when $r \ll c$, that M_2 can indeed be regarded as a material constant.

Since $\mathscr{D}(x)$ vanishes at $x = \pm a$, this model is obviously an equilibrium crack, in Barenblatt's sense. It is also in neutral equilibrium in a constant applied field at infinity if it is assumed that all the work done against the stress σ_1 in the plastic regions is dissipated in plastic deformation without contributing in some way to a fracture mechanism. This question has been discussed in detail by Swinden (1964). The calculations of Yokobori and Ichikawa (1966) simply show for this special case that, if we include the forces of cohesion, then the condition that $\mathscr{D}(\pm a) = 0$, makes the δF of Eq. (130) vanish. As is made clear by Swinden's earlier discussion, their comments are thus not relevant to the use which is made of the model in physical applications, where it is assumed, as in all quasibrittle fracture, that at least some of the work done at the crack tip contributes to that required for the fracture process. As has been shown, if we use the model to represent truly brittle fracture, we obtain the Griffith criterion with $\gamma = \frac{1}{2}\sigma_1\phi(c)$. The unstable equilibrium arising in this low stress limit appears, because r becomes a constant of the material.

Using the same assumptions as Bilby, Cottrell, and Swinden (1963), Smith (1964) has treated two collinear relaxed cracks, and two vertically stacked relaxed cracks (1965a). A periodic sequence of collinear cracks has been discussed by Bilby, Cottrell *et al.* (1964) and by Smith (1965b), and a vertically stacked periodic sequence by Smith (1965a). The solutions are applicable to plane strain, plane stress, and antiplane strain. The periodic collinear array has been used to discuss the size effect in a notched plate of finite width (Smith, 1965b; Bilby and Swinden, 1965). This problem is an example of the type discussed in App. C, where $\mathscr{D}(x)$ and $P(x)$ are periodic. Let a central crack $-c < x < c$ bounded by collinear relaxed regions extending to $x = \pm a$ and with yield stress $\sigma_{zy} = \sigma$, be repeated with period $2h$ along the x axis, and let the applied stress at infinity be $\sigma_{zy} = \sigma$. Then the problem is one of antiplane strain. The solution is readily obtained (Bilby *et al.*, 1964) from that for a single crack by the transformations of variable indicated in App. C. Corresponding to Eq. (145) we find

$$\frac{\sin(\pi c/2h)}{\sin(\pi a/2h)} = \cos(\pi\sigma/2\sigma_1) \tag{153}$$

while, in the special case when $a = h$ (so that the plasticity just spreads between adjacent regions and we reach a state of general yield), $\phi(c)$ may be written (there are errors in (15) of Bilby *et al.*, 1964)

$$\phi(c) = -\left(\frac{4h\sigma_1}{\pi^2 G}\right)\left[2L\left(\frac{\pi c}{2h}\right) + L\left(\frac{\pi}{2} - \frac{\pi c}{h}\right) - L\left(\frac{\pi}{2}\right)\right] \tag{154}$$

where $L(\theta)$ is the Lobachevskiĭ function

$$L(\theta) = -\int_0^\theta \ln(\cos\alpha)\, d\alpha \tag{155}$$

When $a \neq h$, the integral for $\phi(c)$ may be evaluated numerically (Bilby and Swinden, 1965). Since there are no tractions on the surfaces $x = 0$ and $x = h$, the model can be used to represent a plate of thickness h with a surface slit notch of depth c. Assuming that failure will occur when $\phi(c)$ reaches a critical value, ϕ_{crit}, the relation between the parameters σ/σ_1 and c/h at failure can be determined. In this way, for a given size of structure (h), yield stress (σ_1) and loading (σ), a dangerous notch size (c) can be estimated (Bilby and Swinden, 1965; Smith, 1965b). The latter authors have also incorporated work hardening in the model, and have reported preliminary work on a model where the plasticity is relaxed at each tip on two dislocation arrays inclined symmetrically to a crack. Smith (1965b) discusses the role of inclusions in fracture using as a basis the model with two collinear relaxed cracks.

Finite and infinite arrays of parallel and collinear slit and wedge cracks have also been discussed using these methods and the concepts of dislocation theory by Markuzon (1961), Ichikawa *et al.* (1965), Yokobori *et al.* (1965), Yokobori and Ichikawa (1965), Ohashi *et al.* (1965a, 1965b) and Smith (1966d, 1966e). The interaction of slip dislocations, slip bands, and cracks has also been studied recently by Ichikawa and Yokobori (1965), and Yokobori *et al.* (1965b). Louat (1965) for antiplane strain and isotropy and Atkinson (1967) for antiplane and plane strain and for general anisotropy have given, in manageable form, the general solution for the field of a dislocation of arbitrary Burgers vector situated at any point outside or on a slit crack. These solutions can be applied to all two-dimensional problems involving the interaction of dislocation distributions and cracks.

These analyses of the interaction of slip bands, dislocations, and cracks have been used to study not only crack propagation and stability, as has mostly concerned us hitherto, but also to consider the problem of crack initiation. We shall therefore conclude our discussion of applications with a few remarks about this problem. Here, it is entirely appropriate that we should end this chapter by referring to the pioneering work of Zener (1948), Mott (1948, 1953, 1956) and particularly of the late Dr. A. N. Stroh (1954, 1955a, 1955b, 1956, 1957a, 1957b, 1958a, 1958b). This work originated in the idea that the accumulation and interaction of slip dislocations under stress could provide a mechanism for the initiation of cracks in crystalline solids. Stroh (1954, 1955a) treated the formation of the isolated wedge crack from a single slip band, and his discussion of its formation and stability was applied by Cottrell (1957, 1958) to analyze the propagation of wedge cracks and to develop relations between the important parameters in the grain size/lower yield stress relation in iron and mild steel, and the fracture behavior of these materials. Petch (1957, 1959) gave a related discussion.

The initiation of cracks by the stress of accumulated dislocations has been discussed in a number of geometries. Cottrell (1958) proposed a model for the initiation of cracks at intersecting slip planes, and there have been various recent discussions of this model relevant particularly to iron and magnesium oxide mainly to incorporate the effects of dislocations external to the wedge crack (Atkinson and Heald, 1967; Hahn and Rosenfield, 1966) and to allow fully for the change in the equilibrium of the external dislocations as the crack forms (Smith, 1966a; 1966b, Atkinson, 1966). Under appropriate conditions, we expect (Frank and Stroh, 1952) slip bands, cracks, twins, martensite plates, and kink bands all to nucleate one another, since the stress fields they produce are closely related. Other dislocation geometries may nucleate cracks in special circumstances. For example, Stroh (1958b) also treated in detail the proposal of Friedel (1956) that in hexagonal metal crystals, cracks may arise where part of a moving bend plane is held up by an obstacle.

In this chapter, we have limited our discussion to the static theory and have excluded dynamic and inertial questions associated with moving cracks. We have tried to present a unified picture of the mathematical analysis of the initiation and stability of cracks, using the ideas of dislocation theory as a unifying principle, and have indicated in outline how the dislocation theory has developed, particularly since the pioneer work of Stroh, and how it may be related to macroscopic mechanics.

VII. Recommended Research

We have seen how the concepts of continuously distributed crack and slip dislocations provide a general framework within which we can discuss both the deformation and fracture processes and the interaction between them. The comparison of the various models with experiment and the evidence for their validity are discussed in other chapters. Here, we consider possible future developments in the general theory.

An important task is to extend the work on dislocation arrays so that regions of plastic relaxation of finite thickness can be represented. This, of course, forms part of the larger problem of relating the theory of continuous distributions of dislocations to the theory of macroscopic plasticity. Many refinements of the static linear theory presented here are possible. One can, for example, treat moving cracks, taking into account the inertia of the material. Such time-dependent problems are probably less important than problems in which the stress is allowed to be rate-dependent (Snoek effect, delayed yielding, viscoelastic effects in polymers.)

An application of the nonlinear theory of elasticity would be a useful refinement in the case of polymers, though less realistic in the case of metals, since, for them, when Hooke's law is appreciably departed from, one is usually already in a region where the atomic structure must be allowed for.

The ideal to be aimed at is, of course, to remove the principal limitation of any continuum description of the ultimate fracture process and use a theory which takes full account of the forces of interaction between the atoms or other discrete structural units of which all solids are composed. A bridge between the continuum and atomic theories may perhaps be provided by recent developments in generalized continuum mechanics which go some way toward incorporating certain properties of the atomic structure into a continuum theory.

VIII. Summary

In Sects. I and II, the idea of discrete dislocations and of continuous arrays of infinitesimal dislocations is introduced. It is pointed out that there is a close formal relationship between continuous arrays of dislocations and cracks; this is basic to much of what follows. Dislocations thus enter the theory of fracture in two ways: as crystal dislocations which play a part in the physics of fracture; and also as a convenient mathematical or conceptual element in the macroscopic treatment of cracks.

Section III deals with the problem of determining the equilibrium position taken up by a set of parallel and coplanar dislocations under the combined action of their mutual repulsions and the forces exerted on them by an applied stress. One or more of the dislocations may be locked in fixed positions. For a discrete set of dislocations, the problem can, for a variety of applied stresses, be reduced to the solution of a linear second-order differential equation. Some special cases are exhibited, in particular the important case where a number of dislocations are piled up behind a locked dislocation under the action of a uniform applied stress. For a continuous distribution, one is led to a linear integral equation whose solution gives the dislocation density (number of dislocations per unit length) in terms of the applied stress. The solution of the integral equation is derived in an appendix. Instead of treating the continuous case independently, one can also derive results relating to it from the discrete case by passing to the limit of a large number of dislocations.

Section IV is concerned with the close interrelation between cracks and dislocation arrays. With an eye on applications in later sections, Sect. IV,A develops ab initio a number of results in the theory of elasticity for plane cracks. Emphasis is laid on the fact that results for cracks under plane strain conditions (type I and type II deformation) may usually be directly deduced from the results for cracks in a state of antiplane strain (type III deformation) for which the mathematical analysis is much simpler. An expression is derived for the stresses and displacements in the plane of the crack in terms of the externally applied load. In the simplified form which it takes for points near the ends of the crack, this expression plays an important role in succeeding sections. Using the results of Sects. II, III, and IV,A, Sect. IV,B develops the relationship between cracks and continuous arrays of dislocations, and some special cases of interest in fracture mechanics are discussed.

Both in solid-state physics and fracture mechanics it is important to know the variation of the energy of a system with respect to the position of some singularity (dislocation, crack tip, etc.) within it. (By the energy of the system is meant the elastic energy of the solid containing the defect plus the potential energy of the external loading mechanism.) In dislocation theory, this variation of energy has been expressed in terms of the force on a dislocation (or other singularity), and, in fracture mechanics, in terms of the crack extension force. In the language of Sect. IV,B, these two forces are equal in the following sense: the crack extension force is equal to the force on the leading dislocation of the array equivalent to the crack. These and some other matters relating to the energetics of cracks and dislocations are discussed in Sect. V.

In the final section (Sect. VI), the results of the previous sections are applied to some problems in fracture mechanics and dislocation models of crack initiation. In the Griffith model of a crack, the crack extension force (calculated from the elastic solution) is equated with the work done in creating unit length of fresh surface (surface tension, plastic work), and this gives a critical length of crack which may or may not be stable, according to the details of the applied stress and the way the work done in creating the fresh surface varies with crack length. A more detailed model regards the material containing the crack as two semi-infinite slabs held together by cohesive forces acting over the interface between them. To find the equilibrium form of the interface under the combined action of the cohesive forces and applied stresses, one must solve a nonlinear integral equation similar to the Peierls–Nabarro equation of dislocation theory. One can show by inverse methods that cracklike solutions, in fact, exist for suitable laws of force, but detailed solutions are difficult to obtain. It is clear that, over the region where the separation of the crack faces is large compared with the range of cohesive forces, the solution will agree closely with the solution for a purely elastic crack of the type discussed in the previous sections. This fact is exploited in Barenblatt's theory of cracks, which is discussed at some length.

Problems in which plastic flow occurs at the end of a crack may be handled by the methods of macroplasticity. They have also been treated using methods based on the theory of dislocations by Bilby, Cottrell, and Swinden. The section concludes with a discussion of their method and related work, and its application in fracture mechanics.

Appendix A. Elements of the Theory of Elasticity

We use suffix notation in which x_1, x_2, x_3, or, briefly, x_i, $i = 1, 2, 3$, stand for Cartesian coordinates x, y, z. Similarly, the components u_x, u_y, u_z of a vector $\mathbf{u}$ are denoted by u_i, $i = 1, 2, 3$. A repeated suffix is understood to be summed over the values 1, 2, 3, so that, for example,

$$e_{mm} = e_{11} + e_{22} + e_{33}, \qquad \frac{\partial \sigma_{ij}}{\partial x_j} = \frac{\partial \sigma_{i1}}{\partial x_1} + \frac{\partial \sigma_{i2}}{\partial x_2} + \frac{\partial \sigma_{i3}}{\partial x_3} \tag{A.1}$$

The elastic displacement vector u_i is related to the strain tensor e_{ij} by

$$e_{ij} = \frac{1}{2}\left(\frac{\partial u_i}{\partial x_j} + \frac{\partial u_j}{\partial x_i}\right) \tag{A.2}$$

and, in an isotropic material, e_{ij} is related to the stress tensor $\sigma_{ij} = \sigma_{ji}$ by

$$\sigma_{ij} = \lambda e_{mm}\,\delta_{ij} + 2Ge_{ij} \tag{A.3}$$

where δ_{ij} is unity if $i = j$ and zero if $i \neq j$. λ and G are the two Lamé constants. (G is the shear modulus). They are related to the bulk modulus K, Young's modulus E, and Poisson's ratio ν by

$$3(1 - 2\nu)\,K = E = 2(1 + \nu)\,G, \qquad \lambda + \tfrac{2}{3}G = K \tag{A.4}$$

The equation of elastic equilibrium is

$$\partial \sigma_{ij}/\partial x_j = 0 \tag{A.5}$$

In Sect. V and App. B, we make use of the following relations. Let u_i, e_{ij}, σ_{ij} be a set of quantities satisfying Eqs. (A.2), (A.3), and (A.5), and let u'_i, e'_{ij}, σ'_{ij} be a second set satisfying the same relations. Then

$$\sigma_{ij}e'_{ij} = \sigma'_{ij}e_{ij}$$

by Eq. (A.3). Also

$$\sigma_{ij}e'_{ij} = \frac{\partial(\sigma_{ij}u'_i)}{\partial x_j}$$

by Eqs. (A.2) and (A.5) and the fact that $\sigma_{ij} = \sigma_{ji}$.

Finally, if we form the vector

$$v_j = \sigma_{ij}u'_i - \sigma'_{ij}u_i$$

then

$$\partial v_j/\partial x_j = 0 \tag{A.6}$$

that is, the divergence of v_j vanishes. For a surface element of area dS and normal n_j, we write $n_j\,dS = dS_j$. From Eq. (A.6) and Gauss's theorem, it follows that

$$\int_{\Sigma_0} v_j\,dS_j = \int_{\Sigma} v_j\,dS_j$$

where Σ_0 and Σ are two surfaces which can be deformed into one another without passing through singularities of v_j.

When convenient, we use x, y, z for the Cartesian coordinates, u_x, u_y, u_z or u, v, w for the displacement and σ_{xx}, σ_{xy}, and so forth, for the components of the tensor σ_{ij}, and the abbreviated notation $u_{i,j} = \partial u_i/\partial x_j$, $\sigma_{ij,k} = \partial \sigma_{ij}/\partial x_k$. The quantity ϵ_{ijk} is the completely antisymmetrical tensor with $\epsilon_{123} = 1$.

Appendix B. The Force on a Dislocation

In the text, we make much use of the idea of the force on a dislocation. Various derivations of its magnitude are possible, of differing degrees of rigor and plausibility. The following covers the case where the dislocation is simultaneously acted on by stresses produced by external loads and sources of internal stress.

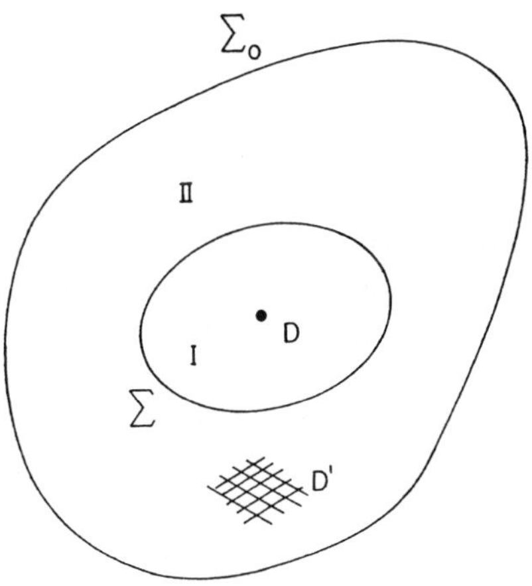

FIG. B.1. To illustrate the force on a dislocation.

Figure B.1 represents a dislocation or other source of internal stress D producing an elastic field u_i^D, σ_{ij}^D in a body bounded by the surface Σ_0. Constant surface loads on Σ_0 and sources of internal stress D' (in particular other dislocations) produce an elastic field u_i^A, e_{ij}^A, σ_{ij}^A. Σ is an arbitrary closed surface which isolates D from the surface and the sources of internal stress D'.

The elastic energy density is $\frac{1}{2}\sigma_{ij}e_{ij}$. We are only interested in that part of the elastic energy which depends on cross terms between A quantities and D quantities. It is

$$E_{\text{el}} = \tfrac{1}{2} \int (\sigma^A_{ij} e^D_{ij} + \sigma^D_{ij} e^A_{ij})\, dv \tag{B.1}$$

taken over the interior of Σ_0. In the sources of internal stress D and D', we cannot, in general, define a displacement, but u_i^D exists in the region II between Σ_0 and Σ, and u_i^A exists in the region I inside Σ. We have $\sigma^A_{ij}\, e^D_{ij} = \sigma^D_{ij}\, e^A_{ij}$, as may be seen if both sides are written out entirely in terms of stress (or of strain) with the help of Hooke's law. Consequently, Eq. (B.1) may be rewritten as

$$E_{\text{el}} = \tfrac{1}{2} \int_{\text{II}} \sigma^A_{ij} u^D_{i,j}\, dv + \tfrac{1}{2} \int_{\text{I}} \sigma^D_{ij} u^A_{i,j}\, dv \tag{B.2}$$

since, for example, $\sigma^A_{ij} e^D_{ij} = \tfrac{1}{2}\sigma^A_{ij}(u^D_{i,j} + u^D_{j,i}) = \sigma^A_{ij} u^D_{i,j}$ because of the symmetry of σ^A_{ij}. We may further put $\sigma^A_{ij} u^D_{ij} = (\sigma^A_{ij} u_i^D)_{,j}$ since $\sigma^A_{ij,j} = 0$ and, similarly, $\sigma^D_{ij} u^A_{i,j} = (\sigma^D_{ij} u_i^A)_{,j}$. Gauss's theorem then converts Eq. (B.2) into

$$E_{\text{el}} = \tfrac{1}{2} \int_{\Sigma_0} \sigma^A_{ij} u_i^D\, dS_j - \tfrac{1}{2} \int_{\Sigma} \sigma^A_{ij} u_i^D\, dS_j + \tfrac{1}{2} \int_{\Sigma} \sigma^D_{ij} u_i^A\, dS_j \tag{B.3}$$

where the minus sign in the second term is correct if, in $dS_j = n_j\, dS$, n_j refers to the outward normal to Σ. The potential energy of the loading mechanism, so far as it depends on cross terms between A and D quantities, may be taken to be

$$E_{\text{ext}} = - \int_{\Sigma_0} \sigma^A_{ij} u_i^D\, dS_j \tag{B.4}$$

because it gives the correct value

$$-\delta E_{\text{ext}} = \int_{\Sigma_0} \sigma^A_{ij}(u_i^D - u_i^{D''})\, dS_j$$

for the work done by the external forces when u_i^D changes to $u_i^{D''}$ as a result of a change in the configuration of D.

From Eqs. (B.3) and (B.4)

$$E_{\text{tot}} = E_{\text{el}} + E_{\text{ext}} = - \tfrac{1}{2} \int_{\Sigma_0} \sigma^A_{ij} u_i^D\, dS_j + \tfrac{1}{2} \int (\sigma^D_{ij} u_i^A - \sigma^A_{ij} u_i^D)\, dS_j$$

In the Σ_0 integral we may replace $\sigma^A_{ij} u_i^D$ by $\sigma^A_{ij} u_i^D - \sigma^D_{ij} u_i^A$ since $\sigma^D_{ij} n_j = 0$

on Σ_0. The divergence of the integrand now vanishes and so, by Gauss's theorem, the integral may be taken over Σ in place of Σ_0. Hence

$$E_{\text{tot}} = \int_{\Sigma} (\sigma_{ij}^D u_i^A - \sigma_{ij}^A u_i^D)\, dS_j \tag{B.5}$$

If D is a dislocation loop bounding a surface C we may take for Σ the positive and negative sides of C together with a narrow tube embracing the dislocation line. The latter contributes nothing in the limit as the tube shrinks onto the dislocation line. In Eq. (B.5), the first term is continuous across C and, in the second term, $\sigma_{ij}^A n_j$ is continuous, while u_i^D suffers a discontinuity b_i. Hence

$$E_{\text{tot}} = -b_i \int_C \sigma_{ij}^A\, dS_j$$

If an element of the dislocation line of length dl and direction s_i moves through a vector distance $d\xi_i$ the area C acquires an extra element

$$dS_j = \epsilon_{jmn}\, dl\, s_m\, d\xi_n$$

and E_{tot} increases by

$$\delta E_{\text{tot}} = -\epsilon_{jmn} b_i \sigma_{ij}^A s_m\, d\xi_n\, dl$$

and we may define the effective force F_n per unit length of dislocation by

$$F_n\, dl\, d\xi_n = -\delta E_{\text{tot}}$$

or

$$F_n = \epsilon_{jmn} b_i \sigma_{ij}^A s_m \tag{B.6}$$

For a pure edge dislocation parallel to the z axis with Burgers vector $(b_x, 0, 0)$ Eq. (B.6) gives

$$F_x = b_x \sigma_{xy}^A, \qquad F_y = -b_x \sigma_{xx}^A \tag{B.7}$$

and for a pure screw dislocation parallel to the z axis with Burgers vector $(0, 0, b_z)$

$$F_x = b_z \sigma_{zy}^A, \qquad F_y = -b_z \sigma_{zx}^A \tag{B.8}$$

Appendix C. Solution of Certain Integral Equations

The solution of

$$\int_{-a}^{a} \frac{\mathscr{D}(x')\, dx'}{x - x'} = -\frac{P(x)}{A}, \qquad -a \leqslant x \leqslant a \tag{C.1}$$

is given by Eq. (C.4) below.

There are several ways of deriving this important result. A "physical" proof is given in Sect. IV,A in connection with the theory of cracks in antiplane strain. More rigorous proofs may involve the theory of functions of a complex variable (Muskhelishvili, 1953) or the general theory of Hilbert transforms (Tricomi, 1957). In all of them, the expressions like $(a^2 - x^2)^{1/2}$ which figure in the solution do not appear naturally but have to be introduced more or less blatantly into the analysis. In the following real-variable proof, they enter through the use of angular variables.

With the change of variables

$$x = a \cos \xi, \qquad x' = a \cos \eta$$

Eq. (C.1) becomes

$$\int_0^\pi \frac{D(\eta) \sin \eta \, d\eta}{\cos \eta - \cos \xi} = p(\xi)$$

where

$$D(\eta) = \mathscr{D}(x'), \qquad p(\xi) = P(x)/A$$

To solve Eq. (C.1), we make use of the relations

$$\int_0^\pi \frac{\cos n\eta \, d\eta}{\cos \eta - \cos \xi} = \pi \frac{\sin n\xi}{\sin \xi}, \qquad n = 0, 1, 2,... \tag{C.2}$$

and

$$\int_0^\pi \frac{\sin n\eta \sin \eta \, d\eta}{\cos \eta - \cos \xi} = -\pi \cos n\eta, \qquad n = 1, 2,... \tag{C.3}$$

Write

$$D(\eta) \sin \eta = \sum_{n=0}^{\infty} a_n \cos n\eta$$

substitute in Eq. (C.1) and use Eq. (C.2) to get

$$p(\xi) = \sum_{n=1}^{\infty} \pi a_n \frac{\sin n\xi}{\sin \xi}$$

Multiply Eq. (C.3) by πa_n and sum over n to get

$$\int_0^\pi \sum_{n=1}^{\infty} a_n \frac{\sin n\eta}{\sin \eta} \frac{\sin^2 \eta \, d\eta}{\cos \eta - \cos \xi} = -\pi^2 \sum_{n=1}^{\infty} a_n \cos n\xi$$

or

$$\int_0^\pi \frac{p(\eta) \sin^2 \eta \, d\eta}{\cos \eta - \cos \xi} = -\pi^2 [D(\xi) \sin \xi - a_0]$$

which, in the original variables, reads

$$\mathscr{D}(x) = \frac{1}{\pi^2 A} \int_{-a}^{a} \left(\frac{a^2 - x'^2}{a^2 - x^2} \right)^{1/2} \frac{P(x')}{x - x'}\, dx' + \frac{C}{\pi(a^2 - x^2)^{1/2}} \tag{C.4}$$

where $C = \pi a a_0$ is undetermined. The reason for this is that the term in C is a solution of the homogeneous equation, that is, of Eq. (C.1) with $P(x) = 0$. Consequently, if $\mathscr{D}(x)$ satisfies Eq. (C.1), so does $\mathscr{D}(x) +$ constant $(a^2 - x^2)^{-1/2}$. If we multiply Eq. (C.4) by dx and integrate from $-a$ to a, the first term on the right gives zero, and we are left with

$$C = \int_{-a}^{a} \mathscr{D}(x)\, dx \tag{C.5}$$

Equation (C.4) can also be derived from the results of Sect. IV,A. If we ask what is the relative displacement of the faces of an antiplane crack subjected to an applied stress $\sigma_{zy}^A(x)$, the answer is given by Eq. (72). On the other hand, if we ask the converse question, what Δw across the crack will annul a stress $\sigma_{zy}^A(x)$ between $x = -a$ and $x = +a$, the answer is the expression Eq. (72) plus any multiple of the Δw of Eq. (75), since the latter gives $\sigma_{zy} = 0$ between $x = -a$ and $x = +a$. Only if we know that there is no net dislocation content between $-a$ and a can this term be omitted. When this solution is, if necessary, generalized to the case of plane strain, as explained at the end of Sect. IV,A and interpreted in terms of a distribution of dislocations (Sect. IV,B) we obtain Eq. (C.4).

In some applications (Sect. VI) it is necessary to consider a generalization of Eq. (C.1) in which the integration extends not only over the interval $-a \leqslant x \leqslant a$ but also over the intervals

$$-a + 2ml \leqslant x \leqslant a + 2ml, \qquad m = \pm 1, \pm 2, \ldots, \pm\infty$$

on the assumption that $\mathscr{D}(x)$, $P(x)$ are periodic with period $2l$, so that

$$\sum_{m=-\infty}^{\infty} \int_{-a}^{a} \frac{\mathscr{D}(x')\, dx'}{x - x' - 2ml} = \frac{\pi}{2l} \int_{-a}^{a} \cot \frac{\pi}{2l}(x - x')\, \mathscr{D}(x')\, dx' = -\frac{P(x)}{A} \tag{C.6}$$

With the help of the identity

$$\cot(\alpha - \beta) = \frac{1}{2} \frac{\cos\alpha + \cos\beta}{\sin\alpha - \sin\beta} + \frac{1}{2} \frac{\cos\alpha - \cos\beta}{\sin\alpha + \sin\beta}$$

13

and a change of variable from x' to $-x'$ in the second term, Eq. (C.6) may be written

$$\frac{\pi}{2l}\int_{-a}^{a}\frac{\mathscr{D}_s(x')\cos(\pi x/2l)+\mathscr{D}_a(x')\cos(\pi x'/2l)}{\sin(\pi x/2l)-\sin(\pi x'/2l)}\,dx' = -\frac{P(x)}{A} \tag{C.7}$$

where

$$\mathscr{D}_s(x') = \tfrac{1}{2}[\mathscr{D}(x')+\mathscr{D}(-x')], \qquad \mathscr{D}_a(x') = \tfrac{1}{2}[\mathscr{D}(x')-\mathscr{D}(-x')]$$

are the even and odd parts of $\mathscr{D}(x')$. It is easy to verify that the part of Eq. (C.7) which depends on $\mathscr{D}_s(\mathscr{D}_a)$ is odd (even) in x, so that, if we write

$$P_s = \tfrac{1}{2}[P(x)+P(-x)] \qquad \text{and} \qquad P_a = \tfrac{1}{2}[P(x)-P(-x)]$$

Eq. (C.6) may be split into

$$\frac{\pi}{2l}\int_{-a}^{a}\frac{\mathscr{D}_s(x')}{\sin(\pi x/2l)-\sin(\pi x'/2l)}\,dx' = -\frac{P_a(x)}{A\cos(\pi x/2l)} \tag{C.8}$$

and

$$\frac{\pi}{2l}\int_{-a}^{a}\frac{\mathscr{D}_a(x')\cos(\pi x'/2l)}{\sin(\pi x/2l)-\sin(\pi x'/2l)}\,dx' = -\frac{P_s(x)}{A} \tag{C.9}$$

The substitutions

$$X = \sin\frac{\pi x}{2l}, \qquad X' = X(x'), \qquad \mathscr{A} = X(a$$

convert Eq. (C.9) into

$$\int_{-\mathscr{A}}^{\mathscr{A}}\frac{\mathscr{D}_a[x'(X')]}{X-X'}\,dX' = -\frac{P_s[x(X)]}{A}$$

which is of the form Eq. (C.1), so that, from Eq. (C.4), we have, with the original variables,

$$\mathscr{D}_a(x) = \frac{1}{2\pi Al}\int_{-a}^{a}\left(\frac{\sin^2\dfrac{\pi a}{2l}-\sin^2\dfrac{\pi x'}{2l}}{\sin^2\dfrac{\pi a}{2l}-\sin^2\dfrac{\pi x}{2l}}\right)^{1/2}\frac{P_s(x')\cos\dfrac{\pi x'}{2l}\,dx'}{\sin\dfrac{\pi x}{2l}-\sin\dfrac{\pi x'}{2l}} \tag{C.10}$$

With the same change of variables, Eq. (C.9) takes the form

$$\int_{-\mathscr{A}}^{\mathscr{A}}\frac{\mathscr{D}_s^*\,dX'}{X-X'} = -\frac{P_a^*}{A}$$

where

$$\mathscr{D}_s^* = \mathscr{D}_s/(1-X'^2)^{1/2}, \qquad P_a^* = P_a/(1-X^2)^{1/2}$$

so that, by Eqs. (C.1) and (C.4)

$$\mathscr{D}_s(x) = \frac{1}{2\pi Al}\int_{-a}^{a}\left(\frac{\sin^2\frac{\pi a}{2l} - \sin^2\frac{\pi x'}{2l}}{\sin^2\frac{\pi a}{2l} - \sin^2\frac{\pi x}{2l}}\right)^{1/2} \frac{P_a(x')\cos\frac{\pi x}{2l}\,dx'}{\sin\frac{\pi x}{2l} - \sin\frac{\pi x'}{2l}} + \frac{\frac{C}{2l}\cos\frac{\pi x}{2l}}{\left(\sin^2\frac{\pi a}{2l} - \sin^2\frac{\pi x}{2l}\right)^{1/2}} \tag{C.11}$$

We have added the complementary solution of the homogeneous equation to $\mathscr{D}_s$ but not to $\mathscr{D}_a$ in order to preserve the stated symmetries; the constant C is chosen so that Eq. (C.5) continues to hold.

Note that Eqs. (C.10) and (C.11) differ only through the appearance of x' or x in the argument of the cosine. With a little manipulation, they may be combined to give

$$\mathscr{D}(x) = \frac{1}{2\pi Al}\int_{-a}^{a}\left(\frac{\sin^2\frac{\pi a}{2l} - \sin^2\frac{\pi x'}{2l}}{\sin^2\frac{\pi a}{2l} - \sin^2\frac{\pi x}{2l}}\right)^{1/2} \frac{P(x')\,dx'}{\sin\frac{\pi}{2l}(x - x')} + \frac{\frac{C}{2l}\cos\frac{\pi x}{2l}}{\left(\sin^2\frac{\pi a}{2l} - \sin^2\frac{\pi x}{2l}\right)^{1/2}} \tag{C.12}$$

Symbols

A	$Gb/2\pi(1 - \nu)$ for edge dislocation; $Gb/2\pi$ for screw dislocation	$H_n(x)$	Hermite polynomial
$\mathbf{b}$, b_i	Burgers vector	I_i	integral associated with crack stress
$\mathscr{D}(x)$	density of infinitesimal dislocations	I_{ij}	integrals associated with crack stress
E	Young's modulus	J_i	integral associated with crack stress
e_{ij}	components of (small) pure strain	$J_n(x)$	Bessel function
f	antiplane strain complex stress function	k	constant in antiplane strain yield criterion
G	shear modulus	K	bulk modulus
G_{ij}	stress due to "cohesive forces"	$K_1^0 = K_{\mathrm{I}}$, $K_2^0 = K_{\mathrm{II}}$, $K_3^0 = K_{\mathrm{III}}$	stress-intensity factors due to all loads other than cohesive forces
$\mathscr{G}_{\mathrm{I}}$, $\mathscr{G}_{\mathrm{II}}$, $\mathscr{G}_{\mathrm{III}}$	crack extension forces		

K_i^c	stress-intensity factors due to cohesive forces	ϵ_{ijk}	permutation tensor
$L(\theta)$	Lobachevskiĭ function	λ	Lamé's constant
$L_n(x)$	Laguerre polynomial	ν	Poisson's ratio
M_i	moduli of cohesion	ϖ_{ij}	components of rotation
$P_n(x)$	Legendre polynomial	σ_1	friction stress
$Q[\phi]$	cohesive stress as function of relative displacement	σ_{ij}	stress tensor
u_i, u, v, w	displacement	σ_{ij}^A	externally applied stress
Y	yield stress	σ_{ij}^C	crack stress
α_{ij}	dislocation tensor	σ_{ij}^D	dislocation stress
β_{ij}	distortion	σ_{ij}^I	stress applied to crack surface
γ	surface energy	Φ, Ψ	conjugate harmonic functions used in plane strain
δ_{ij}	Kronecker delta	φ, ψ	real and imaginary parts of f
		$\phi(x)$	relative displacement

References

Allen, N. P. (1961). Special Report No. 69, p. 419. Iron and Steel Institute.

Amelinckx, S. (1957). *Physica* **23**, 663.

Atkinson, C. (1966). Unpublished work.

Atkinson, C. (1967). *Intern. J. Fracture Mech.* **2**, 567.

Atkinson, C., and Heald, P. T. (1967). *Acta Met.* **15**, 1617.

Barenblatt, G. I. (1959). *Prikl. Mat. Meh.* **23**, 434, 706, 893.

Barenblatt, G. I. (1960). *Prikl. Mat. Meh.* **24**, 316.

Barenblatt, G. I. (1961). "Problems of Continuum Mechanics," p. 21. Soc. Ind. Appl. Math., Philadelphia.

Barenblatt, G. I. (1962). *Advan. Appl. Mech.* **2**, 55.

Barenblatt, G. I. (1964). *Prikl. Mat. Meh.* **28**, 630–643.

Barenblatt, G. I., and Cherepanov, G. P. (1961a). *Prikl. Mat. Meh.* **25**, 752.

Barenblatt, G. I., and Cherepanov, G. P. (1961b). *Prikl. Mat. Meh.* **25**, 46.

Bilby, B. A. (1950). *J. Inst. Metals* **76**, 613.

Bilby, B. A. (1952). *Research (London)* **4**, 387.

Bilby, B. A. (1955). *In* "Report of the Conference on Defects in Crystalline Solids," p. 123. Physical Society, London.

Bilby, B. A. (1960). *In* "Progress in Solid Mechanics" (I.N. Sneddon and R. Hill, eds.), Vol. 1, p. 331. North-Holland, Amsterdam.

Bilby, B. A. (1967). *In* "Modern Theory in the Design of Alloys," p. 3. Iliffe, London.

Bilby, B. A., and Bullough, R. (1954). *Phil. Mag.* **45**, 631.

Bilby, B. A., and Entwisle, A. R. (1954). *Acta Met.* **2**, 15.

Bilby, B. A., and Hewitt, J. (1962). *Acta Met.* **10**, 587.

Bilby, B. A., and Swinden, K. H. (1965). *Proc. Roy. Soc. (London) Ser. A* **285**, 22.

Bilby, B. A., Bullough, R., and Smith, E. (1955). *Proc. Roy. Soc. (London) Ser. A* **231**, 263.

Bilby, B. A., Gardner, L. R. T., and Smith, E. (1958). *Acta Met.* **6**, 29.

Bilby, B. A., Cottrell, A. H., and Swinden, K. H. (1963). *Proc. Roy. Soc. (London) Ser. A* **272**, 304.

Bilby, B. A., Bullough, R., and Grinberg, D. K. (1964). *Discussions Faraday Soc.* **38**, 61.

Bilby, B. A., Cottrell, A. H., Smith, E., and Swinden, K. H. (1964). *Proc. Roy. Soc. (London) Ser. A* **279**, 1.

Blackburn, W. S. (1966a). I.R.D. Research Report No. 66-80. International Research and Development Co., Ltd., Newcastle-upon-Tyne, England.
Blackburn, W. S. (1966b). I.R.D. Research Report No. 66-65. International Research and Development Co., Ltd., Newcastle-upon-Tyne, England.
Bullough, R. (1964). *Phil. Mag.* **9**, 911.
Bullough, R., and Bilby, B. A. (1956). *Proc. Phys. Soc. (London) Ser. B* **69**, 1276.
Burdekin, F. M., Stone, D. E. W., and Wells, A. A. (1965). *In* "Symposium on Fracture Toughness Testing and Its Applications," STP 381, pp. 400–405.
Burgers, J. M. (1939). *Proc. Acad. Sci. Amsterdam* **42**, 293.
Collins, W. D. (1962). *Proc. Roy. Soc. (London) Ser. A* **226**, 359.
Cottrell, A. H. (1957). *In* "Brittleness in Metals Conference," Conf. Brit./P.P.1. Culcheth Laboratories, UKA.E.A.
Cottrell, A. H. (1958). *Trans. AIME*, April, 192.
Cottrell, A. H. (1961). *In* "Symposium on Steels for Reactor Pressure Circuits, 1960," Special Report No. 69, pp. 281–296. Iron and Steel Institute.
Cottrell, A. H. (1962). *In* "Properties of Reactor Materials and the Effects of Radiation Damage" (D. J. Littler, ed.), pp. 5–14. Butterworths, London.
Cottrell, A. H. (1963a). *Proc. Roy. Soc. (London) Ser. A* **276**, 1.
Cottrell, A. H. (1964). *Proc. Roy. Soc. (London) Ser. A* **282**, 2–9.
Cottrell, A. H. (1965a). *In* "Fracture" (C. J. Osborn, ed.), p. 1. Butterworths, London.
Cottrell, A. H. (1965b). *Proc. Roy. Soc. (London) Ser. A* **285**, 10–21.
Dovnorovich, V. I. (1962). *Prikl. Mat. Meh.* **26**, 342.
Dugdale, D. S. (1960). *J. Mech. Phys. Solids* **8**, 100.
Elliot, H. A. (1947). *Proc. Phys. Soc. (London) Ser. A* **59**, 208.
England, A. H. (1963). *Mathematika* **11**, 107.
England, A. H., and Green, A. E. (1963). *Proc. Cambridge Phil. Soc.* **59**, 489.
Eshelby, J. D. (1949). *Phil. Mag.* **40**, 903.
Eshelby, J. D. (1951). *Phil. Trans. Roy. Soc. London, Ser. A* **244**, 87.
Eshelby, J. D. (1956). *Solid State Phys.* **3**, 79.
Eshelby, J. D. (1957). *Proc. Roy. Soc. (London) Ser. A* **241**, 376.
Eshelby, J. D. (1959). *Proc. Roy. Soc. (London) Ser. A* **252**, 561.
Eshelby, J. D. (1960). *In* "Progress in Solid Mechanics" (I.N. Sneddon and R. Hill, eds.), Vol. 2, p. 89. North-Holland, Amsterdam.
Eshelby, J. D. (1963). *Phys. Status Solidi* **3**, 2057.
Eshelby, J. D., Frank, F. C., and Nabarro, F. R. N. (1951). *Phil. Mag.* **42**, 351.
Eshelby, J. D., Read, W. T., and Shockley, W. (1953). *Acta Met.* **1**, 251.
Frank, F. C. (1950). *In* "Symposium on Plastic Deformation of Crystalline Solids," NAVEXOS-P-834, p. 150. U.S. Printing Office, Washington, D.C.
Frank, F. C. (1951). *Phil. Mag.* **42**, 809.
Frank, F. C., and Stroh, A. N. (1952). *Proc. Phys. Soc. (London) Ser. B* **65**, 811–821.
Friedel, J. (1956). "Les Dislocations," p. 214. Gauthier-Villars, Paris.
Friedel, J. (1959). *In* "Fracture" (B. L. Averbach, D. K. Felbeck, G. T. Hahn, and D. A. Thomas, eds.), pp. 498–523. Wiley, New York.
Friedel, J. (1964). "Dislocations," p. 320. Pergamon, London.
Goodier, J. N., and Kanninen, M. F. (1966). "Crack Propagation in a Continuum Model With Nonlinear Atomic Separation Laws," Technical Report No. 165. Stanford University, Stanford, California.
Hahn, G. T., and Rosenfield, A. R. (1965). *Acta Met.* **13**, 293.
Hahn, G. T., and Rosenfield, A. R. (1966). *Acta Met.* **14**, 1815.

Hahn, G. T., and Rosenfield, A. R. (1967). "Sources of Fracture Toughness," Battelle Report. Presented at ASTM Symposium, Los Angeles. ASTM, Philadelphia.
Head, A. K., and Louat, N. (1955). *Australian J. Phys.* **8**, 1.
Head, A. K., and Thomson, P. F. (1962). *Phil. Mag.* **7**, 439.
Hult, J. A. H., and McClintock, F. A. (1957). *In* "Proceedings of the 9th International Congress on Applied Mechanics," Vol. 8, p. 51.
Ichikawa, M., and Yokobori, T. (1965). *Rept. Res. Inst.* **1** (2), 59. Research Institute for Strength and Fracture of Materials, Tohoku University, Sendai, Japan.
Ichikawa, M., Ohashi, M., and Yokobori, T. (1965). *Rept. Res. Inst.* **1** (1), 1. Research Institute for Strength and Fracture of Materials, Tohoku University, Sendai, Japan.
Irwin, G. R. (1948). *In* "Fracturing of Metals," pp. 147–166. ASM, Cleveland.
Irwin, G. R. (1957). *J. Appl. Mech.* **24**, 361.
Jaswon, M. A., and Foreman, A. J. E. (1952). *Phil. Mag.* **43**, 201.
Jeffreys, H., and Jeffreys, B. S. (1962). "Methods of Mathematical Physics," p. 522. Cambridge, London.
Keer, L. M., and Mura, T. (1965). *In* "Proceedings of the 1st International Conference on Fracture, Sendai" (T. Yokobori, T. Kawasaki, and J. L. Swedlow, eds.), Vol. I, p. 99. Japanese Society for Strength and Fracture of Materials, Tokyo.
Kondo, K. (ed.) (1955). "Memoirs of the Unifying Study of Basic Problems in Engineering Sciences by Means of Geometry," Vol. I, pp. 361–373, 458–469. Gakajutsu Bunken Fukyu-kai, Tokyo.
Kondo, K. (ed.) (1958). "Memoirs of the Unifying Study of Basic Problems in Engineering Sciences by Means of Geometry," Vol. II, pp. 202–226. Gakajutsu Bunken Fukyu-kai, Tokyo.
Kondo, K. (ed.) (1962). "Memoirs of the Unifying Study of Basic Problems in Engineering Sciences by Means of Geometry," Vol. III, pp. 82–89. Gakajutsu Bunken Fukyu-kai, Tokyo.
Kröner, E. (1955). *Z. Physik* **142**, 463.
Kröner, E. (1958). "Kontinuumstheorie der Versetzungen und Eigenspannungen." Springer-Verlag, Berlin.
Leibfried, G. (1951). *Z. Physik* **130**, 244.
Leibfried, G. (1954). *Z. Angew. Phys.* **6**, 251.
Leonov, M. Ya., and Onyshko, L. V. (1961). *Vopr. Mashinoved i Proch. v Mashinostr. Akad. Nauk Ukr. SSR.* **7**, 16–25.
Leonov, M. Ya., and Panasyuk, V. V. (1959). *Prikl. Meh.* **5**, 391–401.
Louat, N. (1965). *In* "Proceedings of the 1st International Conference on Fracture, Sendai" (T. Yokobori, T. Kawasaki, and J. L. Swedlow, eds.), Vol. 1, p. 117. Japanese Society for Strength and Fracture of Materials, Tokyo.
Love, A. E. H. (1952). "Theory of Elasticity," p. 204. Cambridge, England.
Lubahn, J. D., and Yukawa, S. (1958). *Proc. ASTM* **58**, 661.
McClintock, F. A., and Irwin, G. R. (1965). *In* "Symposium on Fracture Toughness Testing and Its Applications," STP 381, pp. 84–113. ASTM, Philadelphia.
Markuzon, I. A. (1961). *Prikl. Mat. Meh.* **25**, 356.
Mott, N. F. (1948). *Engineering* **165**, 16.
Mott, N. F. (1953). *Proc. Roy. Soc. (London) Ser. A* **220**, 1.
Mott, N. F. (1956). *J. Iron Steel Inst. (London)* **183**, 233.
Murrell, S. A. F. (1964a). *Brit. J. Appl. Phys.* **15**, 1211.
Murrell, S. A. F. (1964b). *Brit. J. Appl. Phys.* **15**, 1195.
Muskhelishvili, N. I. (1953). "Singular Integral Equations" (J. M. Radok, Trans.). N. V. P. Noordhoff, Groningen, Holland.

Nabarro, F. R. N. (1947). *Proc. Phys. Soc. (London)* **59**, 256.
Nabarro, F. R. N. (1951). *Advan. Phys.* **1**, 269.
Neményi, P. (1931). *Z. Angew. Math. Mech.* **11**, 59.
Nye, J. F. (1953). *Acta Met.* **1**, 153.
Ohashi, M., Ichikawa, M., and Yokobori, T. (1965a). *Rept. Res. Inst.* **1** (2), 79. Research Institute for Strength and Fracture of Materials, Tohoku University, Sendai, Japan.
Ohashi, M., Ichikawa, M., and Yokobori, T. (1965b). *Rept. Res. Inst.* **1** (2), 41. Research Institute for Strength and Fracture of Materials, Tohoku University, Sendai, Japan.
Orowan, E. (1949). *Rept. Progr. Phys.* **12**, 214.
Orowan, E. (1950). *In* "Fatigue and Fracture of Metals" (W. M. Murray, ed.), pp. 139–157. Wiley, New York.
Panasyuk, V. V. (1960). *Nauchn. Zap. Inst. Mashinoved. i Avtomat.* **7**, 114–127.
Paris, P. C., and Sih, G. (1965). *In* "Symposium on Fracture Toughness Testing and Its Applications," STP 381, pp. 30–83. ASTM, Philadelphia.
Peierls, R. E. (1940). *Proc. Phys. Soc. (London)* **52**, 23.
Petch, N. J. (1957). *In* "Brittleness in Metals Conference," Conf. Brit./P.P.2. Culcheth Laboratories, UKA.E.A.
Petch, N. J. (1959). "Fracture" (B. L. Averbach, D. K. Felbeck, G. T. Hahn, and D. A. Thomas, eds.), p. 54. Wiley, New York.
Priestner, R., and Louat, N. (1963). *Acta Met.* **11**, 195.
Sih, G., Paris, P. C., and Irwin, G. R. (1965). *Intern. J. Fracture Mech.* **1**, 189–203.
Smith, E. (1964). *Intern. J. Eng. Sci.* **2**, 379–387.
Smith, E. (1965a). *In* "Proceedings of the 1st International Conference on Fracture, Sendai" (T. Yokobori, T. Kawasaki, and J. L. Swedlow, eds.), Vol. I, p. 133. Japanese Society for Strength and Fracture of Materials, Tokyo.
Smith, E. (1965b). *Proc. Roy. Soc. (London) Ser. A* **285**, 46–57.
Smith, E. (1965c). *Intern. J. Fracture Mech.* **1**, 204.
Smith, E. (1966a). *Acta Met.* **14**, 985–989.
Smith, E. (1966b). *Acta Met.* **14**, 991–996.
Smith, E. (1966c). *In* "Physical Basis of Yield and Fracture" (Conference Series No. 1), pp. 36–46. Institute of Physics & Physical Society, London.
Smith, E. (1966d). *Proc. Roy. Soc. (London) Ser. A* **292**, 134–151.
Smith, E. (1966e). *Intern. J. Eng. Sci.* **4**, 41.
Somigliana, C. (1914). *Rend. Circ. Accad. Lincei* **23** (1), 463.
Somigliana, C. (1915). *Rend. Circ. Accad. Lincei* **24** (1), 655.
Stieltjes, T. J. (1885). *Acta Math.* **6**, 321.
Stroh, A. N. (1954). *Proc. Roy. Soc. (London) Ser. A* **223**, 404.
Stroh, A. N. (1955a). *Proc. Roy. Soc. (London) Ser. A* **232**, 548.
Stroh, A. N. (1955b). *Phil. Mag.* **46**, 198.
Stroh, A. N. (1956). *Phil. Mag.* **1**, 489.
Stroh, A. N. (1957a). *Phil. Mag.* **2**, 1.
Stroh, A. N. (1957b). *Advan. Phys.* **6**, 418.
Stroh, A. N. (1958a). *Phil. Mag.* **3**, 625.
Stroh, A. N. (1958b). *Phil. Mag.* **3**, 597.
Swinden, K. H. (1964). Thesis. University of Sheffield.
Szegö, G. (1939). "Orthogonal Polynomials." (Colloq. Publ. Vol. 33), p. 139. American Mathematical Society, Providence, Rhode Island.
Tetelman, A. S. (1964). *Acta Met.* **12**, 993.

Tricomi, F. G. (1957). "Integral Equations." Wiley (Interscience), New York.
Vitvitskiĭ, P. M., and Leonov, M. Ya. (1960). *Vses. Inst. Nauchn.-Tekhn. Inform. Akad. Nauk SSSR* Pt. 1, p. 14.
Volterra, V. (1907). *Ann. Ecole Norm. Super.* **24**, 400.
Weertman, J. (1964). *Bull. Seismol. Soc. Amer.* **54**, 1035.
Weertman, J. (1966). *Intern. J. Fracture Mech.* **2**, 460.
Wells, A. A. (1963). *Brit. Welding J.*, November, 563–570.
Wells, A. A. (1965). *Proc. Roy. Soc. (London) Ser. A* **285**, 34–45.
Whittaker, E. T., and Watson, G. N. (1927). "Modern Analysis," p. 562. Cambridge, England.
Willis, J. R. (1967). *J. Mech. Phys. Solids* **15**, 151–162.
Wundt, B. M. (1959). ASME Paper No. 59-Met-9. ASME, New York.
Yokobori, T., and Ichikawa, M. (1965). *Rept. Res. Inst.* **1** (2), 47. Research Institute for Strength and Fracture of Materials, Tohoku University, Sendai, Japan.
Yokobori, T., and Ichikawa, M. (1966). *Rept. Res. Inst.* **2** (1), 21. Research Institute for Strength and Fracture of Materials, Tohoku University, Sendai, Japan.
Yokobori, T., Ohashi, M., and Ichikawa, M. (1965a). *Rept. Res. Inst.* **1** (2), 33. Research Institute for Strength and Fracture of Materials, Tohoku University, Sendai, Japan.
Yokobori, T., Ichikawa, M., and Ohashi, M. (1965b). *Rept. Res. Inst.* **1** (2), 69. Research Institute for Strength and Fracture of Materials, Tohoku University, Sendai, Japan.
Zener, C. (1948). *In* "Fracturing of Metals," p. 3. ASM, Cleveland.
Zheltov, Yu. P., and Khristianovich, S. A. (1955). *Izv. Akad. Nauk SSSR, OTN* **5**, 3–41.

J. Mech. Phys. Solids, 1969, Vol. 17, pp. 177 to 199. Pergamon Press. Printed in Great Britain.

THE ELASTIC FIELD OF A CRACK EXTENDING NON-UNIFORMLY UNDER GENERAL ANTI-PLANE LOADING

By J. D. Eshelby
Department of the Theory of Materials, University of Sheffield

(*Received* 12*th November* 1968)

Summary

With the help of a theorem of Bateman's an expression is found for the dynamic elastic field of a crack when one of its tips moves arbitrarily in the plane of the crack, starting from rest. The linear isotropic theory of elasticity is used, and only states of anti-plane strain (mode III deformation) are considered. The crack is initially of finite length and subject to any static anti-plane loading. The solution obtained becomes inaccurate in regions into which disturbances reflected at the other tip have penetrated. The error is estimated for some special cases. The results are used to discuss the equation of motion of a crack tip.

1. Introduction

Several solutions of the equations of linear elasticity have been obtained relating to plane cracks whose tips move with uniform velocity. However, if we wish to know how a crack extends in given circumstances we must first find the elastic field when the tip moves in an arbitrary manner, and then use some physical criterion to pick out the actual motion. In this paper we consider a plane two-dimensional crack, perturbing any prescribed static state of anti-plane strain in a homogeneous isotropic medium, and find the elastic field when one of its two tips begins to move arbitrarily in the plane of the crack.

In a state of anti-plane strain (also called mode III deformation or longitudinal shear) disturbances spread with a single velocity, that of shear waves, and this greatly simplifies the analysis. Although the results we shall obtain for an arbitrarily moving tip represent some gain in generality (Kostrov's (1964) treatment of this problem relates to a different situation; see Section 5), they are subject to an important limitation, illustrated in Fig. 1. If the right-hand tip starts to move at time $t = 0$ with a velocity always less than c, the velocity of shear waves, the static field will remain unaltered outside an expanding circle S of radius ct. To the left of the moving tip the elastic field inside S, which we shall later obtain in Section 4, gives zero stress on the x-axis and a discontinuity in displacement across it. If we extrapolate this solution to $t > l/c$ (where l is the crack length), so that S has expanded to S', the condition that the displacement be continuous across the x-axis to the left of $x = -l$ is violated. It is easy to see that to correct this we must add another solution which is zero outside a second circle S'' which expands with velocity c about $(-l, 0)$, starting when S reaches this point. We shall say that S has been partially reflected at the left-hand tip, though 'diffracted' or 'scattered' would perhaps be more appropriate. When S'' reaches the right-hand tip there is a further reflection, and so on. We shall not try to calculate any of the reflected waves, so that our solution is only valid everywhere for $t < l/c$, but near the moving tip it remains correct up to the time, somewhat exceeding $2l/c$, at which S'' overtakes the tip. For many purposes the elastic field near the tip is all we need to know. We shall also ignore reflections from the surface of the specimen.

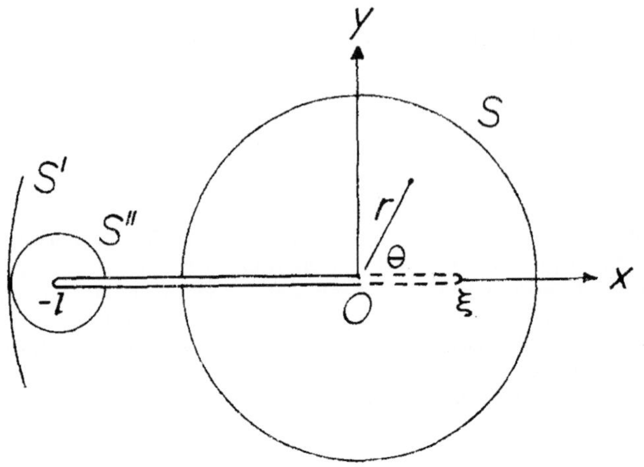

FIG. 1. To illustrate the disturbance produced by motion of the right-hand tip of a crack. The disturbance is zero outside the expanding circle S. When S has reached S' the original solution is modified by reflections at the left-hand tip, but only inside the circle S''.

If both tips start to move, say at the same time, our results for the field at one tip will remain valid up to the time, slightly exceeding l/c rather than $2l/c$, at which disturbances produced (rather than reflected) at the other tip first reach it. Initially the field near one moving tip is not directly affected by the presence of the other. This does not mean that our results are independent of the initial length of the crack, because for a prescribed system of external loading the original static field surrounding the crack depends on its initial length, and the dynamic field depends the static field.

We have been assuming that the crack is extending. If it is contracting, reflections or disturbances from one end will reach the other after a time which falls short of $2l/c$ or l/c. For completeness we shall consider receding as well as advancing tips, even though this requires some distinction of cases in the analysis. In partial justification we recall that cracks in mica can be made to close as well as open.

Section 2 summarizes the theory of static anti-plane cracks. For a semi-infinite crack with a particular type of loading the moving-tip problem can be solved (Section 3) with the help of an elegant result of BATEMAN's (1915). Though his result cannot be applied directly to the case of a finite crack arbitrarily loaded, it nevertheless suggests (Section 4) a successful method of solution. In Section 5 we compare our results with other work and draw some physical conclusions.

2. The Stationary Anti-plane Crack

In a state of anti-plane strain the displacement w is everywhere perpendicular to the xy-plane and satisfies the two dimensional wave equation

$$\frac{\partial^2 w}{\partial x^2} + \frac{\partial^2 w}{\partial y^2} - \frac{1}{c^2}\frac{\partial^2 w}{\partial t^2} = 0, \quad c = (\mu/\rho)^{\frac{1}{2}}, \tag{2.1}$$

where μ is the shear modulus and ρ the density of the medium. The non-vanishing stresses are

$$p_{zx} = \mu \frac{\partial w}{\partial x}, \quad p_{zy} = \mu \frac{\partial w}{\partial y}.$$

If the elastic field is static, w is harmonic, and we may write

$$w = \operatorname{Im} f(z), \tag{2.2}$$

where $f(z)$ is an analytic function of $z = x + iy$. The simplest crack-like solution is obtained by putting $f(z) = Bz^{\frac{1}{2}}$, where B is a real constant. Then

$$w = B \operatorname{Im}(x + iy)^{\frac{1}{2}} = BW_0(x, y), \tag{2.3}$$

where

$$\left.\begin{aligned} W_0(x, y) &= \{\tfrac{1}{2}(x^2 + r^2)^{\frac{1}{2}} - \tfrac{1}{2}x\}^{\frac{1}{2}} \operatorname{sgn} y \\ &= r^{\frac{1}{2}} \sin \tfrac{1}{2}\theta \end{aligned}\right\} \tag{2.4}$$

with the polar co-ordinates shown in Fig. 1.

The displacement (2.4) could be maintained in a circular cylinder slit along the radius $\theta = \pm \pi$ by imposing either a displacement or traction proportional to $\sin \frac{1}{2}\theta$ round its circumference. It will play an important part in what follows. We shall call it *the elementary solution*, and refer to the semi-infinite crack, loaded in the way indicated, as *the elementary crack*.

We now turn to the general case of a finite crack extending, say, from $-l$ to 0 along the x-axis. Throughout we shall assume that the displacement is odd in y. This is not really a restriction. The displacement corresponding to any applied anti-plane stress field can be split into two parts respectively odd and even in y. The former part remains odd when perturbed by the crack, and the latter, since it produces no stress across the x-axis, is unaffected by the presence of the crack.

To make clear what is meant by a crack perturbing a given applied stress it is convenient to imagine that we start with a specimen in which a crack has not yet been made. Let an ' applied ' field w^A, $p_{ij}{}^A$ be induced in it by tractions or displacements imposed on the outer boundary and also, possibly, by sources of internal stress which in anti-plane strain may be taken to be a collection of screw dislocations or a continuous distribution of them. Make a physical cut along the x-axis from $-l$ to 0 and apply a traction $-p_{zy}{}^A(x, 0)$ along the upper surface of the cut and a traction $+p_{zy}{}^A(x, 0)$ along the lower surface*. The traction across the cut is the same as it was across the segment $-l < x < 0$ in the absence of the crack. Consequently the crack does not open, and the elastic field is still w^A, $p_{ij}{}^A$. (Though the word is more appropriate to a crack under tension we shall throughout speak

*Here, and elsewhere, when tractions are applied to the faces of the crack to produce or to annul a prescribed p_{zy} on $y = 0$, the choice of signs is governed by the fact that the outward normals to the upper and lower faces are respectively $(0, -1)$ and $(0, 1)$.

of the crack as *opening* when its faces suffer a relative displacement parallel to the z-axis.)

If we now remove the tractions from the faces of the crack the elastic field becomes

$$w = w^A + w^C, \quad p_{ij} = p_{ij}^A + p_{ij}^C, \tag{2.5}$$

where w^C, p_{ij}^C represent the perturbation due to the presence of the crack. Evidently w^C is the displacement produced by applying tractions $+ p_{zy}^A$ to the upper face of the crack and tractions $- p_{ij}^A$ to the lower face, since these annul the tractions we imposed to prevent it opening. In a finite specimen the actual value of w^C depends not only on p_{zy}^A at the crack, but on the boundary conditions imposed at the outer surface. For example, with the same p_{zy}^A the displacement w^C will differ according as the displacement or the surface traction at the outer surface is held constant while the crack is being created.

In this paper we shall assume that the static problem of finding the perturbation due to the crack has been solved, and that we have to find the field of the moving crack in terms of the total field (2.5). The following is a summary of some properties of the field associated with a static anti-plane crack. (For a simple treatment, see Bilby and Eshelby 1968.) For a crack extending from $x = 0$ to $x = -l$ the displacement may be written in the form

$$w = \operatorname{Im} f(z) = B \operatorname{Im} z^{\frac{1}{2}} g(z). \tag{2.6}$$

By symmetry the function $g(z)$ must behave like $(z + l)^{\frac{1}{2}}$ near the end $(-l, 0)$. About the end $(0, 0)$ it may be expanded in the form

$$g(z) = 1 + \sum_{n=1}^{\infty} a_n z^n, \tag{2.7}$$

where we shall always suppose that the first term is made equal to unity by a suitable choice of B. In simple cases the series (2.7) will converge inside a circle of radius l centred at $(0, 0)$. However, if there is a singularity (dislocation, point force, etc.) of the applied field inside this circle, the radius of convergence will be less than l. If g is made single-valued by a cut along the x-axis from $-l$ to $-\infty$ and $z^{\frac{1}{2}}$ similarly by one from 0 to $-\infty$, then their product is continuous across the x-axis except over the cracked portion $-l < x < 0$. Near the right-hand tip the displacement is the same as that for the elementary crack of equation (2.3) with a suitable value of B. By (2.2) the stress across the x-axis just ahead of the tip is

$$p_{zy}(x, 0) = \tfrac{1}{2}\mu B x^{-\frac{1}{2}}, \quad 0 < x \ll l. \tag{2.8}$$

In fracture mechanics, (2.8) is conventionally written in the form

$$p_{zy}(x, 0) = K (2\pi)^{-\frac{1}{2}} x^{-\frac{1}{2}}, \quad K = (\tfrac{1}{2}\pi)^{\frac{1}{2}} \mu B \tag{2.9}$$

where K is the *stress intensity factor*. We shall usually find it more convenient to use B, which for distinction we shall simply call the *intensity factor*.

If the distance from the crack to every point of the outer surface of the specimen is large (strictly speaking, infinite) compared with the length of the crack, the perturbed field may be written down directly as an integral involving p_{ij}^A along the length of the crack (see, for example, BILBY and ESHELBY 1968). In particular we have

$$B = \frac{2}{\pi\mu l^{\frac{1}{2}}} \int_{-l}^{0} \left(\frac{x'+l}{-x'}\right)^{\frac{1}{2}} P(x')\,dx', \tag{2.10}$$

where $P(x')$ is $p_{zy}{}^{A}(x', 0)$. The argument preceding and following (2.5) shows that (2.8) with (2.10) also gives the stress just ahead of a crack loaded internally with a distribution of force $P(x)$ per unit length of the upper face and an equal and opposite distribution on the lower face.

For a crack perturbing a uniform stress $p_{zy}{}^{A} = \sigma$, $p_{zx}{}^{A} = 0$, we have

$$g(z) = \left(1 + \frac{z}{l}\right)^{\frac{1}{2}} = 1 + \frac{1}{2}\frac{z}{l} - \frac{1}{2.4}\frac{z^2}{l^2} + \frac{1.3}{2.4.6}\frac{z^3}{l^3} - \dots \tag{2.11}$$

and

$$B = l^{\frac{1}{2}}\,\sigma/\mu. \tag{2.12}$$

3. The Moving Elementary Crack

We begin the discussion of what happens when the tip starts to move by considering the case where $g = 1$ and we have the elementary crack of (2.3). This case has been discussed elsewhere (ESHELBY 1969). Here we summarize the analysis, modified so as to be useful in treating the general case.

The problem can be solved with the help of a result due to BATEMAN (1915) which enables one to construct a solution of the two-dimensional wave equation which has a singularity at a moving point whose position as a function of time may be prescribed at will. The solution involves the retarded time associated with the moving point, in the sense in which this term is used in electrodynamics. Suppose that a point (which we shall identify with the moving crack tip) moves along the x-axis according to the arbitrary law

$$x = \xi(t), \tag{3.1}$$

and define the quantity $\tau(x, y, t)$ through the implicit relation

$$c(t - \tau) = +\{[x - \xi(\tau)]^2 + y^2\}^{\frac{1}{2}}. \tag{3.2}$$

The 'retarded time' τ can be given the following interpretation. If the moving point emits signals which propagate with velocity c, a signal received at time t and position (x, y) must have been emitted at time τ when the point was at $(\xi(\tau), 0)$.

Bateman's theorem, somewhat simplified, states that if $\Omega(x, y)$ is homogeneous of degree $\frac{1}{2}$ in x, y and satisfies $\nabla^2 \Omega = 0$ then $\Omega[x - \xi(\tau), y]$ satisfies the wave equation (2.1). It may be verified by differentiation, using (3.2) and Euler's theorem on homogeneous functions.

For Ω we take the elementary solution (2.3) and try to show that

$$w = B\,\mathrm{Im}\{x - \xi(\tau) + iy\}^{\frac{1}{2}} \tag{3.3}$$

represents the displacement associated with a crack whose tip moves according to the law (3.1). Suppose first that the crack is stationary up to $t = 0$ and then starts to move with uniform velocity $v < c$ so that

$$\xi = 0, \quad t < 0; \quad \xi = vt, \quad t > 0.$$

The relation (3.2) then gives

$$\left. \begin{aligned} x - \xi(\tau) &= x, \quad r > ct, \\ x - \xi(\tau) &= \frac{x - vt}{\beta^2} + \frac{v}{\beta c}\left\{\frac{(x-vt)^2}{\beta^2} + y^2\right\}^{\frac{1}{2}}, \quad r < ct, \end{aligned} \right\} \tag{3.4}$$

with $$\beta(v) = (1 - v^2/c^2)^{\frac{1}{2}},$$

which give for the displacement (after some algebra in the case of (3.6) below)

$$w = BW_0(x, y), \quad r > ct, \tag{3.5}$$

$$w = BA(v)\, W_0\left[\frac{x - vt}{\beta(v)},\ y\right], \quad r < ct, \tag{3.6}$$

where $$A(v) = (1 - v/c)^{\frac{1}{4}}/(1 + v/c)^{\frac{1}{4}}. \tag{3.7}$$

As expected, outside an expanding circle $r = ct$ the displacement retains its static value. Inside the circle it has the form of the static solution multiplied by the velocity dependent constant A and with its variables subjected to the 'Lorentz transformation'

$$x \to \frac{x - vt}{\beta}, \quad y \to y.$$

The expression (3.6) thus has all the features required of a moving crack.

When the motion of the tip is arbitrary and (3.2) cannot be solved explicitly, or when $\xi(t)$ is given but the solution of (3.2) would be too cumbersome, it is still possible to establish the crack-like nature of the solution (3.3) and also to find an explicit expression for the elastic field near the tip. We shall suppose that the crack tip, initially at rest, starts to move arbitrarily at $t = 0$, so that

$$\xi = 0, \quad t < 0; \quad \xi = \xi(t), \quad t > 0. \tag{3.8}$$

Throughout we assume that the velocity of the tip is *subsonic*, i.e. $|\dot{\xi}(t)| < c$, and then (3.2) has a single root for τ.

On the x-axis, (3.3) becomes

$$w = \pm B\left\{\tfrac{1}{2}\left|x - \xi(\tau)\right| - \tfrac{1}{2}\left[x - \xi(\tau)\right]\right\}^{\frac{1}{2}} \tag{3.9}$$

(cf. (2.4)). The physical interpretation of (3.2), or a simple graphical construction in the xt-plane, shows that on $y = 0$

$$x - \xi(\tau) \gtrless 0 \text{ according as } x - \xi(t) \gtrless 0 \tag{3.10}$$

so that across the x-axis w is discontinuous to the left of $\xi(t)$ and continuous to the right of it. Again, differentiation of (3.2) shows that

$$\frac{\partial \tau}{\partial x} = 0, \quad \frac{\partial \tau}{\partial y} = 0, \quad \frac{\partial \tau}{\partial t} = 0 \text{ on } y = 0 \tag{3.11}$$

unless $\dot{\xi} = \pm c$, a possibility we have excluded. Hence in finding the stresses and material velocity on $y = 0$ we may treat $\xi(\tau)$ in (3.3) as a constant, and the results will, in view of (3.10), be qualitatively the same as those for a stationary elementary crack with its tip at $\xi(t)$. In particular there is no stress across the x-axis for $x < \xi(t)$.

Across the circle $r = ct$, S in Fig. 1, we expect the relation

$$w \text{ is continuous across } S \tag{3.12}$$

to be satisfied. The dynamic condition

$$c\,(\partial w/\partial n) + \partial w/\partial t \text{ is continuous across } S \tag{3.13}$$

must also be satisfied (LOVE 1934). Here n denotes the normal to S, drawn in the direction in which it is advancing with velocity c. Physically (3.13) ensures that the rate of change of momentum of a 'pill-box' straddling S is entirely accounted for by the tractions on its surfaces without, for example, the help of a layer of body force spread over S. It can be shown that if (3.12) holds along S then so does (3.13) (HILL 1961). The solution (3.3) necessarily satisfies (3.12) and (3.13) even if $\dot{\xi}$ (but not ξ) is discontinuous at $t = 0$. Later, when we construct a more general solution, we shall have to verify that it satisfies these relations.

We shall now find the displacement near the tip correct to order $R^{\frac{1}{2}}$, where

$$R(t) = \{[x - \xi(t)]^2 + y^2\}^{\frac{1}{2}} \tag{3.14}$$

is the distance from the current position of the tip. From it we can find the singular terms, of order $R^{-\frac{1}{2}}$, in the stresses and material velocity, and the energy release rate (Section 5) which does not depend on the higher order terms.

To do this we write (3.2) as

$$\tau = t - R(\tau)/c$$

and carry out a Lagrange expansion of $x - \xi(\tau)$. (Alternatively one can apply this process directly to (3.3), ESHELBY 1969.) We have (WHITTAKER and WATSON 1927, Section 7.32),

$$x - \xi(\tau) = x - \xi(t) + \sum_{n=1}^{\infty} \frac{1}{n!} \left(-\frac{\partial}{c\partial t}\right)^{n-1} \left\{\frac{\dot{\xi}(t)}{c} R^n(t)\right\}. \tag{3.15}$$

To evaluate the nth term we must apply the $n - 1$ operators $\partial/\partial t$ in all possible ways to the $n + 1$ factors $\dot{\xi}(t), R, R, \ldots.$ Suppose that we apply just a single one of the operators to each of $n - 1$ of the factors R, leaving one R and $\dot{\xi}$ undifferentiated. This gives a term of order R. If we apply two of the operators to one R, leaving two factors R and the factor $\dot{\xi}$, undifferentiated, we get a term of order R^2 containing $\ddot{\xi}$ as a factor and a term of order R which does not involve $\ddot{\xi}$. Continuing in this way one can see that terms which only involve $\dot{\xi}$ are of order R, but terms which involve $\ddot{\xi}$ and higher time derivatives of ξ are of order R^2 or higher. Likewise, if we had expended one of the $\partial/\partial t$ on the factor $\dot{\xi}(t)$ which appears in (3.15) we should have obtained terms of order R^2 or higher, each multiplied by $\ddot{\xi}$. Hence the terms of order R may be found by expanding each term in (3.15) by repeated use of Leibnitz's theorem, and in the result rejecting terms which contain $\ddot{\xi}$ and higher derivatives. But what is left is the exact Lagrange expansion for a fictitious crack tip which is moving uniformly, and with which the actual crack tip happens to agree in position and velocity at time t. Hence to order R, $x - \xi(\tau)$ is given by the right-hand side of (3.4) with the replacements

$$x - vt \to x - \xi(t), \quad v \to \dot{\xi}(t) \tag{3.16}$$

and the displacement is given by (3.6) with the same replacements, that is

$$w = A(\dot{\xi})\, B\, W_0 \left[\frac{x - \xi}{\beta(\dot{\xi})},\ y\right] + O(R^{3/2}) \tag{3.17}$$

with

$$A(\dot{\xi}) = (1 - \dot{\xi}/c)^{1/4}/(1 + \dot{\xi}/c)^{1/4}, \quad \beta(\dot{\xi}) = (1 - \dot{\xi}^2/c^2)^{1/2}, \tag{3.18}$$

where

$$\xi = \xi(t), \quad \dot{\xi} = \dot{\xi}(t).$$

If the terms of order R in (3.15) are actually written out it is not at once obvious that they are equal to (3.4) with the replacements (2.28), since the square roots appear in expanded form, but one can show explicitly that they are with the help of a method developed by Schott (1912) for a similar electromagnetic problem.

4. The General Case

We now return to the case where the static field has the general form (2.6), (2.7). We can regard it as made up of terms $z^{1/2}$, $z^{3/2}$, $z^{5/2}$, etc. each induced by a suitable applied stress field. We have seen how we may, so to speak, set the first term in motion with the help of Bateman's theorem, but we cannot use it directly for the others because they are not homogeneous of degree $\frac{1}{2}$. Nevertheless we can use the theorem to build up a solution. The following physical argument is, for the moment, merely intended to provide a motive for seeking the general solution in the form of the integral (4.5) below. Later we shall carry it further.

Suppose that the tip of the elementary crack of Section 3 after remaining at the origin for $t < 0$ moves a distance $\delta\xi$ in time δt and comes to rest once more. From (3.2) and (3.3) we find that $w = B \operatorname{Im} z^{1/2}$ outside $r = ct$ and

$$w = B \operatorname{Im} (z - \delta\xi)^{1/2} = B \operatorname{Im} z^{1/2} - \tfrac{1}{2}\, \delta\xi \operatorname{Im} B\, z^{-1/2} + O(\delta\xi^2)$$

inside a circle of radius $c(t - \delta t)$ centred at $(\delta\xi, 0)$. Between the circles the displacement depends on the details of the motion from 0 to $\delta\xi$. In the limit of small $\delta\xi$ the change in w is thus

$$\delta w = -\tfrac{1}{2}\, \delta\xi \operatorname{Im} B\, z^{-1/2}\, H(ct - r) \tag{4.1}$$

where H is the Heaviside unit step function defined by

$$H(x) = 0, \quad x < 0; \quad H(x) = 1, \quad x > 0.$$

As the circle $r = ct$ sweeps outwards it leaves behind it the new static solution fully established. For a two-dimensional system this simple behaviour is remarkable. Compare, for example, what happens when a screw dislocation is displaced by $\delta\xi$ (Nabarro 1967, p. 488). Here it takes an infinite time for the field inside $r = ct$ to settle down to its new static value.

The displacement

$$w = z^{-1/2}\, H(ct - r) \tag{4.2}$$

which figures in (4.1) is a limiting case of the displacement

$$w = z^{-\frac{1}{2}}\, \varphi\,(ct \pm r) \quad \text{or} \quad w = r^{-\frac{1}{2}} \begin{matrix}\sin\\ \cos\end{matrix}\, (\tfrac{1}{2}\theta)\, \varphi\,(ct \pm r) \tag{4.3}$$

with an arbitrary function φ, which is the only distortionless solution of the two-dimensional wave equation, that is, one which takes the form of the general solution of the one-dimensional wave equation multiplied by an angular factor and by a function of r to allow for attenuation (Heaviside 1903; Eshelby 1969). If $\varphi\,(x)$ has a discontinuity in slope but not in value at $x = 0$, (4.3) (with the lower sign) satisfies (3.12) and (3.13). Whether or not we say that this is still true when φ becomes a step function is a matter of convention. In any case, (4.2) certainly satisfies the spirit of (3.13). If we substitute it into the wave equation in polar coordinates we get a collection of step functions, delta functions and derivatives of delta functions, but they exactly cancel, showing that the disturbance (4.2) is a formal solution of the homogeneous wave equation and requires no layer of body force on S to maintain it. This is not a trivial point; except when $\nu = -\frac{1}{2}$ the displacement $z^{\nu}\, H\,(ct - r)$, which satisfies the wave equation both for $r < ct$ and $r > ct$, does require a layer of body force on $r = ct$ to sustain it. If the type of argument used in this paragraph is not acceptable to the reader it does not matter, because the results we shall now obtain using (4.2) may be verified directly.

If in (4.2), we shift the origin of time and space to t' and $[\xi\,(t'), 0]$, where $\xi\,(t)$ is given by (3.8), multiply by $-\frac{1}{2}F\,[\xi\,(t')]\,\dot{\xi}\,(t')\,dt'$, where F is an arbitrary function, and integrate over t' from 0 to ∞, then we get the more general solution

$$w = -\tfrac{1}{2}\int_0^{\infty} \frac{H\{c\,(t-t') - [(x - \xi\,(t'))^2 + y^2]^{\frac{1}{2}}\}}{[z - \xi\,(t')]^{\frac{1}{2}}}\, F\,[\xi\,(t')]\,\dot{\xi}\,(t')\,dt', \tag{4.4}$$

that is,

$$w = -\tfrac{1}{2}\int_0^{\xi(\tau)} F\,(\xi)\,(z - \xi)^{-\frac{1}{2}}\,d\xi, \tag{4.5}$$

where, as usual, $\xi\,(\tau)$ is defined by (3.2) and (3.8). The time of observation t enters (4.5) through τ in the upper limit along with x and y. One can verify (at the expense of a good deal of algebra) that (4.5) satisfies the wave equation by substituting in (2.1) and using appropriate identities which result when (3.2) is differentiated either once or twice with respect to x, y, t. It also satisfies (3.12), (3.13) across $r = ct$.

Put $F\,(\xi) = c_n\,\xi^n$. Then (4.5) becomes

$$-\tfrac{1}{2}\int_0^{\xi(\tau)} c_n\,\xi^n\,(z - \xi)^{-\frac{1}{2}}\,d\xi = w_n\,(\tau) - w_n\,(0) \tag{4.6}$$

say, where

$$w_n\,(\tau) = c_n \sum_{m=0}^{n} \frac{(-1)^m}{2m+1}\binom{n}{m}\,[z - \xi\,(\tau)]^{m+\frac{1}{2}}\,z^{n-m} \tag{4.7}$$

and

$$w_n\,(0) = c_n \sum_{m=0}^{n} \frac{(-1)^m}{2m+1}\binom{n}{m}\,z^{n+\frac{1}{2}} = \frac{2n\,(2n-1)\ldots 4.2}{(2n+1)\,(2n-1)\ldots 3.1}\,c_n\,z^{n+\frac{1}{2}} \tag{4.8}$$

with
$$\binom{n}{m} = \frac{n(n-1)\ldots(n-m+1)}{m!}.$$

The integral (4.6) is, by construction, a solution of the wave equation, and so is the harmonic function $w_n(0)$. Consequently $w_n(\tau)$ is also a solution. We now choose

$$c_n = \frac{(2n+1)(2n-1)\ldots 3.1}{2n(2n-1)\ldots 4.2} \tag{4.9}$$

so that $w_n(0) = z^{n+\frac{1}{2}}$. If we multiply (4.7) by Ba_n (with $a_0 = 1$), sum over n and take the imaginary part, we get a solution

$$w = B \operatorname{Im} \sum_{n=0}^{\infty} c_n a_n w_n(\tau). \tag{4.10}$$

The retarded time τ is negative for $t < 0$ and any x, y and also for $t > 0$ on and outside the circle $r = ct$. Then $\xi(\tau)$ is zero, $w_n(\tau) = w_n(0)$, and (4.10) becomes

$$w = B \operatorname{Im} z^{\frac{1}{2}} \sum_{n=0}^{\infty} a_n z^n, \tag{4.11}$$

the static field of an arbitrarily loaded crack in the form (2.6), (2.7). Equation (4.10) also satisfies (3.12), (3.13) across $r = ct$, and the kind of argument we applied to (3.3) shows that it actually represents a crack with its tip at $\xi(t)$ and stress-free faces. Hence (4.10) describes what happens when a crack with initial static field (4.11) begins to extend. As in the case of the elementary crack of Section 3 we shall first find the displacement near the tip to order $R^{\frac{1}{2}}$.

All the terms in (4.7) with $m > 0$ are of order $R^{\frac{3}{2}}$ or higher, so that to find the displacement near the tip correct to order $R^{\frac{1}{2}}$ we need only retain the term with $m = 0$ in which we may, with a further error of order $R^{\frac{3}{2}}$, replace z^n by its value $\xi^n(t)$ at the tip. (It is easy to show that after these simplifications w still gives the physically important quantities p_{zx}, p_{zy} or $\dot{w}$ correct to order $R^{-\frac{1}{2}}$ when differentiated with respect to x, y or t.) This gives

$$w = \{\operatorname{Im}[z - \xi(\tau)]^{\frac{1}{2}}\} B(\xi) + O(R^{\frac{3}{2}}) \tag{4.12}$$

where
$$B(\xi) = B \sum_{n=0}^{\infty} a_n c_n \xi^n, \quad \xi = \xi(t). \tag{4.13}$$

We have already evaluated the factor $\{\ldots\}$ in (4.12) when dealing with the elementary crack, equation (3.17). Hence

$$w = A(\xi) B(\xi) W_0 \left[\frac{x - \xi}{\beta(\xi)},\; y\right] + O(R^{\frac{3}{2}}) \tag{4.14}$$

with A and β as in (3.18).

In simple cases the series (4.13) can be summed. For the uniformly loaded crack we have, from (2.11), (4.13) and (4.9),

$$\frac{B(\xi)}{B} = 1 + \frac{1}{1.3}\left(\frac{3}{2}\right)^2 \frac{\xi}{l} - \frac{1}{3.5}\left(\frac{3.5}{2.4}\right)^2 \frac{\xi^2}{l^2} + \frac{1}{5.7}\left(\frac{3.5.7}{2.4.6}\right)^2 \frac{\xi^3}{l^3} - \ldots \tag{4.15}$$

with $B = l^{\frac{1}{2}} \sigma/\mu$, which is reminiscent of the expansion

$$\frac{2}{\pi} E(k_1) = 1 - \frac{1}{2^2} k_1^2 - \frac{1^2\, 3}{2^2\, 4^2} k_1^4 - \frac{3^2\, 5}{2^2\, 4^2\, 6^2} k_1^6 - \ldots \tag{4.16}$$

for $E(k_1)$, the complete elliptic integral of the second kind. If we multiply (4.16 by k_1 and differentiate with respect to k_1, on the left we get $(2/\pi)(2E - K)$ and on the right the series (4.15) with $\xi/l = -k_1^2$, so that

$$B(\xi)/B = (2/\pi)\left[2E(k_1) - K(k_1)\right], \quad k_1 = (-\xi/l)^{\frac{1}{2}}, \tag{4.17}$$

where K is the complete elliptic integral of the first kind. This is satisfactory in the relatively uninteresting case where the tip has receded and $\xi < 0$. If the tip has advanced, $\xi > 0$ and k_1 is imaginary, but (4.17) is in fact still real. A well known transformation changes it into the more palatable form

$$B(\xi)/B = (2/\pi)\left[2E(k)/k' - k' K(k)\right] \tag{4.18}$$

with real modulus $k = \xi^{\frac{1}{2}}/(\xi + l)^{\frac{1}{2}}$.

The square of the complementary modulus, $k'^2 = l/(l + \xi)$, is the ratio of the initial to the current length. Alternatively, the series in (4.13) can be converted into an integral, since it is the term independent of u in

$$\sum_{n=0}^{\infty} a_n u^n \sum_{m=0}^{\infty} c_m \xi^m u^{-m},$$

so that formally

$$\frac{B(\xi)}{B} = \frac{1}{2\pi i} \int_C \sum_{n=0}^{\infty} a_n u^n \sum_{m=0}^{\infty} c_m \xi^m u^{-m} \frac{du}{u}$$

where C encircles $u = 0$ anti-clockwise. The first series is $g(u)$ as defined by (2.7); let its radius of convergence be ρ. Since

$$(1 - \xi/u)^{-\frac{1}{2}} = \sum_{m=0}^{\infty} c_m (\xi/u)^m, \quad |\xi/u| < 1,$$

with c_m as in (4.9), the second (Laurent) series converges outside $|u| = |\xi|$. (We write $|\xi|$ because ξ may be negative for a receding crack.) It is legitimate to multiply the two series together inside their common region of convergence

$$|\xi| < |u| < \rho,$$

and we may take C to run inside this annulus, and replace the series by their sum-functions. Using (2.6) we have

$$B(\xi) = \frac{1}{2\pi i} \int_C f(u)\,(u - \xi)^{-\frac{1}{2}}\, du \tag{4.18a}$$

where, to ensure that the integrand is single-valued in the annulus, we must now make a cut from $u = 0$ to $u = \xi$. Finally we may shrink C onto the cut. If we first perform an integration by parts and use (2.2) we get

$$B(\xi) = \frac{2}{\pi\mu} \int_0^{\xi} p_{zy}{}^{\circ}(\xi', 0)\,(\xi - \xi')^{-\frac{1}{2}}\, d\xi', \quad \xi > 0, \tag{4.19}$$

or
$$B(\xi) = \frac{2}{\pi\mu}\int_0^{\xi} p_{zx}^{\circ}(\xi', 0)(\xi' - \xi)^{-\frac{1}{2}}\, d\xi', \quad \xi < 0, \tag{4.20}$$

where p_{zy}°, p_{zx}° are the stresses for the original static solution (i.e. $p_{ij}^A + p_{ij}^C$ in the notation of (2.5)), and $\xi = \xi(t)$ is the current position of the tip. In (4.20), $p_{zx}^{\circ}(\xi', 0)$ is the limit of $p_{zx}^{\circ}(\xi', y)$ as y approaches zero from above, that is, the value on the upper face of the crack rather than the equal and opposite value on the lower one. With this convention, $p_{zy}^{\circ}(\xi', 0) = -p_{zx}^{\circ}(-\xi', 0)$ for small positive ξ', so that $B(\xi)$ is in fact continuous as ξ passes through zero. If $|\xi| \geqslant \rho$ the proof fails, but, as we shall see, the result ((4.14) with (4.19) or (4.20)) is still valid.

Although, as we have already emphasized, for many purposes it is enough to know only the elastic field near the tip, equation (4.14), it is also possible to find the displacement for arbitrary x, y, t. Rather than try to treat the higher order terms in (4.7) in the same way as the first we shall go back to the general solution (4.5). It is obvious that if we were to multiply (4.6) rather than (4.7) by a_n and sum over n, then we should get an expression for the displacement associated with the moving crack minus the original static displacement. We therefore write

$$w = w_0 + w_d \tag{4.21}$$

where w_0 is the static solution, given by (2.6), and w_d is the disturbance due to the motion of the tip, and look for a solution of the form

$$w_d = -\tfrac{1}{2}\,\mathrm{Im}\int_0^{\xi(\tau)} B(\xi')(z - \xi')^{-\frac{1}{2}}\, d\xi' \tag{4.22}$$

where the notation anticipates the fact that the weighting function is the same as (4.19) or (4.20), as we now show.

Suppose first that the tip has advanced, $\xi(t) > 0$. On the freshly-cracked part of the x-axis the stress must annul the original p_{zy}°:

$$-\tfrac{1}{2}\,\mathrm{Im}\lim_{y\to 0+}\frac{\partial}{\partial y}\int_0^{\xi(\tau)} B(\xi')(z - \xi')^{-\frac{1}{2}}\, d\xi' = -\frac{1}{\mu}\, p_{zy}^{\circ}(x, 0), \quad 0 < x < \xi(t). \tag{4.23}$$

We have seen (equation (3.11)) that $\partial\tau/\partial x = 0$, $\partial\tau/\partial y = 0$ on $y = 0$. Consequently, although the integral in (4.23) is not an analytic function of z (because of the way in which x, y enter τ in the upper limit), it can be treated as such in finding derivatives on the x-axis, and we may replace $\partial/\partial y$ by $i\,\partial/\partial x$ to get

$$\frac{d}{dx}\int_0^{x} B(\xi')(x - \xi')^{-\frac{1}{2}}\, d\xi' = \frac{2}{\mu}\, p_{zy}^{\circ}(x, 0), \quad x > 0, \tag{4.24}$$

of which the solution is (4.19) (WHITTAKER and WATSON 1927, Section 11.8). Normally one considers that (4.24) is the solution of the Abel integral equation (4.19), but the converse is also true. For $x < 0$ the expression (4.23) vanishes, so that p_{zy} remains zero on the original segment of the crack.

If the crack has closed and $\xi(t) < 0$, w_d must annul w_0 on $\xi(t) < x < 0$. If we check that the crack is actually closed at $x = \xi(t)$ it is sufficient to require that $\partial w_d/\partial x$ annuls $\partial w_0/\partial x$, and in place of (4.23) we have

$$-\tfrac{1}{2}\,\mathrm{Im}\lim_{y\to 0+}\frac{\partial}{\partial x}\int_0^{\xi(\tau)} B(\xi')(z-\xi')^{-\frac{1}{2}}\,d\xi' = -\frac{1}{\mu}\,p_{zx}{}^{\circ}(x,0), \quad \xi(t) < x < 0,$$

or

$$\frac{d}{dx}\int_0^{x} B(\xi')(\xi'-x)^{-\frac{1}{2}}\,d\xi' = -\frac{2}{\mu}\,p_{zx}{}^{\circ}(x,0), \quad x < 0,$$

of which the solution is (4.20). When $x < \xi(t)$, (4.23) is zero, so that p_{zy} still vanishes on what is left of the original crack.

In (4.21) with (4.22), and either (4.19) or (4.20) as appropriate, we thus have a solution which meets all requirements: it gives zero stress across the crack to the left of $x = \xi(t)$, reduces to the static solution for $t < 0$ and also for $r > ct > 0$, and satisfies (3.12), (3.13) across $r = ct$. $B(\xi)$ is given directly in terms of the original stresses along the part of the x-axis traversed by the tip since it started to move. As we do not have to assume that $g(z)$ admits the expansion (2.7) there is now no trouble if the applied field has singularities near the tip. To recover (4.14) we integrate (4.22) by parts and easily establish the relation

$$w = \mathrm{Im}\, B[\xi(\tau)][z - \xi(\tau)]^{\frac{1}{2}} + O(R^{\frac{3}{2}}). \tag{4.25}$$

We may, with an error of order R, replace $B[\xi(\tau)]$ by $B[\xi(t)]$, so that (4.25) is equivalent to (4.12), and this leads, as before, to (4.14) with $B(\xi)$ in the form (4.19) or (4.20), but with the limitation $|\xi| \leqslant \rho$ removed.

We now develop further the physical argument which led to the solution (4.5) on which the work of this section has been based. In the end we shall merely obtain our previous results in a slightly different form, but the details are perhaps worth presenting since the same method may be useful in handling other problems.

Equation (4.1) gives the change in field due to a small advance of the tip not only for the elementary crack of Section 3 but also, with the appropriate intensity factor B, for a crack of finite length perturbing an arbitrary applied field, wherever reflections from the other end have not made themselves felt. This becomes clear if we describe the tip movement as follows. Allow the tip to advance to $\delta\xi$ and now, by applying suitable forces to the freshly-formed faces, close it up again between 0 and $\delta\xi$. If these forces are suddenly removed the change of displacement in question is generated. But the response when a specified set of forces is applied or removed near the tip of a crack is obviously independent of the details of the elastic field in which it is immersed and, furthermore, the forces required to close a small region behind the tip depend only on the intensity factor B (cf. (2.10)).

Suppose that the crack advances by a series of small jumps with a pause between each. During each such pause there is a static (i.e. time-independent) field in the neighbourhood of the tip. To see this, note that because of the step function in (4.1) the contribution from the last jump is static inside a circle, S_1 say, of radius $c\Delta t$, where Δt is the time which has elapsed since the last jump, and that the fields contribute by all the previous jumps are static inside similar but larger circles all of which contain S_1. Furthermore, inside S_1 the contribution of all these fields is independent of the space-time path by which the tip arrived at its present position; this is again a consequence of the fact that t enters (4.1) only through the factor $H(\ldots)$.

Let $B^*(\xi')$ be the intensity factor which would be established at the tip if it moved from 0 to ξ' and actually came to rest there. Then the foregoing argument suggests that the elastic field of the arbitrarily-moving crack for any x, y, t should be the sum of the original static field and the integral (4.5) with $F(\xi') = B^*(\xi')$.

At first sight we might suppose that $B^*(\xi')$ is the same as the static intensity factor, $B_{st}(\xi')$ say, for a crack with its ends at $-l$ and ξ' and subject to the prescribed applied field, but this is not true, since we have ignored reflections. The static field established when the tip stops at ξ' is ultimately disturbed by reflections and only after an infinite time does the intensity factor settle down to $B_{st}(\xi')$. $B^*(\xi')$ is, however, the static intensity factor in a modified situation where reflections at the left-hand tip have been artificially suppressed.

The elementary disturbance (4.1) produces a discontinuity in w and zero stress p_{zy} on the negative x-axis. It is reflected at the left-hand tip because for $x < -l$ we have imposed boundary conditions inconsistent with the behaviour of (4.1), namely $w = 0$ and p_{zy} unspecified. To prevent this reflection we must change the boundary conditions to the following: for $x < -l$, $w(x, 0)$ is unspecified, and $p_{zy}(x, 0)$ is independent of time being equal to the value $p_{zy}^\circ(x, 0)$ that it had before the tip started to move. We can construct a physical model which reproduces this behaviour. Start with the static crack with its tips at $-l$ and 0. Slit the x-axis from $-l$ to $-\infty$ and apply the following forces per unit length to the freshly-formed faces:

$$\left.\begin{array}{l} -p_{zy}^\circ(x, 0) \text{ on the upper face,} \\ +p_{zy}(x, 0) \text{ on the lower face,} \end{array}\right\} \quad -\infty < x < -l. \tag{4.26}$$

There is no relaxation and the elastic field is unchanged. Now allow the right-hand tip to move in any way from 0 to ξ'. Then provided we keep the internal loading (4.26) fixed the disturbances which result from the motion will not be reflected. In particular, if the tip stops at ξ' then the value of $B^*(\xi')$ will now remain unaltered indefinitely, so that it may be identified with the static intensity factor of the semi-infinite crack. This can be calculated from

$$B^*(\xi') = \frac{2}{\pi\mu}\int_{-\infty}^{\xi'} (\xi' - x')^{-\frac{1}{2}} P(x')\, dx', \qquad (4.27$$

that is, from (2.10) with the tip at ξ' rather than 0, and l made infinite. The value of $P(x')$ appropriate to the applied field p_{zy}^A and the internal loading (4.26) is

$$P(x') = \begin{cases} p_{zy}^A(x', 0), & -l < x' < \xi', \\ p_{zy}^A(x', 0) - p_{zy}^\circ(x', 0), & -\infty < x' < -l \end{cases}$$

(compare the remarks following (2.10)). There will be no trouble with convergence because $p_{zy}^A(x', 0)$ approaches $p_{zy}^\circ(x', 0)$ for large $|x'|$.

According to the foregoing argument the elastic field of the moving crack is given by (4.21) and (4.22) but with the $B(\xi')$ of (4.19) or (4.20) replaced by $B^*(\xi')$. We now show that $B^*(\xi') = B(\xi')$, so that despite its deficiencies (some of which can be made good) our 'physical' argument does lead to the correct result.

To prove that $B^*(\xi') = B(\xi')$ is a problem purely static in nature. Throughout, the word 'crack' refers to the semi-infinite crack subject to the prescribed applied field and the fixed internal loading (4.26). Suppose first that $\xi' > 0$. Apply a force $-p_{zy}^\circ(x, 0)$ per unit length to the upper surface of the crack between $x = 0$ and $x = \xi'$, and an equal and opposite distribution to the lower face, so as to close the crack between 0 and ξ'. These additional internal forces give an extra contribution to (4.27) which is seen to be equal and opposite to the $B(\xi')$ of (4.20), and so the intensity factor at ξ' becomes $B^*(\xi') - B(\xi')$. But since the crack is now closed between 0 and ξ' its elastic field is the same as that of a crack with its tip at 0 rather than at ξ', so that the intensity factor at ξ' is zero, and consequently $B^*(\xi') = B(\xi')$.

If $\xi' < 0$ the argument needs to be more elaborate. It is convenient to make use of the properties of one-dimensional continuous distributions of dislocations. (Eshelby 1949; Bilby and Eshelby 1968). We need two results. First, a distribution of infinitesimal screw dislocations of density $D(x')$ along the x-axis produces a discontinuity in displacement Δw across the x-axis. With a suitable sign convention we have $\partial(\Delta w)/\partial x = -b\,D(x)$. Secondly, the dislocations lying between x' and $x' + dx'$ produce a stress $p_{zy}(x, 0) = -\mu b\,D(x')/\pi(x - x')$ at $(x, 0)$. The density D refers to the number of infinitesimal dislocations per unit length; their nominal Burgers vector b disappears from the final result.

On the elastic field for the crack with its tip at 0 superimpose the field due to a continuous distribution of dislocations (supposed to be situated in an uncracked medium) whose density is $2p_{zx}{}^{\circ}(x, 0)/\mu b = (2/b)\,\partial w_0(x, 0)/\partial x$ between 0 and ξ' and zero elsewhere. (As in (4.20), $p_{zx}{}^{\circ}(x, y)$ and $\partial w_0(x, 0)/\partial x$ are to be taken on the upper face.) This produces a displacement discontinuity which closes the crack between ξ' and 0, gives $B = 0$ at ξ', (the new tip), and also induces a stress

$$p_{zy}{}^{*}(x, 0) = \frac{1}{\pi}\int_{\xi'}^{0} \frac{p_{zx}{}^{\circ}(x', 0)}{x - x'}\,dx'$$

on the surviving crack surfaces between ξ' and $-\infty$, so that the total field is now partly produced by internal loads $\mp p_{zy}{}^{*}$ on these surfaces. When we remove these unwanted loads we get an intensity factor at ξ' which can be found by setting $P = +p_{zy}{}^{*}(x', 0)$ in (4.27). After a little manipulation the result is found to be the same as the $B(\xi')$ of (4.20).

In the above we have implicitly assumed that the specimen is infinite, otherwise we could not construct the reflectionless semi-infinite crack. However, the static elastic field in a body whose outer boundary Σ is subjected to a traction **T** and a displacement **U** is the same as it would be if the material inside Σ formed part of an infinite medium provided Σ, now merely a surface marked out in the infinite medium, is covered with a layer of body force of surface density **T** and is the seat of a Somigliana dislocation whose variable Burgers vector is equal to **U**; the material outside Σ is undeformed (Gebbia 1902). This result enables us to convert the problem of a crack in a finite body into an equivalent one for a crack in an infinite medium. We now have to require that in addition to the internal loads (4.26) the internal load **T** on Σ and the displacement discontinuity **U** across Σ are both held constant when the tip moves. With these conditions reflections from the surface of the specimen are suppressed, as well as reflections from the other tip.

5. Discussion

It is to be expected that a repetition of the foregoing calculations for states of plane rather than anti-plane deformation will present difficulties because of the existence of *two* velocities of wave-propagation, and the more complicated boundary conditions at the crack faces. Examination of existing solutions for uniformly moving cracks confirms this. In the fracture mechanics of static cracks, anti-plane calculations have often been undertaken not so much for their own sake but rather as a substitute for a plane problem which presents difficulties. In this section we shall take a similar view and suppose that a dynamic anti-plane solution, in addition to describing mode III deformation may also be used as a simplified model for mode I (plane strain tension) or mode II (plane strain shear), and parts of the discussion must be read in this sense. We shall, for example, speak as if velocity-extension curves can be obtained as easily for mode III as for mode I. We shall also attribute a surface energy to a crack which has ‘ opened ’ by the relative sliding of its faces, though this is arguable. However, in the case of a brittle material such as glass (as opposed to mica) in which atomic bonds are permanently ruptured and a crack when once formed does not ‘ heal,’ this is probably legitimate, and the surface energy should be essentially the same as it is for a crack whose faces have actually separated. Again, for mode I, a constant surface energy has often been used to simulate the effect of plastic work at the crack faces, and this approximation, though perhaps never very satisfactory, should be at least as good for the shear modes II and III as it is for the opening mode I.

The solution we have obtained for the field at an arbitrary point and time, (4.21) and (4.22) with (4.19) or (4.20), calls for no special comment, and we shall concentrate on the solution (4.14) for the elastic field near the tip, and try to justify our repeated assertion that it usually gives all that is required. The first term in (4.14) has a very simple structure. It is made up of the elementary static solution $W_0(x', y)$, equation (2.4), with a 'Lorentz-contracted' x-coordinate $x' = (x - \xi)/(1 - \dot{\xi}^2/c^2)^{\frac{1}{2}}$, multiplied by a factor $B(\xi)$ which depends only on the current length $l + \xi$ of the crack and a factor $A(\dot{\xi})$ which depends only on the current velocity of the tip. It does not involve the acceleration or higher derivatives of the velocity, though, as the argument following (3.15) makes clear, the terms denoted by $O(R^{\frac{3}{2}})$ do.

For a static crack a knowledge of the stress intensity factor K, equation (2.9), determines not only the stress ahead of the crack but also the detailed form of the elastic field around the tip. In the dynamic case it is reasonable to define K similarly by

$$p_{zy}(x, 0) = K(\xi, \dot{\xi})(2\pi)^{-\frac{1}{2}}(x - \xi)^{-\frac{1}{2}}, \quad 0 < x - \xi \ll l.$$

Equation (4.14) then gives

$$K(\xi, \dot{\xi}) = (\tfrac{1}{2}\pi)^{\frac{1}{2}}\mu(1 - \dot{\xi}/c)^{\frac{1}{2}} B(\xi). \tag{5.1}$$

To define the field at the tip we must now give both K and $\dot{\xi}$.

An important quantity in fracture mechanics is the *energy release rate* or *crack extension force* G, defined as the amount of energy which 'leaves' the material by way of the tip, calculated per unit length of crack front and per unit advance of the tip. ATKINSON and ESHELBY (1968) have proposed an expression for it which in the case of anti-plane strain takes the form

$$vG = \lim_{S \to 0} \int_S (\mu\dot{w}\, \partial w/\partial n + vE\, n_x)\, dS. \tag{5.2}$$

Here v is the instantaneous velocity of the tip, E is the energy density (elastic plus kinetic), S is a circuit embracing the tip, and (n_x, n_y) the outward normal to S. As S shrinks onto the tip the limit attained must ultimately be independent of the shape as well as the size of S. With (4.14), equation (5.2) gives

$$G(\xi, \dot{\xi}) = \tfrac{1}{4}\pi\,\mu\left(\frac{1 - \dot{\xi}/c}{1 + \dot{\xi}/c}\right)^{\frac{1}{2}} B^2(\xi). \tag{5.3}$$

For the uniformly moving elementary crack described by (3.5) and (3.6), equation (5.2) can be verified directly (ESHELBY 1969). It is only necessary to calculate the total energy inside the circle $r = ct$ at time t and subtract it from the elastic energy inside the same region before the tip started to move. Since the elastic state outside $r = ct$ is unchanged the difference represents the energy which has disappeared from the system up to time t, and, since satisfaction of the equation of motion (2.1) automatically ensures a conservative flow of energy everywhere else, it must have disappeared via the tip. Thus the difference should be, and in fact is, equal to t times the expression (5.3) with $\xi = vt$. As the first term in (4.14) differs only in notation from (3.6) with $B = B(\xi)$ this special case also establishes (5.3) generally, without appeal to (5.2), provided we are prepared to grant that G is unaffected by the addition to the displacement of terms of order $R^{\frac{3}{2}}$ which do not alter the singular terms of order $R^{-\frac{1}{2}}$ in the stress and material velocity. Equation (5.2) may also give an approximate value for the energy flow even if near the tip physical effects occur which are not envisaged by the elementary theory of elasticity. Suppose that we can find a small but finite circuit S on which the displacement and stress satisfy the equations of elasticity to a given degree of accuracy, and inside which the physical state of the medium depends only on $x - vt$ and y to the same accuracy. Then (5.2), with the limit sign ignored, gives the energy flow to this degree of accuracy.

It would be useful to have some idea of the error committed in taking (4.14) to be the solution even after it has been upset by the arrival of reflections. The error can easily be found in the limit of small tip velocities. Our solution then becomes $w = B(\xi)\, W_0(x - \xi, y)$ where $B(\xi)$, given by (4.19) or (4.20), is the intensity factor for the semi-infinite reflectionless crack of Section 4. The correct solution will, at each instant, be very nearly the same as the solution, $w = B_{st}(\xi)\, W_0(x - \xi, y)$ say, for a static crack of the same length. The condition for this to be a good approximation is that the fractional rate of change of crack length shall be small compared with the frequency with which reflections strike the tip, that is $\dot{\xi}/(l + \xi) \ll c/(l + \xi)$ or $\dot{\xi} \ll c$ simply. Even in this quasi-static limit the difference between $B_{st}(\xi)\, W_0$ and $B(\xi)\, W_0$ represents the effect of reflections. We shall only consider the case of uniform loading. Then $B_{st}(\xi)\, W_0 = \sigma(l + \xi)^{\frac{1}{2}}/\mu$ (cf. (2.12)) and $B(\xi)$ is given by (4.18), and the ratio of the incorrect to the correct solution is

$$B(\xi)/B_{st}(\xi) = (2/\pi)\left[2E(k) - k'^2 K(k)\right], \quad k^2 = l/(l + \xi), \tag{5.4}$$

a quantity which increases monotonically from unity to $4/\pi$ as ξ increases from 0 to infinity. The percentage error is 13 when the length of the crack has doubled, 20 when it has increased five-fold, and ultimately reaches $27\frac{1}{4}$. The only exact dynamical solution which is available for comparison is one for a crack under uniform loading with one end fixed and the other moving uniformly, the initial length being zero rather than, as we should have liked, finite. It may be derived from the solution for a crack whose two tips are moving symmetrically in opposite directions. We shall use a suffix 2 (two tips moving) or 1 (one tip moving) to distinguish these solutions.

Broberg (1960) and Craggs (1963) have solved the problem of a crack extending symmetrically under a uniform applied tension. The corresponding anti-plane solution has been obtained by Austwick (1968). The uniform applied stress is $p_{zy}{}^A = \sigma$ and $p_{zx}{}^A = 0$, the tips are at $x = \pm v' t$ for $t > 0$, and for $t < 0$ there is no crack. The displacement at the crack faces,

$$w_2 = \pm \frac{\sigma}{\mu E\left[\beta(v')\right]} (v' t + x)^{\frac{1}{2}} (v' t - x)^{\frac{1}{2}}, \quad y = 0, \quad |x| < v' t, \tag{5.5}$$

has the same elliptical form as for a static crack of length $2v' t$, but its magnitude is reduced by a factor $1/E\left[\beta(v')\right]$; E is an elliptic integral and $\beta(v') = (1 - v'^2/c^2)^{\frac{1}{2}}$. Outside a circle $r = ct$ there is no disturbance, only the uniform applied field σ. Near the right-hand tip the displacement may be written in a form like (4.14):

$$w_2 = A_2(v')\, B_2(v' t)\, W_0\left[\frac{x - v' t}{\beta(v')}, y\right]. \tag{5.6}$$

where $$A_2(v') = \beta^{\frac{1}{2}}(v')/E[\beta(v')]$$

and $$B_2(v'\,t) = \sigma(2v'\,t)^{\frac{1}{2}}/\mu.$$

For comparison with our results we actually need the solution for a crack with one end at rest and the other moving uniformly with velocity v. We can derive it from the solution just given by making the 'Lorentz' transformation

$$x \to \frac{x - v'\,t}{\beta(v')}, \quad y \to y, \quad t \to \frac{t - v'\,x/c^2}{\beta(v')}$$

which changes one solution of the wave equation into another. It turns (5.5) into

$$w_1 = \pm \frac{\sigma}{\mu} \frac{\beta(v')\,\beta^{-\frac{1}{2}}(v)}{E[\beta(v')]} x^{\frac{1}{2}}(vt - x)^{\frac{1}{2}} \tag{5.7}$$

and (5.6) into

$$w_1 = A_1(v)\,B_1(vt)\,W_0\left[\frac{x - vt}{\beta(v)},\ y\right] \tag{5.8}$$

with

$$A_1(v) = \beta(v')/E[\beta(v')], \quad B_1(vt) = \sigma(vt)^{\frac{1}{2}}/\mu,$$

leaves the uniform applied stress σ unchanged, and gives zero stress on the faces of the crack. The velocity v' is now merely a convenient parameter related to v by

$$v = \frac{2v'}{1 + v'^2/c^2}.$$

The separation of the coefficient of W_0 in (5.8) into a velocity factor A_1 and a factor B_1 depending on the current length is unique if we require that, for fixed vt, A_1 shall approach unity for small v, and similarly for A_2, B_2 above.

Before comparing (5.8) with our solution it is interesting to compare (5.8) and (5.6). If we put $v' = v$ in (5.6), but leave v' as a parameter in (5.8), the right-hand tips both move with velocity v, but the left-hand tip moves with velocity $-v$ in the case of w_2 and remains at rest in the case of w_1. The ratio of the displacements near the right-hand tips is

$$w_1/w_2 = (1 + v'^2/c^2)^{\frac{1}{2}}\,E[\beta(v)]/\sqrt{2}\,E[\beta(v')].$$

At low velocities, w_2 is $\sqrt{2}$ times w_1 because crack 2 is twice as long as crack 1, but as v approaches c the effect of one tip on the other becomes unimportant and w_1 and w_2 become equal.

The ratio of our solution (4.14) to (5.8) is $A(v)\,B(\xi)/A_1(v)\,B_1(vt)$. (We suppose, of course, that the tip of our crack only starts to move when the crack described by (5.8) has already grown from zero to a length l, and that thereafter the tips keep pace with each other.) $B(\xi)/B_1(vt)$ is just the expression (5.4) and so lies between 1 and $4/\pi$. The ratio of the velocity factors is

$$A(v)/A_1(v) = E[\beta(v')]/(1 + v'/c)$$

which falls monotonically from 1 to $\pi/4$ as v increases from 0 to c.

It follows that our solution and (5.8) differ by a factor which lies between

$\pi/4$ and $4/\pi$ for any velocity and degree of extension. It is not altogether obvious that the comparison is a sensible one, since we are interested in the error involved in ignoring reflections, and (5.8), of course, shows no trace of them. The nth reflection to arrive at the tip actually represents the starting wave S of Fig. 1 reflected, and consequently attentuated, $2n - 1$ times at one tip or the other. We therefore expect that after an initial period of adjustment, in which the earlier reflections make themselves felt, the field at the tip will become nearly steady. This period of adjustment gets smaller and smaller as the initial length of the crack approaches zero, so that (5.8) shows no discontinuities, although formally it still includes the effects of reflections and, for want of anything better, it is reasonable to use it as our comparison solution.

As far as the writer knows, KOSTROV (1964) is the only other writer who has considered finite cracks in non-uniform motion, but the physical situation which his analysis describes is very different from the one we have treated.

Like ourselves Kostrov only considers states of anti-plane strain. The following is the formal problem which he solves. For $t < 0$ the material is undeformed and at rest, $w = 0$, $\dot{w} = 0$. For $t > 0$ the ends of the crack move according to the laws

$$x = x_1(t), \quad x = x_2(t), \quad x_2 > x_1. \tag{5.9}$$

Starting at $t = 0$ a traction $p(x, t)$ is applied to the upper face of the crack and a traction $-p(x, t)$ to the lower face. Find the resulting disturbance.

The behaviour of the crack before $t = 0$ need not be specified, since the material is then unstressed. We may suppose that the ends remain stationary at $x_1(0)$, $x_2(0)$ or that they move in any way whatsoever. Alternatively we may take the view that the material contained no crack for $t < 0$, and that a crack joining $x_1(0)$ to $x_2(0)$ suddenly appeared at $t = 0$, as if a knife had been plunged into the material.

Kostrov confines himself to calculating the stress p_{zy} on the x-axis, and his solution, like ours, only holds so long as reflections at the ends have not occurred or in regions to which they have not yet penetrated. In contrast to the situation in Fig. 1 the disturbance is confined to a cigar-shaped region bounded by the lines $y = \pm ct$ closed by semi-circles of radius ct centred at $x_1(0)$ and $x_2(0)$, but we may still say that reflection begins when a circle of radius ct about $x_1(0)$ or $x_2(0)$ reaches $x_2(0)$ or $x_1(0)$.

To bring out the differences between Kostrov's solution and our own it will be sufficient to consider the special case where $p(x, t)$ has a constant value, σ say. On Kostrov's solution superimpose a uniform stress $p_{zy} = \sigma$, $p_{zx} = 0$. The resulting field corresponds to the following physical situation. An uncracked body is subject to a uniform stress σ. At $t = 0$ a crack of length $x_2(0) - x_1(0)$ suddenly appears and thereafter extends according to (5.9). For the stress intensity factor at the right-hand tip, Kostrov finds

$$K = 4(2\pi)^{-\frac{1}{2}}\sigma(1 - \dot{\xi}/c)^{\frac{1}{2}}(ct)^{\frac{1}{2}} \tag{5.10}$$

where $\xi(t) = x_2(t)$. Our result for a crack of initial length l is

$$K = (\tfrac{1}{2}\pi)^{\frac{1}{2}}\sigma(1 - \dot{\xi}/c)^{\frac{1}{2}}l^{\frac{1}{2}}\left[1 + \frac{3}{2}\frac{\xi}{l} + \dots\right] \tag{5.11}$$

where the complete contents of $[\dots]$ are given by the right-hand side of (4.15) or (4.18). Both these expressions depend on velocity through a factor $(1 - \dot{\xi}/c)^{\frac{1}{2}}$ and are independent of acceleration, but otherwise they are very different. Our (5.11) involves the initial crack length l because the original static field involves l. When it first appears, Kostrov's crack finds itself in a uniform stress which gives no indication of the initial length $l = x_2(0) - x_1(0)$, and so l does not appear in (5.10), and the solution is the same as it would be for a semi-infinite crack.

No doubt after a sufficient length of time it will not matter whether the crack originally appeared suddenly or started off from a finite length, and the exact solutions of Kostrov's problem and our own should ultimately agree. However, both solutions are limited to times less than about $2l/c$ because each ignores reflections, and so comparison is impossible. As we have seen, at least in some special cases our solution when extrapolated beyond this limit looks not unlike the correct solution, but the same is not true of Kostrov's.

Several other writers have treated the instantaneous formation of a crack (for references, see SIH 1968), but only Kostrov allows the tips to move arbitrarily afterwards. The case of a crack which suddenly appears or, equivalently, grows from zero length with infinite velocity, does not seem to have any very obvious physical application, but Kostrov's results may also be interpreted as the solution of a diffraction or scattering problem.

Suppose, for example, that in the absence of the crack a certain anti-plane disturbance w^A would produce at the x-axis a stress $p_{zy}{}^A = 0$, $t < 0$ and $p_{zy}{}^A = p(x, t)$, $t > 0$. Kostrov's solution with this value for $p(x, t)$ represents the scattered field when the disturbance w^A is incident on a crack extending from $x_1(t)$ to $x_2(t)$. Strictly speaking this is only true if w^A is odd in y, but as in the static case a general w^A may be divided into parts odd and even in y, and the even part is not scattered by the crack. Impulsive loading provides a simple example. If equal and opposite tractions $\pm \frac{1}{2}\sigma$ are suddenly applied at time $t = -Y/c$ to the faces $y = \pm Y$ of a plate, a pair of stress pulses

$$p_{zy} = \tfrac{1}{2}\,\sigma H(ct - y) + \tfrac{1}{2}\,\sigma H(ct + y)$$

are produced, which at $t = 0$ intersect and raise the stress across the x-axis from 0 to σ. If there is a crack along the x-axis the scattered wave is given by the same solution as in the case of the suddenly appearing crack, and, in particular, the stress intensity factor is given by (5.10). A single stress wave

$$p_{zy} = \sigma\, H(ct - y), \quad p_{zx} = 0 \tag{5.13}$$

falling on the crack produces the same K-value, since the displacement corresponding to the difference between (5.12) and (5.13) is even in y.

It is perhaps worth pointing out that Kostrov's solution may be given other physical interpretations. Application of a known analogy between anti-plane elastic and electromagnetic fields (NABARRO 1967, p. 496) converts it into the solution for the scattering of an electromagnetic wave by a conducting ribbon, the field being independent of z and having its magnetic vector parallel to the edges of the ribbon. Or it may be taken to refer to the scattering of sound waves by a rigid strip, a problem considered by, among others, FOX (1948). If the edges of the strip or crack are stationary his equation (165) is equivalent to Kostrov's basic relation (3.2) for the stress intensity factor. Fox's method of dealing with reflections could be used in the elastic problem.

The main reason for calculating the elastic field for an arbitrarily moving tip is to be able to pick out *the actual motion from the ensemble of all possible motions.* The simplest way of doing this is to suppose, along with A. A. Griffith, that the energy released at the tip is all used to provide the surface energy of the faces of the crack. This gives the equation of motion

$$G(\xi, \dot{\xi}) = 2\gamma \tag{5.14}$$

where γ is the surface energy per unit area of either face. For the initial motion of our crack G is, as indicated, a function of ξ and $\dot{\xi}$ only, and the comparisons made above suggest that, with fair accuracy, this will be true throughout the motion; at any rate we shall assume that this is so. Then, since $\ddot{\xi}$ is not involved, the crack tip, regarded as a 'particle,' has no inertia (cf. KÜPPERS 1967). Equation (5.14) may be solved algebraically to give the velocity as a function of extension, $\dot{\xi} = \dot{\xi}(\xi)$, a form in which experimental results are often presented, or it may be integrated to give the trajectory of the crack tip, $\xi = \xi(t)$.

KOSTROV (1964) obtains an equation of motion by equating the stress intensity factor to a material constant, the 'modulus of cohesion' K_c:

$$K(\xi, \dot{\xi}) = K_c \tag{5.15}$$

(actually his modulus of cohesion is $k = (\frac{1}{2}\pi)^{\frac{1}{2}} K_c$). In static fracture theory we have $2\gamma = K_c^2/2\mu$ in anti-plane strain. If γ and K_c are related in this way, (5.14) and (5.15) will give different tip motions, because of the different velocity factors in G (equation (5.3)) and K^2 (equation (5.1)). Physically speaking (5.14) guarantees a motion in which energy (potential plus kinetic plus surface) is conserved, but (5.15) does not. To put it in another way, if (5.15) is to predict the same behaviour as (5.14) with γ independent of velocity, then K_c must depend on velocity, and conversely. Of course the assumption of a constant surface energy will generally be inadequate, though in suitable circumstances plastic work can be simulated by an effective constant γ. In other cases it may be possible to take 2γ in (5.14) to represent the absorption of energy near the tip as a result of processes of which account is not taken in the elementary theory of elasticity. If the state of the material is still fairly adequately described by the elastic solution we may then take γ to be a function of ξ and $\dot{\xi}$ only, since these quantities completely specify the field at the tip, and (5.14) becomes $G(\xi, \dot{\xi}) = 2\gamma(\xi, \dot{\xi})$. In what follows we shall continue to think of γ as a constant.

For uniform loading the general nature of the tip motion predicted by (5.14) is easily made out. G contains a factor which increases with the length of the crack and a factor which decreases with increasing velocity, falling to zero at c with a vertical tangent. Since the product of these factors is to remain constant the velocity increases as the crack lengthens and approaches a terminal value c. Things should be similar for a crack extending under plain strain, either mode I or mode II. Examination of such special solutions as have been obtained shows that for all of them the energy release rate falls to zero at the velocity of Rayleigh waves, c_R. (ATKINSON and ESHELBY (1968) examine an apparent exception.) Consequently the terminal velocity will be c_R for modes I and II: numerically c and c_R are not very different.

Several writers have discussed the equation of motion of a crack using elaborations of a model first proposed by MOTT (1948). The elastic field of the moving crack is supposed to be equal, at each instant, to the static field of a crack of the same length, and from it the potential and kinetic energies of the system are calculated. Throughout the motion the increases in the kinetic and surface energies must be provided by the decrease in potential energy. For a long enough crack the demand for surface energy becomes neglible compared with the demand for kinetic energy, the changes in kinetic and potential energies balance, and the tip reaches a terminal velocity. Mott did not commit himself to its value, but ROBERTS and WELLS (1954) found it to be $0.6c$. A repetition of their calculation, but for the case of anti-plane strain, gives the value $0.8c$.

Some writers using this model (e.g. DULANEY and BRACE 1960) have taken it as obvious that the tip starts with zero velocity, but according to our calculations this is not so. The model effectively assumes that the tip is somehow held fixed while the load is raised until G reaches a value G_0 which exceeds 2γ, and is then released. In these circumstances the initial velocity is found by solving

$A^2(\xi)\, G_0 = 2\gamma$ (cf. (5.14)). If G_0 does not much exceed 2γ the initial velocity is $v = c\,(G_0 - 2\gamma)/2\gamma$. There is nothing improper in a sudden jump in velocity from 0 to v since, as we have seen, the tip exhibits no inertia. If G_0 is less than 2γ then v is negative and, in theory at any rate, the crack begins to close.

Although this model gives a vivid picture of what determines the motion of a crack it cannot be expected to give accurate results since the displacement used does not satisfy the elastic equations of motion. All the evidence available suggests that any accurate calculation based on the elementary theory of elasticity will give a terminal velocity c_R (modes I and II) or c (mode III). Where terminal velocities comparable with Roberts and Wells's estimate are found experimentally it must be that they are determined by some effect not taken account of in the elementary theory.

6. Concluding Remarks

The analysis presented here should be capable of development in several directions. As far as concerns the anti-plane strain case the effects of the successive reflections which we have ignored could no doubt be calculated step by step, at any rate numerically, but it would be gratifying if the complete result could be written down in an intellectually satisfying form, perhaps one involving an expression similar to (4.18a) but with a suitably varied path of integration. The extension to plane deformation may be expected to be complicated, but the solution might present some unexpected features.

At another level it would be useful to develop for a crack a semi-quantitative general picture of its dynamical behaviour such as one already has for a dislocation (Nabarro 1967, ch. 7). A dislocation behaves like a massive particle, but one whose mass is a complicated though rather insensitive functional of its whole previous motion. In Section 5, we have suggested that a crack tip behaves like a massless particle. This may seem surprising since a crack can formally be regarded as a continuous array of dislocations, moving if the crack is extending. However, in the limit of a large number of dislocations, their mutual interactions entirely dominate their individual properties, and so it is not to be expected that the properties of a crack and a dislocation will be very similar. That the tip actually exhibits no inertia is closely related to the rather special properties of the distorsionless solution (4.3). Further examination of some of these points might be rewarding.

References

Atkinson, C. and Eshelby, J. D.	1968	*Int. J. Fracture Mech.* **4**, 3.
Austwick, A.	1968	Master's Dissertation, University of Sheffield.
Bateman, H.	1915	*Electrical and Optical Wave-Motion* p. 138. Cambridge University Press (Reprinted by Dover, N.Y., 1955).
Bilby, B. A. and Eshelby, J. D.	1968	In *Fracture* (Edited by Liebowitz, H.) Academic Press, N.Y.
Broberg, K. B.	1960	*Ark. Fiz.* **18**, 159.
Craggs, J. W.	1963	In *Fracture of Solids* (Edited by Drucker, D. C. and Gilman, J. J.) p. 51. Wiley, N.Y.
Dulaney, E. N. and Brace, W. F.	1960	*J. appl. Phys.* **31**, 2233.

ESHELBY, J. D.	1949	*Phil. Mag.* **40**, 903.
	1969	In *Physics of Strength and Plasticity* (The Orowan 65th Anniversary Volume) (Edited by ARGON A. S.) M.I.T. Press, Cambridge, Mass.
FOX, E. N.	1948	*Phil. Trans.* A **241**, 7.
GEBBIA, M.	1902	*Ann. Mat. pura appl.* **7**, 141.
HEAVISIDE, O.	1903	*Nature, Lond.* **68**, 54. (Electromagnetic Theory. "The Electrician" Printing Publishing Company, London. 1912, **3**, p. 180).
HILL, R.	1961	In *Progress in Solid Mechanics* (Edited by SNEDDON, I. N. and HILL, R.) Vol. 2, p. 247. North-Holland, Amsterdam.
KOSTROV, B. V.	1964	*J. appl. math. Mech.* **28**, 793.
KÜPPERS, H.	1967	*Int. J. Fracture Mech.* **3**, 13.
LOVE, A. E. H.	1934	*The Mathematical Theory of Elasticity* Fourth Edition. Cambridge University Press, §§ 205, 206.
MOTT, N. F.	1948	*Engineering, Lond.* **165**, 16.
NABARRO, F. R. N.	1967	*The Theory of Crystal Dislocations* Oxford University Press.
ROBERTS, D. K. and WELLS, A. A.	1954	*Engineering, Lond.* **178**, 820.
SCHOTT, G. A.	1912	*Electromagnetic Radiation* Cambridge University Press, Appendix C.
SIH, G. C.	1968	*Int. J. Fracture Mech.* **4**, 51.
WHITTAKER, E. T. and WATSON, G. N.	1927	*Modern Analysis*, Fourth Edition. Cambridge University Press.

Axisymmetric Stress Field Around Spheroidal Inclusions and Cavities in a Transversely Isotropic Material[1]

J. D. ESHELBY.[2] In principle one could solve the problem in the way I described in [2] of the author's paper, because the field of a point force in a transversely isotropic medium can be written down explicitly.[3,4] However, I am sure this would be very clumsy, and that Dr. Chen's method is the right way to tackle the problem.

---

[1] By W. T. Chen, published in the December, 1968, issue of the JOURNAL OF APPLIED MECHANICS, Vol. 35, TRANS. ASME, Vol. 90, Series E, pp. 770–773.

[2] Department of the Theory of Materials, University of Sheffield, Sheffield, England.

[3] Lifschitz, Z. M., and Rozentsweig, L. N., *Zhurnal Eksperimental 'noi i Teoreticheskoi Fiziki*, Vol. 17, 1947, p. 17.

[4] Kröner, E., *Zeitschrift fuer Physik*, Vol. 136, 1953, p. 402.

ABSTRACT. A theorem of Bateman's is used to find the elastic field near the tip of an antiplane crack which starts moving in an arbitrary manner. The energy release rate is calculated and found to confirm a general expression previously proposed.

19. The Starting of a Crack

J. D. ESHELBY

19.1. Introduction

An intuitive feeling for the running of cracks, useful in their work, must have been acquired by both the earlier (Baden-Powell 1949) and later (Skertchley 1879) flint-knappers. Modern fracture mechanics is concerned with the prevention of crack propagation rather than with its encouragement. It is the object of this paper to obtain some insight into what goes on near a moving crack tip according to the linear isotropic theory of elasticity. We limit ourselves to states of antiplane strain. It is perhaps worth pointing out that certain interesting effects related to crack branching studied by Orowan and Yoffe (Yoffe 1951) still survive in antiplane strain (McClintock and Sukhatme 1960). It is to be hoped that our type of analysis can also be used for plane strain at the expense of the usual complications consequent on the existence of two velocities of wave propagation.

Elegant mathematical discussions have been given (Broberg 1960, Craggs 1963) of cracks which, starting from zero length, spread with constant velocity. Physically they are perhaps a little unrealistic, not so much because of the limitation to constant velocity, but because according to any reasonable criterion it needs an infinite stress to get a crack of zero length started.

We consider the case where the crack tip starts from rest and moves arbitrarily, the crack being initially of finite length. The analysis is, however, limited to the initial motion in the following sense. In a region close enough to the tip of a static crack, the elastic field, which is independent of the crack length and the details of the loading, is completely characterized by a single parameter, the stress intensity factor. Our results begin to be inaccurate when the disturbance due to the motion of the tip moves out of this region. Also, our analysis ignores the fact that after a finite interval of time the field at the tip will be upset by reflections from the surface of the specimen, and, loosely speaking, reflections from the other end of the crack, and also by disturbances generated at the other tip if it too is in motion. Which of these effects first invalidates our analysis, and after how long, will depend on the details of tip motion, geometry, and loading. Despite these limitations it seems worth working out the problem in detail. It can be solved (Section 19.2) with the help of a result of Bateman's (1915) or by fitting together simple solutions of the wave equation.

The ideas of Griffith (1920), Orowan (1949), and others on the energetics of crack propagation have, in fracture mechanics, become formalized into the concept of crack extension force or energy release rate.

In the static case we may, as Orowan (1955) has emphasized, distinguish two extreme cases. If a specimen containing a crack is deformed by rigid grips, an extension of the crack will obviously reduce the elastic energy. On the other hand, if the specimen is loaded by hanging a weight on it (dead loading), then when the crack extends the specimen becomes more flexible, and the weight descends and does work on the specimen, whose elastic energy consequently increases. However, the weight has lost an amount of potential energy which, as it turns out, is twice the increase of the elastic energy, so that there is again an overall reduction in energy. It is also true that when the rigid and dead loadings happen to produce the same elastic field near the crack tip, the energy reduction per unit extension of the crack is the same. Hence, other things being equal, the energy release rate is a quantity independent of the way in which the load is applied.

For the initial dynamic motion of a crack tip the nature of the loading mechanism is irrelevant until the disturbance produced by the motion of the tip has reached the surface of the specimen. Of the potential energy originally in the region now disturbed, some remains, some is converted into kinetic energy, and some flows out at the crack tip. The outflow, reckoned per unit length of crack front and per unit advance of the tip, is the energy release rate. In Section 19.3 we use the results of Section 19.2 to calculate it, and show that a general expression which has been proposed for the energy release rate gives the correct result.

In Section 19.4 we compare and contrast the elastic field near a moving crack tip with the fields around two other moving elastic singularities which have been studied, the kink and the screw dislocation.

19.2. The Moving Antiplane Strain Crack Tip

In a state of antiplane strain the displacement w is everywhere perpendicular to the xy plane and independent of z. The nonzero stress components are

$$p_{zx} = \mu\, \partial w/\partial x, \qquad p_{zy} = \mu\, \partial w/\partial y, \tag{19.1}$$

where μ is the shear modulus. The displacement must satisfy the wave equation

$$\nabla^2 w - \frac{1}{c^2}\frac{\partial^2 w}{\partial t^2} = 0 \tag{19.2}$$

where

$$c = (\mu/\rho)^{1/2}$$

and ρ is the density of the medium. If w is independent of time, Equation 19.2 becomes $\nabla^2 w = 0$ and solutions can be found by taking w to be the real part of some analytic function of the complex variable $z = x + iy$, which, of course has nothing to do with the z which figures as a suffix in p_{zx} and p_{zy}. If we take $w = \operatorname{Re} f(z)$, then the stresses are given by

$$p_{zx} - ip_{zy} = \mu f'(z).$$

The simplest crack-like solution is obtained by taking w to be a multiple of $z^{1/2}$. In this way we get the basic solution

$$w = W(x, y),$$

where

$$\left.\begin{aligned} W(x, y) &= B \operatorname{Re}(-iz^{1/2}) = B \operatorname{Im} z^{1/2} \\ &= 2^{-1/2} B\{(x^2 + y^2)^{1/2} - x\}^{1/2} \operatorname{sgn} y \\ &= Br^{1/2} \sin \tfrac{1}{2}\theta \end{aligned}\right\} \tag{19.3}$$

with the notation of the figure. The stresses are

$$p_{zx} = -\tfrac{1}{2}\mu Br^{-1/2} \sin \tfrac{1}{2}\theta, \; p_{zy} = \tfrac{1}{2}\mu Br^{-1/2} \cos \tfrac{1}{2}\theta. \tag{19.4}$$

The displacement w is continuous across the x axis ahead of the crack ($\theta = 0$) but it has a discontinuity $\Delta w = 2B(-x)^{1/2}$ behind the tip. The stress is zero on the faces of the crack ($\theta = \pm\pi$). If we put

$$B = (2/\mu^2\pi)^{1/2} K,$$

then

$$p_{zy} = (2\pi)^{-1/2} K x^{-1/2}, \; x > 0 \tag{19.5}$$

is the stress ahead of the crack in the plane of the crack. The constant K defined by Equation 19.5 is the stress intensity factor of fracture mechanics: strictly it should be written K_{III} to show that we are dealing with mode III deformation, that is, antiplane strain.

Across a cylinder $r =$ const. the traction is

$$p_{zr} = \mu \, \partial w / \partial r,$$

so that the displacement Equation 19.3 could be produced in a cylinder with a crack along the radius $\theta = \pi$ by forces proportional to $\sin (1/2)\theta$ distributed around its circumference and directed parallel to the generators. However, Equations 19.3 and 19.4 describe the elastic field near the tip of any crack in antiplane strain provided K is given a suitable value. In particular, if by forces remote from the origin we impose an applied stress p^A_{zx}, p^A_{zy} and then make a straight crack from the origin to $x = -l$, $y = 0$ the displacement is

$$w = (2/\mu^2\pi)^{1/2} K r^{1/2} [\sin \tfrac{1}{2}\theta + a(r/l) \sin \tfrac{3}{2}\theta + b(r/l)^2 \sin \tfrac{5}{2}\theta + \cdots], \tag{19.6}$$

where

$$K = \frac{1}{(\frac{1}{2}\pi l)^{1/2}} \int_{-l}^{0} \left(\frac{x' + l}{-x'} \right)^{1/2} p^A_{zy}(x', 0) \, dx'.$$

Consequently for any type of loading, the elastic field near enough to the tip is the same as that of an elementary crack with a suitably chosen stress-intensity factor K.

In order to set the crack tip in motion we use the following result which is a special case of a somewhat more general theorem due to Bateman (1915). If $W(x, y)$ satisfies $\nabla^2 w = 0$ and is homogeneous of degree 1/2 in x, y, then

$$W[x - \xi(\tau), y]$$

with

$$c\tau = ct - \{[x - \xi(\tau)]^2 + y^2\}^{1/2} \tag{19.7}$$

and

$$\xi = \xi(\tau) \tag{19.8}$$

satisfies the wave equation 19.2. The theorem may be verified by substituting in Equation 19.2, and using the implicit relation Equation 19.7 together with Euler's theorem for homogeneous functions.

The relations 19.7 and 19.8, familiar in electromagnetic theory, can be given the following intuitive interpretation. A point moves along the x axis according to the law $x = \xi(t)$, emitting signals which propagate with velocity c. A signal received at time t and position (x, y) must have been emitted at time τ when the point was at $(\xi(\tau), 0)$.

For W we take the expression 19.3 for the static crack, so that

$$w = B\,\mathrm{Im}\{x - \xi(\tau) + iy\}^{1/2}. \tag{19.9}$$

We now try to show that the solution 19.9 of the wave equation actually represents a crack whose tip is moving according to the law $x = \xi(t)$. When $y = 0$, Equation 19.7 shows that, as the point x approaches the point $\xi(\tau)$ from left or right, $\tau \to t$, $\xi(\tau) \to \xi(t)$. But on the x axis

$$w = \pm B\{|x - \xi(\tau)| - [x - \xi(\tau)]\}^{1/2}$$

(see Equation 19.3) so that w is continuous or discontinuous across the x axis according as x is greater or less than $\xi(t)$. Also, it is not hard to show that for $y = 0$, $p_{zy} \neq 0$ or $p_{zy} = 0$ according as x is greater or less than $\xi(t)$. The main point is that in calculating $p_{zy} = \mu\, \partial w/\partial y$ from Equation 19.1 we have to take into account the variation of $\xi(\tau)$ with y, but that Equation 19.7 shows that $\partial\xi(\tau)/\partial y = 0$ when $y = 0$. Thus Equation 19.7 has the basic properties required of a crack extending from $-\infty$ to $\xi(t)$. It is not hard to convince oneself that for x, y close to the tip $\xi(\tau) \sim \xi(t)$ so that the elastic field near the tip resembles that for a stationary crack. This point is considered more carefully later.

These conclusions are confirmed by considering a special case. Suppose that the tip of a stationary crack starts to move with uniform velocity v at time $t = 0$. Then

$$\left.\begin{aligned} \xi(\tau) &= 0,\ \tau < 0 \\ &= v\tau,\ \tau > 0 \end{aligned}\right\} \tag{19.10}$$

and Equation 19.7 can be solved to give

$$w = W(x, y), \qquad x^2 + y^2 > c^2t^2 \tag{19.11}$$

$$= A(v)W\left[\frac{x - vt}{\beta(v)}, y\right], \qquad x^2 + y^2 < c^2t^2 \tag{19.12}$$

where

$$\beta(v) = (1 - v^2/c^2)^{1/2}$$

and

$$A(v) = \left(\frac{1 - v/c}{1 + v/c}\right)^{1/4}. \tag{19.13}$$

The elastic field is divided into two regions separated by a circle of radius $r = ct$ expanding from the original position of the tip. Outside the circle the static solution remains unaltered. Inside the circle the solution

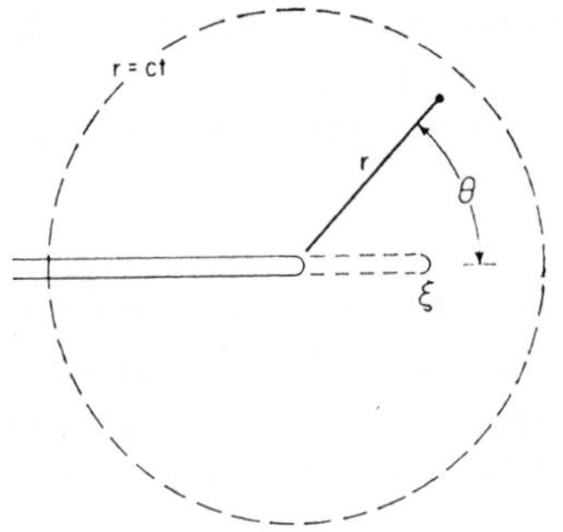

FIGURE 19.1.

takes a form which can be described as the static solution subjected to the Lorentz transformation

$$x \to (x - vt)/\beta(v), \qquad y \to y \tag{19.14}$$

and multiplied by the velocity-dependent constant $A(v)$.

It is worth considering how we might have derived Equations 19.11 and 19.12 if we were unaware of Bateman's theorem. It is obvious that the elastic field will be unaltered outside the circle $r = ct$ and, since the transformation 19.14 is a well-known way of generating moving solutions of the wave equation from stationary ones, we might perhaps have guessed the general form of Equation 19.12, though not the magnitude of A. To fix A we must impose some continuity condition across $r = ct$. If affixed "in," "out" refer to adjacent points just inside and outside the circle, we must obviously have

$$w_{\text{in}} = w_{\text{out}}. \tag{19.15}$$

According to Love (1927), the relation

$$c(\partial w/\partial n)_{\text{in}} + (\partial w/\partial t)_{\text{in}} = c(\partial w/\partial n)_{\text{out}} + (\partial w/\partial t)_{\text{out}} \tag{19.16}$$

must be satisfied across a surface of discontinuity which is advancing with velocity c in the direction of its normal n. Although it is not obvious without some algebra, Equations 19.11 and 19.12 in fact have the same angular dependence round $r = ct$, and Equation 19.15 leads to the value 19.16 for A. Fortunately, Equation 19.16 can also be satisfied all around the circle, and with the same value of A. Actually this is not a coincidence: if Equation 19.15 is valid along a surface of discontinuity which is advancing with velocity c, then 19.16 is also valid. (For a careful discussion, see Hill 1961).

Suppose that after moving for a time t_1 the crack tip increases its velocity to v_1. A new circle of discontinuity will begin to spread with velocity c from the point $(vt_1, 0)$ and outside it (but inside $r = ct$) the displacement will still be Equation 19.12. The previous argument suggests that inside the new circle the solution will be derived from Equation 19.3 by making the substitution

$$x \to [x - vt - v_1(t - t_1)]/\beta(v_1), \qquad y \to y$$

and multiplying by a suitable constant. Calculation shows that if we choose the constant to be $A(v_1)$, that is, Equation 19.13 with v replaced by v_1, then the new solution indeed satisfies Equations 19.15 or 19.16 at its junction with Equation 19.12 across the new circle of discontinuity.

We can go on to the case where the crack tip suffers a succession of changes of velocity. The displacement will evidently be

$$w = A[\dot{\xi}(t)]W\left[\frac{x - \xi(t)}{\beta[\dot{\xi}(t)]}, y\right] + O(R^{3/2}), \tag{19.17}$$

where

$$A[\dot{\xi}(t)] = [c - \dot{\xi}(t)]^{1/4}/[c + \dot{\xi}(t)]^{1/4}.$$

If the point of observation lies within the last discontinuity circle, the first term of Equation 19.17 is exact. If not there is an error, represented by the second term, which increases with distance R from the current position of the tip.

The following argument is supposed to show that, as indicated, the second term in Equation 19.17 is actually of the order $R^{3/2}$. If the reader finds it unacceptable it does not matter, since for the important case of continuous motion the result follows from the discussion of Equation 19.18 below.

Suppose the tip suffers a succession of roughly equal velocity increments at roughly equal intervals of time. If we move a distance R away from the crack tip we cross a number of discontinuity circles proportional to R and arrive at a point where the elastic field is that appropriate to a fictitious crack with a velocity which falls short of that of the actual crack by an amount proportional to R. If there had been no intervening changes of velocity, the fictitious crack would have kept up with the actual one. As things are, it has lagged behind the real crack with a relative velocity of order R for a time about R/c, and so it has fallen behind the real crack by a distance of order R^2. Since w behaves like $R^{1/2}$ near the tip (see Equation 19.3) the first term of Equation 19.17 is in error by an amount of order $R^2\, \partial R^{1/2}/\partial R = \frac{1}{2}R^{3/2}$.

We now return to Bateman's theorem. The problem of a crack moving with successive discontinuous changes of velocity can be solved by an obvious extension of the method which gave Equation 19.12. It is harder to derive Equation 19.17 directly for continuous motion. The first term of Equation 19.17 for unrestricted R (the distance from the current position of the tip) is (see Equation 19.12) the displacement associated with a crack which has been moving uniformly for all time and which at time t happens to coincide in position and velocity with the actual crack, and we have to show that close enough to the crack (Equation 19.9) leads to this expression with an error of order $R^{3/2}$. Equation 19.9 gives the displacement as a function $w[x, y, \tau]$ of position and the retarded time τ defined implicitly by Equation 19.7. What we want is the displacement $w[x, y, \tau(x, y, t)]$ as a function of x, y, t. As in similar electromagnetic problems it can be found with the help of Lagrange's expansion (Whittaker and Watson 1927). The result is

$$w[x, y, \tau(x, y, t)] = w[x, y, t] - \frac{1}{c}\sum_{n=1}^{\infty}\frac{1}{n!}\left(-\frac{1}{c}\frac{\partial}{\partial t}\right)^{n-1}[\dot{\xi}(t)Q(t)\{R(t)\}^n], \tag{19.18}$$

where

$$R(t) = \{[x - \xi(t)]^2 + y^2\}^{1/2}$$

is the distance from the current position of the crack tip, and

$$Q(t) = [\dot{\xi}(t)]^{-1}\,\partial w[x, y, t]/\partial t = O(R^{-1/2}).$$

If we were to pick out the terms of order $R^{1/2}$ in Equation 19.18 we could not easily see that they add up to the first term of Equation 19.17, (supposing that they do); since Equation 19.18 is a power series in c^{-1}, the square roots in Equation 19.17 would appear in an unidentifiable expanded form. However, we can complete the proof by an artifice. In expanding the nth term of Equation 19.18 we have to apply $n - 1$ operators $\partial/\partial t$ to the $n + 2$ factors $\dot{\xi}(t)$, Q, R, R. ... A term of order $R^{1/2}$ results when we leave $\dot{\xi}(t)$ undifferentiated and apply $\partial/\partial t$ to all but one of the remaining factors. Any other arrangement (e.g. with $\dot{\xi}(t)$ differentiated or $\partial/\partial t$ applied twice to Q or one of the factors R) leads to terms of order $R^{3/2}$ or higher, and, further, such terms will contain one or more of $\ddot{\xi}(t)$, $\dddot{\xi}(t)$ ... as a factor. In addition there may be a term of order $R^{1/2}$ which only involves $\dot{\xi}(t)$. So the terms of order $R^{1/2}$ may be picked out by expanding each term of Equation 19.18 and, in the result, throwing out all terms which contain $\ddot{\xi}(t)$, $\dddot{\xi}(t)$.... But what is left is the exact expansion for a fictitious crack which has been moving uniformly for all time and which, at time t, happens to coincide in position and velocity with the real crack, and so its sum must be the first term of Equation 19.17.

19.3. The Energy Release Rate

Atkinson and Eshelby (1968) have argued that the energy release rate for a moving crack may be calculated from

$$vG = \lim_{S\to 0}\int_S (\mathbf{T}\cdot\mathbf{u} + vEn_x)\,dS, \tag{19.19}$$

where S is a circuit surrounding the crack tip, $\mathbf{u}$ is the elastic displacement, $\mathbf{T}$ is the traction across S, (n_x, n_y) is the outward normal to S, E is the total energy density (elastic plus kinetic), and v is the instantaneous velocity of the tip. The limit sign must be taken to mean that S is allowed to shrink on to the tip, and that, for a meaningful result, the integral must ultimately be independent of the shape as well as the size of S. In Equation 19.19, vG is the energy release per unit time and G the energy release per unit advance of the tip.

In the present case, Equation 19.19 reduces to

$$vG = \lim_{S\to 0} \int_S (\mu \dot{w}\, \partial w/\partial n + vEn_x)\, dS \tag{19.20}$$

with

$$E = \tfrac{1}{2}\mu(\partial w/\partial r)^2 + \tfrac{1}{2}\mu(\partial w/r\,\partial\theta)^2 + \tfrac{1}{2}\rho\dot{w}^2 \tag{19.21}$$

With Equation 19.12 the recipe 19.20 gives

$$G = A^2K^2/2\mu, \tag{19.22}$$

which reduces to the accepted static value when $v = 0$.

It is possible to verify directly that, at least in the case of a crack starting off with uniform velocity, Equation 19.19 gives the correct result. Since the elastic state outside $r = ct$ is unchanged and the surface of discontinuity does not carry a finite amount of energy along with it, at time t the energy which has disappeared from the elastic field since the crack started to move is given by the integral

$$\int (E_s - E_m)\, dx\, dy \tag{19.23}$$

extended over the interior of the circle $r = ct$. Here E_s and E_m are the energy densities calculated from Equation 19.21 and the static solution 19.3 or the dynamic solution 19.12, respectively. Since energy conservation is taken care of everywhere else, the energy 19.23 must have disappeared via the tip. If Equation 19.19 is correct, 19.23 should be t times 19.20. A tedious calculation shows that this is so. The following less direct method is instructive and also useful in other connections.

One can verify that for any solution derived from a static solution by the transformation 19.14 the integral 19.19 or 19.20 is independent of the shape and size of S, so that we may ignore the limit sign and, for convenience, take S to be the circle $r = ct$. (Of course we must use the displacement 19.12 rather than 19.11). It is easy to see that E_m is homogeneous of degree -1 in $x - vt$ and y. Consequently,

$$\left[(x - vt)\frac{\partial}{\partial x} + y\frac{\partial}{\partial y}\right] E_m = -E_m$$

or

$$E_m = -\frac{\partial}{\partial x}[(x - vt)E_m] - \frac{\partial}{\partial y}[yE_m],$$

and similarly,

$$E_s = -\frac{\partial}{\partial x}[xE_s] - \frac{\partial}{\partial y}[yE_s].$$

Thus, with the help of Gauss's theorem, 19.23 becomes

$$\int (E_s - E_m)\, dx\, dy = \int_{r=ct} [vtE_m n_x + r(E_s - E_m)]\, dS.$$

But the static and moving solutions confront one another across the circle $r = ct$ and the difference in the energy densities (in the form 19.21) can be found by squaring and rearranging 19.16 and adding $(\partial w/r\,\partial\theta)^2_{\text{in}} = (\partial w/r\,\partial\theta)^2_{\text{out}}$ which follows from 19.15. The result is

$$E_s - E_m = (\mu/c)[\dot{w}\,\partial w/\partial r]_{\text{in}}$$

so that

$$\int (E_s - E_m)\, dx\, dy = t\int_{r=ct} (\mu\dot{w}\,\partial w/\partial n + vEn_x)\, dS,$$

as required.

So far we have simply shown that for a uniformly moving crack, Equation 19.20 gives the correct result. However, near enough to the tip of a crack in arbitrary motion the elastic field, Equation 19.17, is the same, apart from notation, as Equation 19.12, and so the energy release rate must be the same, namely,

$$G = \left[\frac{1 - \dot{\xi}(t)/c}{1 + \dot{\xi}(t)/c}\right]^{1/2} \frac{K^2}{2\mu} \tag{19.24}$$

for a crack tip in arbitrary motion. This expression depends only on the initial K value and the instantaneous velocity of the crack and not, for example, on its acceleration. The fears expressed by Atkinson and Eshelby about the applicability of Equation 19.19 to an accelerating crack thus seem to be unfounded. Of course the acceleration is not entirely without effect. The argument following 19.18 shows that the remainder term in 19.17 contains terms proportional to the acceleration and higher derivatives of the velocity. The larger they are the smaller must be the circuit S in 19.20 before the integral approaches a limiting value. But unless the size of S turns out to be of atomic dimensions, this causes no trouble.

19.4. Discussion

In the ideal case of a solid with surface energy γ a crack tip when released will initially move with a constant velocity found by equating 2γ with the energy release rate G given by Equation 19.22. Note that if K is not large enough A may have to be greater than unity, in which case v is negative and the crack shrinks.

More generally γ may be an equivalent surface energy representing plastic work, or 2γ may be taken to stand for some velocity-dependent dissipative process going on in a small region near the crack tip. Then the equation of motion takes the form

$$G(\dot{\xi}) = 2\gamma(\xi, \dot{\xi})$$

where G is given by Equation 19.24 and γ is a prescribed function of position and velocity.

Since G does not depend on the acceleration, the crack exhibits no inertia. If the tip encounters a region where γ suddenly decreases it immediately speeds up so as to decrease G (see Equation 19.24); there is no question of the tip overshooting, or tripping over its own feet in an attempt to feed more energy into the crack plane than the latter will accept. Closely connected with the absence from G of terms involving the acceleration or higher derivatives of the velocity is the fact that the discontinuity associated with an abrupt change of velocity carries no energy with it.

It is interesting to compare the behavior of a crack tip with that of other moving elastic signularities.

A kink (Eshelby 1962) behaves very much like an electron. It has a mass and is subject to a radiation reaction proportional to the rate of change of acceleration. If it suffers a rapid change of velocity two spheres of discontinuity spread out with the velocities of longitudinal and shear waves, and each carries with it a finite amount of energy.

A screw dislocation (Nabarro 1967) behaves like a moving electrically charged rod. The radiation reaction depends on the whole previous history of the motion so that there is an integral equation of motion. We may say that this is because the problem, though nominally two-dimensional, is really three-dimensional, so that each element of the dislocation is perpetually subjected to disturbances produced by remote elements at earlier times. However, precisely similar effects occur in "really" two-dimensional systems. A smooth heavy particle sliding on a horizontal drumhead has an integral equation of motion similar to that of a charged rod or dislocation. Unfortunately the equations are not quite identical since the particle is coupled to a scalar field whereas the rod or dislocation is coupled to a vector field. As a result the particle does not feel anything analogous to the Lorentz force which acts on the rod, and so this simple model throws no light on the problem (Nabarro 1967) of whether a screw dislocation experiences a Lorentz force or not.

It is not very surprising that a crack tip does not behave much like a kink, but we might have expected that some of the peculiarities of the two-dimensional wave equation which affect the dislocation would also affect the crack. The reason why they do not is that, roughly speaking, the peculiarity of being two dimensional is canceled by the peculiar angular dependence of the elastic field through sines and cosines of half-integral multiples of the angle θ, in the static case, or of the corresponding aberrated (Lorentz-transformed) angle in the moving case. The matter is bound up with the problem of whether the N-dimensional wave equation admits distortionless solutions or not.

A solution of the wave equation is considered to be distortionless if it takes the form

$$w = g(r)f(r \pm ct)\Omega(\theta, \phi \ldots). \tag{19.25}$$

In N dimensions the equation may be written

$$w'' + (N-1)r^{-1}w' + r^{-2}\Lambda w - c^{-2}\ddot{w} = 0, \tag{19.26}$$

where the dash denotes $\partial/\partial r$ and $r^{-2}\Lambda$ is the angular part of the Laplace operator. To get rid of the term in f' we must take $g = r^{-1/2N+1/2}$. Then if we take Ω to be an eigenfunction of Λ, with $\Lambda\Omega = \lambda\Omega$, Equation 19.26 becomes the one-dimensional wave equation only if

$$\lambda = \tfrac{1}{4}(N-1)(N-3).$$

In three dimensions, $g = r^{-1}$, $\lambda = 0$ and there is a distortionless solution, namely,

$$w = r^{-1}f(r \pm ct)$$

only if there is no angular dependence. Otherwise there is no solution of the the form 19.25. For instance, if $\Omega = \cos\theta$, the best we can do is

$$w = [r^{-2}f(r \pm ct) - r^{-1}f'(r \pm ct)]\cos\theta,$$

which is not distortionless in the sense of Equation 19.25; hence the distinction between the induction and radiation zones in electromagnetic and other fields. When $N = 2$,

$$g = r^{-1/2}, \lambda = -\tfrac{1}{4},$$

and

$$w = r^{-1/2}f(r \pm ct)\,{}^{\sin}_{\cos}\,\tfrac{1}{2}\theta, \tag{19.27}$$

and we have a distortionless solution in which, as stated, the usual peculiarities of the two-dimensional wave equation are offset by the peculiar angular dependence.

We can, in fact, use Equation 19.27 to construct our basic solution 19.9. It can be rewritten as

$$w = (x + iy)^{-1/2} f(r \pm ct).$$

For f take the step function $H(ct - r)$, which is unity for $ct > r$ and zero for $ct < r$. Shift the origin of time and space to τ', $\xi(\tau')$, 0, multiply by $-\frac{1}{2}B\dot{\xi}(\tau')d\tau'$, and integrate over all values of the parameter τ'. The result is

$$w = -\frac{1}{2}\int_{-\infty}^{\infty} \frac{BH\{c(t-\tau') - [(x-\xi(\tau'))^2 + y^2]^{1/2}\}}{\{x - \xi(\tau') + iy\}^{1/2}} \dot{\xi}(\tau')\, d\tau'$$

$$= B\{x - \xi(\tau) + iy\}^{1/2},$$

where τ is defined by Equation 19.7. The imaginary part of this expression is 19.9. (The rather offhand treatment of the lower limit can be justified).

The expression 19.27 is the only distortionless solution in the two-dimensional case. The absence of others for angular dependence sin $(3/2)\theta$, sin $(5/2)\theta$ ... is what prevents us treating the higher terms in Equation 19.6 in the same simple way as the first.

REFERENCES

Atkinson, C., and J. D. Eshelby, 1968, *Int. J. Fracture Mech.*, **4**, 3.

Baden-Powell, D. F. W., 1949, *Proc. Prehist. Soc.*, **15**, 38.

Bateman, H., 1915, *Electrical and Optical Wave-Motion*, Cambridge: Cambridge University Press, p. 138.

Broberg, K. B., 1960, *Arkiv för Fiz.*, **18**, 159.

Craggs, J. W., 1963, in *Fracture of Solids*, New York: Interscience, p. 51.

Eshelby, J. D., 1962, *Proc. Roy. Soc.*, **A266**, 222.

Griffith, A. A., 1920, *Phil. Trans. Roy. Soc.*, **A221**, 163.

Hill, R., 1961, in I. N. Sneddon and R. Hill (eds.), *Progress in Solid Mechanics*, New York: Wiley-Interscience, vol. 2, p. 247.

Love, A. E. H., 1927, *The Mathematical Theory of Elasticity*, Cambridge: Cambridge University Press, p. 295.

McClintock, F. A., and S. P. Sukhatme, 1960, *J. Mech. Phys. Solids*, **8**, 187.

Nabarro, F. R. N., 1967, *Theory of Crystal Dislocations*, Oxford: Clarendon Press, p. 496.

Orowan, E., 1949, in *Rep. Prog. Physics*, **12**, London: Physical Society, p. 185.

Orowan, E., 1955, *Welding J.*, **34**, 157s.

Skertchley, S. B. J., 1879, *On the Manufacture of Gunflints*, Mem. Geol. Survey, London: H.M. Stationary Office.

Whittaker, E. T., and G. N. Watson, 1927, *Modern Analysis*, Cambridge: Cambridge University Press, p. 132.

Yoffe, E. H., 1951, *Phil. Mag.*, **42**, 739.

ENERGY RELATIONS AND THE ENERGY-MOMENTUM TENSOR IN CONTINUUM MECHANICS

J. D. Eshelby

Department of the Theory of Materials
University of Sheffield
Sheffield, England

ABSTRACT

The force on a dislocation or point defect, as understood in solid-state physics, and the crack extension force of fracture mechanics are examples of quantities which measure the rate at which the total energy of a physical system varies as some kind of departure from uniformity within it changes its configuration. One may define similarly a force acting on each element of a mobile interface (a phase boundary or martensitic interface, for example).

Methods for calculating such effective forces are reviewed for both quasi-static and dynamic processes, the latter with particular reference to the motion of crack tips. The elastic energy-momentum tensor proves to be a useful tool in such calculations.

1. INTRODUCTION

In solid-state theory, theoretical metallurgy, fracture mechanics, and elsewhere, there are departures from uniformity in a material on various scales which, for want of a better term, we shall call defects. Examples on a microscopic scale are dislocations and point defects in a crystal lattice or their analogs in a continuum theory. On a larger scale there are regions which differ in some way from the bulk of the material, from which they are separated by an interface, for example inclusions of one phase in another, martensitic plates in ferrite, twins and so on. If the two materials on either side of it are in some sense uniform within themselves we may consider the interface itself to be the defect. On a macroscopic level there are cavities and cracks. All these entities can alter their configuration. Dislocations can glide and climb, point defects diffuse, interfaces can migrate, cavities can change their shape, and cracks can expand. The configuration of the defects can be specified by a number, possibly infinite, of parameters. Following the terminology of analytical mechanics and thermodynamics we can call the rate of decrease of the total energy of the system with respect to a parameter the generalized force acting on that parameter, or, in simple cases, on the defect itself. The idea of the force on a lattice defect is now a familiar one (it goes back to an interesting paper by Burton[1]) and the crack extension force of fracture mechanics is a concept of the same kind.

It is always the total energy which is important, the energy of the system we concentrate our attention on, and which contains the defect, plus the energy of the environment with which it interacts, in our case some mechanical loading device. The distinction between the two parts is, in fact, arbitrary. If we strain a test piece in a tensile testing machine an engineer will know where to draw the line between specimen and machine, but an applied mathematician need not. In thermodynamics the matter is handled by introducing enthalpy and Gibbs free energy, quantities which though nominally referring to the system under observation actually relate to the energy of the system plus the energy of its environment.

In this paper, the term elastic energy means, strictly speaking, internal energy under adiabatic conditions and Helmholtz free energy under isothermal ones. We shall rely on the small difference between the two in solids to give us meaningful results in intermediate cases. Of course the difference can sometimes be important, for example, in thermo-elastic effects which can actually contribute to the force on a defect;[2] we shall not consider such effects here.

It is the object of this paper to give some account of these ideas, and their extension to the dynamic case. We shall make extensive use of the properties of the energy-momentum tensor associated with the elastic field. The writer should perhaps admit that this tensor has become an obsession with him since he first noticed its connection with the force on a defect,[3] and no doubt it appears at some points in the argument where one could get along without it. Still, it serves as a convenient thread to tie together the various topics we shall discuss.

2. ELASTICITY IN LAGRANGIAN COORDINATES

For the most part our discussion will apply to finite deformation with a general stress-strain relation. In this section we briefly set out the necessary elastic theory and then in Sec. 3 show how the energy-momentum tensor appears as a formal concept, preparatory to interpreting it in Sec. 4.

We shall use rectangular Cartesian coordinates X_i to label the initial positions of particles of material in the initial unstrained state (Fig. 1*a*). On deformation the particle at X_i suffers a vector displacement $\mathbf{u}$ and its final position is, say, x_i referred to the same coordinate system, so that

$$u_i(X_m) = x_i(X_m) - X_i \tag{1}$$

If we imagine a replica of the coordinate network X_i to be embedded in the material in the initial state, and to deform with it, it becomes the curvilinear network $\tilde{X}_i$ of Fig. 1*b* after deformation. For some purposes it is convenient to refer things to this embedded (convected) coordinate system, but we shall not do so here.

For the stress it will be convenient to use the nominal (Piola-Kirchhoff or Boussinesq) stress p_{ij} defined so that $p_{ij}\,dS$ is the

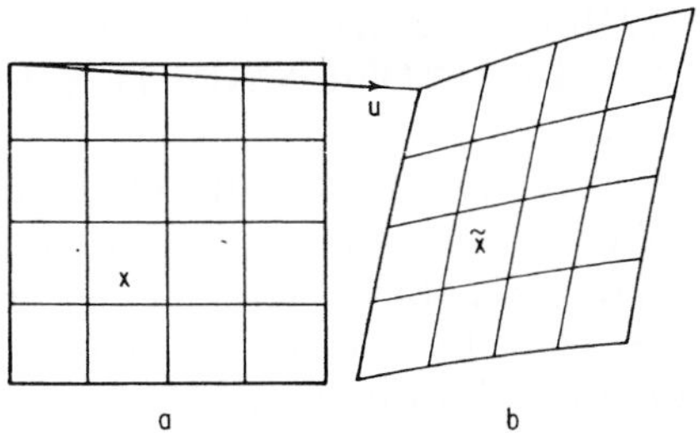

Fig. 1

component parallel to the X_i axis of the force on an element of area which, before deformation, was an element of area dS perpendicular to the X_j axis. If W is the strain energy in unit initial volume,

$$p_{ij} = \frac{\partial W(u_{m,n})}{\partial u_{i,j}} \tag{2}$$

In the absence of body forces the equilibrium condition is

$$\frac{\partial p_{ij}}{\partial X_j} = 0 \tag{3}$$

The p_{ij} are not symmetric, but the fact that W depends only on the final shape and size of an element which was originally cubic, and not on its orientation, gives the relation

$$p_{ij} - p_{ji} = p_{jk}u_{i,k} - p_{ik}u_{j,k} \tag{4}$$

If the material is isotropic there is the additional relation

$$p_{ij} - p_{ji} = p_{ki}u_{k,j} - p_{kj}u_{k,i} \tag{5}$$

The displacement plays a double role: it can be regarded as a simple field vector like, say, the electric field E_i but it also represents a displacement of the material in that the material particle originally at X_m is in fact as $x_i = X_i + u_i(X_m)$. However, if, as we shall, we use Lagrangian coordinates X_m rather than the Eulerian coordinates x_m we may for most purposes regard them as the usual rectangular coordinates of theoretical physics and, ignoring their displacement aspect, treat $u_i(X_m)$ as a vector field of the same kind as $E_i(X_m)$ which does not stick out from X_m in any meaningful sense. This brings the elastic field into line with other physical fields and enables us to apply the methods of general field theory.

3. FORMAL DERIVATION OF THE ENERGY-MOMENTUM TENSOR

We are now in a position to describe the mathematical process which generates the energy-momentum tensor whose physical interpretation we shall discuss in the next section. The matter is usually approached by way of the calculus of variations[4,5] but the writer finds the following method clearer and, in any event, it gives all we shall want. To cover later extensions the calculation is somewhat more general than is necessary to deal with static problems in Lagrangian coordinates. The number of components of u_i need not be the same as the number of X_i; later we shall augment X_1, X_2, X_3 by the time variable $t = X_0$. We shall replace the energy density W (or, rather $-W$) by a general Lagrangian density L, and allow it to depend on the u_i as well as the $u_{i,j}$. Although this is not true in Lagrangian coordinates it is in the Eulerian formulation that we shall briefly glance at later.

Suppose, then, that we have a set of quantities $u_i(X_m)$ depending on the independent variables X_m and a function

$$L = L(u_i, u_{i,j}\ X_m) \tag{6}$$

which depends on u_i and its first derivatives $u_{i,j} = \partial u_i/\partial X_j$ and also explicitly on the X_m. We may regard $L(\cdots, \cdots, X_m)$ as a calculating machine which works out the numerical value of L when the appropriate $u_i, u_{i,j}$ are inserted. The explicit dependence on X_m then means that there is a different machine at each point X_m.

If we regard L as the Lagrangian density and require that its integral over a region of X_m space be an extremum we are led to the Euler equations

$$\frac{\partial}{\partial X_j}\frac{\partial L}{\partial u_{i,j}} - \frac{\partial L}{\partial u_i} = 0 \tag{7}$$

We shall use the notation $\partial L/\partial X_i$ to denote the gradient of L, so that $(\partial L/\partial X_i)\, dX_i$ is, to order dX_i, the numerical value of L at $X_i + dX_i$ minus its numerical value at X_i. From it we must distinguish the explicit partial derivative of L with respect to X_i when its other arguments u_i, u_{ij} and the remaining X_m are held constant. We shall denote it by $(\partial L/\partial X_i)_{\text{exp}}$, so that

$$\left(\frac{\partial L}{\partial X_i}\right)_{\text{exp}} = \left.\frac{\partial L(u_i, u_{i,j}, X_m)}{\partial X_i}\right|_{\substack{u_i,\ u_{i,j}\ \text{const.} \\ X_m\ \text{const.},\ m \neq i}} \tag{8}$$

The components of the gradient of L are thus

$$\frac{\partial L}{\partial X_l} = \frac{\partial L}{\partial u_i} u_{i,l} + \frac{\partial L}{\partial u_{i,j}} \frac{\partial u_{i,j}}{\partial X_l} + \left(\frac{\partial L}{\partial X_l}\right)_{\text{exp}} \tag{9}$$

or noting that $\partial u_{i,j}/\partial X_l = \partial u_i/\partial X_j \partial X_l = \partial u_{i,l}/\partial X_j$ and using the rule for differentiating a product

$$\frac{\partial L}{\partial X_l} = \left(\frac{\partial L}{\partial u_i} - \frac{\partial}{\partial X_j}\frac{\partial L}{\partial u_{i,j}}\right)u_{i,l} + \frac{\partial}{\partial X_j}\left(\frac{\partial L}{\partial u_{i,j}}u_{i,l}\right) + \left(\frac{\partial L}{\partial X_l}\right)_{\text{exp}} \tag{10}$$

So far Eqs. (7) and (9) are purely mathematical relations. We now make the physical assumption that Eq. (7) actually is the governing equation for the field $u_i(X_m)$. Then the first term on the right of Eq. (10) vanishes and we may rewrite Eq. (10) as

$$\frac{\partial P_{lj}}{\partial X_j} = -\left(\frac{\partial L}{\partial X_l}\right)_{\text{exp}} \tag{11}$$

where

$$P_{lj} = \frac{\partial L}{\partial u_{i,j}}u_{i,l} - L\delta_{lj} \tag{12}$$

is the energy-momentum tensor we are seeking.

In most of what follows L will be minus the elastic energy density W. For a uniform material unstressed in the initial state $(\partial W/\partial X_l)$ exp defined as in Eq. (10) is zero. If there is a patch of material where the function W differs from its normal form, but the material is still unstressed in the initial state, we shall say that we have a defect which is a pure inhomogeneity. An example is a small region of abnormally low elastic constants, simulating a lattice vacancy which happens to give rise to no internal stress. Actually we shall be more interested in defects associated with internal stress. One is inclined to think that because internal stress is associated with an incompatible strain there is therefore no displacement function. However, there is one—the displacement associated with Bilby's[6] shape change. One way of generating a state of internal stress is the following.[8,3]

Dice the material into tiny cubes by cuts parallel to the coordinate planes of Fig. 1*a*. Allow each cube to undergo a permanent change of shape and size specified by some suitable finite strain measure, say e_{ij}^T, which is a function of X_m. Apply to each cube surface forces chosen so as to restore its original cubic shape and size. Weld the cubes together again. The coordinate net still has the appearance of Fig. 1*a*, but only because it is held so by a distribution of body force due to the failure of the forces on adjacent cube faces to cancel completely. When these forces are relaxed the network will warp and become, say, the network $\tilde{X}_m$ of Fig. 1*b*. This immediately defines a displacement at X_m, namely the vector joining X_m to the point $\tilde{X}_m$ with the same three coordinate numbers. Naturally there is still a displacement if we deform further by applying external loads. We shall find that dislocations, with their discontinuous displacements, give no trouble.

Wherever $e^T_{ij} = 0$ the energy density is found by inserting $u_{i,j}$ into $W(\ldots, X_m)$. Elsewhere we need to know both $u_{i,j}(X_m)$ and $e_{ij}^T(X_m)$; for example, with the network still held rectangular by the distribution of body force we have a nonzero W at each point, depending on e_{ij}^T even when the $u_{i,j}$ are zero. Whether we absorb the e_{ij}^T – dependence of W into its explicit dependence on X_m, or regard the e_{ij}^T as extra field variables is a matter of choice. We do not need to decide, because all our results will involve integrals taken over surfaces which lie in "good" regions where $e_{ij}^T = 0$. Elsewhere all we need to know is that W exists. In fact we really only need to be assured of the weaker fact that any macroscopic region has a recoverable energy content. The energy does not even necessarily need to be recoverable if we do not call the material's bluff by unloading to see if it is.

4. PHYSICAL INTERPRETATION OF THE STATIC ENERGY-MOMENTUM TENSOR

To get the static energy-momentum tensor associated with an elastic medium we take L in Eq. (12) to be $-W$, where W is the energy density defined in Sec. 2. This gives

$$P_{lj} = W\delta_{lj} - p_{ij}u_{i,l} \qquad (13)$$

Wherever W does not explicitly depend on X_m its divergence vanishes,

$$\frac{\partial P_{li}}{\partial X_i} = 0 \qquad (14)$$

elsewhere the divergence is equal to $(\partial W/\partial X_l)_{\text{exp}}$.

The elastic energy-momentum tensor has been largely ignored, or at any rate its significance has not been appreciated. For example, Morse and Feshbach[8] set up the complete 4×4 array of Sec. 8, give the name "force dyadic" to its spatial part, the dynamic generalization of Eq. (13) (see Eq. (53) below), and then merely wonder if one "can discover a use for the byproduct quantities such as field momentum and force dyadic." However, the energy-momentum tensor is involved in Rice's[9] path-independent integral, which plays a useful role in fracture mechanics.

Sometimes P_{lj} turns up as a convenient auxiliary quantity in ordinary elastic calculations, with no need of interpretation. The following result, which does not seem to have been noticed before, illustrates this.

In the linear theory the elastic energy of a body is given by the surface integral

$$E_{\text{el}} = \frac{1}{2}\int_S p_{ij}\,u_i\,dS_j \tag{15}$$

but this is not valid in the general nonlinear case. But even then there is a surface integral representation of sorts if the material is homogeneous, namely

$$3E_{\text{el}} = \int_S (X_l P_{lj} + u_l p_{lj})\,dS_j \tag{16}$$

To prove this we convert to a volume integral by Gauss' theorem, use Eq. (14) (homogeneity) and Eq. (3) (no body force) and an integrand $3W$ is left.

Our interpretation of the physical meaning of P_{lj} will be the following.[3,10,11] Figure 2*a* represents a body subject to surface

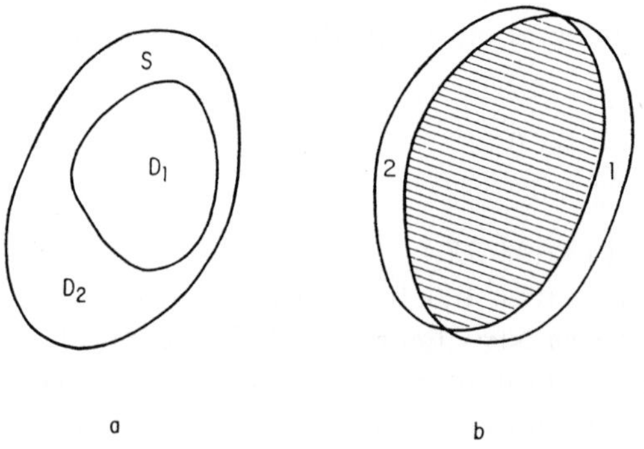

Fig. 2

loading and containing a defect D_1. In addition there may be other defects, symbolized by D_2. We ask what is the change in the total energy of the system (elastic energy plus potential energy of the loading mechanism) if D_1 suffers a small vector displacement $\delta\xi_i$. The answer is that

$$\delta E_{tot} = -\delta\xi_l F_l \tag{17}$$

where

$$F_l = \int_S P_{lj}\, dS_j \tag{18}$$

and S is any surface surrounding D_1 and isolating it from D_2. F_l is thus the force on the defect in the sense explained in the Introduction. Note that by Eq. (14) S may be deformed without altering F_l provided it remains in "good" material where $(\partial W/\partial X_l)_{exp} = 0$.

To establish Eq. (17) we first need to know how to find the elastic field in a body containing a defect which has suffered a vector shift $\delta\xi_i$, given the field before the shift. In the linear infinitesimal theory this is straightforward.[3] To find the elastic field of a defect in a finite body one usually begins with a known expression for the displacement field of the defect when situated in an infinite medium, say $u_i^\infty(X_m)$. This will produce a nonzero traction $p_{ij}^\infty n_j$ at the surface of the body. But we require the surface traction to be zero at the surface of the body when it is not externally loaded, so we must apply a surface traction $-p_{ij}^\infty n_j$ which produces, say, a displacement u_i^I, the "image" displacement. In addition, if there is, in fact, external loading which produces a surface traction $p_{ij}^L n_j$, there will be an additional displacement u_i^L so that the net displacement is $u_i(X_m) = u^\infty(X_m) + u^I(X_m) + u^L(X_m)$. It is now clear how we are to find the displacement after the singularity has been displaced by $\delta\xi_i$. We first replace $u_i^\infty(X_m)$ by $u_i^{\infty'} = u_i^\infty(X_m - \delta\xi_m)$. If the material is elastically inhomogeneous we must also make the change

$$c_{ijkl}(X_m) \to c_{ijkl}(X_m - \delta\xi_m) \tag{19}$$

to ensure that $u_i^{\infty'}$ continues to satisfy the equilibrium equations. The relation Eq. (19) is, of course, just the change we should decide to make if the singularity were a pure inhomogeneity. Next we calculate the new u^I and u^L. Obviously the new u^I differs from the old, and, in addition, the new u^L may differ from the old because of the change in Eq. (19). Indeed, in the case of a defect which is a pure inhomogeneity it is only the change in u^L which signals to the loading mechanism that something has happened inside the material.

A little thought shows that the above process is unnecessarily elaborate. We should get the same result if we shifted the entire original displacement field by $\delta\xi_i$, so that

$$u_i(X_m) \to u_i(X_m - \delta\xi_m) \tag{20}$$

and so forth, coupled with Eq. (19), and then adjusting the surface traction so as to satisfy the original boundary conditions once more. This recipe can be equally well applied to the finite deformation of a solid with any stress-strain relation, and the following will be our prescription for shifting a singularity in the general case.[10,11]

Let $f(X_m)$ denote any quantity associated with the elastic field, in particular the displacement, the stress and the energy density.

Stage (*i*) Make the replacement

$$f(X_m) \to f(X_m - \delta\xi_m) \tag{21}$$

Stage (*ii*) Adjust the surface tractions until the original boundary conditions are satisfied.

It is worth asking what becomes of Eq. (19) in the general case. We now have to shift the functional form of W: $W(\ldots, X_m) \to W(\ldots, X_m - \delta\xi_m)$. But since the arguments $u_{i,j}$ which have to be inserted into the energy density function have themselves suffered the change $u_{i,j}(X_m) \to u_{i,j}(X_m - \delta\xi_m)$ we may just as well write $W[X_m] \to W[X_m - \delta\xi_m]$ where $W[X_m]$ stands for $W(u_{i,j}(X_m), X_m)$ regarded as a simple function of X_m. This change is already covered by Eq. (21).

In the situation of Fig. 2*a* we want to shift D_1 but not D_2. To do this we simply regard S as the surface of a "body" acted on by a complex loading mechanism made up of the material between S and the surface of the body plus the real external loading mechanism, and apply our previous recipe. Of course the adjustment in stage (*ii*) will change the displacement in the material outside S as well as inside it, but one can show that it does not generate a discontinuity across S.

We now have to find how the elastic energy E_{el} of the body and the potential energy E_{ext} of the loading mechanism change during stage (*i*) and stage (*ii*).

In stage (*i*) E_{el} changes by

$$\delta E_{\mathrm{el}}^{(i)} = \int_V \{W[X_m - \delta\xi_m] - W[X_m]\} dV$$

$$= -\delta\xi_l \int_V \frac{\partial W}{\partial X_l} dV + 0(\delta\xi^2) \qquad (22)$$

$$= -\delta\xi_l \int_S W dS_l + 0(\delta\xi^2)$$

by Gauss' theorem. At the beginning of stage (*ii*) the surface traction is $[p_{ij}(X_m) - \delta\xi_l p_{ij,l}(X_m)] n_j + 0(\delta\xi^2)$ and at the end it is $p_{ij}(X_m) n_j$, so that with an error of order $\delta\xi$ it is $p_{ij}(X_m) n_j$ throughout the adjustment. The surface displacement changes from $u_i(X_m) - \delta\xi_l u_{i,l}(X_m) + 0(\delta\xi^2)$ to some final value $u_i^F(X_m)$ which could only be found by detailed calculation. Hence the change in E_{el} during stage (*ii*), equal to the work done on the body during the adjustment, is

$$\delta E_{\mathrm{el}}^{(ii)} = \int_S (u_i^F - u_i + \delta\xi_l u_{i,l}) p_{ij} dS_j + 0(\delta\xi^2) \qquad (23)$$

During stages (*i*) and (*ii*) together the increase in the potential energy of the loading mechanism, being equal to minus the work which it does, is

$$\delta E_{\mathrm{ext}} = \int_S (u_i - u_i^F) p_{ij} dS_j + 0(\delta\xi^2) \qquad (24)$$

Adding Eq. (22) to Eq. (23) to get the change in total energy we arrive at

$$\delta E_{\mathrm{tot}} = -\delta\xi_l \int_S (W\delta_{lj} - p_{ij} u_{i,l}) dS_j + 0(\delta\xi^2) \qquad (25)$$

which is Eq. (17).

The quantity $\delta E_{\mathrm{el}}^{(i)}$ can also be found as follows. In addition to S, the surface of the body, draw the surface S' derived from S by a shift $-\delta\xi_i$ (Fig. 2*b*). Then $\delta E_{\mathrm{el}}^{(i)}$ is the integral of W over the crescent-shaped volumes 1 and 2, the former being given a positive sign and the latter a negative one. (Of course in the case of 2 we have to suppose that the field is slightly extrapolated beyond S.) This gives the last of the surface integrals contained in Eq. (22) at once, and makes it clear that W in the shaded area of Fig. 2*b*, which appears in the first of them and has cancelled from the last, need not appear in the calculation at all. Thus if we do not know, or do not care to specify, W in the neighborhood of the singularity it does not matter. Even formal infinities in the shaded region cancel out.

Again, in calculating $\delta E_{\mathrm{el}}^{(ii)}$ we supposed that the loading mechanism produced a surface traction which was independent of the slight shift of the point of application consequent on the displacement of the singularity by $\delta\xi_i$, that is, in engineering language we assumed dead loading produced by an ideally soft loading machine.

If this is not so the change in the surface traction which occurs during the change of surface displacement from u_i to u_i^F will be of order $\delta\xi$, but this will only alter δE_{ext} by an amount of order $\delta\xi^2$. Consequently Eq. (17) is also valid for an arbitrarily hard or soft loading mechanism. (We might perhaps feel uncomfortable about the extreme case of infinitely hard loading, i.e., rigidly imposed surface displacements, but then $u_i = u_i^F$ at the surface and Eq. (17) is still valid.)

The fact that Eq. (7) is independent of the hardness or softness of the loading mechanism enables us to treat the situation of Fig. 2*a* without further calculation. Draw an internal surface S isolating D_1 from D_2 and the surface of the body. Then, as already mentioned, we may regard the self-stressed material between S and the surface of the body, together with the loading mechanism properly speaking, as constituting a single loading mechanism of unknown hardness or softness acting on the surface S of the "body" bounded by S, and Eq. (17) may be applied.

Thus, generally, F_l in Eq. (18) gives the force on the singularity (or singularities) inside any surface S, due to other singularities outside S, and to imposed surface tractions produced by any type of loading mechanism.

Although we do not in fact need to know the form of the function W inside S, or even that it exists everywhere, when we *do* know it Eq. (11) shows that we may write

$$F_l = \int_v \left(\frac{\partial W}{\partial X_l}\right)_{\mathrm{exp}} dV \tag{26}$$

with the notation of Eq. (8), so that F_l is the same as the energy of a fictitious body with the energy density $(\partial W(\ldots, X_m)/\partial X_l)_{\mathrm{exp}}$ into which we insert the original displacement gradients $u_{i,l}$. It is unnecessary to know how they change during the operations of stages (*i*) and (*ii*). The following is a similar but more general result.

If we make any small change δW_{exp} in the form of the energy function this will induce a change $\delta u_{i,l}$ in the displacement gradients, and the change in the energy density will be

$$\delta W = \delta W_{\mathrm{exp}} + \frac{\partial W}{\partial u_{i,j}} \delta u_{i,j} = \delta W_{\mathrm{exp}} + p_{ij}\delta u_{i,j} \tag{27}$$

and on integration the change in the elastic energy of the body is, since $p_{ij,j} = 0$,

$$\delta E_{\mathrm{el}} = \int_V \delta W_{\mathrm{exp}} dV + \int_S p_{ij}\delta u_i \, dS_j \tag{28}$$

The change in the energy of the loading mechanism is

$$\delta E_{\mathrm{ext}} = -\int_S p_{ij}\delta u_i \, dS_j \tag{29}$$

to the first order, even without dead loading. Hence

$$\delta E_{\mathrm{tot}} = \int_V \delta W_{\mathrm{exp}}(u_{ij})\, dV \tag{30}$$

and so, again, we only need to know the change in the functional form of W and the old $u_{i,j}$; the changes $\delta u_{i,j}$ cancel from δE_{tot}, though not from δE_{el} and δE_{ext} taken separately. This result can be of practical use.[12] There are similar results in other parts of theoretical physics, e.g., the Hellman-Feynman theorem in quantum mechanics.

These conditions are related to the concept of complementary energy. If we derive a quantity ϕ from W by the Legendre transformation

$$-\phi = W - \frac{\partial W}{\partial u_{i,j}} u_{i,j} \tag{31}$$

then ϕ is the complementary energy density. In thermodynamic language it is the enthalpy density for adiabatic processes and the Gibbs free energy density for isothermal ones. For the small change contemplated above

$$-\delta\phi = \delta W_{\text{exp}} - \delta p_{ij} u_{i,j} = \delta W_{\text{exp}} - (\delta p_{ij} u_i)_{,j} \tag{32}$$

the terms $\partial u_{i,j} \partial W/\partial u_{i,j}$ and $-p_{ij}\delta u_{i,j}$ cancelling. Hence if

$$\Phi = \int_V \phi dV$$

is the total complementary energy, comparison with Eq. (32) gives

$$-\delta\Phi = \delta E_{\text{tot}} - \int_S \delta p_{ij} u_i dS_j \tag{33}$$

and for dead loading $\delta p_{ij} n_i = \text{const}$, $\delta E_{\text{tot}} = -\delta\Phi$. For other types of loading $-\delta\Phi$ differs from δE_{tot} by a quantity of the first order, in contrast to Eq. (32).

The application of Eq. (18) to dislocations requires a little discussion because the displacement is discontinuous. On the infinitesimal theory the usual dislocations of physical theory have a discontinuity in u_i across a surface but since $u_{i,l}$ does not, Eq. (18) may be applied to them. The same is not true for general Somigliana dislocations,[3] or even for disclinations[13] where the rotation is discontinuous.

For dislocations with a *finite* constant Burgers vector, Eq. (18) will be valid if $u_{i,l}$ is the same at points of the material which were adjacent across the cut in the initial state. To avoid complication we shall only consider the case where the dislocation is created by cutting the material and sliding the faces without separation or interpenetration. If we identify the undeformed material with Frank's[13] reference crystal we must arrange that points originally separated by $\mathbf{b}$ across the cut are coincident finally. The usual argument of the infinitesimal theory[13,14] may be applied to the finite $u_{i,j}$. It shows that the $u_{i,l}$ (and their gradients $u_{i,lj}$) are the same at points which were opposite one another in the initial state, and this is what we want for Eq. (18) to be applicable, because the integration is with respect to the undeformed state.

The writer has convinced himself that for points *finally* adjacent across the cut $u_{i,j}$ is continuous but $u_{i,jk}$ is not. This fact, if true, should make itself felt in the theory of continuous distributions of dislocations; though there is no trouble with a single infinitesimal dislocation, an isolated bundle of them is, after all, equivalent to a dislocation with a finite Burgers vector.

In the linear theory the displacement may be written $u_i = u_i^\infty + u_i^I + u_i^L + u_i^S$, where, as before, $u_i^\infty + u_i^I$ is the field due to a particular defect divided into its value in an infinite medium and an image term; u_i^L, u_i^S are the displacements produced by external loading and other defects. We can then partition F_l into contributions from external load, other defects, and image effects, say F_l^x, with $x = L, S$ or I. F_l^x is given by the cross-term in Eq. (18) between x-quantities and ∞-quantities. It may be manipulated[10] into the more useful form

$$F_l^x = \int_S (u_i^x p_{ij,l}^\infty - p_{ij}^x u_{i,l}^\infty)\, dS_j \qquad x = L, S, I \tag{34}$$

The quantity which would be denoted by F_l^∞ vanishes if the separation into u_i^∞ and u_i^I has been done properly.

Equation (34) gives the standard results for the forces on defects according to the linear theory. Since most of them can be derived without the help of the energy-momentum tensor this is not very helpful, though once Eq. (18) enabled the writer to decide which of two rival expressions for the interaction between a dislocation and a point defect to attack. The real value of Eq. (18) is that it indicates how far these results are valid. For example, a well-known result in the infinitesimal theory states[13] that the force on a point defect idealized as a center of dilatation is

$$F = -\Delta V \operatorname{grad} p \tag{35}$$

where ΔV is the volume change produced by the introduction of the defect (measured by the expansion of the outer surface of the body) and p is the hydrostatic pressure there would be at the center of the defect if it were not there. Equation (35) can be derived from Eq. (18) by taking S to be a small surface enveloping the defect. In fact the nonlinear theory is grossly inadequate near the defect. What effect does this have on Eq. (35)? If we choose S to be a large enough surface surrounding the defect the quantities which enter P_{lj} will be adequately represented by the linear theory. Since the divergence of P_{lj} is zero on both the nonlinear and linear theories we may contract S to an infinitesimal sphere about the defect and evaluate it using either theory. On the infinitesimal theory the result is Eq. (35) but it holds equally for the nonlinear theory, if p is taken to be not the actual hydrostatic pressure due to the surface loads and other defects (which is meaningless in a nonlinear theory) but rather the hydrostatic pressure which would be calculated from the applied stress and displacement prevailing over the large surface S on the assumption that the linear theory is valid everywhere. Similarly, in other cases results obtained on the linear theory may be used to calculate F_l whenever S in Eq. (18) can in principle be chosen to lie everywhere in regions where the linear theory is adequate.

5. MODIFICATION WHEN THERE ARE BODY FORCES

In deriving Eq. (18) we have supposed that the material inside S is free of body force. However, in ionic crystals a dislocation may carry a charge,[15] and if the crystal is immersed in an electrostatic field there will be a body force acting at its center. There do not seem to be any other sensible cases where a defect is associated with body force, but it is easy to adapt the argument which led to Eq. (18) to cover the general case. The body force inside S is

$$f_i = -\int_S p_{ij}\, dS_j \tag{36}$$

In the shift carried out in stage (i) the electrostatic field (or whatever mechanism is providing the body force) increases its potential energy by $-f_i \delta\xi_l$. In stage (ii) the presence of the body force produces changes in both Eqs. (23) and (24) but they exactly cancel, just as do the terms in $u_i - u_i^F$. So, in all, we have to add a term $+ f_i \delta\xi_i$ to the right-hand side of Eq. (17). Hence Eq. (18) must be modified to read

$$F_l = \int_s P^*_{lj}\, dS_i \tag{37}$$

where
$$P^*_{lj} = P_{lj} - p_{lj} = W\delta_{lj} - P_{ij}(\delta_{il} + u_{i,l})$$
$$= W\delta_{lj} - \frac{\partial W}{\partial\left(\dfrac{\partial x_i}{\partial X_j}\right)} \frac{\partial x_i}{\partial X_l} \tag{38}$$

(see Eq. (1)). The difference in sign between P_{lj} and p_{lj} is merely a matter of convention (see Sec. 8).

It has become traditional to emphasize the difference between the "real" forces acting on the material and the "fictitious" forces acting on defects, but in Eq. (37) the "real" force (Eq. (36)) seems to make a contribution to the "fictitious" force. However, in the case of charged dislocations, the extra term has the required "fictitious" character because when it moves a dislocation does not actually carry marked charges along with it. Rather, atomic rearrangement at its old position eliminates the net charge there, and at its new position atomic rearrangement creates an equal net charge.

One can, of course, replace P_{lj} by P^*_{lj} even when there are no body forces. In an isotropic medium, but not otherwise, P^*_{lj} is, in view of Eq. (5), symmetric, in contrast to P_{lj}.

6. THE FORCE ON AN INTERFACE

So far the energy-momentum tensor has appeared only in an integral over a closed surface. In certain cases it is also possible to give a meaningful interpretation to the force $dF_l = P_{lj}\,dS_j$ on an individual surface element.

Figure 3 represents a specimen in which the material inside a surface S has undergone some kind of transformation. S may be closed or, as indicated by the dotted lines, it may extend to the surface of the specimen. The following are some examples of what we mean by the word "transformation."

In a martensitic transformation region B undergoes a change of natural shape and size. If this is inhibited by the presence of the matrix A, strains are set up inside and outside S. However, adjacent

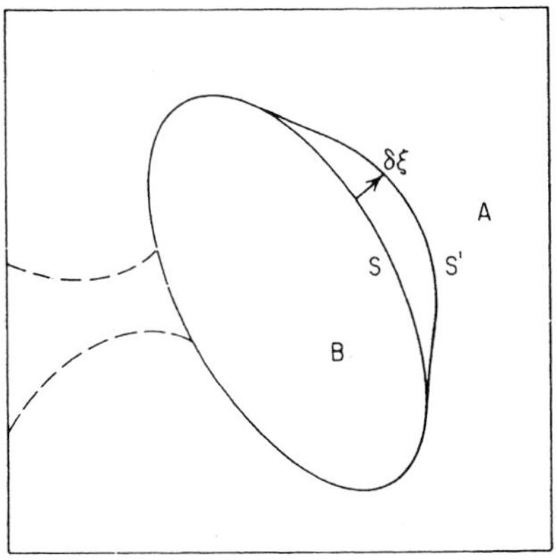

Fig. 3

material particles on opposite sides of S remain adjacent, so that the displacement is continuous. In addition the elastic properties of the material inside B become different from those of A. When a coherent twinned region develops there is also a change of form and elastic properties. Though the material is the same the crystal axes are differently oriented, so that in the linear case, for example, the c_{ijkl} are different in A and B when referred to the same axes. The same is true if S is a grain boundary. In the case of Dauphiné twins[16] in quartz there is a difference of elastic constants in the above sense, but no associated change in form. We may also take S to be the boundary of a cavity, the material inside it having, so to speak, transformed into nothing.

In all these cases the displacement is, or may be taken to be, continuous across S, and we shall assume that this is so in what follows. We also allow an external loading mechanism to act on the outer surface of the specimen.

In the physical cases described, and in others, the boundary may be capable of migrating through the material (growth of martensitic plates, change in the form of a cavity by volume or surface diffusion, and so on). In Fig. 3 migration has made S develop a shallow blister, changing it to S′. The migration may be specified by erecting a small vector $\delta\xi_i$ at each point of S. It is then a sensible question to ask what is the change in the total energy of the system as a result of the migration. The answer is that

$$\delta E_{\text{tot}} = -\int_S \delta\xi_l (P_{lj}{}^A - P_{lj}{}^B)\, dS_j \tag{39}$$

where $P_{lj}{}^A$ and $P_{lj}{}^B$ are the energy-momentum tensors on the A and B sides of S, and the normal in $dS_j = n_j dS$ is directed outwards from B to A. The proof is similar to that of Eq. (17). Cut out and remove the A material which lies between S and S′ in the initial state, and apply suitable tractions to the raw surface to prevent relaxation. The increase in elastic energy is

$$\delta E_{\text{el}}{}^{(i)A} = -\int \delta\xi_l W^A dS_l \tag{40}$$

The displacement on the boundary S′ of A is now $u_i^A + \delta\xi_l u_{i,l}{}^A$ and the traction is $p_{ij}{}^A dS_j + 0(\delta\xi)$. Now alter the displacements to their final values $u_i{}^{FA}$, and the work done on A is

$$\delta E^{(ii)A} = -\int \left(u_i^{FA} - u_i^A - \delta\xi_l u_{i,l}^A\right) p_{ij}^A \, dS_j \tag{41}$$

(Since some of the work expended may have gone towards the potential energy of the part of the loading mechanism associated with A, we cannot label Eq. (41) as purely elastic energy.)

For B we have similarly

$$\delta E_{\rm el}^{(i)B} = \int \delta\xi_l \, W^B dS_l \tag{42}$$

$$\delta E^{(ii)B} = \int \left(u_i^{FB} - u_i^B - \delta\xi_l u_{i,l}^B\right) p_{ij}^B \, dS_j \tag{43}$$

There is no term corresponding to Eq. (24). In a sense A and the part of the actual loading mechanism which acts on its surface provides the loading mechanism for B and conversely. Since u_i is continuous across the interface, we have $u_i^A = u_i^B$, $u_i^{FA} = u_i^{FB}$ for each element dS_j. Also $p_{ij} dS_j$ is continuous across the interface. So, on adding Eq. (40) to Eq. (43) we get Eq. (39).

This expression only gives the change in the elastic energy and the energy of the loading mechanism. If it is to be applied to phase changes we must also include the "chemical" energy, the work required to transform the material which disappeared from A into the material which appeared in B, say

$$\int \delta\xi_l (W_0^B - W_0^A) \, dS_l \tag{44}$$

where $W_0^B - W_0^A$ is the work required to transform a mass of unstressed A into an equal mass of unstressed B, the mass occupying unit volume in the common initial state. More precisely, $W_0^B - W_0^A$ is a change of internal energy in an adiabatic change and of Helmholtz free energy in an isothermal one. The quantity Eq. (44) can be added to Eq. (39) without changing its form by choosing the value of W for $u_{i,j} = 0$, $W(0)$ say, which is so far unspecified, so that $W^A(0) - W^B(0) = W_o^A - W_o^B$. The addition of a constant to W does not, of course, alter Eq. (18), because the extra term is proportional to $\int dS_l$ and so is zero for a closed surface being, by Gauss' theorem, the volume integral of the gradient of unity. It appears that the value ascribed to the $W(0)$ of a particular substance depends on what we propose to transform it into, but in a sense $W(0)$ can be prescribed absolutely since, after all, anything can in principle be transformed into anything else through a suitable intermediary, say the material of a neutron star.

If we use the notation $[Y]$ to denote the discontinuity in Y across S, its value on the side to which the normal points minus its value on the other side, Eq. (39) may be written

$$\delta E_{\text{tot}} = -\int_S \delta\xi_l \left[P_{lj}\right] dS_j$$

$$= -\int_S \delta\xi_l([W]\,\delta_{lj} - p_{ij}[u_{i,l}])dS_j \qquad (45)$$

It is natural to choose the vector $\delta\xi_l$ which describes the shift of the interface everywhere normal to S, but if we do not it makes no difference. The term $[W]\delta_{lj}$ gives a force parallel to the normal, and so does the term $-p_{ij}[u_{i,l}]$ because as $[u_i] = 0$, by supposition it follows that $[u_{i,l}]\delta\xi_l' = 0$ whenever $\delta\xi'$ lies in S. In fact, it is not hard to see[17] that

$$[u_{i,l}] = \left[\frac{\partial \mathbf{u}_i}{\partial n}\right] n_l \qquad (46)$$

where $\partial/\partial n$ denotes differentiation along the normal. Thus Eq. (45) is equivalent to the statement that there is an effective normal force

$$F = [W] - \mathbf{T}\cdot\left[\frac{\partial \mathbf{u}}{\partial n}\right] \qquad (47)$$

per unit area of interface, where $\mathbf{T}$ is the surface traction at the interface.

Equation (45) can be used to find the equilibrium position of phase and twin boundaries in the presence of stresses produced by the transformation itself, or applied externally, or both. Since Eq. (45) must be zero for any small $\delta\xi_l$ the boundary must take up a shape for which Eq. (47) is zero all along it. In the case of a stress-free cavity Eq. (47) becomes the energy density at its surface (a positive quantity). Any increase in its volume leads to a decrease of E_{tot}.[18]

7. RESULTS IN EULERIAN COORDINATES

It is not easy to conduct the argument of Sec. 4 in the Eulerian coordinates x_i of Eq. (1), but we can get the results from the Lagrangian case by introducing a Σ_{lj} defined by $\Sigma_{lj}\,ds_j = P_{lj}\,dS_j$ where ds_j is what dS_j becomes in the final state, so that

$$F_l = \int_S P_{lj}\,dS_j = \int_S \Sigma_{lj}\,ds_j \tag{48}$$

A tedious calculation gives

$$\Sigma_{lj} = w\delta_{lj} - \frac{\partial w}{\partial v_{i,j}} v_{i,l} \tag{49}$$

a direct transcription of P_{lj}. Here $v_i(x_j) = x_i - X_i(x_j)$ is the Eulerian displacement as contrasted with the Lagrangian $u_i(X_j) = x_i(X_j) - X_i$, $v_{i,j}$ stands for $\partial v_i/\partial x_j$ and w is the energy density per unit final volume. Since W depends on X_m, w will depend on v_i as well as $v_{i,j}$ but only through the combination $X_m = x_m - v_m$:

$$w = w(v_{i,j}, x_m - v_m) \tag{50}$$

If, similarly, we define the Eulerian stress through $\sigma_{ij}ds_j = p_{ij}\,dS_j$ we get, surprisingly,

$$\sigma_{ij} = \frac{\partial w}{\partial v_{i,j}} + w\delta_{ij} - \frac{\partial w}{\partial v_{r,j}} v_{r,i} \tag{51}$$

(Actually Eq. (51) is a disguised form of Hamel's[19] expression for σ_{ij}.) The formalism of Sec. 3, with suitable changes in notation, confirms Eqs. (49) and (51). Equation (49) comes directly from Eq. (12). The equilibrium equation corresponding to Eq. (7) can be written as

$$\frac{\partial}{\partial x_i}\frac{\partial w}{\partial v_{i,j}} - \frac{\partial w}{\partial v_i} = -\left(\frac{\partial w}{\partial x_i}\right)_{\text{exp}} = -\frac{\partial \Sigma_{ij}}{\partial x_j} \tag{52}$$

or $\partial\sigma_{ij}/\partial x_j = 0$ with σ_{ij} as in Eq. (51), by Eq. (50) and Eq. (11). Of course this of itself does not actually prove that σ_{ij} is the Eulerian stress.

One consequence of Eq. (51) is that expressions for radiation pressure,[20] an ordinary force, seem to be formed from Σ_{ij} which we associate with the fictitious force on an inhomogeneity. This is because the first term in Eq. (51) has averaged to zero in a periodic process. The analogy drawn by the writer[10] between the lift on an aerofoil and the supposed Lorentz force on a moving dislocation[13] merely illustrates the same sort of thing and has no direct physical content. (Of course in both these cases there are dynamical terms we have ignored.)

8. THE DYNAMIC ENERGY-MOMENTUM TENSOR

If we supplement the spatial coordinates X_1, X_2, X_3 by $X_0 = t$, the time variable, and take for the Lagrangian density $L = T - W$ where $T = \frac{1}{2}\rho\dot{u}_i\dot{u}_i$ is the kinetic energy density per unit original volume, the analysis of Sec. 3 gives a 4×4 energy-momentum tensor with components

$$P_{lj} = (W - T)\delta_{lj} - p_{ij}u_{i,l} \qquad i, j = 1, 2, 3 \tag{53}$$

$$s_j = P_{0j} = -p_{ij}\dot{u}_i \tag{54}$$

$$g_l = -P_{l0} = -\rho\dot{u}_i u_{i,l} \tag{55}$$

$$H = P_{00} = T + W \tag{56}$$

If the material is homogeneous and has time-independent properties, Eq. (11) gives the conservation laws

$$\frac{\partial P_{lj}}{\partial X_j} = \frac{\partial g_l}{\partial t} \tag{57}$$

and

$$-\frac{\partial s_j}{\partial X_i} = \frac{\partial H}{\partial t} \tag{58}$$

The space components of P_{lj} only differ from the static case by the kinetic energy term. The vector s_j is the energy flow vector, and $s_j dS_j$ in the flow of energy through dS_j in the direction of the positive normal.[21] H is the energy density.

The vector g_l is variously called the quasi-momentum, pseudomomentum or field momentum. It is also the crystal momentum of lattice and electron theory, in the continuum limit of zero lattice constant, when umklapp processes are irrelevant.

The quasi-momentum density differs from the ordinary momentum density $G_l = \rho \dot{u}_l$. It can be given various formal interpretations. We could use the curvilinear (and now time-dependent) coordinate network $\tilde{X}_m$ of Fig. 1*b* to describe the motion of a particle moving through the medium, and take the $\tilde{X}_1$, $\tilde{X}_2$, $\tilde{X}_3$ with which it happens to coincide at any instant as its generalized coordinates. Let the particle be a small mesh of side ϵ in Fig. 1*a*. Although its $\tilde{X}_m$ remain constant, it still has a generalized momentum in the sense of analytical mechanics, say $\epsilon^3 \Pi_l$. Then[10] $g_l = G_l - \Pi_l$. In a linear anisotropic medium the ratio S_j/H gives the group velocity of a disturbance, whereas g_l/H is the slowness, a vector whose direction is that of the phase velocity and whose magnitude is the reciprocal of the magnitude of the phase velocity. (It may sound odd to speak of group velocity for a medium governed by a second-order equation, but as the frequency $\omega/2\pi$ depends on the direction of the wave vector k_l the group velocity $\partial\omega/\partial k_l$ differs from the phase velocity ω/k_l.)

Suppose next that the elastic medium is interacting with some other system by way of a potential $U(u_{i,j}, x_m)$ per unit original volume which depends on the deformation and on the spatial position x_m (Eq. (1)) of the volume element. A good example is provided by electrons moving in the solid and interacting with it via a deformation potential.[13] Then one can show[22,10] that for a closed surface S bounding a volume V

$$\frac{d}{dt}\int_V (G_l - g_l)\, dV = \int_S (p_{li} - P_{lj} - U\delta_{lj})\, dS_j \qquad (59)$$

If the surface integral vanishes the total ordinary momentum and the total quasi-momentum inside S increase at the same rate, and one is as good as the other in drawing up a momentum balance sheet. The surface integral will vanish if the disturbance has not reached S, and U is negligible (or a constant) on S. More important, it vanishes if, as usual in theoretical solid-state calculations, we impose periodic boundary conditions; surface elements separated by a period then given contributions which cancel.

It is perhaps worth remarking that a point charge (electron) interacting with a medium through a deformation potential which only depends on the volume change induces a center of dilatation. Two electrons at rest do not interact, because a center of dilatation produces no hydrostatic pressure, and so the force between them is zero by Eq. (35). When they are moving there is a hydrostatic pressure and hence an interaction which, in most cases, is the one responsible for superconductivity.

The considerations above refer to a medium free of defects. It is difficult to fit moving defects rigorously into the P_{lj}, g_l formalism. However, results obtained otherwise can often be reproduced correctly by assuming that the force on a defect and quasi-momentum are related in the same way as ordinary force and momentum. We return to this point in Sec. 10.

We end this section with some remarks on sign. Here, the sign of g_l has been reversed compared with Refs. 3 and 10, so that now the quasi-momentum of a wave packet points roughly in the direction in which it is going. The sign of P_{lj} was originally chosen so that, in the static case, its surface integral gives the force on a defect in the generally accepted sense, rather than, as would perhaps have been more logical, the equal and opposite force, of unspecified origin, which keeps the defect from moving. (Either sign may be found in the field theory literature.) On the other hand, the surface integral of p_{ij} is equal and opposite to the body force inside the surface. This accounts for the difference in sign between P_{lj} and p_{lj} in Eq. (37) and other places where they occur together. In the dynamic case the surface integral of P_{lj} represents the flow of quasi-momentum into the surface, but the integral of the energy-flow vector s_i gives the outflow of energy; hence the difference of sign in Eqs. (57) and (58).

9. STATIC CRACKS

Apart from their obvious importance in fracture mechanics, cracks, because of the weakness of the singularity at their tips, lend themselves nicely to the illustration and extension of some of the results we have reviewed.

When regarded as a very flat hole, a crack is really a planar defect. Nevertheless, in the two-dimensional case at least, each of the tips of a crack can be regarded as a defect or singularity in its own right. The force on the tip, found by integrating P_{lj} round a loop embracing it, is the same as the crack extension force or energy release rate $\mathcal{G}$ of fracture mechanics. We shall begin by reviewing various formal ways of calculating $\mathcal{G}$ by purely elastic methods, and then, rather crudely, see what happens when one allows for the existence of interatomic forces or plasticity near the tip.

We start with the classical argument[23,24] which gives the energy release rate. For mode I deformation the stress p_{22} just ahead of the crack and the opening displacement just behind it are[25]

$$p_{22}(x) = K(2\pi)^{-1/2}x^{-1/2} \qquad \Delta u_2 = \frac{K}{M}(2\pi)^{-1/2}(-x)^{1/2} \tag{60}$$

where K is the stress intensity factor and M is an elastic modulus which is $\mu/(1-\nu)$ for isotropy. Let the tip advance by $\delta\xi$ and apply forces to the new freshly formed surface so as to close it up again. It is clear from the relation between the old p_{22} and the new Δu_2 shown in Fig. 4*a* that the work extracted from the system when these forces are slowly relaxed is

$$\frac{1}{2}\int_0^{\delta\xi} p_{22}(x)\Delta u_2(x-\delta\xi)dx = \frac{K^2}{2M}\delta\xi \tag{61}$$

The crack extension force or energy release rate $\mathcal{G}$ is, by definition, the energy extracted per unit advance, and so

$$\mathcal{G} = \frac{K^2}{2M} \tag{62}$$

The same method can be applied to modes II and III; for mode II Eq. (62) is still correct with $M = \mu/(1-\nu)$ and for Mode III we must put $M = \mu$.

A crack is equivalent to a pileup of dislocations, in the limit where their Burgers vectors are infinitesimal and they are infinitely numerous.[25] Fig. 4*b* illustrates this for mode II. The first dislocation is locked and the rest are free. For mode I the dislocations are still of edge type, but their Burgers vectors are vertical and for mode III they are screws.

For the discrete pileup of Fig. 4*b* the force on the locked dislocation is, as we have seen, given by the integral

$$F_1 = \int_S P_{1j}dS_j \tag{63}$$

taken over the loop $S = S_1$. But as all the other dislocations are in equilibrium (i.e., there is no force on them) we may extend S to embrace any number of them, provided we do not include the locked

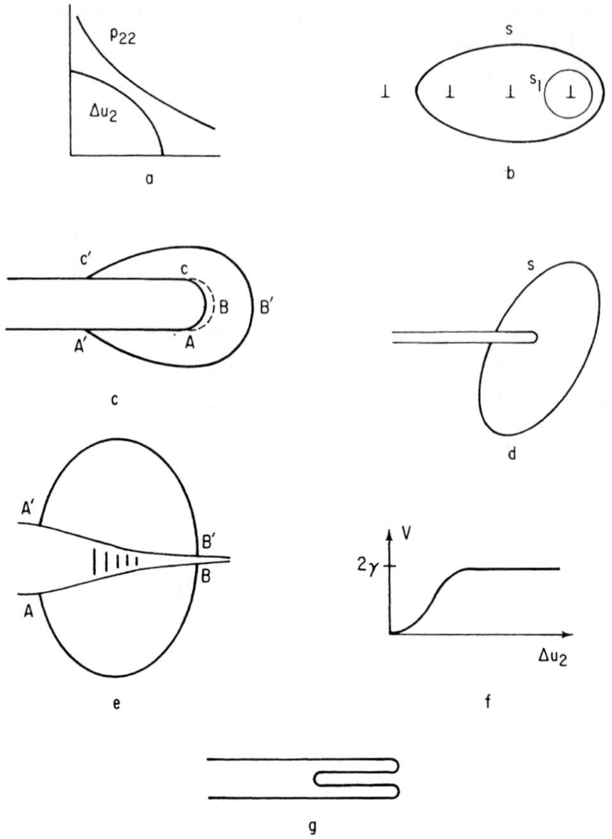

Fig. 4

dislocation at the other end of the crack. We suspect that this result will hold in the continuum limit, so that F_1 is given by Eq. (63) with S any loop embracing the crack tip (Fig. 4*d*). Since F_l and $\mathcal{G}$ both give the energy loss per unit advance we should have $\mathcal{G} = F_1$ One can verify this by trying a few special cases. Actually one can do better.[25] It is possible to write down an expression for the force on the locked dislocation in the discrete case, pass to the limit, and then, using the details of the dislocation-crack analogy show that $F_1 = \mathcal{G}$. Equation (63) is, of course, Rice's[9] path-independent integral expression for $\mathcal{G}$.

One can also get F_1 by regarding the crack as a cavity bounded by flat parallel faces separated by a small but finite distance h and closed by rounded ends (Fig. 4*c*). We can then apply Eq. (45) with $P_{lj}{}^B$ zero. If the boundary moves forward by $\delta\xi$ like a trombone slide, $\delta\xi_1 = (\delta\xi, 0, 0)$ is constant over ABC and zero elsewhere so that Eq. (45) gives

$$F_1 = \mathcal{G} = \int_S P_{1j}\, dS_j \tag{64}$$

taken over ABC. The path may be deformed to $A\,A'B'C'\,C$, and as AA', CC' contribute nothing we are left with an arbitrary loop embracing the tip. We now argue that if the dimensions of S are large compared with h, the elastic field will (as can be verified in simple cases) be almost unaffected by the value of h, and so we can let h tend to zero. The weak point is, of course, that the proof of Eq. (45) involves the removal of material, and the whole point about a hair-line crack is precisely that its advance does not involve the elimination or redistribution of material. The most satisfactory general proof is to repeat word for word for the internal surface S of Fig. 4*d* the argument we applied to the internal surface S of Fig. 2*a*. The fact that S is intersected by the crack leads to no trouble provided the shift made in calculating $\delta E_{el}^{(i)}$ is parallel to the crack. Essentially this argument has been used by Cherepanov[26] and Rice.[27]

It is interesting to see what we get from a crude atomic model, analogous to the Peierls-Nabarro[13] model of a dislocation. In Fig. 4*e* the cracked solid is represented as two elastic blocks pulled open in mode I by externally applied forces and held together by interatomic forces represented by the shading. We suppose that the interatomic forces can be derived from a potential depending on the separation of the blocks, so that, at the faces

$$p_{22} = \frac{\partial V(\Delta u_2)}{\partial(\Delta u_2)} \tag{65}$$

In contrast to the Peierls-Nabarro case, V is not periodic in Δu_1, but has the form shown in Fig. 4*f* with a plateau whose height is 2γ where γ is the surface energy of the material. Let us integrate P_{ij} round the loop S shown in Fig. 4*e*. Because the divergence of P_{ij} is zero the integral can equally well be taken along the upper and lower faces of the crack itself, and we have[9]

$$F_1 = \int_S P_{1j}\, dS_j \tag{66}$$

$$= -\int_A^B p_{22} \frac{\partial}{\partial x_i} (\Delta u_i)\, dX_i \tag{67}$$

$$= \int_B^A p_{22}\, d(\Delta u_i) \tag{68}$$

$$= \int_B^A \frac{\partial V}{\partial \Delta u_i}\, d\Delta u_i \tag{69}$$

$$= V_A - V_B \tag{70}$$

If A lies in the region where the crack is so far open that the interatomic force is practically zero, and B in the region where Δu_i is practically zero, $V_A - V_B$ is 2γ and we have

$$F_1 = \mathcal{G} = 2\gamma \tag{71}$$

or the Griffith condition for brittle cracks.

Equation (68) can also be used in other cases. For example, in the models of Dugdale[28] and of Bilby, Cottrell, and Swinden[29] beyond the crack tip there is a narrow plastic zone in which $p_{22} = \sigma_Y$. If S embraces both tip and plastic zone Eq. (68) gives

$$F_1 = \sigma_Y \Delta u_2(\text{tip}) \tag{72}$$

where Δu_2 (tip) is the so-called crack opening displacement. More generally, if the narrow plastic zone exhibits plastic behavior too complicated to be expressed by a constant yield stress, the integral Eq. (68) is still the plastic work required to extend unit length of the zone to rupture.

The argument which gave Eq. (71) can also be applied locally to the three-dimensional case of a penny-shaped crack, or a similar flat crack whose periphery is not necessarily circular. Take X_3 and X_2 to be tangential to the edge and normal to the plane of the crack, and let S be a closed surface which, so to speak, takes a bite out of the crack surface and intercepts a length δl of its edge. Then the process which led from Eq. (66) to Eq. (71) gives

$$\int_S P_{1j}\, dS_j = 2\gamma\,\delta l \tag{73}$$

which shows that Eq. (62) applies locally all along the edge of a curved crack. This is fortunate since the usual proof is not entirely convincing. For the two-dimensional case we assumed that Δu_2 for the lengthened crack was the same as the original Δu_2 advanced bodily. Although this is wrong one can easily show that the answer is right. But to apply the same argument to a finite length of the edge we must let the crack change its shape by putting out a tongue, and then a certain amount of goodwill is necessary.

10 MOVING CRACKS

Moving cracks sometimes attain speeds not far short of the velocity of transverse elastic waves in the material. Then the static value of $\mathcal{G}$ can no longer be used. In this section we review methods for calculating the dynamic $\mathcal{G}$ first according to pure continuum theory, and then according to the simple Peierls-Nabarro model of the last section. Part of the analysis also applies to defects other than crack tips. Some of the results will then be tried out on a recently published solution for the elastic field near the tip of a crack moving arbitrarily in antiplane strain, and used to give a rather naive treatment of crack branching.

Craggs[30] supposed that the energy release rate per unit time $v\mathcal{G}$ for a crack tip moving with instantaneous velocity v was given by the integral

$$vG = \int_S p_{ij}\,\dot{u}_i\, dS_j \tag{74}$$

taken over a small circle S centered on the tip. Unfortunately, Eq. (74) gives wrong answers in the static limit and, further, its value depends on the shape of S. The right-hand side of Eq. (74) is the integral of minus the normal component of the energy flow vector (Eq. (54)) taken over S, and so gives the rate at which elastic and kinetic energy flows in through S. This energy either leaves via the tip, or goes to increase the energy inside S. So Eq. (74) must be amended to

$$v\mathcal{G} = \int_S p_{ij}\dot{u}_i\,dS_j - \frac{d}{dt}\int_V (W + T)\,dV \tag{75}$$

where V denotes the interior of S. If the displacement has the form

$$u_i = u_i(X_1 - vt, X_2) \tag{76}$$

so that the elastic field is transported rigidly, the energy inside varies only because of the transport of the rigid field into and out of the boundary and

$$\frac{d}{dt}\int (W + T)\,dV = -v\int (W + T)\,dS_1 \tag{77}$$

so that[26,31]

$$v\mathcal{G} = \int_S \left[p_{ij}\dot{u}_i + v(W + T)\delta_{1j}\right] dS_j \tag{78}$$

Alternatively, when Eq. (76) is valid we may replace $\dot{u}_i$ by $-vu_{i,1}$ and cancel v to get

$$\mathcal{G} = \int_S H_{1j}\,dS_j \tag{79}$$

where the tensor

$$H_{lj} = (W + T)\delta_{lj} - p_{ij}u_{i,l} \tag{80}$$

differs from the dynamic P_{lj} only in the sign of the kinetic energy term.

When Eq. (76) does not hold, the displacement can still be written in the form

$$u_i = u_i^0(X_1 - vt_1, X_2) + u_i'(X_1, X_2, t) \tag{81}$$

in an infinity of ways, but in favorable cases it will be possible to arrange that u_i' is negligible compared with u_i^0 close to the point $X_1 = vt, X_2 = 0$. This is so for all moving crack solutions so far published (compare, for example, with Eq. (93)). If this is so Eqs. (78) and (79) will be correct for small enough S, and we have

$$vF_1 = v\mathcal{G} = \lim_{S \to 0} \int_S \left[p_{ij}\dot{u}_i - v(W + T)\delta_{1j}\right] dS_j \qquad (82)$$

or

$$F_1 = \mathcal{G} = \lim_{S \to 0} \int_S H_{1j}\, dS_j \qquad (83)$$

We can also arrive at Eq. (83) and hence also Eq. (82), though rather metaphysically, by arguments based on Sec. 8. In the dynamic case

$$\int_S P_{1j}\, dS_j \qquad (84)$$

should give the force on the crack tip, plus the rate of change of the quasi-momentum inside S, and so

$$F_1 = \int_S P_{1j}\, dS_j - \frac{d}{dt}\int_V g_1\, dv \qquad (85)$$

If Eq. (76) holds the second term is

$$v \int g_1\, dS_1 = \int 2T\, dS_1 \qquad (86)$$

because $g_1 = -\,p\dot{u}_i u_{i,1} = p\dot{u}_i \dot{u}_i$, and we recover Eq. (83). The rest of the argument goes as before.

If the crack tip is moving with a velocity v_l arbitrarily inclined to the X_1-axis, Eq. (83) becomes

$$F_l = \lim_{S \to 0} \int_S H_{lj}\, dS_j \qquad (87)$$

This gives the correct value for the component of force parallel to v_l and, if the displacement has the form

$$u_i = u_i^0(X_m - v_m t) + u'_i(X_m, t) \tag{88}$$

generalizing Eq. (81), the component perpendicular to v_l is zero if u'_i ultimately makes no contribution, and so we need not worry about its significance.

Although in deriving Eq. (81) we have introduced convective terms, the velocity v_l of the singularity does not appear in Eq. (87). This is because v_l is implicitly defined by the field variables themselves. For any nonstatic displacement function we can define a velocity field $v_l(X_m)$ given by the solution of the matrix equation

$$v_l(X_m)\,u_{i,l} = -\,\dot{u}_i \tag{89}$$

If u_i has the form of Eq. (88), $v_l(X_m)$ will take the constant value v_l near the singularity, and in the arguments above we have at several points eliminated $v_l = (v, 0, 0)$, or introduced it and cancelled it out by replacing one side of Eq. (89) by the other.

There seems to be nothing really specific to crack tips in the analysis which led to Eq. (87), and this suggests that it gives the force on a defect generally, in three dimensions as well as two. For static problems H_{lj} reduces to P_{lj}, and, as its divergence vanishes, we can ignore the limit sign and recover our original static results. In the dynamic case the divergence vanishes to the extent that the contribution of u'_i (Eq. (88)) is negligible, so that in the limit Eq. (87) should be independent of the shape as well as size of S.

We can test Eq. (87) on some already worked-out examples. But from the present point of view the results are disappointing because, in fact, a crack tip is almost the only sensible singularity for which the force can be worked out cleanly on the basis of continuum mechanics alone, without additional physical considerations. If one evaluates Eq. (87) for a kink, or a mobile point defect (Frenkel caterpillar), with S a sphere of radius r, or a dislocation with S a cylinder of radius r the results diverge, respectively, like r^{-1}, r^{-3} and $\ln r$. This is because each of them has an effective mass, similar to the electromagnetic mass of an electron, which is proportional, respectively to r_o^{-1},[32] r_o^{-3},[3] and $\ln\,[R(t)/r_o]$ [33] where r_o is an appropriate cutoff radius, different in each case, and $R(t)$ depends on the history of the dislocation's motion. We shall see later that crack tips have no effective mass, and that is why for them the calculation goes through nicely. Whenever one has to introduce a cutoff radius in continuum calculations it is a sign that an atomic treatment is really necessary, and so we shall try to extend the model which led to Eq. (71) and apply it not only to cracks but to other defects as well.

Equations (66) and (71) remain valid in the dynamic case so long as S is taken to run along the crack faces, because the extra dynamic term $-T\delta_{ij}$ contributes nothing. However, when we open S out into the loop of Fig. 4e quasi-momentum terms appear. For example Eq. (71) becomes

$$\int_S P_{1j}\, dS_j - 2\gamma = \frac{d}{dt}\int_V g_1\, dV \tag{90}$$

where V is the area inside S. This says that the rate of increase of quasi-momentum in V is equal to the flux of it through S (the P_{ij} term) plus the quasi-force -2γ which surface tension exerts. It is negative because it tries to close the crack.

Apart from a few trivial changes in suffixes the analysis leading to Eq. (90) applies also to an edge or screw dislocation in the Peierls-Nabarro model, V becoming the Peierls potential. We should again get Eq. (90) with 2γ equal to the difference in Peierls potential at A and B. But if the loop S completely straddles the core of the dislocation V_A and V_B are the same, both being equal to the minimum of the Peierls potential, and we get simply

$$\int_S P_{1j}\, dS_j = \frac{d}{dt}\int_V g_1\, dV \tag{91}$$

We can treat a Frenkel caterpillar (an idealized crowdion), extended along the X_1 axis, in the same way, imagining that the lattice row of atoms which contains the extra atom interacts with the rest of the crystal through a "Peierls potential." We should again get Eq. (91) now regarded as three dimensional so that S is, say, a sphere centered on the defect and large compared with its width. We could also apply a similar argument to a kink free to glide along the X_1 axis, and once more arrive at Eq. (91).

Equation (91) says that the usual quasi-momentum balance equation is not upset by the presence of the defect. We may say that there is no net force on the defect, which is what is to be expected for a freely moving entity. The fact that the net force on the defect is zero does not, of course, stop us calculating one of the partial forces on it which add up to zero. For example,[34,32] if we supposed that a dislocation or kink merely oscillated about a fixed mean position when an elastic wave was incident on it we should find that Eq. (91) was not satisfied. To prevent it shooting away we should have to apply an additional steady force, say by imposing a constant stress, and this steady force, with reversed sign, is, in a sense, the force which the wave exerts on the defect. For the cases we have considered, Eq. (91) justifies the statement[34] that quasi-momentum and the force on a defect are related like ordinary force and momentum.

To conclude this section we shall apply some of the ideas developed to the simple case of a crack tip moving arbitrarily under antiplane strain. As our results will only apply to the linear theory we shall write

$$X_1 = x, X_2 = y, u_3 = w \tag{92}$$

Initially the medium is stationary and the crack runs along the x axis from $-l$ to 0. At $t = 0$ its right-hand tip begins to move according to the arbitrary law $\xi = \xi(t)$. The displacement near the tip is

$$w = A(\dot{\xi})\, B(\xi)\, W_o \left[\frac{x - \xi}{\beta(\dot{\xi})}, y\right] + 0(R\lambda)^{3/2} \tag{93}$$

where

$$W_o(x, y) = \left\{\frac{1}{2}(x^2 + y^2)^{1/2} - \frac{1}{2}x\right\}^{1/2} \tag{94}$$

$$A(\dot{\xi}) = \frac{(1 - \dot{\xi}/c)^{1/4}}{(1 + \dot{\xi}/c)^{1/4}}, \quad \beta(\dot{\xi}) = \left(1 - \frac{\dot{\xi}^2}{c^2}\right)^{1/2} \tag{95}$$

$$B(\xi) = \frac{2}{\pi\mu}\int_o^{\xi} p^o_{zy}(x',0)(\xi - x')^{-1/2}\,dx' \tag{96}$$

R is the distance from the current position of the tip and $p^o_{zy}(x, 0)$ is the stress ahead of the crack before it started to move. Equation (93) ceases to be valid when reflections from the other tip or the surface of the medium overtake the moving tip. Equation (78) gives

$$\mathcal{G}(\xi, \dot{\xi}) = \frac{1}{4}\pi\mu A^2(\dot{\xi})\, B^2(\xi) \tag{97}$$

If we equate this to 2γ we have an equation of motion for the crack tip which, solved for $\dot{\xi}$ and integrated, gives its trajectory under the prescribed loading. The acceleration of the tip only enters the term $O(R^{3/2})$ in Eq. (93) and does not affect $\mathcal{G}$. Consequently we may say that the crack tip has no inertia, as stated earlier. Equation (97) can be verified directly in a simple case.[36] After the tip has been moving for a time t the field outside a circle $r = ct$ is unaltered. So if we calculate the total energy in this circle and subtract it from the elastic energy inside the same circle before the motion started the result should be the time-integral of $\dot{\xi}\mathcal{G}$ from 0 to t. For the case of constant $\dot{\xi}$ and constant B this is found to be so. (B is constant if the external loading happens to make $p_{zy}{}^{o}(x,0)$ strictly proportional to $x^{-1/2}$ and nearly constant in any case at the beginning of the motion.) In a sense this special case establishes Eq. (97) generally, without invoking Eq. (78) if we admit that the term $O(R^{3/2})$ in Eq. (93) does not affect $\mathcal{G}$.

There has been some argument[37] as to whether a crack branches when a critical velocity or, alternatively, a critical $\mathcal{G}$ value is reached. We can use Eq. (97) to give a rather crude discussion. We cannot, of course, really tackle the dynamic problem of a Y-shaped crack (for the static case see a paper by Smith[38]). However, in the case of glass at least, the angle between the branches is often small, and so we consider the limiting case of zero branching angle, Fig. 4*g*. This only differs from the unbranched case by the presence of an unstressed septum extending back from the tip of the crack. After branching Eq. (97) is still valid but now $\mathcal{G}$ must be equal to 4γ rather than 2γ. Since $B(\xi)$ only depends on the current length, the crack can only provide the extra energy by slowing down so as to increase A^2. In fact we must have

$$A^2(v_2) = 2A^2(v_1) \tag{98}$$

if v_1 and v_2 are the velocities just before and just after branching. Since $A(v)$ is unity for $v = 0$ and decreases monotonically to zero at $v = c$, Eq. (98) has no solution unless v exceeds the value v_{br} which is the root of $A^2(v_{br}) = 2$, namely

$$v_{br} = \frac{3}{5}c = 0.6c \tag{99}$$

If v_1 is only marginally greater than v_{br}, $A(v_2)$ is nearly unity and the branches must move very slowly, but if the crack can restrain itself from branching until v_1 is somewhat above v_{br} the branches can move move off with a reasonable velocity. It seems likely that Eq. (93) is valid generally (i.e., when reflections are not ignored and the applied stresses are, possibly, not constant) provided one replaces $A(v)B(\xi)$ by a function $C(t)$ which cannot be separated into velocity and length factors. If this is granted the values of C_1 and C_2 before and after branching can be related by matching the solutions across the circle of radius $c(t - t_1)$ which separates the two solutions if branching occurs at $t = t_1$. This gives[36] $A_2 C_1 = A_1 C_2$, which leads to the same v_{br} as before. According to our argument, therefore, there is a critical *velocity* for branching. It makes no sense to speak of a critical $\mathcal{G}$, because just before branching $\mathcal{G}$ must have the same value, dictated by the material properties, as it would if branching was not just about to occur.

When the appropriate $A(v)$ becomes available we shall be able to calculate v_{br} for the more realistic case of mode I deformation. In the meanwhile we note that two other critical velocities provided by pure continuum theory are nearly the same for mode I as they are for mode III. The first is the absolute upper limit to the tip velocity, about $0.95c$ (the Rayleigh velocity) for mode I and c for mode III. The other is the Yoffe[39,40] velocity above which the radius across which the stress is a maximum no longer lies dead ahead of the crack; it slightly exceeds $0.7c$ in both cases. If it should turn out that the branching velocity v_{br} is also nearly the same for mode I as it is for mode III, it will also be nearly the same as the value $0.6c$ quoted as a common experimental value for the terminal velocity of a crack when there is no question of branching. (This value is usually related to Roberts and Wells'[41] calculated value $0.6c$, but this is an approximation, and the true terminal velocity in their sense is the Rayleigh velocity.) If in fact v_{br} is nearly the same for modes I and III, and if the agreement with the experimental terminal velocity is not simply a coincidence, we may tentatively give the following interpretation. In glass the initial smooth crack surface (mirror) is succeeded by a so-called mist region. Microscopic examination shows that in it small tongues of crack project from the main crack into the material. This suggests that even when macroscopic branching does not occur when v_{br} is reached, microscopic branching may set in and limit the crack velocity.

11. CONCLUDING REMARKS

We have reviewed some of the energy relations in solid mechanics, and emphasized the usefulness of the energy-momentum tensor. A topic we have not touched on is the use of Rice's path-independent integral in contexts where its interpretation as an effective force is not involved.[9,26,42] There are analogs of the energy-momentum tensor in branches of continuum mechanics other than the theory of elasticity. Some of them might repay investigation.

REFERENCES

1. Burton, C. V.: *Phil. Mag.*, (5) **33**:191 (1892).
2. Weiner, J. H.: *J. Appl. Phys.*, **29**:1305 (1958).
3. Eshelby, J. D.: *Phil. Trans.*, **A244**:87 (1951).
4. Hill, E. L.: *Rev. Mod. Phys.*, **23**:253 (1951).
5. Corson, E. M.: "Introduction to Tensors, Spinors, and Relativistic Wave-Equations," p. 65, Blackie and Son, London and Glasgow, 1953.
6. Bilby, B. A.: in I. N. Sneddon and R. Hill (eds.), "Prog. Solid Mech.," vol. 1, p. 331, North-Holland Publishing Company, Amsterdam, 1960.
7. Timoshenko, S., and J. N. Goodier: "Theory of Elasticity," p. 425, McGraw-Hill Book Company, New York, 1951.
8. Morse, P. M., and H. Feshbach: "Methods of Theoretical Physics," p. 323, McGraw-Hill Book Company, New York, 1953.
9. Rice, J. R.: *J. Appl. Mech.*, **35**:379 (1968).
10. Eshelby, J. D.: in F. Seitz and D. Turnbull (eds.), "Prog. Solid State Physics," vol. 3, p. 79, 1956.
11. ——: in E. M. Rassweiler and W. L. Grube (eds.), "Internal Stresses and Fatigue in Metals," p. 41, North-Holland Publishing Company, Amsterdam, 1959.
12. Ardell, A. J., and R. B. Nicholson: *Acta Met.*, **14**:1295 (1966).
13. Nabarro, F. R. N.: "Theory of Crystal Dislocations," Oxford University Press, 1967.
14. Truesdell, C., and R. Toupin: in S. Flügge (ed.), "Encyclopedia of Physics," vol. 3/1, p. 495, Springer-Verlag, Berlin, 1960.
15. Schweinsfeir, R. J., and C. Elbaum: *J. Phys. Chem. Solids*, **28**:597 (1967).
16. Thomas, L. A., and W. A. Wooster: *Proc. Roy. Soc.*, **A208**:43 (1951).
17. Hill, R.: in I. N. Sneddon and R. Hill (eds.), "Prog. Solid Mech.," vol. 2, p. 247, North-Holland Publishing Company, Amsterdam, 1961.
18. Rice, J. R., and D. C. Drucker: *Int. J. Fracture Mech.*, **3**:19 (1967).
19. Truesdell, C.: *J. Rat. Mech. Analysis*, **1**:178 (1952).
20. Post, E. J.: *Phys. Rev.*, **118**:1113 (1960).
21. Love, A. E. H.: "Mathematical Theory of Elasticity," p. 177, Cambridge University Press, 1927.
22. Brenig, W.: *Z. Phys.*, **143**:168 (1955).
23. Irwin, G. R.: *J. Appl. Mech.*, **24**:361 (1957).
24. Barenblatt, G. I.: in H. L. Dryden and Th. von Kármán (eds.), "Adv. Appl. Mech.," vol. 7, p. 56, 1962.
25. Bilby, B. A., and J. D. Eshelby: in H. Liebowitz (ed.), "Fracture," vol. 1, p. 99, Academic Press, Inc., New York, 1968.
26. Cherepanov, G. P.: *Int. J. Solids Structures*, **4**:811 (1968).
27. Rice, J. R.: in H. Liebowitz (ed.), "Fracture," vol. 2, p. 191, Academic Press, Inc., New York, 1968.
28. Dugdale, D. S.: *J. Mech. Phys. Solids*, **8**:100 (1960).
29. Bilby, B. A., A. H. Cottrell, and K. H. Swinden: *Proc. Roy. Soc.*, A272:304 (1963).
30. Craggs, J. W.: "Fracture of Solids," p. 51, Interscience, New York, 1963.
31. Atkinson, C., and J. D. Eshelby: *Int. J. Fracture Mech.*, **4**:3 (1968).
32. Eshelby, J. D.: *Proc. Roy. Soc.*, **A266**:222 (1962).
33. ——: *Phys. Rev.*, **90**:248 (1953).
34. Nabarro, F. R. N.: *Proc. Roy. Soc.*, **A209**:278 (1951).
35. Eshelby, J. D.: *J. Mech. Phys. Solids*, **17**:177 (1969).

36. ——: in A. S. Argon (ed.), "Physics of Strength and Plasticity," Massachusetts Institute of Technology Press, Cambridge, 1969.
37. Congleton, J., and N. J. Petch: *Phil. Mag.*, **16**:749 (1967).
38. Smith, E.: *J. Mech. Phys. Solids*, **16**:329 (1968).
39. Yoffe, E. H.: *Phil. Mag.*, **42**:739 (1951).
40. McClintock, F. A., and S. P. Sukhatame: *J. Mech. Phys. Solids*, **8**:187 (1960).
41. Roberts, D. K., and A. A. Wells: *Engineering*, **178**:820 (1954).
42. Hutchinson, J. W.: *J. Mech. Phys. Solids*, **16**:13 (1968).

DISCUSSION *on Paper by J. D. Eshelby*

C. A. BERG: As you have pointed out, your energy-momentum tensor constructed with the elastic energy density function provides a measure of the force acting on a defect. Can one use the complementary energy density to construct a complementary energy-momentum quantity which would provide an estimate of the displacement of a defect from its equilibrium site when prescribed forces are applied to the body?

J. D. ESHELBY: The natural arguments of the complementary energy are the stresses. To fit the formalism I presented, the stresses would have to be written as the gradient of something. I dare say that if this were done in detail something interesting might come out. But, any connection with displacement would be pretty indirect because the new energy-momentum tensor would still be a stress. My plan to arouse interest in energy-momentum tensors seems to have worked in the case of Professor Berg, and I hope he will follow up the point he has raised.

G. R. IRWIN: When one averages across small scale discontinuities which are essential to progressive crack extension, it is possible to observe that the speed of a running semibrittle (or brittle) crack changes in phase with changes in the crack-extension force. Between the onset of rapid crack extension and the attainment of a limiting or plateau crack speed, the observed behavior is that of a low-inertia, over-damped disturbance. After a limiting velocity has been nearly reached, efforts at crack division (or branching) are observed. These occur with increasing frequency with increase of G and eventually result in successful branching. The limiting velocity prior to branching may be as high as 0.65 C_2 in a nearly pure silicon glass or half as large in a leaded glass with a substantial damping characteristic.

Successful branching always follows the attainment of a plateau velocity and does not appear to depend upon the magnitude of the velocity. The fact that branching is never observed prior to achievement of a limiting speed is readily explained. It is due to the high probability that one of two new crack paths will outrun and stress relieve the other. Only when the effort at crack division produces two new cracks moving fast enough so that their speeds are relatively insensitive to crack size will the branching effort be successful. Because successful branching seems to depend upon achievement of a plateau crack speed, it would seem advisable to center attention first upon explanation of the plateau speed. In some materials the plateau speed may be quite low; 0.2 C_2, for example.

A generalized force concept is somewhat arbitrary and depends upon how we define the crack model and the process speed. Suppose we assume the model contains only one energy reservoir, the elastic stress field. But, we wish to recognize that adiabatic heating due to inelastic strains near the crack may tend to expand the metal and reduce the tensile stresses which are driving the process. After correcting the stress field for the influence of adiabatic heating and by assuming the process speed is nearly constant, it may be possible to obtain a useful force calculation using the path independent integral idea.

International Journal of Fracture Mechanics, Vol. 7, No. 4, December 1971
Wolters-Noordhoff Publishing–Groningen
Printed in the Netherlands

The Fracture Mechanics of Flint-Knapping and Allied Processes

J. G. FONSECA, J. D. ESHELBY AND C. ATKINSON*

Department of the Theory of Materials, University of Sheffield, England

(Received October 2, 1970)

ABSTRACT
The production of flakes of a brittle solid by pressure or impact is studied with the help of a simplified mathematical model.

1. Introduction

Fig. 1 illustrates one of the basic techniques, *flaking*, of flint-knapping, whether for the production of stone tools, or, more recently, gun flints. A suitably prepared lump of flint, the *core*, is subjected to a concentrated force near the edge. If this is done properly a crack runs down parallel to the face of the core, producing a *flake*. The force may be applied steadily (*pressure flaking*) or impulsively (*percussion flaking*). Kerkhof and Müller–Beck [1] have discussed the matter in the context of fracture mechanics. Here we try to give a simplified mathematical treatment of some aspects of the process.

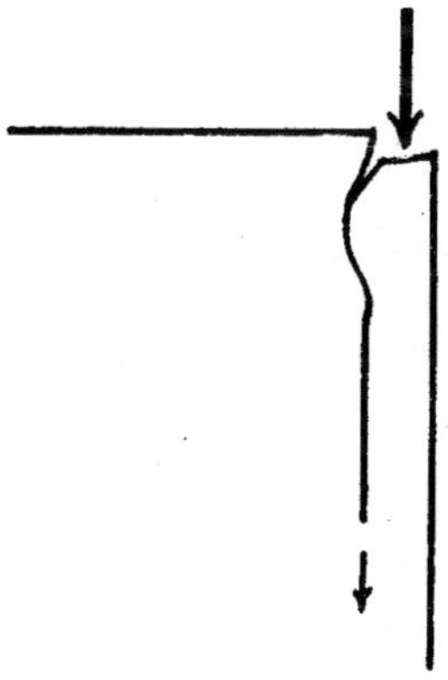

Figure 1. Formation of a flake (After Siret [19]).

The problem we have chosen may, perhaps, seem a frivolous one, but in fact it embodies in a simple form several features which are of general interest in what one may call destructive fracture mechanics. In contrast to, say, the classic situation of a crack perturbing a uniform stress in an infinite medium, the external force producing fracture is highly localized in space, and possibly also in time, and, in addition, because of its closeness to a free surface the crack tip is, throughout its motion strongly influenced by reflections of the disturbance it itself creates. The same features would reappear in, for example, an analysis of the grinding of powders [2], blasting, impact testing of adhesive joints [3], or the explosive shattering of an aircraft canopy so that the pilot can eject safely [4]. In all these cases an important question is, how much of the energy introduced into the material by the external force finds its way into the crack tip? In fracture mechanics the flow of energy into the crack tip is expressed in terms of the energy release rate or crack extension force G defined so that $G\delta l\delta a$ is the energy which disappears

* Now at Department of Mathematics, Imperial College, London.

into a length δl of the crack border when the tip advances by δa. For a brittle solid this energy goes to provide the surface energy of the freshly formed surface of the crack. Evidently, in the situations mentioned above, only a certain fraction of the work done by the applied force will be absorbed by the tip, and it will be one of our principal aims to find, for the particular problem we have chosen to study, what this fraction is, or in other words the efficiency of the flaking process.

To make the problem tractable we shall have to idealize the situation drasticaly. We treat it as a two-dimensional plane problem. Further, since dynamic plane strain problems involving several free surfaces present many difficulties we shall, in the present paper, only consider the

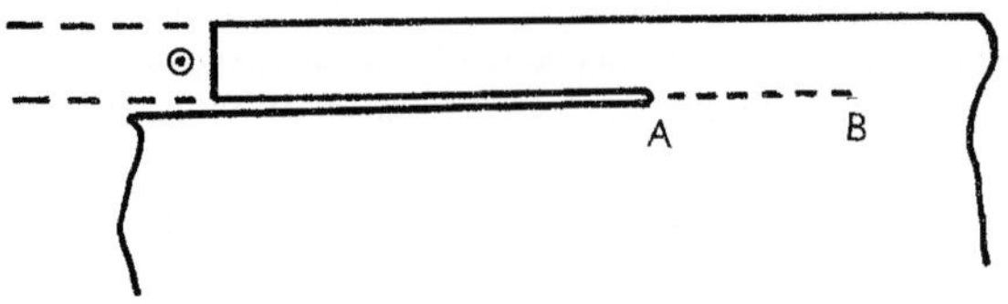

Figure 2. Mode III analogue of fig. 1.

anti-plane analogue of the situation in fig. 1, so that, fig. 2, the force is applied transversely to the end of the flake, as indicated by the conventional sign for an arrow-head pointing towards the reader and the displacement is everywhere normal to the plane of the figure. We do not inquire into the details of the crack initiation process; in section 2 (pressure flaking) we assume that a steady state has been reached, and in section 3 (percussion flaking) we suppose that there is a pre-existing length of crack whose tip starts to move with constant velocity when the pulse transmitted down the flake first strikes it.

2. Pressure Flaking

In this section we consider the case where the crack tip is made to move by applying a force σb to the end of the flake (fig. 2), supposed to be of thickness b. We begin with a static problem. Remote from its junction with the body of the material the stress in the flake becomes uniform and so the displacement will be

$$w(x) = -\sigma(x-x_0)/\mu\,, \tag{2.1}$$

where μ is the shear modulus, provided the point x is a few multiples of b to the left of the tip. This gives a constant stress $\tau_{xz}(x) = -\sigma$, which matches the surface traction σ on the end of the flake. We expect x_0 to be nearly the same as the x-coordinate of the crack tip. Its precise value could be found from the exact elastic solution presented later, but all we shall need is the obvious fact that x_0 will become $x_0+\delta a$ if the crack tip moves to the right by an amount δa. If it does so the work δW done by the applied force σb at the end of the flake is

$$\delta W = \sigma b(\partial w/\partial x_0)\ \delta a = \sigma^2\,\delta a/\mu\,. \tag{2.2}$$

Consider next the increase δE in the elastic energy of the material. After the displacement of the tip the stress, and hence the strain energy density, at the point $(x-\delta a, y)$ in the core or flake is the same as it was at (x, y) before the displacement. So the increase in elastic energy is entirely due to the fact that the flake has lengthened by δa. The elastic energy in the new portion is $\frac{1}{2}\sigma^2 b\delta a/\mu$, and this is δE and equal to one half of δW. The other half of δW provides the energy $G\,\delta a$ which is absorbed by the crack tip, so that

$$G = \sigma^2 b/2\mu\,. \tag{2.3}$$

In the infinitisimal linear theory of elasticity the relation

$$G\,\delta a = \delta W - \delta E = \tfrac{1}{2}\delta W = \delta E \tag{2.4}$$

holds generally for crack extension under dead loading, but in our simple case it is particularly transparent. In anti-plane strain the stress intensity factor K is related to the energy release rate by $G = K^2/2\mu$, so that here

$$K = \sigma b^{\frac{1}{2}} \tag{2.5}$$

Eqn. (2.3) can be derived more formally by evaluating a certain line-integral [5], [6] around a small loop embracing the tip. It is not necessary to know the form of the elastic field near the tip because the integral is path-independent, so that the loop can be opened up until it becomes the curve described by a point which starts anywhere on the lower surface of the crack, moves anti-clockwise along, say, an arc of a circle centered at the crack tip until it reaches a point on the upper surface of the solid to the right of the tip, and thence to the left along this surface until it reaches some point on the upper surface of the flake, and finally crosses the flake to end at any point on the upper surface of the crack. Only the segment across the flake gives a non-vanishing contribution, and it reproduces (2.3).

The same kind of argument will also give G when the crack tip is moving uniformly with a finite velocity c. With suitable origins for x and t, the displacement in the flake is then

$$w = -\sigma(x-ct)/\mu \tag{2.6}$$

provided x is several multiples of b to the left of the tip. This expression satisfies the wave equation and gives the required stress $\tau_{xz} = -\sigma$. In the time δt the crack advances by δa and the load does an amount of work δW which is still given by (2.2). Since the stress pattern moves along rigidly with the tip, it is clear that the increase of energy δE in the material during δt is again equal to the energy contained in a length δa of the flake. This is now made up of elastic energy $\sigma^2 b\delta a/2\mu$ as before, together with kinetic energy

$$\tfrac{1}{2}\rho b\dot{w}^2\,\delta a = \sigma^2\,bc^2\,\delta a/2\mu c_2^2$$

where ρ is the density of the material and $c_2 = (\mu/\rho)^{\frac{1}{2}}$ is the velocity of shear waves in it. Thus in all $\delta E = (\sigma^2/2\mu)b(1+c^2/c_2^2)$. Since $G\delta a = \delta W - \delta E$, we have

$$G = \frac{\sigma^2}{2\mu}\,b\left(1-\frac{c^2}{c_2^2}\right). \tag{2.7}$$

This value is also given by the dynamical extension [5], [7] of the path-independent integral mentioned in connegion with (2.3).

The rate at which the applied force feeds energy into the material is $b\sigma\dot{w} = b\sigma^2 c/\mu$, and the fraction of this which finds its way into the creack tip, that is, the efficiency of the pressure flaking process, is

$$\eta = \frac{cG}{\sigma^2\,c/\mu} = \tfrac{1}{2}\left(1-\frac{c^2}{c_2^2}\right).$$

It falls with increasing velocity, vanishing at $c = c_2$, and for low enough velocities it approaches the value $\frac{1}{2}$ demanded by the general static relation (2.4).

The detailed form of the elastic field, which we have not needed in the above, may, in fact, be found in books on hydrodynamics. To make this clear we replace the problem by a more symmetrical one. Fig. 3 shows two semi-infinite cracks. Suppose that by a force applied to the strip between the cracks far to the left of the tips we induce a uniform stress $\tau_{xz} = -\sigma$ in it. By symmetry $\tau_{yz} = \mu\partial w/\partial y$ is zero all along the plane midway between the cracks. Thus if the material is cut in two along this plane the elastic field is not disturbed and we are left with two replicas of the state of affairs in fig. 2.

Now suppose that fig. 3 refers to a perfect incompressible fluid flowing uniformly along a semi-infinite two-dimensional submerged pipe drawing liquid from or discharging it into an infinite reservoir, the plates which form the sides of the pipe replacing the cracks. The boundary conditions on the velocity potential φ are as follows: $\partial\varphi/\partial y = 0$ on either side of each plate; in the pipe far from the mouth $\partial\varphi/\partial x = \text{const.}$ and $\partial\varphi/\partial y = 0$. Since these are just the conditions which

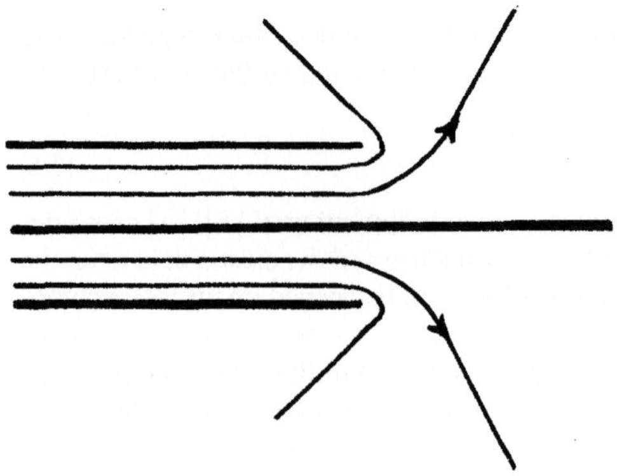

Figure 3. A hydrodynamic analogy.

the displacement w must satisfy in the elastic problem we may, in suitable units, identify w and φ. The solution of the above hydrodynamic problem is contained in the relation [5]

$$z = W + e^{W} \tag{2.9}$$

where $z = x + iy$ and $W = \varphi + i\psi$, where ψ is the stream function conjugate to φ. The transformation (2.9) cannot be inverted analytically to give φ as a function of x and y, but we can easily make out its main features. Well inside the pipe the first term on the right of (2.9) is dominant, so that $w \equiv \varphi = x$. In the body of the liquid remote from the orifice the second term dominates and $w \equiv \varphi = \frac{1}{2} \ln(x^2 + y^2)$. The ends of the walls of the pipe (or the tips of the cracks) are situated at the singularities of the transformation (2.9) where $\partial W/\partial z = \infty$, namely at $z = -1 \pm i\pi$, where $W = \pm i\pi$. Near the lower of these points

$$W = i2^{\frac{1}{2}} z_1^{\frac{1}{2}} + O(z_1^{\frac{3}{2}}) - i\pi \tag{2.10}$$

where $z_1 = z + 1 + i\pi$ is the complex distance from the tip, and so $w \equiv \varphi$ exhibits the square root singularity characteristic of a crack tip.

Evidently with $w = \varphi$, eqn. (2.9) refers to the elastic problem with $b = \pi$, $\sigma = -\mu$ and the general case can be found by suitable scaling. We easily find that if w is given by (2.1) in the flake, then near the crack tip

$$w = \frac{\sigma}{\pi}\left(\frac{2b}{\pi}\right)^{\frac{1}{2}} r^{\frac{1}{2}} \sin \tfrac{1}{2}\theta \tag{2.11}$$

where r, θ are polar coordinates centered at the tip with the crack along $\theta = \pm\pi$. This is the correct expression for an anti-plane crack with the stress intensity factor (2.5).

To get an elastic solution with the tip moving uniformly it is only necessary to make the replacement ("Lorentz transformation"),

$$x \to (x - ct)/\beta\,, \quad y \to y, \quad \beta(1 - c^2/c_2^2)^{\frac{1}{2}} \tag{2.12}$$

in the static solution $w(x, y)$ and then, if we wish the applied load to remain σb, multiply the result by β. In particular, if we define the dynamic stress intensity factor as in the static case, then

$$K(c) = (1 - c^2/c_2^2)^{\frac{1}{4}} K(0)$$

where $K(0) = K$ is the static value (2.5). If, similarly, we introduce a coefficient $L(c)$ describing the relative displacement just behind the tip according to

$$w = \pm \frac{L(c)}{(2\pi)^{\frac{1}{2}}} (ct - x)^{\frac{1}{2}} + O[(ct - x)^{\frac{3}{2}}]$$

we have

$$L(c) = 2(1 - c^2/c_2^2)^{\frac{1}{4}} K(0)/\mu \tag{2.13}$$

with K as given in (2.5). One can show generally that $G=\frac{1}{4}K(c)\,L(c)$ and in the present case (2.13) and (2.14) verify (2.7).

3. Percussion Flaking

We now turn to the case of percussion flaking, where a blow applied to the end of the flake launches a stress pulse which runs along the flake and interacts with the crack tip. In fig. 2 we measure y downwards from the upper surface of the solid, so that the flake is part of the strip $0<y<b$.

Suppose that the end of the flake is at x_1 and that a surface traction $-f'(t)$, independent of y, is applied to it, starting at $t=0$. This produces an anti-plane shear wave with displacement

$$w_a = -(c_2/\mu)f[t-(x-x_1)/c_2]\,, \qquad x>x_1 \tag{3.1}$$

travelling from left to right in the flake. If the head of the pulse has not yet reached to crack tip the displacement everywhere is given correctly by w_0, where

$$\begin{aligned} w_0 &= w_a\,, \qquad 0<y<b \\ w_0 &= 0\,, \qquad b<y<\infty \end{aligned} \tag{3.2}$$

This is trivially true, because w is w_a in the part of the flake occupied by the pulse, and zero elsewhere. But now let us allow the solution (3.2) to evolve in time until the head of the pulse has arrived at B, fig. 2, having overtaken the tip, now at A. Above AB w is finite, and below it zero, so that there is a displacement discontinuity across AB. The correct solution of our problem will be $w=w_0+w'$, where w' has to be found so as to cancel this discontinuity. We must also require that τ_{zy} be continuous across AB; otherwise there would have to be a layer of body force on AB to maintain the motion. In addition, of course, τ_{yz} must be zero on the crack faces and on $y=0$, and w must fall to zero for large y. These conditions, together with the requirement that w reduce to (3.2) before the pulse has reached the crack tip, serve to determine w.

Though it is not necessary to do so it simplifies the analysis if we assume that the flake extends indefinitely to the left of x_1 as shown by the dashed lines, and that in this section there is a disturbance

$$w_a = -(c_2/\mu)f[t+(x-x_1)/c_2]\,, \qquad x<x_1 \tag{3.3}$$

so that w_a is symmetrical about $x=x_1$. Physically, this is the state of affairs which would arise if we applied a force $-2bf'(t)$ at x_1 uniformly spread over the cross-section of the flake. It is clear that if we vary x_1 the sequence of elastic states at the crack tip and beyond it will be the same, measured from a suitably changed time origin. For simplicity we shall take $x_1=0$ so that (3.1) and (3.3) become

$$\left.\begin{aligned} w_a &= -(c_2/\mu)f(t-x/c_2)(t-x/c_2)\,, \qquad x>0 \\ w_a &= -(c_2/\mu)f(t+x/c_2)\,H(t+x/c_2)\,, \qquad x<0 \end{aligned}\right\} \tag{3.4}$$

where the step-functions, defined by

$$H(t)=1,\ t>0\ ; \qquad H(t)=0,\ t<0$$

which are strictly speaking redundant, but emphasise that there is no disturbance for $t<0$ and that the waves have well-defined heads, are inserted for later convenience.

We suppose that the crack starts to move which constant velocity c when the pulse first hits it at $t=0$. The solution would be the same, of course, if it had been moving before this, since its tip would then be unstressed. It will be convenient to use the moving x-coordinate

$$x' = x-ct\,. \tag{3.5}$$

The boundary conditions just enumerated then become

$$\left.\begin{array}{ll}\tau_{yz}(x', b+0, t) = \tau_{yz}(x', b-0, t), & \text{all } x', t\\ \tau_{yz}(x', b\pm 0, t) = 0 & x' < 0, \text{ all } t\\ \tau_{yz}(x', 0, t) = 0 & \text{all } x', t\\ w(x', b+0, t) = w(x', b-0, t), & x' > 0. \text{ all } t\\ w(x', y, t) \to 0 \text{ for } y \to \infty, & \text{all } x', t\end{array}\right\} \tag{3.6}$$

and the equation of motion $\nabla^2 w = \partial^2 w / c_2^2 \partial t^2$ becomes, in terms of x',

$$\frac{\partial^2 w}{\partial x'^2} + \frac{\partial^2 w}{\partial y^2} - \frac{1}{c_2^2}\left(\frac{\partial}{\partial t} - c\frac{\partial}{\partial x'}\right)^2 w = 0\,. \tag{3.7}$$

Transform methods will be used to reduce the partial differential equation (3.7). The first transform, a Laplace transform over time, is denoted by a single bar so that

$$\bar{F} \equiv \bar{F}(x', y, p) = \int_0^\infty \mathrm{e}^{-pt} F(x', y, t)\,\mathrm{d}t \tag{3.8}$$

where the complex number p has a positive real part. The second transform, the Fourier exponential transform over the x coordinate, is denoted by a double bar, for example

$$\bar{\bar{F}} \equiv \bar{\bar{F}}(s, y, p) = (2\pi)^{-\frac{1}{2}} \int_{-\infty}^\infty \mathrm{e}^{isx'} \bar{F}(x', y, p)\,\mathrm{d}x' \tag{3.9}$$

where s is the complex number $s = \sigma + i\tau$.

The differential equation is reduced to

$$\frac{d^2 \bar{\bar{w}}}{\partial y^2} = \gamma^2 \bar{\bar{w}}, \quad \gamma^2 = s^2 + \frac{(p + ics)^2}{c_2^2} \tag{3.10}$$

with the general solution

$$\bar{\bar{w}} = \bar{F}(s, p)\,\mathrm{e}^{\pm\gamma y}\,; \tag{3.11}$$

the behaviour of the exponentials as $y \to \infty$ must be determined. First, the function $\gamma(s, p)$ is factored as follows:

$$\gamma = \gamma_+ \gamma_-, \quad \gamma_\pm = \left[\left(1 \mp \frac{c}{c_2}\right) s \pm i \frac{p}{c_2}\right]^{\frac{1}{2}}, \tag{3.12}$$

where $\gamma_\pm(s, p)$ will denote those branches of the functions which approach positive real numbers when $s \to +\infty$.

Then suppose that

$$\left.\begin{array}{ll}\bar{\bar{w}} = A(s, p)\,\mathrm{e}^{-\gamma y} & \text{for } y \geqq +b\\ \bar{\bar{w}} = B(s, p)\cosh\gamma y + \bar{\bar{w}}_a(s, p) & \text{for } 0 \leqq y \leqq +b\end{array}\right\} \tag{3.13}$$

where $\bar{\bar{w}}_a(s, p)$ is the double transform of w_a, eqn (3.4)

We shall suppose that $\bar{w}$ and $\bar{\tau}_{yz}$ exhibit the following behaviour for large $|x'|$:

$$\left.\begin{array}{ll}|\bar{w}(x', b\pm 0, p)| < M_w \exp(\tau_w^\pm x') & \text{as } x' \to \pm\infty, \ \pm\tau_w^\pm < 0\\ |\tau_{yz}(x', b\pm 0, p)| < M_\tau \exp(\tau_\tau^\pm x') & \text{as } x' \to \pm\infty, \ \pm\tau_\tau^\pm < 0\end{array}\right\} \tag{3.14}$$

where M_w and M_τ are positive real numbers. Define the following four functions of s, p:

$$\left.\begin{array}{l}\varphi_+(b\pm 0) = (2\pi)^{-\frac{1}{2}} \displaystyle\int_0^\infty \mathrm{e}^{isx'} \bar{w}(x', b\pm 0, p)\,\mathrm{d}x'\\ \psi_+(b\pm 0) = (2\pi)^{-\frac{1}{2}} \displaystyle\int_0^\infty \mathrm{e}^{isx'} \bar{\tau}_{yz}(x', b\pm 0, p)\,\mathrm{d}x'\end{array}\right\} \tag{3.15}$$

together with the four functions $\varphi_-(b\pm 0)$, $\psi_-(b\pm 0)$ derived from them by replacing the limits of integration $[0, \infty]$ by $[-\infty, 0]$. The functions $\varphi_+(b\pm 0)$ and $\psi_+(b\pm 0)$ are analytic and vanish at infinity in, respectively, the half planes $\mathrm{Im}\, s > \tau_w^\pm$ and $\mathrm{Im}\, s > \tau_\tau^\pm$, while $\varphi_-(b\pm 0)$ and $\psi_-(b\pm 0)$

are analytic and vanish at infinity in the half-planes Im $s < \tau_w^\pm$ and Im $s < \tau_\tau^\pm$ respectively. Since for both φ and ψ we have $\tau^+ < \tau^-$ the region of analyticity of $\varphi_+(b\pm 0)$ and $\varphi_-(b\pm 0)$ and of $\psi_+(b\pm 0)$ and $\psi_-(b\pm 0)$ overlap severally.

From (3.13) and the definitions (3.15) we have

$$\left.\begin{aligned}\varphi_-(b-0)+\varphi_+(b+0) &= A\,\mathrm{e}^{-\gamma b}\\ \varphi_-(b-0)+\varphi_+(b+0) &= B\cosh\gamma b+\bar{\bar{w}}_a\\ \psi_-(b+0)+\psi_+(b+0) &= -\mu A\gamma\,\mathrm{e}^{-\gamma b}\\ \psi_-(b-0)+\psi_+(b-0) &= \mu B\gamma\sinh\gamma b\end{aligned}\right\}\tag{3.16}$$

Using the conditions (3.6) we obtain

$$\left.\begin{aligned}\psi_\mp(b+0) &= \psi_\mp(b-0)\\ \varphi_+(b+0) &= \varphi_+(b-0)\\ [1+\coth\gamma b]\,\varphi_+(b\pm 0) &= -\varphi_-(b-0)-\coth\gamma b\varphi_-(b+0)+\bar{\bar{w}}_a\\ A &= -B\,\mathrm{e}^{\gamma b}\sinh\gamma b\\ B &= \mathrm{e}^{-\gamma b}\{[\varphi_-(b-0)-\varphi_-(b+0)]-\bar{\bar{w}}_a\}\end{aligned}\right\}\tag{3.17}$$

and substituting back into the last equation in (3.16) we get the Wiener–Hopf equation

$$\psi_-(b\pm 0)+\psi_+(b\pm 0) = \mu\gamma\sinh\gamma b\,\mathrm{e}^{-\gamma b}\{\varphi_-(b-0)-\varphi_-(b+0)-\bar{\bar{w}}_a\}\tag{3.18}$$

By (3.6), in our problem we have $\tau_{yz}(x', b\pm 0, t) = 0$ for all t and $x' < 0$, and so in fact $\psi_-(b\pm 0) = 0$.

The solution $w_a(x, t)$, eqn. (3.4) becomes, in terms of x',

$$\left.\begin{aligned}w_a(x', t) &= -\frac{c_2}{\mu}f\left[\left(1-\frac{c}{c_2}\right)t-\frac{x'}{c_2}\right]H\left[\left(1-\frac{c}{c_2}\right)t-\frac{x'}{c_2}\right]\text{for } x' > -ct\\ w_a(x', t) &= -\frac{c_2}{\mu}f\left[\left(1+\frac{c}{c_2}\right)t+\frac{x'}{c_2}\right]H\left[\left(1+\frac{c}{c_2}\right)t+\frac{x'}{c_2}\right]\text{for } x' < -ct,\end{aligned}\right\}\tag{3.19}$$

whose Laplace–Fourier transform is

$$\bar{\bar{w}}_a(s, p) = -\frac{c_2}{\mu(2\pi)^{\frac{1}{2}}}\left[\frac{\bar{f}\left(\dfrac{c_2 p}{c_2+c}\right)}{(1+c/c_2)\,is+p/c_2}-\frac{\bar{f}\left(\dfrac{c_2 p}{c_2-c}\right)}{(1-c/c_2)\,is-p/c_2}\,.\right]\tag{3.20}$$

Let

$$L(s) = \mathrm{e}^{-\gamma b}\frac{\sinh\gamma b}{\gamma b} = L_+(s)\,L_-(s)\tag{3.21}$$

where $L_\pm(s)$ is regular in Im $s \gtrless \pm\tau^*$, $\tau^* > 0$.

If

$$s = \alpha - \mathrm{i}pc/(c_2^2-c^2)$$

then, as shown in reference [9] the expressions for $L_\pm(s)$ can be written immediately as

$$L_\pm(s) = \exp\{\mp\chi_2(\alpha)-T_\pm(\alpha)\}\prod_{n=1}^{\infty}(1-k^2 b_n^2)^{\frac{1}{2}}\mp\mathrm{i}\alpha b_n\}\,\mathrm{e}^{\pm\mathrm{i}\alpha b_n}\tag{3.22}$$

where

$$\chi_2(\alpha) = -\mathrm{i}\frac{b^*\alpha}{\pi}\left\{1-C+\ln\frac{2\pi}{b^* k}\right\}+\tfrac{1}{2}\alpha b^*$$

$$T_\pm(\alpha) = \frac{\gamma b}{\pi}\cos^{-1}\left(\frac{\alpha}{k}\right)$$

$$k = \mathrm{i}pc_2/(c_2^2-c^2),\quad b_n = b^*/n\pi,\quad b^* = b(1-c^2/c_2^2),$$

C is the Euler constant 0.5772 ..., and the branch of the logarithm used is the one defined by

$$\ln(\xi_1+i\xi_2)=\ln[(\xi_1^2+\xi_2^2)^{\frac{1}{2}}]+i\tan^{-1}(\xi_2/\xi_1) \tag{3.23}$$

with

$$-\pi<\tan^{-1}(\xi_2/\xi_1)\leqq+\pi\,. \tag{3.24}$$

In the appropriate half planes,

$$L_\pm(s)\to\frac{1}{(2b)^{\frac{1}{2}}(1-c^2/c_2^2)^{\frac{1}{4}}(\mp is)^{\frac{1}{2}}}\,,\quad \operatorname{Im} s\gtrless 0 \tag{3.25}$$

for large $|s|$.

We can now re-write the equation (3.18) as

$$\frac{\psi_+(b\pm0)}{L_+(s)[(1-c/c_2)s+ip/c_2]}-\frac{2ic_2bp\bar{f}[c_2p/(c_2-c)]}{(2\pi)^{\frac{1}{2}}(c_2-c)}\frac{L_-[-ip/(c_2-c)]}{(1-c/c_2)is-p/c_2}$$

$$=\mu b[(1+c/c_2)s-ip/c_2]L_-(s)\left[\varphi_-(b-0)-\varphi_-(b+0)+\frac{c_2\bar{f}[c_2p/(c_2+c)]}{\mu(2\pi)^{\frac{1}{2}}[(1+c/c_2)is+p/c_2]}\right.$$

$$\left.-\frac{c_2b\bar{f}[c_2p/(c_2-c)]}{(2\pi)^{\frac{1}{2}}}\left\{\frac{[(1+c/c_2)s-ip/c_2]L_-(s)+2iL_-[-ip/(c_2-c)]p/(c_2-c)}{(1-c/c_2)is-p/c_2}\right\}\right. \tag{3.26}$$

The left side of the above equality is regular in the half-plane $\operatorname{Im} s>\tau^{(1)}$, $\tau^{(1)}=\max\{-\tau^*,\tau_\tau^+\}$ and the right side in the half-plane $\operatorname{Im} s<\tau^{(2)}$, $\tau^{(2)}=\min\{\tau^*,\tau_w^-\}$ and both sides are regular in the strip $\tau^{(1)}<\operatorname{Im} s<\tau^{(2)}$, so both sides can be considered as the analytic continuation of the other and they represent the same function $W(s)$, analytic and single-valued for the entire plane. Furthermore, the fact that $\psi_+(b\pm0)$ and $\varphi_-(b-0)-\bar{\varphi}_-(b+0)$ vanish at infinity in their half-planes coupled with the behaviour of the above-constructed functions $L_+(s)$, $L_-(s)$, provides the following bounds on $W(s)$ for large $|s|$:

$$|W(s)|<M_1|s|^{\alpha_1},\quad s\to\infty \text{ in upper half-plane}$$

$$|W(s)|<M_2|s|^{\alpha_2},\quad s\to\infty \text{ in lower half-plane}\,. \tag{3.27}$$

Hence from the extended Liouville theorem the entire function is a polynomial of degree no greater than the maximum of α_1, α_2. If, as we assume throughout, the crack speed c is less than the shear wave speed, it can be shown that the exponents α_1 and α_2 are, respectively, $-\frac{1}{2}$ and $+\frac{1}{2}$. Then $W(s)$ is identically zero, and we obtain

$$\begin{aligned}\psi_+(b\pm0)&=\frac{2b}{(2\pi)^{\frac{1}{2}}}\frac{c_2p}{c_2-c}\bar{f}\left(\frac{c_2p}{c_2-c}\right)L_-\left(-\frac{ip}{c_2-c}\right)L_+(s)\\ \varphi_-(b-0)-\varphi_-(b+0)&=\bar{\bar{w}}_a+\frac{2}{\mu(2\pi)^{\frac{1}{2}}}\frac{c_2p}{c_2-c}\bar{f}\left(\frac{c_2p}{c_2-c}\right)L_-\left(-\frac{ip}{c_2-c}\right)\gamma^2L_-(s)\end{aligned} \tag{3.28}$$

which can be used in (3.17) and (3.13) to find the final transformed solution.

The stress intensity factor and the crack opening coefficient L (compare (2.14)) can be obtained directly from (3.28) with (3.25):

$$\begin{aligned}K&=(2\pi)^{\frac{1}{2}}\lim_{x'\to0+}x'^{\frac{1}{2}}\tau_{yz}(x',b\pm0,t)=\frac{2b^{\frac{1}{2}}Q(t)}{(1-c^2/c_2^2)^{\frac{1}{4}}}\\ L&=(2\pi)^{\frac{1}{2}}\lim_{x'\to0-}(-x')^{-\frac{1}{2}}\cdot\tfrac{1}{2}[w(x',b+0,t)-w(x',b-0,t)]=\frac{4b^{\frac{1}{2}}Q(t)}{\mu(1-c^2/c_2^2)^{\frac{3}{4}}}\end{aligned} \tag{3.29}$$

where $Q(t)$ is the inverse Laplace transform of

$$\bar{Q}(p)=\frac{c_2p}{c_2-c}\bar{f}\left(\frac{c_2p}{c_2-c}\right)L_-\left(-i\frac{p}{c_2-c}\right). \tag{3.30}$$

Here we have used the results concerning asymptotic relations between functions and their Fourier transforms (cf. theorems in [10], Chap. II) where if the functions $E_+(s)$, $E_-^*(s)$ are analytic in the upper (lower) half-planes Im $s>0$ (Im $s<0$) and are such that

$$E_+(s) \sim (-\mathrm{i}s)^{-\frac{1}{2}},\ E_-^*(s) \sim (\mathrm{i}s)^{-\frac{3}{2}},\ |s| \to \infty$$

then we have for their inverses that

$$e(x) \sim 2^{\frac{3}{2}} x^{-\frac{1}{2}},\quad x \to 0+$$

and

$$e^*(x) \sim -2^{\frac{1}{2}}(-x)^{\frac{1}{2}},\quad x \to 0-.$$

We note in passing that a similar relation [11] for Laplace transforms shows that

$$\lim_{t\to\infty} Q(t) = \lim_{p\to 0} p\bar{Q}(p)$$

and since (3.22) gives

$$\lim_{p\to 0} L_-\left(-\mathrm{i}\frac{p}{c_2-c}\right) = 1$$

we have

$$\lim_{t\to\infty} Q(t) = (1-c/c_2) \lim_{t\to\infty} f'(t) \tag{3.31}$$

As in the static case we have $G = \frac{1}{4}KL$, so that

$$G = \frac{2b}{\mu(1-c/c_2)}[Q(t)]^2. \tag{3.32}$$

We have seen that even in the static case it is not possible to write the displacement explicitly as a function of the coordinates, so we cannot expect to be able to do so in the dynamic case. It is the infinite product in (3.22) which gives trouble. However, there is one important quantity in which it does not appear, namely the total energy, E_{abs} say, absorbed by the crack tip during the whole of its interaction with the crack, given by

$$E_{\mathrm{abs}} = \int_0^\infty cG(t)\mathrm{d}t.$$

From (3.32) and Parseval's theorem for the Laplace transform [12] we have

$$E_{\mathrm{abs}} = \frac{2bc}{\mu(1-c^2/c_2^2)} \frac{1}{2\pi\mathrm{i}} \int_{d-\mathrm{i}\infty}^{d+\mathrm{i}\infty} \bar{Q}(p)\bar{Q}(-p)\mathrm{d}p \tag{3.33}$$

where the path of integration must pass between the singularities of the two factors in the integrand. Using (3.22) and the definition (3.23) we have

$$L_-\left(-\mathrm{i}\frac{p'}{c_2-c}\right)\Big]_{p'=p}\left[L_-\left(-\frac{\mathrm{i}p'}{c_2-c}\right)\right]_{p'=-p} = \exp\left\{\pm\frac{\mathrm{i}pb}{(c_2^2-c^2)^{\frac{1}{2}}}\right\} \tag{3.34}$$

where in the exponent the positive or negative sign must be used according as Im p is positive or negative. These different signs come from the fact that in evaluating $\bar{Q}(p')$ for $p'=-p$ not only in the function evaluated at $-p$ but also the branch cuts etc. of $\bar{Q}(-p)$ are the branch cuts etc. of $\bar{Q}(p)$ inverted through the origin of the p-plane. For example, the function ln $(\xi_1+\mathrm{i}\xi_2)$ defined in (3.23) has a branch cut along the real axis from $-\infty$ to 0, and so the function ln $(\xi_1'+\mathrm{i}\xi_2')$ evaluated for $\xi_1'+\mathrm{i}\xi_2'=-\xi_1-\mathrm{i}\xi_2$ will have a branch cut along the real axis from 0 to $+\infty$, and must be defined as in (3.23) but with (3.24) replaced by

$$0 < \tan^{-1}[(-\xi_2)/(-\xi_1)] \leq 2\pi \tag{3.36}$$

giving

$$[\ln(\xi_1' + i\xi_2')]_{-\xi_1 - i\xi_2} = \ln(\xi_1 + i\xi_2) + i\pi\,\mathrm{sgn}(\xi_1/\xi_2) \tag{3.37}$$

so as to satisfy the inequality (3.24) for all ξ_1, ξ_2. Thus, finally, (3.33) may be re-written in the form

$$E_{\mathrm{abs}} = \frac{2bc}{\pi\mu(1+c/c_2)} \int_0^\infty |i\xi\bar{f}(i\xi)|^2 \exp(-\delta^{\frac{1}{2}}\xi b/c_2)\,d\xi \tag{3.38}$$

where

$$\delta = (1-c/c_2)/(1+c/c_2) \tag{3.39}$$

provided $\bar{f}(p)$ is such that the complex conjugate of $\bar{f}(i\xi)$ is $\bar{f}(-i\xi)$.

Since (3.38) has the form of a Laplace transform, E_{abs} may easily be looked up or calculated for a wide variety of pulse shapes. Here we shall only consider a uniform stress pulse of magnitude σ_a and length a in space or duration a/c_2 in time. For it

$$f'(t) = \sigma_a H(t) - \sigma_a H(t - a/c_2) \tag{3.40}$$

whose Laplace transform is

$$p\bar{f}(p) = \sigma_a[1 - \exp(-pa/c_2)]/p$$

which gives

$$E_{\mathrm{abs}} = \frac{2abc}{c_2 + c}\frac{\sigma_a^2}{\mu} h(a/b\delta^{\frac{1}{2}})$$

where the function

$$h(x) = \frac{2}{\pi}\{\tan^{-1} x - \tfrac{1}{2}x^{-1}\ln(1+x^2)\} \tag{3.41}$$

is equal to x/π for small x and approaches unity for large x.

Some intermediate values are

	$h(x) = 0.28$	0.45	0.67	0.74	0.94
for	$x = 1$	2	5	7	10

If the pulse is a long one, so that $h=1$, it is reasonable to calculate an average G value by dividing E_{abs} by the distance moved by the crack while it is interacting with the pulse. This is c times the time which elapses between the arrival at the crack tip of the head and the tail of the pulse. This is not just ca/c_2 but rather $ca/(c_2 - c)$ because the pulse is overtaking the moving tip. This gives

$$G_{\mathrm{av}} = 2\frac{\sigma_a^2 b}{\mu}\frac{1-c/c_2}{1+c/c_2}. \tag{3.42}$$

This is also the ultimate G value for an infinitely long pulse with $f'(t) = \sigma_a H(t)$, calculated from (3.32) and (3.31). Indeed we may expect that (3.42) will also be the instantaneous value of G throughout the interaction of pulse and tip, except for intervals of order b/c_2 at the beginning and end.

The energy injected into the material by the blow on the end of the flake is the energy, half potential and half kinetic, contained in the pulse (3.40). Its magnitude is $E_{\mathrm{inj}} = \sigma_a^2 ab/\mu$ and so the efficiency of the flaking process is

$$\eta = \frac{E_{\mathrm{abs}}}{E_{\mathrm{inj}}} = \frac{2c/c_2}{1+c/c_2} h(a/b\delta^{\frac{1}{2}}). \tag{3.43}$$

If a/b (and *a fortiori* $a/b\delta^{\frac{1}{2}}$) is large enough for h to be sensibly unity η itself approaches unity at velocities close to c_2; the fall in G at high speeds is compensated by the increased time of interaction between tip and pulse.

4. Discussion

It will be seen from the foregoing that there are some notable differences in the results of our analysis for pressure and for percussion flaking. For example, the efficiency falls with velocity for the former, eqn. (2.8) and rises for the latter, eqn. (3.43). It is not difficult to see qualitatively how these differences arise.

When it interacts with the crack the pulse of section 2 is partly reflected back into the flake, but we implicitly supposed that by the time the reflection reached the end of the flake the whole process was over, and that what happened thereafter was of no interest. However, if we had followed the motion further, maintaining a constant force on the end of the flake, we should have found that the reflected wave was itself reflected at the end of the flake towards the tip where it was again partly reflected back, and so on. After a number of reflections to and fro (experience suggests that two or three would be enough) we should reach a steady state identical with that obtaining in pressure flaking. Under these circumstances the velocity of the end of the flake is controlled by the velocity of the tip, being proportioned to it. Thus the rate at which the load does work and the rate at which the tip absorbs energy go to zero together as the tip velocity falls to zero and the efficiency stays finite. On the other hand in percussion flaking the velocity imparted to the end of the flake by the blow is controlled only by the mechanical impedance of the material; it is actually the same as it would be in pressure flaking with $c=c_2$. Thus the load has to do the same amount of work whatever is the crack speed, and so the efficiency falls at low tip velocities when the crack is absorbing little energy.

It is in fact possible to modify the results for pressure flaking so as to show their relation to those for percussion flaking.

Starting with the uniformly moving elastic field of Section 2 we first add a constant displacement $-\sigma_a(1+c/c_2)/2\mu$ everywhere, so that the displacement (2.6) in the flake becomes

$$w = -\sigma(x-ct)/\mu - \sigma_a(1+c/c_2)/2\mu \tag{4.1}$$

and the complicated field near and beyond the tip is understood to have received the same addition. The reason for this seemingly trivial step, and the particular choice of constant, will appear in a moment. The displacement (4.1) can now be written as $w=w_i'+w_r'$ where

$$w_i' = \sigma_a(c_2t-x-a)/\mu\,, \quad \sigma_a = \tfrac{1}{2}(1+c/c_2)\sigma \tag{4.2}$$

and

$$w_r' = -\sigma_a\delta(c_2t+x)/\mu\,, \quad \delta = (1-c/c_2)/(1+c/c_2)$$

exhibiting it as the sum of an incident wave of constant stress σ_a and a reflected wave of reduced amplitude.

Let us change w_i' and w_r' into

$$w_i = \sigma_a(c_2t-x-a)/\mu\cdot H(a+x-c_2t) \tag{4.3}$$

and

$$w_r = -\sigma_a\delta(c_2t+x)/\mu\cdot H(x+c_2t)$$

which are still solutions of the wave equation. The incident wave is no longer infinite in length but has a tail at $-x=a-c_2t$ behind which there is no disturbance, and, similarly the reflected wave now has a head advancing to the left beyond which the disturbance is zero. We can now see the necessity for adding the displacement $-\sigma_a(1+c/c_2)/2\mu$ everywhere. Without it the linear factor in (4.3) would be c_2t-x which does not vanish at the step in the H-function, and the resulting discontinuity in displacement would produce a delta-function spike in the stress and material velocity, carrying with it an infinite amount of energy.

Suppose that the displacement in the flake is

$$w = w_i + w_r$$

and that the crack tip is at $x=0$ for $t=0$. Then for $t>0$ there is a region of the flake behind the crack tip where the incident and reflected waves overlap and w retains its original value

(2.6) unaffected by our modifications, and joins on to the same complicated field as before near and beyond the crack tip. Further to the left in the flake there is a region where at first the incident and later the reflected wave exists by itself. Evidently we have a rough description of the elastic field when a finite pulse of length a in space or a/c_2 in time launched into the flake is partly reflected at the tip. We must say "rough" because the solution is inaccurate when the head or tail of the pulse is interacting with the tip, but between these two events there will be a nearly steady state which should agree with the results at the end of Section 3. In particular the energy release rate should agree with (3.42). To see that this is so note that G is still given by (2.7) since the field around the tip is unaffected by the modification we have made. But now σ is merely an auxiliary quantity related to the actual stress σ_a by (4.2), and when we express (2.7) in terms of σ_a it indeed becomes (3.42).

If we take the reasonable value $c \sim \frac{1}{2}c_2$ either regime gives a fair efficiency, 40 per cent for (2.8) and 70 per cent for (3.43).

To conclude we return to the situation of fig. 1 where the applied force acts along the flake, so as to produce a compressive stress, σ say, in it. For the static case the energy release rate can be found by the argument of Section 2. It is only necessary to replace 2μ by the appropriate modulus E' for longitudinal compression of the flake, namely E for plane stress or $E/(1-\nu^2)$ for plane strain, where E is Young's modulus and ν Poisson's ratio. Thus

$$G = \frac{\sigma^2}{E'} b \,. \tag{4.4}$$

Intuition suggests that the deformation near the tip will be substantially mode II, but we cannot exclude the possibility of some admixture of mode I. In such a case $G = (K_{\mathrm{I}}^2 + K_{\mathrm{II}}^2)/E'$ and from (4.4) we can only infer the effective stress intensity factor

$$K = (K_I^2 + K_{II}^2)^{\frac{1}{2}} = \sigma b^{\frac{1}{2}}$$

and not K_{I}, K_{II} individually.

The case where the force is applied transversely to the flake so as to peel it off, has been treated by Obreimov [13] [14].

It is tempting to extend (4.4) to the moving case, again following the treatment in Section 2. In place of (2.7) we get

$$G = \frac{\sigma^2}{E'} b \left(1 - \frac{c^2}{c_0^2}\right) \tag{4.5}$$

where $c_0 = (E'/\rho)^{\frac{1}{2}}$ is the velocity of long-wave longitudinal disturbances along the flake. However, this can hardly be correct. The velocity c_0 exceeds c_2 and hence also c_R, the Rayleigh velocity. For all moving mode I or II crack tips so far investigated G becomes zero when the velocity c_R is reached, and, formally at least, it is negative above c_R. Gold'shtein [15] had discussed the peculiarities which appear near the Rayleigh velocity in steady state problems involving free surfaces. Only a detailed solution could show what is wrong with (4.5) and whether it is valid even for low velocities. Similar objections can be raised against other treatments [16], [17], [18] where the dynamic behaviour of cracks is discussed with the help of simple strength of materials arguments. Such objections do not apply to our treatment of the anti-plane case in section 2 since, if challenged, we could exhibit the exact solution.

Acknowledgement

One of us (J.G.F.) would like to thank the British Council for a Fellowship.

RÉSUMÉ
On discute, à l'aide d'un modèle mathematique simplifié, la production d'éclats d'un solide fragile par pression ou choc.

ZUSAMMENFASSUNG
Die durch Druck oder Schlag erfolgte Herstellung von Spänen eines spröden Festkörpers wird mittels eines vereinfachten mathematischen Modells untersucht.

[1] F. Kerkhof and H. Müller–Beck, *Glastechn. Ber.*, 42 (1969) 439.
[2] H. E. Rose, *Dechema-Monogr.* 57 (1966) 27.
[3] J. M. Buist and W. J. S. Naunton, *Trans. Inst. Rubber Ind.*, 25 (1949) 378.
[4] Anon, *Aircr. Engng.*, 42 (1970) 13.
[5] C. Atkinson and J. D. Eshelby, *Int. J, Fracture Mech.*, 4 (1968) 177.
[6] J. R. Rice, *J. Appl. Mech.*, 35 (1968) 379.
[7] J. D. Eshelby, in *Inelastic Behavior of Solids*, R. I. Jaffee and M. F. Kanninen (eds.) Mc.Graw-Hill (1970) 77.
[8] H. Lamb, *Hydrodynamics*, Cambridge University Press (1932) 74.
[9] B. Noble, *Methods Based on the Wiener-Hopf Technique*, Pergamon Press, Oxford (1958) 104.
[10] A. Erdélyi, *Asymptotic Expansions*, Dover Publications, New York (1956).
[11] G. Doetsch, *Theorie und Anwendungen der Laplace-transformation*, Springer-Verlag, Berlin (1937).
[12] I. N. Sneddon, *Fourier Transforms*, McGraw-Hill, New York (1951) 30.
[13] J. S. Obreimov, *Proc. Roy. Soc.*, A 127 (1930) 290.
[14] L. D. Landau and E. M. Lifshitz, *Theory of Elasticity*, Pergamon Press, Oxford (1959) 51.
[15] R. V. Gol'dshtein, *J. Appl. Math. Mech. (PMM)*, 29 (1965) 516.
[16] J. J. Gilman, *Fracture*, B. L. Averbach *et al.* (eds), Wiley, New York (1959) 193.
[17] S. J. Burns and W. W. Webb, *J. Appl. Phys.*, 41 (1970) 2078.
[18] B. Steverding and S. H. Lehnigk, *J. Appl. Phys.*, 41 (1970) 2096.
[19] L. Siret, *Bull. Soc. Anthrop. Brux.*, 43 (1928) 18.

Int. Journ. of Fracture Mech., 7 (1971) 421–433

Sci. Prog., Oxf. (1971) **59,** 161–179.

Fracture mechanics

J. D. Eshelby

The early work of Griffith and others on crack propagation has in recent years been extended and systematized to form the science of Fracture Mechanics. In dealing with stationary cracks its principal theoretical tools are the stress intensity factor which characterizes the elastic field near a crack tip and the energy release rate associated with the extension of a crack. The behaviour of fast-moving cracks presents some intriguing theoretical problems which are only beginning to be solved.

1. Introduction

The scientific study of fracture began with Griffith's[1] work on cracks in brittle solids like glass and its extension by Orowan[2] to metals and other more or less ductile materials. More recently, largely as a result of the work of Irwin,[3] their ideas have been further developed and formalized into the science of Fracture Mechanics, considered by its exponents to be an independent discipline, and now supported by two journals (the *International Journal of Fracture Mechanics* and *Engineering Fracture Mechanics, an International Journal*), a seven-volume treatise,[4] and numerous conferences.

The main task of fracture mechanics is a conservative one, to learn enough about cracks to prevent them from appearing or growing, or at least to be in a position to predict their behaviour. On the other hand, the phenomenon of brittle fracture is, or has been, used constructively in various technical processes, ranging from flint-knapping[5] through blasting and comminution[6] (the grinding of powders) to the shattering of aircraft canopies[7] to allow the pilot to eject. The ideas of fracture mechanics have also been used in geology, in particular in the study of earthquakes; so far this is largely an observational branch of the subject.

The fracture of materials must ultimately be explained in terms of their atomic, or at least microscopic constitution. This topic was the subject of a recent article by Hull[8] in this journal. Here we shall be concerned with some topics drawn from the macroscopic side of fracture mechanics which uses concepts drawn from the mechanics of continua to set up the formal framework within which the detailed physical processes of fracture must act, and also to

Dr Eshelby is a Professor in the Department of the Theory of Materials, University of Sheffield.

provide practical criteria for estimating fracture toughness, the resistance of a material to the propagation of cracks.

The following is an outline of the contents of the present article. Section 2 is concerned with a problem fundamental to fracture mechanics, the determination of the stress field which appears around a crack when the body containing it is subjected to external forces. For most purposes it is enough to know the elastic field near the tip of a crack. This field can be completely specified by three constants, the *stress intensity factors.*

Section 3 is concerned with energetics. Griffith emphasized the importance of the energy flow associated with crack propagation. In modern fracture mechanics it is handled with the help of the *energy release rate* or *crack extension force*, a quantity which measures the rate at which the total energy of the system, made up of the cracked body and the external loading mechanism, varies with the position of a crack tip.

The stress intensity factor and the energy release rate are the two characteristic concepts of modern fracture mechanics and they provide an adequate basis for the discussion of the onset of fracture and the slow propagation of cracks. However, a crack which is actually running may reach a speed which is a sizeable fraction of the velocity of elastic waves in the material, and then the static theory is inadequate. Section 4 is concerned with the theory of moving cracks.

2. Stresses around cracks: the stress intensity factor

When a specimen containing a crack is acted on by external forces a complicated system of stresses is induced in its interior. The stresses are particularly large near the ends of the crack and for most of the purposes of fracture mechanics it is only necessary to know the form of the stress field near the crack tips. In order to give some body to the general discussion, and because stress analysis bulks so large in fracture mechanics, it seems worthwhile outlining the elements of the mathematical theory and carrying it far enough to find the elastic field at a crack tip. (A more detailed, but straightforward account may be found in reference 9.) Only a little of the formal apparatus of the theory of elasticity is required. In fact all we shall really need are the equations of equilibrium[10]

$$\left.\begin{array}{ll}\nabla^2 u+\alpha\partial e/\partial x=0, & \nabla^2 v+\alpha\partial e/\partial y=0\\ \nabla^2 w+\alpha\partial e/\partial z=0 & \text{with } \alpha=1/(1-2\nu)\end{array}\right\}\qquad(2.1)$$

written in terms of the displacement vector **u** with Cartesian components u, v, w and the relation between the stress components σ_{xx}, σ_{xy} . . . . and the displacement, typified by

$$\sigma_{xx}=\lambda e+\mu\partial u/\partial x,\qquad \sigma_{xy}=\mu(\partial u/\partial y+\partial v/\partial x)$$

Here

$$e = \operatorname{div} \mathbf{u} = \partial u/\partial x + \partial v/\partial y + \partial w/\partial z \tag{2.2}$$

is the dilatation (the fractional change of volume on deformation), μ is the shear modulus, ν Poisson's ratio and λ stands for $2\mu\nu/(1-2\nu)$.

It is first necessary to introduce some standard nomenclature. Fig. 1(a) represents a body containing a crack. In Fig. 1(b) the crack has been opened up by a tension T transverse to its length, producing a state of *mode I* deformation. In Fig. 1(c) the shear stress S has made the faces of the crack slide over one another parallel to the plane of the figure and we have *mode II* deformation. In Fig. 1(d) a shear stress τ (represented by the conventional signs for arrow heads and tails) is applied so as to make the crack faces slide over each other

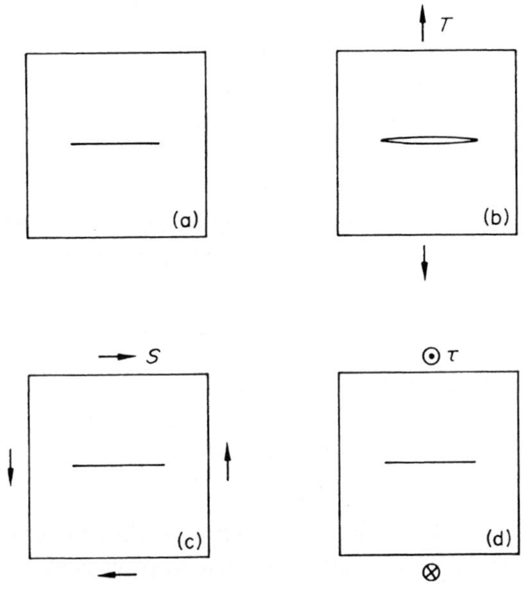

Fig. 1

perpendicular to the plane of the figure producing a state of *mode III* deformation. If the left-hand end of the crack had reached all the way to the surface, the resulting configuration would be roughly what one gets when a telephone directory is torn in half, hence the alternative name tearing mode for mode III. For more complicated loading, even if it is non-uniform, the elastic field near the tip of a crack will still be the sum of simple mode I, II and III deformations.

For the moment we shall only be concerned with two-dimensional states of strain in which the elastic field is independent of the z-coordinate. Modes I and II, for which w is zero and the displacement (u, v) is in the plane of the figure, are examples of *plane strain*. Mode III, with the displacement (0, 0, w) perpendicular to the plane of the figure, is an example of *anti-plane strain* deformation.

Because for it the theory is particularly simple we begin with mode III. Since u and v are zero and w is independent of z, the dilatation (2.2) vanishes and the equations (2.1) reduce to one, namely

$$\nabla^2 w = 0,$$

which says that the surviving displacement component is a harmonic function of x, y. The only non-zero stress components are

$$\sigma_{zx} = \mu \partial w/\partial x, \qquad \sigma_{zy} = \mu \partial w/\partial y.$$

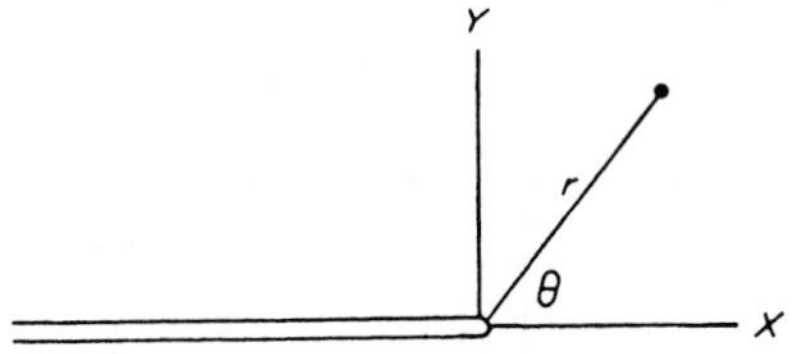

Fig. 2

To deal with the elastic field near a crack tip we set up the Cartesian and polar coordinates shown in Fig. 2. The displacement can then be expanded in terms of the basic set of harmonic functions

$$r^n \sin n\theta, \qquad r^n \cos n\theta \tag{2.3}$$

Obviously w will be odd in θ, and so

$$w = \sum_n a_n r^n \sin n\theta \tag{2.4}$$

We cannot expect n to be integral, for then w would be continuous across the crack, and its faces would not suffer a relative displacement. In fact it is easy to show that if the crack faces are to be free of stress, so that $\sigma_{zy} = 0$ for $\theta = \pm\pi$, then n must be plus or minus half an odd integer. We can exclude negative values of n by an energy argument. Take the term with $n = -\frac{1}{2}$. Then w behaves like $r^{-\frac{1}{2}}$, the stresses like $r^{-3/2}$ and the energy per unit volume, proportional to $(\text{stress})^2/(\text{elastic modulus})$, behaves like r^{-3}. By integration one finds that the elastic energy per unit length of a cylinder of outer and inner radii R, r_0 centred on the crack tip is proportional to $r_0{}^{-1} - R^{-1}$. Letting r_0 tend to zero we find that the total energy in any region embracing the tip is infinite. This is not necessarily mathematically or even physically improper for a permanently existing singularity (a classical point electron for example) but in the present case we start with an energy-free unstressed body and 'induce' a stress field around the crack in it by applying forces to the material. Since these forces would have to do an infinite amount of work to provide the energy round the tip we must exclude $n = -\frac{1}{2}$, and likewise any more negative value.

Close enough to the tip the term with $n = \frac{1}{2}$ will be dominant and, with a rather clumsy notation to be explained in a moment, the displacement is

$$w = \frac{2K_{\text{III}}}{\mu(2\pi)^{\frac{1}{2}}} r^{\frac{1}{2}} \sin \tfrac{1}{2}\theta \tag{2.5}$$

and the stresses are

$$\sigma_{zy} = \frac{K_{\text{III}}}{(2\pi)^{\frac{1}{2}}} r^{-\frac{1}{2}} \cos \tfrac{1}{2}\theta, \quad \sigma_{zx} = -\frac{K_{\text{III}}}{(2\pi)^{\frac{1}{2}}} r^{-\frac{1}{2}} \sin \tfrac{1}{2}\theta \tag{2.6}$$

In particular the stress immediately ahead of the tip ($\theta = 0$) is

$$\sigma_{zy} = \frac{K_{\text{III}}}{(2\pi)^{\frac{1}{2}}} r^{-\frac{1}{2}} \tag{2.7}$$

and the relative displacement of the faces just behind it ($\theta = \pm\pi$) is

$$\Delta w = \frac{4K_{\text{III}}}{\mu(2\pi)^{\frac{1}{2}}} r^{\frac{1}{2}}$$

The quantity K_{III} is by definition the mode III *stress intensity factor*. It is defined only in relation to the stress across the plane of the crack just ahead of the tip, being $(2\pi)^{\frac{1}{2}}$ times the coefficient of the $r^{-\frac{1}{2}}$ singularity, as in (2.7), but as (2.5) and (2.6) make clear a knowledge of its value also enables one to reconstruct the entire elastic field around the tip for small enough r. The factor $(2\pi)^{\frac{1}{2}}$ is conventional but convenient. In a less common definition of the stress intensity factor it is left out.

If the crack is subjected to mode I or mode II loading (Figs. 1b and 1c) instead of mode III the elastic field near the tip is similar though more complicated. It is now governed by the first two of equations (2.1). If we apply $\partial/\partial x$ to the first and $\partial/\partial y$ to the second and add we get $\nabla^2 e = 0$, so that the dilatation is harmonic. If we then apply ∇^2 to the two original equations we get $\nabla^2\nabla^2 u = 0$, $\nabla^2\nabla^2 v = 0$, so that u and v are not harmonic, but rather so-called biharmonic functions.

However, we can stay within the more familiar realm of harmonic functions by a trick. Because e is harmonic one may write $\partial e/\partial x = \nabla^2(\frac{1}{2}xe)$, $\partial e/\partial y = \nabla^2(\frac{1}{2}ye)$, as is easily checked. Consequently the first two of (2.1) can be thrown into the forms

$$\nabla^2[u+\tfrac{1}{2}\alpha xe] = 0, \qquad \nabla^2[v+\tfrac{1}{2}\alpha ye] = 0$$

which state that the bracketed expressions are harmonic functions, U and V say, so that we have

$$u = U - \tfrac{1}{2}\alpha x e, \qquad v = V - \tfrac{1}{2}\alpha y e. \tag{2.8}$$

For a mode I crack tip the harmonic U, V, e can be built up from the set of functions (2.3) which served for mode III. Obviously U and e will contain cosines and V sines, and corresponding to a term in r^n for U, V there will be one in r^{n-1} for e. The demand that the tensile stress σ_{yy} shall vanish on the crack faces and that the total energy near the tip shall be finite requires, as in mode III, that in U and V each n shall be half a positive odd integer. Hence close to the tip, where the terms with $n = \frac{1}{2}$ are dominant, the displacement takes the form (2.8) with

$$e = Ar^{-\frac{1}{2}} \cos \tfrac{1}{2}\theta, \qquad U = Br^{\frac{1}{2}} \cos \tfrac{1}{2}\theta, \qquad V = Cr^{\frac{1}{2}} \sin \tfrac{1}{2}\theta$$

The constants B, C can be related to A by imposing two conditions. One is that the shear stress σ_{xy} shall vanish on the crack faces. The other, less obvious, is that the dilatation e formed from the two expressions (2.8) shall agree with the e already incorporated in them. If the reader cares to carry out this programme he should find that, with a suitable choice for the remaining constant A, the elastic field near the tip takes the form

$$\sigma_{yy} = \frac{K_{\mathrm{I}}}{(2\pi)^{\frac{1}{2}}} r^{-\frac{1}{2}} \frac{1}{4}\left(5 \cos \frac{1}{2}\theta - \cos \frac{5}{2}\theta\right)$$

$$\sigma_{xx} + \sigma_{yy} = \frac{K_{\mathrm{I}}}{(2\pi)^{\frac{1}{2}}} r^{-\frac{1}{2}}. 2 \cos \frac{1}{2}\theta$$

$$\sigma_{xy} = \frac{K_{\mathrm{I}}}{(2\pi)^{\frac{1}{2}}} r^{-\frac{1}{2}}. \frac{1}{4}\left(- \sin \frac{1}{2}\theta + \sin \frac{5}{2}\theta\right)$$

$$\begin{matrix} u \\ v \end{matrix} = \frac{K_{\mathrm{I}}}{(2\pi)^{\frac{1}{2}}} r^{\frac{1}{2}}. \frac{4(1-\nu^2)}{E}\left[1 - \frac{\cos^2 \frac{1}{2}\theta}{2(1-\nu)}\right] \begin{matrix} \cos \\ \sin \end{matrix} \tfrac{1}{2}\theta \tag{2.9}$$

To conform with engineering usage we have expressed the displacements in terms of Young's modulus $E = 2(1+\nu)\mu$.

Two important quantities contained in (2.9) are the stress across the plane of the crack just ahead of the tip and the relative displacement of the crack faces just behind the tip:

$$\sigma_{yy} = \frac{K_{\mathrm{I}}}{(2\pi)^{\frac{1}{2}}} r^{-\frac{1}{2}}, \qquad \Delta v = \frac{K_{\mathrm{I}}}{(2\pi)^{\frac{1}{2}}} r^{\frac{1}{2}}. \frac{8(1-\nu^2)}{E} \tag{2.10}$$

To deal with mode II it is only necessary to interchange sine and cosine in (2.8) and find the new B, C in the same way. We shall only quote the analogue of (2.10):

$$\sigma_{xy} = \frac{K_{\mathrm{II}}}{(2\pi)^{\frac{1}{2}}} r^{-\frac{1}{2}}, \qquad \Delta u = \frac{K_{\mathrm{II}}}{(2\pi)^{\frac{1}{2}}} r^{\frac{1}{2}} \frac{8(1-\nu^2)}{E}$$

The quantities K_{I}, K_{II} are the stress intensity factors for mode I and II loading. As in the case of mode III they are in the first instance defined only in terms of the stress immediately ahead of the crack, but they actually determine the complete elastic field near the tip. Generally speaking the tip field will be a superposition of the three modes; it is then completely specified by K_{I}, K_{II} and K_{III}.

Although this analysis determines the *form* of the tip field it does not tell us its *magnitude*, which is fixed by the stress intensity factors. Their values depend on the details of the external load acting on the specimen, and also on the geometry, on whether, for example, the crack line terminates at the other end in another tip or at the surface and generally on the relation of the crack to external surfaces. Much effort has gone into the calculation of stress intensity factors. Paris & Sih[11] have given a compendium of results. Here we shall merely indicate some of the simpler ones.

Suppose that the crack in Fig. 1 extends along the x-axis from $x = -a$ to $x = a$ and that it is subjected not to one of the simple types of stress indicated in the figure but to a more complicated system of loading which produces stresses $\sigma_{ij}^{\mathrm{A}}(x,y)$ in the absence of the crack. The stress intensity factor at the right-hand tip can be found in terms of the values of σ_{ij}^{A} along the crack from the formula

$$K = \frac{1}{\pi a^{\frac{1}{2}}}\int_{-a}^{a}\left(\frac{x+a}{a-x}\right)^{\frac{1}{2}} \sigma^{\mathrm{A}}(x, 0)\, dx \tag{2.11}$$

To get K_{I}, K_{II}, K_{III}, σ^{A} must be put equal to σ_{yy}^{A}, σ_{xy}^{A}, σ_{zy}^{A} successively.

For the simple types of loading in the figure (2.11) gives

$$K_{\mathrm{I}} = T(\pi a)^{\frac{1}{2}}, \qquad K_{\mathrm{II}} = S(\pi a)^{\frac{1}{2}}, \qquad K_{\mathrm{III}} = \tau(\pi a)^{\frac{1}{2}} \tag{2.12}$$

If internal forces are applied to the crack faces equal and opposite to those produced by the load in the absence of the crack the faces will be pulled together again and the stress will be just $\sigma_{ij}^{\mathrm{A}}(x,y)$ everywhere. This means that the internal forces have cancelled the tip singularities due to the load, so that (2.11) can also be used for *internal* loading. A little puzzling over signs shows that if there is a variable pressure $p(x)$ inside the crack, tending to open it when positive, then the K_{I} it induces is given by (2.11) with $\sigma^{\mathrm{A}}(x,0) = p(x)$. If there is a real gas pressure p in the crack (hydrogen in steel) the result will merely be the first of (2.12) with $T = p$, but the idea of a variable internal pressure (or rather tension) will be useful later.

All we have said so far relates to two-dimensional situations, but the idea of a stress intensity factor carries over into three dimensions. Fig. 3 shows a flat disc-shaped crack. It is assumed that referred to the coordinates indicated the stresses are locally the same as those of a tangential two-dimensional crack suitably loaded, so that the elastic field near the crack border can be specified by three parameters $K_{\rm I}$, $K_{\rm II}$, $K_{\rm III}$ which vary round the periphery. This assumption is confirmed in all cases for which an analytical solution has been obtained; the most elaborate of these is a flat elliptical crack arbitrarily loaded.

The theory of cracks in plates, of particular importance to fracture mechanics, is at present in an unsatisfactory condition. In a state of plane strain there is a tensile stress σ_{zz} perpendicular to the x,y plane, equal to $\nu(\sigma_{xx}+\sigma_{yy})$, which prevents the material expanding or contracting parallel to the z-axis. Thus if we cut a slab out of the cracked solid of Fig. 1(b) to get a plate traversed by a crack, retaining the original loading round the edges, the relaxation of σ_{zz} on the freshly formed faces of the plate will throw the material into a complicated

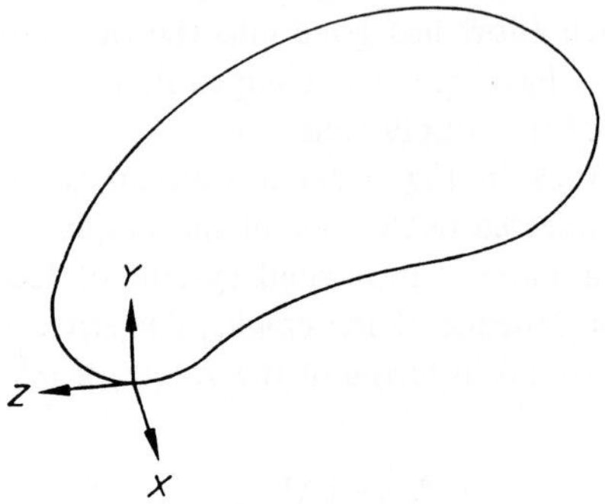

Fig. 3

three-dimensional state of stress. From what has just been said about the disc-shaped crack we might expect that for a crack in a plate under mode I loading the field close enough to the tip would be given by (2.9) with $K_{\rm I} = K_{\rm I}(z)$ now some function of z, though we may suspect that 'close enough' must be interpreted more and more strictly as the faces of the plate are approached. No exact solution has so far been published,* and this is an embarrassment since much of fracture toughness testing is carried out on various forms of cracked plate. In the absence of the exact solution a so-called 'plane stress' approximation is commonly used. It is of doubtful validity and in any case only claims to give the *average* of the field across the thickness of the plate. In particular the averages corresponding with (2.10) are

$$\bar{\sigma}_{yy} = \frac{\bar{K}_{\rm I}}{(2\pi)^{\frac{1}{2}}}\, r^{-\frac{1}{2}}, \quad \Delta\bar{v} = \frac{\bar{K}_{\rm I}}{(2\pi)^{\frac{1}{2}}}\, r^{\frac{1}{2}}.\, \frac{8}{E} \qquad (2.13)$$

where the average $\bar{K}_{\rm I}$ is the same as the plane strain $K_{\rm I}$ for the same type of loading. Because of the suppression of the factor $(1-\nu^2)$ (2.13) is inconsistent with (2.10) with variable $K_{\rm I}$. We return to the matter in the next section.

The importance of the stress intensity factor in fracture mechanics stems from

* Cruse & VanBuren[11a] have recently given some interesting numerical results.

the fact that if at the tip of a certain crack K_I, K_{II}, K_{III} have the same values as they do at the tip of a second crack in identical material then the elastic fields are identical at the two tips despite possible differences of geometry and loading. For brevity we shall suppose that K_{II} and K_{III} are both zero. Then if one crack is found to begin to extend and initiate fracture when K_I reaches a critical value K_{IC} we should expect the second crack to do the same and we have a fracture criterion

$$K_I = K_{IC} \tag{2.14}$$

where K_{IC} is a function which may be tabulated for materials of different chemical composition, crystal structure, grain size, heat treatment and so forth.

The above argument is based on the linear theory of elasticity used to calculate K_I. However, the idea of a critical K_{IC} is still useful when there are departures from linearity or plastic flow near the tip. Imagine that we have a material in which the forces responsible for non-linearity or plasticity can be switched on or off at will. With the forces switched off, load two crack tips to the same value of K_I. Now switch the forces on. As these forces are identical and find themselves acting in identical environments (the identical linear elastic fields) they will, with a certain limitation, produce identical changes in the elastic field, so that physical conditions are still identical at the tips, and they should both begin to extend at the same value of K_I, as calculated from the linear theory. The limitation is that the changes produced by the forces should be confined to a region small compared with the length of the crack and its distance from the surface of the specimen.

Explanations of (2.14) which say that the stress or strain or energy density reach a critical value near the crack tip do not really add anything to the assumption that like causes lead to like effects, but the Russian school of Barenblatt[12] gives a more elaborate interpretation. It is assumed that the crack tip is, so to speak, pulled apart by the applied forces, and held together by cohesive forces acting between the faces just inside the tips. For equilibrium it is required that Δv shall behave smoothly like $r^{3/2}$ near the tip, in place of the usual abrupt $r^{\frac{1}{2}}$. The condition for this is that the K_I produced by the combined action of the external load and the cohesive forces shall be zero. In fact if $K_I = 0$ the $r^{-\frac{1}{2}}$ term in the stresses vanishes, and with it the $r^{\frac{1}{2}}$ term in Δv, and the second term in (2.4) (or rather its analogue for mode I) dominates. We have seen that, for the crack of Fig. 1, the K value can be calculated from the same formula for both internal and external forces, and the same is true for more general geometries. Hence Barenblatt's criterion can be written in the form

$$K_I(\text{app}) + K_I(\text{coh}) = 0 \tag{2.15}$$

It is assumed by him that for small enough loads the cohesive forces are able to adjust themselves so that (2.15) is satisfied, but that ultimately they are unable to do so any longer, so that a stress-singularity develops and the crack begins to extend. It is further assumed that just before this point is reached the shape of the tip, specified by Δv, is a function characteristic of the material, which implies that the cohesive force is a definite function of distance back from the tip. (These assumptions are equivalent to our argument above that with a definite applied K_{I} prescribed non-linear forces produce a unique configuration near the tip.) The upshot of all this is that $K_{\mathrm{I}}(\mathrm{coh})$ in (2.15) is a constant characteristic of the material, negative since the cohesive forces tend to close the crack. If we call it $-K_{\mathrm{IC}}$ (2.15) gives a fracture criterion $K_{\mathrm{I}}(\mathrm{app}) = K_{\mathrm{IC}}$ of the usual type. Barenblatt calls $(\frac{1}{2}\pi)^{\frac{1}{2}}K_{\mathrm{IC}}$ the *modulus of cohesion.*

Other interpretations of the fracture process are based, not on the stress intensity factor, but on a second basic concept of fracture mechanics, the crack extension force or energy release rate, which we introduce in the next section.

3. Energy relations: the crack extension force

If the right-hand tip of the crack in Fig. 1 moves to the right the elastic energy E_{el} of the specimen and the potential energy E_{pot} of the loading mechanism responsible for the applied stress T both change. In his theory of brittle fracture Griffith pointed out that for the crack to be able to extend, the decrease of the total energy $E_{\mathrm{tot}} = E_{\mathrm{el}} + E_{\mathrm{pot}}$ must be equal to or exceed the surface energy of the freshly formed crack surfaces. Suppose that the total energy increases by δE_{tot} when a length δl of the crack border advances by δa. Then we write

$$-\delta E_{\mathrm{tot}} = G\,\delta a\,\delta l \tag{3.1}$$

where, by definition, G is the *energy release rate* or *crack extension force.* As the alternative name suggests we may regard (3.1) as the work done by a fictitious force G per unit length acting along the edge of the crack.

If the applied tension T in Fig. 1(b) is produced by pulling the upper and lower surfaces of the block apart and then clamping them (hard loading) the elastic energy obviously decreases when the crack lengthens, while there is no change in the potential energy of the loading mechanism. On the other hand, if T is produced by, say, a weight (dead loading) the lengthening of the crack makes the specimen more flexible, and the weight descends, losing potential energy, and at the same time doing work on the specimen which thus *increases* its elastic energy. For intermediate types of loading the elastic energy may either increase or decrease. The energy release rate may be calculated without distinguishing between these different cases with the help of the following set of imaginary operations performed at the crack tip.

Fig. 4 shows the stress and displacement near the tip of a crack. Allow the crack tip to advance by δa, but apply to the freshly-formed surfaces forces chosen so as to prevent them separating. The original elastic field remains

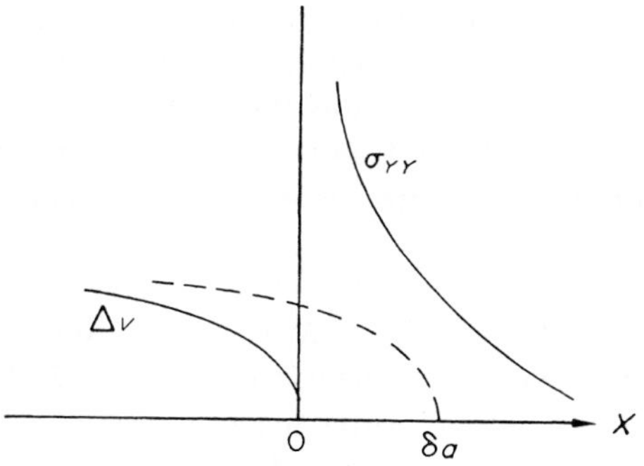

Fig. 4

unchanged. From (2.10) it follows that if these forces are gradually relaxed the stress across the x-axis between 0 and δa falls from

$$\sigma_{yy} = \frac{K_{\mathrm{I}}}{(2\pi)^{\frac{1}{2}}} x^{-\frac{1}{2}}$$

to zero, and the crack opening Δv increases from zero to the value

$$\Delta v = \frac{K_{\mathrm{I}}}{(2\pi)^{\frac{1}{2}}} \frac{8(1-\nu^2)}{E} (-x+\delta a)^{\frac{1}{2}}$$

represented by the dotted curve. The work extracted by the surface forces as they relax is thus

$$\frac{1}{2}\int_0^{\delta a} \sigma_{yy}\, \Delta v\, dx = \frac{1}{2}\frac{K_{\mathrm{I}}^2}{2\pi}\frac{8(1-\nu^2)}{E}\int_0^{\delta a} x^{-\frac{1}{2}}(\delta a - x)^{\frac{1}{2}} dx \tag{3.2}$$

per unit length perpendicular to the plane of the figure. The integral has the value $\frac{1}{2}\pi\, \delta a$, and (3.2) is, by definition, $G_{\mathrm{I}}\, \delta a$, so that we have

$$G_{\mathrm{I}} = \frac{K_{\mathrm{I}}^2}{E}(1-\nu^2) \tag{3.3}$$

Entirely similar calculations for modes II and III give

$$G_{\mathrm{II}} = \frac{K_{\mathrm{II}}^2}{E}(1-\nu^2), \qquad G_{\mathrm{III}} = \frac{K_{\mathrm{III}}^2}{2\mu} \tag{3.4}$$

The formula

$$\bar{G}_{\mathrm{I,II}} = \bar{K}_{\mathrm{I,II}}/E \tag{3.5}$$

is commonly used for the average of G across a crack in a plate. It is obtained by inserting (2.13) in the first member of (3.2). As well as ignoring the uncertain nature of the underlying plane stress solution it in effect assumes that the average of a product is the product of the averages. Later in this section we shall outline an argument which suggests that, nevertheless, (3.5) is correct, though it does not explain precisely how this comes about.

Since G is a definite function of K and the elastic constants we can transcribe the fracture criterion $K_{\text{I}} = K_{\text{IC}}$ into the entirely equivalent one

$$G_{\text{I}} = G_{\text{IC}} \tag{3.6}$$

where $G_{\text{I}} = K_{\text{I}}^{2}(1-\nu^2)/E$ and $G_{\text{IC}}^{2} = K_{\text{IC}}(1-\nu^2)/E$ and similarly for other modes. However, because of its energetic interpretation G_{IC} has a more concrete meaning than does K_{IC}. For a brittle material we may, following Griffith, equate G_{IC} with 2γ, the surface energy of the two faces of the crack. For a ductile material we may write, with Orowan, $G_{\text{IC}} = 2(\gamma+\gamma_{\text{pl}})$ where γ_{pl} is a fictitious surface energy which accounts for the work done in plastic deformation near the tip.

In its role as an effective force on the tip of a crack the energy release rate is closely related to the force on a defect (dislocation, impurity atom, lattice vacancy and so forth) as the term is understood[13] in the theory of lattice defects. This force can be calculated from the formula

$$F_x = \int_S \left(W\, n_x - \frac{\partial \mathbf{u}}{\partial x} \cdot \mathbf{T}\right) dS \tag{3.7}$$

with similar expressions for the y and z components. Here S is a closed surface which envelops the defect we are interested in, and no others, $\mathbf{T}\, dS$ is the force on an element dS of S, (n_x, n_y, n_z) are the components of the normal to dS, W is the elastic energy density and $\mathbf{u}$ is the elastic displacement. With the symbols suitably interpreted, (3.7) is valid not only for infinitesimal but also for finite deformation and an arbitrary stress strain law; $\mathbf{T}dS$ is the force on an element which had area dS and normal (n_x, n_y, n_z) *before* deformation, and W is the energy per unit *undeformed* volume. An important property of the integral (3.7) is that its value is unchanged when S is deformed, provided it does not embrace any other defect.

The physical significance of F_x is that if the defect is displaced by δa parallel to the x-axis $F_x\, \delta a$ gives the decrease in the sum of the elastic energy of the material and the potential energy of the loading mechanism, or in other words F_x is the 'energy release rate' for the defect. One can show that a crack tip qualifies as a singularity in its own right, to which (3.7) may be applied. In a two-dimentional situation we have simply

$$G = F_x \tag{3.8}$$

(supposing that the crack is parallel to the x-axis) where in (3.7) S is any circuit embracing the tip. The resulting integral is identical with Rice's[14] path-independent integral for G. If we need the local G value at some point on a straight or curved crack we must take for S a small closed surface, a sphere say, which, so to speak, takes a bite out of the edge of the crack. If S intercepts a length δl of the crack border then F_x is equal to $G\ \delta l$, provided the axes are oriented as in Fig. 3. There are various ways of making good these assertions;[15] we shall outline one later in connection with the corresponding dynamical problem.

The fact that (3.7) is valid for finite deformation and that it is surface or path-independent make it a useful tool. When the elastic field is known only near the tip one takes for S an infinitesimal circuit. On the other hand, when there are non-linearities near the tip but we believe that the linear solution is adequate far from the tip we take S to be large. We then know that the G so calculated will be the same as the G which would have been obtained from a small circuit and the unknown tip field. If for this case a fracture criterion of the usual form $G_I = G_{IC}$ is found to be adequate an alternative criterion of the form $K_I = K_{IC}$ will also be adequate, where K_I is related to G_I by the linear formula (3.3), even though this formula is inapplicable near the tip. This provides a justification for our earlier assertion that a stress intensity factor based on the linear theory is still significant when the linear theory is inadequate near the tip.

The same kind of argument can be used to justify the plate formula (3.5). The integral of G along the crack is given by (3.7) with S, say, a large circular cylinder with the crack as axis closed by the two circles in which it intersects the faces of the plate. These circular areas actually make no contribution. One can show that the plane stress solution is adequate remote from the tip, the averages given by the theory now being interpreted as constant values across the plate. After inserting the plane stress solution into (3.7) we can contract S on to the tip and use (2.13) to reproduce (3.5). This last step is allowable because the plane stress solution, though not an exact solution of any problem for the actual plate, is in fact the exact solution for a fictitious, but possible, anisotropic plate, and this is enough to make (3.7) independent of S.

4. Fast cracks

Crack tips may propagate through a material with speeds which are comparable with the velocity of elastic waves in it, and then the dynamic equations of elasticity must be used. Various mathematical solutions have been obtained for cracks whose tips move with constant velocity, but since it is found that cracks do not commonly move uniformly what is really wanted is the solution for a crack which is moving arbitrarily.

Ideally one would like to be able to predict the motion of a crack in a given material under given loading. A programme for doing this would go somewhat as follows. (i) Calculate the elastic field near an arbitrarily moving tip. (ii) From it find the energy release rate G. (iii) As in the static case equate G to a quantity

G_C which gives the rate at which energy is absorbed near the tip by non-elastic processes. (iv) Solve the resulting equation of motion. Some progress has already been made along these lines. So far solutions for unsteady crack motion have only been found for anti-plane elastic fields[16, 17] and so we shall have to limit ourselves to mode III, with the hope that for the more important modes I and II things will not be grossly different.

For a non-static state of anti-plane strain the equilibrium equation $\nabla^2 w = 0$ is replaced by the wave-equation

$$\frac{\partial^2 w}{\partial x^2} + \frac{\partial^2 w}{\partial y^2} - \frac{1}{c^2}\frac{\partial^2 w}{\partial t^2} = 0 \tag{4.1}$$

where $c = (\mu/\rho)^{\frac{1}{2}}$ is the velocity of transverse waves (ρ is the density of the medium). We begin with the simplest possible case. Initially the crack is at rest and has the displacement (2.5) not only near the tip but everywhere, that is, all but the first of the terms in the expansion (2.4) are zero. This displacement could, for example, be maintained in a large cylinder slit along the negative x-axis by forces proportional to $\sin \frac{1}{2}\theta$ distributed round the circumference. We re-write (2.5) in the form

$$w = B\, W_0(x, y) \tag{4.2}$$

where B is an abbreviation for the cumbersome coefficient $2K_{\mathrm{III}}/\mu(2\pi)^{\frac{1}{2}}$, and

$$W_0(x, y) = r^{\frac{1}{2}} \sin \frac{1}{2}\theta = \{\tfrac{1}{2}r - \tfrac{1}{2}x\}^{\frac{1}{2}} \tag{4.3}$$

is a basic solution which will play an important part in what follows. We now suppose that at time $t=0$ the tip of the crack starts to move along the x-axis according to the arbitrary law

$$x = \xi(t) \tag{4.4}$$

and ask what the resulting elastic field will be.

The answer is provided by a rather pretty theorem of Bateman's.[18] To explain it we need to introduce the retarded time τ. Suppose that a point moving according to (4.4) is continually emitting signals which propagate with velocity c. A signal received by an observer at x,y at time t must have been emitted at an earlier time, τ say, when the emitter was at $\xi(\tau)$. The retarded time τ is determined as a function of x,y,t by solving the implicit equation

$$c(t-\tau) = \{[x-\xi(\tau)]^2 + y^2\}^{\frac{1}{2}} \tag{4.5}$$

which says that the time of transmission is equal to the distance between the points of emission and reception, divided by the signal velocity. Bateman's theorem states that if $\nabla^2 w(x, y) = 0$ then $w[x-\xi(\tau), y]$ is a solution of the wave-equation (4.1), provided $w(x, y)$ is homogeneous of degree $\frac{1}{2}$. A function

$w(x, y)$ is said to be homogeneous of degree $\frac{1}{2}$ if $w(\lambda x, \lambda y) = \lambda^{\frac{1}{2}} w(x, y)$. The theorem may be verified, though with some difficulty, by substituting in (4.1) and using (4.5). The basic solution (4.2) meets the conditions of the theorem and yields the solution

$$w = BW_0[x - \xi(\tau), y] \tag{4.6}$$

It is easy to verify that (4.6) has the right properties. In particular the stress σ_{zy} is zero on the x-axis to the left of $x = \xi(t)$ and there is a stress singularity of the right type near $x = \xi(t)$. Outside a circle of radius ct about the original position of the tip τ is negative, $\xi(\tau) = 0$ and, as would be expected, the static solution remains undisturbed.

If the tip moves uniformly, so that

$$\xi = 0, t < 0; \quad \xi = vt, t > 0$$

(4.5) can be solved for τ, and after some algebra the displacement is found to be

$$w = B\,A(v)\,W_0\left[\frac{x - vt}{\beta(v)}, y\right] \tag{4.7}$$

inside the circle $r = ct$, with

$$A(v) = \left(\frac{1 - v/c}{1 + v/c}\right)^{\frac{1}{2}}, \ \beta(v) = \left(1 - \frac{v^2}{c^2}\right)^{\frac{1}{2}} \tag{4.8}$$

Equation (4.7) differs from (4.2) by the substitution ('Lorentz transformation')

$$x \to \frac{x - vt}{\beta(v)}, \quad y \to y$$

and the multiplier $A(v)$.

For arbitrary tip motion (4.5) can be solved for τ with the help of Lagrange's expansion[19] to give the displacement

$$w = B\,A(\dot{\xi})\,W_0\left[\frac{x - \xi(t)}{\beta(\dot{\xi})}, y\right] + O(R^{\frac{3}{2}}) \tag{4.9}$$

where R is the distance from the current position of the tip and $\dot{\xi}$ is the current velocity. Comparison with (4.7) shows that close to the tip the elastic field is the same as that of a uniformly-moving fictitious crack which happens to have the same position and velocity as the real crack at the instant of observation.

All this refers to the case where the disturbance spreading from the moving tip is, so to speak, eating into the particularly simple static field (4.2). When this is not so the static field is not homogeneous of degree $\frac{1}{2}$ and Bateman's theorem can only be applied indirectly.[17] An arbitrary tip motion can be analysed into a succession of elementary jumps. It is fairly clear that if the tip jumps from ξ to $\delta\xi$ at time t' the resulting disturbance will be a universal function of $x-\xi$, y and $t-t'$, a Green's function for crack motion in fact, multiplied by $\delta\xi$ and a weight factor depending on the (unknown) conditions near the tip. The Green's function can be found by allowing the jump to occur in the simple stress field to which Bateman's theorem *does* apply. An integral over the jumps will give the total disturbance, and the unknown variable weight factor in the integral can be found from the condition that the disturbance shall annul the original static stress σ_{zy} over the segment of the x-axis so far swept over by the tip. The final result is that (4.9) still holds close to the tip but that now B depends on the current position ξ of the tip. The actual expression for it is

$$B(\xi) = \frac{2}{\pi\mu}\int_0^{\xi} \sigma_{zy}(x', 0)(\xi - x')^{-\frac{1}{2}}\, dx' \tag{4.9a}$$

a weighted average of the static stress σ_{zy} $(x, 0)$ which originally existed ahead of the crack.

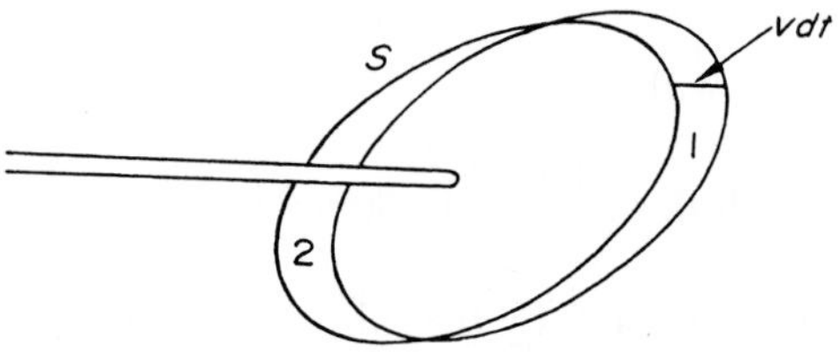

Fig. 5

For a crack of finite length (4.9) is ultimately upset by the arrival at the moving tip of the original disturbance reflected back from the other tip. This is not very restrictive; for example the crack will have already trebled its original length by the time the first reflection overtakes it if its average tip velocity up to this time has the reasonable value $\frac{1}{2}c$.

To find the energy releases rate for a moving crack tip[15, 20] we draw a circuit S around the tip (Fig. 5). (The discussion is no longer confined to mode III.) If $\mathbf{T}.dS$ is the force acting on the surface element dS with outwardly-directed normal the integral

$$I = \int_S \dot{\mathbf{u}} \cdot \mathbf{T}\, dS \tag{4.10}$$

is the rate at which the material outside S does work on the material inside S, that is, I is the rate of flow of energy (elastic plus kinetic) into S. (More formally I is the integral of the normal component of the energy flow vector.[10]) Of the inflow I a part, which by definition is vG, flows out again at the tip, while the rest goes towards increasing the elastic and kinetic energy within S. To get vG we must subtract from (4.10) the rate of increase of the energy contained within S. Suppose for the moment that the elastic field is carried along rigidly with the tip, so that we may write

$$\mathbf{u} = \mathbf{u}(x - vt, y) \tag{4.11}$$

Then in time dt the energy within S is decreased by the energy contained in the crescent-shaped region 1 and increased by the energy in the crescent-shaped region 2. The diagram makes clear that the total rate of change of energy is

$$-\int vE\, n_x\, dS \tag{4.12}$$

where E is the sum of the elastic and kinetic energy densities and n_x is the x-component of the outward normal to dS. Subtracting (4.12) from (4.10) we get

$$vG = \int_S (vE\, n_x + \dot{\mathbf{u}} \cdot \mathbf{T}) dS \tag{4.13}$$

But if (4.11) is true we have $\dot{\mathbf{u}} = -v\partial\mathbf{u}/\partial x$ and (4.13) becomes

$$G = \int_S \left(E\, n_x - \frac{\partial \mathbf{u}}{\partial x} \cdot \mathbf{T} \right) dS \tag{4.14}$$

One can easily show that if (4.11) is satisfied the integrals (4.13) and (4.14) are path-independent, in the sense explained in connection with (3.7), and so we are spared the embarrassing feature of some expressions which have been proposed, that they give different values according as S is, say, a circle or a square.

Of course (4.11) is not strictly satisfied by the elastic field of a moving crack. Nevertheless we can arrange for it to be satisfied to any degree of accuracy by going close enough to the tip, or at least, to be more cautious, this is true for every solution so far published; (4.9) is an example. Consequently (4.4) can be used to calculate G generally, on the understanding that S is to be contracted on to the tip until the integral reaches a constant limiting value. Equation (4.14) can also be used to find the local G value at a point on a crack border in the same way as (3.7), but the appropriate closed surface S must be supposed to shrink on to the crack border.

If the crack is moving slowly enough for the kinetic energy to be ignored (4.4) becomes the static expression (3.7). The integral is now, in fact, path-independent whether (4.11) is satisfied or not and so, as stated in Section 3, (4.14) can be used with a finite circuit S even though it was derived only for an infinitesimal one. In the static case there are, however, more satisfactory ways[15] of deriving (4.14) which start out with a finite circuit.

For the moving anti-plane tip field described by (4.9) and (4.9a) the formula (4.14) gives

$$G(\xi,\dot{\xi}) = \frac{1}{4}\pi\mu\left(\frac{1-\dot{\xi}/c}{1+\dot{\xi}/c}\right)^{\frac{1}{2}} B^2(\xi) \tag{4.15}$$

The velocity factor in (4.15) falls to zero at c. The analogues of (4.15) for modes I and II is as yet unknown but the indications are that they are similar to (4.15) but with a velocity factor which falls to zero at the Rayleigh velocity c_R, about $0{\cdot}9c$.

The expression (4.15) or its unknown analogues represents the rate at which energy flows into the tip in obedience to the equations which govern the elastic field. To find the way in which the tip actually moves we must equate (4.15) to some quantity G_C which specifies how this energy is absorbed near the tip by physical processes not taken account of by the theory of elasticity. The simplest assumption is that G_C has a constant value representing a real or effective surface energy. Since $B(\xi)$ increases with the length of the crack and the velocity factor decreases as the velocity increases, reaching zero at $\dot{\xi} = c$ or $\dot{\xi} = c_R$, the tip would then accelerate up to a limiting velocity c or c_R. This is in disagreement with experiment; limiting velocities in the neighbourhood of $0{\cdot}6c$ are commonly quoted. To reproduce this sort of behaviour we have to suppose that G_C is not constant but increases as the crack lengthens. The tip field and its rate of change are fixed if $K = K_{III}$ and $\dot{\xi}$ are given, and we may perhaps expect G_C to depend mainly on these quantities. This leads to an equation of motion for the tip of the form

$$G(\xi,\dot{\xi}) = G_C(K,\dot{\xi}),$$

where the left-hand side is determined through (4.15) by the applied stresses and the law of tip advance (4.4) while the right-hand side represents a material function. Note that $G(\xi, \dot{\xi})$ does not depend on the acceleration of the tip, so that if we regard the tip as a 'particle' it is one which exhibits no inertia. Küppers[21] drew the same conclusion for mode I from experimental evidence.

In the foregoing we have introduced the basic ideas of fracture mechanics and looked at some of the more interesting theoretical problems, solved and unsolved, connected with them. This may perhaps have conveyed the incorrect impression that fracture mechanics is a somewhat rarified branch of science. In fact, fracture toughness tests, largely based on the ideas we have discussed, are now regularly incorporated into material specifications for high-duty materials, and they have done much to improve the reliability and efficiency of engineering design.

References

1. GRIFFITH A.A. (1921) *Phil. Trans. R. Soc.* A **211**, 180.
2. OROWAN E. (1949) *Rep. Prog. Phys.* **12**, 214.

3. Irwin G.R. (1948) *Trans. Am. Soc. Metals* **40**, 147.
4. Liebowitz H. (1968) *Fracture*. Academic Press, New York.
5. Fonseca J.G., Eshelby J.D. & Atkinson C. (1971) *Int. J. Fracture Mech.* (in press).
6. Rose H.E. (1966) *Dechema-Monogr.* **57**, 27.
7. Anon. (1970) *Aircraft Engng* **42**, 13.
8. Hull D. (1969) *Sci. Prog., Oxf.* **57**, 495.
9. Eshelby J.D. (1968) In *Fracture Toughness*, p. 30. Iron and Steel Institute Publication 121, London.
10. Love A.E.H. (1927) *Theory of Elasticity*, p. 177. Cambridge University Press.
11. Paris P.C. & Sih G.C. (1965) In *Fracture Toughness Testing*, p. 30. American Society for Testing and Materials, Publication No. 381.
11a. Cruse T.A. & VanBuren W. (1971) *Int. J. Fracture Mech.* (in press).
12. Barenblatt G.I. (1959) *Advanc. appl. Mech.* **2**, 55.
13. Nabarro F.R.N. (1968) *Theory of Crystal Dislocations*, p. 562. Cambridge University Press.
14. Rice J.R. (1968) *J. appl. Mech.* **35**, 379.
15. Eshelby J.D. (1970) In *Inelastic Behavior of Solids* (Ed. by M. F. Kanninen), p. 77. McGraw-Hill, New York.
16. Kostrov B.V. (1966) *J. appl. Math. Mech. (PMM)* **30**, 1241.
17. Eshelby J.D. (1969) *J. Mech. Phys. Solids* **17**, 177.
18. Bateman H. (1915) *Electrical and Optical Wave Motion*, p. 138. Cambridge University Press.
19. Whittaker E.T. & Watson G.N. (1927) *Modern Analysis*, p. 132. Cambridge University Press.
20. Atkinson C. & Eshelby J.D. (1968) *Int. J. Fracture Mech.* **4**, 3.
21. Küppers H. (1967) *Int. J. Fracture Mech.* **3**, 13.

Phil. Trans. R. Soc. Lond. A. **274**, 331–338 (1973) [331]
Printed in Great Britain

Dislocation theory for geophysical applications

By J. D. Eshelby
Department of the Theory of Materials, University of Sheffield

A fault plane which has undergone slip over a limited area, a thin intrusion or a crack whose faces have been caused to slide over one another or separate by the action of an applied stress are all physical realizations of a dislocation, that is, an internal surface in an elastic solid across which there is a discontinuity of displacement. Since this discontinuity varies from point to point of the internal surface it is actually a so-called Somigliana dislocation. It can, however, be built up from the more familiar dislocations of crystal physics which have a constant displacement discontinuity.

Methods of finding the elastic displacement field around a dislocation in a solid with free surfaces will be outlined. The field of an infinitesimal dislocated area in a semi-infinite solid can be found quite simply, and from it the field of a general dislocation can be obtained by integration. The energy associated with a dislocation is discussed in connexion with energy release in earthquakes.

1. Introduction: Somigliana dislocations

Figure 1*a* represents a fault plane in which relative slip of the faces is confined to the interior of the curve *C*. Figure 1*b* shows a cross-section of, say, a thin igneous intrusion. Figure 1*c* represents the collapse of a worked-out coal-seam. These are all examples of so-called Somigliana dislocations (Somigliana 1914, 1915; Gebbia 1902). The formal construction of a Somigliana dislocation in an

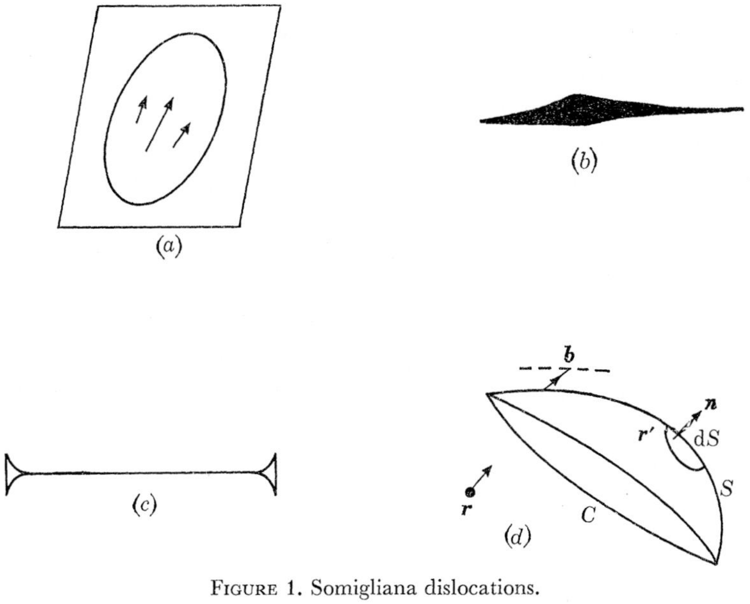

Figure 1. Somigliana dislocations.

elastic continuum goes as follows: make a cut over a surface *S* (figure 1*d*) (not necessarily plane) bounded by a curve *C* and give the two faces of the cut a relative displacement $\boldsymbol{b}(\boldsymbol{r})$ which varies from point to point $\boldsymbol{r}$ of *S*. Where this leaves a gap (as in figure 1*b*) fill it with a layer of the same material, and where there would be interpenetration of matter (as in figure 1*c*) prevent this by scraping away a layer on one face or the other. Finally weld the material together again.

Cracks can be treated in terms of Somigliana dislocations. Suppose that a crack forms in a body under an applied stress, and for simplicity let the faces of the crack slide over one another but not separate. We may say that the applied stress has induced in the medium a Somigliana dislocation of the kind shown in figure 1*a*. If the faces of the crack are freely slipping the discontinuity vector $\boldsymbol{b}$ will adjust itself in such a way that the surface tractions on S due to the applied stress and to the dislocation exactly cancel. If there is friction between the faces the cancellation may be imperfect. In the freely slipping case the dislocation disappears when the applied stress is removed. With friction the surfaces may stick at some stage of the unloading to leave behind a 'fossilized' remnant of the Somigliana dislocation originally produced by the applied stress.

A crack whose faces separate under load may be looked at in the same way; as long as it is held open the elastic field is unaffected if we fail to fill in the gap according to the recipe above.

If the displacement discontinuity $\boldsymbol{b}$ is known as a function of position over S the elastic field (displacement, strain, stress) which it produces, and the energy associated with it, can be calculated, fairly easily if it is remote from free surfaces of the medium and otherwise with more or less complicated corrections for the presence of the free surfaces (§2).

Somigliana dislocations are well suited to describe geophysical discontinuities, but the bulk of modern dislocation theory (Nabarro 1967) considers only the subclass of them, Volterra dislocations, for which $\boldsymbol{b}$ is constant over the discontinuity surface. In §3 we describe some of their properties and indicate how, if necessary, they can be used to synthesize Somigliana dislocations.

2. The elastic field of a Somigliana dislocation

We shall use the following sign convention relating the discontinuity $\boldsymbol{b}$ and the normal $\boldsymbol{n}$ to S (figure 1*d*). Draw an arrow (with head and tail) piercing S and defining the direction of its normal. Then $\boldsymbol{b}$ is the displacement on the head side of S minus the value on the tail side.

To find the elastic displacement produced by a Somigliana dislocation in an infinite medium imagine that, in addition to the dislocation, there is also a concentrated point force acting at the point $\boldsymbol{r}$ at which we wish to find the displacement $u_i(\boldsymbol{r})$ due to the dislocation (figure 1*d*).

If we introduce the dislocation and then apply the force the work done on the medium is simply the sum of the amounts E_{D} and E_{F} of work done when either the dislocation or the force is introduced by itself, because, in the linear theory of elasticity, the response of a body to external loading is unaffected by internal stresses. On the other hand, if first the force and then the dislocation is introduced the work done is $E_{\mathrm{F}}+E_{\mathrm{D}}$ plus two extra terms. One of these is the work done by the point force when its point of application shifts by the displacement which the dislocation produces at $\boldsymbol{r}$. The other is the work done by the surface tractions (produced by the point force) on the two faces of the cut as the faces separate by $\boldsymbol{b}$. But since the final state of the medium is the same whether the dislocation or the point force is introduced first, the sum of these two extra terms must be zero. This leads at once to the relation

$$u_i(\boldsymbol{r}) = \int_S b_j(\boldsymbol{r}')\, p_{jk}^{(i)}(\boldsymbol{r}')\, n_k \,\mathrm{d}S, \tag{2.1}$$

where $p_{jk}^{(i)}(\boldsymbol{r}')$ is the stress produced at a point $\boldsymbol{r}'$ on S by a point force of unit magnitude at $\boldsymbol{r}$ parallel to the x_i axis, and n_k is the normal to S at $\boldsymbol{r}'$. The left-hand side is the work done by the unit force moving through the dislocation displacement and the right-hand side is minus the work done by the surface tractions at the cut. (For the sign of the second term compare the discussion following equation (2.5) below.)

In an isotropic medium with Lamé constants λ and μ and the Poisson ratio σ the Hooke law relation between $p_{jk}^{(i)}$ and the corresponding displacement can be written in the form

$$p_{jk}^{(i)} = \lambda \frac{\partial U_{il}}{\partial x_l'} \delta_{jk} + \mu \frac{\partial U_{ij}}{\partial x_k'} + \mu \frac{\partial U_{ik}}{\partial x_j'} \tag{2.2}$$

where U_{il} is the x_i-component of the displacement at $\boldsymbol{r}(x_1, x_2, x_3)$ due to a unit force acting at $\boldsymbol{r}'(x_1', x_2', x_3')$ parallel to the x_l-axis. If we use the explicit form

$$U_{il} = \frac{1}{8\pi\mu}\left[\delta_{il}\nabla^2 - \frac{1}{2(1-\sigma)}\frac{\partial^2}{\partial x_i \partial x_l}\right]|\boldsymbol{r}-\boldsymbol{r}'| \tag{2.3}$$

(Love 1927), and note that $\partial/\partial x_m'$ is equivalent to $-\partial/\partial x_m$ when acting on $|\boldsymbol{r}-\boldsymbol{r}'|$, equation (2.1) becomes

$$u_i(\boldsymbol{r}) = \frac{1}{8\pi(1-\sigma)} \vartheta_{ijk} I_{jk},$$

with (Eshelby 1961)

$$\vartheta_{ijk} = \frac{\partial^3}{\partial x_i \partial x_j \partial x_k} - \left\{\sigma\delta_{jk}\frac{\partial}{\partial x_i} + (1-\sigma)\,\delta_{ij}\frac{\partial}{\partial x_k} + (1-\sigma)\,\delta_{ik}\frac{\partial}{\partial x_j}\right\}\nabla^2$$

and

$$I_{jk} = \int_S b_j(\boldsymbol{r}')\,|\boldsymbol{r}-\boldsymbol{r}'|\,n_k\,\mathrm{d}S.$$

If the discontinuity surface is flat and happens to lie in a plane $x_3 = $ const. I_{j1} and I_{j2} are zero and one only needs to calculate the three surface integrals

$$I_{j3} = \int_S b_j(\boldsymbol{r}')\,|\boldsymbol{r}-\boldsymbol{r}'|\,\mathrm{d}x_1'\,\mathrm{d}x_2'.$$

The rest of the calculation is mere differentiation. We may regard (2.1) as exhibiting a finite Somigliana dislocation as a mosaic of infinitesimal dislocations of area dS, normal n_k and a discontinuity vector b_j which is, of course, effectively constant all over the small area dS. The displacement produced by one of these elementary dislocations is

$$\mathrm{d}u_i = b_j n_k \mathrm{d}S\, p_{jk}^{(i)}, \tag{2.4}$$

with $p_{jk}^{(i)}$ in the form (2.2). According to (2.3) we have $U_{il}(\boldsymbol{r}, \boldsymbol{r}') = U_{il}(\boldsymbol{r}', \boldsymbol{r})$ and so (2.4) with (2.2) can be given a new interpretation: du_i can be derived from the displacement U_{il} at the point of observation $\boldsymbol{r}$ due to a point force acting not, as hitherto, at $\boldsymbol{r}$, but rather at $\boldsymbol{r}'$. To do this we have, according to (2.2) only to differentiate U_{il} with respect to the x_m', the coordinates of the point of application, and form a suitable linear combination of the derivatives. In other words, du_i is the same as the displacement produced at $\boldsymbol{r}$ by a collection of force-doublets situated at the position $\boldsymbol{r}'$ of the elementary dislocation. If the plane of the dislocation is a plane $x_2 = $ const. and the displacement discontinuity is $(b, 0, 0)$, parallel to the x_1-axis, figure 2*a*, or, equally, the plane is parallel to $x_1 = $ const., and the discontinuity is $(0, b, 0)$, figure 2*b*, we have

$$\mathrm{d}u_i = \mu b\,\mathrm{d}S\left(\frac{\partial U_{i1}}{\partial x_2'} + \frac{\partial U_{i2}}{\partial x_1'}\right),$$

which says that du_i can be produced by a pair of equal and opposite force couples each of moment $\mu b\,\mathrm{d}S$ force–times–distance units, figure 2*c*. If, on the other hand, the discontinuity vector is $(0, b, 0)$ and so normal to the plane $x_2 = $ const. of the dislocation (figure 2*e*) we have

$$\mathrm{d}u_i = b\,\mathrm{d}S\left[\lambda\frac{\partial U_{il}}{\partial x_1'} + \lambda\frac{\partial U_{i3}}{\partial x_3'} + (\lambda+2\mu)\frac{\partial U_{i2}}{\partial x_2'}\right],$$

which is the displacement due to three 'double forces without moment' (Love 1927) which despite their name have, in an obvious sense, a moment (not a couple) which is actually $(\lambda+2\mu)\,b\,\mathrm{d}S$ for the vertical doublet and $\lambda b\,\mathrm{d}S$ for the two horizontal ones (figure $2f$). If we had chosen to discuss the elementary dislocations of figure $2a, b$ in a coordinate system rotated through 45° in the plane of the paper we should have got the forces shown in figure $2d$, made up of two double forces of moment $\pm\mu b\,\mathrm{d}S$ which in fact produce the same displacement as the couples in figure $2c$.

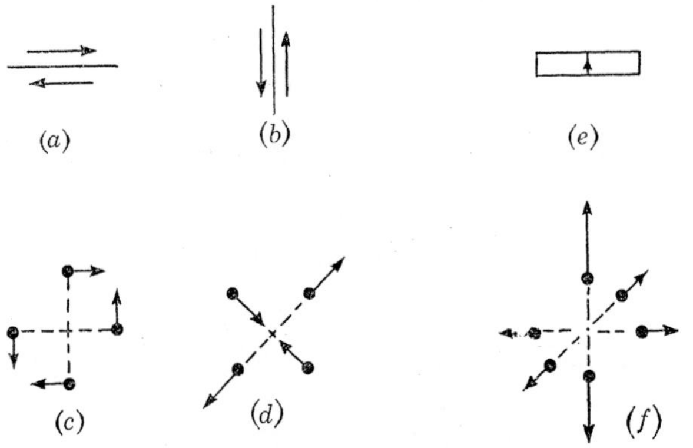

FIGURE 2. Equivalence between elementary dislocated areas and point-force clusters.

The displacement due to a dislocation in a body with stress-free surface Σ, say, can be found from the displacement u_i of the same dislocation in an infinite medium in the following way. Suppose that u_i produces a surface traction $p_{ij}n_j$ on Σ. Calculate the displacement u_i' due to surface tractions $-p_{ij}n_j$ applied to the surface of a finite body bounded by Σ. Then u_i+u_i' is the required displacement since it has the required singularity inside Σ and gives zero traction on Σ.

For the case of a semi-infinite solid bounded by a stress-free plane surface (which should cover many geophysical applications) a simple alternative treatment is possible. We go back to (2.4) which says that the elastic field of an elementary dislocation in an infinite medium is the same as that of a certain collection of point forces operating in an infinite medium. To adapt this expression to give the displacement in a half-space with, say, the surface $x_3 = 0$ stress-free it is evidently only necessary to take U_{il} in (2.4) to stand, as before, for the x_i- component of displacement at $\boldsymbol{r}$ due to a point force acting parallel to the x_l-axis at $\boldsymbol{r}'$, but now in a semi-infinite medium with the plane $x_3 = 0$ stress-free. Formation of doublets by differentiating this new U_{il} with respect to x_m' will not upset the fact that $x_3 = 0$ is stress-free. Integration will then give the field of a finite dislocation in the semi-infinite medium.

The necessary results for the displacement due to a point force in a semi-infinite medium with a stress-free surface have been given with various degrees of explicitness by Mindlin (1936), Westergaard (1952), Lur'e (1964) and Solomon (1968). Mindlin & Cheng (1950) give the field of doublets of the kind shown in figure 2. The result of inserting one of these expressions into (2.2) plus (2.4) is decidedly complicated. Steketee (1958) has written out the displacement for an elementary dislocation like the one in figure $2a$ with the plane of relative slip parallel to the free surface for the special case where the Poisson ratio is $\frac{1}{4}$ $(\lambda = \mu)$.† Even then the result is rather cumbersome (see also Maruyama 1964).

The energy change associated with the appearance or disappearance of a discontinuity surface

† In Steketee's equations (7.18) and (7.20) τ is a misprint for w.

is an important quantity. We begin with the case of the formation of a Somigliana dislocation in an initially stress-free region.

The elastic energy of a body is given by the integral

$$W = \frac{1}{2}\int u_i p_{ij} n_j \,\mathrm{d}S$$

taken over its surface. For a Somigliana dislocation the effective surface is made up of the two faces of the cut on either side of the discontinuity surface S (figure 1 *d*). If p_{ij}^{D} is the stress produced by the dislocation the surface traction $p_{ij}^{\mathrm{D}} n_j$ is continuous across S although the individual components of p_{ij}^{D} are not. Since the two faces of the cut are displaced relatively by $\boldsymbol{b}$ the energy required to establish the dislocation becomes

$$W = -\frac{1}{2}\int_S b_i p_{ij}^{\mathrm{D}} n_j \,\mathrm{d}S. \tag{2.5}$$

This is a positive quantity despite the negative sign which comes about as follows. In figure 1 *d* the face of the cut on, say, the convex side of S has a normal, pointing out of the material, which is in the opposite direction to the normal we have assigned to S. Combined with the sign convention for $\boldsymbol{b}$ (§ 2) this gives the minus sign in (2.5). The quantity (2.5) is also, of course, the energy released if the dislocation disappears.

If the dislocation forms in the presence of a pre-existing stress p_{ij}^{A} we have to take account of the work done at the discontinuity surface against the traction $p_{ij}^{\mathrm{A}} n_j$ and the work done is

$$W = -\int_S (p_{ij}^{\mathrm{A}} + \tfrac{1}{2}p_{ij}^{\mathrm{D}})\, b_i n_j \,\mathrm{d}S, \tag{2.6}$$

which can also be written more symmetrically as

$$W = -\int_S \tfrac{1}{2}(p_{ij}^{\mathrm{I}} + p_{ij}^{\mathrm{F}})\, b_i n_j \,\mathrm{d}S, \tag{2.7}$$

where $p_{ij}^{\mathrm{I}} = p_{ij}^{\mathrm{A}}$ and $p_{ij}^{\mathrm{F}} = p_{ij}^{\mathrm{A}} + p_{ij}^{\mathrm{D}}$ are the initial and final stresses. As explained in § 1, when a freely slipping crack forms a Somigliana dislocation is generated which completely annuls the traction $p_{ij}^{\mathrm{A}} n_{ij}$ over S, so that $p_{ij}^{\mathrm{F}} = 0$, and (2.6) or (2.7) gives a negative energy of formation

$$W = -\frac{1}{2}\int_S p_{ij}^{\mathrm{A}} b_i n_j \,\mathrm{d}S = +\frac{1}{2}\int_S p_{ij}^{\mathrm{D}} b_i n_j \,\mathrm{d}S,$$

so that, comparing with (2.5), there is an energy release when the crack appears which is the same as the energy released when an equivalent dislocation disappears in the absence of an external stress.

Two remarks should perhaps be made about (2.6). The first is that although it represents the work required to form the dislocation in the presence of p_{ij}^{A}, not all this work goes into elastic energy; some of it goes to increase the potential energy of the loading mechanism responsible for p_{ij}^{A}, in geophysical situations ultimately the Earth's gravitational field. The other is that it is assumed that p_{ij}^{A} does not vary appreciably as the dislocation forms (or disappears), a condition which will often, but not always, be fulfilled.

The integrals (2.6) and (2.5) only give the total energy released when a discontinuity appears or disappears. If we actually know how the displacement discontinuity varies with time during these processes the radiation field can in principle be calculated starting from a generalization to time-dependent $\boldsymbol{b}$ of the expression (2.4) for the field of an elementary dislocation (Nabarro 1951; Eshelby 1962; Kosevich 1965).

3. Volterra dislocations

The modern theory of dislocations (Nabarro 1967) confines itself almost entirely to the special case where the displacement discontinuity $\boldsymbol{b}$ is a constant. We shall call such dislocations Volterra dislocations, though strictly speaking they are a combination of Volterra's dislocations of the first, second and third kinds only.

If $\boldsymbol{b}$ is constant (2.1) can be transformed into

$$\boldsymbol{u}(\boldsymbol{r}) = -\frac{\boldsymbol{b}}{4\pi}\Omega - \frac{1}{4\pi}\int_C \frac{\boldsymbol{b}\times \mathrm{d}\boldsymbol{l}}{R} + \frac{1}{4\pi}\frac{1}{2(1-\sigma)}\,\mathrm{grad}\int_C \frac{\boldsymbol{b}\times \mathrm{d}\boldsymbol{l}\cdot \boldsymbol{R}}{R} \tag{3.1}$$

(Burgers 1939). Here

$$\boldsymbol{R} = \boldsymbol{r}-\boldsymbol{r}', \quad R = |\boldsymbol{r}-\boldsymbol{r}'|,$$

$\mathrm{d}\boldsymbol{l}$ is an element of the curve C at $\boldsymbol{r}'$ and

$$\Omega = \int_S \frac{\boldsymbol{R}\cdot\boldsymbol{n}}{R^3}\mathrm{d}S$$

is the solid angle subtended by S at $\boldsymbol{r}$. (To verify that (3.1) agrees with (2.1) for constant $\boldsymbol{b}$ use Stokes's theorem to turn the line integrals into surface integrals.)

It is known that the gradient of Ω is independent of S and consequently also continuous across it. Hence the gradient of $\boldsymbol{u}$, and with it the strain, rotation, and stress around the Volterra dislocation, are independent of S. If this continuous gradient is now integrated to recover the displacement we find that $\boldsymbol{u}$ is a multiple-valued quantity which changes by $\pm\boldsymbol{b}$ each time the curve C is encircled by the path of integration. It is almost equivalent to say that $\boldsymbol{u}$ is only defined modulo $\boldsymbol{b}$, i.e. to within a multiple of $\boldsymbol{b}$. Because of this we cannot observe the discontinuity surface of a Volterra dislocation with discontinuity vector $\boldsymbol{b}$ in a crystal with lattice parameter $\boldsymbol{b}$, because (unless we actually watch the process of displacement) the displacement of one of a lattice of identical atoms can only be measured modulo $\boldsymbol{b}$. It is these facts which make Volterra dislocations so important in crystal physics. Such a dislocation can be regarded as a line singularity (dislocation loop) characterized by a curve C and a constant vector $\boldsymbol{b}$, known in this connection as the Burgers vector.

The simplest kind of Volterra dislocation is one where the curve C (the dislocation line) is an infinite straight line and the discontinuity surface S is a half-plane bonded by C. In figure 3*a* the point C indicates the trace of the curve C and CA is the trace of S. If the Burgers vector $\boldsymbol{b}$ is parallel to the curve C, and so perpendicular to the plane of the figure, we have a *screw* dislocation. The cross products $\boldsymbol{b}\times\mathrm{d}\boldsymbol{l}$ in the line integrals of (3.1) are then zero, and the term in Ω gives a displacement

$$u_3 = \frac{b}{2\pi}\theta, \quad u_1 = 0, \quad u_2 = 0, \tag{3.2}$$

where the angle θ is as indicated in the figure and is limited by $-\pi \leqslant \theta \leqslant \pi$ so as to produce a discontinuity across CD rather than across some other line through C.

The plane AE can be made stress-free by introducing a screw dislocation of opposite sign at the 'image' point C′ so that the displacement due to a screw dislocation at C in a semi-infinite medium bounded by the stress-free plane BE is

$$u_3 = \frac{b}{2\pi}(\theta-\theta'). \tag{3.3}$$

The origin of θ' has been chosen so that the displacement is precisely $\pm\frac{1}{2}b$ on either side of D, and so that if θ', like θ, is limited by $-\pi \leqslant \theta' \leqslant \pi$ the discontinuity surface associated with θ' does not intersect the medium.

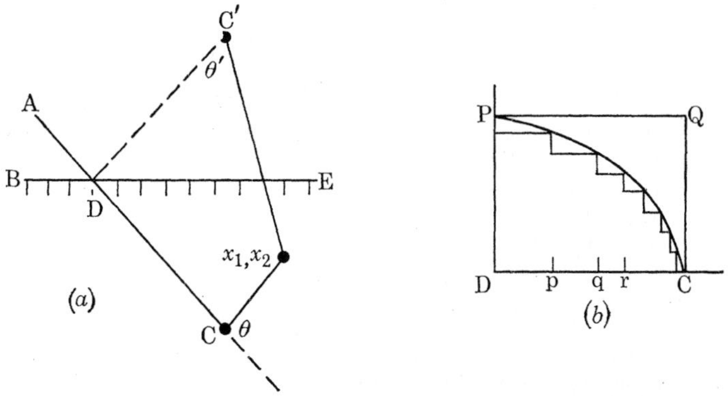

FIGURE 3. (*a*) Screw edge dislocations in a half-space. (*b*) Synthesis of a fault with variable slip from elementary screw and edge dislocations.

In geological terms the situation described by figure 3*a* and equation (3.3) represents a strike-slip fault, but one with the special property that the slip is constant from D to C and zero beyond C. How do we describe a strike-slip fault when the slip varies in some other way? We can, of course, go back to the general Somigliana dislocation. Alternatively, we can build up the fault using Volterra dislocations. Figure 3*b* shows the relative slip plotted as ordinate along DC. The line PQC represents the case we have been considering, constant slip produced by a single dislocation with Burgers vector b at C. Suppose that in fact the slip tapers off as shown by the smooth curve. Introduce n dislocations with equal Burgers vectors $b_0 = b/n$ at points p, q, r, ... Since the slip changes by b_0 each time a dislocation is passed the n dislocations will produce a stepped slip curve, and by choosing the positions of p, q, r, ... suitably the stepped curve may be made to agree with the smooth curve at n points. By increasing n (and thus decreasing b_0) the agreement can be improved indefinitely. In the limit $n \to \infty$, $b_0 \to 0$ we can say that the fault, actually a Somigliana dislocation, has been simulated by a continuous distribution of infinitesimal Voltarra dislocations. (For a general treatment on these lines see Bilby & Eshelby 1968; for a geophysical application see Weertman 1964.)

Suppose next that the relative slip of the two faces of the slip plane DC is in the plane of the figure instead of perpendicular to it, and of course parallel to DC. If the relative slip is constant along the slip plane as far as C and zero beyond it we have an *edge* dislocation. Its field is considerably more complex than (3.2) or (3.3).

In geological terms the edge dislocation is a dip-slip fault for which the relative slip is constant. The more realistic case of non-constant slip can be treated in the way already described for the strike-slip case, except that, in figure 3*b*, p, q, r, ..., must now be edge dislocations.

For three-dimensional situations of the kind shown in figure 1 it may be possible to use a Volterra dislocation to approximate to a Somigliana dislocation. For example (3.1) will give the correct elastic field at *remote* points if $\boldsymbol{b}$ is taken to be a suitable average of $\boldsymbol{b}(\boldsymbol{r})$ over the surface S of the Somigliana dislocation. If something better is needed a Somigliana dislocation can be synthesized from Volterra dislocations. The method illustrated in figure 3*b* can be extended to situations which are not two-dimensional (Leibfried 1954; Eshelby 1963), but it is not very advantageous. It is better to regard the discontinuity surface of the Somigliana dislocation as divided into a mosaic of small areas across each of which b is sensibly constant (cf. the remark preceding (2.4)). Then each elementary area is equivalent to a small dislocation loop to which the appropriate formula may be applied.

REFERENCES (Eshelby)

Bilby, B. A. & Eshelby, J. D. 1968 In *Fracture* (ed. Liebowitz), **1**, 99. New York: Academic Press.

Burgers, J. M. 1939 *Proc. K. ned. Akad. Wet.* **42**, 293.

Eshelby, J. D. 1961 In *Progress in solid mechanics* (ed. Sneddon & Hill), **2**, 89. Amsterdam: North-Holland.

Eshelby, J. D. 1962 *Proc. R. Soc. Lond.* A **266**, 222.

Eshelby, J. D. 1963 *Phys. Stat. Sol.* **3**, 2057.

Gebbia, M. 1902 *Ann. Mat. pura appl.* [3], **7**, 141.

Kosevich, A. M. 1965 *Soviet Phys. Dokl.* **7**, 837.

Leibfried, G. 1954 *Z. angew. Phys.* **6**, 251.

Love, A. E. H. 1927 *Mathematical theory of elasticity*, p. 183. Cambridge University Press.

Lur'e, A. I. 1964 *Three-dimensional problems in the theory of elasticity*, p. 132. New York: Interscience.

Maruyama, T. 1964 *Bull. Earthq. Res. Inst. Tokyo Univ.* **42**, 289.

Mindlin, R. D. 1936 *Physics* **7**, 195.

Mindlin, R. D. & Cheng, D. H. 1950 *J. appl. Phys.* **21**, 926.

Nabarro, F. R. N. 1951 *Phil. Mag.* **42**, 1224.

Nabarro, F. R. N. 1967 *Theory of crystal dislocations.* Oxford: Clarendon Press.

Solomon, L. 1968 *Élasticité linéaire*, p. 541. Paris: Masson.

Somigliana, C. 1914 *Rc. Accad. Lincei* [5], **23** (1), 463.

Somigliana, C. 1915 *Rc. Accad. Lincei* [5], **24** (1), 655.

Steketee, J. A. 1958 *Can. J. Phys.* **36**, 192.

Weertman, J. 1964 *Bull. seism. Soc. Am.* **54**, 1035.

Westergaard, H. M. 1952 *Theory of elasticity and plasticity*, p. 142. Cambridge, Mass.: Harvard University Press.

THE CALCULATION OF ENERGY RELEASE RATES

J. D. Eshelby

Department of the Theory of Materials, University of Sheffield, U.K.

ABSTRACT

The connection between the elastic energy-momentum tensor and the surface or path-independent integral for energy release rates and kindred concepts is recalled and extended to grade 2 materials. For a completely general elastic field there are apparently only the three types of path-independent integrals derived by Günther from Noether's theorem, but if the field is anti-plane (mode III) there is an indefinite number of them. In plane strain the energy-momentum tensor can be re-interpreted as an ordinary compatible stress. This observation, coupled with the elastic reciprocal theorem, leads to an indefinite number of path-independent integrals for this case also. Some other possibilities are also touched on. The paper contains some specimen calculations for specific cases.

NOTATION

The following symbols are used throughout:

F_ℓ	force on a singularity
G	energy release rate
I, L_i, M	path-independent integrals
K	stress intensity factor
n_i	unit normal vector
p_{ij}	nominal stress tensor

$P_{\ell j}$	energy-momentum tensor
dS, dS_j	surface element, $n_j\, dS$
u_i	displacement vector
W	energy density
X_i	initial coordinates
x_i	final coordinates
δ_{ij}	Kronecker delta
ε_{ijk}	is completely antisymmetric with $\varepsilon_{123} = 1$
ξ_ℓ	position of a singularity
$\tilde{\omega}$	rotation

Symbols used in a single context are defined in their place. Repeated suffixes are summed over 1,2,3 in three or 1,2 in two dimensions. Suffixes following a comma denote differentiation with respect to the corresponding X_i or x_i. In the linear theory, where X_i and x_i are identified, x_i is used, with x,y replacing x_1, x_2 when convenient.

INTRODUCTION - THE BASIC PATH-INDEPENDENT INTEGRAL

The matters to be discussed here go back to a famous paper by Emmy Noether [1]. (There is now an English translation.) From it there follows, among other things, the result that if we have a set of quantities $u_i(X_m)$ which depend on the variables X_m (in our case Cartesian coordinates) and which satisfy the equations

$$\partial W/\partial u_i - (\partial W/\partial u_{i,j})_{,j} = 0 , \tag{1}$$

where $W = W(u_i, u_{i,j})$ does not depend explicitly on the X_m, that is, it is the same function of the u_i and $u_{i,j}$ for any X_m, then the tensor

$$P_{\ell j} = W\delta_{\ell j} - (\partial W/\partial u_{\ell,j})\, u_{i,\ell} \tag{2}$$

has zero divergence, that is, $P_{\ell j,j} = 0$. This is easily checked. It then follows from Gauss's theorem that the integral

$$F_\ell = \int_S P_{\ell j}\, dS_j \tag{3}$$

is zero if there are no singularities of the integrand within the surface S, and that, if there are such singularities, F_ℓ has the same value for all surfaces S which can be deformed into one another without passing through these or any other singularities of the integrand. In two dimensions S is a plane circuit and we describe the above property by saying that (3) is path-independent. We shall use this familiar term also in three dimensions in place of the more appropriate 'surface-independent'.

If W is the energy density of an electrostatic field, $P_{\ell j}$ is Maxwell's tensor and F_1, F_2, F_3 give the components of the force on any charge there may be inside S. Similarly, if W is the energy density of an elastic field (3) gives the force on any elastic singularity within S. 'Force' now has the meaning it has in the phrases crack extension force in fracture mechanics, or force on a dislocation, point defect etc. in solid state physics [2]. It is defined so that if the defect is displaced by $\delta\xi_\ell$ then $F_\ell \delta\xi_\ell$ is the decrease in the total energy of the system, made up of the elastic energy of the material containing the singularity and the potential energy of any external loading mechanism. This aspect is emphasized in the synonym energy release rate for crack extension force.

A succession of imaginary physical operations establishes what has just been said. As the writer has presented it several times already [3,4,5] the argument is here cut to the bone, thus begging several questions, but, for variety, it is done for a grade 2 material.

Take Cartesian coordinates X_m in the undeformed material. On deformation a particle of material originally at X_i goes to $x_i(X_m)$ and $u_i(X_m) = x_i(X_m) - X_i$ is the displacement. So far as possible everything is referred to the undeformed state (Lagrangian point of view). In particular, in what follows W is the energy (in the deformed state, of course) per original unit volume and, in the vector surface element $dS_j = n_j dS$, n_j is the normal to and dS the area of an element of a surface S drawn in the undeformed medium.

In a grade 2 material [6] the energy density

$$W = W(u_{i,j}\; u_{i,jk})$$

depends on the first and second derivatives of the displacement with respect to the X_i. The requirement that the volume integral of W be stationary with respect to small variations of the u_i leads to the Euler equation

$$(\partial W/\partial u_{i,j})_{,j} - (\partial W/\partial u_{i,jk})_{,jk} = 0 \; . \qquad (4)$$

We shall suppose that (4) is in fact the equilibrium equation governing the u_i.

As a preliminary we work out the energy change produced by a small change of displacement, equation (5) below. Suppose that the displacement field alters slightly from u_i to $u_i + \delta u_i$ as a result, say, of a change in the external loads or the re-arrangement of sources of internal stress; there is no need to introduce the word 'virtual'. The energy density changes by

$$\delta W = (\partial W/\partial u_{i,j})\, \delta u_{i,j} + (\partial W/\partial u_{i,jk})\, \delta u_{i,jk} \, .$$

The total change of elastic energy inside a closed surface S is this quantity integrated over its interior. With the help of (4), 'differentiation by parts' and Gauss's theorem it may be converted into the integral

$$\delta E = \int \{ [\partial W/\partial u_{i,j} - (\partial W/\partial u_{i,jk})_{,k}] \, \delta u_i + (\partial W/\partial u_{i,jk})\, \delta u_{i,k} \} \, dS_j \qquad (5)$$

To find the force on an elastic singularity we draw a surface S enclosing it and cut out the material inside S from the surrounding material and apply suitable distributions of force and couple over both the freshly formed surfaces so as to prevent relaxation. We next displace the whole elastic field inside S by $\delta\xi_\ell$. This means that at a given material element u_i changes to $u_i + \delta u_i$ with

$$\delta u_i = - \delta\xi_\ell \, u_{i,\ell} \, . \qquad (6)$$

After this change the energy inside S is evidently equal to the energy which was previously inside a surface S' derived from S by giving it a rigid translation $-\delta\xi_\ell$. The volume inside S can be derived from the volume inside S' by adding a crescent-shaped volume to one side of S' and subtracting another from the other side. The change of energy inside S is thus the integral of W over the first crescent-shaped area minus its integral over the second, or in all

$$\delta E(\text{in}) = - \delta\xi_\ell \int W \, dS_\ell \, . \qquad (7)$$

We next change the displacement on the inner surface of the hole from u_i to $u_i + \delta u_i$ (with the δu_i of (6)) so that it matches that of the material inside S. The work required to do this is the expression (5) with the δu_i of (6) and the sign reversed since the outward normal of the piece removed is the inward normal of the hole left. We do not include a similar term for the outer boundary of the material because any energy inflow through it is accompanied by an equal decrease of the potential energy of the loading mechanism, and we are only interested in changes in the total energy of the system. Next weld the piece back into the hole. Although the displacements now match, the surface tractions do not quite do so: there is a layer of body force and couple with surface density proportional to $\delta\xi_\ell$ on the interface which induces a displacement of order $\delta\xi_\ell$. Thus when these unwanted layers are relaxed an amount of work only of order $(\delta\xi_\ell)^2$ is extracted from the system. Hence the total change of energy is (7) minus (5) with the δu_i of (6) so that the force on the singularity is given by the integral (3) with

$$P_{\ell j} = W\delta_{\ell j} - \{\partial W/\partial u_{i,j} - (\partial W/\partial u_{i,jk})_{,k}\} u_{i,\ell} - (\partial W/\partial u_{i,jk}) u_{i,k\ell} . \quad (8)$$

This agrees with the formal field theory expression for the energy-momentum tensor when the Lagrangian depends on the first and second derivatives of the field variables [7,8,9]. A material of grade N can be treated similarly; the appropriate $P_{\ell j}$ agree with the formal results of [8] and [9] for a Lagrangian containing derivatives up to the N th.

The elastic field near a crack tip (or, indeed, some other elastic singularity) in a grade 2 (or N) material may be notably different from what it would be in an ordinary elastic (grade 1) material, but in the body of the material the difference will be negligible. Hence if there is nothing to stop a small surface S surrounding the crack tip from being expanded to macroscopic size, the energy release rate will be the same as it would be for a suitably chosen grade 1 material.

From now on we shall only consider ordinary (grade 1) elastic materials. Then in (3) we may put

$$P_{\ell j} = W\delta_{\ell j} - p_{ij} u_{i,\ell} \quad (9)$$

where

$$p_{ij} = \partial W(u_{m,n})/\partial u_{i,j}$$

is the nominal, Boussinesq or first Piola-Kirchhoff stress tensor which gives the component parallel to the (rectangular) X_i coordinate axis of the force on a surface element which was of unit area and perpendicular to the X_j axis before deformation.

It is not hard to see that in a two-dimensional context a crack tip qualifies as an elastic singularity within the framework of the above argument provided $\delta\xi_\ell$ is parallel to the crack. Equation (3) can also be applied locally to the edge of a three-dimensional crack [5] and to plane stress [10].

When the path is prevented from closing by a mathematical cut representing, for example, a crack or a dislocation, one must distinguish between two kinds of path-independent integrals: in type (a) the path may be deformed provided the points where it meets the cut remain fixed, and in type (b) these points may, in addition, be slid along the cut, not necessarily remaining opposite one another. For a crack parallel to the X_1 axis with stress-free surfaces F_1 is of type (b). However, even in this case, if, moving away from the tip, the crack is at first parallel to the X_1 axis and then deviates from it beyond a certain point, F_1 ceases to be path-independent when its ends are slid along the crack beyond that point. (There is a similar distinction in three dimensions [11].)

In the interval between the appearance of Griffith's energy criterion and the rise of modern fracture mechanics quantities equivalent to the energy release rate had already appeared in the physical and technical literature.

In his 1930 paper on the surface energy of mica Obreimoff [12] gave what is in effect the first formula for the energy release rate for a specific crack configuration. As shown in Fig. 1 a flake, of length ℓ and thickness h say, is made to peel off by

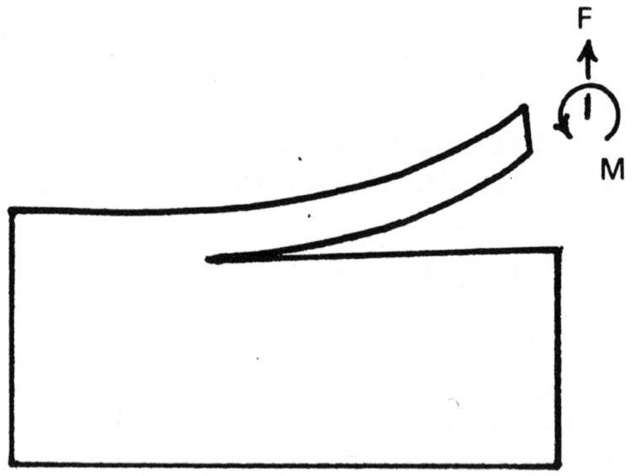

Fig. 1. Obreimoff's method of measuring surface energy.

applying a force F and couple M at the end. Its deformation $y = y(x)$ is calculated by treating it as a beam or plate clamped to a block at the position of the crack tip, say $x = 0$.

As is well known, of the work done by the loading mechanism when the crack extends only half contributes to the energy release rate, the other half being needed to increase the elastic energy. Consequently we have

$$G = 2\gamma = \tfrac{1}{2}F\partial y(\ell)/\partial\ell - \tfrac{1}{2}M\partial y'(\ell)/\partial\ell \tag{10}$$

where the two terms are respectively, half the work done per unit length perpendicular to the paper by the force and by the couple per unit crack extension; the minus sign follows from the usual sign convention for M. Beam theory gives

$$G = \tfrac{1}{2}(M - \ell F)^2/E'I \tag{11}$$

where E' is E or $E/(1-\nu^2)$ according to whether we treat the flake as a beam or a plate, and $I = h^3/12$. If we cut through the beam at x and provide the appropriate local bending moment $M(x)$ and shear force $F(x)$ to the cut end to prevent relaxation the (unchanged) G will be given by (11) with $M - \ell F = M(\ell) - \ell F(\ell)$ replaced by $M(x) - xF(x)$. Beam theory shows that, as expected, this is independent of x. In particular it is equal to $M(0)$, whence

$$G = E'h^3[y''(0)]^2/24 ,$$

the form given by Obreimoff. (Actually by a trivial mistake he got a factor 4 wrong. On the other hand Landau and Lifshitz's [13] rather cryptic derivation gives a value twice too big because of an error of principle; in effect they calculate the work done by the load but forget the fifty percent rake-off exacted by the elastic energy.)

To find G from (3) we surround the tip by a small circuit and then deform it into a circuit which is made up of a large semicircle in the block, centered on the tip, and any cross-section of the flake. (These two segments are actually joined by straight lines along the crack faces and the upper surface of the flake, but they contribute nothing.) On the semicircle the stress is essentially the same as that of a point force applied to a semi-infinite solid so that $p_{ij} \sim 1/r$, $P_{\ell j} \sim 1/r^2$ and it contributes nothing. Hence $G = -F_1$ is the integral of

$$-P_{11} = -W + p_{11}u_{1,1} + p_{12}u_{2,1} \tag{12}$$

across the flake. If we force in the assumptions of elementary beam theory, p_{11} proportional to x_2, $u_{2,1} = y'$ independent of x_2, equation (12) easily gives

$$G = -\tfrac{1}{2}My'' + Fy' \tag{13}$$

evaluated at any cross-section. This is equivalent to (10) or (11). More elegantly, having reduced the integration to one across the beam we may recognize that the beam is a one-dimensional grade 2 medium (Cosserat continuum with constrained rotation) with energy density $W = \frac{1}{2}E'I(y'')^2$. Variation in the body of the integral $\int W dx_1$ gives correctly $y'''' = 0$, the one-dimensional version of (4), and in addition variation at the end points shows that the quantities $M = -E'Iy''$ and $F = -E'Iy'''$ must be interpreted as couple and shearing force, while the one-dimensional version of (8) gives

$$G = -F_1 = \tfrac{1}{2}E'I(y'')^2 - E'Iy'''$$

which is the same as (13).

In 1953 Rivlin and Thomas [14] gave the theory of several tests used on rubber and plastics. An interesting one is the 'trouser test' shown in Fig. 2. The following treatment illustrates how the formalism with nominal stresses and Lagrangian

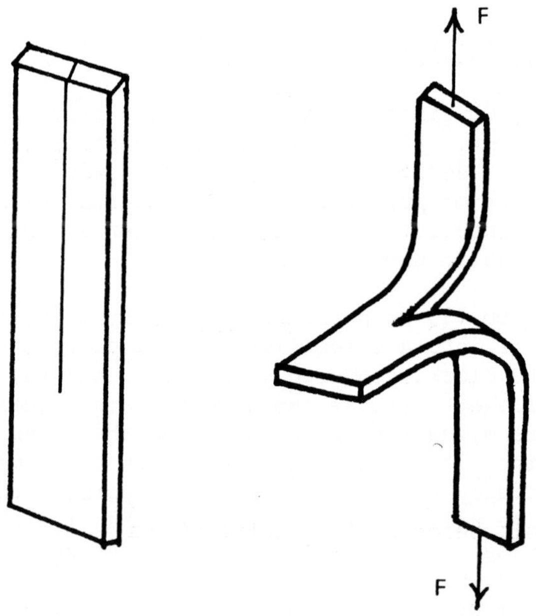

Fig. 2. The trouser test.

coordinates copes neatly with finite deformation. The two legs are extended by forces F and we want to find the force on the tip of the crack (or tear), equal to the total energy release rate Gt (t is the specimen thickness), or in the language of rubber testing Tt, where T is the tearing energy. To get it we must work out the integral (3) where S is a surface embracing the tip of the tear. It can be expanded until it becomes a surface made up of parts of the specimen surface and a cross-section of the unstressed tail, all of which contribute nothing, together with cross-sections of the two legs, one of which contributes

$$\tfrac{1}{2}F = \int (W - p_{i1}\, u_{i,1})\, dS$$

$$= WA - F\partial u_2/\partial X_1 \qquad (14)$$

where A is the unstrained area of one leg. A point on one leg with initial coordinates X_1, $X_2 = 0$ is first swung round to $x_1 = 0$, $x_2 = X_1$ and then stretched vertically by F to the final position $x_1 = 0$, $x_2 = X_1\ell/\ell_o$ where ℓ, ℓ_o are the unstrained and strained lengths of a leg. Hence $u_2 = x_2(X) - X_2 = X_1\ell/\ell_o - X_2$ and so $\partial u_2/\partial X_1 = \ell/\ell_o$ and we have

$$Gt = -F_1 = 2F\ell/\ell_o - 2WA .$$

The minus sign before F_1 merely says that the tear will run to the left. The expression on the right can, of course, be interpreted as the work done by the forces minus the increase in elastic energy when the tear extends by unit length of unstrained material. (Note that this lengthens each leg by ℓ/ℓ_o and that W is the final energy per unit initial volume.) When the forces are small enough for Hooke's law to be valid the second term on the right of (14) is minus one half of the first, just as in the Obreimoff example, and generally on the linear infinitesimal theory. This example shows that a large geometrical deformation alone need not upset the principle.

At the other extreme, far beyond the Hooke's law region, some rubbers reach a final extension ℓ'/ℓ_o which scarcely increases with further loading until the material breaks. In that case the first term can be neglected, $|F_1| = 2F\ell'/\ell_o$, and the force on the cut tip is no longer proportional to F^2 but is, so to speak, equal to the total applied force 2F magnified by the mechanical advantage factor ℓ'/ℓ_o which takes into account the fact that one unit of unstressed tail becomes ℓ'/ℓ_o units of stressed leg.

The formula (14) does not concern itself with the hardness or softness of the loading mechanism and so it applies also when the legs are extended and clamped before the tear is allowed to propagate. $|F_1|$ is then the decrease in internal energy alone per unit increase in tear length since no external work is done. In fact

Rivlin and Thomas treat this as the basic case, defining Tt to be $-(\partial U/\partial c)_\ell$ where U is the strain energy of the specimen and c the (unstrained) tear length. The equation

$$dU = (\partial U/\partial c)_\ell \, dc + (\partial U/\partial \ell)_c \, d\ell$$

combined with the thermodynamic relation $F = (\partial(\frac{1}{2}U)/\partial\ell)_c$ and the geometrical one $(\partial\ell/\partial c)_F = \ell/\ell_0$ then takes us back to (14).

GÜNTHER'S PATH-INDEPENDENT INTEGRALS

It was Günther [15] who first applied Noether's theorem systematically to elastostatics. As well as the F_ℓ of equation (3) he obtained the path-independent integrals

$$L_i = \varepsilon_{ik\ell} \int_S (x_k P_{\ell j} + u_k P_{\ell j}) \, dS_j \tag{15}$$

and

$$M = \int_S (x_\ell P_{\ell j} - \tfrac{1}{2} u_\ell P_{\ell j}) \, dS_j \,. \tag{16}$$

(We follow the notation of Budiansky and Rice [16]). Suppose that we make a picture of an elastic field by drawing the appropriate displacement vector at each point. Then the results that F_ℓ, L_i, M are path-independent depend on the fact that the picture still represents a possible elastic state after it has been, respectively, translated, rotated or enlarged. From this one can read off under what conditions each result is valid. For F_ℓ the deformation may be finite and the material may be non-linear, but it must be homogeneous. For L_i the conditions are the same, except that the material must, in addition, be isotropic. For M the stresses must be linear in the displacement gradients, so we are limited to the usual linear infinitesimal theory, though the material can be anisotropic. The term 'homogeneous' implies that the form of the energy density W is independent of X_1, X_2, X_3. A few relaxations are possible. If the form of W depends on X_2, X_3 but not X_1 then F_1 is still path-independent. (This takes care of the application of (3) to cracks at interfaces.) Similarly L_3 is path-independent if the form of W depends on $r = (x_1^2 + x_2^2)^{\frac{1}{2}}$ and x_3 but not on $\theta = \arctan(x_2/x_1)$ (Curvilinear aeolotropy).

It is hard to think of any operations other than the above three which will convert one picture of a general equilibrium elastic field into another possible one, and in fact Knowles and Sternberg [17] have given an argument to show that F_ℓ, L_i and M are the only path-independent integrals derivable from Noether's theorem for a general field. However, we shall see later that when certain limitations are imposed on the form of the displacement (anti-plane strain, plane strain) many more path-independent integrals appear, some of the form envisaged in Noether's theorem, some not.

In a two-dimensional state of plane strain (16) reduces to the simpler-looking integral

$$M = \int_S x_\ell P_{\ell j} dS_j \qquad (17)$$

taken along a plane curve S with normal (n_1, n_2). Summation over ℓ and j is now, of course, only over 1,2. There is no real difference between (16) and (17). To apply the three-dimensional (16) to the two-dimensional situation the surface of integration can be taken to be the surface of the disk generated by letting the plane area bounded by the curve S sweep through, say, unit distance along the x_3-axis. The two flat faces of the disk make a contribution to (16) which can, with the help of Gauss's theorem, be transformed into an integral over the rim of the disk, where it cancels the second term of (16), leaving (17). The following is a simple application of (17). The G for the right-hand tip of the crack of Fig. 3 is given by the integral F_1 of (3) evaluated over any loop which embraces the right tip but not the left one. It would be nice if we were able to slip the loop over the left-hand tip, because then we could expand the loop to infinity and make use of the simple asymptotic form of the crack field. However, if we do this we actually get G-G', where G' is the energy release rate for the left-hand tip. To see this, contract the large loop back onto the crack to form the keyhole-shaped contour in the figure; the right-hand circle gives G and the

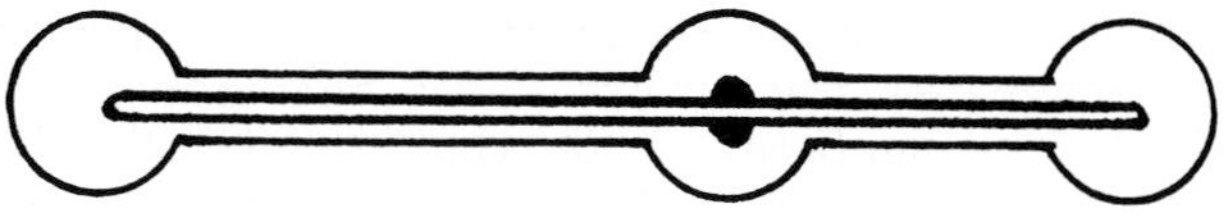

Fig. 3. Contour embracing a crack.

left hand one -G'. (The middle indentations relate to another problem.) To get over this difficulty we can use M instead of F_1, and site the origin of x_ℓ at the left-hand tip. On the keyhole-shaped contour the factor x_ℓ suppresses the contribution from the left-hand circle and on the right-hand circle it can be taken outside the integral to give G x length of crack, and this quantity is also the value of M over a large contour completely enclosing the crack. This argument shows at once, for example, that there is a respectable energy release rate for a crack at an interface between different media, despite the odd behaviour of the elastic field at the tips.

Fig. 4 shows a couple of situations which can be handled with the help of (17). On the left an edge crack of length a is opened

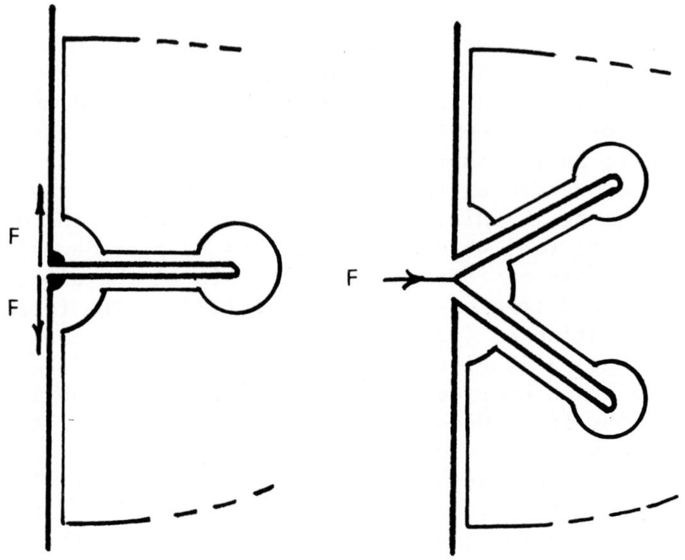

Fig. 4. Corner-loaded edge crack and plane version of conical crack.

in mode I by transverse concentrated forces $\pm$ F applied (perhaps by forcing in a wedge) at the points where the crack faces intersect the vertical free surface. For the circuit shown the straight parts contribute nothing to M. In particular the presence of x_ℓ in (17) gives a factor $x_1 = 0$ on the vertical surface which takes care of the non-zero P_{11} there. The contribution from the tip is a G and from the large semicircle, zero. To evaluate the contribution from one of the small quarter-circles surrounding the forces we can use standard expressions for the field at the tip of an infinite wedge loaded by a point force [18]. The result is

$$G = 2(1-\nu)\,\pi F^2/\mu(\pi^2-4)\,a\,, \quad K = \alpha . 2F/(\pi a)^{\frac{1}{2}}$$

with $\alpha = 1.2967...$; for the analogous mode III situation $\alpha = 1$. Howard was able to use this result in discussing edge cracks under large-scale yielding. The left of the figure shows the two-dimensional analogue of the intractable problem of a conical crack; a knife-edge produces two cracks of length a inclined at an angle β.

In this case 2Ga is equal to the difference of the M-integrals over the throat of the loaded wedge and over an infinite semicircle. This gives

$$G = (1-\nu)(F^2/2\mu a)\{(\beta + \sin\beta)^{-1} - \pi^{-1}\} \qquad (18)$$

OTHER PATH-INDEPENDENT INTEGRALS

For linear anti-plane strain we have

$$P_{11,1} + P_{12,2} = 0 \quad , \quad P_{21,1} + P_{22,2} = 0 \qquad (19)$$

and also

$$P_{11} + P_{22} = 0 \quad , \quad P_{12} = P_{21} \qquad (20)$$

whence, by Gauss's theorem,

$$I = \int \phi_\ell P_{\ell j}\, dS_j \qquad (21)$$

is path independent if $\phi_{1,1} = \phi_{2,2}$ and $\phi_{1,2} = -\phi_{2,i}$ i.e. if $f(z) = \phi_1 + i\phi_2$ is any analytic function of $z = x_1 + ix_2$. In plane strain the first, but not the second, of (20) holds and the nearest analogue of (21) is that, with the same ϕ_1, ϕ_2,

$$I = \int\{\phi_\ell P_{\ell j} + \tfrac{1}{2}(x_1\phi_2)_{,j}(P_{12}-P_{21}) - \tfrac{1}{2}x_1\phi_2(P_{12}-P_{21})_{,j}\}\, dS_j \qquad (22)$$

is path independent. This can be checked with the help of the easily-verified relation

$$\nabla^2(P_{12}-P_{21}) = 0\ . \qquad (23)$$

Equation (21) or (22) can be used to reproduce various known results quite simply. For example, suppose that the crack of Fig. 3 has point forces $\pm F$ applied to the upper (lower) faces at $x_1 = x'$. On putting $f(z)$ equal successively to $1/(z-x')$ and $z-x'$ we get, in mode III,

$$(a-x')\,G + (a+x')\,G' = F^2/\pi\mu$$

and

$$(a-x')^2 G - (a+x')^2 G' = 0$$

whence

$$G = F^2(a+x')/2\pi a\mu(a-x')$$

or $$K = F\{(a+x')/\pi a(a-x')\}^{\frac{1}{2}}\ , \qquad (24)$$

the latter valid on all three modes, and, by superposition, leading at once to the well-known result

$$K = \int_{-a}^{a} \{(a+x')/a-x')\}^{\frac{1}{2}} P(x')\, dx'/(\pi a)^{\frac{1}{2}}$$

for general internal or external loading.

The fact that $P_{12}-P_{21}$ is harmonic, eqn. (23), leads to a more general class of plane-strain path-independent integrals. Write

$$p'_{ij} = \varepsilon_{i\ell 3} P_{\ell j} \;, \quad P_{ij} = - \varepsilon_{i\ell 3} p'_{ij} \;, \quad i,j,\ell = 1,2 \;.$$

Then (19), the first of (20) and (23) show that the $p'_{ij} = p'_{ji}$ satisfy the equilibrium equations without body force and the compatibility equation $\nabla^2(p'_{11} + p'_{22}) = 0$ and are thus ordinary elastic stresses derivable from a displacement u'_i. Choose any other elastic field p^0_{ij}, u^0_i. Then the Rayleigh-Betti reciprocal theorem states that

$$\int (u^0_i\, p'_{ij} - u'_i\, p^0_{ij})\, dS_j \qquad (25)$$

is path-independent. It can be verified that, if, with the traditional Muskhelishvili notation we write

$$2\mu(u_1 + iu_2) = (3-4\nu)\,\phi(z) - z\overline{\phi'(z)} - \overline{\psi(z)}$$

together with the related expressions for the p_{ij}, then u'_i and the p'_{ij} are given by similar expressions with ϕ' and ψ' replaced by

$$\Phi' = (1-\nu)\,\phi'^2/i\mu \;, \quad \Psi' = 4(1-\nu)\,\phi'\psi'/i\mu$$

respectively.

The $P_{\ell j}$ of a rightward-pointing mode I crack along the x_1-axis with its tip at the origin translates into a field p'_{ij}, u'_i which near the tip is summed up in the Airy stress function

$$\chi' = G x\theta/2\pi$$

and thus represents a force (0,G) plus an edge dislocation with Burgers vector

$$b' = (1-2\nu)\; G/2\mu$$

coexisting at the origin. In order to use (25) to find G for some particular geometry and loading we have to try and choose a contour which embraces the crack tip and elsewhere as far as possible runs where $p'_{ij}n_j$ or u'_i are known, and arrange that u^0_i or $p^0_{ij}\, n_j$ vanishes where they are not. The writer has not had much success with (25) so far. An example of what can go wrong: an attempt to find G for an edge crack under uniform tension T gave for it a multiple of T^2 plus a multiple of the square of the relative rotation of the faces of the crack at the corners where they meet the free surface, a quantity which can, presumably, only be found by solving the problem in detail.

REFERENCES

1. Noether, E., Nachr. Ges. Wiss. Göttingen, math.-phys. Klasse, p.235, 1918. (English translation by Tavel, M., Transport Theory and Statistical Physics, 1, 183, 1971.)

2. Nabarro, F. R. N., Theory of Crystal Dislocations, Clarendon Press, Oxford, 1967, 562.

3. Eshelby, J. D., in Solid State Physics, Seitz, F. and Turnbull, D., Eds., Academic Press, New York, 1956, 79.

4. Eshelby, J. D., in Internal Stresses and Fatigue, Rassweiler, G. M., and Grube, W. L., Eds., Elsevier, Amsterdam, 1959, 41.

5. Eshelby, J. D., in Inelastic Behavior of Solids, Kanninen, M. F. et al. Eds., McGraw-Hill, New York, 1970, 77.

6. Toupin, R., Arch. rat. Mech. Anal., 11, 385, 1962.

7. Podolsky, B., and Kikuchi, C., Phys. Rev., 65, 228, 1944.

8. Chang, T. S., Proc. Cant. Phil. Soc., 42, 132, 1946.

9. Thielheim, K. O., Proc. Phys. Soc. Lond., 91, 798, 1967.

10. Eshelby, J. D., Sci. Prog., Oxf., 59, 161, 1971.

11. Eshelby, J. D., Phil. Trans., A244, 87, 1951.

12. Obreimoff, J. W., Proc. Roy. Soc., A127, 290, 1930.

13. Landau, L. D., and Lifshitz, E. M., Theory of Elasticity, Pergamon, Oxford, 1970, 52.

14. Rivlin, R. S., and Thomas, A. G., J. Polym. Sci., 10, 291, 1953.

15. Günther, W., Abh. braunschw. wiss. Ges., 14, 54, 1962.

16. Budiansky, B., and Rice, J. R., J. appl. Mech., 40, 201, 1973.

17. Knowles, J. K. and Sternberg, Eli, Arch. rat. Mech. Anal., 44, 187, 1972.

18. Timoshenko, S. and Goodier, J. N., Theory of Elasticity, McGraw-Hill, New York, 1951, 96.

19. Eshelby, J. D., in Fracture Toughness, I.S.I. Publication 121 The Iron and Steel Institute, London, 1968, 13.

CHAPTER 1

POINT DEFECTS

by J. D. ESHELBY †

1.1 INTRODUCTION

In this chapter we shall mainly be concerned with point defects (vacancies and interstitials) in pure crystals, though substitutional and interstitial impurities will receive some attention, particularly as ideas originally developed to deal with them have subsequently been applied to vacancies and self-interstitials.

Section 1.2 is concerned with the formation of defects by thermal excitation, cold work and irradiation. Part of section 1.3 (on defect mobility) may appear a little out of proportion. However, it has seemed to the writer that many solid state and metallurgical texts introduce the standard jump frequency formula with much less discussion than they devote to related matters, and that therefore a slightly extended treatment at an intermediate level might be useful.

Section 1.4 (Physical properties of defects) is concerned with some of the basic physical properties (energy and volume of formation, effect on electrical resistance) of defects and their measurement. Since the presence of defects has usually to be inferrred from their influence on bulk properties this section could equally well be entitled 'Effect of defects on physical properties'. It is perhaps best to confine this latter phrase to secondary effects consequent on the basic effects mentioned above. Important among them is the interaction between point defects and dislocations. Section 1.5 ('Interaction energies of point defects') serves as a link with other chapters.

1.2 TYPES OF DEFECT. THEIR FORMATION BY THERMAL ACTIVATION, IRRADIATION AND PLASTIC DEFORMATION

Starting from a collection of simple lattice vacancies one can form more elaborate defects by allowing them to aggregate into clusters.

† Dr Eshelby is Professor at the Department of the Theory of Materials, University of Sheffield.

A pair of vacancies on neighbouring sites constitute a divacancy, a dumb-bell shaped configuration which can assume any one of $\frac{1}{2}z$ orientations in a lattice with co-ordination number z. The stable configuration of a more complicated cluster is less obvious. One might guess that a trivacancy in a face-centred cubic lattice would take the form of a triangle of vacancies lying in a {111} plane. However, detailed calculations suggest that the energy will be lowered if a suitable adjacent atom moves in towards the triangle, leaving a configuration which may be described as a tetrahedron of vacancies

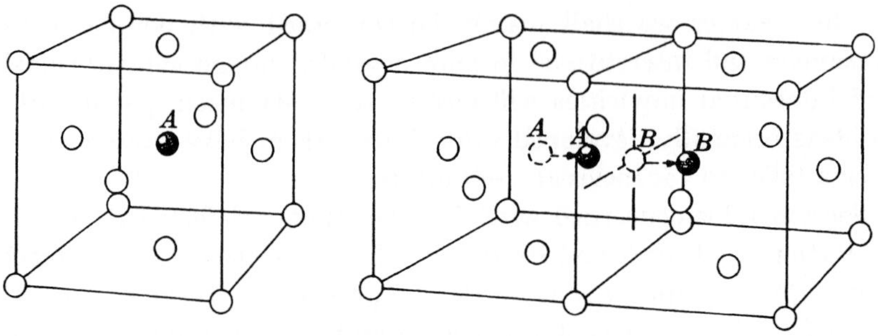

(a) (b)

Fig. 1.1 Body-centred (a) and split (b) interstitial in a face-centred cubic lattice. (Damask and Dienes, 1963)

with an atom in the middle (Damask *et al.*, 1959). Computer calculations of the equilibrium configuration for various clusters of vacancies (and interstitials) have been made by Vineyard (1961) and his associates.

In contrast to the vacancy even a single interstitial can display quite complicated behaviour. Fig. 1.1(*a*) shows an interstitial atom A in a body-centred position in a face-centred cubic lattice. In fig. 1.1(*b*) the atom has moved to the right, displacing the atom B, so that in the resulting dumb-bell configuration one cannot tell whether A or B is 'really' the interstitial. The resulting configuration is called a split interstitial (German *Hantel*). Calculations indicate that its energy is slightly less than that of the simple interstitial of fig. 1.1(*a*). There is, in fact, experimental evidence that split interstitials can exist in several metals. If the atom A in fig. 1.1(*a*) had been displaced vertically we should have got a vertical dumb-bell, and a displacement out of the paper would have given a horizontal dumb-bell, perpendicular to the one in fig. 1.1(*b*). The defect can thus exist in one of three orientations, and can change from one to another. There will evidently be a tetragonal distortion of the lattice near the defect, with its tetragonal axis in the [100], [010] or [001] direction, according to the orientation of the defect. This is very similar to what happens with interstitial carbon (or nitrogen) atoms in body-centred iron. Here the carbon atom can lie at the mid-point of a line joining two iron atoms and oriented in either the [100], [010] or [001] direction. Relaxation due to reorientation under the influence of applied stresses leads to a peak in the internal friction spectrum (Snoek effect), and interaction with Bloch walls affects the magnetic susceptibility. We might expect split interstitials to produce similar effects, and they have in fact been found in nickel and perhaps copper, into which interstitials have been introduced by plastic deformation or irradiation (Seeger and Wagner, 1965).

Point defects can be introduced into a crystal by heating it (thermal excitation) deforming it plastically, or irradiating it with fast particles.

At a finite temperature an otherwise perfect crystal will contain point defects of the types mentioned in section 1.1, and others, in concentrations which are determined by their energies and entropies of formation.

In order to find the equilibrium concentration of a particular type of defect we may start from a result of statistical mechanics which says that the probability of finding a system in a definite state whose Gibbs free energy is G is given by

$$p = C \exp(-G/kT)$$

where k is Boltzmann's constant, T the absolute temperature and C is a normalising constant. With a suitable choice of C, G may be taken to be the excess of the Gibbs free energy over its value in some standard state. G is then the work required to carry the system from the standard to the actual state. If the system is subject to an external pressure, or some more complicated applied stress, part of this work goes towards increasing the potential energy of the loading mechanism. Use of the Gibbs free energy G, rather than the Helmholtz free energy F, automatically allows for this, For a crystal under atmospheric pressure the difference between G and F is negligible.

For definiteness we consider the case of vacancies. Let the standard state be a crystal free of vacancies, and let ΔG be the work required to form one vacancy. The work required to form n vacancies will then be $G = n\Delta G$ if we can neglect the interaction between them. The probability of finding n vacancies at n specified sites is thus $C\exp(-n\Delta G/kT)$. The probability of finding n vacancies at n unspecified sites in a crystal containing N atoms (and hence $N+n$ lattice sites) is thus

$$p(n) = C\frac{(N+n)!}{N!\,n!}\mathrm{e}^{-n\Delta G/kT} \tag{1.1}$$

where the collection of factorials is the number of distinct ways of picking out n unoccupied sites from the $N+n$ (occupied or unoccupied) sites. The most probable value of n is that for which a small change of n leaves $p(n)$ nearly unaltered, so that, say, $p(n) = p(n-1)$. After extensive cancellation this gives

$$n = (N+n)\,\mathrm{e}^{-\Delta G/kT} \tag{1.2}$$

or a most probable concentration (vacant sites divided by total sites)

$$c = \mathrm{e}^{-\Delta G/kT}. \tag{1.3}$$

One can show that (1.2) also gives, very closely, the average value of n,

$$\bar{n} = \frac{\sum_n np(n)}{\sum_n p(n)} = \frac{\sum_n n[(N+n)!/N!n!]x^n}{\sum_n[(N+n)!/N!\,n!]x^n}, \qquad \text{where } x = \mathrm{e}^{-\Delta G/kT}. \tag{1.4}$$

In the summations, terms for which n is not small in comparison with N are inaccurate because of interaction between the defects, or even meaningless (more vacancies than atoms), but they have a negligible effect on the sum for small x and we may carry the summations from 0 to ∞. The denominator in (1.4) is then recognisable as the expansion of $(1-x)^{-(N+1)}$ by the binomial theorem. If we differentiate the denominator with respect to x we get x^{-1} times the numerator. Consequently

$$\bar{n} = \frac{x\,\mathrm{d}(1-x)^{-(N+1)}/\mathrm{d}x}{(1-x)^{-(N+1)}} = (N+1)\frac{x}{1-x}$$

which agrees with (1.2) for large N and small x.

The number of interstitials present in thermal equilibrium may be calculated in the same way. The combinatorial factor in (1.1) may be more complicated if the number of interstitial sites differs from the number of lattice sites, but, as might be expected, the ratio of occupied to total interstitial sites is still given by (1.3) with a suitable ΔG.

More complex defects, in particular clusters of vacancies, may be handled in the same way. For example, a divacancy may be regarded as a dumb-bell which may exist in a number of orientations. Each orientation may be regarded as a separate type of defect and its concentration calculated from (1.3). Multiplication of the number of orientations then gives the total number of defects. However, to get the combinatorial factor in (1.1) right we must take care to choose the number of independent orientations in such a way that two pairs apparently having different orientations and different positions are not in fact the same physical pair. We illustrate this for divacancies in a lattice with arbitrary co-ordination number, and for two types of trivacancy in a face-centred cubic lattice.

An observer looking at the crystal will see divacancies situated at various points in the crystal and with different orientations. If he is not looking along a symmetry element of the crystal lattice he can classify the two vacancies of a pair into a near one and a far one. A particular near vacancy can have its companion in any one of $\frac{1}{2}z$ positions, where z is the number of nearest neighbour sites; if the far vacancy were in one of the remaining $\frac{1}{2}z$ positions it would be the near one. For a particular one of these $\frac{1}{2}z$ orientations the near vacancy can lie at any one of the N lattice sites, so the concentration of them is given by (1.3) with a suitable ΔG. There will be an equal concentration of divacancies with any one of the remaining $\frac{1}{2}z-1$ orientations, so the concentration of divacancies is

$$c = \tfrac{1}{2}z\,\mathrm{e}^{-\Delta G_2/kT}.$$

The most obvious form for a trivacancy in a face-centred cubic lattice is an equilateral triangle of vacancies lying in any one of the four sets of $\{111\}$ planes. An observer can classify them as looking like $\triangle$ or ∇, and can distinguish a 'near' atom in the triangle. The near atom in a $\triangle$ can be situated at any of the N atom positions in the lattice, so that the number of them is given by (1.3) with a suitable ΔG. There is an equal number of ∇s in (111) planes, or in all

$$n = 2N\,\mathrm{e}^{-\Delta G_3/kT} \tag{1.5}$$

trivacancies in (111) planes. There are equal numbers lying in the other three $\{111\}$ planes so that the total concentration of trivacancies is

$$c_3 = 8\,e^{-\Delta G_3/kT}. \tag{1.6}$$

However, calculation (Vineyard, 1961) suggests that a trivacancy in the form of a tetrahedron of vacancies with an atom in the middle will have an energy of formation, $\Delta G_3'$ say, less than ΔG_3. Any tetrahedron will have one face in a (111) plane and it may be a $\triangle$ or a ∇. If the (111) planes are labelled $ABC\ldots$ in the usual way, a $\triangle$ in an A plane can only accept the fourth vacancy in the B position and a ∇ in an A plane can only accept one in the C position, so that there is only one type of tetrahedron associated with a $\triangle$ and one with a ∇. The whole number of tetrahedra in the crystal is thus the same as the number of triangular trivacancies which happen to lie on (111) planes, equation (1.5), apart from the change in ΔG. So

$$c_3' = 2\,e^{-\Delta G_3'/kT} \tag{1.7}$$

is the concentration of tetrahedral trivacancies. Of course the two types co-exist in the proportion corresponding to (1.6), (1.7), but if $\Delta G_3'$ is notably smaller than ΔG_3 the tetrahedral type will predominate at reasonable temperatures and may be regarded as 'the' trivacancy.

It is customary to write an expression like (1.3) for a defect concentration in terms of the energy and entropy of formation rather than in terms of ΔG. It will be convenient to consider this in section 1.4.2.

Theory and experiment both indicate that the energy required to form an interstitial is much greater than the energy required to form a vacancy, so that practically speaking only vacancies are formed thermally. Typically about 1 lattice site in 10^4 may be vacant in a metal just below the melting point.

If a metal is bombarded with high-energy radiation (neutrons, protons, deuterons, α-particles, electrons) atoms are displaced from their lattice sites and ultimately take up interstitial positions, not necessarily near their point of origin. Before settling down they may knock out other atoms, which in turn may displace others. In this way, in contrast to thermal generation, equal numbers of vacancies and interstitials are formed. It is interesting to see how the concentration produced by a given dose of irradiation compares with the number generated thermally. The following is a rather crude estimate for the case of copper irradiated by neutrons.

If the energy of the incident neutron, E_n say, greatly exceeds the energy E_d required to displace the struck atom, the latter can be treated as free. Elementary mechanics then gives for the maximum energy transferred to the atom the value $E_m = 4E_n M_1 M_2/(M_1 + M_2)^2$ where M_1, M_2 are the masses of the colliding particles, or say $4E_n/A$ for a neutron colliding with a particle of atomic weight A. The average energy transferred is $\frac{1}{2}E_m$.

The number of collisions per cm^3 produced by a flux of ϕ neutrons per cm^2 maintained for a time t is $n = \phi t \sigma n_0$, where n_0 is the number of atoms per cm^3 and σ is the collision cross-section. The displaced particle will have an average energy $\frac{1}{2}E_m$ and should thus be able to create about $\frac{1}{2}E_m/E_d$ additional Frenkel pairs; a somewhat more accurate estimate is $\frac{1}{2}E_m/2E_d$. Hence finally the total concentration of pairs is

$$c = \frac{n}{n_0}\frac{E_m}{4E_d} = \frac{\sigma E_n}{AE_d}\phi t.$$

For copper $A = 64$, $E_d \sim 25$ eV and for 1 MeV neutrons $\sigma \sim 1$ barn $= 10^{-24}$ cm^2, so that $c \sim 6 \times 10^{-22}\ \phi t$, and a day in a typical pile flux of 10^{13} neutrons cm^{-2} sec^{-1} can give a pair concentration 6×10^{-4}, comparable with the number of thermal vacancies just below the melting point.

The fact that the atoms are regularly arranged leads to some interesting effects in irradiated materials. If the energy transfer is low a succession of collisions may occur along a close-packed row of atoms. Each atom returns to its equilibrium position, so that energy, but not material, is transported along the chain (focuson). If the conditions of the initiating collision are somewhat different a vacancy may be formed in the chain, and consequently somewhat further along it $n + 1$ atoms must be distributed over the distance normally occupied by n. Such a 'crowdion' configuration (essentially Frenkel and Kontorova's (1938) 'caterpillar') is formally a small loop of edge dislocation with its Burgers vector along the direction of the row, so that its movement transports material along the row, though no individual atom moves a large distance. Individual atoms may travel large distances along channels in the crystal structure.

In the case of plastic deformation, point defects probably mostly arise from the non-conservative motion of jogs on dislocations, produced by their intersection with other dislocations. Seitz (1952) has given the estimate $c \simeq 10^{-4}\ \epsilon$ for the concentration produced by a plastic strain ϵ.

1.3 THE MOBILITY OF POINT DEFECTS

Under the influence of thermal vibrations an interstitial impurity may jump from one interstitial site to the next. An atom adjacent to a vacancy may fall into it; the vacancy has then moved to a new position. These are simple diffusion processes. More complex rearrangements can occur – for example a divacancy or a split interstitial or an interstitial carbon atom in alpha-iron may re-orientate.

It is a problem in kinetic theory (or the theory of rate processes) to find the rate at which such rearrangements occur. It is convenient to start with the highly idealised situation shown in fig. 1.2. There is a

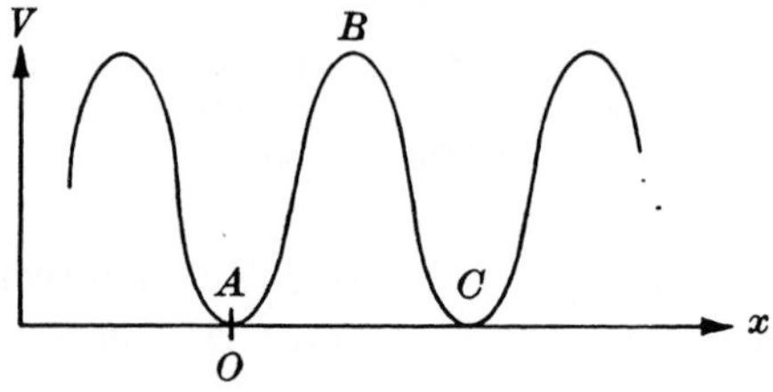

Fig. 1.2.

particle in the well A of the one-dimensional potential $V(x)$. It is required to find the probability that, as a result of thermal agitation, the particle will leave the well by passing over the barrier B. It is convenient to imagine that there is a large number of non-interacting particles in the well as we can then discuss the decrease in the number of particles in A rather than the decrease in the probability that the single particle is still there. This is of course a purely formal device. Fig. 1.2 may be taken as referring to a mental superposition of a large number of wells each containing one or no particles. The collection may refer to a number of actual wells at different points in a crystal, or it may just be an 'ensemble' in the sense of statistical mechanics.

With this understanding we may say that there are n non-interacting particles in A, and that as a result of thermal agitation a number of them will jump over the barrier B. (As many will jump out in the opposite direction, but we are not concerned with them.) It is convenient to specify the rate at which the particles leave by way of a jump frequency f defined by

$$f = -\frac{1}{n}\cdot\frac{1}{2}\cdot\frac{dn}{dt}.$$

Its value is

$$f = \nu \mathrm{e}^{-U/kT} \tag{1.8}$$

where U is the height of the barrier and ν is the frequency with which a particle vibrates in the bottom of the well. The justification usually given for (1.8) is that the exponential gives the probability that a particle has enough energy to surmount the barrier and that ν is the number of attempts it makes in unit time. Though it gives the correct result this argument is, to say the least, rather unconvincing. Since (1.8) and generalisation of it to more realistic situations than the one shown in fig. 1.2 play such an important part in the theory of defect mobility and diffusion it is perhaps worth devoting some space to the matter. We shall give a simple derivation of (1.8) and extend it to motion in a fixed two- or three-dimensional potential. Finally, following Vineyard (1957) we indicate how the method can in principle be extended to the realistic case where we cannot regard the particle as moving in a fixed potential (because the atoms which produce the potential are themselves displaced by the motion of the particle) or where (as in the motion of a vacancy or the re-orientation of a divacancy) the configuration of the defect cannot be described by giving the co-ordinates of a single particle.

To calculate the rate at which particles leave the well A by passing over the barrier B from left to right we only need two results from statistical mechanics. First, Boltzmann's barometric formula states that the number of particles per unit length of the x-axis is const.$\exp[-V(x)/kT]$. Secondly Maxwell's distribution law for velocities states that the average velocity is a constant independent of $V(x)$. This applies to any average velocity of those particles which are moving from left to right, $\bar{v}$, say. (Actually any average velocity is a multiple of $\sqrt{(kT/m)}$ for a particle of mass m, but we do not need to know this.)

Define the flux of particles at x to be the number of particles which in unit time pass the point x moving from left to right. It is equal to the number of particles per unit length at x multiplied by $\bar{v}(x)$. Then we have

$$\frac{\text{flux at } B}{\text{flux at } A} = \frac{\bar{v}(B)\exp[-V(B)/kT]}{\bar{v}(A)\exp[-V(A)/kT]}.$$

The $\bar{v}$s cancel, since, as we have seen, $\bar{v}$ is in fact independent of position, and so with $U = V(B) - V(A)$ we have

$$\text{flux at } B = \text{flux at } A \times \exp(-U/kT).$$

The flux at A can be calculated directly. If ν is the frequency of vibration of a particle in the potential well, each of the n particles at the bottom of the well passes the centre line of the well from left to right ν times a second, and the flux is $n\nu$. So

$$\text{flux at } B = n\nu \exp(-U/kT)$$

and consequently for the jump frequency we have

$$f = \frac{\text{flux at } B}{\text{number in } A} = \nu \exp(-U/kT)$$

in agreement with (1.8).

The artifice which gave us the flux at A becomes rather hard to apply in two or more dimensions, so we re-derive (1.8) by a method which lends itself to extension.

According to Maxwell and Boltzmann the number of particles between x and $x + \mathrm{d}x$ having momentum between p and $p + \mathrm{d}p$ is $N(x, p)\,\mathrm{d}x\,\mathrm{d}p$ where

$$N(x,p) = \text{const.} \exp\{[-V(x) - p^2/2m]/kT\}$$

where m is the mass of a particle. The flux at B is $N(x,p)$ evaluated at B, multiplied by the velocity p/m and integrated with respect to p from 0 to ∞, since we are only interested in particles moving from left to right. The number of particles in the well A is $N(x,p)$ integrated over all momenta and over a range of x which embraces all the particles which are considered to be in the well. Hence f, the ratio of flux to the number of particles in the well, is given by

$$f = \frac{\mathrm{e}^{-U/kT} \int_0^\infty (p/m)\,\mathrm{e}^{-p^2/2mkT}\,\mathrm{d}p}{\int \mathrm{e}^{-V(x)/kT}\,\mathrm{d}x \int_{-\infty}^{\infty} \mathrm{e}^{-p^2/2mkT}\,\mathrm{d}p}.$$

We now suppose that $kT \ll U$. The density of particles is then negligible except at the bottom of the well and we may without noticeable error take the limits of the x-integral in the denominator to be $\pm\infty$. (It is only in such a case that one can make a sensible distinction between particles in the well and particles in transit to the next well.) For the same reason we may replace $V(x)$ by the first non-vanishing term of its Taylor expansion. If A is at $x=0$ this is

$$V(x)=\tfrac{1}{2}V''(0)\,x^2$$

because V' is zero at a minimum and we have chosen $V(0)$ to be zero.

The integrals can be evaluated with the help of

$$\int_0^\infty \mathrm{e}^{-t^2/a^2}\,\mathrm{d}t=\tfrac{1}{2}a\sqrt{\pi},\quad \int_0^\infty t\,\mathrm{e}^{-t^2/a^2}\,\mathrm{d}t=\tfrac{1}{2}a \tag{1.9}$$

to give (1.8) with

$$\nu=\frac{1}{2\pi}\sqrt{\left(\frac{V''(0)}{m}\right)} \tag{1.10}$$

which is precisely the frequency of a particle of mass m vibrating in a potential well with curvature $V''(0)$.

We consider next the two-dimensional situation of fig. 1.3. There are n particles in the well A and it is required to find the rate at which they pass through the saddle-point region B on the way to the well C. Near A we may write the potential as

$$V=\tfrac{1}{2}V''_x\,x^2+\tfrac{1}{2}V''_y\,y^2 \tag{1.11}$$

where x and y are axes chosen parallel to the principal axes of the elliptical equipotentials around A, and near B we may put

$$V=U-\tfrac{1}{2}|V''_{x'}|x'^2+\tfrac{1}{2}V''_{y'}\,y'^2 \tag{1.12}$$

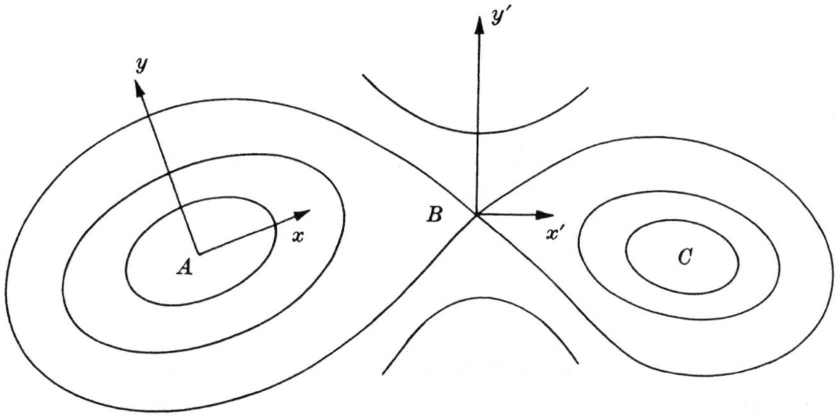

Fig. 1.3..

where x', y' are local axes at B parallel to the principle axes of the hyperbolic equipotentials near B. V_x'' stands for $(\partial^2 V/\partial x^2)_A$, and similarly for the others. The coefficient of x'^2 is negative because, along x', V decreases on either side of B.

The two-dimensional analogue of $N(x,p)$ is given by

$$N(x,y,p_x,p_y)\,\mathrm{d}x\,\mathrm{d}y\,\mathrm{d}p_x\,\mathrm{d}p_y$$
$$= \text{const.} \exp\{[-V(x,y) - (p_x^2 + p_y^2)/2m]/kT\}\,\mathrm{d}x\,\mathrm{d}y\,\mathrm{d}p_x\,\mathrm{d}p_y. \quad (1.13)$$

A particle may be said to have crossed the saddle-point when it has passed over the line $x' = 0$ from left to right. The flux of particles through the saddle-point is thus equal to N (equation (1.13)) multiplied by the velocity $p_{x'}/m$ and integrated along the y' axis and over positive $p_{x'}$. The number of particles in A is N integrated over all momenta and over the area of the well. We may use the same argument as in the one-dimensional case to replace V by the expressions (1.11) and (1.12) and to extend the spatial integrations between $\pm\infty$. The jump frequency, the ratio of flux at B to particles in A, thus takes the form

$$f = \mathrm{e}^{-U/kT}$$
$$\times \frac{\int_{-\infty}^{\infty} \mathrm{e}^{-V_{y'}'' y'^2/2kT}\,\mathrm{d}y' \int_0^{\infty} (p_{x'}/m)\,\mathrm{e}^{-p_{x'}^2/2mkT}\,\mathrm{d}p_{x'}}{\int_{-\infty}^{\infty} \mathrm{e}^{-V_x'' x^2/2kT}\,\mathrm{d}x \int_{-\infty}^{\infty} \mathrm{e}^{-V_y'' y^2/2kT}\,\mathrm{d}y \int_{-\infty}^{\infty} \mathrm{e}^{-p_x^2/2mkT}\,\mathrm{d}p_x \int_{-\infty}^{\infty} \mathrm{e}^{-p_y^2/2mkT}\,\mathrm{d}p_y}. \quad (1.14)$$

Using (1.9) this becomes

$$f = \frac{\nu_x \nu_y}{\nu_{y'}}\,\mathrm{e}^{-U/kT} \quad (1.15)$$

where the ν_x, ν_y, $\nu_{y'}$ are the vibration frequencies of a particle of mass m constrained to move along the x or y axis at A, or the y' axis at B, calculated from the appropriate V'' with the help of (1.11), (1.12). Equation (1.15) is the same as (1.8) with an effective frequency $\nu_x \nu_y/\nu_{y'}$.

Suppose next that fig. 1.3 represents a cross-section of a three-dimensional potential field with a well at A where the equipotentials are ellipsoids, a saddle-point B where they are hyperboloids, and another well at C. Near A and B the potential will be like (1.11) and (1.12) but with extra terms $\frac{1}{2}V_z''$, $\frac{1}{2}V_{z'}''$ where z and z' refer to additional co-ordinate axes erected perpendicular to the paper at A and B.

The distribution of particles over position and momentum is given by an obvious generalisation of (1.13).

The jump frequency is now the flux of particles from left to right across the $y'z'$ plane at B, divided by the number of particles in the well at A. It is easy to see how (1.14) is to be modified. In the numerator (flux) the velocity integral over $p_{x'}$ is unaltered, but since we are integrating over the $y'z'$ plane we must insert an integral over z' similar to the one over y'. In the denominator (number of particles in A) we must insert a z integral similar to the x and y integrals and a p_z integral similar to the p_x and p_y integrals. Applying (1.9) and (1.10) will evidently give

$$f = \frac{\nu_x \nu_y \nu_z}{\nu_{y'} \nu_{z'}} e^{-U/kT} \tag{1.16}$$

with an obvious generalisation of notation.

The foregoing calculations are unrealistic because we are actually interested in the motion of a particle which is moving in a potential which is varying in time because the atoms which produce it are subject to thermal fluctuations and are themselves affected by the movement of the atom we are interested in. Moreover, if the jump process is a complicated one (e.g. the movement of a vacancy, the re-orientation of a split interstitial) it may not be clear how to introduce a co-ordinate which describes its progress. These difficulties are in principle avoided by a treatment on the following lines.

Consider a crystal containing N atoms, not necessarily all the same. Its total potential energy will be a function of the $3N$ variables $x_n, y_n, z_n, n = 1, 2, \ldots N$. Imagine that $x_1, y_1, \ldots z_N$ are taken as $3N$ rectangular co-ordinates in a $3N$-dimensional 'configuration space' and that contours $V = \text{const.}$ are plotted in it. There will be a number of potential wells in it, representing for example, the perfect crystal, the crystal containing a vacancy at a specific site, the crystal containing two vacancies at two specific sites, the crystal containing an interstitial at a specific site, and so on. Between the minima there will be saddle-points. We may take fig. 1.3 as a crude picture of configuration space; the place of the moving particle is taken by a representative point whose movement is dictated by the way the co-ordinates $x_1, y_1, \ldots z_N$ change with time.

A representative point oscillating near the bottom of well A will in general execute a $3N$-dimensional Lissajous figure. However, if it is vibrating along one of the $3N$ principal axes of the hyper-ellipsoidal potential surfaces around A, its motion is simple harmonic. If we rotate our axes to coincide with these directions we get a set of so-called normal co-ordinates. The associated frequencies ν_i, $i = 1, 2, \ldots 3N$ are the possible vibration frequencies of the whole crystal when its representative point is near A. Similarly we can introduce normal co-ordinates at the saddle-point B with associated frequencies ν_i'; $i = 2, 3, \ldots 3N$. One (we have arbitrarily taken it to be the first) is missing because V will decrease, rather than increase, as the associated normal co-ordinate is displaced from B, corresponding to unstable motion (compare x' in fig. 1.3).

To compute the jump frequency for a representative point from well A to well C via B we imagine, as in the case of the single particle, that there is a swarm (statistical ensemble) of such points. The jump frequency is then the flux of points 'from left to right' through B, divided by the number of points in A. The calculation is essentially the same as that for a particle in a two- or three-dimensional potential; the only difficulty is one of visualisation. It is fairly clear that the result will just be what one gets by extending the series (1.8), (1.15), (1.16), namely

$$f = \frac{\nu_1 \nu_2 \ldots \nu_{3N}}{\nu_2' \nu_3' \ldots \nu_{3N}'} \mathrm{e}^{-U/kT} \tag{1.17}$$

where U is the potential difference between B and A.

To make contact with the macroscopic theory of diffusion we suppose that in fig. 1.3 there are n_A defects in A, n_C in C. The net flux of defects across B from left to right is $(n_A - n_C)f$. The diffusion coefficient D is the ratio of the flux of defects per unit area per unit time divided by minus the gradient of the number of defects per unit volume. If the crystal is simple cubic with lattice parameter a each potential well monopolises a volume a^3 and the particles flow across B through an area a^2. Hence the current is $(n_A - n_C)f/a^2$ and the defect density gradient is $(n_C - n_A)/a^4$ which gives

$$D = a^2 f. \tag{1.18}$$

Suppose next that $n_A = n_C = n$ but that a slowly varying potential $\mathscr{V}(x)$ is added to the potential $V(x)$ of fig. 1.3. The height of the barrier from A to B is now $U - \tfrac{1}{2}a\partial\mathscr{V}/\partial x$ and the height of the barrier going from C to B is $U + \tfrac{1}{2}a\,\partial\mathscr{V}/\partial x$. Consequently, even though $n_A = n_C$, there is a flux from left to right over B equal to

$$nf\left[\exp\left(-\frac{a}{2kT}\frac{\partial\mathscr{V}}{\partial x}\right) - \exp\left(\frac{a}{2kT}\frac{\partial\mathscr{V}}{\partial x}\right)\right] \simeq -nf\frac{a}{kT}\frac{\partial\mathscr{V}}{\partial x} \tag{1.19}$$

provided the changes in barrier height are small compared with kT. If we define the drift velocity v of the defects to be the flux of defects per unit area (the quantity (1.19) times a^{-2}) divided by the number of defects per unit volume (n/a^3) we have

$$v = -\frac{D}{kT}\frac{\partial \mathscr{V}}{\partial x}, \tag{1.20}$$

the Einstein relation between mobility and diffusion coefficient.

1.4 THE PHYSICAL PROPERTIES OF DEFECTS

1.4.1 Theory

The theory of the physical properties of point defects presents peculiar difficulties. What goes on near a defect must be found by a detailed calculation which takes into account the displacement of the ions and the redistribution of electrons. On the other hand, if one is interested in, say, the volume change due to a defect or the interaction of a defect with a dislocation some distance from it, it is impossible to take the behaviour of every atom into account and one is driven to use continuum calculations. Often the atomic and continuum approaches must be used simultaneously, suitably patched together. Atomic calculations have to begin with a careful consideration of what approximations are allowable and usually end with extensive numerical computation. They are thus hard to describe briefly and informatively, and their results are difficult to assess and liable to revision. It seems best, therefore, to confine ourselves to a few aspects which can be discussed in fair detail with simple though crude models, and only look rather cursorily at the rest.

The simplest continuum model is the so-called misfitting sphere model (Bilby, 1950). In it the point defect is treated as an elastic sphere inserted into a hole in an elastic matrix, the two being welded together at the interface (the sphere may be too large or too small). The deformation in the matrix is calculated with the help of the linear theory of elasticity, and usually with the assumption of isotropy. The model is obviously not a very good one, though it may have some validity when applied to alloys between metals with nearly the same atomic volume. It would be precarious to apply it to an interstitial, and in the case of a vacancy there is no misfitting atom. Even here, however, it is reasonable to suppose that the distortions round the vacancy are similar to those around a too-small misfitting atom. Despite its shortcomings we shall describe the model in some detail, since it has been, and still is, used in the theory of solid solutions and lattice defects, if only as a method of extrapolating the results of atomistic calculations to points remote from the defect, and because it underlies some of the experiments described in 1.4.2. Further it brings out some points concerning the effect of the free surface of the specimen which might be overlooked in a purely atomistic treatment.

Let $\mathbf{u}(\mathbf{r})$ be the displacement which a point of the elastic medium at $\mathbf{r}$ undergoes when the defect is introduced. According to the theory of elasticity (Love, 1952) it must satisfy the equilibrium equation

$$\mu \nabla^2 \mathbf{u} + (K + \tfrac{1}{3}\mu)\,\mathrm{grad}\,\mathrm{div}\,\mathbf{u} = 0 \qquad (1.21)$$

where μ, K are the shear modulus and bulk modulus of the (isotropic) material. The strains e_{ij} are related to the Cartesian components u_1, u_2, u_3 of $\mathbf{u}$ by

$$e_{ij} = \tfrac{1}{2}(\partial u_i/\partial x_j + \partial u_j/\partial x_i), \quad i, j = 1, 2, 3$$

where x_1, x_2, x_3 stand for x, y, z. The stresses are related to the strains by

$$p_{ij} = (K - \tfrac{2}{3}\mu)\, e\delta_{ij} + 2\mu e_{ij}$$

where

$$e = e_{11} + e_{22} + e_{33} = \mathrm{div}\,\mathbf{u}$$

is the dilatation and δ_{ij} is equal to 1 or 0 according as i and j are or are not equal. In terms of the p_{ij} the equilibrium equation (1.21) takes the form

$$\partial p_{ij}/\partial x_j = 0. \qquad (1.22)$$

(Here and in what follows we use the convention that a repeated suffix is summed over the values 1, 2, 3.)

If we require the displacement outside the defect to be spherically symmetrical and to fall off with r there is only one solution of (1.21), namely

$$\mathbf{u}^{\infty} = c\frac{\mathbf{r}}{r^3} = -c\,\mathrm{grad}\left(\frac{1}{r}\right) \qquad (1.23)$$

where c is a constant which measures the 'strength' of the defect. For many purposes it is sufficient to regard c as a constant determined by experiment, or by fitting (1.23) to the results of an atomic calculation. Later we consider how c is related to the details of the misfitting sphere model.

The solution (1.23) must be modified if we require it to satisfy definite boundary conditions at the surface of the solid containing the defect (see below) and so we have attached the superscript ∞ to $\mathbf{u}$ to indicate that it refers to an infinite medium, or more strictly to one where the inevitable boundary is so far away that we can ignore it, in considering what happens at a finite distance from the defect.

The divergence (and also the curl) of $\mathbf{u}$ is zero, so that any closed surface not embracing the defect suffers no volume change when the defect is introduced, though its shape is distorted. However, since each point of it is displaced radially outwards a surface S enclosing the defect suffers an increase of volume

$$\Delta V^{\infty} = \int \mathbf{u} \cdot \mathbf{n}\, \mathrm{d}S = 4\pi c \tag{1.24}$$

($\mathbf{n}$ is the outward normal to the surface) since the integral is just c times the solid angle subtended by the closed surface at the defect. Because div $\mathbf{u}^{\infty} = 0$ outside the defect all such surfaces suffer the same volume change. It is sometimes useful to look at the matter in the following way. Formally the dilatation $e = \operatorname{div} \mathbf{u}$ is given by

$$e = -c \operatorname{div} \operatorname{grad}\left(\frac{1}{r}\right) = -c\nabla^2\left(\frac{1}{r}\right) = 4\pi c\,\delta(\mathbf{r}), \tag{1.25}$$

so that e vanishes except at the origin where there is a concentrated patch of dilatation of total amount $4\pi c$. An element of volume $\mathrm{d}v$ becomes one of volume $e\,\mathrm{d}v$ after deformation. Consequently the volume change $\int e\,\mathrm{d}v$ associated with any volume V is $4\pi c$ or 0 according as V does or does not include the defect.

Because the divergence and curl of (1.23) both vanish the associated stresses are very simple:

$$p_{ij}^{\infty} = 2\mu \partial u_i^{\infty} / \partial x_j \tag{1.26}$$

and in particular the radial stress is

$$p_{rr}^{\infty} = 4\mu c / r^3. \tag{1.27}$$

As an aside we can derive from (1.23) and (1.27) a result which will be useful later. If we cut out a sphere of radius r about the defect the elastic field outside the resulting hole will be unaltered providing we maintain a pressure p numerically equal to (1.27) inside it. The increase in the volume v of the hole over its value when p (or c) is zero is $\delta v = 4\pi c$. Hence

$$\frac{\delta v}{v} = \frac{p}{K_{\mathrm{h}}}, \quad \text{where } K_{\mathrm{h}} = \tfrac{4}{3}\mu. \tag{1.28}$$

Equation (1.28) is reminiscent of the relation

$$-\frac{\delta v}{v} = \frac{p}{K} \tag{1.29}$$

for the change of volume of a sphere under pressure, and we may think of K_{h} as the bulk modulus for blowing up a spherical hole in an infinite solid, or a small hole in a finite solid, provided it is not too near the surface.

Returning to the main argument we must consider how these results are modified by the fact that the material has a free surface. If in the infinite medium we draw a large surface S surrounding the defect and remove the material outside it, the displacement and stress will only maintain the values (1.23), (1.26) if we apply surface forces

$$T_i = p^{\infty}_{ij} n_j$$

per unit area of S, where n_j is the outward normal to S.

If we remove these forces the displacement and stress will become, say,

$$u_i = u^{\infty}_i + u^{\mathrm{I}}_i, \qquad p_{ij} = p^{\infty}_{ij} + p^{\mathrm{I}}_{ij} \tag{1.30}$$

where u^{I}_i, p^{I}_{ij} is an elastic field free from singularities inside S and so chosen that there are no surface tractions on S, that is

$$p^{\mathrm{I}}_{ij} n_j = -p^{\infty}_{ij} n_j \quad \text{on } S. \tag{1.31}$$

In analogy with electrostatics we may regard u^{I}_i, p^{I}_{ij} as the 'image' elastic field. Consider first the simple case where S is a sphere centred on the defect. Then according to (1.27) $p^{\infty}_{ij} n_j$ reduces to a uniform hydrostatic pressure $p = 16\pi\mu c/3V$ acting over the surface of the sphere, where V is its volume. When this unwanted pressure is removed the sphere expands by an amount $\Delta V^{\mathrm{I}} = pV/K = 4\pi c(4\mu/3\mathrm{K})$ additional to ΔV^{∞} and the whole volume expansion due to the presence of the defect is

$$\Delta V = \Delta V^{\infty} + \Delta V^{\mathrm{I}} = \gamma \Delta V^{\infty} = 4\pi\gamma c \tag{1.32}$$

where

$$\gamma = \frac{3K + 4\mu}{3K} = 3\frac{1-\sigma}{1+\sigma} \tag{1.33}$$

(σ is Poisson's ratio).

For a body of general shape it would be difficult to work out the image field, but it is still possible to show that (1.31) remains valid. There are several ways of doing this; the following method illustrates a useful device. The total volume change is

$$\Delta V = \int_S u_j n_j \, \mathrm{d}S \tag{1.34}$$

where $u_j = u_j^\infty + u_j^{\mathrm{I}}$. To evaluate it we replace the integrand u_j by another vector u_j' which leaves the integral unchanged but which has zero divergence. A suitable choice is

$$u_j' = u_j - p_{ij} x_i / 3K$$

where u_j, p_{ij} are given by (1.30); the second term contributes nothing to the integral by (1.31) and the divergence

$$\frac{\partial u_j'}{\partial x_j} = \frac{\partial u_j}{\partial x_j} - \frac{1}{3K} p_{jj} - \frac{1}{3K}\frac{\partial p_{ij}}{\partial x_j} x_i$$

vanishes by Hooke's law and (1.22). The contribution of the unknown u_j^{I}, p_{ij}^{I} to u_i' has zero divergence throughout the volume inside S and so, by Gauss' theorem, it contributes nothing to the integral. We cannot say the same for the contribution

$$\begin{aligned} u_j^{\infty\prime} &= u_j^\infty - p_{ij}^\infty x_i / 3K \\ &= \frac{c x_j}{r^3} - \frac{2\mu}{3K} x_i \frac{\partial}{\partial x_i}\left(\frac{x_j}{r^3}\right) \end{aligned}$$

from u_j^∞, p_{ij}^∞ because of the singularity at the defect. However, operating on x_j/r^3 with $x_i \partial/\partial x_i$ merely multiplies it by -2 and so $u_j^{\infty\prime} = (1 + 4\mu/3K) u_j^\infty$. Thus the integral (1.34) is γ times the integral (1.24), which establishes (1.32) for any shape of surface and any position of the defect within it.

Since, for example, $\gamma = 1.5$ when $\sigma = \frac{1}{3}$ and $\gamma = 1.8$ when $\sigma = \frac{1}{4}$, the image correction to the expansion is quite important.

Although it is not generally possible to find the details of the change of shape induced in a body with a free surface by a single defect, it is possible to find the macroscopic change of shape when a large number of defects, say n per unit volume, are scattered uniformly through it. The result is that, as is to be expected, it undergoes an increase of volume which is n times the expression (1.32), but that this volume change is *uniform*, that is, though larger, the solid has the same *shape* as it did before the defects were introduced. It is not, perhaps, quite obvious that the complicated changes of shape produced by the individual defects will add up to give this simple result, and it takes a certain amount of calculation to establish it. The following is an outline of one method.

Begin by marking out the surface S of the proposed solid in an infinite medium and distribute the defects uniformly within it. The displacement is now

$$u_i^\infty(\mathbf{r}) = -c\frac{\partial}{\partial x_i}\sum_m \frac{1}{|\mathbf{r}-\mathbf{r}_m|} \tag{1.35}$$

where $\mathbf{r}_m$ is the position vector of the mth defect. In the matrix, not too close to S this may be replaced by

$$u_i^\infty(\mathbf{r}) = -cn\frac{\partial}{\partial x_i}\int \frac{dv'}{|\mathbf{r}-\mathbf{r}'|}\,. \tag{1.36}$$

Close to S the difference between (1.35) and (1.36) will take the form of a ripple with a wave-length of order $n^{-1/3}$ which will, however, by Saint-Venant's principle, only extend into the matrix to a depth which is a few multiples of $n^{-1/3}$. We may say that (1.36) gives the macroscopic displacement in the matrix.

At this stage the volume enclosed by S has been increased by ΔV^∞, for one defect, times the number of defects it contains. It has also become distorted in a way which could be calculated from (1.36) with the help of potential theory. For example, if it was originally a cube it will now be barrel-shaped ($c > 0$) or pincushion-shaped ($c < 0$). To find what happens when S is freed from the matrix we appeal to a related problem.

Mark out a surface S in an infinite solid, make a cut over S, remove the material inside and (by uniform heating or otherwise) make its volume change by a fraction $\Delta V/V$ without any change of shape or elastic constants. Force it back into the hole and weld across the cut so that points originally adjacent across S are again adjacent. It is required to find the displacement in the matrix. The answer to this problem is known (Nabarro, 1940). The displacement is identical with

(1.36) provided we set

$$\Delta V/V = 4\pi\gamma cn. \qquad (1.37)$$

Conversely, our distorted surface S will resume its original shape when freed from the matrix but will have undergone a uniform volume expansion (1.37).

For various purposes (in particular the determination of the number of point defects from simultaneous measurements of dimensions and lattice parameter, section 1.4.2) it is necessary to know the effect of a distribution of defects on the lattice parameter as measured by X-rays.

The information obtainable about a crystal by X-ray diffraction can be summed up in a plot of scattering power in reciprocal space. For a perfect crystal this distribution takes the form of a sharp spot at each reciprocal lattice point. If the crystal is deformed (by external loads or a point defect, for example) the spots will be broadened and displaced but the new positions of their maxima will define a new deformed reciprocal lattice. Suppose that for the perfect crystal the crystal lattice points and the reciprocal lattice points are $L_i\mathbf{a}_i$ and $h_i\mathbf{b}_i$ where $\mathbf{a}_i$ and $\mathbf{b}_i$ are the base vectors of the crystal and reciprocal lattice respectively, and L_i, h_i are integers. If the crystal is deformed the atoms move to $L_i\mathbf{a}_i + \mathbf{u} = (L_i + \Delta L_i)\mathbf{a}_i$ and the points of the reciprocal lattice to $(h_i + \Delta h_i)\mathbf{b}_i$, say, with fractional ΔL_i, Δh_i. Miller and Russell (1952) have shown that for small integral h_i the following relation obtains:

$$\Delta h_i A_{ij} = -h_i B_{ij} \qquad (1.38)$$

where

$$A_{ij} = \sum L_i L_j, \quad B_{ij} = \sum \Delta L_i L_j$$

and the summations are over all atoms in the lattice. In what follows we shall assume for simplicity that the lattice is simple cubic with spacing a and that we can replace the sums (1.38) by integrals. Then

$$A_{ij} = \frac{1}{a^5}\int_V x_i x_j \mathrm{d}v, \quad B_{ij} = \frac{1}{a^5}\int_V x_i u_j \mathrm{d}v. \qquad (1.39)$$

The solution of (1.38) represents a uniform deformation of the reciprocal lattice. The dilatation of the reciprocal lattice is

$$\partial\Delta h_i/\partial h_i = \Delta h_1(100) + \Delta h_2(010) + \Delta h_3(001)$$

and its value can be inferred from measurement of reflections of not too high order. If this were all the information we had we should infer that the crystal had undergone a uniform expansion with $\Delta V/V = -\partial \Delta h_i/\partial h_i$ and we can define

$$\left(\frac{\Delta V}{V}\right)_{\mathrm{X}} = -\frac{\partial \Delta h_i}{\partial h_i} \tag{1.40}$$

to be the volume change of the crystal as measured by X-rays.

It is obvious from the form of B_{ij} that $(\Delta V/V)_{\mathrm{X}}$ depends on a quite different average of the displacements from $\Delta V/V$ which depends on the surface integral (1.34). Consequently if u_i is due, say, to a single defect in the crystal we do not expect $(\Delta V/V)_{\mathrm{X}}$ and $\Delta V/V$ to agree. It turns out, however, that, as might be expected, if there are a large number of defects distributed uniformly through the crystal they do agree, and that, moreover, the detailed deformation of the reciprocal lattice near its origin is what one would have inferred from its macroscopic change of shape.

These points may be shown quite easily for a sphere containing a single defect at $\boldsymbol{\xi}$, so that the displacement is given by (1.23) with $\mathbf{r}$ replaced by $\mathbf{r}-\boldsymbol{\xi}$, together with the image field necessary to give zero surface tractions.

For a sphere of radius R, $A_{ij} = (4\pi R^5/15a^5)\delta_{ij}$ and so

$$\left(\frac{\Delta V}{V}\right)_{\mathrm{X}} = -\frac{\partial \Delta h_i}{\partial h_i} = \frac{15}{4\pi R^5}\int x_i u_i \,\mathrm{d}v.$$

But

$$\int x_i u_i \,\mathrm{d}v = \int_V x_i u_k \frac{\partial x_i}{\partial x_k}\,\mathrm{d}v = \int \left[\frac{\partial}{\partial x_k}(r^2 u_k) - er^2 - x_i u_i\right]\mathrm{d}v$$

so that

$$2\int_V x_i u_i \,\mathrm{d}v = \int_S r^2 u_k n_k \,\mathrm{d}S - \int_V er^2 \,\mathrm{d}v.$$

The first integral is just $R^2 \Delta V = R^2(\Delta V^{\infty} + \Delta V^{\mathrm{I}})$. In the second, e^{∞} contributes $4\pi c\xi^2$ by (1.25), where ξ is the distance of the defect from the centre of the sphere. To find the contribution of e^{I} to the second integral we note that if the defect moves over the sphere $\xi = \text{const.}$ the image field will move round with it but the volume integral of $e^{\mathrm{I}}r^2$ will be unaltered. Hence the defect may be replaced by a uniform distribution of infinitesimal defects over the sphere $\xi = \text{const.}$, provided their total strength is c. But the image dilatation is then, by symmetry, uniform and equal to $\Delta V^{\mathrm{I}}/V$ and this gives an image contribution of $\frac{3}{5}R^2 \Delta V^{\mathrm{I}}$ to the volume integral. Hence finally, using (1.32),

$$\left(\frac{\Delta V}{V}\right)_{\mathrm{X}} = \left[1 + \frac{1}{2\gamma}\left(3 - 5\frac{\xi^2}{R^2}\right)\right]\frac{\Delta V}{V} \tag{1.41}$$

where ΔV is the geometrical volume change, equation (1.32). For a defect at the centre of the sphere the ratio $(\Delta V/V)_{\mathrm{X}}/(\Delta V/V)$ is $1 + 3/2\gamma \sim 2$, and for a defect near the surface it is $1 - 1/\gamma \sim 0.3$. (If we had ignored the image terms the discrepancy would have been greater, with a ratio 2.5 for a defect at the centre and 0 for one near the surface, for any Poisson's ratio.) Consequently, if defects are not distributed uniformly through the crystal the volume changes inferred from X-rays and changes in dimensions or density will not agree. However, if they are distributed uniformly $(\Delta V/V)_{\mathrm{X}}$ is found from (1.41) by multiplying by n, the number of them in unit volume, and replacing ξ^2 by its mean value over the sphere, $\frac{3}{5}R^2$. The square bracket expression then becomes unity, so that the X-ray and geometrical volume expansions are equal. Also, by symmetry the reciprocal lattice will undergo a *uniform* contraction at least near the origin.

By more elaborate calculations it may be shown that for an arbitrary shape of crystal with a uniform distribution of defects $(\Delta V/V)_{\mathrm{X}} = \Delta V/V$ and that the reciprocal lattice is deformed in the way which the macroscopic deformation of the crystal would suggest.

So far, we have not needed to know the relation between the strength c of a defect and the details of the misfitting sphere model but we shall require it in section 1.5. The connection is easily made.

If V_{s}, V_{h} are the volumes of sphere and hole before insertion, we define

$$V_{\mathrm{mis}} = V_{\mathrm{s}} - V_{\mathrm{h}}$$

to be the misfit volume. When the sphere is inserted in the hole V_{s} and V_{h} will change by ΔV_{s} and ΔV_{h} and pressures will be developed on the surfaces of the hole and sphere. Since these must be equal, (1.28) and (1.29) give

$$\tfrac{4}{3}\mu \Delta V_{\mathrm{h}} + K' \Delta V_{\mathrm{s}} = 0 \tag{1.42}$$

where K' is the bulk modulus of the sphere. (We ignore the slight difference between the initial volumes of hole and sphere.) Also, ΔV_{h} and ΔV_{s} must be just such as to eliminate V_{mis}, that is,

$$\Delta V_{\mathrm{h}} - \Delta V_{\mathrm{s}} = V_{\mathrm{mis}}. \tag{1.43}$$

Further, ΔV_{h}, being the volume change of a sphere closely enveloping the defect, is equal to ΔV^{∞}. Equations (1.42) and (1.43) give

$$\Delta V_{\mathrm{h}} = \Delta V^{\infty} = 4\pi c = V_{\mathrm{mis}}/\gamma'$$

where

$$\gamma' = \frac{3K' + 4\mu}{3K'}.$$

From the previous argument, the total volume change is

$$\Delta V = \gamma V_{\mathrm{mis}}/\gamma'. \tag{1.44}$$

If the sphere has the same bulk modulus as the matrix $\Delta V = V_{\mathrm{mis}}$, while if it is rigid ($K' = \infty$) $\Delta V = \gamma V_{\mathrm{mis}}$.

For a vacancy there is nothing in the hole to cause a volume change. In this case we can follow Brooks (1955) and regard the vacancy as a hole with a radius r_{s}, defined by $\frac{4}{3}\pi r_{\mathrm{s}}^3 = \Omega$, pulled inwards by the macroscopic surface tension Γ. The volume change can be found from (1.28) with $p = -2\Gamma/r_{\mathrm{s}}$. This gives ΔV^{∞}, so that the total volume change is given by

$$\Delta V/\Omega = -3\gamma\Gamma/2\mu r_{\mathrm{s}}.$$

The ratio Γ/μ has the dimensions of a length. For a wide range of materials it has a value close to $\frac{1}{4}$Å (Nabarro, 1951). For copper $r_{\mathrm{s}} = 1.4$ Å, giving $\Delta V/\Omega = -0.4$ which is in quite good agreement with the values -0.53 (Tewordt, 1958), -0.29 (Seeger and Mann, 1960) and -0.48 (Johnson and Brown, 1962) obtained by atomistic calculations. However, the same model over-estimates the energy of formation by a factor of about two so it cannot be taken very seriously. In fact, the energy of formation is $4\pi\Gamma r_{\mathrm{s}}^2$ plus the elastic energy in the matrix. The latter is $\frac{1}{2}p\delta v$ calculated from (1.28) with $p = -2\Gamma/r_{\mathrm{s}}$. Its ratio to the surface term is $\Gamma/2\mu r_{\mathrm{s}}$. With $\Gamma/\mu \sim \frac{1}{4}$ Å and $r_{\mathrm{s}} = 1.5$ Å the ratio is about 0.1, so that we may take the energy to be simply $4\pi\Gamma r_{\mathrm{s}}^2$. For copper ($\Gamma = 1400$ dynes cm^{-1}) this gives about 2 eV, which is about twice the value determined from experiment and more respectable calculations.

This crude model actually contains an element of truth. If we cut a large hole in a piece of metal the electrons will tunnel a certain distance into the cavity to an extent governed by the fact that though they thereby reduce their kinetic energy they also increase their potential energy. These ideas can be made the basis of a calculation of surface energy (Huang and Wyllie, 1949). The same principles apply when the hole is reduced to one atomic volume (vacancy). The tunnelling length is then comparable with the dimensions of the cavity; as a result, the increase of energy is less than the macroscopic model would suggest. Detailed calculations on these lines have been made by Fumi (1955) and Seeger and Bross (1956). As the surface tension model suggests, the 'elastic' part of the energy (ion–ion interaction) is not very important.

The electrical resistivity due to point defects can be calculated from the formula (Mott and Jones, 1936)

$$\Delta\rho = \frac{mv}{ne^2} cA$$

where m is the electron mass, v the Fermi velocity, n the number of free electrons per atom, e the electronic charge and c the atom fraction of defects. The quantity A is the scattering cross-section given by

$$A = \int (1 - \cos\theta) f(\theta)\, \mathrm{d}\omega$$

where $f(\theta)$ is the intensity of the scattered wave in the direction θ. If the scattered wave is analysed into a set of products of spherical harmonics of order l, each multiplied by a radial factor, A can be written in terms of the phase shifts η_l of the various components (Mott and Massey, 1965):

$$A = \frac{4\pi}{k^2} \sum_l (l+1) \sin^2(\eta_l - \eta_{l+1})$$

where $k = 2mv/\hbar$.

The phase shifts must satisfy the Friedel (1952) sum rule which requires that the relation

$$\frac{2}{\pi} \sum_l (2l+1)\, \eta_l = Z$$

holds, where Z is the charge required to screen the potential of the defect. This is a powerful check on the results obtained from an assumed potential. Indeed Abelès (1953) was able to obtain a reasonable value for vacancy resistivity simply by taking for the potential a flat-bottomed spherical well of one atomic volume and fixing its depth so that the sum rule was satisfied. In most other calculations a screened Coulomb potential has been used.

The results of a number of calculations suggest that the extra resistance due to one per cent of vacancies in copper, silver or gold is about 1.5 microhm cm, and that the contribution from lattice distortion is small.

1.4.2 Experimental measurement

Individual point defects can be seen by field-ion microscopy (Müller, 1962; Brandon and Wald, 1961). This technique has limitations (in particular, the defects are only observed at the surface) and usually one must study defects indirectly through their effect on some bulk property. An obvious way to do this is to vary the temperature, so that the defect concentration varies according to (1.3), and then measure the change in some bulk property supposedly proportional to the defect concentration. It is, of course, necessary to allow for the changes in the chosen property which would occur even if there were no defects. In the present section we show how this difficulty is overcome in a few typical experiments. First, however, it is necessary to say something about the relation between the free energy, enthalpy and entropy of formation.

Suppose that the concentration of some type of defect is given by

$$c = \mathrm{e}^{-\Delta G/kT} \tag{1.45}$$

and that there is some physical effect proportional to the concentration of defects, so that $X = Ac$ where A is independent of temperature or can be corrected for temperature variation. If the results of a series of measurements of X as a function of temperature are exhibited by plotting $-\ln X$ against $1/kT$, the slope of the resulting curve is not ΔG but rather the quantity ΔH related to it by

$$\Delta H = \frac{\partial}{\partial(1/T)} \cdot \frac{\Delta G}{T} = \Delta G - T\frac{\partial \Delta G}{\partial T}. \tag{1.46}$$

Equation (1.46) is identical with the Gibbs–Helmholtz relation connecting Gibbs free energy and enthalpy, so that ΔH is the enthalpy of formation, or heat of formation. If the material is not under pressure it is just the (internal) energy of formation. Again, according to thermodynamics the negative temperature derivative of a Gibbs free energy is the associated entropy, so that

$$\Delta S = -\frac{\partial \Delta G}{\partial T}$$

is the entropy of formation, and we may write

$$\Delta G = \Delta H - T\Delta S. \qquad (1.47)$$

The entropy of formation can be related to the change which the vibration frequencies of the lattice undergo when it contains a defect. Suppose that in fig. 1.3 (regarded, as before, as a schematic representation of the configuration space describing the whole crystal) A represents the perfect crystal and C represents the state of metastable equilibrium in which there is, say, a vacancy at some specific lattice site. The representative points can pass from A to C or from C to A by way of the saddle-point B. If n_A and n_C are the numbers of representative points in A and C and f_A, f_C are the corresponding jump frequencies we must have $n_A f_A = n_C f_C$ in equilibrium.

The jump frequency f_A is given by (1.17) with $\nu_i = \nu_i^A$ and $U = V_B - V_A$, and f_C by the same expression with $\nu_i = \nu_i^C$ and $U = V_B - V_C$. When we equate $n_A f_A$ with $n_C f_C$ the saddle-point frequencies ν_i and potential V_B cancel out and we are left with

$$n_C/n_A = (\Pi \nu_i^A / \Pi \nu_i^C)\, e^{-V/kT} \qquad (1.48)$$

where V is the difference in potential between C and A, and the products extend over the $3N$ normal vibration frequencies of the perfect and defective crystal. But the ratio n_C/n_A represents the relative probability of finding or not finding a vacancy at the specified site. Comparing (1.48) with (1.45) and (1.47) we see that we must put $V = \Delta H$ and

$$\Delta S = k \ln (\Pi \nu_i^A / \Pi \nu_i^C).$$

The argument, of course, also applies to defects other than vacancies.

If, as we have supposed, the crystal executes purely harmonic vibrations about its perfect and defective states ΔH, ΔS are independent of temperature. (The fact that we have to go through a highly non-linear region to get from one state to the other is irrelevant.) Actually, there are anharmonic and other effects which spoil this, but they are small and tend to cancel (Levinson and Nabarro, 1967) and ΔH, ΔS are commonly treated as constants. We can then write

$$c = e^{\Delta S/k}\, e^{-\Delta H/kT}$$

where $\exp(\Delta S/k)$ is the so-called entropy factor. Then if the proportionality factor A connecting the physical effect X with the concentration c is known both ΔH and ΔS can be found from a logarithmic plot, but if A is not known only ΔH can be determined.

In practice, the variation of X due to the change of defect concentration with temperature will be obscured by the variation of X which would occur if defect formation could somehow be inhibited. For example, the volume change due to defects would be swamped by thermal expansion, and the defect resistivity by the normal change of resistance with temperature.

There are three ways round this difficulty: (i) we may estimate the non-defect part of X by extrapolation from temperatures where the defect contribution can reasonably be neglected; (ii) the non-defect contribution can be eliminated by some special artifice; (iii) we may quench specimens from a number of temperatures down to a single common temperature. If the quench has been fast enough to retain all the defects and if additional defects have not been introduced as a result of quenching stresses we then have a set of specimens containing the defect concentration appropriate to the temperatures from which they were quenched. A plot of X measured at the low temperature against the quenching temperature will then give information about the variation of vacancy concentration with temperature.

An example of the first method is provided by MacDonald's (1954) measurement of the defect contribution to the specific heat of the alkali metals. The presence of defects of concentration c and heat of formation ΔH will give an extra contribution to the specific heat per atom equal to $\partial(\Delta Hc)/\partial T$. According to (1.45) and (1.46) this amounts to

$$\Delta C_p = \frac{(\Delta H)^2}{kT^2}c$$

if we suppose that ΔH is temperature-independent. Fig. (1.4) shows the experimental curve for potassium; it gives $\Delta H = 0.4$ eV.

The second method is illustrated by Simmons and Baluffi's (1960*a*, *b*, 1962, 1963) experiments in which the change in both length and X-ray lattice parameter of a specimen were measured. To explain their method we suppose for the moment that there is no relaxation

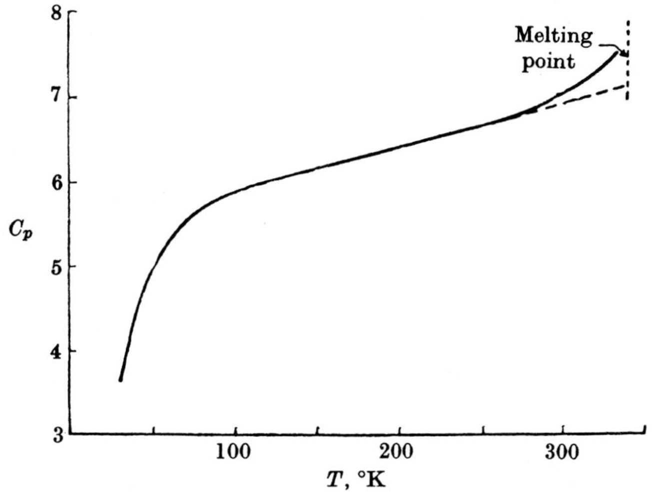

Fig. 1.4. The specific heat of potassium as a function of temperature. (MacDonald, 1954)

of the lattice about a vacancy or interstitial. Then each vacancy increases the volume of the specimen by one atomic volume, and each interstitial decreases it by the same amount, and so a fractional concentration c_V of vacancies and c_i of interstitials produces a fractional volume change $c_V - c_i$. However, if we vary the temperature the volume change due to $c_V - c_i$ will be overshadowed by thermal expansion. On the other hand, if there is no relaxation the X-ray lattice parameter is only affected by thermal expansion and not by the presence of the defects. This is just an extreme case of the fact (Huang, 1947) that altering the scattering power of some of the atoms in a lattice does not affect the X-ray parameter; the scattering power of some lattice points is reduced to zero (vacancies) and the scattering power of some interstitial sites is increased from zero to a finite value (interstitials). Hence if there were no relaxation we should have

$$c_V - c_i = \frac{\Delta V}{V} - \left(\frac{\Delta V}{V}\right)_X = \frac{\Delta V}{V} - 3\frac{\Delta a}{a} \qquad (1.49)$$

where $\Delta a/a$ is the fractional change in lattice parameter. But we have seen in section 1.4 that if relaxation does occur it affects $\Delta V/V$ and $(\Delta V/V)_X$ equally, so that (1.49) is still valid.

Figure 1.5 shows some of Simmons and Baluffi's experimental results for gold. Even a rough estimate suggests that ΔH is much greater

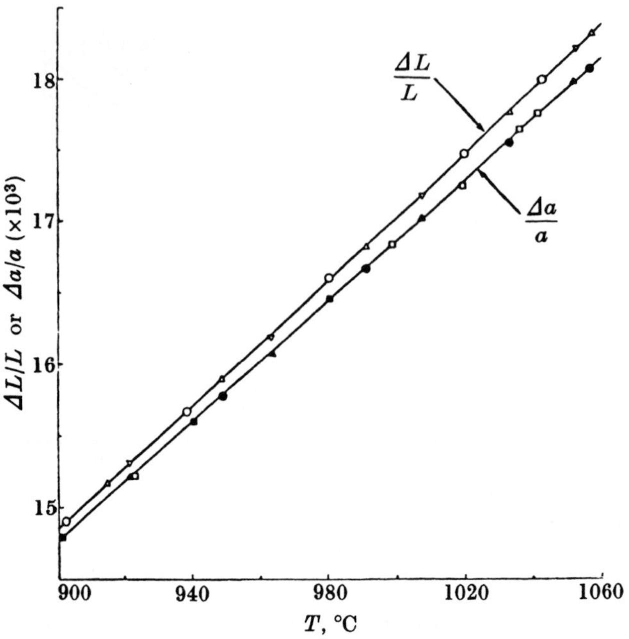

Fig. 1.5. Change of length (L) and lattice parameter (a) of copper with temperature. (Simmons and Baluffi, 1962)

for interstitials than it is for vacancies, so we may expect c_i to be negligible in comparison with c_v. Consequently a plot of the logarithm of $\Delta V/V - 3\Delta a/a$ against $1/kT$ will give both the energy and entropy of formation of vacancies. The following are some values obtained in this way by Simmons and Baluffi:

	Cu	Ag	Au	Al
ΔH (eV)	1.17	1.09	0.94	0.75
$\Delta S/k$	~1.5	~1.5	1	2.4

The estimated error in ΔH is about 10 per cent.

The quenching experiments of Bauerle and Koehler (1957) are a classic example of the third method. Fig. 1.6 shows their results for the quenched-in resistance of gold wires. The energy of formation (0.98 ± 0.03 eV) agrees closely with the value which Simmons and Baluffi obtained from equilibrium measurements. Since Simmons and Baluffi's measurements give the absolute concentration of vacancies they can be combined with Bauerle and Koehler's results to give the resistance increase per vacancy. The result is 1.8 microhm cm. for

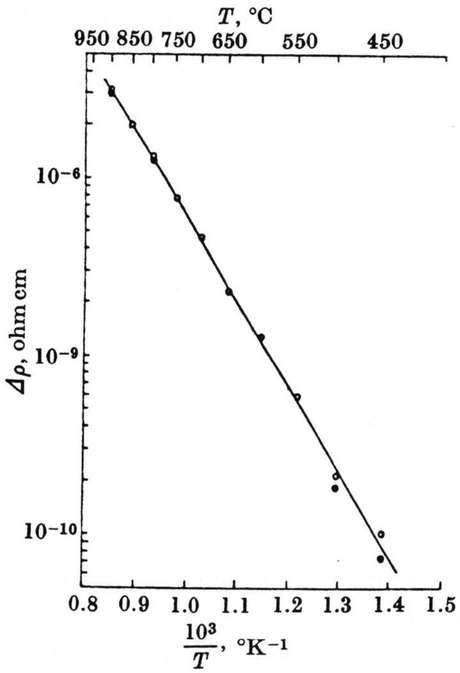

Fig. 1.6. Variation of quenched-in resistivity (ρ) of copper with quenching temperature. (Bauerle and Koehler, 1957)

one per cent vacancies, which agrees reasonably with theoretical estimates.

Huebner and Homan (1963) have found the volume increase associated with the formation of a vacancy in gold. If we apply the thermodynamic relation $\Delta V = \partial \Delta G / \partial p$ to (1.45) we get

$$\frac{\partial \ln c}{\partial p} = -\frac{\Delta V}{kT}.$$

Huebner and Homan subjected gold wires to a temperature of 680 °C and pressures up to 11 000 atmospheres, quenched them and measured the electrical resistance. The slope of the resistance–pressure curve gave for ΔV a fraction 0.53 ± 0.04 of an atomic volume, so that, of the basic increase of one atomic volume per defect, about half is cancelled by relaxation.

If a specimen containing quenched-in vacancies or other point defects is held at a temperature between the temperatures from and to which it was quenched the excess vacancies will gradually disappear until the concentration appropriate to the new temperature is reached. The theory of such isothermal annealing processes is complicated; for a review see Damask and Dienes (1963).

The simplest case is where the excess defects disappear by random migration to a fixed number of unfillable sinks (the surface of the specimen, dislocations). Then, in the language of physical chemistry, we have a first-order reaction and the number of defects varies with time according to the law

$$\frac{dn}{dt} = -\alpha Dn \tag{1.50}$$

where D is the diffusion coefficient for the defect and α is a factor depending on the geometrical arrangement of the sinks. Equation (1.50) has the solution

$$n(t) = n(0)\, e^{-\alpha Dt}$$

so that the logarithm of a physical quantity proportional to n should plot linearly against t. Further, D depends on U, the activation energy for migration (cf. equations (1.18), (1.20)), and so U can be determined from the results of isothermal annealing experiments at several temperatures. (It is not necessary to know the parameter α.)

In this way Bauerle and Koehler (1957) found the migration energy for vacancies in gold by measuring the decay of the extra resistance. They obtained 0.82 ± 0.05 eV. This can be checked by comparison with measurements of the coefficient of self-diffusion. Since self-diffusion occurs by the interchange of vacancies it is proportional to the product of the concentration of vacancies and the diffusion coefficient for vacancies and so contains a factor $\exp[-(U+\Delta H)/kT]$. Diffusion experiments (Makin *et al.*, 1957) give $U+\Delta H = 1.8$ eV which agrees closely with the value 1.80 eV found by combining the $U = 0.82$ eV of Bauerle and Koehler's annealing experiments with the $\Delta H = 0.98$ eV obtained from their quench experiments.

Stored energy measurements can give an absolute value for vacancy concentration. De Sorbo (1960) quenched a series of gold specimens from different temperatures and measured calorimetrically the release of energy as the excess defects (assumed to be vacancies) annealed out. A logarithmic plot gives the energy of formation, and this, since it is the energy released per defect annihilated, can be used to convert the energy measurements into vacancy concentrations.

His result was

$$c = 2\exp(-0.97\,\text{eV}/kT). \tag{1.51}$$

The energy of formation agrees well with the values obtained from equilibrium measurements, though the pre-exponential factor is rather smaller. In conjunction with the resistivity annealing experiments just described Bauerle and Koehler also measured length changes. Combined with (1.51) their results give 0.57 atom volume for the volume increase due to a vacancy, in fair agreement with the value Huebner and Homan obtained from their pressure-dependence measurements.

All the experiments described above refer to thermally generated defects, either in equilibrium or retained by quenching. Similar methods can also be used to study defects introduced by irradiation or plastic deformation. We give a couple of examples.

If irradiation produces equal numbers of vacancies and interstitials we expect, according to (1.49) that the fractional changes in length and X-ray lattice parameter will be equal. This has been verified for copper irradiated with deuterons at low temperatures (Simmons and Baluffi, 1958; Vook and Wert, 1958).

Molenaar and Aarts (1950) measured the rate of change of electrical resistance with deformation in copper at liquid nitrogen temperature. They estimated the number of defects by measuring the resistance decrease on annealing at room temperature. If the defects are assumed to be vacancies producing a resistivity change of 1.5 microhm cm. per atom per cent vacancies, their results give a defect concentration of 4×10^{-6} for one per cent deformation. Henderson and Koehler (1956) measured the stored energy released on annealing copper deformed at a low temperature. With an energy of formation of 1 eV per defect their results give a concentration of 6×10^{-6} defects for 1 per cent deformation. Both these values are comparable with the estimate at the end of section 1.2.

1.5 INTERACTION ENERGIES OF POINT DEFECTS

In addition to producing the macroscopic effects which enable their fundamental parameters to be measured (section 1.4) point defects play an important role in influencing the mechanical properties of metals, chiefly through their interaction with dislocations. In this section we discuss the basic interactions between point defects and the stresses due to dislocations, other defects, or external loading.

Consider a point defect in a crystal in a state of stress, due to other defects, dislocations and externally applied forces. If the point defect is moved about, the total energy of the system, that is, the sum of the internal energy (strictly the Helmholtz free energy) and the potential of the external loading mechanism, varies. We may regard the negative gradient of the total energy as a 'force' acting on the defect. From this force one can calculate at what points the defect will be in equilibrium, and with the help of the Einstein relation (1.20) the drift velocity of a defect can be calculated.

If the problem is treated according to the infinitesimal theory of elasticity it is usually convenient to consider not the total energy but only the cross term in it between the elastic field of the defect and the applied internal stresses which are regarded as exerting a force on it. In this section we treat directly some cases which are important in the theory of the interaction between point defects and dislocations. It is possible to develop a fairly respectable general theory of the forces on lattice defects (see, for example, Eshelby, 1956) which shows that some of the shortcomings of the following treatment are only apparent. For example, equation (1.52) below states that the force on a point defect in the elastic field of a dislocation depends on the value of the hydrostatic pressure due to the latter at the position of the defect, where things are highly non-superposable. The precise treatment shows that for the result to be valid it is only necessary that one can find a closed surface, isolating the defect from the dislocation, on which the elastic fields of dislocation and defect can be superposed with fair accuracy. The appropriate hydrostatic pressure to use is then that found by extrapolating inwards from this surface to the position of the defect, but on the assumption that the defect is absent. Likewise, interaction energies derived on the assumption that the applied stress is uniform remain valid when it is not, provided the applied field near the defect (in the sense just described) does not vary too rapidly. We begin with the simple case of a point defect in a crystal under hydrostatic pressure.

We may define the interaction energy of a defect in a specimen subjected to a hydrostatic pressure p to be the work required to insert the defect in the specimen under pressure p, minus the work required to insert it when $p = 0$. The latter is E_S, the self-energy of the defect. The former is $E_S + p\Delta V$, the second term being simply the work done against p by the increase ΔV (equation (1.32)) in the volume of the specimen. Hence

$$E_{\text{int}} = p\Delta V. \tag{1.52}$$

One might expect that there would be an additional term arising from a cross-term in the elastic energy of the specimen between the elastic field of the defect and the applied stress, but in fact this term is zero. The elastic energy is

$$\tfrac{1}{2}\int (p_{ij}^{\mathrm{D}} + p_{ij}^{\mathrm{A}})(e_{ij}^{\mathrm{D}} + e_{ij}^{\mathrm{A}})\,\mathrm{d}v \tag{1.53}$$

taken over the volume of the specimen. The superscripts A and D refer to the applied field and the defect. The A-A term represents the energy when p is applied in the absence of the defect and the D-D term is the self-energy of the defect. The cross-term is

$$\tfrac{1}{2}\int (p_{ij}^{\mathrm{D}} e_{ij}^{\mathrm{A}} + p_{ij}^{\mathrm{A}} e_{ij}^{\mathrm{D}})\,\mathrm{d}v. \tag{1.54}$$

If the two terms are written out in terms of displacements they are seen to be identical, so that (1.54) may be replaced by twice its first term. From the definition of the strain components, and because $p_{ij} = p_{ji}$, we may write

$$p_{ij}^{\mathrm{D}} e_{ij}^{\mathrm{A}} = \tfrac{1}{2} p_{ij}^{\mathrm{D}}(\partial u_i^{\mathrm{A}}/\partial x_j + \partial u_j^{\mathrm{A}}/\partial x_i) = p_{ij}^{\mathrm{D}}\,\partial u_i^{\mathrm{A}}/\partial x_j$$

and since $\partial p_{ij}^{\mathrm{D}}/\partial x_i = 0$ we may also write it as $\partial(p_{ij}^{\mathrm{D}} u_i^{A})/\partial x_j$. In this form the volume integral can be converted by Gauss' theorem into a surface integral

$$\int p_{ij}^{\mathrm{D}} u_i^{\mathrm{A}} n_j\,\mathrm{d}S \tag{1.55}$$

extended over the surface of the specimen. But $p_{ij}^{\mathrm{D}} n_j$ vanishes at the surface so that the cross-term is zero. Since the displacement is discontinuous across the interface between defect and matrix we should, perhaps, for safety's sake, also extend the integral (1.55) over surfaces just outside and just inside the interface. However, since $p_{ij}^{\mathrm{D}} n_j$ and u_i^{A} (though not u_i^{D}) are continuous across the interface and the two surfaces have oppositely-directed normals they contribute nothing.

If the defect is a misfitting sphere with the same elastic constants as the matrix this concludes the proof that (1.52) is the appropriate interaction energy. With a suitable interpretation it is also correct when the defect is both misfitting and inhomogeneous, with elastic constants differing from those of the matrix. Suppose that at first the sphere is inhomogeneous but perfectly fitting. When the pressure is applied the work done in compressing the solid will not be the same as it would be in the absence of the defect because the presence of the defect makes the specimen harder or softer. There is thus an interaction energy between an applied stress and a perfectly fitting but inhomogeneous defect. It is discussed below and shown to be proportional to p^2. Suppose now that the defect is made to become misfitting. The work required to do this is the sum of E_s, the expression (1.52) and the cross-term in (1.53). In the latter, $\overset{A}{p_{ij}}$, $\overset{A}{e_{ij}}$ now refer to the elastic field set up by the applied forces in the material containing the inhomogeneous but perfectly fitting defect. The cross-term vanishes by the previous argument, which nowhere assumes the material to be homogeneous. Consequently (1.52) gives the interaction energy of the defect in its role of a misfitting inclusion (source of internal stress) over and above its interaction energy as an inhomogeneity. Since the latter quantity is quadratic in the applied stresses, the linear interaction term is always given correctly by (1.52).

Equation (1.52) can be re-written in terms of the variation of lattice parameters a with defect concentration c. Since $\Delta V = (3/a)$ $(\mathrm{d}a/\mathrm{d}c)\Omega$ where Ω is the atomic volume, (1.52) becomes

$$E_{\text{int}} = -\frac{3}{a}\frac{\mathrm{d}a}{\mathrm{d}c}\Omega p. \tag{1.56}$$

This remains true even if we drop the limitation to spherically symmetrical deformation about a defect and isotropic elasticity, provided a uniform distribution of defects produces a volume change in the crystal but no macroscopic change of shape.

A uniform distribution of defects like the one in fig. 1.1(*b*) (or an even less symmetrical one), all with the same orientation, will change both the size and shape of a crystal. The change can be specified by the strain $\overset{T}{e_{ij}}(c)$ which a unit cube suffers when a concentration c is introduced. If the cube is subjected to an applied stress p_{ij} the work done against it when the defects are introduced is $-\overset{T}{e_{ij}}p_{ij}$ and so instead of (1.56) we have for the interaction energy per defect

$$E_{\text{int}} = -\frac{\mathrm{d}\overset{T}{e_{ij}}}{\mathrm{d}c} p_{ij}\Omega; \tag{1.57}$$

the cross-term in the internal energy vanishes by the same argument as before. The difference in sign between (1.56) and (1.57) is merely a result of the convention that for a hydrostatic pressure p the stress components are $p_{11}=p_{22}=p_{33}=-p$, $p_{12}=p_{23}=p_{31}=0$. If we insert these values in (1.57) we get back to (1.56) because $e_{11}^{\mathrm{T}}+e_{22}^{\mathrm{T}}+e_{33}^{\mathrm{T}}$ is just the fractional change of volume. Conversely, if the defects have spherical or cubic symmetry $e_{11}^{\mathrm{T}}=e_{22}^{\mathrm{T}}=e_{33}^{\mathrm{T}}$, $e_{12}^{\mathrm{T}}=e_{23}^{\mathrm{T}}=e_{31}^{\mathrm{T}}=0$ and (1.57) reduces to (1.56) for any state of stress; the defect only pays attention to the hydrostatic component of the p_{ij}. The validity of (1.57) does not depend on the material being isotropic.

An important application of (1.52) is to the interaction between a defect and an edge dislocation. Here the hydrostatic pressure is

$$p=\frac{\mu b}{\pi\gamma}\frac{\sin\theta}{r} \tag{1.58}$$

where r, θ are polar co-ordinates with θ measured from the direction of the Burgers vector b, γ is given by (1.33) and the material is assumed to be isotropic. Equation (1.52) gives

$$E_{\mathrm{int}}=\frac{\mu b}{\pi\gamma}\Delta V\frac{\sin\theta}{r}=\frac{\mu b}{\pi}\Delta V^{\infty}\frac{\sin\theta}{r}. \tag{1.59}$$

Cottrell (1948) first calculated the interaction between a point defect and an edge dislocation by considering the work done in 'blowing up' the defect in the field of the dislocation. He obtained the expression (1.59) multiplied by γ; Bilby (1950) later obtained the correct result. Since Cottrell considered a rigid sphere while Bilby considered one elastically homogeneous with the matrix it is commonly stated that the difference arises from the strain energy inside the sphere. However, we have seen that the linear part of the interaction energy should be the same, for the same ΔV or ΔV^{∞}, whether the sphere is rigid (when there is no strain energy within it) or not. One can also be confused by the fact that the ratio of Cottrell's result to that of Bilby, the ratio of ΔV to ΔV^{∞} for a particular defect, and the ratio of the ΔVs (or ΔV^{∞}s) for two defects with the same volume misfit, the one rigid, the other elastically homogeneous (cf. equation (1.44)), all differ by a factor γ, and that the expression (1.58) itself contains a factor γ^{-1}.

Cottrell's method has the advantage that it is only concerned with what goes on near the defect and so avoids the need to take the not very obvious step of applying (1.52), derived for an externally applied stress, to the case of internal stress. For the former, it is fairly obvious why the total change of volume ΔV appears in (1.52) rather than ΔV^∞, but it is not at all obvious why it should also appear in the interaction between a defect and a dislocation unless we appeal to the anthropomorphic principle that the defect does not know what is producing the pressure. It is therefore perhaps worthwhile to show that, with attention to a few details, Cottrell's method can be made to give the correct result. We limit ourselves to the case of an incompressible sphere. There is nothing specific to the dislocation problem in the following treatment: it applies equally well to any source of internal stress or external loading.

We note first that to correspond with a hole of volume V_h in the unstrained matrix we must cut out a hole of volume

$$V_\mathrm{h}' = V_\mathrm{h}(1 - p/K)$$

in the medium under pressure p, and then put a pressure p in the hole to prevent it collapsing further. Next, as the hole is inflated from the initial volume V_h' to a volume V_s which will enable it to accept the rigid sphere, an additional pressure $(4\mu/3V_\mathrm{h})(V_\mathrm{s} - V_\mathrm{h}')$ will build up, equal to the product of the fractional change in volume and the bulk modulus for blowing up a hole (equation (1.28)). The final pressure is thus

$$p_\mathrm{f} = \frac{4\mu}{3}\cdot\frac{V_\mathrm{s} - V_\mathrm{h}'}{V_\mathrm{h}} + p$$

while the initial pressure is p. The work W done in inserting the sphere is the product of the average pressure and the change of volume:

$$W = \tfrac{1}{2}(p_\mathrm{f} + p)(V_\mathrm{s} - V_\mathrm{h}'). \tag{1.60}$$

When this is worked out the term linear in p is found to be $p\gamma(V_\mathrm{s} - V_\mathrm{h})$ which agrees with (1.52) since $V_\mathrm{s} - V_\mathrm{h} = \Delta V^\infty$ for a rigid sphere.

We can also use (1.60) to find the work required to insert the rigid sphere correct to order p^2. For consistency, however, we must first subtract from it the energy recovered on relaxing the stress in the matrix atom which is removed to make room for the rigid sphere, namely $\tfrac{1}{2}p(V_\mathrm{h} - V_\mathrm{h}')$. The result is

$$E = \tfrac{2}{3}\mu\frac{(\Delta V^\infty)^2}{V_\mathrm{h}} + \gamma\Delta V^\infty p + \tfrac{1}{2}\frac{\gamma V_\mathrm{h}}{K}p^2.$$

The first term, independent of p, is the self-energy of the defect. The second we have already discussed. The third is independent of the degree of misfit and appears because the sphere is an inhomogeneity in the elastic constants. Evidently (at least if $\Delta V^{\infty} = 0$) it will be energetically advantageous for the defect to migrate to a region where p^2 is smaller. There will obviously be a similar though smaller effect if the sphere, though not absolutely rigid, has a greater bulk modulus than the matrix. Conversely, we should expect that vacancies or elastically softer atoms would tend to migrate to positions of high p^2. Also, we should expect similar effects in a more general stress field, including a pure shear stress.

All these effects can be calculated on the inhomogeneous sphere model but, as in the case of (1.56) and (1.57), it is more satisfactory to relate them directly to an associated macroscopic effect, in this case the change of the bulk elastic constants which the presence of defects induces in the solid.

Consider a rod containing a uniform distribution of point defects which, say, reduce its elastic constants. Let one end of it be clamped in a rigid vice so as to impose a fixed strain (not stress). The energy density at the clamped end is proportional to the square of the imposed strain times a suitable elastic constant. So if the impurities migrate to the clamped end and reduce the elastic constants there, the elastic energy will be decreased. Consequently we can say that there is a force urging the individual defects towards the clamped end. Suppose that instead of clamping we impose invariable surface forces (dead loading) chosen so as to produce a region of fixed stress (not strain) at one end of the rod. Then the energy density, being proportional to the square of the imposed stress divided by a suitable elastic constant, will be *increased* if the defects migrate to the stressed end. Nevertheless, it will still be energetically favourable for them to do so because the stressed end becomes elastically softer and this allows the applied forces to do more work on the bar (Le Chatelier's principle). If the stress–strain relation is strictly linear the work done by the applied forces is precisely twice the increase in elastic energy and there is a net reduction of energy.

To make these ideas quantitative suppose that we can at will force a unit cube of material into a rigid container which is nearly a cube, but with angles differing slightly from right angles (so as to induce a shear strain) and which has a volume slightly less than unity (so as to induce a hydrostatic pressure). Insert the defect-free cube of material. The elastic energy is $W_d + W_s$ where W_d and W_s are, respectively, the parts of the elastic energy density which depend on the bulk modulus K and the shear modulus μ. Explicitly

$$W_d = \tfrac{1}{2}K(e_1 + e_2 + e_3)^2 = \frac{1}{18K}(p_1 + p_2 + p_3)^2$$
$$W_s = \mu[e_1^2 + e_2^2 + e_3^2 - \tfrac{1}{3}(e_1 + e_2 + e_3)^2]$$
$$= \frac{1}{4\mu}[p_1^2 + p_2^2 + p_3^2 - \tfrac{1}{3}(p_1 + p_2 + p_3)^2]$$

where e_1, e_2, e_3 are the principal strains and p_1, p_2, p_3 are the principal stresses.

Now force n defects into the strained cube. The work necessary is $n\epsilon$, where ϵ is the heat of solution per defect, plus nE_{int} where E_{int} is the interaction energy we are looking for. (We suppose for the moment that the defects produce no change of lattice parameter, and that their concentration is small enough for us to neglect any interaction between them.) The total energy of the cube is now

$$E = W_d + W_s + n\epsilon + nE_{\text{int}}. \tag{1.61}$$

Start again with the undeformed, defect-free cube. Insert the defects. The energy is $n\epsilon$. Force the cube into the container. Since for fixed strain W_d is proportional to K, and W_s to μ, the total energy of the cube is now

$$E = \frac{K + \Delta K}{K} W_d + \frac{\mu + \Delta\mu}{\mu} W_s + n\epsilon \tag{1.62}$$

where ΔK, $\Delta\mu$ are the changes in elastic constants produced by the presence of the defects. If Ω is the atomic volume and c is the atomic fraction of defects we may write

$$\Delta K = \frac{dK}{dc} n\Omega, \quad \Delta\mu = \frac{d\mu}{dc} n\Omega.$$

The expressions (1.61) and (1.62) refer to the same physical state arrived at in two different ways. Consequently they must be equal, and comparison gives

$$E_{\text{int}} = \frac{1}{K}\frac{dK}{dc} W_d \Omega + \frac{1}{\mu}\frac{d\mu}{dc} W_s \Omega. \tag{1.63}$$

The elastic sphere model gives

$$\frac{1}{K}\frac{\mathrm{d}K}{\mathrm{d}c} = \left[\frac{1+\sigma}{3(1-\sigma)} + \frac{K}{K'-K}\right]^{-1}, \quad \frac{1}{\mu}\frac{\mathrm{d}\mu}{\mathrm{d}c} = \left[\frac{2}{15}\frac{4-5\sigma}{1-\sigma} + \frac{\mu}{\mu'-\mu}\right]^{-1}$$

where K, μ refer to the matrix and K', μ' to the inhomogeneous sphere, and σ is the Poisson's ratio of the matrix. These expressions suggest that 1 per cent of very hard or very soft atoms will respectively increase or decrease either K or μ by about 2 per cent.

It is not hard to justify the following recipe for calculating the analogue of (1.63) for an anisotropic material: write out the energy density in terms of the strain components and the elastic stiffness c_{ij} and replaced each c_{ij} by $\Omega \mathrm{d}c_{ij}/\mathrm{d}c$. Alternatively, write out the energy density in terms of the stresses and the elastic compliances s_{ij} and replace each s_{ij} by $-\Omega\, \mathrm{d}s_{ij}/\mathrm{d}c$.

For the model of a vacancy as a spherical hole pulled in by surface tension the interaction energy is the sum of (1.56), (1.63), plus a further term proportional to p^2, arising from direct interaction between applied stress and surface energy, which merely adds to the first term of (1.63).

The interaction between point defects and dislocations can be studied by annealing experiments in which the rate at which defects migrate to dislocations is measured. If the interaction law is (1.56) (simple misfit) the number of defects which have reached the dislocation at time t after the beginning of the anneal is proportional to $t^{2/3}$. This is the Cottrell–Bilby (1949) law which has been well verified for, for example, interstitial carbon in iron. If the interaction is given by (1.63) (inhomogeneity but no misfit) the interaction energy is proportional to r^{-2} since the dislocation stresses are proportional to r^{-1}. This leads to a $t^{1/2}$ law in place of the $t^{2/3}$ law. Wintenberger's (1957) measurements on the annealing of vacancies in aluminium follow a $t^{1/2}$ law, which suggests that the interaction (1.63) is more important than the interaction (1.56).

According to the isotropic misfitting sphere model the stresses round a point defect are proportional to r^{-3}, so that there is an inhomogeneity interaction energy (1.63) proportional to r^{-6} between two point defects. Because the dilatation is zero there is no misfit interaction (1.56). (We ignore the unimportant image correction.) If the material is not isotropic the dilatation is no longer zero and there is an r^{-3} misfit interaction between two defects. If the crystal is cubic and the defects have cubic symmetry the interaction can still be calculated from (1.56). When the departure from isotropy is small the dilatation due to a defect is (Eshelby, 1956, with an error of sign)

$$e = -\frac{\Delta VK}{c_{11}^2} d \frac{x^4 + y^4 + z^4 - \frac{3}{5}r^4}{r^7}$$

where ΔV is the total volume change it produces, K is the bulk modulus, and $d = c_{11} - c_{12} - 2c_{44}$ is a combination of the stiffness constants c_{ij} which vanishes for isotropy. To find the elastic field of a defect in a crystal which is not nearly isotropic one must know the elastic Green's function, that is, the displacement due to a concentrated force, and this cannot be written in finite form. Mann *et al.* (1961) have calculated it numerically for copper.

Tectonophysics, 28 (1975) 265—274

THE CHANGE OF SHAPE OF A VISCOUS ELLIPSOIDAL REGION EMBEDDED IN A SLOWLY DEFORMING MATRIX HAVING A DIFFERENT VISCOSITY

B.A. BILBY, J.D. ESHELBY and A.K. KUNDU*

Department of the Theory of Materials, University of Sheffield, Sheffield (Great Britain)

(Submitted December 10, 1974; revised version accepted May 30, 1975)

ABSTRACT

Bilby, B.A., Eshelby, J.D. and Kundu, A.K., 1975. The change of shape of a viscous ellipsoidal region embedded in a slowly deforming matrix having a different viscosity. Tectonophysics, 28: 265—274.

The general theory of the elastic fields round ellipsoidal inclusions and inhomogeneities is applied to solve the problem of the slow deformation of a viscous material containing an ellipsoidal inhomogeneity of different viscosity. The theory is used to calculate the two-dimensional finite strain of an elliptic cylinder and of prolate and oblate spheroids under applied pure-strain rates, and attention is drawn to applications in the fields of rock deformation and in the mixing and homogenization of viscous liquids.

INTRODUCTION

It has been pointed out (Eshelby, 1957, 1959) that the general theory of the elastic fields round ellipsoidal inclusions and inhomogeneities may also be used to solve the problem of the deformation of a viscous material containing an ellipsoidal inhomogeneity of different viscosity. This problem is of interest in connection with the theory of the deformation of rocks (Ramsay, 1967; Gay, 1968a, b; Dunnet, 1969; Jaeger, 1969, chapter 5; Wood, 1973; Tan, 1974) and the theory of mixing and homogenization of viscous liquids (Cable, 1968) and so in the present paper we develop the theory of the viscous inhomogeneity in more detail.

We give the name 'inclusion' to a region bounded by a surface S in an elastic solid which has undergone a change of shape and size, which, without the constraint of the surrounding matrix would be an arbitrary homogeneous strain e_{ij}^T. Because of the constraint the inclusion and the material around it

* A.K. Kundu is now at Berhampur College of Technology, West Bengal, India.

are thrown into a state of stress. The resulting elastic field can be found as follows (Eshelby, 1957, 1959; to be referred to as I and II). Cut round the region bounded by S which is to transform and remove it from the matrix. Allow the unconstrained transformation to take place and then restore the region to its original size and shape by applying to it the surface tractions $-p_{ij}^T n_j$, where:

$$p_{ij}^T = \lambda e^T \delta_{ij} + 2\mu e_{ij}^T \tag{1}$$

are the stresses calculated from e_{ij}^T by Hooke's law. We here assume that the material is isotropic with Lamé constants λ and μ; e^T is the dilatation. Now replace the region in the matrix and reweld at the surface S; the strain in the inclusion and matrix is everywhere zero, and there is a distribution of body force $-p_{ij}^T n_j$ over S. Finally, to cancel this, apply a distribution of body force $+p_{ij}^T n_j$ over S, and let this produce the displacement u_i^C everywhere. We can write down u_i^C in terms of the elastic Green's function $U_{ij}(r - r')$, which is the elastic displacement in the i-direction at the point r (coordinates x_k) produced by a unit force applied in the j-direction at the point r' (coordinates $x_k{}'$). $U_{ij}(r - r')$ depends on the modulus $|r - r'|$ of $r - r'$ and is given by (Love, 1927, p. 185):

$$U_{ij}(r - r') = \frac{\delta_{ij}}{4\pi\mu|r - r'|} - \frac{1}{16\pi\mu(1 - \nu)} \frac{\partial^2}{\partial x_i \partial x_j} |r - r'| \tag{2}$$

Here ν is Poisson's ratio, and $\delta_{ij} = 1$ if $i = j$ and is zero otherwise. In terms of $U_{ij}(r - r')$ we have:

$$u_i^C = \int_S U_{ij}(r - r') p_{jk}^T n_k \, dS \tag{3}$$

It is shown in I that if the surface S is an ellipsoid the displacement u_i^C is linear in x_1, x_2, x_3 inside S, so that the strain and stress components in the inclusion, as constrained by the matrix, are uniform. The strain may be written in the form:

$$e_{ij}^C = S_{ijkl} e_{kl}^T \tag{4}$$

The constant coefficients S_{ijkl} are worked out in I; they depend only on Poisson's ratio and the axial ratios b/a, c/a of the ellipsoidal inclusion, which we assume to have semi-axes a, b, c. If p_{ij}^C denotes the stress derived from e_{ij}^C by Hooke's law the (non-uniform) stress in the matrix is p_{ij}^C, but the uniform stress in the inclusion is $p_{ij}^C - p_{ij}^T$, because there is no stress associated with the transformation strain e_{ij}^T.

This useful fact enables us to determine the perturbation caused in a uniform stress field by an ellipsoidal region of differing elastic constants, an ellipsoidal inhomogeneity in the language of I. On the field of the inclusion constrained in the matrix superpose a uniform field e_{ij}^A. It is useful to introduce the deviatoric parts $'e_{ij} = e_{ij} - \frac{1}{3}e\delta_{ij}$, $'p_{ij} = p_{ij} - \frac{1}{3}p\delta_{ij}$ of the strain and stress respectively, where $P = -\frac{1}{3}p = -\frac{1}{3}p_{ii}$ is the hydrostatic pressure. The stress-strain relations then become:

$$p = 3\kappa e, \qquad 'p_{ij} = 2\mu' e_{ij} \tag{5}$$

where $\kappa = \lambda + \frac{2}{3}\mu$ is the bulk modulus. The strain within the ellipsoid is now given by the constant values $e^C + e^A$, $'e^C_{ij} + 'e^A_{ij}$ while the stress in it is $p = 3\kappa(e^C + e^A - e^T)$, $'p_{ij} = 2\mu('e^C_{ij} + 'e^A_{ij} - 'e^T_{ij})$. Thus we could replace the material within S by an ellipsoid of elastic constants κ_1, μ_1, while maintaining continuity of tractions and displacements across S provided that:

$$\kappa_1(e^C + e^A) = \kappa(e^C + e^A - e^T) \tag{6}$$

$$\mu_1('e^C_{ij} + 'e^A_{ij}) = \mu('e^C_{ij} + 'e^A_{ij} - 'e^T_{ij}) \tag{7}$$

The e^C_{ij} can be eliminated from eqs. 6 and 7 with the help of the S_{ijkl}, eq. 4. Then, for given e^A_{ij}, κ_1, μ_1, κ, μ, the quantities e^T_{ij} can be calculated from eqs. 6 and 7 and the field of the inhomogeneity evaluated with the help of eq. 3.

In the sequel we consider only incompressible deformations and flows so that we no longer need to distinguish between e_{ij} and its deviatoric part $'e_{ij}$.

THE VISCOUS INHOMOGENEITY

We can pass to the case of an incompressible elastic medium by letting Poisson's ratio tend to $\frac{1}{2}$ so that $\kappa \to \infty$ and $e \to 0$ in such a way that the hydrostatic pressure $P = -\kappa e$ remains finite. The equilibrium equation, which may be written:

$$\mu\nabla^2 u_i - \frac{3}{2(1+\nu)}\frac{\partial P}{\partial x_i} = 0$$

becomes:

$$\mu\nabla^2 u_i = \partial P/\partial x_i \tag{8}$$

which must be supplemented by the incompressibility condition:

$$\partial u_i/\partial x_i = 0$$

The Green's function (eq. 2) becomes:

$$U_{ij}(r - r') = \frac{\delta_{ij}}{4\pi\mu|r - r'|} - \frac{1}{8\pi\mu}\frac{\partial^2}{\partial x_i \partial x_j}|r - r'| \tag{9}$$

To get the associated pressure P_j we may calculate the hydrostatic pressure associated with eq. 2 and then put $\nu = \frac{1}{2}$ to get:

$$P_j(r-r') = -\frac{1}{4\pi}\frac{\partial}{\partial x_j}\frac{1}{|r-r'|} \tag{10}$$

Alternatively, if eq. 9 is inserted in eq. 8 it is clear that the associated pressure is equal to eq. 10 plus an arbitrary uniform pressure which may be taken to be zero.

Any solution u_i of eq. 8 can be interpreted as the velocity associated with a slow viscous flow in a medium of viscosity μ. The stresses are:

$$p_{ij} = \mu\left(\frac{\partial u_i}{\partial x_j} + \frac{\partial u_j}{\partial x_i}\right) - P\delta_{ij}$$

e_{ij} becomes the rate of strain tensor and eq. 8 becomes the Navier-Stokes equation for slow viscous flow in which inertia terms are negligible. In the viscous context, the field defined by eqs. 9 and 10 is sometimes referred to as a Stokeslet (Batchelor, 1970, p. 240). In particular, the solution for an ellipsoidal elastic inhomogeneity in a uniform strain field e_{ij}^A becomes the solution for an ellipsoidal viscous inhomogeneity in a flow described by the rate of strain tensor e_{ij}^A. We may, if we wish, at the expense of a little artificiality, transcribe the physical argument which led to eq. 3. The interior of an ellipsoidal region in a viscous medium is removed from the surrounding material and is found to be undergoing a spontaneous change of form (due, perhaps, to the progress of some imaginary phase change) specified by the rate of strain tensor e_{ij}^T. This deformation is cancelled by applying tractions $-p_{ij}^T n_j$ which induce an equal and opposite viscous flow and the ellipsoid is returned to the matrix and joined to it. The layer of body force thus built in is relaxed and a uniform rate of strain e_{ij}^A is superimposed overall. The (uniform) rate of strain and the (uniform) stress in the ellipsoid are then connected by eq. 7 with μ replaced by some other value μ_1 for the viscosity.

Less intuitively we may simply start with the flow (eq. 3) representing a certain distribution of body force over S together with a uniform dilatationless flow e_{ij}^A and consider the surface traction on and the deformation of the ellipsoidal surface immediately outside the layer of body force. From the detailed results given in I and II it follows that the traction and deformation are consistent with the presence within the surface of material with a viscosity μ_1 which is related to μ and the shape of the ellipsoid by eq. 7.

To discuss the finite deformation of the viscous inhomogeneity we proceed as follows. For a given deviatoric strain rate e_{ij}^A applied at infinity, eqs. 4 and 7 enable e_{ij}^T to be found in terms of e_{ij}^A. Then the total strain rate within the ellipsoid, which is, by eq. 7:

$$e_{ij}^{\mathrm{inc}} = e_{ij}^C + e_{ij}^A = \frac{\mu}{\mu - \mu_1} e_{ij}^T \tag{11}$$

can also be found in terms of e_{ij}^A. In eq. 8 and in eq. 12 below we have dropped the dash since for each term $e_{mm} = 0$ and so $'e_{ij} = e_{ij}$. Since e_{ij}^{inc} is uniform within S, the ellipsoid always remains an ellipsoid and the analysis can be applied at every instant of the deformation. Integration of the equations for e_{ij}^{inc} thus enables the finite deformation of the inhomogeneity to be followed. This integration is considerably simplified if e_{ij}^A is a pure strain rate whose principal axes coincide initially with those of the ellipsoid, for obviously they will then continue to do so. In the examples below we confine ourselves to applied strain rates of this type. Then in combining eqs. 4 and 7 to determine e_{ij}^T in terms of e_{ij}^A we have only to deal with the three equations:

$$(\mu_1 - \mu)S_{ijkl}e_{kl}^T + \mu e_{ij}^T = (\mu - \mu_1)e_{ij}^A \tag{12}$$

where $ij = 11, 22, 33$. (The incompressible limit of the remaining eq. 6 adds no new information about the change of shape of the ellipsoid and we may ignore it.) The necessary coefficients are given (I 3.7 with $\sigma = \frac{1}{2}$) by:

$$S_{1111} = 1 - (3a^2/4\pi)(I_{ab} + I_{ac})$$

$$S_{1122} = (3b^2/4\pi)I_{ab}$$

and their cyclic counterparts, where:

$$I_{ab} = (I_b - I_a)/3(a^2 - b^2)$$

and I_a, I_b, I_c are the constant coefficients (demagnetizing factors) which appear in the expression:

$$\varphi = \tfrac{1}{2}(a^2 - x^2)I_a + \tfrac{1}{2}(b^2 - y^2)I_b + \tfrac{1}{2}(c^2 - z^2)I_c$$

for the Newtonian potential inside a solid ellipsoid of unit density.

THE DEFORMATION OF AN ELLIPTIC CYLINDER

We take first an elliptic cylinder of viscosity μ_1 with axis parallel to x_3 and cross section ${x_1}^2/a^2 + {x_2}^2/b^2 = 1$ and subject the medium (of viscosity μ) remote from it to a deviatoric pure strain rate with principal axes parallel to those of the ellipse, so that $e_{11}^A = -e_{22}^A > 0$ and $e_{33}^A = 0$ (Fig. 1). Putting $(a + b)^{-1} = k$ we have the results (see I 3.17), $I_a = 4\pi bk$, $I_b = 4\pi ak$, $I_c = 0$, $3I_{ab} = 4\pi k^2$, $I_{aa} = 4\pi/3a^2 - I_{ab}$, $I_{bb} = 4\pi/3b^2 - I_{ab}$, and also, as $c \to \infty$, the limiting values $3c^2 I_{ac} = I_a$, $3c^2 I_{bc} = I_b$, $c^2 I_{cc} = 0$. From these we obtain $S_{1111} = 1 - a^2k^2$, $S_{1122} = b^2k^2$, $S_{1133} = bk$, $S_{3311} = 0$, $S_{3333} = 0$. The other non-vanishing components follow from these by simultaneous interchange of (1,2) and (a,b). For these values of S_{ijkl}, eqs. 12 give $e_{33}^T = 0$, $e_{11}^T = -e_{22}^T$ and, with eq. 11:

$$e_{11}^{\text{inc}} = -e_{22}^{\text{inc}} = \frac{1}{a}\frac{\mathrm{d}a}{\mathrm{d}t} = \frac{\mu(a^2 + {a_0}^2)^2 e_{11}^A}{2a^2a_0^2(\mu_1 - \mu) + \mu(a^2 + {a_0}^2)^2}$$

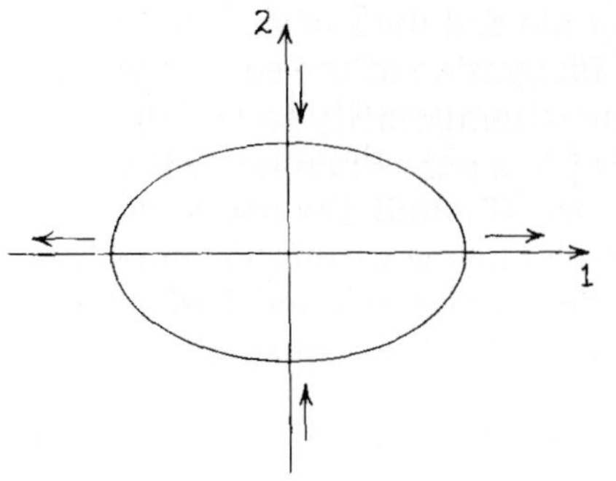

Fig. 1. Deformation of elliptic cylinder by a deviatoric pure strain-rate with principal axes parallel to those of the elliptical cross-section.

where we have set the constant area of the ellipse πab equal to $\pi a_0{}^2$. Integrating with the boundary condition that $t = 0$ when $a = b = a_0$ we have:

$$\ln (a/a_0) + \tfrac{1}{2}\alpha \frac{(a/a_0)^2 - 1}{(a/a_0)^2 + 1} = e^A_{11} t$$

where $\alpha = (\mu_1 - \mu)/\mu$. In terms of the so-called natural strain $S = \ln (a/a_0)$ of the elliptical inhomogeneity and the natural strain $S_H = [\ln(a/a_0)]_H = e^A_{11} t$ applied at infinity, and which would, of course, prevail within the ellipse in the homogeneous case $\alpha = 0$ we have (Cable, 1968):

$$S + \tfrac{1}{2}\alpha \tanh S = S_H = e^A_{11} t \tag{13}$$

Figure 2 shows the general form of this equation, where the curves are labelled not by α but by $R = \mu_1/\mu = 1 + \alpha$. $R = 1$ corresponds to the homogeneous case where $S = S_H$ and $R = 0$ to the deformation of an elliptical hole. Increas-

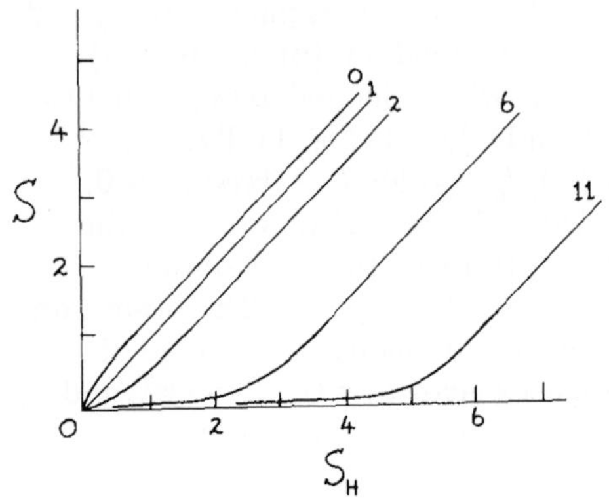

Fig. 2. Total strain $S = \ln(a/a_0)$ in the elliptic cylinder versus applied total strain S_H. The curves are labelled with the viscosity ratio $R = \mu_1/\mu = 1 + \alpha$.

ing values of R represent inhomogeneities of increasing viscosity; $R = \infty$ corresponds to a perfectly rigid inclusion. For large S_H the curves all become parallel to the homogeneous line $R = 1$, but they are translated along the S_H-axis. Thus the stiff inhomogeneity ultimately reaches the same strain as the weak one but after a greater value of the applied strain S_H. Ultimately the increment in S equals the increment in S_H, but the stiff inhomogeneity suffers a smaller strain (and the weak one a larger) when the deformation is in its early stages.

THE DEFORMATION OF PROLATE AND OBLATE SPHEROIDS

We next take a prolate spheroid of viscosity μ_1 with semi-axes $a > b = c$ and subject the medium (of viscosity μ) remote from it to an applied deviatoric pure strain-rate with principal axes parallel to those of the spheroid, chosen so that the 1-axis of the strain-rate is parallel to the a-axis of the spheroid, with $e_{11}^A = -2e_{22}^A = -2e_{33}^A$. If e_{11}^A is positive we expect the spheroid to become increasingly prolate; if e_{11}^A is negative we expect the spheroid to tend to a spherical shape and then to become an oblate spheroid if the deformation is continued.

For the prolate spheroid with $a > b = c$ we have (I 3.16):

$$I_c = I_b = 2\pi v(v^2 - 1)^{-3/2}[v(v^2 - 1)^{1/2} - \cosh^{-1}v] \tag{14}$$

where $v = a/b$, while the quantities S_{ijkl} may be expressed in terms of:

$$\Lambda = (3I_b - 4\pi)/4\pi(a^2 - b^2)$$

as follows; $S_{1111} = 1 - 2\Lambda a^2$, $S_{2222} = S_{3333} = \frac{3}{4}(1 - \Lambda b^2)$, $S_{1122} = S_{1133} = \Lambda b^2$, $S_{2211} = S_{3311} = \Lambda a^2$, $S_{2233} = S_{3322} = \frac{1}{4}(1 - \Lambda b^2)$. Eqs. 12 now show, as we might have anticipated, that $e_{11}^T = -2e_{22}^T = -2e_{33}^T$. When the equations are solved for e_{11}^T and e_{11}^{inc} calculated by eq. 11 we get:

$$e_{11}^{\text{inc}} = \frac{1}{a}\frac{\mathrm{d}a}{\mathrm{d}t} = 4\pi e_{11}^A(a^2 - b^2)[\alpha\{12\pi a^2 - 3I_b(2a^2 + b^2)\} + 4\pi(a^2 - b^2)]^{-1}$$

The volume $4\pi ab^2/3$ of the spheroid is constant and we put $ab^2 = a_0{}^3$. Introducing the variable $(a/a_0)^{3/2} = w$, writing $b^2 = a_0{}^3/a$ and integrating we get:

$$\alpha\left(-\frac{1}{2} + \frac{1}{w^2 - 1} - \frac{w\cosh^{-1}w}{(w^2 - 1)^{3/2}}\right) + \ln w^{2/3} = e_{11}^A t + \text{const.}$$

We put $t = 0$ when $w = 1$ so that at this time we have a sphere of radius a_0. Taking the limit when $w \to 1$ shows the constant to be $-5\alpha/6$ so that finally we have, with $w = (a/a_0)^{3/2}$:

$$\alpha[(1/3) + (w^2 - 1)^{-1}\{1 - (w^2 - 1)^{-1/2}w\cosh^{-1}w\}] = S_H - S \tag{15}$$

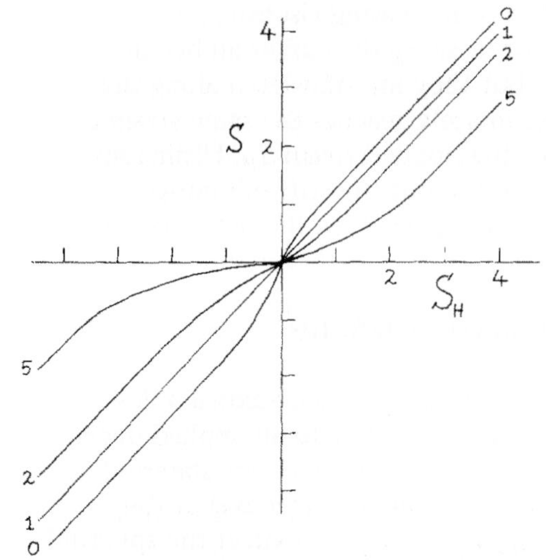

Fig. 3. Total strain $S = \ln(a/a_0)$ in the prolate and oblate spheroids versus applied total strain S_H. The curves are labelled with the viscosity ratio $R = \mu_1/\mu = 1 + \alpha$.

where we have again introduced the natural strain $S = \ln(a/a_0)$ of the inhomogeneity and the natural strain $S_H = e_{11}^A t$ applied at infinity, which is the value of S when $\alpha = 0$.

For the oblate spheroid with $b = c > a$, the expression for $I_c = I_b$ is again given by eq. 14, but with $v < 1$; it is still real because the factors inside and outside the square brackets are both imaginary. A more natural way of writing this expression is then obtained by putting $\cosh^{-1} v = \sinh^{-1}(v^2 - 1)^{1/2}$ and noting that when $v < 1$, $\sinh^{-1}(v^2 - 1)^{1/2} = \sinh^{-1} i(1 - v^2)^{1/2} = i \sin^{-1}(1 - v^2)^{1/2} = i \cos^{-1} v$. Thus eq. 14 becomes for $b = c > a$ (see I):

$$I_c = I_b = 2\pi v(1 - v^2)^{-3/2}[\cos^{-1} v - v(1 - v^2)^{1/2}] \tag{16}$$

for $b = c > a$ (I 3.15). Eq. 15 therefore suffices also to describe the transition of the sphere to an oblate spheroid as w becomes less than 1 and S negative. Once again, we put $\cosh^{-1} w = \sinh^{-1}(w^2 - 1)^{1/2}$ and so obtain the more natural form:

$$\alpha[(1/3) - (1 - w^2)^{-1}\{1 - (1 - w^2)^{-1/2} w \cos^{-1} w\}] = S_H - S \tag{17}$$

as the equation describing the development of the oblate spheroid ($w < 1$) from the sphere ($w = 1$). Eq. 17 may be obtained directly, of course, by repeating the above analysis using eq. 16 instead of eq. 14.

Figure 3 shows S as a function of S_H for various values of α from eq. 15 when $w > 1$ and S, $S_H > 0$, and eq. 17, when $w < 1$ and S, $S_H < 0$.

DISCUSSION

The only previous work on the finite deformation of a viscous inhomogeneity seems to be Gay's (1968) treatment of the problem we have dealt with in the section 'The deformation of an elliptic cylinder'. He finds, in our notation:

$$S = S_H/(1 + \tfrac{2}{5}\alpha) \tag{18}$$

which differs from eq. 13 in being linear in S. Even when S is small enough for eq. 13 to become:

$$S = S_H/(1 + \tfrac{1}{2}\alpha) \tag{19}$$

eq. 18 does not agree with it.

There seem to be several errors in Gay's analysis. Suppose first that S is small, so that the cylindrical inhomogeneity is nearly circular. Gay uses a general three-dimensional solution given by Lamb (1932, p. 595) and used by Taylor (1932) to treat the initial deformation of a spherical viscous inhomogeneity. To find the corresponding two-dimensional solution he simply puts $z = 0$ in Lamb's formulas. Though this reproduces the general shape of the formulas the numerical coefficients are incorrect. Likewise, to get the harmonic functions to insert in the formula he puts $z = 0$ in their three-dimensional counterparts, a process which does not generally generate a two-dimensional harmonic function. When these shortcomings are corrected eq. 18 becomes eq. 19. To treat the general case Gay in effect maps the ellipse back on to the circle and thence deduces that eq. 18 is valid for all S. (With the correction just indicated he would have found instead that eq. 19 is valid for finite S.) It does not seem to be possible to justify this mapping process.

Gay compares his formula with some experimental results but, because of their scatter and the limited range of S_H and α which they cover, it is hardly possible to decide whether they favour any one of eqs. 13, 18 or 19 rather than another.

The results we have obtained have interesting implications for the mixing and homogenizing of viscous liquids, and a preliminary discussion of some of them has already been given by Cable (1968). In mixing theory the important question is how fast is an inhomogeneity stretched and thinned, and so in the present paper we have merely calculated the rate of deformation of the ellipsoid. However, when the appropriate e_{ij}^T have been obtained, as explained in the second section of this paper, the entire flow inside and outside the inhomogeneity can be found from the general theory presented in I and II.

REFERENCES

Batchelor, G.K., 1970. Introduction to Fluid Dynamics. Cambridge University Press, 615 pp.

Cable, M., 1968. The physical chemistry of glassmaking. Proc. Int. Congr. on Glass, London. Society of Glass Technology, pp. 163—178.

Dunnet, D., 1969. A technique of finite-strain analysis using elliptical particles. Tectonophysics, 7: 117—136.

Eshelby, J.D., 1957. The determination of the elastic field of an ellipsoidal inclusion, and related problems. Proc. R. Soc. London, Ser. A, 241: 376—396.

Eshelby, J.D., 1959. The elastic field outside an ellipsoidal inclusion. Proc. R. Soc. London, Ser. A, 252: 561—569.

Gay, N.C., 1968a. Pure-shear and simple-shear deformation of inhomogeneous viscous fluids. 1. Theory. Tectonophysics, 5: 211—234.

Gay, N.C., 1968b. Pure-shear and simple-shear deformation of inhomogeneous viscous fluids. 2. The determination of the total finite strain in a rock from objects such as deformed pebbles. Tectonophysics, 5: 295—302.

Jaeger, J.C., 1969. Elasticity, Fracture and Flow. Methuen, London, 268 pp.

Lamb, H., 1932. Hydrodynamics. Cambridge University Press, 738 pp.

Love, A.E.H., 1927. A Treatise on the Mathematical Theory of Elasticity. Cambridge University Press, 643 pp.

Ramsay, J.G., 1967. Folding and Fracturing of Rocks. McGraw-Hill, New York, 567 pp.

Tan, B.K., 1974. Deformation of particles developed around rigid and deformable nuclei. Tectonophysics, 24: 243—257.

Taylor, G.I., 1932. The viscosity of a fluid containing small drops of another fluid. Proc. R. Soc. London, Ser. A, 138: 41—48.

Wood, D.S., 1973. Patterns and magnitude of natural strain in rocks. Philos. Trans. R. Soc. Lond., Ser. A, 274: 373—382.

Journal of Elasticity, Vol. 5, Nos. 3–4, November 1975
Dedicated to A. E. Green
Noordhoff International Publishing – Leyden
Printed in The Netherlands

The elastic energy-momentum tensor

J. D. ESHELBY
Department of the Theory of Materials, University of Sheffield, Sheffield, U.K.

(Received April 15, 1975)

ABSTRACT

The application to continuum mechanics of the general methods of the classical theory of fields is advocated and illustrated by the example of the static elastic field. The non-linear theory of elasticity is set up in the most convenient form (Lagrangian coordinates and stress tensor). The appropriate energy-momentum tensor is derived, and it is shown that the integral of its normal component over a closed surface gives the force (as the term is used in the theory of solids) on defects and inhomogeneities within the surface. Other topics discussed are Günther's and related integrals, symmetrization of the energy-momentum tensor, and the Eulerian formulation. Some further extensions, existing and potential, are indicated.

RÉSUMÉ

On signale les avantages de l'application à la méchanique des milieux continus des méthodes de la théorie classique des champs. A titre d'example la théorie de l'élasticité est construite sous la forme la plus convenable à cette fin (coordonées de Lagrange, composantes de tension de Boussinesq). On déduit ensuite le tenseur d'énergie-impulsion dont l'intégrale de la composante normale étendue sur une surface fermée donne la force (comme on l'entend en théorie des solides) agissant sur les défaults dans son intérieur. On discute aussi quelques intégrales semblables, entr'elles celles de Günther, la symétrisation du tenseur d'énergie-impulsion et la formulation en coordonées d'Euler. On récapitule par conclure certains autres résultats déjà connus, tout en indiquant quelques extensions possibles.

1. Introduction

In that corner of the theory of solids which deals with lattice defects (dislocations, impurity and interstitial atoms, vacant lattice sites and so forth) which are capable of altering their position or configuration in a crystal, it has been found useful to introduce the concept of the force acting on a defect. If it is enough to specify simply the position of a defect the force is defined to be minus the rate of increase of the total energy of the system with respect to variation of the three position coordinates of the defect. If a larger (possibly infinite) number of parameters is required to specify the configuration of the defect, generalized forces are defined similarly in terms of the variation of the total energy with the parameters in a way familiar in analytical mechanics and thermodynamics. By total energy is meant the energy of the solid being investigated plus the potential energy of any external loading mechanism which is acting on it. The distinction between the two is to some extent arbitrary. In solid state applications the forces are equated to zero to find the equilibrium configuration, and in favourable cases the rate of approach to equilibrium can be estimated from them with the help of the Einstein relation[1] between mobility and diffusion.

In many cases an elastic model of a crystal containing defects will be adequate, or at any rate the only tractable one. The crystal becomes an elastic solid, a crack becomes a crack, a crystal dislocation one of Volterra's elastic dislocations, an oversize impurity atom an elastic sphere forced into a hole too small for it, a lattice vacancy a small region (with or without misfit) whose elastic constants are less than those of the bulk of the medium, and so on. The normal theory of elasticity recognizes nothing which corresponds with the force on a defect. (It has nothing to do with the ordinary body force, of course.) But in fact the appropriate concept has been to hand ever since the appearance of a paper by Noether [1] in 1918, in the form of the energy-momentum tensor[2] which the elastic field possesses in common with every field whose governing equations are derivable from a variational principle, and some for which they are not.

The archetypal energy-momentum tensor is Maxwell's stress tensor in electrostatics ([2], p. 84). The integral of its normal component taken over a closed surface gives the total force on all the electric charges inside that surface (cf. [2], p. 87). The writer, having looked at a book on field theory, felt that the force on a defect ought to be given by a similar expression involving the energy-momentum tensor appropriate to the elastic field. For linear elasticity it proved possible [3] to manipulate an existing expression for the force until it took on the hoped-for form. Later a simpler and less artificial demonstration was devised [4, 5] which led naturally to the desired expression and which, moreover, applied equally well to the case of the finite deformation of a solid with an arbitrary stress-strain relation.

To give a sensible result the integral for the force on a defect must be surface-independent in the sense that its value is unchanged when the surface of integration is deformed in any way, provided it continues to enclose the defect we are interested in and does not embrace any others. The fact that the divergence of the energy-momentum tensor vanishes ensures that this is so. In two dimensions the integral for the force becomes a path-independent one. If the path of integration embraces the tip of a two-dimensional crack the expression for the component of the force parallel to the crack becomes Rice's [6, 7] independently-discovered J-integral (cf. also [8, 9]). In what follows we shall, for brevity, use the expression "path-independent", which has become familiar in fracture mechanics, to stand for "surface-independent or path-independent as appropriate".

Apart from its connection with the theory of lattice defects the energy-momentum tensor and kindred concepts associated with elastic and other material media are of interest for their own sake, but they have received scarcely any attention from applied mathematicians, even during the intensive re-examination and extension of continuum

[1] See, for example, [5] p. 425.

[2] As we shall be almost entirely concerned with static situations this term is not particularly appropriate, but the alternative, stress-energy tensor, is not much better.

mechanics which has been under way for the last couple of decades, perhaps because of the artificial separation which has grown up between applied mathematics and theoretical physics. The papers of Günther [10], Knowles and Sternberg [11] and Green [12] are exceptions.

It is hoped that the present paper may perhaps help to dispel this lack of interest. Apart from anything it may do for the theory itself a treatment of elasticity on the lines of other physical fields furnishes homely (in the British sense) illustrations of the results of the classical part of general field theory. By classical field theory we mean, roughly, the body of doctrine, based on a Lagrangian density depending on a set of field variables and their derivatives, which is often presented as a preliminary to second quantization in books and articles devoted to the quantum theory of fields [13, 14, 15]. As long as one is content with a single-particle description the wave-function of an elementary particle is a perfectly respectable classical field in the sense in which we are using the term [16, 17]. Only when recourse must be had to second quantization is this no longer true. Also relevant are some discussions of the theory of relativity, particularly those in which extra field variables appear alongside the components of the metric tensor which serve as gravitational potentials [18, 19].

As an exercise-ground for field theory, elasticity has the advantage of being non-linear, in contrast to most of the fields commonly considered. One exception is Born's non-linear electrodynamics [16, 20, 21], which in fact has a close connection with the theory of elasticity. There is an exact correspondence between time-dependent linear anti-plane elastic and plane electromagnetic fields [5] and it is not hard to show that there is the same correspondence between the two-dimensional version of Born's theory [22] and the type of non-linear anti-plane elasticity discussed by Neuber [23].

2. Elasticity in Lagrangian coordinates

Unless the contrary is stated our results will relate to the finite deformation of an elastic solid with an arbitrary stress-strain relation. Our coordinate system, measure of deformation and stress tensor are chosen so as to make the description of the elastic field conform with that of other physical fields, though they may not be the best choice for solving specific problems. We introduce fixed rectangular Cartesian coordinates X_m and label each particle of the medium with the X_m appropriate to the position it occupies in the undeformed state (Lagrangian coordinates). After deformation the particle labelled with X_m is at x_i, referred to the same fixed coordinate system, and

$$u_i(X_m) = x_i(X_m) - X_i \tag{2.1}$$

is its displacement vector. If from each point X_m we draw a vector arrow $u_i(X_m)$ with its tail at X_m its tip defines the final position of the particle labelled with X_m. Contrast this with say, the electrostatic field. If we treat the electric field vector $E_i(X_m)$ in the same way as $u_i(X_m)$ the position of its tip has no meaning; indeed it would depend on how many metres we used to denote one volt per metre. If we wish to develop the theory of the elastic field in the same way as for other physical fields we must, in fact, look upon $u_i(X_m)$ in the same way as $E_i(X_m)$, as an unlocated vector associated with the point X_m, though its aspect as a displacement of the material must, of course, be taken into account in the interpretation of any results we obtain.

To match our Lagrangian[3] formulation, we choose our stress tensor so that p_{ij} denotes the force parallel to the X_i-axis on an element of material which was of unit area and perpendicular to the X_j-axis before deformation. Thus p_{ij} is, to use Hill's apt term, the tensor of nominal stresses; it is also known as the Lagrangian, first Piola-Kirchhoff or Boussinesq tensor. The traction on an element of surface embedded in the material is $p_{ij}dS_j$ where, in $dS_j = n_j dS$, dS is the area and n_j the normal of the element before deformation. (We employ the usual summation convention for repeated suffixes.) In the absence of body forces p_{ij} obeys the equilibrium equations

$$\frac{\partial p_{ij}}{\partial X_j} = 0. \tag{2.2}$$

When convenient we shall use a comma followed by suffixes to denote differentiation with respect to the X_m, so that, for example, the last equation could be re-written $p_{ij,j} = 0$, and $u_{i,j}$, $u_{i,jk}$ denote the strain gradient tensor $\partial u_i/\partial X_j$ and its X_k-derivative. If $W(u_{i,j}, X_m)$ is the elastic energy density per unit undeformed volume (in an inhomogeneous medium it may depend on X_m) consideration of the work done in a further small deformation of a material already finitely deformed gives the relation

$$p_{ij} = \frac{\partial W}{\partial u_{i,j}}. \tag{2.3}$$

The p_{ij} are not symmetric, but they obey a relation,

$$p_{li} - p_{jl} = p_{ji}u_{l,i} - p_{li}u_{j,i}, \tag{2.4}$$

which implies that the stress tensor is symmetric when referred to Cartesian final coordinates. The easiest way to derive it is to start with the relation

$$\int_S (x_l p_{ji} - x_j p_{li}) dS_i = 0,$$

which states that the total couple on the material inside S is zero (the final position vector $x_l = X_l + u_l$ rather than X_l is obviously the appropriate lever arm), convert to a volume integral,

$$\int [(\delta_{li} + u_{l,i})p_{ji} - (\delta_{ji} + u_{j,i})p_{li}] dv = 0,$$

and note that, since S is arbitrary, the integrand must vanish.

We shall also need the relation (2.5) below. Consider two elementary cubes of material, the first with edges $\varepsilon \boldsymbol{i}_m$ parallel to the coordinate axes and the second with edges $\varepsilon \boldsymbol{i}'_m$ slightly rotated relative to those of the first, and let the displacement gradient $u'_{l,i}$ impart to the second the same strain (change of shape and size) as $u_{l,i}$ does to the first. Evidently $u'_{l,i}$ must have components $u_{l,i}$ when referred to axes parallel to the $\boldsymbol{i}'_m$, that is,

[3] In the sense of Lagrangian versus Eulerian coordinates. Also, by chance, in the sense of a Lagrangian as against a Hamiltonian formulation of mechanics: for the latter we would have to use a (complementary) energy function depending on stress rather than deformation.

$$u'_{l,i} = u_{r,s}(\delta_{rl}+\omega_{rl})(\delta_{si}+\omega_{si}),$$

where $\omega_{rl} = -\omega_{lr} \ll 1$ specifies the relative orientation of the cubes. In an anisotropic medium, a piece of wood say, the energies of the cubes will be different, for although they have undergone the same changes of shape and size, the grain runs differently in them. But if the material is isotropic the energies are equal, and we must have $\delta u_{l,i}\partial W/\partial u_{l,i} = 0$ with $\delta u_{l,i} = u'_{l,i} - u_{l,i}$, which gives, after a suitable re-naming of suffixes,

$$p_{ij}u_{i,l} - p_{il}u_{i,j} + p_{ji}u_{l,i} - p_{li}u_{j,i} = 0, \tag{2.5}$$

or, with (2.4),

$$p_{lj} - p_{jl} = p_{il}u_{i,j} - p_{ij}u_{i,l}. \tag{2.6}$$

For the linear theory with $p_{lj} = p_{jl}$ neither (2.4) nor (2.6) is true but fortunately (2.5), which then takes the form

$$p_{ij}e_{il} = p_{il}e_{ij},$$

with

$$p_{ij} = \lambda e_{mm}\delta_{ij} + 2\mu e_{ij}, \qquad e_{ij} = \tfrac{1}{2}(u_{i,j}+u_{j,i}),$$

is.

What we have called elastic energy is, more precisely, internal energy under adiabatic conditions and Helmholtz free energy under isothermal conditions [24]. For most solids the difference between them is small enough for us to hope that our results will also apply in intermediate situations.

3. The energy-momentum tensor

In this section we show how an energy-momentum tensor can be generated for (almost) any field. The usual treatment, following Noether, is a variational one, but we shall give a more direct method which will serve our purposes. It is rather more general than would be required for the theory of elasticity as set up in the last section. The X_i can be N-dimensional curvilinear coordinates. The number of u_i need not be the same as the number of X_i and they need not be the components of a vector. They could, for example, be a set of scalars, or $u_1, \ldots u_9$ could be the elements of a 3×3 matrix conventionally numbered.

We suppose that the field is governed by the equations

$$\frac{\partial}{\partial X_j}\frac{\partial W}{\partial u_{i,j}} = \frac{\partial W}{\partial u_i} \tag{3.1}$$

with

$$W = W(u_i, u_{i,j}, X_i). \tag{3.2}$$

Of course (3.1) are the Euler equations which would be found by requiring the integral of (3.2) over a region of X-space to be stationary, but we make no explicit use of this fact.

The W of the last section does not depend on the u_i, but it would do so if the X_i were curvilinear coordinates, for W would then contain covariant derivatives of the u_i which have a term linear in the u_i. Likewise, if the X_i stand for the coordinates x_i of the last section, even Cartesian ones, W may depend on the u_i; see section 7. In any case certain manipulations seem less artificial if W is supposed to contain the u_i even if it does not. In order to generate the energy-momentum tensor we must first distinguish between two X-derivatives of W, namely the ordinary gradient,

$$\frac{\partial W}{\partial X_l} = (\text{grad } W)_l$$

and the derivative

$$\left(\frac{\partial W}{\partial X_l}\right)_{\text{exp.}} = \frac{\partial}{\partial X_l} W(u_i, u_{i,j}, X_m) \Big|_{\substack{u_i,\, u_{i,j} \text{ const.} \\ X_m \text{ const.},\, m \neq l}}$$

which gives the explicit dependence on X_l. This rather clumsy notation, common in field theory, will be adequate since $(\partial W/\partial X_l)_{\text{exp.}}$ will not often appear in what follows. For the u_i there is, of course, no distinction between the two types of derivative. We have

$$\frac{\partial W}{\partial X_l} = \frac{\partial W}{\partial u_i} u_{i,l} + \frac{\partial W}{\partial u_{i,j}} u_{i,jl} + \left(\frac{\partial W}{\partial X_l}\right)_{\text{exp.}} \tag{3.3}$$

If we use (3.1) to replace $\partial W/\partial u_i$ by $\partial(\partial W/\partial u_{i,j})/\partial X_j$ the first two terms on the right of (3.3) combine to give

$$\frac{\partial}{\partial X_j}\left(\frac{\partial W}{\partial u_{i,j}} u_{i,l}\right).$$

Whence, writing $\partial W/\partial X_l = \partial(W\delta_{lj})/\partial X_j$ and re-arranging, we have

$$\frac{\partial P_{lj}}{\partial X_j} = \left(\frac{\partial W}{\partial X_l}\right)_{\text{exp.}}, \tag{3.4}$$

where, by definition,

$$P_{lj} = W\delta_{lj} - \frac{\partial W}{\partial u_{i,j}} u_{i,l} \tag{3.5}$$

is the (canonical) energy-momentum tensor derived from W.

If $W = W(u_i, u_{i,j} u_{i,jk}, X_m)$ depends also on the second derivatives of u_i (grade 2 material) the right-hand sides of (3.1) and (3.3) acquire additional terms

$$-\frac{\partial^2}{\partial X_k \partial X_j} \frac{\partial W}{\partial u_{i,jk}} \quad \text{and} \quad \frac{\partial W}{\partial u_{i,jk}} u_{i,jkl},$$

respectively, and a similar calculation leads to (3.4) with

$$P_{lj} = W\delta_{lj} - \left(\frac{\partial W}{\partial u_{i,j}} - \frac{\partial}{\partial X_k}\frac{\partial W}{\partial u_{i,jk}}\right) u_{i,l} - \frac{\partial W}{\partial u_{i,jk}} u_{i,kl}. \tag{3.6}$$

One can go on to derive the energy-momentum tensor for a material of grade N.

Though it does not play a very important role in field theory the energy-momentum tensor for a field in which the Lagrangian density depends on the first N derivatives of the field variables has been repeatedly re-derived; see, for example, the papers of Podolsky and Kikuchi [25], Chang [26] and Thielheim [27]. Equations (3.5), (3.6) and their higher-order analogues agree with the results of these and other authors.

4. The force on a defect

For the elastic field (2.3) and (3.5) give

$$P_{lj} = W\delta_{lj} - p_{ij}u_{i,l}. \tag{4.1}$$

According to (3.4) the integral

$$F_l = \int_S P_{lj}dS_j \tag{4.2}$$

is zero when taken over a closed surface S within which the material is homogeneous and free of defects so that

$$\left(\frac{\partial W}{\partial X_l}\right)_{\text{exp.}} = 0 \tag{4.3}$$

inside S. If a surface S_1 encloses a second surface S_2 so that (4.3) holds on S_1 and S_2 and in the volume between them, but not within S_2, then (4.2) is not zero, but, since the divergence of the integrand is zero it has the same value over S_1, S_2, and any intermediate surface which is inside S_1 and embraces S_2. In the region between S_1 and S_2 it is thus path-independent in the sense explained in the Introduction. Although in the present paper we do not want to over-emphasize this particular role of the energy-momentum tensor it is perhaps worth sketching the physical argument which shows that (4.2) gives the force on a defect inside S in the sense mentioned in the Introduction.

The left of fig. 1 shows a loaded body containing a defect enclosed by an arbitrary surface S. We shall refer to it as the original system. On the right is an exact replica of it with S marked out and also S', the surface produced by giving S a vector displacement $-\delta\xi_l$ in the *undeformed* state.

To find the energy change associated with a displacement of the singularity by $+\delta\xi_l$ we carry out the following imaginary operations.

(i) In the original system cut out the material inside S and discard it. Apply suitable tractions to the surface of the resulting hole to prevent relaxation.

(ii) Cut out the piece of material inside S' in the replica and apply surface tractions to prevent relaxation. The energy $E_{s'}$ inside S' differs from the energy E_s originally inside S by the addition of the energy in the crescent-shaped region 1 and the removal of the energy in the crescent-shaped region 2, so that

$$E_{s'} - E_s = -\delta\xi_l \int_S W dS_l. \tag{4.4}$$

At this stage there is no change in the energy in the material outside the hole in the original system, or the energy of its loading mechanism.

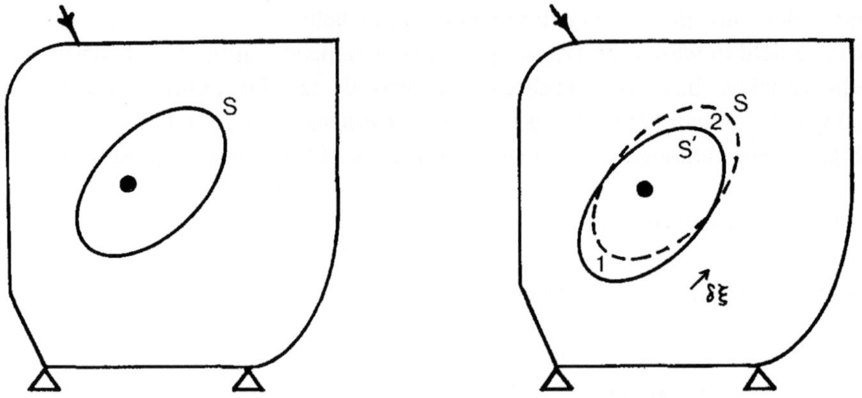

Figure 1. Calculation of the force on a defect.

(iii) We now try to fit the body bounded by S' into the hole S. But since, by construction, S and S' can be made to coincide by a simple translation in the undeformed state they will not usually do so after deformation. In fact after a suitable translation to move it from replica to original the displacement on S' will differ from that on S by

$$\delta u_i = -\delta\xi_l u_{i,l}. \tag{4.5}$$

We therefore give a displacement (4.5) to the surface of the hole, which requires the expenditure of an amount of work

$$\delta W = -\int_S \delta u_i p_{ij} dS_j, \tag{4.6}$$

some of which goes to raise the elastic energy in the material outside S and some to increase the potential energy of the loading mechanism. (The minus sign is correct if, in $dS_j = n_j dS$, n_j is the outward normal to S).

(iv) We can now fit S' into S and weld across the interface. We have not yet quite reached final state, however. Though the displacement matches across the interface the tractions on either side of it will differ by a quantity of order $\delta\xi_l$, and so there is a layer of body force of order $\delta\xi_l$ spread over the interface. As we relax this unwanted force the displacement changes by a quantity of order $\delta\xi_l$ and an amount of energy of order $(\delta\xi_l)^2$ is extracted, which may be neglected in comparison with (4.4) and (4.6).

We are now left with the system as it was to begin with except that the defect has been shifted by $+\delta\xi_l$, as required. The associated change of energy is the sum of (4.4) and (4.6), and if we choose to write it as $-F_l\delta\xi_l$, the effective force F_l is given by (4.2), as claimed. We add a few comments.

The material in the region where the interiors of S and S' overlap (see the right of the figure) plays no direct part in the energy calculation; indeed our set of operations is so devised that we do not even have to say precisely what we mean by a defect in this region. In particular a possible formal infinity in the energy of a singular defect is subtracted out in forming the difference (4.4).

As we apply the displacement δu_i to S in stage (iii) the traction changes from $p_{ij}n_j$ to

$$p_{ij}n_j + 0(\delta\xi_l), \tag{4.7}$$

where the extra term will depend on the exact properties of the medium and on the way the forces exerted by the loading mechanism vary as their points of application move. However, as the extra term in (4.7) only alters (4.6) by a quantity of order $(\delta\xi_l)^2$ our final result is independent of the precise elastic behaviour of the medium and, to use an engineering term, the degree of hardness or softness of the loading mechanism.

Our argument might suggest that in (4.2) the first term of (4.1) gives the change of energy of the material inside S and the second the change in the energy of the loading mechanism and the material outside S, but this is not so. Although the final relaxation in stage iv only changes the total energy by a quantity of order $(\delta\xi_l)^2$ it involves an interchange of energy of order $\delta\xi_l$ across S which, fortunately, we are not required to work out.

It is not hard to see that in the two-dimensional version of the above argument the tip of a crack qualifies as a defect in its own right, even though the crack passes out of the contour S, provided that the part of the crack which lies inside S is straight and $\delta\xi_l$ is parallel to the crack. (For a more careful treatment see [9], and for an extension to three dimensions, [28].) If the crack is parallel to the X_1-axis F_1 is equal to G, the crack extension force or Rice's J-integral. It is also possible to give an interpretation to the other component, F_2. Suppose that the crack begins to spread in a direction inclined at an angle α to its original direction. Then G at the new tip is a function of α, and equation (5.1) below can be used to show that

$$F_2 = \left(\frac{dG(\alpha)}{d\alpha}\right)_{\alpha=0} \tag{4.8}$$

provided that the integral for F_2 is evaluated round a small circuit embracing the tip. Equation (4.8) may be of use in treating the curved path of a crack in a non-uniform stress field.

5. Günther's and related integrals

Günther [10] seems to have been the first to make an explicit application of Noether's theorem to elasticity. In addition to F_l, equation (4.2), he found the path-independent integrals

$$L_{kl} = \int (X_k P_{lj} - X_l P_{kj} + u_k p_{lj} - u_l p_{kj}) dS_j \tag{5.1}$$

and

$$M = \int (X_l P_{lj} - \tfrac{1}{2} u_l p_{lj}) dS_j . \tag{5.2}$$

The L_{kl} are path-independent for a non-linear homogeneous isotropic medium, as may be verified with the help of Gauss' theorem, (2.2) and (2.5). For L_{12} to be path-independent we only need transverse isotropy about the X_3 axis. More generally L_{12} is path-independent if the material has the elastic properties of a treetrunk where, with cylindrical polars ρ, ϕ, X_3, the functional form of W depends on ρ and X_3 but not ϕ, provided it is everywhere referred to the local coordinate directions.

The integral M is path-independent for a homogeneous linear an isotropic medium (use Gauss' theorem, (2.2) and $W = \frac{1}{2}p_{ij}u_{i,j}$) or more generally if with spherical polars r, θ, ϕ, W depends on θ, ϕ but not r. To apply (5.2) to a plane-strain situation we may integrate over a disk with its flat faces in the X_1X_2-plane. Gauss' theorem converts the contributions from the faces into an integral round the rim, where it cancels the second term in (5.2) to leave

$$M = \int X_l P_{lj} dS_j \tag{5.3}$$

in plane strain. For some applications of (5.3) see [29].

We may be led directly to the differential forms of (5.1) and (5.2) by introducing spherical polars r, θ, ϕ and treating $\partial W/\partial\phi$ and $r\partial W/\partial r$ on the lines of $\partial W/\partial X_l$ in (3.3). For a nonlinear homogeneous anisotropic medium $r\partial W/\partial r$ leads to

$$(X_l P_{lj} + u_i \partial W/\partial u_{i,j})_{,j} = 3W + u_i \partial W/\partial u_i, \tag{5.4}$$

the differential form of a result of Green's [12]; for use in a moment we have included a term which only appears if W depends on u_i. For the linear case (5.4) expresses the path-independence of M.

As conservation laws the four-dimensional analogues of F_l and L_{kl} have long been familiar in field theory, but though stated for the electromagnetic field in 1921 [30] the analogue of M has only recently achieved prominence, in connection with high-energy physics (Dilatation or scale symmetry [31]). The W appropriate to an elementary particle of mass m contains a term $\frac{1}{2}m^2 u_i u_i$, and so there is a term $3m^2 u_i u_i$ in (5.4) and it does not express the vanishing of a four-dimensional divergence. (The 3, which stands for δ_{ii}, is now a 4). However, u_i is a wave-like solution of the governing equations with a frequency proportional to the energy of the particle. Thus at high enough energies the other terms in (5.4), because they contain gradients of u_i, swamp the awkward term and there is an approximate conservation law.

Knowles and Sternberg [11] have given an argument to show that in three dimensions F_l, L_{kl} and M are the only path-independent integrals of Noether's type, i.e. of the form

$$\int (\xi_l P_{lj} - \eta_l p_{lj}) dS_j, \tag{5.5}$$

with ξ_l independent of u_i and p_{lj}, and that nothing new appears in plane situations beyond the trivial change from (5.2) to (5.3). However, whether we regard it as three-dimensional or plane, a state of linear isotropic anti-plane strain provides an exception. For it we have $\partial u_3/\partial X_3 = 0$, $u_1 = 0$, $u_2 = 0$, $P_{12} = P_{21}$, $P_{11} + P_{22} = 0$ and Gauss' theorem shows that (5.5) is path-independent if η_l is zero and ξ_1, ξ_2 are any pair of conjugate harmonic functions. There is an analogue of this result for plane strain, though not precisely of the form (5.5). The two may be used to simplify the derivation of some standard results in fracture mechanics [29]. Possibly further integrals may appear when other severe restrictions are placed on the form of u_i.

The canonical energy-momentum tensor P_{lj} is not symmetric. Field theorists have studied the problem of symmetrizing it, that is, finding a symmetric tensor P^S_{lj} such that the divergence of $P^S_{lj} - P_{lj}$ is identically zero. The most elegant method is that of

Rosenfeld [32]. Suppose that the X_i are curvilinear with metric tensor g_{ij} and that W is the energy per unit coordinate mesh, a scalar density. Then

$$g^{lm}g^{jn}P^S_{mn} = \frac{\delta W}{\delta g_{lj}} + \frac{\delta W}{\delta g_{jl}}, \tag{5.6}$$

where

$$\frac{\delta W}{\delta g_{lj}} = \frac{\partial W}{\partial g_{lj}} - \left(\frac{\partial W}{\partial g_{lj,k}}\right)_{,k}$$

is the so-called variational derivative, and g_{lj}, g_{jl} are regarded as independent. Equation (5.6) has the disadvantage that even if we are only interested in Cartesian coordinates we still have to go over to general coordinates, evaluate (5.6) and then set g_{ij} equal to δ_{ij}. Belinfante's [33] recipe is more direct, and leads to the same result as (5.6) [34]. For an isotropic grade one elastic material a simplified version goes as follows.

From the isotropy condition (2.5) and the equilibrium equation (2.2) we have

$$\begin{aligned} P_{lj} - P_{jl} &= p_{il}u_{i,j} - p_{ij}u_{i,l} \\ &= p_{ji}u_{l,i} - p_{li}u_{j,i} \\ &= (p_{ji}u_l - p_{li}u_j)_{,i} \\ &= H_{lji,i} \end{aligned}$$

say, with

$$H_{lji} = p_{ji}u_l - p_{li}u_j. \tag{5.7}$$

Put, tentatively,

$$H_{lji} = S_{jli} - S_{lji}. \tag{5.8}$$

Then

$$P^S_{lj} = P_{lj} + S_{lji,i} \tag{5.9}$$

is symmetric. We also require that the second term in (5.9) shall have no divergence, and this can be secured by making S_{lji} antisymmetric in i, j:

$$S_{lji} = -S_{lij}. \tag{5.10}$$

Between them (5.8) and (5.10) determine the S_{lji}: add to (5.8) the results of applying the permutations $lji \to jil \to lij$ to it and use (5.10) to get

$$S_{jli} = \tfrac{1}{2}(H_{lji} + H_{jil} + H_{lij}).$$

With (5.7) this gives

$$P^S_{lj} = P_{lj} + \tfrac{1}{2}(p_{ij}u_l - p_{ji}u_l + p_{il}u_j - p_{jl}u_i + p_{li}u_j - p_{lj}u_i)_{,i}$$

or for linear isotropy $P^S_{lj} = P_{lj} + (p_{li}u_j - p_{lj}u_i)_{,i}$. An application of Stokes' theorem will show that $\int S_{lji,i}dS_j$ is equal to zero for a closed surface or to a line-integral round the boundary if it is open. Hence $\int P^S_{lj}dS_j$ is equal to (4.2) for a closed surface but not an open one.

In the elastic case there is another symmetric energy-momentum tensor which seems to have no counterpart for other fields [28]. Eqn. (2.6) states that the antisymmetric parts of P_{lj} and p_{lj} are the same, and so

$$P^*_{lj} = P_{lj} - p_{lj} \tag{5.11}$$

is symmetric for an isotropic medium. We may write it as

$$P^*_{lj} = W\delta_{lj} - \frac{\partial W}{\partial\left(\dfrac{\partial x_i}{\partial X_j}\right)} \frac{\partial x_i}{\partial X_l}, \tag{5.12}$$

a form we should have obtained to begin with if, following the often-quoted advice of Kirchhoff, we had taken the $x_i = u_i + X_i$ rather than the u_i as the field variables on which W is supposed to depend. The force on a defect, F_l, eqn. (4.2) will usually be the same as

$$F^*_l = \int P^*_{lj} dS_j, \tag{5.13}$$

since we commonly consider only cases where there is no uncompensated body force associated with the defect and the second term in (5.11) contributes nothing. When, rarely, it is sensible to consider body force as forming part of a defect, (5.13) rather than (4.2) is the correct expression to use [28]. The tensor P^*_{lj} is not symmetric for an anisotropic material, neither is it in the linear isotropic theory where p_{lj} is symmetric; this theory is a consistent field theory, but not a consistent theory of an elastic body to the second order. Nevertheless in these cases too P^*_{lj} is an acceptable substitute for P_{lj}.

6. Eulerian formulation

It is possible to derive an energy-momentum tensor in terms of the rectangular final (Eulerian) coordinates x_i of section 2 with the help of the general formulation of section 3, but in such coordinates it is not easy to repeat the argument of section 4 which shows that it gives the force on a defect. The simplest way is to define the appropriate tensor, Σ_{lj} say, and the ordinary Eulerian stress σ_{lj} by

$$\Sigma_{lj} ds_j = P_{lj} dS_j, \qquad \sigma_{lj} ds_j = p_{lj} dS_j, \tag{6.1}$$

where ds_j is the directed surface element which dS_j transforms into on deformation. Then the integral

$$F_l = \int_s \Sigma_{lj} ds_j$$

over the surface s into which S transforms will agree with (4.2). A tedious calculation gives an expression

$$\Sigma_{lj} = w\delta_{lj} - \frac{\partial w}{\partial u_{i,j}} u_{i,l} \tag{6.2}$$

precisely analogous to P_{lj}; w is the energy per unit final volume and $u_i(x_m) = x_i - X_i(x_m)$ is now a function of the x_m as opposed to our previous $u_i(X_m) = x_i(X_m) - X_i$. On the other hand the second of (6.1) gives

$$\sigma_{lj} = \Sigma_{lj} + \frac{\partial w}{\partial u_{l,j}} = w\delta_{lj} - \frac{\partial w}{\partial\left(\dfrac{\partial X_i}{\partial x_j}\right)} \frac{\partial X_i}{\partial x_l}, \tag{6.3}$$

which is analogous not to p_{lj} but rather to P^*_{lj}, equation (5.13). It is not hard to show that, as required, it is symmetric. Chadwick [35] has independently noted the relationship between σ_{lj} and P^*_{lj}.

It is interesting to see how the formalism of section 3 handles things when left to itself. With suitable changes in notation (3.4) gives

$$\frac{\partial \Sigma_{lj}}{\partial x_j} = \left(\frac{\partial w}{\partial x_l}\right)_{\text{exp.}} \tag{6.4}$$

with the Σ_{lj} of (6.2), but in place of the expected

$$\frac{\partial \sigma_{lj}}{\partial x_j} = 0, \tag{6.5}$$

(3.1) gives

$$\frac{\partial}{\partial x_j} \frac{\partial w}{\partial u_{l,j}} = \frac{\partial w}{\partial u_l}, \tag{6.6}$$

where the quantity whose divergence appears on the left is not one of the more familiar representations of the stress tensor. The second member of (6.6) appears because we must now allow w to depend on u_i since in an inhomogeneous medium the form of the energy function for the particle now at x_i depends on where it originally came from. However, it actually only depends on u_i and x_i through the combination $x_i + u_i = X_i$, which specifies its initial position, and so if we add (6.6) to (6.4) the right-hand members cancel and we are left with (6.5).

The fact that σ_{lj} involves Σ_{lj} has some interesting consequences in dynamical situations. Strictly speaking the form of Σ_{lj} then needs changing but we can make the point intended without working out the details. Suppose that we are interested in the mean force exerted by a wave on an obstacle in a fluid. If the obstacle is fixed we naturally use Eulerian coordinates, and the force on it is the integral of the normal component of σ_{lj} over its surface. If we work to a linear approximation and the process is periodic the time-average of the term $\partial w/\partial u_{l,j}$ in (6.3) is zero and the force appears to be given by the mean value of the energy-momentum tensor Σ_{lj}, or rather its dynamical generalization. This result is useful because of its suggestive analogy with radiation pressure in electromagnetism, but at the same time physically misleading.

7. Closure

The methods of sections 3 and 4 can be applied to continua (see, e.g. [36]) more complicated than a simple elastic solid. We can, for example take (u_1, u_2, u_3) to be the displacement and (u_4, u_5, u_6) a director. If W is known, equation (3.5) or (3.6) or a higher-order analogue provides an energy-momentum tensor. (If there are constraints, incompressibility or constancy of the length of a director say, they need not be explicitly recognised provided that only solutions which conform with them are fed into the final formulas.) An argument on the lines of section 4 will then show that (4.2) with the appropriate P_{lj} gives the force on a defect in the sense outlined in the Introduction. At least, this is what the writer believes, but see reference [37].

The formalism of section 4 can be applied to time-dependent situations by introducing an extra independent variable $X_0 = t$ (u_i still has only three components) and interpreting $-W$ as the density of the appropriate Lagrangian, the kinetic energy minus the elastic energy. Some fairly interesting results may be derived for a medium with or without defects [28]. It ought to be possible to do the same for more complex continua.

Dissipative media present difficulties since the governing equations are not self-adjoint and can only be written in the form (3.1) at the expense of letting W depend on additional dummy field variables. However, the case of steady viscous flow with the inertial terms ignored can be handled with the help of Goodier's [38] analogy. If in a solution of the equations of linear isotropic elasticity for an incompressible medium we interpret displacement, stress and shear modulus as velocity, stress and viscosity we have the solution of a problem in slow viscous flow. The energy density becomes half the rate of dissipation of energy per unit volume. Correspondingly the integral (4.2) gives minus half the rate of change of the total rate of dissipation with respect to the change of position of a defect, now, say, a small rigid sphere perturbing the flow. Jeffery [39] studied the motion of a rigid spheroid in a viscous flow, ignoring the inertial terms. He found that after a long time the details of the motion still depended on initial orientation, whereas experiment showed that spheroids which started off with different orientations all ultimately settled down to one common motion with no memory of their initial conditions, presumably under the influence of the neglected terms. He suggested that the final motion was given by his solution with the initial orientation chosen so as to minimize[4] the final rate of dissipation. A similar principle has been invoked in some studies of the migration of solid particles in viscous flows. Unfortunately there seems to be no respectable physical principle which justifies this procedure. It is equivalent to requiring that for the actual ultimate motion the "force" (4.2) and the "couple" (5.1) be zero for a surface enclosing the solid. Even if the principle is incorrect its consequences might still be salvaged if one could show that those motions of the solid for which F_l and L_{kl} are zero, but not others, are, to the first order, unaltered when the small terms which have been left out are switched on again. The writer has tried hard to devise a proof of this conjecture, but so far without success, possibly because it is not true.

[4] At constant imposed strain rate. This is the same as maximizing it at constant imposed stress [40].

REFERENCES

[1] Noether, E., *Göttinger Nachr.* (*Math.-Phys. Klasse*) (1918) 235 (English translation by M. A. Tavel, *Transport Theory and Statistical Physics* 1 (1971) 183)
[2] Love, A. E. H., *The Mathematical Theory of Elasticity*. Cambridge University Press 1952
[3] Eshelby, J. D., *Phil. Trans.* A244 (1951) 87
[4] Eshelby, J. D. in *Prog. Solid State Physics* (ed. F. Seitz and D. Turnbull) Vol. 3, p. 79. Academic Press, New York 1956
[5] Nabarro, F. R. N., *Theory of Crystal Dislocations*. Oxford University Press 1967
[6] Rice, J. R., *J. appl. Mech.* 35 (1968) 379
[7] Rice, J. R., *Fracture* (ed. H. Liebowitz) Vol. 2, p. 101. Academic Press, New York 1968
[8] Bilby, B. A. and Eshelby, J. D., *Fracture* (ed. H. Liebowitz) Vol. 1 p. 99. Academic Press, New York 1968
[9] Atkinson, C. and Eshelby, J. D., *Int. J. Fracture Mech.* 4 (1968) 3
[10] Günther, W., *Abh. braunschw. wiss. Ges.* 14 (1962) 54
[11] Knowles, J. K. and Sternberg, Eli, *Arch. rat. Mech. Anal.* 44 (1972) 187
[12] Green, A. E., *Arch. rat. Mech. Anal.* 50 (1973) 73
[13] Wenzel, G., *Quantum Theory of Fields*. Interscience, New York 1949
[14] Pauli, W., *Rev. mod. Phys.* 13 (1941) 203
[15] Hund, F., *Materie als Feld*. Springer-Verlag, Berlin 1954
[16] Ivanenko, D. and Sokolov, A., *Klassicheskaya Teoria Polya*. Gostekhizdat, Moscow-Leningrad 1950
[17] Corson, E. M., *Introduction to Tensors, Spinors, and Relativistic Wave-Equations*. Blackie & Son, London and Glasgow 1953
[18] Davies, W. R., *Classical Fields, Particles and the Theory of Relativity*. Gordon and Breach, New York 1970
[19] Mitskevich, N. V., *Fizicheskie Polya v Obshcheĭ Teorii Otnositel'nosti*. Izdatel'stvo "Nauka", Moscow 1969
[20] Born, M. *Atomic Physics*. Blackie & Son, London and Glasgow 1946, p. 286
[21] Born, M., *Ann. Inst. Henri Poincaré* 7 (1937) 155
[22] Pryce, M. H. L., *Proc. Camb. phil. Soc.* 31 (1935) 50
[23] Neuber, H., *Kerbspannungslehre*. Springer-Verlag, Berlin 1958, p. 186
[24] Green, A. E. and Adkins, J. E., *Large Elastic Deformations*. Clarendon Press, Oxford 1960, p. 259
[25] Podolsky, B. and Kikuchi, C., *Phys. Rev.* 65 (1944) 228
[26] Chang, T. S., *Proc. Camb. phil. Soc.* 44 (1948) 76
[27] Thielheim, K. O., *Proc. Phys. Soc. Lond.* 91 (1967) 798
[28] Eshelby, J. D., *Inelastic Behavior of Solids* (ed. M. F. Kanninen et al.) McGraw Hill, New York 1970, p. 77
[29] Eshelby, J. D., *Prospects of Fracture Mechanics* (ed. G. C. Sih et al.) Noordhoff, Leyden 1975, p. 69
[30] Bessel-Hagen, E., *Math. Ann.* 84 (1921) 258
[31] Jackiw, R., *Physics Today*, January 1974, p. 23
[32] Rosenfeld, L., *Acad Roy. Belge*, 18 (1940) No. 6
[33] Belinfante, F. J., *Physica*, 6 (1939) 887
[34] de Wet, J. S., *Proc. Camb. Phil. Soc.* 43 (1947) 511
[35] Chadwick, P., private communication, 1974
[36] Green, A. E. and Rivlin, R. S., *Arch. rat. Mech. Anal.* 17 (1964) 113
[37] Atkinson, C., and Leppington, F. G., *Int. J. Fracture* 10 (1974) 599
[38] Goodier, J. N., *Phil. Mag.* 22 (1936) 678
[39] Jeffery, G. B., *Proc. roy. Soc.* A102 (1922) 161
[40] Eshelby, J. D., *Proc. roy. Soc.* A241 (1957) 376

THE CHANGE OF SHAPE OF A VISCOUS ELLIPSOIDAL REGION EMBEDDED IN A SLOWLY DEFORMING MATRIX HAVING A DIFFERENT VISCOSITY — SOME COMMENTS ON A DISCUSSION BY N.C. GAY

B.A. BILBY, J.D. ESHELBY, M.L. KOLBUSZEWSKI and A.K. KUNDU *
Department of the Theory of Materials, University of Sheffield, Sheffield (United Kingdom)

(Received May 7, 1976)

We are glad to see our results laid out in a handy form, although we would prefer our solution to be described as an exact one rather than one which is "more exact" than Gay's original solution which was, in fact, incorrect.

Gay (1976) has speculated on the behaviour of inhomogeneities in applied flows which contain a component of rotation-rate. We should like to emphasize that the interpretation of observations made in flows of this type requires particular caution. For example, it may be shown (Bilby and Kolbuszewski, unpublished) that in simple shear in two dimensions when $R > 2$, while some inhomogeneities elongate indefinitely and some oscillate between finite values of a/b, those having initially a value of a/b equal to $R/(R-2)$ and lying initially with their a axes along the direction of shear, *retain their original shape and orientation for all time.* Such an inhomogeneity appears *invariant* although the material inside (and outside) is in motion. (It might be possible to detect this motion by looking for changes of structure within the inhomogeneity.) Again, it is found that, for $R > 3.4$ in simple shear, some inhomogeneities may tumble, that is, not only deform but rotate completely, as a rigid one may do. Others exhibit the invariance or oscillatory behaviour already mentioned. There are similar phenomena in three dimensions.

Gay has recalculated some viscosity ratios using a procedure involving the linearized form (for small strains) of our two-dimensional equation (our equation 19). However, while there seems to be no reason why the same type of calculation could not be done using our general equation (13), Gay's use of these two-dimensional equations to make inferences on ellipsoidal inhomogeneities subjected to three-dimensional flows cannot really be justified. A full treatment must evidently await a general discussion of the ellipsoid deforming in three dimensions; some special cases have been considered in unpublished work by Brierley and Howard.

Finally we should like to alert the reader to a point associated with the limit $R \to 0$, where we have, perhaps misleadingly, referred to the inhomogeneity as a hole (Bilby et al., 1975). As is clear from our analysis we have

* A.K. Kundu is now at Berhampur College of Technology, West Bengal, India.

discussed in this paper only flows in which the volume of the inhomogeneity remains constant. Thus when $R \to 0$ we obtain the behaviour of an inhomogeneity consisting of an *incompressible inviscid fluid.* In this there will be, in general, a shape-dependent hydrostatic pressure, and a similar pressure would be required within the hole to which we refer if the equations given are to describe its deformation correctly. It may be shown that, when the inhomogeneity is of constant volume the applied hydrostatic pressure P^{A} at infinity and the hydrostatic pressure $P^{\mathrm{inh.}}$ in the inhomogeneity are related by:

$$p^{\mathrm{inh.}} - p^{\mathrm{A}} = -\frac{\mu}{2\pi}('e_{11}^T I_a + 'e_{22}^T I_b + 'e_{33}^T I_c)$$

where μ is the viscosity of the matrix and $'e_{ij}^T$, I_a, I_b and I_c are defined in the paper (Bilby et al., 1975). For the incompressible inviscid inhomogeneity in the form of an elliptic cylinder in a pure shear two-dimensional flow, the axes of the ellipse being parallel to those of the shear, this reduces to:

$$p^{\mathrm{inh.}} - p^{\mathrm{A}} = 2\mu \frac{a^2 - b^2}{a^2 + b^2} e_{11}^{\mathrm{A}}$$

where e_{11}^{A} ($= -e_{22}^{\mathrm{A}}$) is the applied strain rate along x_1.

REFERENCES

Bilby, B.A., Eshelby, J.D. and Kundu, A.K., 1975. The change of shape of a viscous ellipsoidal region embedded in a slowly deforming matrix having a different viscosity. Tectonophysics, 28: 265—274.

Gay, N.C., 1976. Discussion of paper by B.A. Bilby, J.D. Eshelby and A.K. Kundu. Tectonophysics, 35: 403—407.

Interaction and diffusion of point defects

J. D. Eshelby

Simple derivations are given for some results in the continuum theory of point defects; in particular the interaction between defects, both directly and through image (surface) effects. It is argued that these results have a wider validity than appears at first sight. The jump frequency for particles diffusing from one minimum to another of a rigid one-dimensional potential is derived by a simple argument which shows clearly the origin of the frequency factor, which turns out not to be, in any reasonable sense, a measure of the number of attacks on the barrier per unit time, as commonly stated. The method can quite easily be extended to give the jump frequency of a point defect or vacancy in three dimensions both when the effect of the host lattice is simulated by a rigid set of potential wells and also when, more realistically, it is not (Vineyard formula).

The author is at the University of Sheffield

Before treating, by request, the continuum theory of defects, the author would like to recall in defence of its obvious shortcomings an earlier remark[1] to the effect that 'the theory perhaps suffers from the disadvantage that its limitations are more immediately obvious than are those of other approximate methods which have to be used in dealing with the solid state', and add that a simple treatment at cottage-industry level may sometimes provide a certain degree of insight into some phenomenon or other when an exact calculation, based on a theoretical model which apes reality precisely, will give accurate answers but may be too unsurveyable, or too numerical, to give much insight.

In the next section the elastic interaction of defects is reviewed and in the following section some views are given on the right way to look at jump frequencies.

DEFECT INTERACTIONS

The interaction energy of a defect with a stress field can be worked out in macroscopic terms by considering the result of introducing a uniform distribution of many such defects into a stressed specimen. We suppose that the solution of defects is dilute enough for the interaction between them to be neglected.

Suppose that in a perfectly rigid block of material a badly made cavity has been cut which was meant to be a unit cube but is actually slightly too small and is also slightly parallelepipedal. If we force a unit cube of the elastic material we are interested in into the cavity the resulting contraction and shear will give it an elastic energy

$$W = W_d + W_s \tag{1}$$

Here W_d, the dilatational part of the energy density, is $\frac{1}{2}Ke^2$ where K is the bulk modulus and e the dilatation (actually negative) induced by the too-small cavity. Similarly the shear part of the energy density W_s induced by the small departure from cubic shape is $\frac{1}{2}\mu s^2$ where μ is the shear modulus and s a suitable measure of distortion which we need not write out. Next insert n defects, uniformly scattered, into the deformed cube in the cavity. The work done is nE_s, where E_s is the self-energy of one defect, plus n times the interaction energy E_{int} of a single defect with the imposed stress field.

Now start again with the undeformed cube. Insert the defects in it; the work done is nE_s. Force the cube into the cavity. The work required to do this will differ from (1) for two reasons. First, the introduction of the defects into the unconstrained unit cube changes its volume by, say, e' without producing any corresponding stress (compare the case of thermal expansion), so that now we have to write $W_d = \frac{1}{2}K(e-e')^2$. There would be a corresponding change in W_s if, contrary to our assumption, the defects changed the shape as well as the size of the cube (carbon or nitrogen in ferrite). Secondly, the macroscopic K and μ of the material are changed if the defects are in some sense elastically harder or softer than the matrix. Hence in all W increases by

$$C\frac{d}{dC}W=-CKe\frac{de}{dC}+\frac{1}{2}\frac{dK}{dC}e^2+\frac{1}{2}\frac{d\mu}{dC}s^2$$

$$=CP\frac{de}{dC}+\frac{1}{K}\frac{dK}{dC}W_d+\frac{1}{\mu}\frac{d\mu}{dC}W_s$$

where P is the imposed hydrostatic pressure and C is the concentration of defects. Both this quantity and $nE_{int.}=CE_{int.}/\Omega$ represent the work, over and above nE_s, required to establish the defect-filled cube in the cavity, and they must be equal; otherwise we should be in the energy-making business by loading up the easier way and unloading in the reverse of the harder way. Consequently we must have

$$E_{int.}=E_{mis.}+E_{inh.} \tag{2}$$

where the first term is

$$E_{mis.}=P\frac{de}{dC}\Omega=P\cdot\frac{3}{a}\frac{da}{dC}\Omega=P\,\Delta V \tag{3}$$

and the second is (5) below. The component (3), which is associated with the misfit in the sphere-in-hole model, can be written in terms of the imposed hydrostatic pressure P and either the rate of change of lattice parameter with defect concentration or the volume change ΔV associated with the introduction of a single defect.

This is perhaps the point to comment on the fact that in his original calculation Cottrell[2] found the misfit interaction by calculating the work done in 'blowing up' the defect against the pressure P and arrived at

$$E_{mis.}=P\,\Delta V^{\infty} \tag{4}$$

instead of (3), where ΔV^{∞} is a certain volume less than ΔV and related to it by (7) and (10) below. This is plausible because, as we shall see shortly, ΔV^{∞} is the volume change of any surface *closely* surrounding the defect. The standard calculation of Bilby[3] used a sphere elastically homogeneous with the matrix whereas Cottrell used a rigid sphere, and so the difference between (3) and (4) is commonly said to be due to the elastic energy in the defect. This is untrue. Attention to two quite unsophisticated points, both of which increase the work of insertion, will raise (4) by the necessary factor γ. The first is that the volume of a hole is reduced by a factor $(1-P/K)$ when the body containing it is subjected to an external pressure P, supposing that there is also a pressure P inside the hole. (If not, the volume is even smaller.) The second is that the internal pressure has to be increased somewhat above P to expand the hole so that the (rigid) sphere may be inserted. (The 'bulk modulus for blowing up a hole' is, in fact, $4\mu/3$.)

The second term in (2),

$$E_{inh.}=\frac{1}{K}\frac{dK}{dC}\Omega W_d+\frac{1}{\mu}\frac{d\mu}{dC}\Omega W_s \tag{5}$$

takes account of the defect in its role as elastic inhomogeneity.

To find the interaction between two defects we must calculate the elastic fields which have to be inserted in (3) or (5). If we require the displacement around a defect to be purely radial, spherically symmetric, and to decrease with increasing distance r from the defect, the equations of isotropic linear elasticity force it to be proportional to $1/r^2$,

$$U_R=\frac{c}{r^2}\quad\text{or}\quad u_i=c\frac{x_i}{r^3}=-c\frac{\partial}{\partial x_i}\left(\frac{1}{r}\right) \tag{6}$$

where the constant c measures the strength of the defect. Equation (6) has the same form as the electric field of a point charge and so its divergence is zero, that is to say the dilatation $e=\operatorname{div}\mathbf{u}$ around the defect is zero. However this does not mean that there is no volume change associated with the defect. In fact a sphere of radius r suffers an increase in volume

$$\Delta V^{\infty}=4\pi r^2\,.\,c/r^2=4\pi c \tag{7}$$

when the defect is introduced at its centre. If the spherical surface is deformed into a new surface of arbitrary shape embracing the defect the volume change is still (7) because throughout the volume between the original sphere and the deformed surface the dilatation is zero.

Although (7) is the volume change associated with any surface surrounding the defect it is not the whole of the volume change when the defect is introduced into a finite body with a stress-free surface. At the surface of a sphere round the defect there is a radial strain dU_R/dr accompanied by two equal transverse strains of half this magnitude and opposite sign so as to give zero dilatation. Thus the radial and transverse strains at distance r are

$$-2c/r^3,\qquad c/r^3,\qquad c/r^3 \tag{8}$$

and the corresponding stresses are, by Hooke's law,

$$-4\mu c/r^3,\qquad 2\mu c/r^3,\qquad 2\mu c/r^3 \tag{9}$$

Thus if a sphere of radius r is cut out a normal radial inward force, i.e. a pressure, of amount $4\mu c/r^3$ must be supplied if the original elastic field (6) is to be maintained. If this unwanted pressure is relaxed to give a stress-free surface there will be an additional 'image' volume increase

$$\Delta V^{I}=\frac{4\pi}{3}r^3\,.\,\frac{4\mu c}{r^3}\,.\,\frac{1}{K}=4\pi c\,.\,\frac{4\mu}{3K} \tag{10}$$

or in all a volume change

$$\Delta V=\Delta V^{\infty}+\Delta V^{I}=4\pi c\gamma$$

with

$$\gamma=1+\frac{4\mu}{3K}=3\frac{1-\nu}{1+\nu}$$

where ν is Poisson's ratio. Like (7) the extra volume change (10) is independent of the shape and size of the solid or the position of the defect in it, though this is rather more difficult to show. One way is to use the sphere-in-hole model plus the general result,[4] obvious to some, that the average, or volume integral, of any stress-component over a self-stressed body with a stress-free surface is zero. Applied to the hydrostatic pressure this means that the self-stress produces no change in the volume of the material if Hooke's law holds and the material is elastically homogeneous. Let the body be the sphere plus the solid with the hole to receive it (the theorem does not require the body to be all in one piece). Before assembly the volume of material is the volume inside the outer surface, minus the volume of the hole, plus the volume of the sphere, that is, the volume inside the outer surface plus the misfit volume $V_{mis.}$ by which the sphere exceeds the hole. After the sphere has been forced into the hole the volume of material is just the volume inside the outer surface. As the volume of material has not changed, the increase in the volume inside the outer surface must be equal to $V_{\text{mis.}}$ In addition to showing that ΔV and its parts ΔV^{∞}, ΔV^{I} are independent of the shape and size of the body and the position of the defect within it, this argument shows that $\Delta V = 4\pi c\gamma$ is actually equal to the misfit volume $V_{\text{mis.}}$ For a defect at the centre of a sphere the extra 'image' elastic field associated with ΔV^{I} is a uniform hydrostatic pressure

$$\bar{P} = -K\frac{\Delta V^{\text{I}}}{V} = -16\pi\mu c/3V \tag{11}$$

which is actually a tension for positive c or ΔV, inversely proportional to the volume of the specimen.

In considering the interaction between defects we can usually ignore the modification of the defect elastic fields by image terms (but see below). Then, since each defect produces no hydrostatic pressure the misfit interaction (3) between them is zero, a result apparently first noticed by Bitter.[5] To get the inhomogeneity interaction (5) we need to know the effective elastic constants of a dilute suspension of spheres of elastic constants K', μ' in a matrix with constants K, μ. They have been worked out by many people; the first to arrive at both K and μ correctly seems to have been Dewey.[6] If the Poisson's ratio of the matrix is 1/5 we have the easily memorized results

$$\alpha_K \equiv \frac{1}{K}\frac{dK}{dC} = \frac{K'-K}{\frac{1}{2}(K'+K)}, \qquad \alpha_\mu \equiv \frac{1}{\mu}\frac{d\mu}{dC} = \frac{\mu'-\mu}{\frac{1}{2}(\mu'+\mu)},$$

difference over average; though $\nu = 0{\cdot}20$ is rather small to be typical these expressions are not far out for, say, $\nu = \frac{1}{3}$ or $\nu = \frac{1}{4}$. For the elastic field (6) W_d is zero and W_s is half the sum of the products of the three quantities (8) by the three quantities (9), each to each, which gives

$$W_s = 6\mu c^2/r^6 \tag{12}$$

and hence

$$E_{\text{inh.}} = 6\alpha_\mu \mu c^2 \Omega/r^6$$

for the interaction of one purely misfitting defect with one purely inhomogeneous one. If they are both inhomogeneous and misfitting the result is doubled. Generally, for two different defects 1 and 2 we have

$$E_{\text{inh.}} = 6(\alpha_{\mu 1}c_2^2 + \alpha_{\mu 2}c_1^2)\Omega/r^6 \tag{13}$$

According to the estimate above, α_μ is -2 for a hole and $+2$ for an unshearable ('rigid') inclusion. It is not unreasonable to expect a vacancy to behave like a hole, not necessarily of volume Ω, or at any rate as an elastically soft spot. One might expect interstitials to act as hard spots and put up the bulk modulus (which does not interest us here), but it is not intuitively obvious that they would also put up the shear modulus. Indeed one might perhaps feel that the excessive distortion around the interstitial would allow the neighbouring atoms to as it were slip round it under shear. In that case a more appropriate boundary condition might be a freely slipping interface rather than bonding between sphere and hole. The bulk shear modulus has been worked out for this case, but only for a Poisson's ratio $\frac{1}{2}$.[7] It gives $\alpha_\mu = 1$ in place of the value $\alpha_\mu = 2{\cdot}5$ for bonding, so that, on this model also, an interstitial is still a 'hard' defect. What we have said suggests that vacancies should attract one another, and interstitials repel. For the interaction of an interstitial 1 with a vacancy 2 we should expect to have $\alpha_{\mu 1}$ moderate and positive, c_1^2 large, $\alpha_{\mu 2}$ moderate to large in magnitude and negative, c_2^2 small, leading to attraction between the pair since the second term in (13) will dominate.

The result that the misfit interaction (3) between two defects is zero because each looks for a hydrostatic pressure which its colleague does not provide is peculiar to the isotropic misfitting sphere model and can easily be upset. One way is to make the material anisotropic. Another is to make the defect less symmetrical, a misfitting ellipsoid in a spherical hole say, representing a split interstitial or carbon and nitrogen in alpha-iron. Either modification gives an interaction energy proportional to r^{-3} times the appropriate angular factor. However, we shall only consider here defects with cubic symmetry in a cubic crystal.

Before going over to anisotropy we shall see if anything fresh can be obtained from an isotropic model. As an alternative to the misfitting inclusion model we can imagine that the elastic field of the defect is produced by a small cluster of point forces, with zero resultant and couple, representing, approximately, the force which neighbouring atoms would feel if they refused to move when the defect was introduced. The i-component of displacement produced by unit force at the origin, directed parallel to the x_k-axis is

$$U_{ik} = \frac{1}{4\pi\mu}\frac{\delta_{ik}}{r} - \frac{1}{16\pi\mu(1-\nu)}\frac{\partial^2 r}{\partial x_i \partial x_k} \tag{14}$$

Hence if, say, a cluster of six forces, all of magnitude f, act radially outwards at the face centres of a cube of side a_0, forming three crossed double forces, the displacement for large r/a_0 is

$$u_i = -fa_0\frac{\partial U_{ik}}{\partial x_k} \tag{15}$$

which is identical with (6) if the moments fa_0 of the double forces are related to the defect strength c by

$$fa_0 = 4\pi K\gamma c \tag{16}$$

Equation (15) is strictly valid for all r only if we let the force dipoles become genuine point dipoles, that is, if we make the transition $a_0 \to 0$, $f \to \infty$ in such a way that (16) is satisfied even in the limit. However, if we take the model seriously we ought to let a_0 be finite and of atomic dimensions. As all we shall need is the dilatation we might as well find it directly. The dilatation corresponding to (14) is

$$e = \frac{1}{4\pi K\gamma}\frac{\partial}{\partial x_k}\left(\frac{1}{r}\right)$$

Thus the forces $\pm f$ at $(x_1 \mp \frac{1}{2}a_0, x_2, x_3)$ contribute an amount

$$\frac{f}{4\pi K\gamma}\frac{\partial}{\partial x_1}\{|\mathbf{r} - \tfrac{1}{2}a_0\mathbf{i}|^{-1} - |\mathbf{r} + \tfrac{1}{2}a_0\mathbf{i}|^{-1}\}$$

to the dilatation, where $\mathbf{i}$ is the vector (1, 0, 0). If we expand the quantities in { } in a Taylor series and add the contributions from the other two dipoles we get

$$e = -c\left[\nabla^2 + \frac{1}{24}a_0^2\left(\frac{\partial^4}{\partial x_1^4}+\frac{\partial^4}{\partial x_2^4}+\frac{\partial^4}{\partial x_3^4}\right)+\ldots\right]\frac{1}{r}$$

for the dilatation due to the defect, or, dropping the term $-c\nabla^2(1/r) = 4\pi c\delta(r)$ which just represents a blob of expansion at the origin, and the unwritten higher terms,

$$e = -\frac{35}{8}ca_0^2\Gamma r^{-5} \tag{17}$$

where the angular factor

$$\Gamma(l, m, n) = l^4 + m^4 + n^4 - \tfrac{3}{5} \tag{18}$$

depends on the direction cosines

$$l = x_1/r, \qquad m = x_2/r, \qquad n = x_3/r$$

It is easy to see that any centrally symmetric cluster of forces will give a dilatation proportional to r^{-5}. Since e is harmonic the angular factor Γ must then be a surface harmonic of order 4 and there is only one such which has cubic symmetry, namely (18). The original sphere-in-hole model can be modified to give (17). Replace the misfitting sphere by a water-worn cube or octahedron which has become nearly, but not quite, a sphere. Apply surface forces to make it a sphere the size of the hole, and cement it into the hole. At this stage there is no stress outside the defect, but there is an unwanted cubically symmetric layer of body force at the interface. To remove it we may apply an equal and opposite layer which, as we have seen, induces a dilatation of the form (17) with some value for the constant ca_0^2. Thus (17) gives the first non-vanishing term in the expansion of the dilatation in negative (odd) powers of r for any model of the defect which allows it to manifest its finite size and cubic symmetry. Such a model involves a pair of constants, say the volume change $\Delta V = 4\pi\gamma c$ and ca_0^2 which controls the magnitude of (17), and we can characterize the defect by specifying the pair of numbers $(\Delta V, a_0^2)$. We choose a_0^2 rather than a_0 because in some cases a_0^2 may be negative. An example is a set of forces directed outward from the corners rather than the face centres of the small cube. To see this apply the corner and face forces together. We are approaching the spherically symmetric situation of the misfitting sphere with zero dilatation. Hence the dilatations produced by the corner and face forces must partly cancel. With the $(\Delta V, a_0^2)$ notation our original spherically symmetric defect evidently has to be denoted by $(\Delta V, 0)$.

According to (3) the interaction energy of an ordinary $(\Delta V, 0)$ defect with the dilatation (17) is $-K\Delta V$ times (17). If both defects are of type $(\Delta V, a_0^2)$ each feels the dilatation of the other and the effect is doubled, giving the interaction energy

$$E_{\text{mis.}} = \frac{35}{16\pi}\frac{K}{\gamma}(\Delta V)^2 a_0^2 \frac{\Gamma}{r^5} \tag{19}$$

An interaction energy with this radial and angular dependence and about the same magnitude, was originally presented by Hardy and Bullough[8,9] on the basis of a lattice calculation. As a result the Γ/r^5 interaction is commonly said to be an essentially lattice effect, described variously as involving non-local forces, phonon dispersion curves, or the lattice Green's function. However, to derive (19) the only concession we had to make to the lattice was to admit that the defect was of finite size and had cubic symmetry. To make our calculation comparable with Hardy and Bullough's we should in addition have to replace (14) by the expression for the displacement of one atom due to a force applied to another. According to Siems[10] the effect of doing this would be (for the case considered in ref. 8) to reduce (19) by 25%, so that the Γ/r^5 interaction in a lattice could fairly be described as an essentially continuum effect somewhat reduced by the lattice structure of the crystal.

To finish our discussion of the direct interaction between defects we drop the refinements which led to (19) and ask what effect the introduction of cubic anisotropy has on the original model. For a cubic crystal one of the elastic equilibrium equations is

$$c_{44}\nabla^2 u_1 + (c_{12}+c_{44})\frac{\partial e}{\partial x_1} + d\frac{\partial^2 u_1}{\partial x_1^2} = 0 \tag{20}$$

where

$$d = c_{11} - c_{12} - 2c_{44}$$

is zero for isotropy. If we differentiate (20) with respect to x_1 and add to it the two similar equations in u_2, u_3 we obtain

$$(c_{12}+2c_{44})\nabla^2 e + d\left(\frac{\partial^3 u_1}{\partial x_1^3}+\frac{\partial^3 u_2}{\partial x_2^3}+\frac{\partial^3 u_3}{\partial x_3^3}\right) = 0$$

so that e is no longer necessarily harmonic. If d is small (nearly isotropic material) we may, with an error of order only d^2, use the isotropic value of u_i, $-c\partial(r^{-1})/\partial x_i$ in the present case, in the terms involving d. If in addition we write r^{-1} as $\nabla^2(\frac{1}{2}r)$ we get

$$\nabla^2\left\{(c_{12}+2c_{44})e - \tfrac{1}{2}cd\left(\frac{\partial^4}{\partial x_1^4}+\frac{\partial^4}{\partial x_2^4}+\frac{\partial^4}{\partial x_3^4}\right)r\right\} = 0 \tag{21}$$

which says that { } is harmonic, i.e.

$$e = \frac{\frac{1}{2}cd}{c_{12}+2c_{44}}\left(\frac{\partial^4}{\partial x_1^4}+\frac{\partial^4}{\partial x_2^4}+\frac{\partial^4}{\partial x_3^4}\right)r + h(x_1, x_2, x_3)$$

where the harmonic function h must have no singularities, not even at $r = 0$ (one there would alter the magnitude or nature of the elastic singularity) and so we can leave it out, as merely being some applied field unrelated to the defect. Hence, working out the derivatives,

$$(c_{12}+2c_{44})e = -(15/2)cd\Gamma(l, m, n)r^{-3} \tag{22}$$

with the Γ of equation (18). The angular factors in (17) and (22) are the same not for any deep physical reason, but merely because of the dearth of suitable functions. From (21) one can see that $\nabla^2\nabla^2 e = 0$, i.e. e is a biharmonic function, so that by a standard theorem it has the form $h_1 + r^2 h_2$ with harmonic h_1, h_2. With an r^{-3} dependence we must have $h_1 = A(l, m, n)r^{-3}$, $h_2 = B(l, m, n)r^{-5}$, and we have already seen that, with cubic symmetry, $A = 0$, $B = \text{const.}$ Γ is the only possibility. According to (3) the interaction between two defects via their dilatation fields is

$$E_{\text{mis.}} = 30\pi dc^2\frac{\Gamma}{r^3} = \frac{15}{8\pi\gamma^2}(\Delta V)^2 d\frac{\Gamma}{r^3} \tag{23}$$

If d is not small the interaction is still proportional to r^{-3} and to $(\Delta V)^2$ but $d\Gamma$ is replaced by a power series in d whose successive coefficients are increasingly complicated functions of l, m, n.

The finite-size interaction (19) beats the anisotropy interaction (23) if, roughly,

$$r^2/a_0^2 < d/2K$$

For highly isotropic aluminium with $d/K \sim 0{\cdot}06$ this gives $r < 6a_0$. For most other cubic crystals where d/K is of the order of unity we must have $r < a_0$, so that if a_0 is in fact something like a lattice constant the anisotropy interaction will always win. (Of course we are cheating a little here—for $d/K \sim 1$ equation (23) is not a very good approximation.)

The indirect interaction between defects via their image fields, which we have so far ignored, may become important when there are a number of them, when say the 'defects' are actually the minor constituent in a binary alloy. In that case the image term makes a (negative) contribution to the strain energy per defect which is comparable with the elastic self-energy of a defect. To see this we must first calculate the self-energy. According to (12) the energy in the matrix round a defect is

$$\int_r^\infty \frac{6\mu c^2}{r^6} \cdot 4\pi r^2 \, dr = \frac{8\pi\mu c^2}{r^2} \tag{24}$$

On the other hand, since the matrix expands to partially accommodate it the sphere suffers a volume change $V_{\text{mis.}} - 4\pi c = 4\pi c(\gamma - 1)$, so that the elastic energy of the sphere is

$$\tfrac{1}{2}K[4\pi c(\gamma-1)]^2/\Omega \tag{25}$$

The sum of (24) and (25) gives E_s, the self-energy of a defect:

$$E_s = (8\pi\mu c/3)\,\Delta V$$

By the time a concentration C has been introduced an image pressure (actually a tension for positive c)

$$P(C) = -(16\pi\mu c/3)C \tag{26}$$

equal to (11) times the number of defects in the volume V, will build up linearly with C, so that by (3) the average energy expended in inserting all the defects introduced so far is $E_s + \frac{1}{2}P(C)\,\Delta V$ per defect, which is just $E_s(1-C)$ because of the similarity of the coefficients in E_s and $P(C)$. Consequently the strain energy per atom of alloy takes the simple form

$$E = E_s C(1-C) \tag{27}$$

characteristic of the heat of solution of a regular solution. It is not surprising that (27) looks sensible not only near $C=0$ but also near $C=1$, for there it is just what we should have got from a model in which atoms of the former major constituent were inserted, with minus the former volume misfit, into a matrix of the former minor constituent, but that it also looks sensible near $C=\frac{1}{2}$ is perhaps a matter of luck. However, (27) or some complication of it, has been used to discuss strain energy in alloy theory. One can, for example, 'derive' Hume-Rothery's 15% rule from it.[1] Here we shall just use it to talk about the strain energy associated with non-uniform solute distribution as, for example, in spinodal decomposition. Begin with an AB alloy with a concentration C of A. At first it is homogeneous but then it segregates into, say, cubical regions which are alternatively A-rich and A-lean. If we cut the specimen up along the cube boundaries the A-rich cubes will expand and the others will contract, supposing that da/dC is positive. With sufficient care we could get useful work from these expansions and contractions, and so there is strain energy associated with the non-uniform solute distribution. On the other hand, one can see that it requires no work (beyond the osmotic work done in changing the configurational entropy) to rearrange the solute. In fact as we have seen the defects only interact via their image fields. Clearly the fine-scale rearrangement we have postulated will leave the image hydrostatic tension uniform and unaltered in magnitude in the bulk of the material, and so there is no change in the image interaction. The same is true for any other type of segregation (blobs, stripes etc.) which does not produce macroscopic deformation of the specimen. We can use these facts to avoid the difficulties of a direct calculation of the strain energy. As the energy of the specimen with fluctuating composition and coherent lattice is unaffected by redistributing the solute uniformly, we must have, by (27)

$$E_{\text{coh.}} = E_{\text{hom.}} = E_s\bar{C}(1-\bar{C}) \tag{28}$$

where $\bar{C}$ is the average composition. On the other hand, if we divide the specimen into cubes small enough for the composition to be substantially constant and the stress to be relaxed in each, the energy is

$$E_{\text{relax.}} = E_s\overline{C(1-C)} \tag{29}$$

The elastic strain energy released is (28) minus (29):

$$E_{\text{strain}} = -E_{\text{incoh.}} = E_s(\overline{C^2} - \bar{C}^2) = E_s\overline{(C-\bar{C})^2} \tag{30}$$

On the other hand, the energy required to make an incoherent block (or a collection of loosely assembled infinitesimal cubes) from a uniform block with the same mean composition, or from pure A and B, contains a term equal to (29) minus (28), in its role as $E_{\text{hom.}}$ rather than $E_{\text{coh.}}$, that is, as indicated, (30) with reversed sign. (This decrease is of course exactly cancelled by the strain energy if the blocks are forced to fit together coherently.) Rather paradoxically the energy required to make an inhomogeneous incoherent alloy from pure A and B depends not only on the mean composition $\bar{C}$, but also on its distribution, whereas the energy required to make an inhomogeneous coherent alloy depends only $\bar{C}$.

Since (3) and (5) were derived for the interaction of a defect with a uniform externally applied elastic field it is not entirely obvious that they will work when the 'applied' field is due to another defect, unless we take the anthropomorphic view that the defect cannot distinguish between one field and another. Doubts of this kind can be more or less set at rest by invoking the idea of the force on a defect. If $E_{\text{int.}}$ changes by

$$\delta E_{\text{int.}} = -F_x\,\delta x + 0(\delta x^2) \tag{31}$$

when the defect is shifted by δx parallel to the x-axis, then F_x is, by definition, the x-component of the force on the defect. It is equal to the integral

$$F_x = \int_S \left(W n_x - \frac{\partial \mathbf{u}}{\partial x} \cdot \mathbf{T} \right) dS \tag{32}$$

Here S is a closed surface drawn in the undeformed material and enclosing the site of the defect, n_x is the normal (before deformation) to the element dS, $\mathbf{T}$ dS is the traction on dS and W is the energy density per unit undeformed volume. The writer has devised a number of proofs of (32) of varying generality and simplicity; for the best so far see ref. 11. Evaluation of (32) and similar expressions for F_y, F_z will verify any of the interaction energies we have exhibited, to within an additive constant, of course. However, they will do more. Equation (32) is valid for finite deformation and any non-Hookian behaviour, and, further, it is surface-independent in the sense that it is not affected by deforming S, provided S does not pass through some other singularity. For two defects a distance $2R$ apart evaluate F_x over a sphere S_R of radius R about one of them. If R is large enough linear elasticity will be adequate on S_R. Contract S_R into a small sphere about the defect, and evaluate the integral there. As the theory is linear we can pick out the cross term between the fields due to the two defects, and they turn out to be consistent with (3) and (5). This means that (3) and (5) are valid as long as we can separate the defects by some surface where linear elasticity is reasonably valid; the fact that that theory is hopeless near either defect is irrelevant. It also shows that in, say, (3) P is not the hydrostatic pressure that the second defect produces at the centre of the first (a more or less meaningless concept), but rather the pressure calculated at the site of the first defect from the elastic data on S_R and the linear theory. From its derivation[11] one can see that (31) is still valid when the specimen is a lattice of balls connected by springs; the second term in (32) then just comes from the elements of S which are cut through by springs. The only difficulty is that δx must be an integral number of lattice parameters, so that the term $0(\delta x^2)$ cannot be reduced indefinitely, but this does not matter if F_x varies only slowly with position.

JUMP FREQUENCIES

Figure 1 shows the conventional picture of a particle in a one-dimensional potential well of depth U, and the following equation,

$$f = \nu e^{-U/kT}, \tag{33}$$

gives the usual expression for the jump frequency over the barrier B of height U. Here ν is the frequency of vibration of the particle at the bottom of the well A, and the usual rationalization of (33) is that ν gives the number of attempts in unit time, and the exponential the probability of one particle having enough energy to get to the top. The writer has never felt happy with this argument. The one-dimensional version of the treatment which follows was originally given in an article written in 1966,[12] together with a statement that its extension to two or more dimensions was difficult. In the nine years which have elapsed between then and publication the writer has decided that this is not so, and the argument is here extended in outline to two, three or N dimensions.

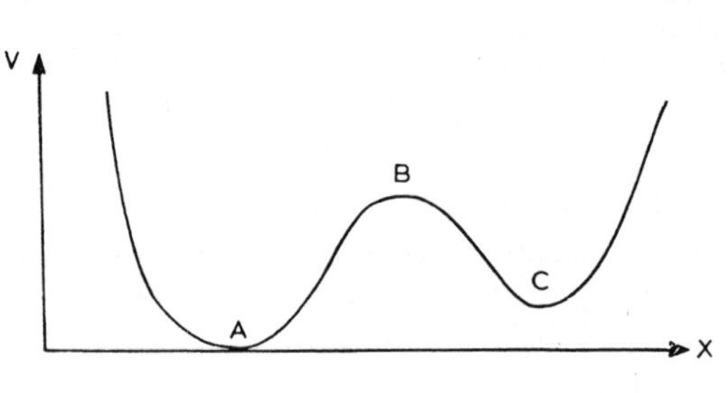

1 For explanation see text

It is a great help if we can imagine that there is a large number n of non-interacting particles in the well A, so that we may talk about the number which leave A via B in unit time rather than about the probability that a single particle leaves. To be able to do this we set up an ensemble in the sense of statistical mechanics, that is, a large number of replicas of Fig. 1 each containing only one particle and all in thermal equilibrium with one another and with a heat bath at temperature T. With the aid of a multiple-headed periscope we then superpose images of the members of the ensemble to get the required picture of a well apparently containing many particles. Or, begging some basic questions, we may take a very long film of the well with a single particle, cut it into many small lengths, print them on top of one another and run the resulting shorter film which will apparently show a well occupied by many, but mutually ignoring, particles.

To derive (33) for one dimension we only need two facts. First, if we write the number of particles between x and $x+dx$ as $\rho(x)\,dx$ then

$$\rho(x) = \text{const. } e^{-V(x)/kT}$$

where $V(x)$ is the potential at x (Boltzmann's barometric formula). Secondly the mean velocity in any given direction, $\bar{v}$ say, is independent of position (Maxwell distribution). We do not actually need to know that, for a particle of mass m, $\bar{v}$ is some multiple of $\surd(kT/m)$.

The flux of particles ϕ at a point in one direction, say left to right, is equal to the linear density of particles moving in the right direction, half the total, times their mean velocity. Evidently the jump frequency we are looking for may be written

$$f = \phi(B)/n \tag{34}$$

where n is the number of particles in the well A and ϕ(B) is the left-to right flux at B. We have

$$\phi(A) = \tfrac{1}{2}\rho(A)\bar{v}(A), \qquad \phi(B) = \tfrac{1}{2}\rho(B)\bar{v}(B)$$

which with $\bar{v}(A) = \bar{v}(B)$ and

$$\rho(B) = \rho(A)\, e^{-U/kT}, \tag{35}$$

gives

$$\phi(B)/\phi(A) = e^{-U/kT} \tag{36}$$

We can get a value for ϕ(A) very simply. If the particles oscillate with frequency ν in the well A each one crosses the bottom of the well from left to right once every $1/\nu$ seconds, so that ϕ(A) is just $n\nu$, and so (36) becomes

$$\phi(B) = \phi(A)\, e^{-U/kT} = n\nu\, e^{-U/kT} \tag{37}$$

which in view of (34) is equivalent to (33). From the writer's point of view (37) gives the correct interpretation of (33): the exp($-$U/kT) is there because the fluxes at B and A are obliged to be in this ratio, and the ν is there because the flux ϕ(A) is proportional to ν as well as to n.

It is easy to see through the trick whereby we avoided the usual integrations of statistical mechanics. The

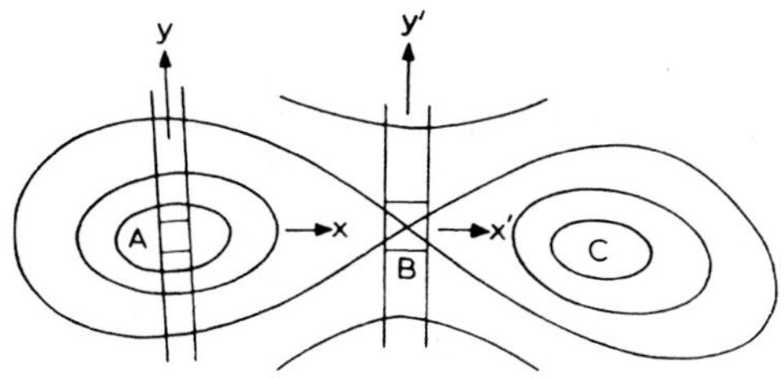

2 For explanation see text

ensemble can be regarded as a set of particles describing undisturbed orbits in phase space, with a certain distribution over energies. Of course particles are continually being thrown out of one orbit by thermal agitation, but as an equal number is being thrown into it this does not matter. Consequently we can combine purely mechanical results such as $\phi(\mathrm{A}) = n\nu$ with statistical results such as (36) in a way which, with luck, will save us trouble.

Figure 2 shows the energy contours for a particle moving in two dimensions, and we want the flux of particles out of well A into well C by way of the saddle point B.

In the figure we have drawn a narrow strip embracing the y-axis. To avoid a large number of dx's we shall take the strip to be of unit width and choose our unit of length so small that nothing varies noticeably across the width of the strip. Also shown is a similar strip about the x-axis, or rather, merely the part of it which forms a square with the first strip, which is all that will concern us.

The x and y coordinates of a particle near the bottom of the well oscillate independently with frequencies ν_x, ν_y. The flux ϕ_x of particles across $x = 0$ can be written in two ways. Each of the n particles in the well crosses $x = 0$ in a given sense once every $1/\nu_x$ seconds. Alternatively ϕ_x is n(strip), the number of particles in the strip, times their mean velocity $\bar{v}$ times $\frac{1}{2}$, since half of them are moving the wrong way. So

$$(\phi_x =)\nu_x n = \tfrac{1}{2}\bar{v}n(\text{strip}) \tag{38}$$

Similarly the flux of particles across the unit length of the x-axis intercepted by the strip is either the number of particles n(square) in the square, times $\frac{1}{2}\bar{v}$, or it is the number of particles in the strip times the frequency ν_y with which they oscillate parallel to the y-axis. Of course a good many of them will drift sideways out of one side of the strip before they reach the x-axis, but as an equal number will drift in through the other side sideways drift is without effect. So we have

$$\nu_y n(\text{strip}) = \tfrac{1}{2}\bar{v}n(\text{square}) = \tfrac{1}{2}\bar{v}\rho(\mathrm{A}) \tag{39}$$

where, because of our choice of unit of length n(square) is equal to the particle density $\rho(\mathrm{A})$ near the bottom of the well. From (38) and (39) we can eliminate n(strip):

$$n = (\tfrac{1}{2}\bar{v})^2\rho(\mathrm{A})/\nu_x\nu_y \tag{40}$$

If we construct similar strips about the y' and x' axes at the saddle-point we get by the same type of argument

$$\phi_{x'} = * = \tfrac{1}{2}\bar{v}n'(\text{strip}) \tag{41}$$

in place of (38). The equations (38) and (41) are quite analogous, but, as the brackets indicate, we shall make no use of the fact that (38) is equal to ϕ_x, and in (41) the * indicates that there is no term $\nu_{x'}n'$ since there is no n' (a saddle-point will not hold water). The analogue of (39) is

$$\nu_{y'}n'(\text{strip}) = \tfrac{1}{2}\bar{v}n'(\text{square}) = \tfrac{1}{2}\bar{v}\rho(\mathrm{B}) \tag{42}$$

From (41) and (42) we get

$$\phi_{x'} = (\tfrac{1}{2}\bar{v})^2\rho(\mathrm{B})/\nu_{y'},$$

from which, using (40) and (35),

$$f = \frac{\phi_{x'}}{n} = \frac{\nu_x\nu_y}{\nu_{y'}}e^{-U/kT} \tag{43}$$

Next take Fig. 2 to represent a section through a three-dimensional potential diagram. Mark out planes $x = \pm\frac{1}{2}$ defining a slab whose trace is the strip about the y-axis, then planes $y = \pm\frac{1}{2}$ which intersect the slab in a square rod whose trace is the square at the centre of the well, and finally planes $z = \pm 1$ which cut out a unit cube from the rod. We can now find relations between n(slab), n(rod), n(cube), and analogous quantities for the saddle-point, just as we related n(strip), n(square) n'(strip), n'(square) in two dimensions. It is fairly obvious that in place of (33) or (43) we shall come up with

$$f = \frac{\nu_x\nu_y\nu_z}{\nu_{y'}\nu_{z'}}e^{-U/kT} \tag{44}$$

(For the details put $N = 3$ in (47), (48), (49) below.)

The model of, say, an interstitial particle moving in a fixed potential $V(x, y, z)$ is not realistic, as the lattice producing V deforms as the particle moves. In other cases, vacancy movement or the re-orientation of a split interstitial for example, it may not even be clear what the appropriate defect coordinate is. We then have to treat the whole lattice together. To conclude we shall outline a derivation of Vineyard's classic equation[13,12] (equation (49) below) by a generalization of the method used for (43).

The potential energy of a crystal containing $\frac{1}{3}N$ atoms may be plotted in N-dimensional space. Near a local minimum linear contributions q_i of the coordinates may be chosen so that the potential and kinetic energies take the forms

$$V + \text{const.} = \tfrac{1}{2}\Sigma k_i q_i^2, \qquad T = \tfrac{1}{2}\Sigma m\dot{q}_i^2 \tag{45}$$

with all the k_i positive, and near a saddle-point

$$V + \text{const.}' = \tfrac{1}{2}\Sigma k_{i'} q_{i'}^2, \qquad T = \tfrac{1}{2}\Sigma m\dot{q}_{i'}^2 \tag{46}$$

with, say, $k_{1'}$ negative and the rest positive. The state of the crystal is described by a representative point wandering about in q-space. To find the rate at which the system jumps out of the minimum via the indicated saddle-point we introduce an ensemble of representative points and carry on as in the derivation of (43). The number of representative points in the q-space volume element $dq = dq_1\,dq_2 \ldots dq_N$ is proportional to $\exp(-V(q)/kT)dq$. Also the rate of change of any of the q_i has a constant mean value, $\bar{v}$ say (Maxwell distribution). This is a result of the fact that we have arranged for the 'mass' coefficients in (45) to be equal to each other and to those in (46). The only difficulty is one of N-dimensional visualization.

After choosing a sufficiently small unit of length in q-space (cf. the two-dimensional case) mark out successively slabs bounded by the planes $q_1 = \pm\frac{1}{2}$, $q_2 = \pm\frac{1}{2}\ldots, q_i = \pm\frac{1}{2}\ldots q_N = \pm\frac{1}{2}$ and let n_i denote the number of representative points in the box R_i formed by the intersection of the first i slabs. Then as an extension of (36) we have

$$\begin{aligned}(\phi_1 =)\nu_1 n &= \tfrac{1}{2}\bar{v}n_1\\ \nu_i n_{i-1} &= \tfrac{1}{2}\bar{v}n_i\\ \nu_N n_{N-1} &= \tfrac{1}{2}\bar{v}n_N = \tfrac{1}{2}\bar{v}\rho(A);\end{aligned} \tag{47}$$

as in two dimensions sideways leakage has no effect. The middle equation says that the flux into R_i through one of its faces $q_i = \pm\frac{1}{2}$ is equal both to the number of points $\frac{1}{2}n_i$ in R_i itself which are moving in the right direction times $\bar{v}$ their mean velocity, and also to the number of points n_{i-1} in R_{i-1}, the previous region in the hierarchy, times the frequency ν_i with which they oscillate parallel to the q_i-axis. Likewise at the saddle-point we have

$$\begin{aligned}\phi_{1'} = * &= \tfrac{1}{2}\bar{v}n_{1'}\\ \nu_{i'} n_{i'-1} &= \tfrac{1}{2}\bar{v}n_{i'}\\ {}_{N'}n_{N'-1} &= \tfrac{1}{2}\bar{v}n_{N'} = \tfrac{1}{2}\bar{v}\rho(B)\end{aligned} \tag{48}$$

from which, eliminating the intermediate n_i from (45) and (46) and using (33) which holds here too, we finally reach the Vineyard formula

$$f = \frac{\phi_{1'}}{n} = \frac{\nu_1\nu_2\ldots\nu_N}{\nu_{2'}\nu_{3'}\ldots\nu_{N'}}e^{-U/kT} \tag{49}$$

REFERENCES

1 J. D. ESHELBY: in 'Solid state physics' (Ed. F. SEITZ AND D. TURNBULL), **3**, 79, 1956, New York, Academic Press
2 A. H. COTTRELL: Report Conf. on the 'Strength of solids', 30, 1948, London, The Physical Society
3 B. A. BILBY: *Proc. Phys. Soc.*, 1950, A**63**, 191
4 G. ALBENGA: *Atti Accad. Sci., Torino, Cl. Sci. fis. mat. nat.*, 1918/19, **54**, 864
5 F. BITTER: *Phys. Rev.*, 1931, **37**, 1526
6 J. M. DEWEY: *J. Appl. Phys.*, 1947, **18**, 579
7 J. D. ESHELBY, *Ann. Physik*, 1958, **1**, 116
8 J. R. HARDY AND R. BULLOUGH: *Phil. Mag.*, 1967, **15**, 1967
9 J. R. HARDY AND R. BULLOUGH: *Phil. Mag.*, 1968, **17**, 833
10 R. SIEMS: *Physica Status Solidi*, 1968, **30**, 645
11 J. D. ESHELBY: *J. Elasticity*, 1975, **5**, 321
12 J. D. ESHELBY: 'The physics of metals', **2**, 1, 1975 (Ed. P. B. HIRSCH), Cambridge University Press
13 G. H. VINEYARD: *J. Phys. Chem. Solids*, 1957, **3**, 121

CHAPTER 3

Boundary Problems

J. D. ESHELBY

Department of the Theory of Materials
University of Sheffield
Sheffield, UK

Dislocations in Solids
Edited by F. R. N. Nabarro

Contents

1. Introduction

The presence of free surfaces can sometimes affect the deformation of an internally stressed solid on an almost macroscopic scale: for instance, crystal whiskers containing an axial screw dislocation exhibit a twist which increases as their cross-section decreases. Such effects can be handled with the help of the theory of elasticity. On a microscopic scale relaxation due to the presence of the free surfaces of the foil affects the image contrast in thin-film electron microscopy. Of course, in this case we are really interested in the deformation of the lattice, but usually for want of anything better we have to treat the atoms as points embedded in an imaginary elastic continuum and deforming with it.

Part of the free surface may take the form of an internal cavity. More generally, if the cavity is not empty but is filled with material having elastic properties different from those of the rest of the solid we have what may be called an elastic inhomogeneity. The interaction between dislocations and inhomogeneities is important in the theory of plastic deformation. These are some of the reasons why the subject-matter of this chapter has received the attention of solid-state theoreticians. The field has also served as a convenient exercise ground for applied mathematicians.

We shall be mainly concerned with the following problem in linear isotropic elasticity: given the elastic field u_i^∞, p_{ij}^∞ due to some source of internal stress (a dislocation loop for example) in an infinite medium, find the elastic field of the same singularity in a finite body with a stress-free boundary S. The solution will take the form $u_i^\infty + u_i^{\text{im}}$, $p_{ij}^\infty + p_{ij}^{\text{im}}$ where the field u_i^{im}, p_{ij}^{im} (which, by analogy with similar electrostatic problems, we may call the "image" field) introduces no new singularities inside S and is such that the surface traction $p_{ij}^{\text{im}} n_j$ cancels the original traction $p_{ij}^\infty n_j$ on S.

Evidently u_i^{im}, p_{ij}^{im} is the field induced in the solid by known tractions $-p_{ij}^\infty n_j$ applied to its surface. This is a standard problem in the theory of elasticity. The difficulties standing in the way of its solution can be made clear with the help of a theorem of Gebbia's [1] which reduces the problem to one relating to an infinite solid. According to Gebbia the elastic field in a body whose outer boundary S is subjected to a traction $\boldsymbol{T}$ and a displacement $\boldsymbol{U}$ is the same as it would be if the material inside S formed part of an infinite medium provided S, now merely a surface marked out in the infinite medium, is covered with a layer of body force of surface density $\boldsymbol{T}$ and is the seat of a Somigliana dislocation whose variable discontinuity vector is equal to $\boldsymbol{U}$.

Figure 1 makes the theorem obvious. Insert the undeformed body into a perfectly-fitting cavity in an infinite matrix. Apply the surface traction $\boldsymbol{T}$. Where the body pulls away from the matrix build it up with a thin layer of material until it meets the matrix, and where it would penetrate the matrix, scrape it away so as just to avoid interference. Finally weld body and matrix together over S. These operations do not affect the elastic field within S and we are left with the situation envisaged in the theorem. The surface tractions have become built in as the layer of body force and the addition and removal of material over S generate the Somigliana dislocation. The latter can be regarded as being made up of a network of small dislocation loops each with a Burgers vector equal to the local value of $\boldsymbol{U}$, and so the displacement inside S can be written down as an integral over S involving $\boldsymbol{T}$, $\boldsymbol{U}$ and the elastic fields of a point force and an elementary dislocation loop acting in an infinite medium. (The field outside S is zero.) Expressed analytically the result is Somigliana's formula for finding the interior field in terms of surface traction and displacement ([2] p. 245). The trouble with this formula is that it requires us to know both $\boldsymbol{T}$ and $\boldsymbol{U}$ on the surface, whereas it is physically clear that a knowledge of $\boldsymbol{T}$ alone (which is all that we have) should be

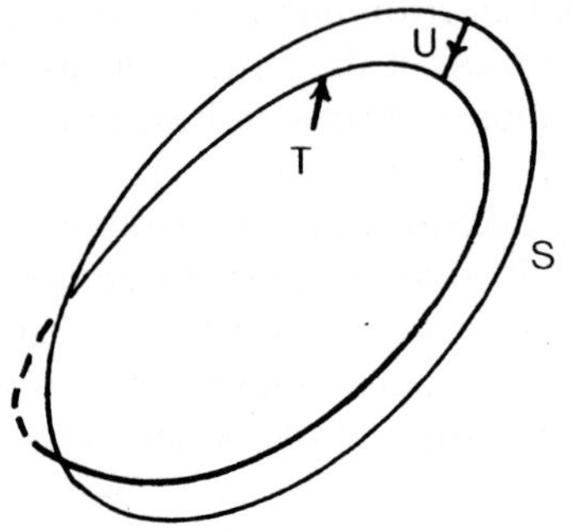

Fig. 1 Gebbia's theorem.

sufficient for determining the interior field. In principle we could use Somigliana's formula to find the displacement at a particular point on the surface of S in terms of the known $\boldsymbol{T}$ and the unknown $\boldsymbol{U}$ at all points of S, solve the resulting integral equation for $\boldsymbol{U}$ and insert the answer into Somigliana's formula to determine the interior field. If the demand for the result is sufficiently urgent the above process could be carried out numerically; Cruse's method [3] seems to be well adapted to such calculations. Consideration of the interaction between a dislocation and an elastic inhomogeneity leads to a related problem. We begin as before with the solution u_i^∞ in an infinite homogeneous medium, allow the elastic constants inside a surface S, which may or may not embrace the singularities of u_i^∞, to change to uniform values different from those of the rest of the material, and ask for the resulting change in the elastic field. By an extension of Gebbia's result the solution may be made to depend on the determination of the field of a point force and of an elementary dislocation loop situated at a general point on the interface S. To see this, suppose that we have two replicas, 1 and 2, of the infinite medium, with the same singularities, but different elastic constants, symbolized collectively by c_1 and c_2, so that $u_i(c_1)$ and $u_i(c_2)$ are the same functions of position with different values of the parameters c. Remove the interior of S in 1 and replace it with the corresponding region of 2, filling, scraping and welding as before. We are left with an inhomogeneous medium with the required singularities, but disturbed by the presence of a layer of body force of known density

$[p_{ij}^{\infty}(c_1) - p_{ij}^{\infty}(c_2)]n_j$ and a Somigliana dislocation with a known misfit vector $u_i^{\infty}(c_1) - u_i^{\infty}(c_2)$ spread over S. The elastic field induced by the removal of these layers can in principle be found in the form of integrals involving the field of a point force and of an elementary loop at the interface.

From the above it is clear that for a direct analytical solution of the free surface problem we should need to know the elastic field of a point force applied at a general point of the free boundary S. (There is a slight formal difficulty if S is a closed surface, e.g. a sphere, for there is no static solution corresponding to an uncompensated force. It may be turned by the device of introducing an auxiliary force and couple which do not affect the final result: see sect. 3.1.) Similarly, for the inhomogeneity problem we should have to know the field of a point force and of an elementary loop at the interface: actually the latter can be deduced from a knowledge of the effect of the force not only on, but also near, the interface. It is only in rather simple situations that these basic fields (Green's functions) can be found. Otherwise some special artifice has to be used for each class of problems.

Early solutions of problems relating to the interaction of dislocations with boundaries were largely the work of physicists, who used a mixture of traditional and improvised methods. Later the subject was taken up by applied mathematicians. Many of the more recent results, analytical, numerical and mixed, cannot easily be treated briefly. We shall mention some of them, but mostly we shall illustrate the discussion only by some of the more presentable special solutions.

2. *Two-dimensional problems*

2.1. Screw dislocations and circular boundaries

The simplest type of two-dimensional elastic field is a state of so-called anti-plane strain. Of the three Cartesian displacement components u, v, w only w is not zero, and it and the associated elastic quantities are all independent of the z-coordinate. Consequently the dilatation $\partial u/\partial x + \partial v/\partial y + \partial w/\partial z$ is zero and the displacement equilibrium equations reduce to Laplace's equation

$$\nabla^2 w = 0.$$

The non-zero stress components are

$$p_{zx} = \mu\, \partial w/\partial x, \qquad p_{zy} = \mu\, \partial w/\partial y,$$

where μ is the shear modulus. The traction on a plane with normal $(n_x, n_y, 0)$ is parallel to the z-axis and has the value

$$T_z = p_{zx}n_x + p_{zy}n_y = \mu\, \partial w/\partial n,$$

where $\partial/\partial n$ denotes differentiation in the direction of the normal.

The displacement associated with a screw dislocation of Burgers vector $(0, 0, b)$ at the origin in an infinite isotropic medium is

$$w = b(\tan^{-1} y/x)/2\pi$$

or in polar coordinates (r, θ)

$$w = b\theta/2\pi. \tag{1}$$

Since the expression (1) is independent of r, the traction, $\mu\, \partial w/\partial n = \mu\, \partial w/\partial r$, is zero on any circle $r = \text{const.}$, so that eq. (1) as it stands already gives the solution for a dislocation with a hollow core of radius a_0 in a concentric cylinder of radius a and infinite length with a stress-free surface. The energy per unit length of the cylinder is

$$E = \tfrac{1}{2}\mu \int_0^{2\pi} \int_{a_0}^{a} (\text{grad } w)^2 r \,\mathrm{d}r\,\mathrm{d}\theta = \frac{\mu b^2}{4\pi} \ln \frac{a}{a_0}. \tag{2}$$

There are, of course, non-zero tractions on the cross-section of the cylinder. If a finite rod is cut from the infinite cylinder suitable forces must be applied to the ends if a state of anti-plane strain is to be maintained. If these forces are not supplied the rod develops a twist: see sect. 3.4.

To treat the case where the dislocation is not at the centre of the circular cross-section it is convenient to have a few geometrical results from the theory of coaxial circles. We summarise them, together with some others which we shall need later.

Figure 2a shows Cartesian coordinates (x, y) and three sets of polar coordinates $(r, \theta), (r_2, \theta_2), (r_1, \theta_1)$ centred at the points O, P_2, P_1 on the x-axis, with x-coordinates

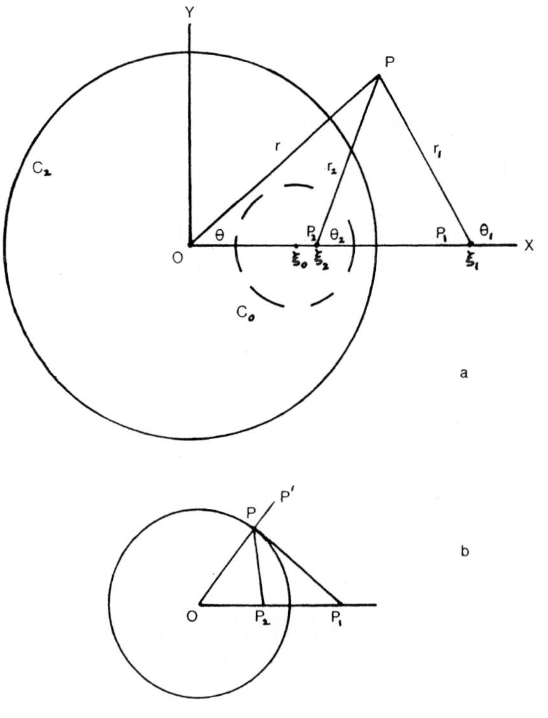

Fig. 2 (a) Coaxial circles, (b) the point P lies on a circle.

0, ξ_2, ξ_1 respectively where ξ_1 and ξ_2 are related by

$$\xi_1\xi_2 = a^2, \tag{3}$$

so that P_1 is the inverse of P_2 with respect to the circle $r = a$ shown in the figure. The point P(x, y) may, of course, lie either inside or outside this circle. Figure 2b refers to the case when P lies on the circle. There are then several simple relations between the θ's and r's which we shall require here and later. In the triangles OPP_2 and OP_1P we have $OP_2/OP = OP/OP_1$ by eq. (3), and they have the angle at O in common. Hence they are similar and the relations

$$r_2/r_1 = a/\xi_1 = \xi_2/a \qquad \text{on } C_2, \tag{4}$$

$$\theta_1 + \theta_2 - \theta = \pi \qquad \text{on } C_2 \tag{5}$$

easily follow.

The changes $\delta\theta_1$, $\delta\theta_2$ in θ_1, θ_2 when P moves out radially by δr to P can be found by projecting the segment PP′ onto directions perpendicular to P_2P and P_1P. This gives $-r_2\,\delta\theta_2 = \delta r \sin P'PP_2 = \delta r \sin OPP_2$ and $-r_1\,\delta\theta_1 = \delta r \sin P'PP_1 = \delta r \sin OP_2P$, or $\delta\theta_1 = \delta\theta_2$ by eq. (4) and the relation $\sin OPP_2/\sin OP_2P = \xi_2/a$. Hence

$$\partial\theta_1/\partial r = \partial\theta_2/\partial r \qquad \text{on } C_2. \tag{6}$$

Later we shall also need the relations

$$(\xi_1 - \xi_2)(r^2 - a^2) = \xi_1 r_2^2 - \xi_2 r_1^2, \tag{7}$$

$$r_2^2 - r_1^2 = (\xi_1 - \xi_2)[(x - \xi_2) + (x - \xi_1)] \tag{8}$$

which follow from

$$r_n^2 = (x - \xi_n)^2 + y^2, \qquad n = 1, 2.$$

Let C_0 in fig. 2a be a second circle for which P_2 and P_1 are inverse points. If its radius is a_0 and its centre is at $(\xi_0, 0)$ we have

$$(\xi_1 - \xi_0)(\xi_2 - \xi_0) = a_0^2 \tag{9}$$

as the analogue of eq. (3), eq. (4) becomes

$$r_2/r_1 = a_0/(\xi_1 - \xi_0) = (\xi_2 - \xi_0)/a_0 \qquad \text{on } C_0, \tag{10}$$

while eqs. (5) and (6) hold also on C_0.

For a screw dislocation at P_2 in the cylinder $r \leqslant a$ with a stress-free surface we have only to put a negative image dislocation at P_1. The resulting displacement [4]

$$w = b(\theta_2 - \theta_1)/2\pi \tag{11}$$

satisfies the condition $\mu\,\partial w/\partial n = \mu\,\partial w/\partial r = 0$ on $r = a$ by virtue of eq. (6).

Equation (11) can equally well be taken to refer to a negative dislocation at P_1 near a circular hole. The change of viewpoint does not of course affect the fact that C_2 is stress-free, but there is a Burgers vector b associated with the hole, and so eq. (11) actually represents a negative dislocation near a positive one with a large hollow core.

To get rid of the dislocation in the hole we may add the displacement $w = b\theta/2\pi$, which leaves the hole stress-free. Then, reversing the sign over all we get, using eq. (5)

$$w = b(\theta_1 - \theta_2 + \theta)/2\pi \tag{12}$$

for a positive dislocation at P_1 near a circular hole.

The dislocation in the cylinder experiences a force due to the image field represented by the second term in eq. (11):

$$F_x = bp_{zy}^{\text{im}} = -\frac{\mu b^2}{2\pi}\left(\frac{\partial\theta_1}{\partial y}\right)_{P_2}$$
$$= \frac{\mu b^2}{2\pi}\frac{1}{\xi_1 - \xi_2} = \frac{\mu b^2}{2\pi}\frac{\xi_2}{a^2 - \xi_2^2}.$$

This may be regarded as the negative gradient

$$F_x = -\partial E/\partial\xi_2$$

of a quantity

$$E = \frac{\mu b^2}{4\pi}\ln(a^2 - \xi_2^2) + \text{const.} \tag{13}$$

which shows how the energy of the dislocation varies with position. If the dislocation has a small hollow core of radius a_0 we may turn the quantity in eq. (13) into an absolute energy by requiring it to agree with eq. (2) for $\xi_2 = 0$:

$$E = \frac{\mu b^2}{4\pi}\ln\frac{a^2 - \xi_2^2}{aa_0}. \tag{14}$$

This is only an approximation, but it is quite easy to find the exact displacement and energy when the singular line is excluded by a stress-free circular hole, not necessarily small. In fact, since eq. (6) is satisfied on C_0 as well as C_2 the solution (11) also leaves c_0 stress-free, and its interior can be removed without upsetting the displacement.

Imagine that the material is slit along the x-axis between C_0 and C_2 and that the two sides of the cut are given a relative displacement b so as to generate the dislocation. The work which must be done in this process, and thus the energy of the dislocation, is given by

$$E = \tfrac{1}{2}b\int p_{zy}\,\mathrm{d}x = \frac{\mu b^2}{4\pi}\int_{\xi_0+a_0}^{a}\left(\frac{1}{x - \xi_2} - \frac{1}{x - \xi_1}\right)\mathrm{d}x.$$

This gives

$$E = \frac{\mu b^2}{4\pi}\ln\frac{(a - \xi_2)(\xi_1 - \xi_0 - a_0)}{(\xi_1 - a)(\xi_0 + a_0 - \xi_2)}, \tag{15}$$

which can be expressed entirely in terms of ξ_0 or ξ_2 with the help of eqs. (3) and (9). Either ξ_0 or ξ_2 is acceptable as a parameter specifying the position of the dislocation, for quite generally it makes no sense to ask whereabouts within its hollow core a dislocation actually lies ([5] p. 570). In fact the difference of the elastic fields for two such supposedly different positions would be characterised by zero traction on the hole and a single-valued displacement round it, and would thus, in the absence of applied loads, be zero. For $a_0 \ll a$ eq. (15) agrees with eq. (14).

We find similarly that for a dislocation in the matrix near a hole the image force due to the terms in θ and θ_2 in eq. (12) is given by $F_x = -\partial E/\partial \xi_1$ with

$$E = \frac{\mu b^2}{4\pi} \ln \frac{\xi_1^2 - a^2}{\xi_1^2} + \text{const.} \tag{16}$$

Here we cannot give a definite value to the additive constant, since the total energy is formally infinite.

Suppose next that the shear modulus is μ_2 inside the circle $r = a$ and μ_1 outside it. The method of images can still be applied [6]. We take w to be a general linear combination of θ_1, θ_2 and θ in both regions, say

$$\begin{aligned} 2\pi w/b &= \alpha\theta + \beta\theta_2 + \gamma(\pi - \theta_1), \qquad r < a \\ &= \alpha'\theta + \beta'\theta_2 + \gamma'(\pi - \theta_1), \qquad r > a. \end{aligned} \tag{17}$$

The use of $\pi - \theta_1$ instead of θ_1 ensures that the displacement has the common value zero on both sides of the interface where it intersects the x-axis. With eq. (6) continuity of the traction $\mu\, \partial w/\partial r$ at the interface gives

$$\mu_2(\beta - \gamma) = \mu_1(\beta' - \gamma').$$

With θ eliminated by eq. (5) continuity of w at the interface gives

$$\beta + \gamma = \beta' + \gamma', \qquad \alpha + \beta = \alpha' + \beta'.$$

If the dislocation is in the cylinder at P_2 we must have $\alpha = 0, \beta = 1, \gamma' = 0$ which gives

$$\begin{aligned} 2\pi w/b &= \theta_2 + K(\pi - \theta_1), \qquad r < a \\ &= (K + 1)\theta_2 - K\theta, \qquad r > a, \end{aligned} \tag{18}$$

with

$$K = (\mu_2 - \mu_1)/(\mu_2 + \mu_1). \tag{19}$$

If the dislocation is in the matrix at P_2 we must put $\alpha = 0, \beta = 0, \gamma' = -1$, which gives [7]

$$\begin{aligned} 2\pi w/b &= (K - 1)(\theta_1 - \pi), \qquad r < a \\ &= K(\theta_2 - \theta) + \theta_1 - \pi, \qquad r > a. \end{aligned} \tag{20}$$

Apart from an additive constant the image displacement, and consequently also the image term near P_2 in eq. (18), is K times what it was in eq. (11), so that the energy corresponding with eq. (14) is

$$E = \frac{\mu_2 b^2}{4\pi} K \ln (a^2 - \xi_2^2) + \text{const.} \tag{21}$$

Similarly, by comparing eq. (20) and eq. (12) we find from eq. (16) that for the dislocation at P_1 in the matrix the energy is

$$E = -\frac{\mu_1}{4\pi} K \ln \frac{\xi_1^2 - a^2}{\xi_1^2} + \text{const.} \tag{22a}$$

With the notation of fig. 3 the displacements in eqs. (11), (12), (18) and (20) apply equally to a dislocation near a plane interface [8]. To adapt eq. (14) to this case replace a^2 by $\xi_1\xi_2$, $\xi_1 - \xi_2$ by 2ξ and note, from fig. 2a that a, ξ_1, ξ_2 become equal for large a. This gives

$$E = \frac{\mu b^2}{4\pi} \ln \frac{2\xi}{a_0} \tag{22b}$$

for a dislocation distant ξ from the stress-free surface of a semi-infinite solid (fig. 3).

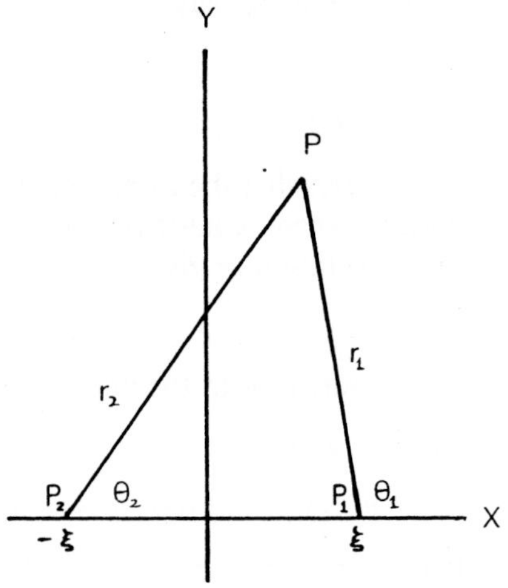

Fig. 3 A dislocation near a plane interface.

The accurate expression, eq. (15), gives

$$E = \frac{\mu b^2}{4\pi} \sinh^{-1} \frac{\xi}{a_0} = \frac{\mu b^2}{4\pi} \cosh^{-1} \frac{\xi_c}{a_0} \tag{23}$$

where 2ξ is the object–image distance, whereas ξ_c is the distance of the centre of the hole of (possibly large) radius a_0 from the free surface.

The relation between eqs. (14) and (21) persists in the limit, and so

$$E = \frac{\mu_2 b^2}{4\pi} K \ln 2\xi + \text{const.} \tag{24}$$

is the energy associated with a screw dislocation at a distance ξ from the surface of a semi-infinite solid of shear modulus μ_2 which is bonded to a second semi-infinite solid of shear modulus μ_1.

2.2. Screw dislocations in cylinders of general cross-section

In this section we shall extend the analysis of the last section to screw dislocations in cylinders of other than circular cross-section with stress-free surfaces. As a justification for doing so we recall that crystal whiskers of, for example, hexagonal or rectangular cross-section have been discovered or grown intentionally, and that in some of them there is an axial dislocation which produces interesting effects. What happens when a finite rod is cut from the infinite cylinder envisaged in the theory of anti-plane strain is discussed in sect. 3.4.

The harmonic anti-plane displacement w may be considered to be the real part of an analytic function of the complex variable $z = x + iy$, say

$$w + i\varphi = g(z).$$

The potential φ conjugate to w is in many respects more useful than w itself [8]. They are related by the Cauchy–Riemann equations

$$\frac{\partial w}{\partial x} = \frac{\partial \varphi}{\partial y} = \frac{p_{zx}}{\mu}, \qquad \frac{\partial w}{\partial y} = -\frac{\partial \varphi}{\partial x} = \frac{p_{zy}}{\mu} \tag{25}$$

which state that the gradients of w and φ are orthogonal. Hence the condition $\partial w/\partial n = 0$ on a free surface becomes $\varphi = $ const. on a free surface. The expression for the force exerted on a screw dislocation by an applied elastic field,

$$F_x = bp_{zy} = \mu b\frac{\partial w}{\partial y}, \qquad F_y = -bp_{zx} = -\mu b\frac{\partial w}{\partial x}$$

becomes a simple gradient in terms of φ:

$$F = -\mu b \operatorname{grad} \varphi. \tag{26}$$

Suppose that in addition to the x–y plane we have a second plane with Cartesian coordinates ξ, η. Then

$$\varphi_\zeta - iw_\zeta = \frac{b}{2\pi}\ln \zeta$$

or

$$\varphi_\zeta = -\frac{b}{2\pi}\ln|\zeta| = -\frac{b}{2\pi}\ln(\xi^2 + \eta^2)^{1/2},$$

$$w_\zeta = \frac{b}{2\pi}\tan^{-1}\frac{\eta}{\xi}, \tag{27}$$

with $\zeta = \xi + i\eta$ represent a screw dislocation at $\zeta = 0$ with the unit circle $|\zeta| = 1$ stress-free, since

$$\varphi_\zeta = 0 \quad \text{or} \quad |\zeta| = 1. \tag{28}$$

The suffix ζ is a reminder that a quantity is considered as a function of ξ, η, not x, y.

Let us relate x, y to ξ, η by

$$\zeta = f(z), \tag{29}$$

where $f(z)$ is analytic. We can then regard φ as a function of x, y, in which case we drop the suffix:

$$\varphi(x, y) = \varphi_\zeta[\xi(x, y), \eta(x, y)].$$

If we write $x + iy = f^{-1}(\zeta)$ we have the Cauchy–Riemann relations $\partial x/\partial \xi = \partial y/\partial \eta$, $\partial x/\partial \eta = -\partial y/\partial \xi$ analogous to eq. (25). Thus if we plot a set of curves $x =$ const., $y =$ const. for small and equal intervals of x and y they will, locally, form a rectangular Cartesian net in the ζ-plane and, referred to it, the equation

$$\frac{\partial^2 \varphi_\zeta}{\partial \xi^2} + \frac{\partial^2 \varphi_\zeta}{\partial \eta^2} = 0$$

satisfied by φ_ζ, being invariant under a rotation or expansion of the coordinate system, becomes

$$\frac{\partial^2 \varphi}{\partial x^2} + \frac{\partial^2 \varphi}{\partial y^2} = 0. \tag{30}$$

(For a more rigorous treatment see, for example, [9].) Suppose next that $f^{-1}(\zeta)$ is chosen so that as ζ wanders over the boundary and interior of the unit circle $|\zeta| = 1$, z wanders over the boundary and interior of some closed curve C in the z-plane, and that $f(z)$ is free of singularities inside C. Let z_0, defined by $f(z_0) = 0$, be the point which corresponds with $\zeta = 0$. Then the potential

$$\varphi(x, y) = -\frac{b}{2\pi} \ln |f(z)| \tag{31}$$

obtained by inserting eq. (29) into eq. (27) is harmonic by eq. (30), satisfies

$$\varphi(x, y) = 0 \quad \text{on } C$$

by eq. (28) and near z_0 takes the form

$$\varphi(x, y) = -\frac{b}{2\pi} \ln |z - z_0| - \frac{b}{2\pi} \ln |f'(z_0)| + \mathrm{O}|z - z_0|$$

with the singularity appropriate to a dislocation at x_0, y_0. Consequently eq. (31) is the potential for a screw dislocation at x_0, y_0 in a cylinder with stress-free boundary C.

To treat the case where the dislocation is not at the point z_0 provided by the chosen $f(z)$ we could replace eq. (27) by the potential function for a dislocation not at the centre of the unit circle in the ζ-plane. Equivalently we can make an intermediate transformation which maps the required point onto $\zeta = 0$. The details may be found in discussions of conformal mapping (e.g. [9]). Here we shall simply quote, and then verify, the result.

If the singularity-free transformation $\zeta = f(z)$ maps the interior of C onto the interior of the unit circle in the ζ-plane so that

$$|f(z)|^2 = f(z)\overline{f(z)} = 1 \quad \text{on } C \tag{32}$$

then

$$\varphi - \mathrm{i}w = -\frac{b}{2\pi} \ln \frac{1 - \overline{f(z')}f(z)}{f(z) - f(z')} \tag{33}$$

represents a screw dislocation at $x', y'(z' = x' + \mathrm{i}y')$ in the cylinder with stress-free contour C. (The bar denotes the complex conjugate.) To verify this, note that by eq. (32) $f(z)$ may be replaced by $1/\overline{f(z)}$ on C, which changes the argument of the logarithm in eq. (33) into the reciprocal of its own complex conjugate, so that its modulus is unity, and that the harmonic function

$$\varphi(x, y;\ \ x', y') = -\frac{b}{2\pi} \ln \left| \frac{1 - \overline{f(z')}f(z)}{f(z) - f(z')} \right| = \varphi(x', y';\ \ x, y) \tag{34}$$

vanishes on C. Also near z' we have

$$\varphi = -\frac{b}{2\pi} \ln |z - z'| + \frac{b}{2\pi} \Phi(x', y') + \mathrm{O}|z - z'| \tag{35}$$

with

$$\Phi(x', y') = \ln \frac{1 - |f(z')|^2}{|f'(z')|}, \tag{36}$$

so that there is the required logarithmic singularity.

The mapping function $f(z)$ is not unique since we can require it to send any point in C into the centre of the unit circle, but this ambiguity does not affect eq. (34). We may take any contour $\varphi =$ const. as the boundary of a stress-free hole excluding the centre of the dislocation. If we needed a hole of another shape we should have to allow $f(z)$ to have singularities inside the hole to pull its boundary into the required form. However, if we are only interested in the conventional small circular stress-free hole of radius a_0 no modification is needed, since according to eq. (35) the contours are nearly circular near enough to the singularity.

In this case the elastic energy is

$$E = \tfrac{1}{2}\mu \int (\operatorname{grad} w)^2 \, \mathrm{d}x \, \mathrm{d}y = \tfrac{1}{2}\mu \int (\operatorname{grad} \varphi)^2 \, \mathrm{d}x \, \mathrm{d}y$$

or by Green's theorem

$$E = \tfrac{1}{2}\mu \int \varphi \frac{\partial \varphi}{\partial n} \, \mathrm{d}S$$

taken over both boundaries with the normals respectively inwards and outwards on the inner and outer boundaries. Since φ is zero on the latter and is given by eq. (35) with $|z - z'| = a_0$ on the former, we have

$$E(x', y') = \frac{\mu b^2}{4\pi} \Phi(x', y') - \frac{\mu b^2}{4\pi} \ln a_0. \tag{37}$$

If we are not interested in the absolute value of E, but only its variation with position, we may omit the term $\ln a_0$, and the result then applies also to the singular case with $a_0 = 0$. The image force on the dislocation is

$$F_x, F_y = \left(\frac{\partial}{\partial x'}, \frac{\partial}{\partial y'}\right) \frac{\mu b^2}{4\pi} \Phi(x', y'). \tag{38}$$

According to eq. (26) we should expect to be able to write

$$F_x, F_y = \left[\left(\frac{\partial}{\partial x'}, \frac{\partial}{\partial y'}\right) b\varphi^{\mathrm{im}}(x, y;\; x', y')\right]_{x=x',\, y=y'}$$

where

$$\varphi^{\mathrm{im}}(x, y;\; x', y') = \varphi + \frac{b}{2\pi} \ln |z - z'|$$

is the potential associated with the image field. This is easily checked if we bear in mind that $b\Phi(x', y')/2\pi$ varies twice as fast with x', y' as does $\varphi^{\mathrm{im}}(x, y;\; x', y')$ with (x, y) held fixed near (x', y').

It is not hard to verify that Φ satisfies the space-charge equation

$$\frac{\partial^2 \Phi}{\partial x'^2} + \frac{\partial^2 \Phi}{\partial y'^2} = -4\,\mathrm{e}^{-2\Phi} \tag{39}$$

([10], [9] p. 203). Thus apart from positions of unstable equilibrium at maxima and saddle points the dislocation is everywhere attracted to the free surface C, near which E goes to negative infinity. This conclusion needs modification when the problem is treated as a three-dimensional one.

These formulas can be given other physical interpretations. For example, φ may be taken to be the electrostatic potential inside a cylindrical condenser (transmission line) with C for outer and a thin wire of radius a_0 for inner conductor. E is the electrostatic energy. In other words the capacity per unit length is const./E. The force on the dislocation becomes the force on the wire. Or φ may be taken to be the stream function and w the velocity potential for a vortex line in a perfect incompressible fluid inside a tube C. E becomes the kinetic energy of the fluid. If the vortex is held fixed by a wire in its core the force on the wire is equal to the force on the dislocation. If the vortex is free the force on the dislocation becomes the drift velocity of the vortex in its own image flow, but turned through a right angle, so that the vortex describes a curve Φ = const., that is, one of constant kinetic energy, as is to be expected. In the above the phrase "in suitable units" must, of course, be understood throughout.

Evidently with the help of a table of conformal transformations [11] or by ransacking textbooks of electrostatics and hydrodynamics we could write down the solutions for an indefinite number of special cases. The following are a few of them.

For a screw dislocation at $(\xi_2, 0)$ in the cylinder $r = a$ we must take $f(z) = z/a$, $z' = \xi_2$. Equations (33) and (37) then give eqs. (11) and (14) at once, together with

$$\varphi = -\frac{b}{2\pi} \ln \frac{ar_2}{r_1}$$

in the notation of fig. 2a.

For a screw dislocation in a plate [12] we may find the mapping function by a physical argument. Figure 4a shows an infinite vertical row of dislocations of alternating sign and spacing d along the y-axis with a positive one at the origin. Evidently

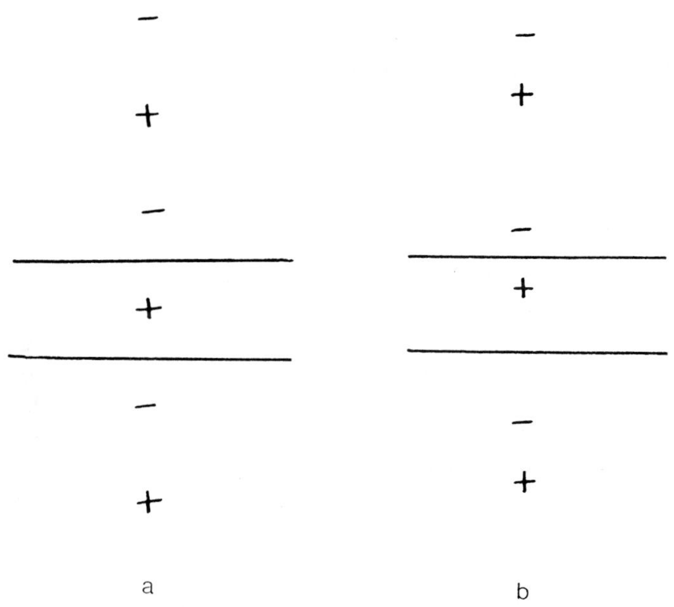

Fig. 4 A dislocation in a plate: (a) in the mid-plane, (b) not in the mid-plane.

either of the planes $y = \pm\frac{1}{2}d$ is straddled by a set of equally-spaced positive–negative pairs whose contributions to φ (or $\partial w/\partial y$) cancel on these planes so that they may be taken to be the stress-free surfaces of a plate with a dislocation at the origin. The argument of the logarithm in eq. (33) must have a zero at each positive dislocation and a pole at each negative one, and this will occur if we put

$$f(z) = \tanh(\pi z/2d) = \frac{\sinh 2X + \mathrm{i} \sin 2Y}{\cosh 2X - \cos 2Y} \tag{40}$$

with

$$X = \pi x/2d, \qquad Y = \pi y/2d.$$

It is easily verified that $|f(z)|^2 = 1$ for $z = x \pm i\frac{1}{2}d$, so that the expression, eq. (40), is the required mapping function. If the dislocation is at the origin,

$$w = \frac{b}{2\pi}\theta \qquad \text{with} \qquad \tan\theta = \frac{\sin(\pi y/d)}{\sinh(\pi x/d)}, \tag{41}$$

and one component of stress is therefore

$$p_{zy} = \mu\frac{\partial w}{\partial y} = \frac{\mu b}{2\pi}\frac{\pi}{2d}\cot\frac{\pi y}{d}\sin\frac{4\pi w}{b}, \tag{42}$$

where we have used the relation $\sin 2\theta = 2\tan\theta/(1 + \tan^2\theta)$. We may compare this with the Peierls law [13]

$$p_{zy}(x, y) = \frac{\mu}{2\pi}\frac{b}{a}\sin\frac{2\pi}{b}[w(x, y) - w(x, -y)]$$

which must be satisfied on the planes $y = \pm\frac{1}{2}a$ across which the Peierls force is supposed to act. The elastic solution of eqs. (41) and (42) already satisfies the Peierls condition across the planes

$$y = \pm\frac{d}{\pi}\tan^{-1}\frac{\pi a}{2d} = \pm\tfrac{1}{2}a\left[1 - \tfrac{1}{3}\left(\frac{\pi a}{2d}\right)^2 + \cdots\right].$$

Thus if we cut out the material between these planes and move them together until their spacing is a the elastic solution provides just the right stresses to support the Peierls force when it is switched on in the gap.

If the dislocation is at $(0, y')$, eq. (33) gives

$$\varphi = -\frac{b}{2\pi}\tfrac{1}{2}\ln\frac{\cosh[\pi x/d] - \cos[\pi(y - y')/d]}{\cosh[\pi x/d] + \cos[\pi(y + y')/d]},$$

$$w = \frac{b}{2\pi}\tan^{-1}\frac{\tan[\pi(y - y')/2d]}{\tanh[\pi x/2d]} - \frac{b}{2\pi}\tan^{-1}\frac{\tan[\pi(y + y' + d)/2d]}{\tanh[\pi x/2d]},$$

$$\Phi(y') = \ln\left[\frac{2d}{\pi}\cos\frac{\pi y'}{d}\right]. \tag{43}$$

The first term in w represents the array of positive dislocations shifted so that the dislocation within the plate is at $(0, y')$ and the second the negative dislocations shifted by as much in the opposite direction (fig. 4b). Each of the planes $y = \pm\frac{1}{2}d$ is still straddled by compensating pairs.

With x replaced by $x(c_{55}/c_{44})^{1/2}$, w becomes the displacement in an anisotropic plate provided the dislocation axis and the normal to the plate are both two-fold axes of the crystal [14].

A dislocation in a stress-free rectangle can be dealt with by repeated reflection of the array of fig. 4b in a pair of vertical lines [4], by a suitable mapping [15] or by lifting the results from Greenhill's [16] treatment of a vortex line in a rectangular tube. The function which maps the rectangle with corners $x = \pm\frac{1}{2}a$, $y = \pm\frac{1}{2}d$ into the unit circle, centre to centre, is

$$f(z) = \mathrm{sn}\,\lambda z\,\mathrm{dn}\,\lambda z/\mathrm{cn}\,\lambda z \tag{44}$$

where λ and the modulus k of the elliptic functions are fixed by

$$K(k)/a = K'(k)/d = \lambda$$

with the usual notation [17]. If the dislocation is at the centre of the rectangle this gives

$$\varphi = \frac{b}{2\pi}\tanh^{-1}(\mathrm{cn}\,2\lambda x\,\mathrm{cn}\,2\lambda y),$$

$$w = \frac{b}{2\pi}\tan^{-1}(\mathrm{sd}\,2\lambda x\,\mathrm{ds}\,2\lambda y),$$

where the elliptic functions of x have modulus k and those of y modulus k'. We shall not give φ and w for a general position of the dislocation, but only the function Φ which determines the energy:

$$\Phi(x', y') = -\tfrac{1}{2}\ln\lambda[\mathrm{dc}^2(2\lambda x', k) + \mathrm{dc}^2(2\lambda y', k') - 1]$$

(cf. Greenhill [18]; our relation (39), evaluated at the origin, can be used to restore the additive constant which he rejects).

2.3. Edge dislocations and circular boundaries

In this section we try to do for edge dislocations what was done for screw dislocations in sect. 2.1. It is assumed that the material, infinite in the z-direction, is in a state of plane strain so that w, the z-component of the displacement, is zero and the other two components, u and v, are independent of z. (For a remark on the status of plane stress and generalized plane stress in dislocation theory see sect. 3.2.)

We begin by recalling the theory of plane strain as developed in terms of the Airy stress function, and introduce some of the stress functions which will be needed.

The details of a plane strain elastic field are all contained directly or implicitly in its Airy stress function χ in terms of which the stresses are

$$p_{xx} = \frac{\partial^2\chi}{\partial y^2}, \qquad p_{yy} = \frac{\partial^2\chi}{\partial x^2}, \qquad p_{xy} = -\frac{\partial^2\chi}{\partial x\,\partial y},$$

$$p_{zx} = 0, \qquad p_{zy} = 0, \qquad p_{zz} = \nu(p_{xx} + p_{yy}); \tag{45}$$

the last relation comes from the requirement that e_{zz} be zero (ν is Poisson's ratio). The stress function is biharmonic, that is, it satisfies

$$\nabla^2\nabla^2\chi = 0. \tag{46}$$

It is useful to remember that, if h is harmonic and r_n is the distance from some fixed point, then

$$h, \quad xh, \quad yh, \quad r_n^2 h \tag{47}$$

are all biharmonic.

The displacement components, if required, can be calculated from χ in the following way. They may be written in the form

$$u = \varphi - \frac{1}{2\mu}\frac{\partial \chi}{\partial x}, \qquad v = \psi - \frac{1}{2\mu}\frac{\partial \chi}{\partial y} \tag{48}$$

where φ, ψ are a pair of conjugate harmonic functions. To find them from χ write $\nabla^2\chi = P$ and let Q be the harmonic function conjugate to P so that we have

$$\frac{\partial P}{\partial x} = \frac{\partial Q}{\partial y}, \qquad \frac{\partial P}{\partial y} = -\frac{\partial Q}{\partial x} \tag{49}$$

and $F(z) = P + \mathrm{i}Q$ is an analytic function of the complex variable $z = x + \mathrm{i}y$. Then

$$\varphi + \mathrm{i}\psi = \frac{1-\nu}{2\mu}\int_{z_0}^{z} F(z')\,\mathrm{d}z' \tag{50}$$

with arbitrary z_0. If Q cannot be found from P by inspection insert a factor $1 = \mathrm{d}(z' - z)/\mathrm{d}z'$ in the integrand of eq. (50), integrate by parts and use the relation $\mathrm{d}F/\mathrm{d}z = \partial F/\partial x$ and eq. (49) to get

$$\varphi + \mathrm{i}\psi = \frac{1-\nu}{2\mu}(z - z_0)F(z_0) + \frac{1-\nu}{2\mu}\int_{z_0}^{z}(z - z')\left(\frac{\partial P}{\partial x'} - \mathrm{i}\frac{\partial P}{\partial y'}\right)\mathrm{d}z', \tag{51}$$

where the first term, a rigid-body displacement, may be dropped. Equations (48) and (51) combined are equivalent to the Cesàro line-integrals [19] for u, v in terms of strains and strain gradients.

The traction on a curve C has x and y components $-\mathrm{d}(\partial\chi/\partial y)/\mathrm{d}s$ and $\mathrm{d}(\partial\chi/\partial x)/\mathrm{d}s$ where $\mathrm{d}/\mathrm{d}s$ denotes differentiation along the arc of C. The condition that C be free of traction is thus

$$-\frac{\mathrm{d}}{\mathrm{d}s}\left(\frac{\partial\chi}{\partial y}\right) = 0, \qquad \frac{\mathrm{d}}{\mathrm{d}s}\left(\frac{\partial\chi}{\partial x}\right) = 0 \quad \text{on } C. \tag{52}$$

The vanishing of these two expressions implies that the gradient of χ is constant along C, so that if we plot $\chi = \chi(x, y)$ as a surface over a horizontal x–y plane it is tangential to some plane

$$\chi = q + lx + my \tag{53}$$

around a curve whose projection onto the x–y plane is C. Since the addition of a linear function of x and y to χ does not affect the stresses we can arrange that the plane represented by eq. (53) is of zero slope and height, so that the boundary conditions take the simplified form [4]

$$\chi = 0, \qquad \partial\chi/\partial n = 0 \quad \text{on } C, \tag{54}$$

where $\partial/\partial n$ denotes differentiation along the normal to C. We shall find it simplest to use eq. (54) wherever possible. However, if there are several separate stress-free boundaries we can use the condition of eq. (54) for one of them, but on the others we must require χ to behave like eq. (53) on and near the boundary, with, in general, different constants q, l, m for each of them.

The following are some stress-functions we shall encounter. For an edge dislocation with Burgers vector $(b, 0)$ the stress function is

$$\chi = -Dy \ln r \tag{55}$$

with

$$D = \mu b/2\pi(1 - \nu), \tag{56}$$

and the displacements are

$$u = \frac{b}{2\pi}\theta + \frac{b}{2\pi}\frac{1}{2(1-\nu)}\frac{xy}{r^2}, \qquad v = \frac{b}{2\pi}\frac{1-2\nu}{2(1-\nu)}\ln\frac{1}{r} + \frac{b}{2\pi}\frac{1}{2(1-\nu)}\frac{y^2}{r^2}. \tag{57}$$

If the Burgers vector is $(0, b)$ in the sense that u is single-valued while v contains a term $+b\theta/2\pi$ we have

$$\chi = +Dx \ln r. \tag{58}$$

The simple displacement

$$u = \frac{b}{2\pi}\theta, \qquad v = \frac{b}{2\pi}\ln r \tag{59}$$

has the harmonic stress function

$$\chi = -\frac{\mu b}{\pi}(x\theta + y \ln r). \tag{60}$$

It describes a compound defect made up of a dislocation $(b, 0)$ and a concentrated force $(0, -2\mu b)$. The x and y derivatives of eq. (60) give

$$\chi = \text{const.}\ \theta \qquad \text{and} \qquad \chi = \text{const.}\ \ln r, \tag{61}$$

representing, respectively, a concentrated couple and a centre (or rather line) of dilatation.

From the difference of eqs. (55) and (60) we get

$$\chi = \frac{1-2\nu}{4\pi(1-\nu)}y \ln r + \frac{1}{2\pi}x\theta \tag{62}$$

for the unit force $(0, 1)$.

We begin by considering the case of an edge dislocation in a cylindrical rod with its lateral surface free of stress. The solutions we shall obtain for this case will apply to a finite rod only if suitable forces are applied to the ends so as to maintain the state of plane strain. However, we shall see in sect. 3.4. that the end-forces have no resultant or resultant moment, so that their removal produces no drastic change in deformation, in contrast with what happens in the case of screw dislocations in rods. We also treat the complementary case of a dislocation in the neighbourhood of a circular stress-free hole, and then look at the more general problem of a circular elastic inhomogeneity.

Volterra [20, 21] himself dealt with an edge dislocation in a tube; some of his results are reproduced by Love ([2], section 156A). To find the Airy stress function for an edge dislocation interacting with free cylindrical surfaces concentric with the dislocation line we add to the basic

$$\chi = -Dr \sin\theta \ln r$$

of eq. (55) multiples of the three other biharmonic functions

$$r^3 \sin\theta, \qquad r\sin\theta, \qquad r^{-1}\sin\theta \tag{63}$$

which have the same angular dependence. For a solid cylinder of radius a with the dislocation along its axis we need only the first two of them to satisfy eqs. (54):

$$\chi = -Dy\left[\ln\frac{r}{a} - \tfrac{1}{2}\left(\frac{r^2}{a^2} - 1\right)\right]. \tag{64a}$$

For a dislocation in an infinite medium with a hollow core of radius a we need only the last two:

$$\chi = -Dy\left[\ln\frac{r}{a} + \tfrac{1}{2}\left(\frac{a^2}{r^2} - 1\right)\right],$$

or, rejecting terms which give no stress, simply

$$\chi = -Dy[\ln r + \tfrac{1}{2}a^2/r^2], \tag{64b}$$

which satisfies eq. (52) rather than eq. (54).

For a dislocation in a cylindrical tube with inner and outer radii a_0, a, we need all three of the functions (63). If we impose the condition of eq. (54) on, say, the inner surface we must require the χ-surface to be tangential to the plane, eq. (53), at the outer surface, or, considering the symmetry, $\chi = Ca\sin\theta$, $\partial\chi/\partial r = C\sin\theta$ on $r = a$. The result is

$$\chi = -Dy\left[\ln\frac{r}{a_0} - \frac{1}{2}\frac{a^2 + r^2}{a^2 + a_0^2}\left(1 - \frac{a_0^2}{r^2}\right)\right]. \tag{65}$$

Photoelastic pictures of the associated stress field have been given by Corbino [22].

The elastic energy for the last case can be found as the work required to establish the dislocation by sliding the two surfaces $y = \pm 0$, $a_0 < x < a$ through a distance b:

$$\begin{aligned} E &= \tfrac{1}{2}b\int_{a_0}^{a} p_{xy}(x, 0)\,\mathrm{d}x = \tfrac{1}{2}b[\partial\chi/\partial y]_{a_0}^{a} \\ &= \tfrac{1}{2}bD\left[\ln\frac{a}{a_0} - \frac{a^2 - a_0^2}{a^2 + a_0^2}\right] \end{aligned} \tag{66}$$

or for $a \gg a_0$

$$E = \frac{\mu b^2}{4\pi(1-\nu)} \ln\left(\frac{a}{ea_0}\right), \qquad e = 2.718\ldots. \tag{67}$$

If the dislocation in the tube has Burgers vector $(0, b)$ the stress function is evidently the function in (65) with the factor $-Dy$ replaced by $+Dx$. The stresses and displacements for this case are given by Love ([2] section 156A).

We shall also need the stress function

$$\chi = \ln\frac{r}{a} - \frac{1}{2}\left(\frac{r^2}{a^2} - 1\right) \tag{68}$$

obtained by striking out $-Dy$ in eq. (64a). It is made up of a two-dimensional hydrostatic pressure and a centre of dilatation, eq. (65), combined so as to make the circle $r = a$ stress-free. For $r > a$ it represents the field in an infinite solid subjected to a uniform pressure perturbed by a circular hole. The centre of dilatation takes account of the inward movement of the matrix when a plug is removed from the compressed material to leave the hole.

Suppose next that the edge dislocation is excentric, say at $P_2(\xi_2, 0)$ in fig. 2a. To find the stress function for a dislocation in a cylinder, circular or not, with stress-free boundary C we have to subtract from the stress function for an infinite medium another biharmonic function which has the same value and normal derivative on C, so satisfying eqs. (54), but which is free of singularities inside C. If C is a circle there is in fact a formula, similar to Poisson's integral formula for a harmonic function, which determines interior values of a biharmonic function from the boundary value and normal derivative [23], but we shall not make use of it. We start with the case where the Burgers vector $(b, 0)$ is parallel to the x-axis.

Following our treatment of the screw dislocation we begin by adding together the stress functions $-Dy \ln r_2$, $+Dy \ln r_1$ for a real dislocation at P_2 and an equal and opposite image one at P_1 to form the first tentative solution

$$\chi_0 = -Dy \ln t \tag{69}$$

with

$$t = r_2 a / r_1 \xi_2, \qquad t^2 = r_2^2 \xi_1 / r_1^2 \xi_2. \tag{70}$$

The extra factor $a/\xi_2 = \xi_1/a$ which we have inserted in t ensures that t is unity in the circle $r = a$, so that χ_0 satisfies the first of the boundary conditions (54), but not the second. Near $t = 1$, $\ln t$ can be expanded in powers of $t^2 - 1$:

$$\ln t = \tfrac{1}{2}\ln t^2 = \tfrac{1}{2}\ln\left[1 + (t^2 - 1)\right] = \tfrac{1}{2}(t^2 - 1) + \ldots.$$

The first term of the expansion reproduces the value and gradient of $\ln t$ correctly on the circle. Hence if we subtract it from $\ln t$ we get a function

$$X(t) = \ln t - \tfrac{1}{2}(t^2 - 1)$$

which satisfies both of the conditions of eq. (54) on the circle. The same will be true of the product of $X(t)$ and a function of x and y. Hence

$$\chi = -DyX(t) \tag{71}$$

solves the problem, provided it is biharmonic. Fortunately it is, having the form of the fourth of the quantities (47), with $h = y/r_1^2$.

Dropping a term const. y, which gives no stress, χ becomes [24–26]

$$\chi = -Dy \ln \frac{r_2}{r_1} + \tfrac{1}{2}Dy \frac{\xi_1}{\xi_2}\frac{r_2^2}{r_1^2}. \tag{72}$$

Or again, replacement of r_2^2 in the second term by, say, $r_2^2 - r_1^2$ likewise merely adds a term const. y and so, by eq. (8), eq. (72) is also equivalent to

$$\chi = -Dy \ln \frac{r^2}{r_1} + D(\xi_1 - \xi_2)\frac{\xi^1}{\xi_2}(x - \bar{\xi})\frac{y}{r_1^2}, \tag{73}$$

with

$$\bar{\xi} = \tfrac{1}{2}(\xi_1 + \xi_2).$$

The stress functions, eqs. (72) and (73), satisfy the general boundary condition eq. (52) rather than the special form eq. (54). The image stress function is made up of the simple image term $Dy \ln r_1$ together with the last term in eq. (73), which can be written in the form

$$D\frac{\xi_1}{\xi_2}\frac{\partial}{\partial \xi_1}(y \ln r_1) + \tfrac{1}{2}D\frac{\xi_1}{\xi_2}(\xi_1 - \xi_2)\frac{\partial}{\partial y}(\ln r_1), \tag{74}$$

representing a dislocation dipole and a doublet of centres of dilatation (eq. (61)) supplementing the image dislocation at P_1.

The image stress is derived from eq. (72) with the term $-Dy \ln r_2$ omitted. The non-logarithmic term contributes nothing to the image stress at P_2, though it does so elsewhere on the x-axis. We may say that the image force is the same as that due to a simple negative image dislocation at P_1 represented by a term $Dy \ln r_1$, the additional singularities separated out in (74) having no effect. The image force is thus $1/(1 - \nu)$ times what it is for a screw dislocation in the same situation, eq. (14), and so the energy is

$$E = \tfrac{1}{2}bD \ln \frac{a^2 - \xi_2^2}{eaa_0}, \tag{75}$$

where the additive constant has been chosen so as to agree with eq. (67) for $\xi_2 = 0$.

In the limit of large a, $\xi_1 - \xi_2$ and $x - \bar{\xi}$ become 2ξ and x in the notation of fig. 3, and we get

$$\chi = -Dy \ln \frac{r_2}{r_1} + 2D\xi \frac{xy}{r_1^2} \tag{76}$$

and

$$E = \tfrac{1}{2}bD \ln \frac{2\xi}{ea_0} \tag{77}$$

for a dislocation with Burgers vector $(b, 0)$ at $(-\xi, 0)$ in the semi-infinite solid $x < 0$.

When the Burgers vector is $(0, b)$ instead of $(b, 0)$ the calculation is more troublesome. Suppressing the factor D we start with the simple object-image stress function $(x - \xi_2) \ln r_2 - (x - \xi_1) \ln r_1$ (cf. eq. (58)) and simplify it to

$$\chi_0 = (x - \xi_2) \ln t \tag{78}$$

by adding a harmless centre of dilatation, eq. (61), at the image point.

From the previous argument it is clear that $(x - \xi_2)X(t)$ satisfies the boundary conditions and has the right singularity, but unfortunately it is not biharmonic, since $(x - \xi_2)/r_1^2$ is not harmonic. We therefore write

$$\chi = \chi_0 - \chi_1$$

and try to choose χ_1 so that it is biharmonic and singularity-free inside C and has a value and derivative which coincide with those of χ_0 on C.

We look for χ_1 in the form

$$\chi_1 = g + \tfrac{1}{2}(r^2 - a^2)h \tag{79}$$

with harmonic g and h. (Equation (79) can actually generate any biharmonic function, but we make no use of this fact.) As χ_0 is already zero on C, g is zero everywhere. Since $\partial/\partial n$ is equivalent to $\partial/\partial r$ on C we have

$$\frac{1}{a}\frac{\partial \chi_1}{\partial n} = h \quad \text{on } C_2 \tag{80}$$

and

$$\frac{\partial \chi_0}{\partial n} = (x - \xi_2)\left[\left(\frac{x - \xi_2}{r_2^2} - \frac{x - \xi_1}{r_1^2}\right)\frac{x}{r} + \left(\frac{y}{r_2^2} - \frac{y}{r_1^2}\right)\frac{y}{r}\right] \quad \text{on } C_2$$

or

$$\frac{1}{a}\frac{\partial \chi_0}{\partial n} = (x - \xi_2)(r_2^{-2} - r_1^{-2}) \quad \text{on } C_2, \tag{81}$$

where we have used the facts that $r = a$ and that $r_1^2 \xi_2 = r_2^2 \xi_1$ on C_2, by eq. (4).

Before eq. (81) can be compared with eq. (80) it must be cleaned up with the help of the various identities which hold on C_2. First we must use eq. (4) to replace r_2^{-2} by $(\xi_1/\xi_2)r_1^{-2}$; otherwise there would be an unwanted extra singularity at $r_2 = 0$. Then we must eliminate $(x - \xi_2)$ in favour of $(x - \xi_1)$, since $(x - \xi_1)/r_1^2$ is harmonic, but $(x - \xi_2)/r_1^2$ is not. This can be done with the help of the relation in (8), which holds generally. This gives

$$\frac{1}{a}\frac{\partial \chi_0}{\partial n} = -(\xi_1 - \xi_2)\left(\frac{x - \xi_1}{\xi_2 r_1^2} + \frac{1}{a^2}\right) \quad \text{on } C_2. \tag{82}$$

If we now require h, eq. (80), to agree with the right-hand side of eq. (82) not only on C but throughout its interior, the χ_1 of eq. (79) satisfies the required conditions. Hence finally we have

$$\chi = D(x - \xi_2)\ln t + \tfrac{1}{2}D(r^2 - a^2)(\xi_1 - \xi_2)\left\{\frac{x - \xi_1}{\xi_2 r_1^2} + \frac{1}{a^2}\right\}. \tag{83}$$

It may be manipulated into the form

$$\chi = D(x - \xi_2)\ln\frac{r_2}{r_1} + D(\xi_1 - \xi_2)(x - \xi)\frac{x - \xi_1}{r_1^2} + \tfrac{1}{2}D(\xi_1 - \xi_2)\frac{r^2}{a^2} \tag{84}$$

which like its analogue eq. (73) satisfies the conditions of eq. (52) rather than of eq. (54). In the last term, r^2 may, of course, be replaced by r_1^2 or r_2^2. Equation (84) can be further manipulated so as to agree with the results of Leibfried and Dietze (quoted by Seeger [26]) and Dundurs and Sendeckyj [27]. The image (climb) force, now equal to the product of b and the yy-component of the image stress at P_2, is easily found to be

$$F_x = \frac{Db}{\xi_1 - \xi_2} - \frac{Db}{\xi_1}, \tag{85}$$

where the first term corresponds to a simple image at P_1 and the second to the non-logarithmic terms in eq. (84), comprising a two-dimensional hydrostatic pressure together with additional terms singular at P_1, which can be interpreted on the lines of eq. (74) as a centre of dilatation, a dislocation dipole, and a doublet of centres of dilatation. To find the elastic energy we replace ξ_1 by a^2/ξ_2 and integrate with respect to ξ_2, fixing the additive constant as in eq. (75):

$$E = \tfrac{1}{2}bD\left(\ln\frac{a^2 - \xi_2^2}{eaa_0} + \frac{\xi_2^2}{a^2}\right). \tag{86}$$

To find the elastic field of a dislocation with Burgers vector $(b, 0)$ at P_1 near the hole $r \leqslant a$ we start by changing t into t^{-1} in eq. (71). The function $X(t^{-1})$ shares with $X(t)$ the property of having zero value and derivative at $t = 1$. Hence

$$\chi = -DyX(t^{-1}) \tag{87}$$

satisfies the boundary conditions of eq. (54) on C_2 and it is, in fact, biharmonic. The term $-Dy\ln r_1$ represents a dislocation $(b, 0)$ at P_1, so that the expression (87) is the stress function for a dislocation near a stress-free circular hole $r = a$. However, because of the term $+Dy\ln r_2$ there is a Burgers vector $(-b, 0)$ associated with the hole, and eq. (87) actually represents a dislocation at P_1 near a dislocation of opposite sign with a large hollow core. We may cancel the dislocation in the hole by adding the stress function of eq. (64b). With stressless terms rejected the result is clearly just eq. (72) with the suffixes 1 and 2 interchanged, plus (64b), that is

$$\chi = -Dy\left[\ln\frac{rr_1}{r_2} + \tfrac{1}{2}a^2\left(\frac{1}{r^2} - \frac{r_1^2}{\xi_1^2 r_2^2}\right)\right]. \tag{88}$$

The elastic energy corresponding to eq. (87) is evidently eq. (75) with 1 and 2 interchanged. When the field (64b) is added there is an extra term

$$bD/\xi_1 - bDa^2/\xi_1^3 \tag{89}$$

in the image force which is equivalent to a term

$$-bD \ln \xi_1 - \tfrac{1}{2}bDa^2/\xi_1^2 \tag{90}$$

in the energy. This gives

$$E = \tfrac{1}{2}bD\left[\ln \frac{\xi_1^2 - a^2}{\xi_1^2} - \frac{a^2}{\xi_1^2}\right] + \text{const.} \tag{91}$$

The additive constant cannot be given a definite value, since in an infinite matrix the energy is formally infinitive even if we cut out a stress-free hole around P_1.

Similarly, if the Burgers vector in the situation just considered is changed from $(b, 0)$ to $(0, b)$ we start by interchanging 1 and 2 in eq. (84) and adding a term

$$\chi = Dx\left[\ln r + \tfrac{1}{2}\frac{a^2}{r^2}\right] \tag{92}$$

to get rid of the dislocation in the hole. However, this result is still not quite the correct solution. The last term in eq. (84) becomes $-\frac{1}{2}D(\xi_1 - \xi_2)r^2/a^2$ and represents a uniform hydrostatic pressure in the matrix. To get rid of it while still leaving the hole stress-free we add $-D(\xi_1 - \xi_2)$ times the expression (68). The energy is the expression (86) with 1 and 2 interchanged plus corrections from the added terms. It takes the simple form

$$E = \tfrac{1}{2}bD \ln \frac{\xi_1^2 - a^2}{\xi_1^2} + \text{const.} \tag{93}$$

By combining the stress functions (71) and (87) we can find the formula which replaces eq. (71) when the inner circle C_0 is also stress-free, and thus test the accuracy of eq. (75). If the combination is chosen so that the coefficient of ln r is $-Dy$ it will take the form

$$\begin{aligned}\chi &= Dy\{AX(t) + (1 - A)[-X(t^{-1})]\} \\ &= -Dy[\ln t - \tfrac{1}{2}A(t^2 - 1) + \tfrac{1}{2}(1 - A)(t^{-2} - 1)] \\ &= -DyY(t)\end{aligned}$$

say, with

$$t^2 = r_2^2\xi_1/r_1^2\xi_2, \tag{94}$$

which satisfies eqs. (54) and leaves C_2 stress-free for any A. On C_0 its value $-DyY(t_0)$ is of the form of eq. (53) and its gradient

$$\frac{\partial\chi}{\partial y} = -DY(t_0) - Dy\left[Y'(t)\frac{\partial t}{\partial y}\right]_{t_0}$$

will agree with the gradient of eq. (53) provided the second term, or simply $Y'(t_0)$, is zero. This gives

$$A = 1/(1 + t_0^2).$$

(If the value of χ and the y-component of its gradient agree with eq. (53) there is no need to test the x-component.)

To get the associated energy we can evaluate the integral in eq. (66) between limits corresponding to $t = t_0$, $t = 1$:

$$E = \tfrac{1}{2}b\left[\frac{\partial \chi}{\partial y}\right]_{t=t_0}^{t=1} = \tfrac{1}{2}b[Y(1) - Y(t_0)]$$

$$= \tfrac{1}{2}bD\left(\ln\frac{1}{t_0} - \frac{1 - t_0^2}{1 + t_0^2}\right). \tag{95}$$

This agrees with eq. (75) for small a_0.

To find the field of an edge dislocation with a general Burgers vector $b_x = b \cos \varphi$, $b_y = b \sin \varphi$ the above results may be combined linearly. For example, the stress function for such a dislocation in a stress-free cylinder is $\cos \varphi$ times eq. (73) plus $\sin \varphi$ times eq. (84). The energy is $\cos^2 \varphi$ times eq. (75) plus $\sin^2 \varphi$ times eq. (86) since the image field due to b_x does not exert any force on the component b_y of the Burgers vector, and conversely, Leibfried and Dietze, quoted by Seeger [26], state that eq. (75) gives the energy for any orientation of the Burgers vector if a_0 is much less than a. This is correct if we are interested in the *value* of E, but it is not good enough if we are interested in its *gradient*; the non-logarithmic term in eq. (86) gives an additional (negative) term in the force of eq. (85) which is comparable in magnitude with the contribution from the logarithmic term.

We next suppose that fig. 2a represents a circular cylindrical inclusion (region 2) of elastic constants μ_2, ν_2 in a matrix (region 1) with constants μ_1, ν_1, and that an edge dislocation is introduced at P_1 or P_2.

The various cases have been worked out by Dundurs and his associates (Dundurs and Mura [28], dislocations in the matrix; Dundurs and Sendeckyj [27], dislocations in the inclusion; Dundurs [7], general review of this and related work). List [29] has given a unified treatment using the method of Muskhelishvili. Aderogba [30] has given a formula which might be useful in this kind of problem. It gives the perturbation due to the introduction of a circular cylindrical inhomogeneity into any initial elastic field.

In a sense there are now *two* elastic fields associated with each region, its true field and the image field, the field of the other region extrapolated into it. We might hope to find the solutions in the way which served for the screw dislocation, namely by taking a linear combination of all the image singularities already encountered in connexion with the stress-free cylinder and cylindrical hole and then adjusting the constants to give continuity of traction and displacement at the interface.

To realize this programme it is actually necessary to introduce additional singularities in the form of concentrated forces coincident with the image dislocations and directed at right angles to their Burgers vectors, and also, in some cases, concentrated couples as well.

As a specimen we present Dundurs and Sendeckyj's [27] stress function for a dislocation with Burgers vector $(b, 0)$ at P_1 in a matrix with elastic constants μ_1, ν_1 near the cylindrical inclusion C_2 with constants μ_2, ν_2.

The stress function is

$$\chi = -D_1 y\left[\ln r_1 - \tfrac{1}{2}(P+Q)\ln\frac{r_2}{r} + \tfrac{1}{2}Pa^2\left(\frac{1}{r^2} - \frac{r_1^2}{\xi_1^2 r_2^2}\right)\right]$$
$$+ \tfrac{1}{2}D_1(Q-P)(x-\xi_2)(\theta_2-\theta) \tag{96}$$

in the matrix and

$$\chi = -D_1(1 - \tfrac{1}{2}P - \tfrac{1}{2}Q)y\ln r_1 + \tfrac{1}{2}D_1(P-Q)[(x-\xi_1)\theta_1 - (\xi_1-\xi_2)\theta_1]$$

in the inclusion, and the elastic energy is

$$E = \tfrac{1}{2}bD_1\left[\tfrac{1}{2}(P+Q)\ln\frac{\xi_1^2 - a^2}{\xi_1^2} - \tfrac{1}{2}(3P-Q)\frac{a^2}{\xi_1^2}\right] + \text{const.} \tag{97}$$

with

$$P = \frac{\mu_1 - \mu_2}{\mu_1 + \mu_2\kappa_1}, \qquad Q = \frac{\mu_1\kappa_2 - \mu_2\kappa_1}{\mu_2 + \mu_1\kappa_2}$$

$$\kappa_1 = 3 - 4\nu_1, \qquad \kappa_2 = 3 - 4\nu_2, \qquad D_1 = \frac{\mu_1 b}{2\pi(1-\nu_1)}.$$

In eq. (96) the new terms are those in θ, θ_2; eqs. (7) and (8) have been used to make the other terms resemble eq. (88). From eqs. (62) and (55) it is clear that the term $(x - \xi_2)\theta_2$ represents a force transverse to the x-axis combined with a dislocation whose stress function might have been absorbed into the first line. The term $-x\theta$ represents an opposite force and dislocation at O and the term $\xi_2\theta$ a couple at O which compensates the moment of the two forces. If the Burgers vector changes to $(0, b)$ the forces are directed along the x-axis and the couple is not needed; its place is actually taken by a centre of dilatation.

For a plane interface Head [31] devised an original method which deserves mention, though we shall have to refer to the original paper for the rather lengthy details.

In plane strain one of the Michell–Beltrami compatibility equations reads $\nabla^2 p_{xx} + \partial^2 P/\partial x^2 = 0$ which, since $P = p_{xx} + p_{yy}$ is harmonic, may be written as $\nabla^2 T = 0$ with

$$T = p_{xx} + \tfrac{1}{2}xP$$

and so, if we take $x = 0$ for the interface, continuity of p_{xx} is equivalent to continuity of the harmonic function T. If we arrange that p_{xy} is continuous at one point of the interface it is enough to require continuity of $\partial p_{xy}/\partial y$, or equally well of $\partial p_{xx}/\partial x$ elsewhere on it, by one of the equilibrium equations. This quantity can be written as a linear combination of $\partial T/\partial x$ and $\partial P/\partial x$, plus a term which vanishes on $x = 0$. Similarly continuity of displacement may be replaced by continuity of $\partial v/\partial y$ and $\partial^2 u/\partial y^2$ and these quantities may be written, respectively, as a linear combination of T and P and of $\partial T/\partial x$ and $\partial P/\partial x$, plus, in each case, a term which vanishes on $x = 0$. The boundary conditions are thus reduced to the continuity of the four harmonic functions

$$T \tag{98}$$

$$\frac{1}{2\mu} T - \frac{1-\nu}{2\mu} P \tag{99}$$

$$\frac{\partial}{\partial x}(T - \tfrac{1}{2}P) \tag{100}$$

$$\frac{\partial}{\partial x}\left(\frac{1}{2\mu} T + \frac{1-2\nu}{4\mu} P\right) \tag{101}$$

across $x = 0$. Head now forms the linear combination

$$\left(\alpha + \frac{1}{2\mu}\right) T - \frac{1-\nu}{2\mu} P \tag{102}$$

of the first pair and

$$\frac{\partial}{\partial x}\left[\left(\beta + \frac{1}{2\mu}\right) T + \left(\frac{1-2\nu}{4\mu} - \tfrac{1}{2}\beta P\right)\right] \tag{103}$$

of the second, and requires α, β to be chosen so that (103) is a multiple of the x-derivative of (102). If β is fixed this can be done by varying α, but α would be different in the two media. However if both α and β are allowed to vary they may be chosen to be the same in the two media. This leads to a quadratic equation for α, with roots α', α'' say. The corresponding quantities (102), say

$$V' = A'T + B'P, \qquad V'' = A''T + B''P \tag{104}$$

are then harmonic, continuous across $x = 0$ and, because of (103), the quantities $K'\ \partial V'/\partial x$ and $K''\ \partial V''/\partial x$ are also continuous across $x = 0$ where K', K'' depend on α', α'' and the elastic constants. These are the boundary conditions for the potential at the interface between two dielectrics, so that we expect that point singularities can be dealt with by imaging.

For a dislocation with, say, Burgers vector $(b, 0)$ at P_1 (fig. 3) in a homogeneous medium we have

$$V' = C' \frac{y}{r_1^2} + D' \frac{2\xi y(x-\xi)}{r_1^4}. \tag{105}$$

The two terms may be dealt with separately. The electrostatic analogy suggests that when the elastic constants in the region $x < 0$ change eq. (105) should be replaced by

$$C' \frac{y}{r_1^2} + C'^* \frac{y}{r_2^2}, \qquad x > 0$$

$$C'^{**} \frac{y}{r_1^2}, \qquad x < 0.$$

The constant C' must stay unchanged to give the correct singularity at P_1 and the boundary conditions fix the other two. The second term in eq. (105) can be modified similarly, and then V'' may be treated in the same way.

In this way we end up with two known combinations, eqs. (104), of T and P from which p_{xx} and p_{yy} can be extracted, and hence also p_{xy} by integration of $\partial p_{xy}/\partial Y = -\partial p_{xx}/\partial x$. The numerical coefficients in the result depend on the elastic constants and on the roots α', α''. Head shows that if the two Poisson's ratios are the same only $\alpha' + \alpha''$ and $\alpha'\alpha''$ are involved, so that the coefficients are free of radicals. Comparison with Dundurs' results [for example eq. (96) above] suggests that the same is true in the general case. Indeed this must be so since the T and P found by solving the simultaneous equations (104) are rational symmetric functions of α' and α'' [32] though it is difficult to make out the details.

Dislocations near plane interfaces in anisotropic media have been treated by Pastur et al. [33], Head [34], Gemperlova [35] and Tucker [36].

2.4. Other solutions for edge dislocations

There has not been much analytical progress in extending the results of the last two sections to non-circular boundaries, because of difficulties in handling biharmonic boundary value problems. Since the same difficulties also apply to other physical phenomena governed by the same equations there is not, as there was in the case of anti-plane strain, an extensive store-house of results to draw on.

Equations (46) and (54) also govern the transverse deflection of a plate clamped around C. If the plate is deflected by a point force at ξ the displacement is

$$w = \text{const.}\,(\boldsymbol{r} - \boldsymbol{\xi})^2 \ln |\boldsymbol{r} - \boldsymbol{\xi}|^2 - w_1(\boldsymbol{r}),$$

where on C w_1 has the same value and normal derivative as the first, singular, term. When regarded as a stress function w represents a wedge dislocation (disclination) made by cutting out a narrow wedge of material with its apex at $\boldsymbol{\xi}$ and closing and welding together the faces of the gap. If w is differentiated with respect to ξ_x or ξ_y the resulting stress function has a singularity of the form const. $(x - \xi_x) \ln |\boldsymbol{r} - \boldsymbol{\xi}|$ or const. $(y - \xi_y) \ln |\boldsymbol{r} - \boldsymbol{\xi}|$ and so represents an edge dislocation [4, 37]. A number of solutions have been given for clamped plates deflected by a point force at a single point of high symmetry. They can be used as the solution for a disclination but, as they cannot be appropriately differentiated they are of no help in the dislocation problems. Indeed the only useful solution seems to be Michell's for a force at an arbitrary point in a clamped circular plate [2]. It can be used to reproduce some of the formulas of the last section.

Seeger [26], quoting from the work of Leibfried and Dietze, gives the value

$$E = \frac{\mu b^2}{4\pi(1-\nu)} \ln\left(\frac{2d}{\pi a_0} \cos\frac{\pi y'}{d}\right) \tag{106}$$

for the energy of an edge dislocation, with arbitrarily oriented Burgers vector, distant y' from the mid-plane of a plate of thickness d. This result is remarkably simple when one recalls the difficulty of solving elastic problems relating to slabs, but the promised details have not been published.

If in fig. 4b the array of screw dislocations becomes an array of edge dislocations with Burgers vectors $(0, \pm b)$ the appropriate Airy stress function is easily seen to be

$$\chi = -\frac{x}{1-\nu}\varphi \tag{107}$$

with the φ of eqs. (43). The forces between the dislocations increase by a factor $1/(1-\nu)$ and so the potential energy of the real dislocation in the field of the images is $1/(1-\nu)$ times the energy eq. (37) with the Φ of eqs. (43), and this agrees with eq. (106).

However, eq. (107) only satisfies the first of the boundary conditions of eqs. (54), and the normal but not the tangential traction is zero at the faces of the plate. To liquidate the tractions completely we should have to carry out the usual procedure in which by repeated imaging the boundary conditions are alternately satisfied on one of the planes at the expense of a certain, but ultimately vanishing, violation of them on the other. At each imaging an image dislocation acquires a family of singularities of the type of eq. (74), but in addition its previous family acquires a family of its own. Hence finally we have an array of image dislocations each accompanied by a group of singularities which increases in complexity as we move outwards from the real dislocation in the plate. The implication of eq. (106) is that these extra singularities, either group by group or collectively, exert no force on the real dislocation. Equation (106) is perhaps subject to the limitation already noticed in connection with eq. (75); see the discussion following eq. (95). Lee and Dundurs' [38] comparison of eq. (106) with their own numerical results also suggests this.

The energy of a dislocation near a crack, an infinitely flat elliptical hole, is of interest in fracture mechanics and can easily be found with the help of the formalism developed in that subject. Suppose a straight crack has its ends at (0, 0) and $(l', 0)$. First calculate the three stress intensity factors at the tip $(l', 0)$ in the form of suitably weighted integrals of the unperturbed dislocation stresses along the proposed site of the crack, and from them find the energy release rate $G(l')$. (See, for example, ref. [39].) By definition $G(l')\,\mathrm{d}l'$ is the reduction in total energy when the tip extends by $\mathrm{d}l'$. (The total energy is the sum of the elastic energy and, though it does not concern us here, the potential energy of the loading mechanism if any.) Thus if the crack establishes itself by growing from zero to a finite length l, the end (0, 0) staying fixed, the energy of the dislocation falls from E_0 to

$$E = E_0 - \int_0^l G(l')\,\mathrm{d}l'.$$

The interaction energy between a dislocation and an inclusion of any shape can be found very easily with the help of a simple general result [40] provided the difference in elastic constants is small. Suppose that in an initially homogeneous body the Lamé constants μ, λ change to $\mu + \delta\mu$, $\lambda + \delta\lambda$ where $\delta\mu$ and $\delta\lambda$ may possibly vary continuously from point to point in addition to being discontinuous across interfaces. Then an estimate for the increase in elastic energy is the integral

$$E = \tfrac{1}{2}\int(\delta\lambda e^2 + 2\delta\mu e_{ij}e_{ij})\,\mathrm{d}v \tag{108}$$

taken over the region where $\delta\mu$ and $\delta\lambda$ are not zero, using the original strains e_{ij}. The error in eq. (108) is of order $(\delta\mu)^2$ even though the strains are altered by quantities of order $\delta\mu$. (For brevity we shall suppose that $\delta\mu$ and $\delta\lambda$ are of the same order.) In eq. (108) it is assumed that the strengths of the sources of internal stress (or rather strain) are unaffected by the change in the elastic constants; in the present case this means that the Burgers vectors of the dislocations are not altered. If the stresses are partly produced by external loads eq. (108) gives the change of the total energy, made up of the elastic energy and the potential energy of the loading mechanism. There is a similar result for anisotropy.

To verify these statements suppose that despite the change in elastic constants the original e_{ij} remain unchanged. The change of energy is then exactly the quantity of eq. (108), even for finite $\delta\mu$, $\delta\lambda$. To maintain these "wrong" strains it is necessary to impose a certain volume distribution of body force and also suitable layers of force on the interfaces across which the elastic constants change abruptly. They could be found by substituting the old strains into the new equilibrium equations with body force and the new interface boundary conditions, but we know in any case that they are of order $\delta\mu$. If they are now relaxed the change of displacement is of order $\delta\mu$ and an amount of work E_1 of order $(\delta\mu)^2$ is extracted from the system, and so eq. (108) is still correct apart from being too large by an amount E_1 of order $(\delta\mu)^2$. Equation (108) is an example of a general class of theorems of which the Hellman–Feynman theorem [41] in quantum mechanics is the best known. The correction $-E_1$ which makes eq. (108) correct apart from an error of order $(\delta\mu)^3$ can also be calculated without solving a boundary value problem since with sufficient accuracy the displacement induced by the restraining body forces can be calculated using the elastic Green's function for the original uniform medium, eq. (112) below.

Equation (108) can give not too bad results even if $\delta\mu$, $\delta\lambda$ are not particularly small. For example it reproduces the functional form of the two terms in eq. (97) correctly, and with, say, $\nu_1 = \nu_2 = \frac{1}{4}$, $\mu_1 = 1.25\,\mu_2$ it gives 0.10 and 0.03 for their coefficients $\frac{1}{2}(P + Q)$ and $\frac{1}{2}(3P - Q)$ in place of the correct 0.11 and 0.04. If the angle between the Burgers vector and the x-axis is Θ rather than zero eq. (97) becomes

$$E = \tfrac{1}{2}bD_1\left(L \ln \frac{\xi_1^2 - a^2}{\xi_1^2} + M\frac{a^2}{\xi_1^2}\cos 2\Theta + N\frac{a^2}{\xi_1^2}\right) \tag{109}$$

where for small $\delta\mu$, L and M are of order $\delta\mu$ and N is of order $(\delta\mu)^2$. Equation (108) reproduces the terms in L and M but, naturally it fails to give the term in N, which, however, appears in the correction $-E_1$.

Reference [7] contains an extensive list of papers on two-dimensional inhomogeneity problems, to which may be added the following. The problem of an edge dislocation in a semi-infinite solid with a finite plane inhomogeneous surface layer is treated in ref. [42] for isotropic and in ref. [43] for anisotropic media. Reference [38] deals with the case when the dislocation is in the surface layer.

3. *Three-dimensional problems*

3.1. Dislocations in a semi-infinite medium

A problem of importance in, for example, electron microscopy and geophysics is the determination of the effect of a free surface on the elastic field of a dislocation loop. In the geophysical applications it will usually be enough to consider a semi-infinite solid (half-space) and this case may also be adequate in the electron microscopy application, or at least it represents the first step towards the solution for a dislocation in a plate.

In dealing with any elastic problem concerning a half-space it is useful to introduce what may be called a geometrical image field. A physical picture of an original displacement u_i^∞ may be produced by drawing an arrow at every point of space to represent the displacement there. Reflect the whole pattern of arrows in the horizontal plane $x_3 = 0$ and interpret the reflected pattern as a picture of the geometrical image displacement field u_i^{G}. We evidently have

$$u_i^{\mathrm{G}}(x_1, x_2, x_3) = u_i^\infty(x_1, x_2, -x_3), \qquad i = 1, 2$$

and

$$u_3^{\mathrm{G}}(x_1, x_2, x_3) = -u_3^{\mathrm{G}}(x_1, x_2, -x_3). \tag{110}$$

A few sketches combined with eq. (110) will show that the original and image loops are identical for horizontal or vertical prismatic loops and for vertical shear loops with a horizontal Burgers vector, and equal and opposite for horizontal shear loops and vertical shear loops with a vertical Burgers vector.

From eq. (110) we easily get

$$p_{33}^{\mathrm{G}} = p_{33}^\infty, \qquad p_{31}^{\mathrm{G}} = -p_{31}^{\mathrm{G}}, \qquad p_{23}^{\mathrm{G}} = -p_{23}^\infty \quad \text{on} \quad x_3 = 0.$$

Hence if we add the G-field to the ∞-field the shear force on x_3 is annulled and the normal force is doubled. The effect of removing the normal forces can then be calculated, say from Boussinesq's expression ([2] p. 192) for the field due to an arbitrary normal load applied to the surface of a semi-infinite solid. In this way Steketee [44] found the field of an elementary dislocation loop in an infinite solid. (In his equations (7.18) and (7.20) τ is a misprint for w.) He actually used a Fourier method to annul the normal tractions (cf. also Maruyama [45]). Alternatively we could annul the normal force and double the tangential forces by subtracting $u_i^{\rm G}$, and complete the calculation with the help of the expressions for a tangential surface force. Though this is less convenient the final result may look simpler when written in terms of $u_i^\infty - u_i^{\rm G}$ rather than $u_i^\infty + u_i^{\rm G}$ (see below).

The solution for an infinitesimal loop or other singularity in a semi-infinite medium may also be made to depend upon known expressions for the elastic field of a point force in a semi-infinite medium as given in refs. [46–50] and in two dimensions in [51–53].

The field of an infinitesimal loop at the origin in an infinite medium with Burgers vector b_1, normal n_1 and area $\mathrm{d}S$ can be written in the form [54]

$$\mathrm{d}u_i^\infty = \mu b_j n_k \,\mathrm{d}S\left(\frac{\partial U_{ij}}{\partial x_k} + \frac{\partial U_{ik}}{\partial x_j} + \frac{2\nu}{1-2\nu}\delta_{jk}\frac{\partial U_{im}}{\partial x_m}\right), \qquad (111)$$

where

$$U_{ij} = [(3-4\nu)\,\delta_{ij}/r + x_i x_j/r^3]/16\pi\mu(1-\nu) \qquad (112)$$

is the i-component of the displacement produced by a unit concentrated force at the origin of an infinite medium, directed parallel to the x_j-axis ([2] p. 183), or explicitly

$$\mathrm{d}u_i^\infty = -b_j n_k \,\mathrm{d}S[(1-2\nu)(\delta_{ij}l_k + \delta_{ik}l_j - \delta_{jk}l_i) + 3l_i l_j l_k]/8\pi(1-\nu)r^2$$

with $l_i = x_i/r$.

If the loop is at position ξ_i eq. (111) becomes

$$\mathrm{d}u_i = -\mu b_j n_k \,\mathrm{d}S\left[\delta_{jl}\frac{\partial}{\partial \xi_k} + \delta_{kl}\frac{\partial}{\partial \xi_j} + \frac{2\nu}{1-2\nu}\delta_{jk}\frac{\partial}{\partial \xi_l}\right] U_{il}(x_s, \xi_s), \qquad (113)$$

where $U_{il}(x_s, \xi_s)$ denotes the displacement of eq. (111) translated by ξ_i. We have taken advantage of the fact that it depends only on $x_s - \xi_s$ to replace ∂/∂_{x_i} by $-\partial/\partial\xi_i$, and to save later re-writing the affix ∞ has been dropped. Since ξ_i denotes the point of application of the point force, eq. (113) states that $\mathrm{d}u_i$ is the same as the displacement field due to a certain distribution of force doublets at ξ_i, with zero resultant and, since the expression (111) is symmetric in jk, no resultant moment. Let us next give a new meaning to $U_{il}(x_s, \xi_s)$ in eq. (113) and allow it to stand for the displacement due to a point force at ξ_i in a semi-infinite medium with a stress-free surface. (It is of course no longer a function of $x_s - \xi_s$ only.) The new U_{il} is made up of the old one plus an image field with no new singularities, and so eq. (113) now gives the field of the loop in a semi-infinite medium. More generally we may take $U_{il}(x_s, \xi_s)$ to refer to a point force in any body with a partly stress-free surface provided it extends to infinity in some direction, so that it is legitimate to think of an apparently uncompensated point force acting on it, though in fact it is balanced by forces of order r^{-2} at infinity. Even for a closed surface, say a sphere, where this is not allowable, we can save eq. (113) by re-defining $U_{il}(x_s, \xi_s)$. Let it stand for the i-component of the displacement produced by a point force parallel to the x-axis at ξ_1 in a finite body with a stress-free surface, together with an equal and opposite force at an arbitrary fixed point P (independent of ξ_i) and a point couple at P whose moment balances that of the two forces. Then eq. (113) gives the field of the elementary loop alone since the resultants of the forces and couples at P associated with the force-cluster at ξ_i are zero.

Using Mindlin's results and eq. (113) Bacon and Groves [54, 55] have derived the displacement field of an elementary loop in a semi-infinite medium and have presented their results in a remarkably compact form. The loop is at (0, 0, c) in the semi-infinite solid $x > 0$. They write

$$du_i = du_i^{\infty} + du_i^{\mathrm{I}} + du_i^{\mathrm{S}} \tag{114}$$

where du_i^{∞} is the field in an infinite medium, du_i^{I} is the displacement of a loop of opposite sign at the image point (0, 0, $-c$), so that if

$$du_i^{\infty} = F_i(x_1, x_2, x_3 - c)$$

then

$$du_i^{\mathrm{I}} = -F_i(x_1, x_2, x_3 + c),$$

and du_i^{S} is the additional displacement required to complete the cancellation of tractions on $x_3 = 0$. A loop of area dS, Burgers vector b_i and normal n_i is specified by the tensor $b_i n_i \, dS$ and its elastic field is the sum of the fields of a set of elementary loops of the type $b_1 n_1 \, dS$, $b_1 n_2 \, dS$ and so on in which all but one of the tensor components vanish, and so it is enough to exhibit the following special cases in which the Burgers vector and normal are parallel to one or other of the unit vectors $\boldsymbol{i}_1, \boldsymbol{i}_2, \boldsymbol{i}_3$ of the coordinate system:

(i) prismatic loop: $\boldsymbol{b} = b\boldsymbol{i}_3$, $\quad \boldsymbol{n} = \boldsymbol{i}_3$:

$$du_i^{\mathrm{S}} = Kc[A_{i3}(R^{-1})_{,i3} - (x_3 R^{-1})_{,i33}], \tag{115a}$$

(ii) prismatic loop: $\boldsymbol{b} = b\boldsymbol{i}_j$, $\quad \boldsymbol{n} = \boldsymbol{i}_j$, $\quad j = 1$ or 2:

$$du_i^{\mathrm{S}} = K\{(A_{ij} - 2\nu A_{i3})(R^{-1})_{,i} - BR_{,ijj} + 2\nu(x_3 R^{-1})_{,i3} - c(x_3 R^{-1})_{ijj} + 2Bc\,\delta_{i3}(R^{-1})_{,jj} + D[x_j(R + x_3 + c)^{-1}]_{,ij}\}, \tag{115b}$$

(iii) shear loop: $\boldsymbol{b} = b\boldsymbol{i}_j$, $\quad \boldsymbol{n} = \boldsymbol{i}_3$ $\quad$ or $\boldsymbol{b} = b\boldsymbol{i}_3$, $\quad \boldsymbol{n} = \boldsymbol{i}_j$, $\quad j = 1$ or 2:

$$du_i^{\mathrm{S}} = Kc[-A_{i3}(R^{-1})_{,ij} + (x_3 R^{-1})_{,ij3}], \tag{115c}$$

(iv) shear loop: $\boldsymbol{b} = b\boldsymbol{i}_1$, $\quad \boldsymbol{n} = \boldsymbol{i}_2$ $\quad$ or $\boldsymbol{b} = b\boldsymbol{i}_2$, $\quad \boldsymbol{n} = \boldsymbol{i}_1$:

$$du_i^{\mathrm{S}} = K\{B[\delta_{i1}(R^{-1})_{,2} + \delta_{i2}(R^{-1})_{,1} + 2c\,\delta_{i3}(R^{-1})_{,12} - R_{,i12} - c(x_3 R^{-1})_{,i12} + D[x_1(R + x_3 + c)^{-1}]_{,i2}\}, \tag{115d}$$

where

$$A_{ij} = 2\nu + 2B\,\delta_{ij}, \qquad B = 2(1 - \nu), \qquad D = B(1 - 2\nu)(1 - 2\,\delta_{i3}),$$
$$K = b\,dS/4\pi(1 - \nu), \qquad R^2 = x_1^2 + x_2^2 + (x_3 + c)^2,$$

$(\ )_{,i}$, $(\ )_{,i1}$ denote $\partial/\partial x_1$, $\partial^2/\partial x_i\,\partial x_1$ and so forth, and repeated suffixes are not to be summed.

In (iii) Groves and Bacon's image field and the geometrical image field defined by eq. (110) are equal and opposite, and for (i), (ii) and (iv) they are identical. The appropriate choice of sign is responsible for the simplicity of eqs. (115).

A general finite loop in the half-space can be dealt with by dividing its discontinuity surface, any conveniently chosen surface spanning the loop, into a network of infinitesimal loops each specified by the tensor $b_i n_j\,dS$. This tensor, in turn, can be decomposed into nine tensors in each of which only a single element is not zero, so that its elastic field is the same as that of one of the nine elementary loops displayed in eqs. (115a–d). The field of the complete finite loop can be built up by summation and integration.

There is some simplification for a finite plane prismatic loop parallel to the surface [56]. In this case we have $j = 3$, $k = 3$ in eq. (111) and it is not difficult to see that du_i^∞ only involves derivatives of r^{-1}. Likewise only derivatives of R^{-1} occur in the appropriate du_i^S, eq. (115a), terms in $(R + x_3 + c)^{-1}$ being absent. Consequently the integration reduces to finding the Newtonian potential

$$\int |\boldsymbol{r} - \boldsymbol{r}'|^{-1}\,dx_1'\,dx_2'$$

of a uniform disk bounded by the dislocation line or its image. In the general case one also has to find the corresponding biharmonic (or direct) potential in which $|\boldsymbol{r} - \boldsymbol{r}'|^{-1}$ is replaced by $|\boldsymbol{r} - \boldsymbol{r}'|$ in order to find u_i^∞ and u_i^I [57], and to get u_i^S one must evaluate Boussinesq's first logarithmic potential ([2] p. 192) as well; for a single attracting point it has the form of eq. (118) below, and is responsible for the appearance of $(R + x_3 + c)^{-1}$ in eq. (115). (The expression in eq. (119) is Boussinesq's second logarithmic potential.) Baštecká [58] has calculated the elastic field of a circular prismatic loop with its plane parallel to the free surface.

Division into elementary loops is not always the most convenient way of dealing with dislocations in a half-space. In particular, straight dislocations can be handled more directly. We begin with a screw dislocation running normal to a free surface [59].

In cylindrical polar coordinates r, θ, z the field of a screw dislocation along the z-axis of an infinite medium is

$$u_z^\infty = \frac{b}{2\pi}\theta, \qquad p_{z\theta} = \frac{\mu b}{2\pi}\frac{1}{r}; \tag{116}$$

the other components are zero. There is a couple $\frac{1}{2}\mu b a^2$ about the z-axis on a circle of radius a in $z = 0$. This suggests that the image field for the semi-infinite solid $z > 0$ may be constructed by introducing a distribution of couples along the negative z-axis.

A couple at the origin gives the displacement ([2] p. 187)

$$\boldsymbol{u} = -\operatorname{curl}(0, 0, \varphi)$$

where $\varphi = C/r$. This equation gives a solution of the elastic equations with any harmonic φ. If, further, φ depends on r and z but not on θ we have

$$u_\theta = \partial\varphi/\partial r, \qquad p_{\theta z} = \mu(\partial u_\theta/\partial z), \qquad p_{\theta r} = \mu((\partial u_\theta/\partial r) - u_\theta/r) \tag{117}$$

with the remaining components zero. If we put $\varphi = c/r$, the potential of a point charge, $p_{\theta z}$ is proportional to r^{-3} on $z = 0$ and so does not cancel the stress in eq. (116) on $z = 0$. If we integrate c/r along the negative z-axis it becomes

$$\varphi = c \ln (R + z) \tag{118}$$

with

$$R^2 = r^2 + z^2 = x^2 + y^2 + z^2.$$

This is the potential of a uniformly charged wire along the negative z-axis. With eq. (118), eq. (117) gives a stress proportional to r^{-2} on $z = 0$. We therefore integrate once more to get

$$\varphi = c[z \ln (R + z) - R], \tag{119}$$

the potential of a semi-infinite wire whose charge is proportional to $|z|$. With eq. (119), eq. (117) gives $p_{\theta z} = c\mu/r$ on $z = 0$ which cancels the $p_{\theta z}$ of eq. (116) with $c = -b/2\pi$, as required, and the image field is

$$u_\theta^{\text{im}} = -\frac{b}{2\pi}\frac{r}{R + z}, \qquad p_{\theta z}^{\text{im}} = -\frac{\mu b}{2\pi}\frac{r}{R(R + z)},$$

$$p_{\theta r}^{\text{im}} = \frac{\mu b}{2\pi}\left(\frac{z^2}{R^3} + \frac{1}{R + z}\right). \tag{120}$$

Note that $1/(R + z)$ may also be written in the form $R/r^2 - z/r^2$. The complete field is the sum of the expressions (116) and (120).

With the help of her theory of angular dislocations Yoffe [60] has given an elegant treatment of the general case where a straight dislocation, of arbitrary Burgers vector, meets the surface of a semi-infinite solid at any angle.

Figure 5a shows a dislocation meeting the free surface of the semi-infinite solid $x_3 < 0$. We cannot allow it to terminate in the infinite solid with which we have to start the calculation, and so we let it continue along the negative x_2-axis so as to form an angular dislocation. (In physical terms the dislocation produces a growth step on the free surface, say along the negative x_2-axis. When a second semi-infinite solid is welded onto the first the step gets built in as a dislocation.)

The angular dislocation can be analysed into a network of infinitesimal loops, one of which, together with its geometrical image, as defined by eq. (110), is shown in the figure. We suppose for the moment that the Burgers vector is horizontal. Then the argument following eq. (110) shows that if the original and image elementary loops are described in the same sense their Burgers vectors are the same. Thus if an image dislocation is synthesized from the image loops its horizontal arm will cancel that of the original dislocation, and the remaining parts of original and image may be joined up to form a single angular dislocation whose field is known (Yoffe [61]). It produces a purely normal traction on $x_3 = 0$, since, as we have seen, each elementary object–image loop pair does so. Yoffe cancels this normal traction by adding the elastic field

$$2\mu u_i = x_3 \frac{\partial^2 \varphi}{\partial x_i\, \partial x_3} + (1 - 2\nu)\frac{\partial \varphi}{\partial x_i}, \qquad i = 1, 2$$

$$2\mu u_3 = x_3 \frac{\partial^2 \varphi}{\partial x_3^2} - 2(1 - \nu)\frac{\partial \varphi}{\partial x_3} \tag{121}$$

with harmonic φ. The stresses which contribute to the traction on $x_3 = 0$ are

$$p_{13} = x_3 \frac{\partial^2 \varphi}{\partial x_1\, \partial x_3^2}, \qquad p_{23} = x_3 \frac{\partial^2 \varphi}{\partial x_2\, \partial x_3^2}$$

and

$$p_{33} = x_3 \frac{\partial^3 \varphi}{\partial x_3^3} - \frac{\partial^2 \varphi}{\partial x_3^2},$$

so that the surface will be stress-free if $\partial^2\varphi/\partial x_3^2$ is chosen to be equal in value to p_{33} for the angular dislocation on $x_3 = 0$.

We refer to the original paper for the general expressions for φ and the field of the angular dislocation and only consider the case where the dislocation meets the surface normally. The angular dislocation then becomes a straight dislocation along the x-axis, with the elastic field of eq. (57), supposing that its Burgers vector is parallel to the x-axis. The appropriate φ is then

$$\varphi = \frac{\nu\mu b}{2\pi(1 - \nu)}\left[x_3 \ln(R - x_3) - \frac{x_2 x_3}{R - x_3}\right]$$

with

$$R^2 = x_1^2 + x_2^2 + x_3^2.$$

The total shear stress p_{12} in $x_3 = 0$ has the same functional form as for the infinite dislocation, but multiplied by a factor $1 + \nu(1 - 2\nu)$, or say about 1.1. Two like

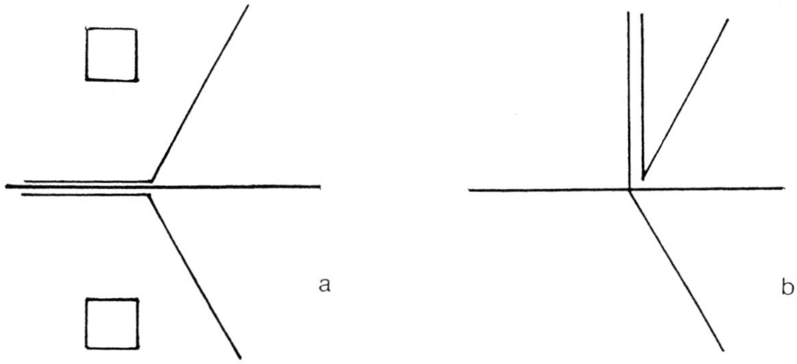

Fig. 5 (a) An angular dislocation, an elementary loop, and their images in a free surface. (b) An acute-angled and an obtuse-angled dislocation combine to form a straight dislocation and its image.

parallel dislocations will thus experience a more than normal repulsion near the surface and should splay apart there. Actually the contrary behaviour is sometimes found in ionic crystals; evidently elastic image effects will not explain it [62]. It is possibly due to an electrostatic surface effect [63].

If next the Burgers vector of the dislocation is vertical the Burgers vectors of the elementary loops in fig. 5a are opposite, the horizontal arms of the original and image dislocations do not cancel but reinforce one another and their oblique arms cannot be joined up to form a single angular dislocation. The configuration still produces, by our general argument, a purely normal traction on $x_3 = 0$, but the function φ required to annul it would, so to speak, have to waste a great deal of effort in cancelling the singularity along the negative x_3-axis. To avoid this we swing the horizontal arms round until they lie along the positive x_3-axis to give a pair of angular dislocations (fig. 5b) one acute, the other obtuse. Yoffe finds that they produce the same shear stresses on $x_3 = 0$ as would an infinite screw dislocation along the x-axis, and these can be cancelled by a field like that of eq. (120) but reflected in $x_3 = 0$, since the solid now occupies the half-space $x_3 < 0$. They also give a normal stress which can be cancelled by the field of eq. (121) with a suitable φ. We again refer to the original paper for the details.

3.2. Dislocations in plates and disks

The elastic field for a screw dislocation traversing a plate may be found by an extension of the analysis for the half-space [59]. We assume that the image field is of the same form as eq. (117). Then both u_θ^{im} and $p_{z\theta}^{\text{im}}$ satisfy the single equilibrium equation

$$\frac{\partial^2 f}{\partial r^2} + \frac{1}{r}\frac{\partial f}{\partial r} - \frac{f}{r^2} + \frac{\partial^2 f}{\partial z^2} = 0 \tag{122}$$

which has the basic solutions

$$\mathrm{e}^{\pm kz} J_1(kr), \qquad \mathrm{e}^{\pm kz} Y_1(kr), \qquad \mathrm{e}^{\pm ikz} I_1(kr), \qquad \mathrm{e}^{\pm ikz} K_1(kr), \tag{123}$$

where k is arbitrary and J_1, Y_1, I_1, K_1 are the Bessel functions usually so denoted.

On a free surface $z =$ const. we must have

$$p_{z\theta}^{\text{im}} = -p_{z\theta}^{\infty} = -\frac{\mu b}{2\pi}\frac{1}{r}, \tag{124}$$

which may also be written as

$$p_{z\theta}^{\text{im}} = -\frac{\mu b}{2\pi}\int_0^\infty J_1(kr)\,\mathrm{d}k \tag{125}$$

by using the standard results $J_1 = -J_0'$, $J_0(0) = 1$, $J_1(\infty) = 0$.

To extend the expression in eq. (125) into the interior of the semi-infinite solid $z > 0$ we insert a factor e^{-kz} into the integrand, so making it a solution of eq. (122)

which falls off with increasing z and agrees with eq. (124) for $z = 0$. From eq. (117) the corresponding displacement is

$$u_\theta^{\text{im}} = -\frac{b}{2\pi}\int_0^\infty e^{-kz}J_1(kr)\frac{dk}{k},$$

which by a standard result reproduces the displacement of eq. (120).

Similarly, for a plate with stress-free surfaces $z = \pm d$ we insert a factor $\cos kz/\cos kd$ into the integrand of eq. (125). The corresponding displacement is now

$$u_\theta^{\text{im}} = -\frac{b}{2\pi}\int_0^\infty \frac{\sinh kz}{\cosh kd}J_1(kr)\frac{dk}{k}. \tag{126}$$

The expansion

$$\operatorname{sech} kd = 2\sum_{n=0}^{\infty}(-1)^n \exp[-(2n+1)kd]$$

converts eq. (126) into a sum of simple image terms like eq. (120); for each of them the origin of z is displaced and that part of the z-axis which is the seat of couples lies outside the plate:

$$u_\theta^{\text{im}} = \frac{b}{2\pi}\sum_{n=0}^{\infty}(-1)^n \times\left\{\frac{r}{d_n+z+[(d_n+z)^2+r^2]^{1/2}} - \frac{r}{d_n-z+[(d_n-z)^2+r^2]^{1/2}}\right\}$$

with

$$d_n = (2n+1)d.$$

Alternatively eq. (126) can be expressed as a series of modified Bessel functions by the theory of residues. The complete field, original plus image, is

$$u_z = \frac{b}{2\pi}\theta, \qquad u_\theta = -\frac{b}{2\pi}\frac{z}{r} + \frac{b}{\pi}\sum(\sin\tfrac{1}{2}n\pi)\frac{2}{n\pi}K_1\left(\frac{n\pi r}{2d}\right)\sin\frac{n\pi z}{2d},$$

$$p_{\theta z} = \frac{\mu b}{\pi d}\sum(\sin\tfrac{1}{2}n\pi)K_1\left(\frac{n\pi r}{2d}\right)\cos\frac{n\pi z}{2d},$$

$$p_{r\theta} = \frac{\mu b}{\pi}\frac{z}{r^2} - \frac{\mu b}{\pi d}\sum(\sin\tfrac{1}{2}n\pi)K_2\left(\frac{n\pi r}{2d}\right)\sin\frac{n\pi z}{2d}, \tag{127}$$

$$u_r = 0, \qquad p_{rr} = p_{\theta\theta} = p_{zz} = p_{rz} = 0 \tag{128}$$

with summation over integral n (terms with even n are actually zero).

The solution of eq. (127) can be verified without reference to the method by which it was found. By eq. (123) each term of u_θ satisfies eq. (122), and in $p_{z\theta}$ each term vanishes on $z = \pm d$. Since $K_n(x)$ falls off rapidly as x increases the Bessel function terms are unimportant a few multiples of d away from the dislocation axis, and the field components reduce to

$$u_z = \frac{b}{2\pi}\theta, \qquad u_\theta = -\frac{b}{2\pi}\frac{z}{r}, \qquad p_{\theta z} = 0, \qquad p_{r\theta} = \frac{\mu b}{\pi}\frac{z}{r^2}, \tag{129}$$

together with eq. (128). This simple field in fact satisfies the equilibrium equations and the boundary conditions on $z = \pm d$ and there seems to be no need for the Bessel function terms. However, the field of eq. (129) is evidently the average of that of eq. (120) and of the same field reflected in $z = 0$ but with couples of opposite hand. There is thus a distribution of couples along the z-axis not only in the image but also in the plate itself. It is the task of the Bessel function terms to cancel the part of the couple distribution inside the plate without upsetting the boundary conditions at the surfaces. That they do get rid of the extra axial singularity may be verified by expanding the term $-bz/2\pi r$ of the complete expression for u_θ (eq. (127)), as a Fourier series in sin $(n\pi z/2d)$, and combining each term with a corresponding Bessel function term. The result is the original series for u_θ with the term $-bz/2\pi r$ deleted and each $K_1(x)$ replaced by $K_1(x) - x^{-1}$, which is non-singular at $x = 0$.

A screw dislocation in a plate exerts a total force

$$F = b' \int_{-d}^{d} p_{\theta z}\, \mathrm{d}z = \frac{4bb'}{\pi^2} \sum_{n \text{ odd}} \frac{1}{n} K_1\left(\frac{n\pi r}{2d}\right) \tag{130}$$

on another screw dislocation with Burgers vector b' distant r from it. If r is much less than d the force has a value

$$F = \frac{bb'}{2\pi r}\cdot 2d,$$

which is the same as the force on a length $2d$ of one of a pair of dislocations in an infinite medium. As r grows beyond d the force falls rapidly: for $r = d$ it is about one third of the value for an infinite medium, for $r = 2d$ about one twentieth. Thus like dislocations in a thin film can be made to approach each other to within a distance of about d under the influence of very moderate forces.

In electron micrographs a collection of like screw dislocations in a foil often seem to form a fairly regular lattice with a spacing of the order of the film thickness [64]. This is what one would expect of a set of entities with short-range hard repulsions pushed together by some external force. The elastic stresses which can exist in a deformed or non-uniformly heated foil will not usually be such as to provide this force, but there will evidently be an effective force tending to make the dislocations congregate wherever there is a local thinning of the foil, since there they will be shorter and their self-energies will be less.

It is not hard to verify that the stress on a plane parallel to the axis of a screw dislocation in a plate can be reduced to zero by introducing into the plate a second dislocation of opposite sign which is the geometrical image of the first with respect to the plane. Consequently eq. (130) also gives the image force on a dislocation distant $\frac{1}{2}r$ from the edge of a semi-infinite plate.

The displacements and stresses of eq. (127) can be adapted so as to give the elastic field in a circular disk of outer radius a and inner radius a_0, the latter representing the hollow core of the dislocation. Consider the Bessel-function terms for one particular value of n taken by themselves. They satisfy the governing equations and the condition that the plate surfaces be stress-free. By eq. (123) this will still be true if we replace K_1 in u_θ by a linear combination $A_n I_1 + B_n K_1$ of itself and the second solution, $I_1(x)$, of the equation satisfied by K_1, with corresponding changes in the other terms. If we treat the terms for all n in this way the boundary conditions at the surfaces of the plate are not upset, and we can choose the A_n, B_n so that the cylindrical surfaces $r = a$, $r = a_0$ are also stress-free. The final result is

$$u_z = \frac{b}{2\pi}\theta, \qquad u_\theta = -\frac{b}{2\pi}\frac{z}{r} + \sum\left[A_n I_1\left(\frac{n\pi r}{2d}\right) + B_n K_1\left(\frac{n\pi r}{2d}\right)\right]\sin\frac{n\pi z}{2d},$$

$$p_{\theta z} = \frac{\mu\pi}{2d}\sum n\left[A_n I_1\left(\frac{n\pi r}{2d}\right) + B_n K_1\left(\frac{n\pi r}{2d}\right)\right]\cos\frac{n\pi z}{2d},$$

$$p_{r\theta} = \frac{\mu b}{\pi}\frac{z}{r^2} + \frac{\mu\pi}{2d}\sum\left[A_n I_2\left(\frac{n\pi r}{2d}\right) - B_n K_2\left(\frac{n\pi r}{2d}\right)\right]\sin\frac{n\pi z}{2d},$$

$$u_r = 0, \qquad p_{rr} = p_{\theta\theta} = p_{zz} = p_{rz} = 0, \tag{131}$$

where

$$A_n = \frac{4b}{\pi^2}\frac{1}{n^3}\sin\tfrac{1}{2}n\pi\left[\left(\frac{2d}{n\pi a_0}\right)^2 K_2\left(\frac{n\pi a}{2d}\right) - \left(\frac{2d}{n\pi a}\right)^2 K_2\left(\frac{n\pi a_0}{2d}\right)\right]$$
$$\times\left[I_2\left(\frac{n\pi a}{2d}\right)K_2\left(\frac{n\pi a_0}{2d}\right) - I_2\left(\frac{n\pi a_0}{2d}\right)K_2\left(\frac{n\pi a}{2d}\right)\right]^{-1}$$

and B_n is the same expression with I_2 in place of K_2 in the numerator.

The energy E required to form the dislocated disk can be found by integrating $\frac{1}{2}bp_{\theta z}$ over the area in which a plane through its axis intersects the disk. The result is simple if the outer radius is large and the inner radius either large or small compared with the thickness:

$$E = \frac{\mu b^2}{4\pi}\cdot 2d\cdot\tfrac{2}{3}\left(\frac{d}{a_0}\right)^2, \qquad a_0 \gg d$$
$$= \frac{\mu b^2}{4\pi}\cdot 2d\cdot\ln\left(\frac{d}{2.24a_0}\right), \qquad a_0 \ll d.$$

The expressions (131) have the complicated appearance one expects of the solution of a boundary-value problem for a finite cylinder. The fact that it could be obtained fairly easily stems from the circumstance that the dilatation is zero, so that we had to deal with harmonic rather than biharmonic functions. It would be much more difficult to find the corresponding solution for an edge dislocation. However, the case of an infinite plate has been discussed, though in a disguised form.

The problem of finding the elastic field of an edge dislocation passing through a plate perpendicular to its stress-free surfaces is equivalent to the same problem for a line-force traversing the plate in the same way. To see this note that the stress function of eq. (60) for a dislocation (0, b) plus a force (0, $-2\mu b$) is harmonic, so that by the last expression of eq. (45), p_{zz}, p_{zx} and p_{zy}, are zero everywhere. Hence any plane z = const. is stress-free and the plane strain solution is itself the solution for a plate. This means that the image fields of the two singularities cancel, or that the image fields of a dislocation (b, 0) and a line-force (0, $2\mu b$) are identical. Green and Willmore [65] have studied the line-force case, and their paper makes clear how difficult an analytical solution of the problem is. They give some curves of the image field.

Remote from the dislocation the elastic field in the plate will be given closely by the generalized plane stress solution ([2] p. 207). This is derived from the plane solution, eqs. (45), (48) by making the changes $\mu \to \mu$, $\nu \to \nu/(1 + \nu)$, putting p_{zz} equal to zero, and interpreting p_{xx}, p_{yy}, p_{xy}, u and v as denoting merely averages across the thickness of the plate, though in fact their values will be nearly constant across it. Very close to the dislocation line, the generalized plane stress solution, interpreted as an average or otherwise, is incorrect: for example v will have nearly the value given by eq. (57) with the original, not the modified, Poisson's ratio. From the plane strain solution one can also derive a so-called plane stress solution ([2] p. 206) which satisfies the equilibrium equations and leaves the surfaces of the plate stress-free. However, for the edge dislocation it is useless because it contains unwanted stresses of order r^{-3} near the singular line.

The discussion in the last paragraph suggests that the total force exerted by one edge dislocation in a plate on another is the same as it would have exerted on a length $2d$ of the latter if they were both in an infinite medium. When the dislocations are very close or very far apart compared with the plate thickness the result follows from the assumed validity of the plane strain solution close to and the generalized plane stress solution remote from one of the dislocations, but strictly speaking one can say nothing precise about the case when their distance apart is of the order of the thickness of the plate.

The stress field of the dislocation, and with it the interaction between two dislocations, may be changed radically if the plate can relieve its stresses by buckling. The problem has been studied by Mitchell and Head [66]. They conclude that for a circular plate the condition for buckling is $R > 10t^2/b$ for a dislocation of Burgers vector b at the centre of a circular disk of radius R and thickness t.

3.3. Other three-dimensional solutions

In this section we refer to some three-dimensional results which do not fall under the head of sects. 3.1. or 3.2.

Salamon and Dundurs [56] have calculated the field of an elementary loop in a pair of bonded half-spaces, that is, an infinite solid with elastic constants μ_1, ν_1 for $z > 0$ and μ_2, ν_2 for $z < 0$. The method of solution is similar to Groves and Bacon's [55] (see sect. 3.1.) for the stress-free half-space, and starts from known expressions [67, 68] for the field of a point force in such a composite solid. Hsieh and Dundurs

[69] have considered continuous distributions of dislocations in the same composite solid. As an example they work out the field of a screw dislocation along the z-axis. With the notation of eq. (120) and the K of eq. (19) the displacement is

$$u_\theta = -K\frac{b}{2\pi}\frac{r}{R+r} \quad \text{for} \quad z > 0 \quad \text{where} \quad \mu = \mu_1$$

$$= -K\frac{b}{2\pi}\frac{r}{R-r} \quad \text{for} \quad z < 0 \quad \text{where} \quad \mu = \mu_2$$

$$u_z = \frac{b}{2\pi}\theta$$

in both regions.

Some work has been done on the interaction of dislocations with spherical boundaries. Nabarro [70] gave the elastic field of an infinitesimal shear loop at the centre of a sphere with a free surface, and Coulomb [71] found an approximate solution for a spherical cavity embraced by a concentric circular shear loop.

Weeks et al. [72] found the elastic field for a straight screw dislocation threading a spherical cavity along a diameter. The change in energy on bringing the dislocation from infinity to this position is

$$E = -\frac{\mu b^2 a}{2\pi}\left(\frac{\pi^2}{12} + \ln\frac{a}{a_0}\right),$$

where a is the radius of the sphere and a_0 is the usual core radius. They also gave the interaction energy of an edge or screw dislocation with a general spherical inhomogeneity in the limit where the ratio of the radius of the inclusion to its distance from the dislocation is small (see also [73]).

Willis and Bullough [74] have calculated the field of a circular prismatic loop lying outside a spherical cavity, with its axis passing through the centre of the sphere, and worked out the attractive force between loop and cavity. Willis et al. [75] treated the case of a screw dislocation near a spherical cavity. They give a formula for the variation of the interaction energy with distance and study in detail the dilatation (which controls the motion of point defects) induced by the sphere in the otherwise dilatationless stress field of the dislocation.

In these last two papers the problem was solved with the help of a series expansion about the centre of the sphere of the type presented by Love [2] in ch. XI of his book. The work of Collins [76–78] and Blokh [79] might perhaps be exploited in this kind of problem. They have given a general method for finding the change of elastic field when either a spherical cavity or rigid inclusion is introduced into a given field (see also [80]). The perturbed field can be written down at once as an integral involving the original field, but in evaluating it one may, of course, be reduced to using a series expansion. If the differences between the elastic constants of matrix and inclusion are small we may use eq. (108) to find the interaction energy with an inhomogeneity of any shape. The energy given by eq. (108) may be written in the form

$$E = \int [W(\mu_2, \nu_2) - W(\mu_1, \nu_1)]\, \mathrm{d}v, \tag{132}$$

where $W(\mu, \nu)$ is the energy density in a homogeneous material with elastic constants μ, ν, and the integration extends over the inclusion. For an edge dislocation with the field of eq. (57) we have

$$W(\mu, \nu) = \tfrac{1}{2}bD \frac{1}{2\pi}\left(\frac{1}{r^2} + \frac{\nu}{1-\nu}\frac{\cos 2\theta}{r^2}\right). \tag{133}$$

To find its interaction with an inhomogeneous sphere of radius a centred at (R, Θ), in the plane $z = 0$ we have to work out the integrals of $1/r^2$ and $\cos 2\theta/r^2$ over the sphere. The second is easy; since $\cos 2\theta/r^2$ is harmonic its volume average over the sphere is equal to the value at the centre, and so we have at once

$$\int \frac{\cos 2\theta}{r^2}\,\mathrm{d}v = \frac{4\pi}{3}a^3\frac{\cos 2\Theta}{R^2}.$$

For the other term we set up spherical polar coordinates ρ, ϑ, φ and also cylindrical polars $\rho' = \rho \sin\vartheta$, z, φ at the centre of the sphere, with the polar axis parallel to the dislocation line and φ measured from the plane passing through the dislocation and the centre of the sphere. For r^{-2} we use the two-dimensional expansion

$$\begin{aligned}\frac{1}{r^2} &= \frac{1}{R^2 + \rho'^2 + 2R\rho'\cos\varphi}\\ &= \frac{1}{R^2 - \rho'^2}\left(1 - 2\frac{\rho'}{R}\cos\varphi + 2\frac{\rho'^2}{R^2}\cos 2\varphi - \dots\right)\end{aligned}$$

of which only the first term survives the φ-integration, and so we have simply

$$\begin{aligned}\int \frac{\mathrm{d}v}{r^2} &= 2\pi \int_0^\pi \int_0^a \frac{\sin\vartheta\rho^2\,\mathrm{d}\rho\,\mathrm{d}\vartheta}{R^2 - \rho^2\sin^2\vartheta}\\ &= 4\pi a F(R/a)\end{aligned}$$

with

$$F = 1 - x^{-1}(1 - x^2)^{1/2}\sin^{-1} x.$$

Hence the interaction energy takes the form

$$E = \tfrac{1}{2}bD_1 \cdot 2a \cdot \left[LF\left(\frac{R}{a}\right) + M\frac{a^2}{R^2}\cos 2\Theta\right], \tag{134}$$

where the values of L and M can be inferred from eqs. (133) and (132). If, for example, matrix and sphere have shear moduli μ_1, μ_2 and the same Poisson's ratio ν we have

$$L = \frac{\mu_2 - \mu_1}{\mu_1}, \qquad M = \frac{\nu}{3(1-\nu)}L.$$

For a screw dislocation eq. (133) is replaced by

$$W = \frac{\mu b^2}{8\pi^2} \frac{1}{r^2}$$

and the interaction energy with the sphere is, similarly, with the same F,

$$E = \frac{b^2 a}{2\pi} (\mu_2 - \mu_1) F\left(\frac{a}{R}\right) \tag{135}$$

with an error of order $(\mu_2 - \mu_1)^2$; any difference in the Poisson's ratios only makes itself felt in the next approximation. Equation (135) ought to agree with Willis et al.'s [75] series solution if in it we put $\mu_1 = \mu$ everywhere except in the over-all multiplier $\mu - \mu_1$, our $\mu_1 - \mu_2$, but it is not easy to check this. At large distances, where $F(a/R) = a^2/3R^2$, eqs. (134) and (135) agree with the results of Weeks et al. [72] to the stated accuracy.

3.4. Approximate methods

In most of the cases we have considered, the problem of annulling the stresses on all the surfaces of a finite solid is shirked by allowing the material to extend to infinity in one or more directions. The exception eq. (131), suggests that in any case such a complete solution would not be very informative. However, useful results can be obtained for certain finite bodies if we are prepared to tolerate the degree of approximation associated with the theory of the strength of materials and use the intuitive methods employed in it.

A useful tool [81, 82] in such calculations is Colonnetti's theorem [13]. Stated in words it says that the work done in introducing a dislocation into a stressed body is equal to the work done by the loads in moving through the displacements produced by the dislocation.

As a first example consider an edge dislocation passing through a rectangular rod

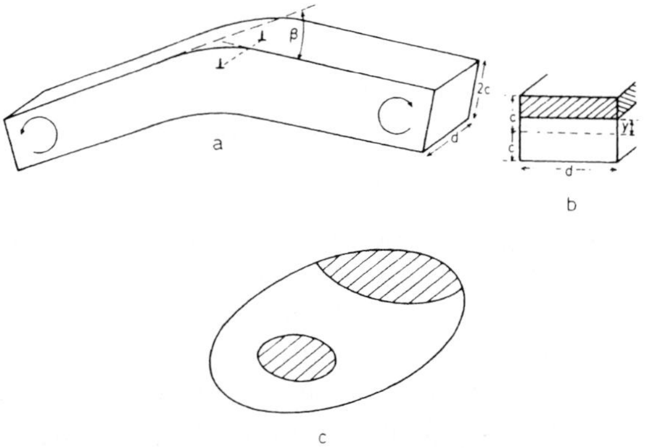

Fig. 6 (a) An edge dislocation passes through a rod with couples applied to its ends. (b) The area over which the product of fibre stress and Burgers vector must be integrated. (c) The dislocation is not straight, or forms a closed loop.

(or plate) as shown in fig. 6a, b. Obviously the rod will be unstrained except near the dislocation, but because of the local deformation there the ends will be rotated through some angle β. To find it bend the rod by couples M applied to the ends. The work done by the dislocation, if it is introduced from the top, is the fibre stress induced by the couples, integrated over the shaded area in fig. 6b and multiplied by the magnitude of the Burgers vector, which we suppose parallel to the axis of the rod. Elementary beam theory gives the value $3bM(c^2 - y^2)/4c^3$ for the work, where $2c$ is the depth of the beam and y is the height of the dislocation above the mid-plane. The work done by the couples is βM, the product of the couple and the relative rotation of the ends. Equating these two quantities we have

$$\beta = \tfrac{3}{4}b(c^2 - y^2)/c^3$$

(Kroupa [83], Siems et al. [84]). Some similar results have been given by Siems et al. [82].

The same method applies if the cross-section is not rectangular and the dislocation is not straight, or if it forms a closed loop, fig. 6c. There will then be bending about each of the principal axes of the cross-section and they may be found by applying a corresponding bending couple and equating βM to b times the integral of the fibre stress over the appropriate shaded area. If the dislocation deforms by gliding on a cylinder with the shaded area as base it does no work against the applied stresses and β is unaltered. Thus an arbitrary dislocation in the rod produces the same total bending as its projection on a cross-section of the rod. For example, suppose that half the dislocation in fig. 6b glides towards the reader so as to form an axial screw dislocation joined to the surface by edge segments at its ends. The total bending is still the same, but it now occurs in two equal instalments at the exit points. The two planes of bending will not usually be parallel because of the twist induced by the intervening screw segment (see below). A component of the Burgers vector not parallel to the axis of the rod does not affect β.

In the anti-plane solutions for a screw dislocation discussed in sect. 2.2. there is a net couple, M say, about the z-axis acting on each cross-section of the material. Consequently if a finite rod is cut out of the infinite cylinder to which the solution applies the rod will develop a twist unless appropriate couples are provided. For a screw dislocation at the centre of a circular cylinder we find $M = \frac{1}{2}\mu b a^2$ and so, dividing by the torsional rigidity $\frac{1}{2}\pi\mu a^4$, we get the value

$$\alpha = b/\pi a^2 \tag{136}$$

radians per unit length for the twist (Mann [85]). There will, of course, be end effects which can in principle be made out from eq. (127).

The relation between the signs of the dislocation and the twist can be described as follows. On the surface of the undislocated cylinder scribe a generator and a set of circles with spacing b. When the dislocation is introduced the circles join up to form a helix and the generator is twisted into another helix, of much coarser pitch. The two helices are of opposite hand. A similar rule applies to the cases which follow.

Twists of the order predicted by eq. (136) have actually been observed in crystal whiskers containing an axial screw dislocation, but the whisker cross-section is not usually circular, and the dislocation is not necessarily at the centre. We therefore need to know the twist due to a dislocation at (x, y) in a whisker whose cross-section is bounded by an arbitrary closed curve C. It can be found at once if the appropriate torsion function $\Psi(x, y)$ is known.

The torsion function [19] satisfies

$$\nabla^2\Psi = -2 \quad \text{inside } C$$

$$\Psi = 0 \quad \text{on } C$$

and in terms of it the torsional rigidity is the integral

$$D = 2\mu \int \Psi \, dx \, dy$$

taken over the interior of C. If the whisker is given a twist α_1 by applying end couples $M = \alpha_1 D$ the stresses in it are

$$p_{zx} = \mu\alpha_1 \, \partial\Psi/\partial y, \qquad p_{zy} = -\mu\alpha_1 \, \partial\Psi/\partial x$$

and they exert a force

$$F_x = bp_{zy} = -\mu b\alpha_1 \, \partial\Psi/\partial x,$$

$$F_y = -bp_{zx} = -\mu b\alpha_1 \, \partial\Psi/\partial y$$

on an axial screw dislocation in the rod, so that Ψ acts as a potential function for the force. If the dislocation moves in from the boundary to (x, y) the work done on it is

$$-\int (F_x \, dx + F_y \, dy) = b\alpha_1\mu \int \frac{\partial\Psi}{\partial s} \, ds$$

per unit length of the whisker, taken along any path, or simply $b\alpha_1\mu\Psi(x, y)$ since Ψ is zero on the boundary. The work done by the end couples is $M_1\alpha$, where α is the twist actually produced by the presence of the dislocation and M_1 is the couple associated with the externally applied twist α_1. By Colonnetti's theorem these two amounts of work are equal and we have

$$\alpha(x, y) = \Psi(x, y)\mu b/D. \tag{137}$$

If we write

$$\alpha = \kappa \frac{\text{Burgers vector}}{\text{area of cross-section}}$$

κ is close to unity for a dislocation near the centre of a reasonably equiaxed cross-section. For a dislocation at the centre of an equilateral triangle, a square or a regular hexagon it has the values 10/9, 1.048, 1.015 respectively. It is also unity for a dislocation at the centre of an ellipse of any excentricity. In contrast it has the value $\frac{3}{4}$ for a long thin rectangle with the dislocation anywhere on the centre line not too near the ends; this follows from the torsion function,

$$\Psi = (\tfrac{1}{2}d)^2 - y^2 \tag{138}$$

for a narrow rectangle with its long sides at $y = \pm\frac{1}{2}d$. If the dislocation has a hollow core we must use the torsion function for a suitable hollow rod. Equation (137) still applies, with Ψ given its (constant) value on the inner boundary. (We may imagine that a singular dislocation moves in from the surface until it falls into a ready-prepared hole.) If the hollow core is nearly as large as the cross-section, so that the whisker becomes a thin tube, not necessarily of constant thickness, the torsional rigidity is very nearly

$$D = \mu\Psi_{\mathrm{I}}(A_{\mathrm{O}} + A_{\mathrm{I}}),$$

where A_{O}, A_{I} are the areas within the outer and inner boundaries of the tube and Ψ_{I} is the value of Ψ at the inner boundary [19]. If we were interested in the torsion problems we should have to estimate Ψ_{I}, but it cancels from eq. (137) to give

$$\alpha = b/(A_{\mathrm{O}} + A_{\mathrm{I}}) \tag{139}$$

or, since A_{I} must be nearly as large as A_{O} for eq. (139) to apply, say simply

$$\alpha = \tfrac{1}{2}b/A_{\mathrm{O}}.$$

(Actually the more elaborate expression of eq. (139) is exact for a tube bounded by any pair of concentric ellipses with parallel axes and the same excentricity.) Thus a solid dislocated whisker with κ of order unity loses about half its twist when hollowed out to form a tube.

As long as the whisker is maintained in a state of anti-plane strain by suitable end couples its elastic energy per unit length is given by eq. (37). When the couples are relaxed the energy is decreased by $\frac{1}{2}\alpha^2 D$, where α is the twist due to the dislocation and eq. (37) has to be replaced by

$$E(x, y) = \frac{\mu b^2}{4\pi}\left[\ln \Phi(x, y) - \frac{2\pi}{S}\Psi^2(x, y)\right], \tag{140}$$

where $S = D/\mu$ is the geometrical torsional rigidity.

For a circle the original energy is given by eq. (14), Ψ is $\frac{1}{2}(a^2 - r^2)$ and eq. (140) gives

$$E = \frac{\mu b^2}{4\pi}[\ln(a^2 - \xi_2^2) - (a^2 - \xi_2^2)^2/a^4], \tag{141}$$

which is plotted in fig. 7. Surprisingly the dislocation is in metastable equilibrium at the centre, being trapped in a potential well of depth $\mu b^2 \ln(\frac{1}{2}\mathrm{e})/8\pi$ and radius $[1 - \frac{1}{2}(2)^{1/2}]^{1/2}a = 0.54a$. It can be dislodged by applying a couple large enough to undo half the twist, or by clamping one end and deflecting the other ([13] p. 273).

The $E(x, y)$ surface is similar to that of eq. (141) for other reasonably equiaxed cross-sections, and one can show that the dislocation is stable at the centre of any regular polygon. From eqs. (43) and (138) it can be shown that it is unstable in a long thin rectangle; as the cross-section elongates from a square instability probably starts when the ratio of the sides is about 2:1. It has been suggested, by analogy with the foregoing, that a rod containing an edge dislocation parallel to its axis will exhibit a curvature when the end couples necessary to maintain a state of plane strain are relaxed, but this is not so because in fact there are no such couples ([13] p. 274). To see directly that there is no curvature, bend the whisker by end couples and introduce the edge dislocation. The fibre stresses due to the couples exert no forces on it, it does no work and, by Colonnetti's theorem, neither do the couples. Hence the ends remain parallel and there is no curvature. By applying simple tension instead of a bending moment we can show similarly that the edge dislocation does not change the length of the rod. The energy per unit length of a long rod containing an edge dislocation is therefore given by the unmodified plane strain formula. For a circular rod to which eqs. (75) or (86) or a suitable combination of them applies there is no stable position for an edge dislocation. For a mixed dislocation free to glide along a diameter of a

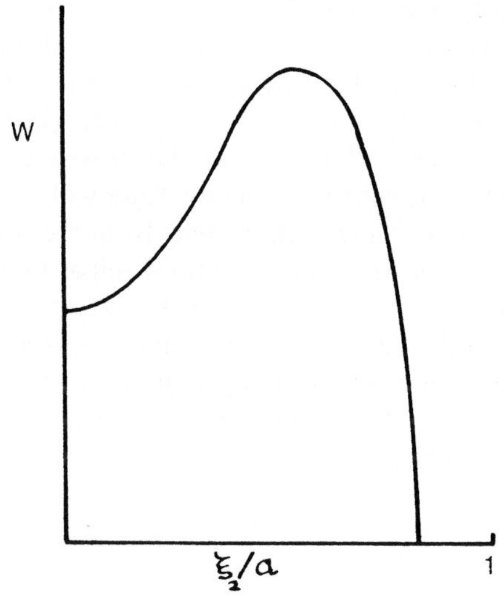

Fig. 7 Energy of a screw dislocation in a circular cylinder as a function of its distance from the axis.

circular cylinder and having its Burgers vector inclined at an angle θ to the cylinder axis the energy is given by $\cos^2\theta$ times that of eq. (141) plus $\sin^2\theta$ times eq. (75). The potential well of fig. 7 becomes shallower with increasing θ and disappears when $\tan\theta$ exceeds $(1 - \nu)^{1/2}$ [86].

Gomer [87] observed that when a whisker twisted by a screw dislocation was stretched the ends rotated relative to one another. The writer [88] gave a theory of the effect. The angle of rotation, $\delta\theta$ say, is given by

$$\delta\theta = \frac{\alpha_0 l_0(\lambda^2 - \mu^2)(S - I)}{\mu^2 S + (\lambda^2 - \mu^2)I} \tag{142}$$

where α_0 is the initial twist due to the screw dislocation, λ is the ratio of the final length to the initial length l_0 of the whisker, μ^2 is the ratio of the final to the initial area of the cross-section, while S is the torsional rigidity for unit shear modulus and I

the polar moment of inertia about the centroid. Equation (142) was derived from the theory of small deformations superposed on large, so that although the dislocation strains are small λ and μ are finite. The result does not depend on the details of the finite stress–strain relations, except in as far as they determine the relation between λ and μ. If Hooke's law can be applied to the extension eq. (142) becomes

$$\delta\theta = -2\alpha_0(1+\nu)(1 - I/S)\,\delta l \tag{143}$$

where δl is the increase in the length of the whisker.

The writer claimed [89], incorrectly, that eq. (143) showed that one could detect the presence of the internal stresses set up by the dislocation by measuring the response of the whisker to external forces, contrary to a general result in the linear theory of elasticity [90]. The presence of the dislocation gives a prismatic whisker a certain observable initial twisted shape. A rod, cast, forged or turned from the solid to the same twisted form, and free of internal stress, would also show the effect predicted by eq. (143). Thus nothing can be learned from the tensile test which could not be inferred directly from the external form of the whisker. (In the only case where the initial twist cannot be observed, that of a circular cross-section, S is equal to I and there is no effect.) To see this, without using the theory of the extension of rods with initial twist, we may start with the fact, evident from the analysis of ref [89], that eq. (143) applies equally well when α_0 is the total initial twist produced by a set of dislocations distributed in any way over the cross-section. If, in particular, they are infinitesimal and uniformly distributed over the cross-section we have simply a plastically twisted rod with no macroscopic internal stresses but with the same response to axial loading as it had when it contained a single dislocation which produced the same initial twist α_0.

Addendum (1976)

Gavazza and Barnett [91] have looked at an interesting general point relating to image forces. The Peach–Koehler formula gives [13]

$$f_m = -\varepsilon_{jmn} b_i p_{ij} t_n \tag{144}$$

for the force per unit length acting on an element of a dislocation loop, with Burgers vector b_i at a point where its tangent is the unit vector t_n, due to its interaction with an externally-applied stress p_{ij}. It is commonly assumed that if for a dislocation in a finite solid with stress-free surfaces we wish to find the force which its own image stress p^{I}_{ij} exerts on it we have only to put $p_{ij} = p^{\mathrm{I}}_{ij}$ in eq. (144). This is plausible on the anthropomorphic grounds that the dislocation does not discriminate between externally-exerted and self-produced stresses in its neighbourhood, but it is not hard to raise doubts.

The part of the elastic energy which depends on the image field is

$$E_{\mathrm{im}} = \tfrac{1}{2} b_i \int_C p^{\mathrm{I}}_{ij}\,\mathrm{d}S_j \tag{145}$$

where C is a suitable cap bounded by the dislocation line L. Let the loop be changed to a slightly different one L' in which every element $\mathrm{d}L$ of L has undergone a small displacement $\delta\xi_m$ which adds (or subtracts) a surface element $\mathrm{d}S_j = \varepsilon_{jmn}\,\delta\xi_m t_n\,\mathrm{d}L$ to the edge of C, converting it to the cap C' bounded by L'. The change in E_{im} results from the change in the region of integration and from the change in p^{I}_{ij}:

$$\begin{aligned}\delta E_{\mathrm{im}} &= \tfrac{1}{2}b_i \int_{C'-C} p^{\mathrm{I}}_{ij}\,\mathrm{d}S_j + \tfrac{1}{2}b_i \int_C \delta p^{\mathrm{I}}_{ij}\,\mathrm{d}S_j \\ &= \tfrac{1}{2}b_i \int_L p^{\mathrm{I}}_{ij}\varepsilon_{jmn}\,\delta\xi_m t_n\,\mathrm{d}L + \tfrac{1}{2}b_i \int_C \delta p^{\mathrm{I}}_{ij}\,\mathrm{d}S_j. \qquad (146)\end{aligned}$$

The first term in eq. (146) evidently gives half the force predicted by eq. (144). The question is, does the second term provide the other half? The following is a compressed and somewhat distorted version of Gavazza and Barnett's careful argument to show that the second term in eq. (146) is indeed equal to the first. It may be checked with the help of Gauss's theorem, the equilibrium equations and Hooke's law that the integral

$$\int_S (p_{ij}u'_i - p'_{ij}u_i)\,\mathrm{d}S_j \qquad (147)$$

involving two elastic fields u_i, p_{ij} and u'_i, p'_{ij} has the same value on $S = S_1$ as it has on $S = S_2$ if S_1 can be deformed into S_2 without passing through singularities. Take for the dashed and undashed quantities the elastic fields of the final and initial loops L', L; then by expanding S until it becomes the outer free surface where $p_{ij}n_i$ and $p'_{ij}n_j$ are both zero we see that the integral term (147) vanishes. If we split the two fields into image and infinite-medium terms we get

$$\begin{aligned}&\int_S (u'^{\infty}_i p^{\infty}_{ij} - u^{\infty}_i p'^{\infty}_{ij})\,\mathrm{d}S_j + \int_S (u'^{\mathrm{I}}_i p^{\infty}_{ij} - u^{\mathrm{I}}_i p'^{\infty}_{ij})\,\mathrm{d}S_j \\ &\qquad + \int_S (u'_i p^{\mathrm{I}}_{ij} - u_i p'^{\mathrm{I}}_{ij})\,\mathrm{d}S_j = 0. \qquad (148)\end{aligned}$$

The first integral is of just the same form as (147) and so its surface may be deformed into a large sphere of radius r; this is allowed since the infinity quantities are supposed to exist everywhere. As the remote displacement and stress of a loop are proportional to r^{-2} and r^{-3} the integral is proportional to r^{-3} and yet independent of r, and so zero. The second and third integrals taken together are of the form of the term (147) and can be contracted to a jacket enveloping a surface C'' which contains both C and C'. Ultimately the jacket becomes the two sides of C'' plus two narrow tubes containing L and L'. It can be shown that the tubes contribute nothing because any volume element traversed by a dislocation is in static equilibrium. As the integrand is continuous across C'' the rest of the second integral is also zero. In the final integral, as u_i is continuous except across C and h'_i except across C', and the discontinuities are both b_i we have

$$b_i \int_{C'} p^{\mathrm{I}}_{ij}\, \mathrm{d}S_j - b_i \int_{C} p'^{\mathrm{I}}_{ij}\, \mathrm{d}S_j = 0.$$

Putting $p'^{\mathrm{I}}_{ij} = p^{\mathrm{I}}_{ij} + \delta p^{\mathrm{I}}_{ij}$ and dividing by two we get

$$\tfrac{1}{2}b_i \int_{C} \delta p^{\mathrm{I}}_{ij}\, \mathrm{d}S_j = \tfrac{1}{2}b_i \int_{C'-C} p^{\mathrm{I}}_{ij}\, \mathrm{d}S_j = \tfrac{1}{2}b_i \int_{L} p^{\mathrm{I}}_{ij}\, \varepsilon_{jmn}\, \delta\xi_m t_n\, \mathrm{d}S_j,$$

as required.

Generalizing one of the results in sect. 2.3., Vitek [92] has given the Muskhelishvili complex stress functions for the elastic field of an edge dislocation near an elliptical hole in an isotropic medium. Of special importance is the case where one axis of the ellipse is allowed to dwindle to zero, leaving a dislocation in the neighbourhood of a crack. This configuration, already touched on in sect. 2.4, is important in connection with crack-tip plasticity (Riedel [93] gives useful references to the fracture mechanics literature). Vitek [94] gives the crack limit of his ellipse solution, and the same problem has also been treated by Hirth and Wagoner [95] who allow their dislocation to have both screw and edge components, and by Rice and Thompson [96] who give the image force on the dislocation. It has also been treated for an anisotropic medium. Many years ago Stroh [97] gave expressions for the elastic field of a crack in an anisotropic medium in the form of integrals involving the traction which the applied forces produced along the proposed site of the crack before it was introduced. It has been applied to the dislocation case by Atkinson [98], Atkinson and Clements [99] and Solovev [100]. The complementary case of a dislocation near what in fracture mechanics would be called an external crack has been discussed by Tomate [101]. Here instead of, say, the segment $|x| < a$ of the x-axis being cracked it is the only part of the x-axis which is not cracked; in other words cracks extend inwards from $-\infty$ to $-a$ and from ∞ to a. Tomate only considers a screw dislocation, but he allows the half spaces $y > 0$ and $y < 0$ to have different elastic constants, since his real interest is in imperfectly bonded composites.

A useful addition to the stock of three-dimensional special solutions is Comninou and Dundurs' [102] expression for the elastic field of one of Yoffe's [61] angular dislocations when it is situated in a semi-infinite solid with a stress-free surface. The vertex is an arbitrary distance below the surface, the two infinite arms of the angle point away from the surface and one of them is perpendicular to it. This is not really a restriction, for if we add together the elastic fields of a pair of such angular dislocations with a common vertex and the same Burgers vector the arms which are perpendicular to the surface will be coincident but described in opposite senses, so that their contributions to the elastic field cancel, leaving the field of a single angular dislocation with no non-geometrical limitations on the orientations of its arms with respect to the free surface or each other. Similarly if we add Yoffe's [60] solution for a straight dislocation in a half-space, discussed in sect. 3.1., we can cancel one arm of the angle and arrive at an angular dislocation with an arm which reaches the surface. By repeating this manoeuvre we can arrange that both arms reach the surface. As explained by Yoffe [61], angular dislocations can be added together to produce polygonal loops, dislocation tetrahedra and so forth. The authors reach their solution by introducing an image angular dislocation which at the site of the proposed free surface cancels the shear stresses and doubles the normal stress (compare the argument following eq. (110)) and then liquidate the latter by adding the field, eq. (121) with a suitably chosen harmonic φ.

A problem mentioned in sect. 3.3., the interaction between a screw dislocation and a spherical inhomogeneity, has now also been treated by Gavazza and Barnett [103].

References

[1] M. Gebbia, Annali Mat. pura appl. [3] 7 (1902) 141.
[2] A. E. H. Love, A Treatise on the Mathematical Theory of Elasticity (Cambridge University Press, Cambridge, 1927).
[3] T. A. Cruse, Int. J. Solids Struct. 5 (1969) 1259.
[4] J. D. Eshelby, Phil. Trans. Roy. Soc. A244 (1951) 87.
[5] P. Coulomb and J. Friedel, in Dislocations and Mechanical Properties of Crystals (J. C. Fisher et al., eds.) (Wiley, New York, 1957, p. 555).
[6] A. K. Head, Phil. Mag. 44 (1953) 92.
[7] J. Dundurs, in Mathematical Theory of Dislocations (T. Mura, ed.) (A.S.M.E., New York, 1969) p. 70.
[8] J. D. Eshelby, Phil. Mag. 3 (1958) 440.
[9] Z. Nehari, Conformal Mapping (McGraw-Hill, New York, 1952).
[10] M. v. Laue, J. Radioakt. Elektronik 15 (1918) 205.
[11] H. Kober, Dictionary of Conformal Representations (Dover, New York, 1957).
[12] G. Leibfried and H.-D. Dietze, Z. Phys. 126 (1949) 790.
[13] F. R. N. Nabarro, Theory of Crystal Dislocations (Clarendon Press, Oxford, 1967).
[14] G. B. Spence, J. Appl. Phys. 33 (1962) 729.
[15] R. Siems, Phys. kondens. Materie 2 (1964) 1.
[16] A. G. Greenhill, Q. Jl. pure appl. Math. 15 (1878) 10.
[17] L. M. Milne-Thomson, Jacobian Elliptic Function Tables (Dover, New York, 1950, p. 35).
[18] A. G. Greenhill, The Applications of Elliptic Functions (Macmillan, London, 1892, p. 292).
[19] I. S. Sokolnikoff, Mathematical Theory of Elasticity (McGraw-Hill, New York, 1956).
[20] V. Volterra, Annls scient. Éc. norm. sup., Paris [3] 24 (1907) 401.
[21] V. Volterra and E. Volterra, Sur les Distortions des Corps Elastiques (Mémorial des Sciences Mathematiques, fasc. 147) (Gauthier-Villars, Paris, 1960).
[22] O. M. Corbino, Nature (Lond.) 90 (1910) 540.
[23] Ph. Frank and R. v. Mises, Die Differential- und Integralgleichungen der Mechanik und Physik (Riemann–Weber), Part I (Vieweg, Brunswick, 1925) p. 635.
[24] W. F. Brown, Phys. Rev. 60 (1941) 139.
[25] J. S. Koehler, Phys. Rev. 60 (1941) 397.
[26] A. Seeger, in Encyclopedia of Physics (S. Flügge, ed.) vol. 7 part 1 (Springer, Berlin, 1955) pp. 560–3.
[27] J. Dundurs and G. P. Sendeckyj, J. Mech. Phys. Solids 13 (1965) 141.
[28] J. Dundurs and T. Mura, J. Mech. Phys. Solids 12 (1964) 177.
[29] R. D. List, Proc. Camb. Phil. Soc. 65 (1969) 823.
[30] K. Aderogba, Proc. Camb. Phil. Soc. 73 (1972) 269.
[31] A. K. Head, Proc. Phys. Soc. (Lond.) B 66 (1953) 793.
[32] W. S. Burnside and A. W. Panton, The Theory of Equations (Longmans, London, 1912) p. 167.
[33] L. A. Pastur, E. P. Fel'dman, A. M. Kosevich and V. M. Kosevich, Sov. Phys. Solid State 4 (1963) 1896.
[34] A. K. Head, Phys. Stat. solidi 10 (1965) 481.
[35] J. Gemperlova, Phys. Stat. solidi 30 (1968) 261.

[36] M. O. Tucker, Phil. Mag. 19 (1969) 1141.
[37] J. D. Eshelby, Br. J. appl. Phys. 17 (1966) 1131.
[38] M.-S. Lee and J. Dundurs, Int. J. Engng. Sci. 11 (1973) 87.
[39] B. A. Bilby and J. D. Eshelby, in Fracture (H. Liebowitz, ed.) (vol. 1, Academic, New York, 1969) p. 99.
[40] A. J. Ardell and R. B. Nicholson, with an Appendix by J. D. Eshelby, Acta Metall. 14 (1966) 1295.
[41] J. I. Musher, Am. J. Phys. 34 (1966) 267.
[42] R. Weeks, J. Dundurs and M. Stippes, Int. J. Engng. Sci. 6 (1968) 365.
[43] T. Kurihara, Int. J. Engng. Sci. 11 (1973) 891.
[44] J. A. Steketee, Can. J. Phys. 36 (1958) 192.
[45] T. Maruyama, Bull. Earthquake Res. Inst. 42 (1964) 289.
[46] R. D. Mindlin, Physics 7 (1936) 195.
[47] R. D. Mindlin and D. H. Cheng, J. appl. Phys. 21 (1950) 926.
[48] H. M. Westergaard, Theory of Elasticity and Plasticity (Harvard, Cambridge, Mass., 1952) p. 142.
[49] A. I. Lur'e, Three-dimensional Problems in the Theory of Elasticity (Interscience, New York, 1964) p. 132.
[50] L. Solomon, Élasticité linéaire (Masson, Paris, 1968) p. 541.
[51] E. Melan, Z. angew. Math. Mech. 12 (1932) 343.
[52] E. Melan, Z. angew. Math. Mech. 20 (1940) 368.
[53] T. K. Tung and T. H. Lin, J. appl. Mech. 29 (1966) 363.
[54] D. J. Bacon and P. P. Groves, in Fundamental Aspects of Dislocation Theory (J. A. Simmons et al., eds.) (Spec. Pub. 317, vol. 1, Nat. Bur. Standards, Washington, 1970) p. 35.
[55] P. P. Groves and D. J. Bacon, Phil. Mag. 22 (1970) 83.
[56] N. J. Salamon and J. Dundurs, J. Elasticity 1 (1971) 153.
[57] J. D. Eshelby, in Progress in Solid Mechanics (I. N. Sneddon and R. Hill, eds.) vol. 2 (North-Holland, Amsterdam, 1961) p. 119.
[58] J. Baštecká, Czech. J. Phys. B14 (1970) 702.
[59] J. D. Eshelby and A. N. Stroh, Phil. Mag. 42 (1951) 1401.
[60] E. H. Yoffe, Phil. Mag. 6 (1961) 1147.
[61] E. H. Yoffe, Phil. Mag. 5 (1960) 161.
[62] T. Vreeland and J. D. Eshelby, in Dislocations and Mechanical Properties of Crystals (J. C. Fisher et al., eds.) (Wiley, New York, 1957) p. 91.
[63] J. D. Eshelby, C. W. A. Newey, P. L. Pratt and A. B. Lidiard, Phil. Mag. 3 (1958) 75.
[64] W. J. Tunstall, P. B. Hirsch and J. Steeds, Phil. Mag. 4 (1964) 511.
[65] A. E. Green and T. J. Willmore, Proc. Roy. Soc. A193 (1948) 229.
[66] L. H. Mitchell and A. K. Head, J. Mech. Phys. Solids 9 (1961) 131.
[67] L. Rongved, Proc. Second Midwestern Conference on Solid Mechanics (1953) 1–13.
[68] J. Dundurs and M. Hetényi, J. appl. Mech. 32 (1965) 671.
[69] C. F. Hsieh and J. Dundurs, Int. J. Engng. Sci. 11 (1973) 933.
[70] F. R. N. Nabarro, Phil. Mag. 42 (1951) 1224.
[71] P. Coulomb, Acta metall. 5 (1957) 538.
[72] R. W. Weeks, S. R. Pati, M. F. Ashby and P. Barrand, Acta metall. 17 (1969) 1403.
[73] R. Bullough and R. C. Newman, Phil. Mag. 7 (1962) 529.
[74] J. R. Willis and R. Bullough, in Proc. Brit. Nuc. Energy Soc. Conf. Reading (S. F. Pugh et al., eds.) (A.E.R.E., Harwell, 1971) p. 133.
[75] J. R. Willis, M. R. Hayns and R. Bullough, Proc. Roy. Soc. A329 (1972) 121.
[76] W. D. Collins, Proc. Lond. Math. Soc. 3 (1959) 9.
[77] W. D. Collins, J. Lond. Math. Soc. 34 (1959) 343.
[78] W. D. Collins, Z. angew. Math. Phys. 11 (1960) 1.
[79] V. I. Blokh, Teoriya Uprugosti (University Press, Kharkov, 1964) p. 450.
[80] J. H. Bramble, Z. angew. Math. Phys. 12 (1961) 1.
[81] J. D. Eshelby, Phys. Stat. solidi 2 (1962) 1021.
[82] R. Siems, P. Delavignette and S. Amelinckx, Phys. Stat. solidi 3 (1963) 872.
[83] F. Kroupa, Czech. J. Phys., B9 (1959) 332, 488.

[84] R. Siems, P. Delavignette and S. Amelinckx, Phys. Stat. solidi 2 (1962) 421.
[85] E. H. Mann, Proc. Roy. Soc. A199 (1949) 376.
[86] J. P. Hirth and F. C. Frank, Phil. Mag. 3 (1958) 1110.
[87] R. Gomer, J. Chem. Phys. 28 (1958) 457.
[88] J. D. Eshelby, in Growth and Perfection of Crystals (R. H. Doremus et al., eds.) (Wiley, New York, 1958) p. 130.
[89] J. D. Eshelby, in Proc. Symp. Internal Stresses and Fatigue in Metals (G. M. Rassweiler et al., eds.) (Elsevier, Amsterdam, 1959) p. 41.
[90] R. V. Southwell, Theory of Elasticity (Clarendon Press, Oxford, 1936) p. 77.
[91] S. D. Gavazza and D. M. Barnett, Scripta Metall. 9 (1975) 1263.
[92] V. Vitek, J. Mech. Phys. Solids 24 (1976) 67.
[93] H. Riedel, J. Mech. Phys. Solids 24 (1976) 277.
[94] V. Vitek, J. Mech. Phys. Solids 24 (1976) 263.
[95] J. P. Hirth and R. H. Wagoner, Int. J. Solids Structures 12 (1976) 117.
[96] J. R. Rice and R. M. Thompson, Phil. Mag. 29 (1974) 73.
[97] A. N. Stroh, Phil. Mag. 3 (1958) 625.
[98] C. Atkinson, Int. J. Fracture Mech. 2 (1966) 567.
[99] C. Atkinson and D. L. Clements, Acta Metall. 21 (1973) 55.
[100] V. A. Solovev, Phys. Stat. Sol. 65b (1974) 857.
[101] O. Tomate, Int. J. Fracture Mech. 4 (1968) 357.
[102] M. Comninou and J. Dundurs, J. Elasticity 5 (1975) 203.
[103] S. D. Gavazza and D. M. Barnett, Int. J. Engng. Sci. 12 (1974) 1025.

PHILOSOPHICAL MAGAZINE A, 1980, VOL. 42, NO. 3, 359–367

The force on a disclination in a liquid crystal

By J. D. ESHELBY
Department of the Theory of Materials, University of Sheffield,
Sheffield S1 3JD, England

[Received 20 December 1979 and accepted 29 January 1980]

ABSTRACT

It is argued that the Peach–Koehler force on a disclination in a nematic liquid crystal is, unlike its namesake for a dislocation in a solid, not a fictitious configurational force, but a real force which, for equilibrium, must be balanced by an external force applied to the singular line. Formally this is a consequence of the near identity of the Ericksen stress and the appropriate energy–momentum tensor.

§ 1. INTRODUCTION

The force on a segment of dislocation in an elastic solid (Nabarro 1967) is defined so that the energy which could in principle be extracted by letting the segment undergo a small displacement is the scalar product of the force and the displacement. A force acting on other entities in a solid, for example a point defect, an inter-phase interface or a crack tip, can be similarly defined. It can be expressed as the integral of the elastic energy–momentum tensor over a surface embracing the defect, or an appropriate part of it. A convenient name is needed to distinguish such a force from an ordinary force which can be directly balanced by a weight or spring. Read (1953) calls it the force on a configuration while Roitburd (1971) speaks of a configurational force; we shall adopt the latter term. The term configurational force has also been applied by Kléman (1977) to the apparently similar force on disclinations in a nematic liquid crystal which forms the main subject of the present paper.

Recently the writer had to prepare a paper (Eshelby 1980) about configurational forces on defects in the generalized continua, of greater or less physical relevance, which have been devised and studied by applied mathematicians. In it the force on a disclination in a nematic liquid crystal was to have served as a realistic example. However, the writer was rather disconcerted to find that the appropriate energy–momentum tensor is, to within an unimportant constant hydrostatic pressure, the same as the Ericksen stress tensor which gives the traction that the fluid exerts on a surface element in it. This implies that the supposedly configurational force on a disclination in a nematic is in fact a real force exerted on the core of the disclination by the surrounding medium. In § 2 we establish this for a simple case, helped by an obvious but somewhat misleading analogy with screw dislocations in solids. The extension to the general case (§ 3) is then easy. Section 4 contains some comments.

§ 2. A simple case

We begin with a particularly simple situation often used for illustration (de Gennes 1974) and specified by

$$\left.\begin{array}{lll} n_1 = \cos\phi, & n_2 = \sin\phi, & n_3 = 0 \\ n_{i,3} = 0, & W = \tfrac{1}{2}k_{33}\phi_{,i}\,\phi_{,i} & \end{array}\right\}. \tag{2.1}$$

That is, the unit director n_i lies in the x_1x_2 plane, makes an angle ϕ with the x_1 axis and is independent of x_3, and the curvature–elasticity constants k_{22} and k_{33} are made equal, so that the general Oseen–Frank energy density W takes the simple form shown. (A suffix following a comma denotes differentiation and a repeated suffix is to be summed over 1, 2 in this section and 1, 2, 3 in the others.) The generally non-linear equilibrium equation takes the linear form

$$\nabla^2\phi = 0 \tag{2.2}$$

not, of course, to be confused with a linear approximation. The solution

$$\phi = m\tan^{-1}\frac{x_2}{x_1-\xi} - m\tan^{-1}\frac{x_2}{x_1-\eta} \tag{2.3}$$

represents a disclination of strength m at $(\xi, 0)$ and one of strength $-m$ at $(\eta, 0)$. When we calculate the total energy E of the field and differentiate with respect to defect position, we find that there is an effective force

$$F_1 = -\frac{\partial E}{\partial \xi} = -\frac{2\pi k_{33}m^2}{\xi-\eta}$$

on the one at $(\xi, 0)$ and an equal and opposite force on the other; that is, there is an attractive force

$$F = 2\pi k_{33}m^2/R \tag{2.4}$$

tending to reduce the distance R between them.

With the alternative notation

$$\phi = u_3, \quad k_{33} = \mu \tag{2.5}$$

these equations relate to a state of anti-plane strain with displacement $(0, 0, u_3)$ in a linear isotropic elastic medium of shear modulus μ. That the quantities equated in eqn. (2.5) are dimensionally inconsistent does not matter, since the transliterated energy density $W = \tfrac{1}{2}\mu\, u_{3,i}\, u_{3,i}$ has the right dimensions, which is all that is necessary here. Equation (2.2) is the elastic equilibrium equation, eqn. (2.3) the displacement field of a pair of screw dislocations with Burgers vectors $\pm 2\pi m$, and eqn. (2.4) is the orthodox expression for the force between them.

It is shown in the theory of lattice defects (Nabarro 1967) that the configurational force on a defect is given by the integral

$$F_l = \int_S P_{lj}\, dS_j, \tag{2.6}$$

where S is a surface—or in two dimensions a circuit—enclosing the defect,

dS_j is the product of the surface element dS and its unit normal ν_j, and

$$P_{lj} = W\delta_{lj} - p_{ij}u_{i,l}$$

is the energy–momentum tensor, in which p_{ij} is the ordinary stress tensor and u_i the displacement vector. The divergence $P_{lj,j}$ vanishes, so that eqn. (2.6) is not affected by deforming S so long as it continues to embrace only the defect we are interested in. (We use the convention that it is the second suffix in P_{lj} which is contracted with the surface element in eqn. (2.6) or in forming the divergence, and similarly for t_{lj} below.)

For anti-plane strain, eqn. (2.6) becomes

$$F_l = \int_S (W\delta_{lj} - k_{33}\phi_{,j}\phi_{,l})\, dS_j. \tag{2.7}$$

To avoid later rewriting we have used eqn. (2.5) to translate back into the nematic notation. Equation (2.7) reproduces eqn. (2.4). More generally, for a dislocation at the origin interacting with an applied displacement ϕ^{A}, so that

$$\phi = m \tan^{-1}\frac{x_2}{x_1} + \phi^{\mathrm{A}}(x_1, x_2) \tag{2.8}$$

it gives, correctly,

$$F_1 = 2\pi m k_{33}\phi_{,2}{}^{\mathrm{A}}(0,0), \quad F_2 = -2\pi m k_{33}\phi_{,1}{}^{\mathrm{A}}(0,0) \tag{2.9}$$

for the so-called Peach–Koehler force on the dislocation.

If we now take eqn. (2.8) to refer to a disclination interacting with an imposed disturbance ϕ^{A} of the director field, the energy density will be the same, point for point, as it was on the elastic interpretation, both before and after a small shift of the defect, so we expect that the configurational force will still be given by eqn. (2.7). In fact, Dafermos (1970) has already given an integral for the force on a disclination in a tube which agrees with eqn. (2.7) after correction and simplification (see the Appendix). The force (2.9) calculated from eqn. (2.7) agrees with the Peach–Koehler force on a disclination, already so christened by Kléman (1977) and calculated by him for a more general geometry.

The use of the integral (2.7) to find the force (2.9) may seem unnecessary since eqn. (2.9) can be derived without it for dislocations in solids (Nabarro 1967) and also for disclinations in nematics, either by appeal to the identity of the nematic and elastic energy densities or directly in the manner of Kléman (1977). We have nevertheless introduced eqn. (2.7) so as to be able to make an important point in the following paragraphs.

In the theory of dislocations in solids it is emphasized that the Peach–Koehler force is a fictitious configurational force not to be confused with ordinary body force. The nematic–elastic solid analogy suggests that the same is true of the similarly named force on a disclination in a nematic, but this is not in fact so. The theory of nematics shows that on a surface S in the fluid there is a total traction equal to the integral

$$F_l^{\mathrm{E}} = \int_S t_{lj}\, dS_j, \tag{2.10}$$

where t_{lj} is the Ericksen stress tensor given in the general case by

$$t_{lj} = (-p_0 + W)\delta_{lj} - \frac{\partial W}{\partial n_{i,j}} n_{i,l} \tag{2.11}$$

(Ericksen 1976), where p_0 is a uniform hydrostatic pressure. (For the relation of eqn. (2.11) to a form more commonly quoted, see the discussion following eqn. (4.4) below.) When the conditions (2.1) apply, eqn. (2.11) takes the form

$$t_{lj} = (-p_0 + W)\delta_{lj} - k_{33}\phi_{,j}\phi_{,l}. \tag{2.12}$$

If S embraces a disclination, eqn. (2.10) gives the force exerted by the medium outside S on the medium inside it. Since the divergence $t_{lj,j}$ of t_{lj} is zero S can be made as small as we like without affecting eqn. (2.10), and so F_l^{E} is actually a force exerted by the medium on the disclination core.

Since the term $-p_0\delta_{lj}$ makes no contribution to the integral over a closed circuit, the real force (2.10) is equal to the supposedly configurational force given by the integral (2.7), and consequently to the Peach–Koehler force (2.9), and to secure equilibrium an equal and opposite force must be applied to the core from outside. So, for example, eqn. (2.3) does not simply describe a pair of opposite disclinations, but instead such a pair plus outwardly directed concentrated forces of magnitude (2.4) applied at $(\xi, 0)$ and $(\eta, 0)$.

The mathematical analogy with anti-plane elasticity which gives a purely configurational force on a dislocation is therefore physically misleading. A better analogue is steady flow in a perfect incompressible fluid. Equation (2.8) gives the velocity potential for a vortex fixed at the origin in a velocity field specified by ϕ^{A}, and (2.9) is the real lift force on a small aerofoil inserted in the core of the vortex to stop it drifting in the flow ϕ^{A}.

§ 3. The general case

These results are easily extended to the case of general three-dimensional geometry and the full Oseen–Frank energy density, or any allowable generalization of it. The Ericksen stress tensor is given by eqn. (2.11). Thus, on any element ds of a disclination line there is a force, $f_i^{\mathrm{E}}\, ds$ say, equal to the value of the integral (2.10)—now three-dimensional—taken over any surface intercepting the element ds from the line, for example a small sphere with ds as diameter. We shall call f_i^{E} the Ericksen force. The example of § 2 makes it clear that f_i^{E} is not usually zero. Thus a disclination whose shape is prescribed at will can only exist in static equilibrium if a balancing force $-f_i^{\mathrm{E}}\, ds$ is applied to each element from outside, say by inserting a suitably shaped stiff wire into the core which will automatically provide the required reaction at each point. We are not, of course, thinking of a real wire which would upset the director pattern (Williams, Cladis and Kléman 1973), but an idealized one which would not. We might, perhaps, impose the artificial boundary condition that in each element of fluid in contact with it the wire induces the same director orientation as the mathematical solution prescribes there in the absence of the wire, and then let the radius of the wire tend to zero. Without the artificial support of the external force the line will deform towards the shape for which the force is not needed.

If we happen to have overlooked the existence of the Ericksen force we can still introduce an apparently configurational Peach–Koehler force f_i^{PK}, defined so that $\delta\xi_i f_i^{\mathrm{PK}}\, ds$ is the decrease in the energy of the system (excluding, of course, the potential energy of the mechanism applying the unrecognized balancing force) when a segment ds of the disclination is given a small displacement $\delta\xi_i$, and then try to express it as an integral of the type (2.6)—now three-dimensional—taken over a surface S which intercepts the element ds from the line. The appropriate energy–momentum tensor turns out to be

$$P_{lj} = W\delta_{lj} - \frac{\partial W}{\partial n_{i,j}} n_{i,l}. \tag{3.1}$$

The writer has sketched a formal proof of this (Eshelby 1980), but once the existence of the Ericksen force has been appreciated there is no need for it. In fact the line integral

$$\int \delta\xi_i(s) f_i^{\mathrm{E}}(s)\, ds$$

gives the energy extracted by the mechanism applying the external force balancing the Ericksen force when each point of the line is given a small displacement $\delta\xi_i(s)$. Since this energy comes from the curvature–elastic energy of the medium and from the potential energy of the loading mechanism (if any) with which we impose couples and forces on the outer surface of the medium, it follows trivially that f_i^{PK} is equal to the Ericksen force f_i^{E}, so that P_{lj} differs from t_{lj} by, at most, a tensor which contributes nothing to eqn. (2.6). For neatness we take the difference to be $p_0\delta_{lj}$, so as to make eqn. (3.1) agree with the general formalism (Eshelby 1980).

In the last paragraph we have used the term Peach–Koehler force even though one cannot write down a Peach–Koehler formula like eqn. (2.9) for it in the general three-dimensional (or even two-dimensional) non-linear case. There is, however, a class of three-dimensional solutions in which the limitation $n_{i,3} = 0$ is dropped from (2.1) and the linear equation (2.2), now with the three-dimensional Laplace operator, remains valid. For it Kléman (1977) has given the three-dimensional extension of eqn. (2.9). Our $f_i^{\mathrm{PK}} = f_i^{\mathrm{E}}$ should agree with it, and in fact does so.

§ 4. Discussion

The fact that the apparently configurational force on a disclination in a nematic is actually a true force produced by the Ericksen stress seems not to have been noticed hitherto. For the situation described by the conditions (2.1), several authors have added together the solutions for two single disclinations, as in eqn. (2.3), taking advantage of the fact that the governing equation (2.2) is linear. However, they all apparently overlooked the fact that the Ericksen stresses are not additive, since, unlike eqn. (2.2), eqn. (2.12) is not linear, in consequence of which the resultant of the Ericksen stress on a circuit around one of the defects is made up of a term from its own field, which is zero by symmetry, one from the field of the other defect, also zero, and a cross-term between the two fields which is not zero. One of these authors went so far as to check that there was no net couple on a circuit about a defect, but did not go on to ask whether there was a net force on it. If Dafermos' integral, eqn. (A 3) of our Appendix, had not contained an error, its similarity to the integral of the Ericksen stress might have prompted discovery of the relation (A 4) which shows that the similarity is in fact an identity.

If ϕ satisfies the equilibrium equation (2.2) then, as one can check, the divergence $t_{lj,j}$ of the Ericksen stress (2.12) is automatically zero, and t_{lj} is in static equilibrium. Also, for a given solution of eqn. (2.2), t_{lj} is fixed, apart from p_0, and outside our control. It is perhaps the fact that in this sense the Ericksen stress looks after itself and can apparently be ignored which led to the oversight mentioned in the last paragraph.

The argument of § 2 can be clarified if we adopt some sort of model for the centres of the defects. For definiteness imagine, with Dafermos and others, that the core of a disclination is occupied by a pool of isotropic fluid under a hydrostatic pressure p_1. If we solve eqn. (2.2) with definite boundary conditions at the outer surface of the medium and the nematic–isotropic interface, supposed at first to be of fixed shape and position, the tractions $t_{lj}\nu_j$ and $-p_1\nu_l$ will not, except by chance, match point by point across the interface, and an externally applied layer of body force equal to their difference will have to be spread over it to maintain equilibrium. Even the resultant of the Ericksen traction, given by eqn. (2.10), will not usually be zero. However, if we now allow ourselves to displace the interface bodily, we may be able to find a location for it which makes (2.10) zero. Then, as an added refinement, we could possibly deform the interface so as to have $t_{lj}\nu_j = -p_1\nu_l$ all round it. (We ignore such realities as interfacial tension.)

Things are similar for three dimensions and an arbitrary energy density. If we choose a solution n_i of the Euler equations

$$\frac{\partial W}{\partial n_i} - \left(\frac{\partial W}{\partial n_{i,j}}\right)_{,j} = \lambda n_i, \tag{4.1}$$

then the Ericksen stress t_{lj} of eqn. (2.11) is fixed except for p_0, and

$$t_{lj,j} = 0 \tag{4.2}$$

is satisfied automatically. In fact, if in

$$t_{lj,j} = W_{,l} - \left(\frac{\partial W}{\partial n_{i,j}}\right)_{,j} n_{i,l} - \frac{\partial W}{\partial n_{i,j}} n_{i,lj}$$

we use eqn. (4.1) to eliminate $(\partial W/\partial n_{i,j})_{,j}$ we have

$$t_{lj,j} = W_{,l} - G_l + \tfrac{1}{2}\lambda(n_j n_j)_{,l} \tag{4.3}$$

with

$$G_l = \frac{\partial W}{\partial n_i} n_{i,l} + \frac{\partial W}{\partial n_{i,j}} n_{i,jl}.$$

The term involving the Lagrange multiplier λ vanishes since $n_i n_i = 1$. The quantity G_l is simply $W_{,l}$ if, as we have been tacitly assuming, the medium is homogeneous, so that W does not depend on x_l explicitly. Hence $t_{lj,j}$ is zero, as stated. It is the connection between eqns. (4.1) and (4.2) which saves the static Ericksen–Leslie equations from being, as suggested by Lee and Eringen (1974), over-determined.

We note in passing that in place of eqn. (2.11), t_{lj} is more often written in the less explicit form

$$t_{lj} = -p\delta_{lj} - \frac{\partial W}{\partial n_{i,j}} n_{i,l}, \tag{4.4}$$

where the non-uniform hydrostatic pressure p originates as a Lagrange multiplier to take care of incompressibility. Equation (4.2) becomes a condition to be imposed on p, and gives, using eqn. (4.1) again as in deriving eqn. (4.3),

$$0 = t_{lj,j} = -p,_l - G_l = -(p+W),_l,$$

so that $p+W$ is a constant, p_0 say, and eqn. (4.4) agrees with eqn. (2.11), as required. It is also perhaps worth noting that the term in Frank's energy density which involves the constant k_{24}, and is usually omitted on the grounds that it is a divergence and so does not affect eqn. (4.1), actually contributes to t_{lj} and hence to the Ericksen force. The geometry excludes it in the situation of § 2.

When we treat a curvilinear disclination as a mathematical line singularity, we have to forego a detailed core adjustment of the kind discussed above for the two-dimensional case, and to discover the equilibrium shape of, say, a disclination segment anchored at the ends we have merely to find the form which liquidates the Ericksen force on each element of it. In this problem the only use we make of the force is to reduce it to zero, and so it does not much matter whether we believe that it is real, or, incorrectly, that it is similar to the configurational force on a lattice defect. The point of view adopted will, however, influence our attitude to the approach to equilibrium. A comparison with dislocations in solids will make this clear.

We are not surprised to find in a crystal a dislocation segment with an obviously non-zero Peach–Koehler force acting on it, because we can suppose that its movement to a form of minimum energy has been blocked, or at any rate slowed down, by, say, lattice friction and inadequate diffusion which inhibit glide and climb. In other words it ' tends ' to move under the action of the Peach–Koehler force, but may do so only slowly, or not at all, because of effects not taken account of in the continuum theory.

There are many passages in the liquid-crystal literature which seem to suggest that, similarly, a segment of disclination in a nematic can exist, at least temporarily, in any configuration, from which it may tend to the minimum-energy form. But such a view cannot be taken when we have realized that an arbitrary shape can be maintained only by the artificial application of an external force distribution along the singular line. If the external force is removed the Ericksen force which it balanced disappears as well and in compensation a flow develops which, because there is nothing to stop it, inexorably convects the disclination towards the minimum energy configuration, at a rate controlled by generalized viscosity and inertia. (There seems to be no need to suppose, with Imura and Okano (1973), that the director pattern migrates *through* the medium.) However, suppose that a pair of opposite disclinations traverse a layer of nematic and are anchored to the slide and cover-slip which confine it. If they are not too close they will at first simply bulge towards each other, but, given time, they may be able to drag their anchors and get nearer together. It is perhaps in this way that one should interpret statements that disclinations of opposite sign tend to group together, and so forth.

The conclusions of this paper should also apply to forces on point defects in nematics. In addition they should apply to defects in cholesteric liquid crystals, since although their energy density differs in form from that of nematics, the tensor t_{lj} is still given by eqn. (2.11). Things are, however, different for a smectic A liquid crystal. For example, under some circumstances its energy density can be taken (Kléman 1977) to be

$$W=\tfrac{1}{2}K_1(u_{3,11}+u_{3,22})^2+\tfrac{1}{2}K_3[(u_{3,31})^2+(u_{3,32})^2]+\tfrac{1}{2}B(u_{3,3})^2,$$

where u_3 is the small transverse displacement of the molecular layers, supposed to be originally parallel to the x_1x_2 plane. This expression is the same as the energy density of a particular type of linear anisotropic grade-two elastic solid. The writer has given the energy–momentum tensor appropriate to such a material (Eshelby 1975, 1980). With it the integral (2.6) can be made to reproduce the expression of Kléman (1977) for the force on a dislocation in a smetic A. It is a genuine configurational force of the kind familiar in the theory of dislocations in solids, and not a body force acting at the centre of the dislocation.

ACKNOWLEDGMENTS

The writer would like to thank Professor J. L. Ericksen and Sir Charles Frank for valuable correspondence.

APPENDIX

Dafermos (1970) treats the problem of a disclination at $x_i=y_i$ in a tube of nematic with outer boundary S_0 under the conditions (2.1) with the orientation ϕ prescribed around the tube :

$$\phi(x_i)=\phi_0(x_i) \quad \text{on } S_0. \tag{A 1}$$

He writes

$$\phi=\psi+\theta,$$

where

$$\psi=I\tan^{-1}(x_2-y_2)/(x_1-y_1) \tag{A 2}$$

is the field of the defect in an infinite medium and θ is the image term needed to satisfy eqn. (A 1). His expression (4.43) for the positive gradient of the energy with respect to y_i, reversed in sign to give the force, and with a further apparent change of sign to reconcile his inward with our outward normal, may be written as

$$F_l=\int_{S_0}\left[W\delta_{lj}+k_{33}\frac{\partial\phi}{\partial x_j}\left(\frac{\partial\theta}{\partial x_l}+\frac{\partial\theta}{\partial y_l}\right)\right]dS_j. \tag{A 3}$$

We have restored a general value to the constant k_{33}, which he normalizes to 2.

The W term in eqn. (A 3) comes from Dafermos' (4.42) and the term in k_{33} from the subsequent unnumbered equation. Both involve the use of

Gauss' theorem. Examination of the way he applies it shows that the normal is directed inwards in (4.42) but outwards in the other equation, though in constructing his eqn. (4.43)—our eqn. (A 3)—Dafermos appears to assume that it is inwards for both of them. Hence the sign of the k_{33} term in eqn. (A 3) must be reversed.

From eqn. (A 1) we have

$$\frac{\partial\psi}{\partial y_i}+\frac{\partial\theta}{\partial y_i}=0 \quad \text{on } S_0$$

and from eqn. (A 2)

$$\frac{\partial\psi}{\partial x_i}+\frac{\partial\psi}{\partial y_i}=0 \quad \text{everywhere,}$$

so that $\partial\theta/\partial y_i=\partial\psi/\partial x_i$ on S_0 and consequently

$$\frac{\partial\theta}{\partial x_i}+\frac{\partial\theta}{\partial y_i}=\frac{\partial\psi}{\partial x_i} \quad \text{on } S_0. \tag{A 4}$$

With eqn. (A 4) and the above-mentioned change in sign, eqn. (A 3) becomes

$$F_l=\int_{S_0}(W\delta_{lj}-k_{33}\phi_{,j}\phi_{,l})\,dS_j$$

in agreement with eqn. (2.7); as the divergence of the integrand is zero the contour S_0 can be freed from the tube.

Earlier in his paper Dafermos derives otherwise an explicit expression for the force on a disclination interacting with others or with an image field. It differs in sign from the results of other authors. In comparing, say, his eqn. (4.29) with our eqn. (2.9), note that $k_{33}=2$ and that he considers, incorrectly, that the force is the positive, not the negative, gradient of the energy with respect to y_i—see the text below his eqn. (4.3). This discrepancy comes, perhaps, from a wrongly directed normal in his integral (4.3).

References

Dafermos, C. M., 1970, *Q. Jl Mech. appl. Math.*, **23**, S49.

Ericksen, J. L., 1976, *Advances in Liquid Crystals*, Vol. 2, edited by G. Brown (New York: Academic Press), p. 233.

Eshelby, J. D., 1975, *Prospects of Fracture Mechanics*, edited by G. C. Sih, H. C. van Elst and D. Broek (Noordhoff), p. 69; 1980, *Proceedings of the Third International Symposium on Continuum Models of Discrete Systems* (in the press).

de Gennes, P. G., 1974, *The Physics of Liquid Crystals* (Oxford University Press).

Imura, H., and Okano, K., 1973, *Physics Lett.* A, **42,** 403.

Kléman, M., 1977, *Points. Lignes. Parois*, Vol. 1 (Paris: Les Éditions de Physique).

Lee, J. D., and Eringen, A. C., 1974, *Liquid Crystals and Ordered Fluids*, edited by J. F. Johnson and R. S. Porter (New York: Plenum Press), p. 315.

Nabarro, F. R. N., 1967, *Theory of Crystal Dislocations* (Oxford University Press).

Read, W. T., 1953, *Dislocations in Crystals* (New York: McGraw-Hill).

Roitburd, 1971, *Soviet Phys. Doklady*, **16,** 305.

Williams, C. E., Cladis, P. E., and Kléman, M., 1973, *Molec. Crystals liquid Crystals*, **21,** 355.

CONTINUUM MODELS OF DISCRETE SYSTEMS
University of Waterloo Press, 1980

Freudenstadt, Germany
June 24-30, 1979

THE ENERGY-MOMENTUM TENSOR OF COMPLEX CONTINUA

J.D. Eshelby

Department of the Theory of Materials
University of Sheffield
Sheffield, U.K.

1. INTRODUCTION

In the physics of solids and in fracture mechanics the idea of an effective force acting on some kind of defect in an elastic solid (e.g. a lattice vacancy or other point defect, a dislocation or a crack tip) has become familiar. (For a review see [1] or [2]). If the configuration of the defect can be adequately specified by giving its position ξ_1, ξ_2, ξ_3 the force is defined to be

$$F_\ell = -\frac{\partial}{\partial \xi_\ell} E_{tot} = -\frac{\partial}{\partial \xi_\ell} (E_{e\ell} + E_{ext}) \qquad (1)$$

where E_{tot} is the total energy of the system, made up of the elastic energy $E_{e\ell}$ of the body containing the defect and the potential energy E_{ext} of any external loading mechanism acting on the body.

Expressions such as effective force, fictitious force, energy gradient, or 'force' in quotation marks have been used to distinguish the force so defined from ordinary mechanical forces. Recently several authors have referred to it as a configurational force, and we shall adopt this term.

For a defect in a finitely deformed elastic solid with an arbitrary stress-strain law we may write

$$F_\ell = \int_S P_{\ell j} dS_j \tag{2}$$

where S is a surface enclosing the defect or, in fracture mechanics, a contour embracing the crack tip, and $P_{\ell j}$ is a certain tensor, the static energy-momentum tensor which the general theory of fields prescribes for the elastic field. (We use the customary summation convention; dS_j is $n_j dS$, where dS is an element of S and n_j is its outwardly-directed normal). The divergence of $P_{\ell j}$ vanishes, and so F_ℓ is unchanged (is path-independent, in the language of fracture mechanics) if S is deformed in any way, provided it continues to enclose only the defect we are interested in.

It is the object of this paper to try to extend (2) to continua more complex than a simple elastic solid in a way which requires as little discussion as possible of the details of the theory of the generalized continuum being studied.

2. THE ENERGY-MOMENTUM TENSOR

We use rectangular Cartesian coordinates X_m and denote differentiation with respect to them by a comma:

$$\phi^\alpha_{,i} = \partial\phi^\alpha/\partial X_i \quad , \quad \phi^\alpha_{,ijk} = \partial^3\phi^\alpha/\partial X_i \partial X_j \partial X_k \quad \text{etc.}$$

Unless otherwise stated X_1, X_2, X_3 will be Lagrangian coordinates labelling points of the medium in an undeformed reference state.

We shall assume that there is an energy density, per unit undeformed volume, of the form

$$W = W(\phi^\alpha, \phi^\alpha_{,i} \cdots \phi^\alpha_{,ij\ldots t} , X_m) \tag{3}$$

which depends on a set of field variables ϕ^α and their derivations $\phi^\alpha_{,ij\ldots t}$ up to some finite order and also, if the material is inhomogeneous, explicitly on the X_m. Three of the ϕ^α will usually be the (finite) displacement components u_1, u_2, u_3. The rest may be, for example, the components of one or more directors or the like. A repeated α is to be summed over its whole range, a repeated Latin suffix over 1, 2, 3 or, in a two-dimensional situation, over 1, 2.

It is convenient to begin by defining the variational derivative

$$\frac{\delta W}{\delta\phi^\alpha} = \frac{\partial W}{\partial\phi^\alpha} - \left(\frac{\partial W}{\partial\phi^\alpha_{,j}}\right)_{,j} + \left(\frac{\partial W}{\partial\phi^\alpha_{,aj}}\right)_{,aj} - \cdots \\ \cdots \pm \left(\frac{\partial W}{\partial\phi^\alpha_{,ab\ldots tj}}\right)_{,ab\ldots tj} \mp \cdots \tag{4}$$

with the sign of the general term + or - according as ab...tj contains an even or odd number of symbols. The variational derivative

$$\frac{\delta W}{\delta\phi^\alpha_{,pq\ldots r}}$$

is simply the expression (4) with ϕ^α replaced by $\phi^\alpha_{,pq\ldots r}$. Inspection shows that the sum of the terms on the right of (4), excluding the first, is just minus the X_j-derivative of $\delta W/\delta\phi^\alpha_{,j}$, and so we arrive at the useful identity

$$\frac{\partial W}{\partial\Phi} = \frac{\delta W}{\delta\Phi} + \left(\frac{\delta W}{\delta\Phi_{,j}}\right)_{,j} \tag{5}$$

where Φ could, for example, be ϕ^α, $\phi^\alpha_{,a}$ or $\phi^\alpha_{,ab..t}$.

The small change in W induced by a small change $\delta\phi^\alpha$ of the field variables is

$$\delta W = \frac{\partial W}{\partial\phi^\alpha}\delta\phi^\alpha + \frac{\partial W}{\partial\phi^\alpha_{,a}}\delta\phi^\alpha_{,a} + \frac{\partial W}{\partial\phi^\alpha_{,ab}}\delta\phi^\alpha_{,ab} + \cdots \\ \cdots + \frac{\partial W}{\partial\phi^\alpha_{,ab\ldots st}}\delta\phi^\alpha_{,ab\ldots st} + \cdots \tag{6}$$

or, applying (5) to each term,

$$\delta W = \frac{\delta W}{\delta\phi^\alpha}\delta\phi^\alpha + \frac{\delta W}{\delta\phi^\alpha_{,a}}\delta\phi^\alpha_{,a} + \ldots + \frac{\delta W}{\delta\phi^\alpha_{,ab\ldots st}}\delta\phi^\alpha_{,ab\ldots st}\ \cdots$$

$$+ \left(\frac{\delta W}{\delta\phi^\alpha_{,j}}\right)_{,j}\delta\phi^\alpha + \ldots + \left(\frac{\delta W}{\delta\phi^\alpha_{,ab\ldots sj}}\right)_{,j}\delta\phi^\alpha_{,ab\ldots s}\ \cdots \tag{7}$$

When the last dummy suffix in each term in the top row is replaced by j each vertically aligned pair of terms is recognisable as the two terms in the differentiation of a product with respect to X_j, and we have

$$\delta W = \frac{\delta W}{\delta\phi^\alpha}\delta\phi^\alpha + (Q_{\alpha j}[\delta\phi^\alpha])_{,j} \tag{8}$$

with

$$Q_{\alpha j}[\delta\phi^\alpha] = \frac{\delta W}{\delta\phi^\alpha_{,j}}\delta\phi^\alpha + \frac{\delta W}{\delta\phi^\alpha_{,aj}}\delta\phi^\alpha_{,a} + \ldots$$

$$\ldots + \frac{\delta W}{\delta\phi^\alpha_{,ab\ldots sj}}\delta\phi^\alpha_{,ab\ldots s} + \ldots \tag{9}$$

If we assume that the equilibrium of the medium is governed by the Euler equations

$$\delta W/\delta\phi^\alpha = 0 \tag{10}$$

integration of (8) over a volume V bounded by the surface S, followed by the use of Gauss' theorem gives the relation

$$\delta\int_V W\,dV = \int_S Q_{\alpha j}[\delta\phi^\alpha]dS_j \tag{11}$$

which expresses the change in the energy of the material inside S when small changes $\delta\phi^\alpha$, $\delta\phi^\alpha_{,i}\ldots$ are made in the values of the ϕ^α and their derivatives on S.

The gradient of W has components

$$\frac{\partial W}{\partial X_\ell} = \frac{\delta W}{\delta\phi^\alpha}\phi^\alpha_{,\ell} + (Q_{\alpha j}[\phi^\alpha_{,\ell}])_{,j} + \left(\frac{\partial W}{\partial X_\ell}\right)_{\text{exp}} \tag{12}$$

that is, what one gets by putting $\delta\phi^\alpha = \varepsilon\phi^\alpha_{,\ell}$ in (9), dividing by ε and adding the explicit derivative of (3) with respect to X_ℓ, in which ϕ^α, $\phi^\alpha_{,ab\ldots}$ ... and X_m, $m\neq\ell$ are all held constant. Equation (12) can be re-written in the form

$$\frac{\partial P_{\ell j}}{\partial X_j} = \frac{\delta W}{\delta\phi^\alpha}\phi^\alpha_{,\ell} + \left(\frac{\partial W}{\partial X_\ell}\right)_{\text{exp}} \tag{13}$$

where

$$P_{\ell j} = W\,\delta_{\ell j} - \frac{\delta W}{\delta\phi^\alpha_{,j}}\phi^\alpha_{,\ell} - \frac{\delta W}{\delta\phi^\alpha_{,aj}}\phi^\alpha_{,a\ell} - \ldots$$

$$\ldots - \frac{\delta W}{\delta\phi^\alpha_{,ab\ldots sj}}\phi^\alpha_{,ab\ldots s\ell} - \ldots \tag{14}$$

is the (canonical) energy-momentum tensor.

To avoid trouble when (14) is worked out for a specific case some convention on repeated suffixes following a comma is needed (see the remarks on (29) below). We adopt the following. Imagine that in (3) the suffixes are written out in some arbitrary fixed order, and treat, for example $\phi^\alpha_{,122}$, $\phi^\alpha_{,212}$, $\phi^\alpha_{,221}$ as distinct in carrying out the differentiation of W. For this to work the order of the suffixes must be preserved in the manipulations leading from (6) to (14).

3. THE FORCE ON A DEFECT

If (10) holds, and in addition the material properties are independent of X_ℓ, though not necessarily of X_m, $m \neq \ell$, so that

$$(\partial W/\partial X_\ell)_{exp} = 0 \tag{15}$$

then

$$\partial P_{\ell j}/\partial X_j = 0 \tag{16}$$

and the integral (1) is zero provided there are no singularities inside the closed surface S. If within S there is a surface S_0 inside which the material is inhomogeneous, so that (15) does not hold, and/or there is a singularity, then F_ℓ is not in general zero, but, in view of (16), its value is not changed if S is deformed, so long as it does not intersect S_0.

We now try to show that F_ℓ represents the force on the defect, in the sense discussed in the Introduction, that is, that if the defect is given a small displacement $\delta\xi_\ell$ then $F_\ell \delta\xi_\ell$ is minus the increase of the total energy of the system, made up of the energy of the medium, both inside and outside S, and the potential energy of any loading mechanism acting on the body. We shall use an extension of the method of [3] as being less precarious than the short-cut method of [1] and [2].

The left of Figure 1 shows the material in a standard uniform reference state, D is the defect, or, more precisely, the site marked out in the undeformed material for the proposed defect, S a surface surrounding it and S' the same surface after a small displacement $-\delta\xi_\ell$. On the right is a replica of the system on the left, the original, derived from it by a uniform translation. We now carry out the following succession of imaginary operations.

As a preliminary give the defect a displacement $\delta\xi_\ell$ and record the new values, say

$$\phi^{\alpha 1}, \ \phi^{\alpha 1}_{,a}, \ \ldots \ \phi^{\alpha 1}_{,abc\ldots}, \tag{17}$$

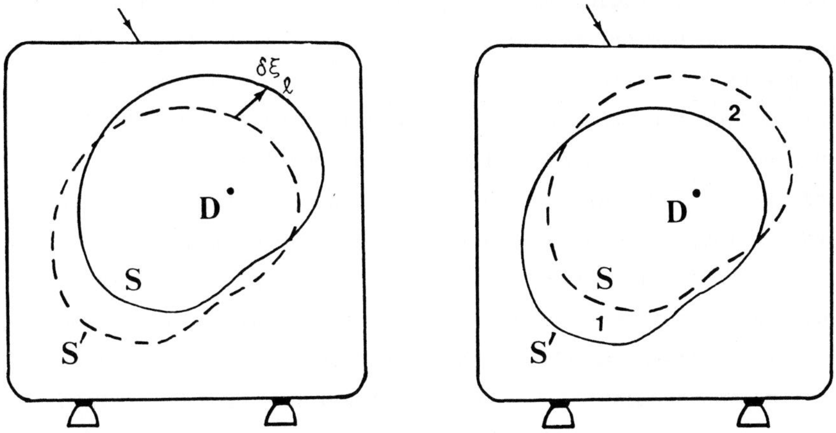

Figure 1 - Calculation of the force on a defect

of the field variables and their derivatives on S. Restore the defect to its original position.

In the original system cut out the material inside S and discard it. Apply suitable generalized forces to the surface of the resulting hole so as to preserve the original fields ϕ^α in the material exterior to S.

Similarly, cut out the piece of material inside S' in the replica and apply generalized forces to its surface so as to keep the ϕ^α unchanged in its interior. The energy $E_{S'}$ inside the new inclusion S' differs from the energy E_S in the discarded inclusion bounded by S by the addition of the energy in region 1 and the removal of the energy in region 2, so that

$$E_{S'} - E_S = -\delta\xi_\ell \int_S W\, dS_\ell \tag{18}$$

The region of overlap of the interiors of S and S' contributes nothing; even a formal infinity in the energy density is 'subtracted out'. At this stage there is no change in the energy of the material outside the hole in the original system, nor in the energy of the loading mechanism attached to its outer surface.

We now try to fit the body bounded by S' into the hole S by giving S' a translation the reverse of the one used to derive the replica from the original. Generally speaking we shall not succeed, since S and S', which, by construction, coincide in size and shape before deformation, will not do so after deformation. Even if they did, the ϕ^{α} other than the displacement components, or some of their derivatives, would be discontinuous across the interface, so that an unwanted layer of generalized force would have to be distributed over the interface to maintain local equilibrium. Consequently, before inserting the inclusion S' into the hole S we must adjust the ϕ^{α} and their various derivatives on S and S' to the common final values (17) originally noted down. That is, we must impose a change

$$\delta\phi^{\alpha}(S) = \phi^{\alpha 1} - \phi^{\alpha} \tag{19}$$

and similarly for the derivatives, on S, the surface of the hole. The corresponding change on the surface of the inclusion is

$$\delta\phi^{\alpha}(S') = \phi^{\alpha 1} - (\phi^{\alpha} - \delta\xi_{\ell}\, \phi^{\alpha}_{,\ell}) \ , \tag{20}$$

since before the final adjustment the displacement on S' was $\phi^{\alpha} - \delta\xi_{\ell}\, \phi^{\alpha}_{,\ell}$ not ϕ^{α}.

According to (11) the change (20) introduces an amount of energy

$$\delta E_{S'} = \int Q_{\alpha j}[\delta\phi^{\alpha}(S')]dS_j \tag{21}$$

into the inclusion. Similarly, during the adjustment (19) an amount of work

$$\delta E_S = - \int Q_{\alpha j}[\delta\phi^{\alpha}(S)]dS_j \tag{22}$$

is done on the surface of the hole; the minus sign is correct if the normal implicit in $dS_j = n_j dS$ is directed out of the inclusion. There is an integral similar to (22) which gives the flow of energy out of the material at its surface. However, it is exactly equal to the energy which flows into the loading mechanism, and so cancels from the total energy change, which is the quantity we actually require.

The total increase of energy is the sum of (18), (21) and (22). Fortunately, because of the linearity of $Q_{\alpha j}[\ldots]$ in its arguments the terms involving the unknown quantities (17) cancel out in the sum, and if we choose to write the total energy increase in the form $-F_\ell \delta\xi_\ell$, then F_ℓ is given by (2) with (14). Since the replacement of the interior of S by the interior of S' has the effect of displacing the defect by $+\delta\xi_\ell$ we have established that F_ℓ is the configurational force on it.

4. DISCUSSION

Here we apply the general result of the last section to a few specific cases, including some rather artificial examples from plate theory. Where possible we compare with earlier work.

The simplest case is where W has the form

$$W = W(\phi^\alpha, \phi^\alpha_{,i}) \tag{23}$$

so that (14) becomes

$$P_{\ell j} = W\delta_{\ell j} - (\partial W/\partial\phi^\alpha_{,j})\phi^\alpha_{,\ell} \quad . \tag{24}$$

For a finitely deformed grade 1 elastic solid we identify ϕ^1, ϕ^2, ϕ^3 with the displacement components u_1, u_2, u_3, and (14) becomes

$$P_{\ell j} = W\delta_{\ell j} - p_{ij}\, u_{i,\ell} \tag{25}$$

where p_{ij} is the first Piola-Kirchhoff stress tensor. Some applications of (25) to crystal defects may be found in [4] and [1]. For the elastic case in two dimensions the argument based on Figure 1 goes through as before if D is a crack tip so long as the part of the crack intercepted by S is straight and parallel to $\delta\xi_\ell$. So for a traction-free crack lying (before deformation) along the X_1-axis the integral

$$F_1 = \int_A^B P_{1j}\, d\, S_j \tag{26}$$

taken along a curve which starts at A on the lower surface, rounds the tip and ends at B on the upper surface is equal to Irwin's G [5], the energy release rate or crack extension force, or Rice's [6] J-integral.

Equation (26) will be valid for a generalized continuum if 'traction-free crack' is taken to mean that the integral (11) taken along any short segment of either crack face is zero for any $\delta\phi^\alpha$.

Atkinson and Leppington [7], using a method of their own, have given the analogue of (26) for a simple type of micropolar continuum to which (24) applies. Our result agrees with theirs.

Equations (23), (24) also cover a crack in a plate treated according to Reissner's [8] theory. The energy per unit area depends on the first gradients with respect to X_1, X_2 of five field variables, the in-plane displacement of the mid-plane, $u_i = u_i(X_1,X_2,0)$, the in-plane displacements $u_i(X_1,X_2,h)$ (2h is the thickness), introduced through the variables $\beta_i = \{u_i(X_1,X_2,h) - u_i(X_1,X_2,0)\}/h$ and the transverse displacement w. Equation (24) gives

$$P_{\ell j} = W\delta_{\ell j} - N_{ij}\, u_{i,\ell} - M_{ij}\beta_{i,\ell} - Q_j w_{,\ell} \tag{27}$$

with the usual notation. Inserted in (26), equation (27) gives the total energy release rate, i.e. the result of integrating the variable $G(X_3)$ across the thickness. Bergez [9] has already given this result.

If W has the form

$$W = W(\phi^\alpha, \phi^\alpha_{,i}, \phi^\alpha_{,ij}) \tag{28}$$

equation (14) gives

$$P_{\ell j} = W\,\delta_{\ell j} - \left[\frac{\partial W}{\partial \phi^\alpha_{,j}} - \left(\frac{\partial W}{\partial \phi^\alpha_{,jk}}\right)_{,k}\right]\phi^\alpha_{,\ell} - \left\{\frac{\partial W}{\partial \phi^\alpha_{,kj}}\right\}\phi^\alpha_{,k\ell}. \tag{29}$$

Note that in [..] $\phi^\alpha_{,jk}$ appears in the denominator of the second term, in contrast to the $\phi^\alpha_{,kj}$ in {..}. This is a consequence of the need to preserve the order of suffixes, as explained in section 2. In [10] and [2] the writer had not appreciated this point and wrote $\phi^\alpha_{,jk}$ in both positions.

When ϕ^1, ϕ^2, ϕ^3 are identified with the elastic displacement components (29) applies to a grade 2 material. The quantities [..] and {..} are actually the stress and hyperstress in it [11], though the success of our analysis does not depend on our noticing this. Atkinson and Leppington [7] have given the analogue of (26) for a crack in a linear isotropic Cosserat continuum with constrained rotation. Equation (29) reproduces their result though, as they note, the incorrect version in [10] does not.

In the elementary Kirchhoff theory of the bending of plates the energy per unit area,

$$W = \frac{1}{2}\nu D(\nabla^2 w)^2 + \frac{1}{2}(1-\nu)Dw_{,ij}\,w_{,ij} \tag{30}$$

where D is the plate modulus, ν Poisson's ratio and w the transverse displacement, is formally the energy-density of a grade 2 material with the single field variable $\phi^1 = w$. Equation (29) gives appropriate to a strongly anisotropic linear grade 2 elastic solid. Our equation (29) can be made to reproduce Kléman's [12] expression for the configurational force on a dislocation in such a medium.

For a defect in a nematic liquid crystal things are different. Here, assuming incompressibility, W depends on a director n_i of unit length and its derivatives $\partial n_i/\partial x_j$ with respect to Eulerian coordinates x_i. With our Lagrangian X_i it therefore depends formally on n_i, $\partial n_i/\partial X_j$ and also $\partial u_i/\partial X_j$. If we put $\phi^1 \ldots \phi^6 = u_i, n_i$ in (24) and transform to the x_i we get simply

$$P_{\ell j} = W\,\delta_{\ell j} - (\partial W/\partial n_{i,j})n_{i,\ell} \tag{31}$$

where $n_{i,j}$ stands for $\partial n_i/\partial x_j$, not $\partial n_i/\partial X_j$, a result we could, in this case, have got directly by conducting the argument of section 3 in Eulerian coordinates from the start.

With a suitable choice of elastic constants, and for certain geometries, the non-linear equation (10) governing n_i can be made to take the linear form

$$\nabla^2\phi = 0 \tag{32}$$

(not to be confused with a linear approximation) by the substitution $n_1 = \cos\phi$, $n_2 = \sin\phi$, $n_3 = 0$. We can then use (31) to derive known results explicitly. In particular, by taking S in (2) to be a narrow tube embracing a short segment of a curved disclination we can reproduce Kléman's [12] expression for the Peach-Koehler force on it.

Kléman calls this force configurational. It is, on the contrary, or perhaps one should say in addition, real. The theory of nematics shows that there is a traction on a surface element in the fluid given by the Ericksen stress tensor (see, for example, [13], equation (3.11))

(27) with $N_{ij} = 0$, $\beta_{i,\ell} = w,_{i\ell}$.

The energy density for a linear isotropic state of plane strain written in terms of stresses, which are then themselves written in terms of an Airy stress function χ,

$$W = -\nu(\nabla^2\chi)^2/4\mu + \chi,_{ij}\chi,_{ij}/4\mu$$

(μ is the shear modulus) is similar to (30) but does not really fit into the formalism of section 3. Among other anomalies the Euler equation (10) with $\phi^1 = \chi$ gives not the equation for equilibrium (which is taken care of by the use of χ) but the compatibility equation $\nabla^2\nabla^2\chi = 0$. If, nevertheless, we work out the tensor, $P_{\ell j}(\chi)$ say, furnished by (29) and compare it with the ordinary linear elastic $P_{\ell j}$ we get the surprising result

$$\int_A^B P_{\ell j}(\chi)\, dS_j = -\int_A^B P_{\ell j} dS_j + \frac{\mu}{1-\nu}(u_{2,1} - u_{1,2})\chi,_{\ell}\Big|_A^B .$$

The integrated-out term is zero for a stress-free crack parallel to the X_ℓ-axis or for a closed circuit embracing a dislocation (with continuous $u_{2,1} - u_{1,2}$) or other line defect. We may perhaps rationalise the minus sign by saying that as we are working in terms of stress rather than deformation we ought really to use the appropriate Legendre transform

$$W - P_{ij}\,\partial W/\partial P_{ij} = W - \chi,_{ij}\,\partial W/\partial\chi,_{ij} = -W$$

instead of W itself.

Under certain circumstances the energy density of a smectic-A liquid crystal can be taken to have the form [12]

$$W = \frac{1}{2}K_1(u_{3,11} + u_{3,22})^2 + \frac{1}{2}K_2[(u_{3,31})^2 + (u_{3,32})^2] + \frac{1}{2}B(u_{3,3})^2$$

$$t_{\ell j} = (-p_0+W)\delta_{\ell j} - (\partial W/\partial n_{i,j})n_{i,\ell} \tag{33}$$

which only differs from the $P_{\ell j}$ of (31) by a constant hydrostatic pressure p_0. As the latter contributes nothing to an integral over a closed surface the integral which gives the Peach-Koehler force actually represents a real force which the fluid exerts on the core of the disclination, and to maintain the configuration an equal and opposite force must be provided from outside, say by embedding a stiff wire in the core to provide the necessary reaction. To find the equilibrium shape of a disclination line anchored at the ends we may bend the wire until the fluid exerts no force on any element of it, and then remove it. Alternatively, and equivalently, we could, in ignorance of the force, find the shape for which the integral (2) is zero for a surface embracing any element of the line. The energy of the system is then stationary for any small deformation of the line. The equality expressed by (31) and (33) simply says that in a deformation of the disclination (or the displacement of a point defect) the work done by the externally-applied balancing force is stored as deformation energy of the fluid and the potential energy of the loading mechanism, if any, attached to its outer surface.

A number of authors have, in two dimensions, added together solutions of (32) for several disclinations without, apparently, noticing that although $t_{\ell j}$ exerts no net force on a circuit round one disclination in isolation, nevertheless, being, unlike (32), non-linear, it does exert such a force when the solutions are combined, and moreover, one which is equal to the supposedly configurational force which they calculate by integrating the energy and differentiating with respect to defect position. This equality, expressed by $t_{\ell j} = -p_0\delta_{\ell j} + P_{\ell j}$, is not, of course, confined to two dimensions or cases where (32) is valid.

The basic reason for the different status of the force on a dislocation in a solid and a disclination in a nematic liquid crystal is that the former moves through the medium, the latter with it.

REFERENCES

[1] ESHELBY, J.D., "Energy Relations and the Energy-Moment Tensor in Continuum Mechanics," in *Inelastic Behavior of Solids*, McGraw-Hill, New York, 1970, p. 77.

[2] ESHELBY, J.D., "The Elastic Energy-Momentum Tensor," *Journal of Elasticity*, Vol. 5, 1975, p. 321.

[3] ESHELBY, J.D., "The Continuum Theory of Lattice Defects," in *Solid State Physics*, Vol. 3, Academic Press, New York, 1956, p. 79.

[4] ESHELBY, J.D., "The Force on an Elastic Singularity," *Philosophical Transactions*, Vol. A 244, 1951, p. 87.

[5] BILBY, B.A. and ESHELBY, J.D., "Dislocations and Fracture," in *Fracture*, Vol. 1, Academic Press, New York, 1968, p. 99.

[6] RICE, J.R., "Mathematical Analysis in the Mechanics of Fracture," in *Fracture*, Vol. 2, Academic Press, New York, 1968, p. 191.

[7] ATKINSON, C. and LEPPINGTON, F.G., "Some Calculations of the Energy-Release Rate G for Cracks in Micropolar and Couple-Stress Elastic Media," *International Journal of Fracture*, Vol. 10. 1974, p. 599.

[8] REISSNER, E., "The Effect of Transverse Shear Deformation on the Bending of Elastic Plates," *Journal of Applied Mechanics*, Vol. 12, 1945, p. A-69.

[9] BERGEZ, D., "La Rupture des Plaques et des Coques Fissurées," Thèse, Université de Paris VI, Paris, 1974.

[10] ESHELBY, J.D., "The Calculation of Energy Release Rates," in *Prospects of Fracture Mechanics*, Noordhoff, Leyden, 1975, p. 69.

[11] TOUPIN, R.A., "Theories of Elasticity with Couple-Stress," *Archive for Rational Mechanics and Analysis*, Vol. 17, 1964, p. 85.

[12] KLÉMAN, M., *Points. Lignes. Parois*, Vol. 1, Les Éditions de Physique, Paris, 1977.

[13] DAFERMOS, C.M., "Disinclinations in Liquid Crystals," *Quarterly Journal of Mechanics and Applied Mathematics*, Vol. 23, 1970, p. S49.

Aspects of the Theory of Dislocations

by J. D. Eshelby

Department of the Theory of Materials, University of Sheffield

Summary

Various topics in the theory of dislocations in linear elastic continua are reviewed using, wherever possible, simple physical arguments to derive the mathematical results. The elastic field of a general Somigliana dislocation is derived and specialized to the physically important case where its discontinuity vector is constant. A general account of the effect of free surfaces is illustrated by some specific examples. In discussing dislocation energetics it is emphasized that once the Peach–Koehler expression describing the interaction with an externally applied stress field has been established, its extension to the case of internal stresses and image stresses follows without further mathematics. In the final Section it is demonstrated that certain solutions for uniformly and non-uniformly moving dislocations can be derived rather simply from the corresponding static solutions.

1. Introduction

The modern mathematical theory of dislocations really begins with G. I. Taylor's (1934) recognition of the relationship between a particular one of the various kinds of elastic dislocation treated by Volterra (1907) and the line defect postulated by himself, Orowan (1934), and Polanyi (1934) to explain the plasticity of crystals. Volterra's dislocations are a special class of what have come to be called Somigliana (1914, 1915) dislocations, though they seem to have been first introduced into the theory of elasticity by Gebbia (1891) under the rather uninformative name of deformations of the third kind. To create a Somigliana dislocation we start (see Fig. 2.1) by making a cut over a surface S, bounded by a curve C, or possibly closed, in an elastic solid and giving the two faces of the cut a small relative displacement $\mathbf{b}$ which varies from point to point of the surface. A component of displacement normal to the cut will give rise to gaps and interference, and, when these have been made good by filling and scraping, the faces of the cut are welded together again, leaving the whole specimen in a state of internal stress.

Macroscopic examples of Somigliana dislocations are provided by a worked-out coal-seam whose roof and floor have been forced into contact by overburden pressure or a thin igneous intrusion between bedding planes. Or again, if, having made a cut, we apply suitable tractions to the outer surface of the specimen the faces of the cut, or crack, will separate, and we have generated a Somigliana dislocation. So long as we maintain the external stress it makes no difference whether or not we fill the gap according to prescription, but if we do so the dislocation will survive when the applied stress is removed. If the cut is plane a suitably chosen applied shear stress will make the faces of the crack slide over one another so as to make the traction on the crack zero, thus adding to the applied field the elastic field of a Somigliana dislocation which happens to have its discontinuity vector everywhere in the plane of the cut. If the faces can slip freely the dislocation will disappear when the applied stress is removed; we may say that it is merely "induced" by the applied stress. However, if we have previously smeared the crack with glue we can remove the applied stress when the glue has set, and be left with a permanent fossilized version of the dislocation and its stress field. More realistically, if slip has occurred over a finite area of a fault plane, friction and normal pressure may wholly or partially inhibit slipping back when the applied shear stress no longer acts.

To apply the theory of elasticity to the study of deformed crystal lattices we imagine that the undeformed continuum is a transparent jelly in which is marked a regular array of points to represent the centres of the atoms, and that the points move with the medium when it deforms. Suppose that in this medium we make a cut bounded by a curve C and create a Somigliana dislocation whose discontinuity vector, instead of being variable, is constant and equal to a lattice vector. Then an observer who has not watched us construct the dislocation—and who cannot distinguish one atom from another, so that for him the position of an atom is only defined modulo a lattice vector—will not notice any discontinuity at the cut and it is only the large distortion of the lattice near the bounding curve C which will tell him that something has happened.

Suppose that in the undeformed continuum we have drawn a closed circuit going from point to point of the lattice of dots and embracing the curve C. Then when we have made the dislocation the circuit will be disrupted by a gap. The vector required to close the gap is, with a suitable sign convention, the (local) Burgers vector of the dislocation. The observer who has not watched us make the dislocation can also be led to the conclusion that there is a dislocation, and can find its Burgers vector, without reference to the undeformed lattice, for if he tries to form a closed circuit, not too small, about C by going, say, ten lattice spacings north, ten west, ten south, and then ten east, he will end up one lattice spacing from his starting point. We can also look at things more physically. Suppose that, as above, we create a Somigliana dislocation in a continuum dotted with points to represent the atoms, making sure that the discontinuity exceeds a few lattice spacings over a fair fraction of the surface. Now replace the dots by atoms, at the same time abolishing the supporting elastic continuum. We shall find that over most of the former cut surface the interatomic forces will pull the atoms into perfect registry, squeezing the misfit into the neighbourhood of a set of closed curves, with each of which is associated a Burgers vector which is a lattice vector. Subject to certain constraints these lines, or, rather, narrow regions of misfit, are able to move through the lattice, and the equilibrium form, if any, which they will take up as a result of their mutual interaction and the effect of applied and internal stresses can, in simple cases, be worked out by continuum theory. The assertion that the regions of misfit, or dislocation cores, are narrow, and that they are mobile, is qualitatively confirmed by the model of Peierls and Nabarro, which uses a mixture of continuum and atomic theory.

The prime reference for the theory of dislocations is still the book by Nabarro (1967) now supplemented by a five-volume collective work under his editorship (Nabarro, 1980). The books by Hirth and Lothe (1968) and Lardner (1974) contain more mathematical details; the latter also deals with the connection between dislocations and the theory of cracks. Seeger's (1955) *Handbuch* article is still well worth reading. A recent review (Bacon *et al.*, 1979) is devoted to the application of anisotropic linear elasticity to dislocations and other lattice defects. An article by Kosevich (1979) gives references to Soviet work.

So long as the dislocation was purely a theoretical construct a fairly elementary treatment of its properties was sufficient, and was mostly supplied by physicists and metallurgists. Later they were joined by applied mathematicians, who found in dislocation theory, and also in the allied subject of fracture mechanics, a welcome addition to the rather limited stock of worthwhile topics to which the theory of elasticity could be applied. Later, the actual observation of dislocations in the electron microscope called for, and the availability of computers made possible, more elaborate calculations. The present article is pitched at the earlier level, and is confined to the linear theory. We shall concentrate on some of the concepts which are more or less foreign to the traditional theory of elasticity and have to be added to it to build up the theory of dislocations. We start (Section 2) with the calculation of the elastic field induced by establishing a displacement discontinuity, in particular a constant one, across a surface in a solid, as explained above, beginning with a medium of infinite extent. In a finite body the presence of free surfaces modifies the field of a dislocation ("image" effects, Section 3) and, in addition by deforming the surface of the body the dislocation, signals its presence to any external loading mechanism. If the dislocation deforms, the energy of the system varies, in a way which depends on the presence of other sources of internal stress, the form of the surface and the coupling to external loads. The energy changes can be picturesquely described (Section 4) in terms of a configurational or Peach–Koehler force acting on the dislocation. In this context the dislocation is supposed to deform by passing through a succession of neighbouring static configurations. However, if its speed is comparable with the velocity of elastic waves its elastic field has to satisfy the dynamical elastic equations, and in Section 5 we present a few dynamical solutions which can be deduced from static solutions without too much trouble.

2. The Elastic Field of a Dislocation

2.1. *Preliminaries*

Figure 2.1 illustrates the construction of a Somigliana dislocation in the way outlined in the Introduction. In (a) we have made a cut in the unstressed material, bounded by the curve C, and have given the labels S, S' to the two faces of the cut, at present coincident. In (b) we have given S' a small displacement $\mathbf{b}(\mathbf{r}')$ which, generally speaking, varies as $\mathbf{r}'$ moves about on S. Where there is a gap we fill it up, and where there would be interpenetration we scrape material away to avoid it. Finally, we weld the material together again. We are left with the situation of Fig. 2.1(c), a surface S with a discontinuity vector $\mathbf{b}(\mathbf{r}')$ associated with each point $\mathbf{r}'$ on it. We evidently need some sort of sign convention. Draw an arrow to represent the normal $\mathbf{n}$ at each point, and label with $+$ and $-$ the sides of S on, respectively, the head and tail sides of the arrow. Then we define $\mathbf{b}$ by

$$\mathbf{b}(\mathbf{r}') = \mathbf{u}^{+}(\mathbf{r}') - \mathbf{u}^{-}(\mathbf{r}'), \tag{2.1}$$

where $\mathbf{u}^{+}(\mathbf{r}')$ and $\mathbf{u}^{-}(\mathbf{r}')$ are the displacements, on the sides indicated, of two points originally coincident at $\mathbf{r}'$. The opposite convention is also used.

For brevity we shall often talk as if there was a gap everywhere in Fig. 2.1(b), leaving the reader to make the adjustments necessary to cover the case of interpenetration.

In the next Section we shall discuss the determination of the elastic field produced by a Somigliana dislocation. Here we discuss some results which depend on geometry assisted by only a minimum of the theory of elasticity—to connect (2.2), (2.3), and (2.4) below we need Hooke's law, and in connection with Fig. 2.2 we need the fact that an elastic body resumes its unstressed form when the load is removed and that a rigid body motion induces no stress in it.

Before dislocations had become associated with the physics of crystals, mathematicians had set themselves the task, one could almost say had invented the game, of determining what conditions have to be imposed on the discontinuity **b** to make the site of the cut become in some sense undetectable. In modern dislocation theory the limitation imposed on **b** comes from crystallography rather than from the investigations just mentioned, and so we shall only present a simplified intuitive account of them. (On the topic of discontinuities in general, see Hill, 1961.)

In a featureless linear isotropic continuum it is natural to start by requiring the six strain components e_{ij} to be continuous across the cut. In Fig. 2.1 let us take at each point of the cut

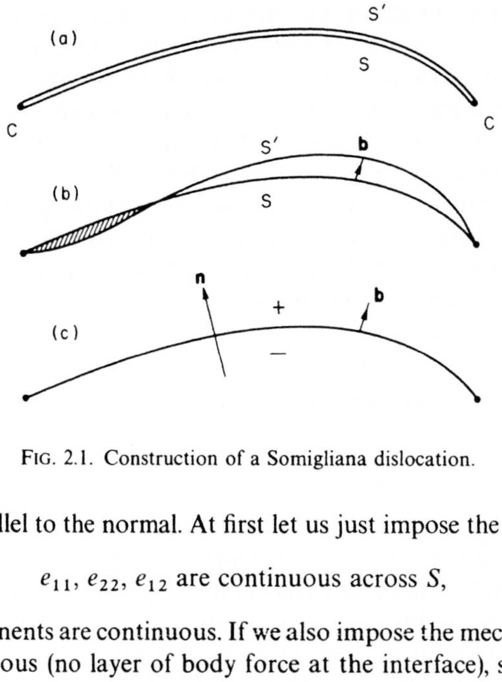

FIG. 2.1. Construction of a Somigliana dislocation.

local axes with x_3 parallel to the normal. At first let us just impose the geometrical condition

$$e_{11},\ e_{22},\ e_{12} \text{ are continuous across } S, \tag{2.2}$$

i.e. the in-plane components are continuous. If we also impose the mechanical condition that the traction is continuous (no layer of body force at the interface), so that

$$\sigma_{13},\ \sigma_{23},\ \sigma_{33} \text{ are continuous across } S, \tag{2.3}$$

(2.2) and (2.3) combined with Hooke's law then show that in addition the remaining strain components

$$e_{13},\ e_{23},\ e_{33} \text{ are continuous across } S, \tag{2.4}$$

as are also $\sigma_{11}, \sigma_{22}, \sigma_{12}$. If at this point we were to allow the material to be anisotropic, (2.2) and (2.3) would not imply (2.4); as there is in general a relative rotation of the material on the two sides of S the components of the modulus tensor C_{ijkl}, when referred to common axes, suffer a jump across S which must be compensated for by discontinuities of non-negligible order in e_{13}, e_{23}, e_{33}.

The requirement (2.2) imposes a simple geometrical constraint on the discontinuity **b** (Weingarten, 1901). We can imagine constructing the Somigliana dislocation in two stages. Stage 1 is essentially geometrical. Starting from the two surfaces of the crack S and S' in contact in the unstrained material, we keep S fixed with a set of infinitesimal clamps and enforce the displacement $\mathbf{b}(\mathbf{r}')$ on the points $\mathbf{r}'$ of S'. With the local axes used above we have, with an obvious notation, $e_{\alpha\beta}(S) = 0$ (where here and below α, β only take the values 1, 2) and, say, $e_{\alpha\beta}(S') = e^0_{\alpha\beta}$. After filling and scraping we weld the body together again. In stage 2 we remove the constraints and the body relaxes into a state of equilibrium. During this relaxation S and S', now joined together, each acquire the same additional strain, $e'_{\alpha\beta}$ say, so that now we have $e_{\alpha\beta}(S) = e'_{\alpha\beta}$, $e_{\alpha\beta}(S') = e^0_{\alpha\beta} + e'_{\alpha\beta}$. But according to (2.2) we must now have $e_{\alpha\beta}(S') = e_{\alpha\beta}(S)$, i.e. $e^0_{\alpha\beta}(S') = 0$. Hence the deformation which **b** induces in S' during stage 1 must produce no strain in its surface. Hence, in the language of the theory of surfaces, S' is applicable to S. This is Weingarten's result.

According to Somigliana, Weingarten's result is incomplete, and the e_{ij} would also be continuous if **b** were everywhere normal to S. His claim is refuted by the simple example of a sphere inserted into a hole slightly too small for it. We may suppose that the sphere was originally a perfect fit but was then fitted with a thin spherical jacket, so that we do in fact have a Somigliana dislocation. Also, as a sphere is not applicable to one of different radius we have the required violation of Weingarten's condition. Since the sphere is compressed and the hole is expanded, meridian circles of the two respectively decrease and increase in length so that the hoop strain, $e_{\theta\theta} = \partial u_r/r\partial\theta + u_r/r$ in the usual notation, is discontinuous. What Somigliana's analysis actually shows is that $\partial u_r/r\partial\theta$ is continuous, and similarly for more general curved surfaces. When S is flat what he says is true, but this case is already covered by Weingarten's result since, to the accuracy of the linear theory, for a plane a small variable normal displacement induces no stretching or contraction in the surface.

The simplest displacement **b** which meets the condition of applicability is the rigid-body displacement

$$\mathbf{b} = \mathbf{a} + \boldsymbol{\omega} \times \mathbf{r} \tag{2.5}$$

with constant small **a** and $\boldsymbol{\omega}$, the case studied at length by Volterra (1907); for a modern summary see Volterra and Volterra (1960). Figure 2.2 shows two surfaces S_1, S_2 bounded by the same curve C. The material in a toroidal hole containing C is removed, so rendering the body multiply-connected.

Make a cut over S_1 and remove material so as to leave a gap specified by (2.5). (The reader

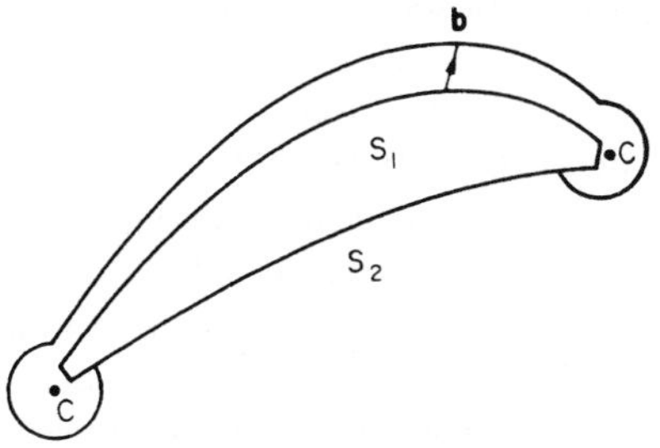

FIG. 2.2. A Volterra dislocation.

will easily make the adjustments required to cover the case where the "gap" is wholly or partly an interpenetration.) If we make a cut over S_2 and close the gap at S_1 the material between S_1 and S_2 will undergo a rigid-body motion specified by the right-hand side of (2.5) and a gap will open up over S_2, again given by (2.5). Having noted this we start again, with S_2 uncut and a gap (2.5) at S_1. If we now close this gap and weld the faces of S_1 together, the material is put into a state of stress, both between S_1 and S_2 and also outside them. If we next make a cut over S_2 this stress disappears and a gap (2.5) opens up at S_2; all we have done, in a roundabout way, is to give the piece of material between S_1 and S_2 a rigid-body displacement. If we now close the gap at S_2 the state of stress just relaxed is, naturally, restored. This shows, if it is not already obvious, that creating and closing the gap (2.5) over S_2 produces the same state of stress as creating and closing the gap (2.5) over S_1 does; the strains in the two cases are also, of course, the same, though the displacements differ by the right-hand side of (2.5) between S_1 and S_2 and are identical outside them.

It is also clear that closing a gap at S_1 will induce no discontinuity in the e_{ij}, or in their derivatives of all orders in the neighbourhood of S_2 and conversely, so that the site of the cut is undetectable by investigating the strain or its derivatives of various orders, in contrast to the situation governed by (2.2) and (2.3), where the e_{ij} are continuous, but their first derivatives are usually not.

As explained in the Introduction, when we use the continuum to simulate a crystal lattice we are limited to the case where **b** is constant and equal to a lattice vector, i.e. to (2.5) with $\boldsymbol{\omega} = \mathbf{0}$ and a suitably chosen **a**.

2.2. *The elastic field of a Somigliana dislocation*

The first person to have published an expression for the displacement field of a Somigliana dislocation seems to have been Gebbia (1902). In vector notation, and with our sign convention, it takes the form

$$\mathbf{u}(\mathbf{r}) = -\frac{1}{4\pi}\int_S \mathbf{b}\frac{\mathbf{R}\cdot d\mathbf{S}}{R^3} + \frac{1}{4\pi}\int_S \mathbf{b}\times\left(d\mathbf{S}\times\nabla'\frac{1}{R}\right)$$
$$+\frac{1}{4\pi}\frac{1}{2(1-\nu)}\nabla\int_S\left[\mathbf{b}\left(\nabla'\cdot\frac{\mathbf{R}}{R}\right)-(\mathbf{b}\cdot\nabla')\frac{\mathbf{R}}{R}\right]\cdot d\mathbf{S}, \tag{2.6}$$

where $\mathbf{R} = \mathbf{r}-\mathbf{r}'$, $R = |\mathbf{r}-\mathbf{r}'|$, ∇' denotes the gradient with respect to $\mathbf{r}'$, and $\mathbf{b}$, $d\mathbf{S}$ stand for $\mathbf{b}(\mathbf{r}')$, $d\mathbf{S}(\mathbf{r}')$, and the material is supposed to be isotropic with Poisson's ratio ν.

No notice seems to have been taken of (2.6) in the dislocation literature, probably because Gebbia regarded it as simply a tool for use in his method of solving elastic boundary value problems.

Gebbia's proof of (2.6) is decidedly involved. His result, and also its analogue for an anisotropic medium, can be derived, or at any rate arrived at, by the following intuitive argument (Read, 1953). Consider the interaction of the dislocation with a point force applied at **r**, of unit magnitude, and parallel to the x_i-axis, as in Fig. 2.3. (The surface S' is to be ignored for the moment.). Let E_D, E_F be the amounts of work expended in introducing either the dislocation or the force by itself. If we introduce first the dislocation, then the force, the work done is simply $E_D + E_F$, since, on the linear theory, the response of the material to the force is unaffected by the prestress due to the dislocation. If, alternatively, we introduce the force first and then the dislocation, the work required will be $E_D + E_F$ plus two new terms. One is the

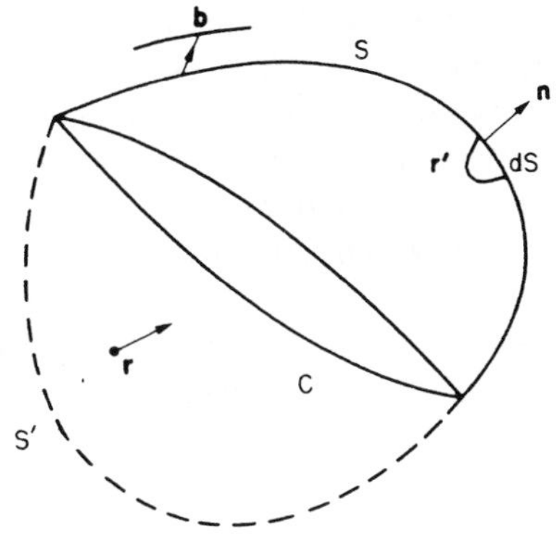

FIG. 2.3. Calculation of the elastic field of a Somigliana dislocation.

work done by the force as its point of application is shifted through the displacement induced by the dislocation. The other is the work done on the two faces of the cut by the tractions due to the point force as the faces suffer the relative displacement **b**. But these extra terms must cancel; otherwise one method of loading would require less work than the other, and by loading in the easier way and then unloading in the reverse of the harder way we should be able to create energy. This gives at once

$$u_i(\mathbf{r}) = \int_S b_j(\mathbf{r}')\sigma_{jk}^{(i)}(\mathbf{r}')n_k\,dS, \tag{2.7}$$

where $\sigma_{jk}^{(i)}$ is the stress produced by the point force. The left-hand side is the work done by the force and the right-hand side is minus the work done by the tractions at the cut; the minus sign comes from the convention (2.5) combined with the fact that the outward normal to the material is $-\mathbf{n}$ on the positive side of the discontinuity surface.

This heuristic argument can be made respectable by using the Rayleigh–Betti reciprocal theorem which says that the integral

$$\int_S (u_i\sigma'_{ij} - u'_i\sigma_{ij})dS_j \tag{2.8}$$

taken over the closed surface S vanishes for any two elastic fields u_i, σ_{ij} and u'_i, σ'_{ij} provided S does not embrace any body forces or singularities. The proof of (2.8) simply consists in using Gauss' theorem to convert to a volume integral over the interior of S, using the equilibrium conditions $\sigma_{ij,j} = 0, \sigma'_{ij,j} = 0$, and noting that we have $u_{i,j}\sigma'_{ij} = u_{i,j}C_{ijkl}u'_{k,l} = u'_{i,j}\sigma_{ij}$ in virtue of the symmetry relation $C_{ijkl} = C_{klij}$ satisfied by the elastic moduli. Let u_i be the dislocation displacement and u'_i the displacement due to a point force, as above. For S we may take a large surface S_0 enclosing the dislocation and force together with a small sphere excluding the force and a closely fitting jacket embracing the discontinuity surface of the dislocation. These latter surfaces reproduce the terms in (2.7), but on the right we have to add minus the expression (2.7) taken over S_0. In some presentations with pretensions to rigour, this term is simply ignored. In others it is stated to go to zero as S_0 becomes indefinitely large, but this requires a knowledge of the behaviour of u_i which we do not have until the problem is solved. There is, however, a simple way out of the difficulty. The extra term, or indeed just the contribution of a single element dS to it, is, as is easily checked, itself a solution of the elastic equilibrium equations, free of singularities inside S_0. Thus the contribution from S_0 represents a displacement produced by some type of loading applied outside S_0, and its omission does not affect the essential properties of the solution which is, in fact, undefined to within an arbitrary singularity-free solution of the elastic equations.

In an isotropic medium the displacement due to a unit point force $\mathbf{e}_i$ in the u_i-direction is

$$\mathbf{u}^{(i)}(\mathbf{r}) = \frac{\mathbf{e}_i}{4\pi\mu R} - \frac{1}{16\pi\mu(1-\nu)}(\mathbf{e}_i \cdot \nabla')\nabla' R \tag{2.9}$$

or in suffix notation

$$u_j^{(i)} = G_{ij}(\mathbf{R}) = \frac{\delta_{ij}}{4\pi\mu R} - \frac{1}{16\pi\mu(1-\nu)} \frac{\partial^2 R}{\partial x_i' \partial x_j'}, \tag{2.10}$$

where G_{ij} is the so-called elastic Green's tensor. If we calculate the stress $\sigma_{jk}^{(i)}$ from (2.10) and insert it in (2.7), the result can be manipulated to agree with (2.6).

2.3. *The field of a crystal dislocation*

For a crystal dislocation with a constant discontinuity $\mathbf{b}$ we may add to the expression in the square brackets in the last integral of (2.6) the vanishing quantity $(\mathbf{R}/R) \cdot \nabla' \mathbf{b} - (\mathbf{R}/R)(\nabla' \cdot \mathbf{b})$ which turns it into what can be recognised as the expansion of $\nabla' \times (\mathbf{b} \times \mathbf{R}/R)$, and then with the help of Stokes' theorem this term may be converted into a line integral round the curve C bounding S. The second term can be converted to a line integral directly by taking the $\mathbf{b}\times$ outside the integral and using

$$\int_S d\mathbf{S} \times \nabla'(\ldots) = \int_C d\mathbf{x}'(\ldots),$$

an alternative form of Stokes' theorem. In this way we get

$$\mathbf{u}(\mathbf{r}) = -\frac{\mathbf{b}}{4\pi}\Omega - \frac{1}{4\pi}\int_C \frac{\mathbf{b} \times d\mathbf{x}'}{R} + \frac{1}{4\pi}\frac{1}{2(1-\nu)}\nabla \int_C \frac{\mathbf{b} \times d\mathbf{x}' \cdot \mathbf{R}}{R}, \tag{2.11}$$

where

$$\Omega(\mathbf{r}) = \int_S \frac{\mathbf{R} \cdot d\mathbf{S}}{R^3} \tag{2.12}$$

is the solid angle subtended by S at $\mathbf{r}$.

This important result, due to Burgers (1939), exhibits the displacement as the sum of single-valued terms and one, the first, which produces the required cyclic behaviour. Peach and Koehler (1950) were, rather surprisingly, able to turn even the solid angle (2.12) into a line integral. It contains

$$\frac{\mathbf{R}}{R^3} = \nabla' \frac{1}{R}.$$

If we could replace the gradient by the curl of something we could use Stokes' theorem to convert (2.12) to a line integral. This is, in fact, possible (Madelung, 1950). Equation (2.12) gives the field of, say, a magnetic monopole as the gradient of a scalar potential. But the magnetic field is the same as that of a very fine semi-infinite solenoid if we choose to overlook the concentrated return field inside the solenoid. Hence we ought to be able to express the field as the curl of the vector potential of the solenoid. This leads to

$$\frac{\mathbf{R}}{R^3} = -\nabla' \times \mathbf{A} \tag{2.13}$$

with

$$\mathbf{A}(\mathbf{R}) = \frac{\mathbf{m} \times \mathbf{R}}{R(R + \mathbf{m}\cdot\mathbf{R})}, \tag{2.14}$$

where $\mathbf{m}$ is any unit vector. Equation (2.13) is easily checked for, say, $\mathbf{m} = (0, 0, 1)$. Stokes' theorem then gives

$$\Omega(\mathbf{r}) = -\int_C \mathbf{A}(\mathbf{r} - \mathbf{r}')\cdot d\mathbf{r}'. \tag{2.15}$$

This integral has the necessary discontinuity of 4π across the cylindrical surface $S_\mathbf{m}$ generated by drawing a semi-infinite straight line from each point of C in the direction $-\mathbf{m}$; in other words $S_\mathbf{m}$ is the shadow surface thrown (in a dusty atmosphere) by the loop C when it is illuminated by a lamp at a remote point in the direction of $\mathbf{m}$. It is not hard to show that the contribution of a single element $d\mathbf{r}'$ to (2.15) is the solid angle subtended at $\mathbf{r}$ by the hairpin made up of $d\mathbf{r}'$ and a pair of lines drawn from its extremities to infinity in the direction of $-\mathbf{m}$. The representation of Ω as a line integral is, therefore, perhaps a little fraudulent; it is simply (2.12) with the arbitrary cap closing the loop drawn out into the cylindrical shadow surface $S_\mathbf{m}$. Peach and Koehler's expression actually contains the average of (2.14) over the six vectors $\mathbf{m} = (\pm 1, 0, 0)$, $(0, \pm 1, 0)$, $(0, 0, \pm 1)$. Some readers may find illuminating Amar's (1980) derivation of (2.15) from (2.12) with the help of the theory of exterior forms.

For anisotropy $\sigma_{jk}^{(i)}$ is the stress derived from $u_j^{(i)} = G_{ij}$, where now $G_{ij}(\mathbf{r} - \mathbf{r}')$ is the elastic Green's function for an anisotropic medium, giving, as before, the x_j-component of the displacement at $\mathbf{r}$ produced by a unit point force at $\mathbf{r}'$ parallel to the x_i-axis. The property

$$G_{ij}(\mathbf{r} - \mathbf{r}') = G_{ji}(\mathbf{r}' - \mathbf{r}) \tag{2.16}$$

expresses the fact that if we apply forces F_i at $\mathbf{r}$ and F_i' at $\mathbf{r}'$ the total work done must be the same whether we apply F_i first, then F_i', or, conversely, so that the cross-terms in the energy, $F_i G_{ij}(\mathbf{r} - \mathbf{r}')F_j'$ and $F_i' G_{ij}(\mathbf{r}' - \mathbf{r})F_j$, must be equal. The relation $G_{ij}(\mathbf{r}) = G_{ji}(-\mathbf{r})$ which with (2.16) also implies

$$G_{ij}(\mathbf{r}) = G_{ji}(\mathbf{r}) \tag{2.17}$$

or $u_j^{(i)} = u_i^{(j)}$ is equivalent to the statement that equal and opposite point forces produce zero total displacement at a point halfway between them. To establish it we have to use the fact that the anisotropic elastic equations are invariant under the inversion $\mathbf{r} \to -\mathbf{r}$.

It is easy enough to find a formal solution of the equation

$$C_{rsjk} G_{ij,ks}(\mathbf{r}) + \delta_{ir}\delta(\mathbf{r}) = 0$$

governing Green's function by taking its Fourier transform and writing the solution in the form of a triple integral over the transform variables (k_1, k_2, k_3); see, for example, Eshelby

(1951). To evaluate the integral explicitly one would have to be able to write down explicitly the solution of a certain sextic equation whose coefficients involve the k_i. Gebbia (1904) discussed some cases in which this is possible. (This remarkable paper contains, among other things, an anticipation of Feynman's (1949) method for dealing with integrals involving products.) Head (1979) has also discussed this point. For a hexagonal (transversely isotropic) material the factorization can be carried out and so G_{ij} can be given explicitly (Kröner, 1953). For the other case of most interest in metal physics, that of cubic symmetry, factorization is, unfortunately, not usually possible. An exception is the case where, with the usual abbreviated notation, $c_{12}+c_{44}=0$. The equations for u_1, u_2, u_3 are then uncoupled, and, after a suitable stretching of one of the coordinate axes, each equation becomes Laplace's equation. This is the continuum limit of the Rosenstock–Newell crystal, in which the three displacement equations are similarly decoupled by a suitable choice of the interatomic force law. Though the condition $c_{12} = -c_{44} < 0$ is not actually ruled out by stability considerations, no known crystal satisfies it even approximately, so that this attractively simple case is not really of much value. However, Dederichs and Leibfried (1969) have examined the possibility of a perturbation treatment for the cubic case, starting from the isotropic or Rosenstock–Newell case.

When the stress $\sigma_{jk}^{(i)}$ in (2.7) with constant b_j has to be derived from the general anisotropic Green's function, we cannot isolate a term like (2.12) from the dislocation displacement, but we can show that (2.7) has the right discontinuity across S by a pretty argument due to Burgers (1939). In Fig. 2.3 let S' be a second surface bounded by C, so that $S+S'$ forms a closed surface. With the factor b_j suppressed the integral (2.7) taken not over S but over $S+S'$ is the x_j-component of the total traction due to the point force, namely zero if **r** is outside $S+S'$ and $-\delta_{ij}$ if it is inside, minus because the traction balances the force. Hence the integral over $S+S'$ increases by δ_{ij} as **r** breaches S from inside $S+S'$. But obviously there is no sudden change in the contribution from S', so that the integral over S alone suffers the jump. Hence, restoring the factor b_j, u_i jumps by b_i as **r** passes through S in the direction of the positive normal, as required. Burgers extended this argument to give a proof of (2.7). It is first necessary to verify that (2.7) actually satisfies the equilibrium equations written in terms of displacements. In fact $\sigma_{jk}^{(i)}$ by itself, when regarded as a displacement $u_i^{(jk)}$, does so, as is easily checked if we recall the symmetry relation (2.17). Burgers also felt obliged to check that (2.7) contained no unwanted singularities along C. In the energy argument we first used to derive (2.7) there is no need for this. If there had been, say, a distribution of body force along C and a line of dilatation coinciding with it, the cross-terms between these singularities and the field of the point force would have appeared in (2.7) as a line integral along C involving the displacement and displacement gradient of the field of the point force. By not allowing for such terms we unconsciously ensured that the corresponding singularities were absent.

Suppose that in the above argument we keep the point of observation **r** fixed and deform the surface S. Then except for a jump of $\pm b_i$ when S sweeps over **r**, the integral over $S+S'$ is unaffected by the deformation, and so, since the integral over S', as before, does not vary, it follows that, provided it spans C, the exact form of S is, the jump apart, unobservable and irrelevant. This is Volterra's fundamental result, though here derived by a mechanical rather than a geometrical argument.

Though the displacement cannot, the displacement gradient, and hence also the strain and stress, can be written as a line integral in the anisotropic case. We shall find it convenient to write $x_{k'}$ for x'_k, $dS_{k'}$ for $dS_k(\mathbf{r}')$, $(\ldots)_{,k'} = \partial(\ldots)/\partial x_{k'}$, $(\ldots)_{,k} = \partial(\ldots)/\partial x_k$.

Differentiate (2.7) with respect to x_l and use the fact that $\sigma_{jk}^{(i)}$ depends only on $x_m - x_{m'}$ to get

$$u_{i,l} = b_j \int_S \sigma^{(i)}_{jk,l}\, dS_{k'} = -b_j \int_S \sigma^{(i)}_{jk,l'}\, dS_{k'}. \tag{2.18}$$

Apply Stokes' theorem in the form

$$\int_S \left(dS_{k'} \frac{\partial}{\partial x_{l'}} - dS_{l'} \frac{\partial}{\partial x_{k'}} \right) (\ldots) = \int_C \varepsilon_{klq}\, dx_{q'} (\ldots) \tag{2.19}$$

to get

$$u_{i,l} = -b_j \varepsilon_{klq} \int_C \sigma^{(i)}_{jk}\, dx_{q'} \tag{2.20}$$

plus the surface integral (2.18) with l', k' interchanged. This, however, is zero because we have

$$\sigma^{(i)}_{jk,k'} = -\sigma^{(i)}_{jk,k} = 0 \tag{2.21}$$

in equilibrium, and so the integral (2.20), taken round the loop, is the complete expression. Alternatively, in the first form of (2.18) write $dS_{k'}$ as $-(x_r - x_{r'})_{,k'} dS_{r'}$. By (2.21) this can be written as

$$u_{i,l} = -b_j \int_S [\sigma^{(i)}_{jk,l} (x_r - x_{r'})]_{,k'}\, dS_{r'}. \tag{2.22}$$

Stokes' theorem (2.19) converts this into the line integral

$$u_{i,l} = -b_j \varepsilon_{qkr} \int_C \sigma^{(i)}_{jk,l} (x_r - x_{r'})\, dx_{q'} \tag{2.23}$$

together with the integral (2.22) with r', k' interchanged. The latter vanishes since it contains

$$\left[3 + (x_r - x_{r'}) \frac{\partial}{\partial x_r} \right] \sigma^{(i)}_{jk,l}$$

and $\sigma^{(i)}_{jk,l}$ is homogeneous of degree -3 in the $x_r - x_{r'}$ (Indenbom and Orlov, 1967). As we shall see, the special virtue of (2.23) is that $\varepsilon_{qkr}(x_r - x_{r'})\, dx_{q'}$ is the x_k-component of $\mathbf{R} \times d\mathbf{x}'$.

2.4. *Straight dislocations*

The elastic fields of infinite straight dislocations may be found from the three-dimensional formulae, but it seems worthwhile to treat them directly.

We begin with isotropy, and go over to the elementary notation in which (x_1, x_2, x_3) and (u_1, u_2, u_3) become (x, y, z) and (u, v, w). The displacement of a screw dislocation with Burgers vector $(0, 0, b)$ has the simple form

$$w = \frac{b}{2\pi} \theta = \frac{b}{2\pi} \tan^{-1} \frac{y}{x}, \quad u = v = 0, \tag{2.24}$$

and since the dilatation div $\mathbf{u}$ is zero the non-vanishing stress components are

$$\sigma_{zx} = \mu \frac{\partial w}{\partial x} = -\frac{\mu b}{2\pi} \frac{\sin\theta}{r}, \quad \sigma_{zy} = \mu \frac{\partial w}{\partial y} = \frac{\mu b}{2\pi} \frac{\cos\theta}{r}. \tag{2.25}$$

Equation (2.24) hardly needs proving. With div $\mathbf{u} = 0$ the equilibrium equation

$$(1-2\nu)\nabla^2\mathbf{u} + \text{grad div } \mathbf{u} = 0 \tag{2.26}$$

reduces to $\nabla^2 w = 0$, so that w must be a harmonic function which changes by b when the dislocation is encircled once, and this leads directly to (2.24).

The expressions for the displacement $(u, v, 0)$ around an edge dislocation with Burgers vector $(b, 0, 0)$ are considerably more complicated:

$$\begin{aligned} u &= \frac{b}{2\pi}\theta + \frac{1}{2(1-\nu)}\frac{b}{2\pi}\frac{xy}{r^2}, \\ v &= \frac{1-2\nu}{2(1-\nu)}\frac{b}{2\pi}\ln\frac{1}{r} + \frac{1}{2(1-\nu)}\frac{b}{2\pi}\frac{y^2}{r^2}. \end{aligned} \tag{2.27}$$

The stress components σ_{zx}, σ_{zy} are zero and the rest can be derived from the simple Airy stress function

$$\chi = -\frac{\mu b}{2\pi(1-\nu)} y \ln r \tag{2.28}$$

according to the normal rule

$$\begin{aligned} \sigma_{xx} &= \partial^2\chi/\partial y^2, \quad \sigma_{yy} = \partial^2\chi/\partial x^2, \\ \sigma_{xy} &= -\partial^2\chi/\partial x\,\partial y, \quad \sigma_{zz} = \nu\nabla^2\chi. \end{aligned} \tag{2.29}$$

Newcomers to dislocation theory are sometimes disconcerted to find that in some references the y^2/r^2 in the second term of v is replaced by $-x^2/r^2$. However, the difference is a multiple of $(y^2+x^2)/r^2 = 1$, and so amounts only to an unimportant translation.

The displacements (2.27) and the associated stresses play an important part in even elementary treatments of dislocation theory, but are a good deal more complicated-looking than other formulae encountered at that level. The writer has therefore tried to provide simple proofs involving a minimum of the theory of elasticity, and in which each step of the argument appears to be reasonably inevitable. One way is to begin by first solving the simpler problem of finding the cylindrically symmetrical stress field which is set up when a narrow wedge-shaped gap is cut out and then closed up. The field of an edge dislocation can then be found from it by an essentially geometrical argument (Eshelby, 1966). Or one can start from a relationship which connects certain states of plane strain with their anti-plane analogues, and derive (2.27) from (2.24) (Bilby and Eshelby, 1969). Here we give a third method, originally devised to give a simple derivation of a standard result in fracture mechanics (Eshelby, 1971).

We start from the fact that as the dilatation

$$e = \frac{\partial u}{\partial x} + \frac{\partial v}{\partial y} \tag{2.30}$$

is harmonic we have $\nabla^2(xe) = 2\partial e/\partial x$, $\nabla^2(ye) = 2\partial e/\partial y$ so that the equilibrium equation (2.26), which becomes

$$\nabla^2 u + \frac{1}{1-2\nu}\frac{\partial e}{\partial x} = 0, \quad \nabla^2 v + \frac{1}{1-2\nu}\frac{\partial e}{\partial y} = 0,$$

may be re-written as

$$\nabla^2 u' = 0, \quad \nabla^2 v' = 0$$

with

$$u' = u + \alpha x e, \quad v' = v + \alpha y e, \quad \alpha = 1/2(1-2\nu).$$

For an edge dislocation with Burgers vector $(b, 0)$ the strains, being dimensionless and proportional to b, must be proportional to r^{-1}. The dilatation must therefore be

$$e = \frac{b}{2\pi} A \frac{y}{r^2} = -\frac{b}{2\pi} A \frac{\partial \theta}{\partial x}, \tag{2.31}$$

the only harmonic function proportional to r^{-1} and with the right symmetry. Likewise u', v' must each be a harmonic function whose gradient is proportional to r^{-1}, i.e. a linear combination of θ and $\ln r$. The geometry shows that u' must be $b\theta/2\pi$, while v' may be a multiple of $\ln r$. Hence

$$\frac{2\pi}{b} u = \theta - A\alpha \frac{xy}{r^2}, \quad \frac{2\pi}{b} v = B \ln r - A\alpha \frac{y^2}{r^2}. \tag{2.32}$$

The relation

$$B = 1 + (1+\alpha)A \tag{2.33}$$

between the constants A, B follows from the requirement that the e calculated from (2.30) and the displacements (2.32) agree with the e of (2.31) already incorporated in them. An easy calculation gives

$$\sigma_{yy} = \frac{\mu b}{\pi} \frac{\partial}{\partial x} \left[\left(\frac{1}{2} A - B \right) \theta + \frac{xy}{r^2} \right]. \tag{2.34}$$

Hence, integrating with respect to x between $\pm\infty$, there is a total traction $-\mu b(\frac{1}{2}A - B)$ on both the faces of any slab bounded by planes y = constant above and below the dislocation, so that there must be an unwanted concentrated vertical force

$$F = 2\mu b(\tfrac{1}{2}A - B)$$

lurking at the singular point (0, 0). To get rid of it we put $\frac{1}{2}A = B$, which with (2.33) turns (2.32) into (2.27).

Two other cases are worth looking at. With $A = 0$, (2.32) gives Taylor's (1934) original solution

$$u = \frac{b}{2\pi}\theta, \quad v = \frac{b}{2\pi} \ln r \tag{2.35}$$

which actually represents the required dislocation with Burgers vector $(b, 0)$ plus an unwanted concentrated force $(0, -2\mu b)$. Because the dilatation (2.31) is zero, (2.35) fails to reproduce some of the properties of an edge dislocation (e.g. its interaction with point defects) even qualitatively. However, as not only the dilatation but also the rotation is zero (2.35) is a useful starting point for constructing the field of a moving dislocation (Section 5).

The case $A = -2$, $F = 2\mu b/(1-2\nu)$ is also interesting. Both σ_{yy} and σ_{xy} vanish on the x-axis and so nothing happens if we cut the material along, say, the negative x-axis. This suggests the following simple experiment. Place a block of foam rubber on a table, make a horizontal slit from the centre to the edge and cover the cut faces with a suitable glue. Pass a rod through the block at the head of the cut and press down on the ends. The two faces will slide over one another with a relative displacement which is fairly convincingly constant along the length of the cut. When the glue has set the rod can be removed to leave a permanent edge dislocation. This solution is in fact just what one gets by taking the well-known expression for a normal point load, actually $\mu b/(1-2\nu)$, acting on the edge of the half space $y>0$ and then allowing x, y to range over the whole xy-plane provided with the (mathematical) cut needed to make u single-valued.

The solution of plane elastic problems in an anisotropic medium is in a sense simpler than it is for an isotropic one. For the latter the field quantities are biharmonic functions of x_1, x_2, whereas the former involve harmonic functions of suitably chosen linear continuations of x_1 and x_2 (Lekhnitskii, 1963). To see this assume tentatively, following Eshelby *et al.* (1953),

$$u_k = A_k f(p_1 x_1 + p_2 x_2) \tag{2.36}$$

and substitute into the equilibrium equation

$$C_{ijkl} u_{k,lj} = 0 \tag{2.37}$$

to get, after cancelling out f'', the three equations

$$C_{i\beta k\alpha} p_\alpha p_\beta A_k = 0, \quad i = 1, 2, 3, \tag{2.38}$$

with α, β summed only over 1, 2. These equations can be solved for A_1, A_2, A_3 only if the determinant of the coefficients is zero.

The argument already used in the isotropic case shows that the strains behave like r^{-1}, and so we must take the function f to be $\ln(p_1 x_1 + p_2 x_2)$ or, rejecting a constant displacement proportional to $\ln p_1$, $\ln(x_1 + p x_2)$ where we have written p for p_2/p_1. Hence in the present case nothing is lost by putting $p_1 = 1, p_2 = p$ simply, and the equation $\det(C_{i\beta k\alpha} p_\alpha p_\beta) = 0$ is a sextic in p. Let its roots be $p_{(l)}$, $l = 1, 2, \ldots, 6$, and $A_{(l)i}$ the corresponding sets of coefficients furnished by (2.38). The remainder of the analysis depends on the $p_{(l)}$ all being complex, and never purely real. To see that this is so we compare the energy density

$$\tfrac{1}{2} C_{i\beta k\alpha} u_{i,\beta} u_{k,\alpha}$$

with the result of multiplying (2.38) by $\tfrac{1}{2} A_i$:

$$\tfrac{1}{2} C_{i\beta k\alpha} A_i p_\beta A_k p_\alpha = 0, \quad p_1 = 1, p_2 = p.$$

If p were real the choice $u_{k,1} = A_k p_1 = A_k$, $u_{k,2} = A_k p_2 = A_k p$ would make the energy density zero for a non-zero deformation, which is impossible in a stable elastic solid. Hence p is complex, as required.

Each of the complex displacements

$$u_i = A_{(l)i} \ln(x_1 + p_{(l)} x_2)$$

satisfies (2.36) and so does the real linear combination of them

$$u_i = \sum_{l=1,2,3} \frac{D_{(l)}}{2\pi} A_{(l)i} \ln(x_1 + p_{(l)} x_2) + \text{c.c.} \tag{2.39}$$

with constant complex coefficients $D_{(l)}$, and where c.c. denotes the complex conjugate of what precedes it, provided $p_{(1)}, p_{(2)}, p_{(3)}$ are, say, the three roots with positive real parts, so that no two of them are complex conjugates. Like the Taylor solution (2.35), the displacement (2.39) actually represents a mixture of a dislocation with Burgers vector (b_1, b_2, b_3) and the only other line singularity with stresses behaving like r^{-1}, namely a line force of say, intensity (F_1, F_2, F_3) per unit length of the x_3-axis. The six real constants contained in the three complex $D_{(l)}$ can be adjusted to make the b_i and F_i whatever we choose; in our application the b_i are dictated by the crystallography and the F_i must be zero.

The editor of Lekhnitskii's (1963) book implies that the above type of analysis uses the theory of analytic functions of several complex variables in the same way as does quantum field theory, but the comparison is quite misleading. Here we simply use a sum of functions each of which is an analytic function of a different complex variable, but quantum field theory uses something much more sophisticated, a single function of several complex variables, e.g. an analytic function of $x+iy$ which contains parameters which are themselves analytic functions of $x'+iy'$.

It is easy to see from (2.39), without working out the details, that u_i takes the form

$$u_i(r, \theta) = f_i(\theta) + C_i \ln r \tag{2.40}$$

with constant C_i, and that the associated stresses take the form

$$\sigma_{ij}(r, \theta) = \Sigma_{ij}(\theta)/r. \tag{2.41}$$

We can use this to find a quantity we shall need later, the energy E contained between two circles (strictly, cylinders of unit length) of radii $R_1, R_2 > R_1$ centred on the dislocation line. It is equal to the integral

$$E = \frac{1}{2}\int u_i \sigma_{ij} dS_j$$

taken along the circles and the two sides of the cut needed to make u_i single-valued. The contributions involving $f_1(\theta)$ in (2.40) are equal and opposite for the two circles, and the contributions from the logarithmic term are separately zero for each circle because there is no net traction on either of them. This leaves only the contribution from the cut. If we take it to be the radial plane θ = constant we get

$$E(\mathbf{t}) = E_0(\mathbf{t}) \ln (R_2/R_1) \tag{2.42}$$

with the so-called pre-logarithmic factor

$$E_0(\mathbf{t}) = \frac{1}{2} b_i \Sigma_{ij}(\theta) n_i(\theta), \tag{2.43}$$

where $n_i = (-\sin\theta, \cos\theta, 0)$ is the normal to the cut with, naturally, a sign convention which makes the result positive. As the position of the cut is irrelevant E_0 must, in fact, be independent of θ; that is, although the individual $\sigma_{ij}(r, \theta)$ are fairly complicated functions of θ, the traction on any radial plane has a component parallel to the Burgers vector which is independent of the inclination of the plane. (For later use we have written $E(\mathbf{t})$, $E_0(\mathbf{t})$ to emphasize that the energy (2.42) is a function of the direction $\mathbf{t}$ of the dislocation line relative to the crystal axes.)

A number of authors, beginning with Stroh (1958), have improved and extended the theory of straight dislocations in anisotropic media. We refer the reader to the recent article by Bacon *et al.* (1979) and turn to an interesting connection between the field of an infinite straight dislocation and the field of a dislocation loop, associated with the names of Lothe (1967), Brown (1967), Indenbom and Orlov (1967), and others.

There is nothing to stop us evaluating the integral (2.20) or (2.23) over an unclosed length of dislocation, with an apparent violation of Burgers vector conservation at the ends. Actually the violation is only apparent (Eshelby and Laub, 1967); the segment is surrounded by a continuous distribution of infinitesimal dislocations with their total Burgers vector equal to the Burgers vector of the segment, which fan out at one cut end, curl round and enter the other end, so ensuring a formal conservation of the Burgers vector. When several such segments, with a common Burgers vector, are joined end to end, the inflows and outflows at the junctions cancel. Although picturesque, this result is not very helpful, though it emphasizes the point that one need have no compunction in considering terminated dislocation segments so long as this is merely a preparation for fitting them together to form a closed loop, or a line going from infinity to infinity.

Figure 2.4 shows a loop in the plane $x_3 = 0$. We shall study the contribution, $du_{i,j}$ say, of a single element $dx_{q'}$ of the integral (2.23) to the displacement gradient $u_{i,j}(i, j = 1, 2, 3)$ at a point $(x_1, x_2, 0)$ in the plane of the loop. In (2.23) the factor $\varepsilon_{qkr}(x_r - x_{r'})dx_{q'}$ is the vector product of $x_r - x_{r'}$ and $dx_{q'}$ and in the present case only has an x_3-component, equal to $r \sin(\theta - \alpha)dS$ in the notation of the figure. The remaining factor $b_j\sigma^{(i)}_{jk,l}$ is the product of r^{-3} and a function of direction in the x_1x_2-plane. Hence we can write

$$u_{i,j}(x_1, x_2) = \int du_{i,j} \tag{2.44}$$

with

$$\begin{aligned} du_{i,j} &= (ds/r^2)\sin(\theta - \alpha)\Theta_{ij}(\theta) && (2.45)\\ &= (d\theta/r)\Theta_{ij}(\theta) && (2.46)\\ &= (d\theta/p)\sin(\theta - \alpha)\Theta_{ij}(\theta); && (2.47)\end{aligned}$$

for the second and third forms we use the fact that $dS \sin(\theta - \alpha) = r\,d\theta$ and that p, the perpendicular from (x_1, x_2) onto the tangent AB to the loop at (x_1, x_2), is $r \sin(\theta - \alpha)$. Suppose that instead of merely being the tangent to a curved dislocation the line AB actually

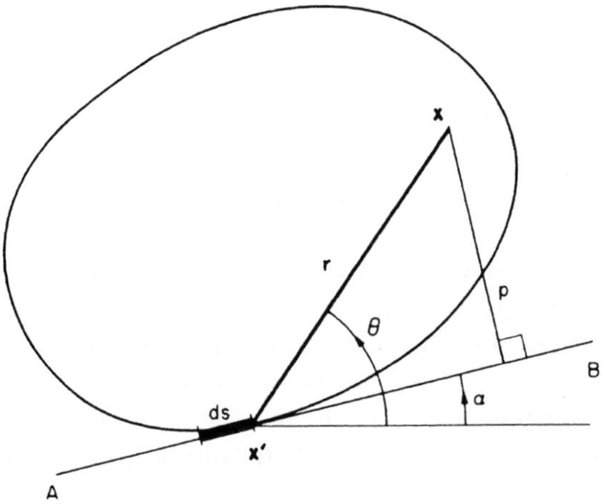

FIG. 2.4. To illustrate the Lothe–Brown formula.

represents an infinite straight dislocation. As p is constant along it we have

$$u_{i,j} = \int du_{i,j} = \frac{1}{p}\int_{\alpha}^{\alpha+\pi} \sin(\theta-\alpha)\Theta_{ij}(\theta)\,d\theta.$$

But from the two-dimensional theory we know that at a distance p from the straight dislocation its displacement gradient has the form $U_{ij}(\alpha)/p$, where $U_{ij}(\alpha)$ depends only on its orientation α, and so we have

$$\int_{\alpha}^{\alpha+\pi} \sin(\theta-\alpha)\Theta_{ij}(\theta)\,d\theta = U_{ij}(\alpha).$$

Differentiate twice with respect to α:

$$\Theta_{ij}(\alpha)+\Theta_{ij}(\alpha+\pi) - \int_{\alpha}^{\alpha+\pi} \sin(\theta-\alpha)\Theta_{ij}(\theta)\,d\theta = U''_{ij}(\theta).$$

Since G_{ij} is an even function of the vector $\mathbf{x}-\mathbf{x}'$ so is $\sigma^{(i)}_{jk,l}$ in (2.23), which is derived from it by two differentiations, and consequently $\Theta_{ij}(\alpha+\pi)$ is equal to $\Theta_{ij}(\alpha)$ so that we have

$$\Theta_{ij}(\theta) = \tfrac{1}{2}[U_{ij}(\theta)+U''_{ij}(\theta)]. \tag{2.48}$$

If we insert (2.46) into (2.44), (2.44) into (2.42), and multiply both sides by C_{klij} to obtain the stress from the displacement gradient, we get the remarkable result that the stress due to the loop, in the plane of the loop, is given by

$$\sigma_{kl}(x_1,x_2) = \frac{1}{2}\int \frac{\Sigma_{kl}(\theta)+\Sigma''_{kl}(\theta)}{|\mathbf{x}-\mathbf{x}'|}\,d\theta, \tag{2.49}$$

where $\Sigma_{kl}(\alpha)/p$ is the stress at a distance p from an infinite straight dislocation whose orientation is specified by α and which has the same Burgers vector as the loop we are interested in. (The notation is that of (2.41), but the angular argument of Σ_{kl} here has a different meaning.) Note that the associated straight dislocation is directed not, as one might perhaps have guessed, along the tangent at $x_{i'}$, but along the radius vector from ds to the point of observation. For some developments of this result see the article by Bacon *et al.* (1979).

2.5. *Treatment of the singular line*

According to the foregoing calculations the stresses at a distance r from a dislocation line behave like r^{-1}. In certain situations this physically unrealistic singularity is tolerable and in some others it can be dealt with by introducing an inner cut-off radius r_0 to the exact choice of which the results are insensitive. In particular it may enter through a factor $\ln(R/r_0)$ with $R \gg r_0$, as in (2.42). The singularity can be removed by using nonlinear elasticity (Gairola, 1979). However, the region where nonlinearity becomes important is also more or less the same as the region where one can no longer neglect the atomic structure of the material. There have been several calculations aimed at using the theory of generalized continua to tone down the singularity without straying out of the realm of linear continuum mechanics. Cohen (1966) and Knésel and Semela (1972) treat the solid as a grade 2 material (Toupin, 1964) in which the energy density depends on strain and strain gradient. This work is an elastic counterpart of Podolsky's (1942) use of "grade 2" electrostatics to moderate the field near a point charge. Eringen (1977) used his non-local elasticity. In it the stress at a point depends linearly on the strain everywhere through an integral with a suitable kernel. Following a well-known method in theoretical physics we may say, equivalently, that the Fourier components $\sigma_{ij}(\mathbf{k}, \omega)$, $e_{kl}(\mathbf{k}, \omega)$ of stress and strain are connected by elastic constants $C_{ijkl}(\mathbf{k}, \omega)$ which themselves depend on wave number $\mathbf{k}$ and frequency $\omega/2\pi$. The application to a dislocation has a close analogue in McManus's (1949) treatment of an electron by non-local electrodynamics. Though non-local the treatment is non-localized in the sense that it cannot give an account of properties which depend on the periodicity associated with the underlying lattice. The same is, of course, the case for any true continuum theory.

The model of Peierls (1940) and Nabarro (1947) attempts to give some account of the nonlinear behaviour close to the centre of the dislocation, and of lattice periodicity. We shall give a rather oversimplified sketch of the method, treating a screw rather than, as they did, an edge dislocation. The analysis is simpler, and in addition the fact that for the screw but not the edge the atoms move at right angles to the xy-plane, so that, as always in states of anti-plane strain, x and y are indifferently Eulerian or Lagrangian coordinates, allows us to evade certain difficulties.

The model treats the crystal as two elastic blocks separated by a gap $y = \pm\frac{1}{2}a$ across which interatomic forces act, the relation between a and b depending on the crystallography. The tractions which the interatomic forces exert on the blocks are some prescribed function of the relative displacement

$$\Delta w = w(x, \tfrac{1}{2}a) - w(x, -\tfrac{1}{2}a)$$

of the blocks. Rather than set up the general formalism we shall simply ask what force law is required to maintain the simple elastic solution (2.24) outside the gap. On $y = \frac{1}{2}a$ we have

$$\sigma_{zy} = \frac{\mu b}{2\pi}\frac{\cos\theta}{r}\cdot\frac{r\sin\theta}{y} = \frac{\mu b}{2\pi a}\sin 2\theta \tag{2.50}$$

and if we eliminate θ between (2.50) and

$$\Delta w = \frac{b\theta}{\pi} = \frac{b}{\pi}\tan^{-1}\frac{a}{2x}, \tag{2.51}$$

we get

$$\sigma_{zy} = \frac{\mu b}{2\pi a}\sin\frac{2\pi}{b}\Delta w, \tag{2.52}$$

which is, in fact, the Peierls–Nabarro law, fixed by the requirements that it shall be periodic, and for simplicity sinusoidal, with period b and that it shall reduce to Hooke's law for small Δw. Thus the elastic solution itself provides the solution of (2.49). The width 2ζ of the dislocation is conventionally taken to be the distance between the points at which the displacement discontinuity attains $\frac{1}{4}b$ and $\frac{3}{4}b$ as it varies from 0 to b along the slip plane. From (2.51), $2\zeta = a$ for a screw dislocation; for an edge the result is $2\zeta = a/(1-\nu)$. These small values confirm the assertion in the Introduction that the atomic misfit is concentrated into a small region, though one cannot have much confidence in their exact values.

So far we have simulated the atomic structure by a layer of material with a non-linear periodic deformation law, but, because its properties are independent of x, we have not managed to localize the atoms. That is, our results would have been the same if we had replaced (2.24) and (2.51) by

$$w = \frac{b}{2\pi}\tan^{-1}\frac{y}{x-\xi} \quad \text{and} \quad \Delta w = \frac{b}{\pi}\tan^{-1}\frac{a}{2(x-\xi)}$$

in the elastic blocks and across the slip plane. But one would expect that if there is a solution with all atoms in stable equilibrium for, say, $\xi = 0$, there would be precisely similar solutions for $\xi = b, 2b, \ldots$, representing the same dislocation displaced by one, two, ... atomic distances, and between them solutions with $\xi = \frac{1}{2}b, \frac{3}{2}b, \ldots$, in which the atoms are in unstable equilibrium. For other values of ξ one would expect no solutions, except, perhaps, in the presence of an applied stress. In particular we should expect the energy associated with the interatomic forces to depend on ξ. According to the argument so far the energy in the gap at x is

$$\int_0^{\Delta w(x)} \sigma_{zy}\, d(\Delta w) = \frac{\mu b^2}{4\pi^2}\frac{\frac{1}{2}a}{(x-\xi)^2 + (\frac{1}{2}a)^2}$$

per unit area, and so the total energy in the gap is

$$\frac{\mu b^2}{4\pi^2}\int_{-\infty}^{\infty}\frac{\frac{1}{2}a}{(x-\xi)^2 + (\frac{1}{2}a)^2}\,dx, \tag{2.53}$$

which is obviously independent of ξ and is in fact equal to $\mu b^2/4\pi$. To get over this we suppose that the atom at x is acted on by a force $b\sigma_{zy}(x)$ and that the elastic field in the blocks is translated rigidly with ξ, so that the elastic energy in them, actually also $\mu b^2/4\pi$, is independent of ξ, and can be ignored in what follows. Suppose first that pairs of atoms face each other across the slip plane. Then we must modify the integral (2.52) by replacing $x-\xi$ by $nb-\xi$ and $\int \ldots dx$ by $\Sigma_n \ldots b$:

$$E_{\mathrm{PN}} = \frac{\mu b^3}{4\pi^2}\sum_{n=-\infty}^{\infty}\frac{\frac{1}{2}a}{(nb-\xi)^2 + (\frac{1}{2}a)^2}. \tag{2.54}$$

The summation is the vertical electric field due to a grid of charged wires at a point whose coordinates are $x = \xi$, $y = \frac{1}{2}a$ relative to the nearest wire, and so is made up of the constant field of a smeared-out sheet with the same overall charge plus a fluctuating term which falls off more or less exponentially with y. Its actual value (see Bromwich, 1955, or books on electrostatics and hydrodynamics) is

$$\frac{\pi\sinh(\pi a/b)}{b\cosh(\pi a/b) - b\cos(2\pi\xi/b)}.$$

If a is of the order of b the hyperbolic sine and cosine are large and nearly equal to one another and to $\frac{1}{2}\exp(\pi a/b)$ and we have

$$E_{\mathrm{PN}} = \frac{\mu b^2}{4\pi}[1 + 2\exp(-\pi a/b)\cos(2\pi\xi/b)] \tag{2.55}$$

very nearly. Suppose that there is a uniform applied stress $\sigma_{zy} = \sigma$ acting on the material. The work which it does on the gap when ξ changes by $\delta\xi$ is the integral of $\delta\xi\sigma\partial\Delta w(\xi)/\partial\xi$ along the x-axis, i.e. $\delta\xi b\sigma$, which is equivalent to adding $E' = \xi b\sigma$ to (2.55). For the dislocation to be able to move freely through the lattice, σ must be large enough to make the curve $E_{\mathrm{PN}}(\xi) + \xi b\sigma$

monotonic decreasing everywhere, which gives $\sigma = \mu \exp(-2\pi\zeta/b)$ with $\zeta = \frac{1}{2}a$. For $a \sim b$ this is small compared with the shear modulus, though not so small as we might like. If we suppose that the atoms instead of being in a square array in the xy-plane are piled triangularly we ought to sample the integrand of (2.53) at $\xi + nb$ in the upper plane and $\xi + (n+\frac{1}{2})b$ in the lower. To find E_{PN} we have only to add to (2.54) the same expression with ξ replaced by $\xi + \frac{1}{2}b$ and divide by 2. This doubles the argument of both exponential and cosine in (2.55), giving $\sigma = 2\mu \exp(-4\pi\zeta/b)$, which is satisfyingly smaller, but with the embarrassment that E_{PN} has period $\frac{1}{2}b$ rather than, or in addition to, the expected b. For an edge dislocation one can obtain, for example, $\sigma = [\mu/(1-\nu)] \exp(-2\pi\zeta/b)$, or $\sigma = [\mu/(1-\nu)] \exp(-4\pi\zeta/b)$, $\zeta = a/2(1-\nu)$, according to the details of the model (Huntington, 1955).

3. The Effect of Free Surfaces

3.1. *General*

The formulae we have discussed so far refer to a dislocation in an infinite medium. To emphasize this we may write u_i^∞, σ_{ij}^∞ for the elastic field. Mark out a surface S in the infinite medium and make a cut over S. We are left with a dislocation in a finite body with a stress-free surface S. The elastic field will now be, say,

$$u_i^D = u_i^\infty + u_i^{\mathrm{im}}, \quad \sigma_{ij}^D = \sigma_{ij}^\infty + \sigma_{ij}^{\mathrm{im}}, \tag{3.1}$$

where the additional field u_i^{im}, $\sigma_{ij}^{\mathrm{im}}$ may conveniently be called the "image" field by analogy with the terminology of electrostatics. Since the total traction $(\sigma_{ij}^\infty + \sigma_{ij}^{\mathrm{im}})n_j$ must be zero on S, the image field is simply the field induced in a body bounded by S when surface tractions $-\sigma_{ij}^\infty n_j$ are applied to its surface. The determination of the image field thus reduces to a standard problem of elastostatics, to find the field inside a body from given surface tractions.

Quite apart from its application to the dislocation image problem the problem of finding elastic fields from surface data is itself connected, in a way which the writer finds illuminating, with the theory of Somigliana dislocations through what we shall call Gebbia's (1891, 1902) theorem. According to the theorem the elastic field of a body whose outer boundary S is subjected to a traction **T** and a displacement **U** is the same as it would be if the material inside S formed part of an infinite medium provided S, now simply a surface marked out in the infinite medium, is covered with a layer of body force of surface density **T** and is the site of a Somigliana dislocation whose variable discontinuity vector is **U**. (For the sign to agree with (2.1) the normal must be directed into the body.) The part of the infinite medium outside S is actually stress-free. To see this start with the undeformed body inserted, with perfect fit, in a hole in an infinite matrix. Impose **T** and **U** and suppose, for the moment, that the surface of the solid is everywhere pulled away from the hole in the matrix, leaving a gap. Fill the gap with a thin layer of material and cement it to the body and matrix. We now have the situation envisaged in the theorem: the surface traction **T** has become built in as a layer of body force and the inserted layer constitutes the Somigliana dislocation. Also, as no traction has been applied to the surface of the hole the surrounding matrix is stress-free, as asserted. It only remains to make the usual adjustment to cover interpenetration. According to this argument the displacement at a point **r** inside S is

$$u_i(\mathbf{r}) = \int_S [u_j^{(i)}(\mathbf{r}-\mathbf{r}')T_j(\mathbf{r}') - U_j(\mathbf{r}')\sigma_{jk}^{(i)}(\mathbf{r}-\mathbf{r}')n_k(\mathbf{r}')]\, dS(\mathbf{r}'). \tag{3.2}$$

The first term gives the contribution of the layer of body force and the second the contribution of the Somigliana dislocation in (2.7) with $b_j = -U_j$; the minus is necessary as the normal is now supposed to be directed outwards from S. Equation (3.2) is just Somigliana's formula as given, e.g. by Love (1927); we have supposed there are no body forces. For more details see Jaswon and Symm's (1977) book, which will also provide formal justification for the arguments which follow. If $\mathbf{r}$ is outside S, the left-hand side of (3.2) is zero, as we have seen. If $\mathbf{r}$ is actually on S it is $\frac{1}{2}u_i(\mathbf{r})$.

Suppose that instead of filling up the gap between solid and hole we apply tractions $\mathbf{T}'\,dS$ to the surface of the hole, chosen so that each point of the hole is drawn into contact with the point on the surface of the solid with which it coincided in the undeformed uncut infinite medium. If we cement the cut together we have the same elastic field inside S as before (though the field in the matrix is no longer zero), no dislocation and a layer of body force of density $\mathbf{T}+\mathbf{T}'$. That is, we can drop the second term in (3.2) at the expense of replacing the known T_j by the unknown $T_j + T_j'$ in the first. Conversely, we could apply a force $-\mathbf{T}\,dS$ to each element of the surface of the hole, thus imposing a displacement, $\mathbf{U}'$ say, on it and then fill up the gap with the modified discontinuity $\mathbf{U}-\mathbf{U}'$. We also have a layer of force doublets formed by the built-in tractions $\pm\mathbf{T}\,dS$, but they are separated by a vector distance $\mathbf{U}-\mathbf{U}'$ which is itself of order $\mathbf{T}$ so that the effect of the layer of doublets is of second order in the loads and so has to be ignored in the linear theory. The upshot is that the first term in (3.2) may be dropped at the expense of replacing U_j by $U_j - U_j'$ in the second.

Things are particularly simple when the body is, so to speak, deformed from inside by a dislocation loop, so that there is a $\mathbf{U}$ but automatically no $\mathbf{T}$ in Gebbia's construction. In fact if we start with the loop in an infinite medium and make a cut over S a gap will open up leaving the body bounded externally by S with a traction-free surface, and with the material outside S stress-free throughout. This state of stress can be regarded as being the sum of the infinite-medium field of the loop plus that of the Somigliana dislocation with the unknown discontinuity vector $\mathbf{U}(\mathbf{r}')$. One can in principle set up an integral equation for $\mathbf{U}$ by requiring it to produce a traction which cancels the traction $\sigma_{ij}^{\infty} n_j$ at each point of S. This is in fact practicable in two dimensions, where the Somigliana dislocation can be simulated by a distribution of crystal dislocations. When we have done this we are left with an infinite block in which the material outside S is stress-free, and its removal to give a dislocation in a finite solid with a traction-free surface is then a mere formality. This rather attractive way of looking at the image problem has been strongly advocated by Marcinkowski (1979). On the other hand, the old-fashioned method of images, in the limited range of cases where it works, also has its appeal, and in Section 3.2 we consider a few simple examples.

3.2. *Some examples*

As usual it is simplest to start with a screw dislocation in an isotropic medium. When dealing with a state of anti-plane strain it is often convenient to associate with w the harmonic function ϕ conjugate to it so that $w + i\phi$ is an analytic function of the complex variable $z = x + iy$, not to be confused with the z which features as a suffix in (2.25). As, according to the Cauchy–Riemann relations, the gradients of w and ϕ are orthogonal, the traction $(0, 0, T_z)$ across a curve at a point where its normal is $(n_x, n_y, 0)$ is given by

$$T_z = \mu\frac{\partial w}{\partial x}n_x + \mu\frac{\partial w}{\partial y}n_y = \mu\frac{\partial w}{\partial n} = \mu\frac{\partial \phi}{\partial s},$$

where $\partial/\partial s$ denotes differentiation along the curve. In particular, ϕ is constant along a stress-free contour.

With (2.24) is associated

$$\phi = -\frac{b}{2\pi}\ln r, \tag{3.3}$$

and if we wish to find the field for a dislocation in an infinite rod bounded by a curve C we must add to (3.3) a harmonic function chosen so that the resulting ϕ is constant on C. The expression (3.3) can also be regarded as the scalar potential of an electrostatic line charge, the stream function of a vortex, or the magnetic vector potential of a current filament, so that we can obtain solutions by plundering other areas of mathematical physics.

For example, if we have a screw dislocation situated eccentrically at $(\xi_1, 0)$ in a circular cylinder with its circumference $x^2 + y^2 = a^2$ stress-free, plane electrostatics suggests that we should introduce a negative image at the inverse point

$$(\xi_2, 0), \quad \xi_2 = a^2/\xi_1.$$

For a point (x, y) on the circumference the quantity

$$t = r_1\xi_2^{1/2}/r_2\xi_1^{1/2} = r_1 a/r_2\xi_1 \tag{3.4}$$

(where r_1, r_2 are radii drawn from $(\xi_1, 0)$ and $(\xi_2, 0)$) has the value unity. (Consider the two similar triangles formed by connecting the ends $(0, 0)$, (x, y) of the radius vector to either the object or image.) Hence

$$\phi = -\frac{b}{2\pi}\ln t \tag{3.5}$$

has the right singularity at the object point, has a constant value (zero) on the circumference, and so solves the problem. The corresponding displacement is, of course,

$$w = \frac{b}{2\pi}\tan^{-1}\frac{y}{x-\xi_1} - \frac{b}{2\pi}\tan^{-1}\frac{y}{x-a^2/\xi_1}. \tag{3.6}$$

Actually t is constant, though not unity, on any circle with respect to which ξ_1 and ξ_2 are inverse points, so that (3.5) and (3.6) also apply to the case where the core of the dislocation is excluded by a circular hole of radius a_0, say. We might perhaps ask, with certain authors, whether the centre of the dislocation is at $(\xi_1, 0)$ or at the centre $(\xi_0, 0)$ of the hole, with which it does not exactly coincide or, perhaps, elsewhere. However, it is meaningless to ask where a dislocation is situated in a hole, circular or otherwise. The difference of the solutions for two positions would give zero stress at the hole, and there would be no Burgers vector associated with it and the difference field would be zero.

Suppose, more generally, that the dislocation is at (x', y') inside the arbitrary closed contour C and that the conformal transformation

$$\zeta = f(z)$$

maps C onto the unit circle $|\zeta| = 1$ so that

$$\overline{f(z)} = 1/f(z) \quad \text{on} \quad C. \tag{3.7}$$

Then w and ϕ are given by

$$\phi - iw = \frac{b}{2\pi} \ln F(z, z')$$

with

$$F(z, z') = \frac{1 - f(z')f(z)}{f(z) - f(z')}.$$

To see this it is only necessary to check that $\phi = (b/2\pi) \ln |F|$ behaves like $-(b/2\pi) \ln |z - z'|$ near z' and to note that $|F| = 1$ on C, in view of (3.7).

We return to the case of a screw dislocation off-centre in a circular cylinder. The energy per unit length of the cylinder can be found most easily by slitting the x-axis from the hole at the core to the outer circumference and calculating the work required to establish the displacement difference b across the cut as the stress σ_{zy} builds up to its final value:

$$E = \frac{1}{2} b \int_{\xi_0 + a_0}^{a} \sigma_{zy}\, dx = \frac{1}{2} b \int_{\xi_0 + a_0}^{a} \frac{\partial \phi}{\partial x} dx,$$

that is, $-\mu b^2/2\pi$ times the value of $\ln t$ at the lower limit. (At the upper limit it is zero.) If a_0 is comparable with a, this needs a little geometry, but for $a_0 \ll a$ we have $r_1 \sim a_0$, $r_2 \sim \xi_2 - \xi_1$ and

$$E = \frac{\mu b^2}{4\pi} \ln \frac{a^2 - \xi_1^2}{a_0 a}. \tag{3.8}$$

To find the plane strain field of an edge dislocation in a cylinder it is simplest to work in terms of the Airy stress function in (2.29).

If the curve C bounding the cross-section of the cylinder is stress-free the x- and y-components of traction, $-d(\partial\chi/\partial y)/ds$, $d(\partial\chi/\partial x)/ds$, are zero, which means that the gradient of χ is zero along C. If we add $q + lx + my$ to any χ the stresses are unaltered. So by choosing q, l, m suitably we can arrange for χ and its gradient to be zero on C, and so take

$$\chi = 0, \quad \partial\chi/\partial n = 0 \tag{3.9}$$

for the boundary conditions. (The first, of course, implies $\partial\chi/\partial s = 0$.)

For the edge dislocation $(b, 0)$ at $(\xi_1, 0)$ in the circular cylinder $x^2 + y^2 = a^2$ we may start with the function (3.5) we used for a screw dislocation in the same situation. Multiplied by y and $1/(1 - \nu)$ it gives the trial function

$$\chi = -\frac{b}{2\pi(1 - \nu)} y \ln t \tag{3.10}$$

which is biharmonic, has the right singularity near the dislocation (see (2.28)), and satisfies the first of (3.9) but not the second. To remedy this we note that since

$$\ln t = \tfrac{1}{2} \ln t^2 = \tfrac{1}{2} \ln [1 + (t^2 - 1)] = \tfrac{1}{2}(t^2 - 1) + \ldots$$

the function

$$L(t) = \ln t - \tfrac{1}{2}(t^2 - 1) \tag{3.11}$$

and its gradient with respect to t itself, or the x, y implicit in t, vanish on the circle $t = 1$, so that

$$\chi = -\frac{b}{2\pi(1-\nu)} y L(t) \tag{3.12}$$

satisfies both of (3.9), and is the required solution provided it is biharmonic. Dropping a term constant times y which does not affect the stresses, we can write

$$\chi = -\frac{b}{2\pi(1-\nu)} y \left\{ \ln\frac{r_1}{r_2} + \frac{1}{2}\frac{r_1^2 \xi_1}{r_2^2 \xi_2} \right\}. \tag{3.13}$$

The extra term is indeed of a standard biharmonic form, being the product of a harmonic function, here a multiple of y/r_2^2, and a squared radius vector, here r_1^2. It represents a collection of doublets of force doublets located at the image point which can be resolved, perhaps not very informatively, into a doublet of the simple image term $y \ln r_2$ plus a doublet of centres of dilatation. At the actual site $(\xi_1, 0)$ of the dislocation, though not elsewhere, the extra terms contribute nothing to the stresses, so that, so to speak, the dislocation thinks that, as in the screw case, its image field is just that of a dislocation of opposite sign at the image point.

The solutions (3.5) and (3.12) relate to a screw or edge dislocation in an infinite cylinder in a state of anti-plane or plane strain. If we cut out a finite rod from the cylinder the elastic state described by (3.5) or (3.12) will be maintained so long as appropriate tractions are applied to the cut ends, but if we remove them to give a rod with stress-free ends the solutions will be modified. For a long rod the new fields can be found by elementary means everywhere except in regions of the order of a extending in from the ends. On any cross-section of the circular cylinder containing the screw dislocation there is a couple about the z-axis of magnitude

$$M = \int (x\sigma_{zy} - y\sigma_{zx})\, dx\, dy = \mu \int \left[\frac{\partial}{\partial y}(xw) - \frac{\partial}{\partial x}(yw) \right] dx\, dy$$

or

$$M = \mu \int_s w(xn_y - yn_x)\, ds$$

by Gauss' theorem, where the contour s must embrace the whole area of the cross-section and respect the discontinuity in w, so that it can be taken to be made up of the circumference of the cylinder and of the hole at the core, joined by lines along the two sides of the cut used in calculating the energy (3.7). The outer circumference obviously contributes nothing and so, too, though less obviously, does the circumference of the hole. The contribution from the cut gives

$$M = \mu b \int_{\xi_1}^{a} x\, dx = \frac{1}{2}\mu b(a^2 - \xi_1^2) \tag{3.14}$$

if we neglect the radius of the hole in comparison with a.

To maintain a state of anti-plane strain in a finite rod couples $\pm M$ must be applied to the ends. If they are not the rod will develop a twist α (Mann, 1949) given, in radians per unit length, by (3.14) divided by the torsional rigidity, $\frac{1}{2}\pi\mu a^4$, of the rod:

$$\alpha = \frac{b}{\pi a^2}\left(1 - \frac{\xi_1^2}{a^2}\right). \tag{3.15}$$

Its sign is given by the following rule. When the dislocation is introduced, circles with spacing b on the surface of the rod join up to make a helix with a certain sense; α twists a generator of the cylinder into a helix of opposite hand (and much coarser pitch). Such a twist has been observed in crystal whiskers containing dislocations.

When the couples maintaining anti-plane strain are removed an amount of work $\frac{1}{2}M\alpha$ per unit length is extracted from the rod so that (3.7) must be replaced by

$$E = \frac{\mu b^2}{4\pi}\left[\ln\frac{a^2 - \xi_1^2}{a_0 a} - \frac{(a^2 - \xi_1^2)^2}{a^4}\right].$$

As a function of ξ_1, E has a minimum at $\xi_1 = 0$, a maximum at $\xi_1 = (1 - 2^{-\frac{1}{2}})^{\frac{1}{2}}\, a$ at a height $(\mu b^2/8\pi)(1 + \ln\frac{1}{2})$ above the minimum and falls to minus infinity as ξ_1 approaches a. Hence, rather surprisingly, the dislocation is in stable (strictly, metastable) equilibrium at the centre of the rod.

For a long rod the elastic field will deviate from the sum of the anti-plane solution (3.5) and the simple torsion specified by (3.15) only in the neighbourhood of the ends. However, if the rod is short enough to be called a disc the elastic field will be, so to speak, all end effect. Eshelby and Stroh (1951) have given the solution for a screw dislocation at the centre of a circular disc, with or without a cylindrical hole at the core. It appears to be the only solution for a dislocation in a finite body with the tractions exactly liquidated on *all* surfaces. We shall only quote their expression for a dislocation in an infinite plate of thickness $2d$ without a hole at the centre. In cylindrical polars (r, θ, z) the non-vanishing displacement and stress components are

$$w = u_z = \frac{b}{2\pi}\theta, \quad u_\theta = -\frac{b}{2\pi}\frac{z}{r} + \sum_n \frac{2b\sin\frac{1}{2}n\pi}{\pi^2 n} K_1(nR)\sin nZ,$$

$$\sigma_{r\theta} = \frac{\mu b_z}{\pi r^2} - \sum_n \frac{\mu b \sin\frac{1}{2}n\pi}{\pi d} K_2(nR)\cos nZ, \tag{3.16}$$

$$\sigma_{\theta z} = \sum_n \frac{\mu b\sin\frac{1}{2}n\pi}{\pi d} K_1(nR)\sin nZ, \quad R = \frac{\pi r}{2d}, \quad Z = \frac{\pi z}{2d}.$$

The summations are over positive integral n, though the terms for odd n are zero. If the terms involving the modified Bessel functions K_m are omitted, the remaining terms give zero stress on the surfaces $z = \pm d$ and the correct behaviour of w, but there is an unwanted singularity at $r = 0$, traceable to a distribution of couples along the dislocation line, which it is the business of the Bessel functions to cancel.

The dislocation interacts repulsively with a similar one in the plate through the stress $\sigma_{\theta z}$ (see Section 4). As $K_m(nR)$ falls rapidly to zero for $nR > 1$ the repulsion is negligible until the distance between them falls to about d and then rises rapidly. A set of like screw dislocations subject to some influence (e.g. local thinning of the plate) which drives them together should thus form a more or less regular lattice with a spacing of the order of the plate thickness. Something of the sort may sometimes be seen in thin film electron micrographs of metallic foils.

In contrast to the case of a screw dislocation there is no net couple (or force) on the cross-section of an infinite cylinder, circular or otherwise, containing an edge dislocation (Nabarro, 1967, p. 274) and so, apart from end effects, the field is unaffected when a finite rod is cut out. The image field of an edge dislocation $(b, 0)$ in a plate with faces parallel to the xy-plane is actually the same as that of a line force $(0, 2\mu b)$ distributed along the dislocation line. To see this note that for the solution (2.35) the out-of-plane stress components $\sigma_{zz}, \sigma_{zx}, \sigma_{zy}$ are zero. Consequently the solution represents a state of both plane strain and plane stress, and there is no relaxation when a plate is cut out. Hence the image fields of the dislocation and the line force $(0, -2\mu b)$ which the solution represents must cancel. Green and Willmore's (1948) investigation of the line force problem shows that there is no possibility of obtaining an explicit solution analogous to (3.15).

4. Energetics of Dislocations

4.1. *The force on a dislocation*

A dislocation in a crystal can, subject to certain constraints, migrate through the lattice, and we must allow the same freedom to the elastic dislocation which we use to simulate it, even though this property does not flow naturally from its definition in terms of the cutting and welding of a continuum. The dislocation will tend to adopt a configuration which minimizes the energy of the system, made up of the elastic energy of the body containing the dislocation and the potential energy of any external loading mechanism attached to the body.

The elastic field in the body can be regarded as being made up of the field of the dislocation u_i^D, σ_{ij}^D, which can be split into an infinite-medium field $u_i^\infty, \sigma_{ij}^\infty$ and an image field $u_i^{\mathrm{im}}, \sigma_{ij}^{\mathrm{im}}$, the externally applied field u_i^A, σ_{ij}^A, and possibly also a field, u_i^O, σ_{ij}^O, say, due to other sources of internal stress. In the energy we can distinguish terms quadratic in these fields individually and also cross-terms between them. In this Section we concentrate on the interaction between the dislocation field and the externally applied field. We therefore omit u_i^O, σ_{ij}^O for the moment. We cannot also drop the image field because with an infinite medium there would be nowhere to apply the external load. The total energy is made up of the following terms:

$$E = E_s + E_{\mathrm{im}} + \frac{1}{2}\int_S u_i^A \sigma_{ij}^A \, dS_j - \int_S (u_i^A + u_i^D)\sigma_{ij}^D \, dS_j. \tag{4.1}$$

The first is the elastic self-energy of the dislocation loop by itself in an infinite medium. The second is the change in this energy when a closed surface S is marked out in the infinite medium, a cut is made over S, and the material outside it is removed, allowing the traction on S to relax to zero. As already explained, the elastic displacement and stress then change from u_i^∞, σ_{ij}^∞ to u_i^D, σ_{ij}^D as defined in (3.1).

The first integral in (4.1) is the additional energy which enters the body when it is loaded by the applied surface tractions $\sigma_{ij}^A n_j$. Here we make explicit use of the fact that, according to the linear theory, the response of a body to external forces is unaffected by the presence of internal stresses in it (Southwell, 1936). In the elastic energy of the body there is therefore no coupling cross-term between internally and externally generated stresses. The necessary coupling is provided by the second integral. It represents the potential energy of the loading mechanism which applies the traction $\sigma_{ij}^A n_j$. This integral is minus the work done by the loading mechanism as each surface element moves to its final position $x_i + u_i^A + u_i^D$ from some standard reference configuration which we take to be the unloaded and undislocated specimen. For the moment we suppose that, in engineering language, we have dead loading, i.e. the traction which the loading mechanism exerts on a particular element dS does not change when dS is displaced.

Though the point does not directly affect the present argument, one might perhaps wonder why, in the absence of the dislocation, $u_i^D = 0$, the work done by the load is not equal to the work done on the body, in which case the two integrals in (4.1) would cancel. Consider the operation of attaching a weight to a spring. When we hook the weight on to the unstretched spring we have to support the whole weight with our hand. When the weight has reached its final position it has developed a tension which is just able to support the weight and the hand can be removed; halfway between it supports half the weight, the hand the rest. Hence, of the potential energy loss of the weight, only part, actually a half, is stored in the stretched spring. The same argument can be applied to the action of the loading mechanism on each surface element dS.

The only term in (4.1) which contains the applied field and the field of the dislocation is

$$E(A, D) = -\int_S u_i^D \sigma_{ij}^A \, dS_j \tag{4.2}$$

and we wish to know how it changes when the form of the loop varies. The integral can be related more closely to the shape of the loop by a trick. Re-write (4.2) as

$$E(A, D) = -\int (u_i^D \sigma_{ij}^A - u_i^A \sigma_{ij}^D) \, dS_j. \tag{4.3}$$

The value is unchanged because the dislocation produces no traction on S, and so the added term is zero. But the integrand has now acquired the form (2.8) and so has zero divergence. Hence S can be replaced by any other surface enclosing the loop, in particular the surface shown in Fig. 4.1 made up of a tube with the loop as axis and the top and bottom surfaces of the (arbitrary) discontinuity surface.

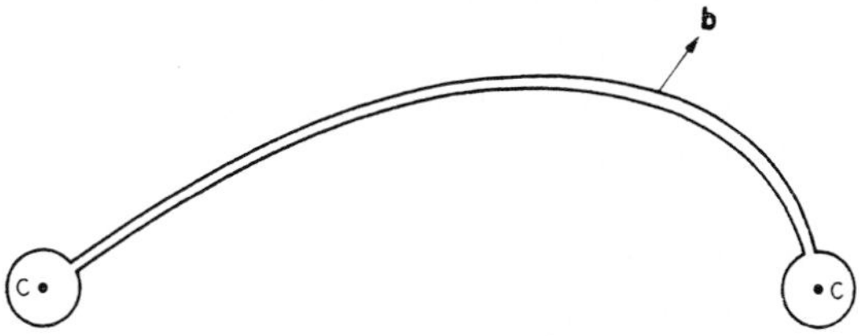

FIG. 4.1. Surface used in finding the Peach–Koehler force.

We now have to try to show that the integral over the tube is zero. In the literature this is sometimes done by supposing that the tube is actually hollow with a stress-free surface, and then shrinking it to zero, or by assuming that its cross-section is of atomic dimensions and allowing it to move with the loop when it deforms. We prefer, in the spirit of theoretical physics, to try to outface the singularity. Suppose the cross-section of the tube is everywhere a circle of radius a, though this is not really necessary. Divide the tube into a large number of short cylindrical beads threaded by the dislocation. In carrying out the integration of the second term in (4.3) over one bead we may take the slowly varying u_i^A outside the integral. As σ_{ij}^D is proportional to r^{-1}, the distance from the axis, the total traction on the curved surface of the cylinder appears to be of order $a.a^{-1}$, which does not go to zero with a. However, since the total traction on the circular ends is of order $a^2.a^{-1} = a$ and each bead is in static equilibrium, the total traction on the cylindrical surface must also be of order a, and so the second term in (4.3) goes to zero with a, as required. There is no trouble with the first term since u_i^P is only of order $\ln r$. We are left with the integral over the two sides of the cut. The second term in (4.3) is continuous across the cut and in the first u_i^P has a discontinuity b_i across it, so that with our previous sign convention we have

$$E(A, D) = -b_i \int \sigma_{ij}^A dS_j \tag{4.4}$$

taken over the positive side of the cut. Note that the image field does not affect (4.4).

If the dislocation suffers a small change of shape the change in $E(A, D)$ is just the integral (4.4) taken over the freshly formed portion of the cut. Suppose that the change of shape is carried out by giving each element dx_k of the loop a small displacement $\delta\xi_l$. Each element then sweeps out an elementary area $dS_j = \varepsilon_{jkl} dx_k \delta\xi_l$ and so we have

$$\delta E(A, D) = -\int \delta\xi_l F_l ds \tag{4.5}$$

with

$$F_l = \varepsilon_{jkl} b_i \sigma_{ij} t_k, \tag{4.6}$$

where $t_k = dx_k/ds$ is the local tangent vector to the loop and σ_{ij} is equal to σ_{ij}^A. (We do not actually write σ_{ij}^A in (4.6) because, as we shall see, it is of more general application.)

Consequently we shall get the correct energy change if we suppose that a force (4.6), the Peach–Koehler force (Peach and Koehler, 1950), acts on the dislocation. We might, perhaps, feel uncomfortable because (4.6) seems to involve the value of the applied stress at the dislocation where, in our treatment, there is a singularity due to the dislocation, or, from a more physical point of view, where there is a strong departure from linearity. However, in principle we could have calculated the energy change (4.5) directly from (4.2) as

$$\delta E(A, D) = -\int_S \delta u_i^P \sigma_{ij}^A dS_j,$$

which only involves the σ_{ij}^A on S, remote from the dislocation, at the expense of solving a boundary value problem to determine the δu_i^P induced by the deformation of the loop. Equation (4.6) is merely a convenient transformation of the result of such a hypothetical calculation, and what one actually has to insert in (4.6) is the value of σ_{ij}^A at the site of the proposed dislocation before it is actually introduced, calculated from the surface traction $\sigma_{ij}^A n_j$ on the assumption of linearity.

The component of (4.6), $F_l t_l$, parallel to the dislocation line, which would be more or less meaningless, is in fact zero. Let us take local coordinates with x_3 parallel to the dislocation line and x_2 perpendicular to the plane, the slip plane, defined by b_1 and t_k. Then if the dislocation is of pure-edge type with Burgers vector $(b_1, 0, 0)$ (4.6) gives

$$F_1 = b_1\sigma_{12}, \quad F_2 = -b_1\sigma_{11}. \tag{4.7}$$

Motion in the direction of F_1, i.e. parallel to the Burgers vector, is referred to as "glide" and in the direction of F_2 as "climb". Evidently when climb occurs the construction of the necessary extension to the discontinuity will involve filling a gap or scraping to avoid interpenetration, and our analysis assumes that the necessary material is supplied or removed. In a real crystal, as opposed to its continuum representation, this can occur by the emission or absorption of lattice vacancies. Consequently a discussion of the energetics of climb involves, for example, consideration of the work done in injecting the emitted vacancies against the osmotic pressure of the vacancies already present in the lattice. As such ideas are outside the competence of continuum mechanics we shall not pursue the matter. For a screw dislocation with Burgers vector $(0, 0, b_3)$ we have

$$F_1 = b_3\sigma_{32}, \quad F_2 = -b_3\sigma_{31}. \tag{4.8}$$

As b_i and t_i are now parallel, every plane is a potential glide plane as far as continuum theory is concerned.

Returning to the general geometry and Burgers vector of (4.6), the glide force takes the form

$$F_l^{\text{glide}} = Fm_l, \quad \text{with} \quad F = b_i\sigma_{ij}n_j, \quad m_l = \varepsilon_{lrs}n_r t_s. \tag{4.9}$$

F is the scalar product of the Burgers vector and the traction exerted by the applied stress on the glide plane (with normal n_j) and m_l is a unit vector in the glide plane and normal to the dislocation line. If we subtract (4.9) from (4.6) we are left with the climb force.

4.2. *Image force and force due to internal stresses*

It is not essential to the validity of (4.6) that the load applied to each surface element of the body should be independent of the displacement of the element. In engineering language the part of the external mechanism for loading a particular element need not be perfectly soft, but may have an arbitrary degree of stiffness. In fact, with arbitrary stiffness, when u_i^p in (4.1) suffers a change of order $\delta\xi_l$ as a result of the deformation of the loop specified by $\delta\xi_l$ the force on an element will be $\sigma_{ij}^A dS_j$ at the beginning of the change and $\sigma_{ij}^A dS_j + O(\delta\xi_l)$ during the course of it, so that the work done by the loading mechanism will only differ from its value for dead loading by a quantity of order $(\delta\xi_l)^2$, and so (4.5) is not affected.

We shall now try to show that the irrelevance of the stiffness of the loading mechanism enables us to extend the Peach–Koehler formula (4.6) to cover the image field and the internal stress field due to other defects without the use of additional mathematics. We begin with the image force. Let us start with the loop in an infinite medium. Call this state (i). Then the only non-zero term in (4.1) is E_S. Make a small change in the form of the loop, note the change $\delta E_S^{(i)}$ in E_S and restore the loop to its original form. Make a cut over S, discard the material outside it, and apply surface tractions $\sigma_{ij}^\infty n_j$ to S. The field inside S is unchanged. Call this state (ii). In state (i) we can regard the material outside S as a loading mechanism of unknown stiffness acting on S. In state (ii) we have replaced it by externally applied forces which produce the same traction on S. According to our argument this makes no difference to the change of total energy when the loop suffers a small change of form. (It is perhaps easier to see this for the reverse change from (ii) to (i).) Hence if we make the same change in the form of the loop as we did in stage (i), we shall get the same energy change $\delta E_S^{(i)}$ as before. But, in fact in stage (ii) the dislocation is acted on by two elastic fields, the image field and the field produced by applied tractions $\sigma_{ij}^A n_j = \sigma_{ij}^\infty n_j$. Yet despite this the energy change is what it would be in an infinite medium. We must conclude that the effects of these two fields cancel. In other words the image field produces the same effect as applied tractions $-\sigma_{ij}^\infty n_j$. But, by definition, it is precisely these tractions which generate the image stress σ_{ij}^{im} in the body. Hence the image stress produces a Peach–Koehler force given by (4.6) with $\sigma_{ij} = \sigma_{ij}^{\text{im}}$. (For a proof of this by an entirely different method see a paper by Gavazza and Barnett, 1975.)

We can handle in the same way a case not covered by (4.1), namely where there are other sources of internal stress in the body in addition to the dislocation loop we are interested in, for example, other dislocations, point defects, or misfitting inclusions representing various metallurgical irregularities. Suppose that they produce a stress σ_{ij}^{O} in the body. Draw a surface S_0 embracing the loop we are discussing and isolating it from the other sources of stress. We may now regard the interior of S_0 as our body and the self-stressed material between S_0 and S as a loading mechanism, of unknown but irrelevant stiffness, acting on S_0. It follows that the force on the dislocation, so far as it depends on the internal stress σ_{ij}^{O}, is given by (4.6) with $\sigma_{ij} = \sigma_{ij}^{O}$. In particular, we can use (4.6) to study the interaction of dislocations with one another.

An analogy is often drawn between the interactions of dislocations and current-carrying wires. The Peach–Koehler formula (4.6) is evidently roughly analogous to the expression for the force on a current element, the stress taking the role of the applied magnetic field. If we are interested in the interaction of two dislocation loops the appropriate stress is given by the integral (2.20) turned into a stress with the help of Hooke's law, and this integral is roughly similar to the Biot–Savart integral for the magnetic field. In some cases the analogy is numerically exact, e.g. if the two loops are coplanar with their Burgers vectors parallel (or anti-parallel) and in the plane of the loops, and if the material is isotropic with zero Poisson's ratio (Hart, 1953). However, since two like dislocations repel whereas two like currents attract, two loops with the same Burgers vector must be simulated by loops carrying opposite currents, and this prompts the Zen-like question, What are three like dislocation loops simulated by? The answer, based on an unpublished comment of Nabarro's is: three like loops each carrying a current $i = \sqrt{(-1)}$. For this apparently flippant idea (which can be a useful mnemonic) there is a respectable precedent in the now outdated regularization procedure of quantum electrodynamics (Jauch and Rohrlich, 1955), in which some of the coupling constants are made purely imaginary so as to change the sign of part of the interaction.

It is clear that the Peach–Koehler force (4.6), now extended to cover internal and image stresses as well as externally applied stresses, is not a straightforward force of the kind that can be balanced by a weight, but, that it simply provides a picturesque description of the energy changes associated with deformation of the dislocation. In particular, as is emphasized in the literature, it is not to be confused with a distribution of body force acting along the singular line. In a strict elastic continuum, where a dislocation is made by cutting and re-welding, and so is immobile, there is nothing objectionable about there being a non-zero Peach–Koehler force acting on it. When we use the dislocated continuum to simulate a dislocation in a crystal lattice and, so to speak, switch on crystallinity, the dislocation will try to reduce the energy by moving in the direction indicated by the Peach–Koehler force, which thus becomes a useful tool for determining equilibrium configurations.

The name configurational force is sometimes given to fictitious forces of the kind typified by the Peach–Koehler force. Recently Kléman (1977) has applied the same term to an apparently similar force acting on disclinations in nematic liquid crystals. In certain circumstances it is given by an equation like (4.6) with b_i replaced by the rotation associated with the disclination, and σ_{ij} by the torque per unit area. However, in addition to a torque there is a traction on an element of the medium given by the Ericksen stress tensor, and one can show (Eshelby, 1980a, b) that the supposedly configurational force on an element of the disclination is equal to the traction which the Ericksen stress exerts on a surface embracing the element, and the element can only be in equilibrium if an equal and opposite real force is applied to it from outside. Hence to call this force a configurational or Peach–Koehler force (Kléman, 1977) seems inappropriate. It is true that it, like the force on a disclination in an elastic continuum, is a measure of the variation of the energy of the system made up of the medium and any loading mechanism attached to its outer boundary, but this is just because the work done by or on the external balancing force is added to or extracted from the system, a state of affairs with no parallel for the dislocation. We might equally well argue that the force exerted on a charge in an electric field was configurational on the grounds that the mechanical work done in pushing it against the field is stored up in the electrostatic energy of the field. It seems to the writer that to be worthy of the name configurational a force must not be balanceable by an ordinary force and, for preference, there should also exist another, "real", force for it not to be confused with.

4.3. *The self-energy of a dislocation*

Returning to (4.1) it still remains to deal with E_S, the self-energy of the loop. Ideally we should calculate E_S and then find its variation for a small change of form. If the result could be thrown into the form (4.5), then (4.6) could be interpreted as the Peach–Koehler force which the loop exerts on itself. The problem is a difficult one, and attempts to improve existing treatments (Bacon *et al.*, 1979) are still under way. Here we shall only present a highly oversimplified treatment, based on an idea of Schoeck and Kirchner (1978).

Let $E_S(L)$ be the self-energy of a loop of a certain size, specified by some characteristic length L, say its maximum diameter. We suppose that E_S is prevented from diverging by omitting from the integration of the energy density the interior of a tube of radius r_0 embracing the singular line. Next make a magnified version of the loop in which L has become L' and associate the same strain with corresponding points $\mathbf{r}$ in the original and $\mathbf{r}' = L'\mathbf{r}/L$ in the enlarged version, so that the stresses in the enlarged version will be in static equilibrium if they were in the original. Because corresponding volume elements are in the ratio $L'^3{:}L^3$ the enlarged configuration will have the energy $E_S(L)L'^3/L^3$. We have enlarged the Burgers vector $\mathbf{b}L'/L$, but in fact we wish it to remain the same, so we must divide the above value by $(L'/L)^2$, since the energy is proportional to the square of the Burgers vector. The resulting energy $E_S(L)L'/L$ is still not quite what we want, since we have also increased the cut-off radius to r_0L'/L. To remedy this we must add the elastic energy E_{tube} within a curved tube of outer and inner radii $L'r_0/L$ and r_0 embracing the singular line. If r_0 is small compared with any characteristic lengths which enter the problem, it is highly appealing to associate with each element ds of the tube an energy $E(\mathbf{t})ds$ where $E(\mathbf{t})$ is the expression (2.42) for an infinite straight dislocation of the appropriate direction $\mathbf{t}$ and with the argument of the logarithm chosen to be $(r_0L'/L)/r_0 = L'/L$. If we admit this the elastic energy in the tube will be

$$E_{\text{tube}} = \ln(L'/L)\int E_0(\mathbf{t})\,ds = CL\ln(L'/L), \tag{4.10}$$

where

$$C = \int E_0(\mathbf{t})\,ds/L$$

is independent of L, being in fact the value of the integral in (4.10) when L happens to be unity. If we now let $L' = L + dL$ differ only slightly from L we have

$$dE_S = E_S(L') - E_S(L) = E_S(L)dL/L + C\ln(1 + dL/L)$$

or

$$\frac{d}{dL}\left(\frac{E_S}{L}\right) = \frac{C}{L}$$

which integrates to give

$$E_S = LC \ln (L/L_0)$$

with an unknown constant of integration L_0. We may equally well write the logarithm as $\ln (R/r_0)$ with prescribed r_0 and unknown R, and then we have

$$E_S = \int E(\mathbf{t}) ds \tag{4.11}$$

with $E(\mathbf{t})$ as in (2.42) and with the logarithmic factor $\ln(R_2/R_1) = \ln(R/r_0)$ undetermined but independent of $\mathbf{t}$ and of position along the loop. For isotropy, (2.42) gives

$$E(\mathbf{t}) = \frac{\mu b^2}{4\pi}\left(\cos^2\theta + \frac{1}{1-\nu}\sin^2\theta\right)\ln\frac{R}{r_0}, \tag{4.12}$$

where θ is the angle between the Burgers vector and the dislocation line.

The shortcoming in the above argument is, of course, the identification of each element of the integral (4.10) with the corresponding element of that of a tangential infinite straight tube. If a straight dislocation is bent into a curve the elastic field is, so to speak, compressed on the inside of the curve and expanded on the outside of the curve. The resulting increase and decrease in energy density do not precisely cancel, and as a result $E(\mathbf{t})$ depends on the local curvature of the dislocation line. Indeed, strictly speaking, it is a functional which depends on the whole shape of the loop and on the point on the loop to which it applies (Stroh, 1954). The above model is, however, widely used. Exact calculations for specific cases suggest that R in the logarithmic factor is of the order of the dimensions of the loop. If we admit the validity of (4.11) we can go on to discuss the equilibrium form of a dislocation in an applied stress field.

Consider a dislocation segment confined to the xy-plane, with its ends locked at (0,0) and $(a, 0)$ and between them following the curve $y = y(x)$. If it is acted upon by an applied stress $\sigma_{ij}(x, y)$ the energy (4.1) of the system is

$$\int_0^a E(\tan^{-1} y')(1+y'^2)^{1/2} dx - \int_0^a dx \int_0^{y(x)} b\sigma(x, y) dy = \int_0^a \mathscr{E}(y, y') dx \tag{4.13}$$

say. The first integral in (4.13) is the self-energy and the second is the integral (4.4) extended over the area bounded by the dislocation above and closed by some fixed arbitrary curve below; we have taken it to be the x-axis. In the second integral $b\sigma$ is an abbreviation for $b_i\sigma_{ij}n_j$. The equilibrium shape which minimizes the total energy must satisfy the Euler equation

$$\frac{\partial \mathscr{E}}{\partial y} - \frac{d}{dx}\frac{\partial \mathscr{E}}{\partial y'} = 0$$

prescribed by the calculus of variations. It gives

$$\{E(\theta) + E''(\theta)\} y''/(1+y'^2)^{3/2} + b\sigma(y) = 0,$$

which may be written in the form

$$T(\theta)/\rho + b\sigma = 0, \tag{4.14}$$

where ρ is the radius of curvature of the dislocation and

$$T(\theta) = E(\theta) + E''(\theta). \tag{4.15}$$

The second term in (4.14) is the Peach–Koehler force due to the applied stress, and so the first term can evidently be interpreted as the Peach–Koehler force which the dislocation exerts on itself at each point. It takes the form of the product of a line tension T (de Wit and Koehler, 1959) and the curvature $1/\rho$ and is directed towards the centre of curvature so as to tend to straighten out the dislocation, supposing that T is positive. However, in an anisotropic medium it is possible for $E''(\theta)$ to be negative and exceed $E(\theta)$ in magnitude for some orientations. In that case a dislocation with fixed end points and, say, $b\sigma = 0$ may prefer to take a zigzag rather than a straight course between them. For isotropy, (4.15) applied to (4.12) gives

$$T(\theta) = \frac{\mu b^2}{4\pi}\left(\frac{1+v}{1-v}\cos^2\theta + \frac{1-2v}{1-v}\sin^2\theta\right)\ln\frac{R}{r_0}. \tag{4.16}$$

However, because the assumption that E depends only on θ but not on the form of the dislocation is not a very good one, it does not follow that the best choice of the logarithmic factor will be the same for (4.12) and (4.16). For example, Laub and Eshelby (1966) found that for a long straight dislocation deformed into a shallow sine curve the R in (4.16) was equal to the wavelength divided by π, which is comparable with the distance at which the fields of the straight and deformed dislocations become indistinguishable.

A related problem is the determination of the form of a dislocation node. The crystallographic theory of dislocations shows that several dislocations may meet at a point (node) provided that the Burgers vector observed on entering the junction along one of them is equal to the sum of the Burgers vectors observed on exiting along each of the others. One may expect that there will be some condition controlling them similar to that governing joined strings under tension in elementary statics. It may be established by a variational calculation, or by comparison with Herring's (1951, 1953) treatment of the analogous problem for grain boundaries. (We suppose that the dislocations are constrained to lie in one plane.) The result is that on the imaginary cut end of a dislocation there acts a "vector line tension" $\mathbf{T}$ given by

$$\mathbf{T} = \mathbf{t}E(\theta) + \mathbf{n}E'(\theta), \tag{4.17}$$

where, as before, E is the line energy, $\mathbf{t}$ is the tangent, and $\mathbf{n}$ is the normal to the dislocation. In equilibrium the sum of the $\mathbf{T}$'s of the dislocations terminating at a node must be zero. Equation (4.17) may also be applied to a curved dislocation. On an element ds the sum of the forces (4.17) on the ends will be $ds\, d\mathbf{T}/ds$, and from the relations

$$ds/\rho = d\theta = \mathbf{n}\cdot d\mathbf{t} = -\mathbf{t}\cdot d\mathbf{n}$$

which relate arc length s and inclination θ to $\mathbf{n}$ and $\mathbf{t}$ we easily find that its normal component is $T(\theta)ds/\rho$, which takes us back to (4.13) (Chou and Eshelby, 1962).

5. Dislocations in Motion

5.1. *Uniform motion*

So far, when we have spoken of a dislocation moving we have merely meant that it passed through a succession of neighbouring static configurations. However, if while deforming a dislocation line, or part of it, moves with a velocity comparable with the speed of elastic waves, its elastic field must be a solution of the dynamic equations of elasticity. Such solutions will not usually be very relevant to the actual behaviour of moving dislocations because they take no account of the strong retarding effect of interactions with, for example, point defects, electrons and phonons, and, indeed, with the lattice itself, regarded as a perturbation of the continuum. Some of these effects can be incorporated into the continuum description by ascribing suitable viscoelastic properties to the medium (Kosevich and Natsik, 1966; Eshelby, 1961). Here we shall simply show how certain results for moving infinite straight dislocations can be derived rather easily, starting from simple static solutions.

The substitution

$$x \to x'/\beta_n = (x - Vt)/\beta_n, \quad y \to y, \quad \beta_n = (1 - V^2/c_n^2)^{1/2}$$

converts a solution of

$$\nabla^2 f(x, y) = 0$$

into a uniformly moving solution of the wave equation

$$\{c_n^2 \nabla^2 - \partial^2/\partial t^2\} f(x'/\beta_n, y) = 0.$$

Consequently for a uniformly moving screw dislocation we have, from (2.24),

$$w = \frac{b}{2\pi} \tan^{-1} \frac{\beta_2 y}{x - Vt}, \tag{5.1}$$

where $c_2 = (\mu/\rho)^{1/2}$ is the velocity of shear waves in an isotropic medium of density ρ and shear modulus μ. One can check that (5.1), like its static counterpart, satisfies the Peierls relation (2.52).

To solve the same problem for an edge dislocation we may start from the dilatationless and irrotational solution (2.35) for a dislocation plus a vertical force and the fact that the elastic equations of motion reduce to the wave equation with c_n equal to c_2 or c_1 if, respectively, the dilatation or the rotation happens to be zero. The two quantities

$$u^{(2)}(x, y) = u_0(x'/\beta_2, y), \quad v^{(2)}(x, y) = \beta_2^{-1} v_0(x'/\beta_2, y) \tag{5.2}$$

satisfy the wave equation with $c_n = c_2$ and, thanks to the factor β_2^{-1} multiplying v_0, the vector $(u^{(2)}, v^{(2)})$, like the static (u_0, v_0), has zero dilatation, and so is a solution of the equations of motion. So too is the displacement

$$u^{(1)}(x, y) = u_0(x'/\beta_1, y), \quad v^{(1)}(x, y) = \beta_1 v_0(x'/\beta_1, y), \tag{5.3}$$

for it satisfies the wave equation with $c_n = c_1 = [(2-2\nu)\mu/(1-2\nu)\rho]^{\frac{1}{2}}$, the velocity of irrotational waves and, like (u_0, v_0), has zero rotation because of the factor β_1 before v_0.

The unwanted vertical force $-2\mu b$ of the static solution (2.35) remains unchanged for the moving version (5.3) but becomes $-2\mu b\alpha^2$, with

$$\alpha^2 = \tfrac{1}{2}(1 + \beta_2^2) = 1 - V^2/2c_2^2$$

for (5.1). So if we multiply (5.3) by $-\alpha^2$, add it to (5.2), and divide by $1 + (-\alpha^2) = V^2/2c_2^2$, we get a moving solution with the right Burgers vector $(b, 0)$ and no force, namely

$$u = \frac{b}{2\pi}\frac{2c_2^2}{V^2}\left[\tan^{-1}\frac{\beta_1 y}{x - Vt} - \alpha^2 \tan^{-1}\frac{\beta_2 y}{x - Vt}\right],$$

$$v = \frac{b}{2\pi}\frac{2c_2^2}{V^2}\left[\beta_1 \ln\{(x-Vt)^2+(\beta_1 y)^2\}^{1/2} - \frac{\alpha^2}{\beta^2}\ln\{(x-Vt)^2+(\beta_2 y)^2\}^{1/2}\right] \tag{5.4}$$

which reduces to (2.27) in the limit $V = 0$, as a rather tedious calculation will show.

The elementary solutions (5.2) and (5.3) may also be used to construct an image field which when added to (5.4) turns it into the solution for a dislocation moving parallel to the stress-free plane $y = Y$. Introduction of the geometrical image obtained by displacing (5.4) upwards so that its singular point is at a height Y above $y = Y$ is not adequate since it cancels σ_{yy} but doubles σ_{xy} on that plane, or the reverse if we change its sign. The relevant stresses corresponding to (5.2) and (5.3) are

$$\sigma_{yy}^{(2)} = \frac{b}{2\pi}\frac{2\mu\beta_2}{x'^2+(\beta_2 y)^2}, \quad \sigma_{xy}^{(2)} = \frac{b}{2\pi}\frac{2\mu\alpha^2 x'/\beta_2}{x'^2+(\beta_2 y)^2}, \tag{5.5}$$

$$\sigma_{yy}^{(1)} = \frac{b}{2\pi}\frac{2\mu\beta_1\alpha^2 y}{x'^2+(\beta_1 y)^2}, \quad \sigma_{xy}^{(1)} = \frac{b}{2\pi}\frac{2\mu\beta_1 x'}{x'^2+(\beta_1 y)^2}, \tag{5.6}$$

in terms of which the stresses associated with (5.4) are

$$\sigma_{ij} = (2c_2^2/V^2)(\sigma_{ij}^{(i)} - \alpha^2\sigma_{ij}^{(2)}). \tag{5.7}$$

If we displace, say, the (2)-solution so that its singular point is at a height $\beta_1 Y/\beta_2$ above $y = Y$ then on that plane the denominators in (5.5) will have the same form as the denominators of the undisplaced field (5.6), which occur both in the original field (5.7) and its geometrical image. Similarly, we may displace the (1)-solution so that the denominators of (5.6) agree with those of (5.5). In this way, we can arrange that not only another dilatationless field but also an irrotational one is available to help cancel the stresses due to the dilatationless part of the original field (5.4), (5.7), and similarly for its irrotational part, so adding to the richness of the solution and the chance of satisfying the boundary conditions. This suggests that we try the following form for the image field:

$$\begin{aligned}\mathbf{u}^{\text{im}} &= A\mathbf{u}^{(2)}(x, y-2Y) + B\mathbf{u}^{(1)}(x, y-2Y) + C\mathbf{u}^{(2)}(x, y-Y-\beta_1 Y/\beta_2)\\ &\quad + D\mathbf{u}^{(1)}(x, y-Y-\beta_2 Y/\beta_1)\end{aligned} \tag{5.8}$$

with $\mathbf{u}^{(n)} = (u^{(n)}, v^{(n)})$ in the notation of (5.2), (5.3). The first and second terms are ordinary images of (5.2) and (5.3) while the last two are "extraordinary" images with their singularities shifted to points at distances $\beta_1 Y/\beta_2$ and $\beta_2 Y/\beta_1$ above $y = Y$. The four constants are fixed by the requirement that for the sum of (5.7) and the above image field, evaluated on $y = Y$, the individual parts of σ_{yy} which contain the denominators $x'^2 + (\beta_2 Y)^2$ and $x'^2 + (\beta_1 Y)^2$ (there are no others) shall vanish separately, and similarly for σ_{xy}. This gives

$$\frac{A}{\alpha^2} = B = \frac{2c_2^2}{V^2}\frac{\alpha^4+\beta_1\beta_2}{\alpha^4-\beta_1\beta_2},$$

$$\frac{C}{\beta_1\beta_2} = \frac{D}{\alpha^2} = -\frac{2c_2^2}{V^2}\frac{2\alpha^2}{\alpha^4-\beta_1\beta_2},$$

which completes the solution. The image terms become very large near the Rayleigh velocity where $\alpha^4 - \beta_1\beta_2$ vanishes.

The same device can be used to deal with a screw dislocation moving along the x-axis parallel to the interface $y = Y$ between media with parameters μ', c_2' above it and μ, c_2 below. For the undashed medium we take the field to be the sum of (5.1) and an ordinary image, of unknown strength, at the geometrical image point and for the dashed medium an image of unknown strength at a depth below the interface which makes the denominator in its stress components agree on $y = Y$ with that of the real dislocation, i.e. we now choose its position so as to compensate for the difference between the coefficients β_2 and β_2' in the two different media, instead of, as in the previous example, the difference between β_2 and β_1 in a single medium. The displacement is thus given by

$$2\pi w/b = \tan^{-1}[\beta_2 y/x'] + A \tan^{-1}[\beta_2(y-2Y)/x']$$

in the lower medium and

$$2\pi w/b = B \tan^{-1}[\beta_2'(y - Y + \beta_2 Y/\beta_2')/x']$$

in the upper. On matching stress and displacement at the interface we get

$$A = \frac{\mu'\beta_2' - \mu\beta_2}{\mu'\beta_2' + \mu\beta_2}, \quad B = \frac{2\mu\beta_2}{\mu'\beta_2' + \mu\beta_2}.$$

The analysis fails if the dislocation is supersonic with respect to either medium, so to go over to the case, $\mu' = 0$, of a stress-free surface (Hirth and Lothe, 1968) we must let ρ' tend to zero along with μ', in such a way that c_2' stays greater than V. The result is $A = -1$, $B = 2$, so that, as one might have guessed, there is a simple negative image. The term in B is without effect so long as we regard the upper medium as empty space, but if we think of it as a material of negligible elastic constants and density but with a definite value for c_2', the B-term gives its deformation.

The same method will handle an edge dislocation gliding parallel to an interface. Starting with the moving solution (5.4), (5.7) we use the image field (5.8) again for the lower medium, but with different A, B, C, D and for the upper medium a similar image expression involving the dashed parameters μ', c_2', c_1', and with the depths of the four image dislocations adjusted so that for each of them the denominator in its stress components agrees on $y = Y$ with one or other of the two denominators which appear in (5.5) and (5.6). The eight constants so introduced are enough to satisfy the four boundary conditions for each denominator separately, and so, fortunately, the eight simultaneous equations fall into two groups of four. It hardly seems worthwhile to write out the results in detail.

5.2. *Non-uniform motion*

More ambitiously, returning to an infinite isotropic solid, we may try to construct the elastic field of a screw dislocation moving non-uniformly along the x-axis by starting from the special case where it executes small oscillations

$$\xi = \xi_0 \exp(i\omega t) \tag{5.9}$$

about the origin. Much closer than the wavelength $2\pi c_2/\omega$ of the disturbance emitted the dislocation simply carries its static field rigidly about with it so that we have

$$w = \frac{b}{2\pi}\tan^{-1}\frac{y}{x - \xi(t)} = \frac{b}{2\pi}\theta + \delta w$$

with

$$\delta w = \frac{b}{2\pi}\frac{\sin\theta}{r}\xi_0 \exp(i\omega t) \tag{5.10}$$

provided we are several multiples of ξ_0 away from the origin. The solution of the two-dimensional wave equation of which (5.10) is the limiting form for small r must likewise be proportional to $\sin\theta$ and to $\xi_0 \exp(i\omega t)$ and must represent outgoing waves. The solution must therefore be (5.10) with $1/r$ replaced by some multiple of the Bessel function of the third kind (Hankel function) $H_1^{(2)}(kr)$, $k = \omega/c_2$, which represents an outgoing wave when multiplied by $\exp(i\omega t)$. The appropriate multiple can be read off from the standard expansion

$$-\frac{1}{2}i\pi k H_1^{(2)}(kr) = \frac{1}{r} - \frac{1}{2}\left[\ln(0.54kr) + \frac{1}{2}i\pi\right]k^2 r \tag{5.11}$$

for small kr (the coefficient 0.54 actually stands for $\frac{1}{2}\exp(\gamma - \frac{1}{2}) = 0.5401\ldots$, where γ is Euler's constant $0.5772\ldots$), giving

$$\delta w = -\frac{1}{4}ibk\xi_0 \sin\theta\, H_1^{(2)}(kr)\exp(i\omega t). \tag{5.12}$$

For later use we note that for (5.12) to apply within a range of a few multiples of b from the centre of the dislocation, where our interest will lie, the velocity of the dislocation must be much less than c_2. In fact we must, on the one hand, have $\xi \ll b$ for (5.10) to apply, and, on the other, the wavelength of the disturbance emitted, $c_2/2\pi\omega$, must be large compared with the lattice spacing b, and these two conditions taken together imply $\dot{\xi} \ll c_2$.

If we apply to (5.9) the operator

$$\frac{1}{2\pi i}\int_C (\ldots)\frac{d\omega}{\omega},$$

where C is the real ω-axis indented below the origin, it changes the oscillation (5.9) into the step-function

$$\xi(t) = \xi_0 H(t), \quad H(t) = 1, \quad t > 0, \quad H(t) = 0, \quad t < 0, \tag{5.13}$$

representing a sudden small shift of the dislocation's position by ξ_0. The same operator applied to (5.12) will give the corresponding change of displacement. The result,

$$\delta w = \frac{b\xi_0}{2\pi r}\sin\theta \int_1^\infty \frac{\delta(v - c_2 t/r)}{(v^2-1)^{1/2}} v\,dv = \frac{b\xi_0}{2\pi}\frac{y}{r^2}\frac{c_2 t}{(c_2^2 t^2 - r^2)^{1/2}} H(c_2 t - r), \tag{5.14}$$

follows easily if precariously when we use the integral representations

$$H_1^{(2)}(kr) = -\frac{2}{\pi}\int_1^\infty \frac{\exp(-ikrv)}{(v^2-1)^{1/2}} v\,dv$$

for the Hankel function (Watson, 1945) and

$$\delta(z) = \frac{1}{2\pi}\int_C \exp(i\omega z)\,d\omega$$

for Dirac's delta function and interchange the order of integration.

An arbitrary motion

$$\xi = \xi(t) \tag{5.15}$$

along the x-axis can be considered to be made up of jumps of amount $\dot{\xi}\,dt$ in successive time intervals dt. Hence replacing t, $r^2 = x^2 + y^2$, ξ_0 in (5.11) by τ, $[x - \xi(\tau)]^2 + y^2$, $\dot{\xi}(\tau)\,d\tau$ and integrating over all τ up to t, we get

$$w = \frac{b}{2\pi}\int_{-\infty}^{\tau_0} \frac{y}{[x-\xi(\tau)]^2 + y^2}\,\frac{c_2(t-\tau)}{s}\,\dot{\xi}(\tau)\,d\tau \tag{5.16}$$

with

$$s^2 = c_2^2(t-\tau)^2 - [x - \xi(\tau)]^2 - y^2$$

and where τ_0 is the root of $s^2(\tau) = 0$ which is less than t. For another derivation see Nabarro (1951a).

According to (5.14) the field near the dislocation depends on the entire previous history of its motion. This is a result of the fact that, for example, even after the jump (5.13) has been completed, disturbances from remote parts of the z-axis continue to arrive at the origin. As a result the dynamic behaviour of the dislocation is somewhat peculiar.

To conclude, we return to (5.12) and try to use it to find how the dislocation responds when acted on by a small alternating applied stress, using a crude version of Nabarro's (1951b) argument. Suppose that at first we only retain the term $1/r$ in (5.11). Then the displacement is the sum of (5.10) and $b\theta/2\pi$, the field of a static dislocation rigidly transported to and fro according to the law (5.9). Because, as we have seen, we are limited to small velocities (see the remarks below (5.12)) and hence to β_2 nearly unity in (5.1), we could just as well say that the field of the oscillating dislocation agrees at each instant with that of a dislocation with the same instantaneous position and velocity, but which has been moving uniformly for all time. This means that the displacement and stress obey the Peierls relation (2.52) near the centre of the oscillating dislocation. All this is on the assumption that the second term in (5.11) has been suppressed. When it is restored there is an extra term

$$\delta w' = -\frac{b}{4\pi}k^2 y\left[\ln(0.54\,kr) + \frac{1}{2}i\pi\right]\xi_0 \exp(i\omega t) \tag{5.17}$$

which upsets the agreement with the Peierls law. We can put this right by imposing a displacement equal and opposite to (5.17) from outside. The corresponding applied stress should be

$$\sigma_{zy}^{\text{app}} = -\mu\frac{\partial \delta w'}{\partial y} = \frac{\mu b^2}{4\pi}k^2\left[\frac{y^2}{r^2} + \ln(1.47kr) + \frac{1}{2}i\pi\right]\xi_0 \exp(i\omega t). \tag{5.18}$$

Of course we cannot actually produce a non-uniform stress like (5.18) with the help of remotely applied tractions. However, because of the condition $kb \ll 1$ the logarithm is large (and negative) for $r \sim b$ and the term y^2/r^2 is unimportant in comparison with it (for neatness we shall replace it by $\frac{1}{2}$, its average value, absorb it into the logarithm and round off the coefficient) and so we may simplify (5.18) to

$$\sigma_{zy}^{\text{app}} = \frac{\mu b^2}{4\pi}\frac{\omega^2}{c_2^2}\left[-\ln\left(\frac{c_2}{\omega r}\right) + \frac{1}{2}i\pi\right]\xi_0 \exp(i\omega t). \tag{5.19}$$

Then, as the logarithm varies only slowly with r this stress can be adequately simulated over the region which matters by a uniform applied stress equal to (5.19) with r replaced by some constant value $r_0 \sim b$. The Peierls–Nabarro model suggests $r_0 = \frac{1}{2}b$, but the choice is not critical. If we multiply the stress by b to convert it to a force according to (4.8) we get the equation of motion

$$F_x = [-m(\omega) + iR]\,\omega^2\,\xi$$

which connects applied force F_x and dislocation position ξ when both vary like $\exp(i\omega t)$. The coefficient

$$m(\omega) = \rho\frac{b^2}{4\pi}\ln\frac{\omega_0}{\omega}, \quad \omega_0 = \frac{c_2}{r_0},$$

can be interpreted as a mass per unit length which depends weakly on frequency. The resistive term involving

$$R = \frac{1}{8}\rho b^2,$$

however, since it involves an odd power of i and an even power of ω, is not even approximately proportional to some time derivative of ξ. The best we can do is to say, with Nabarro (1951b), that the mass has a small imaginary component.

Our treatment of the oscillating dislocation is an example of the kind of argument, which may appear somewhat uncouth to an applied mathematician, that one often has to use to extract physical results from the continuum theory. It also illustrates a fortunate fact, which we also found in discussing line energy and tension, and which is true elsewhere in the theory of dislocations, that, even when the elastic solution has to be modified near the core, the shortcomings of the continuum model are often muffled by a merely logarithmic dependence on the details.

References

Amar, H. (1980) Elementary application of differential forms to electromagnetism, *Am. J. Phys.* **48,** 252–253.

Bacon, D. J., Barnett, D. M., and Scattergood, R. O. (1979) Anisotropic continuum theory of lattice defects, *Progress in Materials Science* **23,** 51–262.

Bilby, B. A. and Eshelby, J. D. (1969) Dislocations and fracture, *Fracture* **1,** ed. by H. Liebowitz, 99–182, Academic Press, New York.

Bromwich, T. J. I'A. (1955) *An Introduction to the Theory of Infinite Series*, 2nd edn, 314, Macmillan, London.

Brown, L. M. (1967) A proof of Lothe's theorem, *Phil. Mag.* **15,** 363–370.

Burgers, J. M. (1939) Some considerations of the field of stress connected with dislocations in a regular crystal lattice, *Proc. Konikl. Nederl. Acad. Wetenschap.* **42,** 293–325, 378–399.

Chou, Y. T. and Eshelby, J. D. (1962) The energy and line tension of a dislocation in a hexagonal crystal, *J. Mech. Phys. Solids* **10,** 27–34.

Cohen, H. (1966) Dislocations in couple-stress elasticity, *J. Math. Phys.* **45,** 35–44.

Dederichs, P. H. and Leibfried, G. (1969) Elastic Green's function for anisotropic cubic crystals, *Phys. Rev.* **188,** 1175–1183.

Eringen, C. (1977) Edge dislocation in nonlocal elasticity, *Int. J. Engng Sci.* **15,** 177–183.

Eshelby, J. D. (1951) The force on an elastic singularity, *Phil. Trans. Roy. Soc. Lond.* A **244,** 87–112.

Eshelby, J. D. (1961) Dislocations in visco-elastic materials, *Phil. Mag.* **6,** 953–963.

Eshelby, J. D. (1966) A simple derivation of the elastic field of an edge dislocation, *Br. J. Appl. Phys.* **17,** 1131–1135.

Eshelby, J. D. (1971) Fracture mechanics, *Sci. Prog. Oxf.* **59,** 161–179.

Eshelby, J. D. (1980a) The energy-momentum tensor of complex continua, *Proceedings of the Third International Symposium on Continuum Models of Discrete Systems, Freudenstadt*, ed. by E. Kröner and K.-H. Anthony, 651–665, University of Waterloo Press.

Eshelby, J. D. (1980b) The force on a disclination in a liquid crystal, *Phil. Mag.* A **42,** 359–367.
Eshelby, J. D. and Laub, T. (1967) The interpretation of terminating dislocations, *Can. J. Phys.* **45,** 887–892.
Eshelby, J. D., Read, W. T., and Shockley, W. (1953) Anisotropic elasticity with applications to dislocation theory, *Acta Metall.* **1,** 251–259.
Eshelby, J. D. and Stroh, A. N. (1951) Dislocations in thin plates, *Phil. Mag.* **42,** 1401–1405.
Feynman, R. P. (1949) Space–time approach to quantum electrodynamics, *Phys. Rev.* **76,** 769–789 (Appendix, 785).
Gairola, B. K. D. (1979) Nonlinear elastic problems, *Dislocations in Solids* **1,** ed. by F. R. N. Nabarro, 225–342, North-Holland, Amsterdam.
Gavazza, S. D. and Barnett, D. M. (1975) The image force on a dislocation loop in a bounded elastic medium, *Scripta Metall.* **9,** 1263–1265.
Gebbia, M. (1891) Formule fondamentali della statica dei corpi elastici, *Rend. Circ. Matem. di Palermo* **5,** 320–323.
Gebbia, M. (1902) Le deformazioni tipiche dei corpi solidi elastici, *Ann. Mat. pura appl.* (3) **7,** 141–230.
Gebbia, M. (1904) Le deformazioni tipiche dei solidi elastici, *Ann. Mat. pura appl.* (3) **10,** 157–200.
Green, A. E. and Willmore, T. J. (1948) Three-dimensional stress systems in isotropic plates, II, *Proc. R. Soc. Lond.* A **193,** 229–248.
Hart, E. W. (1953) A magnetic analog for the interaction of dislocation loops, *J. Appl. Phys.* **24,** 224–225.
Head, A. K. (1979) The Galois unsolvability of the sextic equation of anisotropic elasticity, *J. Elasticity* **9,** 9–20.
Herring, C. (1951) Surface tension as a motive for sintering, *The Physics of Powder Metallurgy*, ed. by W. E. Kingston, 143–179, McGraw-Hill, New York.
Herring, C. (1953) The use of macroscopic concepts in surface-energy problems, *Structure and Properties of Solid Surfaces*, ed. by R. Gomer and C. S. Smith, 5–72, University of Chicago Press.
Hill, R. (1961) Discontinuity relations in mechanics of solids, *Progress in Solid Mechanics* **2,** ed. by I. N. Sneddon and R. Hill, 248–276, North-Holland, Amsterdam.
Hirth, J. P. and Lothe, J. (1968) *Theory of Dislocations*, McGraw-Hill, New York.
Huntington, H. B. (1955) Modification of the Peierls–Nabarro model for edge dislocation core, *Proc. Phys. Soc. Lond.* B **68,** 1043–1048.
Indenbom, V. L., and Orlov, S. S. (1967) Dislocations in an anisotropic medium, *Sov. Phys. JETP Letters* **6,** 274–277.
Jaswon, M. A. and Symm, G. T. (1977) *Integral Equation Methods in Potential Theory and Elastostatics*, Academic Press, New York.
Jauch, J. M. and Rohrlich, F. (1955) *The Theory of Photons and Electrons*, 223, Addison-Wesley, Cambridge, Mass.
Kléman, M. (1977) *Points. Lignes. Parois dans les Fluides Anisotropes et les Solides Cristallins*, Les Éditions de Physique, Orsay, France.
Knesel, Z. and Semela, F. (1972) The influence of couple-stresses on the elastic properties of an edge dislocation, *Int. J. Engng Sci.* **10,** 83–91.
Kosevich, A. M. (1979) Crystal dislocations and the theory of elasticity, *Dislocations in Solids* **1,** ed. by F. R. N. Nabarro, 37–141, North-Holland, Amsterdam.
Kosevich, A. M. and Natsik, V. D. (1966) Dislocation damping in a medium having dispersion of the elastic moduli, *Sov. Phys. Solid State* **8,** 993–999.
Kröner, E. (1953) Das Fundamentalintegral der anisotropen elastischen Differentialgleichungen, *Z. Phys.* **136,** 402–410.
Lardner, R. W. (1974) *Mathematical Theory of Dislocations and Fracture*, University of Toronto Press.
Laub, T. and Eshelby, J. D. (1966) The velocity of a wave along a dislocation, *Phil. Mag.* **14,** 1285–1293.
Lekhnitskii, S. G. (1963) *Theory of Elasticity of an Anisotropic Elastic Body*, Holden-Day, San Francisco.
Lothe, J. (1967) Dislocation bends in anisotropic media, *Phil. Mag.* **15,** 353–362.
Love, A. E. H. (1927) *A Treatise on the Mathematical Theory of Elasticity*, 4th edn, Cambridge University Press.
McManus, H. (1949) Classical electrodynamics without singularities, *Proc. R. Soc. Lond.* A **195,** 323–336.
Madelung, E. (1950) *Die Mathematischen Hilfsmittel des Physikers*, 181, Springer-Verlag, Berlin.
Mann, E. H. (1949) An elastic theory of dislocations, *Proc. R. Soc. Lond.* A **199,** 376–394.
Marcinkowski, M. J. (1979) *Unified Theory of the Mechanical Behavior of Matter*, Wiley, New York.
Nabarro, F. R. N. (1947) Dislocations in a simple cubic lattice, *Proc. Phys. Soc. Lond.* **59,** 256–272.
Nabarro, F. R. N. (1951a) The synthesis of elastic dislocation fields, *Phil. Mag.* **42,** 1224–1231.
Nabarro, F. R. N. (1951b) The interaction of screw dislocations and sound waves, *Proc. R. Soc. Lond.* A **209,** 278–290.
Nabarro, F. R. N. (1967) *Theory of Crystal Dislocations*, Clarendon Press, Oxford.
Nabarro, F. R. N. (1980) *Dislocations in Solids* (editor, volumes 1–5), North-Holland, Amsterdam.
Orowan, E. (1934) Zur Kristallplastizität III: Über den Mechanismus des Gleitvorganges, *Z. Phys.* **89,** 634–659.
Peach, M. and Koehler, J. S. (1950) The forces exerted on dislocations and the stress fields produced by them, *Phys. Rev.* **80,** 436–439.
Peierls, R. (1940) Size of a dislocation, *Proc. Phys. Soc. Lond.* **53,** 34–37.
Podolsky, B. (1942) A generalized electrodynamics, Part I: Non-quantum, *Phys. Rev.* **62,** 68–71.
Polanyi, M. (1934) Über eine Art Gitterstörung, die einen Kristall plastisch machen könnte, *Z. Phys.* **89,** 660–664.

Read, Jr., W. T. (1953) *Dislocations in Crystals*, 123, McGraw-Hill, New York.

Schoeck, G. and Kirchner, H. O. K. (1978) The elastic energy of dislocation loops in anisotropic media, *J. Phys.* F. **8**, L43–L46.

Seeger, A. (1955) Theorie der Gitterfehlstellen, *Handbuch der Physik* VII/2, ed. by S. Flügge, 383–665, Springer-Verlag, Berlin.

Somigliana, C. (1914) Sulla teoria delle distorsioni elastiche: Nota I, *Rend. R. Accad. Lincei* $[5]_1$ **23**, 463–472.

Somigliana, C. (1915) Sulla teoria delle distorsioni elastiche: Nota II, *Rend. R. Accad. Lincei* $[5]_1$ **24**, 655–666.

Southwell, R. V. (1936) *An Introduction to the Theory of Elasticity for Engineers and Physicists*, 78, Clarendon Press, Oxford.

Stroh, A. N. (1954) Constrictions and jogs in extended dislocations, *Proc. Phys. Soc. Lond.* B **67**, 427–436.

Stroh, A. N. (1958) Dislocations and cracks in anisotropic elasticity, *Phil. Mag.* **3**, 625–647.

Taylor, G. I. (1934) The mechanism of plastic deformation of crystals, *Proc. R. Soc. Lond.* A **145**, 362–387.

Toupin, R. A. (1964) Theories of elasticity with couple-stress, *Arch. Rat. Mech. Anal.* **17**, 85–112.

Volterra, V. (1907) Sur l'équilibre des corps élastiques multiplement connexes, *Ani. Sci. Éc. Norm. Sup.* [3] **24**, 401–518.

Volterra, V. and Volterra, E. (1960) Sur les distorsions des corps élastiques (Théorie et applications), *Mémorial des Sciences Mathématiques*, 147, Gauthier-Villars, Paris.

Watson, G. N. (1945) *A Treatise on the Theory of Bessel Functions*, 180, Cambridge University Press.

Weingarten, G. (1901) Sulle superficie di discontinuità nella teoria della elasticità dei corpi solidi, *Atti Accad. naz. Lincei Rc.* [5] **10**, 57–60.

Wit, G. de and Koehler, J. S. (1959) Interaction of dislocations with an applied stress in anisotropic crystals, *Phys. Rev.* **116**, 1113–1125.

Engineering Fracture Mechanics Vol. 16, No. 3, pp. 453–455, 1982
Printed in Great Britain.

0013-7944/82/030453-03$03.00/0

TECHNICAL NOTE

THE STRESSES ON AND IN A THIN INEXTENSIBLE FIBRE IN A STRETCHED ELASTIC MEDIUM

CHEREPANOV[1] has considered the problem of a finite thin inextensible fibre embedded in an elastic solid which is stretched by a uniform stress σ_∞ parallel to the fibre. He introduces a distribution of body force of amount $X(x)$ per unit length spread over that part of the x-axis, $-l<x<l$, which is to be occupied by a fibre of length $2l$. The force distribution has to be chosen so that it produces a displacement along the x-axis which for $|x|<l$ annuls the applied displacement $u_x=\sigma_\infty x/E$, (E is Young's modulus for the solid) and elsewhere is unspecified.

Simple statics shows that the shear traction τ across the fibre-matrix interface is

$$\tau(x)=-X(x)/2\pi r_0 \tag{1}$$

where r_0 is the radius of the fibre, and that the stress in the fibre, averaged over its cross-section, is

$$\sigma(x)=-\frac{2}{r_0}\int_{-l}^{x}\tau(x)\,\mathrm{d}x. \tag{2}$$

Cherepanov takes the displacement at $(x,0,0)$ due to a unit force at $(t,0,0)$ directed parallel to the x-axis to have the x-component

$$u_x=\frac{1+\nu}{2\pi E(t-x)}, \tag{3}$$

(ν is Poisson's ratio for the solid), and so is led to the integral equation

$$\int_{-l}^{l}\frac{f(t)\,\mathrm{d}t}{t-x}=\frac{2\pi}{1+\nu}\sigma_\infty x \tag{4}$$

in which, taking into account the evident fact that $X(t)$ is odd in t, and implicitly assuming that X changes sign only at $t=0$, he writes

$$\begin{aligned} f(t)&=|X(t)|,\ -l<x<0\\ &=-|X(t)|,\ 0<x<l. \end{aligned} \tag{5}$$

Spelled out in detail what (5) says is that either $f(t)$ is equal to $X(t)$ over the whole range $-l<x<l$ and $X(t)$ happens to be negative for positive t or, alternatively, that $f(t)$ is equal to $-X(t)$ over the whole range $-l<x<l$ and $X(t)$ happens to be positive for positive t.

Any particular solution of (4) is even in x and to it may be added a multiple of the complementary solution $(l^2-x^2)^{-1/2}$ [2]. The solution which vanishes at $x=0$ is

$$f(x)=\frac{2\sigma_\infty x^2}{(1-\nu)(l^2-x^2)^{1/2}}. \tag{6}$$

But the derivation of (6) was based on the assumption that $f(x)$ is odd in x, so that we have a contradiction and (6) must be incorrect.

In the resulting expression for the shear traction,

$$\tau=\frac{\sigma_\infty x|x|}{\pi\gamma_0(1-\nu)(l^2-x^2)^{1/2}} \tag{7}$$

the x^2 of (6) has been replaced by $x|x|$, thus making τ odd in x, as required, apparently because of a misinterpretation of (5).

Since the elegant expression (7) is, unhappily, incorrect, it is perhaps worth presenting the following treatment of the problem which though approximate indicates the general form of $\tau(x)$ and $\sigma(x)$. Unfortunately the solution is especially unreliable near the ends of the cylinder, a feature it has in common with solutions of other field equations relating to thin cylinders for which the radius r_0 can "almost" be ignored, and only makes itself felt through $\ln(l/r_0)$, the logarithm of the aspect ratio.

The reason why (6) is incorrect is that (3) should actually read

$$u_x=\frac{1+\nu}{2\pi E|t-x|}; \tag{8}$$

along its line of action the force pushes material in front of it forwards and drags along material behind it. With $t-x$ replaced by $|t-x|$ in eqn (4) we cannot rely on a Cauchy principle value to see us over the singularity at $x=t$, and, indeed the equation with $|t-x|$ has no solution.

To get round this we must take into account the finite radius, $r_0 \ll l$, of the fibre, and choose the force distribution along the x-axis so as to cancel the applied displacement $\sigma_\infty x/E$ not on the x-axis but on the surface of the cylinder, or, since we suppose it is circular, along one of its generators. Off the x-axis (8) has to be replaced by the more complicated expression[3]

$$u_x = 4\alpha(1-\nu)/R - \alpha r_0^2/R^3$$

with

$$R^2 = (x-t)^2 + r_0^2, \quad \alpha = 1/16\pi\mu(1-\nu)$$

which leads to the integral equation

$$\alpha \int_{-l}^{l} \left[\frac{4(1-\nu)}{R} - \frac{r_0^2}{R^3}\right] X(t)\,\mathrm{d}t + \frac{\sigma_\infty x}{E} = 0. \tag{9}$$

To avoid a full discussion of this type of equation, familiar in, for example, antenna theory and the flow of fluids past slender bodies we shall at first lean heavily on Landau and Lifshitz'[4] rather rough-and-ready treatment of the closely similar problem of a conducting cylinder perturbing a uniform electric field. We begin by writing $X(t)-X(x)+X(x)$ for $X(t)$ in (9) so getting

$$X(x)\int_{-l}^{l} \left[\frac{4(1-\nu)}{R} - \frac{r_0^2}{R^3}\right] \mathrm{d}t + I = -\frac{\sigma_\infty x}{E\alpha} \tag{10}$$

with

$$I = \int_{-l}^{l} [X(t)-X(x)] \left[\frac{3-4\nu}{R} + \frac{(t-x)^2}{R^3}\right] \mathrm{d}t.$$

Except close to the ends of the cylinder the integral multiplying $X(x)$ in (10) is

$$4(1-\nu)\ln[4(l^2-x^2)/r_0^2] + 2. \tag{11}$$

We now argue, with Landau and Lifshitz, first that in I, because the factor $X(t)-X(x)$ cancels the singularity at $t=x$, we may replace R by $|t-x|$ so that

$$\frac{I}{4(1-\nu)} = \int_{-l}^{l} \frac{X(t)-X(x)}{|t-x|}\,\mathrm{d}t \tag{12}$$

and secondly that $X(t)$ is almost proportional to t, so that the integral (12) is approximately $-2X(x)$. (Put $X(t)=Ct$ and the integral gives $-2Cx=-2X(x)$). This gives, using (1),

$$\tau(x) = \frac{\sigma_\infty}{(1-\nu)\gamma_0} \frac{x}{\ln(l^2/r_0^2) + \ln(1-x^2/l^2) + \epsilon} \tag{13}$$

with

$$\epsilon = 2\ln 2 - 2 + 1/((1/2)-(1/2)\nu).$$

If we suppose the aspect ratio l/γ_0 to be large enough for $\ln(l^2/\gamma_0^2)$ to be considered large as well we may expand the denominator in (13) by the binomial theorem to get

$$\tau(x) = \frac{\sigma_\infty x}{(1+\nu)r_0} \frac{1}{\ln(l^2/r_0^2)} \left[1 - \frac{\ln(1-x^2/l^2)+\epsilon}{\ln(l^2/r_0^2)} + \cdots\right] \tag{14}$$

If we only retain the first term in the square brackets (2) gives

$$\sigma(x) = \frac{\sigma_\infty}{2(1+\nu)} \frac{l^2-x^2}{\gamma_0^2} \frac{1}{\ln(l^2/r_0^2)}$$

for the stress in the fibre, with a maximum

$$\sigma_{\max} = \sigma(0) = \frac{\sigma_\infty}{2(1+\nu)} \frac{l^2}{r_0^2} \frac{1}{\ln(l^2/r_0^2)}$$

which is smaller than the value given by Cherepanov by a factor $1/\ln(l^2 r_0^2)$.

Since the calculation forces us to introduce the radius r_0 of the cylinder we should strictly allow for the fact that the fibre will also inhibit the radial displacement $-\nu\sigma_\infty\gamma_0/E$ produced by σ_∞, but the corrections are unimportant. It is not hard to bolster up the above somewhat unsatisfactory analysis. Van Dyke, in an appendix to a paper by Taylor[5] has treated the same electrostatic problem as Landau and Lifshitz, and if we follow his method we are led to (14). Hallén[6] has given a lengthy treatment of a charged cylinder in an otherwise field-free space. As he himself points out, his method can be modified to treat the case of a cylinder in a uniform field. If we follow his method with the necessary modifications we are again led to (14). As one might expect, the higher, unwritten, terms in the expansion (14) so obtained do not agree with the higher terms in the expansion of the denominator in (13).

It ought perhaps to be emphasized again that the present calculation, like even the more sophisticated versions of its electrostatic prototypes, gives inaccurate results near the ends of the cylinder, so that it fails to give a firm answer, even for an inextensible fibre, to various questions of interest in the study of fibre reinforcement.

Department of the Theory of Materials
University of Sheffield
Mappin Street
Sheffield S1 3JD, England

J. D. ESHELBY

REFERENCES

1. G. P. Cherepanov, Invariant Gamma Integrals. *Engng Fracture Mech.* **14**, 39–58 (1981).
2. G. Leibfried, *Z. Phys.* **130**, 214–226 (1951).
3. A. E. H. Love, *Mathematical Theory of Elasticity*, p. 183. Cambridge University Press (1924).
4. L. D. Landau and E. M. Lifshitz, *Electrodynamics of Continuous Media*, p. 18. Pergamon Press, Oxford (1960).
5. G. I. Taylor, *Proc. R. Soc.* A291, 145–158 (1966).
6. E. Hallén, *Arkiv för Mat., Astron. och Fysik.* 21A, No. 22, 1–44 (1929).

Lectures on

The Elastic Energy-Momentum Tensor

J. D. Eshelby

University of Sheffield.

References

J.D. Eshelby

Phil. Trans. A 244 87 (1951)

Prog. Sol. State Phys. 3 ed Seitz, Academic (1956)

Symp. Int. Stress & Fatigue, ed Rassweiler, Elsevier (1959)

Inelastic Behavior of Solids, ed. Kanninen, McGraw-Hill (1970)

Prospects in Fracture Mechanics, ed. Sih, Elsevier (1975)

J. Elasticity 5 321 (1975)

C. Atkinson & J.D. Eshelby

Int. J. Fracture Mech. 4 3 (1968)

B.A. Bilby & J.D. Eshelby

Fracture 1, ed Liebowitz, Acad. (1968)

Pp. 15, 16, 17, 19, 23 unpublished.

Force on a defect.

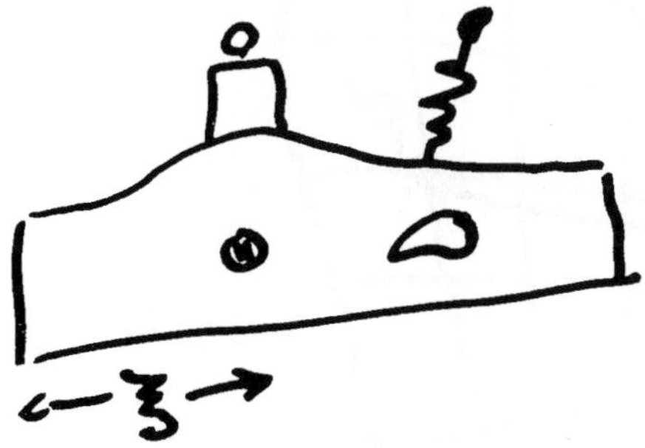

$$F_\xi = -\frac{\partial}{\partial \xi}(E_{el} + E_{load})$$

Point defect, $\underset{\sim}{F} = -\Delta V \operatorname{grad} p$

Dislocation $(b,0,0)$ $F_x = b\, p_{xy}$

$F_y = -b\, p_{xx}$

Crack tip

$F = \mathcal{G} = J$

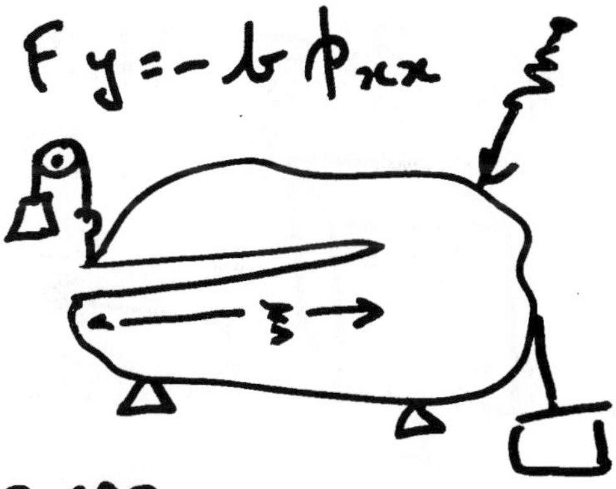

Maxwell, e.s. charge

$$F_l = \int_S \left(E_l E_j - \tfrac{1}{2} E_i E_i \delta_{lj}\right) dS_j$$

$$= \int_V (\quad)_{,j}\, dv = \int_V dv \left(E_{j,j} E_l + \cancel{E_{l,j} E_j - E_i E_{i,l}}\right)$$

j,l

$$= \int_V \rho E_l \, dv \qquad \operatorname{curl} \underset{\sim}{E} = 0$$

$$\operatorname{div} \underset{\sim}{E} = \rho$$

$$F_l = \int P_{lj}\, dS_j, \quad P_{lj} = E_l E_j - \tfrac{1}{2} E_i E_i \delta_{lj}$$

Lagrangian coordinates & stress

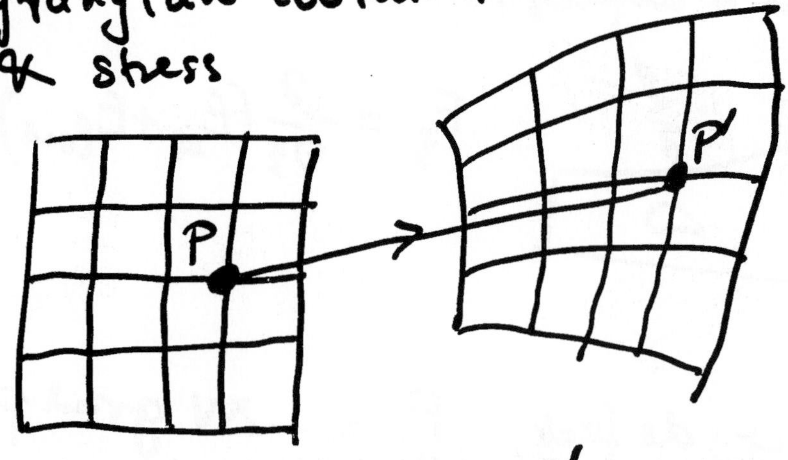

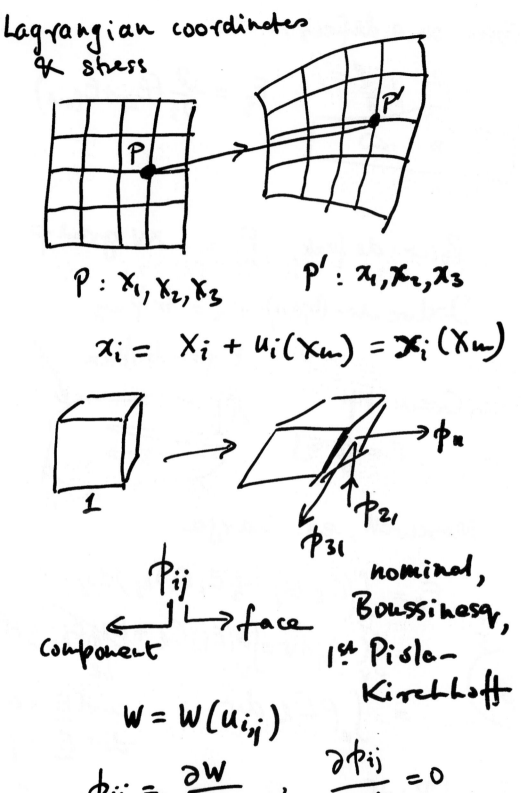

$P : X_1, X_2, X_3$ $\qquad$ $P' : x_1, x_2, x_3$

$$x_i = X_i + u_i(X_m) = x_i(X_m)$$

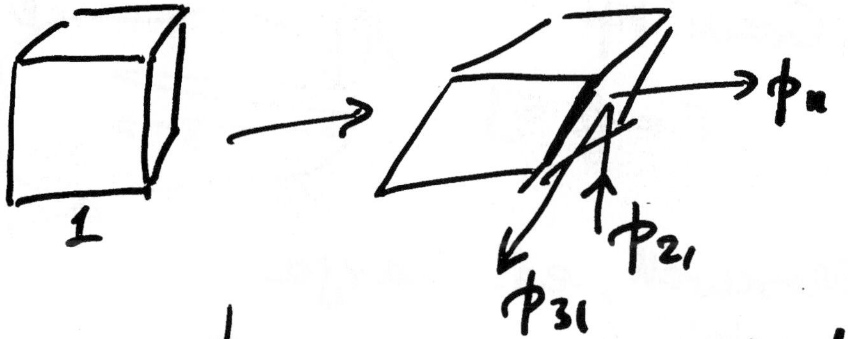

ϕ_{ij} — i → component, j → face

nominal, Boussinesq, 1st Piola-Kirchhoff

$$W = W(u_{i,j})$$

$$\phi_{ij} = \frac{\partial W}{\partial u_{i,j}}, \qquad \frac{\partial \phi_{ij}}{\partial X_j} = 0$$

Symmetry relations

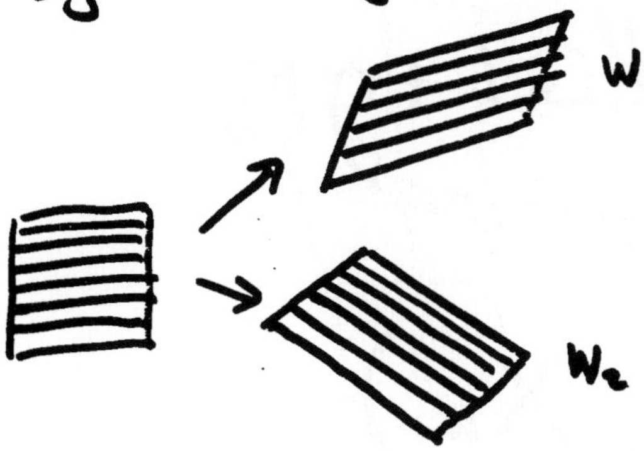

$$W_1 = W_2$$

$$\frac{\partial W}{\partial u_{j,i}} x_{l,i} = \frac{\partial W}{\partial u_{l,i}} x_{j,i}$$

$$p_{lj} - p_{jl} = p_{ji} u_{l,i} - p_{li} u_{j,i} \qquad (1)$$

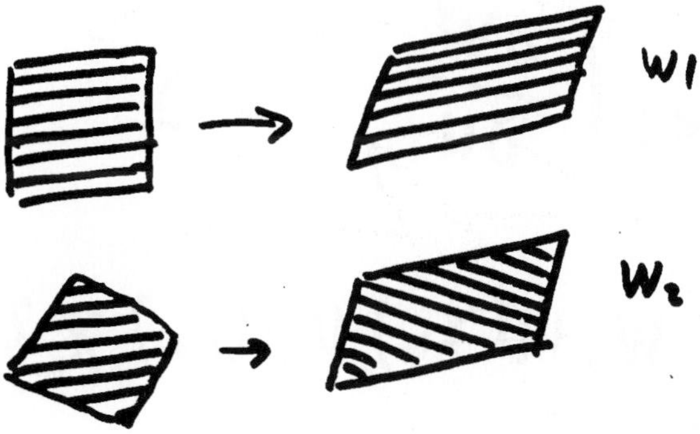

$W_1 \neq W_2$ except for isotropy; then

$$\frac{\partial W}{\partial u_{i,j}} x_{i,l} = \frac{\partial W}{\partial u_{i,l}} x_{i,j}$$

i.e. $$p_{lj} - p_{jl} = p_{il} u_{i,j} - p_{ij} u_{i,l} \qquad (2)$$

or, by (1) $$p_{ij} u_{i,l} - p_{il} u_{i,j} = p_{li} u_{j,i} - p_{ji} u_{l,i} \qquad (3)$$

$$= (p_{li} u_j - p_{ji} u_l)_{,i}$$

For linear theory (1), (2) invalid, (3) O.K.

Ordinary and explicit derivatives

Emmy Noether 1918

$$W = W(u_i,\ u_{i,j},\ x_m)$$

$$u_{i,j} = \frac{\partial u_i}{\partial x_j}$$

$$\frac{\partial W}{\partial u_i} - \frac{\partial}{\partial x_j}\left(\frac{\partial W}{\partial u_{i,j}}\right) = 0$$

$$\frac{\partial W}{\partial x_l} = \text{grad}_l\, W = \frac{\partial}{\partial x_l} W\{u_i(x_m), u_{i,j}(x_m), x_m\}$$

$$\left(\frac{\partial W}{\partial x_l}\right)_{\text{exp.}} = \frac{\partial}{\partial x_l} W(u_i, u_{i,j}, x_m)\Big|_{\substack{u_i,\ u_{i,j} \\ x_m,\ m \neq l \\ \text{all const.}}}$$

$$\frac{\partial W}{\partial u_i} = \frac{\partial}{\partial x_j}\left(\frac{\partial W}{\partial u_{i,j}}\right)$$

$$\frac{\partial W}{\partial x_l} = \frac{\partial W}{\partial u_i}\frac{\partial u_i}{\partial x_l} + \frac{\partial W}{\partial u_{i,j}}\frac{\partial u_{i,j}}{\partial x_l} + \left(\frac{\partial W}{\partial x_l}\right)_{\text{exp.}}$$

$$\frac{\partial}{\partial x_j}\left(\frac{\partial W}{\partial u_{i,j}}\right)u_{i,l} + \frac{\partial W}{\partial u_{i,j}}\frac{\partial}{\partial x_j}u_{i,l}$$

$$\frac{\partial}{\partial x_j}(W\delta_{lj}) = \frac{\partial}{\partial x_j}\left(\frac{\partial W}{\partial u_{i,j}}u_{i,l}\right) \qquad + \left(\frac{\partial W}{\partial x_l}\right)_{\text{exp.}}$$

$$\frac{\partial}{\partial x_j}\left(W\delta_{lj} - \frac{\partial W}{\partial u_{i,j}}u_{i,l}\right) = \left(\frac{\partial W}{\partial x_l}\right)_{\text{exp.}}$$

$$\frac{\partial P_{lj}}{\partial x_j} = \left(\frac{\partial W}{\partial x_l}\right)_{\text{exp}}$$

$$\boxed{P_{lj} = W\delta_{lj} - \frac{\partial W}{\partial u_{i,j}}u_{i,l}}$$

$$\int_S P_{lj}\, dS_j = \int_V \left(\frac{\partial W}{\partial x_l}\right)_{\text{exp}} dv \quad \text{generally}$$

$$= 0 \quad \text{for homogeneity}$$

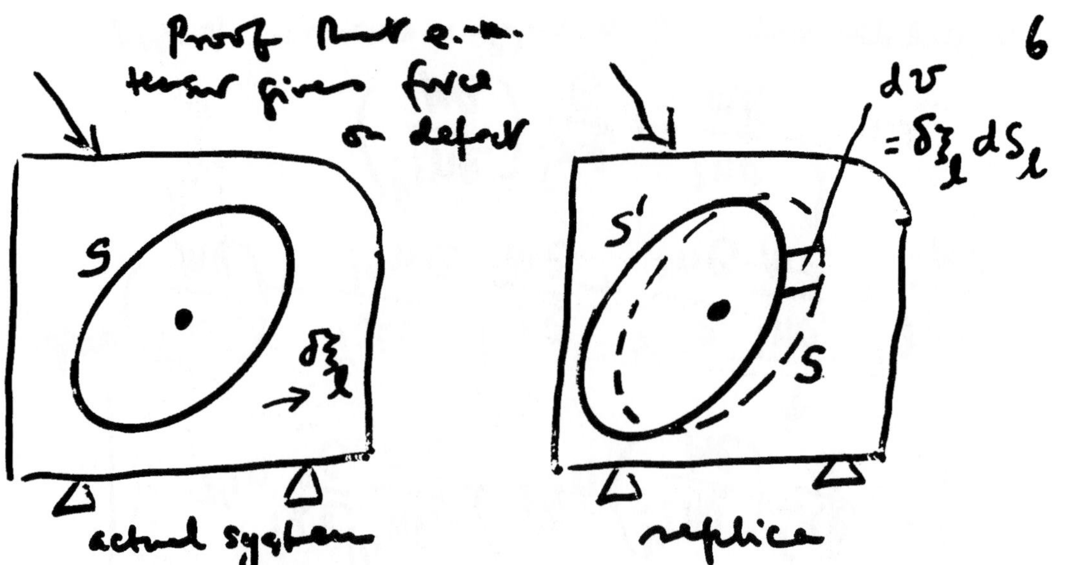

$$\delta E_{I} = -\delta\xi_l \int_S W\, dS_l + O(\delta\xi^2)$$

$$\delta E_{II} = -\int_S \delta \underset{\sim}{u} \cdot \underset{\sim}{T}\, dS + O(\delta\xi^2)$$

$$\left(\delta \underset{\sim}{u} = -\frac{\partial \underset{\sim}{u}}{\partial x_l} \delta\xi_l \right)$$

$$\left[\delta E_{III} = O(\delta\xi^2) \right]$$

$$\delta E_{tot} = \delta E_{I} + \delta E_{II} + \delta E_{III}$$

$$= -\delta\xi_l F_l + O(\delta\xi^2)$$

$$F_l = \int_S \left(W n_l - \frac{\partial \underset{\sim}{u}}{\partial x_l} \cdot \underset{\sim}{T} \right) dS$$

$$= \int_S P_{lj}\, dS_j ,$$

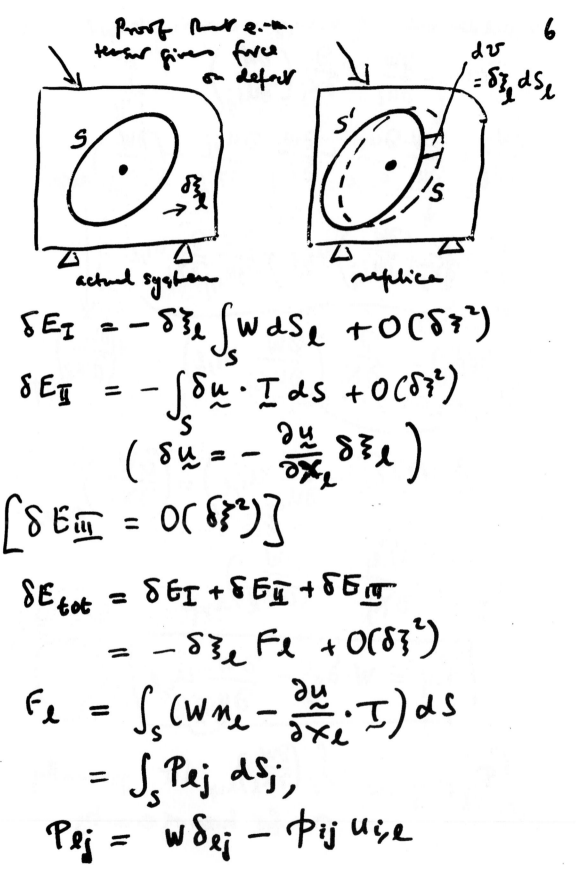

$$P_{lj} = W \delta_{lj} - p_{ij} u_{i,l}$$

1

Trouser test

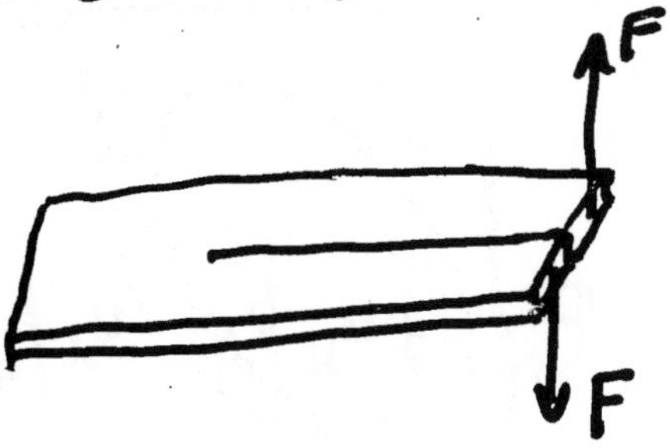

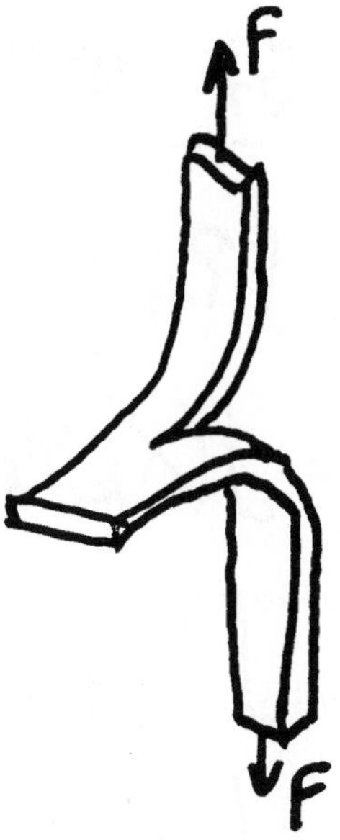

$$\tfrac{1}{2}F_1 = \int (W - p_{i1}\, u_{i,1})\, dS$$

$$= WA - F\,\frac{\partial u_2}{\partial x_1}$$

$$(X_1, X_2 = 0) \to (x_1 = 0, X_2 = x_1) \to (x_1 = 0, x_2 = X_1 \tfrac{l}{l_0})$$

$$u_2 = x_2 - X_2 = \frac{l}{l_0} X_1 - X_2$$

$$\mathcal{G} t = -F_1 = 2F \frac{l}{l_0} - 2WA$$

$$t\mathcal{G} = \frac{M^2(0)}{EI}$$ Buckling in compression

$$= 8(P - P_{Eul.})$$

Günther 1962

$$f_l = \int P_{lj}\, dS_j$$

$$L_{kl} = \int (x_k P_{lj} - x_l P_{kj} + u_k p_{lj} - u_l p_{kj})\, dS_j$$

$$M = \int (x_l P_{lj} - \tfrac{1}{2} u_l p_{lj})\, dS_j \qquad 3D$$

$$= \int x_l P_{lj}\, dS_j \qquad 2D$$

$$F_2 = \left.\frac{\partial F_1}{\partial \alpha}\right|_{\alpha=0} = \left.\frac{\partial \mathcal{G}}{\partial \alpha}\right|_{\alpha=0}$$

$$\mathcal{G} = \frac{1-\nu}{2\mu} \frac{P^2}{a} \left[\frac{1}{\beta + \sin\beta} - \frac{1}{\pi} \right]$$

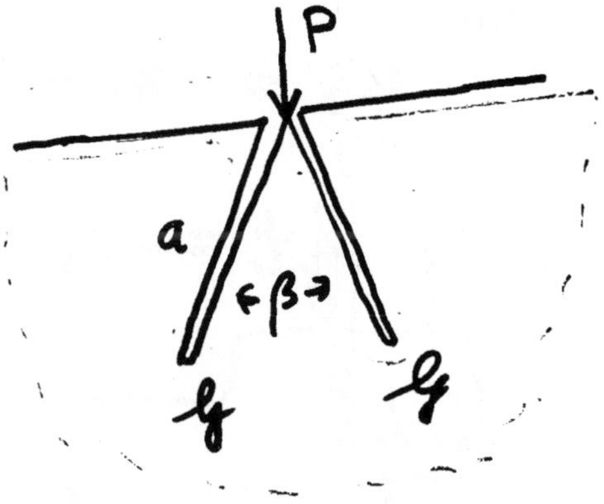

$$\mathcal{G} = \frac{1-\nu}{2\mu} K^2$$

$$K = \frac{2}{\sqrt{a}} \sqrt{\frac{\pi}{\pi^2 - 4}} \left(Q - \frac{2}{\pi} P\right)$$

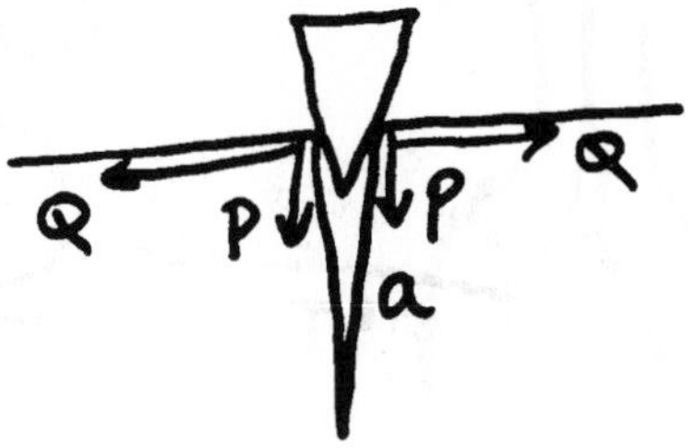

Treatment of previous problem by beam theory

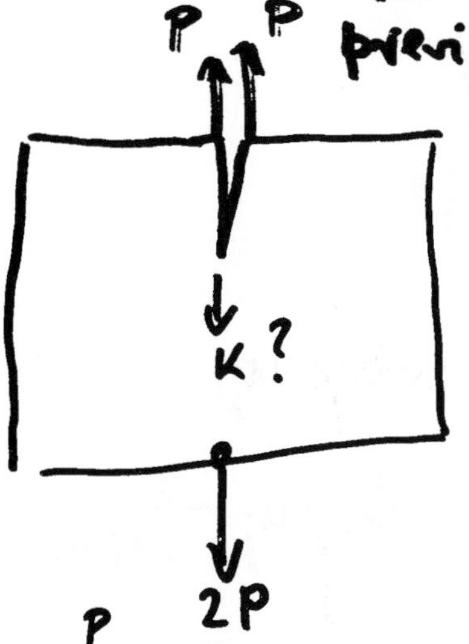

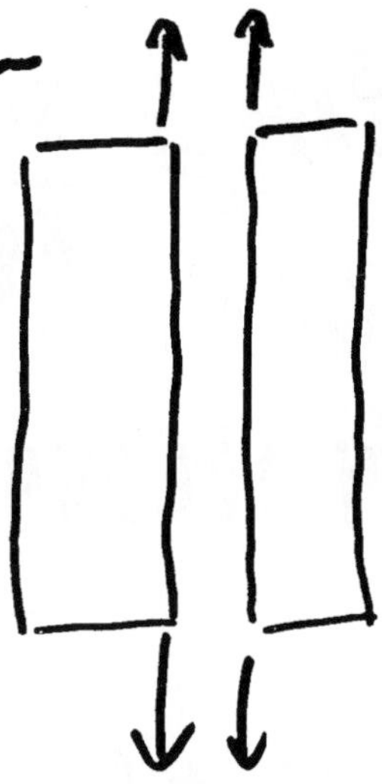

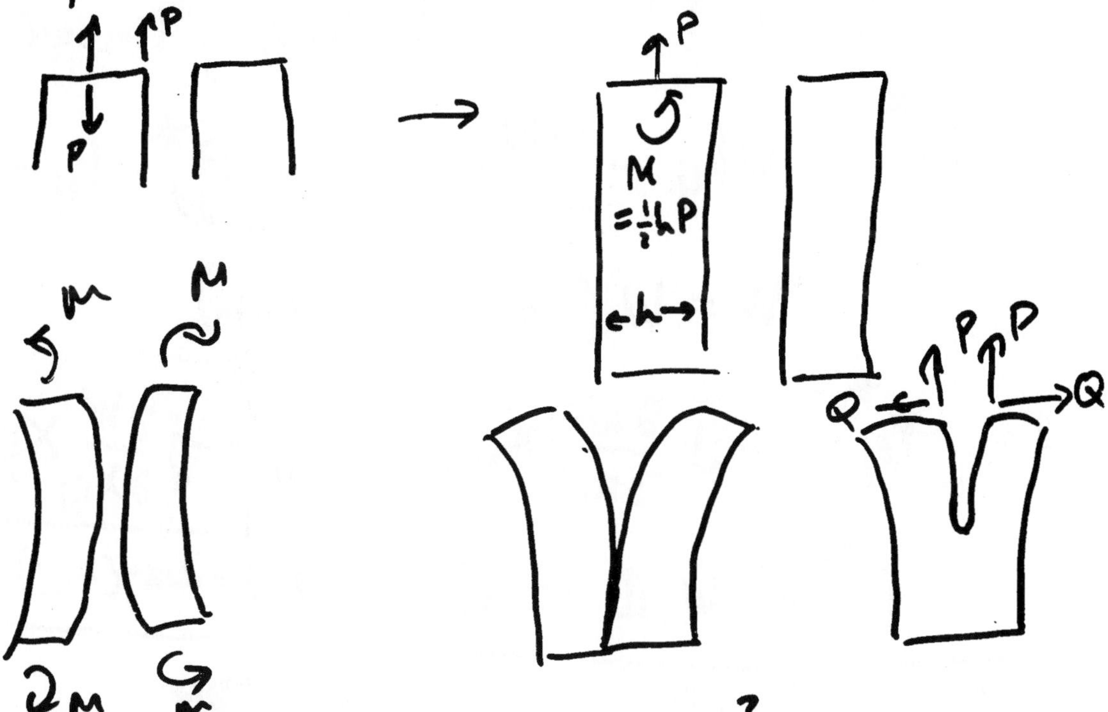

$$g = \frac{M^2(0)}{E'I} = \frac{\left(Qa + \frac{h}{2}P\right)^2}{E' \cdot \frac{1}{12}h^3} = \frac{K^2}{E'}$$

$$K = 2\sqrt{3}\left(\frac{a}{h}\right)^{3/2}\frac{1}{a^{1/2}}\left(Q + \frac{h}{2a}P\right)$$

Symmetrization
η e.-m.
tensor

$$p^{S\,lj} = \frac{\delta W}{\delta g_{lj}} + \frac{\delta W}{\delta g_{jl}}$$

$$p^{S}_{lj} = P_{lj} + \tfrac{1}{2}(\phi_{ij}u_i - \phi_{ji}u_i + \phi_{il}u_j - \phi_{jl}u_i + \phi_{li}u_j - \phi_{lj}u_i)_{,i}$$

$$= P_{lj} + (\phi_{li}u_j - \phi_{lj}u_i)_{,i}, \quad \text{linear}$$

$$P^{*}_{lj} = W\delta_{lj} - \frac{\partial W}{\partial x_{i,j}} x_{i,l} = P^{*}_{jl} = P_{lj} - \phi_{lj}$$

$$W = W(x_i, x_{i,j}, X_m)$$

$$P^{*}_{lj} = -J\frac{\partial w}{\partial x_{i,j}} x_{i,l} \quad ; \quad \sigma_{lj} = -j\frac{\partial W}{\partial X_{i,j}} X_{i,l}$$

$$P^{*} = W\,\mathbb{I} - C\cdot S \qquad J = \det(x_{i,j}) = j^{-1}$$

see p. 13

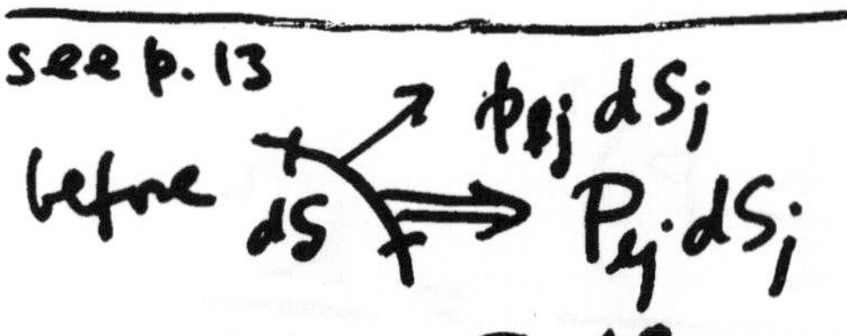

$$\Sigma_{lj} = w\delta_{lj} - \frac{\partial w}{\partial u_{i,j}} u_{i,l} \sim P_{lj}$$

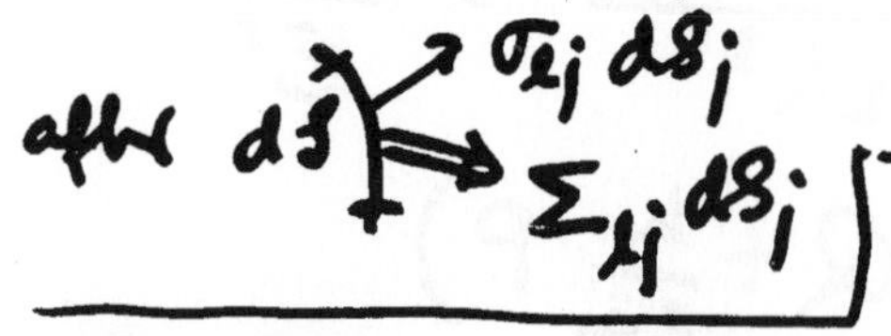

$$\sigma_{lj} = \frac{\partial w}{\partial u_{l,j}} + \Sigma_{lj} \sim P^{*}_{lj}$$

$$\frac{\partial}{\partial x_j}\left(\frac{\partial w}{\partial u_{l,j}}\right) = \frac{\partial w}{\partial u_l} \; ; \quad \frac{\partial \Sigma_{lj}}{\partial x_j} = \left(\frac{\partial w}{\partial x_l}\right)_{exp} \; ; \quad \left(\frac{\partial w}{\partial x_l}\right)_{exp} + \frac{\partial w}{\partial u_l} = 0$$

A more elaborate Günther-type integral

$$\int (X_k X_l - \tfrac{1}{2} X_i X_i \, \delta_{kl}) T_{lj} \, dS_j$$

$$= \int \left[X_l (T_{lk} - T_{kl}) + X_k T_{ll} + (X_k X_l - \tfrac{1}{2} X_i X_i) \frac{\partial T_{lj}}{\partial x_j} \right] dv$$

$= 0$ if

ⓐ $\frac{\partial T_{lj}}{\partial x_j} = 0$, ⓑ $T_{lj} = T_{jl}$, ⓒ $T_{ll} = 0$

For isotropy $P^*_{lj} = W \delta_{lj} - \frac{\partial W}{\partial x_{i,j}} x_{i,l}$

satisfies ⓐ, ⓑ; also ⓒ if

$$W \delta_{ll} = \frac{\partial W}{\partial x_{i,j}} x_{i,j}$$

i.e. if W is homogeneous

of degree $\frac{3}{2}$ in $\frac{3}{2}$ dimensions.

Equivalence of P_{ij} to an ordinary stress in 2D infinitesimal isotropic elasticity (plane strain)

$$\frac{\partial P_{ij}}{\partial x_j} = 0, \quad P_{11} + P_{22} = 0, \quad P_{12} \neq P_{21}$$

$$\nabla^2 (P_{12} - P_{21}) = 0$$

$$p'_{ij} = \varepsilon_{il3} P_{lj}, \quad P_{ij} = -\varepsilon_{il3} p'_{lj}$$

$$p'_{11} = P_{21}, \quad p'_{22} = -P_{12}$$

$$p'_{12} = P_{22}, \quad p'_{21} = -P_{11}$$

$$\frac{\partial p'_{ij}}{\partial x_j} = 0, \quad p'_{ij} = p'_{ji}, \quad \nabla^2 (p'_{11} + p'_{22}) = 0$$

So p'_{ij} is an equilibrated compatible ordinary stress.

$$2\mu (u_1 + i u_2) = (3 - 4\nu) \varphi(z) - z \overline{\varphi'(z)} - \overline{\psi(z)}$$

For $p'_{ij} \sim P_{lj}$ $\quad \varphi, \psi \rightarrow \Phi, \Psi$:

$$\Phi' = (1-\nu) \frac{\varphi'^2}{i\mu}, \quad \Psi' = a(1-\nu) \frac{\psi' \varphi'}{i\mu}$$

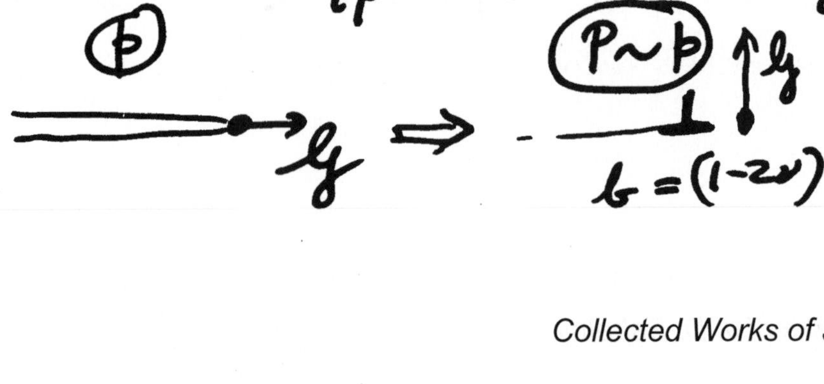

$b = (1-2\nu) g / 2\mu$

Euler coordinates & stress

$$w = w(u_i, u_{i,j}, x_i)$$

$$u_i = x_i - X_i = x_i - X_i(x_k), \text{ not } x_i(X_k) - X_i$$

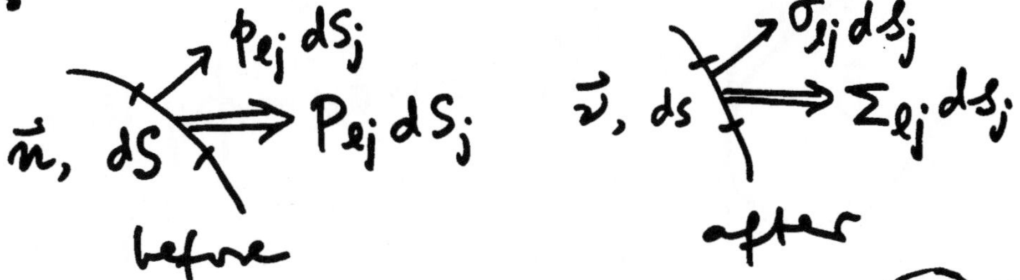

$$F_l = \int_S P_{lj}\,dS_j = \int_s \Sigma_{lj}\,ds_j$$

$$\Sigma_{lj} = w\delta_{lj} - \frac{\partial w}{\partial u_{i,j}} u_{i,l} \sim P_{lj}$$

$$\sigma_{lj} = \frac{\partial w}{\partial u_{l,j}} + \Sigma_{lj} \sim P^*_{lj} \; ; \; \sigma_{lj} = w\delta_{lj} - \frac{\partial w}{\partial X_{i,j}} X_{i,l}$$

$$\frac{\partial \Sigma_{lj}}{\partial x_j} = \left(\frac{\partial w}{\partial x_l}\right)_{exp}$$

$$\frac{\partial}{\partial x_j}\left(\frac{\partial w}{\partial u_{l,j}}\right) = \frac{\partial w}{\partial u_l}$$

$$\frac{\partial}{\partial x_j}\left(\frac{\partial w}{\partial u_{l,j}} + \Sigma_{lj}\right) = \frac{\partial}{\partial x_j}(\sigma_{lj}) = \left(\frac{\partial w}{\partial x_l}\right)_{exp} + \frac{\partial w}{\partial u_l} = 0$$

because W depends on x_l, u_l only thru $X_l = x_i - u_i$

Effective force on an interface

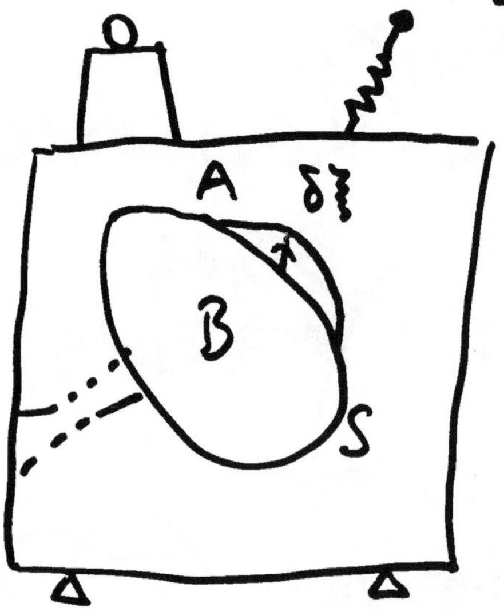

$$\delta E_{tot} = -\int_S \delta\xi_l \, [P_{lj}] \, dS_j$$

$$P_{lj} = W\delta_{lj} - p_{ij} u_{i,l}$$

$$[\;] = (\;)_A - (\;)_B$$

$$dF_l = [P_{lj}] n_j \, dS$$

$$[u_{i,l}] = \left[\frac{\partial u_i}{\partial n}\right] n_l \,, \quad [p_{ij} n_j] = 0$$

so

$$F_l = F \cdot n_l$$

$$F = \left[W - \underset{\sim}{T} \cdot \frac{\partial \underset{\sim}{u}}{\partial n} \right]$$

$$F = \tfrac{1}{2}\left(p_{ij}^{out} + p_{ij}^{in}\right) e^T_{ij}$$

Gavazza's formula

Case where W depends on higher derivatives of field quantities A.

$$W = W(A, A_{,i}, A_{,ij} \dots x)$$

$$\frac{\delta W}{\delta A} \overset{\text{def}}{=} \frac{\partial W}{\partial A} - \left(\frac{\partial W}{\partial A_{,j}}\right)_{,j} + \left(\frac{\partial W}{\partial A_{,jk}}\right)_{,jk} \cdots \pm \left(\frac{\partial W}{\partial A_{,jk\dots m}}\right)_{,jk\dots m} \mp \dots$$

variational derivative

$$= \frac{\partial W}{\partial A} - \left(\frac{\delta W}{\delta A_{,j}}\right)_{,j}$$

$$\boxed{\frac{\partial W}{\partial A} = \frac{\delta W}{\delta A} + \left(\frac{\delta W}{\delta A_{,j}}\right)_{,j}}$$

$$\delta W = \frac{\partial W}{\partial A}\delta A + \frac{\partial W}{\partial A_{,j}}\delta A_{,j} + \frac{\partial W}{\partial A_{,jk}}\delta A_{,jk} + \cdots$$

$$= \frac{\delta W}{\delta A}\delta A + \left(\frac{\delta W}{\delta A_{,j}}\right)\delta A_{,j} + \frac{\delta W}{\partial A_{,jk}}\delta A_{,jk}$$

$$+ \left(\frac{\delta W}{\delta A_{,j}}\right)_{,j}\delta A + \left(\frac{\delta W}{\delta A_{,jk}}\right)_{,k}\delta A_{,j} + \left(\frac{\delta W}{\delta A_{,jkl}}\right)_{,l}\delta A_{,jk}$$

so $\delta W = \frac{\delta W}{\delta A}\delta A + \left(Q_j[\delta A]\right)_{,j}$

where

$$Q_j[\delta A] = \frac{\delta W}{\delta A_{,j}}\delta A + \frac{\delta W}{\delta A_{,jk}}\delta A_{,k} + \frac{\delta W}{\delta A_{,jkm}}\delta A_{,km} + \cdots$$

and

$$\delta\int W\,dv = \int \delta W\,dv$$

$$= \int \frac{\delta W}{\delta A}\delta A\,dv + \int Q_j[\delta A]\,dS_j$$

$$\varepsilon\frac{\partial W}{\partial x_l} = \delta W[\delta A = \varepsilon A_{,l}] + \varepsilon\left(\frac{\partial W}{\partial x_l}\right)_{exp}$$

$$= \frac{\delta W}{\delta A}\varepsilon A_{,l} + Q_{j,j}[\delta A = \varepsilon A_{,l}]$$

$$\frac{\partial P_{lj}}{\partial x_j} = \frac{\delta W}{\delta A}A_{,l} + \left(\frac{\partial W}{\partial x_l}\right)_{exp}$$

Analog of argument on p. 6 gives

$$P_{lj} = W\delta_{lj} - Q_j\,[\delta A = A_{,l}]$$

$$= W\delta_{lj} - \frac{\delta W}{\delta A_{,j}} A_{,l} - \frac{\delta W}{\delta A_{,jk}} A_{,lk}$$

$$\cdots - \frac{\delta W}{\delta A_{,jk\cdots p}} A_{,lk\cdots p} \cdots$$

$$\boxed{A \to u_i}$$

$$P_{lj} = W\delta_{lj} - \frac{\delta W}{\delta u_{i,j}} u_{i,l} - \frac{\delta W}{\delta u_{i,pj}} u_{i,pl}$$

$$\cdots - \frac{\delta W}{\delta u_{i,p\cdots sj}} u_{i,p\cdots sl} \cdots$$

$$\frac{\partial P_{lj}}{\partial x_j} = \frac{\delta W}{\delta u_i} u_{i,l} + \left(\frac{\partial W}{\partial x_l}\right)_{exp}$$

Examples of
Grade 2 situation

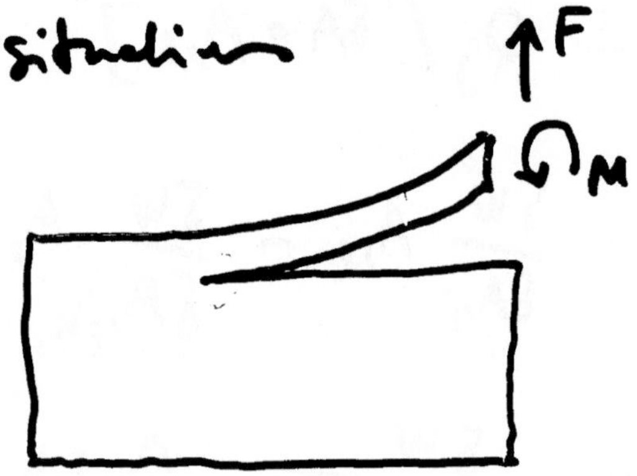

$y = u_2$

$y' = u_{2,1}$

$y'' = u_{2,11}$

...

Obreimoff $\quad 2\gamma = \mathcal{G} = \dfrac{M^2(0)}{2E'I}$

J-integral $\quad \mathcal{G} = -\frac{1}{2} M y'' + F y'$

Beam as Cosserat continuum with constrained rotation:

$$W = \tfrac{1}{2} E' I (y'')^2$$

$$\mathcal{G} = \tfrac{1}{2} E' I (y'')^2 - E' I y''' y' \,;$$

all of these are the same.

Plate flexure

$A \to w$

$$W = \frac{Eh^3}{24(1-\nu^2)}\left[(\nabla^2 w)^2 + 2(1-\nu)\left(\left(\frac{\partial^2 w}{\partial x \partial y}\right)^2 - \frac{\partial^2 w}{\partial x^2}\frac{\partial^2 w}{\partial y^2}\right)\right]$$

$F_1 = \int P_{1j}\, dS_j$ = force on crack tip

$= \int \mathcal{G}(z)\, dz$

Plane strain

$A \to \chi$

$$W = \frac{1-\nu}{4\mu}\left[(\nabla^2 \chi)^2 + \frac{2}{1-\nu}\left(\left(\frac{\partial^2 \chi}{\partial x \partial y}\right)^2 - \frac{\partial^2 \chi}{\partial x^2}\frac{\partial^2 \chi}{\partial y^2}\right)\right]$$

$F_1 = \int P_{1j}\, dS_j = ?$ because we're not working with a displacement.

4D generalization of e.-m. tensor

$$X_1, X_2, X_3, X_0 = t, \quad (u_1, u_2, u_3), \quad u_{i,0} = \dot{u}_i$$

$$L = T - W(u_{i,j}), \quad T = \rho \dot{u}_i \dot{u}_i$$

$$\frac{\partial}{\partial x_j}\left(\frac{\partial L}{\partial u_{i,j}}\right) + \frac{\partial}{\partial t}\left(\frac{\partial L}{\partial \dot{u}_i}\right) = 0$$

	0	1 2 3
0	H	s_j
1, 2, 3	$-g_l$	P_{lj}

$$P_{lj} = (W - T)\delta_{lj} - p_{ij} u_{i,l}; \quad l, j = 1, 2, 3$$

$$s_j = P_{0j} = -p_{ij}\dot{u}_i, \quad H = W + T$$

$$g_l = -P_{l0} = -\rho \dot{u}_i u_{i,l} \neq G_l = \rho \dot{u}_l$$

$$g_l = G_l - \pi_l, \quad \pi_l = \frac{\partial L}{\partial \dot{u}_l}$$

g_l ↓ quasi-momentum; G_l → ordinary momentum; π_l → canonical momentum

$$\frac{\partial \omega}{\partial k_j} = \frac{s_j}{H} = (v_{g})_j, \quad \frac{\omega}{k_l} g_l = \text{"}\frac{1}{(v_{ph})_j}\text{"} = \frac{(v_{ph})_j}{|v_{ph}|^2} = \text{slowness}$$

$$\vec{g} = \frac{H}{c_l}\hat{k}, \quad \vec{g} = \frac{H}{c_t}\hat{k}, \text{ pure } l \text{ or } t.$$

↑ for anisotropic linear solid (for $\frac{\partial \omega}{\partial k_j}$)

↑ for linear isotropic solid (for $\vec{g} = \frac{H}{c_l}\hat{k}$, $\vec{g} = \frac{H}{c_t}\hat{k}$)

$$\frac{\partial p_{lj}}{\partial x_j} - \frac{\partial G_l}{\partial t} = -f_l \qquad \vec{f}, \text{ external force}$$

$$\frac{\partial P_{lj}}{\partial x_j} - \frac{\partial g_l}{\partial t} = f_n u_{n,l}$$

$$\frac{\partial (P_{lj} - p_{lj})}{\partial x_j} + \frac{\partial}{\partial t}(G_l - g_l) = f_n \cdot (\delta_{nl} + u_{n,l})$$

$$= f_n \frac{\partial x_n}{\partial X_l}$$

$$= \frac{\partial U}{\partial X_l}$$

if $U = U[u_{r,s}(x_i), x_i]$

e.g. $U = C \operatorname{div} \vec{u} \cdot \rho(\vec{x})$, $\rho = \psi^* \psi$

↖ electron interacting with crystal via 'deformation potential'

Then

$$\frac{d}{dt} \int (G_l - g_l) \, dv = \int (p_{lj} - P_{lj} + U \delta_{lj}) \, dS_j$$

and $\frac{d}{dt} \int G_l \, dv = \frac{d}{dt} \int g_l \, dv$

for, e.g., periodic boundary conditions.

$$\frac{\partial P_{1j}}{\partial x_j} = \frac{\partial g_1}{\partial t} + \left(\frac{\partial W}{\partial x_1}\right)_{exp}$$

$$\int P_{1j}\, dS_j - \frac{d}{dt}\int g_1\, dv = \int \left(\frac{\partial W}{\partial x_1}\right)_{exp} dv \overset{?}{\doteq} F_1$$

cf $\boxed{v\frac{\partial}{\partial x_1} \simeq -\frac{\partial}{\partial t}}$ can absorb g_1 - term into $\int \cdots dS_j$;

$$\frac{\partial g_1}{\partial t} \simeq -v\frac{\partial g_1}{\partial x_1} = -v\frac{\partial}{\partial x_1}(-\rho \dot{u}_i\, u_{i,1})$$

$$\simeq -\frac{\partial}{\partial x_1}(\rho \dot{u}_i \dot{u}_i) = -\frac{\partial}{\partial x_1}(2T)$$

$$-\frac{d}{dt}\int g_1\, dv = -\int \frac{\partial}{\partial t} g_1\, dv = \int 2T\, dS_1$$

$$F_1 \overset{?}{\doteq} \int [(W - T + 2T)\delta_{1j} - p_{ij}\, u_{i,1}]\, dS_j$$

$$\overset{?\checkmark}{\simeq} \int \left(E\, n_1 - \vec{T}\cdot\frac{\partial \vec{u}}{\partial x_1}\right) dS$$

$$v F_1 \simeq \int (v E\, n_1 + \vec{T}\cdot\dot{\vec{u}})\, dS$$

23

$\Delta\dot{v} = vl(-x)^{-1/2}$

$\sigma = kx^{-1/2}$

Alternate way to find $\mathcal{G}$ for moving crack

$$v\mathcal{G} = \int \Delta\dot{v}\,\sigma\,dx$$

Smear technique

$$\sigma = \int f(\xi)\,k\,(x-\xi)^{-1/2}\,d\xi$$

$$\Delta\dot{v} = \int f(\xi')\,vl\,(-x+\xi')\,d\xi'$$

$$\int f(\xi)\,d\xi = 1\,,\quad f(\xi) \to \delta(\xi)$$

$$v\mathcal{G} = \int f(\xi)d\xi \int f(\xi')d\xi' \underbrace{\int_{-\xi}^{\xi'} dx(-x+\xi')^{-1/2}(x+\xi)^{-1/2}}_{\pi} \cdot vkl$$

$$v\mathcal{G} = \pi vkl \underbrace{\int f(\xi)\,d\xi}_{1} \underbrace{\int f(\xi')d\xi'}_{1}$$

✓

$= \pi vkl$

Mechanics

SOLID MECHANICS AND ITS APPLICATIONS

Series Editor: G.M.L. Gladwell

Aims and Scope of the Series

The fundamental questions arising in mechanics are: *Why?*, *How?*, and *How much?* The aim of this series is to provide lucid accounts written by authoritative researchers giving vision and insight in answering these questions on the subject of mechanics as it relates to solids. The scope of the series covers the entire spectrum of solid mechanics. Thus it includes the foundation of mechanics; variational formulations; computational mechanics; statics, kinematics and dynamics of rigid and elastic bodies; vibrations of solids and structures; dynamical systems and chaos; the theories of elasticity, plasticity and viscoelasticity; composite materials; rods, beams, shells and membranes; structural control and stability; soils, rocks and geomechanics; fracture; tribology; experimental mechanics; biomechanics and machine design.

1. R.T. Haftka, Z. Gürdal and M.P. Kamat: *Elements of Structural Optimization.* 2nd rev.ed., 1990 ISBN 0-7923-0608-2
2. J.J. Kalker: *Three-Dimensional Elastic Bodies in Rolling Contact.* 1990 ISBN 0-7923-0712-7
3. P. Karasudhi: *Foundations of Solid Mechanics.* 1991 ISBN 0-7923-0772-0
4. *Not published*
5. *Not published.*
6. J.F. Doyle: *Static and Dynamic Analysis of Structures.* With an Emphasis on Mechanics and Computer Matrix Methods. 1991 ISBN 0-7923-1124-8; Pb 0-7923-1208-2
7. O.O. Ochoa and J.N. Reddy: *Finite Element Analysis of Composite Laminates.* ISBN 0-7923-1125-6
8. M.H. Aliabadi and D.P. Rooke: *Numerical Fracture Mechanics.* ISBN 0-7923-1175-2
9. J. Angeles and C.S. López-Cajún: *Optimization of Cam Mechanisms.* 1991 ISBN 0-7923-1355-0
10. D.E. Grierson, A. Franchi and P. Riva (eds.): *Progress in Structural Engineering.* 1991 ISBN 0-7923-1396-8
11. R.T. Haftka and Z. Gürdal: *Elements of Structural Optimization.* 3rd rev. and exp. ed. 1992 ISBN 0-7923-1504-9; Pb 0-7923-1505-7
12. J.R. Barber: *Elasticity.* 1992 ISBN 0-7923-1609-6; Pb 0-7923-1610-X
13. H.S. Tzou and G.L. Anderson (eds.): *Intelligent Structural Systems.* 1992 ISBN 0-7923-1920-6
14. E.E. Gdoutos: *Fracture Mechanics.* An Introduction. 1993 ISBN 0-7923-1932-X
15. J.P. Ward: *Solid Mechanics.* An Introduction. 1992 ISBN 0-7923-1949-4
16. M. Farshad: *Design and Analysis of Shell Structures.* 1992 ISBN 0-7923-1950-8
17. H.S. Tzou and T. Fukuda (eds.): *Precision Sensors, Actuators and Systems.* 1992 ISBN 0-7923-2015-8
18. J.R. Vinson: *The Behavior of Shells Composed of Isotropic and Composite Materials.* 1993 ISBN 0-7923-2113-8
19. H.S. Tzou: *Piezoelectric Shells.* Distributed Sensing and Control of Continua. 1993 ISBN 0-7923-2186-3
20. W. Schiehlen (ed.): *Advanced Multibody System Dynamics.* Simulation and Software Tools. 1993 ISBN 0-7923-2192-8
21. C.-W. Lee: *Vibration Analysis of Rotors.* 1993 ISBN 0-7923-2300-9
22. D.R. Smith: *An Introduction to Continuum Mechanics.* 1993 ISBN 0-7923-2454-4
23. G.M.L. Gladwell: *Inverse Problems in Scattering.* An Introduction. 1993 ISBN 0-7923-2478-1

Mechanics

SOLID MECHANICS AND ITS APPLICATIONS

Series Editor: G.M.L. Gladwell

24. G. Prathap: *The Finite Element Method in Structural Mechanics*. 1993 ISBN 0-7923-2492-7
25. J. Herskovits (ed.): *Advances in Structural Optimization*. 1995 ISBN 0-7923-2510-9
26. M.A. González-Palacios and J. Angeles: *Cam Synthesis*. 1993 ISBN 0-7923-2536-2
27. W.S. Hall: *The Boundary Element Method*. 1993 ISBN 0-7923-2580-X
28. J. Angeles, G. Hommel and P. Kovács (eds.): *Computational Kinematics*. 1993 ISBN 0-7923-2585-0
29. A. Curnier: *Computational Methods in Solid Mechanics*. 1994 ISBN 0-7923-2761-6
30. D.A. Hills and D. Nowell: *Mechanics of Fretting Fatigue*. 1994 ISBN 0-7923-2866-3
31. B. Tabarrok and F.P.J. Rimrott: *Variational Methods and Complementary Formulations in Dynamics*. 1994 ISBN 0-7923-2923-6
32. E.H. Dowell (ed.), E.F. Crawley, H.C. Curtiss Jr., D.A. Peters, R. H. Scanlan and F. Sisto: *A Modern Course in Aeroelasticity*. Third Revised and Enlarged Edition. 1995 ISBN 0-7923-2788-8; Pb: 0-7923-2789-6
33. A. Preumont: *Random Vibration and Spectral Analysis*. 1994 ISBN 0-7923-3036-6
34. J.N. Reddy (ed.): *Mechanics of Composite Materials*. Selected works of Nicholas J. Pagano. 1994 ISBN 0-7923-3041-2
35. A.P.S. Selvadurai (ed.): *Mechanics of Poroelastic Media*. 1996 ISBN 0-7923-3329-2
36. Z. Mróz, D. Weichert, S. Dorosz (eds.): *Inelastic Behaviour of Structures under Variable Loads*. 1995 ISBN 0-7923-3397-7
37. R. Pyrz (ed.): *IUTAM Symposium on Microstructure-Property Interactions in Composite Materials*. Proceedings of the IUTAM Symposium held in Aalborg, Denmark. 1995 ISBN 0-7923-3427-2
38. M.I. Friswell and J.E. Mottershead: *Finite Element Model Updating in Structural Dynamics*. 1995 ISBN 0-7923-3431-0
39. D.F. Parker and A.H. England (eds.): *IUTAM Symposium on Anisotropy, Inhomogeneity and Nonlinearity in Solid Mechanics*. Proceedings of the IUTAM Symposium held in Nottingham, U.K. 1995 ISBN 0-7923-3594-5
40. J.-P. Merlet and B. Ravani (eds.): *Computational Kinematics '95*. 1995 ISBN 0-7923-3673-9
41. L.P. Lebedev, I.I. Vorovich and G.M.L. Gladwell: *Functional Analysis*. Applications in Mechanics and Inverse Problems. 1996 ISBN 0-7923-3849-9
42. J. Menčik: *Mechanics of Components with Treated or Coated Surfaces*. 1996 ISBN 0-7923-3700-X
43. D. Bestle and W. Schiehlen (eds.): *IUTAM Symposium on Optimization of Mechanical Systems*. Proceedings of the IUTAM Symposium held in Stuttgart, Germany. 1996 ISBN 0-7923-3830-8
44. D.A. Hills, P.A. Kelly, D.N. Dai and A.M. Korsunsky: *Solution of Crack Problems*. The Distributed Dislocation Technique. 1996 ISBN 0-7923-3848-0
45. V.A. Squire, R.J. Hosking, A.D. Kerr and P.J. Langhorne: *Moving Loads on Ice Plates*. 1996 ISBN 0-7923-3953-3
46. A. Pineau and A. Zaoui (eds.): *IUTAM Symposium on Micromechanics of Plasticity and Damage of Multiphase Materials*. Proceedings of the IUTAM Symposium held in Sèvres, Paris, France. 1996 ISBN 0-7923-4188-0
47. A. Naess and S. Krenk (eds.): *IUTAM Symposium on Advances in Nonlinear Stochastic Mechanics*. Proceedings of the IUTAM Symposium held in Trondheim, Norway. 1996 ISBN 0-7923-4193-7
48. D. Ieşan and A. Scalia: *Thermoelastic Deformations*. 1996 ISBN 0-7923-4230-5

Mechanics

SOLID MECHANICS AND ITS APPLICATIONS

Series Editor: G.M.L. Gladwell

49. J.R. Willis (ed.): *IUTAM Symposium on Nonlinear Analysis of Fracture*. Proceedings of the IUTAM Symposium held in Cambridge, U.K. 1997 ISBN 0-7923-4378-6
50. A. Preumont: *Vibration Control of Active Structures*. An Introduction. 1997 ISBN 0-7923-4392-1
51. G.P. Cherepanov: *Methods of Fracture Mechanics: Solid Matter Physics*. 1997 ISBN 0-7923-4408-1
52. D.H. van Campen (ed.): *IUTAM Symposium on Interaction between Dynamics and Control in Advanced Mechanical Systems*. Proceedings of the IUTAM Symposium held in Eindhoven, The Netherlands. 1997 ISBN 0-7923-4429-4
53. N.A. Fleck and A.C.F. Cocks (eds.): *IUTAM Symposium on Mechanics of Granular and Porous Materials*. Proceedings of the IUTAM Symposium held in Cambridge, U.K. 1997 ISBN 0-7923-4553-3
54. J. Roorda and N.K. Srivastava (eds.): *Trends in Structural Mechanics*. Theory, Practice, Education. 1997 ISBN 0-7923-4603-3
55. Yu.A. Mitropolskii and N. Van Dao: *Applied Asymptotic Methods in Nonlinear Oscillations*. 1997 ISBN 0-7923-4605-X
56. C. Guedes Soares (ed.): *Probabilistic Methods for Structural Design*. 1997 ISBN 0-7923-4670-X
57. D. François, A. Pineau and A. Zaoui: *Mechanical Behaviour of Materials*. Volume I: Elasticity and Plasticity. 1998 ISBN 0-7923-4894-X
58. D. François, A. Pineau and A. Zaoui: *Mechanical Behaviour of Materials*. Volume II: Viscoplasticity, Damage, Fracture and Contact Mechanics. 1998 ISBN 0-7923-4895-8
59. L.T. Tenek and J. Argyris: *Finite Element Analysis for Composite Structures*. 1998 ISBN 0-7923-4899-0
60. Y.A. Bahei-El-Din and G.J. Dvorak (eds.): *IUTAM Symposium on Transformation Problems in Composite and Active Materials*. Proceedings of the IUTAM Symposium held in Cairo, Egypt. 1998 ISBN 0-7923-5122-3
61. I.G. Goryacheva: *Contact Mechanics in Tribology*. 1998 ISBN 0-7923-5257-2
62. O.T. Bruhns and E. Stein (eds.): *IUTAM Symposium on Micro- and Macrostructural Aspects of Thermoplasticity*. Proceedings of the IUTAM Symposium held in Bochum, Germany. 1999 ISBN 0-7923-5265-3
63. F.C. Moon: *IUTAM Symposium on New Applications of Nonlinear and Chaotic Dynamics in Mechanics*. Proceedings of the IUTAM Symposium held in Ithaca, NY, USA. 1998 ISBN 0-7923-5276-9
64. R. Wang: *IUTAM Symposium on Rheology of Bodies with Defects*. Proceedings of the IUTAM Symposium held in Beijing, China. 1999 ISBN 0-7923-5297-1
65. Yu.I. Dimitrienko: *Thermomechanics of Composites under High Temperatures*. 1999 ISBN 0-7923-4899-0
66. P. Argoul, M. Frémond and Q.S. Nguyen (eds.): *IUTAM Symposium on Variations of Domains and Free-Boundary Problems in Solid Mechanics*. Proceedings of the IUTAM Symposium held in Paris, France. 1999 ISBN 0-7923-5450-8
67. F.J. Fahy and W.G. Price (eds.): *IUTAM Symposium on Statistical Energy Analysis*. Proceedings of the IUTAM Symposium held in Southampton, U.K. 1999 ISBN 0-7923-5457-5
68. H.A. Mang and F.G. Rammerstorfer (eds.): *IUTAM Symposium on Discretization Methods in Structural Mechanics*. Proceedings of the IUTAM Symposium held in Vienna, Austria. 1999 ISBN 0-7923-5591-1

Mechanics

SOLID MECHANICS AND ITS APPLICATIONS

Series Editor: G.M.L. Gladwell

69. P. Pedersen and M.P. Bendsøe (eds.): *IUTAM Symposium on Synthesis in Bio Solid Mechanics*. Proceedings of the IUTAM Symposium held in Copenhagen, Denmark. 1999 ISBN 0-7923-5615-2
70. S.K. Agrawal and B.C. Fabien: *Optimization of Dynamic Systems*. 1999 ISBN 0-7923-5681-0
71. A. Carpinteri: *Nonlinear Crack Models for Nonmetallic Materials*. 1999 ISBN 0-7923-5750-7
72. F. Pfeifer (ed.): *IUTAM Symposium on Unilateral Multibody Contacts*. Proceedings of the IUTAM Symposium held in Munich, Germany. 1999 ISBN 0-7923-6030-3
73. E. Lavendelis and M. Zakrzhevsky (eds.): *IUTAM/IFToMM Symposium on Synthesis of Nonlinear Dynamical Systems*. Proceedings of the IUTAM/IFToMM Symposium held in Riga, Latvia. 2000 ISBN 0-7923-6106-7
74. J.-P. Merlet: *Parallel Robots*. 2000 ISBN 0-7923-6308-6
75. J.T. Pindera: *Techniques of Tomographic Isodyne Stress Analysis*. 2000 ISBN 0-7923-6388-4
76. G.A. Maugin, R. Drouot and F. Sidoroff (eds.): *Continuum Thermomechanics*. The Art and Science of Modelling Material Behaviour. 2000 ISBN 0-7923-6407-4
77. N. Van Dao and E.J. Kreuzer (eds.): *IUTAM Symposium on Recent Developments in Non-linear Oscillations of Mechanical Systems*. 2000 ISBN 0-7923-6470-8
78. S.D. Akbarov and A.N. Guz: *Mechanics of Curved Composites*. 2000 ISBN 0-7923-6477-5
79. M.B. Rubin: *Cosserat Theories: Shells, Rods and Points*. 2000 ISBN 0-7923-6489-9
80. S. Pellegrino and S.D. Guest (eds.): *IUTAM-IASS Symposium on Deployable Structures: Theory and Applications*. Proceedings of the IUTAM-IASS Symposium held in Cambridge, U.K., 6–9 September 1998. 2000 ISBN 0-7923-6516-X
81. A.D. Rosato and D.L. Blackmore (eds.): *IUTAM Symposium on Segregation in Granular Flows*. Proceedings of the IUTAM Symposium held in Cape May, NJ, U.S.A., June 5–10, 1999. 2000 ISBN 0-7923-6547-X
82. A. Lagarde (ed.): *IUTAM Symposium on Advanced Optical Methods and Applications in Solid Mechanics*. Proceedings of the IUTAM Symposium held in Futuroscope, Poitiers, France, August 31–September 4, 1998. 2000 ISBN 0-7923-6604-2
83. D. Weichert and G. Maier (eds.): *Inelastic Analysis of Structures under Variable Loads*. Theory and Engineering Applications. 2000 ISBN 0-7923-6645-X
84. T.-J. Chuang and J.W. Rudnicki (eds.): *Multiscale Deformation and Fracture in Materials and Structures*. The James R. Rice 60th Anniversary Volume. 2001 ISBN 0-7923-6718-9
85. S. Narayanan and R.N. Iyengar (eds.): *IUTAM Symposium on Nonlinearity and Stochastic Structural Dynamics*. Proceedings of the IUTAM Symposium held in Madras, Chennai, India, 4–8 January 1999 ISBN 0-7923-6733-2
86. S. Murakami and N. Ohno (eds.): *IUTAM Symposium on Creep in Structures*. Proceedings of the IUTAM Symposium held in Nagoya, Japan, 3-7 April 2000. 2001 ISBN 0-7923-6737-5
87. W. Ehlers (ed.): *IUTAM Symposium on Theoretical and Numerical Methods in Continuum Mechanics of Porous Materials*. Proceedings of the IUTAM Symposium held at the University of Stuttgart, Germany, September 5-10, 1999. 2001 ISBN 0-7923-6766-9
88. D. Durban, D. Givoli and J.G. Simmonds (eds.): *Advances in the Mechanis of Plates and Shells The Avinoam Libai Anniversary Volume*. 2001 ISBN 0-7923-6785-5
89. U. Gabbert and H.-S. Tzou (eds.): *IUTAM Symposium on Smart Structures and Structonic Systems*. Proceedings of the IUTAM Symposium held in Magdeburg, Germany, 26–29 September 2000. 2001 ISBN 0-7923-6968-8

Mechanics

SOLID MECHANICS AND ITS APPLICATIONS

Series Editor: G.M.L. Gladwell

90. Y. Ivanov, V. Cheshkov and M. Natova: *Polymer Composite Materials – Interface Phenomena & Processes*. 2001 ISBN 0-7923-7008-2
91. R.C. McPhedran, L.C. Botten and N.A. Nicorovici (eds.): *IUTAM Symposium on Mechanical and Electromagnetic Waves in Structured Media*. Proceedings of the IUTAM Symposium held in Sydney, NSW, Australia, 18-22 Januari 1999. 2001 ISBN 0-7923-7038-4
92. D.A. Sotiropoulos (ed.): *IUTAM Symposium on Mechanical Waves for Composite Structures Characterization*. Proceedings of the IUTAM Symposium held in Chania, Crete, Greece, June 14-17, 2000. 2001 ISBN 0-7923-7164-X
93. V.M. Alexandrov and D.A. Pozharskii: *Three-Dimensional Contact Problems*. 2001 ISBN 0-7923-7165-8
94. J.P. Dempsey and H.H. Shen (eds.): *IUTAM Symposium on Scaling Laws in Ice Mechanics and Ice Dynamics*. Proceedings of the IUTAM Symposium held in Fairbanks, Alaska, U.S.A., 13-16 June 2000. 2001 ISBN 1-4020-0171-1
95. U. Kirsch: *Design-Oriented Analysis of Structures*. A Unified Approach. 2002 ISBN 1-4020-0443-5
96. A. Preumont: *Vibration Control of Active Structures*. An Introduction (2nd Edition). 2002 ISBN 1-4020-0496-6
97. B.L. Karihaloo (ed.): *IUTAM Symposium on Analytical and Computational Fracture Mechanics of Non-Homogeneous Materials*. Proceedings of the IUTAM Symposium held in Cardiff, U.K., 18-22 June 2001. 2002 ISBN 1-4020-0510-5
98. S.M. Han and H. Benaroya: *Nonlinear and Stochastic Dynamics of Compliant Offshore Structures*. 2002 ISBN 1-4020-0573-3
99. A.M. Linkov: *Boundary Integral Equations in Elasticity Theory*. 2002 ISBN 1-4020-0574-1
100. L.P. Lebedev, I.I. Vorovich and G.M.L. Gladwell: *Functional Analysis*. Applications in Mechanics and Inverse Problems (2nd Edition). 2002 ISBN 1-4020-0667-5; Pb: 1-4020-0756-6
101. Q.P. Sun (ed.): *IUTAM Symposium on Mechanics of Martensitic Phase Transformation in Solids*. Proceedings of the IUTAM Symposium held in Hong Kong, China, 11-15 June 2001. 2002 ISBN 1-4020-0741-8
102. M.L. Munjal (ed.): *IUTAM Symposium on Designing for Quietness*. Proceedings of the IUTAM Symposium held in Bangkok, India, 12-14 December 2000. 2002 ISBN 1-4020-0765-5
103. J.A.C. Martins and M.D.P. Monteiro Marques (eds.): *Contact Mechanics*. Proceedings of the 3rd Contact Mechanics International Symposium, Praia da Consolação, Peniche, Portugal, 17-21 June 2001. 2002 ISBN 1-4020-0811-2
104. H.R. Drew and S. Pellegrino (eds.): *New Approaches to Structural Mechanics, Shells and Biological Structures*. 2002 ISBN 1-4020-0862-7
105. J.R. Vinson and R.L. Sierakowski: *The Behavior of Structures Composed of Composite Materials*. Second Edition. 2002 ISBN 1-4020-0904-6
106. Not yet published.
107. J.R. Barber: *Elasticity*. Second Edition. 2002 ISBN Hb 1-4020-0964-X; Pb 1-4020-0966-6
108. C. Miehe (ed.): *IUTAM Symposium on Computational Mechanics of Solid Materials at Large Strains*. Proceedings of the IUTAM Symposium held in Stuttgart, Germany, 20-24 August 2001. 2003 ISBN 1-4020-1170-9

Mechanics

SOLID MECHANICS AND ITS APPLICATIONS

Series Editor: G.M.L. Gladwell

109. P. Ståhle and K.G. Sundin (eds.): *IUTAM Symposium on Field Analyses for Determination of Material Parameters – Experimental and Numerical Aspects.* Proceedings of the IUTAM Symposium held in Abisko National Park, Kiruna, Sweden, July 31 – August 4, 2000. 2003 ISBN 1-4020-1283-7
110. N. Sri Namachchivaya and Y.K. Lin (eds.): *IUTAM Symposium on Nonlinear Stochastic Dynamics.* Proceedings of the IUTAM Symposium held in Monticello, IL, USA, 26 – 30 August, 2000. 2003 ISBN 1-4020-1471-6
111. H. Sobieckzky (ed.): *IUTAM Symposium Transsonicum IV.* Proceedings of the IUTAM Symposium held in Göttingen, Germany, 2–6 September 2002, 2003 ISBN 1-4020-1608-5
112. J.-C. Samin and P. Fisette: *Symbolic Modeling of Multibody Systems.* 2003 ISBN 1-4020-1629-8
113. A.B. Movchan (ed.): *IUTAM Symposium on Asymptotics, Singularities and Homogenisation in Problems of Mechanics.* Proceedings of the IUTAM Symposium held in Liverpool, United Kingdom, 8-11 July 2002. 2003 ISBN 1-4020-1780-4
114. S. Ahzi, M. Cherkaoui, M.A. Khaleel, H.M. Zbib, M.A. Zikry and B. LaMatina (eds.): *IUTAM Symposium on Multiscale Modeling and Characterization of Elastic-Inelastic Behavior of Engineering Materials.* Proceedings of the IUTAM Symposium held in Marrakech, Morocco, 20-25 October 2002. 2004 ISBN 1-4020-1861-4
115. H. Kitagawa and Y. Shibutani (eds.): *IUTAM Symposium on Mesoscopic Dynamics of Fracture Process and Materials Strength.* Proceedings of the IUTAM Symposium held in Osaka, Japan, 6-11 July 2003. Volume in celebration of Professor Kitagawa's retirement. 2004 ISBN 1-4020-2037-6
116. E.H. Dowell, R.L. Clark, D. Cox, H.C. Curtiss, Jr., K.C. Hall, D.A. Peters, R.H. Scanlan, E. Simiu, F. Sisto and D. Tang: *A Modern Course in Aeroelasticity.* 4th Edition, 2004 ISBN 1-4020-2039-2
117. T. Burczyński and A. Osyczka (eds.): *IUTAM Symposium on Evolutionary Methods in Mechanics.* Proceedings of the IUTAM Symposium held in Cracow, Poland, 24-27 September 2002. 2004 ISBN 1-4020-2266-2
118. D. Ieşan: *Thermoelastic Models of Continua.* 2004 ISBN 1-4020-2309-X
119. G.M.L. Gladwell: *Inverse Problems in Vibration.* Second Edition. 2004 ISBN 1-4020-2670-6
120. J.R. Vinson: *Plate and Panel Structures of Isotropic, Composite and Piezoelectric Materials, Including Sandwich Construction.* 2005 ISBN 1-4020-3110-6
121. *Forthcoming*
122. G. Rega and F. Vestroni (eds.): *IUTAM Symposium on Chaotic Dynamics and Control of Systems and Processes in Mechanics.* Proceedings of the IUTAM Symposium held in Rome, Italy, 8–13 June 2003. 2005 ISBN 1-4020-3267-6
123. E.E. Gdoutos: *Fracture Mechanics. An Introduction. 2nd edition.* 2005 ISBN 1-4020-3267-6
124. M.D. Gilchrist (ed.): *IUTAM Symposium on Impact Biomechanics from Fundamental Insights to Applications.* 2005 ISBN 1-4020-3795-3
125. J.M. Huyghe, P.A.C. Raats and S. C. Cowin (eds.): *IUTAM Symposium on Physicochemical and Electromechanical Interactions in Porous Media.* 2005 ISBN 1-4020-3864-X
126. H. Ding, W. Chen and L. Zhang: *Elasticity of Transversely Isotropic Materials.* 2005 ISBN 1-4020-4033-4
127. W. Yang (ed): *IUTAM Symposium on Mechanics and Reliability of Actuating Materials.* Proceedings of the IUTAM Symposium held in Beijing, China, 1–3 September 2004. 2005 ISBN 1-4020-4131-6